Agriculture in the United States

A Documentary History

WAYNE D. RASMUSSEN is an Agricultural Historian in the Economic Research Service, United States Department of Agriculture, Washington, D.C. He received his Ph.D. in 1950 from George Washington University. In 1973, Dr. Rasmussen was awarded the Distinguished Service award of the United States Department of Agriculture. He is the editor of *Readings in the History of American Agriculture* (1960), and co-author of *Century of Service: The First 100 Years of the United States Department of Agriculture* (1963), and *The Department of Agriculture* (1972).

Agriculture in the United States

A Documentary History

Edited by

Wayne D. Rasmussen

National Economic Analysis Division
United States Department of Agriculture

Volume 1

Random House New York

Reference Series Editors:

William P. Hansen
Fred L. Israel

FIRST EDITION

9 8 7 6 5 4 3 2 1

*Agriculture in the United States: A Documentary
History* is now exclusively published and distributed
by *Greenwood Press, Inc.*
  51 Riverside Avenue
  Westport, Ct. 06880
          ISBN 0-313-20147-1 Set
          0-313-20148-X v.1
          0-313-20149-8 v.2
          0-313-20150-1 v.3
          0-313-20151-X v.4

1. Agriculture — United States — History — Sources.
I. Title.
S441.R33   630′.973   74–9643
ISBN 0-394-49976-X Vol. 1

Preface

Agriculture or farming — the conscious production, directly or indirectly from the soil, of food, fiber and related products used by man — is one of our most fundamental occupations. Possible food shortages concentrated attention on agriculture in the late 1970's. If the world is to meet its food needs in the future, agricultural production must be increased in both the developed and the developing nations.

It is a major goal of the compiler of these volumes to show how change took place in farming in the United States. When this nation won its independence, over 90 percent of the population had to farm to meet the needs of the nation and to provide a surplus to pay for at least part of our imports. In our Bicentennial year, only about 4 percent of the population was meeting those same needs. Two major revolutions in agricultural production have taken place. The first, the change from hand labor to horse power and from self-sufficient agriculture to commercial agriculture throughout the nation, was stimulated by the Civil War. The second, the completion of the change from horse power to machine power and the adoption of a package of improved practices, was stimulated by World War II.

The agricultural revolutions came in distinctive periods of stress but were based upon a national land policy, which was unique when adopted; other Federal policies to encourage agriculture; invention, experimentation, and education; and the willingness of American farmers, when convinced it was to their economic advantage, to adopt new farm practices. These volumes are divided into sections covering these topics, within seven chronological periods, which have significance to agricultural as well as to American history.

Each section includes documents, generally in chronological order, illustrative of that topic and period. Legislation and other key documents are usually reprinted in full, omitting footnotes, while many others, representing trends or containing limited specific material, are reprinted in part.

The documents are illustrative of the major types of original material available for research in agricultural history. They include legislation, travel accounts, farm manuals, correspondence, diaries, articles from farm journals, advertisements, and reports of the Federal and State Departments of Agriculture, agricultural colleges, and experiment stations. Representative writings of outstanding national leaders, scientists, educators, and administrators are found under appropriate headings. Most accounts tend to be optimistic about agricultural progress, although some critical accounts and reports of failures are included.

Throughout the compilation of these volumes, I have been aided by colleagues in the Department of Agriculture, fellow members of the agricultural historical profession and many others. Karen E. Rasmussen worked as my irreplaceable research assistant. Gladys L. Baker gave generously of her advice. Vivian B. Whitehead assisted on every aspect of the project. Marion F. Rasmussen helped from beginning to end. The staff of the National Agricultural Library proved helpful beyond the usual call of duty. Other libraries whose resources contributed to the volumes include the Huntington Library, the Widener Library of Harvard University, the Library and staff of the Nebraska State Historical Society, and the Library of the University of North Carolina. Others who helped me with particular problems include: Mark T. Allen, Michigan State University; Leonard J. Arrington, Church of Jesus Christ of Latter Day Saints; Betty Baxtresser, National Agricultural Library; Ray A. Billington, Huntington Library; James T. Bonnen, Michigan State University; Douglas E. Bowers, U.S. Department of Agriculture; Harold F. Breimyer, University of Missouri; David E. Brewster, U.S. Department of Agriculture; Emerson M. Brooks, Silver Spring, Maryland; Melba M. Bruno, National Agricultural Library; Vernon Carstensen, University of Washington; Rosalie E. R. Cherry, U.S. Department of Agriculture; Robert G. Dunbar, Montana State University; Gilbert C. Fite, Eastern Illinois University; Tom Fulton, U.S. Department of Agriculture; Alan Fusonie, National Agricultural Library; G.E. Fussell, Sudbury, Suffolk, England; Mary W. M. Hargreaves, University of Kentucky; Cecil L. Harvey, U.S. Department of Agriculture; Darwin P. Kelsey, Old Sturbridge Village; Frances Loomis, Washington, D.C.; Mildred G. Mosby, U.S. Department of Agriculture; Gerald R. Ogden, U.S. Department of Agriculture; Don Paarlberg, U.S. Department of Agriculture; William D. Pennington, William Woods College; Charles S. Peterson, Utah State University; Harold T. Pinkett, National Archives; Jane M. Porter, U.S. Department of Agriculture; James E. Potter, Nebraska State Historical Society; Leroy C. Quance, U.S. Department of Agriculture; Linda M. Rasmussen, Cambridge, Massachusetts; Joe D. Reid, Jr., University of Pennsylvania; Andrea Rhodin, Washington, D.C.; Earl M. Rogers, University of Iowa; Howard S. Russell, Wayland, Massachusetts; Mark D. Schmitz, University of Delaware; Joel A. Schor, U.S. Department of Agriculture; Morgan B. Sherwood, University of California, Davis; James H. Shideler, University of California, Davis; Maryanna S. Smith, U.S. Department of Agriculture; Homer E. Socolofsky, Kansas State University; Paul Stone, Arlington, Massachusetts; Vivian D. Wiser, U.S. Department of Agriculture.

Contents

VOLUME TWO: Changes in Farming (Continued)

VOLUME THREE

VOLUME FOUR

Colonial America

Introduction

The phrase "Colonial America" may well call to mind that historic day in 1607 when a group of English adventurers landed on the banks of the James River, there to establish the first permanent settlement by Englishmen in America. Yet to the south, the Spaniards had long since planted permanent settlements, introduced the horse and other animals, and acclimated European plants. Even there we have not reached the beginnings of agriculture in what is now the United States, for the American Indians had been practicing agriculture for several centuries.

The major staple crops developed by the Indians include corn or maize, the white potato, and tobacco. The mainstay of the world's cotton crop is a variety, *Gossypium hirsutum* L., that was cultivated by the Indians of Mexico. The peanut, a native of South America, is important in American agriculture. Other food plants first used by the Indians and now cultivated commercially in the United States include avocados, kidney and lima beans, chilipeppers, pumpkins, squashes, sweetpotatoes, and tomatoes. The North American Indians in some areas collected and boiled maple sap for syrup and sugar.

Domesticated animals were the greatest lack in Indian agriculture. Only the dog, practically universal, and the turkey in the Southwest had been domesticated in the present-day United States. Perhaps, considering the state of civilization and tiny population of the New World and its abundant wild life, the lack of domestic animals did not mean a shortage of protein food and animal fiber, although such a lack would have become imporant as population increased. The major shortage and a basic hindrance to the development of agriculture was the virtual non-availability of animal power.

All agriculture was of necessity carried on entirely by hand. Here the North American Indian was handicapped in that he had not yet learned the use of metals. His crude tools were of wood, stone, bone, or shell. Thus, he had none of the saws, axes, or other tools of metal that proved so useful to our frontiersmen in clearing the forests. Instead, he killed trees by girdling them or by scorching their roots. As the trees died, the ground was exposed to sunlight and was planted. Later, the dead trunks and stumps of the trees were burned.

The corn was usually planted in hills. The Indian farmers, most often the women, dug holes with sticks, dropped two or four seeds in each, and covered them with heaped-up earth. Beans squashes, pumpkins, and even sunflower might be planted in the same hill or between the hills. In general, the hills and ground immediately around them would be kept free of weeds, the soil between the hills not being broken.

Thus, the American Indian developed a number of plants which are among our most important staples; he left clearings which were used by the first European settlers and taught the settlers how to clear land; he developed methods of planting corn which are used, in greatly modified form, to the present; he taught the New Englanders the value of fish as fertilizer; he practiced land rotation instead of crop rotation; and he used intertillage, a practice that was adopted by the new settlers.

The first Europeans to move into the Indian's world and make permanent settlement in what is now the United States were the Spaniards. Since they followed a systematic pattern in introducing plants and animals into the regions they settled, it seems evident that they introduced many of the crops of Spain into Florida and later into the Southwest. Some of their introductions are well known, others may be presumed. It was not, however, until a number of years had passed that the Spanish introductions could have influenced the English settlements. Perhaps the transfer of the potato to Europe, from whence it was introduced into North America, and the transfer of "Orinoco" tobacco seed by one means or another to the Jamestown colonists were the most important Spanish contributions to agriculture in the English colonies.

French settlers in Canada were carrying on a limited agriculture before any permanent English settlement was made in America. For example, wheat, rye, barley, oats, hemp, flax, turnips, radishes, cabbages, and other seeds were sown at Port Royal in 1606. One Louis Hebert, a physician, is usually credited with being the moving spirit in this work. It was continued until 1613, when the settlement was destroyed by the English and Hebert returned to France. He was later persuaded to return to the New World and settled in Quebec in 1617, where he lived until his death in 1627.

The years 1609 and 1610 are sometimes remembered as the "starving time" of Jamestown, the first permanent settlement by Englishmen in America. Neither the Spanish settlements far to the south nor the French holdings to the north could be called upon for assistance. Indeed, they were considered a menace. The English settlers did not, at the same time, realize they they would have to learn new methods of growing crops from the Indians. Instead, while they begged, borrowed, and stole food from their neighbors, they attempted to carry out English farming practices and to grow the familiar crops of the old world in the untamed soils of a vast wilderness.

The farming practices that the first settlers brought with them were but a step removed from those of the Middle Ages. Even the more knowledgeable farmers tended to folow rules of thumb, such as those laid down by Thomas Tusser or Stevens and Liebault. While England was upon the threshhold of far-reaching improvements in her agriculture as the overseas movement started, such improvements were still in the future insofar as the Jamestown and Plymouth settlers were concerned and, in fact, were to have little influence upon agriculture in colonial America.

The two most important changes that took place in English agriculture during the 17th century were the enclosure of former open-field farms and the conversion of arable land into pasture. Enclosure led to more efficient and more profitable farming — at least more profitable for the landlords and the freeholders or yeomen who owned sizeable plots. Common rights, that is, the right to use certain meadow, wood, and other lands in common with fellow villagers, were extinguished, by force if not by mutual consent. This abolition of common rights, with enclosures and the growing wool trade, enabled the landholders to breed better livestock.

In most cases, prior to the enclosure movement, the land of the manor would be farmed under the three-field system. The arable land would be divided into three great fields. These in turn would be divided into the strips assigned to each man. Every year one of the great fields lay fallow; one was planted in wheat or rye; and other was planted to barley, oats, vetches, beans, or peas. After harvest, the livestock, which had been grazing on the commons or waste land, were turned into the arable fields.

The tools used on the farm were crude. The plows, large and cumbersome, were usually drawn by oxen. After the soil was broken iron- or wooden-toothed harrows were pulled over the land. All grain crops were seeded broadcast. Harvesting was done by hand, using scythes or reaping hooks; threshing was done with flails. Hoes, mattocks, and spades completed the list of implements. The change to horse-drawn implements was not to come for over a century, after Jethro Tull emphasized horse-drawn cultivators in a book first published in 1733.

The village farmer usually made and repaired his own tools except for a bit of metal work. This of course, was in accordance with the general pattern of life. The fields provided food. The women spun wool and hemp; men tanned their own leather. Each village had its mill and those of some size had a smith and a carpenter. However, most farmers carved their own household utensils and made their own furniture. Thus, the villagers who left England for America were accustomed to supplying their own needs. If they were not entirely self-sufficient, and of course they were not, they were sufficient enough to survive in the wilderness.

The gentlemen and sons of gentlemen, however, who seem to have been predominant among the first settlers of Jamestown, had not learned the skills of the villagers. Thus, they were handicapped in undertaking their new life. Too, they had been directed by the King and the London Council for Virginia to make a special effort to find mineral wealth. This is readily understood when one remembers that the gold, silver, pearls, and precious stones found in the New World by Pizzaro, Cortes, and other adventurers had enriched Spain and its rulers almost beyond imagination.

Some of the adventurers were interested in agriculture, and the first expedition had brought swine, chickens, and seeds. One of the first accounts of the new settlement speaks of the pineapples, oranges, and tree cotton from the West Indies that had been planted in addition to the potatoes (perhaps sweet potatoes), pumpkins, melons, and garden seeds. At one time and another several of the settlers were assigned to agricultural work but such efforts as were put forth for the first two years were without effect.

The starving time of 1609–1610 was followed in the summer of 1610 by a number of events which lent hope to the surviving colonists. A well-equipped expedition, which included milk cows, arrived in June. During the summer, the colonists took control of several Indian fields, including some located in healthier sites than the peninsula. Even more to the point, the colonists began to adopt Indian methods of growing corn. During the starving time some of the settlers had lived with friendly Indians and had learned much of their agriculture. Two Indian prisoners had, a little earlier, taught some of the colonists how to grow maize and this knowledge was now put to use.

Under the energetic leadership of Sir Thomas Dale in 1611, more cleared land was taken from the Indians. Dale had brought more livestock and issued strict orders that none, not even a chicken, was to be killed without his permission. The result was a fairly rapid increase in the livestock population. Fishing was placed upon an organized basis, with the result of an additional increase in the supply of food. Dale even organized a new settlement on higher and healthier land at Henrico and there established a number of the settlers.

By 1612, many kinds of agricultural products had been tried in the colony. Mention has been made of the oranges, cotton, and pineapples. Most vegetables common to English gardens and the ordinary English fruits had been planted, as had most cereals. Other plants included European grapes, flax, hemp figs, indigo, and rice, and some effort was made from 1613 on to establish a silk industry. However, interest in such crops was overshadowed by the rapid rise of tobacco as the staple crop of Virginia.

When the first colonists landed, they found the native Americans using tobacco. The settlers began growing that type of tobacco, *Nicotiana rustica,* although it grew poorly and was weak and bitter compared with the *Nicotiana tabacum* grown in the Spanish colonies. One of the settlers, probably John Rolfe, apparently secured seed of the Spanish type from Trinidad. At any rate, in 1612 Rolfe was growing the ''Orinoco'' tobacco and made a shipment of it to England, where it was favorably received. The large returns from this first shipment and those following served as a stimulus to tobacco planting. The Virginia farmer, and the Virginia Company itself, now had a staple that could be readily grown and commanded a good price. So much effort naturally went into tobacco that Sir Thomas Dale, in 1616, ordered that each person responsible for his own maintenance must plant two acres of corn for himself and each man servant as a condition to raising any tobacco, under pain of forfeiture of his tobacco crop. The crop of 1618, sent to England in 1619, amounted to 20,000 pounds.

The earlier elements which were to lead to the development of a free landholding class and the establishment of the plantation system were further developed by 1619. By that time, the Virginia Company and several joint stock plantations were employing a considerable number of white indentured servants and other laborers. Generally, an indentured servant was a person who had agreed to work a number of years in return for his passage. Early in its history, the Virginia Company had recognized that one could acquire a share in the company by venturing one's self as a colonist instead of venturing money. This concept was extended in 1618 to grant 50 acres of land to proprietors for each person transported to the colony. Settlers paying their own way were also to receive 50 acres. These rights to land called headrights, were subsequently extended freely and were often granted in varying amounts to the transported person himself, particularly to an indentured servant at the end of his indenture, and were granted to those bringing slaves into the country.

As these circumstances might suggest, land was freely available while labor was exceedingly scarce. The first representative assembly of the New World met at Jamestown in 1619, and, seeing the importance of labor, recognized indenture. In an endeavor to keep its men in the colony and to provide for the future, the Company took a practical step the same year when it sent a shipload of marriageable women to the colony. It required only that each husband of those wooed and won pay, in tobacco, the costs of bringing his wife to the colony. However, another event having to do with labor and the plantation system far exceeded these in long-range significance. Late in August a Dutch privateer sold 20 Africans, captured from a slaver, to Jamestown colonists. These Negroes were sold, apparently, as indentured servants with an indefinite term of service. First by custom and then by law Negroes sold under such conditions soon became slaves. Thus, a year before colonists arrived at Plymouth, the Jamestown settlers were upon the road to a plantation agriculture dominated by staple crops and slavery.

On September 16, 1620, a group of 102 religious dissenters set sail for the New World. This group, the Pilgrims, and their ship, the *Mayflower*, were embarking upon the best known of all colonizing expeditions to America. The Pilgrims were mainly farmers who had formed a joint stock company with some London merchants to establish a colony in the region of Chesapeake Bay. The Pilgrims were to settle the country; the merchants were to finance the expedition. Funds for financing the settlement were raised by the sale of shares of stock in the company at ten pounds each. Every migrant over 16 years of age was to receive a share for going to America and an extra share if he furnished cash or provisions to the amount of ten pounds. The company was to be in existence seven years, during which time all profits from trade, fishing, and other work of the colonists was to go into a common fund. The colonists were to be supported from this fund. At the end of seven years, the fund and all land and other property were to be divided among settlers and financiers in accordance with the numbers of shares held by each individual. This communal system was similar to the one first established in Jamestown.

The Mayflower, driven off course, made land at what is now Provincetown, Massachusetts, on November 21, 1620. The colonists then explored the bay area and settled upon Plymouth as a proper place for the new settlement, landing there at the end of December. The site had a good harbor and a number of cleared fields. Their previous Indian occupants had succumbed a few years earlier to some virulent disease. The coastal plain was very narrow at Plymouth and sandy, while the neighboring upland was hilly and rocky. The climate was rigorous and not particularly favorable to farming. It appears that settlement was made in spite of these factors because the intent of the company was to have the settlers give more attention to cutting lumber, fishing, and trading with the Indians, and the site seemed suitable for these activities.

The colonists had arrived in the midst of winter and were faced by two immediate problems, shelter and food. They set to work to build cottages, grouped closely together for protection against the Indians. Not the log cabins of tradition, the cottages were built of poles and logs set upright or took the form of dugouts. Unfortunately, these crude shelters did not provide sufficient protection against the elements. The exposure brought on a disease which killed nearly half of the group.

Perhaps the strange food also encouraged disease, for arriving as they did in mid-winter, the colonists had to rely partly upon native game and fish and upon such Indian foods as they could obtain. Although there were no Indians with whom to trade, the Pilgrims did find a quantity of corn and beans that had been stored by Indians on Cape Cod.

With the coming of spring in 1621, the Pilgrims turned to farming. They worked entirely by hand, since no animals had been brought over on the *Mayflower*. The land was broken with hoes and mattocks, and the other farm operations were carried on with essentially the same kinds of tools that the first settlers in Jamestown used. The Pilgrims did not have even a plow until 1632. However, the work was easier than it had been in Jamestown since the Pilgrims were working with land that the Indians had cleared. The Pilgrims were fortunate in that, during their very first spring, Squanto, an Indian who had been kidnapped, had lived in England and learned English, and had then been returned to the Massachusetts Bay region, came to the colonists and taught them how to grow corn. He also supplied them with game and was likely responsible for the friendly attitude of the chief of his tribe, Massasoit, toward the Pilgrims.

Under Squanto's tutelage, the Pilgrims planted about 20 acres of corn fertilized with fish and five acres of English grain in the spring of 1621. As reported in a letter by Edward Winslow, the corn succeeded; the English grain failed. It was not until some years had passed that wheat, barley, and similar crops were cultivated successfully.

The system for working and sharing the results in common, which had been part of the agreement with the London partners, proved unsatisfactory. In 1623, Governor William Bradford abandoned the system and assigned each family a separate tract of land to plant. The following year, each family was assigned one acre of land to be kept from year to year. At the end of the seven year period of the contract with the London merchants, the settlers arranged to buy out their interests and then apportioned 20 acres of land to each family. The meadow land was still held in common and assigned to families in accordance with livestock held.

The first cattle, three heifers and a bull, were brought to Plymouth in 1624. Goats were also imported at an early date. Horses and sheep were not brought in until later.

Ten years after the Pilgrims had settled at Plymouth, the Puritans, carrying with them their charter as the "Governor and Company of Massachusetts Bay," made their great migration. Earlier groups had established settlements at Gloucester in 1623, which failed, and at Salem. John Endicott had been sent as governor to the Salem colony in 1628, and had carried to it seeds of English grains and vegetables. The next year, the Salem settlers purchased hogs from the Plymouth settlers and brought goats and possibly other livestock from England.

During 1630, the year of the "great migration," about 2000 migrants, under the leadership of the governor of the Company, John Winthrop, arrived in Massachusetts and soon established settlements at Charlestown, Boston, Roxbury, Dorchester, Watertown, Medford, and Lynn. These Puritans, profiting by the experiences of the Jamestown, Plymouth, and Salem settlers, and aided by their comparative wealth, came better equipped than the earlier groups to wrest a living from the New World. Thought had been given as to what might be necessary to insure success and efforts were made to meet the needs. Thus, in 1629, the governing body of the Company decreed that past and prospective settlers be allocated 200 acres of land each "for the p' sent accomodacon of the people lately gone thither, as well to build them houses as to inclose & manure, & feede their cattle on." In accordance with this planning, considerable numbers of livestock were brought into the colony in 1630. The settlers were also well provided with tools of the type being used in England. The emphasis was upon hand tools; in 1636 there were only 30 plows in the Massachusetts Bay Colony.

There was never any doubt after the first year, when the Puritans adopted the Pilgrim custom of celebrating a day of thanksgiving, that the settlers could feed themselves. On the other hand, the sandy soil gave no promise of surpluses for export and many of the colonists turned to livestock raising. The cattle pastured upon the native grass, which was thin and sparse in summer, and browsed upon the trees and were fed marsh hay in the winter. When possible the settlers would fence a neck of land on a river, or a peninsula, where the cattle might be kept at night as a protection against the wolves and against their straying. Where conditions did not permit such simple expedients, the colonists would erect stockades. During the day, boys or older men would look after the cattle.

The emphasis upon livestock in Massachusetts Bay, Plymouth as well, led to local regulations of a type that are more often associated with the far west of 250 years later. Thus, the Council at Plymouth, on November 15, 1636, ordered that every man register the brand by which he marked his cattle and mark all of his cattle with that brand. Another Plymouth regulation of the same year allowed persons whose crops were injured when livestock broke into "sufficient enclosures" to impound the trespassing livestock until damages were paid.

The New England town usually obtained, by grant or preemption, title to between six and eight square miles of land. The area was occupied jointly by a group of settlers, and was laid out in home plots, tillage land, meadow and grazing land, and woodland. The home plots, ranging in size from one-half acre to as many as 20 acres were grouped together about an area known as the town common and were often granted by lot. The farm land was then granted to families. There was variation in method; some towns assigned single tracts to individuals, in others the grants were broken into noncontiguous tracts. There were often restraints on aliena- tion of these grants, but they were usually free of quitrents, the fixed fee paid in commutation of personal services due a feudal lord.

The New England land grants were characterized by small grants to actual settlers. This led to the development of the family farm rather than the manor or plantation. The important place of the town on the land system led to comparatively close settlement within the area granted to the town. This made the development of democratic local government, schools, religious institutions, and, eventually, in- dustry, simpler than in the plantation areas. The tendency in granting townsites to make them adjacent to previous grants made for solid, concentrated settlement, a factor sometimes mentioned as being important in the American Revolution. The total influence of the New England land system upon the United States of the present day cannot be determined; that it was an important influence is evident.

The end of the colonial period saw the agricultural patterns of the Northern colonies rather firmly fixed. New England had developed no staple crops and the products of her agriculture, except for livestock, went to supply her own needs. The future of New England obviously lay in trade, shipping, fishing, and manufacturing rather than in an agriculture that would produce surpluses for export.

New York was unique among the northern colonies in that large estates determined her development. Western New York had not been settled; large holdings in the eastern part of the colony were delaying agricultural progress. New Jersey and Pennsylvania were producing surpluses of farm products and exporting large quantities of food. Pennsylvania was leading the colonies in the production of wheat and was to continue to do so for some time.

Labor was scarce and land was cheap. Thus, intensive farming was not practical anywhere in the colonies. There was little free labor for hire and slavery had been tried and found uneconomical. Indentures supplied most of the non-family labor that was available on farms. The family farm operated by family labor was the prevalent farm and labor unit.

Corn was the most important grain crop, except in Pennsylvania where wheat ruled. Rye, oats, various vegetables, and fruit trees were widely grown as was flax and hemp. Livestock was generally unimproved even though it furnished the basis for a considerable export trade.

The life of the frontier farmer, constantly battling the forest and often the Indian, was hard. That of his wife, burdened by numerous children and constant, monotonous tasks carried out in a dingy cabin in a gloomy forest clearing, was drab, and all too short. However, in well settled areas rural social life was developing.

The general pattern of agricultural development in the southern colonies of Maryland, Virginia, North Carolina, South Carolina, and Georgia during the colonial period followed that discernable in Virginia as early as 1619. While the New England colonies and the middle colonies of New Jersey, Pennsylvania and Delaware were developing small farms operated by family labor, the southern colonies were developing plantations operated by slave labor. There were, of course, large holdings in New England and the middle colonies, particularly in New York, just as there were numerous small holdings in all of the southern colonies. Nevertheless, the family farm dominated the north and the plantation dominated the south. In each case, climate and soil were determining factors even though religion and other social factors were not without influence.

Much of the land in the southern colonies was held in large estates, although many of the colonies also granted "headrights," similar to those of Virginia. Some of the lands granted as headrights, as well as some of the large grants, were subject to small annual payments known as quitrents. An anonymous writer, describing Maryland in 1635, wrote of the land grants and the quitrents. William Penn, too, established a system which included large grants and headrights, both subject to quitrents. Later, John Lawson, in writing of North Carolina stated "our Land pays to the Lords but an easy Quit-Rent, or yearly Acknowledgement." Even though these payments were small, they were a constant irritant to colonists subject to them.

Tobacco continued to be a major crop, but two additional plantation-grown crops, rice and indigo, developed in South Carolina. Rice had been planted when the Ashley River was first settled, but it did not become an export crop until the turn of the century. Better varieties and better methods of cultivation increased production per acre, but rice nevertheless required a great deal of hard, unhealthy labor. The work was done by Negro slaves toiling ankle deep in mud and water. The mosquitoes infesting the swamps infected many workers with malaria. Thus the rice swamps early acquired a reputation for causing a high mortality among the slaves. By the Revolution, a slave was able to cultivate about three acres of rice yielding 75 bushels, and take care of his food crops.

The British government from time to time regulated the trade by specifying the areas to which rice might be shipped and by levying duties on it. The general trend of prices was up and the last 15 years of the colonial period were exceedingly prosperous.

Like rice, indigo was planted in the very first years of settlement, but its large-scale, commercial cultivation awaited a labor supply, a better market, and a bounty. The best indigo was produced in the Spanish and French colonies, but war was interferred every so often with these supplies. Thus, the British government passed legislation from time to time to encourage its cultivation in America. Finally, in 1748, a bounty was granted on all indigo produced in the British American colonies, provided it reached certain standards. This bounty, continued until 1770 and then reduced, provided a tremendous impetus to the industry.

A young lady, Eliza Lucas (later Pinckney), is creditied with establishing indigo on a commercial basis in South Carolina. Miss Lucas was in charge of her father's plantation while he served as governor of the island of Antigua. During the 1740's she experimented with several crops, and decided that indigo offered unusual possibilities. After considerable experimentation, she succeded in producing a commercial crop, and, encouraged by the bounty of 1748, many other planters soon followed her example.

Maize was the most important crop of the south, but more and more of it was being used on the plantation. Cattle raising was of major importance on the piedmont, but the industry was a transitional one, useful until the land came under the domination of planter and farmer. Self-sufficiency was declining, and there was a greater dependence upon imports, including food products, by the Revolution. At the same time, the economy was dependent upon the export of its staples, a factor that was to distinguish the south from the rest of the Nation until the present time. Small, nonslaveholding farmers were numerous in every southern colony but the economic and political direction of society was in the hands of the planter.

European Background

Thomas Tusser on Good Husbandrie, 1580

From Thomas Tusser, *Five Hundred Points of Good Husbandrie* (London, 1580), pp. 30–52, 115–16, 121–28.

Septembers Husbandrie

At Mihelmas lightly new fermer comes in,
  new husbandrie forceth him new to begin
Old fermer, still taking the time to him given,
  makes August to last untill Mihelmas even.

New fermer may enter (as champions say)
  on all that is fallow, at Lent ladie day:
In Wood and and, old fermer to that will not yeeld,
  for loosing of pasture, and feede of his feeld.

Provide against Mihelmas, oargaine to make,
  for ferme to give over, to keepe or to take:
Booming of either, let wit beare a stroke,
  For bueing or selling of a pig in a poke.

Good ferme and well stored, good housing and drie,
  good corne and good dairie, good market and nie:
Good shepheard, good tilman, good Jack and good Gil,
  makes husband and huswife their cofers to fil.

Let pasture be stored, and fenced about,
  and tillage set forward as needeth without:
  fore ye doe open your purse to begin,
  with anything dooing for fancie within.

No storing of pasture with beggedglie tit,
    with ragged, with aged and evil athit:
Let carren and barren be shifted awaie,
    for best is the best, whatsoever ye paie.

Horse, Oxen, plough, tumbrel, cart, waggon, & waine,
    the lighter and stronger, the greater thy game.
The soile and the seede, with the sheafe and the purse,
    the lighter in substance, for profite the wurse.

To borow to daie and to-morrow to mis,
    for lender and borower, noiance it is:
Then have of thine owne, without lending unspilt,
    What followeth needfull, here learne if thou wilt.

A Digression to
Husbandlie Furniture

Barne locked, gofe ladder, short pitchforke and long,
    flaile, strawforke and rake, with a fan that is strong:
Wing, cartnave and bushel, peck, strike readie hand,
    get casting sholve, broome, and a sack with a band.

A stable wel planked, with key and a lock,
    walles stronglie wel lyned, to beare off a knock:
A rack and a manger, good litter and haie,
    sweete chaffe and some provender everie daie.

A pitchfork, a doongfork, seeve, skep and a bin,
    a broome and a paile to put water therein:
A handbarow, wheelebarow, sholve and a spade,
    a currie combe, mainecombe, and whip for a Jade.

A buttrice and pincers, a hammer and naile,
    an aperne and siszers for head and for taile:
Hole bridle and saddle, whit lether and nall,
    with collers and harneis, for thiller and all.

A panel and wantey, packsaddle and ped,
    A line to fetch litter, and halters for hed.
With crotchis and pinnes, to hang trinkets theron,
    and stable fast chained, that nothing be gon.

Strong exeltred cart, that is clouted and shod,
    cart ladder and wimble, with percer and pod:
Wheele ladder for harvest, light pitchfork and tough,
    shave, whiplash wel knotted, and cartrope ynough.

Ten sacks, whereof everie one holdeth a coome,
    a pulling hooke handsome, for bushes and broome:
Light tumbrel and doong crone, for easing sir wag,
    sholve, pickax, and mattock, with bottle and bag.

A grinstone, a whetstone, a hatchet and bil,
    with hamer and english naile, sorted with skil:
A frower of iron, for cleaving of lath,
    with roule for a sawpit, good husbandrie hath.

A short saw and long saw, to cut a too logs,
    an ax and a nads, to make troffe for thy hogs:
A Dovercourt beetle, and wedges with steele,
    strong lever to raise up the block fro the wheele.

Two ploughs and a plough chein, ij culters, iij shares,
    with ground cloutes & side clouts for soile that so tares:
With ox bowes and oxyokes, and other things mo,
    for oxteeme and horseteeme, in plough for to go.

A plough beetle, ploughstaff, to further the plough,
    great clod to a sunder that breaketh so rough;
A sled for a plough, and another for blocks,
    for chimney in winter, to burne up their docks.

Sedge collers for ploughhorse, for lightnes of neck,
    good seede and good sower, and also seede peck:
Strong oxen and horses, wel shod and wel clad,
    wel meated and used, for making thee sad.

A barlie rake toothed, with yron and steele,
    like paier of harrowes, and roler doth weele:
A sling for a moether, a bowe for a boy.
    a whip for a carter, is hoigh de la roy.

A brush sithe and grasse sithe, with rifle to stand,
    a cradle for barlie, with rubstone and sand:
Sharpe sikle and weeding hooke, haie fork and rake,
    a meake for the pease, and to swinge up the brake.

Short rakes for to gather up barlie to binde,
    and greater to rake up such leavings behinde:
A rake for to hale up the fitchis that lie,
    a pike for to pike them up handsom to drie.

A skuttle or skreine, to rid soile fro the corne,
    and sharing sheares readie for sheepe to be shorne:
A fork and a hooke, to be tampring in claie,
    a lath hammer, trowel, a hod, or a traie.

Strong yoke for a hog, with a twicher and rings,
    with tar in a tarpot, for dangerous things:
A sheepe marke, a tar kettle, little or mitch,
    two pottles of tar to a pottle of pitch.

Long ladder to hang al along by the wal,
    to reach for a neede to the top of thy hal:
Beame, scales, with the weights, that be sealed and true,
    sharp moulspare with barbs, that the mowles do so rue.

Sharpe cutting spade, for the deviding of mow,
    with skuppat and skavel, that marsh men alow:
A sickle to cut with, a didall and crome
    for draining of ditches, that noies thee at home.

A clavestock and rabetstock, carpenters crave,
    and seasoned timber, for pinwood to have:
A Jack for to saw upon fewell for fier,
    for sparing of firewood, and sticks fro the mier.

Soles, fetters, and shackles, with horselock and pad,
    a cow house for winter, so meete to be had:
A stie for a bore, and a hogscote for hog,
    a roost for thy hennes, and a couch for thy dog.

                    Here endeth husbandlie furniture.

Thresh seed and to fanning, September doth crie,
    get plough to the field, and be sowing of rie:
To harrow the rydgis, er ever ye strike,
    is one peece of husbandrie Suffolk doth like.

Sowe timely thy whitewheat, sowe rie in the dust,
    let seede have his longing, let soile have hir lust:
Let rie be partaker of Mihelmas spring,
    to beare out the hardnes that winter doth bring.

Some mixeth to miller the rie with the wheat,
    Temmes lofe on his table to have for to eate:
But so we it not mixed, to growe so on land,
    least rie tarie wheat, till it shed as it stand.

If soile doe desire to have rie with the wheat,
    by growing togither, for safetie more great,
Let white wheat be ton, be it deere, be it cheape,
    the sooner to ripe, for the sickle to reape.

Though beanes be in sowing but scattered in,
    yet wheat, rie, and peason, I love not too thin:
Sowe barlie and dredge, with a plentifull hand,
    least weede, steed of seede, over groweth thy land.

No sooner a sowing, but out by and by,
    with mother or boy that Alarum can cry:
And let them be armed with sling or with bowe,
    to skare away piggen, the rooke and the crowe.

Seed sowen, draw a forrough, the water to draine,
    and dike up such ends as in harmes doe remaine:
For driving of cattell or roving that waie,
    which being prevented, ye hinder their praie.

Saint Mihel doth bid thee amend the marsh wal,
    the brecke and the crab hole, the foreland and al:
One noble in season bestowed thereon,
    may save thee a hundred er winter be gon.

Now geld with the gelder the ram and the bul,
    sew ponds, amend dammes, and sel webster thy wul:
Out fruit go and gather, but not in the deaw,
    with crab and the walnut, for feare of a shreaw

The Moone in the wane, gather fruit for to last,
    but winter fruit gather when Mihel is past:
Though michers that love not to buy nor to crave,
    makes some gather sooner, else few for to have.

Fruit gathred too timely wil taste of the wood,
    wil shrink and be bitter, and seldome proove good:
So fruit that is shaken, or beat off a tree,
    with brusing in falling, soone faultie wil bee.

Now burne up the bees that ye mind for to drive,
    at Midsomer drive them and save them alive:
Place hive in good ayer, set southly and warme,
    and take in due season wax, honie, and swarme.

Set hive on a plank, (not too low by the ground)
    where herbe with the flowers may compas it round:
And boordes to defend it from north and north east,
    from showers and rubbish, from vermin and beast.

At Mihelmas safely go stie up thy Bore,
    least straying abrode, ye doo see him no more:
The sooner the better for Halontide nie,
    and better he brawneth if hard he doo lie.

Shift bore (for il aire) as best ye do thinke,
    and twise a day give him fresh vittle and drinke:
And diligent Cislye, my dayrie good wench,
    make cleanly his cabben, for measling and stench.

Now pluck up thy hempe, and go beat out the seed,
    and afterward water it as ye see need:
But not in the river where cattle should drinke,
    for poisoning them and the people with stinke.

Hempe huswifely used lookes cleerely and bright,
    and selleth it selfe by the colour so whight:
Some useth to water it, some do it not,
    be skilful in dooing, for feare it do rot.

Wife, into thy garden, and set me a plot,
    with strawbery rootes, of the best to be got:
Such growing abroade, among thornes in the wood,
    wel chosen and picked proove excellent good.

The Barbery, Respis, and Goosebery too,
    looke now to be planted as other things doo:
The Goosebery, Respis, and Roses, al three,
    with Strawberies under them trimly agree.

To gather some mast, it shal stand thee upon,
    with servant and children, er mast be al gon:
Some left among bushes shal pleasure thy swine,
    for feare of a mischiefe keepe acrons fro kine.

For rooting of pasture ring hog ye had neede,
    which being wel ringled the better do feede:
Though yong with their elders wil lightly keepe best,
    yet spare not to ringle both great and the rest.

Yoke seldom thy swine while the shacktime doth last,
    for divers misfortunes that happen too fast:
Or if ye do fancie whole eare of the hog,
    give eie to il neighbour and eare to his dog.

Keepe hog I advise thee from medow and corne,
    for out aloude crying that ere he was borne:
Such lawles, so haunting, both often and long,
    if dog set him chaunting he doth thee no wrong.

Where love among neighbors do beare any stroke,
    while shacktime indureth men use not to yoke:
Yet surely ringling is needeful and good,
    til frost do enuite them to brakes in the wood.

Get home with thy brakes, er an sommer be gon,
    for teddered cattle to sit there upon:
To cover thy hovel, to brewe and to bake,
    to lie in the bottome, where hovel ye make.

Now sawe out thy timber, for boord and for pale,
    to have it unshaken, and ready to sale:
Bestowe it and stick it, and lay it aright,
    to find it in March, to be ready in plight.

Save slab of thy timber for stable and stie,
    for horse and for hog the more clenly to lie:
Save sawe dust, and brick dust, and ashes so fine,
    for alley to walke in, with neighbour of thine.

Keepe safely and warely thine uttermost fence,
    with ope gap and brake hedge do seldome dispence:
Such runabout prowlers, by night and by day,
    see punished justly for prowling away.

At noone if it bloweth, at night if it shine,
    out trudgeth Hew make shift, with hooke & with line:
Whiles Gillet, his blouse, is a milking thy cow,
    Sir Hew is a rigging thy gate or the plow.

Such walke with a black or a red little cur,
    that open wil quickly, if anything stur;
Then squatteth the master, or trudgeth away,
    and after dog runneth as fast as he may.

Some prowleth for fewel, and some away rig
    fat goose, and the capon, duck, hen, and the pig:
Some prowleth for acornes, to fat up their swine,
    for corne and for apples, and al that is thine.

Thus endeth Septembers husbandrie.

Octobers Husbandrie

Now lay up thy barley land, drie as ye can,
    when ever ye sowe it so looke for it than:
Get daily aforehand, be never behinde;
    least winter preventing do alter thy minde.

Who laieth up fallow too soone or too wet,
    with noiances many doth barley beset.
For weede and the water so soketh and sucks,
    that goodnes from either it utterly plucks.

Greene rie in September when timely thou hast,
    October for wheat sowing calleth as fast.
If weather will suffer, this counsell I give,
    Leave sowing of wheat before Hallomas eve.

Where wheat upon edish ye mind to bestowe,
  let that be the first of the wheat ye do sowe:
He seemeth to hart it and comfort to bring,
  that giveth it comfort of Mihelmas spring.

White wheat upon peaseetch doth grow as he wold,
  but fallow is best, if we did as we shold:
Yet where, how, and when, ye entend to begin,
  let ever the finest be first sowen in.

Who soweth in raine, he shall reape it with teares,
  who soweth in harmes, he is ever in feares,
Who soweth ill seede or defraudeth his land,
  hath eie sore abroode, with a coresie at hand.

Seede husbandly sowen, water furrow thy ground,
  that raine when it commeth may run away round,
Then stir about Nicoll, with arrow and bowe,
  take penie for killing of everie crowe.

A Digression to the Usage of Divers Countries

Each soile hath no liking of everie graine,
  nor barlie and wheat is for everie vaine:
Yet knowe I no countrie so barren of soile
  but some kind of corne may be gotten with toile.

In Brantham, where rie but no barlie did growe,
  good barlie I had, as a meany did knowe:
Five seame of an aker I truely was paid,
  for thirtie lode muck of each aker so laid.

In Suffolke againe, where as wheat never grew,
  good husbandrie used good wheat land I knew:
This Proverbe experience long ago gave,
  that nothing who practiseth nothing shall have.

As gravell and sand is for rie and not wheat,
  (or yeeldeth hir burden to tone the more great,)
So peason and barlie delight not in sand,
  but rather in claie or in rottener land.

Wheat somtime is steelie or burnt as it growes,
  for pride or for povertie practise so knowes.
Too lustie of courage for wheat doth not well,
  nor after sir peeler he looveth to dwell.

Much wetnes, hog rooting, and land out of hart,
  makes thistles a number foorthwith to upstart.
If thistles so growing proove lustie and long,
  it signifieth land to be hartie and strong.

As land full of tilth and in hartie good plight,
    yeelds blade to a length and encreaseth in might,
So crop upon crop, upon whose courage we doubt,
    yeelds blade for a brag, but it holdeth not out.

The straw and the eare to have bignes and length,
    betokeneth land to be good and in strength.
If eare be but short, and the straw be but small,
    it signifieth barenes and barren withall.

White wheat or else red, red rivet or whight,
    far passeth all other, for land that is light.
White pollard or red, that so richly is set,
    for land that is heavie is best ye can get.

Maine wheat that is mixed with white and with red
    is next to the best in the market mans hed:
So Turkey or Purkey wheat many doe love,
    because it is flourie, as others above.

Graie wheat is the grosest, yet good for the clay,
    though woorst for the market, as fermer may say.
Much like unto rie be his properties found,
    coorse flower, much bran, and a peeler of ground.

Otes, rie, or else barlie, and wheat that is gray,
    brings land out of comfort, and soone to decay:
One after another, no comfort betweene,
    is crop upon crop, as will quickly be seene.

Still crop upon crop many fermers do take,
    and reape little profit for greedines sake.
Though breadcorne & drinkcorn such croppers do stand:
    count peason or brank, as a comfort to land.

Good land that is severall, crops may have three,
    in champion countrie it may not so bee:
Ton taketh his season, as commoners may,
    the tother with reason may otherwise say.

Some useth at first a good fallow to make,
    to sowe thereon barlie, the better to take.
Next that to sowe pease, and of that to sowe wheat,
    then fallow againe, or lie lay for thy neat.

First rie, and then barlie, the champion saies,
    or wheat before barlie be champion waies:
But drinke before bread corne with Middlesex men,
    then lay on more compas, and fallow agen.

Where barlie ye sowe, after rie or else wheat,
  if land be unlustie, the crop is not great,
So lose ye your cost, to your coresie and smart,
  and land (overburdened) is cleane out of hart.

Exceptions take of the champion land,
  from lieng alonge from that at thy hand.
(Just by) ye may comfort with compas at will,
  far off ye must comfort with favor and skill.

Where rie or else wheat either barlie ye sowe,
  let codware be next, thereupon for to growe:
Thus having two crops, whereof codware is ton,
  thou hast the lesse neede, to lay cost thereupon.

Some far fro the market delight not in pease,
  for that ery chapman they seeme not to please.
If vent of the market place serve thee not well,
  set hogs up a fatting, to drover to sell.

Two crops of a fallow enricheth the plough,
  though tone be of pease, it is land good ynough:
One crop and a fallow some soile will abide,
  where if ye go furder lay profit aside.

Where peason ye had and a fallow thereon,
  sowe wheat ye may well without doong thereupon:
New broken upland, or with water opprest,
  or over much doonged, for wheat is not best.

Where water all winter annoieth too much,
  bestowe not thy wheat upon land that is such:
But rather sowe otes, or else bullimong there,
  gray peason, or runcivals, fitches, or tere.

Sowe acornes ye owners, that timber doe loove,
  powe hawe and rie with them the better to proove;
If cattel or cunnie may enter to crop,
  yong oke is in daunger of loosing his top.

Who pescods delighteth to have with the furst,
  if now he do sowe them, I thinke it not wurst.
The greener thy peason and warmer the roome,
  more lusty the layer, more plenty they come.

Go plow up or delve up, advised with skill,
  the bredth of a ridge, and in length as you will.
Where speedy quickset for a fence ye wil drawe,
  to sowe in the seede of the bremble and hawe.

Through plenty of acornes, the porkling to fat,
  not taken in season, may perish by that,
If ratling or swelling get once to the throte,
  thou loosest thy porkling, a crowne to a grote.

What ever thing fat is, againe if it fall,
  thou ventrest the thing and the fatnes withall,
The fatter the better, to sell or to kil,
  but not to continue, make proofe if ye wil.

What ever thing dieth, go burie or burne,
  for tainting of ground, or a woorser il turne.
Such pestilent smell of a carrenly thing,
  to cattle and people great peril may bring.

Thy measeled bacon, hog, sow, or thy bore,
  shut up for to heale, for infecting thy store:
Or kill it for bacon, or sowce it to sell,
  for Flemming, that loves it so deintily well.

With strawisp and peasebolt, with ferne & the brake,
  for sparing of fewel, some brewe and do bake,
And heateth their copper, for seething of graines:
  good servant rewarded, refuseth no paines.

Good breadcorne and drinkcorne, full xx weekes kept,
  is better then new, that at harvest is rept:
But foisty the breadcorne and bowd eaten malt,
  for health or for profit, find noysome thou shalt.

By thend of October, go gather up sloes,
  have thou in a readines plentie of thoes,
And keepe them in bedstraw, or still on the bow,
  to staie both the flixe of thyselfe and thy cow.

Seeith water and plump therein plenty of sloes,
  mix chalke that is dried in powder with thoes.
Which so, if ye give, with the water and chalke,
  thou makest the laxe fro thy cow away walke.

Be sure of vergis (a gallond at least)
  so good for the kitchen, so needfull for beast,
It helpeth thy cattel, so feeble and faint,
  if timely such cattle with it thou acquaint.

Thus endeth Octobers husbandrie.

Novembers Husbandrie

At Hallontide, slaughter time entereth in,
   and then doth the husbandmans feasting begin:
From thence unto shroftide kill now and then some,
   their offal for houshold the better wil come.

Thy dredge and thy barley go thresh out to malt,
   let malster be cunning, else lose it thou shalt:
Thencrease of a seame is a bushel for store,
   bad else is the barley, or huswife much more.

Some useth to winnow, some useth to fan,
   some useth to cast it as cleane as they can:
For seede goe and cast it, for malting not so,
   but get out the cockle, and then let it go.

Thresh barlie as yet but as neede shal require,
   fresh threshed for stoover thy cattel desire:
And therefore that threshing forbeare as ye may,
   till Candelmas comming, for sparing of hay.

Such wheat as ye keepe for the baker to buie,
   unthreshed til March in the sheafe let it lie,
Least foistnes take it if sooner yee thresh it,
   although by oft turning ye seeme to refresh it.

Save chaffe of the barlie, of wheate, and of rie,
   from feathers and foistines, where it doth lie,
Which mixed with corne, being sifted of dust,
   go give to they cattel, when serve them ye must.

Greene peason or hastings at Hallontide sowe,
   in hartie good soile he requireth to growe:
Graie peason or runcivals cheerely to stand,
   at Candlemas sowe, with a plentifull hand.

Leave latewardly rering, keepe now no more swine,
   but such as thou maist, with the offal of thine:
Except ye have wherewith to fat them away,
   the fewer thou keepest, keepe better yee may.

To rere up much pultrie, and want the barne doore,
   is naught for the pulter and woorse for the poore.
So, now to keepe hogs and to sterve them for meate,
   is as to keepe dogs for to bawle in the streate.

As cat a good mouser is needfull in house,
  because for hir commons she killeth the mouse,
So ravening curres, as a meany doo keepe,
  makes master want meat, and his dog to kill sheepe.

(For Easter) at Martilmas hang up a beefe,
  for stalfed and pease fed plaie pickpurse the theefe:
With that and the like, er an grasse biefe come in,
  thy folke shal looke cheerelie when others looke thin.

Set garlike and beanes, at S. Edmond the king,
  the moone in the wane, thereon hangeth a thing:
Thencrease of a pottle (wel prooved of some)
    shal pleasure thy household er peskod time come.

When raine is a let to thy dooings abrode,
  set threshers a threshing to laie on good lode:
Thresh cleane ye must bid them, though lesser they yarn,
  and looking to thrive, have an eie to thy barne.

Take heede to thy man in his furie and heate,
  with ploughstaff and whipstock, for maiming thy neate:
To thresher for hurting of cow with his flaile,
  or making thy hen to plaie tapple up taile.

Some pilfering thresher will walke with a staffe,
  will carrie home corne as it is in the chaffe,
And some in his bottle of leather so great
  will carry home daily both barlie and wheat.

If houseroome will serve thee, lay stover up drie,
  and everie sort by it selfe for to lie.
Or stack it for litter, if roome be too poore,
  and thatch out the residue noieng thy doore.

Cause weekly thy thresher to make up his flower,
  though slothfull and pilferer thereat doo lower:
Take tub for a season, take sack for a shift,
  yet garner for graine is the better for thrift.

All maner of strawe that is scattered in yard,
  good husbandlie husbands have daily regard,
In pit full of water the same to bestowe,
  where lieng to rot, thereof profit may growe.

Now plough up thy hedlond, or delve it with spade,
  where otherwise profit but little is made:
And cast it up high, upon hillocks to stand,
  that winter may rot it, to compas thy land.

If garden requier it, now trench it ye may,
    one trench not a yard from another go lay:
Which being well filled with muck by and by,
    go cover with mould for a season to ly.

Foule privies are now to be clensed and fide,
    let night be appointed such baggage to hide:
Which buried in garden, in trenches alowe,
    shall make very many things better to growe.

The chimney all sootie would now be made cleene,
    for feare of mischances, too oftentimes seene:
Old chimney and sootie, if fier once take,
    by burning and breaking, soone mischeefe may make.

When ploughing is ended, and pasture not great,
    then stable thy horses, and tend them with meat:
Let season be drie when ye take them to house,
    for danger of nittes, or for feare of a louse.

Lay compas up handsomly, round on a hill,
    to walke in thy yard at thy pleasure and will,
More compas it maketh and handsom the plot,
    if horsekeeper daily forgetteth it not.

Make hillocks of molehils, in field thorough out,
    and so to remaine, till the yeere go about.
Make also the like whereas plots be too hie,
    all winter a rotting for compas to lie.

Thus endeth Novembers husbandrie.

*   *   *

Julies Husbandrie

Go muster thy servants, be captaine thy selfe,
    providing them weapon and other like pelfe.
Get bottles and walletts, keepe field in the heat,
    the feare is as much, as the danger is great.

With tossing and raking and setting on cox,
    grasse latelie in swathes is hay for an ox:
That done, go and cart it and have it away,
    the battel is fought, ye have gotten the day.

Pay justly thy tithes whatsoever thou bee,
    that God may in blessing send foison to thee.
Though Vicar be bad, or the Parson as evill,
    go not for thy tithing thy selfe to the Devill.

Let hay be well made, or avise else avouse,
　　for molding in goef, or of firing the house.
Lay coursest aside for the ox and the cow,
　　the finest for sheepe and thy gelding alow.

Then downe with the hedlonds, that groweth about,
　　leave never a dallop unmowne and had out.
Though grasse be but thin, about barlie and pease,
　　yet picked up cleane ye shall find therein ease.

Thry fallow betime, for destroieng of weede,
　　least thistle and duck fall a blooming and seede,
Such season may chance, it shall stand thee upon,
　　to till it againe, er an Sommer be gon.

Not rent off, but cut off, ripe beane with a knife,
　　for hindering stalke of hir vegetive life.
So gather the lowest, and leaving the top,
　　shall teach thee a trick, for to double thy crop.

Wife, pluck fro thy seed hemp the fiemble hemp clene,
　　this looketh more yellow, the other more grene:
Use ton for thy spinning, leave Mihel the tother,
　　for shoo thred and halter, for rope and such other.

Now pluck up thy flax, for the maidens to spin,
　　first see it dried, and timelie got in.
And mowe up thy branke, and away with it drie,
　　and howse it up close, out of danger to lie.

While wormwood hath seed, get a handful or twaine,
　　to save against March to make flea to refraine:
Where chamber is sweeped, and wormwood is strowne,
　　no flea for his life dare abide to be knowne.

What saver is better (if physick be true),
　　for places infected, than wormwood and rue.
It is as a comfort for hart and the braine,
　　and therefore to have it, it is not in vaine.

Get grist to the mill, to have plentie in store,
　　least miller lack water, as many doo more.
The meale the more yeeldeth, if servant be true,
　　and miller that tolleth, take none but his due.

Thus endeth Julies husbandrie.

*　*　*

Augusts Husbandrie

Thry fallow once ended, go strike by and by,
    both wheat land and barlie, and so let it ly.
And as ye have leisure, go compas the same,
    when up ye doo lay it, more fruitfull to frame.

Get downe with thy brakes, er an showers doo come,
    that cattle the better may pasture have some.
In June and in August, as well doth appeere,
    is best to mowe brakes, of all times in the yeere.

Pare saffron betweene the two S. Maries daies,
    or set or go shift it, that knowest the waies.
What yeere shall I doo it (more profit to yeeld?)
    the fourth in garden, the third in the feeld.

In having but fortie foote workmanly dight,
    take saffron ynough for a Lord and a knight.
All winter time alter as practise doth teach,
    what plot have ye better, for linnen to bleach.

Maides, mustard seede gather, for being too ripe,
    and weather it well, er ye give it a stripe:
Then dresse it and laie it in soller up sweete,
    least foistines make it for table unmeete.

Good huswifes in sommer will save their owne seedes,
    against the next yeere, as occasion needes.
One seede for another, to make an exchange,
    with fellowlie neighbourhood seemeth not strange.

Make sure of reapers, get harvest in hand,
    the corne that is ripe, doo but shed as it stand.
Be thankfull to God, for his benefits sent,
    and willing to save it with earnest intent.

To let out thy harvest, by great or by day,
    let this by experience leade thee a way.
By great will deceive thee, with lingring it out,
    by day will dispatch, and put all out of dout.

Grant harvest lord more by a penie or twoo,
    to call on his fellowes the better to doo:
Give gloves to thy reapers, a larges to crie,
    and dailie to loiterers have a good eie.

Reape wel, scatter not, gather cleane that is shorne,
    binde fast, shock apace, have an eie to thy corne.
Lode safe, carrie home, follow time being faire,
    gove just in the barne, it is out of despaire.

Tithe dulie and trulie, with hartie good will,
    that God and his blessing may dwell with thee still:
Though Parson neglecteth his dutie for this,
    thanke thou thy Lord God, and give erie man his.

Corne tithed (sir Parson) to gather go get,
    and cause it on shocks to be by and by set:
Not leaving it scattering abrode on the ground,
    nor long in the field, but away with it round.

To cart gap and barne, set a guide to looke weele,
    and hoy out (sir carter) the hog fro thy wheele:
Least greedie of feeding, in following cart,
    it noieth or perisheth, spight of thy hart.

In champion countrie a pleasure they take,
    to mowe up their hawme, for to brew and to bake.
And also it stands them in steade of their thack,
    which being well inned, they cannot well lack.

The hawme is the strawe of the wheat or the rie,
    which once being reaped, they mowe by and bie:
For feare of destroieng with cattle or raine,
    the sooner ye lode it, more profit ye gaine.

The mowing of barlie, if barlie doo stand,
    is cheapest and best, for to rid out of hand:
Some mowe it and rake it, and sets it on cocks,
    some mowe it and binds it, and sets in on shocks.

Of barlie the longest and greenest ye find,
    leave standing by dallops, till time ye doo bind:
Then early in morning (while deaw is thereon),
    to making of bands till the deaw be all gon.

One spreadeth those bands, so in order to ly,
    as barlie (in swatches) may fill it thereby:
Which gathered up, with the rake and the hand,
    the follower after them bindeth in band.

Where barlie is raked (if dealing be true),
    the tenth of such raking to Parson is due:
Where scatring of barlie is seene to be much,
    there custome nor conscience tithing should gruch.

Corne being had downe (any way ye alow),
    should wither as needeth, for burning in mow:
Such skill appertaineth to harvest mans art,
    and taken in time is a husbandly part.

No turning of peason till carrege ye make,
    nor turne in no more, than ye mind for to take:
Least beaten with showers so turned to drie,
    by turning and tossing they shed as they lie.

If weather be faire, and tidie thy graine,
    make speedily carrege, for feare of a raine:
For tempest and showers deceiveth a menie,
    and lingering lubbers loose many a penie.

In goving at harvest, learne skilfully how
    ech graine for to laie, by it selfe on a mow:
Seede barlie the purest, gove out of the way,
    all other nigh hand gove as just as ye may.

Stack pease upon hovell abrode in the yard,
    to cover it quicklie, let owner regard:
Least Dove and the cadow, there finding a smack,
    with ill stormie weather doo perish thy stack.

Corne carred, let such as be poore go and gleane,
    and after, thy cattle to mowth it up cleane.
Then spare it for rowen, till Mihel be past,
    to lengthen thy dairie no better thou hast.

In harvest time, harvest folke, servants and all,
    should make all togither good cheere in the hall:
And fill out the black boule of bleith to their song,
    and let them be merie all harvest time long.

Once ended thy harvest, let none be begilde,
    please such as did helpe thee, man, woman, and childe.
Thus dooing, with alway such helpe as they can,
    thou winnest the praise of the labouring man.

Now looke up to Godward, let tong never cease
    in thanking of him, for his mightie encrease:
Accept my good will, for a proofe go and trie:
    the better thou thrivest, the gladder am I.

Works After Harvest

Now carrie out compas, when harvest is donne,
    where barlie thou sowest, my champion sonne:
Or laie it on heape, in the field as ye may,
    till carriage be faire, to have it away.

Whose compas is rotten and carried in time,
   and spred as it should be, thrifts ladder may clime.
Whose compas is paltrie and carried too late,
   such husbandrie useth that many doo hate.

Er winter preventeth, while weather is good,
   for galling of pasture get home with thy wood.
And carrie out gravell to fill up a hole:
   both timber and furzen, the turfe and the cole.

Howse charcole and sedge, chip and cole of the land,
   pile tallwood and billet, stacke all that hath band.
Blocks, rootes, pole and bough, set upright to the thetch:
   the neerer more handsome in winter to fetch.

In stacking of baven, and piling of logs,
   make under thy baven a hovell for hogs,
And warmelie enclose it, all saving the mouth,
   and that to stand open, and full to the south.

Once harvest dispatched, get wenches and boies,
   and into the barne, afore all other toies.
Choised seede to be picked and trimlie well fide,
   for seede may no longer from threshing abide.

Get seede aforehand, in a readines had,
   or better provide, if thine owne be too bad.
Be carefull of seede, or else such as ye sowe,
   be sure at harvest, to reape or to mowe.

When harvest is ended, take shipping or ride,
   Ling, Saltfish and Herring, for Lent to provide.
To buie it at first, as it commeth to rode,
   shall paie for thy charges thou spendest abrode.

Choose skilfullie Saltfish, not burnt at the stone,
   buie such as be good, or else let it alone.
Get home that is bought, and goe stack it up drie,
   with peasestrawe betweene it, the safer to lie.

Er ever ye jornie, cause servant with speede
   to compas thy barlie land where it is neede.
One aker well compassed, passeth some three,
   thy barne shall at harvest declare it to thee.

This lesson is learned by riding about,
   the prices of vittels, the yeere thorough out.
Both what to be selling and what to refraine,
   and what to be buieng, to bring in againe.

Though buieng and selling doth woonderfull well,
    to such as have skill how to buie and to sell:
Yet chopping and changing I cannot commend,
    with theefe and his marrow, for feare of ill end.

The rich in his bargaining needes not be tought,
    of buier and seller full far is he sought.
Yet herein consisteth a part of my text,
    who buieth at first hand, and who at the next.

At first hand he buieth that paieth all downe,
    at second, that hath not so much in the towne,
At third hand he buieth that buieth of trust,
    at his hand who buieth shall paie for his lust.

As oft as ye bargaine, for better or wurse,
    to buie it the cheaper, have chinkes in thy purse:
Touch kept is commended, yet credit to keepe,
    is paie and dispatch him, er ever ye sleepe.

Be mindfull abrode of Mihelmas spring,
    for thereon dependeth a husbandlie thing:
Though some have a pleasure, with hauke upon hand,
    good husbands get treasure, to purchase their land.

Thy market dispatched, turne home againe round,
    least gaping for penie, thou loosest a pound:
Provide for thy wife, or else looke to be shent,
    good milch cow for winter, another for Lent.

In traveling homeward, buie fort ie good crones,
    and fat up the bodies of those seelie bones.
Leave milking and drie up old mulley thy cow,
    the crooked and aged, to fatting put now.

At Bartilmewtide, or at Sturbridge faire,
    buie that as is needfull, thy house to repaire:
Then sell to thy profit, both butter and cheese,
    who buieth it sooner, the more he shall leese.

If hops doo looke brownish, then are ye too slowe,
    if longer ye suffer those hops for to growe.
Now sooner ye gather, more profit is found,
    if weather be faire and deaw of a ground.

Not breake off, but cut off, from hop the hop string,
    leave growing a little againe for to spring.
Whose hill about pared, and therewith new clad,
    shall nourish more sets against March to be had.

Hop hillock discharged of everie let,
    see then without breaking, ech pole ye out get.
Which being untangled above in the tops,
    go carrie to such as are plucking of hops.

Take soutage or haier (that covers the kell),
    set like to a manger and fastened well:
With poles upon crotchis as high as thy brest,
    for saving and riddance is husbandrie best.

Hops had, the hop poles that are likelie preserve,
    (from breaking and rotting) againe for to serve:
And plant ye with alders or willowes a plot,
    where yeerelie as needeth mo poles may be got.

Some skilfullie drieth their hops on a kell,
    and some on a soller, oft turning them well.
Kell dried will abide, foule weather or faire,
    where drieng and lieng in loft doo dispaire.

Some close them up drie in a hogshed or fat,
    yet canvas or soutage is better than that:
By drieng and lieng they quickly be spilt:
    thus much have I shewed, doo now as thou wilt.

Old fermer is forced long August to make,
    his goodes at more leisure away for to take.
New fermer he thinketh ech houre a day,
    untill the old fermer be packing away.

> *Thus endeth and holdeth out Augusts husbandrie,*
> *till Mihelmas Eve.*

The Countrie Farme, 1616

From Charles Stevens and John Liebault, *Maison Rustique, or, The Country Farme*, translated from the French by Richard Surflet and revised by Gervase Markham (London, 1616. Copy in James M. Gwin Collection, National Agricultural Library), pp. 35–39, 532–33.

The Particular Workes That a Husbandman Must be Carefull to Doe Everie Moneth in the Yeare

Furthermore, to the end that his people may not live idle, and that they may not loose one small minute of time; which being imployed about some one or other worke, he shall dispose of his workes so, as that they may everie one have his certaine time, and he shall know at his fingers ends what things is to be done everie moneth and time of the yeare. Yet thus ever to governe his memorie, that there labours following being more naturall to the Kingdome of France than to any of her neighbours, they shall, for their satisfaction, because the Booke is now intended generall, returne to the sixt Chapter, and there behold the convenient labours fit for colder Countries, as is the Island of Great Britaine, Ireland, and the Low Countries.

In the moneth of Ianuarie, chiefely toward the end, hee shall cut downe his Wood which hee appointeth for Building, or other Worke, when the Moone is vnder the Earth: for the brightnesse of the Moone maketh the Wood more tender, and the Wood which shall be cut at such time will endure a long time without rotting. He shall dung the Fruit-trees, not letting the dung touch their roots. He shall graft all such great and little Trees which bud betimes, as Rose-trees, Damaske, Plum-trees, Apricock-trees, Almond-trees, and Cherry-trees. He shall digge the Earth for the casting in of Nuts, Almonds, and the kernels of Apricockes, Peaches, and Plums, and such others, in grounds that are cold and moist, in the two first quarters of the Moone. Hee shall cut his Vine in faire and beautifull Weather. Hee shall plough the grounds that are drie, light, white, leane, sandie, full of roots and great hearbes, and which were not eared in October. Hee shall give the second care vnto those his grounds that are most barren, and scatter vpon them the chaffe of Beanes, Wheat, or Barley. Hee shall cut downe the boughes of the Willowes for Railes for Vines and Stakes for Hedges. Hee shall prepare props and thicke square Laths to vnderset his Vines. Hee shall cut and take away the superfluous boughes of the Trees, the Moone decreasing. Hee shall turne the vppermost of all the dung made since S. *Martins* day vndermost, and contrarily, to the end it may be well rotten when hee shall carrie it out to spread it vpon his Field and Medow. Hee shall furnish afresh or make new his Carts, Tumbrels, Ploughes, and other his Instruments necessarie for his Husbandrie. Hee shall make prouisions of verie sharpe yron tooles to cut and cleanse his Trees and Vines. Aboue all things, let them beware of Sowing, because the Earth as then is too open, heauie, full of vapours, and like vnto Wooll not well carded.

In Februarie in the new Moone he shall transplant Vines of two or three yeare, which shall now alreadie haue taken good root, but he shall not touch them of one yeare, which will not be remoued because of the small strength which they haue as yet got. He shall carrie dung out into his Corne-fields, Vineyards, Medowes, and Gardens. Hee shall cast trenches for the planting of new Vines. Hee shall cut the roots of the Vines, and set square Laths or Props for the defending of them. Hee shall prune and cleanse the Trees of whatsoeuer is superfluous: Hee shall cleanse them from wormes, filthinesse, and wormeeatings, canker, and rottennesse, which are to be found in the drie leaues. Hee shall make readie his Garden-grounds to sow and set therein all manner of hearbes. Hee shall giue the Earth her second earing for the receiving of Beanes, Barley, Oates, Hempe, Millet, and such other Seed of small Pulse. He shall ouerlooke his Vines, especially those which he knoweth to be weake and tender. He shall repaire the Hedges of his Gardens. He shall plant woods for Timber-trees aud Talwood. He shall also plant the slips of Oliue trees, Pomegranate trees, Quince trees, Figge trees, Poplar trees, Willow trees, Elme trees, Osiers, and others, as well Fruit Trees as wild ones, which haue roots. Hee shall cleanse the Douehouse, Henne-house, and place where the Peacocks and Geese make their haunt, because that these Cattell in the end of this moneth begin to be hot, and to tread. Hee shall ouer-looke his Warren, to stoare it anew, and to handsome vp the Earths. Hee shall buy Bees: he shall make cleane their Hiues verie carefully, and kill their Kings. Hee shall buy Faulcons, Sparrow-hawkes, and other Birds of the prey, which he shall put into Mue in the end of this moneth.

In March, euen in the beginning of it, he shall sow Lyne, Woad (if it were not sowne in Februarie) Oates, Barley, Millet, Pannick, Hempe, Peason, Lentils, Tare euerlasting, Lupines, small kinds of Corne, as the Fetch, Fasels, and other such like bitter kinds of small Pulse. He shall giue a second earder vnto new plowed fallowes, which are now by this time well amended and dunged, so as that he may make them readie to sow. He shall weed his Corne: he shall get Grifts to graft, when the sap beginneth to climbe the Trees, and before that they put forth any buds. He shall plant these Fruits, great Nuts, Chesnuts, Almonds, small Nuts, Filberds, and the stones of Oliues and Apricocks, and diuers other Fruits. He shall sow diuers Nurseries with the kernels of Apples, Peares, Mulberries, and such other like Fruits. He shall plant such Hearbes as are set low and close by the ground, as the slips of Artichokes, Thistles necessarie for vse, Sage, Lauander, Rosemarie, Strawberrie, Gooseberrie-bush, Roses, Lillies, Citruls Cucumers, Melons and Pompions. He shall trim vp his Gardens as well for the Kitchin or commoditie, as that which is drawne into quarters, or for pleasure, and shall sow therein whatsoeuer necessarie Seeds. He shall cut and vncouer the roots of Vines and Fruit-trees, to the end they may bring forth more fruit. He shall put dung to the roots of the Trees: he shall gather vp the loppings to make Fuell of.

In Aprill, about S. *George* his day, you shall set abroad your Citron and Orenge Trees, as also all such other Trees as you had kept within house from S. *Martins* day, from which he shall remoue the earth from foot to foot, taking from them such roots as are put forth towards the vppermost part of the earth, as also all superfluous boughes, not suffering any one branch to exceed another either in breadth or height. He shall plant, if he haue not alreadie done it, Oliue trees, Pomegranate trees, Citron trees, and Mulberrie trees, and shall prune them carefully. He shall graft the Figge tree, Chesnut tree, Cherrie tree, and Orenge tree. He shall cut the new Vine, for at this time it endureth best to be cut. He shall be carefull to feed his Pigeons, because at this time they find but little in the fields. He shall put Horse to his Mares, the hee-Asse, to the shee-Asse, and Rams to the Ewes. He shall make cleane the Hiues of the Honey-flies, and shall kill the Butterflies, which abound when Mallowes are in flower.

In May hee shall water the Trees that are newly planted: hee shall sheare his Sheepe, fill vp his Wines, gather great store of Butter, and make much Cheese, geld his Calues, and begin to looke to his Bees and Silkewormes, of which he shall gather together a great number. He shall weed his Corne, cast the earth off his Vines the second time, vncouering and freeing their roots from the earth about them, to the end that the heat may not hurt them: he shall take away all the greene branches and tender boughes which beare no fruit: he shall crop the ouer-ranke boughes of Trees, he shall graft such Oliue trees as must be grafted in the bud.

In Iune hee shall make readie his Threshing floore, and cause it to be thorowly cleansed of straw, durt, and dust: he shall cut downe his Medowes, mow his Barley, crop his Vines, thresh his Corne to sow in Seed time.

In Iuly hee shall mow his Wheat and other graine vsed to make pottage of: hee shall graft in the bud: hee shall gather from Apple-trees and Peare-trees the faultie Apples and Peares, and those which doe ouer-charge the Trees: he shall digge his Vines againe the second time, and plucke vp from them the Grasse called Dogstooth: he shall lay eeuen and fill vp the earth where it is any where cleft or broken, to the end that the Sunne may not burne before hand the Vine: He shall cut downe such Wood as shall serue for his Fuell all the yeare long.

In August he shall pull his Line and Hempe; gather such fruits from off the Trees as he meaneth to preserue. Hee shall take away the leaues from about such Grapes as are slow and backward, to the end they may receiue and reape the more heat from the Sunne. He shall make his Veriuice. He shall digge the Earth to make Wells, or to find the heads of Fountaines, if he haue need. He shall thinke vpon making readie his Wine vessels and other things necessarie for his Vintage.

In September he shall giue his land that commeth to be tilled againe, after it hath beene fallow, the last earder. He shall sow his Wheat, Masling, Rye, and such like Corne. Hee shall gather his Vintage: beat downe Nuts: cut downe late Medow grounds, to haue the after-Crop. Hee shall gather stubble for the thatching of his house, and for fuell to the Ouen all the yeare. He shall cut away the branches of Madder, and gather the Seed to sow in the beginning of the March following. Hee shall gather the leaues of Woad, and order them in such sort, as that they may be made vp into balls, and he shall cause them to be dried in the Sunne, or at a fire not verie hot. He shall cut downe Rice and Millet.

In October he shall make his Wines, and turne them into Vessels. He shall bestow his Orenge, Ceron, and Pomegranate Trees in some couered place, to avoid the danger of the dominent Cold. He shall make his Honey and Waxe, and driue the old flees.

In Nouember he shall couch his Wines in his Cellar. He shall gather Acornes to feed Swine. Hee shall gather Chesnuts small and great, and such Garden-fruits as will keepe. He shall take Radish out of the Earth, taking off their leaues, and putting them vnder the Sand, to keepe them from the Frost. Hee shall lay bare the rootes of Artichokes, and couer them againe verie well, that the Frost may not perish them. Hee shall make Oyles. Hee shall make Hiues for Bees, Panniers, Dung-pous, and Baskets of Osier. Hee shall cut Willowes for to make Frames to beare vp Vines, and shall bind the Vines, and draw the climbing Poles from the Vines.

In December hee shall oftentimes visite his Fields, thereby to let out the water which may stand in them after great Raine. Hee shall cause water to runne through the old Medowes, and dung them if need be. Hee shall make prouision of Dung to manure his fallowes that are broken vp and tilled. Hee shall couer with dung the rootes of the Trees and Hearbes which he intendeth to keepe vnto the Spring. Hee shall cut off the boughes and heads of Willowes, Poplars, Saplings, and other Trees, to the end that their boughes may more speedily put forth and grow so soone as Winter shall be past. Hee shall cut downe his Wood as well to build withall as to make his fire with. Hee shall make readier his Nets to catch Birds, and to beset the Hares, when as the Fields shall be ycie, or couered with Snow, or ouerflowne with Waters in such sort, as that a man can doe no worke in them. Hee shall also occupie himselfe (as long as he pleaseth) in making a thousand pretie Instruments and necessarie things of Wood, as are Platters, Trenchers, Spindles, Bathing-Tubs, Dishes, and other things requisite for household store: as also Harrowes, Rakes, and Handles for these Tooles. He shall repaire his Teames, Yokes, Ploughes, and all other Instruments necessarie for the fitting and garnishing of Cattell going to Cart or Plough, to the end that all may be in good order when they are to goe to labour. He shall also make prouision of Spades, Shouels, Pickaxes, Peeles, Hatchets, Wedges, Sawes, and other furniture fit for a Countrie house store.

The Condition and State of a Huswife

I doe not find the state or place of a Huswife or Dairie-woman to be of lesse care and diligence than the office of her Husband, vnderstood alwayes, that the woman is acquited of Field matters, in as much as shee is tyed to matters within the House and base Court (the Horses excepted) as the husband is tyed to doe what concerneth him, euen all the businesses of the Field. Likewise, according to our custome of France, Countrie Women looke vnto the things necessarie and requisite about Kine, Calues, Hogges, Pigges, Pigeons, Geese, Duckes, Peacockes, Hennes, Fesants, and other sorts of Beasts, as well for the feeding of them as for the milking of them: making of Butter and Cheese: and the keeping of Lard to dresse the labouring men their victuals withall. Yea, furthermore they haue the charge of the Ouen and Cellar: and we leaue the handling of Hempe vnto them likewise; as also the care of making Webs, of looking to the clipping of Sheepe, of keeping their Fleeces, of spinning and combing of Wooll to make Cloth to cloath the familie, of ordering of the Kitchin Garden, and keeping of the Fruits, Hearbes, Rootes, and Seeds: and moreouer, of watching and attending the Bees. It is true, that the buying and felling of Cattell belongeth vnto the man, as also the disposing and laying out of money, together with the hyring and paying of seruants wages: But the surplusage to be employed and layed out in pettie matters, as in Linnens, Clothes for the household, and all necessaries of household furniture, that of a certainetie belongeth vnto the woman. I meane also that she must be such a one as is obedient vnto God and to her husband, giuen to store vp, to lay vp and keepe things sure vnder locke and key, painefull, peaceable, not louing to stirre from home, mild vnto such as are vnder her when there is need, and sharpe and seuere when occasion requireth: not contentious, full of words, toyish, tatling; nor drowsie-headed. Let her dispose of her stuffe and implements vnder her hand in such fort, as that euerie thing may haue his certaine place, and that in good order, to the end that when they be to be vsed, they may be found and easily come by and deliuered. Let her alwayes haue her eye vpon her maids: and let her be alwayes first at worke, and last from it, the first vp, and the last in bed. Let her not suffer to be lost or purloyned, no not the least trifle that is. Let her not grumble at any time for any seruice done to the Lord of the Farme: for the value of the least crum of Bread denyed, or vnwillingly graunted or giuen vnto him or his, may loose the quantitie of a whole Loafe afterward. Let her not trouble her braine with the reports & speeches of others, but let her acquaint her husband with them in good fort and manner. Let her gratifie her neighbors willingly, neuer attempting to inueagle or draw away any of their men seruants or maids from them: neither let her keepe companie with them, except when shee may doe them good, or helpe them, or when she maketh some marriage, or assemblies of great companie. Let her not suffer her daughters to gad and wander abroad vpon the Sabbath, except they be in such companie as is faithfull, or that she her selfe be present with them. Let her compell her sonnes to be formost at worke, and let her shew them the example of their father, that this may be as a double spurre vnto the men seruants. Let her not endure them to vtter or speake any vnchast word, oath, or blasphemie in her house: and let her cause Tale-bearers to be silent, and not to trouble themselues with other folkes matters. Let her keepe close vp her Stubble and lopping of Trees for fuell for the Ouen. Let her not suffer the stalkes of her Beanes, Peason, Fetches, Thistles, Danewort, the refuse of pressed things, and other vnprofitable hearbes, to be lost, for in winter they being burnt into ashes, will affoord prouision to lay Bucks withall, or else be fold by little and little vnto the Towne. Let her giue good account vnto the Mistresse or Lord of the Egges and young ones, as well of Birds as of other Beasts. Let her be skilfull in naturall Physicke, for the benefite of her owne folke and others when they shall fall out to be ill: and so in like manner in things good for Kine, Swine, and Fowls: for to haue a Physition alwayes, when there is not verie vrgent occasion and great necessitie, is not for the profit of the house. Let her keepe all them of her house in friendly good will one toward another, not suffering them to beare malice one against another. Let her gouerne her Bread so well, as that no one be suffered to vse it otherwise than in temperate fort: and in the time of Dearth, let her cause to be ground amongst her Corne, Beanes, Pease, Fetches, or Sarrasins Corne, in some small quantitie; for this mingling of these flowers raiseth the paste, maketh the Bread light, and to be of a greater bulke. At the same time she shall reserue the drosse of the Grapes shee presseth, affoording them some little corner, for the imploying of them in the defraying of some part of charge for the seruants Drinke, that so the Wine may serue for her husband and extraordinarie commers. But the naturall remedies which shee shall acquaint her selfe withall for the succour of her folke in their sicknesses, may be those, or such as those are, which I shall set downe by writing, in manner of a Countrie Dispensatorie, leauing the other more exquisite Remedies to bee vsed by the professed Physitions of the great Townes and Cities.

*　*　*

The Plow Mans Instruments and Tooles

The carefull and diligent plow-man, long time before he be to begin to eare his ground, shall take good heed, and see that all his tooles and implements, for to be vsed in plowing time, be readie and vvell appointed, that so he may haue them for his vse vvhen need shall be: as namely a waggon or two, according to the greatnesse of the firme, and those of a reasonable good bigge size, and handsome to handle, vvell furnished vvith wheels, vvhich must be finely bound and nayled, and of a good height, but more behind than before: one or two carres, vvhich may be made longer or shorter, according as the matter, vvhich shall be layed vpon them, shall require: one light and swift carr, the bodie layed vvith plankes, and sufficient strong to beare corne, vvine, vvood, stones, and other matters that are of great vveight: a plow furnished vvith a sharpe culture, and other parts: tumbrills to carrie his dung out into his grounds: wheele-barrowes and dung-pots to lade and carrie out dung in: strong and stout forkes to load and lay vpon heapes the corne-sheaues: pick-axes to breake small the thicke clods: the roller to breake the little clods: rakes, pick-axes, and mattockes, or other instruments to plucke vp vveeds that are strong and vnprofitable: harrowes and rakes with yron or woodden teeth, to couer the feed with earth: sickles to sheare or cut down haruest: flailes to thresh the corne: fannes and sieues to make cleane the good corne, and to separate it from the chaffe, dust, and other filth.

And because the plow is of all instruments belonging to the arable field the principallest, and varieth the oftest according to the variation of climats, I vvill here giue you a little touch of the seuerall plows for euerie seuerall soyle; and first to speake of the composition of plows, it consisteth vpon the beame, the skeath, the head, the hales the spindles, the rest, the shelboard, the plow-foot, the culture, and the share; then the slipe to keepe the plow from wearing, and the arker-staffe to cleanse the plow vvhen it shall be loaden vvith earth or other vild matter. The plow vvhich is most proper for the stiffe blacke clay, would be long, large, and broad, with a deepe head, and a square shelboard, so as it may turne vp a great furrow, the culture vvould be long and little or nothing bending, and the share would haue a verie large wing; as for the foot it vvould be long and broad, & so set as it may giue vvay to a great furrow. The plow for the vvhite, blew, or gray clay, vvould not be so large as that for the blacke clay, onely it vvould be somewhat broader in the britch; it hath most commonly but one hale, and that belonging to the left hand, yet it may haue two at your pleasure, the culture vvould be long, and bending, and the share narrow, vvith a vving comming vp to arme and defend the shelboard from vvearing. The plow for the red sand, would be lesse than any before spoken of, more light and more nimble; the culture would be made circular, or much bending like that for the white clay, yet much thinner, and the share vvould be made as it were with a halfe vving, neither so large as that for the black clay, nor so narrow as that for the white clay, but in a meane between both. The plow for the white sand differs nothing from that of the red sand, only it oft hath one addition more, that is, at the further end of the beame there is a paire of round wheeles which bearing the beame, vpon a loose mouing axle-tree, being just the length of two furrowes and no more, doth so certainely guide the plow to his true furrow that it can neuer loose land by swaruing, nor take too much land by the greedinesse of the yrons: the culture and share for this plow are like those for the red sand, onely they are a little lesse, the culture being not fully so long, nor so much bent, nor the share so broad, but a little sharper pointed, and this plow also serueth for the grauell howsoeuer mixt, whether with peeble, flint or otherwise. The plow for blacke clay mixt with red sand, and the white clay mixt with white sand, would be made of a middle size betwixt that for the blacke clay, and that for the red sand, being not so huge as the first, nor so slender as the later, but of a meane and competent greatnesse; and so also the culture and share must be made answerable, neither so bigge and streight as the greatest, nor so sharpe and long as the smallest. Lastly, the blacke clay mixt with white sand, and the white clay mixt with red sand, would haue a plow in all points like that for the red sand simple, onely the culture would be more sharpe, long, and bending, and the share so narrow, sharpe, and small, that it should be like a round pike, onely bigge at the setting on.

Thus you see the diuersitie of plowes, and how they serue for euerie seuerall soyle: now it is meet to know the implements belonging to their draught, vvhich if it be Oxen, then there is but the plow cleuise, the teames, the yoakes, and beeles; but if it be Horse, then they are two-fold, as single or double; single, as vvhen they draw in length one horse after another, and then there is needfull but the plow cleuise, and swingle-tree, treates, collers, harnesse, and cart bridles; or double, when they draw two and two together in the beare geares, and then there is needfull the plow, cleuise, and teame, the toastred, the swingle-trees, the treates, the harnesse, the collars, the round withs, or bearing geares, bellie-bands, back-bands, and bridles. Also, there be of harrowes two kinds, one with vvoodden teeth, the other vvith yron teeth: the vvoodden are for all simple clayes, or such as easily breake, and the yron for sands, mixt grounds, or any binding earth, and for new broken swarthes, or such earths as are subject to weeds, or quicke growth: for sleighting tooles, the barkeharrowes vvill serue loose grounds, and the roller those vvhich bind.

Jethro Tull on Husbandry, 1733

From Jethro Tull, *Horse-Hoeing Husbandry: or, An Essay on the Principles of Vegetation and Tillage* (London, 1751), pp. 36–73, 254–77.

Of Tillage

Tillage is breaking and dividing the Ground by Spade, Plough, Hoe, or other Instruments, which divide by a fort of Attrition (or Contusion) as Dung does by Fermentation.

By Dung we are limited to the Quantity of it we can procure, which in most Places is too scanty: But by Tillage, we can inlarge our Field of subterranean Pasture without Limitation, tho' the external Surface of it be confin'd within narrow Bounds: Tillage may extend the Earth's internal Superficies, in proportion to the Division of its Parts; and as Division is infinite, so may that Superficies be.

Every time the Earth is broken by any sort of Tillage, or Division, there must arise some new Superficies of the broken Parts, which never has been open before: For when the Parts of Earth are once united and incorporated together, 'tis morally impossible, that they, or any of them, should be broken again, only in the same Places; for to do that, such Parts must have again the same numerical

Figures and Dimensions they had before such Breaking, which even by an infinite Division could never be likely to happen: As the Letters of a Distichon, cut out and mixt, if they should be thrown up never so often, would never be likely to fall into the same Order and Position with one another, so as to recompose the same Distich.

Although the internal Superficies may have been drain'd by a preceding Crop, and the next Plowing may move many of the before divided Parts, without new-breaking them; yet such as are new-broken, have, at such Places where they are so broken, a new Superficies, which never was, or did exist before; because we cannot reasonably suppose, that any of those Parts can have in all Places (if in any Places) the same Figure and Dimensions twice.

For as Matter is divisible *ad infinitum,* the Places or Lines whereat 'tis so divisible, must be, in relation to Number, infinite, that is to say, without Number; and must have at every Division Superficies of Parts of infinite Variety in Figure and Dimensions.

And because 'tis morally impossible, the same Figure and Dimensions should happen twice to any one Part, we need not wonder, how the Earth, every time of Tilling, should afford a new internal Superficies (or artificial Pasture); and that the till'd Soil has in it an inexhaustible Fund, which by a sufficient Division (being capable of an infinite one) may be produc'd.

Tillage (as well as Dung) is beneficial to all Sorts of Land. Light Land, being naturally hollow, has larger Pores, which are the Cause of its Lightness: This, when it is by any means sufficiently divided, the Parts being brought nearer together, becomes, for a time, Bulk for Bulk, heavier; *i.e.* The same Quantity will be contain'd in less Room, and so is made to partake of the Nature and Benefits of strong Land, *viz.* to keep out too much Heat and Cold, and the like.

But strong Land, being naturally less porous, is made for a time lighter (as well as richer) by a good Division; the Separation of its Parts makes it more porous, and causes it to take up more room than it does in its natural State; and then it partakes of all the Benefits of lighter Land.

When strong Land is plow'd, and not sufficiently, so that the Parts remain gross, 'tis said to be rough, and it has not the Benefit of Tillage; because most of the artificial Pores (or Interstices) are too large; and then it partakes of the Inconveniencies of the hollow Land untill'd.

For when the light Land is plow'd but once, that is not sufficient to diminish its natural Hollowness (or Pores); and, for want of more Tillage, the Parts into which 'tis divided by that once (or perhaps twice) Plowing, remain too large; and consequently the artificial Pores are large also, and, in that respect, are like the ill-till'd strong Land.

Light-land, having naturally less internal Superficies, seems to require the more Tillage or Dung to enrich it; as when the poor, hollow, thin Downs have their upper Part (which is the best) burnt, whereby all (except a *Caput Mortuum*) is carried away; yet the Salts of this spread upon that barren Part of the Staple, which is unburnt, divide it into so very minute Particles that their Pasture, will nourish Two or Three good Crops of Corn: But then the Plough, even with a considerable Quantity of Dung, is never able afterwards to make a Division equal to what those Salts have done; and therefore such burnt Land remains barren.

Artificial Pores cannot be too small, because Roots may the more easily enter the Soil that has them, quite contrary to natural Pores; for these may be, and generally are, too small, and too hard for the Entrance of all weak Roots, and for the free Entrance of strong Roots.

Insufficient Tillage leaves strong Land with its natural Pores too small, and its artificial ones too large. It leaves Light-land, with its natural and artificial Pores both too large.

Pores that are too small in hard Ground, will not easily permit Roots to enter them.

Pores that are too large in any sort of Land, can be of little other Use to Roots, but only to give them Passage to other Cavities more proper for them; and if in any Place they lie open to the Air, they are dry'd up, and spoil'd, before they reach them.

For fibrous Roots (which alone maintain the Plant; the other Roots serve for receiving the Chyle from them, and convey it to the Stem) can take in no Nourishment from any Cavity, unless they come into Contact with, and press against, all the Superficies of that Cavity, which includes them; for it dispenses the Food to their Lacteals by such Pressure only: But a fibrous Root is not so press'd by the Superficies of a Cavity whose Diameter is greater than that of the Root.

The Surfaces of great Clods form Declivities on every Side of them, and large Cavities, which are as Sinks to convey, what Rain and Dew bring, too quickly downwards to below the plow'd Part.

The First and Second plowings with common Ploughs scarce deserve the Name of Tillage; they rather serve to prepare the Land for Tillage.

The Third, Fourth, and every subsequent plowing, may be of more Benefit, and less Expence, than any of the preceding ones.

But the last Plowings will be more advantageously perform'd by way of Hoeing, as in the following Chapters will appear.

For the finer Land is made by Tillage, the richer will it become, and the more Plants it will maintain.

It has been often observ'd, that when Part of a Ground has been better till'd than the rest, and the whole Ground constantly manag'd alike afterwards for Six or Seven Years successively; this Part that was but once better till'd, always produc'd a better Crop than the rest, and the Difference remain'd very visible every Harvest.

One Part being once made finer, the Dews did more enrich it; for they penetrate within and beyond the Superficies, whereto the Roots are able to enter: The fine Parts of the Earth are impregnate, throughout their whole Substance, with some of the Riches carried in by the Dews, and there reposited; until, by new Tillage, the Insides of those fine Parts become Superficies; and as the Corn drains them, they are again supply'd as before; but the rough large Parts cannot have that Benefit; the Dews not penetrating to their Centres, they remain poorer.

I think nothing can be said more strongly to confirm the Truth of this, than what is related by the Authors quoted by Mr. *Evelyn*, to this Effect, *viz.*

Take of the most barren Earth you can find, pulverize it well, and expose it abroad for a Year, incessantly agitated; it will become so fertile as to receive an exotic Plant from the furthest *Indies;* and to cause all Vegetables to prosper in the most exalted Degree, and to bear their Fruit as kindly with us as in their natural Climates.

This artificial Dust, he says, will entertain Plants which refuse Dung, and other violent Applications; and that it has a more nutritive Power than any artificial Dungs or Compost whatsoever; And further, that by this Toil of pulverizing, 'tis found, that Soil may be so strangely alter'd from its former Nature, as to render the harsh and most uncivil Clay obsequious to the Husbandmen, and to bring forth Roots and Plants, which otherwise require the lightest and hollowest Mould.

'Tis to be suppos'd, that the *Indian* Plants had their due Degrees of Heat and Moisture given them; and I should not chuse to bestow this Toil upon the poorest of Earth in a Field or Garden, tho' that be the most sure wherein to make the Experiment.

I never myself try'd this way of pounding or grinding, because impracticable in the Fields.

But I have had the Experience of a Multitude of Instances, which confirm it so far, that I am in no Doubt, that any Soil (be it rich or poor) can ever be made too fine by Tillage.

For 'tis without dispute, that one cubical Foot of this minute Powder may have more internal Superficies, than a thousand cubical Feet of the same, or any other Earth till'd in the common Manner; and, I believe, no two arable Earths in the World do exceed one another in their natural Richness Twenty times; that is, One cubical Foot of the richest is not able to produce an equal Quantity of Vegetables, *cæteris paribus,* to Twenty cubical Feet of the poorest; therefore 'tis not strange, that the poorest, when by pulverizing it has obtain'd One hundred times the internal Superficies of the rich untill'd Land, it should exceed it in Fertility; or, if a Foot of the poorest was made to have Twenty times the Supersicies of a Foot of such rich Land, the poorest might produce an equal Quantity of Vegetables with the rich. Besides, there is another extraordinary Advantage, when a Soil has a larger internal Superficies in a very little Compass; for then the Roots of Plants in it are better supply'd with Nourishment, being nearer to them on all Sides within Reach, than it can be when the Soil is less fine, as in common Tillage; and the Roots in the one must extend much further than in the other, to reach an equal Quantity of Nourishment: They must range and fill perhaps above Twenty times more Space to collect the same Quantity of Food.

But in this fine Soil, the most weak and tender Roots have free Passage to the utmost of their Extent, and have also an easy, due, and equal Pressure everywhere, as in Water.

Hard Ground makes a too great Resistance, as Air makes a too little Resistance, to the Superficies of Roots.

Farmers, just when they have brought their Land into a Condition fit to be further till'd to much greater Advantage, leave off, supposing the Soil to be fine enough, when, with the Help of Harrows, they can cover the Seed; and afterwards with a Roller they break the Clods; to the end that, if a Crop succeed, they may be able to mow it, without being hinder'd by those Clods: By what I could ever find, this Instrument, call'd a Roller, is seldom beneficial to good

Husbands; it rather untills the Land, and anticipates the subsiding of the Ground, which in strong Land happens too soon of itself.

But more to blame are they, who neglect to give their Land due Plowing, trusting to the Harrow to make it fine; and when they have thrown in their Seed, go over it Twenty times with the Harrows till the Horses have trodden it almost as hard as a Highway, which in moist Weather spoils the Crop; but on the contrary, the very Horses, when the Earth is moist, ought all to tread in the Furrows only, as in plowing with a Hoe-plough they always do, when they use it instead of a common Plough.

Of Hoeing

Hoeing is the breaking or dividing the Soil by Tillage, whilst the Corn or other Plants are growing thereon.

It differs from common Tillage (which is always perform'd before the Corn or Plants are sown or planted) in the times of performing it; 'tis much more beneficial; and 'tis perform'd by different Instruments.

Land that is before Sowing tilled never so much (tho' the more 'tis till'd the more it will produce) will have some Weeds, and they will come in along with the Crop for a Share of the Benefit of the Tillage, greater or less, according to their Number, and what Species they are of.

But what is most to be regarded is, that as soon as the Ploughman has done his Work of plowing and harrowing, the Soil begins to undo it, inclining towards, and endeavouring to regain, its natural specific Gravity; the broken Parts by little and little coalesce, unite, and lose some of their Surfaces; many of the Pores or Interstices close up during the Seed's Incubation and Hatching in the Ground; and, as the Plants grow up, they require an Increase of Food proportionable to their increasing Bulk; but on the contrary, instead thereof, that internal Superficies, which is their artificial Pasture, gradually decreases.

The Earth is so unjust to Plants, her own Off-Spring, as to shut up her Stores in proportion to their Wants; that is, to give them less Nourishment when they have need of more: Therefore Man, for whose Use they are chiefly design'd, ought to bring in his reasonable Aid for their Relief, and force open her Magazines with the Hoe, which will thence procure them at all times Provision in abundance, and also free them from Intruders; I mean, their spurious Kindred, the Weeds, that robb'd them of their too scanty Allowance.

There's no doubt, but that One-third Part of the Nourishment raised by Dung and Tillage, given to Plants or Corn at many proper Seasons, and apportion'd to the different Times of their Exigencies, will be of more Benefit to a Crop, than the Whole apply'd, as it commonly is, only at the time of Sowing. This old Method is almost as unreasonable as if Treble the full Stock of Leaves, necessary to maintain Silkworms till they had finished their Spinning, should be given them before they are hatched, and no more afterwards.

Next to Hoeing, and something like it, is Transplanting, but much inferior; both because it requires a so much greater Number of Hands, that by no Contrivance can it ever become general, nor does it succeed, if often repeated; but Hoeing will maintain any Plant in the greatest Vigour 'tis capable of, even unto the utmost Period of Age. Besides, there is Danger in removing a whole Plant, and Loss of Time before the Plant can take Root again, all the former Roots being broken off at the Ends in taking up (for 'tis impossible to do it without), and so must wait until by the Strength and Virtue of its own Sap (which by a continual Perspiration is daily enfeebled) new Roots are form'd, which, unless the Earth continue moist, are so long in forming, that they not only find a more difficult Reception into the closing Pores; but many times the Plant languishes and dies of an Atrophy, being starv'd in the midst of Plenty; but whilst this is thus decaying, the hoed Plant obtains a more flourishing State than ever, without removing from the same Soil that produc'd it.

'Tis observ'd that some Plants are the worse for Transplanting. *Fenochia* removed is never so good and tender as that which is not, it receives such a Check in Transplanting in its Infancy; which, like the Rickets, leaves Knots that indurate the Parts of the Fennel, and spoil it from being a Dainty.

Hoeing has most of the Benefits without any Inconveniencies of Transplanting; because it removes the Roots by little and little, and at different times; some of the Roots remaining undisturb'd, always supply the moved Roots with Moisture, and the whole Plant with Nourishment sufficient to keep it from fainting, until the moved Roots can enjoy the Benefit of their new Pasture, which is very soon.

Another extraordinary Benefit of the new Hoing Husbandry is, that it keeps Plants moist in dry Weather, and this upon a double Account.

First, as they are better nourished by Hoeing, they require less Moisture, as appears by Dr. *Woodward's* Experiment, that those Plants which receive the greatest Increase, having most terrestrial Nourishment, carry off the least Water in proportion to their Augment: So Barley or Oats, being sown on a Part of a Ground very well divided by Dung and Tillage, will come up and grow vigorously without Rain, when the same Grains, sown at the same time, on the other Part, not thus enriched, will scarce come up; or, if they do, will not thrive till Rain comes.

Secondly, The Hoe, I mean the Horse-hoe (the other goes not deep enough), procures Moisture to the Roots from the Dews, which fall most in dry Weather; and those Dews (by what Mr. *Thomas Henshaw* has observ'd) seem to be the richest Present the Atmosphere gives to the Earth; having, when putrefy'd in a Vessel, a black Sediment like Mud at the Bottom. This seems to cause the darkish Colour to the upper Part of the Ground. And the Sulphur, which is found in the Sediment of the Dew, may be the chief Ingredient of the Cement of the Earth; Sulphur being very glutinous, as Nitre is dissolvent. Dew has both these.

These enter in proportion to the Fineness and Freshness of the Soil, and to the Quantity that is so made fine and fresh by the Hoe. How this comes to pass, and the Reason of it, are shewn in the Chapter of Tillage.

To demonstrate that Dews moisten the Land when fine, dig a Hole in the hard dry Ground, in the driest Weather, as deep as the Plough ought to reach: Beat the Earth very fine, and fill the Hole therewith; and, after a few Nights· Dews, you'll find this fine Earth become moist at the Bottom, and the hard Ground all round will continue dry.

Till a Field in Lands; make one Land very fine, by frequent deep Plowings; and let another be rough, by insufficient Tillage, alternately; then plow the whole Field cross-ways in the driest Weather, which has continued long; and you will perceive, by the Colour of the Earth, that every fine Land will be turn'd up moist; but every rough Land will be dry as Powder, from Top to Bottom.

Altho' hard Ground, when thoroughly soak'd with Rain, will continue wet longer than fine till'd Land adjoining to it; yet this Water serves rather to chill, than nourish the Plants standing therein, and to keep out the other Benefits of the Atmosphere, leaving the Ground still harder when 'tis thence exhaled; and being at last once become dry, it can admit no more Moisture, unless from a long-continued Deluge of Rain, which seldom falls till Winter, which is not the season for Vegetation.

As fine hoed Ground is not so long soaked by Rain, so the Dews never suffer it to become perfectly dry: This appears by the Plants, which flourish and grow fat in this, whilst those in the hard Ground are starved, except such of them, which stand near enough to the hoed Earth, for the Roots to borrow Moisture and Nourishment from it.

And I have been informed by some Persons, that they have often made the like Observations; that, in the driest of Weather, good Hoeing procures Moisture to Roots; tho' the Ignorant and Incurious fansy, it lets in the Drought; and therefore are afraid to hoe their Plants at such times, when, unless they water them, they are spoil'd for want of it.

There is yet one more Benefit Hoing gives to Plants, which by no Art can possibly be given to Animals: For all that can be done in feeding an Animal is, what has been here already said of Hoing; that is, to give it sufficient Food, Meat and Drink, at the times it has occasion for them; if you give an Animal any more, 'tis to no manner of purpose, unless you could give it more Mouths, which is impossible; but in hoeing a Plant, the additional Nourishment thereby given, enables it to send out innumerable additional Fibres and Roots, as in one of the Glasses with a Mint in it, is seen; which fully demonstrates, that a Plant increaseth its Mouths, in some Proportion to the Increase of Food given to it: So that Hoeing, by the new Pasture it raises, furnishes both Food and Mouths to Plants; and 'tis for want of Hoing, that so few are brought to their Growth and Perfection.

In what manner the Sarrition of the Antients was performed in their Corn, is not very clear: This seems to have been their Method; *viz.* When the Plants were some time come up, they harrowed the Ground, and pull'd out the Weeds by Hand. The Process of this appears in *Columella,* where he directs the Planting of *Medica* to be but a sort of Harrowing or Raking amongst the young Plants, that the Weeds might come out the more easily: *Ligneis Rastris statim jacta Semina obruantur. Post sationem Ligneis Rastris sarriendus, & identidem runcandus est Ager, ne alterius generis Herba invalidam Medicam perimat.*

They harrowed and hoed *Rastris*; so that their *Occatio* and *Sarritio* were performed with much the same sort of Instrument, and differed chiefly in the Time: The first was at Seed-time, to cover the Seed, or level the Ground; the other was to move the Ground after the Plants were up.

One sort of their Sarrition was, *Segetes permota Terra debere adobrui, ut fruticare possint.* Another sort was thus: *In Locis autem frigidis sassiri nec adobrui, sed plana Sarritione Terram permoveri.*

For the better understanding of these Two sorts of Sarrition, we must consider, that the Antients sowed their Corn under Furrow; that is, when they had harrowed the Ground, to break the Clods, and make it level, they sowed the Seed, and then plowed it in: This left the Ground very uneven, and the Corn came up (as we see it does here in the same case) mostly in the lowest Places betwixt the Furrows, which always lay higher: This appears by *Virgil's Cum Sulcos æquant Sata.* Now, when they used *Plana Sarritio,* they harrowed Length-ways of the Furrows, which being somewhat harden'd, there could be little Earth thrown down thence upon the young Corn.

But the other sort of Sarrition, whereby the Corn is said *Adobrui,* to be cover'd, seems to be perform'd by Harrowing cross the Furrows; which must needs throw down much Earth from the Furrows, which necessarily fell upon the Corn.

How this did contribute to make the Corn *fruticare,* is another Question: I am in no doubt to say, it was not from covering any Part of it (so I see that has a contrary Effect), but from moving much Ground, which gave a new Pasture to the Roots: This appears by the Observation of the extraordinary Frutication of Wheat ho'd without being cover'd; and by the Injury it receives by not being uncover'd when any Earth falls on the Rows.

The same Author saith, *Faba, & cætera Legumina, cum quatuer Digitis à Terra extiterint, recte farrientur, excepto tamen Lupino, cujus Semini contraria est Sarritio; quoniam unam Radicem habet, quæ sive Ferro succisa feu vulnerata est, totus Frutex emoritur.*

If they had ho'd it only betwixt Rows, there had been no Danger of killing the Lupine, which is a Plant most proper for Hoeing. What he says of the Lupine's having no need of Sarrition, because it is able of itself to kill Weeds, shews the Antients were Ignorant of the chief Use of Hoeing; *viz.* to raise new Nourishment by dividing the Earth, and making a new internal Superficies in it.

Sarrition scratched and broke so small a Part of the Earth's Surface, amongst the Corn and Weeds, without Distinction, or favouring one any more than the other, that it was a Dispute, whether the Good it did in facilitating the Runcation (or Hand-weeding) was greater, than the Injury it did by bruising and tearing the Corn: And many of the Antients chose rather to content themselves with the Use of Runcation only, and totally to omit all Sarrition of their Corn.

But Hoeing is an Action very different from that of Sarrition, and is every way beneficial, no-way injurious to Corn, tho' destructive to Weeds. Therefore sone modern Authors shew a profound Ignorance, in translating *Sarritio,* Hoeing: They give an Idea very different from the true one: For the Antients truly hoed their Vineyards, but not their Corn; neither did they plant their Corn in Rows, without which they could not give it the Vineyard-hoeing: Their Sarculation was used but amongst small Quantities of sown Corn, and is yet in Use for Flax; for I have seen the *Sarculum* (which is a sort of a very narrow Hoe) used amongst the Plants of Flax standing irregularly: But this Operation is too tedious, and too chargeable, to be apply'd to great Quantities of irregular Corn.

If they ho'd their Crops sown at random, one would think they should have made mad Work of it; since they were not at the Pains to plant in Rows, and hoe betwixt them with their Bidens; being the Instrument with which they tilled many of their Vineyards, and enters as deep as the Plough, and is much better than the *English* Hoe, which indeed seems, at the first Invention of it, to be designed rather to scrape Chimneys, than to till the Ground.

The highest and lowest Vineyards are ho'd by the Plough; first the high Vineyards, where the Vines grow (almost like Ivy) upon great Trees, such as Elms, Maples, Cherry-trees, &c. These are constantly kept in Tillage, and produce good Crops of Corn, besides what the Trees do yield; and also these great and constant Products of the Vines are owing to this sort of Hoe-tillage; because neither in Meadow or Pasture Grounds can Vines be made to prosper; tho' the Land be much richer, and yet have a less Quantity of Grass taken off it, than the Arable has Corn carried from that.

The Vines of low Vineyards, ho'd by the Plough, have their Heads just above the Ground, standing all in a most regular Order, and are constantly plowed in the proper Season: These have no other Assistance, but by Hoeing; because their Head and Roots are so near together, that Dung would spoil the Taste of the Wine they produce, in hot Countries.

All Vineyards must be ho'd one way or other, or else they will produce nothing of Value; but Corn-fields without Hoeing do produce something, tho' nothing in Comparison to what they would do with it.

Mr. *Evelyn* says, that when the Soil, wherein Fruit-trees are planted, is constantly kept in Tillage, they grow up to be an Orchard in half the Time they would do, if the Soil were not till'd; and this keeping an Orchard-Soil in Arable, is Horse-hoing it.

In some Places in *Berkshire* they have used, for a long time, to Hand-hoe most sorts of Corn, with very great Success; and I may say this, that I myself never knew, or heard, that ever any Crop of Corn was properly so ho'd, but what very well answer'd the Expence, even of this Hand-work; but be this never so profitable, there are not a Number of Hands to use it in great Quantities; which possibly was one Reason the Antients were not able to introduce it into their Corn-fields to any purpose; tho' they should not have been ignorant of the Effect of it, from what they saw it do in their Vineyards and Gardens.

In the next Place I shall give some general Directions, which by Experience I have found necessary to be known, in order to the Practice of this Hoing-Husbandry.

    I. *Concerning the Depth to Plant at.*
   II. *The Quantity of Seed to plant.*
  III. *And the Distance of the Rows.*

I. 'Tis necessary to know how deep we may plant our Seed, without Danger of burying it; for so 'tis said to be, when laid at a Depth below what 'tis able to come up at.

Different Sorts of Seeds come up at different Depths; some at six Inches, or more; some at not more than half an Inch: The way to know for certain the Depth any sort will come up at is, to make Gauges in this manner: Saw off 12 Sticks of about 3 Inches Diameter: Bore a Hole in the End of each Stick, and drive into it a taper Peg; let the first Peg be half an Inch long, the next an Inch, and so on; every Peg to be half an Inch longer than the former, till the last Peg be Six Inches long; then in that sort of Ground where you intend to plant, make a Row of Twenty Holes with the Half-inch Gauge; put therein Twenty good Seeds; cover them up, and stick the Gauge at the End of that Row; then do the like with all the other Eleven Gauges: This will determine the Depth, at which the most Seeds will come up.

When the Depth is known, wherein the Seed is sure to come up, we may easily discover whether the Seed be good or not, by observing how many will fail: For in some sorts of Seeds the Goodness cannot be known by the Eye; and there has been often great Loss by bad Seed, as well as by burying good Seed; both which Misfortunes might be prevented by this little Trouble; besides, 'tis not convenient to plant some sorts of Seed at the utmost Depth they will come up at; for it may be so deep, as that the Wet may rot or chill the first Root, as in Wheat in moist Land.

The Nature of the Land, the Manner how it is laid, either flat, or in Ridges, and the Season of Planting, with the Experience of the Planter, acquired by such Trials, must determine the proper Depths for different Sorts of Seeds.

II. The proper Quantity of Seed to be drill'd on an Acre, is much less than must be sown in the common way; not because Hoeing will not maintain as many Plants as the other; for, on the contrary, Experience shews it will, *cæteris paribus,* maintain more; but the Difference is upon many other Accounts: As that 'tis impossible to sow it so even by Hand, as the Drill will do; for let the Hand spread it never so exactly (which is difficult to do some Seeds, especially in windy Weather), yet the Unevenness of the Ground will alter the Situation of the Seed; the greatest Part rebounding into the Holes, and lowest Places; or else the Harrows, in Covering, draw it down thither; and tho' these low Places may have Ten times too much, the high Places may have little or none of it: This Inequality lessens, in effect, the Quantity of the Seed; because Fifty Seeds, in room on One, will not produce so much as One will do; and where they are too thick, they cannot be well nourished, their Roots not spreading to near their natural Extent, for want of Hoeing to open the Earth. Some Seed is buried (by which is meant the laying them so deep, that they are never able to come up, as *Columella* cautions, *Ut absque ulla Resurectionis Spe sepeliantur*): Some lies naked above the Ground; which, with more uncovered by the first Rain, feeds the Birds and Vermin.

Farmers know not the Depth that is enough to bury their Seed, neither do they make much Difference in the Quantity they sow on a rough, or a fine Acre; tho' the same that is too little for the one, is too much for the other; 'tis all mere Chance-work, and they put their whole Trust in good Ground, and much Dung, to cover their Errors.

The greatest Quantity of Seed I ever heard of to be usually sown, is in *Wiltshire,* where I am informed by the Owners themselves, that on some sorts of Land they sow Eight Bushels of Barley to an Acre; so that if it produce Four Quarters to an Acre, there are but Four Grains for One that is sown, and is a very Poor Increase, tho' a good Crop; this is on Land plowed once, and then double-dung'd, the Seed only harrow'd into the stale and hard Ground, 'tis like not Two Bushels of the Eight will enter it to grow; and I have heard, that in a dry Summer an Acre of this scarce produces Four Bushels at Harvest.

But, in Drilling, Seed lies all the same just Depth, none deeper, nor shallower, than the rest; here's no Danger of the Accidents of burying, or being uncover'd, and therefore no Allowance must be made for them; but Allowance must be made for other Accidents, where the sort of Seed is liable to them; such as Grub, Fly, Worm, Frost, &c.

Next, when a Man unexperienc'd in this Method has proved the Goodness of his Seed, and Depth to plant it at, he ought to calculate what Number of Seeds a Bushel, or other Measure or Weight, contains: For one Bushel, or one Pound of small Seed, may contain double the number of Seeds, of a Bushel, or a Pound, of large Seed of the same Species.

This Calculation is made by weighing an Ounce, and counting the Number of Seeds therein; then weighing a Bushel of it, and multiplying the Number of Seeds of the Ounce, by the Number of Ounces of the Bushel's Weight; the Product will shew the Number of Seeds of a Bushel near enough: Then, by the Rule of Three, apportion them to the square Feet of an Acre; or else it may be done, by divideing the Seeds of the Bushel by the square Feet of an Acre; the Quotient will give the Number of Seeds for every Foot: Also consider how near you intend to plant the Rows, and whether Single, Double, Treble, or Quadruple; for the more Rows, the more Seed will be required.

Examine what is the Produce of one middle-siz'd Plant of the Annual, but the Produce of the best and largest of the perennial Sort; because that by Hoeing will be brought to its utmost Perfection: Proportion the Seed of both to the reasonable Product; and, when 'tis worth while, adjust the Plants to their competent Number with the Hand-hoe, after they are up; and plant Perennials generally in single Rows: Lastly, Plant some Rows of the Annual thicker than others, which will soon give you Experience (better than any other Rule) to know the exact Quantity of Seed to drill.

III. The Distances of the Rows are one of the most material Points, wherein we shall find many apparent Objections against the Truth; of which tho' full Experience be the most infallible Proof, yet the World is by false Notions so prejudiced against wide Spaces between Rows, that unless these common (and I wish I could say, only vulgar) Objections be first answer'd, perhaps no-body will venture so far out of the old Road, as is necessary to gain the Experience; without it be such as have seen it.

I formerly was at much Pains, and at some Charge, in improving my Drills, for planting the Rows at very near Distances; and had brought them to such Perfection, that One Horse would draw a Drill with Eleven Shares, making the Rows at Three Inches and half Distance from one another; and at the same time sow in them Three very different Sorts of Seeds, which did not mix; and these too, at different Depths; as the Barley-Rows were Seven Inches asunder, the Barley lay Four Inches deep; a little more than Three Inches above that, in the same Chanels, was Clover; betwixt every Two of these Rows was a Row of St. Foin, cover'd half an Inch deep.

I had a good Crop of Barley the first Year; the next Year, Two Crops of Broad-Clover, where that was sown; and where Hop-Clover was sown, a mix'd Crop of That and St. Foin; and every Year afterwards a Crop of St. Foin; but

I am since, by Experience, so fully convinced of the Folly of these, or any other such mix'd Crops, and more especially of narrow Spaces, that I have demolish'd these Instruments (in their full Perfection) as a vain Curiosity, the Drift and Use of them being contrary to the true Principles and Practice of Horse-Hoeing.

Altho' I am satisfied, that every one, who shall have seen as much of it as I have, will be of my Mind in this Matter; yet I am aware, that what I am going to advance, will seem shocking to them, before they have made Trials.

I lay it down as a Rule (to myself) that every Row of Vegetables, to be Horse-ho'd, ought to have an empty Space or Interval of Thirty Inches on one Side of it at least, and of near Five Feet in all Sorts of Corn.

In Hand-hoeing there is always less Seed, fewer Plants, and a greater Crop, *cæteris paribus*, than in the common Sowing: Yet there the Rows must be much nearer together, than in Horse-hoeing; because as the Hand moves many times less Earth than the Horse, the Roots will be sent out in like Proportion; and if the Spaces or Intervals, where the Hand-hoe only scratches a little of the upper Surface of them, should be wide, they would be so hard and stale underneath, that the Roots of perennial Plants would be long in running thro' them; and the Roots of many annual Plants would never be able to do it.

An Instance which shews something of the Difference between Hand-hoeing and Deep-hoeing is, That a certain poor Man is observ'd to have his Cabbages vastly bigger than any-body's else, tho' their Ground be richer, and better dung'd: His Neighbours were amaz'd at it, till the Secret at length came out, and was only this: As other People ho'd their Cabbages with a Hand-hoe, he instead thereof dug his with a Spade: And nothing can more nearly equal the Use of the Horse-hoe than the Spade does.

And when Plants have never so much *Pabulum* near them, their fibrous Roots cannot reach it all, before the Earth naturally excludes them from it; for, to reach it all, they must fill all the Pores, which is impossible: So far otherwise it is, that we shall find it probable, that they can only reach the least Part of it, unless the Roots could remove themselves from Place to Place, to leave such Pores as they had exhausted, and apply themselves to such as were unexhausted; but they not being endow'd with Parts necessary for local Motion (as Animals are), the Hoe-Plough supplies their want of Feet; and both conveys them to their Food, and their Food to them, as well as provides it for them; for by transplanting the Roots, it gives them Change of the Pasture, which it increases by the very Act of changing them from one Situation to another, if the Intervals be wide enough for this Hoeing Operations to be properly perform'd.

The Objections most likely to prepossess Peoples Minds, and prevent their making Trials of this Husbandry, are these:

First, They will be apt to think, that these wide, naked Spaces, not being cover'd by the Plants, will not be sufficient to make a good Crop.

For Answer, we must consider, that tho' Corn, standing irregular and *Sparsim,* may seem to cover the Ground better than when it stands regular in Rows, this Appearance is a mere *Deceptio visus;* for Stalks are never so thick on any Part of the Ground as where many come out of one Plant, or as when they stand in a Row; and a ho'd Plant of Corn will have Twenty or Thirty Stalks, in the same Quantity of Ground where an un-ho'd Plant, being equally single, will have only Two or Three Stalks. These tillered ho'd Stalks, if they were planted *Sparsim* all over the Interval, it might seem well cover'd, and perhaps thicker than the sown Crop commonly is; so that tho' these ho'd Rows seem to contain a less Crop, they may contain, in reality, a greater Crop than the sown, that seems to exceed it; and 'tis only the different placing that makes one seem greater, and the other less, than it really is; and this is only when both Crops are young.

The next Objection is, That the Space or Interval not being *planted,* much of the Benefit of that Ground will be lost; and therefore the Crop must be less than if it were planted all over.

I answer, It might be so, if not Horse-ho'd; but if well Horse-ho'd, the Roots can run through the Intervals; and, having more Nourishment, make a greater Crop.

The too great Number of Plants, plac'd all over the Ground in common sowing, have, whilst it is open, an Opportunity of *wasting,* when they are very young, that Stock of Provision, for want of which the greatest Part of them are afterwards starv'd; for their irregular Standing prevents their being relieved with fresh Supplies from the Hoe: Hence it is, that the old Method exhausting the Earth to no purpose, produces a less Crop; and yet leaves less *Pabulum* behind for a succeeding one, contrary to the Hoeing-Husbandry, wherein Plants are manag'd in all respects by a quite different Oeconomy.

In a large Ground of Wheat it was prov'd, that the widest ho'd Intervals brought the greatest Crop of all: Dung without Hoeing did not equal Hoeing without Dung. And, what was most remarkable, amonst Twelve Differences of wider and narrower Spaces, more and less ho'd, dung'd and undung'd, the Hand-sow'd was considerably the worst of all; tho' all the Winter, and Beginning of the Spring, that made infinitely the most promising Appearance; but at Harvest yielded but about One-fifth Part of Wheat of that which was most hoed; there was some of the most hoed, which yielded Eighteen Ounces of clean Wheat in a Yard in Length of a double Row, the Intervals being Thirty Inches, and the Partition Six Inches.

A Third Objection like the two former is, that so small a Part of the Ground, as that whereon the Row stands, cannot contain Plants or Stalks sufficient for a full Crop.

This some Authors endeavour to support by Arguments taken from the perpendicular Growth of Vegetables, and the Room they require to stand on; both which having answer'd elsewhere, I need not say much of them here; only I may add, that if Plants could be brought to as great Perfection, and so to stand as thick all over the Land, as they do in the ho'd Rows, there might be produc'd, at once, many of the greatest Crops of Corn that ever grew.

But since Plants thrive, and make their Produce, in proportion to the Nourishment they have within the Ground, not to the Room they have to stand upon it, one very narrow Row may contain more Plants than a wide Interval can nourish, and bring to their full Perfection, by all the Art that can be used; and 'tis impossible a Crop should be lost for want of room to stand above the Ground, tho' it were less than a Tenth-part of the Surface.

In wide Intervals there is another Advantage of Hoeing, I mean Horse-hoeing (the other being more like Scratching and Scraping than Hoeing): There is room for many Hoeings which must not come very near the Bodies of some annual Plants, except whilst they are young; but, in narrow Intervals, this cannot be avoided at every Hoeing: 'Tis true, that in the last Hoeings, even in the middle of a large Interval, many of the Roots may be broken off by the Hoe-plough, at some considerable Distance from the Bodies; but yet this is no Damage, for they send out a greater Number of Roots than before; as in the Mint, in Chap. I. appears.

In wide Intervals, those Roots are broken off only where they are small; for tho' they are capable of running out to more than the Length of the external Parts of a Plant; yet 'tis not necessary they should always do so; if they can have sufficient Food nearer to the Bodies of the Plants.

And these new, young, multiply'd Roots are fuller of Lacteal Mouths than the older ones; which makes it no Wonder, that Plants should thrive faster by having some of their Roots broken off by the Hoe; for as Roots do not enter every Pore of the Earth, but miss great Part of the Pasture, which is left unexhausted, so when new Roots strike out from the broken Parts of the old, they meet with that Pasture, which their Predecessors miss'd, besides that new Pasture which the Hoe raises for them; and those Roots which the Hoe pulls out without breaking, and covers again, are turn'd into a fresh Pasture; some broken, and some unbroken: All together invigorant the Plants.

Besides, the Plants of sown Corn, being treble in Number to those of the drill'd, and of equal Strength and Bulk, whilst they are very young, must exhaust the Earth whilst it is open, thrice as much as the drill'd Plants do; and before the sown Plants grow large, the Pores of the Earth are shut against them, and against the Benefit of the Atmosphere; but for the drill'd, the Hoe gives constant Admission to that Benefit; and if the Hoe procures them (by dividing the Earth) Four times the Pasture of the sown during their Lives, and the Roots devour but one half of that, then tho' the ho'd Crop should be double to the sown, yet it might leave twice as much *Pabulum* for a succeeding Crop. 'Tis impossible to bring these Calculations to Mathematical Rules; but this is certain in Practice, that a sown Crop, succeeding a large undung'd ho'd Crop, is much better than a sown Crop, that succeeds a small dung'd sown Crop. And I have the Experience of poor, worn out Heath-ground, that, having produc'd Four successive good ho'd Crops of Potatotes (the last still best), is become tolerable good Ground.

In a very poor Field were planted Potatotes, and, in the very worst Part of it, several Lands had them in Squares a Yard asunder; these were plowed four ways at different times: Some other Lands adjoining to them, of the very same Ground, were very well dung'd and till'd; but the Potatoes came irregularly, in some Places thicker, and in others thinner: These were not ho'd, and yet, at first coming up, looked blacker and stronger than those in Squares not dung'd, either that Year, or ever, that I know of; yet these Lands brought a good Crop of the largest Potatoes, and very few small ones amongst them; but in the dung'd Lands, for want of Hoeing, the Potatoes were not worth the taking up; which proves, that in those Plants that are planted so as to leave Spaces wide enough for Repetitions of Hoeing, that Instrument can raise more Nourishment to them, than a good Coat of Dung with common Tillage.

Another thing I have more particularly observ'd, *viz*. That the more successive Crops are planted in wide Intervals, and often ho'd, the better the Ground does maintain them; the last Crop is still the best, without Dung, or changing the sort of Plant; and this is visible in Parts of the same Field, where some Part has a first, some other Part a second, the rest a third Crop growing all together at the same time; which seems to prove, that as the Earth is made by this Operation to dispense or distribute her Wealth to Plants, in proportion to the Increase of her inner Superficies (which is the Pasture of Plants); so the Atmosphere, by the Riches in Rain and Dews, does annually reimburse her in proportion to the same Superficies, with an Overplus for Interest: But if that Superficies be not increased to a competent degree, and, by frequent Repetitions of Hoeing, kept increasing (which never happens in common Husbandry) this Advantage is lost; and, without often repeated Stercoration, every Year's Crop grows worse; and it has been made evident by Trials, which admit of no Dispute, that Hoeing, without Dung or Fallow, can make such Plants as stand in wide Intervals, more vigorous in the same Ground, than both common Dunging and Fallowing can do without Hoeing.

This sort of Hoeing has in truth every Year the Effect of a Summer-fallow; tho' it yearly produce a good Crop.

This is one Reason of the different Effects Plants have upon the Soil; some are said to enrich it, others to burn it, *i. e.* to impoverish it; but I think it may be observed, that all those Plants, which are usually ho'd, are reckoned among the Enrichers; and tho' it be certain, that some Species of Plants are, by the Heat of their Constitution, greater Devourers than those of another Species of equal Bulk; yet there is Reason to believe, that were the most cormorant Plant of them all to be commonly ho'd, it would gain the Reputation of an Enricher or Improver of the Soil; except it should be such, as might occasion Trouble, by filling it full of its shatter'd Seeds, which might do the Injury of Weeds to the next Crop; and except such Plants, which have a vast Bulk to be maintained a long time, as Turnep-seed.

The wider the Intervals are, the more Earth may be divided; for the Row takes up the same Room with a wide, or a narrow Interval; and therefore with the wide, the unho'd Part bears a less Proportion to the ho'd Part than in the narrow.

And 'tis to no purpose to hoe, where there is not Earth to be ho'd, or room to hoe it in.

There are many ways of Hoeing with the Hoe-Plough; but there is not room to turn Two deep clean Furrows in an Interval that is narrower than Four Feet Eight Inches; for if it want much of this Breadth, one, at least, of these Furrows, will reach, and fall upon the next Row, which will be very injurious to the Plants; except of grown St. Foin, and such other Plants, that can bear to have the Earth pull'd off them by Harrows.

Thus much of Hoeing in general may suffice: And different sorts of Plants, requiring different Management; that may more properly be described in the Chapter, where particular Vegetables are treated of.

It may not be amiss to add, that all sorts of Land are not equally proper for Hoeing: I take it, that a dry friable Soil is the best. Intractable wet Clays, and such Hills that are too steep for Cattle to draw a Plough up and down them, are the most improper.

That 'tis not so beneficial to hoe in Common-fields, is not in respect of the Soil, but to the old Principles, which have bound the Owners to unreasonable Customs of changing the Species of Corn, and make it necessary to fallow every Second, Third, or Fourth Year at farthest.

* * *

Of Differences between the Old and the New Husbandry

In order to make a Comparison between the Hoeing-Husbandry, and the old Way, there are Four Things; whereof the Differences ought to be very well considered.

> I. *The Expence* ⎫
> II. *The Goodness* ⎬ of a Crop.
> III. *The Certainty* ⎭
> IV. *The Condition in which the Land is left after a Crop.*

The Profit or Loss arising from Land, is not to be computed, only from the Value of the Crop it produces; but from its Value, after all Expences of Seed, Tillage, &c. are deducted.

Thus, when an Acre brings a Crop worth *Four Pounds,* and the Expences thereof amount to *Five Pounds,* the Owner's Loss is *One Pound;* and when an Acre brings a Crop which yields *Thirty Shillings,* and the Expence amounts to no more than *Ten Shillings,* the Owner receives *One Pound,* clear Profit, from this Acre's very small Crop, as the other loses *One Pound* by his greater Crop.

The usual Expences of an Acre of Wheat, sown in the old Husbandry, *in the Country where I live, is, in some Places, for Two Bushels and an half of Seed; in other Places Four Bushels and an half; the least of these Quantities at* Three Shillings per *Bushel, being the present Price, is* Seven Shillings *and* Six-pence. *For Three Plowings, Harrowing, and Sowing,* Sixteen Shillings; *but if plow'd Four times, which is better,* One Pound. *For Thirty Load of Dung, to a Statute Acre, is* Two Pounds Five Shillings. *For Carriage of the Dung, according to the Distance, from* Two Shillings *to* Six-pence *the Load,* One Shilling *being the Price most common, is* One Pound Ten Shillings. *The Price for Weeding is very uncertain; it has sometimes cost* Twelve Shillings, *sometimes* Two Shillings per *Acre.*

	l.	*s.*	*d.*
In Seed and Tillage, nothing can be abated of	01	03	06
For the Weeding, one Year with another, is more than	00	02	00
For the Rent of the Year's Fallow	00	10	00
For the Dung; 'tis in some Places a little cheaper, neither do they always lay on quite so much; therefore abating 15s. in that Article, we may well set Dung and Carriage at	02	10	00
Reaping commonly 5s. sometimes less	00	04	06
Total	04	10	00

Folding of Land with Sheep is reckoned abundantly cheaper than Cart-dung; but this is to be questioned, because much Land must lie still for keeping a Flock (unless there be Downs); and for their whole Year's keeping, with both Grass and Hay, there are but Three Months of the Twelve wherein the Fold is of any considerable Value; this makes the Price of their Manure quadruple to what it would be, if equally good all the Year, like Cart-dung: And folding Sheep yield little Profit, besides their Dung; because the Wool of a Flock, except it be a large one, will scarce pay the Shepherd and the Shearers. But there is another thing yet, which more inhances the Price of Sheep-dung; and that is, the dunging the Land with their Bodies, when they all die of the Rot, which happens too frequently in many Places; and then the whole Crop of Corn must go to purchase another Flock, which may have the same Fate the ensuing Year, if the Summer prove wet; and so may the Farmer be served for several more successive Years, unless he should break, and another take his Place, or that dry Summers come in time to prevent it. To avoid this Misfortune, he would be glad to purchase Cart-dung at the highest Price, for supplying the Place of his Fold; but 'tis only near Cities, and great Towns, that a sufficient Quantity can be procured.

But, supposing the Price of Dunging to be only Two Pounds Ten Shillings, and the general Expence of an Acre of Wheat, when sown, at Three Shillings per Bushel, to be Four Pounds Ten Shillings, with the Year's Rent of the Fallow;

The Expences of planting an Acre of Wheat in the Hoeing-Husbandry, is Three Pecks of Seed; at *Three Shillings per* Bushel, is *Two Shillings* and *Three-pence.* The whole Tillage, if done by Horses, would be *Eight Shillings;* because our Two Plowings, and Six Hoeings, are equal to Two Plowings; the common Price whereof is *Four Shillings* each; but this we diminish half, when done by Oxen kept on St. Foin, in this manner; *viz.* Land worth *Thirty Shillings* Rent, drill'd with St. Foin, will well maintain an Ox a Year, and sometimes Hay will be left to pay for the Making: We cannot therefore allow more than *One Shilling* a Week for his Work, because his keeping comes but to *Seven-pence* a Week round the Year.

In plain Plowing, Six Feet contains Eight Furrows; but we plow a Six-feet Ridge at Four Furrows, because in this there are Two Furrows cover'd in the Middle of it, and one on each Side of it lies open. Now what we call one Hoeing, is only Two Furrows of this Ridge, which is equal to a Fourth Part of one plain Plowing; so that the Hoeing of Four Acres requires an equal Number of Furrows with One Acre, that is plow'd plain, and equal time to do it in (except that the Land, that is kept in Hoeing, works much easier than that which is not).

All the Tillage we ever bestow upon a Crop of Wheat that follows a ho'd Crop, is equal to Eight Hoeings; Two of which may require Four Oxen each, One of them Three Oxen, and the other Five Hoeings Two Oxen each. However, allow Three Oxen to each single Hoeing, taking them all one with another, which is Three Oxen more than it comes to in the Whole.

Begin at Five in the Morning, and in about Six Hours you may hoe Three Acres, being equal in Furrows to Three Rood; *i. e.* Three Quarters of an Acre. Then turn the Oxen to Grass, and after resting, eating, and drinking, Two Hours and an half, with another Set of Oxen begin Hoeing again; and by or before half an Hour after Seven at Night, another like Quantity may be ho'd. These are the Hours the Statute has appointed all Labourers to work, during the Summer Half-year.

To hoe these Six Acres a Day, each Set of Oxen draw the Plough only Eight Miles and a Quarter, which they may very well do in Five Hours; and then the Holder and Driver will be at their Work of Plowing Ten Hours, and will have Four Hours and an half to rest, &c.

The Expence then of hoeing Six Acres in a Day, in this manner, may be accounted, at *One Shilling* the Man that holds the Plough, *Six-pence* the Boy that drives the Plough, *One Shilling* for the Six Oxen, and *Six-pence* for keeping the Tackle in Repair. The whole Sum for hoeing these Six Acres is *Three Shillings,* being *Six-pence per Acre.*

They who follow the old Husbandry cannot keep Oxen so cheap, because they can do nothing without the Fold, and Store-sheep will spoil the St. Foin. They may almost as well keep Foxes and Geese together, as Store-sheep and good St. Foin. Besides, the sowed St. Foin cost Ten times as much the Planting as drill'd St. Foin does, and must be frequently manured, or else it will soon decay; especially upon all sorts of chalky Land, whereon 'tis most commonly sown.

The Expence of Drilling cannot be much; for as we can hoe Six Acres a Day, at Two Furrows on each Six-feet Ridge, so we may drill Twenty-four Acres a Day, with a Drill that plants Two of those Ridges at once; and this we may reckon a *Peny Halfpeny* an Acre. But because we find it less Trouble to drill single Ridges, we will set the Drilling, at most, *Six-pence per* Acre.

As every successive Crop (if well managed) is more free from Weeds than the preceding Crop; I will set it all together at *Six-pence* an Acre for Weeding.

For a Boy or a Woman to follow the Hoe-plough, to uncover the young Wheat, when any Clods of Earth happen to fall on it, for which Trouble there is seldom necessary above once to a Crop, *Two-pence* an Acre. *One Peny* is too much for Brine and Lime for an Acre.

Reaping this Wheat is not worth above half as much as the Reaping of a sown Crop of equal Value; because the drill'd standing upon about a Sixth Part of the Ground, a Reaper may cut almost as much of the Row at one Stroke,

The whole Expense of an Acre of drilled Wheat

	l.	s.	d.
For Seed	00	02	03
For Tillage	00	04	00
For Drilling	00	00	06
For Weeding	00	00	06
For Uncovering	00	00	02
For Brine and Lime	00	00	01
For Reaping	00	02	06
Total	00	10	00
The Expence of an Acre of sowed Wheat is	04	00	00
To which must be added, for the Year's Rent of the Fallow	00	10	00
Total	04	10	00

at he could at Six, if the same stood dispersed all over the Ground, as the sowed does; and because he who reaps sowed Wheat, must reap the Weeds along with the Wheat; but the drilled has no Weeds; and besides, there go a greater Quantity of Straw, and more Sheaves, to a Bushel of the sowed, than of the drilled. And since some Hundred Acres of drilled Wheat have been reaped at *Two Shillings* and *Six-pence per* Acre, I will count that to be the Price.

If I have reckon'd the Expence of the drilled at the lowest Price, to bring it to an even Sum; I have also abated in the other more than the whole Expence of the drilled amounts unto.

And thus the Expence of a drilled Crop of Wheat is but the Ninth Part of the Expence of a Crop sown in the common Manner.

'Tis also some Advantage, that less Stock is required where no Store-sheep are used.

Of the different Goodness of a Crop.

The Goodness of a Crop consists in the Quality of it, as well as the Quantity; and Wheat being the most useful Grain, a Crop of this is better than a Crop of any other Corn, and the ho'd Wheat has larger Ears (and a fuller Body) than sow'd Wheat. We can have more of it, because the same Land will produce it every Year, and even Land, which, by the Old Husbandry, would not be made to bear Wheat at all: So that, in many Places, the New Husbandry can raise Ten Acres of Wheat for One that the Old can do; because where Land is poor, they sow but a Tenth Part of it with Wheat.

We do not pretend, that we have always greater Crops, or so great as some sown Crops are, especially if those mention'd by Mr. *Houghton* be not mistaken.

The greatest Produce I ever had from a single Yard in Length of a double Row, was Eighteen Ounces: The Partition of this being Six Inches, and the Interval Thirty Inches, was, by Computation, Ten Quarters (or Eighty Bushels) to an Acre.

I had also Twenty Ounces to a like Yard of a Third successive Crop of Wheat; but this being a treble Row, and the Partitions and Interval being wider, and supposed to be in all Six Feet, was computed to Six Quarters to an Acre. And if these Rows had been better order'd than they were, and the Earth richer, and more pulveriz'd, more Stalks would have tillered out, and more Ears would have attained their full Size, and have equall'd the best, which must have made a much greater Crop than either of these were.

But to compare the different Profit, we may proceed thus: The Rent and Expence of a drill'd Acre being One Pound, and of a sow'd Acre Five Pounds; One Quarter of Corn, produced by the drill'd, bears an equal Proportion in Profit to the One Pound, as Five Quarters, produced by the other, do to the Five Pounds. As suppose it be of Wheat, at Two Shillings and Six-pence a Bushel, there is neither Gain nor Loss in the one nor the other Acre, though the former yield but One Quarter, and the other Five; but if the drill'd Acre yield Two Quarters, and the sow'd Acre Four Quarters at the same Price, the drill'd brings the Farmer One Pound clear Profit, and the sown, by its Four Quarters, brings the other One Pound Loss. Likewise suppose the drilling Farmer to have his Five Pounds laid out on Five Acres of Wheat, and the other to have his Five Pounds laid out on One dung'd Acre; then let the Wheat they produce be at what Price it will, if the Five Acres have an equal Crop to the One Acre, the Gain or Loss must be equal: But when Wheat is cheap, as we say it is when sold at Two and Six-pence a Bushel, then if the Farmer, who follows the old Method, has Five Quarters on his Acre, he must sell it all to pay his Rent and Expence; but the other having Five Quarters on each of his Five Acres, the Crop of One of them will pay the Rent and Expence of all his Five Acres, and he may keep the remaining Twenty Quarters, till he can sell them at Five Shillings a Bushel, which amounts to Forty Pounds, wherewith he may be able to buy Four of his Five Acres at Twenty Years Purchase, out of One Year's Crop, whilst the Farmer who pursues the old Method, must be content to have only his Labour for his Travel; or if he pretends to keep his Wheat till he sells it at Five Shillings a Bushel, he commonly runs in Debt to his Neighbours, and in Arrear of his Rent; and if the Markets do not rise in time, or if his Crops fail in the Interim, his Landlord seizes on his Stock, and then he knows not how it may be sold; Actions are brought against him; the Bailiffs and Attorneys pull him to Pieces; and then he is undone.

The Certainty of a Crop.

The Certainty of a Crop is much to be regarded, it being better to be secure of a moderate Crop, than to have but a mere Hazard of a great one. The Farmer who adheres to the old Method is often deceiv'd in his Expectation, when his Crop at coming into Ear is very big, as well as when 'tis in Danger of being too little. Our hoeing Farmer is much less liable to the Hazard of either of those Extremes; for when his Wheat is big, 'tis not apt to lodge or fall down, which Accident is usually the utter Ruin of the other; he is free from the Causes which make the contrary Crop too little.

A very effectual Means to prevent the failing of a Crop of Wheat, is to plow the pulveriz'd Earth for Seed early, and when 'tis dry. The early Season also is more likely to be dry than the latter Season is.

The Advocate for the old Method is commonly late in his sowing; because he can't fallow his Ground early, for fear of killing the Couch, and other Grass that maintains his folding Sheep, which are so necessary to his Husbandry: And when 'tis sow'd late, it must not be sow'd dry, for then the Winter might kill the young Wheat. Neither can he at that time plow dry, and sow wet, because he commonly sows under Furrow; that is, sows the Seed first, and plows it in as fast as 'tis sown. If he sows early (as he may if he will) in light Land, he must not sow dry, for fear the Poppies and other Weeds should grow, and devour his Crop; and if his Land be strong, let it be sown early, wet or dry (tho' wet is worst), 'tis apt to grow so stale and hara by the Spring, that his Crop is in Danger of starving, unless the Land be very rich, or much dung'd; and then the Winter and Spring proving kind, it may not be in less Danger of being so big as to fall down, and be spoil'd. Another thing is, that though he had no other Impediment against plowing dry, and sowing wet, 'tis seldom that he has time to do it in; for he must plow all his Ground, which is Eight Furrows in Six Feet; and, whilst it is wet, must lie still with his Plough. When he sows under Furrow, he fears to plow deep, lest he bury too much of his Seed; and if he plows shallow, his Crop loses the Benefit of deep plowing, which is very great. When he sows upon Furrow (that is after 'tis plow'd) he must harrow the Ground level to cover the Seed; and that exposes the Wheat the more to the cold Winds, and suffers the Snow to be blown off it, and the Water to lie longer on it; all which are great Injuries to it.

Our Hoeing Husbandry is different in all of the fore-mentioned Particulars.

1. We can plow the Two Furrows whereon the next Crop is to stand, immediately after the present Crop is off.

2. We have no Use of the Fold; because our Ground has annually a Crop growing on it, and it must lie still a Year, if we would fold it, and that Crop would be lost; and all the Good the Fold could do to the Land, would be only to help to pulverize it for one single Crop; its Benefit not lasting to the Second Year. And so we should be certain of losing one Crop for the very uncertain Hopes of procuring one the ensuing Year by the Fold; when 'tis manifest by the adjoining Crops, that we can have a much better Crop every Year, without a Fold, or any other Manure.

3. We can plow dry, and drill wet, without any manner of Inconvenience.

4. He fears the Weeds will grow, and destroy his Crop: We hope they will grow, to the end we may destroy them.

5. We do not fear to plant our Wheat early (so that we plow dry), because we can help the Hardness or Staleness of the Land by Hoeing.

6. The Two Furrows of every Ridge whereon the Rows are to be drill'd, we plow dry; and if the Weather prove wet before these are all finish'd, we can plow the other Two Furrows up to them, until it be dry enough to return to our plowing the first Two Furrows; and after finishing them, let the Weather be wet or dry, we can plow the last Two Furrows. We can plow our Two Furrows in the Fourth Part of the Time they can plow their Eight, which they must plow dry all of them, in every Six Feet; for they can't plow part dry, and the rest when 'tis wet, as we can.

7. We never plant our Seed under Furrow, but place it just at the Depth which we judge most proper; and that is pretty shallow, about Two Inches deep; and then there is no Danger of burying it.

8. We not only plow a deep Furrow, but also plow to the Depth of Two Furrows; that is, we trench-plow where the Land will allow it; and we have the greatest Convenience imaginable for doing this, because there are Two of our Four Furrows always lying open; and Two plow'd Furrows (that is, one plow'd under another) are as much more advantageous for the nourishing a Crop, as Two Bushels of Oats are better than One for nourishing an Horse: Or if the Staple of the Land be too thin or shallow, we can help it by raising the Ridges prepar'd for the Rows the higher above the Level.

9. We also raise an high Ridge in the Middle of each Interval above the Wheat before Winter, to protect it from the cold Winds, and to prevent the Snow from being driven away by them. And the Furrows or Trenches, from whence the Earth of these Ridges is taken, serve to drain off the Water from the Wheat, so that, being drier, it must be warmer than the harrow'd Wheat, which has neither Furrows to keep it dry, nor Ridges to shelter it, as every Row of ours has on both Sides of it.

The Condition in which the Land is left after a Crop.

The different Condition the Land is left in after a Crop, by the one and the other Husbandry, is not less considerable than the different Profit of the Crop.

A Piece of Eleven Acres of a poor, thin, chalky Hill was sown with Barley in the common Manner, after a ho'd Crop of Wheat; and produced full

Five Quarters and an half to each Acre (reckoning the Tythe); which was much more than any Land in all the Neighbourhood yielded the same Year; tho' some of it be so rich, as that One Acre is woth Three Acres of this Land: And no Man living can remember, that ever this produced above half such a Crop before, even when the best of the common Management has been bestow'd upon it.

A Field, that is a sort of an Heath-ground, used to bring such poor Crops of Corn, that heretofore the Parson carried away a whole Crop of Oats from it, believing it had been only his Tythe. The best Management that ever they did or could bestow upon it, was to let it rest Two or Three Years, and then fallow and dung it, and sow it with Wheat, next to that with Barley and Clover, and then let it rest again; but I cannot hear of any good Crop that it ever produced by this or any other of their Methods; 'twas still reckoned so poor, that nobody cared to rent it. They said Dung and Labour were thrown away upon it; then immediately after Two sown Crops of black Oats had been taken off it, the last of which was scarce worth the mowing, it was put into the Hoeing Management; and when Three ho'd Crops had been taken from it, it was sown with Barley, and brought a very good Crop, much better than ever it was known to yield before; and then a good Crop of ho'd Wheat succeeded the Barley, and then it was again sown with Barley, upon the Wheat-stubble; and that also was better than the Barley it used to produce.

Now all the Farmers of the Neighbourhood affirm, that it is impossible but that this must be very rich Ground, because they have seen it produce Six Crops in Six Years, without Dung or Fallow, and never one of them fail. But, alas! this different Reputation they give to the Land, does not at all belong to it, but to the different Sorts of Husbandry; for the Nature of it cannot be alter'd but by that, the Crops being all carried off it, and nothing added to supply the Substance those Crops take from it, except (what Mr. *Evelyn* calls) the celestial Influences; and that these are received by the Earth, in proportion to the Degrees of its Pulveration.

A Field was drilled with Barley after an ho'd Crop; and another adjoining to it on the same Side of the same poor Hill, and exactly the same sort of Land, was drill'd with Barley also, Part of it after the sown Crop, the same Day with the other; there was only this Difference in the Soil, that the former of these had no manner of Compost on it for many Years before, and the latter was dunged the Year before: Yet its Crop was not near so good as that which followed the ho'd Crop; tho' the latter had twice the Plowing that the former had before drilling, and the same Hoeings afterwards; *viz.* Each was ho'd Three times.

A Field of about Seventeen Acres was Summer-fallowed, and drilled with Wheat; and with the Hoeing brought a very good Crop (except Part of it, which being eaten by trespassing Sheep in the Winter, was somewhat blighted); the *Michaelmas* after that was taken off, the same Field was drilled again with Wheat, upon the Stubble of the former, and ho'd: This Second Crop was a good one, scarce any in the Neighbourhood better. A piece of Wheat adjoining to it, on the very same sort of Land (except that this latter was always reckoned better, being thicker in Mould above the Chalk), sown at the same time on dung'd Fallows, and the Ground always dung'd once in Three Years; yet this Crop fail'd so much, as to be judged, by some Farmers, not to exceed the Tythe of the other: That the ho'd Field has receiv'd no Dung or Manure for many Years past, is because it lies out of the Reach for carrying of Cart-Dung, and no Fold being kept on my Farm: But I cannot say, I think there was quite so much Odds betwixt this Second undung'd ho'd Crop and the sown; yet this is certain, that the former is a good, and the latter a very bad Crop.

I could give many more Instances of the same Kind, where ho'd Crops and sown Crops have succeeded better after ho'd Crops than after sown Crops, and never yet have seen the contrary; and therefore am convinced, that the Hoeing (if it be duly performed) enriches the Soil more than Dung and Fallows, and leaves the Land in a much better Condition for a succeeding Crop. The Reason I take to be very obvious: The artificial Pasture of Plants is made and increased by Pulveration only; and nothing else there is in our Power to enrich our Ground, but to pulverize it, and keep it from being exhausted by Vegetables. Superinductions of Earth are an Addition of more Ground, or changing it, and are more properly purchasing than cultivating.

Their One Year's Tillage, which is but Two Plowings before Seed-time, commonly makes but little Dust; and that which it does make, has but a short time to lie exposed for Impregnation; and after the Wheat is sown, the Land lies unmoved for near Twelve Months, all the while gradually losing its Pasture, by subsiding, and by being continually exhausted in feeding a treble Stock of Wheat-plants, and a Stock of Weeds, which are sometimes a greater Stock. This puts the Advocates for the old Method upon a Necessity of using of Dung, which is, at best, but a *Succedaneum* of the Hoe; for it depends chiefly on the Weather, and other Accidents, whether it may prove sufficient by Fermentation to pulverize in the Spring, or no: And it is a Question whether it will equal Two additional Hoeings, or but one; tho', as I have computed it, one Dunging costs the Price of One hundred Hoeings.

When they have done all they can, the Pasture they raise, is generally too little for the Stock that is to be maintain'd upon it, and much the greatest Part of the Wheat plants are starved; for from Twenty Gallons of Seed they sow on an Acre, they receive commonly no more than Twenty Bushels of Wheat in their Crop, which is but an Increase of Eight Grains for one: Now, considering how many Grains there are in one good Ear, and how many Ears on one Plant, we find, that there is not One Plant in Ten that lives till Harvest, even when there has not been Frost in the Winter sufficient to kill any of them; or if we count the Number of Plants that come up on a certain Measure of Ground, and count them again in the Spring, and likewise at Harvest, we shall be satisfied, that most or all of the Plants that are missing, could die by no other Accident than want of Nourishment.

They are obliged to sow this great Quantity of Seed, to the end that the Wheat, by the great Number of Plants, may be the better able to contend with the Weeds; and yet, too often, at Harvest, we see the great Crop of Weeds, and very little Wheat among them. Therefore this Pasture, being insufficient to maintain the present Crop, without starving the greatest Part of its Plants, is likely to be less able to maintain a subsequent Crop, than that Pasture which is not so much exhausted.

When their Crop of Wheat is much less than ours, their Vacancies, if computed all together, may be greater than those of our Partitions and Intervals; theirs, by being irregular, serve chiefly for the Protection of Weeds; for they cannot be plow'd out, without destroying the Corn, any more than Cannons firing at a Breach, whereon both Sides are contending, can kill Enemies, and not Friends.

Their Plants stand on the Ground in a confused manner, like a Rabble; ours like a disciplin'd Army: We make the most of our Ground; for we can, if we please, cleanse the Partitions with a Hand-hoe; and for the rest, if the Soil be deep enough to be drill'd on the Level, in treble Rows, the Partitions at Six Inches, the Intervals Five Feet; Five Parts in Six of the whole Field may be pulveriz'd every Year, and at proper times all round the Year.

The Partitions being one Sixth-part for the Crop to stand on, and to be nourished in the Winter, one other Sixth-part being well pulveriz'd, may be sufficient to nourish it from thence till Harvest; the Remainder, being Two-thirds of the Whole, may be kept unexhausted, the One-third for one Year, and the other Third of it Two Years; all kept open for the Reception of the Benefits descending from above, during so long a time; whilst the sowed Land is shut against them every Summer, except the little time in which it is fallow'd, once in Three Years, and a little, perhaps, whilst they plow it for Barley in the Winter, which is a Season seldom proper for pulverizing the Ground.

Their Land must have been exhausted as well by those supernumerary Plants of Wheat, while they lived, as by those that remain for the Crop, and by the Weeds. Our Land must be much less exhausted, when it has never above one Third-part of the Wheat-plants to nourish that they have, and generally no Weeds; so that our ho'd Land having much more vegetable Pasture made, and continually renewed, to so much a less Stock of Plants, must needs be left, by every Crop, in a much better Condition than theirs is left in by any one of their sown Crops, altho' our Crops of Corn at Harvest be better than theirs.

They object against us, saying, That sometimes the Hoeing makes Wheat too strong and gross, whereby it becomes the more liable to the Blacks (or Blight or Insects): But this is the Fault of the Hoer; for he may choose whether he will make it too strong, because he may apply his Hoeings at proper times only, and apportion the Nourishment to the Number and Bulk of his Plants. However, by this Objection they allow, that the Hoe can give Nourishment enough, and therefore they cannot maintain, that there is a Necessity of Dung in the Hoeing-Husbandry; and that, if our Crops of Wheat should happen to suffer, by being too strong, our Loss will be less than theirs, when that is too strong, since it will cost them Nine times our Expence to make it so.

A Second Objection is, That as Hoeing makes poor Land become rich enough to bear good Crops of Wheat for several Years successively, the same must needs make very good Land become too rich for Wheat. I answer, That if possibly it should so happen, there are Two Remedies to be used in such a Case; the one is to plant it with Beans, or some other Vegetables, which cannot be over-nourished, as Turneps, Carrots, Cabbages, and such-like, which are excellent Food for fatting of Cattle; or else they may make use of the other infallible Remedy, when that rich Land, by producing Crops every Year in the Hoeing-Husbandry, is grown too vigorous and resty, they may soon take down its Mettle, by sowing it a few Years in their old Husbandry, which will fill it again with a new Stock of Weeds, that will suck it out of Heart, and exhaust more of its Vigour, than the Dung, that helps to produce them, can restore.

There is a Third Objection, and that is, That the Benefit of some Ground is lost where the Hoe-plough turns at each End of the Lands: But this cannot be much, if any, Damage; because about Four-square Perch to a Statute Acre is sufficient for this Purpose; and that, at the Rate of *Ten Shillings* Rent, comes to but Three-pence, tho' this varies, according as the Piece is longer or shorter; and supposing the most to be Eight Perch, that is but *Six-pence per* Acre; and that is not lost neither; for whether it be of natural or artificial Grass, the Hoe-plough, in turning on it, will scratch it, and leave some Earth on it, which will enrich it so much, that it may be worth its Rent for Baiting of Horses or Oxen upon it. And besides, these Ends are commonly near Quick-hedges or Trees, which do so exhaust it, that when no Cattle come there to manure it, 'tis not worth the Labour of plowing it.

American Indian Agriculture

John Smith's Description of Virginia, 1612

From John Smith, *The Voyages and Discoveries of Captain John Smith in Virginia* (Oxford, 1612), pp. 56–63.

Of Such Things Which are Naturall in Virginia and How They Vse Them

Virginia doth afford many excellent vegitables and liuing Creatures, yet grasse there is little or none but what groweth in lowe Marishes: for all the Countrey is overgrowne with trees, whose droppings continually turneth their grasse to weedes, by reason of the rancknesse of the ground; which would soone be amended by good husbandry. The wood that is most common is Oke and Walnut; many of their Okes are so tall and straight, that they will beare two foot and a halfe square of good timber for 20 yards long. Of this wood there is 2 or 3 seuerall kinds. The Acornes of one kind, whose barke is more white than the other, is somewhat sweetish; which being boyled halfe a day in severall waters, at last afford a sweete oyle, which they keep in goards to annoint their heads and ioints. The fruit they eate, made in bread or otherwise.

There is also some Elme, some black walnut tree, and some Ash: of Ash and Elme they make sope Ashes. If the trees be very great, the ashes will be good, and melt to hard lumps: but if they be small, it will be but powder, and not so good as the other.

Of walnuts there is 2 or 3 kindes: there is a kinde of wood we called Cypres, because both the wood, the fruit, and leafe did most resemble it; and of those trees there are some neere 3 fadome about at the root, very straight, and 50, 60, or 80 foot without a braunch.

By the dwelling of the *Savages* are some great Mulbery trees; and in some parts of the Countrey, they are found growing naturally in prettie groues. There was an assay made to manage silke, and surely the wormes prospered excellent well, till the master workeman fell sicke: at which time they were eaten with rats.

In some parts, were found some Chestnuts whose wild fruit equalize the best in *France, Spaine, Germany,* or *Italy,* to their tast[e]s that had tasted them all.

Plumbs there are of 3 sorts. The red and white are like our hedge plumbs: but the other, which they call *Putchamins,* grow as high as a *Palmeta.* The fruit is like a medler; it is first greene, then yellow, and red when it is ripe: if it be not ripe it will drawe a man's mouth awrie with much torment; but when it is ripe, it is as delicious as an Apricock.

They haue Cherries, and those are much like a Damsen; but for their tastes and colour, we called them Cherries. We see some few Crabs, but very small and bitter.

Of vines, [there is] great abundance in many parts, that climbe the toppes of the highest trees in some places, but these beare but fewe grapes. But by the riuers and Savage[s] habitations where they are not overshadowed from the sunne, they are covered with fruit, though never pruined nor manured. Of those hedge grapes, wee made neere 20 gallons of wine, which was neare as good as your French Brittish wine, but certainly they would proue good were they well manured.

There is another sort of grape neere as great as a Cherry, this they call *Messaminnes*; they bee fatte, and the iuyce thicke: neither doth the tast so well please when they are made in wine.

They haue a small fruit growing on little trees, husked like a Chesnut, but the fruit most like a very small acorne. This they call *Chechinquamins,* which they esteeme a great daintie. They haue a berry much like our gooseberry, in greatnesse, colour, and tast; those they call *Rawcomenes,* and doe eat them raw or boyled.

Of these naturall fruits they liue a great part of the yeare, which they vse in this manner. *The walnuts, Chesnuts, Acornes, and Chechinquamens* are dryed to keepe. When they need them, they breake them betweene two stones, yet some part of the walnut shels will cleaue to the fruit. Then doe they dry them againe vpon a mat ouer a hurdle. After, they put it into a morter of wood, and beat it very small: that done, they mix it with water, that the shels may sinke to the bottome. This water will be coloured as milke; which they cal *Pawcohiscora,* and keepe it for their vse.

The fruit like medlers, they call *Putchamins,* they cast vppon hurdles on a mat, and preserue them as Pruines. *Of their Chesnuts* and *Chechinquamens* boyled 4 houres, they make both broath and bread for their chiefe men, or at their greatest feasts.

Besides those fruit trees, there is *a white populer,* and another tree like vnto it, that yeeldeth a very cleere and an odoriferous *Gumme like Turpentine, which some called Balsom.* There are also *Cedars* and *Saxafras trees.* They also yeeld gummes in a small proportion of themselues. Wee tryed conclusions to extract it out of the wood, but nature afforded more than our arts.

In the wat[e]ry valleyes groweth *a berry,* which they call *Ocoughtanamnis,* very much like vnto Capers. These they dry in sommer. When they will eat them, they boile them neare halfe a day; for otherwise they differ not much from poyson. *Mattoume* groweth as our bents do in meddows. The seede is not much vnlike to rie, though much smaller. This they vse for a dainty bread buttered with deare suet.

During Sommer there are either *strawberries* which ripen in April; or mulberries which ripen in May and Iune. Raspises hurres; or a fruit that the Inhabitants call *Maracocks,* which is a pleasant wholsome fruit much like a lemond.

Many *hearbes* in the spring time there are commonly dispersed throughout the woods, good for brothes and sallets, as Violets, Purslin, Sorrell, &c. Besides many we vsed whose names we know not.

The chiefe roote they haue for foode is called *Tockawhoughe.* It groweth like a flagge in low muddy freshes. In one day a *Savage* will gather sufficient for a weeke. These rootes are much of the greatnes and taste of *Potatoes.* They vse to couer a great many of them with oke leaues and ferne, and then couer all with earth in the manner of a colepit; over it, on each side, they continue a great fire 24 houres before they dare eat it. Raw it is no better then poison, and being roasted, except it be tender and the heat abated, or sliced and dried in the sun, mixed with sorrell and meale or such like, it will prickle and torment the throat extreamely, and yet in sommer they vse this ordinarily for bread.

They haue an other roote which they call *wighsacan:* as th[e] other feedeth the body, so this cureth their hurts and diseases. It is a small root which they bruise and apply to the wound. *Pocones* is a small roote that groweth in the mountaines, which being dryed and beate in powder turneth red: and this they vse for swellings, aches, annointing their ioints, painting their heads and garments. They account it very pretious, and of much worth. *Musquaspenne* is a roote of the bignesse of a finger, and as red as bloud. In drying, it will wither almost to nothing. This they vse to paint their Mattes, Targets, and such like.

There is also *Pellitory of Spaine, Sasafrage,* and diuers other simples, which the Apothecaries gathered, and commended to be good and medicinable.

In the low Marishes, *growe plots of Onyons* containing an acre of ground or more in many places; but they are small, not past the bignesse of the Toppe of ones Thumbe.

Of beastes the chiefe are Deare, nothing differing from ours. In the deserts towards the heads of the riuers, ther[e] are many, but amongst the riuers few.

There is a beast they call Aroughcun, much like a badger, but vseth to liue on trees as Squirrels doe. *Their Squirrels* some as neare as greate as our smallest sort of wilde rabbits; some blackish or blacke and white, but the most are gray.

A small beast they haue, they call Assapanick, but we call them flying squirrels, because spreading their legs, and so stretching the largenesse of their skins that they haue bin seene to fly 30 or 40 yards. *An Opassom hath an head like a Swine,* and a taile like a *Rat,* and is of the bignes of a Cat. Vnder her belly shee hath a bagge, wherein shee lodgeth, carrieth, and sucketh her young. *Mussascus* is a beast of the forme and nature of our water *Rats,* but many of them smell exceeding strong of muske. Their Hares [are] no bigger then our Conies, and few of them to be found.

Their *Beares* are very little in comparison of those of *Muscovia* and *Tartaria.* The *Beaver* is as bigge as an ordinary water dogge, but his legges exceeding short. His fore feete like a dogs, his hinder feet like a Swans. His taile somewhat like the forme of a Racket bare without haire; which to eate, the *Savages* esteeme a great delicate. *They haue many Otters,* which, as the Beavers, they take with snares, and esteeme the skinnes great ornaments; and of all those beasts they vse to feede, when they catch them.

There is also a beast they call Vetchunquoyes in the forme of a wilde Cat. *Their Foxes* are like our siluer haired Conies, of a small proportion, and not smelling like those in England. *Their Dogges* of that country are like their Woloues, and cannot barke but howle; and their woloues [are] not much bigger then our English Foxes. *Martins, Powlecats, weessels and Minkes* we know they haue, because we haue seen many of their skinnes, though very seldome any of them aliue.

But one thing is strange, that we could never perceiue their vermine *destroy our hennes, egges, nor chickens,* nor do any hurt: nor their flyes nor serpents [to be] anie waie pernitious; where [as] in the South parts of *America,* they are alwaies dangerous and often deadly.

Of birds, the Eagle is the greatest devourer. Hawkes there be of diuerse sorts as our Falconers called them, *Sparowhawkes, Lanarets, Goshawkes, Falcons* and *Osperayes;* but they all pray most vpon fish. *Patrridges* there are little bigger than our Quailes, wilde Turkies are as bigge as our tame. There are wooosels or blackbirds with red shoulders, thrushes, and diuerse sorts of small birds, some red, some blew, scarce so bigge as a wrenne, but few in Sommer. In winter there are great plenty of Swans, Craynes gray and white with blacke wings, Herons, Geese, Brants, Ducke, Wigeon, Dotterell, Oxeies, Parrats, and Pigeons. Of all those sorts great abundance, and some other strange kinds, to vs vnknowne by name. But in sommer not any, or a very few to be seene.

Of fish we were best acquainted with *Sturgeon, Grampus, Porpus, Seales, Stingraies* whose tailes are very dangerous. Brettes, mullets, white Salmonds, Trowts, Soles, Plaice, Herrings, Conyfish, Rockfish, Eeles, Lampreyes, Catfish, Shades, Pearch of 3 sorts, Crabs, Shrimps, Creuises, Oysters, Cocles, and Muscles. But the most strange fish is a smal one so like the picture of *S.* George his Dragon, as possible can be, except his legs and wings: and the To[a]defish which will swell till it be like to brust, when it commeth into the aire.

Concerning the entrailes of the earth little can be saide for certainty. There wanted good Refiners: for these that tooke vpon them to haue skill this way, tooke vp the washings from the mouneteaines and some moskered shining stones and spangles which the waters brought down; flattering themselues in their own vaine conceits to haue bin supposed that they were not, by the meanes of that ore, if it proued as their arts and iudgements expected. Only this is certaine, that many regions lying in the same latitude, afford mines very rich of diuerse natures. The crust also of these rockes would easily perswade a man to beleeue there are other mines then yron and steele, if there were but meanes and men of experience that knew the mine from *spare.*

Of their Planted fruits in Virginia and how they vse them.

They diuide the yeare into 5. seasons. Their winter some call *Popanow,* the spring *Cattapeuk,* the sommer *Cohattayough,* the earing of their Corne *Nepinough,* the haruest and fall of leafe *Taquitock.* From September vntill the midst of Nouember are the chiefe Feasts and sacrifice. Then haue they plenty of fruits as well planted as naturall, as corne greene and ripe, fish, fowle, and wilde beastes exceeding fat.

The greatest labour they take, is in planting their corne, for the country naturally is ouergrowne with wood. To prepare the ground they bruise the barke of the trees neare the root, then do they scortch the roots with fire that they grow no more.

The next yeare with a crooked peece of wood, they beat vp the woodes by the rootes; and in that [those] moulds, they plant their corne. Their manner is this. They make a hole in the earth with a sticke, and into it they put 4 graines of wheat and 2 of beanes. These holes they make 4 foote one from another. Their women and children do continually keepe it with weeding, and when it is growne midle high, they hill it about like a hop-yard.

In Aprill they begin to plant, but their chiefe plantation is in May, and so they continue till the midst of Iune. What they plant in Aprill they reape in August, for May in September, for Iune in October. Every stalke of their corne commonly beareth two eares, some 3, seldome any 4, many but one, and some none. Every eare ordinarily hath betwixt 200 and 500 graines. The stalke being green hath a sweet iuice in it, somewhat like a suger Cane, which is the cause that when they gather their corne greene, they sucke the stalkes: for as wee gather greene pease, so doe they their corne being greene, which excelleth their old.

They plant also pease they cal *Assentamens,* which are the same they cal in Italye, *Fagioli.* Their Beanes are the same the Turkes call *Garnanses,* but these they much esteeme for dainties.

Their corne they rost in the eare greene, and bruising it in a morter with a Polt, lappe it in rowles in the leaues of their corne, and so boyle it for a daintie. They also reserue that corne late planted that will not ripe[n], by roasting it in hot ashes, the heat thereof drying it. In winter they esteeme it being boyled with beans for a rare dish, they call *Pausarowmena.* Their old wheat they first steep a night in hot water, in the morning pounding it in a morter. They vse a small basket for their Temmes, then pound againe the great, and so separating by dashing their hand in the basket, receaue the flower in a platter made of wood scraped to that forme with burning and shels. Tempering this flower with water, they make it either in cakes, couering them with ashes till they bee baked, and then washing them in faire water, they drie presently with their owne heat: or else boyle them in water eating the broth with the bread which they call *Ponap.*

The grouts and peeces of the cornes remaining, by fanning in a Platter or in the wind away the branne, they boile 3 or 4 houres with water; which is an ordinary food they call *Vstatahamen.* But some more thrifty then cleanly, doe burne the core of the eare to powder which they call *Pungnough,* mingling that in their meale; but it never tasted well in bread, nor broth.

Their fish and flesh they boyle either very tenderly, or broyle it so long on hurdles over the fire; or else, after the *Spanish* fashion, putting it on a spit, they turne first the one side, then the other, til it be as drie as their ferkin beefe in the west *Indies,* that they may keepe it a month or more without putrifying. The broth of fish or flesh they eate as commonly as the meat.

In May also amongst their corne, they plant Pumpeons, and a fruit like vnto a muske millen, but lesse and worse; which they call *Macocks.* These increase exceedingly, and ripen in the beginning of Iuly, and continue vntil September. They plant also *Maracocks* a wild fruit like a lemmon, which also increase infinitely: they begin to ripe[n] in September and continue till the end of October.

When all their fruits be gathered, little els they plant, and this is done by their women and children; neither doth this long suffice them: for neere 3 parts of the yeare, they only obserue times and seasons, and liue of what the Country naturally affordeth from hand to mouth, &c.

Indian Customs, 1634

From William Wood, *New England's Prospect* (London, 1634), pp. 75–77, 105–08.

Of [Indian] dyet, cookery, meale-times, and hospitality at their kettles

Having done with their most needfull cloathings and ornamentall deckings; may it please you to feast your eyes with their belly-timbers, which I suppose would be but *stibium* to weake stomacks as they cooke it, though never so good of it selfe. In Winter-time they have all manner of fowles of the water and of the land, & beasts of the land and water, pond-fish, with Catharres and other rootes, *Indian* beanes and Clamms. In the Summer they have all manner of Sea-fish, with all sorts of Berries. For the ordering of their victuals, they boile or roast them, having large Kettles which they traded for with the *French* long since, and doe still buy of the *English* as their neede requires, before they had substantiall earthen pots of their owne making. Their spits are no other than cloven sticks sharped at one end to thrust into the ground; into these cloven sticks they thrust the flesh or fish they would have rosted, behemming a round fire with a dozen of spits at a time, turning them as they see occasion. Some of their scullerie having dressed these homely cates, presents it to his guests, dishing it up in a rude manner, placing it on the verdent carpet of the earth which Nature spreads them, without either trenchers napkins, or knives, upon which their hunger-sawced stomacks impatient of delayes, fals aboard without scrupling at unwashed hands, without bread, salt, or beere: lolling on the Turkish fashion, not ceasing till their full bellies leave nothing but emptie platters: they seldome or never make bread of their *Indian* corne, but feeth it whole like beanes, eating three or foure cornes with a mouthfull of fish or flesh, sometimes eating meate first, and cornes after, filling chinkes with their broth. In Summer, when their corne is spent, Isquouter-squashes is their best bread, a fruite like a young Pumpion. To say, and to speake paradoxically, they be great eaters, and yet little meate-men; when they visit our *English,* being invited to eate, they are very moderate, whether it be to shew their manners, or for shamefastnesse, I know not; but at home they will eate till their bellies stand fouth, ready to split with fulnesse; it being their fashion to eate all at some times, and sometimes nothing at all in two or three dayes, wise Providence being a stranger to their wilder wayes: they be right Infidels, neither caring for the morrow, or providing for their owne families; but as all are fellowes at foot-ball, so they all meete friends at the kettle, saving their Wives, that dance a Spaniell-like attendance at their backes for their bony fragments. If their imperious occasions cause them to travell, the best of their victuals for their journey is *Nocake,* (as they call it) which is nothing but *Indian* Corne parched in the hot ashes; the ashes being sifted from it, it is afterward beaten to powder, and put into a long leatherne bag, trussed at their backe like a knapsacke; out of which they take thrice three spoonefulls a day, dividing it into three meales. If it be in Winter, and Snow be on the ground, they can eate when they please, stopping Snow after their dusty victuals, which otherwise would feed them little better than a Tiburne halter. In Summer they must stay till they meete with a Spring or Brooke, where they may have water to prevent the imminent danger of choaking. with this strange *viaticum* they will travell foure or five daies together, with loads fitter for Elephants than men. But though they can fare so hardly abroad, at home their chaps must walke night and day as long as they have it. They keepe no set meales, their store being spent, they champe on the bit, till they meete with fresh supplies, either from their owne endeavours, or their wives indus- try, who trudge to the *Clam-bankes* when all other meanes faile. Though they be sometimes scanted, yet are they as free as Emperours, both to their Country-men and *English,* be he stranger, or neare acquaintance; counting it a great discourtesie, not to eate of their high-conceited delicates, and sup of their un-oat-meal'd broth, made thicke with Fishes, Fowles, and Beasts boyled all together; some remaining raw, the rest converted by over-much seething to a loathed mash, not halfe so good as *Irish Boniclapper.*

* * *

Of their women, their dispositions, employments, usage by their husbands, their Apparell, and Modesty.

To satisfie the curious eye of women-readers, who otherwise might thinke their sex forgotten, or not worthy a record, let them peruse these few lines, wherein they may see their owne happinesse, if weighed in the womans ballance of these ruder *Indians,* who scorne the tuterings of their wives, or to admit them as their equals, though their qualities and industrious deservings may justly claime the preheminence, and command better usage and more conjugall esteeme, their per- sons and features being every way correspondent, their qualifications more excel- lent, being more loving, pittifull, and modest, milde, provident, and laborious than their lazie husbands. Their employments be many: First their building of houses, whose frames are formed like our garden-arbours, something more round, very strong and handsome, covered with close-wrought mats of their owne weav- ing, which deny entrance to any drop of raine, though it come both fierce and long, neither can the piercing North winde finde a crannie, through which he can conveigh his cooling breath, they be warmer than our *English* houses; at the top is a square hole for the smoakes evacuation, which in rainy weather is covered with a pluver; these bee such smoakie dwellings, that when there is good fires, they are not able to stand upright, but lie all along under the smoake, never using any stooles or chaires, it being as rare to see an *Indian* sit on a stoole at home, as it is strange to see an *English* man sit on his heeles abroad. Their houses are smaller in the Summer, when their families be dispersed, by reason of heate and occasions. In Winter they make some fiftie or threescore foote long, fortie or fiftie men being inmates under one roofe; and as is their husbands occasion these poore tectonists are often troubled like snailes, to carrie their houses on their backs sometime to fishing-places, other times to hunting-places, after that to a planting place, where it abides the longest: an other work is their planting of corne, wherein they exceede our *English* husband-men, keeping it so cleare with their Clamme shell-hooes, as if it were a garden rather than a corne-field, not suffering a choaking weede to advance his audacious head above their infant corne, or an undermining worme to spoile his spurnes. Their corne being ripe, they gather it, and drying it hard in the Sunne, conveigh it to their barnes, which be great holes digged in the ground in forme of a brasse pot, seeled with rinds of trees, wherein they put their corne, covering it from the inquisitive search of their gurmandizing husbands, who would eate up both their allowed portion, and reserved feede, if they knew where to finde it. But our hogges having found a way to unhindge their barne doores, and robbe their garners, they are glad to implore their husbands helpe to roule the bodies of trees over their holes, to prevent those pioners, whose theeverie they as much hate as their flesh. An other of their employments is their Summer processions to get Lobsters for their husbands, wherewith they baite their hookes when they goe a fishing for Basse or Codfish. This is an every dayes walke, be the weather cold or hot, the waters rough or calme, they must dive sometimes over head and eares for a Lobster, which often shakes them by their hands with a churlish nippe, and bids them adiew. The tide being spent, they trudge home two or three miles, with a hundred weight of Lobsters at their backs, and if none, a hundred scoules meete them at home, and a hungry belly for two dayes after. Their husbands having caught any fish, they bring it in their boates as farre as they can by water, and there leave it; as it was their care to catch it, so it must be their wives paines to fetch it home, or fast: which done, they must dresse it and cooke it, dish it, and present it, see it eaten over their shoulders; and their loggerships having filled their paunches, their sweete lullabies scramble for their scrappes. In the Summer these *Indian* women when Lobsters be in their plenty and prime, they drie them to keepe for Winter, erecting scaffolds in the hot sun-shine, making fires likewise underneath them, by whose smoake the flies are expelled, till the substance remain hard and drie. In this manner they drie Basse and other fishes without salt, cutting them very thinne to dry suddainely, before the flies spoile them, or the raine moist them, having a speciall care to hang them in their smoakie houses, in the night and dankish weather.

In Summer they gather flagges, of which they make Matts for houses, and Hempe and Rushes, with dying stuffe of which they make curious baskets with intermixed colours and protractures of antique Imagerie: these baskets be of all sizes from a quart to a quarter, in which they carry their luggage.

Indian Agriculture, 1755

From James Adair, *The History of the American Indians; Particularly Those Nations adjoining to the Mississippi, East and West Florida, Georgia, South and North Carolina, and Virginia* (London, 1755), pp. 405–10.

The Indians formerly had stone axes, which in form commonly resembled a smith's chisel. Each weighed from one to two, or three pounds weight — They were made of flinty kind of stone: I have seen several, which chanced to escape being buried with their owners, and were carefully preserved by the old people, as respectable remains of antiquity. They twisted two or three tough hiccory slips, of about two feet long, round the notched head of the axe; and by means of this simple and obvious invention, they deadened the trees by cutting through the bark, and burned them, when they either fell by decay, or became thoroughly dry. With these trees they always kept up their annual holy fire; and they reckon it unlawful, and productive of many temporal evils, to extinguish even the culinary fire with water. In the time of a storm, when I have done it, the kindly women were in pain for me, through fear of the ill consequences attending so criminal an act. I never saw them to damp the fire, only when they hung up a brand in the appointed place, with a twisted grape-vine, as a threatening symbol of torture and death to the enemy; or when their kinsman dies. In the last case, a father or brother of the deceased, takes a fire-brand, and brandishing it two or three times round his head, with lamenting words, he with his right hand dips it into the water, and lets it sink down.

By the aforesaid difficult method of deadening the trees, and clearing the woods, the contented natives got convenient fields in process of time. And their tradition says they did not live straggling in the American woods, as do the Arabians, and rambling Tartars; for they made houses with the branches and bark of trees, for the summer-season; and warm mud-walls, mixt with soft dry grass, against the bleak winter, according to their present plan of building, which I shall presently describe. Now, in the first clearing of their plantations, they only bark the large timber, cut down the sapplings and underwood, and burn them in heaps; as the suckers shoot up, they chop them off close by the stump, of which they make fires to deaden the roots, till in time they decay. Though to a stranger, this may seem to be a lazy method of clearing the wood-lands; yet it is the most expeditious method they could have pitched upon, under their circumstances, as a common hoe and a small hatchet are all their implements for clearing and planting.

Every dwelling-house has a small field pretty close to it: and, as soon as the spring of the year admits, there they plant a variety of large and small beans, peas, and the smaller sort of Indian corn, which usually ripens in two months, from the time it is planted; though it is called by the English, the six weeks corn. Around this small farm, they fasten stakes in the ground, and tie a couple of long split hiccory, or white oak-sapplings, at proper distances to keep off the horses: though they cannot leap fences, yet many of the old horses will creep through these enclosures, almost as readily as swine, to the great regret of the women, who scold and give them ill names, calling them ugly mad horses, and bidding them "go along, and be sure to keep away, otherwise their hearts will hang sharp within them, and set them on to spoil them, if envy and covetousness lead them back." Thus they argue with them, and they are usually as good as their word, by striking a tomohawk into the horse, if he does not observe the friendly caution they gave him at the last parting. Their large fields lie quite open with regard to fencing, and they believe it to be agreeable to the best rules of oeconomy; because, as they say, they can cultivate the best of their land here and there, as it suits their conveniency, without wasting their time in fences and childishly confining their improvements, as if the crop would eat itself. The women however tether the horses with tough young bark-ropes, and confine the swine in convenient penns, from the time the provisions are planted, till they are gathered in — the men improve this time, either in killing plenty of wild game, or coursing against the common enemy, and thereby secure the women and girls, and get their own temples surrounded with swan-feathered cap. In this manner, the Indians have to me, excused their long-contracted habit and practice.

The chief part of the Indians begin to plant their out-fields, when the wild fruit is so ripe, as to draw off the birds from picking up the grain. This is their general rule, which is in the beginning of May, about the time the traders set off for the English settlements. Among several nations of Indians, each town usually works together. Previous thereto, an old beloved man warns the inhabitants to be ready to plant on a prefixed day. At the dawn of it, one by order goes aloft, and whoops to them with shrill calls, "that the new year is far advanced, — that he who expects to eat, must work, — and that he who will not work, must expect to pay the fine according to old custom, or leave the town, as they will not sweat themselves for an healthy idle waster." At such times, may be seen many war-chieftains working in common with the people, though as great emperors, as those the Spaniards bestowed on the old simple Mexicans and Peruvians, and equal in power, (i. e. persuasive force) with the imperial and puissant Powhatan of Virginia, whom our generous writers raised to that prodigious pitch of power and grandeur, to rival the Spanish accounts. About an hour after sun-rise, they enter the field agreed on by lot, and fall to work with great cheerfulness; sometimes one of their orators cheers them with jests and humorous old tales, and sings several of their most agreeable wild tunes, beating also with a stick in his right hand, on the top of an earthern pot covered with a wet and well-stretched deerskin: thus they proceed from field to field, till their seed is sown.

Corn is their chief produce, and main dependance. Of this they have three sorts; one of which hath been already mentioned. The second sort is yellow and flinty, which they call "hommony-corn." The third is the largest, of a very white and soft grain, termed "bread-corn." In July, when the chesnuts and corn are green and full grown, they half boil the former, and take off the rind; and having sliced the milky, swelled, long rows of the latter, the women pound it in a large wooden mortar, which is wide at the mouth, and gradually narrows to the bottom: then they knead both together, wrap them up in green corn-blades of various sizes, about an inch-thick, and boil them well, as they do every kind of seethed food. This sort of bread is very tempting to the taste, and reckoned most delicious to their strong palates. They have another sort of boiled bread, which is mixed with beans, or potatoes; they put on the soft corn till it begins to boil, and pound it sufficiently fine; — their invention does not reach to the use of any kind of milk. When the flour is stirred, and dried by the heat of the sun or fire, they sift it with sieves of different sizes, curiously made of the coarser or finer cane-splinters. The thin cakes mixt with bear's oil, were formerly baked on thin broad stones placed over a fire, or on broad earthen bottoms fit for such a use: but now they use kettles. When they intend to bake great loaves, they make a strong blazing fire, with short dry split wood, on the hearth. When it is burnt down to coals, they carefully rake them off to each side, sweep away the remaining ashes: then they put their well-kneeded broad loaf, first steeped in hot water, over the hearth, and an earthen bason above it, with the embers and coals a-top. This method of baking is as clean and efficacious as could possibly be done in any oven; when they take it off, they wash the loaf with warm water, and it soon becomes firm, and very white. It is likewise very wholesome, and well-tasted to any except the vitiated palate of an Epicure.

The French of West-Florida, and the English colonists, got from the Indians different sorts of beans and peas, with which they were before entirely unac-quainted. And they plant a sort of small tobacco, which the French and English have not. All the Indian nations we have any acquaintance with, frequently use it on the most religious occasions. The women plant also pompions, and different sorts of melons, in separate fields, at a considerable distance from the town, where each owner raises an high scaffold, to over-look this favourite part of their vegetable possessions: and though the enemy sometimes kills them in this their strict watch duty, yet it is a very rare thing to pass by those fields, without seeing them there at watch. This usually is the duty of the old women, who fret at the very shadow of a crow, when he chances to pass on his wide survey of the fields; but if pinching hunger should excite him to descend, they soon frighten him away with their screeches. When the pompions are ripe, they cut them into long circling slices, which they barbacue, or dry with a slow heat. And when they have half boiled the larger sort of potatoes, they likewise dry them over a moderate fire, and chiefly use them in the spring-season, mixt with their favourite bear's oil. As soon as the larger sort of corn is full-eared, they half-boil it too, and dry it either by the sun, or over a slow fire; which might be done, as well, in a moderately hot oven, if the heat was renewed as occasion required. This they boil with venison, or any other unsalted flesh. They commonly have pretty good crops, which is owing to the richness of the soil; for they often let the weeds out-grow the corn, before they begin to be in earnest with their work, owing to their laziness and unskilfulness in planting: and this method is general through all those nations that work separately in their own fields, which in a great measure checks the growth of their crops. Besides, they are so desirous of having *multum in parvo,* without much sweating, that they plant the corn-hills so close, as to thereby choak up the field. — They plant their corn in straight rows, putting five or six grains into one hole, about two inches distant — They cover them with clay in the form of a small hill. Each row is a yard asunder, and in the vacant ground they plant pumpkins, water-melons, marsh-mallows, sun-flowers, and sun-dry sorts of beans and peas, the last two of which yield a large increase.

They have a great deal of fruit, and they dry such kinds as will bear it. At the fall of the leaf, they gather a number of hiccory-nuts, which they pound with a round stone, upon a stone, thick and hollowed for the purpose. When they are beat fine enough, they mix them with cold water, in a clay bason, where the shells subside. The other part is an oily, tough, thick, white substance, called by the traders hiccory milk, and by the Indians the flesh, or fat of hiccory-nuts, with which they eat their bread. A hearty stranger would be as apt to dip into the sediments as I did, the first time this vegetable thick milk was set before me. As ranging the woods has given me a keen appetite, I was the more readily tempted to believe they only tantalized me for their diversion, when they laughed heartily at my supposed ignorance. But luckily when the bason was in danger, the bread was brought in piping hot, and the good-natured landlady being informed of my simplicity, shewed me the right way to use the vegetable liquid. It is surprising to see the great variety of dishes they make out of wild flesh, corn, beans, peas, potatoes, pompions, dried fruits, herbs and roots. They can diversify their courses, as much as the English, or perhaps the French cooks: and in either of the ways they dress their food, it is grateful to a wholesome stomach.

Their old fields abound with larger strawberries than I have seen in any part of the world; insomuch, that in the proper season, one may gather a hat-full, in the space of two or three yards square. They have a sort of wild potatoes, which grow plentifully in their rich low lands, from South-Carolina to the Mississippi, and partly serve them instead of bread, either in the woods a hunting, or at home when the foregoing summer's crop fails them. They have a small vine, which twines, chiefly round the watry alder; and the hogs "feed" often upon the grapes. Their surface is uneven, yet inclining to a round figure. They are large, of a coarse grain, well-tasted, and very wholesome; in the woods, they are a very aggreeable repast. There grows a long flag, in shallow ponds, and on the edges of running waters, with an ever-green, broad, round leaf, a little indented where it joins the stalk; it bears only one leaf, that always floats on the surface of the water, and affords plenty of cooling small nuts, which make a sweet-tasted, and favourite bread, when mixed with Indian corn flour. It is a sort of marsh-mallows, and reckoned a speedy cure for burning maladies, either outward or inward, — for the former, by an outward application of the leaf; and for the latter, by a decoction of it drank plentifully. The Choktah so highly esteem this vegetable, that they call one of their head-towns, by its name.

Colonial Settlement – Virginia

John Rolfe's Description of Virginia, 1616

From John Rolfe, "Virginia in 1616," *Virginia Historical Register and Literary Advertiser*, (July, 1848), pp. 101–11, 113.

We copy the following article from the *Southern Literary Messenger* for June 1839, where it is introduced with a short preface in these words: "We derive the subjoined interesting historical paper from so high a source, that we do not hesitate to vouch its authenticity. It appears that it was carefully transcribed from the Royal MSS. in the British Museum, and is entitled in Casley's Catalogue of those MSS., 'John Rolf's Relation of the State of Virginia, 17th Century.' The remark in the tract itself, 'the estate of this colony, as it remained in *May last,* when Sir Thomas Dale left the same,' proves that it must have been written a year after May, 1616 — as the governor left the colony, and returned to England at that time; and the expression, 'both *here* and in Virginia,' established the fact that the paper was written in England. Rolf, the narrator, had been married to the celebrated Pocahontas, daughter of Powhatan, a few year before. She and her husband accompanied Sir Thomas Dale on his return to England, and arrived in Plymouth on the 12th June, 1616. The narrative itself, independent of the fact that it sustains and corroborates most of the accounts which have been preserved of the early state of the colony, will be read with interest, as the production of Mr. Rolf, the chosen partner of her who has been emphatically styled the guardian angel of the colony, and the ancestor of some of the most respectable and distinguished families of Virginia. We give the text *verbatim et literatim.*"

MAY IT PLEASE YOUR HIGHNESS:

There have been of late divulged many impressions, judicially and truly penned; partlie to take away the ignominie, scandalls and maledictions wherewith this action hath ben branded, and partlie to satisfie all, (especially the best) with the manner of the late proceedings and the prosperitie likely to ensue. How happily and plenteously the good blessings of God have fallen upon the people and colony since the las impression, faithfully written by a gent. of good merit, Mr. Ralph Hamor, (some tyme an actuall member in the Plantation, even then departing when the foundacoun and ground worke was new laid of their now thrift and happines,) of the earthie and worldly man is scarcely believed, but of heavenlier minds they are most easilie discerned, for they daily attend and marke how those blessings, (though sometimes restrayned for a tyme,) in the end, are poured upon the servants of the Lord. Shall your Ma'tie, with pietie and pittie — with pietie, being zealous for God's glory, and with pittie, (mourning the defects,) vouchsafe to reade thus much of the estate of this colony, as it remained in May last, when Sir Thomas Dale left the same, I shall deeme my selfe most happie in your gracious acceptance, and most readilie offer to your approved judgment, whether this cause, so much despised and disgraced, doe not wrongfully suffer many imputacions.

First, to meete with an objection commonly used amongst many men, who search truthes no farther than by common reports, namely, how is it possible Virginia can now be so good, so fertile a countrey, so plentifullie stored with food and other commodities? It is not the same still it was when men pined with famine? Can the earth now bring forth such a plentifull increase? Were there not governors, men and meanes to have wrought this heretofore? And can it now, on the suddaine, be so fruitfull? Surely, say they, these are rather bates to catch and intrapp more men into woe and miserie, then otherwise can be imagined. These, with many as frivolous, I have heard instigated, and even reproachfullie spoken against Virginia. To answeare whom, (the most parte of them incredulous worldlings — such as believe not, unless they feele the goodness of the Lord sensiblie to touch them,) though it be not much materiall, yet let them know, 'tis true, Virginia is the same it was, I meane for the goodnes of the seate, and fertileness of the land, and will no doubt so contynue to the world's end, — a countrey as worthy good report, as can be declared by the pen of the best writer. A countrey spacious and wide, capable of many hundred thousands of inhabitants. For the soil most fertile to plant in, for ayre fresh and temperate, somewhat hotter in summer, and not altogether so cold in winter as in England, yet so agreeable it is to our constitutions, that now 'tis more rare to heare of man's death then in England amongst so many people as are there resident. For water, most wholesome and verie plentifull, and for fayre navigable rivers and good harbours, no countrey in christendom, in so small a circuite, is so well stored. For matter fit for buildings and fortifications, and for building of shipping, with everie thing thereto apperteyning, I may boldly avouch scarce anie or no countrey knowne to man is of itself more abundantly furnished. Theis things (may some say,) are of great consequence toward the settling of a plantation, but where are the beasts and cattle to feede and cloth the people? I confesse this is a mayne want; yet some there are already, as neate cattle, horses, mares and gotes, which are carefullie preserved for increase. The nomber whereof, hereafter shalbe sett downe in a particular note by themselves. There are also great store of hoggs, both wild and tame, and poultrie great plentie, which every man, if they will, themselves may keepe. But the greatest want of all is least thought on, and that is good and sufficient men, as well of birth and qualitie, to command soldiers, to march, discover and defend the countrey from invasions, as also artificers, laborers, and husbandmen, with whom, were the colony well provided, then might tryall be made what lyeth hidden in the wombe of the ground. The land might yearlie abound with corne and other provisions for man's sustentation — buildings, fortifications and shipping might be reared, wrought and framed — commodities of divers kinds might be yearly reaped and sought after, and many things (God's blessinge contynuing,) might come with ease to establish a firme and perfect common weale.

But to come again to the matter, from which I have a little straied, and to give a more full answeare to the objectors, may you please to take notice, that the beginning of this plantation was governed by a president and councell, aristocratically. The president yearlie chosen out of the councell, which consisted of twelve persons. This government lasted about two years, in which tyme such envie, dissentions and jarres were daily sowne amongst them, that they choaked the seed and blasted the fruits of all men's labors. If one were well disposed and gave good advisement to proceed in the business — others, out of the malice of their hearts, would contradict, interdict, withstand and dash all. Some rung out and sent home too loud praises of the riches and fertilness of the country, before they assayed to plant, to reape or search the same; others said nothing, nor did any thing thereunto; all would be *keisars,* none inferior to other. Some drew forward, more backward — the vulgar sort looked for supplie out of England — neglected husbandry — some wrote — some said there was want of food, yet sought for none — others that would have sought could not be suffered; in which confusion much confusion yearlie befell them, and in this government happened all the miserie. Afterward a more absolute government was graunted, monarchially, wherein it still contynueth, and although for some few years it stood at a stay, especially in the manuring and tilling of ground, yet men spent not their tyme idely nor improfitably, for they were daily employed in palazadoing and building of townes, impaling grounds and other needful businesses, which is now both beneficiall to keepe the cattle from ranging, and preserveth the corne safe from their spoile. Being thus fitted and prepared to sow corne, and to plant other seeds and fruits in all the places of our habitations, — one thing, notwithstanding, much troubled our governor, namely, enmitie with the Indians; for, however well we could defend ourselves, townes and seates from any assaulte of the natives, yet our cattle and corne lay too open to their courtesies, and too subject to their mercies: whereupon a peace was concluded, which still continueth so firme, that our people yearely plant and reape quietly, and travell in the woods a fowling and a hunting as freely and securely from feare of danger or treacherie as in England. The great blessings of God have followed this peace, and it, next under him, hath bredd our plentie — everie man sitting under his fig tree in safety, gathering and reaping the fruits of their labors with much joy and comfort. But a question may be demanded what these fruits are — for such as the countrey affordeth naturally (for varietie and goodnes) are comparable to the best in christendom, (growing wild as they doe,) — I pass them over, other discourses having largely manifested them to the view of the world. But for the people's present labors they have Indian wheate, called mays in the West Indies, pease and beanes, English wheate, peas, barley, turnips, cabbages, pumpions, West Indian and others, carretts, parsnips, and such like, besides hearbs and flowers, all of our English seede, both for pleasure and for the kitchen, so good, so fruitful, so pleasant and profitable, as the best made ground in England can yield. And that your Ma'tie may know what two men's labor, with spade and shalve only, can manure in one year, fiftie pounds in money was offered for their cropp, which they refused to take; for hempe and flax, none better in England or Holland — silkewormes, some of their labors, and tasts of other good and vendible commodities were now brought home. Likewise tobacco, (though an esteemed weed,) very commodious, which there thriveth so well, that no doubt but after a little more triall and expense in the curing thereof, it will compare with the best in the West Indies. For fish and fowle, deere and other beasts, reports and writinge have rather been too sparing then prodigall.

About two years since, Sir Thomas Dale, (whose worth and name, in concluding this peace, and managing the affairs of this colony, will out last the standing of this plantation,) found out two seasons in the year to catch fish, namely, the spring and the fall. He himself tooke no small paines in the tryall, and at one hall with a scryne caught five thousand three hundred of them, as bigg as codd. The least of the residue or kind of salmon trout, two foote long; yet durst he not adventure on the mayne skull for breaking his nett. Likewise, two men with axes and such like weapons, have taken and kild neere the shoare and brought home fortie as great as codd in two or three hours space, so that now there is not so great plentie of victualls in anie one of the forenamed kind yearlie with small paines to be gotten in any part of England amongst so few people as are there resident. And, whereas, heretofore we were constrayned yearly to go to the Indians and intreate them to sell us corne, which made them esteeme verie basely of us — now the case is altered; they seeke to us — come to our townes, sell their skins from their shoulders, which is their best garments, to buy corne — yea, some of their pettie kings have this last yeare borrowed four or five hundred bushells of wheate, for payment whereof, this harvest they have mortgaged their whole countries, some of them not much less in quantitie then a shire in England. By this meanes plentie and prosperitie dwelleth amonst them, and the feare and danger of famine is clean taken away, wherewith the action hath a long time suffered injurious defamations.

Now that your highnes may with the more ease understand in what condition the colony standeth, I have briefly sett downe the manner of all men's several imployments, the number of them, and the several places of their aboad, which places or seates are all our owne ground, not so much by conquest, which the Indians hold a just and lawfull title, but purchased of them freely, and they verie willingly selling it.

The places which are now possessed and inhabited are sixe.

1. Henrico and the lymitts
2. Bermuda Nether ⎫ Hundreds.
3. West and Sherley ⎭
4. James Towne
5. Kequoughtan
6. Dales-Gift

Members belonging to ye Bermuda Towne, a place so called there, by reason of the strength of the situation, were it indifferently fortified.

The generall mayne body of the planters are divided into

1. Officers.
2. Laborers.
3. Farmors.

The officers have the charge and care as well over the farmors as laborers generallie — that they watch and ward for their preservacions; and that both the one and the other's busines may be daily followed to the performance of those imployments, which from the one are required, and the other by covenant are bound unto. These officers are bound to maintayne themselves and families with food and rayment by their owne and their servants' industrie.

The laborers are of two sorts. Some employed only in the generall works, who are fedd and clothed out of the store — others, specially artificers, as smiths, carpenters, shoemakers, taylors, tanners, &c., doe worke in their professions for the colony, and maintayne themselves with food and apparrell, having time lymitted them to till and manure their ground.

The farmors live at most ease — yet by their good endeavours bring yearlie much plentie to the plantation. They are bound by covenant, both for themselves and servants, to maintaine your Ma'tie's right and title in that kingdom, against all foreigne and domestique enemies. To watch and ward in the townes where they are resident. To do thirty-one dayes service for the colony, when they shalbe called thereunto — yet not at all tymes, but when their owne busines can best spare them. To maintayne themselves and families with food and rayment — and every farmor to pay yearlie into the magazine, for himself and every man servant, two barrells and a half a piece of their best Indian wheate, which amounteth to twelve bushells and a halfe of English measure.

Thus briefly have I sett downe every man's particular imployment and manner of living; albeit, lest the people — who generallie are bent to covett after gaine, especially having tasted of the sweete of their labors — should spend too much of their tyme and labor in planting tobacco, knowne to them to be verie vendible in England, and so neglect their tillage of corne, and fall into want thereof, it is provided for — by the providence and care of Sir Thomas Dale — that no farmor or other — who must maintayne themselves — shall plant any tobacco, unless he shall yearely manure, set and maintayne for himself and every man servant two acres of ground with corne, which doing they may plant as much tobacco as they will, els all their tobacco shalbe forfeite to the colony — by which meanes the magazin shall yearely be sure to receave their rent of corne; to maintayne those who are fedd thereout, being but a few, and manie others, if need be; they themselves will be well stored to keepe their families with overplus, and reape tobacco enough to buy clothes and such other necessaries as are needeful for themselves and household. For an easie laborer will keepe and tend two acres of corne, and cure a good store of tobacco — being yet the principall commoditie the colony for the present yieldeth. For which, as for other commodities, the councell and company for Virginia have already sent a ship thither, furnished with all manner of clothing, houshold stuff and such necessaries, to establish a magazin there, which the people shall buy at easie rates for their commodities — they selling them at such prices that the adventurers may be no loosers. This magazin shalbe yearelie supplied to furnish them, if they will endeavor, by their labor, to maintayne it — which wilbe much beneficiall to the planters and adventurers, by interchanging their commodities, and will add much encouragement to them and others to persevere and follow the action with a constant resolution to uphold the same.

The people which inhabite the said six severall places are disposed as followeth:

At Henrico, and in the precincte, (which is seated on the north side of the river, ninety odd myles from the mouth thereof, and within fifteen or sixteen myles of the falls or head of that river, being our furthest habitation within the land,) are thirty-eight men and boyes, whereof twenty-two are farmors, the rest officers and others, all whom maintayne themselves with food and apparrell. Of this towne one capten Smaley hath the command in the absence of capten James Davis. Mr. Wm. Wickham minister there, who, in his life and doctrine, give good examples and godly instructions to the people.

At Bermuda Nether Hundred, (seated on the south side of the river, crossing it and going by land, five myles lower than Henrico by water,) are one hundred and nineteen — which seate conteyneth a good circuite of ground — the river running round, so that a pale running cross a neck of land from one parte of the river to the other, maketh it a peninsula. The houses and dwellings of the people are sett round about by the river, and all along the pale, so farr distant one from the other, that upon anie alarme, they can succor and second one the other. These people are injoyned by a charter, (being incorporated to the Bermuda towne, which is made a corporacoun,) to effect and performe such duties and services whereunto they are bound for a certain tyme, and then to have their freedome. This corporacoun admit no farmors, unles they procure of the governor some of the colony men to be their servants, for whom (being no members of the corporacoun,) they are to pay rent corne as other farmors of this kind — these are about seventeen. Others also comprehended in the said number of one hundred and nineteen there, are resident, who labor generallie for the colonie; amongst whom some make pitch and tarr, potashes, charcole and other works, and are maintayned by the magazin — but are not of the corporacoun. At this place (for the most part) liveth capten *Peacdley*, deputy marshal and deputy governor. Mr. Alexander Whitaker, (sonne to the reverend and famous divine, Dr. Whitaker,) a good divine, hath the ministerial charge here.

At West and Sherley Hundred (seated on the north side of the river, lower then the Bermudas three or four myles,) are twenty-five, commanded by capten Maddeson — who are imployed onely in planting and curing tobacco, — with the profitt thereof to clothe themselves and all those who labor about the generall business.

At James Towne (seated on the north side of the river, from West and Sherley Hundred lower down about thirty-seven myles,) and fifty, under the command of lieutenant Sharpe, in the absence of capten Francis West, Esq., brother to the right ho'ble the Le. Lawarre, — whereof thirty-one are farmors; all theis maintayne themselves with food and rayment. Mr. Richard Burd minister there — a verie good preacher.

At Kequoughtan (being not farr from the mouth of the river, thirty-seven miles below James Towne on the same side,) are twenty — whereof eleven are farmors; all those also maintayne themselves as the former. Capten George Webb commander. Mr. Wm. Mays minister there.

At Dales-Gift (being upon the sea, neere unto Cape Charles, about thirty myles from Kequoughtan,) are seventeen, under the command of one lieutenant Cradock; all these are fedd and maintayned by the colony. Their labor is to make salt and catch fish at the two seasons aforementioned.

So the number of officers and laborers are two hundred and five. The farmors 81; besides woemen and children, in everie place some — which in all amounteth to three hundred and fifty-one persons — a small number to advance so great a worke.

Theis severall places are not thus weakly man'd, as capable of no greater nomber, (for they will maintayne many hundreds more,) — but because no one can be forsaken without losse and detriment to all. If then so few people, thus united, ordered and governed, doe live so happily, every one partaking of the others labor, can keepe in possession so much ground as will feed a far greater nomber in the same or better condition; and seeing too, too many poore farmors in England worke all the yeare, rising early and going to bed late, live penuriously, and much adoe to pay their landlord's rent, besides a daily karking and caring to feed themselves and families, what happines might they enjoy in Virginia, were men sensible of theis things, where they may have ground for nothing, more than they can manure; reape more fruits and profitts with half the labor, void of many cares and vexacions, and for their rent a matter of small or no moment, I leave to your singular judgment and consideracoun, nothing doubting, but He (who, by his infinite goodnes, with so small means, hath settled these poore and weake beginnings so happily,) will animate, stirr up and encourage manie others cheerefully to undertake this worke, and will assuredly add a daily strength to uphold and maintayne what he hath already begun.

Seeing then this languishing action is now brought to this forwardness and strength, no person but is provided for, either by their owne or others labors, to subsist themselves for food, and to be able to rayse commodities for clothing and other necessaries, envy it selfe, poysoned with the venom of aspes, cannot wound it. . . .

YOUR HIGHNES' MOST FAITHFUL AND LOYALL SUBJECT,

JOHN ROLF.

The nomber of neate cattle, horses and goates, which were alive in Virginia at Sir Thomas Dale's departure thence:

Cowes,			
Heifers,	} 83		
Cow calves,		in all } 144	
Steeres	41		
Bulles,	20		

Memorand: 20 of the cowes were great with calfe at his departure.

Horses,	3 } in all } 6.		
Mares,	3		
Goates and Kidds,	} male and female, in all } 216		

Hoggs, wild and tame, not to be nombred.
Poultry, great plenty.

Regulating Tobacco in Virginia, 1618

From *Colonial Records of Virginia*, State Senate, Extra (Richmond, 1874), pp. 23–24.

All the general Assembly by voices concluded not only the acceptance and observation of this order, but of the Instruction also to Sir George Yeardley next preceding the same. Provided first, that the Cape Marchant do accepte of the Tobacco of all and everie the Planters here in Virginia, either for Goods or upon billes of Exchange at three shillings the pounde the beste, and 18d the second sorte. Provided also that the billes be only payde in Englande. Provided, in the third place, that if any other besides the Magazin have at any time any necessary comodity wch the Magazine doth wante, it shall and may be lawfull for any of the Colony to buye the said necessary comodity of the said party, but upon the termes of the Magazin viz: allowing no more gaine then 25 in the hundred, and that with the leave of the Governour. Provided lastly, that it may be lawfull for the Govern^r to give leave to any Mariner, or any other person, that shall have any suche necessary comodity wanting to the Magazin to carrie home for England so muche Tobacco or other naturall comodities of the Country as his Customers shall pay him for the said necessary comodity or comodities. And to the ende we may not only persuade and incite men, but inforce them also thoroughly and loyally to aire their Tobacco before they bring it to the Magazine, be it enacted, and by these presents we doe enacte, that if upon the Judgement of power sufficient even of any incorporation where the Magazine shall reside, (having first taken their oaths to give true fentence, twoe whereof to be chosen by the Cape Marchant and twoe by the Incorporation,) any Tobacco whatsoever shall not proove vendible at the second price, that it shall there immediately be burnt before the owner's face. Hitherto suche lawes as were drawn out of the Instructions.

Instructions to George Yeardley from the Virginia Company, 1618

From Susan M. Kingsbury, ed., *The Records of the Virginia Company of London* (Washington, 1933), III, pp. 90–109.

The Treasurer and Companie of Adventurers and Planters of the City of London for the first Colony in Virginia. To Captain George Yeardly Elect Governor of Virginia and to the Council of State there being or to be Greeting.

Our former cares and Endeavours have been chiefly bent to the procuring and sending people to plant in Virginia so to prepare a way and to lay a foundation whereon A flourishing State might in process of time by the blessing of Almighty

God be raised. Now our trust being that under the Government of you Captain Yeardly with the advice and Assistance of the said Council of State such public provisions of Corn and Cattle will again be raised as may draw on those Multitudes who in great Abundance from diverse parts of the Realm were preparing to remove thither if by the late decay of the said public Store their hopes had not been made frustrate and their minds thereby clene discouraged   We have thought good to bend our present cares and Consultations according to the Authority granted unto us from his Majesty under his great Seal to the setling there of A laudable form of Government by Majestracy and just Laws for the happy guiding and governing of the people there inhabiting like as we have already done for the well ordering of our Courts here and of our Officers and accions for the behoof of that plantation. And because our intent is to Ease all the Inhabitants of Virginia forever of all taxes and public burthens as much as may be and to take away all occasion of oppression and corruption we have thought fit to begin (according to the laudable Example of the most famous Common Wealthes both past and present) to alot and lay out A Convenient portion of public lands for the maintenance and support as well of Magistracy and officers as of other public charges both here and there from time to time arising   We therefore the said treasurer and Company upon a solemn treaty and resolution and with the advice consent and assent of his Majesties Council here of Virginia being Assembled in A great and general Court of the Council and Company of Adventurers for Virginia require you the said Governor and Council of Estate to put in Execution with all convenient Speed a former order of Our Courts (which had been commended also to Captain Argal at his making Deputy Governor) for the laying and seting out by bounds and metes of three thousand Acres of land in the best and most convenient place of the territory of James town in Virginia and next adjoining to the said town to be the seat and land of the Governor of Virginia for the time being and his Successors and to be called by the name of the Governors Land which Governors Land shall be of the freed grounds by the common labor of the people sent thither at the Companies Charges And of the Lands formerly conquer'd or purchased of the Paspeheies and of other grounds next adjoining   In like sort we require you to set and lay out by bounds and Metes other three thousand Acres of good land within the territory of James town which shall be convenient and in such place or places as in your discretions you shall find meet which latter three thousand Acres shall be and so called the Companies Land   And we require you Captain Yeardley that immediately upon your arrival you take unto you the Guard assigned to Captain Argal at his going Deputy Governor or sithence by him assumed to be of your guard [for the better defence] of your Government and that as well the said guard as also fifty other persons now sent and transported with you you place as tennants on the said Governors land and that all other persons heretofore transported at the Common Charge of the Company since the coming away of Sr. Thomas Dale Knight late Deputy Governor be placed as Tennants on the said Companies Lands   And we will and ordain that all the said Tennants on the Governors and Companies Lands shall occupy the same to the half part of the profits of the said Lands so as the one half to be and belong to the said Tennants themselves and the other half respectively to the said Governor and to us the said Treasurer and Company and our Successors   And we further will and ordain that of the half profits arising out of the said Companies Lands and belonging to us the said Treasurer and Company the one Moiety be imploied for the Entertainment of the said Councel of Estate there residing and of other public officers of the general Colony and plantation (besides the Governor) according to the proportion as hereafter we shall Express and in the mean time as you in your discretions shall think meet   And the other moiety be carefully gathered kept and ship'd for England for the public use of us the said Treasurer and Company and our Successors   And we further will and ordain that out of the half profits of the said Companies Lands to us belonging one fifth part be deducted and alotted for the Wages of the Bailiffs and other Officers which shall have the oversight and Government of the said Tenants and Lands and the dividing gathering keeping or shiping of the particular moiety of the profits belonging Either to the said Council and Officer there or to us the said treasurer and Company and our Successors as aforesaid   *Provided* alwaies that out of the said Companies Land A Sufficient part be exempted and reserved for the securing and Wintering of all sorts of Cattle which are or shall be the public Stock and Store of the said Company   And forasmuch as our intent is to Establish one Equal [*blank of several lines*] Plantations, whereof we shall speak afterwards, be reduced into four Cities or Burroughs *Namely* the cheif City called James town Charles City Henrico and the Burrough of Kiccowtan   And that in all these foresaid Cities or Burroughs the ancient Adventurers and Planters which [were] transported thither with intent to inhabit at their own costs and charges before the coming away of S^R Thomas Dale Knight and have so continued during the space of three years shall have upon a first division to be afterward by us augmented one hundred Acres of land for their personal Adventure and as much for every single share of twelve pound ten Shillings paid [for such share] allotted and set out to be held by them their heirs and assigns forever   And that for all such Planters as were brought thither at the Companies Charge to inhabit there before the coming away of the said S^R Thomas Dale after the time of their Service to the Company on the common Land agreed shall be expired there be set out One hundred Acres of Land for each of their personal Adventurers to be held by them their heirs and Assigns for ever. Paying for every fifty Acres the yearly free Rent of one Shilling to the said treasurer and Company and their Successors at one Entire payment on the feast day of S Michael the archangel for ever   And in regard that by the singular industry and virtue of the said S^R Thomas Dale the former difficulties and dangers were in greatest part overcome to the great ease and security of such as have been since that time transported thither   We do therefore hereby ordain that all such persons as sithence the coming away of the said S^R Thomas Dale have at their own charges been transported thither to inhabit and so continued as aforesaid there be allotted and set out upon a first division fifty acres of land to them and their heirs for ever for their personal Adventure paying a free rent of one Shilling yearly in manner aforesaid And that all persons which since the going away of the said S^R Thomas Dale have been transported thither at the Companies charges or which hereafter shall be so transported be placed as tenants on the Companies lands for term of seven years occupy the same to the half part of the profits as is abovesaid   We therefore will and ordain that other three thousand Acres of Land be set out in the fields and territory of Charles City and other three thousand Acres of Land in the fields and territories of Henrico And other three thousand Acres of land in the fields and territory of Kiccowtan all which to be and be called the Companies lands and to be occupied by the Companies Tenants for half profits as afore said   And that the profits belonging to the Company be disposed by their several moieties in the same manner as before set down touching the Companies lands in the territory of James town with like allowance to the Bailies and reservation of ground for the common Store of Cattle in those several places as is there set down And our will is that such of the Companies tenants as already inhabite in those several Cities or Burroughs be not removed to any other City or Burrough but placed on the Companies Lands belonging to those Cities or Burroughs where they now inhabite   *Provided* alwaies that if any private person without fraud or injurious intent to the public at his own charges have freed any of the said Lands formerly appointed to the Governor he may continue and inhabite there till a valuable recompence be made him for his said Charges   And we do hereby ordain that the Governors house in James town first built by S^R Thomas Gates Knight at the charges and by the Servants of the Company and since enlarged by others by the very same means be and continue for ever the Governors house any pretended undue Grant made by misinformation and not in a general and quarter Court to the contrary in anywise notwithstanding. And to the intent that godly learned and painful Ministers may be placed there for the service of Almighty God & for the spiritual benefit and comfort of the people   We further will and ordain that in every of those Cities or Burroughs the several quantity of one hundred Acres of Land be set out in quality of Glebe land toward the maintenance of the several Ministers of the parishes to be there limited and for a further supply of their maintenance there be raised a yearly standing and certain contribution out of the profits growing or renuing within the several farmes of the said parish and so as to make the living of every Minister two hundred pounds Sterling ᴘ annum or more as hereafter there shall be cause   And for a further Ease to the Inhabitants of all taxes and Contributions for the Support and Entertainment of the particular magistrates and Officers and of other charges to the said Citys and Burroughs respectively belonging   We likewise will and ordain that within the precincts or territories of the said Cities and Burroughs shall be set out and alotted the several Quantities of fifteen hundred Acres of Land to be the common Land of the said Citie Or Burrough for the uses aforesaid and to be known and called by the name of the Cities Or Burroughs Land *And Whereas* by a special Grant and licence from his Majesty a general Contribution over this Realm hath been made for the building and planting of a college for the training up of the Children of those Infidels in true Religion moral virtue and Civility and for other godly uses   We do therefore according to a former Grant and order hereby ratifie confirm and ordain that a convenient place be chosen and set out for the planting of a University at the said Henrico in time to come and that in the mean time preparation be there made for the building of the said College for the Children of the Infidels according to such Instructions as we shall deliver And we will and ordain that ten thousand acres partly of the Lands they impaled and partly of other Land within the territory of the said Henrico be alotted and set out for the endowing of the said University and College with convenient possessions   Whereas also we have heretofore by order of Court in consideration of the long good and faithful Service done by you Captain George Yeardley in our said Colony and plantation of Virginia and in reward thereof as also in regard of two single shares in money paid into our treasury granted unto you the said Captain Yeardley all that parcel of Marsh ground called Weynock and also one other peice and parcel of Land adjoining to the same Marsh called by the Natives *Konwan* one parcel whereof abutteth upon a Creek there called Mapscock towards Creek on the West and extendeth in breadth to landward from the head of the said creek called Mapscock up to the head of the said Creek called Queens Creek (which creek called Queens Creek is opposite to that point there which is now called the Tobacco point and abutteth south upon the River and North to the Landward) all which several Lands are or shall be henceforward accounted to be lying within the territory of the said Charles City and exceed not the quantity of two thousand and two hundred acres   We therefore the said Treasurer and Company do hereby again grant ratifie and Confirm unto you the said Captain George Yeardley the said several Grounds and Lands to have and to hold the said Grounds and Lands to you the said Captain George Yeardley your heirs and Assigns for Ever   And for the better Encouragement of all sorts of neccessary and laudable trades to be set up and exercised within the said four Cities or Burroughs   We do hereby ordain that if any artizans or tradesmen shall be desirous rather to follow his particular Art or trade then to be imploied in husbandry or other rural business It shall be lawful for you the said Governor and Councel to alot and set out within any of the precincts aforesaid One dwelling house with four Acres of Land adjoining and held in fee simple to every said tradsman his heirs and Assigns for ever upon condition that the said tradesman his heirs and Assigns do continue and exercise his trade in the said house paying only a free rent of four pence by the year to us the said Treasurer and Company and our Successors at the feast of St Michael the Archangel for ever   And touching all other particular Plantations set out or like to be set out in convenient Multitudes either by divers of the ancient Adventurers Associating themselves together (as the Society of Smiths hundred and Martins hundred) or by some ancient Adventurer or Planter associating others unto him (as the plantation of Captain Samuel Argall and Captain John Martin and that by the late Lord La warre advanced) or by some new Adventurers joining themselves under one head (as the plantation of Christopher Lawne Gentleman and others now in providing) Our Intent being according to the Rules of Justice and good government to alot unto every one his due yet so as neither to breed Disturbance to the Right of others nor to interrupt the good form of Government intended for the benefit of the people and strength of the Colony We do therefore will and ordain that of the said particular plantations none be placed within five Miles of the said former Cities and Boroughs And that if any man out of his own presumption or pleasure without special direction from us hath heretofore done otherwise a convenient time be Assigned him and then by your Discretions to remove to Some farther place by themsleves to be chosen with the Allowance and Assent of the Governor for the time being and the Council of Estate   And that the Inhabitants of the said City or Burrough too near unto which he or they were placed make him and them a valuable recompense for their Charges and expence of time in freeing of Grounds and building within those precincts   In like sort we ordain that no latter particular plantation shall at any time hereafter be seated within ten Miles of a former   We also will and ordain that no particular plantation be or shall be placed straglingly in divers places to the weakening of them but be united together in one seat and territory that so also they may be incorporated by us into one body corporate and live under Equal and like Law and orders with the rest of the Colony   We will and ordain also for the preventing of all fraud in abusing of our grants contrary to the intent and just meaning of them That all such person or persons as have procured or hereafter shall procure grants from us in general Words unto themselves and their Associates or to like Effect shall within one year after the date hereof deliver up to us in writing under their hands and seals as also unto you the said Governor and Councel what be or were the names of those their first Associates   And if they be of the Adventurers of us the Company which have paid into our treasury money for their shares that then they express in that their writing for how many shares they join in the said particular Plantation to the End a Due proportion of Land may be set out unto them and we the said Treasurer and Company be not defrauded of Our due And if they be not of the Adventurers of the Company which have paid into our treasury money for their shares yet are gone to inhabit there and so continue for three years there be allotted and set out fifty Acres of Land for every such person paying a free rent of twelve pence the year in manner aforesaid and All such persons having been planted there since the coming away of S^R Thomas Dale   And forasmuch as we understand that certain persons having procured such Grants in general Words to themselves and their Associates or to like Effect have corruptly of late endeavoured for gain and Worse respects to draw many of the ancient Planters of the said four Cities or Burroughs to take grants also of them and thereby to become associated unto them with intent also by Such means to overstrengthen their party   And thereupon have adventured on divers Enormous Courses tending to the great hurt and hindrance of the Colony Yea and have also made Grants of like Association to Masters of Ships and Mariners never intending there to inhabit, thereby to defraud his Majesty of the Customs due unto him   We to remedy and prevent such unlawful and greedy Courses tending also directly to faction and sedition Do hereby ordain that it shall not be lawful for the Grantees of such Grants to associate to any other unto them then such as were their Associates from the first time of the said grants without express licence of us the s^d Treasurer and Company in a great General and Quarter Court under our seal obtained   And that all such after or under Grants of Association made or to be made by the said Grantees shall be to all intents and purposes utterly void. And for as much as we understand that divers particular persons (not members of our Company) with their Companies have provided or are in providing to remove into Virginia with intent (as appeareth) by way of Association to shroud themselves under the General Grants last aforesaid which may tend to the Great disorder of our Colony and hinderance of the good Govern-. ment which we desire to Establish. We do therefore hereby ordain that all such persons as of their own Voluntary Will and authority shall remove into Virginia without any Grant from us in a great general and Quarter Court in writing under our seal shall be deemed (as they are) to be occupiers of our Land that is to say of the Common Lands of us the said Treasurer and Company   And shall yearly pay unto us for the said occupying of our Land one full fourth part of the profits thereof till such time as the same shall be granted unto them by us in manner aforesaid   And touching all such as being Members of our Company and Adventurers by their monies paid into our Treasury shall either in their own person or by their agents Tennants or Servants set up in Virginia any such particular Plantation tho with the privity of us the said Treasurer and Company yet without any grant in Writing made in our said General Quarter Courts as is requisite   We will and ordain that the said Adventurers or Planters shall within two year after the arrival of them or their Company in Virginia procure our grant in writing to be made, in Our General Quarter Court and under our seal, of the Lands by them possessed or occupied or from thenceforth shall be deemed only Occupiers of the Common Land As is aforesaid till such times as our said grant shall be obtained. We also not more intending the reformation of the Errors of the said   [1] than for advancing of them into good Courses and therein to assist them by all good means We further hereby ordain that to all such of the said particular   [1] as shall truly fully observe the orders Afore and hereafter specified there be alotted and set out over and above Our former Grants One hundred Acres of glebe land for the Minister of every   [1] and fifteen hundred Acres of Burough Land for the public use of the said Plantation. Not intending yet hereby either to abridge or enlarge such grant of glebe or common Land as shall be made in any of our grants in writing to any of the said particular plantations   We also will and ordain that the like proportion of maintenance out of the   [1] and profits of the Earth be made for the several Ministers of the said particular Plantations as have been before set down for the Ministers of the said former Cities and burroughs   We will and ordain that the Governor for the time being and the said Council of Estate do justly perform or cause to be performed all such grants Covenants and Articles as have or shall be in writing in Our great and general Quarter Courts to any of the said particular Plantations Declaring all other grants of Lands in Virginia not made in one of our great and general Quarter Courts by force of his Majesties Letters patents to be void   And to the End aforesaid we will and ordain that all our grants in writing under our seal made in our great and general Quarter Courts be Entered into your records to be kept there in Virginia   Yet directly forbiding that a Charter of Land granted to Captain Samuel Argal and his Associates bearing date the twentieth of March 1616 be entered in your Records or otherwise at all respected forasmuch as the same was obtained slight and cunning   And afterwards upon suffering him to go Governor of Virginia was by his own voluntary act left in our Custody to be cancelled upon Grant of a new Charter which.   [1] We do also hereby declare that heretofore in one of our said general and Quarter Courts we have ordained and enacted and in this present Court have ratified and Confirmed these orders and laws following. That all Grants of Lands privileges and liberties in Virginia hereafter to be made be passed by Indenture A Counterpart whereof to be sealed by the Grantees and to be kept   [1] the Companies   Evidences And that the Secretary of the Company have the Engrossing of all such Indentures That no patents or Indentures of Grants of Land in Virginia be made and sealed but in a full General and Quarter Court the same having been first thoroughly perused and Approved *under* the hands of A Select Committee for that purpose.   [1] That all Grants of   [1] in Virginia to such Adventurers as have heretofore brought in their money here to the treasury for their several shares being of twelve pounds ten shillings the share be of one hundred Acres the share upon the first division and of as many more upon A Second Division when the land of the first division shall be Sufficiently peopled   And for Every person which they shall transport thither within seven years after Midusmmer Day One thousand six hundred and Eighteen if he continue there three years or dye in the mean time after he is Shiped it be of fifty Acres the person upon the first Division and fifty more upon a second Division the land of the first being Sufficiently peopled *without* paying any rent to the Company for the one or the Other   And that in all such Grants the names of the said Adventurers and the several Number of Each of their Shares be Expressed *Provided* alwaies and it is ordained that if the said Adventurers or any of them do not truly and Effectually within One Year next after the Sealing of the said Grant pay and discharge all such Sums of money wherein by subscription (or otherwise upon notice thereof given from the Auditors) they stand indebted to the Company or if the said Adventurers or any of them having not lawful Right either by purchase from the Company or by Assignment from some other former Adventurers within one year after the said Grant or by Special Gift of the Company upon merit preceding in A full Quarter Court to so many shares as he or they pretend Do not within one year after the said Grant satisfie and pay to the said Treasurer and Company for every share so wanting after the rate of twelve pounds ten shillings the share That then the said Grant for so much as concerneth the   [1] part and all the shares of the said person so behind and not satisfying as aforesaid shall be utterly void *Provided* also and it is ordained that the Grantees shall from time to time during the said seven years make a true Certificate to the said Treasurer Councel and Company from the Chief .Officer or Officers of the places respectively of the Number names ages sex trades and conditions of every such person so transported or shiped to be entered by the Secretary into a Register book for that purpose to be made   That for all persons not comprised in the order next before which during the next seven years after Midsummer day 1618 shall go into Virginia with intent there to Inhabite   If they continue there three years or dye after they are shiped there shall be a grant made of fifty acres for every person upon A first division and as many more upon a second division (the first being peopled) which grants to be made respectively to such persons and their heirs at whose charges the said persons going to Inhavite in Virginia shall be transported with reservation of twelve pence yearly Rent for every fifty acres to be answered to the said treasurer and Company and their Successors for ever after the first seven years of every such Grant   In which Grants a provisoe to be inserted that the Grantees shall from time to time during the said Seven years make A true Certificate to the said Treasurer Councel and Company from the Chief Officer or Officers of places Respectively of the Number names ages sex trades and Conditions of every such person so transported or shiped to be entred by the Secretary into a Register book for that purpose to be made that all Grants as well of one sort as the other respectively be made with equal favours and grants of like Liberties and immunities as near as may be to the End that all Complaint of partiality [or] differencie may be prevented   All which said orders we hereby will and

ordain to be firmly and unvoilably kept and observed And that the Inhabitants
of Virginia have notice of them for their use and benefit   *Lastly* we do hereby
require and Authorize you the said Captain George Yeardley and the said Council
of Etats Associating with you such other as you shall there find meet to Survey
or cause to be Survey'd all the Lands and territories in Virginia above mentioned
and the same to set out by bounds and metes especially so as that the territories
of the said Several Cities and Buroughs and other particular plantations may be
conveniently divided and known the one from the other   *Each* survey to be set
down distinctly in writing and returned to us under your hands and seals *In Witness*
whereof we have hereunto set our Common Seal   *Given* in a great and general
Court of the Council and Company of Adventurers of Virginia held the Eighteenth
Day of November 1618   And in the years of the Reign of Our Soverain Lord
James by the grace of God King of England Scotland France and Ireland Defender
of the Faith &c Vizt of England France and Ireland the Sixteenth and of Scotland
the two and fiftieth.

Nov^R 18. 1618.

The Governor of Virginia Explains Land Division, 1619

From Susan M. Kingsbury, ed., *The Records of the Virginia Company of London* (Washington, 1906),
I, pp. 249–50.

He began wth the second reason as being fresh in memory, & reading the
Orders in the title of Graunts, he shewed, that all Lande were to be graunted
either to Planters in Virginia by their persons, or to Adventurers by their purses,
or by extraordinary merritts of service. That the Adventurers by their purses,
were they onely and their Assignees, who paid in their severall shares
of — 12 — 10 — to the Common Treasure for the charges of transporting men
to the private Lands of the Adventurers, there was also allowance made to them
of 50 Acres the person. But noe further allowance for any such private expences
as was now demaunded.

Then he came to the first reason, and shewed that these Orders were not
newly devised, but taken out of the Lawes Pattente, namely the second & third;
divers passages of wch he there openly read, importing that the Lande in Virginia
were to be devided amongst the Adventurers by mony, or service and the Planters
in person, and that he is to be reputed an Adventurer by money, who payeth
it into the Companies Treasury, insomuch that if any man be admitted for an
Adventurer, and have paid in noe mony to the Common Treasury, he is to be
compelled thereto by suite of Lawe, yea though he never subscribed to any such
payment, as is expresly sett downe in the Third Lawes Pattente.

Thirdly he shewed, it was not beneficiall to them of Martins hundred in
point of advantaging their perticuler Plantacon, for the benefitt grew not by a

bare title to Land, but by cultivating & peopling it so to reape profit, now of such Land it was in every Adventurers power to have asmuch as he pleased wth out any other payment: for if an Adventurer (for examples sake) who had but one share of One hundreth Acres would send over Twenty men to inhabite & occupy it, fewer at this day will not doe it, he was by the Orders allready established to have for these 20 an addition of — 1000 — Acres of Land upon a first devision and as much more upon a second; And if then he would also people his — 1000 — Acres wth Ten score men more, he were to have another addition of — 10000 — Acres more upon a first devision and asmuch upon a second, & so forward to what extent of Land himselfe should desire. On the other side to enlarge a mans right unto new Land, and not to make use and proffitt of the Old, were to increase a matter of opinion, rather then of realty, & a shadowe rather then a substance.

Tobacco Laws of Virginia, 1631–32

From William Waller Hening, ed., *The Statutes at Large; A Collection of All the Laws of Virginia*. I (New York, 1823), p. 164.

Act XXI.

BE it also further ordered, That no planter or mayster of a famylie shall plante or cause to be planted above two thousand plants per pol, and that those that shall not plante or be otherwise imployed shall not transferr or make over theire right of plantinge unto any other; and to prevent any greater quantities, every planter or mayster of a famylie plantinge a cropp of tobacco, more or lesse, shall by tyed to procure one of his neighbours or some sufficient man to come and nomber his or theire plants of tobacco, who will uppon his oath declare and testifie unto the commander of that place, before the tenth day of July, that he hath counted and nombred the sayd plants, and shall say in his conscience the iust and true nomber of them, which thinge yf the sayd planter or mayster of a famylie shall neglect, or that the nomber of the plants is found to exceede the proportion of 2000 per pol, then the commander is hereby to present it to the next mounthlie cort, and the commissioners thereof shall give present order to have all that whole cropp of tobacco cutt down under payne of imprisonment and censure of the governor and counsell and grand assembly yf they neglect the execution thereof. Also uppon the neglect of the commander, he shall be censured in like manner.

Act XXII.

IT is likewise enacted, That no person shall tend, or cause to be tended, above 14 leaves, nor gather or cause to be gathered above 9 leaves uppon a plant of tobacco; and the several commanders shall hereby have power to examine the truth thereof; and yf any offend, to punish the servants by whippinge, and to bind over the mayster unto the next quarter cort at James Citty to be censured by the governor and counsell.

Act XXIII.

IT is ordered and ordeyned, That no person shall tend, or cause to be tended any slipps of old stalkes of tobacco, or any of the second cropps, upon the forfeiture of the whole cropp, whereof halfe to be to the informer, and halfe to publique uses as aforesayd.

Colonial Settlement – New England

Settlers at Plymouth Learn to Plant Corn — William Bradford's Account, 1621

From William Bradford, *History of Plymouth Plantation* (Boston, 1856), pp. 100, 105.

Afterwards they (as many as were able) began to plant ther corne, in which servise Squanto stood them in great stead, showing them both the maner how to set it, and after how to dress & tend it. Also he tould them, excepte they gott fish & set with it (in these old grounds) it would come to nothing, and he showed them that in the midle of Aprill [1621] they should have store enough come up the brooke, by which they begane to build, and taught them how to take it, and wher to get other provisions necessary for them; all which they found true by traill & experience. Some English seed they sew, as wheat & pease, but it came not to good, eather by the badness of the seed, or lateness of the season or both, or some other defecte.

* * *

They begane now to gather in the small harvest they had, and to fitte up their houses and dwellings against winter, being all well recovered in health & strength, and had all things in good plenty; for as some were thus imployed in affairs abroad, others were excersised in fishing, aboute codd & bass & other fish, of which they tooke good store, of which every family had their portion. All the somer ther was no wante. And now begane to come in store of foule, as winter aproached, of which this place did abound when they came first (but afterward decreased by degrees). And besids water foule ther was great store of wild turkies, of which they tooke many, besids vension, etc. Besids they had aboute a peck a meale a weeke to a person, or now since harvest, Indean corn to that proportion. Which made many afterwards write so largly of their plenty hear to their friends in England, which were not fained but true reports.

A Description of Plymouth Colony, 1621

Letter from Edward Winslow to George Morton, December 11, 1621, in Alexander Young, *Chronicles of the Pilgrim Fathers of the Colony of Plymouth, 1620–1625* (Boston, 1841), pp. 230–38.

LOVING AND OLD FRIEND,

Although I received no letter from you by this ship, yet forasmuch as I know you expect the performance of my promise, which was, to write unto you truly and faithfully of all things, I have therefore at this time sent unto you accordingly, referring you for further satisfaction to our more large Relations.

You shall understand that in this little time that a few of us have been here, we have built seven dwelling-houses and four for the use of the plantation, and have made preparation for divers others. We set the last spring some twenty acres of Indian corn, and sowed some six acres of barley and pease; and according to the manner of the Indians, we manured our ground with herrings, or rather shads, which we have in great abundance, and take with great ease at our doors. Our corn did prove well; and, God be praised, we had a good increase of Indian corn, and our barley indifferent good, but our pease not worth the gathering, for we feared they were too late sown. They came up very well, and blossomed; but the sun parched them in the blossom.

Our harvest being gotten in, our governor [William Bradford] sent four men on fowling, that so we might, after a special manner, rejoice together after we had gathered the fruit of our labors [The First Thanksgiving]. They four in one day killed as much fowl as, with a little help beside, served the company almost a week. At which time, amongst other recreations, we exercised our arms, many of the Indians coming amongst us, and among the rest their greatest king, Massasoyt, with some ninety men, whom for three days we entertained and feasted; and they went out and killed five deer, which they brought to the plantation, and bestowed on our governor, and upon the captain and others. And although it be not always so plentiful as it was at this time with us, yet by the goodness of God we are so far from want, that we often wish you partakers of our plenty.

We have found the Indians very faithful in their covenant of peace with us, very loving, and ready to pleasure us. We often go to them, and they come to us. Some of us have been fifty miles by land in the country with them, the occasions and relations whereof you shall understand by our general and more full declaration of such things as are worth the noting. Yea, it hath pleased God so to possess the Indians with a fear of us and love unto us, that not only the greatest king amongst them, called Massasoyt, but also all the princes and peoples round about us, have either made suit unto us, or been glad of any occasion to make peace with us; so that seven of them at once have sent their messengers to us to that end. Yea, an isle at sea, which we never saw, hath also, together with the former, yielded willingly to be under the protection and subject to our sovereign lord King James. So that there is now great peace amongst the Indians themselves, which was not formerly, neither would have been but for us; and we, for our parts, walk as peaceably and safely in the wood as in the highways in England. We entertain them familiarly in our houses, and they as friendly bestowing their venison on us. They are a people without any religion or knowledge of any God, yet very trusty, quick of apprehension, ripe-witted, just. The men and women go naked, only a skin about their middles.

For the temper of the air here, it agreeth well with that in England; and if there be any difference at all, this is somewhat hotter in summer. Some think it to be colder in winter; but I cannot out of experience so say. The air is very clear, and not foggy, as hath been reported. I never in my life remember a more seasonable year than we have here enjoyed; and if we have once but kine, horses, and sheep, I make no question but men might live as contented here as in any part of the world. For fish and fowl, we have great abundance. Fresh cod in the summer is but coarse meat with us. Our bay is full of lobsters all the summer, and affordeth variety of other fish. In September we can take a hogshead of eels in a night, with small labor, and can dig shellfish out of their beds all the winter. We have muscles and othus at our doors. Oysters we have none near, but we can have them brought by the Indians when we will. All the spring-time the earth sendeth forth naturally very good sallet herbs. Here are grapes, white and red, and very sweet and strong also; strawberries, gooseberries, raspas [rasberries], &c.; plums of three sorts, with black, and red, being almost as good as a damson; abundance of roses, white, red and damask; single, but very sweet indeed. The country wanteth only industrious men to employ; for it would grieve your hearts if, as I, you had seen so many miles together by goodly rivers uninhabited; and withal, to consider those parts of the world wherein you live to be even greatly burthened with abundance of people. These things I thought good to let you understand, being the truth of things as near as I could experimentally take knowledge of, and that you might on our behalf give God thanks, who hath dealt so favorably with us.

Our supply of men from you came the 9th of November, 1621, putting in at Cape Cod, some eight or ten leagues from us. The Indians that dwell thereabout were they who were owners of the corn which we found in caves, for which we have given them full content, and are in great league with them. They sent us word there was a ship near unto them, but thought it to be a Frenchman; and indeed for ourselves we expected not a friend so soon. But when we perceived that she made for our bay, the governor commanded a great piece to be shot off, to call home such as were abroad at work. Whereupon every man, yea boy, that could handle a gun, were ready, with full resolution that, if she were an enemy, we would stand in our just defence, not fearing them. But God provided better for us than we supposed. These came all in health, not any being sick by the way, otherwise than by sea-sickness, and so continue at this time, by the blessing of God. The good-wife Ford was delivered of a son the first night she landed, and both of them are very well.

When it pleaseth God we are settled and fitted for the fishing business and other trading, I doubt not but by the blessing of God the gain will give content to all. In the mean time, that we have gotten we have sent by this ship; and though it be not much, yet it will witness for us that we have not been idle, considering the smallness of our number all this summer. We hope the merchants will accept of it, and be encouraged to furnish us with things needful for further employment, which will also encourage us to put forth ourselves to the uttermost.

Now because I expect your coming unto us, with other of our friends, whose company we much desire, I thought good to advertise you a few things needful. Be careful to have a very good bread-room to put your biscuits in. Let your cask for beer and water be iron-bound, for the first tire, if not more. Let not your meat be dry salted; none can better do it than the sailors. Let your meal be so hard trod in your cask that you shall need an adz or hatchet to work it out with. Trust not too much on us for corn at this time, for by reason of this last company that came, depending wholly upon us, we shall have little enough till harvest. Be careful to come by some of your meal to spend by the way; it will much refresh you. Build your cabins as open as you can, and bring good store of clothes and bedding with you. Bring every man a musket or fowling-piece. Let your piece be long in the barrel, and fear not the weight of it, for most of our shooting is from stands. Bring juice of lemons, and take it fasting; it is of good use. For hot waters, aniseed water is the best; but use it sparingly. If you bring any thing for comfort in the country, butter or sallet oil, or both, is very good. Our Indian corn, even the coarsest, maketh as pleasant meat as rice; therefore spare that, unless to spend by the way. Bring paper and linseed oil for your windows [oiled paper was used until glass windows were imported], with cotton yarn for your lamps. Let your shot be most for big fowls, and bring store of powder and shot. I forbear further to write for the present, hoping to see you by the next return. So I take my leave, commending you to the Lord for a safe conduct unto us, resting in him,

YOUR LOVING FRIEND,
E.W.

The Nature of the New England Soil, 1634

From William Wood, *New England's Prospect* (London, 1634), pp. 11–15.

The Soyle is for the generall a warme kinde of earth, there being little cold-spewing land, no Morish Fennes, no Quagmires, the lowest grounds be the Marshes, over which every full and change the Sea flowes: these Marshes be rich ground, and bring plenty of Hay, of which the Cattle feed and like, as if they were fed with the best up-land Hay in *New England;* of which likewise there is great store which growes commonly between the Marshes and the Woods. This Medow ground lies higher than the Marshes, whereby it is freed from the over-flowing of the Seas; and besides this, in many places where the Tres grow thinne, there is good fodder to be got amongst the Woods. There be likewise in divers places neare the plantations great broad Medowes, wherein grow neither shrub nor Tree, lying low, in which Plaines growes as much grasse, as may be throwne out with a Sithe, thicke and long, as high as a mans middle; some as high as the shoulders, so that a good mower may cut three loads in a day.

But many obiect, this is but a course fodder: True it is, that it is not so fine to the eye as *English* grasse, but it is not sowre, though it grow thus ranke; but being made into Hay, the Cattle eate it as well as it were Lea-hay and like as well with it; I doe not thinke *England* can shew fairer Cattle either in Winter, or Summer, than is in those parts both Winter and Summer; being generally larger and better of milch, and bring forth young as ordinarily as Cattle doe in *England,* and have hitherto beene free from many diseases that are incident to Cattle in *England.*

To returne to the Subject in hand, there is so much hayground in the Countrey, as the richest voyagers that shall venture thither, neede not feare want of fodder, though his Heard increase into thousands, there being thousands of Acres that yet was never medled with. And whereas it hath beene reported, that some hath mowne a day for halfe of a loade of Hay: I doe not say, but it may be true, a man may doe as much, and get as little in *England,* on *Salisbury* Plaine, or in other places where Grasse cannot be expected: So Hay-ground is not in all places in *New England:* Wherefore it shall behoue every man according to his calling and estate, to looke for a fit situation at the first; and if hee be one that intends to live on his stocke, to choose the grassie Vallies before the woody Mountaines. Furthermore, whereas it hath beene generally reported in many places of *England,* that the Grasse growes not in those places where it was cut the fore-going yeares, it is a meere falshood; for it growes as well the ensuing Spring as it did before, and is more spiery and thicke, like our *English* Grasse: and in such places where the Cattle use to graze, the ground is much improved in the Woods, growing more grasse, and lesse weedy. The worst that can be sayd against the meddow-grounds, is because there is little edish or after-pasture, which may proceede from the late mowing, more than from any thing else; but though the edish be not worth much, yet is there such plenty of other Grasse and feeding, that there is no want of Winter-fodder till *December,* at which time men beginne to house their milch-cattle and Calves: Some, notwithstanding the cold of the Winter, have their young Cattle without doores, giving them meate at morning and evening. For the more upland grounds, there be different kinds, in some places clay, some gravell, some a red sand; all which are covered with a black mould, in some places above a foote deepe, in other places not so deepe. There be very few that have the experience of the ground, that can condemne it of barrennesse; although many deeme it barren, because the *English* use to manure their land with fish which they doe not because the land could not bring corne without it, but because it brings more with it; the land likewise being kept in hart the longer: besides, the plenty of fish which they have for little or nothing, is better so used, than cast away; but to argue the goodnesse of the ground, the *Indians* who are too lazie to catch fish, plant corne eight or ten yeares in one place without it, having very good crops. Such is the rankenesse of the ground that it must bee sowne the first yeare with *Indian* Corne, which is a soaking graine, before it will be fit for to receive *English* feede. In a word, as there is no ground so purely good, as the long forced and improoved grounds of *England,* so is there none so extreamely bad as in many places of *England,* that as yet have not beene manured and improved; the woods of *New England* being accounted better ground than the Forrests of *England* or woodland ground, or heathy plaines.

For the naturall soyle, I preferre it before the countrey of Surry, or Middlesex, which if they were not inriched with continuall manurings, would be lesse fertile than the meanest ground in *New England;* wherefore it is neyther impossible, nor much improbable, that upon improvements the soile may be as good in time as *England*. And whereas some gather the ground to be naught, and soone out of hart, because *Plimouth* men remove from their old habitations, I answer, they do no more remove from their habitation, than the Citizen wch hath one house in the Citty & another in the Countrey, for his pleasure, health, & profit. For although they have taken new plots of ground, and built houses upon them, yet doe they retaine their old houses still, and repaire to them every Sabbath day; neyther doe they esteeme their old lots worse than when they first tooke them: what if they doe not plant on them every yeare? I hope it is no ill husbandry to rest the land, nor is alwayes that the worst that lies sometimes fallow. If any man doubt of the goodnesse of the ground, let him comfort himselfe with the cheapenesse of it; such bad land in *England* I am sure wil bring in store of good monie. This ground is in some places of a soft mould, and easie to plow; in other places so tough and hard, that I have seene ten Oxen toyled, their Iron chaines broken, and their Shares and Coulters much strained: but after the first breaking up it is so ease, that two Oxen and a Horse may plow it; there hath as good *English* Corne growne there, as could be desired; especially Rie and Oates, and Barly: there hath beene no great tryall as yet of Wheate, and Beanes; onely thus much I affirme, that these two graines grow well in Gardens, therefore it is not improbable, but when they can gather feede of that which is sowne in the countrey, it may grow as well as any other Graine: but commonly the seede that commeth out of *England* is heated at Sea, and therefore cannot thrive at land.

The ground affoards very good kitchin Gardens, for Turneps, Parsnips, Carrots, Radishes, and Pumpions, Muskmillions, Isquouterquashes, Coucumbers, Onyons, and whatsoever growes well in *England*, growes as well there, many things being better and larger: there is likewise growing all manner of Hearbes for meate, and medicine, and that not onely in planted Gardens, but in the Woods, without eyther the art or the helpe of man, as sweet Marjoran, Purselane, Sorrell, Peneriall, Yarrow, Mirtle, Saxifarilla, Bayes, &c. There is likewise Strawberries in abundance, very large ones, some being two inches about; one may gather halfe a bushell in forenoone: In other seasons there bee Gooseberries, Bilberies, Resberies, Treackleberies, Hurtleberries, Currants; which being dryed in the Sunne are little inferiour to those that our Grocers sell in *England:* This land likewise affoards Hempe and Flax, some naturally, and some planted by the *English,* with Rapes if they bee well managed.

New Haven Agriculture Regulations, 1640–1641

From Charles J. Hoadley, ed., *Records of the Colony and Plantation of New Haven from 1638 to 1649*, (Hartford, 1857), p. 26–27, 49.

It is agreed by the towne and accordingly ordered by the court that the Neck shall be planted or sowen for the tearme of seaven yeares, and that John Brockett shall goe about laying it out forthw^{t}h, and all differences . . . shall be arbitrated by indifferent men w^{c}h shall be chosen to that end.

It is ordered thatt Mr. Davenports quarter, Mr. Eatons, Mr. Newmans and Mr. Tenches quarters shall have their first division of upland to begin at the sea side after the small lotts are layd out, and so goe on to the cow pasture, and to have their meaddow in the east meaddowes. And Mr. Evance quarter, Mr. Fowlers, Mr. Gregsons, Mr. Lambertons and the subburbs, are to begin w^{t}h their lands att the oyster poynt, and so come on to the oxe pasture in order, and to have their meaddow in the west meadowes, in the meadowes called Mr. Malbon meadow, on the Indian side, and in the sollitary cove. Allso that the cow pasture shall begin on the hither side of the Beever ponds, and the oxe pasture on the farre side of the Bever pond, and the way to them both to begin att Mr. Tenches corner.

It is ordered that no planter or planters shall make purchase of any lands or plantation from the Indians or others for their owne private use or advantage, but in the name & for the use of the whole plantation.

Itt is ordered that some speedy course shall be taken to keepe hogs out of the neck.

It is ordered that a convenient way to the hay place be left comon for all the towne.

It is ordered that no cattell belonging to this towne shall goe w^{t}hout a keeper after the first of May next.

It is ordered thatt those thatt kill wolves and foxes shall have for every wolfe head 15^s and for every foxe head 2^s 6^d, and if any be setting guns or traps shall hap to kill any hogs or other cattell, the towne is to bear the damage till some other course be determined. . . .

It was agreed that every planter in the towne shall have a proportion of land according to the proportion of estate w^{c}h he hath given in, and number of heads in his famyly, (viz) in the first division of upland & meadow 5 acres for every hundred pound, [an]d 5 acres for every two heads, of upland, but halfe an acre of meadow to a head [and] in the necke an acre to every hundred pound, and halfe an acre to every head.

* * *

It is agreed that the small lotts shall begin at the great rock on the farre side of the mill river, and so come downe, towards the sea, and then begin att the lower end of the farre side of the iland in the East river, and so come downe againe on the hither side, and if there shall fall out to be some small proportion wch will not amount to the quantity of him whose lott falls last in the Mill river, it shall be in his choyce w^ther he will have itt, yea or no.

It is allso agreed thatt in the East meadowes the first lott shall begin at the neck on the hither side of the river, and so goe on in order to the uper end so farr as there is meadow, and then begin at pasto^{rs} farme, and so goe up againe on the other side so farr as there is meadow, and whosoeu^r by lott falls next to the farmes thatt are layd out by choyce, if there be not their proportion there, they must take the rest where it falls next in order beyond them.

It is ordered thatt Thomas Fugill shall have the Iland in the mill river for his proportion, he being willing to have it when others refused it because it was bad.

Colonial Settlement – New York

Agriculture in New Netherlands, 1625

From "Wassenaers Historie Van Europa," in E. B. O'Callaghan, ed., *The Documentary History of the State of New York* (Albany, 1850), pp. 37–41.

Agriculture progresses in New Netherland in this wise. It is very pleasant, all products being in abundance, though wild. Grapes are of very good flavour, but will be, henceforward better cultivated by our people. Cherries are not found there. There are all sorts of fowls, both in the water and in the air. Swans, geese, ducks, bitterns, abound. The men scarcely ever labour, except to provide some game, either fowl or other description, for cooking, and then they have provided every thing. The women must attend to the remainder, tilling the soil, &c. When our people arrived there, they were busy cleaning up and planting. Before this vessel had left, the harvest was far advanced. It excites little attention if any one [of the Indians] abandon his wife; in case she have children, they usually follow her. Their summers are fine, but the days there are shorter than with us here. The winters are severe, but there is plenty of fuel, as the country is well wooded and it is at the service of whoever wants it.

There is some respect paid to those in authority amongst them; but these are no wise richer than others. There is always so much ado about them that the chief is feared and obeyed as long as he is near, but he must shift for himself like others. There is nothing seen in his house more than in those of the rest.

As regards the prosperity of New Netherland, we learn by the arrival of the ship whereof Jan May of Hoorn, was skipper, that every thing there was in good condition. The colony began to advance bravely and continues in friendship with the natives. The fur, or other trade, remains in the West India company, others being forbidden to trade there. Rich beavers, otters, martins and foxes are found there. This cargo consists of five hundred otter skins, and fifteen hundred beavers, and a few other things, which were in four parcels, for twenty-eight thousand, some hundred guilders. . . .

Good care having been taken by the directors of the West India company, in the spring to provide everything for the colony in Virginia, near the *Maykans* on the river *Mauritius,* by us called New Netherland, special attention was directed this month, (April [1625]) to reinforce it, as follows:

As the country is well adapted for agriculture and the raising of every thing that is produced here, the aforesaid Lords resolved to take advantage of the circumstance, and to provide the place with many necessaries, through the Honble. Pieter Evertsen Hulst, who undertook to ship thither, at his risk whatever was requisite, to wit; one hundred and three head of cattle; stallions, mares, steers and cows, for breeding and multiplying, besides all the hogs and sheep that might be thought expedient to send thither; and to distribute these in two ships of one hundred and forty lasts, in such a manner that they should be well feddered and attended to. Each animal had its own stall, with a floor of three feet of sand; fixed as comfortably as any stall here. Each animal had its respective servant who attended to it and knew its wants, so as to preserve its health, together with all suitable forage, such as oats, hay and straw, &c. In addition to these, country people take with them all furniture proper for the dairy; all sorts of Seed, ploughs and agricultural implements, so that nothing is wanting. What is most remarkable is, that nobody in the two ships can discover where the water is stowed for these cattle. As it was necessary to have another [ship] on that account, I shall here add:—the above parties caused a deck to be constructed on board. Beneath this were stowed in each ship three hundred tons of fresh water which was pumped up and thus distributed among the cattle. On this deck lay the ballast and thereupon stood the horses and steers, and thus there was no waste. He added the third ship so that, should the voyage continue longer, nothing may be wanting to the success of the expedition. In the eyes of the far seeing, the plan of this colony, which lay right beside the Spanish passage from the West Indies, was well laid.

In company with these, goes a fast sailing vessel at the risk of the Directors. In these aforesaid vessels also go six complete families with some freemen, so that forty five new comers or inhabitants are taken out, to remain there. The natives of New Netherland are very well disposed so long as no injury is done them. But if any wrong be committed against them they think it long till they be revenged and should any one against whom they have a grudge, be peaceably walking in the woods or going along in his sloop, even after a lapse of time, they will slay him, though they are sure it will cost them their lives on the spot, so highly prized is vengeance among them.

In our previous discourses, mention is made of New Netherland. Here is additional information: On further enquiry it is found, that they have a chief in time of war, named *Sacjama,* [Sachem] but above him is a greater *Sacjama* (pointing to Heaven) who rules the sun and moon. When they wage war against each other, they fortify their tribe or nation with palisades, serving them for a Fort, and sally out the one against the other. They have a tree in the centre, on which they place sentinels to observe the enemy and discharge arrows. None are exempt in war, but the Priests, and the women who carry their husband's arrows and food. The meat they eat consists of game and fish; but the bread is cakes baked fore-father's fashion, in the ashes; they almost all eat that in war. They are a wicked, bad people, very fierce in arms. Thir dogs are small. When the Honble Lambrecht van Twenhuyzen, once a skipper, had given them a big dog, and it was presented to them on ship-board, they were very much afraid of it; calling it, also, a Sachem of dogs, being the biggest. The dog, tied with a rope on board, was very furious against them, they being clad like beasts with skins, for he thought they were game; but when they gave him some of their bread made of Indian corn, which grows there, he learned to distinguish them, that they were men.

There are oaks of very close grain; yea, harder than any in this country, as thick as three or four men. There is Red-wood which being burned, smells very agreeably; when men sit by the fire on benches made from it, the whole house is perfumed by it. When they keep watch by night against their enemies, then they place it [the fire] in the centre of their huts, to warm their feet by it; they do not sit, then, up in the tree, but make a hole in the roof, and keep watch there, to prevent attacks.

Poisonous plants have been found there, which should be studied by those who have a fancy to cultivate land Hendrick Christiaensen carried thither, by order of his employers, Bucks, and Goats, also Rabbits, but they were found poisoned by the herbs. The Directors intend to send thither this spring voyage, [1625] a quantity of hogs which will be of great service to the colony; to be followed by cows, with young calves.

Very large oysters, sea fish and river fish are in such great abundance there, that they cannot be sold; and in rivers so deep, as to be navigated upwards with large ships. . . .

Chastity appears, on further enquiry, to hold a place among them, they being unwilling to cohabit with ours, through fear of their husbands. But those who are single, evince every friendly disposition. Further information is necessary. Whatever else is of value in the country, such as mines and other ores shall by time and further exploration be made known to us. Much profit is to be expected from good management.

A Description of Manhattan Island, 1626

From E. B. O'Callaghan, ed., *Documents Relative to the Colonial History of the State of New York*, I (Albany: Weed, Parsons, & Co., 1856), pp. 37–38.

HIGH AND MIGHTY LORDS:

Yesterday, arrived here the Ship the Arms of Amsterdam, which sailed from New Netherland, out of the River Mauritius, on the 23rd September. They report that our people are in good heart and live in peace there; the Women also have borne some children there. They have purchased the Island Manhattes from the Indians for the value of 60 guilders; 'tis 11,000 morgens in size. They had all their grain sowed by the middle of May, and reaped by the middle of August. They send thence samples of summer grain; such as wheat, rye, barley, oats, buckwheat, canary seed, beans and flax.

The cargo of the aforesaid ship is· — 7246 Beaver skins.

178½ Otter skins.

675 Otter skins.

48 Minck skins.

36 Wild cat skins.

33 Mincks.

34 Rat skins.

Considerable Oak timber and Hickory.

Herewith, High and Mighty Lords, be commended to the mercy of the Almighty.

*In Amsterdam, the 5*th *November, A*^d *1626.*
YOUR HIGH MIGHTINESSES' OBEDIENT,
(SIGNED)   P. SCHAGEN.

High and Mighty Lords,
My Lords the States General
at the Hague.

A Description of New Netherlands, 1646

From Isaac Jogues, "Novum Belgium [1646]," in J. F. Jameson, ed., *Narratives of New Netherlands 1609–1664* (New York, 1909), pp. 259–63.

New Holland, which the Dutch call in Latin *Novum Belgium* — in their own language, *Nieuw Nederland,* that is to say, New Low Countries — is situated between Virginia and New England. The mouth of the river, which some people call Nassau, or the Great North River, to distinguish it from another which they call the South River, and which I think is called Maurice River on some maps that I have recently seen, is at 40 deg. 30 min. The channel is deep, fit for the largest ships, which ascend to Manhattes Island, which is seven leagues in circuit, and on which there is a fort to serve as the commencement of a town to be built here, and to be called New Amsterdam.

This fort, which is at the point of the island, about five or six leagues from the [river's] mouth, is called Fort Amsterdam; it has four regular bastions, mounted with several pieces of artillery. All these bastions and the curtains were, in 1643, but mounds, most of which had crumbled away, so that one entered the fort on all sides. There were no ditches. For the garrison of the said fort, and another which they had built still further up against the incursions of the savages, their enemies, there were sixty soldiers. They were beginning to face the gates and bastions with stone. Within the fort there was a pretty large stone church, the house of the Governor, whom they call Director General, quite neatly built of brick, the storehouses and barracks.

On the island of Manhate, and in its environs, there may well be four or five hundred men of different sects and nations: the Director General told me that there were men of eighteen different languages; they are scattered here and there on the river, above and below, as the beauty and convenience of the spot has invited each to settle: some mechanics however, who ply their trade, are ranged under the fort; all the others are exposed to the incursions of the natives, who in the year 1643, while I was there, actually killed some two score Hollanders, and burnt many houses and barns full of wheat.

The river, which is very straight, and runs due north and south, is at least a league broad before the fort. Ships lie at anchor in a bay which forms the other side of the island, and can be defended by the fort.

Shortly before I arrived there, three large ships of 300 tons each had come to load wheat; they found cargoes, the third could not be loaded, because the savages had burnt a part of the grain. These ships had come from the West Indies, where the West India Company usually keeps up seventeen ships of war.

No religion is publicly exercised but the Calvinist, and orders are to admit none but Calvinists, but this is not observed; for besides the Calvinists there are in the colony Catholics, English Puritans, Lutherans, Anabaptists, here called Mnistes, etc.

When any one comes to settle in the country, they lend him horses, cows, etc.; they give him provisions, all which he returns as soon as he is at ease; and as to the land, after ten years he pays to the West India Company the tenth of the produce which he reaps.

This country is bounded on the New England side by a river which they call the Fresche [Connecticut] River, which serves as a boundary between them and the English. The English, however, come very near to them, choosing to hold lands under the Hollanders, who ask nothing, rather than depend on the English Milords, who exact rents, and would fain be absolute. On the other side, southward, towards Virginia, its limits are the river which they call the South River, on which there is also a Dutch settlement, but the Swedes have one at its mouth extremely well supplied with cannons and men. It is believed that these Swedes are maintained by some Amsterdam merchants, who are not satisfied that the West India Company should alone enjoy all the commerce of these parts. It is near this river that a gold mine is reported to have been found.

See in the work of the Sieur de Laet of Antwerp, the table and chapter on New Belgium, as he sometimes calls it, or the map "Nova Anglia, Novum Belgium et Virginia."

It is about fifty years [sic] since the Hollanders came to these parts. The fort was begun in the year 1615; they began to settle about twenty years ago, and there is already some little commerce with Virginia and New England.

The first comers found lands fit for use, deserted by the savages, who formerly had fields here. Those who came later have cleared the woods, which are mostly oak. The soil is good. Deer hunting is abundant in the fall. There are some houses built of stone; lime they make of oyster shells, great heaps of which are found here, made formerly by the savages, who subsist in part by that fishery.

The climate is very mild. Lying at 40⅔° there are many European fruits, as apples, pears, cherries. I reached there in October, and found even then a considerable quantity of peaches.

Ascending the river to the 43d degree, you meet the second [Dutch] settlement, which the tide reaches but does not pass. Ships of a hundred and a hundred and twenty tons can come up to it.

There are two things in this settlement (which is called Renselaerswick, as if to say, settlement of Renselaers, who is a rich Amsterdam merchant) — first, a miserable little fort called Fort Orenge, built of logs, with four or five pieces of Breteuil cannon, and as many pedereros. This has been reserved and is maintained by the West India Company. This fort was formerly on an island in the river; it is now on the mainland, towards the Hiroquois, a little above the said island.

Secondly, a colony sent here by this Renselaers, who is the patron. This colony is composed of about a hundred persons, who reside in some twenty-five or thirty houses built along the river, as each found most convenient. In the principal house lives the patron's agent; the minister has his apart, in which service is performed. There is also a kind of bailiff here, whom they call the seneschal, who administers justice. All their houses are merely of boards and thatched, with no mason work except the chimneys. The forest furnishing many large pines, they make boards by means of their mills, which they have here for the purpose.

They found some pieces of ground all ready, which the savages had formerly cleared, and in which they sow wheat and oats for beer, and for their horses, of which they have great numbers. There is little land fit for tillage, being hemmed in by hills, which are poor soil. This obliges them to separate, and they already occupy two or three leagues of country.

Trade is free to all; this gives the Indians all things cheap, each of the Hollanders outbidding his neighbor, and being satisfied provided he can gain some little profit.

This settlement is not more than twenty leagues from the Agniehronons [Mohawks], who can be reached by land or water, as the river on which the Iroquois lie [Mohawk River], falls into that which passes by the Dutch; but there are many low rapids, and a fall of a short half league, where the canoe must be carried.

There are many nations between the two Dutch settlements, which are about thirty German leagues apart, that is, about fifty or sixty French leagues [150 miles]. The Wolves, whom the Iroquois call Agotsaganens, are the nearest to the settlement of Renselaerswick and to Fort Orange. War breaking out some years ago between the Iroquois and the Wolves, the Dutch joined the latter against the former; but four men having been taken and burnt, they made peace. Since then some nations near the sea having killed some Hollanders of the most distant settlement, the Hollanders killed one hundred and fifty Indians, men, women and children, they having, at divers times, killed forty Hollanders, burnt many houses, and committed ravages, estimated at the time that I was there at 200,000 l. (two hundred thousand livres.) Troops were raised in New England. Accordingly, in the beginning of winter, the grass being trampled down and some snow on the ground, they gave them chase with six hundred men, keeping two hundred always on the move and constantly relieving one another; so that the Indians, shut up in a large island, and unable to flee easily, on account of their women and children, were cut to pieces to the number of sixteen hundred, including women and children. This obliged the rest of the Indians to make peace, which still continues. This occurred in 1643 and 1644.

From Three Rivers in New France, August 3, 1646.

Colonial Settlement – Maryland

A Description of Maryland, 1635

From "A Relation of Maryland, 1635," in Clayton Colman Hall, ed., *Narratives of Early Maryland, 1633–1684* (New York, 1910), pp. 76–77, 79–83, 91–97.

Their comming thus to seate upon an Indian Towne, where they found ground cleered to their hands, gave them opportunity (although they came late in the yeere) to plant some Corne, and to make them gardens, which they sowed with English seeds of all sorts, and they prospered exceeding well. They also made what haste they could to finish their houses; but before they could accomplish all these things, one Captaine Cleyborne (who had a desire to appropriate the trade of those parts unto himselfe) began to cast out words amongst the Indians, saying, That those of Yoacomaco were Spaniards and his enemies; and by this meanes endeavoured to alienate the mindes of the Natives from them, so that they did not receive them so friendly as formerly they had done. This caused them to lay aside all other workes, and to finish their Fort, which they did within the space of one moneth; where they mounted some Ordnance, and furnished it with some murtherers [small cannons], and such other meanes of defence as they thought fit for their safeties: which being done, they proceeded with their Houses and finished them, with convenient accommodations belonging thereto: And although they had thus put themselves in safety, yet they ceased not to procure to put these jealousies out of the Natives minds, by treating and using them in the most courteous manner they could, and at last prevailed therein, and setled a very firme peace and friendship with them. They procured from Virginia, Hogges, Poultrey, and some Cowes, and some male cattell, which hath given them a foundation for breed and increase; and whoso desires it, may furnish himselfe with store of Cattell from thence, but the hogges and Poultrey are already increased in Maryland, to a great stocke, sufficient to serve the Colonie very plentifully. They have also set up a Water-mill for the grinding of Corne, adjoyning to the Towne. [Mill Creek which was on the north side of St. Mary's River].

Thus within the space of six moneths, was laid the foundation of the Colonie in Maryland; and whosoever intends now to goe thither, shall finde the way so troden, that hee may proceed with much more ease and confidence then these first adventurers could, who were ignorant both of Place, People, and all things else, and could expect to find nothing but what nature produced: besides, they could not in reason but thinke, the Natives would oppose them; whereas now the Countrey is discovered, and friendship with the natives is assured, houses built, and many other accommodations, as Cattell, Hogges, Poultry, Fruits and the like brought thither from England, Virginea, and other places, which are usefull, both for profit and Pleasure: and without boasting it may be said, that this Colony hath arived to more in six moneths, then Virginia did in as many yeeres. If any man say, they are beholding to Virginea for so speedy a supply of many of those things which they of Virginia were forced to fetch from England and other remote places, they will confess it, and acknowledge themselves glad that Virginea is so neere a neighbour, and that it is so well stored of all necessaries for to make those parts happy, and the people to live as plentifully as in any other part of the world, only they wish that they would be content their neighbours might live in peace by them, and then no doubt they should find a great comfort each in other.

* * *

This Countrey affords naturally, many excellent things for Physicke and Surgery, the perfect use of which, the English cannot yet learne from the Natives: They have a roote which is an excellent preservative against Poyson, called by the English, the Snake roote. Other herbes and rootes they have, wherewith they cure all manner of woundes; also Saxafras, Gummes, and Balsum. An Indian seeing one of the English, much troubled with the tooth-ake, fetched of the roote of a tree, and gave the party some of it to hold in his mouth, and it eased the paine presently. They have other rootes fit for dyes, wherewith they make colours to paint themselves.

The Timber of these parts is very good, and in aboundance, it is usefull for building of houses, and shippes; the white Oake is good for Pipe-staves, the red Oake for wainescot. There is also Walnut, Cedar, Pine and Cipresse, Chestnut, Elme, Ashe, and Popler, all which are for Building, and Husbandry. Also there are divers sorts of Fruit-trees, as Mulberries, Persimons, with severall other kind of Plummes, and Vines, in great aboundance. The Mast and the Chesnuts, and what rootes they find in the woods, doe feede the Swine very fat, and will breede great store, both for their owne provision, or for merchandise, and such as is not inferior to the Bacon of Westphalia.

Of Strawberries, there is plenty, which are ripe in Aprill: Mulberries in May; and Raspices in June; Maracocks which is somewhat like a Limon, are ripe in August.

In the Spring, there are severall sorts of herbes, as Cornsallet, Violets, Sorrell, Purslaine, all which are very good and wholsome, and by the English, used for sallets, and in broth.

In the upper parts of the Countrey, there are Bufeloes, Elkes, Lions, Beares, Wolves, and Deare there are in great store, in all places that are not too much frequented, as also Beavers, Foxes, Otters, and many other sorts of Beasts.

Of Birds, there is the Eagle, Goshawke, Falcon, Lanner, Sparrow-hawke, and Merlin, also wild Turkeys in great aboundance, whereof many weigh 50. pounds, and upwards; and of Partridge plenty: There are likewise sundry sorts of Birds which sing, whereof some are red, some blew, others blacke and yellow, some like our Black-birds, others like Thrushes, but not of the same kind, with many more, for which wee know no names.

In Winter there is great plenty of Swannes, Cranes, Geese, Herons, Ducke, Teale, Widgeon, Brants, and Pidgeons, with other sorts, whereof there are none in England.

The Sea, the Bayes of Chesopeack, and Delaware, and generally all the Rivers, doe abound with Fish of severall sorts; for many of them we have no English names: There are Whales, Sturgeons very large and good, and in great aboundance; Grampuses, Porpuses, Mullets, Trouts, Soules, Place, Mackerell, Perch, Crabs, Oysters, Cockles, and Mussles; But above all these, the fish that have no English names, are the best except the Sturgeons: There is also a fish like the Thornebacke in England, which hath a taile a yard long, wherein are sharpe prickles, with which if it strike a man, it will put him to much paine and torment, but it is very good meate: also the Tode-fish, which will swell till it be ready to burst, if it be taken out of the water.

The Mineralls have not yet beene much searched after, yet there is discovered Iron Oare; and Earth fitt to make Allum, *Terra lemnia,* and a red soile like Bolearmonicke, with sundry other sorts of Mineralls, which wee have not yet beene able to make any tryall of.

The soil generally is very rich, like that which is about Cheesweeke neere London, where it is worth 20. shillings an Acre yeerely to Tillage in the Common-fields, and in very many places, you shall have two foote of blacke rich mould, wherein you shall scarce find a stone, it is like a sifted Garden-mould, and is so rich that if it be not first planted with Indian corne, Tobacco, Hempe, or some such thing that may take off the ranknesse thereof, it will not be fit for any English graine; and under that, there is found good loame, whereof wee have made as good bricke as many in England; there is great store of Marish ground also, that with good husbandry, will make as rich Medow, as any in the world: There is store of Marle, both blue, and white and in many places, excellent clay for pots, and tyles; and to conclude, there is nothing that can be reasonably expected in a place lying in the latitude which this doth, but you shall either find it here to grow naturally: or Industry, and good husbandry will produce it.

He that well considers the situation of this Countrey, and findes it placed betweene Virginia and New-England, cannot but, by his owne reason, conclude that it must needs participate of the naturall commodities of both places, and be capable of those which industry brings into either, the distances being so small betweene them: you shall find in the Southerne parts of Maryland, all that Virginia hath naturally; and in the Northerne parts, what New-England produceth: and he that reades Captaine John Smith shall see at large discoursed what is in Virginia, and in Master William Wood, who this yeere hath written a treatise of New-England, he may know what is there to be expected.

In the first place I name Corne, as the thing most necessary to sustaine man; That which the Natives use in the Countrey, makes very good bread, and also a meate which they call Omene [Hominy] it's like our Furmety, and is very savory and wholesome; it will Mault and make good Beere; Also the Natives have a sort of Pulse, which we call Pease and Beanes, that are very good. This Corne yeelds a great increase, so doth the Pease and Beanes: One man may in a season, well plant so much as will yeeld a hundred bushells of this Corne, 20 bushells of Beanes and Pease, and yet attend a crop of Tobacco: which according to the goodnesse of the ground may be more or lesse, but in ordinarily accompted betweene 800 and 1000 pound weight.

They have made tryall of English Pease, and they grow very well, also Musk-mellons, Water-mellons, Cow-cumbers, with all sorts of garden Roots and Herbes, as Carrots, Parsenips, Turnips, Cabbages, Radish with many more; and in Virginia they have sowed English Wheate and Barley, and it yeelds twise as much increase as in England; and although there be not many that doe apply themselves to plant Gardens and Orchards, yet those that doe it, find much profit and pleasure thereby: They have Peares, Apples, and severall sorts of Plummes, Peaches in abundance, and as good as those of Italy; so are the Mellons and Pumpions: Apricocks, Figgs and Pomegranates prosper exceedingly; they have lately planted Orange and Limon trees which thrive very wel: and in fine, there is scarce any fruit that growes in England, France, Spaine or Italy, but hath been tryed there, and prospers well. You may there also have hemp and Flax, Pitch and Tarre, with little labour; it's apt for Rapeseed, and Annis-seed, Woad, Madder, Saffron, etc. There may be had Silke-wormes, the Countrey being stored with Mulberries: and the superfluity of wood will produce Potashes.

And for Wine, there is no doubt but it will be made there in plenty, for the ground doth naturally bring foorth Vines, in such aboundance, that they are as frequent there, as Brambles are here. Iron may be made there with little charge; Brave ships may be built, without requiring any materialls from other parts: Clab-board, Wainscott, Pipe-staves and Masts for ships the woods will afford plentifully. In fine, Butter and Cheese, Porke and Bacon, to transport to other countrys will be no small commodity, which by industry may be quickly had there in great plenty, etc. And if there were no other staple commodities to be hoped for, but Silke and Linnen (the materialls of which, apparantly will grow there) it were sufficient to enrich the inhabitants.

* * *

What person soever, subject to our soveraigne Lord the King of England, shal be at the charge to transport into the Province of Maryland, himselfe or his deputy, with any number of able men, betweene the ages of 16 and 50, each man being provided in all things necessary for a Plantatio (which, together with their transportation, will amount to about 20 *l*. a man, as by an aestimate hereafter following may appeare) there shalbe assigned unto every such adventurer, for every five men which he shall so transport thither, a proportion of good land within the said Province, containing in quantity 1000 acres of English measure, which shall be erected into a Mannor, and be conveyed to him, his heires, and assignes for ever, with all such royalties and priviledges, as are usually belonging to Mannors in England; rendring and paying yerely unto his Lordship, and his heires for every such Mannor, a quit rent of 20 shillings, (to be paid in the Commodities of the Countrey) and such other services as shall be generally agreed upon for publike uses, and the common good.

What person soever, as aforesaid, shall transport himselfe, or any less number of servants than five, (aged, and provided as aforesaid) he shall have assigned to him, his heires and assignes for ever, for himselfe, 100 acres of good land within the said Province; and for and in respect of every such servant, 100 acres more, he be holden of his Lordship in freehold, paying therefore, a yeerely quit rent of 2 shillings for every hundred acres, in the Commodities of the Countrey.

Any married man that shall transport himselfe, his wife and children; shall have assigned unto him, his heires and assignes for ever, in freehold, (as aforesaid) for himselfe 100 acres; and for his wife 100 acres; and for every child that he shall carry over, under the age of 16 yeers, 50 acres; paying for a quit rent 12 pence for every fifty acres.

Any woman that shall transport herselfe or any children, under the age of six yeeres, shall have the like Conditions as aforesaid.

Any woman that shall carry over any women servants, under the age of fourty yeeres, shall have for and in respect of every such woman servant, 50 acres; paying onely a quit rent as aforesaid.

Instructions and advertisements, for such as shall intend to goe, or send, to plant in Maryland.

This Countrey of Maryland, lieth from England to the Southwest, about 1200 leagues by Sea: the voyage is sometimes performed thither in 5 or 6 weeks, but ordinarily it is two moneths voyage, and oftner, within that time then beyond it. The returne from thence to England, is ordinarily made in a moneth, and seldome exceeds six weekes.

The best time of the yeere for going thither is to be there by Michaelmas, or at furthest by Christmas, for he that comes by that time shall have time enough to build him a house, and to prepare ground sufficient to plant in the spring following. But there is conveniency of passage thither in most moneths of the yeere; and any one that will send unto Mr. Peasleys, or Master Morgans house, may there be informed of the certaine time when any of his Lordships company is to goe away, and so save the charge of unnecessary attendance here in London.

A Particular of Such Necessary Provisions as Every Adventurer Must Carry, According to the Number of His Servants: Together With an Estimate of Their Prices.

In Victualls.

For one man, for a yeere,

	l—s—d
Imprimis, eight bushells of meale	2—8—0
Item, two bushells of Oatmeale	0—9—0
Item, one bushell of Pease	0—4—0
Item, one gallon of Oyle	0—3—6
Item, two gallons of Vinegar	0—2—0
Item, one gallon of Aquavitæ	0—2—6
Item, one bushell of Bay-salt	0—2—0
Item, in Sugar, Spice and Fruit	0—6—8
Summ.	3–17—8

In Apparrell.

For one man,

	l—s—d
Item, two Munmoth caps or hats	0—4—0
Item, three falling Bands	0—1—3
Item, three shirts	0—7—6
Item, one Wastcoate	0—2—2
Item, one suite of Canvas	0—7—6
Item, one suite of Frize	0–10—0
Item, one suite of Cloth	0–16—0
Item, one course cloth, or frize coate	0–15—0
Item, three paire of stockings	0—4—0
Item, sixe paire of shooes	0–13—0
Item, Inkle [laces] for garters	0—0—2
Item, one dozen of points	0—0—3
Summ.	4—0-10

In Bedding.

For two men,

	l—s—d
Item, two paire of Canvas sheets	0—16—0
Item, seven ells of Canvas to make a bed and boulster to be fill'd in the country	0—8—0
Item, one Rugg for a bed	0—8—0
Item, five ells of course Canvas to make a bed at Sea, to bee fill'd with straw	0—4—0
Item, one course Rugg at Sea	0—6—0
Summ.	2—2—0
whereof one mans part is,	1—1—0

In Armes.

For one man,

	l—s—d
Item, one musket	1—0—0
Item, 10 pound of Powder	0–11—0
Item, 40 pound of Lead, Bullets, Pistoll and Goose shot, of each sort some.	0—4—0
Item, one sword	0—5—0
Item, one belt	0—1—0
Item, one bandeleere and flaske	0—2—o
Item, in Match	0—2—6
Summ.	2—5—6

In Tooles.

For five persons, and so after the rate for more or lesse.

	l—s—d
Item, 5 broad Howes, at 2 s. a piece	0—10—0
Item, 5 narrow Howes, at 16 d. a piece	0—6—8
Item, 2 broad Axes, at 3 s. 8 d. a piece	0—7—4
Item, 5 felling Axes, at 1 s. 6 d. a piece	0—7—6
Item, 2 steele Hand-sawes, at 1 s. 4 d.	0—2—8
Item, Two-handsawes at 5 s.	0—10—0
Item, a Whip-saw set and filed, with boxe, file and wrest	0—10—0
Item, 2 Hammers, at 12 d.	0—2—0
Item, 3 Shovells, at 1 s. 6 d.	0—4—6
Item, 3 Spades, at 1 s. 6 d.	0—4—6
Item, 2 Awgurs, at 6 d.	0—1—0
Item, 6 Chissells at 6 d.	0—3—0
Item, 2 Piercers stocked, at 4 d.	0—0—8
Item, 3 Gimlets, at 2 d.	0—0—6
Item, 2 Hatchets, at 1 s. 9 d.	0—3—6
Item, 2 Frowes to cleave Pales, at 1 s. 6 d.	0—3—0
Item, 2 Hand-bills, at 1 s. 8 d.	0—3—4
Item, one Grindstone	0—4—0
Item, Nailes of all sorts	2—0—0
Item, 2 Pickaxes, at 1 s. 6 d.	0—3—0
Summ.	6—7—2
whereof one mans part is,	1—5—8

Houshold Implements
For 6 persons, and so after the rate, for more

	l—s—d
Item, one Iron pot	0—7—0
Item, one Iron kettle	0—6—0
Item, one large Frying-pan	0—2—6
Item, one Gridiron	0—1—6
Item, two Skillets	0—5—0
Item, one Spit	0—2—0
Item, Platters, Dishes, and spoones of wood	0—4—0
Summ.	1—8—0
whereof one mans part is,	0—4—8

An estimate of the whole charge of transporting one servant, and providing him of all necessaries for one yeere.

	l—s—d
Inprimis, In Victualls	3–17—8
Item, In apparell	4—0-10
Item, In bedding	1—1—0
Item, In Armes	2—5—6
Item, In tooles	1—5—8
Item, In houshold Implements	0—4—8
Item, Caske to put his goods in	0-10—0
Item, fraight for his goods at halfe a tunne	1-10—0
Item, For his Victuall, and passage by Sea	6—0—0
	20.–15.–4

Of which charge, the Adventurer having the greatest part of it in provision and goods; in case any servant die by the way, or shortly after his comming thither, the goods of that servant being sold in the Countrey, will returne all his charge againe, with advantage.

A Computation of a servants labour, and the profit that may arise by it, by instance in some particulars, which may be put in practise the first yeere.

One man may at the season plant so much corne, as ordinarily yeelds
    of Wheate 100. bushels, worth upon the place, at Two shillings   l—s—d
    a Bushell, ...10—0—0
Of Beanes and Pease, 20. bushels, worth at three shillings a bushell, ...3—0—0
The same man will plant of Tobacco, betweene 800. and a 1000.
    weight, which at the lowest rate, at two pound 10. shil. the
    hundred, is worth, ...20—0—0
The same man may within the same yere, in the winter, make 4000.
    of Pipe-staves, worth upon the place foure pound the thousand. .....16—0—0
                                     49.—00—00.

Beside all their other labours in building, fencing, clearing of ground, raising of Cattell, gardening, etc.

If a mans labour be imployed in Hempe and Flaxe, it will yeeld him as much profit, as Tobacco at this rate; and so in many other Commodities, whereof this Countrey is capable.

No man neede to doubt of the vent of these Commodities, for Merchants send shipping to those parts, who will buy off these Commodities at the aforesaid rates, in as great a quantitie, as they shalbe able to make ready for them; because they yeeld a great encrease of profit in other Countreys, which the Planters themselves may make advantage of to themselves, if they have shipping, and thinke fit to deale in such a kind of trade. As for instance, a 1000. of Pipe-staves, which are rated upon the place at foure pound, being carried to the Canaries, will yeeld 15. or 20. *l.* Where likewise, and at the Westerne Islands, the Indian Corne will yeeld a great increase of benefit. The benefit also which may be raised by trade out of Swine onely, may easily be conceived to be very great, seeing they multiplie exceedingly, aske little tendance, and lesse charge of keeping in that Countrey, so abounding with Mast, Chestnuts, etc. For Porke being transported into Spaine, or the Westerne Islands will yeeld about 6. pence a pound, and Bacon, 8. pence. or 9. pence.

Colonial Settlement – Pennsylvania

William Penn's Description of Land Division in Pennsylvania, 1681

From William Penn, "Some Account of the Province of Pennsylvania [1681]," in Albert Cook Myers, ed., *Narratives of Early Pennsylvania, West New Jersey and Delaware 1630–1707* (New York, 1912), pp. 208–10.

The Conditions.

My Conditions will relate to three sorts of People: 1st. Those that will buy: 2dly. Those that take up Land upon Rent: 3dly. Servants. To the first, the Shares I sell shall be certain as to number of Acres; that is to say, every one shall contain Five thousand Acres, free from any Indian incumbrance, the price a hundred pounds, and for the Quit rent but one English shilling or the value of it yearly for a hundred Acres; and the said Quit-Rent not to begin to be paid till 1684. To the second sort, that take up Land upon Rent, they shall have liberty so to do, paying yearly one peny per Acre, not exceeding Two hundred Acres. To the third sort, to wit, Servants that are carried over, Fifty Acres shall be allowed to the Master for every Head, and Fifty Acres to every Servant when their time is expired. And because some engage with me that may not be disposed to go, it were very advisable for every three Adventurers to send an Overseer with their Servants, which would well pay the Cost.

The Divident may be thus; if the persons concern'd please, a Tract of Land shall be survey'd; say Fifty thousand Acres to a hundred Adventurers; in which some of the best shall be set out for Towns or Cities; and there shall be so much Ground allotted to each in those Towns as may maintain some Cattel and produce some Corn; then the remainder of the fifty thousand Acres shall be shar'd among the said Adventurers (casting up the Barren for Commons, and allowing for the same) whereby every Adventurer will have a considerable quantity of Land together; likewise every one a proportion by a Navigable River, and then backward into the Country. The manner of divident I shall not be strict in; we can but speak roughly of the matter here; but let men skilful in Plantations be consulted, and I shall leave it to the majority of votes among Adventurers when it shall please God we come there, how to fix it to their own content.

These Persons that Providence Seems to Have Most Fitted for Plantations Are,

1st. Industrious Husbandmen and Day-Labourers, that are hardly able (with extreme Labour) to maintain their Families and portion their Children.

2dly. Laborious Handicrafts, especially Carpenters, Masons, Smiths, Weavers, Taylors, Tanners, Shoemakers, Shipwrights, etc. where they may be spared or are low in the World: And as they shall want no encouragement, so their Labour is worth more there than here, and there provision cheaper.

3dly. A Plantation seems a fit place for those Ingenious Spirits that being low in the World, are much clogg'd and oppress'd about a Livelyhood, for the means of subsisting being easie there, they may have time and opportunity to gratify their inclinations, and thereby improve Science and help Nurseries of people.

4thly. A fourth sort of men to whom a Plantation would be proper, takes in those that are younger Brothers of small Inheritances; yet because they would live in sight of their Kindred in some proportion to their Quality, and can't do it without a labour that looks like Farming, their condition is too strait for them; and if married, their Children are often too numerous for the Estate, and are frequently bred up to no Trades, but are a kind of Hangers on or Retainers to the elder Brothers Table and Charity: which is a mischief, as in it self to be lamented, so here to be remedied; For Land they have for next to nothing, which with moderate Labour produces plenty of all things neccessary for Life, and such an increase as by Traffique may supply them with all conveniencies.

Lastly, There are another sort of persons, not only fit for, but necessary in Plantations, and that is, Men of universal Spirits, that have an eye to the Good of Posterity, and that both understand and delight to promote good Discipline and just Government among a plain and well intending people; such persons may find Room in Colonies for their good Counsel and Contrivance, who are shut out from being of much use or service to great Nations under settl'd Customs: These men deserve much esteem, and would be harken'd to. Doubtless 'twas this (as I observ'd before) that put some of the famous Greeks and Romans upon Transplanting and Regulating Colonies of People in divers parts of the World; whose Names, for giving so great proof of their Wisdom, Virtue, Labour and Constancy, are with Justice honourably delivered down by story to the praise of our own times; though the World, after all its higher pretences of Religion, barbarously errs from their excellent Example.

A Contemporary Description of Pennsylvania, 1683

From "Letter of Thomas Paschall, 1638," in Albert Cook Myers, ed., *Narratives of Early Pennsylvania, West New Jersey and Delaware 1630–1707* (New York, 1912), pp. 250–54.

My kind love remembred unto Thee, and thy wife, and to all the rest of thy Family, hoping that you are all in good health, as through the goodness of God we all are at this present writing, Excepting one of my servants, who was a Carpenter, and a stout young man, he died on board the Ship, on our Voyage. I thank God I, and my Wife, have not been sick at all, but continued rather better than in England; and I do not find but the Country is healthfull, for there was a Ship that came the same day with us into the river, that lost but one Passenger in the Voyage, and that was their Doctor, who was ill when he came on board, and those people that came in since continue well. William Penn and those of the Society are arrived. W. P. is well approved of, he hath been since at New Yorke, and was extraordinarily entertained, and he behaved himself as Noble. Here is a place called Philadelphia, where is a Market kept, as also at Upland [Chester]. I was at Bridlington [Burlington, N.J.]-fair, where I saw most sorts of goods to be sold, and a great resort of people; Where I saw English goods sold at very reasonable rates; The Country is full of goods, Brass and Pewter lieth upon hand. That which sells best, is Linnen cloath, trading Cloath for the Indians; I bought Kersey and it doth not sell, Broad Cloath is wanting, and Perniston, and Iron-potts; and as for the Swedes, they use but little Iron in Building, for they will build, and hardly use any other toole but an Ax; They will cut down a Tree, and cut him off when down, sooner then two men can saw him, and rend him into planks or what they please; only with the Ax and Wooden wedges, they use no Iron; They are generaly very ingenous people, lives well, they have lived here 40 Years, and have lived much at ease, having great plenty of all sorts of provisions, but then they weer but ordinarily Cloathd; but since the English came, they have gotten fine Cloaths, and are going proud. Let all people know that have any mind to come hither, that they provide Comfortable things for their passage, and also some provitions to serve them here, for although things are to be had at reasonable rates here, yet it is so far to fetch, that it spends much time, so that it's better to come provided for half a Year then to want one day, I thank God we have not wanted, but have fared well beyond what we did in England.

The River is taken up all along, by the Sweads, and Finns and some Dutch, before the English came, near eight score miles, and the Englishmen some of them, buy their Plantations, and get roome by the great River-side, and the rest get into Creeks, and small rivers that run into it, and some go into the Woods seven or eight Miles; Thomas Colborne is three miles in the Woods, he is well to pass, and hath about fourteen Acres of Corne now growing, and hath gotten between 30 and 40 *li.* by his Trade, in this short time. I have hired a House for my Family for the Winter, and I have gotten a little House in my Land for my servants, and have cleared Land about six Acres; and this I can say, I never wisht my self at Bristol again since my departure. I live in the Schoolkill Creek, near Philadelphia, about 100 Miles up the River. Here have been 24 Ships with Passengers within this Year, so that provisions are somewhat hard to come by in some places, though at no dear rate, there is yet enough in the River, but it is far to fetch, and suddainly there will be an Order taken for continuall supply. Now I shall give you an impartial account of the Country as I find it, as followeth. When we came into Delawarebay we saw an infinite number of small fish in sholes, also large fish leaping in the Water; The River is a brave pleasant River as can be desired, affording divers sorts of fish in great plenty, it's planted all along the Shoare, and in some Creeks, especially in Pensilvania side, mostly by Sweads, Finns, and Dutch, and now at last, English throng in among them, and have filed all the Rivers and Creeks a great way in the Woods, and have settled about 160 Miles up the great River; some English that are above the falls, have sowed this Year 30 or 40 bushels of Wheat, and have great stocks of Cattel; Most of the Sweads, and Finns are ingeneous people, they speak English, Swead, Finn, Dutch and the Indian; They plant but little Indian corne, nor Tobacco; their Women make most of the Linnen cloath they wear, they Spinn and Weave it and make fine Linnen, and are many of them curious housewives: The people generally eat Rye bread, being approved of best by them, not but that here is good Wheat, for I have eaten as good bread and drank as good drink as ever I did in England, as also very good butter and cheese, as most in England. Here is 3 sorts of Wheat, as Winter, Summer, and Buck Wheat; the Winter Wheat they sow at the fall, the Summer Wheat in March, these two sorts are ripe in June; then having taken in this, they plow the same land, and sow Buck Wheat, which is ripe in September: I have not given above 2s. 6d per skipple, (which is 3 English pecks) for the best Wheat and that in goods which cost little more then half so much in England, here is very good Rye at 2s per skipple, also Barly of 2 sorts, as Winter, and Summer, at 4 Guilders per skipple; also Oats, and 3 sorts of Indian Corne, (two of which sorts they can Malt and make good bear of as of Barley,) at four Guilders per Skiple, a Guilder is four pence halfpenney. I have bought good Beef, Porke, and Mutton at two pence per pound and some cheaper, also Turkeys and Wild-geese at the value of two or three Pound of Shot apeice, and Ducks at one Pound of Shot, or the like value, and in great plenty: here is great store of poultry, but for Curlews, Pidgons, and Phesants, they will hardly bestow a shot upon them. I have Venison of the Indians very cheap, although they formerly sold it as cheap again to the Sweads: I have four Dear for two yards of trading cloath, which cost five shillings, and most times I purchase it cheaper: We had Bearsflesh this fall for little or nothing, it is good food, tasting much like Beef; There have been many Horses sold of late to Barbadoes, and here is plenty of Rum, Sugar, Ginger, and Melasses. I was lately at Bridlington-fair, where were a great resort of people, with Cattle and all sorts of Goods, sold at very reasonable rates.

Here are Gardens with all sorts of Herbs, and some more then in England, also Goose-beries and Roasetrees, but what other Flowers I know not yet: Turnips, Parsnips, and Cabbages, beyond Compare. Here are Peaches in abundance of three sorts I have seen rott on the Ground, and the Hogs eate them, they make good Spirits from them, also from Corne and Cheries, and a sort of wild Plums and Grapes, and most people have Stills of Copper for that use. Here are Apples, and Pears, of several sorts, Cheries both Black and Red, and Plums, and Quinces; in some places Peach Stones grow up to bear in three Years: the Woods are full of Oakes, many very high and streight, many of them about two foot through, and some bigger, but very many less; A Swead will fell twelve of the bigger in a day; Here are brave Poplar, Beach, Ash, Lymetrees, Gum-trees, Hickary-trees, Sasafras, Wallnuts, and Chesnuts, Hazel, and Mullberies: Here growes in the Woods abundance of Wortle-beries or Whorts, Strawberies and Blackberies, better then in England, as also three sorts of Grapes and Plums; Here is but few Pine-trees, and Ceder; Here is good Firestone plenty enough in most places: and the Woods are full of runs of water. I have lately seen some Salt, very good to salt meat with, brought by an Indian out of the Woods: they say there is enough of it: but for Minnerals or Mettals, I have not seen any, except it be Marcasite, such as they make Vitriol or Copperis with in England. Here are Beavers, Rackoons, Woolves, Bears, a sort of Lyons, Polecatts, Mushratts, Elks, Mincks, Squirills of several sorts and other small Creatures, but none of these hurt unless surprised: also Rattle Snakes and black Snakes, but the Rattle Snakes I have not seen, though I have rambled the Woods much these three Months, since the beginning of September. The Indians are very quiet and peaceable, having their understandings, and qualifications, and when abused will seek revenge, they live much better since the English came; getting necessarys as cheap again as formerly, and many of them begin to speake English, I have heard one say *Swead no good, Dutch men no good,* but *Englishman good.* William Penn is settling people in Towns. There are Markets kept in two Towns viz. Philadelphia, being Chiefest, Chester, formerly called Upland. To write of the Seasons of the Year I cannot, but since I came it hath been very pleasant weather. The Land is generally good and yet there is some but ordinary and barren ground. Here are Swamps which the Sweads prize much, and many people will want: And one thing more I shall tell you, I know a man together with two or three more, that have happened upon a piece of Land of some Hundred Acres, that is all cleare, without Trees, Bushes, stumps, that may be Plowed without let, the farther a man goes in the Country the more such Land they find. There is also good Land, full of Large and small Trees, and some good Land, but few Trees on it. The Winter is sharp and the Cattel are hard to keep. The people that come must work and know Country affairs; They must be provided with some provisions for some time in the Country, and also some to help along on Board the Ship. I have more to write, but am shortned in time. *Vale.*

THOMAS PASKELL.

Pennsilvania, the last of January, 1682/3.

William Penn's Description of Settlements and Farming in Pennsylvania, 1685

From William Penn, "A Further Account of the Province of Pennsylvania [1685]," in Albert Cook Myers, ed., *Narratives of Early Pennsylvania, West New Jersey and Delaware 1630–1707* (New York, 1912), pp. 263–69, 274–76.

Of Country Settlements.

1. We do settle in the way of Townships or Villages, each of which contains 5,000 acres, in square, and at least Ten Families; the regulation of the Country being a family to each five hundred Acres. Some Townships have more, where the Interests of the People is less than that quantity, which often falls out.

2. Many that had right to more Land were at first covetous to have their whole quantity without regard to this way of settlement, tho' by such Wilderness vacancies they had ruin'd the Country, and then our interest of course. I had in my view Society, Assistance, Busy Commerce, Instruction of Youth, Government of Peoples manners, Conveniency of Religious Assembling, Encouragement of Mechanicks, distinct and beaten Roads, and it has answered in all those respects, I think, to an Universall Content.

3. Our Townships lie square; generally the Village in the Center; the Houses either opposit, or else opposit to the middle, betwixt two houses over the way, for near neighborhood. We have another Method, that tho the Village be in the Center, yet after a different manner: Five hundred Acres are allotted for the Village, which, among ten families, comes to fifty Acres each: This lies square, and on the outside of the square stand the Houses, with their fifty Acres running back, where ends meeting make the Center of the 500 Acres as they are to the whole. Before the Doors of the Houses lies the high way, and cross it, every man's 450 Acres of Land that makes up his Complement of 500, so that the Conveniency of Neighbourhood is made agreeable with that of the Land.

4. I said nothing in my last of any number of Townships, but there are at least Fifty settled before my leaving those parts, which was in the moneth called August, 1684.

5. I visitted many of them, and found them much advanced in their Improvements. Houses over their heads and Garden plots, Coverts for their Cattle, an encrease of stock, and several Enclosures in Corn, especially the first Commers; and I may say of some Poor men was the beginnings of an Estate; the difference of labouring for themselves and for others, of an Inheritance and a Rack Lease, being never better understood.

Of the Produce of the Earth.

1. The Earth, by God's blessing, has more than answered our expectation; the poorest places in our Judgment producing large Crops of Garden Stuff and Grain. And though our Ground has not generally the symptoms of the fat Necks that lie upon salt Waters in Provinces southern of us, our Grain is thought to

excell and our Crops to be as large. We have had the mark of the good Ground amongst us from Thirty to Sixty fold of English Corn.

2. The Land requires less seed: Three pecks of Wheat sow an acre, a Bushel at most, and some have had the increase I have mention'd.

3. Upon Tryal we find that the Corn and Roots that grow in England thrive very well there, as Wheat, Barly, Rye, Oats, Buck-Wheat, Pease, Beans, Cabbages, Turnips, Carrets, Parsnups, Colleflowers, Asparagus, Onions, Charlots, Garlick and Irish Potatos; we have also the Spanish and very good Rice, which do not grow here.

4. Our low lands are excellent for Rape and Hemp and Flax. A Tryal has been made, and of the two last there is a considerable quantity Dress'd Yearly.

5. The Weeds of our Woods feed our Cattle to the Market as well as Dary. I have seen fat Bullocks brought thence to Market before Mid Summer. Our Swamps or Marshes yeeld us course Hay for the Winter.

6. English Grass Seed takes well, which will give us fatting Hay in time. Of this I made an Experiment in my own Court Yard, upon sand that was dug out of my Cellar, with seed that had lain in a Cask open to the weather two Winters and a Summer; I caus'd it to be sown in the beginning of the month called April, and a fortnight before Midsummer it was fit to Mow. It grew very thick: But I ordered it to be fed, being in the nature of a Grass Plott, on purpose to see if the Roots lay firm: And though it had been meer sand, cast out of the Cellar but a Year before, the seed took such Root and held the earth so fast, and fastened itself so well in the Earth, that it held fast and fed like old English Ground. I mention this, to confute the Objections that lie against those Parts, as of that, first, English Grass would not grow; next, not enough to mow; and, lastly, not firm enough to feed, from the Levity of the Mould.

7. All sorts of English fruits that have been tryed take mighty well for the time: The Peach Excellent on standers, and in great quantities: They sun dry them, and lay them up in lofts, as we do roots here, and stew them with Meat in Winter time. Musmellons and Water Mellons are raised there, with as little care as Pumpkins in England. The Vine especially, prevails, which grows every where; and upon experience of some French People from Rochel and the Isle of Rhee, Good Wine may be made there, especially when the Earth and Stem are fin'd and civiliz'd by culture. We hope that good skill in our most Southern Parts will yield us several of the Straights Commodities, especially Oyle, Dates, Figgs, Almonds, Raisins and Currans.

Of the Produce of our Waters.

1. Mighty Whales roll upon the Coast, near the Mouth of the Bay of Delaware. Eleven caught and workt into Oyl one Season. We justly hope a considerable profit by a Whalery; they being so numerous and the Shore so suitable.

2. Sturgeon play continually in our Rivers in Summer: And though the way of cureing them be not generally known, yet by a Receipt I had of one Collins, that related to the Company of the Royal Fishery, I did so well preserve some, that I had them good there three months of the Summer, and brought some of the same so for England.

3. Alloes, as they call them in France, the Jews Allice, and our Ignorants, Shads, are excellent Fish and of the Bigness of our largest Carp: They are so Plentiful, that Captain Smyth's Overseer at the Skulkil, drew 600 and odd at one Draught; 300 is no wonder; 100 familiarly. They are excellent Pickled or Smokt'd, as well as boyld fresh: They are caught be nets only.

4. Rock are somewhat Rounder and larger, also a whiter fish, little inferior in rellish to our Mallet. We have them almost in the like plenty. These are often Barrell'd like Cod, and not much inferior for their spending. Of both these the Inhabitants increase their Winter store: These are caught by Nets, Hooks and Speers.

5. The Sheepshead, so called, from the resemblance of its Mouth and Nose to a Sheep, is a fish much preferr'd by some, but they keep in salt Water; they are like a Roach in fashion, but as thick as a Salmon, not so long. We have also the Drum, a large and noble fish, commended equal to the Sheepshead, not unlike to a Newfoundland Cod, but larger of the two. Tis so call'd from a noise it makes in its Belly, when it is taken, resembling a Drum. There are three sorts of them, the Black, Red and Gold colour. The Black is fat in the Spring, the Red in the Fall, and the Gold colour believed to be the Black, grown old, because it is observ'd that young ones of that colour have not been taken. They generally ketch them by Hook and Line, as Cod are, and they save like it, where the People are skilful. There are abundance of lesser fish to be caught of pleasure, but they quit not cost, as those I have mentioned, neither in Magnitude nor Number, except the Herring, which swarm in such shoales that it is hardly Credible; in little Creeks, they almost shovel them up in their tubs. There is the Catfish, or Flathead, Lampry, Eale, Trout, Perch, black and white, Smelt, Sunfish, etc.; also Oysters, Cockles, Cunks, Crabs, Mussles, Mannanoses.

Of Provision in General.

1. It has been often said we were starv'd for want of food; some were apt to suggest their fears, others to insinuate their prejudices, and when this was contradicted, and they assur'd we had plenty, both of Bread, Fish and Flesh, then 'twas objected that we were forc't to fetch it from other places at great Charges: but neither is all this true, tho all the World will think we must either carry Provision with us, or get it of the Neighbourhood till we had gotten Houses over our heads and a little Land in tillage, We fetcht none, nor were we wholly helpt by Neighbours; The Old Inhabitants supplied us with most of the Corn we wanted, and a good share of Pork and Beef: 'tis true New York, New England, and Road Island did with their provisions fetch our Goods and Money, but at such Rates, that some sold for almost what they gave, and others carried their provisions back, expecting a better Market neerer, which showed no scarcity, and that we were not totally destitute on our own River. But if my advice be of any Value I would have them to buy still, and not weaken their Herds, by Killing their Young Stock too soon.

2. But the right measure of information must be the proportion of Value of Provisions there, to what they are in more planted and mature Colonies. Beef

is commonly sold at the rate of two pence per Pound; and Pork for two pence half penny; Veal and Mutton at three pence or three pence half penny, that Country mony; an English Shilling going for fifteen pence. Grain sells by the Bushel; Wheat at four shillings; Rye, and excellent good, at three shillings; Barly, two shillings six pence; Indian Corn, two shillings six pence; Oats, two shillings, in that money still, which in a new Country, where Grain is so much wanted for feed, as for food, cannot be called dear, and especially if we consider the Consumption of the many new Commers.

3. There is so great an encrease of Grain by the dilligent application of People to Husbandry that, within three Years, some Plantations have got Twenty Acres in Corn, some Forty, some Fifty.

4. They are very careful to encrease their stock, and get into Daries as fast as they can. They already make good Butter and Cheese. A good Cow and Calf by her side may be worth three pounds sterling, in goods at first Cost. A pare of Working Oxen, eight pounds: a pare of fat ones, Little more, and a plain Breeding Mare about five pounds sterl.

5. For Fish, it is brought to the Door, both fresh and salt. Six Alloes or Rocks for twelve pence; and salt fish at three fardings per pound, Oysters at 2s. per bushel.

6. Our Drink has been Beer and Punch, made of Rum and Water: Our Beer was mostly made of Molosses, which well boyld, with Sassafras or Pine infused into it, makes very tollerable drink; but now they make Mault, and Mault Drink begins to be common, especially at the Ordinaries and the Houses of the more substantial People. In our great Town there is an able Man, that has set up a large Brew House, in order to furnish the People with good Drink, both there, and up and down the River. Having said this of the Country, for the time I was there, I shall add one of the many Letters that have come to my hand, because brief and full, and that he is known to be a Person of an extraordinary Caution as well as Truth, in what he is wont to Write or Speak:

PHILADELPHIA, the 3d of the 6th month [August] 1685. *Governour.*

Having an opportunity by a Ship from this River, (out of which several have gone this Year) I thought fit to give a short account of proceedings, as to settlements here, and the Improvements both in Town and Country. As to the Country, the Improvements are large, and settlements very throng by way of Townships and Villages. Great inclinations to Planting Orchards, which are easily raised, and some brought to perfection. Much Hayseed sown, and much Planting of Corn this Year, and great produce, said to be, both of Wheat, Rye and Rise; Barly and Oates prove very well, besides Indian Corn and Pease of several sorts; also Kidny Beans and English Pease of several kinds, I have had in my own Ground, with English Roots, Turnaps, Parsnaps, Carrets, Onions, Leeks, Radishes and Cabbidges, with abundance of sorts of Herbs and Flowers. I have but few seeds that have mist except Rosemary seed, and being English might be old. Also I have such plenty of pumpkins, Musmellons, Water Mellons, Squashes, Coshaws, Bucks-hens, Cowcumbers and Simnells of Divers kinds; admired at by new Commers that the Earth should so plentifully cast forth, especially the first Years breaking up; and on that which is counted the Worst Sort of Sandy Land. I am satisfied, and many more, that the Earth is very fertil, and the Lord has done his part, if Man use but a moderate Dilligence. Grapes, Mulberies and many wilde Fruits and natural Plums, in abundance, this year have I seen and eat of. A brave Orchard and Nursery have I planted, and thrive mightily, and Fruit the first Year. I endeavor choice of Fruits and Seeds from many parts; also Hay Seed; and have sowed a field this spring for tryall. First, I burned the leaves, then had it Grub'd, not the Field but the small Roots up, then sowed great and small Clover, with a little old Grass seed, and had it only raked over, not Plowed nor Harrowed, and it grows exceedingly; also for experience I sowed some patches of the same sort in my Garden and Dunged some, and that grows worst. I have planted the Irish Potatoes, and hope to have a brave increase to Transplant next Year. Captain Rapel (the Frenchman) saith he made good Wine of the grapes (of the country) last Year, and Transported some, but intends to make more this Year. Also a French man in this Town intends the same, for Grapes are very plentiful.

* * *

1. It is agreed on all hands, that the Poor are the Hands and Feet of the Rich. It is their labour that Improves Countries; and to encourage them, is to promote the real benefit of the publick. Now as there are abundance of these people in many parts of Europe, extreamly desirous of going to America; so the way of helping them thither, or when there, and the return thereof to the Disbursers, will prove what I say to be true.

2. There are two sorts, such as are able to transport themselves and Families, but have nothing to begin with there: and those that want so much as to transport themselves and Families thither.

3. The first of these may be entertained in this manner. Say I have 5000 Acres, I will settle Ten Families upon them, in way of Village, and built each an house, an out house for Cattle, furnish every Family with Stock, as four Cows, two Sows, a couple of Mares, and a yoke of Oxen, with a Town Horse, Bull and Boar; I find them with Tools, and give each their first Ground-seed. They shall continue Seven Year, or more, as we agree, at half encrease, being bound to leave the Houses in repair, and a Garden and Orchard, I paying for the Trees and at least twenty Acres of Land within Fence and improved to corn and grass; the charge will come to about sixty pounds English for each Family: At the seven years end, the Improvement will be worth, as things go now, 120 *l.* besides the value of the encrease of the Stock, which may be neer as much more, allowing for casualties; especially, if the People are honest and careful, or a man be upon the spot himself, or have an Overseer sometimes to inspect them. The charge in the whole is 832 *l.* And the value of stock and improvements 2400 *l.* I think I have been modest in my computation. These Farms are afterwarde fit for Leases at full rent, or how else the Owner shall please to dispose of them. Also the People will by this time be skilled in the Country, and well provided to settle themselves with stock upon their own Land.

4. The other sort of poor people may be very beneficially transported upon these terms: Say I have 5000 Acres I should settle as before, I will give to each Family 100 Acres which in the whole makes 1000; and to each Family thirty pounds English, half in hand, and half there, which in the whole comes to 300 *l.* After four years are expired, in which time they may be easie, and in a good condition, they shall each of them pay five pounds, and so yearly for ever, as a Fee-farm rent; which in the whole comes to 50 *l.* a Year. Thus a man that buys 5000 Acres may secure and settle his 4000 by the gift of one, and in a way that hazard and interest allowed for, amounts to at least ten per cent. upon Land security, besides the value it puts upon the rest of the 5000 Acres. I propose

that there be at least two working hands besides the wife; whether son or servant; and that they oblige what they carry; and for further security bind themselves as servants for some time, that they will settle the said land accordingly and when they are once seated their improvements are security for the Rent.

5. There is yet another expedient, and that is, give to ten Families 1000 Acres for ever, at a small acknowledgement, and settle them in way of Village, as afore; by their seating thus, the Land taken up is secured from others, because the method of the Country is answered, and the value such a settlement gives to the rest reserved, is not inconsiderable; I mean, the 4000 Acres; especially that which is Contiguous: For their Children when grown up, and Handicrafts will soon covet to fix next them, and such after settlements to begin at an Improved Rent in Fee, or for long Leases, or small Acknowledgements, and good Improvements, must advance the whole considerably. I conceive any of these methods to issue in a sufficient advantage to Adventurers, and they all give good encouragement to feeble and poor Families.

6. That which is most advisable for People, intended thither, to carry with them, is in short all things relating to Apparel, Building, Housholdstuf, Husbandry, Fowling and Fishing. Some Spice, Spirits and double ear, at first were not a miss: But I advise all to proportion their Estates thus; one-third in Money, and two thirds in Goods. Upon pieces of eight, there will be almost a third gotten, for they go at 6 *s.* and by goods well bought, at least fifty pounds sterl. for every hundred pounds; so that a man worth 400 *l.* here, is worth 600 *l.* there, without sweating.

A Contemporary Description of Pennsylvania, 1686

From "Letter of Doctor More," in Albert Cook Myers, ed., *Narratives of Early Pennsylvania, West New Jersey and Delaware 1630–1707* (New York, 1912), pp. 292–93.

In a Letter, of October last [1686], from Thomas Holmes Surveyor General.

We have made three Purchases of the Indians, which, added unto the six former Sales they made us, will, I believe, be Land enough for Planters for this Age; they were at first High, and upon their Distances; but when we told them of the Kindness our Governour had always shown them; that the Price we offer'd far exceeded former Rates, and that they offered us the Land before we sought them, they agreed to our last Offer, which is something under three hundred Pounds sterling. The Kings salute our Governour; they hardly ever see any of us, but they ask, with much affection when he will come to them again; we are upon very good terms with them. I intend to send the Draughts for a Map by the first ——

In a Letter from James Claypole Merchant in Philadelphia and One of the Councel.

I have never seen brighter and better Corn then in these parts, especially in the County of Chester. Provisions very cheap; Pork at two Pence, and good fat fresh Beef at three half-pence the Pound, in our Market. Fish is plentiful; Corn cheap; Wheat three and six pence a Bushel; Rye half a Crown; Indian Corn two Shillings, of this Money: And it is without doubt that we shall have as good Wine as France produces. Here is great appearance of a Trade, and if we had small Money for Exchange, we should not want Returns. The Whale-Fishery is considerable; several Companies out to ketch them: There is one caught that its thought will make several hundred Barrels of Oyle. This besides Tobacco and Skins, and Furs, we have for Commerce.

A Contemporary Description of Pennsylvania, 1700

From Francis Daniel Pastorius, "Circumstantial Geographical Description of Pennsylvania [1700],"
in Albert Cook Myers, ed., *Narratives of Early Pennsylvania, West New Jersey and Delaware 1630–1707*
(New York, 1912), pp. 374, 397–98.

THE MANNER AND METHOD IN WHICH WILLIAM PENN SOUGHT TO PROCURE SETTLERS FOR THE UNINHABITED PROVINCE WHICH HE RECEIVED AS A GIFT. THE OFFER FOR SALE.

1. He sent out a notice to the purchasers that they should send in their names to certain places in London and enter into agreements, [and] there he sold 3000 acres of land (Dutch measure) for 100 pounds sterling, with the reservation of a perpetual yearly payment therefor of an English shilling for each 100 acres. The money should be paid down for the receipt in London, and upon its presentation the amount of land would be measured out for the purchaser.

2. To each person who has the necessary money for the voyage, but has no means to establish himself upon his arrival, and to buy land, William Penn gives fifty acres, with a perpetual yearly fee of a penny for each acre. And this fee shall give them as valid a claim as if they had purchased the land for themselves and their heirs forever.

3. To the servants and children (to encourage them to greater industry and obedience) he gives full permission to take perpetual possession of a field of fifty acres, so soon as they shall have worked out their stipulated time, and to pay for each [acre] a yearly fee of only half a penny, and thus become their own masters. Hereupon the book and register of the purchasers was begun at the appointed bargain-place, and the German Company, or Society, was the first to enter into an agreement, and in the beginning paid down the money in London for twenty thousand acres, upon the receipt of an order of acquittance.

4. It is to be remarked that William Penn did not drive forth the naked native inhabitants of the land with military authority, but brought with him upon his arrival especial clothing and hats for the principal Indians, and thereby secured their goodwill, and purchased their land (and territory) to the extent of twenty

leagues, and they, thereupon, withdrew that much farther back into the wild forests. . . .

Of the nature of the land I can write with certainty only after one or more years of experience. The Swedes and Low Dutch who have occupied it for twenty years and more are in this as in most other things of divided opinions; *laudatur ab his, culpatur ab illis.* Certain it is that the soil does not lack fruitfulness and will reward the labor of our hands as well as in Europe if one will duly work and manure it, both which things are for the most part lacking. For the above mentioned old inhabitants are poor agriculturists. Some of them have neither barns nor stables, and leave their grain for years together unthreshed and lying in the open air, and allow their cattle, horses, cows, swine, etc., to run in the woods summer and winter, so that they derive little profit from them. Certainly the penance with which God punished the disobedience of Adam, that he should eat his bread in the sweat of his brow, extends also to his posterity in these lands, and those who think to spare their hands may remain where they are. *Hic opus, hic labor est,* and it is not enough to bring money hither, without the inclination to work, for it slips out of one's hands, and I may well say with Solomon: It has wings. Inasmuch as in the past year very many people came hither both out of England and Ireland and also from Barbadoes and other American islands, and this province does not yet produce sufficient provisions for such a multitude, therefore all victuals are somewhat dear, and almost all the money goes out of the land to pay for them. Yet we hope in time to have a greater abundance of both things, because William Penn will coin money and agriculture will be better managed. Working people and husbandmen are in the greatest demand here, and I certainly wish that I had a dozen strong Tyrolese to cut down the thick oak trees, for in whatever direction one turns, one may say: "We go into the primitive forest." It is nothing but forest, and very few cleared places are to be found, in which respect as also in some others the hope I had previously formed is deceived, namely, that in these wild orchards no apples or pears are to be found, and this winter (which indeed has been very cold) no deer, turkeys, etc., were to be had. The wild grapes are very small and better suited to make into vinegar than into wine. The walnuts have very thick shells, and few thick kernels within, so that they are scarcely worth the trouble of cracking. The chestnuts, however, and hazelnuts are somewhat more palatable; also the peaches, apples and pears are very good, no fault is to be found with them, except that there are not so many of them as some desire. On the other hand there are more rattlesnakes (whose bite is fatal) in the land than is agreeable to us. I must also add this, *tanquam testis oculatus,* that on October 16 I found fine (March) violets in the bushes; also that, after I had on October 24 laid out the town of Germantown, and on the 25th had gone back there with seven others, we on the way found a wild grape-vine, running over a tree, on which were some four hundred clusters of grapes; wherefore we then hewed down the tree and satisfied all eight of us, and took home with us a hatfull apiece besides. Also as I on August 25 was dining with William Penn, a single root of barley was brought in which had grown in a garden here and had fifty grains upon it. But all grains do not bear so much and it is as we say in the proverb, one swallow does not make a summer. Yet I doubt not that in the future more fruitful examples of this sort will present themselves, when we shall put the plow to the land in good earnest. I lament the vines which I brought with me, for when we were already in Delaware Bay they were drenched with seawater and all but two were spoiled. The abovementioned William Penn has a fine vineyard of French vines planted; its growth is a pleasure to behold and brought into my reflections, as I looked upon it, the fifteenth chapter of John, "I am the true vine, and my Father is the husbandman."

Colonial Settlement – Georgia

James Oglethorpe's Description of Georgia, 1733

From James Edward Oglethorpe, "A New and Accurate Account of the Provinces of South Carolina and Georgia," (London, 1733), in W. B. Stevens, ed., *Collections of the Georgia Historical Society* (Savannah, 1840), I, pp. 58–72.

England Will Grow Rich by Sending Her Poor Abroad. Of Refugees, Indians, Roman Colonies.

Besides the persons described in the preceding chapter, there are others whom it may be proper to send abroad for the reasons hereafter given, which reasons will also shew at whose expense these other sorts of indigent people ought to be removed. I think it may be laid down for a rule, that we may well spare all those, who have neither income, nor industry, equal to their necessities, are forced to live upon the fortunes, or labors of others; and that they who now are an heavy rent-charge upon the public, may be made an immense revenue to it, and this by an happy exchange of their poverty for an affluence.

Believing it will be granted that the people described in the last chapter ought in prudence to go abroad; and that we are bound in humanity and charity to send them: there arises a question, whether our aiding their departure be consistent with good policy? I raise this objection on purpose to answer it, because some who mean very well to the public have fancied that our numbers absolutely taken, without a distinction, are real wealth to a nation. Upon a little examination, this will appear to be a mistaken notion. It arises from a misapplication of Sir William Petty's Political Arithemtic, and of Sir William Temple's Observations on the united Netherlands. But when these great men esteem people as the wealth of a nation, surely they can only mean such as labor, and by their industry add yearly to the capital stock of their country, at the same time, that they provide the necessaries or comforts of life for themselves. Perhaps the rasp-houses may be reckoned part of the riches of Holland, because the drones are made to work in them: but is an infirmary of incurables wealth to a community? Or (which is worse, because it is remediable and is not remedied) are hundreds of prisons filled with thousands of English debtors, are they a glory, or a reproach, a benefit, or a burthen, to the nation? Who can be so absurd as to say that we should be enriched by the importation of a multitude of cripples, who might be able perhaps to earn a fourth part of what is necessary to sustain them? If ten thousand of these would be an addition to our wealth, ten millions of them must add a thousand times as much to it. Did the fire of London add to the wealth of the nation? I am sure it gave abundance of employment to the poor, just as people are employed in trade to feed and cloth the inhabitants of prisons. But these are also a slow fire, an hectic fever to consume the vitals of the state. The true state of national wealth is like that of private wealth, it is comparative. The nation, as well as individuals, must work to save and not to spend. If I work hard all day and at night give my wages to the next cripple I see, it may be profitable to my soul, but my worldly fortune is in the same condition as if I had stood idle. If the produce of the nation be in movables, land and labor fifty millions in a year, and only forty-eight millions are expended to maintain the people: now has the nation added two millions to its capital, but if it spends fifty-one millions, then is that to be made good by sinking part of the personal estate, or mortgaging the real. And upon a par, plus a million, and minus a million in earnings and expenses will operate nothing towards increasing the national wealth, if you proceed in *infinitum,* it is only impoverishing the rich to maintain the poor; it seems indeed to have something of leveling in it; to prevent which, I think our men of fortune would act wisely once for all; to put these poor people on a footing of their own, and shake off the perpetual incumbrance by a single act of prudent beneficence.

One of the gentlemen would have Scotland, Ireland and Wales sunk under water, but all the people saved and settled in England. He certainly deceived himself with a view of the artificial strength of the Dutch, when their fishery was at the highest pitch, and when they were carriers for mankind. But they have not been able to preserve these branches of trade entire, and their numbers must decrease as do the means of maintaining them. Therefore instead of taking it for granted, that numbers of people necessarily create a traffic; we may invert the proposition, and safely hold, that an extensive traffic will infallibly be attended with sufficient numbers of people.

And yet these unhappy people, who are not able to earn above a fourth part of their sustenance at home, and as we have shown are a load on the fortunes and industry of others, may in the new province of Georgia well provide by their labor a decent maintenance, and at the same time enrich their mother country.

Upon what has been said, the reader may be desirous to see a state of difference (with respect to the interests of the industrious and wealthy part of the nation,) between a poor person here, earning but half his sustenance, and the same person settled in a freehold, of a fertile soil without tithes or taxes: and in this computation let us remember that of the many thousands of poor debtors, who fill our prisons, few earn any thing at present; and this colony is chiefly intended for the unfortunate, there being no danger of the departure of such as are able to maintain themselves here.

A man who is equal in ability, only to the fourth part of a laborer, (and many such there are,) we will suppose to earn four-pence per diem, five pounds per annum, in London; his wife and a child of above seven years old four-pence per diem more: upon a fair supposition (because it is the common case) he has another child too young to earn any thing. These live but wretchedly at an expense of twenty pounds per annum, to defray which they earn ten pounds; so that they are a loss to the rich and industrious part of the nation of ten pounds per annum, for there are but three general methods of supplying the defect of their ability. Whatever they consume more than they earn, must be furnished, first either by the bounty, or charity of others; or secondly, by frauds, as by running in debt to the ruin of the industrious, &c., or, thirdly, by what our law calls force and felony, as theft and robbery, &c. They must be supplied at some of these rates, therefore (as I said before,) this family is a loss to the rich and industrious of ten pounds per annum, and if the particulars of their consumption, or an equivalent for them could have brought ten pounds from any foreign market, then has the whole community lost ten pounds by this family.

Now this very family in Georgia, by raising rice and corn sufficient for its occasions, and by attending the care of their cattle and land (which almost every one is able to do for himself in some tolerable degree) will easily produce in the gross value, the sum of sixty pounds per annum, nor is this to be wondered at, because of the valuable assistance it has from a fertile soil and a stock given gratis, which must always be remembered in this calculation.

The lots to be assigned to each family, as it is said, will be about fifty acres. The usual wages of a common laborer in Carolina is three shillings per diem, English value, or twenty shillings of their money. Therefore our poor man, (who is only equal to the fourth part of a man,) at about nine pence per diem, earns about twelve pounds per annum, his care of his stock on his land in his hours of resting from labor, (amounting to one half of each day) is worth also twelve pounds per annum, his wife and eldest child may easily between them earn as much as the man; so that the sum remaining to be raised by the wealth of the soil and the stock thereon (abstracted from the care and labor of the husbandman) is only twelve pounds per annum, it must be observed that though this family, when in London, was dieted but meanly, yet it could afford very little for clothes out of the twenty pounds it then expended, but now it will fare much better in Georgia, at the same expense, because provisions will be cheap, and it will also pay forty pounds a year to England for apparel, furniture and utensils of the manufacture of this kingdom. Behold then the benefit the common weal receives by relieving her famishing sons. Take it stated only upon one hundred such families as follows,

In London an hundred men earn	500 *l.*
An hundred women and an hundred children,	500 *l.*
	Total, 1000 *l.*

In Georgia an hundred families earn,

An hundred men for labor	1200 *l.*
Ditto for care,	1200 *l.*
An hundred women and a hundred children,	2400 *l.*
Land and stock in themselves,	1200 *l.*
	Total, 6000 *l.*

In London an hundred families consume,	2000 *l.*
Supplied by their labor,	1000 *l.*
By the wealth of others,	1000 *l.*

| In Georgia an hundred families consume of their own produce, | 2000 *l.* |
| Of English produce, | 4000 *l.* |

Thus taking it that we gained one thousand pounds per annum, (which was the value of their labor) before their removal, that we now gain four thousand pounds, and we have got an addition of three thousand pounds per annum to our income; but if, (as the truth is) we formerly lost one thousand pounds per annum, and the nation now gains four thousand pounds per annum, the rich and industrious are now profited to the value of five thousand pounds per annum. I might also shew other great advantages in the increase of our customs, our shipping, and our seamen. It is plain that these hundred families, thus removed, employ near two hundred families here to work for them, and thus by their absence they increase the people of Great Britain, for hands will not be long wanting where employment is to be had; if we can find business that will feed them, what between the encouragement and increase of propagation on the one hand, and the preservation of those who now perish for want on the other: we should quickly find we had strengthened our hive by sending a swarm away to provide for themselves.

It is also highly for the honor and advancement of our holy religion to assign a new country to the poor Germans, who have left their own for the sake of truth. It will be a powerful encouragement to martyrs and confessors of this kind to hold fast their integrity, when they know their case not to be desperate in this world. Nor need we fear that the King of Prussia will be able to engross them all, we shall have a share of them if we contribute cheerfully to their removal. The Society for the Propagation of the Gospel in Foreign Parts have gloriously exerted themselves on this occasion: they have resolved to advance such a sum of money to the Trustees for the colony of Georgia, as will enable them to provide for seven hundred poor Salzburghers. This is laying a foundation for the conversion of the heathen, at the same time they snatch a great number of poor Christians out of the danger of apostacy. It is to be hoped this laudable example will be followed by private persons, who may thus at once do much for the glory of God, and for the wealth and trade of Great Britain. Subjects thus acquired by the impolitic persecutions, by the superstitious barbarities of the neighboring princes, are a noble addition to the capital stock of the British Empire. If our people be ten millions, and we were to have an access of ten thousand useful refugees, every stock-jobber in Exchange-alley must allow that this would increase our wealth and figure in the world, as one added to a thousand, or, as one-tenth per cent. This would be the proportion of our growth compared with our neighbors, who have not been the persecutors; but as against the persecutor, the increase of our strength would be in a double ratio, compounded as well of negative as of positive quantity. Thus if A and B are worth one thousand pounds each, and a third person gives twenty shillings to A, and A is become richer than B by one-tenth per cent., but if A gains twenty shillings from B, then A is become richer than B by two-tenths or one-fifth per cent., for A is worth one thousand and one pounds, and B is worth only nine hundred and ninety-nine pounds.

The increase of our people, on this fruitful continent, will probably, in due time, have a good effect on the natives, if we do not shamefully neglect their conversion: if we were moderately attentive to our duty on this head, we have no reason to doubt of success. The Spaniard has at this day as many Christians as he has subjects in America, negroes excepted. We may more reasonably hope to make converts and good subjects of the Indians in amity with us, by using them well, when we grow numerous in their neighborhood, than the Spaniards could have expected to have done by their inexpressible cruelties, which raised the utmost aversion in the minds of the poor Indians against them and their religion together. One of their own friars who had not relinquished his humanity, tells us of an Indian prince, who just as the Spaniards were about to murder him, was importuned by one of their Religious to become a Christian; the priest told him much of heaven and hell, of joy and misery eternal; the prince desired to be informed which of the two places was allotted for the Spaniards? Heaven, quoth the priest; says the prince, I'm resolved not to go there. How different from this was the reflection of an Indian chief in Pennsylvania: what is the matter, says he, with us that we are thus sick in our own air, and these strangers well? It is as if they were sent hither to inherit our land in our steads; but the reason is plain, they love the great God and we do not. Was not this Indian almost become a Christian? New England has many convert Indians, who are very good subjects, though no other colony had such long and cruel wars with its Indian neighbors.

The pious benefactions of the people of England have in all ages equalled, if not surpassed, all instances of the kind in other countries. The mistaken piety of our ancestors gave a third part of the kingdom to the church. Their intentions were right though they erred in the object. Since the statutes against *mortmain* and superstitious uses, our great and numerous foundations of hospitals and almshouses are the wonder of foreigners. Some of these, especially of the largest, are doubtless of great use, and excellently administered. And yet, if the numbers in this nation, who feel the woes of others and would contribute to relieve them, did but consider the cases of the people described in the last chapter, of the German emigrants, and even of the poor Indians; they would be apt to conclude that there ought to be a blessing in store for these also. About eight pounds allowed to an indigent person here, may poorly support him, and this must be repeated yearly; but a little more, than double that sum, relieves him for life, sends him to our new world, gives plenty there to him and his posterity; putting them in possession of a good estate, of which, they may be their own stewards.

But this is not all, that sum which settles one poor family in the colony does not end there; it in truth purchases an estate to be applied to like uses, in all future times. The author of these pages is credibly informed that the Trustees will reserve to themselves square lots of ground interspersed at proper distances among the lands, which shall be given away. As the country fills with people, these lots will become valuable, and at moderate rents will be a growing fund to provide for those whose melancholy cases may require assistance hereafter. Thus the settlement of five hundred persons will open the way to settle a thousand more afterwards with equal facility. Nor is this advance of the value of these lots of land a chimerical notion; it will happen certainly and suddenly. All the lands within fifty miles of Charlestown have within these seven years increased near fourfold in their value, so that you must pay three or four hundred pounds for a plantation, which seven years ago you could have bought for a hundred pounds, and it is certain that fifty years ago you might have purchased at Charlestown for five shillings a spot of land which the owner would not sell at this day for two hundred pounds sterling.

The legislature is only able to take a proper course for the transportation of small offenders, if it shall seem best, when the wisdom of the nation is assembled; I mean only those who are but novices in iniquity. Prevention is better than the punishment of crimes, it may reform such to make them servants to such planters as were reduced from a good condition. The manners and habits of very young offenders would meliorate in a country not populous enough to encourage a profligate course of life, but a country where discipline will easily be preserved. These might supply the place of negroes, and yet (because their servitude is only to be temporary) they might upon occasion be found useful against the French, or Spaniards; indeed, as the proportion of negroes now stands, that country would be in great danger of being lost, in case of a war with either of those powers. The present wealth of the planters in their slaves too probably threatens their future ruin, if proper measures be not taken to strengthen their neighborhood with large supplies of free-men. I would not here be understood to advance that our common run of Old-Baily transports would be a proper beginning in the infancy of Georgia. No, they would be too hard for our young planters, they ought never to be sent any where but to the sugar islands, unless we had mines to employ them.

The property of the public, with regard to its immense debt, and the anticipation of taxes attending that debt, will probably be a reason to many worthy patrons, not to afford a large pecuniary assistance in parliament, though they give all other furtherance to this settlement, and yet powerful reasons might be offered why the commons of Great Britain, with justice to those that sent them, might apply a large sum of public money to this occasion. Let us suppose that twenty-five thousand of the most helpless people in Great Britain were settled there at an expense of half a million of money; the easiness of the labor in winding off the silk and tending the silk worm would agree with the most of those who throughout the kingdom are chargeable to the parishes. That labor with the benefit of land stocked for them gratis, would well subsist them, and save our parishes near two hundred thousand pounds a year directly in their annual payments; not to compute would also be saved indirectly, by the unwillingness of many pretended invalids to go the voyage, who would then betake themselves to industrious courses to gain a livelihood.

I shall consider the benefit of employing them in raising silk when I come in the fifth chapter, to treat of the commerce of Carolina. I shall only here observe that the number of poor last mentioned, being thus disposed of, would send us goods, at least to the value of five hundred thousand pounds annually, to pay for their English necessaries; and that would be somewhat better than our being obliged to maintain them at the rate of two hundred thousand pounds a year here at home.

I cannot dismiss this inquiry concerning the proper persons to plant this colony, without observing that the wisdom of the Roman state discharged not only its ungovernable distressed multitude, but also its emeriti, its soldiers, which had served long and well in war, into colonies upon the frontiers of their empire. It was by this policy that they elbowed all the nations round them. Their military hospital went a progress, we can trace its stages northward from the Tyber to the Po, to the Rhone, to the Rhine, to the Thames: the like advances they made on all sides round them, and their soldiers were at least as fond of the estates thus settled on them as ours can be of their pensions.

What I said before in this chapter, with regard to the increasing fund, to arise by reserved lots of gound interspersed among the lands that will be distributed to the planters, will hold good in the same manner in such settlements as might be made at a national expense, so that twenty thousand people, well settled, will raise the value of the reserved lands, in such measure as will bring Great Britain to resemble the present Carolina in one happy instance, viz. that there is not a beggar, or very poor person in the whole country. Then should we have no going to decay, no complaining in our streets.

Of the Present and (Probable) Future Trade of South Carolina and Georgia. Rice, Silk, Cotton, Wine, &c.

The present state of South Carolina and its commerce may give us an idea of the condition of the early settlements in the new colony of Georgia. The first essays in trade and husbandry will doubtless be in imitation of their nearest neighbors. We shall therefore consider these colonies together, the difference in their air and soil being hardly discernible, and the same traffic being proper for them both.

We are not to imagine that either the present branches of trade in that country, will be perpetual, or that there is not room to introduce others of more importance than any they have hitherto been acquainted with. Thus it will necessarily fall out that their present exports of lumber and deer skins will decrease, or rather wholly cease when the country grows populous: and this for an obvious reason, the land will be better employed, it will be disafforrested, and no longer left vacant to the growth of great woods, and the sustenance of wild herds of deer. But the very reason why these branches of trade will cease will also be the cause of their taking up others, or improving them to such a degree, as must put these colonies in a condition to vie with the most flourishing countries of Europe and Asia: and that without prejudice to their dependence on Great Britain. We shall by their growth in people and commerce have the navigation and dominion of the ocean established in us more firmly than ever. We shall be their market for great quantities of raw silk, and perhaps for wine, oil, cotton, drugs, dying-stuffs, and many other lesser commodities. They have already tried the vine and the silk-worm, and have all imaginable encouragement to expect that these will prove most valuable staple commodities to them. And I have been credibly informed, that the Trustees for Georgia furnish proper expenses for a skilful botanist to collect the seeds of drugs and dying-stuffs in other countries in the same climate, in order to cultivate such of them as shall be found to thrive well in Georgia. This gentleman could not be expected to proceed at his own charges, but he is the only person belonging to the management of that trust who does not serve gratis.

The raw silk, which Great Britain and Ireland are able to consume, will employ forty or fifty thousand persons in that country, nor need they be the strongest, or most industrious part of mankind; it must be a weak hand indeed that cannot earn bread where silk-worms and white mulberry trees are so plenty. Most of the poor in Great Britain, who are maintained by charity, are capable of this, though not of harder labor: and the planters may be certain of selling their raw silk to the utmost extent of the British demand for that commodity; because a British Parliament will not fail to encourage the importation of it from thence, rather than from aliens, that the planters may be able to make large demands upon us for our home commodities: for this will be the consequence of their employing all their people in producing a commodity, which is so far from rivalling, that it will supply a rich manufacture to their mother country.

The present medium of our importation of silk will not be the measure hereafter of that branch of trade when the Georgians shall enter into the management of the silk-worm. Great Britain will then be able to sell silk manufactures cheaper than all Europe besides, because the Georgians may grow rich, and yet afford their raw silk for less than half the price that we now pay for that of Piedmont: the peasant of Piedmont, after he has tended the worm, and wound off the silk, pays half of it for the rent of the mulberry trees, and the eggs of the silk-worm: but in Georgia the working hand will have the benefit of all his labor. This is fifty in a hundred, or cent per cent difference in favor of the Georgians, which receives a great addition from another consideration, viz. the Georgian will have his provisions incomparably cheaper than the Piedmontese, because he pays no rent for the land that produces them; he lives upon his own estate. But there is still another reason why Great Britain should quickly and effectually encourage the production of silk in Georgia; for, in effect, it will cost us nothing; it will be purchased by the several manufactures of Great Britain, and this, I fear, is not our present case with respect to Piedmont: especially (if as we have been lately told) they have prohibited the importation of woollen goods into that principality.

That this little treatise may be the more satisfactory to the reader, I could wish I had been minutely informed of the present state of our silk trade; of the medium value of silk per pound; to what amount it is imported; of its duty, freight, commission and insurance; and lastly, by what returns in commerce it is purchased. I am persuaded, these estimates would afford plentiful matter for observations in favor of this position, viz. that Great Britain ought vigorously to attempt to get this trade into her own hands. I shall however aim at a computation, upon my memory of facts, which I have heard from those who understand that commerce.

1. Great Britain imports silk from Piedmont, near the yearly value of three hundred thousand pounds.

2. The medium price is about twelve shillings per pound in Piedmont.

3. The duty here is about four shillings per pound.

4. The price of raw silk in London, is generally more than half of the price of the wrought goods in their fullest perfection.

1st Observ. If the Piedmontese paid no rent for the mulberry-tree and silk worm, he might afford silk at six shillings per pound.

2d Observ. If silk were bought in Piedmont at six shillings per pound, and imported duty free, it might be sold in London at seven shillings per pound. For, the commission, insurance and exchange, or interest of money would be but half what they are at present, and there must be some allowance for the interest of the money that was usually applied to pay the duty.

3d Observ. Therefore Great Britain, by encouraging the growth of silk in Georgia, may save above a hundred thousand pound per annum of what she lays out in Piedmont.

4th Observ. The Georgian (without taking the cheapness of his provisions into question) may enable Great Britain to undersell all her rivals in Europe in the silk manufacture in a proportion resembling what follows.

		l.	s.	d.
France,	Raw-silk, one pound weight,	0	14	0
	Workmanship,	0	16	0
	Total,	1	10	0
Great Britain,	Raw-Silk, one pound weight,	0	7	0
	Workmanship,	0	16	0
	Total,	1	3	0

The difference of these is seven pence in thirty, which is near twenty-five pound in an hundred, and is about thirty per cent. The reader is desired to consider these computations as stated by guess. But the same reasoning will hold in a considerable degree upon the exact state of the several values.

Rice is another growth of this province that doth not interfere with Great Britain. But we reap their harvests; for when they have sold the rice in a foreign market, they lay out the money in our manufactures to carry home with them. They have already made an handsome progress in Carolina, in cultivating this grain. They have exported above ten thousand tons of it by weight in a year already, all produced in a few years from so small a quantity as was carried thither in a bag, fit to hold only a hundred pound sterling in silver; they have sold cargoes of it in Turkey. They have all the world for their market. A market not easily glutted.

The indulgence of the British Legislature to Carolina in this branch of their trade, shows our new Georgians what encouragement they may expect from that august body, as soon as they shall learn the management of the silk-worm. The law for the ease of the rice trade, is alone sufficient to enrich whole provinces: they are now at liberty to proceed in their voyages directly to any part of Europe, south of Cape Fenesterre, or to Asia and Africk before they touch at Great Britain. The difference of the charge of freight is not half the benefit they receive from this act of Parliament; they arrive at the desired ports time enough to forestall the markets of Spain, Portugal, and the Levant. It now frequently happens that cargoes arrive safe, which, as the law stood formerly, would have been lost at sea, by means of the deviation. This new law, in a manner, forces them into the Spanish, Portuguese, and Levant trades, and gives them two returns of commerce instead of one. They may now dispose of their American grain in the first place, and then come laden to Great Britain with the most profitable wares of the countries where they traded; and lastly, buy for ready money such British manufactures as they have occasion to carry home.

When I speak of the future trade of these happy provinces, I might expatiate upon many valuable branches of it besides the silk and rice; branches which it must enjoy as certainly as nature shall hold her course in the production of vegetables, and the revolution of seasons. But because I would not swell this treaties to too expensive a bulk, I shall content myself with acquainting the reader that they have no doubt of the kindly growth of cotton, almonds, olives, &c. And in short, of every vegetable that can be found in the best countries under the same latitude.

I foresee an objection against what is here laid down: it may be said that all the countries under the same latitude do not produce the same commodities; that some of them are incapable of raising choice vegetables, which others of them nourish with the utmost facility. For answer to this objection, what was said in the second chapter should be considered: the intemperate heats of Barbary, Egypt and Arabia are there accounted for, from the vicinity of boundless sandy deserts; on the other hand, near Mount Caucasus in Asia, and particularly in the kingdom of Kaschmere, or Kasimere, (which is entirely surrounded by prodigious mountains) their seasons are almost as cold as ours in England, though they lie in the same latitude with Tangier, or Gibraltar.

These instances of the temperature in countries equidistant from the Equator, are very opposite to each other, the medium between them is the happy portion of Georgia; which therefore must be productive of most of the valuable commodities in the vegetable world.

A Voyage to Georgia, 1735

From Francis Moore, "A Voyage to Georgia, Begun in the year 1735," in W. B. Stevens, ed., *Collections of the Georgia Historical Society* (Savannah, 1840), I, pp. 98–100.

Beyond the villages, commence lots of five hundred acres: these are granted upon terms of keeping ten servants, &c. Several gentlemen who have settled on such grants have succeeded very well, and have been of great service to the colony. Above the town is a parcel of land called Indian lands; these are those reserved by king Toma Chi Chi for his people. There is near the town, to the east, a garden belonging to the Trustees, consisting of ten acres; the situation is delightful, one half of it is upon the top of the hill, the foot of which the river Savannah washes, and from it you see the woody islands in the sea. The remainder of the garden is the side and some plain low ground at the foot of the hill, where several fine springs break out. In the garden is variety of soils; the top is sandy and dry, the sides of the hill are clay, and the bottom is a black, rich garden mould well watered. On the north part of the garden is left standing a grove of part of the old wood, as it was before the arrival of the colony there. The trees in the grove are mostly bay, sassafras, evergreen oak, pellitory, hickory, American ash, and the alurel tulip. This last is looked upon as one of the most beautiful trees in the world; it grows straight-bodied to forty or fifty foot high; the bark smooth and whitish, the top spreads regular like an orange tree in English gardens, only larger; the leaf is like that of a common laurel, but bigger, and the under side of a greenish brown; it blooms about the month of June; the flowers are white, fragrant like the orange, and perfume all the air around it; the flower is round, eight or ten inches diameter, thick like the orange flower, and a little yellow near the heart. As the flowers drop, the fruit which is a cone with red berries succeeds them. There are also some bay trees that have flowers like the laurel, only less.

The garden is laid out with cross-walks planted with orange trees, but the last winter, a good deal of snow having fallen, had killed those upon the top of the hill down to their roots, but they being cut down sprouted again, as I saw when I returned to Savannah. In the squares between the walks, were vast quantities of mulberry trees, this being a nursery for all the province, and every planter that desires it has young trees given him gratis from this nursery. These white mulberry trees were planted in order to raise silk, for which purpose several Italians were brought at the Trustees' expense, from Piedmont by Mr. Amatis; they have fed worms, and wound silk to as great perfection as any that ever came out of Italy; but the Italians falling out, one of them stole away the machines for winding, broke the coppers and spoiled all the eggs which he could not steal, and fled to South Carolina. The others who continued faithful, had saved but a few eggs when Mr. Oglethorpe arrived; therefore he forbade any silk should be wound, but that all the worms should be suffered to eat through their balls, in order to have more eggs against next year. The Italian women are obliged to take English girls apprentices, whom they teach to wind and feed; and the men have taught our English gardeners to tend the mulberry trees, and our joiners have learned how to make the machines for winding. As the mulberry trees increase, there will be a great quantity of silk made here.

Besides the mulberry trees, there are in some of the quarters in the coldest part of the garden all kinds of fruit trees usual in England, such as apples, pears, &c. In another quarter are olives, figs, vines, pomegranates and such fruits as are natural to the warmest parts of Europe. At the bottom of the hill, well sheltered from the north wind and in the warmest part of the garden, there was a collection of West India plants and trees, some coffee, some cocoa-nuts, cotton, Palma-christi, and several West Indian physical plants, some sent up by Mr. Eveleigh, a public-spirited merchant at Charlestown, and some by Dr. Houstoun, from the Spanish West Indies, where he was sent at the expense of a collection raised by that curious physician, Sir Hans Sloan, for to collect and send them to Georgia, where the climate was capable of making a garden which might contain all kinds of plants, to which design, his Grace, the Duke of Richmond, the Earl of Derby, the Lord Peters, and the Apothecary's Company contributed very generously; as did Sir Hans himself. The quarrels among the Italians proved fatal to most of these plants, and they were laboring to repair that loss when I was there, Mr. Miller being employed in the room of Dr. Houstoun, who died in Jamaica.

We heard he had wrote an account of his having obtained the plant from whence the true Balsamum Capivi is drawn; and that he was in hopes of getting that from whence the Jesuits' Bark is taken, he designing for that purpose to send to the Spanish West Indies.

There is a plant of Bamboo cane brought from the East Indies, and sent over by Mr. Towers which thrives well. There was also some tea seeds, which came from the same place; but the latter, though great care was taken, did not grow.

Three miles from Savannah, within land, that is to say to the south, are two pretty villages, Hampstead and Highgate, where the planters are very forward, having built neat huts, and cleared and planted a great deal of land. Up the river also there are several other villages and two towns, not much better than villages, on the Georgia side, the one called Joseph's town, which some Scotch gentlemen are building at their own expense, and where they have already cleared a great deal of ground. Above that is Ebenezer, a town of the Saltzburghers. On the Carolina side is Purysburgh, chiefly inhabited by Swiss. There are also a party of rangers under the command of Capt. McPherson, and another under the command of Capt. Æneas M'Intosh; the one lying upon the Savannah river, the other upon the Ogeechee. These are horsemen and patrol the woods to see that no enemy Indians, nor other lawless persons, shelter themselves there.

There were no public buildings in the town, besides a storehouse; for the courts were held in a hut thirty foot long, and twelve foot wide, made of split boards, and erected on Mr. Oglethorpe's first arrival in the colony. In this hut also divine service was performed; but upon his arrival this time Mr. Oglethorpe ordered a house to be erected in the upper square, which might serve for a court house, and for divine service till a church could be built, and a work house over against it; for as yet there was no prison here.

Importation of Negro Slaves to Georgia, 1750

From Allen P. Chandler, ed., *Colonial Records of Georgia*, (Atlanta, 1904), I, 56–62.

May it please Your Majesty.

The Trustees for establishing the Colony of Georgia in America in pursuance of the Powers and in Obedience to the Directions to them given by Your Majesty's most Gracious Charter humbly lay before Your Majesty the following Law Statute and Ordinance which they being for that purpose assembled have prepared as fit and necessary for the Government of the said Colony and which They most humbly present under their Common Seal to Your most Sacred Majesty in Council for your Majesty's most Gracious Approbation and Allowance.

An Act for repealing an Act Intituled (An Act for rendering the Colony of Georgia more a defensible by prohibiting the Importation and Use of Black Slaves or Negroes into the same) & for permitting the Importation and Use of them in the Colony under proper Restrictions and Regulations, and for other Purposes therein memtioned.

Whereas an Act was passed by his Majesty in Council in the Eighth Year of his Reign Intituled (an Act for rendering the Colony of Georgia more defensible by prohibiting the Importation and Use of Black Slaves or Negroes into the same) by which Act the Importation and Use of Black Slaves or Negroes in the said Colony was absolutely prohibited and forbid under the Penalty therein mentioned AND WHEREAS at the time of passing the said Act the said Colony of Georgia being in its Infancy the Introduction of Black Slaves or Negroes would have been of dangerous Consequence but at present it may be a Benefit to the said Colony and a Convenience and Encouragement to the Inhabitants thereof to permit the Importation and Use of them into the said Colony under proper Restrictions and Regulations without Danger to the said Colony as the late War hath been happily concluded and a General Peace established. THEREFORE WE the Trustees for establishing the Colony of Georgia in America humbly beseech Your Majesty that it may be ENACTED AND be it ENACTED That the said Act and every Clause and Article therein contained be from henceforth repealed and made void and of non Effect AND be it FURTHER ENACTED that from and after the first day of January in the Year of Our Lord One thousand seven hundred and fifty it shall and may be lawful to import or bring Black Slaves or Negroes into the Province of Georgia in America and to keep and use the same therein under the Restrictions and Regulations hereinafter mentioned and directed to be observed concerning the same AND for that purpose be it FURTHER ENACTED that from and after the said first day of January in the Year of Our Lord One thousand seven hundred and fifty it shall and may be lawful for every Person inhabiting and holding and cultivating Lands within the said Province of Georgia and having and constantly keeping one white Man Servant on his own Lands capable of bearing Arms and aged between sixteen and sixty five Years to have and keep four Male Negroes or Blacks upon his Plantation there and for every Person inhabiting and holding and cultivating Lands within the said Province of Georgia and having and constantly keeping two white Men Servants capable of bearing Arms and aged between sixteen and sixty five Years to have and keep eight Male Negroes or Blacks upon his Plantation there and so in Proportion to the Number of such white Men Servants capable of bearing Arms and of such Age as aforesaid as shall be kept by every Person within the said Province AND BE IT FURTHER ENACTED that every Person who shall from and after the said first day of January in the Year of Our Lord One thousand seven hundred and fifty have and keep more than four Male Negroes or Blacks to every such Male Servant as aforesaid contrary to the Intent and true Meaning of this Act shall forfeit the Sum of Ten pounds Sterling Money of Great Britain for every such Male Negroe or Black which he shall have and keep above the said Number and shall also forfeit the further Sum of Five pounds of like Money for each Month after during which he shall retain and keep such Male Negroe or Black the said several Sums of Ten pounds and Five pounds to be recovered and applyed in such manner as is hereinafter mentioned AND BE IT FURTHER ENACTED that no Artificer within the said Province of Georgia (Coopers only excepted) shall take any Negroe or Black as an Apprentice nor shall any Planter or Planters within the said Province lend or let out to any other Planter or Planters within the same any Negroe or Negroes Black or Blacks to be employed otherwise than in manuring and cultivating their Plantations in the Country AND BE IT FURTHER ENACTED that if any Proprietor or Proprietors of Negroes or Blacks which shall be imported or brought into or used within the said Province shall inflict any Chastisement endangering the Limb of a Negroe or Black shall for the first Offence forfeit not less than the Sum of Five pounds Sterling Money of Great Britain and for the second Offence not less than the Sum of Ten pounds of like Money to be recovered and applied in such manner as is hereinafter mentioned; But in the Case of Murder of a Negroe or Black the Criminal to be tried according to the Laws of Great Britain AND BE IT FURTHER ENACTED that all and every Negroe and Negroes Black and Blacks which shall be imported into or born within the said Province of Georgia shall be registered in a proper Office or Offices to be kept for that Purpose within the said Province and that no Sale of any such Negroe or Negroes Black or Blacks shall be good or valid unless the same be duly registered as aforesaid And that Inquisitions shall be made and taken once in every Year (or oftner if need shall be) into the several Registers thereof by Juries to be impannelled for that purpose within the several Districts of the said Province who shall immediately after such Inquisition make their several Reports and Returns to the President and Magistrates of the said Province AND WHEREAS the permitting Ships with Negroes or Blacks to send them on Shore when ill of contagious Distempers (particularly the Yellow Fever) must be of the most dangerous Consequence Therefore for the Prevention of so great a Calamity BE IT FURTHER ENACTED that no Ship which shall bring any Negroes or Blacks to the said Province shall land any Negroe or Negroes Black or Blacks within the said Province until such Ship shall have been visited by the proper Officer or Officers of the said Province for that purpose and shall have obtained a Certificate of Health And that no Ship which shall come to the said Province with Negroes or Blacks shall come nearer to the said Province than Cockspur at the Mouth of the River Savannah but that every such Ship shall first anchor and remain there until such Ship shall have been visited by the proper Officer or Officers And if upon Inspection any such Ship shall be found to be infected such Ship shall perform such Quarantine in Tybee Creek in the River Savannah as by the President and Assistants of the said Province shall be from time to time order and directed And to the End that due Care may be taken of the Crews of sucn infected Ships and of the Negroes brought therein BE IT FURTHER ENACTED that a Lazaretto be forthwith built within the said Province under the Direction and Inspection of the President and Magistrates thereof on the West Side of Tybee Island in the said River Savannah for the Use and Convenience of the said Colony where the whole Crews of such infected Ships and the Negroes brought therein may be conveniently lodged and assisted with Medicines and accommodated with Refreshments for their more speedy Recovery such Medicines and Refreshments to be provided at the Expence of the Captain of the Ship And in Case any Master of a Ship shall attempt to land any Negroes in any other Part of the Colony except as aforesaid he shall for the said Offence forfeit the Sum of Five hundred pounds Sterling Money of Great Britain And in Case he shall land any Negroes before his Ship is visited and the proper Certificate of Health obtained or not perform the full Quarantine directed he shall for the said Offence not only forfeit the like Sum of Five hundred pounds but also the Negroes on board the said Ship The said Forfeitures to be recovered and applied in such manner as is herein after mentioned AND BE IT FURTHER ENACTED that if any Person or Persons shall not permit or even oblige his or their Negroe or Negroes Black or Blacks to attend at some time on the Lords Day for Instruction in the Christian Religion in such Place and Places as the Protestant Ministers of the Gospel within the said Province shall be able to attend them contiguous to the Residence of such Negroe or Negroes Black or Blacks such Person or Persons shall for every such Offence forfeit the Sum of Ten pounds Sterling Money of Great Britain to be recovered and applied in such manner as is herein after mentioned AND BE IT FURTHER ENACTED that all and every Intermarriage and Intermarriages between the white People and the Negroes or Blacks within the said Province shall be deemed unlawful Marriages and the same are hereby declared to be absolutely null and void And that if any white Man shall be convicted of lying with a Female Negroe or Black Or if any white Woman shall be convicted of lying with a Male Negroe or Black such white Man or Woman so offending shall on every such Conviction forfeit the Sum of Ten pounds Sterling Money of Great Britain to be recovered and applied in such manner as is herein after mentioned Or otherwise such white Man or Woman so convicted shall receive such Corporal Punishment as the Court before whom such Conviction shall be shall judge proper to inflict and such Male or Female Black or Negroe shall also receive such Corporal Punishment as the said Court shall order and direct. AND WHEREAS great Advantages may arise to the Inhabitants of the said Colony of Georgia and to the British Nation by the Culture and raising of Silk within the said Province BE IT THEREFORE FURTHER ENACTED that every Planter within the said Colony who shall at any time hereafter have or keep any Male Negroes or Blacks shall have and keep for every four Male Negroes or Blacks one Female Negroe or Black and so in proportion to such greater Number of Male Negroes or Blacks as every such Planter shall keep   And that every such Planter shall instruct such their Female Negroes or Blacks or cause them to be well instructed in the Art of winding or reeling of Silk from the Silk Balls or Cocoons and shall at the proper Season in every Year send such their Female Negroes or Blacks to Savannah in Order to learn the said Art or to such other Place or Places within the said Province as the President and Magistrates thereof shall from time to time appoint for that purpose   AND BE IT FURTHER ENACTED that every Planter within the said Province who shall not at all times hereafter keep and instruct or cause to be instructed such Female Negroe or Black Negroes or Blacks in proportion to the Number of Male Negroes or Blacks which they shall keep as aforesaid shall for every such Female Negroe or Black which they ought to keep and instruct or cause to be instructed and shalt not so keep and instruct or cause to be instructed as aforesaid pursuant to the true Intent and Meaning of this Act forfeit the Sum of Ten pounds Sterling Money of Great Britain to be recovered and applied in such manner as is herein after mentioned   AND for the more effectual carrying on the Culture and raising of Silk within the said Province of Georgia in America BE IT FURTHER ENACTED that every Planter within the said Province shall plant five hundred Mulberry Trees on every five hundred Acres of Land which they shall hold within the said Province and so in proportion to a lesser Number of Acres of Land and shall well and sufficiently fence in such Mulberry Trees so as to defend and protect them against Cattle and shall from time to time keep up the same Number and cultivate them according to the best of their Skill and Judgment Or in Default thereof shall forfeit the Sum of Ten pounds Sterling Money of Great Britain to be recovered and applied in such manner as is herein after mentioned AND to the End that a sufficient Fund may be raised for the future Maintenance of the Minister or Ministers of the Gospel and of the several Officers of the Civil Government within the said Province of Georgia and for the building and repairing of the Church, the Wharf, the Prison, the Lazaretto and the other necessary Publick Buildings within the said Province and for the providing of Pilots and Pilot Boats for the Use thereof BE IT FURTHER ENACTED that a Duty of fifteen Shillings Sterling Money of Great Britain shall be paid for every Negroe or Black of the Age of Twelve Years or upwards which now are in the said Colony or shall at any time hereafter be imported or otherwise brought into the said Colony for Sale by the Person or Persons who shall import or bring the same into the said Colony And that a Duty of One Shilling a Year like Money per Head shall be paid for all Negroes or Blacks of the said Age which now do or shall hereafter inhabit within the said Province by the Owner or Owners of such Negroes or Blacks And that the said several Duties shall be collected in such manner and paid to such Persons as by the Common Council of the said Trustees for establishing the Colony of Georgia in America shall be from time to time ordered and directed And that such Duties shall from time to time be applied by the said Common Council of the said Trustees for the Purposes aforesaid AND BE IT FURTHER ENACTED that all Fines Forfeitures and Penalties which shall be incurred and become forfeited by Virtue of this Act shall and may be recovered in any Court of Record in the said Province of Georgia by Action of Debt Bill Plaint or Information. one half to the said Trustees for establishing the Colony of Georgia in America to be applied for the Benefit of the said Colony as the Common Council of the said Trustees or the Major Part of them for that purpose present and assembled shall think fit and proper and the other half to such Person or Persons as shall sue for the same And if any Action or Suit shall be brought or prosecuted against any Person or Persons for what he or they shall do in or about the Prosecution or putting in Execution of this Act or any of the Powers therein contained it shall and may be lawful to and for the Defendant or Defendants in such Action or Suit to plead the General Issue and to give the special Matter in Evidence And if Judgment shall be given against the Plaintiff or Plaintiffs in such Suit then the Defendant or Defendants shall by the Court before whom the Cause shall be tried be allowed double the Costs and Charges he or they shall have been put unto by such Suit or Suits And the said Court are hereby empowered to levy the said Costs by Distress and Sale of any Goods of the said Plaintiff or Plaintiffs returning the Overplus (if any) to such Plaintiff or Plaintiffs AND BE IT FURTHER ENACTED that the said Common Council of the said Trustees shall have full Power and Authority to lessen abate and mitigate all Forfeitures and Penalties and all such Share of such Forfeitures and Penalties as shall become due to the said Trustees by this Act or by any Clause or Clauses therein contained as the said Common Council or the Major part of them who shall be present and assembled for that purpose shall think fit and proper.

To which the Common Seal was affixed the Eighth day of August 1750.

Encouragement of the Silk Industry in Georgia, 1755

From Allen P. Chandler, ed., *Colonial Records of Georgia* (Atlanta, 1904), VII, pp. 114–15.

At a Council Held in the Council Chamber on Monday the Seventeenth Day of February 1755—The Minutes of the preceeding Board were read and approved

The Board taking into Consideration that the Season for feeding Silk Worms is approaching and being desirous to give all due Encouragement to the Culture of Silk, they came to the following Resolution

RESOLVED—That three Shillings Sterling be given for every Pound of Cocoons fit for use, which shall be raised in this Province, and brought to the Publick Filature in Savannah this Year, provided that in taking them down from the Brooms, that Care is taken to keep the spotted ones separate, which are to be paid for according to their Quality

Development of Colonial Agriculture

A Contemporary Description of Maryland, 1656

From John Hammond, "Leah and Rachel or, The Two Fruitful Sisters, Virginia and Maryland (1656)," in Clayton Colman Hall, *Narratives of Early Maryland 1633–1684* (New York, 1910), pp. 290–93.

The Country is as I said of a temperate nature, the dayes, in summer not so long as in England, in winter longer; it is somewhat hotter in June, July and August then here, but that heat sweetly allayed by a continual breaze of winde, which never failes to cool and refresh the labourer and traveller; the cold seldom approaches sencibly untill about Christmas, (although the last winter was hard and the worst I or any living there knew) and when winter comes, (which is such and no worse then is in England), it continues two monthes, seldom longer, often not so long and in that time although here seldom hard-weather keep men from labour, yet there no work is done all winter except dressing their own victuals and making of fires.

The labour servants are put to, is not so hard nor of such continuance as Husbandmen, nor Handecraftmen are kept at in England, as I said little or nothing is done in winter time, none ever work before sun rising nor after sun set, in the summer they rest, sleep or exercise themselves five houres in the heat of the day, Saturdayes afternoon is alwayes their own, the old Holidayes are observed and the Sabboath spent in good exercises.

The Women are not (as is reported) put into the ground to worke, but occupie such domestique imployments and houswifery as in England, that is dressing victuals, righting up the house, milking, imployed about dayries, washing, beastly and not fit to be so imployed are put into the ground, for reason tells sowing, etc. and both men and women have times of recreations, as much or more than in any part of the world besides, yet som wenches that are nasty, us, they must not at charge be transported and then mantained for nothing, but those that prove so aukward are rather burthensome then servants desirable or usefull.

The Country is fruitfull, apt for all and more then England can or does produce. The usuall diet is such as in England, for the rivers afford innumerable sortes of choyce fish, (if they will take the paines to make wyers or hier the Natives, who for a small matter will undertake it), winter and summer, and that in many places sufficient to serve the use of man, and to fatten hoggs. Water-fowle of all sortes are (with admiration to be spoken of) plentifull and easie to be killed, yet by many degrees more plentifull in some places then in othersome. Deare all over the Country, and in many places so many that venison is accounted a tiresom meat; wilde Turkeys are frequent, and so large that I have seen some weigh neer threescore pounds; other beasts there are whose flesh is wholsom and savourie, such are unknowne to us; and therefore I will not stuffe my book with superfluous relation of their names; huge Oysters and store in all parts where the salt-water comes.

The Country is exceedingly replenished with Neat cattle, Hoggs, Goats and Tame-fowle, but not many sheep; so that mutton is somwhat scarce, but that defect is supplied with store of Venison, other flesh and fowle. The Country is full of gallant Orchards, and the fruit generally more luscious and delightfull then here, witnesse the Peach and Quince, the latter may be eaten raw savourily, the former differs and as much exceeds ours as the best relished apple we have doth the crabb, and of both most excellent and comfortable drinks are made. Grapes in infinite manners grow wilde, so do Walnuts, Smalnuts, Chesnuts and abundance of excellent fruits, Plums and Berries, not growing or known in England; graine we have, both English and Indian for bread and Bear, and Pease besides English of ten several sorts, all exceeding ours in England; the gallant root of Potatoes are common, and so are all sorts of rootes, herbes and Garden stuffe.

It must needs follow then that diet cannot be scarce, since both rivers and woods affords it, and that such plenty of Cattle and Hogs are every where, which yeeld beef, veal, milk, butter, cheese and other made dishes, porke, bacon, and pigs, and that as sweet and savoury meat as the world affords; these with the help of Orchards and Gardens, Oysters, Fish, Fowle and Venison, certainly cannot but be sufficient for a good diet and wholsom accommodation, considering how plentifully they are, and how easie with industry to be had.

Beare is indeed in some place constantly drunken, in other some, nothing but Water or Milk and Water or Beverige, and that is where the goodwives (if I may so call them) are negligent and idle; for it is not for want of Corn to make Malt with (for the Country affords enough) but because they are sloathfull and carelesse: but I hope this Item will shame them out of those humours, that they will be adjudged by their drink, what kinde of Housewives they are.

Those Servants that will be industrious may in their time of service gain a competent estate before their Freedomes, which is usually done by many, and they gaine esteeme and assistance that appear so industrious. There is no Master almost but will allow his Servant a parcell of clear ground to plant some Tobacco in for himself, which he may husband at those many idle times he hath allowed him and not prejudice, but rejoyce his Master to see it, which in time of Shipping he may lay out for commodities, and in Summer sell them again with advantage, and get a Sow-Pig or two, which any body almost will give him, and his Master suffer him to keep them with his own, which will be no charge to his Master, and with one years increase of them may purchase a Cow Calf or two, and by that time he is for himself, he may have Cattel, Hogs and Tobacco of his own, and come to live gallantly; but this must be gained (as I said) by Industry and affability, not by sloth nor churlish behaviour.

And whereas it is rumoured that Servants have no lodging other then on boards, or by the Fire side, it is contrary to reason to believe it: First, as we are Christians; next as people living under a law, which compels as well the Master as the Servant to perform his duty; nor can true labour be either expected or exacted without sufficient cloathing, diet, and lodging; all which both their Indentures (which must inviolably be observed) and the Justice of the Country requires.

A Contemporary Description of North Carolina, 1709

From John Lawson, *Lawson's History of North Carolina* [1709] (Richmond, 1952), pp. 75–84, 173–78.

The Wheat of this Place is very good, seldom yielding less than thirty fold, provided the Land is good where it is sown; Not but that there has been Sixty-six Increase for one measure sown in Piny-Land, which we account the meanest Sort. And I have been informed by People of Credit, that Wheat which was planted in a very rich Piece of Land, brought a hundred and odd Pecks for one. If our Planters, when they found such great Increase, would be so curious as to make nice Observations of the Soil and other remarkable Accidents, they would soon be acquainted with the Nature of the Earth and Climate and be better qualified to manage their Agriculture to more Certainty and greater Advantage, whereby they might arrive to the Crops and Harvests of Babylon, and those other fruitful Countries so much talked of. For I must confess I never saw one Acre of Land managed as it ought to be in Carolina since I knew it; and were they as negligent in their Husbandry in Europe as they are in Carolina, their Land would produce nothing but Weeds and Straw.

They have tried Rye, and it thrives very well; but having such Plenty of Maize, they do not regard it, because it makes black Bread, unless very curiously handled.

Barley has been sowed in small quantities, and does better than can be expected; because that Grain requires the ground to be very well worked with repeated Ploughings, which our general Way of breaking the Earth with Hoes, can, by no means perform, though in several Places we have a light rich, deep, black mould, which is the particular Soil in which Barley best thrives.

The naked Oats thrive extraordinary well; and the other would prove a very bold Grain; but the Plenty of other Grains makes them not much coveted.

The Indian Corn, or Maize, proves the most useful Grain in the World; and had it not been for the Fruitfulness of this Species, it would have proved very difficult to have settled some of the Plantations in America. It is very nourishing, whether in Bread, sodden, or otherwise; and those poor Christian Servants in Virginia, Maryland, and the other northerly Plantations that have been forced to live wholly upon it, do manifestly prove that it is the most nourishing Grain for a Man to subsist on, without any other Victuals. And this Assertion is made good by the Negro-Slaves, who, in many Places eat nothing but this Indian Corn and Salt. Pigs and Poultry fed with this Grain, eat the sweetest of all others. It refuses no Ground, unless the barren Sands, and when planted in good Ground, will repay the Planter, seven or eight hundred fold; besides the Stalks bruised and boiled, make very pleasant Beer, being sweet like the Sugar Cane.

There are several sorts of Rice, some bearded, others not, besides the red and the white, But the white Rice is the best. Yet there is a sort of perfumed Rice in the East Indies, which gives a curious Flavor in the Dressing. And with this sort America is not yet acquainted; neither can I learn that any of it has been brought over to Europe, the Rice of Carolina being esteemed the best that comes to that Quarter of the World. It is of great Increase, yielding from eight hundred to a thousand fold, and thrives best in wild Land that has never been broken up before.

Buck-Wheat is of great Increase in Carolina; but we make no other use of it, than instead of Maiz, to feed Hogs and Poultry; and Guinea Corn, which thrives well here, serves for the same use.

Of the Pulse-kind, we have many sorts. The first is the Bushel-Bean, which is a spontaneous Product. They are so called, because they bring a Bushel of Beans for one that is planted. They are set in the Spring, round Arbors, or at the Feet of Poles, up which they will climb and cover the Wattling, making a very pretty Shade to sit under. They continue flowering, budding and ripening all the Summer long, till the Frost approaches when they forbear their Fruit and die. The Stalks they grow on come to the Thickness of a Man's Thumb; and the Bean is white and mottled, with a purple Figure on each side it, like an Ear. They are very flat, and are eaten as the Windsor-Bean is, being an extraordinary well relished Pulse, either by themselves or with Meat.

We have the Indian Rounceval, or Miraculous Peas, so called from their long Pods, and great Increase. These are later Peas, and require a pretty long Summer to ripen in. They are very good; and so are the Bonavis, Calavancies, Nanticokes, and abundance of other Pulse, too tedious here to name, which we found the Indians possessed of, when first we settled in America, some of which sorts afford us two Crops in one Year; as the Bonavis and Colavancies, besides several others of that kind.

Now I am launched into a Discourse of the Pulse, I must acquaint you that the European Bean planted here, will, in time, degenerate into a dwarfish sort, if not prevented by a yearly Supply of foreign Seed, and an extravagant rich Soil; yet these Pigmy-Beans are the sweetest of that kind I have ever met withal.

As for all the sorts of English Peas that we have yet made tryal of, they thrive very well in Carolina. Particularly the white and gray Rouncival, the common Field-Peas, and Sickle-Peas, yield very well, and are of a good Relish. As for the other sorts, I have not seen made any tryal of as yet, but question not their coming to great Perfection with us.

The Kidney-Beans were here before the English came, being very plentiful in the Indian Corn-Fields.

The Garden-Roots that thrive well in Carolina are, Carrots, Leeks, Parsnips, Turneps, Potatoes of several delicate sorts, Ground Artichokes, Radishes, Horse-Radish, Beet, both sorts, Onions, Shallot, Garlick, Cives, and Wild-Onions.

The Sallads are, the Lettice, Curled, Red Cabbage and Savoy. The Spinage, round and prickley, Fennel, sweet and the common Sort, Samphire, in the Marshes excellent, so is the Dock, or Wild-Rhubarb, Rocket, Sorrell, French and English, Cresses, of several Sorts, Purslain wild, and that of a larger Size which grows in the Gardens; for this Plant is never met withal in the Indian Plantations, and is, therefore, supposed to proceed from Cow-Dung, which Beast they keep not. Parsley, two Sorts, Asparagus thrives to a Miracle, without hot Beds or dunging the Land, White-Cabbage from European, or New England Seed, for the People are negligent and unskilful, and dont care to provide Seed of their own. The Colly-Flower we have not yet had an Opportunity to make tryal of, nor has the Artichoke ever appeared amongst us, that I can learn. Coleworts, plain and curved, Savoys; besides the Water-Melons of several Sorts, very good, which should have gone amongst the Fruits. Of Musk-Melons we have very large and good, and several sorts, as the Golden, Green, Guinea, and Orange. Cucumbers, long, short and prickley, all these from the Natural Ground, and great Increase, without any help of Dung or Reflection; Pompions, yellow and very large, Burmillions, Cashaws, and excellent Fruit boiled; Squashes, Simnals, Horns, and Gourds, besides many other Species of less Value, too tedious to name.

Our Pot herbs and others of use, which we already possess, are Angelica, wild and tame, Balm, Bagloss, Borage, Burnett, Clary, Marigold, Pot-Marjoram, and other Marijorams, Summer and Winter Savory, Columbines, Tansey, Wormwood, Nep, Mallows, several sorts, Drage, red and white, Lambs Quarters, Thyme, Hyssop, of very large Growth, sweet Bazzil, Rosemary, Lavender. The more Physical are Carduns, Benedictus, the Scurvy-grass of America, I never here met any of the European sort; Tobacco of Many sorts, Dill, Carawa, Cummin, Anise, Coriander, all sorts of Plantain of England, and two sorts spontaneous, good Vulneraries, Elecampane, Comfrey, Needle, the Seed from England, none Native; Monks-Rhubarb, Burdock, Asarum, wild in the Woods, reckoned one of the Snake-Roots; Poppies in the Garden, none wild yet discovered; Wormseed, Feverfew, Rue, Ground-Ivey, spontaneous but very small and scarce, Aurea-Virga, four sorts of Snake-Root, besides the common Species, which are great Antidotes against that Serpent's Bite, and are easily raised in the Garden; Mint, James-Town weed, so called from Virginia, the Seed it bears is very like that of an Onion, it is excellent for curing Burns, and assuaging Inflamations, but taken inwardly brings on a sort of drunken Madness. One of our Marsh-Weeds, like a Dock, has the same Effect, and possesses the Party with Fear and Watchings. The Red-Root, whose Leaf is like Spear-Mint, is good for Thrushes and sore Mouths, Camomil, but it must be kept in the Shade, otherwise it will not thrive; Housleek, first from England; Vervin, Night-Shade, several kinds; Harts-Tongue, Yarrow abundance, Mullein the same, both of the Country; Sarsparilla, and abundance more I could name, yet not the hundreth part of what remains, a Catalogue of which is a Work of many Years, and without any other Subject, would swell to a large Volume, and requires the Abilities of a skillful Botanist. Had not the ingenious Mr. Banister (the greatest Virtuoso we ever had on the Continent) been unfortunately taken out of this World, he would have given the best Account of the Plants of America, of any that ever yet made such an Attempt in these parts. Not but we are satisfied, the Species of Vegetable in Carolina are so numerous that it requires more than one Man's Age to bring the chiefest Part of them into regular Classes; the Country being so different in its Situation and Soil, that what one place plentifully affords, another is absolutely a Stranger to; yet we generally observe that the greatest Variety is found in the Low Grounds, and Savannas.

The Flower-Garden in Carolina is as yet arrived but to a very poor and jejune Perfection. We have only two sorts of Roses; the Clov-July-Flowers, Violets, Prince Feather and Tres Colores. There has been nothing more cultivated in the Flower-Garden, which at present occurs to my Memory; but as for the wild spontaneous Flowers of this Country, Nature has been so liberal that I cannot name one tenth part of the valuable ones; And since, to give Specimens would only swell the Volume, and give little Satisfaction to the Reader, I shall therefore proceed to the Present State of Carolina, and refer the Shrubs and other Vegetables of larger growth till hereafter, and then shall deliver them and the other Species in their Order.

When we consider the Latitude and convenient Situation of Carolina, had we no farther Confirmation thereof, our Reason would inform us that such a Place lay fairly to be a delicious Country, being placed in that Girdle of the World which affords Wine, Oil, Fruit, Grain and Silk, with other rich Commodities, besides a sweet Air, moderate Climate and fertile Soil; these are the Blessings, (under Heaven's Protection) that spin out the Thread of Life to its utmost Extent, and crown our Days with the Sweets of Health and Plenty, which, when joined with Content, renders the Possessors the Happiest Race of Men upon Earth.

The Inhabitants of Carolina, through the Richness of the Soil, live an easy and pleasant Life. The Land being of several sorts of Compost, some stiff, others light, some marl, others rich, black Mould, here barren of Pine, but affording Pitch, Tar and Masts; there vastly rich, especially on the Freshes of the Rivers, one part bearing great Timbers others being Savannas or natural Meads, where no Trees grow for several Miles, adorned by Nature with a pleasant Verdure, and beautiful Flowers, frequent in no other Places, yielding abundance of Herbage for Cattle, Sheep, and Horses. The Country in general affords pleasant Seats, the Land, (except in some few Places,) being dry and high Banks, parcelled out into most convenient Necks, (by the Creeks,) easy to be fenced in for securing their Stocks to more strict Boundaries, whereby, with a small trouble of fencing, almost every man may enjoy, to himself, an entire Plantation, or rather Park. These, with the other Benefits of Plenty of Fish, Wild-Fowl, Venison, and the other Conveniences which this Summer-Country naturally furnishes, has induced a great many Families to leave the more Northerly Plantations and sit down under one of the mildest Governments in the World; in a Country that, with moderate

Industry, will afford all the Necessaries of Life. We have yearly abundance of Strangers come among us, who chiefly strive to go Southerly to settle, because there is a vast Tract of rich Land betwixt the Place we are seated in and Cape Fair, and upon that River, and more Southerly which is inhabited by none but a few Indians, who are at this time well affected to the English, and very desirous of their coming to live among them. The more Southerly the milder Winters, with the Advantages of purchasing the Lords Land at the most easy and moderate Rate of any Lands in America, nay, (allowing all Advantages thereto annexed,) I may say the Universe does not afford such another; Besides, Men have a great Advantage of choosing good and commodious Tracts of Land at the first Seating of a Country or River, whereas the Later Settlers are forced to purchase smaller Dividends of the old Standers, and sometimes at very considerable Rates; as now in Virginia and Maryland where a thousand Acres of good Land cannot be bought under twenty shillings an Acre, besides two Shillings yearly Acknowledgement for every hundred Acres; which Sum, be it more or less, will serve to put the Merchant or Planter here into a good posture of Buildings, Slaves, and other Necessaries, when the Purchase of his Land comes to him on such easy Terms. And as our Grain and Pulse thrives with us to admiration, no less do our Stocks of Cattle, Horses, Sheep and Swine multiply.

The Beef of Carolina equalizes the best that our neighboring Colonies afford; the Oxen are of a great size when they are suffered to live to a fit Age. I have seen fat and good Beef at all times of the Year, but October and the cool Months are the Seasons we kill our Beeves in, when we intend them for Salting or Exportation; for then they are in their prime of Flesh, all coming from Grass, we never using any other Food for our Cattle. The Heifers bring Calves at eighteen or twenty Months old, which makes such a wonderful Increase, that many of our Planters, from the very mean Beginnings, have raised themselves, and are now Masters of hundreds of fat Beeves and other Cattle.

The Veal is very good and white, so is the Milk very pleasant and rich, there being at present considerable Quantities of Butter and Cheese made that is very good, not only serving our own Necessities, but we send out a great deal among our Neighbors.

The Sheep thrive very well at present, having most commonly two Lambs at one yeaning. As the Country comes to be open, they prove still better, Change of Pasture being agreeable to that useful Creature. Mutton is (generally) exceeding Fat and of a good Relish; their Wool is very fine and proves a good Staple.

The Horses are well shaped and swift; the best of them would sell for ten or twelve Pounds in England. They prove excellent Drudges and will travel incredible Journeys. They are troubled with very few Distempers, neither do the cloudy faced grey Horses go blind here as in Europe. As for Spavins, Splints and Ring-Bones, they are here never met withal, as I can learn. Were we to have our Stallions and choice of Mares from England, or any other of a good Sort, and careful to keep them on the Highlands, we could not fail of a good Breed; but having been supplied with our first Horses from the neighboring Plantations, which were but mean, they do not as yet come up to the Excellency of the English Horses; though we generally find that the Colt exceeds in Beauty and Strength, its Sire and Dam.

The Pork exceeds any in Europe; the great Diversity and goodness of the Acorns and Nuts which the Woods afford, making that Flesh of an excellent Taste and produces great Quantities; so that Carolina, (if not the chief,) is not inferior in this one Commodity to any Colony in the hands of the English.

As for Goats, they have been found to thrive and increase well, but being mischievous to Orchards and other Trees, makes People decline keeping them.

Our Produce for Exportation to Europe and the Islands in America, are Beef, Pork, Tallow, Hides, Deer-Skins, Furs, Pitch, Tar, Wheat, Indian-Corn, Peas, Masts, Staves, Heading, Boards and all sorts of Timber and Lumber for Madera and the West-Indies, Rozin, Turpentine and several sorts of Gums and Tears, with some medicinal Drugs, are here produced; Besides Rice and several other foreign Grains, which thrive very well. Good Bricks and Tiles are made and several sorts of useful Earths, as Bole, Fuller's-Earth, Oaker and Tobacco-pipe-Clay, in great plenty; Earths for the Potter's Trade and fine Sand for the Glass-Makers. In building with Brick, we make our Lime of Oyster-Shells, though we have great Store of Lime-stone towards the Heads of our Rivers, where are Stones of all sorts that are useful, besides vast Quantities of excellent Marble. Iron-Stone we have plenty of, both in the Low-Grounds and on the Hills. Lead and Copper has been found, so has Antimony heretofore; But no Endeavors have been used to discover those Subteraneous Species; otherwise we might in all proba-bility, find out the best Minerals, which are not wanting in Carolina. Hot Baths we have an account of from the Indians that frequent the Hill-Country, where a great likelihood appears of making Salt-peter, because the Earth in many places, is strongly mixed with a nitrous Salt, which is much coveted by the Beasts, who come at some Seasons in great Droves and Herds, and by their much licking of this Earth, make great Holes in those Banks, which sometimes lie at the heads of great precipices, where their Eagerness after this salt hastens their End by falling down the high Banks, so that they are dashed in Pieces. It must be confessed that the most noble and sweetest Part of this Country is not inhabited by any but the Savages; and a great deal of the richest Part thereof, has no Inhabitants but the Beasts of the Wilderness; For, the Indians are not inclinable to settle in the richest Land, because the Timbers are too large for them to cut down, and too much burthened with Wood for their Laborers to make Plantations of; besides, the Healthfulness of those Hills is apparent by the Gigantick Stature and Grey-Heads so common amongst the Savages that dwell near the Mountains. The great Creator of all things having most wisely diffused his Blessings, by parceling out the Vintages of the World into such Lots as his wonderful Foresight saw most proper, requisite and convenient for the Habitations of his Creatures. Towards the Sea we have the Conveniency of Trade Transportation and other Helps the Water affords; but oftentimes those Advantages are attended with indif-ferent Land, a thick Air, and other Inconveniences; when backwards, near the Mountains, you meet with the richest Soil, a sweet, thin Air, dry Roads, pleasant small murmuring Streams, and several beneficial Productions and Species, which are unknown in the European World. One Part of this Country affords what the other is wholly a Stranger to.

We have Chalybeate Waters of several Tastes and different Qualities, some purge, others work by the other Enunctories. We have amongst the Inhabitants, a Water that is inwardly, a great Apersive, and outwardly, cures Ulcers, Tetters and Sores by washing therewith.

* * *

And, indeed, all the Experiments that have been made in Carolina, of the Fertility and natural Advantages of the Country, have exceeded all Expectations as affording some Commodities which other Places, in the same Latitude, do not. As for Minerals as they are subterraneous Products, so in all new Countries, they are the Species that are last discovered; and especially in Carolina, where the Indians never look for any thing lower than the Superficies of the Earth, being a Race of Men the least addicted to delving of any People that inhabit so fine a Country as Carolina is. As good if not better Mines than those the Spaniards possess in America, lie full West from us; and I am certain we have as Mountainous Land and as great Probability of having rich Minerals in Carolina as any of those Parts that are already found to be so rich therein. But waving this Subject till some other Opportunity, I shall now give you some Observations in general, concerning Carolina; which are, first, that it lies as convenient for Trade as any of the Plantations in America; that we have Plenty of Pitch, Tar, Skins of Deer, and Beeves, Furs, Rice, Wheat, Rie, Indian Grain, sundry sorts of Pulse, Turpentine, Rosin, Masts, Yards, Planks and Boards, Staves and Lumber. Timber of many common sorts, fit for any Uses; Hemp, Flax, Barley, Oats, Buck-Wheat, Beef, Pork, Tallow, Hides, Whale-Bone and Oil, Wax, Cheese, Butter, &c., besides Drugs, Dyes, Fruit, Silk, Cotton, Indico,, Oil and Wine that we need not doubt of as soon as we make a regular Essay, the Country being adorned with Pleasant Meadows, Rivers, Mountains, Valleys, Hills, and rich Pastures, and blessed with wholesome, pure Air; especially a little backwards from the Sea, where the wild Beasts inhabit, none of which are voracious. The Men are active, the Women fruitful to Admiration, every House being full of Children, and several Women that have come hither barren, having presently proved fruitful. There cannot be a richer Soil, no Place abounding more in Flesh and Fowl, both wild and tame, besides, Fish, Fruit, Grain, Cider, and many other pleasant Liquors, together with several other Necessaries for Life and Trade, that are daily found out, as new Discoveries are made. The Stone and Gout seldom trouble us; the Consumption we are wholly Strangers to, no Place affording a better Remedy for that Distemper than Carolina. For Trade we lie so near to Virginia that we have the Advantage of their Convoys; as also Letters from thence in two or three Days at most, in some Places in as few Hours. Add to this the great Number of Ships which come within those Capes, for Virginia and Maryland take off our Provisions and give us Bills of Exchange for England, which is Sterling Money. The Planters in Virginia and Maryland are forced to do the same, the great Quantities of Tobacco that are planted there, Making Provisions scarce; and Tobacco is a Commodity oftentimes so low as to bring nothing, whereas Provisions and Naval Stores never fail of a Market. Besides, where these are raised in such Plenty as in Carolina, there always appears good Housekeeping, and Plenty of all manner of delicate Eatables. For Instance, the Pork of Carolina is very good, the younger Hogs fed on Peaches, Maiz, and such other natural

Produce; being some of the sweetest Meat that the World affords, as is acknowledged by all Strangers that have been there. And as for the Beef in Pamticough and the Southward Parts, it proves extraordinary. We have not only Provisions plentiful, but Cloaths of our own Manufactures, which are made and daily increase; Cotton, Wool, Hemp, and Flax being of our own Growth; and the Women to be highly commended for their Industry in Spinning and ordering their Housewifery to so great Advantage as they generally do, which is much more easy by reason this happy Climate, visited with so mild Winters, is much warmer than the Northern Plantations, which saves abundance of Cloaths, fewer serving our Necessities and those of our Servants. But this is not all, for we can go out with our Commodities to any other Part of the West Indies, or elsewhere, in the Depth of Winter; whereas, those in New England, New York, Pennsylvania, and the Colonies to the Northward of us cannot stir for Ice, but are fast locked into their Harbours. Besides we can trade with South Carolina, and pay no Duties or Customs no more than their own Vessels both North and South being under the Same Lords Proprietors. We have, as I observed before, another great Advantage, in not being a Frontier, and so continually alarmed by the Enemy; and what has been accounted a Detriment to us, proves one of the greatest Advantages any People could wish, which is, our Country's being faced with a Sound near ten Leagues over in some Places, through which, although there be Water enough for as large Ships to come in at, as in any part hitherto seated in both Carolinas; yet the Difficulty of that Sound to Strangers, hinders them from attempting any Hostilities against us; and at the same time, if we consider the Advantages thereof, nothing can appear to be a better Situation, than to be fronted with such a Bulwark, which secures us from our Enemies. Furthermore, our Distance from the Sea rids us of two Curses, which attend most other Parts of America, viz: Muskeetos and the Worm-biting, which eat Ships Bottoms out; whereas at Bath-Town, there is no such thing known; and as for Muskeetos, they hinder us of as little Rest as they do you in England. Add to this, the unaccountable Quantities of Fish this great Water or Sound, supplies us withal, whenever we take the Pains to fish for them; Advantages I have no where met withal in America, except here. As for the Climate, we enjoy a very wholesome and serene Sky, and a pure and thin Air, the Sun seldom missing to give us his daily Blessing, unless now and then on a Winter's Day, which is not often; and when cloudy, the first Appearance of a North-West-Wind clears the Horizon, and restores the Light of the Sun. The Weather, in summer, is very pleasant. The hotter Months being refreshed with continual Breezes of cool reviving Air; and the Spring being as pleasant and beautiful, as in any Place I ever was in. The Winter, most commonly, is so mild, that it looks like an Autum, being now and then attended with clear and thin North-West Winds, that are sharp enough to regulate English Constitutions, and free them from a great many dangerous Distempers, that a continual Summer afflicts them withal, nothing being wanting as to the natural Ornaments and Blessings of a Country, that conduce to make reasonable Men happy. And for those that are otherwise, they are so much their own Enemies, where they are, that they will scarce ever be any one's Friends or their own, when they are transplanted so, it is much better for all sides, that they remain as they are. Not but that there are several good People that, upon just Grounds, may be uneasy under their present Burdens; and such I would advise to remove to the Place I have been treating of, where they may enjoy their Liberty and Religion, and peaceably eat the Fruits of their Labour, and drink the Wine of their own Vineyards, without the Alarms of a troublesome wordly Life. If a Man be a Botanist, here is a plentiful Field of Plants to divert him in. If he be a Gardner, and delight in that pleasant and happy Life, he will meet with a Climate and Soil that will further and promote his Designs, in as great a Measure, as any Man can wish for; and as for the Constitution of this Government, it is so mild and easy, in respect to the Properties and Liberties of a Subject, that without rehearsing the Particulars, I say once for all, it is the mildest and best established Government in the World, and the Place where any Man may peaceably enjoy his own without being invaded by another; Rank and Superiority ever giving Place to Justice and Equity, which is the Golden Rule that every Government ought to be built upon, and regulated by. Besides it is worthy our Notice, that this Province has been settled, and continued the most free from the Insults and Barbarities of the Indians of any Colony that was ever yet seated in America, which must be esteemed as a particular Providence of God, handed down from Heaven to these People, especially when we consider how irregularly they settled North-Carolina, and yet how undisturbed they have ever remained, free from any foreign Danger or Loss, even to this very Day. And what may well be looked upon for as great a Miracle, this is, a Place where no Malefactors are found deserving Death, or even a Prison for Debtors, there being no more than two Persons, as far as I have been able to learn, ever suffered as Criminals, although it has been a Settlement near sixty Years; One of whom was a Turk that committed Murder, the other an old Woman, for Witchcraft. These tis true, were on the Stage and acted many Years before I knew the Place, but as for the last, I wish it had been undone to this day, although they give a great many Arguments to justify the Deed which I had rather they should have had a Hand in than myself; seeing I could never approve of taking Life away upon such Accusations, the Justice whereof I could never yet understand.

But to return to the Subject in Hand, we there make extraordinary good Bricks throughout the Settlement. All sorts of Handicrafts, as Carpenters, Joiners, Masons, Plaisters, Shoemakers, Tanners, Taylors, Weavers, and most others, may with small Beginnings, and God's Blessing, thrive very well in this Place, and provide Estates for their Children, Land being sold at a much cheaper Rate there than in any other Place in America, and may, as I suppose, be purchased of the Lords Proprietors here in England, or of the Governour there for the time being, by any that shall have a mind to transport themselves to that Country. The Farmers that go thither (for which sort of Men it is a very thriving Place) should take with them some particular Seeds of Grass, as Trefoil, Clover-grass all sorts, Sanfoin, and Common Grass, or that which is a Rarity in Europe, especially, what has sprung and rose first from a warm Climate, and will endure the Sun without flinching. Likewise, if there by any extraordinary sort of Grain for Increase or Hardiness, and some Fruit-Trees of choice kinds, they will be both profitable and pleasant to have with you, where you may see the Fruits of your Labor in Perfection, in a few Years. The necessary Instruments of Husbandry I need not acquaint the Husbandman withal. Hoes of all sorts and Axes, must be had, with Saws, Wedges, Augurs, Nails, Hammers, and what other Things may be necessary for building with Brick, or Stone, which sort your Inclination and Conveniency lead you to.

For, after having looked over this Treatise, you must needs be acquainted with the Nature of the Country, and therefore cannot but be Judges, what it is that you will chiefly want. As for Land, none need want it for taking up, even in the Places there seated on the Navigable Creeks, Rivers, and Harbours, without being driven into remoter Holes and Corners of the Country for Settlements, which all are forced to do, who, at this day, settle in most or all of the other English Plantations in America; which are already become so populous that a New-Comer cannot get a beneficial and commodious Seat, unless he purchases, when, in most Places in Virginia and Maryland, a thousand Acres of good Land, seated on a Navigable Water, will cost a thousand Pounds; whereas, with us, it is at present obtained for the fiftieth Part of the Money. Besides our Land pays to the Lords but an easy Quit-Rent, or yearly Acknowledgement; and the other Settlements pay two Shillings per hundred. All these things duly weighed, any rational Man that has a mind to purchase Land in the Plantations for a Settlement of himself and Family, will soon discover the Advantages that attend the Settlers and Purchasers of Land in Carolina above all other Colonies in the English Dominions in America. And as there is a free Exercise of all Persuasions amongst Christians, the Lord's Proprietors to encourage Ministers of the Church of England have given free Land towards the Maintenance of a Church, and especially for the Parish of S. Thomas in Pampticough, over-against the Town is already laid out for a Glebe of two hundred and twenty-three Acres of rich well situated Land, that a Parsonage-House may be built upon.

Agriculture in South Carolina, 1718

From Thomas Nairne, *A Letter from South Carolina* (London, 1718), pp. 10–11.

We have Pompions, Melons, Cucumbers, Squashes, and other Vine-Fruits, which ripen, and are eat all the Summer, from the middle of *June* to the first of *October*. Fig-Trees bear two Crops a Year, one ripe at the End of *June*, the other all *August*. By so great variety of Peaches, Melocotons, and Nectarines, there is this Advantage, that we have them in Season from the 20th of *June* to the End of *September*, for during all that Time, one Kind or another of them is in Perfection.

Rice is clean'd by Mills, turn'd with Oxen or Horses. 'Tis very much sow'd here, not only because it is a vendible Commondity, but thriving best in low moist Lands, it inclines People to improve that sort of Ground, which being planted a few Years with Rice, and then laid by, turns to the best Pasture.

Silk-worms with us are hatch'd from the Eggs about the 6th of *March*, Nature having wisely ordain'd them to enter into this new Form of Being, at the same time that the Mulberry-Leaves, which are their Food, begin to open.

Being attended and fed six Weeks, they eat no more, but have small Bushes set up for them to spin themselves into Balls, which thrown into warm Water, are wound off into raw Silk.

Rosin, Tar and Pitch, are all produc'd from the Pine-Trees; Rosin, by cutting Channels in the standing green Trees, that meet in the Point at the Foot of the Tree, where two or three small Pieces of Board are fitted to receive it. The Channels are cut as high as one can reach with an Axe, and the Bark is peeled off from all those Parts of the Tree that are expos'd to the Sun, that the Heat of it may the more easily force out the Turpentine, which falling upon the Boards placed at the Root, is gathered and laid in Heaps, which melted in great Kettles, becomes Rosin.

Tar is made thus: First they prepare a circular Floor of Clay, declining a little towards the Center, from which is laid a Pipe of Wood, whose upper Part is even with the Floor, and reaches two Foot without the Circumference; under this End the Earth is dug away, and Barrels placed to receive the Tar as it runs. Upon the Floor is built up a large Pile of dry Pine-Wood, split in Pieces, and surrounded with a Wall of Earth, which covers it all over, only a little at the Top, where the Fire is first kindled. After the Fire begins to burn, they cover that likewise with Earth, to the End there may be no Flame, but only Heat sufficient to force the Tar downward into the Floor. They temper the Heat as they please, by thrusting a Stick through the Earth, and letting the Air in at as many Places as they see convenient.

Pitch is made either by boiling Tar in large Iron Kettles, set in Furnaces, or by burning it in round Clay-holes, made in the Earth.

Rice Growing in South Carolina, 1731

From Mark Catesby, *Natural History of Carolina* . . . . (London, 1731), I, p. xvii.

This beneficial Grain was first planted in *Carolina*, about the Year 1688, by Sir *Nathaniel Johnson*, then Governor of that Province, but it being a small unprofitable Kind little Progress was made in its Increase. In the Year 1696 a Ship touched there from *Madagascar* by Accident, and brought from thence about Half a Bushel of a much fairer and larger Kind, from which small Stock it is increased as at present.

The first Kind is bearded, is a small Grain, and requires to grow wholly in Water. The other is larger, and brighter, of a greater Increase, and will grow both in wet and tolerable dry Land. Besides these two Kinds, there are none in *Carolina* materially different, except small Changes occasioned by different Soils, or Degeneracy by successive sowing one Kind in the same Land, which will cause it to turn red.

In *March* and *April* it is sown in shallow Trenches made by the Hough, and good Crops have been made without any further Culture than dropping the Seeds on the bare Ground and covering it with Earth, or in little Holes made to receive it without any further Management. It agrees best with a rich and moist Soil, which is usually two Feet under Water, at least two Months in the Year. It requires several Weedings till it is upward of two Feet high, not only with a Hough, but with the Assistance of Fingers. About the middle of *September* it is cut down and housed, or made into Stacks till it is thresh'd, with Flails, or trod out by Horses or Cattle; then to get off the outer Coat or Husk, they use a Hand-Mill, yet there remains an inner Film which clouds the Brightness of the Grain, to get off which it is beat in large wooden Mortars, and Pestles of the same, by *Negro* Slaves, which is very laborious and tedious. But as the late Governor *Johnson* (as he told me) had procured from *Spain* a Machine which facilitates the Work with more Expedition, the Trouble and Expence ('tis hoped) will be much mitigated by his Example.

Navigation Act, 1660

From Statutes of the Realm, V, pp. 246–50.

AN ACT FOR THE ENCOURAGING AND INCREASING OF SHIPPING AND NAVIGATION.

For *the increase of shipping and encouragement of the navigation of this nation, wherein, under the good providence and protection of God, the wealth, safety and strength of this kingdom is so much concerned;* (2) be it enacted by the King's most excellent majesty, and by the lords and commons in this present parliament assembled, and by the authority thereof, That from and after the first day of *December* one thousand six hundred and sixty, and from thenceforward, no goods or commodities whatsoever shall be imported into or exported out of any lands, islands, plantations or territories to his Majesty belonging or in his possession, or which may hereafter belong unto or be in the possession of his Majesty, his heirs and successors, in *Asia, Africa* or *America,* in any other ship or ships, vessel or vessels whatsoever, but in such ships or vessels as do truly and without fraud belong only to the people of *England* or *Ireland,* dominion of *Wales* or town of *Berwick* upon *Tweed,* or are of the built of and belonging to any the said lands, islands, plantations or territories, as the proprietors and right owners thereof, and whereof the master and three fourths of the mariners at least are *English;* (3) under the penalty of the forfeiture and loss of all the goods and commodities which shall be imported into or exported out of any the aforesaid places in any other ship or vessel, as also of the ship or vessel, with all its guns, furniture, tackle, ammunition and apparel; one third part thereof to his Majesty, his heirs and successors; one third part to the governor of such land, plantation, island or territory where such default shall be committed, in case the said ship or goods be there seized, or otherwise that third part also to his Majesty, his heirs and successors; and the other third part to him or them who shall seize, inform or sue for the same in any court of record, by bill, information, plaint or other action, wherein no essoin, protection or wager of law shall be allowed;

(4) and all admirals and other commanders at sea of any the ships of war or other ship having commission from his Majesty or from his heirs or successors, are hereby authorized and strictly required to seize and bring in as prize all such ships or vessels as shall have offended contrary hereunto, and deliver them to the court of admiralty, there to be proceeded against; and in case of condemnation, one moiety of such forfeitures shall be to the use of such admirals or commanders and their companies, to be divided and proportioned amongst them according to the rules and orders of the sea in case of ships taken prize; and the other moiety to the use of his Majesty, his heirs and successors.

II. And be it enacted, That no alien or person not born within the allegiance of our sovereign lord the King, his heirs and successors, or naturalized, or made a free denizen, shall from and after the first day of *February,* which will be in the year of our Lord one thousand six hundred sixty-one, exercise the trade or occupation of a merchant or factor in any the said places; (2) upon pain of the forfeiture and loss of all his goods and chattels, or which are in his possession; one third to his Majesty, his heirs and successors; one third to the governor of the plantation where such person shall so offend; and the other third to him or them that shall inform or sue for the same in any of his Majesty's courts in the plantation where such offence shall be committed; (3) and all governors of the said lands, islands, plantations or territories, and every of them, are hereby strictly required and commanded, and all who hereafter shall be made governors of any such islands, plantations or territories, by his Majesty, his heirs or successors, shall before their entrance into their government take a solemn oath, to do their utmost, that every the aforementioned clauses, and all the matters and things therein contained, shall be punctually and *bona fide* observed according to the true intent and meaning thereof: (4) and upon complaint and proof made before his Majesty, his heirs or successors, or such as shall be by him or them thereunto authorized and appointed, that any the said governors have been willingly and wittingly negligent in doing their duty accordingly, that the said governor so offending shall be removed from his government.

III. And it is further enacted by the authority aforesaid, That no goods or commodities whatsoever, of the growth, production or manufacture of *Africa, Asia* or *America,* or of any part thereof, or which are described or laid down in the usual maps or cards of those places, be imported into *England, Ireland* or *Wales,* islands of *Guernsey* and *Jersey,* or town of *Berwick* upon *Tweed,* in any other ship or ships, vessel or vessels whatsoever, but in such as do truly and without fraud belong only to the people of *England* or *Ireland,* dominion of *Wales,* or town of *Berwick* upon *Tweed,* or the lands, islands, plantations or territories in *Asia, Africa* or *America,* to his Majesty belonging, as the proprietors and right owners thereof, and whereof the master, and three fourths at least of the mariners are *English;* (2) under the penalty of the forfeiture of all such goods and commodities, and of the ship or vessel in which they were imported, with all her guns, tackle, furniture, ammunition and apparel; one moiety to his Majesty, his heirs and successors; and the other moiety to him or them who shall seize, inform or sue for the same in any court of record, by bill, information, plaint or other action, wherein no essoin, protection or wager of law shall be allowed.

IV. And it is further enacted by the authority aforesaid, That no goods or commodities that are of foreign growth, production or manufacture, and which are to be brought into *England, Ireland, Wales,* the islands of *Guernsey* and *Jersey,* or town of *Berwick* upon *Tweed,* in *English*-built shipping, or other shipping belonging to some of the aforesaid places, and navigated by *English* mariners, as aforesaid, shall be shipped or brought from any other place or places, country or countries, but only from those of the said growth, production or manufacture, or from those ports where the said goods and commodities can only, or are, or usually have been, first shipped for transportation, and from none other places or countries; (2) under the penalty of the forfeiture of all such of the aforesaid goods as shall be imported from any other place or country contrary to the true intent and meaning hereof, as also of the ship in which they were imported, with all her guns, furniture, ammunition, tackle and apparel; one moiety to his Majesty, his heirs and successors, and the other moiety to him or them that shall seize, inform or sue for the same in any court of record, to be recovered as is before exprest.

V. And it is further enacted by the authority aforesaid, That any sort of ling, stock-fish, pilchard, or any other kind of dried or salted fish, usually fished for and caught by the people of *England, Ireland, Wales,* or town of *Berwick* upon *Tweed;* or any sort of cod-fish or herring, or any oil or blubber made or that shall be made of any kind of fish whatsoever, or any whale-fins or whale-bones, which shall be imported into *England, Ireland, Wales,* or town of *Berwick* upon *Tweed,* not having been caught in vessels truly and properly belonging thereunto as proprietors and right owners thereof, and the said fish-cured saved and dried, and the oil and blubber aforesaid (which shall be accounted and pay as oil) not made by the people thereof, and shall be imported into *England, Ireland* or *Wales,* or town of *Berwick* upon *Tweed,* shall pay double aliens custom.

VI. And be it further enacted by the authority aforesaid, That from henceforth it shall not be lawful to any person or persons whatsoever, to load or cause to be loaden and carried in any bottom or bottoms, ship or ships, vessel or vessels whatsoever, whereof any stranger or strangers-born (unless such as shall be denizens or naturalized) be owners, part-owners or master, and whereof three fourths of the mariners at least shall not be *English*, any fish, victual, wares, goods, commodities or things, of what kind or nature soever the same shall be, from one port or creek of *England, Ireland, Wales,* islands of *Guernsey* or *Jersey,* or town of *Berwick* upon *Tweed,* to another port or creek of the same, or of any of them; under penalty for every one that shall offend contrary to the true meaning of this branch of this present act, to forfeit all such goods as shall be loaden and carried in any such ship or vessel, together with the ship or vessel, and all her guns, ammunition, tackle, furniture and apparel; one moiety to his Majesty, his heirs and successors, and the other moiety to him or them that shall inform, seize or sue for the same in any court of record, to be recovered in manner aforesaid.

VII. And it is further enacted by the authority aforesaid, That where any ease, abatement or privilege is given in the book of rates to goods or commodities imported or exported in *English*-built shipping, that is to say, shipping built in *England, Ireland, Wales,* islands of *Guernsey* or *Jersey,* or town of *Berwick* upon *Tweed,* or in any the lands, islands, dominions and territories to his Majesty in *Africa, Asia,* or *America,* belonging, or in his possession, that it is always to be understood and provided, that the master and three fourths of the mariners of the said ships at least be also *English;* (2) and that where it is required that the master and three fourths of the mariners be *English,* that the true intent and meaning thereof is, that they should be such during the whole voyage, unless in case of sickness, death, or being taken prisoners in the voyage, to be proved by the oath of the master or other chief officer of such ships.

VIII. And it is further enacted by the authority aforesaid, That no goods or commodities of the growth, production or manufacture of *Muscovy,* or to any the countries, dominions or territories to the great duke or emperor of *Muscovy or Russia* belonging, as also that no sort of masts, timber or boards, no foreign salt, pitch, tar, rosin, hemp or flax, raisins, figs, prunes, olive-oils, no sorts of corn or grain, sugar, pot-ashes, wines, vinegar, or spirits called *aqua-vitae,* or brandy-wine, shall from and after the first day of *April,* which shall be in the year of our Lord one thousand six hundred sixty-one, be imported into *England, Ireland, Wales,* or town of *Berwick* upon *Tweed,* in any ship or ships, vessel or vessels whatsoever, but in such as do truly and without fraud belong to the people thereof, or some of them, as the true owners and proprietors thereof, and whereof the master and three fourths of the mariners at least are *English:* and that no currans nor commodities of the growth, production or manufacture of any the countries, islands, dominions or territories to the *Othoman* or *Turkish* empire belonging, shall from and after the first day of *September,* which shall be in the year of our Lord one thousand six hundred sixty-one, be imported into any the afore-mentioned places in any ship or vessel, but which is of *English*-built, and navigated, as aforesaid, and in no other, except only such foreign ships and vessels as are of the built of that country or place of which the said goods are the growth, production or manufacture respectively, or of such port where the said goods can only be, or most usually are, first shipped for transportation, and whereof the master and three fourths of the mariners at least are of the said country or place, under the penalty and forfeiture of ship and goods, to be disposed and recovered as in the foregoing clause.

IX. Provided always, and be it hereby enacted by the authority aforesaid, That for the prevention of the great frauds daily used in colouring and concealing of aliens goods, all wines of the growth of *France* or *Germany,* which from and after the twentieth day of *October* one thousand six hundred and sixty shall be imported into any the ports or places aforesaid, in any other ship or vessel than which doth truly and without fraud belong to *England, Ireland, Wales,* or town of *Berwick* upon *Tweed,* and navigated with the mariners thereof, as aforesaid, shall be deemed aliens goods, and pay all strangers customs and duties to his Majesty, his heirs and successors, as also to the town or port into which they shall be imported; (2) and that all sorts of masts, timber or boards, as also all foreign salt, pitch, tar, rosin, hemp, flax, raisins, figs, prunes, olive-oils, all sorts of corn or grain, sugar, pot-ashes, spirits commonly called brandy-wine, or *aqua-vitae,* wines of the growth of *Spain,* the islands of the *Canaries* or *Portugal, Madera,* or western islands; (3) and all the goods of the growth, production or manufacture of *Muscovy* or *Russia,* which from and after the first day of *April,* which shall be in the year of our Lord one thousand six hundred sixty-one, shall be imported into any the aforesaid places in any other than such shipping, and so navigated; (4) and all currans and *Turkey* commodities which from and after the first day of *September* one thousand six hundred sixty-one, shall be imported into any the places aforesaid, in any other than *English*-built shipping, and navigated as aforesaid, (5) shall be deemed aliens goods, and pay accordingly to his Majesty, his heirs and successors, and to the town or port into which they shall be imported.

X. And for prevention of all frauds which may be used in colouring or buying of foreign ships, be it enacted by the authority aforesaid, and it is hereby enacted, That from and after the first day of *April,* which shall be in the year of our Lord one thousand six hundred sixty-one, no foreign-built ship or vessel whatsoever shall be deemed or pass as a ship to *England, Ireland, Wales,* or town of *Berwick,* or any of them belonging, or enjoy the benefit or privilege of such a ship or vessel, until such time that he or they claiming the said ship or vessel to be theirs, shall make appear to the chief officer or officers of the customs in the port next to the place of his or their abode, that he or they are not aliens, and shall have taken an oath before such chief officer or officers, who are hereby authorized to administer the same, that such ship or vessel was *bona fide* and without fraud by him or them bought for a valuable consideration, expressing the sum, as also the time, place and persons from whom it was bought, and who are his part-owners (if he have any); (2) all which part-owners shall be liable to take the said oath before the chief officer or officers of the custom-house of the port next to the place of their abode, and that no foreigner directly or indirectly hath any part, interest or share therein; and that upon such oath he or they shall receive a certificate under the hand and seal of the said chief officer or officers of the port where such person or persons so making oath do reside, whereby such ship or vessel may for the future pass and be deemed as a ship belonging to the said port, and enjoy the privilege of such a ship or vessel; and the said officer or officers shall keep a register of all such certificates as he or they shall so give, and return a duplicate thereof to the chief officers of the customs at *London,* for such as shall be granted in *England, Wales,* and *Berwick,* and to the chief officers of the customs at *Dublin,* for such as shall be given in *Ireland,* together with the names of the person or persons from whom such ship was bought, and the sum of money which was paid for her, as also the names of all such persons who are part-owners of her, if any such be.

XI. And be it further enacted by the authority aforesaid, That if any officers of the customs shall from and after the said first day of *April* allow the privilege of being a ship or vessel to *England, Ireland, Wales,* or town of *Berwick,* or any of them belonging, to any foreign-built ship or vessel, until such certificate be before them produced, or such proof and oath taken before them; (2) or if any officer of the customs shall allow the privilege of an *English*-built ship, or other ship to any the aforesaid places belonging, to any *English* or foreign-built ship coming into any port, and making entry of any goods, until examination whether the master and three fourths of the mariners be *English;* (3) or shall allow to any foreign-built ship bringing in the commodities of the growth of the country where it was built, the privilege by this act of such ship given, until examination and proof whether it be a ship of the built of that country, and that the master and three fourths of the mariners are of that country; (4) or if any person who is or shall be made governor of any lands, islands, plantations or territories in *Africa, Asia* or *America,* by his Majesty, his heirs or successors, shall suffer any foreign-built ship or vessel to load or unload any goods or commodities within the precincts of their governments, until such certificate be produced before them, or such as shall be by them appointed to view the same, and examination whether the master and three fourths of the mariners at least be *English;* (5) that for the first offence such officer of the customs and governors shall be put out of their places, offices or governments.

XII. Provided always, That this act, or any thing therein contained, extend not, or be meant, to restrain and prohibit the importation of any the commodities of the *Streights* or *Levant-Seas,* loaden in *English* built shipping, and whereof the master and three fourths of the mariners at least are *English,* from the usual ports or places for lading of them heretofore within the said *Streights* or *Levant-Seas,* though the said commodities be not of the very growth of the said places.

XIII. Provided also, That this act or any thing therein contained, extend not, or be meant, to restrain the importing of any *East-India* commodities loaden in *English* built shipping, and whereof the master and three fourths of the mariners at least are *English,* from the usual place or places for lading of them in any part of those seas, to the southward and eastward of *Cabo bona Esperanza,* although the said ports be not the very places of their growth.

XIV. Provided also, That it shall and may be lawful to and for any of the people of *England, Ireland, Wales,* islands of *Guernsey* or *Jersey,* or town of *Berwick* upon *Tweed,* in vessels or ships to them belonging, and whereof the master and three fourths of the mariners at least are *English,* to load and bring in from any of the ports of *Spain* or *Portugal,* or western islands, commonly called *Azores,* or *Madera* or *Canary* islands, all sorts of goods or commodities of the growth production or manufacture of the plantations or dominions of either of them respectively.

XV. Provided, That this act, or any thing therein contained, extend not to bullion, nor yet to any goods taken, or that shall be *bona fide* taken, by way of reprisal by any ship or ships belonging to *England, Ireland* or *Wales,* islands of *Guernsey* or *Jersey,* or town of *Berwick* upon *Tweed,* and whereof the master and three fourths of the mariners at least are *English,* having commission from his Majesty, his heirs or successors.

XVI. Provided always, That this act, or any thing therein contained, shall not extend, or be construed to extend, to lay aliens duties upon any corn of the growth of *Scotland,* or to any salt made in *Scotland,* nor to any fish caught, saved and cured by the people of *Scotland,* and imported directly from *Scotland* in *Scotch* built ships, and whereof the master and three fourths of the mariners are of his Majesty's subjects; (2) nor to any seal-oil of *Russia,* imported from thence into *England, Ireland, Wales,* or town of *Berwick* upon *Tweed,* in shipping *bona fide* to some of the said places belonging, and whereof the master and three fourths of the mariners at least are *English.*

XVII. Provided also, and it is hereby enacted, That every ship or vessel belonging to any of the subjects of the *French* King, which from and after the twentieth day of *October* in the year of our Lord one thousand six hundred and sixty shall come into any port, creek, harbour or road of *England, Ireland, Wales,* or town of *Berwick* upon *Tweed,* and shall there lade or unlade any goods or commodities, or take in or set on shore any passengers, shall pay to the collector of his Majesty's customs in such port, creek, harbour or road, for every ton of which the said ship or vessel is of burthen, to be computed by such officer of the customs as shall be thereunto appointed, the sum of five shillings current money of *England:* (2) And that no such ship or vessel be suffered to depart out of such port, creek, harbour or road, until the said duty be fully paid: (3) And that this duty shall continue to be collected, levied and paid, for such time as a certain duty of fifty solls *per* ton, lately imposed by the *French* King, or any part thereof, shall continue to be collected upon the shipping of *England* lading in *France,* and three months after and no longer.

XVIII. And it is further enacted by the authority aforesaid, That from and after the first day of *April,* which shall be in the year of our Lord one thousand six hundred sixty-one, no sugars, tobacco, cotton-wool, indicoes, ginger, fustick, or other dying wood, of the growth, production or manufacture of any *English* plantations in *America, Asia* or *Africa,* shall be shipped, carried, conveyed or transported from any of the said *English* plantations to any land, island, territory, dominion, port or place whatsoever, other than to such other *English* plantations as do belong to his Majesty, his heirs and successors, or to the kingdom of *England* or *Ireland,* or principality of *Wales,* or town of *Berwick* upon *Tweed,* there to be laid on shore, (2) under the penalty of the forfeiture of the said goods, or the full value thereof, as also of the ship, with all her guns, tackle, apparel, ammunition and furniture; the one moiety to the King's majesty, his heirs and successors, and the other moiety to him or them that shall seize, inform or sue for the same in any court of record, by bill, plaint, or information, wherein no essoin, protection or wager of law shall be allowed.

XIX. And be it further enacted by the authority aforesaid, That for every ship or vessel, which from and after the five and twentieth day of *December* in the year of our Lord one thousand six hundred and sixty shall set sail out of or from *England, Ireland, Wales,* or town of *Berwick* upon *Tweed,* for any *English* plantation in *America, Asia* or *Africa,* sufficient bond shall be given with one surety to the chief officers of the custom-house of such port or place from whence the said ship shall set sail, to the value of one thousand pounds, if the ship be of less burthen than one hundred tons; and of the sum of two thousand pounds, if the ship shall be of greater burthen; that in case the said ship or vessel shall load any of the said commodities at any of the said *English* plantations, that the same commodities shall be by the said ship brought to some port of *England, Ireland, Wales,* or to the port or town of *Berwick* upon *Tweed,* and shall there unload and put on shore the same, the danger of the seas only excepted: (2) And for all ships coming from any other port or place to any of the aforesaid plantations, who by this act are permitted to trade there, that the governor of such *English* plantations shall before the said ship or vessel be permitted to load on board any of the said commodities, take bond in manner and to the value aforesaid, for each respective ship or vessel, that such ship or vessel shall carry all the aforesaid goods that shall be laden on board in the said ship to some other of his Majesty's *English* plantations, or to *England, Ireland, Wales,* or town of *Berwick* upon *Tweed:* (3) And that every ship or vessel which shall load or take on board any of the aforesaid goods, until such bond given to the said governor, or certificate produced from the officers of any custom-house of *England, Ireland, Wales,* or of the town of *Berwick,* that such bonds have been there duly given, shall be forfeited with all her guns, tackle, apparel and furniture, to be imployed and recovered in manner as aforesaid; and the said governors and every of them shall twice in every year after the first day of *January* one thousand six hundred and sixty, return true copies of all such bonds by him so taken, to the chief officers of the custom in *London. Confirmed by* 13 *Car.* 2. *stat.* I. *c.* 14.

A Contemporary Description of Virginia, 1724

From Hugh Jones, *The Present State of Virginia* (London, 1724), pp. 34–42, 54–57, 59–61.

The most Southerly of these Rivers is called *James River,* and the next *York River,* the Land in the Latitude between these Rivers seeming most nicely adapted for *sweet scented,* or the finest *Tobacco;* for 'tis observed that the Goodness decreaseth the farther you go to the Northward of the one, and the Southward of the other; but this may be (I believe) attributed in some Measure to the Seed and Management, as well as to the Land and Latitude: For on *York River* in a small Tract of Land called *Digges's Neck,* which is poorer than a great deal of other Land in the same Latitude, by a particular Seed and Management, is made the famous Crop known by the Name of the *E Dees,* remarkable for its mild Taste and fine Smell.

The next great River is *Rappahannock,* and the fourth is *Potowmack,* which divides *Virginia* from the *Province* of *Maryland.*

These are supplied by several lesser Rivers, such as *Chickahommony* and others, *navigable* for Vessels of great Burthen.

Into these Rivers run abundance of great *Creeks* or short Rivers, navigable for *Sloops, Shallops, Long-Boats, Flats, Canoes* and *Periaguas.*

These *Creeks* are supplied with the *Tide,* (which indeed does not rise so high as in *Europe,* so prevents their making good *Docks*) and also with fresh-Water-runs, replenished with *Branches* issuing from the *Springs,* and soaking through the *Swamps;* so that no Country is better watered, for the Conveniency of which most Houses are built near some Landing-Place; so that any Thing may be delivered to a Gentleman there from *London, Bristol,* &c. with less Trouble and Cost, than to one living five Miles in the *Country* in *England;* for you pay no Freight for Goods from *London,* and but little from *Bristol;* only the Party to whom the Goods belong, is in Gratitude engaged to freight *Tobacco* upon the Ship consigned to her Owners in *England.*

Because of this Convenience, and for the Goodness of the Land, and for the sake of Fish, Fowl, &c. Gentlemen and Planters love to build near the Water; though it be not altogether so healthy as the *Uplands* and *Barrens,* which serve for *Ranges for Stock.*

In the *Uplands* near the Ridge generally run the *main Roads,* in a pleasant, dry, sandy Soil, free from Stones and Dirt, and shaded and sheltered chiefly by Trees; in some Places being not unlike the Walks in *Greenwich Park.*

Thus neither the Interest nor Inclinations of the *Virginians* induce them to cohabit in Towns; so that they are not forward in contributing their Assistance towards the making of particular Places, every Plantation affording the Owner the Provision of a little Market; wherefore they most commonly build upon some convenient Spot or Neck of Land in their own Plantation, though Towns are laid out and establish'd in each County; the best of which (next *Williamsburgh*) are *York, Glocester, Hampton, Elizabeth Town,* and *Urbanna.*

The Colony now is encreased to *twenty nine Counties,* naturally bounded (near as much as may be) one with another about as big as *Kent;* but the frontier Counties are of vast Extent, though not thick seated as yet.

The whole Country is a perfect Forest, except where the Woods are cleared for Plantations, and old Fields, and where have been formerly *Indian Towns,* and *poisoned Fields* and *Meadows,* where the Timber has been burnt down in Fire-Hunting of otherwise; and about the Creeks and Rivers are large rank *Morasses* or *Marshes,* and up the Country are poor *Savannahs.*

The Gentlemen's Seats are of late built for the most Part of good Brick, and many of Timber very handsom, commodious, and capacious; and likewise the common Planters live in pretty Timber Houses, neater than the *Farm Houses* are generally in *England:* With Timber also are built Houses for the *Overseers* and *Out-Houses;* among which is the Kitchen apart from the *Dwelling House,* because of the Smell of hot Victuals, offensive in hot Weather.

Of the **Negroes,** *with the Planting and Management of* **Indian Corn,** *Tobacco, &c. and of their Timber, Stock, Fruits, Provision, and Habitations, &c.*

The *Negroes* live in small Cottages called *Quarters,* in about six in a *Gang,* under the Direction of an *Overseer* or *Bailiff;* who takes Care that they *tend* such Land as the Owner allots and orders, upon which they raise *Hogs* and *Cattle,* and plant *Indian Corn* (or *Maize*) and *Tobacco* for the Use of their Master; out of which the *Overseer* has a Dividend (or Share) in Proportion to the Number of *Hands* including himself; this with several Privileges is his Salary, and is an ample Recompence for his Pains, and Encouragement of his industrious Care, as to the Labour, Health, and Provision of the *Negroes.*

The *Negroes* are very numerous, some Gentlemen having Hundreds of them of all Sorts, to whom they bring great Profit; for the Sake of which they are obliged to keep them well, and not over-work, starve, or famish them, besides other Inducements to favour them; which is done in a *great Degree,* to such especially that are laborious, careful, and honest; tho' indeed some Masters, careless of their own Interest or Reputation, are too cruel and negligent.

The *Negroes* are not only encreased by fresh Supplies from *Africa* and the *West India* Islands, but also are very prolifick among themselves; and they that are born there talk *good English,* and affect our Language, Habits, and Customs; and tho' they be naturally of a barbarous and cruel Temper, yet are they kept under by severe Discipline upon Occasin, and by good Laws are prevented from running away, injuring the *English,* or neglecting their Business.

Their Work (or Chimerical hard Slavery) is not very laborious; their greatest Hardship consisting in that they and their Posterity are not at their own Liberty or Disposal, but are the Property of their Owners; and when they are free, they know not how to provide so well for themselves generally; neither did they live so plentifully nor (many of them) so easily in their own Country, where they are made Slaves to one another, or taken Captive by their Enemies.

The Children belong to the Master of the Woman that bears them; and such as are born of a *Negroe* and an *European* are called *Molattoes;* but such as are born of an *Indian* and *Negroe* are called *Mustees.*

Their Work is to take Care of the *Stock,* and plant *Corn, Tobacco, Fruits,* &c. which is not harder than *Thrashing, Hedging,* or *Ditching;* besides, tho' they are out in the violent Heat, wherein they delight, yet in wet or cold Weather there is little Occasion for their working in the Fields, in which few will let them be abroad, lest by this means they might get sick or die, which would prove a great Loss to their Owners, a good *Negroe* being sometimes worth three (nay four) Score Pounds Sterling, if he be a Tradesman; so that upon this (if upon no other Account) they are obliged not to overwork them, but to cloath and feed them sufficiently, and take Care of their Health.

Several of them are taught to be *Sawyers, Carpenters, Smiths, Coopers,* &c. and though for the most Part they be none of the aptest or nicest; yet they are by Nature cut out for hard Labour and Fatigue, and will perform tolerably well; though they fall much short of an *Indian,* that has learn'd and seen the same Things; and *those Negroes* make the best Servants, that have been *Slaves* in their *own Country;* for they that have been *Kings* and *great Men* there are generally lazy, haughty, and obstinate; whereas the others are sharper, better humoured, and more laborious.

The *Languages* of the *new Negroes* are various harsh *Jargons,* and their *Religions* and *Customs* such as are best described by Mr. *Bosman* in his Book intitled (I think) *A Description of the Coasts of* Africa.

The *Virginia* Planters readily learn to become good *Mechanicks* in Building, wherein most are capable of directing their Servants and Slaves.

As for Timber they abound with excellent good; having about eight Sorts of *Oak,* several Kinds of *Walnut-Tree,* and *Hickory* and *Pignut, Pine, Cedar,* and *Cypress* for *Shingles;* which Covering is lighter than *Tiles,* and being nailed down, are not easily blown off in any Tempest or *Gust.*

The Oak, &c. is of quick Growth, consequently will not last so long as ours; though it has a good *Grain,* and is freer from Knots, and will last long enough for Shipping, and ordinary Uses.

When a *Tract of Land is seated,* they *clear* it by selling the Trees about a Yard from the Ground, lest they should shoot again. What Wood they have Occasion for they carry off, and burn the rest, or let it lie and rot upon the Ground.

The Land between the Logs and Stumps they *how* up, planting *Tobacco* there in the Spring, inclosing it with a slight *Fence* of cleft Rails. This will last for *Tobacco* some Years, if the Land be good; as it is where *fine Timber*, or *Grape Vines* grow.

Land when hired is *forced* to bear *Tobacco* by penning their Cattle upon it; but *Cowpen Tobacco* tastes strong, and that planted in wet marshy Land is called *Nonburning Tobacco,* which smoaks in the Pipe like Leather, unless it be of a good Age.

When Land is tired of *Tobacco,* it will bear *Indian Corn* or *English Wheat,* or any other *European Grain* or *Seed,* with wonderful Increase.

Tobacco and *Indian Corn* are planted in *Hills* as Hops, and secured by *Wormfences,* which are made of Rails supporting one another very firmly in a particular Manner.

Tobacco requires a great deal of Skill and Trouble in the right Management of it.

They raise the Plants in *Beds,* as we do Cabbage Plants; which they *transplant* and *replant* upon Occasion after a Shower of Rain, which they call a *Season.*

When it is grown up they *top* it, or nip off the Head, *succour* it, or cut off the Ground Leaves, *weed* it, *hill* it; and when ripe, they *cut* it down about six or eight Leaves on a Stalk, which they carry into airy *Tobacco Houses;* after it is withered a little in the Sun, there it is hung to dry on *Sticks,* as Paper at the Paper-Mills; when it is in proper Case, (as they call it) and the Air neither too moist, nor too dry, they *strike* it, or take it down, then cover it up in *Bulk,* or a great Heap, where it lies till they have Leisure or Occasion to *stem* it (that is pull the Leaves from the Stalk) or *strip* it (that is take out the great Fibers) and *tie* it up in *Hands,* or *streight lay it;* and so by Degrees *prize* or press it with proper Engines into great Hogsheads, containing from about six to eleven hundred Pounds; four of which Hogsheads make a *Tun,* by Dimension, not by Weight; then it is ready for Sale or Shipping.

There are two Sorts of *Tobacco,* viz. *Oroonoko* the stronger, and *Sweet-scented* the milder; the first with a sharper Leaf like a Fox's Ear, and the other rounder and with finer Fibres: But each of these are varied into several Sorts, much as Apples and Pears are; and I have been informed by the *Indian Traders,* that the *Inland Indians* have Sorts of *Tobacco* much differing from any planted or used by the *Europeans.*

The *Indian Corn* is planted in Hills, and weeded much as *Tobacco.*

This Grain is of great Increase and most general Use; for with this is made good *Bread, Cakes, Mush,* and *Hommony* for the *Negroes,* which with good *Pork* and *Potatoes* (red and white, very nice and different from ours) with other *Roots* and *Pulse,* are their general Food.

Indian Corn is the best Food for *Cattle, Hogs, Sheep* and *Horses;* and the *Blades* and *Tops* are excellent *Fodder,* when well cured, which is commonly used, though many raise good *Clover* and *Oats;* and some have planted Sanfoin, &c.

In the *Marshes,* and *Woods,* and *old Fields* is good *Range* for *Stock in the Spring, Summer, and Fall; and the Hogs* will run fat with certain Roots of Flags and *Reeds,* which abounding in the *Marshes* they root up and eat.

Besides, at the *Plantations* are standard *Peach-Trees,* and *Apple-Trees,* planted out in *Orchards,* on Purpose almost for the *Hogs.*

The *Peaches* abound, and are of a delicious Taste, and *Apple-Trees* are raised from the *Seeds* very soon, which kind of Kernel Fruit needs no grafting, and is diversify'd into numberless Sorts, and makes, with good Management, an excellent *Cyder,* not much inferior to that of *Herefordshire,* when kept to a good Age; which is rarely done, the *Planters* being good *Companions* and *Guests* whilst the *Cyder* lasts. Here *Cherries* thrive much better (I think) than in *England;* tho' the *Fruit-Trees* soon decay, yet they are soon raised to great Perfection.

As for *Wool,* I have had near as good as any near *Leominster;* and it might be much improved if the *Sheep* were housed every Night, and foddered and littered as in *Urchinfield,* where they have by such Means the finest *Wool;* but to do this, would be of little Use, since it is contrary to the Interest of *Great Britain* to allow them Exportation of their Woollen Manufactures; and what little Woollen is there made might be nearly had as cheap, and better from *England.*

As for *Provision,* there is Variety of excellent, *Fish* in great Plenty easily taken; especially *Oysters, Sheepsheads, Rocks, large Trouts, Crabs, Drums, Sturgeons,* &c.

They have the same tame Fowl as in *England,* only they propagate better; but these exceed in *wild Geese* and *Ducks, Cohoncks, Blew-Wings, Teal, Swans,* and *Mallard.*

Their *Beef* and *Veal* is small, sweet, and fat enough; their *Pork* is famous, whole *Virginia Shoots* being frequently *barbacued* in *England;* their *Bacon* is excellent, the *Hams* being scarce to be distinguished from those of *Westphalia;* but their *Mutton* and *Lamb* some Folks don't like, though others extol it. Their *Butter* is good and plentiful enough. Their *Venison* in the lower Parts of the Country is not so plentiful as it has been, tho' there be enough and tolerably good; but in the *Frontier Counties* they abound with *Venison, wild Turkies,* &c. where the common People sometimes dress *Bears,* whose Flesh, they say, is not to be well distinguished from good *Pork* or *Bacon.*

They pull the *Down* of their living *Geese* and wild and tame *Ducks,* wherewith they make the softest and sweetest *Beds.*

The *Houses* stand sometimes two or three together; and in other Places a Quarter, half a Mile, or a Mile, or two, asunder, much as in the *Country* in *England.*

* * *

Employment and Trade.

The Plenty of the Country, and the good Wages given to Work-Folks occasion very few Poor, who are supported by the Parish, being such as are lame, sick, or decrepit through Age, Distempers, Accidents, or some Infirmities; for where there is a numerous Family of poor Children the Vestry takes Care to bind them out Apprentices, till they are able to maintain themselves by their own Labour; by which Means they are never tormented with Vagrant, and Vagabond Beggars, there being a Reward for taking up Runaways, that are at a small Distance from their Home; if they are not known, or are without a Pass from their Master, and can give no good Account of themselves, especially Negroes.

In all convenient Places are kept Stores or Ware-Houses of all Sorts of Goods, managed by Store-Keepers or Factors, either for themselves or others in the Country, or in *Great Britain.*

This Trade is carried on in the fairest and genteelest Way of Merchandize, by a great Number of Gentlemen of Worth and Fortune; who with the Commanders of their Ships, and several *Virginians* (who come over through Business of Curiosity, or often to take Possession of Estates, which every Year fall here to some or other of them) make as considerable and handsom a Figure, and drive as great and advantageous a Trade for the Advancement of the Publick Good, as most Merchants upon the *Royal-Exchange.*

At the Stores in *Virginia,* the Planters, &c. may be supplied with what *English* Commodities they want.

The Merchants, Factors, or Store-Keepers in *Virginia* buy up the Tobacco of the Planters, either for Goods or current *Spanish* Money, or with *Sterling* Bills payable in *Great Britain.*

The Tobacco is rolled, drawn by Horses, or carted to convenient Rolling Houses, whence it is conveyed on Board the Ships in Flats or Sloops, &c.

Some Years ago there was made an Act to oblige all Tobacco to be sent to convenient Ware-Houses, to the Custody and Management of proper Officers, who were by Oath to refuse all bad Tobacco, and gave printed Bills as Receipts for each Parcel or Hogshead; which Quantity was to be delivered according to Order upon Return of those Bills; and for their Trouble and Care in viewing, weighing, and stamping, the Officers were allowed 5s. *per* Hogshead.

The Intent of this Law was to improve the Commodity, prevent Frauds in publick Payments; and for Ease of the common Planters, and Expedition and Conveniency of Shipping.

But though the first Design was for publick Tobacco only, yet the private Crops of Gentlemen being included in the Law, was esteemed a great Grievance; and occasioned Complaints, which destroyed a Law, that with small Amendments might have proved most advantageous.

The Abrogation of this Law reduced the Sailors to their old Slavery of rolling the Tobacco in some Places; where they draw it for some Miles, as Gardeners draw a Roller, which makes them frequently curse the Country, and thro' Prejudice give it a very vile Character.

The Tobacco purchased by the Factors or Store-Keepers, is sent Home to their Employers, or consign'd to their correspondent Merchants in *Great Britain.*

But most Gentlemen, and such as are beforehand in the World, lodge Money in their Merchant's Hands here, to whom they send their Crop of Tobacco, or the greatest Part of it.

This Money is employed according to the Planter's Orders; chiefly in sending over yearly such Goods, Apparel, Liquors, &c. as they write for, for the Use of themselves, their Families, Slaves and Plantations; by which Means they have every Thing at the best Hand, and the best of its Kind.

Besides *English* Goods, several Merchants in *Virginia* import from the *West-Indies* great Quantities of Rum, Sugar, Molossus, &c. and Salt very cheap from the *Salt Islands,* which Things they purchase with Money, or generally with Pork, Beef, Wheat, *Indian-Corn,* and the like.

In some of the poorer Parts of the Country abounding in Pine, do they gather up the *Lightwood,* or Knots of the old Trees, which will not decay, being piled up (as a Pit of Wood to be burnt to Charcoal) and encompassed with a Trench, and covered with Earth, is set on Fire; whereby the Tar is melted out, and running into the Trench is taken up, and filled into Barrels; and being boiled to a greater Consistency becomes Pitch.

Of Pitch and Tar they send Home great Quantities, though not near so much as *North Carolina,* which formerly was the *South* Part of *Virginia;* but has long since been given away to Proprietors, tho' the Bounds between the Colony of *Virginia,* and the Government of *North Carolina* are disputed; so that there is a very long *List* of Land fifteen Miles broad between both Colonies (called the *disputed Bounds*) in due Subjection to neither; which is an *Asylum* for the Runagates of both Countries.

* * *

Agriculture in the "Uplands."

Beyond *Col. Spotswood'* s Furnace above the Falls of *Rappahannock* River, within View of the vast *Mountains,* he has founded a Town called *Germanna,* from some *Germans* sent over thither by *Queen Anne,* who are now removed up farther: Here he has Servants and Workmen of most handycraft Trades; and he is building a Church, Court-House and Dwelling-House for himself; and with his Servants and Negroes he has cleared Plantations about it, proposing great Encouragement for People to come and settle in that uninhabited Part of the World, lately divided into a County.

Beyond this are seated the Colony of *Germans* or *Palatines,* with Allowance of good Quantities of rich Land, at easy or no Rates, who thrive very well, and live happily, and entertain generously.

These are encouraged to make Wines, which by the Experience (particularly) of the late *Col. Robert Beverly,* who wrote the *History of Virginia,* was done easily and in large Quantities in those Parts; not only from the Cultivation of the wild Grapes, which grow plentifully and naturally in all the good Lands thereabouts, and in the other Parts of the Country; but also from the *Spanish, French, Italian,* and *German* Wines, which have been found to thrive there to Admiration.

Besides this, these Uplands seem very good for *Hemp* and *Flax,* if the Manufacture thereof was but encouraged and promoted thereabouts; which might prove of wonderful Advantage in our *Naval Stores* and Linens.

Here may likewise be found as good *Clapboards,* and *Pipe-Staves, Deals, Masts, Yards, Planks,* &c. for Shipping, as we are supplied with from several other Countries, not in his Majesty's Dominions.

As for *Trees, Grain, Pults, Fruits, Herbs, Planks, Flowers,* and *Roots,* I know of none in *England* either for Pleasure or Use, but what are very common there, and thrive as well or better in that Soil and Climate than this for the generality; for though they cannot brag of Gooseberries and Currants, yet they may of Cherries, Strawberries, &c. in which they excel: Besides they have the

Advantage of several from other Parts of *America,* there being Heat and Cold sufficient for any; expect such as require a continual Heat, as Lemons and Oranges, Pine-Apples, and the like, which however may be raised there with Art and Care.

The worst Thing in their Gardens, that I know, is the Artichoak; but this I attribute to Want of Skill and good Management.

Mulberry Trees and Silkworms thrive there to Admiration, and Experience has proved that the Silk Manufacture might be carried on to great Advantage.

There is Coal enough in the Country, but good Fire Wood being so plentiful that it encumbers the Land, they have no Necessity for the Trouble and Expence of digging up the Bowels of the Earth, and conveying them afterwards to their several Habitations.

There grows Plenty of *Sumack,* so very useful in the Dying Trade.

The Land is taken up in Tracts, and is Freehold by Patent under the King, paying two Shillings as a yearly *Quit-Rent* for hundred Acres.

Most Land has been long since *taken up* and *seated,* except it be high up in the Country.

For surveying of Land, when any is *taken up,* bought, exchanged, or the Right contested, there is appointed a *Surveyor* in each County, nominated and examined by the *Governors of the College,* in whose Gift those Places are under the *Surveyor General.*

But of this I may be more particular upon another Occasion; only I shall here observe, that every five or seven Years all People are obliged to go a *Procession* round their own Bounds, and renew their Landmarks by cutting fresh *Notches* in the boundary Trees.

Sometimes whole Plantations are sold, and at other Times small Habitations and Lands are let; but this is not very common, most having Land of their own; and they that have not may make more Profit by turning Overseers, or by some other better Ways, than by *Farming.*

Agriculture in Pennsylvania and New Jersey, 1748

From Peter Kalm, *Travels Into North America* **(London, 1772), I, pp. 39, 56–59, 110, 133, 173–75, 271–75.**

Philadelphia reaps the greatest profits from its trade to the *West Indies.* For thither the inhabitants ship almost every day a quantity of flour, butter, flesh and other victuals; timber, plank, and the like. In return they receive either sugar, molasses, rum, indigo, mahogany, and other goods, or ready money. The true mahogany, which grows in *Jamaica*, is at present almost all cut down.

They send both *West India* goods, and their own productions to *England,* the latter are all sorts of woods, especially black walnut, and oak planks for ships; ships ready built, iron, hides, and tar. Yet this latter is properly bought in *New Jersey,* the forests of which province are consequently more ruined than any others.

Ready money is likewise sent over to *England;* from whence in return they get all sorts of goods there manufactured, viz. fine and coarse cloth, linen, iron ware, and other wrought metals, and *East India* goods. For it is to be observed, that *England* supplies *Philadelphia* with almost all stuffs and manufactured goods which are wanted here.

A great quantity of linseed goes annually to *Ireland*, together with many of the ships which are built here. *Portugal* gets wheat, corn, flour, and maize which is not ground. *Spain* sometimes takes some corn. But all the money, which is got in these several countries, must immediately be sent to *England*, in payment for the goods which are got from thence, and yet those sums are not sufficient to pay all the debts. . . .

* * *

Sept. 19th. As I walked this morning into the fields, I observed that a copious dew was fallen; for the grass was as wet as if it had rained. The leaves of the plants and trees had contracted so much moisture, that the drops ran down. I found on this occasion that the dew was not only on the superior, but likewise on the inferior side of the leaves. I therefore carefully considered many leaves both of trees and of other plants; both of those which are more above, and of those which are nearer to the ground. But I found in all of them, that both sides of the leaves were equally bedewed, except those of the *Verbascum Thapsus,* or *great Mullein,* which, though their superior side was pretty well covered with the dew, yet their inferior had but a little.

Every countryman, even a common peasant, has commonly an orchard near his house, in which all sorts of fruit, such as peaches, apples, pears, cherries, and others, are in plenty. The peaches were now almost ripe. They are rare in *Europe*, particularly in *Sweden;* for in that country hardly any people besides the rich taste them. But here every countryman had an orchard full of peach trees, which were covered with such quantities of fruit, that we could scarcely walk in the orchard, without treading upon those peaches which were fallen off; many of which were usually left on the ground, and only part of them sold in town, and the rest was consumed by the family and strangers. Nay, this fine fruit was frequently given to the swine.

This fruit is however sometimes kept for winter use, and prepared in the following manner. The fruit is cut into four parts, the stone thrown away, and the fruit put upon a thread, on which they are exposed to the sunshine in the open air, till they are sufficiently dry. They are then put into a vessel for winter. But this manner of drying them is not very good, because the rain of this season very easily spoils and putrifies them, whilst they hang in the open air. For this reason a different method is followed by others, which is by far the most eligible. The peaches are as before cut into four parts, are then either put upon a thread, or laid upon a board, and so hung up in the air when the sun shines. Being dried in some measure, or having lost their juice by this means, they are put into an oven, out of which the bread has but just been taken, and are left in it for a while. But they are soon taken out and brought into the fresh air; and after that they are again put into the oven, and this is repeated several times, till they are as dry as they ought to be. For it they were dired up at once in the oven, they would shrivel up too much, and lose part of their flavour. They are then put up and kept for the winter. They are either baked into tarts and pyes, or boiled and prepared as dried apples and pears are in *Sweden*. Several people here dry and preserve their apples in the same manner as their peaches.

The peach trees were, as I am told, first planted here by the *Europeans*. But at present they succeed very well, and require even less care, than our apple and pear trees.

The orchards have seldom other fruit than apples and peaches. Pear trees are scarce in this province. They have cherry trees in the orchards, but commonly on the sides of them towards the house, or along the enclosures. Mulberry trees are planted on some hillocks near the house, and sometimes even in the court-yards of the house. The black walnut trees, or *Juglans nigra*, grow partly on hills, and in fields near the farm-houses, and partly along the enclosures; but most commonly in the forests. No other trees of this kind are made use of here. The chesnuts are left in the fields; here and there is one in a dry field, or in a wood.

The *Hibiscus esculentus,* or *Okra*, is a plant which grows wild in the *West Indies,* but is planted in the gardens here. The fruit, which is a long pod, is cut whilst it is green, and boiled in soups, which thereby become as thick as pulse. This dish is reckoned a dainty by some people, and especially by the negroes.

Capsicum *annuum*, or *Guinea pepper*, is likewise planted in gardens. When the fruit is ripe it is almost entirely red, it is put to a roasted or boiled piece of meat, a little of it being strewed upon it, or mixed with the broth. Besides this, cucumbers are pickled with it. Or the pods are pounded whilst they are yet tender, and being mixed with salt are preserved in a bottle; and this spice is strewed over roasted or boiled meat, or fried fish, and gives them a very fine taste. But the fruit by itself is as biting as common pepper.

* * *

Sept. 30th. Wheat and rye are sown in autumn about this time, and commonly reaped towards the end of *June*, or in the beginning of *July*. These kinds of corn, however, are sometimes ready to be reaped in the middle of *June*, and there are even examples that they have been mown in the beginning of that month. Barley and oats are sown in *April*, and they commonly begin to grow ripe towards the end of *July*. Buck-wheat is sown in the middle or at the end of *July*, and is about this time or somewhat later, ready to be reaped. If it be sown before the above-mentioned time, as in *May*, or in *June*, it only gives flowers, and little or no corn.

Mr. *Bartram* and other people assured me, that most of the cows, which the *English* have here, are the offspring of those which they bought of the *Swedes*, when they were masters of the country. The *English* themselves are said to have brought over but few. The *Swedes* either brought their cattle from home, or bought them of the *Dutch*, who were then settled here.

* * *

A soil like this in *New Jersey*, one might be led to think, could produce nothing, because it is so dry and poor. Yet the maize, which is planted on it, grows extremely well, and we saw many fields filled with it. The earth is of that kind in which tobacco commonly succeeds, but it is not near so rich. The stalks of maize are commonly eight feet high, more or less, and are full of leaves. The maize is planted, as usual, in rows, in little squares, so that there is a space of five feet and six inches between each square, both in length and breadth; on each of these little hills three or four stalks come up, which were not yet cut for the cattle; each stalk again has from one to four ears, which are large and full of corn. A sandy ground could never have been better employed. In some places the ground between the maize is ploughed, and rye sown in it, so that when the maize is cut, the rye remains upon the field.

We frequently saw *Asparagus* growing near the enclosures, in a loose soil, on uncultivated sandy fields. It is likewise plentiful between the maize, and was at present full of berries, but I cannot tell whether the seeds are carried by the wind to the places where I saw them; it is however certain, that I have likewise seen it growing wild in other parts of *America*.

* * *

Oct. 28th. We continued our journey in the morning; the country through which we passed was for the greatest part level, though sometimes there were some long hills; some parts were covered with trees, but far the greater part of the country was without woods; on the other hand, I never saw any place in *America*, the towns excepted, so well peopled. An old man, who lived in this neighbourhood, and accompanied us for some part of the road, however assured me, that he could well remember the time, when between *Trenton* and *New Brunswick* there were not above three farms, and he reckoned it was about fifty and some odd years ago. During the greater part of the day, we had very extensive corn-fields on both sides of the road, and commonly towards the south the country had a great declivity. Near almost every farm was a spacious orchard full of peaches and apple trees, and in some of them the fruit was fallen from the trees in such quantities, as to cover nearly the whole surface. Part of it they left to rot, since they could not take it all in and consume it. Wherever we passed by, we were always welcome to go into the fine orchards, and gather our pockets full of the choicest fruit, without the possessor's so much as looking after it. Cherry trees were planted near the farms, on the roads, &c.

The *barns* had a peculiar kind of construction hereabouts, which I will give a concise description of. The whole building was very great, so as almost to equal a small church; the roof was pretty high, covered with wooden shingles, declining on both sides, but not steep: the walls which support it were not much higher than a full grown man; but, on the other hand, the breadth of the building was the more considerable: in the middle was the threshing floor, and above it, or in the loft or garret, they put the corn which was not yet threshed, the straw, or any thing else, according to the season: on one side were stables for the horses, and on the other for the cows. And the small cattle had likewise their particular stables or styes; on both ends of the buildings were great gates, so that one could come in with a cart and horses through one of them, and go out at the other: here was therefore under one roof the threshing floor, the barn, the stables, the hay loft, the coach house, &c. This kind of buildings is chiefly made use of by the *Dutch* and *Germans;* for it is to be observed, that the country between *Trenton* and *New York* is inhabited by few *Englishmen*, but, instead of them, by *Germans* or *Dutch*, the latter of which especially are numerous.

* * *

Nov. 21st. The *Swedes* and all the other inhabitants of the country plant great quantities of maize, both for themselves and for their cattle. It was asserted that it is the best food for hogs, because it makes them very fat, and gives their flesh an agreeable flavour, preferable to all other meat. I have given in two dissertations upon this kind of corn to the *Swedish Royal Academy of Sciences,* which stand in their *Memoirs* for 1751 and 1752.

The wheels of the carts which are here made use of, are composed of two different kinds of wood. The felloes were made of what is called the *Spanish* oak, and the spokes of the white oak.

The *Sassafras* tree grows every where in this place I have already observed several particulars in regard to it and intend to add a few more here. On throwing some of the wood into the fire, it causes a crackling as salt does. The wood is made use of for posts belonging to the enclosures, for it is said to last a long time in the ground: but it is likewise said, that there is hardly any kind of wood, which is more attacked by worms than this, when it is exposed to the air without cover; and that in a short time it is quite worm-eatin through and through. The *Swedes* related that the *Indians,* who formerly inhabited these parts, made bowls of it. On cutting some part of the sassafras tree, or its shoots, and holding it to the nose, it has a strong but pleasant smell. Some people peel the root, and boil the peel with the beer which they are brewing, because they believe it wholesome. For the same reason, the peel is put into brandy, either whilst it is distilling, or after it is made.

An old *Swede* remembered that his mother cured many people of the dropsy, by a decoction of the root of sassafras in water, drank every morning: but she used, at the same time, to cup the patient on the feet. The old man assured me, he had often seen people cured by this means, who had been brought to his mother wrapped up in sheets.

* * *

Nov. 23d. [Raccoon, N. J.] Several kinds of gourds and melons are cultivated here: they have partly been originally cultivated by the Indians, and partly brought over by *Europeans*. Of the gourds there was a kind which were crooked at the end, and oblong in general, and therefore they were called *crooked necks* (Crocknacks;) they keep almost all winter. There is yet another species of gourds which have the same quality: others again are cut in pieces or slips, drawn upon thread, and dried; they keep all the year long, and are then boiled or stewed. All sorts of gourds are prepared for eating in different manners, as is likewise customary in *Sweden*. Many farmers have a whole field of gourds.

Squashes are a kind of gourds, which the *Europeans* got from the Indians, and I have already mentioned them before. They are eaten boiled, either with flesh or by themselves. In the first case, they are put on the edge of the dish round the meat; they require little care, for into whatever ground they are sown, they grow in it and succeed well. If the seed is put into the fields in autumn, it brings squashes next spring, though during winter it has suffered from frost, snow, and wet.

The *Calabashes* are likewise gourds, which are planted in quantities by the *Swedes* and other inhabitants, but they are not fit for eating, and are made use of for making all sorts of vessels; they are more tender than the squashes, for they do not always ripen here, and only when the weather is very warm. In order to make vessels of them, they are first dried well; the seeds, together with the pulpy and spungy matter in which they lie, are afterwards taken out and thrown away; the shells are scraped very clean within, and then great spoons or ladles, funnels, bowls, dishes, and the like, may be made of them: they are particularly fit for keeping seeds of plants in, which are to be sent over sea, for they keep their power of vegetating much longer, if they be put in calabashes, than by any other means. Some people scrape the outside of the calabashes before they are opened, dry them afterwards, and then clean them within; this makes them as hard as bones: they are sometimes washed, so that they always keep their white colour.

Most of the farmers in this country, sow *Buck-wheat,* in the middle of *July;* it must not be sown later, for in that case the frost ruins it; but if it be sown before *July*, it flowers all the summer long, but the flowers drop, and no seed is generated. Some people plough the ground twice where they intend to sow buck-wheat; other plough it only once, about two weeks before they sow it. As soon as it is sown the field is harrowed. It has been found by experience, that in a wet year buck-wheat has been most likely to succeed: it stands on the fields till the frost comes on. When the crop is favourable, they get twenty, thirty, and even forty bushels from one. The *Swedish* churchwarden *Ragnilson*, in whose house we were at this time, had got such a crop: they make buck-wheat cakes and pudding. The cakes are commonly made in the morning, and are baked in a frying pan, or on a stone: are buttered and then eaten with tea or coffee, instead of toasted bread with butter, or toast, which the *English* commonly eat at breakfast. The buck-wheat cakes are very good, and are likewise usual at *Philadelphia* and in other *English* colonies, especially in winter. Buck-wheat is an excellent food for fowls; they eat it greedily, and lay more eggs, than they do with other food; hogs are likewise fattened with it. Buck-wheat straw is of no use; it is therefore left upon the field, in the places where it has been thrashed, or it is scattered in the orchards, in order to serve as a manure by putrifying. Neither cattle nor any other animal will eat of it, except in the greatest necessity, when the snow covers the ground, and nothing else is to be met with. But though buck-wheat is so common in the *English* colonies, yet the *French* had no right notion of it in *Canada,* and it was never cultivated among them. . . .

Rye, wheat, and buck-wheat are cut with the sickle, but oats are mown with a scythe. The sickles which are here made use of are long and narrow, and their sharp edges have close teeth on the inner side. The field lies fallow during a year, and in that time the cattle may graze on it.

All the inhabitants of this place, from the highest to the lowest, have each their orchard, which is greater or less according to their wealth. The trees in it are chiefly peach trees, apple trees, and cherry trees: compare with this what I have already said upon this subject before.

Farming in Pennsylvania, 1752–1756

Excerpt from "Memorandum in Husbandry on my Own Plantation," by William Logan, tenant of Matthew Potter, near Germantown, Pennsylvania, 1748–58. Manuscript in National Agricultural Library.

1752, August and September: I put up a new fence round the Field called Simpson's Field with cedar rails and white oak post. N.B. Such of the post that have their bark on were just fell and cut in the woods and such as are barked were the preceeding spring. . . .

1753: This spring I put up a new fence on the back of the Vandeveer Orchard next to the garlic field, and also at the foot of it with oak post and rails split in the fall.

1753, July or 7th month: I cleared the hill between the rye field and the spring house meadow and sowed it with turnips. N.B. This ground being not sufficiently plowed for want of time and the weather very dry after sowing they did not grow well. Sowed after well dunging it my fruit pasture with clover and timothy seed but a heavy rain coming on immediately after sowing it and continuing to rain for several days I did not harrow it in, being afterwards fearful of hurting the seed if it should be sprouted. N.B. Very little of this seed came up, and what did, the fowls and geese ate it.

1754, January: In the wane of this month I fell 300 white oak posts for railed fence and 200 white oak post for 3 railed fence. N.B. The 3 railed were put up in a ditch between the brick kiln field behind the barn going towards Germantown Woods.

1754, 3rd month: I planted about 300 hills of hops in the small field near Potters at the end of the Vandeveer Orchard where the rabbit shed stands. N.B. They grew well but I did not chuse to pole them the first year.

1754, 8 month: I plowed in the stubble of the wheat which Matthew Potter reaped this year in the field on the right hand side of the lane going down to the gate which I called the Poplar Field. Being about 5 acres. . . .

1754, 9 month, 15th: Harrowed it well and put about 50 bushels of stone lime in small heaps to an acre which 10 month 1st was obliged to slack with water hauled out in hog heads being a very dry lime and from the buckwheat ground about two poles wide towards Neagle's side from the barns northward I laid on wet soap ashes about 40 to 50 bushels to the acre or more.

1754, 10th month 3–7: I spread the lime thin and sowed the ground with wheat . . . plowing it in directly with the lime. N.B. The weather was so very dry for near two months before this wheat was sown that caused it being delayed so late waiting for rain. . . .

1754: This fall I hauled a quantity of dung and soap ashes from Philadelphia and put in heaps on that part of Poplar Field (where the Buckwheat was) from the barn to the chestnut trees and by the side of the orchard and in the spring spread it and sowed it with oats and clover-grass and timothy seed except a strip opposite the barn about two poles wide I sowed with vetches.

1755, 3 month: N.B. This spring was so very dry, and for so very long a time, that the grass seeds never came up or perished soon after, and the vetches came up well, but all died by the excessive drought. The oats were but poor also. I dunged about 1 and ½ acres of this Poplar Field above the chestnut trees and sowed it with flax. This spring I made a new fence with the oak posts I split in January 1754 down the side of the lane next to the garlic field from the orchard to the gate on the Germantown road.

1755, 4 month: Also, from the corner of the Vandeveer Orchard going down to Matthew Potters also from my garden corner along the old orchard side going to Potters.

1755, 6 month: Also a new wormed fence to divide the Great Field beyond the stone house running towards the York Road from the stone house meadow. N.B. This fence was made with rails split on the spot this spring and barked by Jacob Neaglee.

1755. In the Poplar Field I plowed down all my last years stubble and sowed that half. . . . where the oats and flax was with wheat. But first limed the lower corner about one acre next to the brick kiln and sowed all those 4 acres where the wheat was with rye. I had a most extraordinary crop of grain of this land, but touched as many other places were with mildew. Had 220 shocks of wheat and 170 shocks of rye.

1755, 12 month: I hauled out my dung being about 110 loads and spread it partly on the north side of Germantown meadow and on the great meadow opposite my house and partly on the part called the Timothy Patch.

1756, 6 month: As this ground was constantly kept watered and the weather mostly cool the grass did not come forward. . . .

1756, 1st month, 28th to 30th: I fell a quantity of white oak post timber sufficient to make about 400 posts and a quantity of black Spanish oaks sufficient to make about 1,000 rails. Plowed the small pasture where the spruce trees grow and soon after my fruit pasture and soon after my young orchard.

1756, 2nd month: This month I cleared the pieces of woods in Germantown meadow on both sides and plowed it and put a post and rail fence up at the bottom of the meadow.

1756, 3rd month: I put up the fence from the barns on the York Road down toward the corner of my field going to the mill with cedar posts and chestnut rails. Hauled out and spread about 100 bushels of dry ashes and spread part of it on Germantown meadow, and partly on the great meadow opposite my house.

4th month: Harrowed and sowed the spruce pasture behind the house with flax seed. . . .

7th: N.B. This produced a very good crop of flax, but what was sowed on my place where the fruit tree grow was blasted and little worth. Plowed the beginning of the 4th month the piece of ground I cleared last winter on the side of the race near the mill and sowed it down with oats and the foul meadow grass seed I had from Jared Elliot in New England I harrowed and sowed the new cleared ground on both sides of Germantown meadow. That is the south with oats and rye grass seed and the north side with turnip seed. Had a fine rain after two days. N.B. This was bad seed being of Dutch turnip and the grass overpowered it so that I plowed it up and plowed it down again and sowed it with turnips.

1756, 8–9 month: Plowed down all my stubble in Poplar Field where the wheat and rye grew.

1756: This spring I sowed down all of the Poplar Field with buckwheat, it had been plowed down last fall and again this spring and when in blossom I plowed it down and plowed it in. I covered it over with lime about 50 bushels to an acre and sowed and plowed it in all together. N.B. It was exceedingly dry time and had been for near 3 months nor was there any rain until the 8th of the next month.

1756: Matthew Potter says we have had for my own family's use since last harvest for and towards my half of the crop raised last year:

23½ bushels of wheat which I compute at the current price
    of 5 shillings ....£ 5.17.6
45½ bushels of rye at 3 shillings 6.16.6
61 bushels of buckwheat at ²/₆ shillings..................... 7.11.3
24 bushels of Indian corn at ²/₆ shillings.................. 3. - -

I raised and sold 326 pounds of hops at 10 d of
    which my half is £6.15.-
    Sold also 4 cart loads of apples .... 6.- -

In the 11 month I killed 6 hogs and more afterwards
    in the 12th month as per account 12.18.3

Agriculture in Georgia, 1739–1760

From H. P. Holbrook, ed., *Journal and Letters of Eliza Lucas Pinckney* (Wormsloe, Ga., 1850), pp. 5, 10, 12–13, 15–18, 23, 26–27.

July 1. 1739. Wrote my father a long letter on his plantation affairs. and on his change of commissions with Major Heron. on the Augustine Expedition. on the pains I had taken to bring the Indigo, Ginger, Cotton and Lucern and Casada to perfection and had greater hopes from the Indigo (if I could have the seed earlier next year from the West Indies) than any of ye rest of ye things I had try'd.

* * *

Dr Miss Bartlett

The contents of your last concerns us much as it informs us of the accident to Col° Pinckney I hope Mrs Pinckney don't apprehend any further danger from the fall than its spoiling him for a horseman. If it only prevents him riding that dancing beauty Chickasaw for the future I think 'tis not much to be lamented for he has as many tricks and airs as a dancing bear. Won't you laugh at me if I tell you I am so busy in providing for Posterity I hardly allow myself time to sleep or eat and can but just snatch a moment to write to you and a friend or two more. I am making a large plantation of oaks wc^h I look upon as my own property whether my father gives me the land or not and therefore I design many yeer hence when oaks are more valuable than they are now wc^h you know they will be when we come to build fleets I intend I say 2 thirds of the produce of my oaks for a charrity (I'll tell you my scheme another time) and the other third for those that shall have the trouble of putting my design in execution. I sopose according to custom you will read this to your Uncle and Aunt. she is good girl says Mrs Pinckney she is never Idle and she always means well. tell the little Visionary says your Uncle come to town and partake of some of the amusements suitable to her time of life. pray tell him I think these so and what he may now think whims and projects may turn out well by and by. out of many surely one may hitt.

* * *

June 4, 1741

Hon'd Sir

Never were letters more welcome than yours of feb^y 19th and 20th and March ye 11th and 21st wch came almost together. it was near 6 months since we had the pleasure of a line from you. Our fears increased apace and we dreaded some fatal accident had befallen. but hearing of y^r recovery from a dangerous fitt of Illness has more than equalled great as it was our Sorrow and Anxiety nor shall we ever think ourselves sufficiently grateful to Almighty God for the continuance of so great a blessing. I sympathize most sincerely with y^r Inhabitance of Antigua in so great a Calamity as a scarcity of provisions and the want of the necessarys of life to y^r poorer sort. We shall send all we can get of all sorts of provisions particularly what you write for. I wrote this day to Starrat for a bar of butter. We expect the boat from Garden Hill when I shall be able to give you an account of affairs there. The Cotton Guinea corn. and most of the Ginger planted there was cutt off by a frost. I wrote you in a former letter we had a good crop of Indigo upon the ground. I make no doubt this will prove a very valuable Commodity in time if we could have the seed early enough to plant the end of March that the seed might be dry enough to gather before our frost.

* * *

We hear Carthagena is taken — M Wallis is dead Capt Norberry was lately killed in a duel by Capt Dobbins whose life is despaird of by the wounds he received. He is much blamed for quarreling with such a brawling Man as Norberry who was disregarded by every body. Norberry has left a wife and 3 or 4 children in very bad circumstances to lament his rashness.

Mama tenders you her affections and Polly joyns in Duty with

My D^{r}Papa
y^r m obed and ever D D
ELIZA LUCAS.

* * *

Septr 20. 1741. Wrote to my father on plantation business and Concerning a planter's importing negroes for his own use. Col^o Pinckney thinks not — but thinks twas proposed in the assembly and rejected — promised to look over the act and let me know. also informed my father of the alteration tis Soposed there will be in the value of our money occationed by a late Act of Parliament that Extends to all America wch is to disolve all private banks I think by the 30th of last Month or be liable to lose their Estates and put themselves out of the King's protection. informed him of the Tyranical Gov^n at Georgia.

Octr 29. 1971. Wrote to my father acknowledging the receipt of a p^s of rich yellow Lutstring consisting of 19 yards for myself d^o of blue for my Mama. also for a p^s of Holland and Cambrick received from London at the same time. Tell him we have had a moderate and healthy summer and preparing for the King's birth day next day. Tell him shall send the rice by Bullard.

Novr 11. 1741. Wrote to Mr Murray to send down a boat load of white oak staves, bacon and salted beef for the West Indies. sent up at the same time a bar^l salt ½ w^t salt peter. some brown sugar for the bacon. Vinegar and a couple bottles Wine for Mrs Murray and desire he will send down all the butter and hogs lard.

Jany 1741–2. Wrote my father about the Exchange with Col^o Heron.   the purchasing his house at Georgia. The Tyranical Gov^n at Georgia. there went home last year a petition from a great number of sufferers for redress, with their causes and proofs. they got safe and were only answered that the Magistrates that did the Injustice must make Satisfaction. The people are now sending home an Agent to apply for redress in another way. Returned my father thanks for a present I received from him by Capt. Sutherland of twenty pistols.   and for the sweetmeats by Capt Gregory. Shall send the preserved fruit as they come in season Begged the favour of him to send to England for Dr Popashes Cantatas. Wildens Anthems. Knellers Rules for tuning. about the Jerusalem Thorn. shall try different soils for the Lucern grass this year. The ginger turns out but poorly. We want a supply of Indigo Seed. Sent by this Vessel a waiter of my own Japaning my first Essay. Sent also the Rice and beef. Sent Govr Thomas of Philadelphia' Daughter a tea chest of my own doing also Congratulate my father on my brother's recovery from the small pox and having a Commission

*   *   *

May 22^d 1742

I am now set down my dear Brother to obey your Commands and give you a short description of the part of the World I now inhabit — S° Carolina then is an Extensive Country near the Sea. Most of the settled part of it is upon a flatt. the Soil near Charles Town sandy but further distant. clay and swamp lands. It abounds with fine navigable rivers and great quanties of fine timber — The Country at a great distance that is to say about a hundred and fifty mile from C^{rs} Town very hilly The soil in general very fertile and there are few European or American fruits or grain but what grow here the Country abounds with wild fowl Venison and fish Beef Veal and Mutton are here in much greater perfection than in the Islands tho' not equal to that of England — Fruit extreamly good and in profusion, and the oranges exceed any I ever tasted in the West Indies or from Spain or Portugal. The people in general hospitable and honest and the better sort add to these a polite gentile behaviour. The poorer sort are the most indolent people in the world, or they would never be wretched in so plentiful a country as this. The winters here are fine and pleasant but 4 months in the year are extreamly disagreeable excessive hott much thunder and lightening and musketoes and sand flies in abundance C^{rs} Town the Metropolis, is a neat pretty place the inhabitants polite and live a very gentile manner the streets and houses regularly built. the ladies and gentleman gay in their dress. upon the whole you will find as many agreeable people of both sexes for the size of the place as almost any where St Phillip's Church in C^{rs} Town is a very Elegant one and much frequented. there are sever. more places of publick Worship in the town and the generality of people of a religious turn of mind.

I began in haste and have observed no method or I should have told you before I came to Summer, that we have a most charming Spring in this Country especially for those who travel through the Country for the Scent of the young Myrtle and yellow Jessamine with which the woods abound is delightful. The staple commodity here is rice and the only thing they export to Europe. Beef, Pork and Lumber they send to the West Indies.

Pray Inform me how my good friend Mrs Boddicott my cousin Bartholomew and all my old acquaintance doe Mama and Polly joyn in love

with Dr Brother
your's affec^{tely}

To George Lucas Esqr

*   *   *

Feby 3^d 1743–4.

Wrote to my Aunt concerning Mr Smith. The same day to my father   Thanks for all the things he has sent. Sent by the return of the Vessel 2 bar^s Rice — d° Corn. 3 d° pickled Pork 2 Keggs oysters — one of Eggs by way of Experiment. put up in salt. in case they answer my scheme is to supply my Fathers refining house in Antigua with Eggs from Carolina.

* * *

March 15th 1760.

With how much pleasure I receive your letters my Dear Friend I won't attempt to say and the comfort I have at hearing my dear Children are well your own maternal heart can better conceive than I express, so far I can with great truth affirm that 'tis the greatest felicity I have upon earth

* * *

A great cloud seems at present to hang over this province   we are continually insulted by the Indians in our back settlements and a violent kind of small pox rages in C^hrs Town that almost puts a stop to all business   sev.^l of those I have to transact business with are fled into the country but by the Divine Grace I hope a month or two will change the prospect, we expect shortly troops from Gen'l Amherst wc^h I trust will be able to manage these savage enemies and ye small pox as it does not spread in ye Country must soon be over for want of subjects   I am now at Belmont to keep my people out of the way of ye violent distember for the poor blacks have died very fast even by inoculation but ye people in Ch^rs Town were inoculation mad I think I may well call it and rushed into it with such precipitation y^t I think it impossible they could have had either a proper preparation or attendance had there been 10 Doct^rs in town to 1 — the Doct^rs could not help it the people would not be said nay. We lose with this fleet our good Govr Lyttleton   he goes home in the Trent man of warr before he goes to his Gov^t at Jamaica. My sincere thanks to M^r & Mrs Watson

Poor John Motte who was inocculated in England is now very bad with ye small pox   it could never never have taken then to be sure.

April 23. 1760. The small pox is, I thank God, much abated. those that now have it, have it favorably. indeed few have it now but by inocculation. Our Indian affairs are much in the same situation as 4 weeks ago when the Trent sailed.

The Tilling of Land, 1760

From Jared Eliot, *Essays Upon Field Husbandry in New England, as it is or May be Ordered* (Boston, 1760). Reprinted: Harry J. Carman, ed. (New York, 1934), pp. 98–126.

THE FIFTH ESSAY

For your Sakes no Doubt this is written; that he that plougheth should plough in Hope, and that he that thresheth in Hope, should be Partaker of his Hope.

I Cor. ix 10.

The Sluggard will not plough by Reason of the Cold, therefore shall he beg in Harvest, and have Nothing.

Prov. xx. 4.

Wealth or Riches may be considered as nominal or real, natural or artificial: Nominal or artificial are those Things which derive all, or the greatest Part of their Value, from Opinion, Custom, common Consent, or a Stamp of Authority, by which a Value is set; such as Silver, Gold, Pearls, precious Stones, Pictures, Bills of Credit. Some of these Things have a Degree of intrinsick Value in them, but not in any Proportion to the Value to which they are raised by Custom or Consent: For Instance, Silver and Gold have a certain Degree of intrinsick Worth, but nothing equal to Iron in the necessary Service of Life, either for Instrument or Medicine. A Diamond hath an intrinsick Worth from its Hardness; but as to many other precious Stones, a Load-Stone, a Mill-Stone, or a Grind-Stone, is of much more real Worth and Use to Mankind. Pearls are prescribed in Medicine for great People: but is not of Use but as a testacous Powder; and for that Use an Oyster Shell will do as well: But many Things in high Esteem have no intrinsick Worth at all.

Natural or real Wealth are such Things as supply the Necessities or Conveniences of Life: These are obtained from the Earth, or the Sea; such as Corn, Flesh, or Fish, Fruits, Food and Raiment.

Husbandry and Navigation are the true Source of natural or real Wealth, Without Husbandry, even Navigation cannot be carried on; without it we should want many of the Comforts and Conveniences of Life. Husbandry then is a Subject of great Importance, without which all Commerce and Communication must come to an End, all social Advantages cease, Comfort and earthly Pleasure be no more. Nay, this is the very Basis and Foundation of all nominal or artificial Wealth and Riches. This rises and falls, lives or dies, just in Proportion to the Plenty or Scarcity of real Riches or natural Wealth, We have a pregnant Proof of this, 2 *Kings* vi. 25. And there was a great Famine in Samaria; *and behold they besieged it until an Asses Head was sold for four Score Pieces of Silver, and the fourth Part of a Kab of Doves Dung for five Pieces of Silver*. With Submission, I rather think it should be rendered, the Contents of the Dove's Crop. The Dove's returning home from the Field with Crops full of Pease and other Grain, would, when extracted, be a welcome Entertainment to the hungry Inhabitants, and sell for a good Price; whereas the proper Excrements or Dung, especially of such Animals who void no Urine, is so loathsome, and so destitute of Nourishment, as to be unfit for Food, even in Times of greatest Extremity.

If second hand Food has been so high in the Market, how valuable are the clean Productions of the Earth? Husbandry is the true Mine from whence are drawn true Riches and real Wealth. As Dung and other Manure is of such Advantage in raising Corn, in the foregoing Essay the Reader had set before him the *Norfolk* Husbandry, where we find that Clay answers all the Ends and Purposes of Dung, and for Duration much exceeds it: As also divers Ways of making, and Methods to increase the Quantity of Dung or Manure.

In this Essay, I design to shew how Land may be tilled, and the Dung so applied as that a little Dung shall extend as far, and do as much to promote and produce a Crop of Corn, as six Times so much Dung applied in the Common Way. The old worn out Land is to be tilled in such a Manner that affords a Prospect, that the same Land in two or three Years, shall produce Crops without Dung, or any Sort of Manure, in some Measure agreeable to the Method of the excellent and truly learned Mr. [Jethro] *Tull;* a Summary of whose Principles or Doctrine, I here present to the Reader in his own Words.

The only Way we have to inrich the Earth, is to divide it into many Parts, by Manure or by Tillage, or by both: This is called *Pulveration.* The Salt of Dung divide or pulverize the Soil by *Fermentation,* Tillage by the Attrition or *Contusion* of Instruments, of which the Plough is the Chief. The Superficies or Surfaces of those divided Parts of the Earth, is the Artificial Pasture of Plants, and affords the Vegetable Pabulum to such Roots as come into contact with it. There is no Way to exhaust the Earth of this Pabulum, but by the Roots of the Plants, and Plants are now proved to extend their Roots more than was formerly thought they did. Division is infinite, and the more Parts the Soil is divided into, the more of that Superficies or vegetable Pasture must it have, and more of those Benefits which descend from the Atmosphere will it receive. Therefore if the Earth be divided, if it be by Tillage, it answers the same End as if it had been performed by Dung.

In the fore-cited Passage, Mr. *Tull* has had but little Regard to the Capacity of his Reader: Nor will it be much better understood than if it had been wrote in an unknown Tongue, there being so many Words used by him which common Farmers do not understand; and therefore that Book has not been so useful as otherwise it might have been. That excellent Writer seems to me to have entered deeper into the true Principles of Husbandry, than any Author I have ever read. Had he taken Pains to accommodate himself to the Unlearned, his Book would have been much more useful than now it is.

I am very sensible, that the low Stile, the Plainness and Simplicity of these Essays, has exposed them to the Centure of those who do not well consider for whom they are intended and written.

It is much easier to let the Pen run forward in a pompous Parade of Learning, than to bring it into such subjection as to convey and communicate important Truths in such Words as shall be understood, and to use such Plainness and Simplicity as will bring all down to the Level of the most inferior Capacity.

It was a learned Man of the Age, instructed in the School at *Tarsus,* who compleated his Studies in the famous College at *Jerusalem*, under the Tuition of *Gamaliel*, the illustrious President of that renowned Seat of Learning: He was the very Man who said, 1. *Cor.* xiv. 19. *I thank God I speak with Tongues more than you all; yet in the Church I had rather speak five Words with my Understanding, that I might teach others also, than ten thousand Words in an unknown Tongue.*

I purpose to proceed in the same plain simple Manner, to set before the Reader the Way of mending our poor Land, and raising Crops, either without any Dung at all, or if any be applied, it shall be in such a small Quantity, that the Expence will be but little compared with the common Way of Husbandry.

In this Undertaking, I pretend to no other Merit than that,

I. To explain the Doctrine or Principles of Mr. *Tull* in such a Manner as to be open to any common Understanding.

II. To offer such Reasons and Proofs for the Support of these Principles, as will naturally occur.

III. To direct to the Performance of the Work with Instruments less intricate, more plain, cheap and commodious, than those used and described by Mr. *Tull*.

Under these three Heads may, I think, be comprehended all that I design at present to say of this Method of Husbandry, 'till Time and Experience shall enable me to write farther upon this important Subject: For if I succeed according to my Expectation and Desire, I apprehend Husbandry, in the Tillage of Land, will stand upon a good Footing.

The only Way we have to inrich the Land, is by Dung, or by Tillage separately, or by both of them together: It is performed by dividing the Earth into many Parts, or as the common Way of speaking, it is done by making the Ground mellow and soft, so that the Roots may freely pass and find their proper Nourishment. The more mellow and fine the Earth is made, the more Roots will be sent out, from Corn or whatever is sowed or planted in such mellow Land; and the more soft and mellow the Ground is made, there will be not only more Roots, but they will be longer and extend farther: so that the Corn, Turnip, Carrot, or whatever Plant it is, will receive so much the more Nourishment, and consequently grow so much the bigger and better. Dung, or any other Measure, divides the Ground, sets the Parts at a Distance, and so gives a free Passage to the Roots of Plants. In this Action the Salts in Dung hath much the same Operation and Effect as Leaven, or Emptyings hath on Dough; it makes it rise, makes it light, that is, sets the Parts at a Distance. If nothing be done to divide the Parts, and make the Ground mellow by ploughing, or Dung, or both, no Crop can be expected. Sow or plant upon untill'd Land, which is hard and uncultivated, no Corn will grow. If the Earth can be as well divided, and made as mellow by ploughing, digging or howing, why should not Tillage do without Dung; provided the Tillage be equal, or in Proportion to Dung? To do this in the common Way of repeated plain Ploughing and Harrowing, would be too much Charge and Labour: For Mr. *Tull* said, that three Times plain ploughing did only prepare the Land for Tillage. There is a Way of Tillage alone, without Dung, to make the Ground fine and mellow; and this Way is cheap and effectual; is done in the following manner,

First plough your Ground plain, and plough it deep; if you have no Dung, you must have the more loose mellow Earth: When it is thus ploughed, harrow it well with an Iron-Tooth Harrow; let it lye a Fortnight exposed to the Sun, Air and Dews, then plough it into Ridges; to every Ridge there must be eight

Furrows of the Plain-ploughing, two Furrows covered, four ploughed, and two left open; so that in Ridge-ploughing the Team and Plough travels but half so far as in Plain-ploughing: Ridge ploughing will cost but half so much as Plain-ploughing.

I suppose I need not give any particular Directions concerning ploughing the Land into Ridges, every Ploughman understands this, or if he doth not, he may soon learn it of them that do. When it is thus ploughed into Ridges, it is prepared to plant with Wheat, or Cabbages, Carrots, or what else you see fit to plant. In what Manner, and with what Instruments the Seeds of Wheat, Turnips, or Cabbages are to be planted, I shall describe under the third Head, when I come to speak of the Instruments by which it is performed. I shall only add in this Place that the Wheat is to be planted in two Rows on the Middle of the Ridge, the Rows to be at ten Inches Distance; the Cabbages and Turnips in one Row on the Middle of the Ridge, the Turnips at six Inches Distance from each other, Cabbages at a Foot and Half, or two Foot Distance; Carrots are to be planted in two Rows at ten Inches distance, that is, the Space between the Rows is to be ten Inches, the Carrots to be planted at six Inches Distance one from the other, as they stand in the Line or Row.

The Reader will observe, that as yet there is no more Tillage applied to the Land than what is common and usual in our ordinary Way of Husbandry. Now, what follows, is that in which the Art and Mistery doth consist and when it is described and set before you, will appear so simple, so little, so mean, that it will be to you as *go wash in Jordan* was to *Naaman* the Syrian. Suppose it be Turnips, Cabbages or Carrots planted in the Spring, (for as to what relates to Wheat the golden Grain, I purpose to treat of that distinctly by itself) as soon as your Cabbages and Turnips can be seen, weed them with a small hand Hoe. The Carrots for the first Time must be weeded with the Fingers; this is tedious Work: When this is done, and the Plants a little grown so as to be plainly seen, then take one Yoke of Oxen, a long Yoke so long that one Ox may go in one Furrow, and the other Ox in the other, and the Ridge between, in the same Manner as we plough Indian Corn: and with a common Ox Plow, turn of a Furrow from the Ridge, coming as close to the Plants as you can, and not plough them up; you may come within two or three Inches, if the Oxen and Plough are good. Thus take off a Furrow from each Side of every Ridge till all is ploughed; let it lye in this State a Fortnight or three Weeks, then with the Plough turn up the two Furrows to the Ridge; stay about as long as before, and turn the two Furrows off from the Ridge again; the oftner this is repeated so much the better: We ordinarily do it but four Times; but seven times will do better. When the Plants grow larger, you must keep the Plough at a greater Distance; for if you plough as near the Plants as when they are small, you will cut off too many Roots.

You must hoe between the Rows of Carrots with a narrow hand Hoe, to kill the Weeds; and to till the Ground between the Rows, you must mind to dig deep.

Turnips, and whatever is planted in a single Line or Row, must be tended with a hand Hoe, while the Plants are young, and 'till all the Weeds are destroyed so that you may use the Plough. I have been obliged to enter into the practical

Part of this Sort of Husbandry, without which I should not be able to explain the Principles, or doctrinal Part, as I proposed under the first Head.

1. This Way of tilling Land makes it exceeding fine, soft and mellow, beyond what you would imagine: This, we have shewed already, is one Thing requisite and needful.

2. By this Tillage we open such Clods and Parts of Earth as never were opened before, and consequently never was touched by any Root; its whole nourishing Virtue remains intire: In short it is new Land. Every one knows what new Land will do before its native and original Strength and Vigour is consumed and exhausted by the Roots of Corn and other Plants. Thus this Sort of Tillage doth, in a Degree furnish us with new Land. In this Way old Things become new.

3. In this Way of Tillage we intirely destroy and extirpate all Weeds and *Grass* yea, even that stubborn Grass called *Blue Grass,* which is so hurtful to Corn; by which a whole Crop is frequently almost destroyed. This Grass by many is called *Dutch Grass*; and probably that Grass in *England* there called *Couch Grass,* may be the same, and miscalled here *Dutch*, from a Resemblance or Likeness of Sound; their Farmers making the same Complaint of it as ours do here. The Destruction of Weeds and Grass is of great Advantage in Tillage. Weeds very much exhaust the Land, hinder and damnify the Crop: The more these Robbers are destroyed, the more Nourishment there is for Corn.

This Method not only destroys the Weeds for the present, but for the future also; for ploughing stirs up the latent Seeds of Weeds, sets them a growing, and then destroys them when they are come up. The Seeds of Weeds are numerous and hardy, they will lye many Years in the Ground, and when by the Plough are properly situated for Growth, they will come up very plentifully: *Charlock* [Wild Mustard] commonly called *Terrify,* which cannot be subdued in the common Way of Tillage, I suppose in this Way, may be effectually conquered.

That the Destruction of Weeds is one Design we have in View when we till Land, is what is allowed by all; nay, many think that this is the only End, and at least they act and conduct as if they thought so: If it were not so, why do they neglect to hoe and plough if there be no Weeds? And why do they aim at going no deeper than just to cut up the Weeds? But there are other great Advantages to be had by Tillage, besides killing Weeds, as has been said already, and will further appear.

4. This Way of repeated ploughing keeps the Land from going out of Tillage. If Land be never so much ploughed and harrowed, and made ever so light and mellow, yet in a Year's Time the Tillage is spent in a great Degree. The Weight of great Rains, and the natural Weight of the Earth, settles it down so that it is daily growing closer and harder; there is less and less Room for the Roots to extend and spread, find their Food and get Nourishment; for the Roots in Plants are as the Mouth is to Man and Beast; the more Roots the more growth. When Land, by the Law of Gravitation, is thus continually sinking down, closeing together, and so going out of Tillage, we then plough it once in a Month, or oftener, if there be need. Thus the Tillage is kept up in the same State as at first. I find that a great heavy Rain if it fall soon after the Land has been ploughed, it will need ploughing again: In dry Weather it will continue in a State

of Tillage much longer. Our Indian Corn has this repeated Tillage; but our Wheat suffers much for Want of after Tillage: We sow one Year and reap the next, so that from sowing Time 'till Harvest, is ten or eleven Months.

5. There is in Land a twofold and opposite State which renders Tillage absolutely necessary: This repeated plowing answers for both. In the common and ordinary State of Land, it is too hard and close, the parts are so nigh that there are no Holes or Passages left for the Roots to spread downwards, and side ways; or at least these Pores, Holes or Passages, are too small and too few to give Room for the Roots: Often and repeated ploughing sets the Particles of Earth at such a Distance, and so enlargeth these Pores or Holes in the Earth, that the Growth of Plants is by this Means greatly promoted.

Although this be the ordinary State of Land which makes Tillage necessary; yet there is some Land in a State just the Reverse: it is too light, its Parts are at too great a Distance, the Pores and Passages are too wide, so that the Roots are not big enough to fill the Pores, or Holes. If the Roots do not touch the Earth it cannot get Nourishment: The Root should be inclosed on all Sides by the Earth. Every one knows that Roots above Ground in the open Air, can do the Plant no good. All the Difference between Roots under Ground, which do not touch the Earth, and some Roots above Ground, is, that one is shaded, and the other is exposed to the Sun and Wind: But as Roots in the most hollow and light Land, touch the Earth in some Places, so they get some Nourishment and keep alive, yet the Plant makes but a poor Progress.

I have a Piece of Summer Wheat in a drained Swamp, that almost died of this Disease: The Land was so new that it would not bear a Team, so that it could not be ploughed; the Top Earth was exceeding light and puffy; the Seed was howed in, it came up and grew well, so long as the Blade could live upon the Milk of the Wheat Kernel; but when that Store was spent, and the Time was come that it must live by Nourishment obtained by the Roots, it turned yellow, and the Tops died: One of my Sons told me the Wheat would all die; but an heavy Rain fell, which so closed and pressed this light Earth together, and so lessened the Pores, that the Roots were inclosed on all Sides, the Corn recovered its Colour, grew vigorously and well, and put up good large Ears. This Land as much required ploughing as hard heavy Land would have done. Repeated ploughing in Land that is too light, and the Pores too large, will settle it down and close it together, contract and lessen the Pores, as well as raise the heavy Land, and enlarge its Pores. This seeming Contradiction, this blowing Hot and Cold out the same Mouth, may be well enough reconciled, and accounted for in a philosophical Manner: but so long as Experience shews that all this is true, it will be to no Advantage to the Farmer to say any more about it: Nor should I have entered so far into the Philosophy of Tillage as I have done, were it not necessary for a practical Farmer to understand it so far as to make a Judgment, and see into the Reason of this new Kind of Tillage and Farming: And this is the more needful, as there is a Prejudice in Men's Minds against what is new, or at least what Men suppose to be new.

6. This Method or Way of repeated ploughing, fits and prepares the Land to receive and retain all the Benefits of the Atmosphere: It is now open to receive the floating Particles of Sulphur, and the nitrous Salts of the Air, the Benefit

of the Sun's Rays, which, when accompanied with a sufficient Degree of Moisture, enlivens and invigorates all Nature. When the Winter hath brought a universal Gloom upon the Face of the vegitable Creation, Paleness and Death appears on all Sides: The Psalmist saith of it, *Thou hidest thy Face they are troubled.* Then speaking of the Sun, *thou sendest forth thy Spirit they are created, and thou renewest the Face of the Earth.*

But above all this, we are hereby put in Possession of the Dews, which is one of the rich Treasures of the *Atmosphere;* when Land is made fine a good Depth, it is prepared with open Mouth, to drink in and retain the Dews: When the Dew falls upon Land that is untilled, or but poorly tilled, the Ground being hard it doth not sink deep, so the next Day's Sun carries it all off again. It is the same if Land be too light and loose; there is not a sufficient Connection of Parts to convey the Dew from one Particle of Earth to another: I apprehend, that the Moisture of the Dew passeth down in well prepared Land as Water is conveyed through a Rag in Filtration, if the Rag hath large Holes in it the Water will stop: But let this be as it will, it is certain, and known to every observing Farmer, that the best tilled Land in a dry Time, always is moister, and bears the Drought much better than the same Sort of Land which is but poorly tilled; that *Indian* Corn, which is the best ploughed and hoed, will always bear the Drought best. And what is the Reason? Because the Land is prepared to receive and retain the Dew. Mr. *Evelin* made the following Experiment; he dug a Hole in the Ground a good Depth, reduced the Earth to fine Powder, and filled up the Hole with it: a Drought came on, this powdered Earth was moist to the Bottom, when the adjoining Land was exceeding hard and dry. Another Experiment was made thus, a Gallon of Rain-Water was put into a Bowl, and a Gallon of Dew-Water in another Vessel, and set them to dry away in the Sun; the Consequence was, the Sediment or Settlings of the Dew-Water was more in Quantity, blacker and richer than that of the Rain-Water. The Dews and the Salts of the Air, is all by which the Land is inriched; for the other Advantages of Ploughing are but transient. The Advantage this Way is so much, that Mr. *Tull* saith, that Land he hath improved this Way, by this kind of Husbandry, going into another Hand, who used it in the common Way of Husbandry, that Part of the Field was so much inriched by the new Tillage, that there was a visible Difference for the better seven Years after. I suppose, that it is this alone which changed the Colour of my Land in six Months; for having ploughed very deep, and turned up much Fox-colour'd dead Earth, it soon became of a good brown Colour; so that this Kind of Tillage seems likely to put us in Possession of *Joseph's* Blessing: of which we have an Account, *Deut.* xxxiii. 13, 14, and of Joseph he said; Blessed of the Lord be his Land, for the precious Things of Heaven, for the Dew, and for the Deep that coucheth beneath; and for the precious Fruits brought forth by the Sun, and for the precious Things put forth by the Moon. Some understand by the Deep that coucheth beneath, to be the Springs and subterraneous Waters: but it seems more likely, to intend the Riches of the under Earth which coucheth beneath; which, like a couching Lion, must be roused and raised up by a proper Tillage, in order to exert its full Strength.

Thus I have explained the Principles of this kind of Husbandry, the Foundation and Reason of it, in as plain and easy a Manner as I can.

Before I took any Step or Pace towards this Sort of Tillage, I read all I could find upon the Subject with Care, thought and studied on it with Attention; wrote to my good Friend, Mr. *John Bartram,* a Farmer in *Pennsylvania,* a Man of Worth, to know his Opinion of it. He judiciously observ'd, that *England,* where it had been practised with Success, was an Island, having the Sea on all Sides, the Air must be filled with more Vapours and larger Dews, than what we enjoy upon the Continent; their *Atmosphere* being much more replete with Riches for the Earth, than what is to be expected in our dry thin Air. Notwithstanding all this, it ran strongly in my Head to try; for I considered, that, *as God had not left himself without Witness, in that he had given us Rain and fruitful Seasons,* so, in some Degree, he hath given us the other Benefits of the Atmosphere, *to fill our Hearts with Food and Gladness;* therefore thought it our Duty to take all the Advantage of it that we can; and that we would try the Method as far as we could, without the proper Instruments, how much there was of Truth in the Doctrine or Principles, if used and applied in this Climate; and so proceed, or forbear to get the Drill Plough, and other Instruments, as we should find Encouragement: Having made some Trials one Year, this leads me to the second Thing.

2. To offer such Reasons and Proofs for the Support of those Principles, as did occur upon the one Year's Trial which I made. After my Land was prepared and ploughed into Ridges, it was planted with Cabbages, Carrots, Turnips, Onions and Beets, and a Furrow ploughed off from each Side of the Ridge, and then ploughed on; and this being repeated four or five Times from Spring to Fall, the Event was, the Weeds were killed, the Ground grew fine and mellow, Clods and Knots broken and reduced to Dust; the Plants put out numerous Roots, spread and grew very finely; all the Ground was mellow, not only the Furrows which were ploughed, but also the Comb or Ridge in the Middle, as it was narrow and so exposed to the Air and Dew on three Sides, it was struck through, grew mellow, and received as much Advantage by the Tillage, as that Part of the Ridge which was plowed off and on. The Land being ploughed deep, there was a great Quantity of fine Earth prepared to receive the Dews and Salts of the Air, and sufficient Room for the Roots to spread and branch out on all Sides, so that every Thing grew a-pace, and were large, although there was no Dung applied; the same Land would produce in the ordinary Way, Carrots no bigger than a common Candle; in this there were many, 8, 10, and some 12 Inches of Circumference; they were so large, that three Ridges of fifteen Rods long each, two Rows on a Ridge, produced more than twenty Bushels: So an whole Acre's Product, yielding in the same Proportion, would be two hundred and thirty Bushels: had the three Ridges yielded no more than twenty Bushels, besides the greater Increase of the Crop, it is done cheap and with more Ease, as the Horse-Plough performs the Work with more Expedition than it can be done by Hand, so it is done much better for the present Crop, and also mends and enricheth the Land, and prepares it for future Improvement. It is easier this Way, to raise five Bushels of Carrots than one in the common Way. I also tried this Method of Tillage with Turnips planted in a single Row; by the Middle of *June* they were surprizingly large: as I did not weigh or measure them, I am not able to give a perfect Account of them.

In a former Essay, I made mention of a Society in *Scotland,* consisting of three hundred Members, many of them Noblemen of the first Rank; this Society was erected to promote Husbandry and Manufactures; they published a Book of their Transactions; by the Favour of Mr. *Collinson* of *London,* I had an Opportunity to read it, and find in their Fallow Year, instead of the old chargeable Way of Summer Fallow, they plough into Ridges, then plant Cabbages and Turnips; their Cabbages and the early sort are ripe before the Time of sowing Wheat: With frequent Horse-ploughing they grow large, and the Land in fine order for sowing Wheat in the common Way. By this Means, they sometimes raise a Crop of great Value, and have their Land in better order for Wheat than in the old Way of fallowing their Land. The Lord *Rea,* observed to the Society, that he expected to see that Part of his Wheat which grew where the Rows of Cabbages grew before to be poor, but was surprized to find, that in the very Line where the Cabbages grew, in that Range was the biggest Wheat. One would expect that the Land would have been exhausted by so many large Plants. The true Reason of what appeared so strange, was this, the broad Leaves of Cabbage made a large Shade, and within that Shade there would be a swift Undulation of the Air, and consequently a Stream of the nitrous and sulphurous Particles of the Air, would be drawn in and lodged there; I suppose by this Means, that Part of the Land became more inriched than the open Part of the Field.

Pease are found to make Land mellow, to inrich, and so well to prepare it for Wheat, that I have many Times known Farmers to invite others, who had Pease, to sow their Land without paying any Rent, meerly for the Advantage it would be to their Crop of Wheat. Pease make a Shade; where the Land is shaded the Air will be condensed, and, consequently, make room for the rushing in of more Air, so that in this Shade there will be a greater Lodgment of the nitrous Salts, and consequently the Land will be made rich. The same is found by Experience to be true of Potatoes, and therefore, it is accounted to be an inricher of Land. It has been found that Potatoes may be successively planted without Dung, and have good Crops.

It will be asked, if so, why do not Weeds, which make Shades, inrich the Ground? The Reason is plain, because the Land is not tilled, and so prepared to receive and retain the Dews and Salts of the Air: So turn it and set it in every Light, we shall see and find, that Tillage tends to inrich Land, and fits it to bring forth Fruit. My Carrots put forth such Numbers of small fibrous Roots, for the Nourishment of the main Root, that when the Time came to pull them up, they were, comparatively, hairy like a Rat. Roots are to Plants as the Mouth is to Animals; therefore, in feeding Plants we have the greater Advantage: An Horse, Ox or Sheep, has but one Mouth; provide as much Hay and Provender as you will, he can eat but such a Proportion; if you give *Benjamin's* Mess, five Times more than he can eat, it will do no good. But it is otherwise with Plants, the more Provision you make for them of good rich Mold, the more Roots will they put forth, take in so much the more Food, and consequently, grow so much the larger.

Another Proof of the Truth of the Doctrine, or Principles, laid down as the Ground-Work or Basis of this new Husbandry, I shall borrow from the old Husbandry, in the Manner of raising Indian Corn. The Land being previously prepared, the Land Planted, and Corn come up, we plough a Furrow off from the Corn on each side, then hoe it; the next Time plough up to the Corn; so that this Tillage is nearly the same with what is now proposed for Wheat, or what ever we would plant: Only by the Way, I would observe, that the ploughing between the Rows is so shallow, as though they had nothing else in View and Design, but only to kill the Grass and Weeds; whereas it is found by Experience, that if there be no Grass or Weeds, the ploughing and hoeing will make the Corn grow; it is also found true by Experience, that the better the Land is ploughed and hoed, the better and longer will it bear the Drought, and better Crop there will be: Nay, what is still more remarkable, if the Indian Corn be well tilled, the next Crop, whether it be Oats or Flax, so much the bigger and better will that succeeding Crop be, so that the Land must have gained Strength and Riches: If it were not so, why did not the Indian Crop exhaust and spend the Strength of the Land, especially when we consider how large the Corn is made to grow by the good Tillage? But we find the Contrary, the better the Crop of Indian, the better will be the Crop of Oats. There is no sort of Husbandry, wherein the superior Force and Virtue of Tillage doth so evidently appear, as in raising Indian Corn; for if you should plough and harrow the best of Land, and sow or plant the Corn, and never do any Thing more to it, there will be less Corn than if you should plant poor Land, and tend it well; the poor Land well ploughed and hoed, shall bring more Corn than the rich Land; so that by this, we may see the Efficacy and Advantage of this repeated Tillage, which falls in successively, according to the Exigency and Want of the Plant in its several Degrees of Growth: This keeps the Land in a State of Tillage. It is hard to find a Reason why it should not have the same Effect upon Wheat, and every other Plant that is capable of the like Culture; for every one knows, that without this Indian Corn, in good Land, will produce very little, and in poor Land, nothing at all.

We have seen and experienced the Effect of this Kind of Tillage in Indian Corn all our Life, and yet never thought of applying the same Method to other Plants; for we generally go on by Tradition, and do not enter into the Reason of Procedure.

It is natural for Mankind to admire and be pleas'd with new Things, without Reason, and to despise others without Sense or Judgment. The useless Tricks which Horses or Dogs are taught, are admired and valued, and the Instructor is looked upon as little less than a Conjurer; whereas we daily see an Horse or Ox, with little Pains taught, when made fast to a Plough, to keep the Furrow without Variation; and at the End of the Work, at a Word's speaking, come about and return into his Work again. As this is ten Times more useful, so it is more worthy of Admiration. What we see often we little regard.

The Culture of Indian Corn, to a Man of Consideration and Reflection, holds forth much useful Instruction, and is a good Proof of those Principles we have now under Consideration.

Having gone through the Consideration of the Proofs that do occur for the Support of the Doctrine or Principles on which we design to make Tryals,

Third, I come now to direct the Performance of the Work with Instruments less intricate, more plain, cheap and commodious, than those directed to and described by Mr. *Tull.*

Having found by Experience the Advantage of planting Seeds in Rows, and also finding that to plant by Hand is a slow and chargeable Way; therefore I designed to use it no longer than was necessary to find, that it was likely the Method would answer the Design proposed: Being satisfied in that Point, the next Thing was to get Instruments suitable to the Work.

The Instruments peculiar to this Husbandry, are Drill Ploughs. By a Drill I mean an Instrument that will make one Channel or more, upon a Ridge, and drop in the Seed at due Distances, and in a just Proportion: This is what it will do in better Order than Men can possibly do it with their Fingers, and will do more in one Day, than One Hundred Men can do by Hand. There is not much Reason to call it a Plough, for there is no Affinity or Likeness between them but only in this, the Drill has two Coulters by which the Channels are cut.

There are in Use several Sorts of Drills; there is the Wheat Drill, the Turnip Drill, and divers others; but these named are the Chief; to which I have added a Dung Drill, by which Dung, Ashes, or any other Manure, may be conveyed into the Channels where the Seed is to be dropped. Mr. *Tull's* Wheat Drill is a wonderful Invention, but it being the first invented of that Kind, no Wonder if it be intricate, as indeed it is, and consists of more Wheels, and other Parts, than there is really any Need of. This I was very sensible of all along, but knew not how to mend it, therefore applied myself to the Reverend Mr. *Clap,* President of *Yale* College, and desired him for the Regard which he had to the Publick and to me, that he would apply his mathematical Learning, and mechanical Genius, in that Affair; which he did to so good Purpose, that this new modelled Drill can be made with a fourth Part of what Mr. *Tull's* will cost. This [I] look upon as a great Improvement, and take this Opportunity to make my Acknowledgements for the Favour. When this Drill came home, I found the Wheels were too low for our Ridges, therefore it must be mounted upon new Wheels. The next Thing I wanted in order to compass my Design, was a Dung Drill; this is an Invention intirely new, for which there was no Precedent or Model. For this I applied myself to *Benoni Hylliard,* a very ingenious Man of this Town, a Wheel-Wright by Trade. I told him what I wanted, and desired him to make one. At first we could think of no Way but to make it as a distinct Instrument: But at length his Ingenuity led him to set this and the Wheat Drill upon one Frame, so that it became one Instrument. Mr. *Tull,* it is true, might think this Drill not to be needful; for he tells us, that he tried applying of Dung by Hand to the Channels, but found that this Assistance of Dung was not necessary: For he writes, that, to his great Surprize, he found that the Want of Dung might be supplied by repeated Horse-ploughings, and that Two Shillings in Horse-ploughing would do more than *Forty Shillings* in Dung. I should be glad, if in our Climate the One-half of this would prove true. The Land which I design to make use of, is so low and poor, that I shall have need enough of my Dung-Drill, at least, when I first begin with this kind of Husbandry. I hope that in Time, the Land may be so inriched by Tillage, that this may prove needless.

The Dung-Drill, exhibits or sheds into the Channel eighty Bushels of Dung to the Acre, which is about two Cart Loads; the Board on the Fore-side of the Drill-Box, is made fast only by a Spring, so that if any clod, Lump or Stone, cannot pass through the Drill, the Fore-board opens and lets it out, and the Spring shuts it again: Thus the Danger of stopping or breaking the Drill is prevented.

Wheat is planted in two Rows, but Turnips in one Row on the Middle of the Ridge. The Engine is so contrived, that the Wheat-Drill may be taken off, and a Turnip-Drill be put on; and then the Dung-Drill can be so ordered, that the Dung shall be conveyed into that one Channel, either so much Dung as was shed into the two Channels, or half so much more or less as we please.

The Hopper of the Wheat-Drill, holds about a Peck, and the Dung-Hopper two Bushels and an half. Before we plant either Wheat or Turnips, the Tops of the Ridges must be harrowed and made smooth; to do this, Mr. *Tull* used two Harrows at once, one upon one Ridge, and the other upon the next Ridge; a Pole from the Out-side of each Harrow held them together, an Horse made fast to the Middle of the Pole, drew both Harrows: But instead of all this, we have a small Harrow on the Fore-part of the Frame, which first harrows the Ridge; after the Harrow comes two Coulters, which makes the two Channels at ten Inches Distance; The Dung-Drill fills these Channels with Dung; then comes the two Coulters belonging to the Wheat-Drill, and opens the two Channels, and the Wheat-Drill drops in the Wheat Seed, half a Bushel to an Acre; after this follows a small Harrow, which covers the Seed. There is a Tongue or Neb to go between the Oxen; a long Yoak is used for this Work, so that one Ox travels in one Furrow, and the other in the next, with the Ridge between: One Horse might draw it with Ease, could we find any Way to do it, and the Horse travel in the Furrow; if the Horse walk upon the Ridge it would be hurtful. Mr. *Tull's* Wheat-Drill, required two Pair of Wheels: We have two Drills fastened upon a Frame two Foot eight Inches square, and two Harrows, each performing its respective Part of Work at one Movement; and to the Whole but one Pair of Wheels: The Shaft of the Dung-Drill carries round the Shaft of the Wheat-Drill by a Cogg-Wheel; the several Parts are all plain Work, open and easy to the Understanding; this I esteem a compendious Instrument. It has cost me a great deal more than it will to make another, Imitation being so much more easy than Invention.

The next Instrument that was thought necessary for this Kind of Husbandry, is the Hoe-Plough, of which Mr. *Tull* has given us a Draught, which I shewed to our best Plough-Wrights, but they could not understand it, so that I was almost discouraged: But at length I found a Way to do well enough, without any such strange-built Hoe-Plough: Nor is there any Manner of Difficulty about it, for the Furrows may be ploughed from, and up to the Ridges, with a common Plough, a Yoak of Oxen in a long Yoak, so that one Ox may go in one Furrow, and the other Ox in the other Furrow, and the Ridge between. Let the Plants be what they will, we can come as near to them as is needful; or it may be done with one Horse, with an Horse-Plough; but the Way with Oxen I like best, because there is sufficient Strength to plough deep, which is of great Importance, in order to raise a great deal of Mould, for the Purposes above-mentioned. What will be the Success of raising Wheat in this Method, will be left to Experience, and the History of that Trial to be communicated in another Essay. It is high Time something were done; our old Towns raising very little Wheat, it is purchased at the new Towns, and these new Towns will be old in Time; and then what shall we do unless some better Way can be found to manage our old Land which is plain and smooth? For any Man's Reason will tell him, that stony, rocky, rough Land, is by no Means fit for this Sort of drilling Husbandry; there is enough of such plain Land to produce a vast deal of Corn, could there be found any Way to make it bring forth good Crops.

Mr. *Tull* saith, that the Wheat planted in this Manner is not subject to blast, therefore it is a Method that may enable those Parts of *New-England,* to raise Wheat, who never could, in ordinary, attain to it; of this we can have no Certainty but by Trial.

Another Instrument necessary in this Sort of Husbandry, is the *Turnip-Drill:* This is an Instrument which drops a single Turnip-Seed into a Channel cut for that Purpose on the Middle of the Ridge, at six inches Distance; but mine is made to drop one at three Inches Distance, lest the Fly should destroy any of them, or any Seeds should fail coming up; They should be six Inches Distance; if they should be too thick, it is easy to cut them out. They raise Turnips in Abundance in *England* to feed Cattle; some do it in the random Way of sowing, then where they are too thick hoe them up, till they are thinned to a proper Proportion; but then they grow so close together, that it might be difficult and chargeable Work to do it: It is found by Experience, that this Way of Drilling, and tending them with the Horse-Plough, is, by far, the cheapest and most profitable Way.

Mr. *Tull* saith, that his Turnips drilled and well ploughed, weighed from six Pounds to fourteen, did produce Six Hundred and Forty Bushels to the Acre: I should be very glad of half that Quantity: As to the Ease in raising them in this Method, by ploughing the Furrows off and on, I am satisfied by Experience, and that they will grow larger; what I tried were Spring Turnips.

The Usefulness of Turnips for Cows when they calve, for Winter Milch-Cows, is known to all those who have tried.

To make a Turnip-Drill that will drop a single Seed and no more, is a nice Piece of Work. Any Thing farther relating to drilling and ploughing of Wheat and Turnips, must be referred to another Essay, when Time and Experience shall enable me.

Summer Wheat standing so short a Time upon the Ground, to drill that will not be of Advantage. The raising Summer Wheat is a new Part of Husbandry, which obtains greatly of late; and indeed it is Wisdom to have two Strings to the Bow, as was intimated in the *Third* Essay.

As Summer Wheat requires so much Dung, which we cannot well spare, I thought it might do well in the drained Swamps; I have tried this Year, and it looks promising: There is no Part of Husbandry affords me a greater Satisfaction than this; indeed it looks strange to see Wheat growing where not long since there was Flags and Mire; this I suppose to be the first Wheat that ever was raised in such Land in *New-England:* I design to sow a considerable Piece the next Year. Most of the Summer Wheat which was raised last Year, was sold for Seed at a greater Price than Winter Wheat.

But to return from this Digression. There are two Things which may be objected against the Theory or Principle laid down; it may be objected, that if repeated Ploughing will inrich Land; whence is it then that Land is spent and impoverished by Indian Corn?

1. Indian Corn seldom is tended as it ought to be; if there be any ploughing between the Rows it is shallow, just so as to kill Weeds, but not so as to make a great Quantity of soft mellow Earth.

2. It is succeeded by Oats generally, which is a great Spunger, and this without Rest or Relief.

3. The Plants are set too thick for Benefit of the Land, and many Times for the Corn too.

Look upon that Plant in Blossom Time, when it is in its full Pomp and Pride, observe its Height, its Breadth, its Verdure, that deep green shews it to be replete with rich Sap. A Man that spends more than his Income, altho' that be very great, yet he will grow poor; so in Land, if the Exhaustion be more than the Assistance it receives by Dung or Tillage, the Land will not gain but grow poor: That which is called hoeing scarce deserves the Name of Tillage, for really it is but scraping. I have had Thoughts of trying to plant Corn at a great Distance every Way, and plough deep every Time; this, doubtless, would be better for the Land; if the Crop were less, it may be, we should have a Compensation in the Advance of the Land. One of my Sons, upon this Sort of Reasoning, has planted his Corn this Year much thinner than ordinary, and ploughed it deeper than what is common, so that will serve to make some Discovery.

Another Objection is, that sandy Land seems to have all the needful Properties or Qualifications; it is light, mellow, the Parts at due Distance, and there is Room for a Passage to the Roots; and yet it is found that Sands are barren; but this is where it is perfect Sand; for it is found by Experience, that where Loam or Clay is mixed with Sand in a good Proportion, it proves the best of Land; whether this Mixture be by Art, as in the *Norfolk* Husbandry, or if it be by a natural Mixture.

No Fertility is to be expected from perfect Sand, for every Grain of Sand is a pebble Stone, and surely none can reasonably expect Corn from Stones alone, although these Stones lye in never such good Order: That a Grain of Sand is a Pebble, appears by being viewed in the Microscope or Magnifying Glass; as also, that Sand is one of the Ingredients in the making Glass.

I have thought of these and other Objections, and have helped myself to get over them; whether I have done so for the Reader, he can best tell?

I have two Things of great Importance to communicate, with which I shall conclude this Essay. In the *Fourth* Essay, I informed the Reader, I was in Hopes, that I had found certain Times for cutting Bushes, which would be more effectual for their Destruction than any yet discovered; that if I found it so, I would give Notice of it in my next; am glad I am able to perform that Promise; the Times are in the Months of *June, July* and *August; in the old Moon that Day the Sign is in the Heart:* It will not always happen every Month; it happens so but once this Year, and that proves to be on Sunday. Last Year in *June* or *July,* I forgot which, I sent a Man to make Trial; in going to the Place, some of the Neighbours understanding by him the Business he was going about, and the Reason of his going at that Point of Time, they also went to their Land, and cut Bushes also on that Day; their's were tall Bushes that had never been cut; mine were short Bushes such as had been often cut, but to no Purpose, without it was to increase their Number: The Consequence was, that in ever Place it killed so universally, that there is not left alive, scarce one in a hundred; the Trial was made in three or four Places on that same Day. In *July* or *August,* on the critical Day, another Swamp was cut, the Brush was, the greatest Part of it, Swamp Button Wood, the most difficult to subdue of any Wood I know; I have been lately to see it, and find the Destruction of these Bushes are not so universal as among Alders and other Sorts of Growth; it is hard to say how many remain alive, it may be one third or a quarter Part; all that I can say, with Certainty, is that they are now few, compared with what there was last Year: I did not know but that those which are alive, might be such as came up since; but upon Examination, I found the last year's Stumps, and could plainly see where they had been cut off; this was not because the Season was better when there was such Success; for in this last mentioned Piece of Swamp, there were sundry Spots of Alders and other Sorts of Bushes, they seem to be as universally killed as those before mentioned: The Reason why there was not the same Success attending the cutting these Button Bushes as the other Sorts, I suppose to be from the stubborn Nature of this Kind, which would yield to no cutting; the ordinary Way has been to dig or plough it up by the Roots; so that considering the Nature of this Bush, I have had great Success; the Ground being very boggy, those who mowed them, were obliged to cut them very high, which was another Disadvantage.

To shew such a Regard to the Signs, may incur the Imputation of Ignorance or Superstition; for the Learned know well enough, that the Division of the Zodiac into Twelve Signs, and the appropriating these to the several Parts of the animal Body, is not the Work of Nature, but of Art, contrived by Astronomers for Convenience. It is also as well known, that the Moon's Attraction hath great Influence on all Fluids.

It is also well known to Farmers, that there are Times when Bushes, if cut at such a Time, will universally die. A Regard to the Sign, as it serveth to point out and direct to the proper Time, so it becomes worthy of Observation.

If Farmers attend the Time with Care, and employ Hands on those Days, they will find their Account in it. This Rule attended to, may save the Country many Thousand Days Work. A Farmer of good Credit told me, that he had found by Experience, that Bushes cut with a sharp Tool, would die more than when cut with a dull one. This looks agreeable to Reason, for the sharp Scythe leaves the Mouths of the Sap Vessels all open by which Means they bleed more plentifully: The dull Instrument bruises the Part, and in a Degree doth close up the Wound.

Another important Article, is concerning red Clover Seed. It hath been the prevailing Opinion of Farmers in this Country, that Clover Seed must be laid very shallow in the Ground; and by the Books of Husbandry, their Opinion and Practice is the same in England. It hath been said, that no Harrow must be used after the Seed is sown: Nay it has been thought, that if the Ground was mellow, that the Seed would sink too deep in the Earth, and never come up. The constant Lesson was, take *Heed* you do not bury your Clover Seed too deep. But we have gone upon mistaken Principles: By Experience it is found, that the best Way is to plough in the Seed, that it will come up at full Furrow Depth; and this is the Practice in those Towns where they have raised Clover with great Success, and sold the Seed with such great Profit and Advantage.

A Farmer in this Town sows and ploughs in his Clover Seed in this Manner; has upon the Ground sundry Acres ordered in this Manner; it was sowed this last Spring, finds no Difficulty attending to it as to its coming up. Shallow sowing has been very detrimental, attended with great Loss when a dry Season follows soon upon it; whereas this Danger is prevented by sowing deep: I have sustained Loss by sowing shallow, which, according to the old Rule, was tho't necessary: There is the Loss of Seed, and Profit of the Land, which is more than the Loss of a Crop of Corn.

Besides the Advantage we have by deep sowing of Clover Seed, to secure it in Time of early Drought, there is another great Benefit arising from it: The Seed being lodged so deep, it will be well fed and nourished, it hath its Provision all round on every Side, meerly wallows in Wealth; so that the Grass grows strong and large. I suppose this to be the Reason that the Clover Seed which we have from those Parts where they bury their Seed so deep, is so much better than ours, which springs from shallow sowed Clover: Their Seed is a larger Seed than what we raise, and the Grass is larger too; Their Seed looks plump and well fed.

The supreme Ruler of the Universe takes Care of the whole Race of Mankind. His Goodness, in a special Manner, meets us with Instruction, and lays it in our Way, that we may find it in our ordinary Vocation, for our Profit and Improvements, whether it be Merchandise or Farming. *The Kingdom of Heaven is like unto a Merchant-Man seeking goodly Pearls, who, when he had found one Pearl of great Price, he went and sold all that he had and bought it.* Mat. xiii. 45, 46. Judah *shall plough, and* Jacob *shall break his Clods. Sow to yourselves Righteousness, and reap Mercy, break up your fallow ground, for it is Time to seek the Lord, till he shall come to rain Righteousness upon you.* Hosea x. 11, 12.

Commercial Advertisements, 1773

From *Boston Gazette and Country Journal,* November 22, 29; December 6, 1773.

Strayed or Stolen from the Subscriber on the Evening of the 12th lost, a sorrell Mare, with a Blaze in her Face, about 9 Years old, about 14 and half Hands high, Trots and Paces, and carries herself well, she had both her Fore Hoofs crack'd, she's something poor, and lame in one of her Fore-feet. Whoever shall take up said Thief and Mare, and return them to *Samuel Hide* of Newtown, shall have TEN DOLLARS Reward, upon Conviction of Theft: — Or if said Mare only is return'd, they shall have TWO DOLLARS Reward; and all necessary Expences allow'd, by Newton, Nov. 13, 1773. ELIPHALET MURDOCK

* * *

RAISINS

A few Casks of choice Raisins, just arrived and free from Sand,

TO BE SOLD
By Charles Miller,

At his Store in King-Street — ALSO — New Superfine and common Philadelphia FLOUR, Barr Iron, a few Casks Old Brandy, Lisbon Wine, a few Bales Cotton, of superfine Quality, &c. &c. &c.

* * *

TO BE SOLD

A likely Negro Girl about 8 Years old strong and hearty. Also a strong, hearty Negro Girl about 4 Months old to be given away.

Inquire Edes & Gill

A Survey of Agriculture, c.1769

From Anonymous, *American Husbandry* (London, 1775), Reprinted: Harry J. Carman, ed. (New York, 1939), pp. 52–54, 70–85, 89–96, 335–53.

Labor in New England.

Respecting the lower classes in New England, there is scarcely any part of the world in which they are better off. The price of labour is very high, and they have with this advantage another no less valuable, of being able to take up a tract of land whenever they are able to settle it. In Britain a servant or labourer may be master of thirty or forty pounds without having it in their power to lay it out in one useful or advantageous purpose; it must be a much larger sum to enable them to hire a farm, but in New England there is no such thing as a man procuring such a sum of money by his industry without his taking a farm and settling upon it. The daily instances of this give an emulation to all the lower classes, and make them point their endeavours with peculiar industry to gain an end which they all esteem so particularly flattering.

This great ease of gaining a farm, renders the lower class of people very industrious; which, with the high price of labour, banishes every thing that has the least appearance of begging, or that wondering, destitute state of poverty, which we see so common in England. A traveller might pass half through the colony without finding, from the appearance of the people, that there was such a thing as a want of money among them. The condition of labourers in England is far from being comfortable, if compared with their American brethren, for they may work with no slight diligence and industry, and yet, if their families are large, be able to lay up nothing against old age: indeed the poor laws are very destructive of any such provident conduct. Those laws have the effect of destroying prudence without giving an adequate recompense; the condition of the aged or diseased poor who depend on their support is in many cases lamentable; or at least much inferior to what their own previous industry would have procured them had they not been seduced by the idea of this worse than no dependance. And without extending our reflections to this part of their lives we may determine that the pay they receive for their work does not rise proportionably with the price of all their necessaries; the consequence of which is to them great oppression. On the contrary, the New England poor have no delusive poor laws to depend on: they aim at saving money enough to fix them into a settlement; their industry rarely fails of its end, so that the evening of an industrious life is universally that of a little planter in the midst of all necessaries. The public consequence of this may be easily deduced; it is a very high price of labour, and an amazing increase of people; since marriages must abound greatly in a country where a family, instead of being a burden, is an advantage.

I have more than once mentioned the high price of labour: this article depends on the circumstance I have now named; where families are so far from being burthensome, men marry very young, and where land is in such plenty, men very soon become farmers, however low they set out in life. Where this is the case, it must at once be evident that the price of labour must be very dear; nothing but a high price will induce men to labour at all, and at the same time it presently puts a conclusion to it by so soon enabling them to take a piece of waste land. By day labourers, which are not common in the colonies, one shilling will do as much in England as half a crown in New England. This makes it necessary to depend principally on servants and on labourers who article themselves to serve three, five, or seven years, which is always the case with new comers who are in poverty.

* * *

New York: The Climate, Soil, Productions, Husbandry, and Exports.

The colony of New York lies between latitude 41° and 44°, which tho' partly the same parallel as New England, yet is it attended with a different climate in some respects; but in every circumstance superior, since there are productions that will not thrive in New England, which do admirably here; not owing to the greater heat (for New England is as hot as New York) but to a better and more salubrious air. The spring in New York is earlier, and the autumn late; the summer is long and warm; indeed sometimes the heat is great, but rarely oppressive; the winter is severe but short; it is not so sharp as in New England, and they have in general a clear bright sky. In winter the snow lies deep, and for two or three months; and they travel on it in sledges both here and in New England, in the manner that is common in the northern parts of Europe.

Sometimes indeed the cold is extraordinary great; of which Dr. Mitchel gives an instance. By the observations, says he, made, in January 1765, by the masters of the college at New York, Fahrenheit's thermometer fell 6 degrees below 0, which is 21 degrees below 15, the greatest cold in England. Water then froze instantly, and even strong liquors in a very short time. And we are told it is not uncommon there to see a glass of water set upon the table, in a warm room freeze before you can drink it, &c.

The soil of the province is in general very good; on the coast it is sandy but backwards, they have noble tracks of rich black mold, red loam, and friable clays, with mixtures of these soils in great varieties; at some miles distance from the sea, the country swells into fine hills and ridges, which are all covered with forest trees, and the soil on many of these is rich and deep, an advantage not common in poor countries. The river Hudson which is navigable to Albany, and of such a breadth and depth as to carry large sloops, with its branches on both sides, intersect the whole country, and render it both pleasant and convenient. The banks of this great river have a prodigious variety; in some places there are gently swelling hills covered with plantations and farms; in others towering mountains spread over with thick forests: here you have nothing but abrupt rocks of vast magnitude, which seem shivered in two to let the river pass the immense clefts; there you see cultivated vales, bounded by hanging forests, and the distant view completed by the *Blue Mountains* [Catskill Mountains] raising their heads above the clouds. In the midst of this variety of scenery of such grand and expressive character, the river Hudson flows, equal in many places to the Thames at London, and in some much broader. The shores of the American rivers are too often a line of swamps and marshes; that of Hudson is not without them, but in general it passes through a fine, high, dry, and bold country, which is equally beautiful and wholesome.

In general the soil of this province exceeds that of New England: besides the varieties I have already mentioned, there is on Long Island sands that are made quite fertile with oyster-shells, a fish caught there in prodigious quantities: they have the effect of shell marle in Scotland. The productions of New York are the same in general as those of New England, with an exception of some fruits that will not thrive in the latter country; but almost every article is of a superior quality: this is very striking in wheat, of which they raise in New England, as I have already observed, but little that is good, whereas in New York their wheat is equal to any in America, or indeed in the world, and they export immense quantities of it; whereas New England can hardly supply her own consumption.

They sow their wheat in autumn, with better success than in spring: this custom they pursue even about Albany, in the northern parts of the province, where the winters are very severe. The ice there in the river Hudson is commonly three or four feet thick. When professor Kalm was here, the inhabitants of Albany crossed in the third of April with six pair of horses. The ice commonly dissolves at that place about the end of March or the beginning of April. On the 16th of November the yachts are put up, and about the beginning or middle of April they are in motion again. If wheat will do here in autumn, where the ground is sometimes frozen four feet deep, one would apprehend it would succeed even more to the north.

Wheat in many parts of the province yields a larger produce than is common in England: upon good lands about Albany, where the climate is the coldest in the country, they sow two bushels and better upon one acre, and reap from 20 to 40; the latter quantity however is not often had; but from 20 to 30 bushels are common, and this with such bad husbandry as would not yield the like in

England, and much less in Scotland. This is owing to the richness and freshness of the soil. In other parts of the province, particularly adjoining to New Jersey and Pennsylvania, the culture is better and the country more generally settled. Though there are large tracts of waste land within twenty miles of the city of New York.

Rye is a common crop upon the inferior lands, and the sort they produce is pretty good, though not equal to the rye of England. The crops of it are not so great in produce as those of wheat on the better lands.

Maize is sown generally throughout the province, and they get vast crops of it. They chuse the loose, hollow loams, or sandy lands for it, not reckoning the stiff or clayey ones will do at all for it: half a bushel will seed two acres and yield an hundred bushels in return: about Albany, where they have frosts in the summer, maize suits them particularly, because tho' the shoots are damaged or even killed by the frost, yet the roots send forth fresh ones. Maize, from the greatness of the produce, may easily be supposed a rich article of culture, and especially in a province that has so fine an inland navigation through it as New York. It is also of great advantage in affording a vast produce of food for cattle in the winter, which in this country is a matter of great consequence, where they are obliged to keep all their cattle housed from November till the end of March — with exception indeed of unprovident farmers, who trust some out the chief of the winter, to their great hazard.

Barley is much sown in all the southern parts of the province; and the crops they sometimes get of it are very great, but the grain is not of a quality equal to that of Europe. They make much malt and brew large quantities of beer from it at New York, which serves the home consumption, and affords some also for exportation. Pease are a common article of culture here, and though uncertain in their produce, yet are they reckoned very profitable; and the straw is valued as winter food. Thirty bushels per acre they consider as a large crop, but sometimes they get scarcely a third of that. Oats they sow in common, and the products are generally large; sixty bushels an acre have been known on land of but moderate fertility. Buckwheat is every where sown, and few crops are supposed to pay the farmer better, at the same time that they find it does very little prejudice to the ground, in which it resembles pease.

Potatoes are not common in New England, but in New York many are planted; and upon the black, loose, fresh woodland they get very great crops, nor does any pay them better if so well, for a the city of New York there is a constant and ready market for them; I have been assured that from five to eight hundred bushels have been often gained on an acre.

There are many very rich meadows and pastures in all parts of the province; and upon the brooks and rivers, the watered ones (for they are well acquainted with that branch of husbandry) are mown twice and yield large crops of hay. In their marshes they get large crops also, but it is a coarse bad sort; not however to a degree as to make cattle refuse it; on the contrary, the farmers find it of great use in the winter support of their lean cattle, young stock, and cows.

The timber of this province consists chiefly of oak, ash, beech, chestnut, cedar, walnut, cypress, hickory, sassafras, and the pine; nor is there any preceptible difference in their value of the wood here and in New England; though it declines, for ship building, when you get further to the south; with some exceptions however, for there are other species of trees even in the most southern colonies that are equal to any for that purpose. New York not being near so much settled as New England, timber is much more plentiful, so that the planters and new settlers make great profit by their lumber. Upon most of the streams that fall into the river Hudson, there are many saw mills for the mere purpose of sawing boards, planks, and other sorts of lumber, which goes down in immense quantities to New York, from whence it is shipped for the West Indies. We shall by and by see that this is a very great article in the profit of every planter. Among all the woods of this province, are found immense numbers of vines of several species, and quite different from those of Europe, some of the grapes resembling currants rather than ours. Wine has been, and is, commonly, made of them, but of a sort too bad to become an article of export.

Hemp is cultivated in all parts of the province, but not to a greater amount than their own consumption: flax is however a great article in the exports; it succeeds extremely well, and pays the farmer a considerable profit. Linseed oil is another article of export, the seed for which is raised by the planters; but more is exported unmanufactured. Turneps also are grown in large quantities, and by some planters upon a system much improved of late years. The fruits in this province are much superior to those in New England; and they have some, as peaches and nectarines, which will not thrive there. Immense quantities of melons and water-melons are cultivated in the fields near New York, where they come to as great perfection as in Spain and Italy; nor can it well be conceived how much of these fruits and peaches, &c. all ranks of people eat here, and without receiving any ill consequence from the practice. This is an agreeableness far superior to any thing we have in England; and indeed, the same superiority runs through all their fruits, and several articles of the kitchen garden, which are here raised without trouble, and in profusion. Every planter and even the smallest farmers have all an orchard near their house of some acres, by means of which they command a great quantity of cyder, and export apples by ship-loads to the West Indies. Nor is this an improper place to observe that the rivers in this province and the sea upon the coast are richly furnished with excellent fish: oysters and lobsters are no where in greater plenty than in New York. I am of the opinon they are more plentiful than at any other place on the globe; for very many poor families have no other subsistence than oysters and bread. Nor is this the only instance of the natural plenty that distinguishes this country: the woods are full of game, and wild turkies are very plentiful; in these particulars New York much exceeds New England.

These upon the whole are circumstances which contribute much to the plenty and happiness of living in this country; and among other causes, contribute very greatly to the plenty and general welfare of all ranks of the people, nor should I here omit making some observations on the state of the settlers and other inhabitants.

To what causes it is I know not, but New York is much less populous than New England to the north, and Pennsylvania to the south: there is no circumstance that results from nature, or from the government of the province that can account for this; but to whatever cause it may be owing, certain it is, that we ought to esteem it as fortunate for such persons as now chuse to settle there. There are vast tracts of unpatented land yet remaining on the river Hudson and its branches, which abound in every beneficial circumstance that can render a new country desirable to settle in.

This however, will not, in all probability last long, for the new settlements increase every day; so that in a few years there will not be many such spaces abounding in wood and navigable water unoccupied.

But there is one mistake made by most new settlers, especially on the river Hudson; they have in general an idea that the only good soils are the deep black loam, or clays; and accordingly reject all the tracts that consist of a thin reddish loam on rock; but I have been assured by some intelligent gentlemen, that experiment has proved this soil, though so thin, fertile to a great degree in most of the productions which are common in the whole province: they have mentioned particularly, barley, pease, potatoes, turneps, clover, and even wheat. And as a confirmation that this opinion was just, I was favoured with the following particular of the produce of a new field of this soil, which having been rejected by several new settlers, was planted by the person to whom I am obliged for this intelligence. The piece of land contained sixteen acres, the soil a light thin loam, of a reddish colour, on a lime-stone rock.

First Year
Grubbed, ploughed, and prepared for potatoes, and planted without dung: produce 11,000 bushels, which were sold at 10d. per bushel, which is 453£.

Second Year
Ploughed once, and sown with wheat, produce 512 bushels, which sold for 85£.

Third Year
Planted again with potatoes, produced 8,496 bushels, which sold for 10d. a bushel, or 354£.

Fourth Year
Sown with wheat again, produce 600 bushels, which sold for 120£.

Fifth Year
Sown with barley, produce 730 bushels, which sold for 73£.

Sixth Year
Ploughed once, and sown with pease, the produce 630 bushels, which sold for 53£.

With this crop of pease clover was sown, and left an excellent pasture, which was reckoned as profitable as any other piece of land in the whole plantation.

First year ....£.	453
Second do.	85
Third do.	354
Fourth do.	120
Fifth do.	73
Sixth do.	53
Total £.	1,138

Which is near £. 11 15 0 per acre per annum.

Now upon this account I have several remarks to make, which I think important, as it shews what may be done in this country, by good husbandry, even when no manure is used. The reader doubtless observes, that the *system* of management in this field ran upon the principle of an intervening crop of potatoes or pease between every two of wheat and barley. This is the husbandry which I would always recommend, but which is diametrically opposite to the practice of the New York planters; who make not the least scruple of taking six or eight crops successively of maize, wheat, rye, barley, or oats, without ever thinking of the least necessity of introducing pease, buckwheat, turneps, clover, or any other plant which in its nature or culture would prove a preparation for corn. The idea exemplified in the preceding sketch shews quite a different conduct.

* * *

The same gentleman to whom I am indebted for the preceding account, gave me another of the expences and product of a considerable plantation on the river Hudson. This I shall insert with pleasure; for such accounts are what I have most aimed at gaining for all the colonies, not always with success indeed; but it is only from such that we can form a just idea of the advantages of American husbandry. Such accounts of agriculture in Europe are common in numerous books, while the management and state of the agriculture of the colonies has been little attended to, for which I am clear no good reason can be assigned.

The plantations in question consisted of 1,600 acres, situated partly on the banks of the river Hudson and partly on each side of a small river that runs into it; the purchaser was not the first settler, for the land was marked out, a house built, and some offices, with a small tract of land cleared: nothing, however, was done either expensively, or with good judgment; and the place was in a state of neglect when purchased. The price was 370£.

A small saw mill, and additional offices were built on it immediately, which with some other improvements, of no great amount, came to 260£.

Eight hundred acres were grubbed, and the trees sawn and rived into plank, board, shingles, and staves: the whole expence of which was 1162£. Many of the trees were oak and elm, of great size; also some limes of extraordinary growth.

Eight new inclosures were made, the fences, posts, and rails and ditching, with all expences, came to 32£.

The stock fixed on the plantation was as follows:

Eight negroes, at 34 £ ....£.	272
Four indented servants, at 11£. each, for 3 years....................	132
Two hired by the year at New York, 12£.............................	72
Three German emigrants, at 9£.......................................	81
Servants, provisions, and cloathing for negroes, besides	
what produced..	56
Implements of husbandry, expences of, exclusive of timber........	87
Salary of overseer, 3 years	110
Seed for the first crop..	90
Sundry expences	113
Cattle	230
Provisions, &c. for 3 years..	300

£. 1,543

The produce of the three years, in various articles, came to the following sums.

Lumber

	£.	s.	d.
17,000 feet of boards, at 5£. 2s. 6d. per 1,000...........	87	2	6
970 plank, at 3s. 8d.	177	16	0
220,000 singles, at 12s. per 1,000	132	0	0
60,000 staves, at 4£. 10s. per 1,000.........................	270	0	0
260 pieces of timber, at 7s. 6d.	97	0	0
Sundry articles of various kinds................................	187	10	0
	951	8	6

Recapitulation

	£.	s.	d.
Purchase	370	0	0
Saw mill, &c.	260	0	0
Clearing, 800 acres	1,162	0	0
Eight inclosures..	32	0	0
Stock	1,243	0	0
Provisions	300	0	0
Total	3,367	0	0

The annual after expence was:

	£.	s.	d.
Interest of capital..	168	7	0
Repairs...	12	0	0
Inclosing..	10	0	0
Negroes	16	0	0
Servants wages..	135	0	0
Implements..	13	0	0
Sundry expences..	100	0	0
	454	7	0

Produce 1st year

	£.	s.	d.
4 acres of potatoes, 260 bushels per acre, 1,040 bushels at 8d.	34	13	0
82 acres Indian corn, 30 bush. per acre 240 bushels at 1s. 6d..	184	10	0
10 acres pease, failed...	0	0	0
22 of wheat, 22 bushels per acre, at 3s.........................	72	12	0
	291	15	0

Produce 2d year

	£.	s.	d.
6 acres potatoes, 200 bushels per acre,			
1,200 bushels, at 10d.	50	0	0
135 acres Indian corn, 32 bush. per acre,			
4,320 bush. at 1s. 6d.	324	0	0
90 acres wheat, 20 bush. per acre 1,800 bush. at 3s.	270	0	0
40 acres pease, 15 bush. per acre 600 bush. at 1s. 3d.	37	10	0
40 acres barley ⎫			
2 do. potatoes ⎪			
16 turneps ⎬ for plantation	0	0	0
35 oats ⎪			
32 clover ⎪			
20 Indian corn ⎭			

416 acres in culture.   £.   681   10   0

Produce 3d year

	£.	s.	d.
8 acres potatoes, 300 bush. per acre,			
2,400 bush. at 10d.	50	0	0
170 acres of Indian corn, 35 bush. per acre,			
5,950 bush. at 2s.	595	0	0
60 acres wheat, 16 bush. per acre, 960 bush. at 3s.	144	0	0
80 acres pease; 40 failed, 40 at 10 bush.			
400 bush. at 1s. 3d.	25	0	0
Cattle	87	10	0
150 acres clover ⎫			
2 potatoes ⎪			
20 barley ⎪			
20 oats ⎬ for plantation	0	0	0
10 Indian corn ⎪			
2 wheat ⎪			
38 turneps ⎭			

560 acres in culture.   £   901   10   0

	£.	s.	d.
First year	291	15	0
Second	681	10	0
Third	901	10	0
Lumber	951	8	6
	2826	3	6

	£.	s.	d.
Capital	3,367	0	0
Product first three years	2,826	3	6

Remains...	540	16	6
Three years interest...	504	0	0

£. 1,044 16 6

800 acres were soon in culture, which were usually employed in the product of

8 acres potatoes...£.	50
100 do. Indian corn	300
100 do. wheat..	300
40 do. pease	60
400 do. clover	
20 do. barley	
20 do. oats	
80 do. turneps	
30 do. sundries, including orchard and yielding	
Cattle	200
Fruit...	25
Annual lumber..	60
	995
Expences	454

Profit £. 541

This profit is besides the annual improvement of waste from which the lumber is cut, and also the advantage of the surrounding wastes, which are granted as fast the family increases, but which will not admit of calculation, because wastes are converted to profit merely in proportion to the ability, that is the money, of the planter.

The first observation I shall make on this account, is the lumber paying nearly the expence of clearing, which is an high advantage, and certainly owing to the expedition of the saw mill: in many parts of the northern provinces, where a saw mill is not used, the expence of clearing is infinitely the greatest part of a new settler's work. But it is plain from every article in this account, that the great advantages in settling is the command of a large sum of money, that the planter may go spiritedly to work and make his ground produce him something considerable immediately, which can never be done if he has not money enough to clear away the woods speedily. Half this capital I am clear would not yield a proportionable profit; on the contrary, it might not afford half such interest for the amount. It is by means of this advantage that near two thirds of the whole expenditure is repaid by the product of the three first years, which would be far enough from the case, if the sum of money at the beginning of the undertaking had been much less. If the planter's time and trouble, for three years, be not reckoned, as indeed it need not in reason be, then the sum of 1,044£. might be reckoned the original capital, which would make the annual profit on the undertaking immensely great.

But the great superiority of the account of this improvement, over those that can be made in the cultivated parts of Europe, is the *increase* of cultivation.

The account here is stated at 800 cultivated acres, and 60£. a year from lumber; but this does not include the annual increase of the cleared land, which may be carried on as fast as the planter's money will allow. Instead of 6£. a year in lumber, it might be 2 or 300£. by having hands enough, and the land, when cleared, all brought into culture at the same profit as the first 800 acres; the quantity of land to be had does not stop, let him be as able and industrious as he will. This advantage I think greater even than the making the profit above stated.

*   *   *

The account of a settlement given above, is not to be supposed a picture of the profit which every one makes by going to New York; I would on no account have it imagined that this is the case: this was executed by means of a large sum of money, for so 3,000£. must be reckoned in America; and not only by a sum of money, but also by the exertion of much better husbandry than is common in the colonies. So far from every settler making a profit like this, not one in forty equals the proportion of it. In general, the settlers come with a small sum of money, very many of them with none at all, depending on their labour for three, five, or seven years to gain them a sum sufficient for taking a plantation, which is the common case of the foreign emigrants of all sorts. It is common to see men demand, and have, grants of land, who have no substance to fix themselves further than cash for the fees of taking up the land; a gun, some powder and shot, a few tools, and a plough; they maintain themselves the first year, like the Indians, with their guns, and nets, and afterwards by the same means with the assistance of their lands; the labour of their farms they perform themselves, even to being their own carpenters and smiths: by this means, people who may be said to have no fortunes are enabled to live, and in a few years to maintain themselves and families comfortably. But such people are not to be supposed to make a profit in cash of many years, nor do they want, or think of it. And as to the planters who begin their undertakings with small sums of money, though they do better, and even make a considerable profit by their business, yet are they very far from equalling what I have now described; this is for want of money, for I might add that not one new settler in a thousand is possessed of a clear three thousand pounds.

The conclusion which I deduce from these particulars is that new settlements in New York are undertaken to good advantage, profit in money considered, only by those who have a good sum of money ready to expend; and by this term, I mean particularly men who have from two to five thousand pounds clear; in Britain such people cannot from the amount of their fortune get into any valuable trade or manufacture, unless it is by mere interest, or being related to persons already in trade. But it is evident that in New York they may, with such a sum of money, take, clear, stock, and plant a tract of land that shall not only amply support them in all the necessaries of life, but at the same time yield a neat profit sufficient for the acquisition of a considerable fortune.

I shall next lay before the reader the exports of this province as taken on an average of three years since the peace.

Flour and biscuit 250,000 barrels, at 20s.	£.	250,000
Wheat 70,000 qrs.		70,000
Beans, pease, oats, Indian corn and other grains		40,000
Salt beef, pork, hams, bacon, and venison		18,000
Bees wax 30,000 lb. at 1s.		1,500
Tongues, butter, and cheese		8,000
Flax-seed, 7,000 hhds. at 40s.		14,000
Horses and live stock		17,000
		————
Product of cultivated lands		418,500
Timber planks, masts, boards, staves, and shingles		25,000
Pot ash, 7,000 hhds.		14,000
Ships built for sale, 20, at £. 700		14,000
Copper ore, and iron in bars and pigs		20,000
		————
	£.	526,000

Let me upon this table observe that far the greater part of this export is the produce of the lands including timber; and even the metals may be reckoned in the same class; this shews us that agriculture in New York is of such importance as to support the most considerable part of the province without the assistance of either the fishery or of commerce; not that the city of New York has not traded largely, perhaps equal to Boston, but the effects of that trade have been chiefly the introduction of money by the means of barter, besides the exportation of their own products: whereas New England's exports consist five parts in six of fish, and the other products of the fishery; a strong proof that agriculture is far more profitable in one country than in the other; for settlers in colonies will never take to the sea, in a country whose agriculture yields well; but in very bad climates, and such as destroy instead of cherishing the products of the earth, any branch of industry pays better than cultivating the earth. This is a distinction that ought to be decisive with those who have a choice to make which of these colonies they will go to; for men do not usually settle themselves in countries where they are to make their livelihood by encountering a boisterous sea and leading a life of perpetual hardships and violent labour: This is very different from the employment of those who support themselves in so fine a country as New York, by agriculture.

The rural management in most parts of this province is miserable: seduced by the fertility of the soil on first settling, the farmers think only of exhausting it as soon as possible, without attending to their own interest in a future day: this is a degree of blindness which in sensible people one may fairly call astonishing. The general system is to crop their fields with corn till they are absolutely exhausted; then they leave them what they call fallow, that is, to run to weeds for several years, till they think the soil has recovered somewhat of its fertility, when they begin again with corn, in succession, as long as it will bear any, leaving it afterwards to a fallow of weeds. If no spontaneous growth came, but such as cattle would freely eat, the evil would not be great, because then the land would not have more to support than it would gain by the dung, &c. of the stock supported. But the contrary is the case: an infinite quantity of rubbish comes which no beast will touch, this seeds the land in so constant a succession that the soil is never without a large crop on it. The extent to which this practice is carried would astonish any person used to better husbandry; it is owing to the plenty of land; the farmers, instead of keeping all their grounds in good order, and a due succession of valuable crops, depend on new land for every thing, and are regardless of such management as would make their old fields equal the value of the new ones.

Instead of this, the New York farmers should imitate the conduct of those of Britain: they should never *exhaust* their lands; and when they were only *out of order* they should give them what ought to be esteemed the most beneficial fallow; that is, crops which, while growing, receive great culture, at the same time that they do not much exhaust the soil; such as all sorts of roots, and pulse, and every kind of leguminous plant, with the various kinds of clovers. By introducing these in proper succession, the land is never exhausted. In the remarkable instance given of a plantation manged on this system, we find a crop of this nature introduced between every two of maize, wheat, barley, or oats; and in every round of the system, several years under clover, which is instead of the fallow of weeds of the generality of the New York farmers.

The benefit of pursuing this plan is very great; for the lands, when laid down to clovers, maintain more cattle on fifty acres than with weeds they would on four hundred; this quantity of cattle improves the ground by the summer feeding and enables the farmer to raise great store of manure in the winter, by which means his crops of corn, &c. are by much more abundant. It further keeps the whole plantation in a state of profit; whereas in the common method only a part, and that not the largest, is valuable at once, his dependance for product being only on the new broken uplands.

Another part of husbandry, in which the New York farmers are very defective, is the management of their meadows and pastures: they make it a rule to mow every acre that is possible for hay; and as long as they get a tolerable *quantity,* they are strangely inattentive to the quality; weeds, rushes, flags, and all sorts of rubbish they call good hay, and suppose their cattle have not more sense in distinguishing than themselves. This is owing also to their grasping mere extent of land, and caring but little for the good husbandry of it. Many of their meadows are marshes, which, with little trouble, might be drained, and at once improved prodigiously, yet are such undertakings very seldom set about: others of the upland sort are equally filled with various weeds, from the slovenly manner in which they are laid down, or left to clothe themselves; but the appearance of these do not at all startle men whose ideas of agriculture are so little polished.

In respect to the management of cattle, and the raising manure, the farmers of New York are equally inattentive with their neighbours of New England.

I before observed, that vines of several sorts grew spontaneously in all the woods of this province; and that wine, though bad, had been made of them: their being bad has no weight with me, since wild vines in no part of the world produce good wine; but if they would plant vineyards of them, and cultivate them with the same care as is taken in wine countries, I have no doubt but they would produce excellent wine. Some endeavours have been made in this branch, by several patriotic persons in this province; but they have all been on the scent of bringing vines from other countries, scarce any of which ever thrive, and some of them will not live: the frosts are so excessive cold in winter that these foreign vines, used to so different a climate, either come to nothing or produce a grape very different from what they do in their own country. The instance of the great success with which the Dutch planted French vines at the Cape of Good Hope proves nothing in this case, because the climate in general, at that Cape, is not only one of the finest in the world, but the winters are mild, and in every respect different from the peculiar climate of North America.

But good culture and a proper choice of a high dry situation, of which there are plenty in this province, and even rocky ones, would in all probability be attended with success, and make these native grapes yield a wine that would add infinitely to the value of the exports of the province. This is an object of too much importance to be left to the dilatory proceedings of the planters themselves; they are in general engaged in a plain line of husbandry, from which most of them have not the capacity or knowledge to deviate, and the rest want money for it. But the government should order a vineyard to be planted under the direction of an overseer skilled in this branch of agriculture, and also by the same means take care that it was cultivated in perfection. The expence of this would not be great. . . .

* * *

Georgia: The Climate, Soil, Productions, and Exports.

Georgia is in many respects the same country as Carolina, differing very little in climate, but generally in favour of it. Upon the coast it is not above sixty or seventy miles from north to south; but in the internal country the distance is upwards of one hundred and fifty miles. The climate upon the coast is hot, damp, and unwholesome, like Carolina, though there are hilly spots which form strong exceptions. The flat country extends in general about two hundred miles from the sea, and the interior tract, which reaches from thence to the Apalachian mountains and is about one hundred miles broad, ranks in every respect among the very finest in all America. The climate at the same time that it is hot enough to produce the most valuable staples is healthy and agreeable to an extraordinary degree; free from those sudden changes and violent extremes that are felt in the maritime part of the province and which are so pernicious to health wherever found. In this country the soil is of a fertility that even exceeds the back parts of South Carolina, especially on the river Savannah and its branches to the west and north-west of Augusta, and indeed all round that town: no flat lands are found, no swamps, no marshes, but high, dry tracts, waving in gentle hills, and the vales watered with numerous streams. The soil [is] a deep black loam, so rich that there is scarcely any exhausting its fertility: it was for a long time unknown that such a country existed here; but upon first settling Georgia, and for many years afterwards, the flat sandy coast was the only part of the province attended to or known; and as long as this has been the case with any of our colonies to the south of New York, they have languished, and emerged from their languor as soon as they penetrated into the rich and healthy part of the country. Georgia was a very inconsiderable province as long as the people confined themselves to the coast; but the efforts since made here have been by means of removing backwards, where silk, indigo, and other commodities of great value are cultivated with a success far greater than was ever found in any of the maritime parts of our colonies. But however clear the excellencies of these interior parts of Georgia may appear, to such as have viewed them with an understanding eye, yet are they not one tenth peopled: it is but a few years since any attention at all has been given to this province by American or European settlers; but after the arrangement of the American governments in 1763 had confined all the colonies to as narrow bounds as the encroachments of the French before the war and their operations in it, they then found good land, unpatented, scarce; which pushed them upon a more industrious search: this was the cause of Georgia receiving since the peace such an accession of people that settled in numbers in the back parts of it; and it was the same cause that contributed to the peopling several districts of North Carolina, which had been long neglected.

The soil and face of the country in the maritime part of the province resembles South Carolina; it consists of a flat territory, very sandy, and in general either pine barrens or swamps; the slips of oakland are not large or numerous: the swamps are inferior to those of Carolina in the production of rice, and in general the country is not so good for the whole breadth of the flat part; but this inferiority is not great; all the sea coast of America, from Jersey to Florida, has a strong similarity.

The vegetable productions both of trees, shrubs, roots, flowers, &c. are the same as those of South Carolina; nor is there much difference in the growth, for though Georgia lies to the south of that province, yet is the climate not hotter than that of Carolina; and there are some parts of the latter, particularly Charles Town, much hotter than most in Georiga. Relative to a further account of the soil, climate, and products of this province, I shall here insert an extract from a letter written by a planter who went from England and settled not far from Augusta, and has resided there eight years.

I must freely own, that in some instances I was much disappointed in my expectations of this country — I thought the soil had been more generally good, and the climate I was taught to imagine was more agreeable to an English constitution; but in summer I find the heat very oppressive, and gives one, for two or three hours in the afternoon, a languor which I never experienced in England even in the hottest days; going out is then disagreeable, and the only way to be tolerably at ease is to keep one's self perfectly quiet, to sit still in rooms that admit much air but no sun, and to be cautious in diet: this season lasts through July, August, and most of September. The way to enjoy the agreeable parts of any avocation during these months, is to rise early in the morning and to transact whatever business requires your being abroad, by eleven o'clock, or at most by twelve (unless the days are cloudy), and then to keep the house till five in the afternoon; in the evening the air is cool enough to render the fields pleasant. What I have now told you is not general with all constitutions; I have a servant who came from England with me that feels no more inconvenience from being exposed through the heat of the day to the sun than the very negroes themselves, who generally delight in the most meridian beams; and among my neighbours, I know two or three who are of the same temperament. But my own constitution is very different, for a fever would be the least consequence I believe — and indeed I partly know it from experience — which would ensue from my using any fatiguing exercise from one to three o'clock in the afternoon in summer, when the sky is clear; for with a south wind the sun's beams are so intensely hot that the only pleasure I feel is to be perfectly at rest.

But at the same time, Sir, that I describe these inconveniencies, let me remark that I should use very different terms if I lived near the coast: I have been often at Savannah, when I have longed ardently to be at home; the climate is there beyond comparison worse than at Augusta, and the farther west we go, the better it becomes. This will doubtless appear very strange to you, as there is so little difference in the latitudes of these places; but that is a circumstance which has little to do with climate in this part of the world. I attribute the great contrast there is between the sea coast and the western part of the province, to the flatness of one, and the varied surface of the other — also to one being full of swamps and marches, and the other being entirely free from them. Flat countries have always less wind and agitations in the air, which render it far more pure and wholesome to breathe in, as they carry off speedily every noxious quality; there is scarce an instance on the globe, of a hilly or a mountainous country being unhealthy; even under the line or the tropics such are always inhabited by a hardy and robust race of people. The other circumstance is of yet more consequence; the effluvia of stagnant waters in a hot climate, and especially of such as rice-swamps — which are shallow, sometimes fields of mud, at others thinly covered with water — cannot but prove prodigiously injurious to the health of the human body, at the same time that it renders the heat not only burning, but close and suffocating: such a thick heavy atmosphere, in a country so flat as not to be windy, must necessarily make the maritime part of this province far more hot than the internal part, creating a difference greater than what many degrees of latitude could occasion.

In the observations have made on the climate's being uncommonly hot, I confine myself entirely to the hottest part of the summer, July, August, and part of September, and perhaps, but not always, a week the latter end of June. As to the rest of the year, you have no idea in England of the charms of this climate at a distance from the sea. March, April, May, and June, are a warm spring, in which scarce a day offends you: the sky is a clear expanse, clouds rarely to be seen, and the heat nothing offensive; the beauty of our country is then enjoyed every hour of the day — in short, no season in any part of the world can hardly be more agreeable than these months in the back country of Georgia. The latter part of September and October are also perfectly agreeable, in being sufficiently warm without a melting heat. But this is not all I have to say in our favour, for to me the winter is a most pleasing season here; the degree of heat is that of a warm spring, with some days as hot as a common summer; but in some months in the latter country I have felt days as hot as are generally experienced in Jamaica: ideas of heat should not therefore be taken from the height to which the thermometer rises in certain days, but to the mean height when every day is registered. In the winter season, and also in spring, we have extreme cold winds, particularly the north-west; and also sharp frosts; but the sudden changes from heat to cold, which are so much complained of on the coast, are rarely felt with us in any such degree as is common there. I have heard many of my neighbours complain of these frosts and cold winds, but to an European constitution they are natural, and are certainly wholesome whenever the changes are not sudden from heat to cold; and even in that case they are better than constant heat, if any caution is used in dress. I must say for my own part, that neither frost nor wind have ever proved disagreeable to me; and that upon the whole, I much prefer the climate to any in which I have lived before; and yet I have resided at Cadiz, Naples, and the West Indies, not to speak of Boston and England. But whatever I have mentioned on this head is relative only to the country to the west of Augusta, that is, the western half of Georgia; for the other part does not by many degrees enjoy so good a climate.

And another letter was written as follows.

The soil in this neighborhood is good; in general, we have very little that is bad; and none that will not produce some useful crop or other. Like you in England we have wastes uncultivated, which spread almost over the whole province; but our wastes are such only for want of people to accept their property, whereas yours are such from being of a poor and almost worthless soil. I have travelled over parts of Scotland, and even the northern counties of England, which carry such an aspect of barrenness, and are so dreary and waste, as nothing in all this country can be opposed to them. All land here that is uncultivated is either a very rich and valuable forest or a meadow, which in its natural state would be worth ten or twelve shillings an acre in England. On the coast they have swamps which produce nothing, tho' not many, but here swamps are rare; the low grounds on some of the rivers are more properly marshes; they are small, and such as are found in the best and most beautiful counties of England; low meadows on rivers, wherever they are found, were (in a state of nature) marches: we have none but what might easily be drained, and would then be the richest meadows in the world, especially if kept as watered ones. Even these marches are with us found full of tall and beautiful cedars and cypresses.

Our flat tracts, or more properly the surface of gentle waves of country, rather than levels or hills, are of a rich loamy soil; the surface from twelve to eighteen inches deep of a fine light, black, sandy loam, which as the appearance of being the earth which has been formed by the rotting of vegetables; and yet, which is extraordinary, we have this soil where no trees are found. Under this loam we find another of a reddish brown colour, three, four, or five feet deep, and then meet with clay and in some places rock; this under-stratum of loam has the appearance of being admirable land. Other tracts of this sort, and especially the sides of hills, are covered with a reddish loam with many stones in it, from one to two feet deep, and under it rock: the appearance of this soil is not so good, but on experience we find it to be very fertile. In some vales, between gentle hills, we find the black loam three or four feet deep, a soil which I am persuaded might be applicable to any purpose in the world. A true clay on the surface is scarcely ever found in these tracts; but in the low lands on the river sides the soil is a very strong loam, near the clay. Some of the rivers, however, run among the hills, with high rocky shores. For 150 miles from the sea the country abounds with what they call a *pine barren,* which is a light white sand, very poor, covered with pines; it is reckoned the worst part of that country; we are not entirely without it; here and there is a pine barren, but they are rare. There are other varieties of soil, but not in considerable quantity; we have sandy tracts, which though light are very rich, and of a nature entirely different from the pine barren sand. Some spots on the rivers and the highest hills are rocky, and so rough as not to admit of culture; but these are covered with forest trees, and add very much to the beauty of the country.

Uncultivated tracts of country in this part of America are very different from such in other parts of the world; the plenty of the finest timber is astonishing to an European upon his first arrival. We have several sorts of oak which come to a prodigious size, twice or thrice as large as oaks in England; and some of these are much more excellent for ship-building than is commonly imagined: an injudicious choice of the sort of oak was for some years the cause of this idea; but later trials have diffused a more correct knowledge of the value of our timber, for some has been found superior in duration even to British oak. This wood is also cut in various articles of lumber, which are exported to the West Indies; pine, cypress, and cedar are likewise appropriated to the same use: this a vast advantage annexed to those parts of the country which have a good water-carriage, since these sorts of woods converted into lumber will pay the expences of clearing the thickest forests in this country, even if a proportion of the timber be of other sorts and not used in building: when this is the case therefore, a man enters not only into the possession of an estate without expence, but even an estate that is ready to cultivate.

Our forests are generally open, consisting of large trees, growing so thin that you may generally ride through every part of them, rarely having any under-wood, and in some tracts they are wide enough for waggons to pass every where: the labour, therefore, of clearing, when the wood is not of a proper sort for lumber, is not great. We have immense numbers of wild mulberry trees, upon the leaves of which we feed our silk-worms, without forming any plantations for that purpose. Walnuts and hickories are also very plentiful upon the best lands, and grow to a very great size.

A third letter contained the following particulars.

My plantation is situated on a small but navigable creek, which falls into the river Savannah, about thirty miles west of Augusta; when I first came, I had a very large tract of country to chuse in, for the settlements to the west of that town at any distance from it were not numerous; had I then been as well acquainted with agriculture as I am now, I could have made choice of a plantation, consisting more entirely of rich land; but as it is, I have no great reason to complain, and what I lose in soil, I gain in the extreme beauty of the situation.

My house is on the side of a hill; behind it is a fine spreading wood of oak, walnut, hickory, &c. before it a large tract of grass which I have cleared, and which is bounded by the river, whose course I command from almost every window, for three miles on each side: on the other side of it, and all round the lands adjacent to my house, are the fields which I have in culture. The whole plantation, which is my property, consists of 6,340 acres, at least, in the rough manner in which the surveyor-general's people reported the survey, that is, the quantity registered; the only fence which surrounded it for some years was trees marked with a hatchet, or crosses dug in the meadows, with here and there a post set up; but other settlers having since fixed near me, who have taken up small grants of land, their fences have been made in some places in my boundary line, which have saved me the trouble. In some places I am yet open to the country not granted away; for this tract of land containing a large proportion of excellent soil, I paid not quite one hundred pounds, including every charge and fee incurred in order to procure it.

The method here taken is for the person who wants land to fix upon a spot and take what he likes, under condition of peopling it in a given number of years: I had twenty allowed me, but they are now giving only ten or fifteen years. It is not common to see people fixing by each other; they generally plant themselves at a distance, for the sake of having an uncultivated country around them for their cattle to range in: all the country not granted away belongs to the king, and is [held in] common for every man to turn his cattle upon, but not in the manner such right is enjoyed in England, where the same thing is done not by permission but by right; for here every new comer has a liberty of fixing in this common part of the country, and inclosing his property immediately if he pleases; so that the lands on which we turn our cattle farther than our own bounds, are continually decreasing: the consequence is, the planters who on this account have not range enough for their large stocks, take up new grants of small quantities of land farther to the westward for the sake of sending their cattle thither, by which means they are enabled to keep very great stocks, even so far as a thousand head. I have four hundred and forty head of cows, oxen, bulls, heifers, &c. but their value is far from what it would be in England.

The plenty of timber in this country is a great advantage to new settlers, in rendering their buildings and many of their utensils of no other expence than that of labour, tools, and a little iron. My house, a barn, a stable, and some other conveniencies cost me no more at first than one hundred and seventy-four pounds in cash and the labor of ten negroes during three months; this was done by hiring carpenters and paying them by the month; and two of the slaves learnt so much of the art in that time, that by working since with them occasionally, they are become good carpenters enough to raise a shed, or build any plain outhouse, such as you see common in England in little farmyards: our woods is of so little value, that their making waste is of no consequence. I have made many additions to my house since, at a small expence, so that it is now a very convenient and agreeable habitation.

When I consider that for one hundred pounds a man may in this rich and plentiful country buy an extensive tract of land; that for two more he may raise a good house and offices; that he may buy slaves for thirty or forty pounds apiece, or hire white labour a very little dearer than in England; and that he may settle himself with as few as he pleases, and increase them as he can; when this is considered, it surprises me to think that more people of small fortunes do not come among us, but that they should prefer the narrow way in which they must live in Europe. The plenty of this country is much greater than you can think of; a little planter, that is a good gun-man himself, or has a slave that is so, may in half a day kill much more game than two families will eat in a week; and in parts of the country where it is comparatively scarce, an easy walk will yield him a day or two's subsistence of this sort for a moderate table. By game we here understand deer, rabbits, wild kine, and wild hogs, turkies, geese, ducks, pigeons, partridges, teal, &c. Our rivers are equally abounding with excellent fish, which is an advantage not inferior to the other; and the two together in hunting, shooting, and fishing, affording a diversion equal to what is met with in any part of the world, and superior to most. Sporting here is carried on with unlimited freedom, and in a style far superior to what I have any where else met with; and whoever keeps house in this country must presently find the immense advantages attending the great plenty of these articles, which reduce all expences of this sort very low. And now [that] I am giving you the information you want on this head, I shall add that our great plenty of fruit is another point in which this country is very fortunate; we have melons, cucumbers, water-melons, peaches, pears, apples, plums, &c. &c. in any quantity we please, almost without trouble or culture. The climate is so favorable that to plant them is all the attention requisite. Upon a new settler fixing, one of his first works is to inclose and plant a large orchard. Peaches are the most plentiful of any kind of fruit: a stone set, becomes a bearing tree in three years, and the fruit that drops from it rises in young trees; so that a single tree would become a wood of peaches in a few years, if they were not grubbed up.

You see, Sir, from this account, upon the truth of which you may absolutely relie, that you want for nothing in this country that nature can give us. Rich land is plentiful; building no where so cheap; game, fish, flesh, fowl, and fruit, and in the utmost profusion; labour by slaves very cheap, by servants not dear. And to this may be added a government mild and equal, in which more liberty is no where to be found; taxes too trifling to be mentioned, and where neither tythes nor poor's rates are to be found: may I not therefore conclude, that if mere living well, plentifully, and at ease, be considered, no country can exceed, perhaps not equal, this. In respect of society, we are deficient; but this is made up to a studious person, or one who does not dislike retirement, by the amusements of the field, by the employments of agriculture, and by reading; not however that we are without company; I have eight or nine neighbours within twenty miles of me, with whom I visit; and some of them are families in which a rational conversation is by no means wanting!

In a successive letter the same person gave the following particulars.

Respecting the agriculture we pursue, about which you enquire, I shall give you the best account I am able: the objects most attended to are Indian corn, wheat, and provisions, which long occupied this country chiefly; but for a few years past silk and indigo have made great strides among us. The country near the sea is not near so fertile in corn and provisions as that about us; we therefore not only send great quantities for the West India export, but also to feed the towns and rice plantations in part. I am much in doubt whether the common husbandry of raising corn and provisions be not as profitable as that of indigo or rice; the best planters we have, and it is the same in Carolina, do not reckon they make in total product above 20£. or 25£. each for their working hands: I have exceeded this some years by Indian corn, wheat, &c. and some of my neighbours have carried it much farther than me, by more skill and closer application.

The first business in our husbandry is clearing the ground, which is for corn generally done by grubbing the trees up by the roots in order for the plough to go: this method I have followed in all the land I have cleared: the expence is small, from the ease of stirring the light soil; and after raising my house, offices, and negroe camp, with lengths and posts for fences, &c. the residue I have sent down the river in several sorts of lumber, as boards, planks, staves, pieces, casks, &c. to Augusta and Savannah, but not with such advantage as others lower down on the river, who have not such a distance: after clearing, I have planted the land with Indian corn, for three and four years successively, and got from thirty-five to sixty bushels an acre, and at the same time from twenty-five to fifty bushels of Indian pease an acre. Of wheat my crops are not so great; but from thirty to forty bushels an acre is my usual crop. Barley we also sow, usually after wheat or Indian corn; I get the same quantity as of wheat; I have some fields the soil of which is so rich that I have got for six years successively crops of these kinds of grain, and all equally good, and I do not now find that the soil has much abated of its fertility;

but in some of my fields, it begins to wear out; but fine grass will come, and soon yield nearly as good profit as middling crops of corn. I do not hink our farmers in England grow quite so many successive crops of corn as we do here, yet I imagine our products to be much the largest. I have never laid on any dung or other manure for corn.

My black deep loams, which were covered with wood, yield kitchen plants of very fine flavour and an extraordinary size; I allotted a piece of it near my house for a garden, in which several articles of common product much exceed what I remember seeing in England, and yet I have never manured it: this, however, is much owing to climate. I have raised cabbages of 60 lb. weight, and turneps of 25 lb. Potatoes thrive astonishingly in it; I have had 300 bushels from a bed which in size did not exceed a quarter of an acre; and several of my neighbours having found their great increase in this soil have begun to go pretty much into them as an article of sale: they find a ready market at Savannah for the West Indies. I design taking the same hint, and believe they will be as profitable as any other article.

There is one circumstance in this country which is very valuable in planting; it is the warmth of the climate rendering it unnecessary to house or otherwise attend cattle in winter more than in summer; they find their subsistence in the woods and natural meadows, and return home of nights only for the sake of food given them, not so much through necessity, as to induce them to be regular. Our swine fare yet better, for the woods abound greatly with mast and fruit of various sorts, which they are greedy in finding, and keep themselves fat on; but we use them to come home in the same manner as the kine. The number of hogs kept in this country by every planter is very great; they who begin only with a sow or two, in a few years are masters of fourscore, or an hundred head; I have above three hundred of all sorts and sizes, and in a few years, if the country does not settle very fast, shall have twice or thrice as many. Pork and beef barrelled make a considerable article of our product; and hides are not the most inconsiderable part of the product of our cow kine.

Besides the articles I have mentioned, we cultivate indigo, silk, tobacco, and hemp, but not in such large quantities (silk excepted) as they do in Carolina; I have them all upon my plantations; indigo and tobacco require the same soil, which is the richest and deepest we can give them, but it must be dry; tobacco is but just coming it, but we make as good, if not better than any in Virginia; and I am of opinion, since the price has risen, it will be as beneficial as any article we can go upon. Hemp is sown in the low lands on the rivers and in drained marshes, where the soil is a stiff loam, upon clay: we do not reckon it so profitable as either indigo or tobacco; but as the land which suits it is not the right sort for those crops, it is cultivated in small quantities. I cannot speak with any precision of the products which my neighbours gain of these commodities, but I can tell you pretty accurately what I have done myself. When I had increased my ten negroes to twenty, from that number eight of whom were women (who I should observe do as much field-work as the men), they made me, with the help of five white labourers,

	£.	s.	d.
36 acres of Indian corn, the produce of which sold for in sterling money..	146	13	8
26 acres of wheat, the produce of which sold, sold for in ditto	93	8	0
12 acres barley, the product of which sold, yielded.......	18	10	0
40 barrels of pork at 1£. 12s.	64	0	0
26 ditto of beef at 1£. 2s..	28	12	0
16 acres tobacco yielded 12 hogsheads........................	96	0	0
33 pounds silk...	30	0	0
420 pounds of indigo from 4 acres and a half	52	10	0
Hides, live stock sold, and small articles.....................	47	18	0
	£. 577	11	8

This product from 25 hands is above 23£ each but the two following years I had not such good success. You see, by this account, that I made 33 pounds of silk: this is an article which deserves more attention than has been given it, either by the inhabitants of this province, or in encouragement from the mother country: the number of hands who made this 33 lb. in a season was but eleven, which is 3 lb. a head. This appears to be a very considerable object; among them were four of my children (who did it by way of amusement, and indeed a very rational one it is), and three women; they did not employ seven weeks in it, and I need not tell you that in this business it is but a part of the day that is employed. Georgia has, proportioned to its inhabitants, made greater progress in feeding silk-worms than any of the other colonies, but yet her people do not make near the quantities they might; supposing they made but two pounds a head, and less than this I have never made for every person I have employed in it, this would be a vast acquisition for all the women, many of the children, the old men, and disabled persons, with proper assistance from others; and in Carolina, where the people are so much more numerous, the importance of the object would in a national light be still greater. This is so favourite a theme with me that were I to trouble you with all I could say on the subject I should go near to exhaust your patience.

Hemp has been too inconsiderable an article with me to come to market, but I hope next year to send two or three tons down to Savannah. It may be to your satisfaction also to know that I have made some trials of wine, not yet as an export, but have used several casks in my own family, which have proved better than I expected. Four years ago I planted about a quarter of an acre of a dry rocky spot of land, hanging on the side of a hill, to the south of which I thought promised well, I used setts of our native grapes, having no others; these are so plentiful all over the country, that you can scarce go an hundred yards without meeting with numbers. This little plantation, which, for want of better knowledge than I could gain from one or two treatises on the vineyard culture, was not managed near so well as it ought, has turned out much better than I expected. I have made wine from the same grapes growing wild, and have the satisfaction to find that the produce of my cultivated ones is beyond comparison of a finer flavour, which shews that we have much to hope from attending well to our native vines; were they managed with the skill that is exerted in the wine countries of Europe, they would perhaps turn out among the profitable articles of our husbandry. One of the neighbouring planters, a Frenchman by birth, has written to France to a relation to send a vigneron, well experienced, with some setts of the Burgundy grape: this is to be done at my expence, and I have good expectations from the scheme; though my neighbour is quite of a different opinion, and thinks that (farther than our own consumption) if I make it succeed, the culture will not prove near so profitable as indigo: how far this will prove true, I am not yet a judge. Madeira, which is the only wine we import, comes very dear to us. If we could make a sort sufficiently good to be substituted for that, it would be a great acquisition. I shall give you the account of my sale in a year since the former, being nearly what my present produce is.

	£.	s.	d.
50 acres of Indian corn, produce in sterling money	187	14	0
35 acres of wheat produce in ditto	132	0	0
20 acres of barley ditto	35	0	0
50 barrels of pork	82	0	0
40 ditto of beef	50	0	0
Hides	24	10	0
Live stock sold	30	0	0
Lumber	36	0	0
47 pounds of silk	47	0	0
16 acres of tobacco, 11 hogsheads	88	0	0
Indigo	87	10	0
£.	799	14	0

The number of negroes 25, and 5 white labourers, 30 in all; the total divided gives 26£. a head; and this I believe may generally be equalled by those who have any luck in fixing on tolerable land, without possessing great skill in choosing the best.

This, Sir, is a very important part of the information you request me to give, as it explains to you what is to be expected from settling in this country; I have little doubt of your friend being able to make from 20£. to 30£. a head annual produce, sold at market, for all the working hands he employs, besides supplying the plantation with all the provisions consumed, both by the family, slaves, cattle, poultry, &c.; this however is not profit, for negroes cost at present from 40 £ to 60 £ a head, if good ones are bought. Cloathing, physic, attendance, &c. come to something and distempers will now and then break out among them which prove very destructive, though in general the increase will keep up the number. Implements, tools, furniture, manufacturers, &c. &c. are all much dearer than in England, except the articles which are made of wood. Wine, tea, sugar, and spices, are some of them dearer, and none of them cheaper, than in England. Repairs of buildings is an article of some expence. Negroes must have an overseer, at the annual expence of from 40 £ to 60 £ All these and some other articles are deductions from the planter's profit; nor should I omit to add, Sir, that the nature of the country, as it prevents many of the expences which are common in England, so it brings on us others of which you know nothing. Hospitality, to a degree totally unknown in Europe, is the virtue of all America; and a man can hardly through inclination, but especially from example, be niggardly on any occasions that call for it; his great expence will be wine, rum, and a few other articles of housekeeping; not that this amounts to any thing very considerable.

In general, our planters are very much on the thriving hand, yet few are rich, though I have heard of some large fortunes in Carolina; we have scarce one in the province whose circumstances do not improve. As to making a fortune, I believe no part of the world is better adapted for it, provided the planter is skilled in land, has good degree of knowledge in matters of agriculture, and is very industrious and attentive to his business: you may easily suppose that all these qualifications are necessary; indeed I know of no business in which money is to be made without knowledge and industry: I must also add, that he ought to have his capital, especially if it is not considerable, free; for interest here is 8, 9, and 10 per cent. and money even on such terms very difficult to be got; he should therefore possess, or raise whatever he wants, in England, for no dependance is to be had on getting any here.

There are very great advantages in the husbandry which is carried on in this country, of a nature not general in others, especially in Europe; the quantity of land to be had by any person that pleases is a circumstance no where else to be met with on the globe, at least in countries where the religious and civil liberty of mankind is secured; and among all our colonies, none has better land or a more favorable climate than the back parts of Georgia. The expences of living are low, and particularly the necessaries of life so very plentiful that subsistence is no where easier gained; on the contrary, the articles of merchandise produced here yield a large price, as it is the nature of mankind to rate luxuries much higher than necessaries, and not to let the value of one depend on that of the other. Thus the planter feeds and subsists himself and family very cheap, while he sells his produce very dear: silk, indigo, wine, hemp, tobacco, &c. are by no means necessaries, but their value is much greater; this is a circumstance of great value to the planter; how far you enjoy in the same in certain articles produced in England, I am not a good judge; luxurious articles may be very dear, but the necessaries which support the planters, and their workmen are dear also.

These extracts, from pretty long correspondence carried on with a view of settling a relation in Georgia (which is since done), I am happy in being allowed to insert: it is true they principally concern only one plantation, but they abound with many valuable circumstances that concern the whole province, and as such could not but be deemed highly worthy of insertion.

The following is a state of the exports of Georgia,
upon an average of three years since the peace

18,000 barrels of rice, at 40s..	£ 36,000
Indigo, 17,000 lb. at 2s..	1,700
Silk, 2,500 lb. at 20s..	2,500
Deer and other skins...	17,000
Boards, staves, &c..	11,000
Tortoise-shell, drugs, cattle, &c..	6,000
	£ 74,200

But since this account was published the articles are most of them very much increased; the rice is raised to 23,000 barrels, and the price to 3 £ 10s. so that this article alone makes more than the whole of the above articles; the indigo is proportionally increased, but the silk is declined: the Indian trade at Augusta is also thriven very much of late.

Before I take my leave of this province it will be proper to mention the large tract of country lately acquired by the government of the Cherokees, containing by estimation about seven millions of acres. This country lies to the westward of Augusta, and is bounded on one side by the Savannah's branches; from the description I have heard of it, I apprehend the plantation described in the above letters very nearly resembles it: the soil is as rich as any part of America; every article of spontaneous vegetation, luxuriant in the highest degree: the climate, like that of the western part of all our southern colonies which bounds upon the Ap[p]alachian mountains, as desirable as any in the world, both for the production of profitable staples and healthiness. It is further said to be as well watered with streams and rivers as can be wished, with three or four of them navigable for large canoes. Many are the people who have given in petitions for grants of land in it, so that it is expected in Georgia the whole tract, large as it is, will be settled in a few years: the articles of culture upon which the planters will go are particularly indigo, hemp, flax, cotton, tobacco, vines, and silk, not to speak of Indian corn, wheat, and other provisions: for all these, it is said, no part of America can be better adapted. . . .

The American Revolution
and Agriculture

Introduction

Land that was seemingly unlimited in extent and available to very European immigrant characterized the original 13 colonies and was the greatest distinguishing factor, in the economic sphere, between the Old World and the New. In the Old World, one was born to the land or never had it. In the New World, one acquired land merely by coming to a new colony or by working a few years until the indenture given for one's passage was paid. Such was the dream, and it was attainable for most people, a point made by the anonymous author of the famous *American Husbandry,* published in London in 1775. When this hope and dream seemed to be threatened, the American was willing to fight to preserve it.

At the end of the French and Indian War, the British government, anxious to preserve peace with the western Indians and to protect the fur trade, forbade settlement west of the Allegheny Mountains. This boundary line, established by the Proclamation of 1763, was known as the Proclamation Line. No land could be purchased from the Indians and no settlements could be made west of the line without the permission of imperial authorities, while settlers then living west of the line were directed to leave.

With this one regulation the British government antagonized three groups of people. The frontiersmen who were west of the line were naturally bitter. Assurances by the British that eventually land would be purchased from the Indians and opened to settlement was no comfort to one ordered to leave land he had claimed. The governing officials of over one-half of the colonies comprised the second group antagonized by the Proclamation Line. A number of the colonies had been founded under charters that extended their east-west boundaries from sea to sea. The officials of such colonies felt that their charters were being invaded when the British exercised authority over their western claims.

In some cases, the officials were identified with the third group antagonized, the land company speculators. The first such company, the Ohio Company, owned by prominent Virginians, was granted 200,000 acres on the Ohio in 1748. Many other companies were formed before and after 1763, and their conflicting claims led to rivalries and antagonisms. Generally, those organizing and claiming rights to land before 1763 regarded the Proclamation Line as a costly barrier to their speculative activities.

The problem of land settlement in the West was not the only source of friction over land in America just prior to the Revolution. As settlers acquired land from proprietors and the crown, they found that their titles were subject to certain limitations imposed by the proprietors and the crown in accordance with English law and tradition. One such limitation, found in all of the colonies except New England, and there in certain cases, was quitrent. Quitrent was a yearly fee on land held. A remnant of the feudal system, it had been established as a fixed fee that might be paid the lord of the manor in lieu of performing for him certain services or doing customary work.

Quitrents varied widely between colonies and from time to time, but the Maryland plan as found in "A Relation of Maryland, 1635," and in Penn's and Pastorius' descriptions of Pennsylvania are rather typical. Since the colonies were competing for settlers, the quitrents in any one could not become too burdensome, but generally they were all that the proprietors or crown could hope to collect. In every colony and throughout the colonial period, however, the settlers were reluctant to pay quitrents and the authorities responsible for collection often could collect only a fraction of the amount due. In some areas, considerable feeling was aroused against the government when it attempted to collect. It appears that such feeling, whether that colony was owned by proprietors or the crown, reflected against the British government. While the feeling against quitrent could hardly be advanced as a major cause of the Revolution, it is notable that the new states abolished quitrents and forbade them for the future.

Restrictions upon land settlement and land alienation caused resentment against authority, generally identified with Great Britain, but restrictions upon trade, particularly when attempts were made to enforce them, seem to have been more prolific sources of discontent. Very generally, by 1775, New England was dependent of its income chiefly upon fisheries and trade; the Middle Colonies upon trade and food crops; and the South upon the plantation staples of tobacco, rice, and indigo. In each colony, cattle raising and small farming characterized the frontier, which, in 1763, included the back county of New England, the Mohawk Valley, the Great Valley of Pennsylvania, the Shenandoah Valley, and the southern piedmont lying between the fall line of the rivers and the Alleghenies. Commercial farming was carried on to a limited extent in the eastern parts of the Middle Colonies, and plantation agriculture characterized the coastal plain of the South.

The difficulty that New England and the Middle Colonies faced in developing agricultural exports was that they grew no great staple for which there was a demand in Europe. Instead, the products that they grew were the same as those grown in northern Europe. Some wheat and livestock products could be marketed in Europe in times of great scarcity. The West Indies furnished a limited market for provisions, the best available for products of northern farms.

Regulations affecting the sale of northern farm products took two forms: one, the imposition of heavy duties on colonial products; and, two, restrictions on trade with the French, Dutch, and Spanish West Indies. The latter restrictions were imposed most often in time of war and were frequently violated by the New England traders.

While trade restrictions had little direct influence on the frontiersmen, they were of vital importance to the southern planters. As early as 1621 the British government required that all tobacco be shipped to England. While the rule was not always enforced, it was in effect until the Revolution and the taxes levied upon tobacco increased steadily. The problem of a market and of price became more acute as new lands were placed in cultivation, creating a surplus, while, at the same time, yields per acre and per worker were becoming less in the original Virginia and Maryland areas as fertility declined and soil erosion made cultivation more difficult. As an eminent historian has said: "To what extent the people laid the blame on the government or in how far they vented their wrath at the hapless agents of the system, such as merchant or shipper, we do not know. But certain it is that agrarian disaster was ever present to form a great background of discontentment and unrest upon which the more immediate disputes developed." Certainly the Virginia planter class furnished much of the leadership for the Revolution.

The restrictions on colonial trade varied from time to time from 1621 on, but the Navigation Act of 1660 may be regarded as representative since it repeated earlier restrictions and added new. Under its terms, ships engaging in the colonial trade must be British (including colonial) owned and built, and at least three-fourths of the crews must be British subjects. More important to the planters, all sugar, tobacco, cotton, indigo, ginger, fustic and other woods used for dyes exported from the colonies should be sent directly to England. In 1706, molasses, rice, and naval stores were added to the list.

The key to the trade question was that within a short time after the colonies were founded farmers began producing surpluses for which there was no market in America—a continuing problem for American farmers in spite of seeming shortages in the 1970's. At the outbreak of the Revolution, farmers were utterly dependent, therefore, for their growth and prosperity, on the sale of farm produce in overseas markets.

Even before independence was declared, the question of exports was a matter of controversy in the colonies, particularly after the Continental Congress resolved that both imports and exports should be subject to regulation. A New Yorker, Samuel Seabury, argued in 1774 that other nations would supply the farm products needed in England and particularly denounced the ban on exporting sheep. The Virginia General Assembly, the next year, resolved that no sheep under four years old be slaughtered for meat, that woolen, cotton, and flax be cultivated, and that the brewing of malt liquors be encouraged. A Virginian encouraged the growing of dye plants, so the colony would be self-sufficient in that area. The South Carolina Provincial Congress agreed that since rice could legally be exported, other commodities should be valued in relation to it.

Restrictions upon trade in agricultural products had an effect in helping bring about the Revolution. After the war started, the farms supplied most of the men in the patriot armies since over 90 percent of the population lived on farms. The southern plantations furnished many leaders for the military forces, such as George Washington and Henry Lee, and many leaders of the Revolutionary government, such as Thomas Jefferson and Henry Laurens.

On the whole, agricultural life went on as usual during the war and, if anything, was stimulated rather than injured. Many farmers seemed quite indifferent to the progress of the war except when their immediate interests were threatened. New

England was, for the most part, free of invasion. The middle colonies were the scene of extensive military operations, but many farmers profited by charging high prices for food sold to the British, French, and American armies. In the South, blockade runners carried tobacco to Europe. Rice cultivation and export went on with some interruption and some indigo continued to be exported.

One effect of the war, with imports of British cloth practically nonexistent, was to encourage the raising of sheep and the growing of flax and hemp. The New York Committee of Safety made such a strong recommendation to farmers in 1776. In all three cases, the comparative prosperity of the industry was due to the wartime demand and there was a decline after the war. A second effect was to encourage the growing of food crops, particularly in New England and the plantation South, since both areas had been importing food.

While food was generally abundant during the war, the American army experienced shortages from time to time. As armed men gathered around Boston after April 19, 1775, when hostilities began at Lexington and Concord, the Massachusetts Provincial Congress found that it had more than it could do to feed them, and urgent requests were sent to New Hampshire, Connecticut, and Rhode Island to send their troops to Boston well supplied with provisions. In October 1775, the County of Providence, Rhode Island, urged a system of certificates to stop shipments of food to British troops. This problem continued throughout the war in several of the States of the new nation.

Much time and attention was given to the problem of food supply. At times shortages threatened to hamper military operations. The problem was dramatized by the severe suffering of the Continental army at Valley Forge during the winter of 1777-1778, although the problem of food for the army appears again and again in Washington's correspondence. There appears to have been no real shortage of food in the states, or even in Pennsylvania, at the time, but transportation and organization difficulties kept some supplies from reaching the troops. The greatest difficulty lay, however, in the fact that the farmers were unwilling to sell their produce for Continental currency. By the winter of 1779–1780, this problem became so great that Congress abandoned its attempts at direct procurement, and, on December 2, 1779, decided to submit requisitions to the various states. The new system was inefficient and expensive in many respects. On the other hand, the state agents seemed to be better able to persuade the farmers to part with supplies. They were aided a great deal in this respect after France became active in the war, since the French made coin available through purchases for their own forces and through loans to the Continental Congress.

In general, the farmers fed the American, French, and many of the British forces and made money doing it. At the same time, the farming population furnished most of the soldiers. Few farmers were Tories; many of those not actively engaged in the war were neutral, some because they were pacifists, others because they felt that their interests were not involved. At times, particularly on the frontier, these groups would become aroused and render valuable service to the Revolutionary cause. Farmers' problems, particularly restrictions on land and on shipments of agricultural products, had helped bring about the Revolution, and the farmers themselves had been a vital key in winning independence.

Proclamation of 1763

From Great Britain, *Annual Register*, 1763, pp. 208–13.

WHEREAS we have taken into our royal consideration the extensive and valuable acquisitions in America, secured to our crown by the late definitive treaty of peace concluded at Paris the 10th day of February last; and being desirous that all our loving subjects, as well of our kingdoms as of our colonies in America, may avail themselves, with all convenient speed, of the great benefits and advantages which must accrue therefrom to their commerce, manufactures, and navigation; we have thought fit, with the advice of our privy council, to issue this our royal proclamation, hereby to publish and declare to all our loving subjects, that we have, with the advice of our said privy council, granted our letters patent under our great seal of Great Britain, to erect within the countries and islands, ceded and confirmed to us by the said treaty, four distinct and separate governments, stiled and called by the names of Quebec, East Florida, West Florida, and Grenada, and limited and bounded as follows, viz.

First, the government of Quebec, bounded on the Labrador coast by the river St. John, and from thence by a line drawn from the head of that river, through the lake St. John, to the South end of the lake Nipissim; from whence the said line, crossing the river St. Lawrence and the lake Champlain in 45 degrees of North latitude, passes along the High Lands, which divide the rivers that empty themselves into the said river St. Lawrence, from those which fall into the sea; and also along the North coast of the Bayes des Chaleurs, and the coast of the Gulph of St. Lawrence to Cape Rosieres, and from thence crossing the mouth of the river St. Lawrence by the West end of the island of Anticosti, terminates at the aforesaid river St. John.

Secondly, The government of East Florida, bounded to the Westward by the Gulph of Mexico and the Apalachicola river; to the Northward, by a line drawn from that part of the said river where the Catahoochee and Flint rivers meet, to the source of St. Mary's river, and by the course of the said river to the Atlantic Ocean; and to the East and South by the Atlantic Ocean, and the Gulph of Florida, including all islands within six leagues of the sea coast.

Thirdly, The government of West Florida, bounded to the Southward by the Gulph of Mexico, including all islands within six leagues of the coast from the river Apalachicola to lake Pontchartrain; to the Westward by the said lake, the lake Maurepas, and the river Mississippi; to the Northward, by a line drawn due East from that part of the river Mississippi which lies in thirty-one degrees North latitude, to the river Apalachicola, or Catahoochee; and to the Eastward by the said river.

Fourthly, The government of Grenada, comprehending the island of that name, together with the Grenadines, and the islands of Dominico, St. Vincent, and Tobago.

And to the end that the open and free fishery of our subjects may be extended to, and carried on upon the coast of Labrador and the adjacent islands, we have thought fit, with the advice of our said privy council, to put all that coast, from the river St. John's to Hudson's Streights, together with the islands of Anticosti and Madelane, and all other smaller islands lying upon the said coast, under the care and inspection of our governor of Newfoundland.

We have also, with the advice of our privy council, thought fit to annex the islands of St. John and Cape Breton, or Isle Royale, with the lesser islands adjacent thereto, to our government of Nova Scotia.

We have also, with the advice of our privy council aforesaid, annexed to our province of Georgia, all the lands lying between the rivers Attamaha and St. Mary's.

And whereas it will greatly contribute to the speedy settling our said new governments, that our loving subjects should be informed of our paternal care for the security of the liberty and properties of those who are, and shall become inhabitants thereof: we have thought fit to publish and declare, by this our proclamation, that we have, in the letters patent under our great seal of Great Britain, by which the said governments are constituted, given express power and direction to our governors of our said colonies respectively, that so soon as the state and circumstances of the said colonies will admit thereof, they shall, with the advice and consent of the members of our council, summon and call general assemblies within the said governments respectively, in such manner and form as is used and directed in those colonies and provinces in America, which are under our immediate government; and we have also given power to the said governors, with the consent of our said councils, and the representatives of the people, so to be summoned as aforesaid, to make, constitute, and ordain laws, statutes, and ordinances for the public peace, welfare, and good government of our said colonies, and of the people and inhabitants thereof, as near as may be, agreeable to the laws of England, and under such regulations and restrictions as are used in other colonies; and in the mean time, and until such assemblies can be called as aforesaid, all persons inhabiting in, or resorting to, our said colonies, may confide in our royal protection for the enjoyment of the benefit of the laws of our realm of England: for which purpose we have given power under our great seal to the governors of our said colonies respectively, to erect and constitute, with the advice of our said councils respectively, courts of judicature and public justice within our said colonies, for the hearing and determining all causes as well criminal as civil, according to law and equity, and as near as may be, agreeable to the laws of England, with liberty to all persons who may think themselves aggrieved by the sentence of such courts, in all civil cases, to appeal, under the usual limitations and restrictions, to us, in our privy council.

We have also thought fit, with the advice of our privy council as aforesaid, to give unto the governors and councils of our said three new colonies upon the continent, full power and authority to settle and agree with the inhabitants of our said new colonies, or to any other person who shall resort thereto, for such lands, tenements, and hereditaments, as are now, or hereafter shall be, in our power to dispose of, and them to grant to any such person or persons, upon such terms, and under such moderate quit rents, services, and acknowledgments, as have been appointed and settled in other colonies, and under such other conditions as shall appear to us to be necessary and expedient for the advantage of the grantees, and the improvement and settlement of our said colonies.

And whereas we are desirous, upon all occasions, to testify our royal sense and approbation of the conduct and bravery of the officers and soldiers of our armies, and to reward the same, we do hereby command and impower our governors of our said three new colonies, and other our governors of our several provinces on the continent of North America, to grant, without fee or reward, to such reduced officers as have served in North America during the late war, and are actually residing there, and shall personally apply for the same, the following quantities of land, subject, at the expiration of ten years, to the same quit rents as other lands are subject to in the province within which they are granted, as also subject to the same conditions of cultivation and improvement, viz.

To every person having the rank of a field officer, 5000 acres.

To every captain, 3000 acres.

To every subaltern or staff officer, 2000 acres.

To every non-commission officer, 200 acres.

To every private man 50 acres.

We do likewise authorise and require the governors and commanders in chief of all our said colonies upon the continent of North America to grant the like quantities of land, and upon the same conditions, to such reduced officers of our navy of like rank, as served on board our ships of war in North America at the times of the reduction of Louisbourg and Quebec in the late war, and who shall personally apply to our respective governors for such grants.

And whereas it is just and reasonable, and essential to our interest, and the security of our colonies, that the several nations or tribes of Indians, with whom we are connected, and who live under our protection, should not be molested or disturbed in the possession of such parts of our dominions and territories as, not having been ceded to, or purchased by us, are reserved to them, or any of them, as their hunting grounds; we do therefore, with the advice of our privy council, declare it to be our royal will and pleasure, that no governor, or commander in chief, in any of our colonies of Quebec, East Florida, or West Florida, do presume, upon any pretence whatever, to grant warrants of survey, or pass any patents for lands beyond the bounds of their respective governments, as described in their commissions: as also that no governor or commander in chief of our other colonies or plantations in America, do presume for the present, and until our further pleasure be known, to grant warrant of survey, or pass patents for any lands beyond the heads or sources of any of the rivers which fall into the Altantic Ocean from the west or north-west; or upon any lands whatever, which not having been ceded to, or purchased by us, as aforesaid, are reserved to the said Indians, or any of them.

And we do further declare it to be our royal will and pleasure, for the present, as aforesaid, to reserve under our sovereignty, protection, and dominion, for the use of the said Indians, all the land and territories not included within the limits of our said three new governments, or within the limits of the territory granted to the Hudson's Bay company; as also all the land and territories lying to the westward of the sources of the rivers which fall into the sea from the west and north-west as aforesaid; and we do hereby strictly forbid, on pain of our displeasure, all our loving subjects from making any purchases or settlements whatever, or taking possession of any of the lands above reserved, without our special leave and licence for that purpose first obtained.

And we do further strictly enjoin and require all persons whatever, who have either willfully or inadvertently seated themselves upon any lands with the countries above described, or upon any other lands, which not having been ceded to, or purchased by us, are still reserved to the said Indians as aforesaid, forthwith to remove themselves from such settlements.

And whereas great frauds and abuses have been committed in the purchasing lands of the Indians, to the great prejudice of our interests, and to the great dissatisfaction of the said Indians; in order, therefore, to prevent such irregularities for the future, and to the end that the Indians may be convinced of our justice and determined resolution to remove all reasonable cause of discontent, we do, with the advice of our privy council, strictly enjoin and require, that no private person do presume to make any purchase from the said Indians of any lands reserved to the said Indians within those parts of our colonies where we have thought proper to allow settlement; but that if at any time any of the said Indians should be inclined to dispose of the said lands, the same shall be purchased only for us, in our name, at some public meeting or assembly of the said Indians, to be held for that purpose by the governor or commander in chief of our colony respectively within which they shall lie: and in case they shall lie within the limits of any proprietaries, conformable to such directions and instructions as we or they shall think proper to give for that purpose: and we do, by the advice of our privy council, declare and enjoin, that the trade with the said Indians shall be free and open to all our subjects whatever, provided that every person who may incline to trade with the said Indians, do take out a licence for carrying on such trade, from the governor or commander in chief of any of our colonies respectively, where such person shall reside, and also give security to observe such regulations as we shall at any time think fit, by ourselves or commissaries, to be appointed for this purpose, to direct and appoint for the benefit of the said trade: and we do hereby authorise, enjoin, and require the governors and commanders in chief of all our colonies respectively, as well those under our immediate government, as those under the government and direction of proprietaries, to grant such licences without fee or reward, taking especial care to insert therein a condition that such licence shall be void, and the security forfeited, in case the person, to whom the same is granted, shall refuse or neglect to observe such regulations as we shall think proper to prescribe as aforesaid.

And we do further expressly enjoin and require all officers whatever, as well military as those employed in the management and direction of Indian affairs within the territories reserved, as aforesaid, for the use of the said Indians, to seize and apprehend all persons whatever, who standing charged with treasons, misprisions of treasons, murders, or other felonies or misdemeanors, shall fly from justice and take refuge in the said territory, and to send them under a proper guard to the colony where the crime was committed of which they shall stand accused, in order to take their trial for the same.

Given at our Court at *St. James's.* the seventh day of *October,* one thousand seven hundred and sixty-three, in the third year of our reign.

GOD save the KING.

Samuel Seabury Addresses the Farmers
of New York, 1774

From Samuel Seabury, *Letters of a Westchester Farmer* (1774–1775), Clarence H. Vance, ed.; Reprinted (White Plains: Westchester County Historical Society, 1930), pp. 43–47, 62–65.

Permit me to address you upon a subject, which, next to your eternal welfare in a future world, demands your most serious and dispassionate consideration. The American Colonies are unhappily involved in a scene of confusion and discord. The bands of civil society are broken; the authority of government weakened, and in some instances taken away: Individuals are deprived of their liberty; their property is frequently invaded by violence, and not a single Magistrate has had courage or virtue enough to interpose. From this distressed situation it was hoped, that the wisdom and prudence of the Congress lately assembled at Philadelphia, would have delivered us. The eyes of all men were turned to them. We ardently expected that some prudent scheme of accommodating our unhappy disputes with the Mother-Country, would have been adopted and pursued. But alas! they are broken up without even attempting it: they have taken no one step that tended to peace: they have gone on from bad to worse, and have either ignorantly misunderstood, carelessly neglected, or basely betrayed the interests of all the Colonies.

I shall in this, and some future publication, support this charge against the Congress, by incontestible facts: But my first business shall be to point out to you some of the consequences that will probably follow from the Non-importation, Non-exportation, and Non-consumption Agreements, which they have adopted, and which they have ordered to be enforced in the most arbitrary manner, and under the severest penalties. On this subject, I choose to address myself principally to You the *Farmers* of the Province of New-York, because I am most nearly connected with you, being one of your number, and having no interest in the country but in common with you; and also, because the interest of the farmers in general will be more sensibly affected, and more deeply injured by these agreements, than the interest of any other body of people on the continent. Another reason why I choose to address myself to you is, because the Farmers are of the greatest benefit to the state, of any people in it: They furnish food for the merchant, and mechanic; the raw materials for most manufacturers, the staple exports of the country, are the produce of their industry: be then convinced of your own importance, and think and act accordingly.

The Non-importation Agreement adopted by the Congress is to take place the first day of December next; after which no goods, wares, or merchandize, are to be imported from Great-Britain or Ireland; no East-India Tea from any part of the world; no molasses, syrups, paneles, coffee, or pimento, from our islands in the West-Indies; no wine from Madeira, or the Western-Islands; no foreign indigo.

The Non-Exportation Agreement is to take effect on the tenth day of September next; after which we are not to export, directly or indirectly, any merchandize or commodity whatsoever to Great-Britain, Ireland, or the West-Indies, except RICE to Europe, — unless the several acts and parts of acts of the British Parliament, referred to by the fourth article of Association, be repealed.

The Non-consumption Agreement is to be in force the first day of March next; after which we are not to purchase or use any East-India Tea whatsoever; nor any goods, wares, or merchandize from Great-Britain or Ireland, imported after the first of December, nor molasses, &c. from the West-Indies; nor wine from Madeira, or the Western Islands, nor foreign indigo.

Let us now consider the probable consequences of these agreements, supposing they should take place, and be exactly adhered to. The first I shall mention is, clamours, discord, confusion, mobs, riots, insurrections, rebellions, in Great-Britain, Ireland, and the West-Indies. This consequence does not indeed immediately affect You, the Farmers of New-York; nor do I think it a probable one: But the Congress certainly intended it should happen in some degree, or the effect they propose from these agreements cannot possibly take place. They intend to distress the manufacturers in Great-Britain, by depriving them of employment — to distress the inhabitants of Ireland, by depriving them of flax-seed, and of a vent for their linens, — to distress the West-India people, by with-holding provisions and lumber from them, and by stopping the market for their produce. And they hope, by these means, to force them all to join their clamours with ours, to get the acts complained of, repealed. This was the undoubted design of the Congress when these agreements were framed; and this is the avowed design of their warm supporters and partizans, in common conversation.

But where is the justice, where is the policy of this proceedure? The manufacturers of Great-Britain, the inhabitants of Ireland, and of the West-Indies, have done us no injury. They have been no ways instrumental in bringing our distresses upon us. Shall we then revenge ourselves upon them? Shall we endeavour to starve them into a compliance with our humours? Shall we, without any provacation, tempt or force them into riots and insurrections, which must be attended with the ruin of many — probably with the death of some of them? Shall we attempt to unsettle the whole British Government — to throw all into confusion, because our self-will is not complied with? Because the ill-projected, ill-conducted, abominable scheme of some of the colonists, to form a republican government independent of Great-Britain, cannot otherwise succeed? — Good God! can we look forward to the ruin, destruction, and desolution of the whole British Empire, without one relenting thought? Can we contemplate it with pleasure; and promote it with all our might and vigour, and at the same time call ourselves *his Majesty's most dutiful and loyal subjects?* Whatever the Gentlemen of the Congress may think of the matter, the spirit that dictated such a measure, was not the spirit of humanity. . . .

Holland, the Baltic, and the river St. Lawrence, would afford the Irish a sufficient supply of flax-seed. If they look out in time they cannot be disappointed. Canada produces no inconsiderable quantity already.

I have been well informed, that many bushels have been bought up there at a low price, brought to New-York, and sold to the Irish factors at a great advance. Are the Irish such novices in navigation, that they cannot find the way to Quebec? Or are they so blind to their own interest, as to continue giving a high price for flax-seed at New-York, when they might have a considerable supply from Canada, at a much more reasonable rate?

You will say that as soon as the Irish send their ships to Quebec for seed, the price will rise till it comes to an equality with ours. I know it. I know also, that the more the price rises, the more Canadians will be encouraged to raise it. I know also, that the more they raise and sell, the less demand there will be for ours, and the less price it will fetch at market.

Nor should we distress the inhabitants of the West-Indies so much as at first sight we may imagine. Those islands produce now many of the necessaries of life. The quantity may easily be encreased. Canada would furnish them with many articles they now take from us; flour, lumber, horses, &c. Georgia, the Floridas, and the Mississippi abound in lumber; Nova Scotia in fish. All these countries would be enriched by our folly, and would laugh at it.

* * *

There is one article more of the Association, which exhibits such a striking instance of the ignorance, or inattention of the Congress to the Farmers interest, that I must take notice of it to you; especially as it will give me an opportunity of mentioning as striking an instance of the arbitrary, illegal, and tyrannical procedure of the Committee of Correspondence in New-York.

The article I mean, is the seventh, relative to the encreasing of the number and improving of the breed of sheep. No sheep of any kind are to be exported to the West-Indies, or elsewhere. Why, for Gods sake, were *weathers* included in this prohibition? Will weathers encrease the number, or improve the breed of sheep? I with the Gentlemen of the Congress, and the Committee-men of New-York, would try the experiment. Let them buy a score of weathers, and feed, and nurse them for a twelve-month; and then publish an account of the number of lambs they have produced, their enormous size, with the quantity & fineness of their wool; that we may know in what manner the number and breed of sheep may be encreased, and improved, by keeping weathers. But let this account be under oath, or I shall not believe that they have succeeded, either in encreasing the number, or improving the breed. I solemnly declare I never had one lamb produced from a weather in my whole life; and have always been so ignorant, that I should no more expect a lamb from a weather, than a calf from an ox.

But it may be said, that weathers will produce wool, and that it is for the sake of the wool that their exportation is prevented. I readily own that weathers will produce wool, though not lambs. But let me ask you, my brother farmers, which of you would keep a flock of sheep barely for the sake of their wool? Not one of you. If you cannot sell your sheep to advantage at a certain age, you cannot keep them to any profit. An Ewe should not be kept after she is six years old, nor a weather after he is four: few of you choose to keep them so long. What now must be done with our sheep when they become so old that we can keep them no longer with advantage? We are ordered to kill them sparingly: a queer phrase; however, let it pass. If it is not *classical,* it is *congressional;* and that's enough. And after having killed them *sparingly,* if we have any to *spare,* we must *spare* them to our poor neighbours. But supposing that after *killing them sparingly, and sparing* as many to my poor neighbours as they want, I should, by reason of *killing them sparingly,* have still more to *spare* — what shall I do with them? Exported they must not be. Why! fat them well, and sell them to the New-Yorkers: The deuce take them for a set of gundy gutted fellows — will they let us export nothing? Do they intend to eat all our wheat, and rye, and corn, and beef, and pork, and mutton, and butter, and cheese, and turkeys, and geese, and ducks, and fowls, and chickens and eggs, &c? the devil is in't if their bellies are not filled. And yet see their ill-nature and malice against us farmers. — After having furnished them with all this good cheer, which they must have at their own price too, they will not in return let us have a dish of tea to please our wives, nor a glass of Madeira to cheer our spirits, nor even a spoonful of Molasses to sweeten our butter-milk. To be serious —

Had the Congress attended in the least to the farmers interest, they never would have prohibited the exportation of sheep, after they came to a certain age. It is the exportation that keeps up the price of sheep; it is the advantageous price that encourages the farmer to feed them: Take away the profit of selling them, and the farmer will keep but very few. For they are not, and I am confident never will be in this country, worth keeping for their wool alone.

However, right or wrong, the Congress have passed the decree. *Thou shalt not export sheep,* was pronounced at Philadelphia; and, right or wrong, the Committee of New-York are determined to put it in execution: And *thou shalt not export sheep,* is echoed back from New-York.

How this decree is to be supported in New-York, may be learned from the following affair. A Gentleman, an officer in the King's service, had purchased a number of sheep to carry with him to St. Vincent's: Mr. GAIN's news-paper says eighteen. The New-Yorkers, probably affraid that they should loose their share of the mutton, assembled on the dock, sent for the Committee, and in open violation of the laws of their country, obliged the merchant to whom the vessel had been consigned, to have the sheep landed; the sheep were committed to safe durance till the vessel sailed, and then were delivered to the proprietor — I suppose to the person who had sold them to the officer: Though how he could be the proprietor after he had sold them, I cannot see. Had I been the person, I would have had nothing to do with them; the Committee might have done what they pleased with them — *killed them sparingly,* or *spared* them to their poor neighbours. But had there been law or justice in the government, I would have been paid for them: though, now I think of it, I would have made a present of them to the Committee, upon condition that they should make the experiment how far the number and breed of sheep can be increased and improved by keeping weathers; for I have been positively assured, that these same sheep, which made all this bustle, were nothing more.

Here now, my friends, is a flagrant instance of injustice and cruelty committed by a riotous mob; — for a number of people, be they Committee-men, or who you please, assembled to do an unlawful action, especially in the night, deserve no better name, — against both the buyer and seller of the sheep, in open violation of the laws of the government in which we live, and of the rights of the city in which it was perpetrated; and not a single magistrate had virtue or courage enough to interpose. O shame to humanity! Hold up your heads, ye Committee-men of New-York! Deny the charge if ye can. But remember, the instant ye deny it, ye forfeit all pretensions to truth or conscience.

Shipment of Sheep, 1774

From: "U. S. Continental Congress, Correspondence, Proceedings, etc.," in Peter Force, ed., *American Archives:* Fourth Series (Washington, 1837), I, p. 963.

NEW-YORK, NOVEMBER 6, 1774.

A discovery being made the eighteen Sheep were on board a Sloop in this Harbour, bound for the *West Indies;* a number of citizens waited on the Captain, and informed him that the exportation of Sheep was contrary to a Resolution of the Continental Congress, and thereupon obtained his promise that they should be re-landed, and not carried out of the Harbour. The people were satisfied, and patiently waited till evening, when a report prevailing that the vessel was to sail that night, about two hundred inhabitants assembled on the wharf, appointed and sent four persons to wait on the Committee of Correspondence, and request their advice concerning the measures proper to be taken. By their advice, the Merchant to whom the vessel came consigned, was sent for, and desired to cause the Sheep to be landed, and delivered to one of the Committee appointed on this occasion by the people, which person gave his promise to return the Sheep as soon as the vessel had sailed. Accordingly the Sheep were landed, delivered, and soon after the vessel was sailed, returned to the proprietor; on which the people, being well satisfied, peaceably dispersed.

Proceedings of the Virginia Convention, 1775

From "Proceedings, Virginia-Convention," Peter Force, ed., *American Archives:* Fourth Series (Washington, 1839), II, pp. 170–72.

MONDAY, MARCH 27, 1775.

The Committee appointed to prepare a plan for the encouragement of Arts and Manufactures, reported the following Resolutions; which, being severally read, were unanimously agreed to.

Whereas, it hath been judged necessary for the preservation of the just rights and liberties of *America*, firmly to associate against Importations; and as the freedom, happiness, and prosperity of a State greatly depend on providing within itself a supply of articles necessary for subsistence, clothing, and defence; and whereas, it is judged essential, at this critical juncture, to form a proper plan for employing the different inhabitants of this Colony, providing for the poor, and restraining vagrants and other disorderly persons, who are nuisances to every society; a regard for our Country, as well as common prudence, call upon us to encourage Agriculture, Manufactures, economy, and the utmost industry: Therefore, this Convention doth Resolve as follows:

Resolved unanimously, That it be earnestly recommended to the different Magistrates, Vestries and Church-wardens throughout this Colony, that they pay a proper attention, and strict regard to the several Acts of Assembly made for the restraint of vagrants and the better employing and maintaining the poor.

Resolved unanimously, That from and after the first day of *May* next, no person or persons whatever ought to use, in his or their families, unless in case of necessity, and on no account sell to butchers; or kill for market, any Sheep under four years old; and where there is a necessity for using any mutton in his, her, or their families, it is recommended to kill such only as are least profitable to be kept.

Resolved unanimously, That the setting up and promoting Woollen, Cotton, and Linen Manufactures ought to be encouraged in as many different branches as possible, especially Coating, Flannel, Blankets, Rugs, or Coverlids, Hosiery, and coarse Cloths, both borad and narrow.

Resolved unanimously, That all persons having proper lands for the purpose, ought to cultivate and raise a quantity of Flax, Hemp, and Cotton, sufficient not only for the use of his or her own family, but also to spare to others on moderate terms.

Resolved unanimously, As Salt is a daily and indispensable necessary of life, and the making of it amongst ourselves must be deemed a valuable acquisition, it is therefore recommended that the utmost endeavours be used to establish Salt Works, and that proper encouragement be given to Mr. *James Tait,* who hath made proposals, and offered a scheme to the publick, for so desirable a purpose.

Resolved unanimously, That Saltpetre and Sulphur, being articles of great and necessary use, the making, collecting, and refining them to the utmost extent, be recommended, the Convention being of opinion that it may be done to great advantage.

Resolved unanimously, That the making of Gunpowder be recommended.

Resolved unanimously, That the manufacturing of iron into Nails and Wire, and other necessary articles, be recommended.

Resolved unanimously, That the making of Steel ought to be largely encouraged, as there will be a great demand for this article.

Resolved unanimously, That the making of different kinds of Paper ought to be encouraged; and as the success of this branch depends on a supply of old Linen and Woollen Rags, the inhabitants of this Colony are desired, in their respective families, to preserve these articles.

Resolved unanimously, That whereas Wood Combs, Cotton and Wool Cards, Hemp and Flax Heckles, have been for some time made to advantage in some of the neighbouring Colonies, and are necessary for carrying on Linen and Woollen Manufactures, the establishing such Manufactures be recommended.

Resolved unanimously, That the erecting Fulling Mills and mills for breaking, swingling, and softening Hemp and Flax, and also that the making Grindstones be recommended.

Resolved unanimously, That the brewing Malt Liquors in this Colony would tend to render the consumption of foreign Liquors less necessary. It is therefore recommended that proper attention be given to the cultivation of Hops and Barley.

Resolved unanimously, That it be recommended to all the inhabitants of this Colony, that they use, as the Convention engageth to do, our own Manufactures, and those of other Colonies, in preference to all others.

Resolved unanimously, That for the more speedily and effectually carrying these Resolutions into execution, it be earnestly recommended that Societies be formed in different parts of this Colony; and it is the opinion of this Convention, that proper Premiums ought to be offered in the several Counties and Corporations, to such persons as shall excel in the several branches of Manufactures, and it is recommended to the several Committees of the different Counties and Corporations, to promote and encourage the same to the utmost of their power.

The Members of the Convention then, in order to encourage Mr. *James Tait,* who is about to erect Salt Works, undertook, for their respective Counties, to pay the sum of Ten Pounds to *Robert Carter Nicholas,* Esquire, for the use of the said *James Tait,* on or before the 10th day of *May* next.

His Excellency the Governor having, by Proclamation bearing date the 21st day of *March*, in the present year, declared that His Majesty hath given orders, that all vacant Lands within this Colony shall be put up in lots at publick sale, and that the highest bidder for such lots shall be the purchaser thereof, and shall hold the same subject to a reservation of one-half penny sterling per acre, by way of annual quitrent, and of all Mines of gold, silver, and precious stones, which terms are an innovation on the established usage of granting Lands within this Colony:

Resolved, That a Committee be appointed to inquire whether His Majesty may, of right, advance the terms of granting Lands in this Colony, and make report thereof to the next General Assembly or Convention; and that, in the mean time, it be recommended to all persons whatever to forbear purchasing or accepting grants of Lands on the conditions before-mentioned; and that *Patrick Henry, Richard Bland, Thomas Jefferson, Robert Carter Nicholas,* and *Edmund Pendleton,* Esquires, be appointed of the said Committee.

Resolved, That the Delegates from the several Counties in this Colony, as also from the City of *Williamsburgh,* and Borough of *Norfolk*, do, without delay, apply to their respective Counties and Corporations for Fifteen Pounds, current money, and transmit the same, so soon as collected, to *Robert Carter Nicholas,* Esquire, for the use of the Deputies sent from this Colony to the General Congress.

On a motion made,

Resolved, That *Thomas Jefferson,* Esquire, be appointed a Deputy to represent this Colony in General Congress, in the room of the Honourable *Peyton Randolph,* Esquire, in case of the non-attendance of the said *Peyton Randolph,* Esquire.

Resolved, That the said Deputies, or any four of them, be a sufficient number to represent this Colony in General Congress.

Resolved, That the thanks of this Convention be presented to the Rev. Mr. *Selden,* for performing Divine Service, and for his seasonable and excellent Sermon yesterday.

Resolved, That the thanks of this Convention are justly due to the Town of *Richmond* and the neighbourhood, for their polite reception and entertainment of the Delegates.

Mr. *Alexander Purdie* having offered to print the proceedings of this Convention, for the use of the Members thereof, it is ordered, that the Clerk deliver him a copy of the said proceedings for that purpose.

Resolved, That this Convention doth consider the delegation of its members as now at an end; and that it be recommended to the People of this Colony to choose Delegates to represent them in Convention for one year, as soon as they conveniently can.

PEYTON RANDOLPH, *President.*
JOHN TAZEWELL, *Clerk of the Convention.*

Provincial Congress, 1775

From Proceedings of the South Carolina Provincial Congress, 1775," John Drayton, ed., *Memoirs of the American Revolution (Charleston, 1821),* I, pp. 173–74.

The next day, being Sunday, the Congress was opened with the celebration of divine service, by the Rev. Mr. Turquand: after which, a Committee was appointed to adjust some mode of compensation. For, as rice was to be exported, the indigo planters chose to have some mode of compensation, however little satisfactory the same might be; and the other party, sensible the same would never be executed, as either there would not be any occasion for it; or, the hostile situation of affairs would render the compensation a dead letter, very readily agreed to indulge them in the most feasible manner. Upon this, resolutions were passed on the subject; rice being assumed as the basis of valuation, at fifty-five shillings currency the hundred: and as that rose or fell in price, so the other commodities were to rise or fall likewise. Indigo, was valued at 30s. the pound. Hemp at £8 the hundred weight. Corn, at 12s. 6d. the bushel. Flour, of the best sort, at £4 10s.; and of the second sort, at £4, the hundred weight. Lumber, inch pine boards, per 1000 feet, at £20, in Charlestown; and £15, in Beaufort and Georgetown; and other plank and scantling in proportion. Pork, at £13, the barrel. Butter, at £3s. the pound. Resolutions also were passed confirming powers, and appointing Committees to effect the different exchanges of the above commodities, as the conveniences of parties should require; either in kind, or in money.

Provisions for the Continental Army, 1775

From "U. S. Continental Congress, Correspondence, Proceedings, etc.," Peter Force, ed., *American Archives:* Fourth Series (Washington, 1840), III, pp. 45–47.

Whereas, I, the underwritten, *Philip Van Rensselaer,* of the City of *Albany,* and Province of *New-York,* merchant, being appointed by *Walter Livingston,* Esquire, for supplying the Troops under the command of General *Schuyler:* and as a large quantity of barrelled pork will be wanted for supplying the said Troops, and at present finding a great scarcity of that article in this Province, and am informed, cannot possibly be supplied with a sufficient quantity requisite and necessary for said Troops, without having assistance from the Province of *Connecticut;* in consequence of which begs leave to request, that the honourable Provincial Congress, or Committee now sitting for the said Province of *New-York,* would be pleased to take the same into their consideration; and humbly conceives, that were they to write to Governour *Trumbull,* or any other person or persons, which they may think proper, requesting that leave may be given to ship four hundred and fifty barrels good merchantable pork, they might be supplied, and consign the same to the care of Messrs. *Dennis* and *Dawson* of said City of *New-York,* merchants, who have my particular directions to receive the said pork on my account, provided leave can be obtained.

PHILIP VAN RENSSELAER.

In Provincial Congress, New York, August 8, 1775.

Ordered, That Governour *Trumbull* be requested, and he is hereby requested, by this Congress, to permit the quantity of four hundred and fifty barrels of good merchantable Pork to be sent to *New-York,* to the care of Messrs. *Dennis* and *Dawson,* to be forwarded for the use of the Continental Army in the northern parts of this Colony.

A true copy from the Minutes:

JOHN MCKESSON, *Secretary.*

David Welsh to Governour Trumbull.

TICONDEROGA, *August 5, 1775.*

SIR:

The men at this place, belonging to the Colony of *Connecticut,* think they are not well used, as they were promised several things, they don't think there are any steps taken to fulfil it, our Commissaries being superseded by Commissaries in New-York Government; and they avow the principles, that if soldiers have bread and pork, it is enough; and Captain *Phelps* has wrote to the Colonel, that they tell him expressly that he has no business to buy any thing, but only to forward provisions. I heard a few days ago, that he is dismissed, but I don't know the certainty of that.

Several of the companies have no brass kettles to this day. About a week ago I got one for my company, and don't think I shall have any more this year. Pails and bottles we can't get as yet; and not more than one tenth part of the bowls that we were to have.

Some things commanded in the act of Assembly for the soldiers can't all be got here, but the chief of them might be got as well here as at *Boston,* but they would cost something more; and if some things can't be had, there is the more reason for having others; and if there had not been a shifting of Commissaries, I believe we should have been better provided. Several companies have no frying-pans. I have afore notified our Commissaries and then at *Albany,* that we want these things.

Our water here is very bad and unwholesome, and great part of the time there is nothing else for the Troops. At the present we have some beer, but it wont last long, and if our Commissaries do not get some, I don't think any body else will. Rum and molasses are wanted. The rum that comes, as far as I have seen, is worse than none. We expected to have had books and paper, but have not had one book, and but four quires of paper. I think there has not been one pound of soap brought for the Army. A small matter of coffee and chocolate was bought about two weeks ago, so that the sick have a small matter, but none for them that can keep about. Only one barrel of vinegar here and one at *Crown Point* has arrived, and that, all said, was not worth any thing. One barrel of sugar came here, and one to *Crown Point,* and that goes only to the sick. Since the Troops arrived here, it would take about half of their wages to make them live as well as they were to be provided by the act of the Assembly.

There are five companies from *York* Government at *Lake George,* and they have their complement of officers. They were to have forty or fifty men each, but I am well informed that they have not above eighty men on the ground. There is not a soldier from *York* come over *Lake George* to stay, and I don't think we shall have any before the middle of *September* or *October,* unless something is done more than a *New-York* Congress will do. If you and our Colony rely on them to fulfil the engagements of our Colony, I assure you, Sir, that they are determined never to do it. They have not a soldier on this side of *Lake George,* that I know of, except a few sailors. Several officers are arrived, and more expected; and why all the places of profit should be filled up with men in *York* Government, I don't know, and our people be obliged to do all the drudgery. Commissaries' places are profitable, and commanders on board of the vessels profitable. And why should they have all the places of profit? Is it because we have no man capable of any thing but drudgery? Sir, unless you or somebody sees to it, I don't think we shall have one hundred and fifty men here by the middle of *September* or *October* from *New-York* Government. The advantage of their situation is such that it will make them rich. Are we to be wholly ruled by the Committee of *New-York?* Is it for their unfaithfulness in the common cause? Have they not been till very lately, a great part of them, as strong set against the common cause? Neither have I any reason to think that there is a thorough change in them. Why should Mr. *Halsey* be dismissed from the service to make way for a *Yorker,* when every man says he did well? Are our men fit for nothing but privates? If there is not a check put on them, you, Sir, will be put to it to raise men another year. They have a number of carpenters, and the building of batteaus, &c., goes on well; but upon a par, I suppose, it will take six of our men to replace as much money as one of them on a par, and one of our men will do as much as six of them.

I am informed that the Continental Congress are to give out commissions; if they do, unless it is well looked to, there will be a great number of officers, and but a few soliders. Sir, you may rely upon it that the *New-York* Commissaries will not attempt to fulfill what the Colony of *Connecticut* have engaged; and unless they take some steps to do it themselves, I don't see how they can answer it, to promise great things, and not to take suitable care to fulfil.

Sir, please to excuse me for my troubling you with this letter.

I am, Sir, your most obedient and humble servant, &c.

DAVID WELSH.

To *Jonathan Trumbull,* Esq., Governour of *Connecticut,* &c.

A Virginian on Dye Plant, 1775

From "U. S. Continental Congress, Correspondence, Proceedings, Etc.," Peter Force, ed., *American Archives:* Fourth Series (Washington, 1840), III, p. 716.

James Stewart to the People of Virginia.

WILLIAMSBURGH, *September 15, 1775.*

The subscriber, who is an inhabitant of Virginia, and just returned from *England* (where he has been for these eighteen months past, on purpose to make himself acquainted with the culture and preparation of several dyes) has brought in with him the seeds and roots of madder, woad, and welde, (commonly called dyer's weed,) which are the fundamental dyes of all colours, either in the linen, cotton, or woollen manufactures, with a view to propagate them, and makes no doubt of being able to afford them full as cheap as they are sold in *England.* He has likewise brought in the seeds and roots of the aranatto, which dyes yellow and pompadour colours; also, the genuine rhubarb and licorice plants, with some thriving olive trees, &c., &c. But as the cultivation of them all is too much for him to undertake, he offers to supply any gentleman, or company of gentlemen, in *Virginia,* with seeds and roots, and to instruct them how to prepare them for the manufacturers; and as the utensils for preparing the different articles for market are to be had in the Country, at a small expense, nothing else is required but the labour of one hand for every five acres. Madder sells in *England,* according to the quality, from ten pence to two shillings and five shillings per pound; woad from eighteen pounds to twenty pounds a ton, four or five crops of which may be made yearly in *Virginia;* and welde is worth five shillings a sheaf, but, for the convenience of exportation, it is intended to manufacture it as they do indigo.

He also offers to instruct one or two ingenious spinning wheel makers, that may be appointed by any County Committee, to make a machine, or wheel, for spinning cotton, with which one hand may spin from fifteen to thirty threads at a time; and he expects no further recompense than as the merit of the machine may appear to deserve.

All persons who intend applying to him must be expeditious, as the land for the cultivation of the above articles ought to be prepared this fall. He may be spoke with at Mrs. *Vobe's,* for these eight or ten days; afterwards at *Winchester,* in *Frederick* County; and all letters for him may be left at the constitutional post-office in this City, directed to the care of Mr. *Alexander Wodrow,* merchant in *Falmouth.*

JAMES STEWART.

Food Shipments to Baltimore, 1775

From "Baltimore County Committee," in Peter Force, ed., *American Archives:* Fourth Series (Washington, 1843), IV, pp. 1732–33.

At a meeting of the Committee, *November 27, 1775:*

Application being made to this Committee by Mr. *Joshua Hilton*, to load his Sloop with Flour, &c., for *New-England*, saying that they were in great want of such Provisions,

Resolved, That the Committee of Correspondence do immediately write to the *New-England* Delegates to know whether Provisions are really scarce in those Colonies.

A copy from the Minutes:

GEORGE LUX, *Secretary.*

N. B. *David McMechen* attended at the meetings of the Committee from *November* 6 to *December* 18, inclusive, *George Lux* being absent out of Town.

At a meeting of the Committee, *December 4, 1775:*

Application having been made to the Committee by *Lawrence Callohoun* and *Robert Clark*, for liberty to transport to *Annapolis*, by water, some Beef and Sheep for the use of the inhabitants of that city during the sitting of the Convention, a certificate was granted for eight quarters of Beef and twelve Sheep to the said *Callohoun* and *Clark*, to transport the same, they, and the skipper of the boat, making oath that they will land the same at *Annapolis*, and procure a certificate of such landing from the Committee of that place, to be returned to this Committee at a future meeting.

The Committee having received from the Committee of *Philadelphia*, a Letter from *Daniel Chamier,* Esq., Commissary of Stores at *Boston*, with a Commission and other papers directed for *Robert Long*, appointing him Preventive Officer, Weigher, and Guager, for *Baltimore* Town; on consideration thereof,

It was *Resolved*, To transmit the said Commission and Papers to the Convention at *Annapolis*, to know their opinion whether the said Commission should be delivered to the said *Robert Long*.

Ordered, That the copies of the Enrollments of the Militia Companies that have been delivered in to the Committee, be transmitted to the Provincial Convention.

The Committee appointed Messrs. *Samuel Purviance, John Boyd,* and *Darby Lux*, to prepare, against *Monday* next, a Memorial, representing to the Convention the necessity of fortifying *Baltimore*.

GEORGE LUX, *Secretary.*

At a meeting of the Committee, *Monday, December 11, 1775:*

The following is a copy of a Letter from the *Massachusetts-Bay* Delegates in Congress to this Committee, on the subject of shipping Flour to the *New-England* Governments:

"Philadelphia, December 4, 1775.

SIR: We acknowledge the receipt of your letter of the 27th *November*, wrote by order of your Committee, upon the subject of permits for shipping provisions to *New-England*. In reply to which, we observe, that the *New-England* Colonies stand constantly in need of supplies of bread, flour, and corn, from your country, more especially the Colony of the *Massachusetts-Bay*, where the Continental Army are now fixed. The exportation of the articles before-mentioned ought to be under a very strict regulation, to prevent any misapplication of what may be intended for our friends there. We submit it to your consideration, whether it would not be advisable to require either a certificate from some Committee of Inspection in those Governments, and, where such certificate cannot be conveniently obtained, and the people applying for such permits are residents of *New-England*, or of your Colony, to require the shipper, or the master of the vessel in which the provisions are to be exported, to give bond that they shall be landed or delivered to our friends in those Governments, and to oblige the master to make oath, that he will use his best endeavours, that they shall be so landed and delivered.

We are, with great respect, your most humble servants,

"JOHN HANCOCK,
"THOMAS CUSHING,
"SAMUEL ADAMS,
"JOHN ADAMS.

Rhode Island Attempts to Stop Sales to Great Britain, 1775

From "U. S. Continental Congress, Correspondence, Proceedings, etc.," Peter Force, ed., *American Archives:* Fourth Series, (Washington, 1849), III, pp. 974–75.

At a Meeting of the Committee of Inspection of the Several Towns in the County of Providence, on *Friday, the 6th of October, 1775:*

Whereas there has been great suspicion among the inhabitants in this County, as well as in the Towns of the Colony of *Massachusetts-Bay*, that our cruel and unnatural enemies, from time to time, receive supplies of butter, cheese, and other provisions, by reason of the large quantities carried to the Town of *Newport*, &c., under pretence of supplying that place, *Nantucket*, &c.: And whereas, whilst it is our duty to have our friends, wherever they may be, reasonably supplied, it is equally our duty to prevent our enemies from receiving succours of any kind. It is therefore the opinion of this Committee, that no Butter, Cheese, or any other articles of Provisions, be hereafter transported, either by land or water, to *Rhode-Island*, except under the following regulations, viz:

Those persons who may have the beforementioned articles to dispose of, shall, before they proceed to market, procure a certificate or certificates from the Committees of the respective places where such articles are laden, of the quantities of each article they may carry, and that they are persons friendly to the cause of *American* freedom, which shall serve as a pass through the respective Towns to the market: and provided that the said Provisions are destined for *Rhode-Island*, that they carry the same, together with the certificate, to *John Collins*,

Esquire, Chairman of the Committee of Inspection in the Town of *Newport,* or to some person by him to be appointed; whereupon, they may dispose of their Butter or other articles, to such persons as the said *John Collins*, or his substitute, may recommend to be friends to their Country, and to none other. And upon their return from the market they shall deliver certificates, signed by the said *John Collins*, Esq., or his substitute, as aforesaid, of the quantity disposed of, and to whom, unto such persons from whom they received their certificates; and the same shall discharge them from all cause of suspicion with their Town and countrymen. And all persons travelling with Butter, Cheese, &c., in any considerable quantities, without such certificate and certificates, shall be liable to be detained, and their goods, &c., stopped, until, at their expense, such certificates are procured, or until the order of the Committee of Inspection of such District where the same may be stopped, be taken there-upon.

And whereas the honourable General Assembly of the *Massachusetts-Bay* have taken particular order with respect to the transportation of Provisions to *Nantucket;* and it is altogether unnecessary, and may be very prejudicial, to attempt to supply that island from any part of this Colony, by water; therefore, it is

Resolved, That no Provisions of any kind be suffered to pass through any sea-port Town of this Colony, under pretence of sending them to *Nantucket;* but all such Provisions shall be stopped, until the matter may be inquired into by the Committee of Inspection of the District where they may be stopped, and such order taken thereon as they shall see fit.

Resolved, That all persons who shall endeavour to elude these Resolutions, or in anywise counteract them, shall, upon conviction, have their names published in the Newspapers, in order that they may be avoided, as enemies to their Country.

Voted, That the above and foregoing Resolutions be immediately printed in the *Providence Gazette.*

Voted, That Captain *Solomon Owen*, Messrs. *John Brown, Joseph Russell, Job Manchester*, and *Noah Mathewson*, be a Committee to receive any complaints that may be exhibited to them, and lay the same before this Committee at their next meeting, which stands adjourned to the house of the Widow *Waterman*, in *Smithfield,* on the second *Tuesday* in *November* next, at ten o'clock, A. M.

Whereas there have been many complaints of such as are venders of Goods and Merchandise in this County, for selling them at higher prices than settled by the Association Agreement of the Continental Congress, under pretence of buying them at a higher rate, which we deem a breach and violation of said Association: We do hereby forewarn all persons from selling any Goods at a higher price than they were usually sold at before said Association took place, on any pretence whatever, as they will thereby incur the just censure of this Committee; and their names will be published to the world accordingly.

CALEB HARRIS,
Clerk of the County Committee.

Virginia Ordinance Establishing Tobacco Payments, 1776

From "Virginia Convention," in Peter Force, ed., *American Archives:* Fourth Series (Washington, 1843), V, pp. 142–43.

AN ORDINANCE FOR ESTABLISHING TOBACCO PAYMENTS DURING THE DISCONTINUANCE OF THE INSPECTION LAW, AND FOR OTHER PURPOSES THEREIN MENTIONED. [JANUARY 1776]

Whereas, by reason of the expiration of an Act of General Assembly, for improving the staple of Tobacco, and preventing frauds in his Majesty's Customs, the people in this Colony may be subjected to great difficulties for want of a certain mode of making Tobacco Payments for levies or other debts, and sundry disputes may arise between them and the officers or creditors, which may increase the confusions in the Colony, already too much convulsed by the unhappy disputes with *Great Britain:*

For remedy whereof, *Be it ordained, by the Delegates and Representatives of the people of this Colony, now met in General Convention, and by the authority of the same,* That it shall and may be lawful for any person who shall be indebted for levies or other demands, payable in Tobacco, to discharge the same by good, sound, and merchantable Tobacco, leaf or stemmed, tied up in bundles, and clear of trash and dirt; and if the collector or creditor shall refuse to receive Tobacco tendered in such payment, on account of its not being clean, sound, or merchantable, it shall be referred to two judicious neighbours, to be chosen one by each of them, (or if one shall refuse to nominate, the other may choose them both,) who, being sworn to give an impartial judgment, shall determine the point between the parties; and if they disagree, they shall choose a third person, who shall be sworn in like manner, and his judgment shall be final; the payments of levies and rents to be made on the plantation of the debtor, with a reasonable allowance, in cases of rents, for the charge of carrying the same to the next inspection; in other cases, at the place appointed by the contract.

And be it further ordained, That where the Vestries shall not have compounded with their Ministers for his receiving money in lieu of Tobacco for his salary, according to a late Act of Assembly, in such case the Collector of the Parish Levy shall convey the Tobacco, so to be received for levies, to the house of the Minister, who shall receive the Tobacco, so brought from time to time, until his full salary of sixteen thousand pounds of Tobacco, with the allowance of four per cent. for cask, and four per cent. for shrinkage, with the usual expense for transporting the same to the nearest publick landing, on some navigable river, is fully paid; and the residue of the Tobacco, so to be received, shall be by the Collector carefully prized up into hogsheads, and sold, according to the directions of the last mentioned act; but this is not to extend to or affect such Counties or Parishes where by law the inhabitants are allowed to pay their levies at a certain price in money.

And be it further declared and ordained, That the several Vestries shall be empowered to levy for the Collector of their several levies, such additional allowance for his trouble in collecting the Tobacco in manner aforesaid, as to them shall seem reasonable, according to the extent of the Parish; and shall also allow the Minister two Shillings and six Pence for every thousand pounds of Tobacco by him received for his salary as aforesaid, for prizing up the same.

Recommendations of the New York Committee of Safety, 1776

From "New York Committee of Safety," in Peter Force, ed., *American Archives:* Fourth Series, (Washington, 1884), V, pp. 1457–58.

In Committee of Safety, New-York, April 19, 1776.

The Continental Congress have, by Resolve, on the 21st of *March* last, recommended it to the several Assemblies, Conventions, and Councils or Committees of Safety, and Committees of Correspondence and Inspection, that they exert their utmost endeavours to promote, among other things, the culture of Hemp, Flax, and Cotton, and the growth of Wool in the United Colonies.

It is therefore most earnestly recommended to the inhabitants of this Colony to attend to the following considerations, which, exclusive of the best regard to the publick good, must engage every Farmer, from a just attention to his own private interest, to the increase of the staples of Hemp, Flax, and Wool.

By the restraints which the Parliament of *Great Britain* have most tyrannically imposed on our commerce, and the danger thence arising on the exportation of Provisions, the husbandman has no inducement to employ his skill and industry in the produce of those articles, beyond the necessity of home consumption. It is therefore evident that the culture of grain must, under our present circumstances, naturally diminish. This diminution ought, from motives both of private interest and publick utility, to be compensated for by the improvement of our lands, in such a way as will most infallibly be attended with great profit to the landholder. As, by the danger to which our exports are exposed, the usual growth of grain will necessarily be discouraged, it will be laudable economy to devote a larger part of our lands than usual to the culture of Hemp and Flax, and the pasturing of Sheep.

The present great scarcity of Linen and Woollen Goods will be increased to a distressing degree by the continuance of arbitrary Parliamentary restraints on our Trade, and prudent Continental restrictions on our imports; the latter of which would, indeed, necessarily fail, were there no other reasons for it than the want of sufficient exports to support them. From this scarcity we may, in a great measure, be relieved, without loss to the publick, and with manifest profit to the farmer, by improving our lands in such a manner as will furnish both clothing and employment for our inhabitants. Our soil and climate naturally invite us to it; but our necessities, and the prosperity of the husbandman, clearly command it. It is doubtless the most advantageous use to which, in ordinary times, we can apply our lands; but, at this juncture, the advantage will be so highly improved by the enhanced prices of Hemp, Flax, and Wool, that every farmer who neglects to take uncommon pains for the increase of those necessary articles, will be most culpably inattentive to the general weal, and his own private interest. Little, therefore, woul need to be said on this head, were we not urged to it by its vast importance. For this reason, this Committee do earnestly recommend it to every farmer in this Colony to exert himself to raise large quantities of Hemp and Flax, and to increase his stock and improve his breed of Sheep. And in order to increase the breed of Sheep, it is earnestly recommended to the inhabitants of this Colony not to kill any Lambs nor any Ewes under four years old, until further order of the Provincial Congress. And lest any farmers or butchers should be so far lost to all sense of publick virtue as to disregard this recommendation concerning the increase of our flocks of Sheep, it is most ardently recommended to the inhabitants of this City totally to abstain from the purchase of Lamb, and of such Ewe mutton as they shall discover to have been, at the time of killing, under four years old; and are requested to report to this Committee, or the Provincial Congress, all such persons as shall attempt to sell or purchase any Lamb or Mutton contrary to this recommendation.

Ordered, That the foregoing Recommendations be forthwith published in all the publick Newspapers in this Colony.

The Committee adjourned to four o'clock, in the afternoon.

Resolutions on Former Indian Lands in Virginia, 1776

From Peter Force, ed., *American Archives:* Fourth Series (Washington, 1843), VI, p. 1044.

In Convention, Virginia, June 24, 1776.
Whereas divers petitions from the inhabitants on the Western frontiers have been presented to this Convention, complaining of exorbitant demands made on them for Lands claimed by persons pretending to derive titles from *Indian* deeds and purchases:

Resolved, That all person actually settled on any of the said Lands ought to hold the same, without paying any pecuniary or other consideration whatever to any private person or persons, until the said petitions, as well as the validity of the titles under such *Indian* deeds and purchases, shall have been considered and determined on by the Legislature of this country; and that all persons who are now actually settled on any unlocated or unappropriated Lands in *Virginia*, to which there is no other just claim, shall have the pre-emption or preference in the grants of such Lands.

Resolved, That no purchases of Lands within the chartered limits of *Virginia* shall be made, under any pretence whatever, from any *Indian* tribe or nation, without the approbation of the *Virginia* Legislature.

EDMUND PENDLETON *President.*

Crops in Rhode Island, 1776

From Thomas Vernon, *Diary* (Providence, 1881), p. 44

Wednesday, July 31, 1776. [Glocester, R. I.]
Arose ten minutes before five. A very clear, calm, fine morning. A great dew last night. We had radishes with our tea for breakfast. Afterwards walked down to the river. . . . Afterwards we all took a walk to the northward. Can't help noticing that the farmers in the town are much backward in their work. The grass and grain suffer greatly for not being cut in season, and the corn for not being hoed. The reason is plain that there are not people left to do the necessary work.

Land Bounties for Army Men, 1776

From Edmund C. Burnett, ed., *Letters of Members of the Continental Congress* (Washington, 1923), II, p. 103.

Elbridge Gerry to Joseph Trumbull

Philadelphia Sepr 26th 1776

DEAR SIR

I have only Time to advise You that General Schuyler finds himself so uneasy at the Northward from the Reflections of the people that he proposes to resign, from which We have Reason to hope that Harmony will ensue. the Army is to consist the ensuing Year of eighty eight full Batalions to be enlisted for the War, the Officers to have Bounties of Land, the Men one hundred Acres each and twenty Dollars. the Express is Waiting. my Respects to all Friends and believe me to be

YOURS SINCERELY
E GERRY

Colo Trumbull

Provisions for Feeding the Army, 1776

From "U. S. Continental Congress, Proceedings, December 26, 1776-December 28, 1776" in Peter Force, ed., *American Archives: Fifth Series* (Washington, 1853), III, pp. 1612, 1614.

Resolved, That *J. Trumbull,* Esq., Commissary-General, be empowered to import, at the Continental risk, from *Virginia* and *Maryland,* and the other Southern States, such quantities of Flour and other provisions as he may judge necessary for the support of the Army.

Resolved, That the Delegates of *Virginia* be empowered and directed to write to the Governour and Council of their State, and request them to contract with proper persons for the delivery of 10,000 barrels of Flour on *James, York, Rappahannock,* and *Potomack* Rivers, to the order of *Joseph Trumbull,* Commissary-General of the Continental Army, or of a larger quantity should he require a further supply; the said *Joseph Trumbull* to send Vessels to take it in, and to pay for, or draw orders on the President of Congress for the payment of the same. . . .

This Congress, having maturely considered the present crisis, and having perfect reliance on the wisdom, vigour, and uprightness of General *Washington,* do, hereby,

Resolve, That General *Washington* shall be, and he is hereby, vested with full, ample, and complete powers to raise and collect together, in the most speedy and effectual manner, from any or all of these *United States,* sixteen Battalions of Infantry, in addition to those already voted by Congress; to appoint Officers for the said Battalions; to raise, officer, and equip three thousand Light-Horse, three Regiments of Artillery, and a Corps of Engineers, and to establish their pay; to apply to any of the States for such aid of the Militia as he shall judge necessary; to form such Magazines of Provisions, and in such places as he shall think proper; to displace and appoint all officers under the rank of Brigadier-General, and to fill up all vacancies in every other department in the *American* Army; to take wherever he may be, whatever he may want for the use of the Army, if the inhabitants will not sell it, allowing a reasonable price for the same; to arrest and confine persons who refuse to take the Continental currency, or are otherwise disaffected to the *American* cause; and return to the States of which they are citizens their names, and the nature of their offences, together with the witnesses to prove them:

That the foregoing powers be vested in General *Washington,* for and during the term of six months from the date hereof, unless sooner determined by Congress. . . .

Saturday, December 28, 1776.

The Committee to whom the Report of the gentlemen who were directed to repair to *Toconderoga,* and the Papers therein mentioned, were referred, brought in a Report, which was taken into consideration: Whereupon,

Resolved, That in the opinion of Congress, the Northern Army may be supplied more advantageously and conveniently, in the present mode of the

Commissary-General's governing himself by such regulations as have been, and may be, from time to time, ordained by the General or Commander-in-Chief, than by Contractors:

That the Commissary-General be directed, without fail, to supply the Northern Army this winter with Vegetables twice in every week, or more frequently, if possible, and to take effectual measures that they be well supplied with Vinegar: . . .

Maryland's Plans for Feeding the Army, 1778

From "Journal and Correspondence of the Council of Maryland, 1777–1778," in William Hand Browne, ed., *Archives of Maryland* (Baltimore, 1897), XVI, pp. 456–57.

Wednesday 7th January 1778

William Winder Junr Ephraim Stevens and Levin Woolford are appointed Purchasers of Cattle for Somerset County. John Postley, John Richardson and William Wise for Worcester County and Copies of the following Instructions were sent

Whereas the Board of War has requested of this State as well as of other States an immediate Supply of Provisions for the Use of the Continental Army. This Board desirous of complying with the said Requisition and being fully convinced that it will much conduce to the general Interest as well as the Interest of the Proprietors of Cattle at places from whence they may be readily taken away by the Enemy that such of those Cattle as are fit for Beef should be immediately collected at a full and just price and applied to the Subsistance of the Army.

Therefore Messrs William Winder jun^r Ephraim Stevens and Levin Woolford of Somerset County or any or either of them is hereby appointed to purchase and Collect the Cattle in the said County which are fit for Slaughter for the use of the said Army and he is to procure the same on Contract if the Proprietors will sell them for a just price. But if the proprietors of such Cattle along the sound and on the Water sides from whence they may readily taken by our Enemies on board the Men of War refuse to dispose of the same then the said William Winder Junr Ephraim Stevens and Levin Woolford or any or either of them is impowered and required to Seize such Cattle for the use and Subsistance of the Army as aforesaid leaving if necessary sufficient for the Subsistance of the Owner and his Family and paying the value of the Cattle so seized at the rate as near as can be estimated of one Shilling per pound for the neat Beef of good fatted Cattle and nine pence per pound for inferior and adding one fourth part of the sum for the fifth Quarter. And the said William Winder jun^r Ephraim Stevens and Levin Woolford as soon as convenient are to have the weight of the Cattle so seized estimated by an Honest man or two who are good Judges thereof if the proprietor of the Cattle and the person seizing the same shall disagree therein and the said William Winder Junr Ephraim Stevens and Levin Woolford are also desired to send all Cattle by them collected as soon as may be to the Camp under twenty Drovers to the Commissary of Purchases together with an Account of the cost of the Cattle and to send a Copy of the said Account to this Board.

Continental Congress Limits Exports, 1778

From W. C. Ford, ed., *Journals of the Congressional Congress* (Washington, 1908), IX, pp. 569–70.

Whereas, it hath been found by Experience that Limitations upon the Prices of Commodities are not only ineffectual for the Purposes proposed, but likewise productive of very evil Consequences to the great Detriment of the public Service and grievous Oppression of Individuals;

Resolved, That it be recommended to the several States to repeal or suspend all Laws or Resolutions within the said States respectively limiting, regulating, or restraining the Price of any Article, Manufacture or Commodity.

Whereas, the Practice of exporting Wheat, Rice, Rye, Indian Corn, Flour, Bread, Beef, Pork, Bacon, live Stock, and other Provisions hath been attended with the pernicious Consequences not only of raising the Price of such Articles and streighthening the Armies of these States for Subsistence, but also of affording Supplies to their Enemies, thereby enabling them more effectually to prosecute the present unjust War,

Resolved, That it be recommended to the several States to take effectual Measures for preventing the Exportation of the said Articles, or any of them, excepting so much as may be necessary for the Crews of Ships or Vessels of War, or of such as may be laden with other Merchandizes untill the   Day of   next, and for punishing all Persons who under Color thereof may

Resolved, That the Governors of Virginia and Maryland be requested to forward immediately by Water to the Head of Elk, the Provisions purchased for the Use of the continental Army, within those States, and which lie contiguous, or convenient to Navigation, and to take such Measures for that Purpose as they shall deem most expedient.

Whereas, by a change of circumstances in the commerce of these states, the regulation of prices lately recommended by Congress may be unnecessary; and the measure not being yet adopted by all the states: therefore,

Resolved, That it be recommended to the legislatures of the several states that have adopted it, to suspend or repeal their laws made for that purpose.

Letters on Illegal Exports

From Edmund C. Burnett, ed., *Letters of Members of the Continental Congress* (Washington, 1926), III, p. 411.

Henry Laurens to the President of South Carolina (Rawlins Lowndes)

15th September [1778.]

DEAR SIR

I had the honor of addressing Your Excellency on the 10th Instant by Mr. Frisch. on that or the next day Congress framed a Resolve earnestly recommending to the Government of Maryland to take Measures for preventing infractions on the General Embargo, but yesterday we received repeated Intelligence of illicit exportations of Provisions from that State. it appears to be an indubitable truth, that refraining on our part from Exportation of Provisions would be a more sure means of driving the Enemy from our Coast, than all our troops. If we export freely the Enemy will be supported and our Army starved. . . .

Francis Lewis to the Governor of Maryland (Thomas Johnson, Jr.)

SIR

. . . Various complaints have been lately made to Congress, purporting that some Traders at Baltimore are engrossing the flour and other provision, and shipping off the same in direct violation of the Embargo, which if not immediately remedy'd will tend to ruin our Armies and the Fleet of our Allies. . . .

Letter to Henry Laurens Regarding Food Supplies

From: "Letters of Joseph Clay," in Georgia Historical Society, *Collections* (Savannah, Ga., 1913), VIII, pp. 75–78.

Savannah, May 30th, 1778.

SIR

I shoud have wrote you before this informing you with the State of your concerns, here so far as they had come to my Knowledge, had I not expected to have been with you before this when I shou'd have had the pleasure of acquainting you with them in person & which nothing has prevented but the very distressed condition of our State the Continual depredations from E. Florida of almost every kind both by Land & Water the Machinations of our Enemies amongst ourselves supported & Encouraged from that country, w'ch has too evidently shewed itself amongst us lately, particularly in the back parts of your State held so dismal a prospect to View as rendered the Idea of leaving a Wife & Eight Children (the Eldest of whom between 13 & 14 years of age) intolerable, & I may say to the highest degree imprudent at this juncture as they woud in case of any Invasion or Public Calamity during my absence have been intirely helpless — We are now with Assistance of your State carrying on an Expedition against that Country which if Successfull will remove many of our distresses, give Security to our State & promote the general Cause Necessity has Reduced & drove us to this Step, we coud not debate on the subject, matters were Ripening so fast that either we must reduce them or they would subdue us, the Daily increase of Men they were obtaining from the defection of the back Country people woud have made them so formidable in a Short time that they woud have over run us, & rendered our Situation so very uncomfortable that we must have either quitted the State or Submitted to them Whereas if Succeed every thing will go on well with us in a short time Our Trade will increase w'ch has been greatly annoyed from Augustine Our Lands will be better cultivated, & we shall have more time to attend to & regulate our internal Police & fall on some means to pay off & Fund our Debts the depredation of the Floridians to the So'ward for some time past preceeding our Troops marching that way made it extremely dangerous the attempting to fetch any Rice from Broughton Island a party of them were once or twice on the Island though I do not learn that they did any considerable damage there, since I wrote you last, we have had 1066 Bushels brought to Sunburry & sold there a 2/ £ Bushel & Mr. Gervais I make no doubt, informd you that he had purchased a Boat for the purpose of bringing Rice from the Island & that he had received some from there w'ch produced a better price than it woud have done here as well as gave a considerable advantage from the difference of Money the Army now Marching to the So'ward have also had some of the Rice, what Quantity we do not know. Mr. Baillie sets off this Day for Broughton Island by Water in Order to bring some more away I get an Account of what has been taken away for the Use of the Army there has always been a White person at the plantation till within some time past, the last person that lived there was killed by some of the Floridians between Broughton Island & Yekly where he had been or was going on some Acc't or other I have been Acting or rather Endeavouring to Act as D. P. M. G. to the Army in this State for between 2 & three Months past, but the want of Money has rendered it impracticable for me to afford any affectual service in that Department. I have at this time Drafts on me drawn by Gen'l Howe for several thousand pounds & not a shilling in hand to answer any part of them I have advanced a considerable sum myself & the Agent, Mr. Wereat, has done every thing in his power, & has supplyd the Gen'l with all the Cash either of his own or the United States that he coud lay his hands on, & I believe the Gen'l has also borrowed of Individuals this is a very distressing Situation & must be attended to or the Consequences may be very fatal I believe I might have borrowed Money at 8 £ Ct. Interest or I coud have obtained Cash for Bills on Congress had I any Authority to draw or Borrow, But I have none that I know of, indeed I never received any Instructions or line on any matter within my Department except one Dated the   March last from the Auditor Gen'l Mr. Gibson inclosing me the Draft for 202,423 Doll's on the Commissioners of the loan Office I have mentioned these matters to Mr.

Telfair, on of our Delegates who I hope will lay them before Congress in Order that some remedy may be apply'd to prevent the Army being at any time hereafter reduced to the same Situation the State has Voted a sum for the Use of the Army say £60,000 if the same shall be necesary part of w'ch is now Striking off & Signing & in 4 or 5 Days I expect to receive a part of it, this will afford a supply for sometime but some provision must be made for the future tis with the utmost reluctance that the State advances a shilling to the Army not so much for want of the will as the means they have very little dependence on raising any sum adequate to its wants But by Borrowing at Eight £Ct. Interest or limitting Paper Bills of Credit When the Money the Congress sent here Mr. Wood arrived we were paying Interest for near £140,000 our Curr'y Borrowed principally for Army Uses And we have Emitted so much in paper Bills of Credit, that the Value of it is reduced to nothing & every Article of produce (Rice excepted) and Merchandise is risen to an extravagant height in price Indico 20/£ lb Corn 12 a 15/£ Bushel Beef 2/£ lb Butter 6/ & 7/£ & Rum 60/£ Gall. Osnabr'gs 12/ a 15/£ yd. & so on the Prices of Goods have been raised from Complicated causes, two principally the one the large Emission of Paper Bills Credit, the other a Spirit of Extortion nursed & kept up by Jews & others worse than Jews who are contin'ly buying up every Species of Goods they can lay their hands on & selling them out again at advanced Prices these kinds of dealers in my opinion are greater Enemies to the United States & do them more injury than the Fleets & Armies of G. Britian. It is with equal reluctance the Army receive their pay in the Curr'y of this State & they complain of it overmuch & w'th some reason as their whole pay for a year if paid to them in this Curr'y will not maintain them at this juncture three Months whereas if it was paid them in the Curr'y of the United States the Case woud be very different as they coud purchase every Article they stand in need of for near half the price they can do w'th our Curr'y. The Continental Troops in this State have never received one farthing in any other Money than the paper Bills of Credit Emitted by this State from the first Hour they came into it to this minute w'ch has been equally prejudicial to the State. I hope you will excuse my trespassing on your time with these very disagreeable Subjects but they Engage my attention so much that when I touch on them I hardly know when to leave off. I can hardly expect the favour of a line from you as very moment of your time I know is & must be devoted to matters of the highest moment & consequence to the United States.

I am w'h respects and regard

D'R SIR YOUR MOST OBED'T SERV'T
JOSEPH CLAY

John Jay on the Destruction of Crops, 1778

From Edmund C. Burnett, ed., *Letters of Members of the Continental Congress* (Washington, 1926), III, pp. 540–41.

The President of Congress (John Jay) to the President of South Carolina (Rawlins Lowndes)

Philadelphia 18th: Decr: 1778.

SIR,

I have the Honor of transmitting to you the enclosed Act of Congress of the 16th: Inst: It was passed at the Instance of the Minister of France to enable him to purchase and export from your State 6,000 Barrels of Rice for the Use of the French-Fleet.

The Middle and Eastern States cannot supply more Wheat this year, than the Inhabitants and Army will consume. New-York, New-Jersey and Pennsylvania have been so much embarrassed and injured by military Operations, as to afford at present but a small proportion of their usual Supplies. The Crops now in the ground indeed are great and promise plenty the next Season. The wheat in Maryland and Virginia and I may add North-Carolina, has been so destroyed or spoiled by a Fly which infests those Countries that but little Flour and that in general of a bad Quality can be procured there.

For these and other reasons the Congress think the proposed measure expedient. Unless there should be some weighty and at present unforeseen Objection to the Exportation in question, I flatter myself South-Carolina will chearfully give our Ally this proof of her disposition and determination to sustain and succour Armaments sent by him to defend the American cause.

I have the Honor to be with great Respect

YOUR EXCELLENCY'S MOST OBEDT: AND HBLE SERVT:

J. J.

The Army Faces A Flour Shortage, 1779

From Edmund C. Burnett, ed., *Letters of Members of the Continental Congress* (Washington, 1928), IV, p. 476.

Jesse Root to Jeremiah Wadsworth

Philadelphia Octr. 6th 1779

SIR

I recd. your favour of the 28th ulto. Supplying the army with bread is and has been a very Serious business   every exertion has been made in these States to procure and forward flour to the army for some time past with but little

Success, and our dependance has been on the Contracts you entered into with the people of New York to exchange Salt Sugar etc: for flour. You will before this reaches you receive the resolutions of Congress on that head. the Com'ttee early reported and urged the necessity of the reports being taken up, but our foreign affairs pressing, it was delayed more than a fortnight. although some Members not rightly understanding the business at first blamed the Measure, yet the Committee was able to vindicate your Conduct and I beleive you will find by the resolution passed no blame is thrown upon you in this Instance. our foreign affairs, the business of Finance, and the daily orders necessary to be passed, prevent the attending to the appointment of a Successor in your office, So early as otherways they would   I beleive, Sir, your early attachment to the Cause of your Country, the zeal and activity you have shown through the whole of the time, the Signal Services you have rendered in your present office, being Called to it in the worst of times, will not be forgotten by Congress nor rewarded by delivering you up to be sacrificed. it is the wish of Congress you should Continue to Serve, but it is beleived you are in earnest to resign and they Expect to part with you. . . .

Problems in Feeding the Army, 1779

William Hand Browne, ed., *Archives of Maryland*, (Baltimores, 1897), XXI, pp. 547–49.
George Washington to John Jay

Head Quarters, West point 4 Octob 1779

Sir

I had the honor of receiveing your Excellencys Letters of the 26th & 27th Ultimo at half after twelve O'Clock yesterday. Immediately upon the Receipt of it, I set about concerting the measures necessary for a cooperation with his Excellency the Count D Estaing agreeable to the powers vested in me by the Resolve of Congress of the 26th Ult[o]

I have called upon the State of Massachusetts for 2000 Militia Connecticut for 4000 New York for 2500, New Jersey for 2000 and Pennsylvania for 1500, the last is below the Quota that she ought to furnish, in proportion to her strength, but I was induced to make a requisition of that number only, upon a consideration, that we shall be obliged to call largely upon that State for the means of transportation of provisions and supplies of all kinds, I have also taken the liberty, to press the States above mentioned to use the most vigorous exertions in procuring supplies of provisions especially of Flour, for the want of which, I fear we shall be much embarrassed, should we draw such a head of men together as will be necessary to give our operations a tolerable prospect of success.

I have not heard from Gen[l] Sullivan but by report since the 30th August, I have however dispatched an Express to him (upon a supposition that he has compleated the object of his expedition and is upon his return) desiring him to hasten his march, and directing him to leave as few Men as he possibly can in the frontier Garrisons, I have also wrote to Gen¹ Gates desiring him to hold all the Continental troops under his command ready to march this way, should the Count D'Estaing, upon settling a plan of operations determine upon an attempt against New York. But there is a possibility that he may, upon being made acquainted with the numbers and situation of the enemy prefer an attack upon Rhode Island, I have desired Gen¹ Gates to be looking towards and preparing for such an event, I had, upon the first report of the Counts standing towards this Coast, stationed Major Lee in Monmouth with a letter for him, to be carried on board upon his first appearance in which, I informed him of the enemys force by Sea & Land, their position at that time, and pointed out to him the measures which I thought it would be most advantagious for him to pursue upon his arrival I am preparing fresh letters for him, in which, I shall inform him fully of all posterior Events, and the measures I am taking for a cooperation.   I am also engaging and sending down proper pilots to him.

I have taken the liberty to countermand the march of Col Clarke with the two Regiments of North Carolina, upon a presumption, that from the favorable aspect of affairs to the Southward, I shall stand justifiable for such a measure

I observe by a Resolve of Congress lately transmitted to me, that three of the Continental Frigates were ordered to South Carolina. I do not know the views of Congress in making this disposition but should they have no particular object in contemplation I would venture to recommend their being ordered to join the Counts Fleet, which in my opinion woud be much benefitted by an additional number of Frigates, especially for the Navigation of the North River and the Sound. I think it would be also well, should the Marine Committee be directed to turn their attention to the transportation of Flour from the Delaware and Chesapeak by Water, Should we obtain the command of the Sea, Vessels may, without the least danger, be introduced into the Hook, thence to Amboy, from whence their Cargoes may be easily conveyed in Boats to Newark Bay, or should some of them run round into the Sound, it would be equally, nay more convenient Should we operate to the Eastward, measures of this kind will be indispensibly necessary, as the length and difficulty of Land Carriage will render the support of any considerable Body of Men almost impossible.

The Wheat of Maryland being in more forwardness for grinding than any other, I could wish that Governor Johnson may be requested to push the purchases within that State. The Commissary General gives the fullest encouragement on the score of Beef, but of Flour he continues to express his fears

I Have the Honor To Be Yours &c

G. Washington

Copy

N. B. The Frigates were Countermanded & ordered to be in Sound before the Gen¹ˢ Orders was received

His Excellency John Jay Esqʳ

Inflation of Food Prices, 1779

From Edmund C. Burnett, ed., *Letters of Members of the Continental Congress* (Washington, 1928), IV, p. 544.

William Floyd to the Governor of New York (George Clinton)

Philadelphia, December 21st 1779

SIR,

. . . Congress have Resolved to Remove from this place the 1st of May next But have not yet Determined on the place where they will Remove to.

You will not be surprized at their wish to Quit this City, when you are informed of the amazing Expense of Living here; Beef in the market Current at 3 Doll's pr. lb; pork four; wood 100 lb pr. Cord; flour 100 lb pr. hundred w't, and other things in proportion; it Seems as if the Devil was with all his Emmisaries let loose in this State to Ruin our money, and they, the authorities of this State are So Slow in the Collecting their taxes, that it will have but little Effect towards preventing it.

However Critical and Difficult our Situation may be, yet it Cannot be improper you Should know it. Long Since Congress Resolved to Stop the farther Emition of money; Relying on the Taxes to be Raised in the Several States for Money to Carry on the war, with the Small Sums which they Expected to get by loaning.

But alass what is our Situation! Our Treasury nearly Exhausted. Every Department out of Cash, no Magazines of provision laid up, our army Starving for want of Bread, on the Brink of a General Mutiny, and the prospect of a Spedy Supply is very Small. This is a melancholly Situation and would give our Enemies great pleasure if they knew it.

Under these Circumstances the grand Difficulty is to know what can be done for Relief. And it appears to me that if the Several States does not take on themselves to draw forth the Supplies for the use of the army, by a tax on the necessary articles, or Some other mode that may be in their power, God only knows what will become of us next Campaign; our army Cannot be kept together. But I'll Conclude the disagreable tale and Subscribe my Self your most Obedt. and humble Servt.

WM. FLOYD.

Taxes to Support the Army, 1780

From Edmund C. Burnett, ed., *Letters of Members of the Continental Congress* (Washington, 1931), V, p. 9.

Oliver Ellsworth to the Governor of Connecticut (Jonathan Trumbull)

Philadelphia, Jany. 14, 1780.

SIR,

Congress have no late intelligence from Europe, nor do they yet learn the destination of the troops lately embarked from New York. . . .

Much credit is due and given to Connecticut for the supply of beef said to be coming on from thence to the army, whose distress has been great and situation yet remains critical for want of provisions. Eight dollars have been given in its vicinity by the soldiers for a quart of meal and half a dollar for an ear of corn. Flour and grain are procured sufficient, it is said, for some months and now forwarding as fast as may be; and every attention will in future be paid to furnish money for the beef department.

The failure of that great resource, the press, gives as was expectable a violent shock, but it is hoped will prove a salutary one. The system of taxation urged by necessity is now establishing itself fast. All the States in the Union, so far as I can be informed, are now levying and collecting, pursuant to the requisitions of Congress, tho' in some of them their Assemblies have not yet been together to consider of the quotas last required. Maryland Assembly has indeed lately been together and adjourned without making provision therefor, owing to their not being able to obtain a vote in the Senate for the sale of British and Tory property. But as they have called on their people at large to shew their sense on that question, and are soon to meet again, there is no doubt of that obstacle's being removed and that Maryland will chearfully and fully furnish her quota. She is now making every exertion to supply the army with bread.

Greater unanimity has at no time perhaps prevailed in Congress than at present, or ever been more necessary.

Debate on Public Land Sales, 1782

From Edmund C. Burnett, ed., *Letters of Members of the Continental Congress* (Washington, 1933), VI, pp. 516–19.

The North Carolina Delegates to the Governor of North Carolina (Alexander Martin)

Philadelphia, 22nd October, 1782.

DEAR SIR:

. . . The great subject that has for some months drawn the attention of Congress is the means of supporting Public Credit and raising a supply for the expences of the next year and paying the deficiencies of the present year.

We have called the subject of supply a great one, because it is emphatically such with us, since we have little to fear from the Military force of the enemy who are decidedly inferior in the field and must continue so while we can pay, feed and cloath our Army, and never was an army worse paid than they have been and are, to say nothing of the cloaths and rations.

No Citizen of North America will be surprised at this assertion when he looks over the estimate for the expences of the year and considers how small a portion of that sum we have actually paid or borrowed.

Congress have agreed for the next year to attempt to borrow on France or Holland Five Millions of Dollars. We expect to borrow at five per cent. per annum. The Brokerage and other expences will raise the Interest per year to 6 per cent. Some Gentlemen think that we should attempt to raise larger Taxes at home in order to be relieved from the expence of Foreign Interest. We admit that, for our parts, we have been decidedly in favour of borrowing. The peculiar quality of our staple and the scarcity of money in North Carolina, we think, will justify our conduct. We contend that we are sufficiently rich to deserve Credit, tho' we are at present unable to pay Taxes because Tar, Pitch, Turpentine, Lumber, Indian Corn and the other general staples of North Carolina are too bulky to bear transportation in the time of War. In time of Peace all those articles may be very acceptable to either of the Nations that may lend us money. We apprehend that in time of Peace it would be more convenient for the Citizens of North Carolina in general to pay one Guinea than at present to pay one Dollar. We hope that our Constituents will at least approve the loan, knowing how difficult it would be for them to pay a Specie Tax of the twentieth part of their Quota for the annual expences of the War.

It is a fact, not to be dissembled, that the different States are not fully agreed as to the best and most eligible means of funding and paying off the annual supply. The smaller States and some of the larger ones are extremely desirous to do something with the back-lands. Their eyes are so eagerly fixed on those forests that they seem to stumble over the more obvious and productive subjects of Revenue which are nearer at hand. To this hour the State of Rhode Island have not adopted the 5 per cent. duty. They object that such a Tax is not according to the Confederation, while they contend for a participation of the Western lands which is contrary to the express terms of Confederation.

In our public Letter No. 6, August 18th, we mentioned the Western Lands and expressed our opinion that these lands might enable us to pay off a great part of the State debt or such debts as have been contracted for Militia service in the State. As no mention is made of the Western Lands in any of the requisitions that are now forwarded to the General Assembly, it becomes necessary that we should explain why we turned your attention to those Lands.

The subject of Revenue was at first submitted to a Committee of thirteen, being a member from each State, and after much deliberation and debate they reported:

1st. That the Western Lands, if ceded to the United States, might contribute towards a fund for paying the debts of those States.

2nd. That it be recommended to the several States to impose a Land Tax of one Dollar on everyone hundred Acres of Land. That they also impose a Poll Tax of half a Dollar on all male slaves from 16 to 60 and all free men from 16 to 21 and a Dollar on all free men from 21 to 60 except those in the Federal Army, etc.

4th. That an excise be laid on all distilled Spiritous Liquors of one-eighth of a Dollar for every gallon.

We had no such high expectations as some of the Committee concerning the productive value of the Western Lands, nor did we think that the States at large had any claims on them. The Land Tax of one Dollar per hundred Acres was, in our view, insufferably unequal. The vast tracts of sandy barren land in North Carolina can never be measured with the same scale as the uniformly fertile Lands in some of the Northern States. We shall make no remark on the other articles. There was a clear majority of the Committee in favor of the report, on which occasion we wrote the Letter referred to.

However, when the report was taken up by Congress, every part of it which respected the subjects of Taxation was rejected. The Southern States on this occasion were more fortunate than in Committee, for three of the minor States having but one Delegate in Congress were entitled to no vote when the report was brought forward; hence the question was lost by these who had been a majority, for they could produce but 5 or 6 votes at most. In the meanwhile it is our duty to tell you that the debate concerning Back-Lands is only dismissed for the present; it will certainly be revived whenever the Northern and Minor States are better represented. And as those lands are likely to prove the subject of warm and obstinate contention, it may be proper to consider whether there is any middle path by which the State may equally consult its honor, its interest and the public Peace.

It is expressly provided by the 8th Article of Confederation which is our Magna Charter, that in paying the expences of the present War the Quota of each State shall be fixed according to the Value of all lands in the several States "granted to, or surveyed for any person together with the buildings or improvements thereon" to be estimated from time to time as Congress shall direct. The enemy being in the Country is the reason given why this mode of fixing the quota has hitherto been neglected. For our parts we are determined to endeavour, by every possible means, to put this mode of fixing the Quota in a Train of Execution. Our State cannot, in prudence, desert that measure. It is more favorable for us than for most other States in the Union. Other States must pay for their large Towns and land highly cultivated, while we have few Towns and much wood land. But this mode, when the Western lands are included, appears to be less favorable. It is very certain that should we keep those lands, should we open the land office and sell them out, the clear revenue arising from the sale would be very small; they would bring little money into the Treasury, but they would render our Quota of the National Debt near double of what it is at present. For in fixing the Quota every State must be charged with its located lands, and as Land Jobbers are not a very popular set of men in any Country, and as the Lands are probably to be valued by indifferent people we may be assured that the Western Lands, which are located, but not improved, will be rated at their full value, and we suspect that our Western Lands on this side of Ohio are nearly double of those Lands already located in the State.

Connecticut, which has a very extraordinary claim to some Western Lands, also New York and Virginia, have made cessions of part of their claims to the United States, but they have not yet ceded all that is required, nor are the terms, especially those of Virginia, acceptable to Congress.

If North Carolina should be induced to give up any part of her Western Territory, we presume she will at least require the following preliminaries: . . .

On the general subject of supplies we need hardly inform you that our Army is extremely clamorous, we cannot pay them — we can hardly feed them. There is no money in the Treasury and we are obliged to draw upon the Foreign Loans before they are perfected.

We know that our State will not be able to raise the Quota that is assigned to her for the year 1783, but we are confident that she will do all in her power. We have attempted to fix her Quota as low as possible, and from a Paper we have sent you will perceive that we have deducted something from the Quota of last year, though our circumstances are now more favorable. . . .

The enemy are extremely desirous to get their Soldiers, who are prisoners, out of our hands, but they do not offer a *quid pro quo*. Congress have determined not to make any exchange 'til the Commissioner, who shall be sent to negotiate a general Cartel, shall produce Powers expressly derived from the King of Great Britain, and shall have paid or secured payment for a large balance due the United States for the sustenance of Prisoners. Attempts have been made in the meanwhile to secure a partial exchange. Congress have been zealously pressed to confirm the exchange of Lord Cornwallis for Mr. Laurens. This proposition is the more extraordinary as the enemy at first pretended to liberate Mr. Laurens unconditionally. The finesse was certainly curious and original. However, Congress have not adopted the example in giving up Lord Cornwallis unconditionally. Such an Officer is rated at 1,200 men, which is too large a gift to be made to a Nation that never wishes to give us anything except hard measures. The Southern States who are best acquainted with his Lordship's good qualifications are determined, if possible, that he shall be indulged a few months longer in the repose of his native Country unless he should choose to be paroled in South Carolina.

We confess that in this debate we have thought it our particular duty to give Lord Cornwallis credit for the numberless murders he has committed in the Carolinas and Georgia.

Negotiations in Paris for Peace move on with such a steady pace that they are hardly progressive. . . .

Farming

In the New Nation,

1783-1840

Introduction

As the United States entered upon its new destiny as a free, independent nation with the Treaty of Paris of 1783 marking the end of the Revolution, it faced a number of farm-related problems. These included land disposition, foreign restrictions on American farm products, the question of slave labor, and the need for improvements in farming methods which would bring higher yields and better products. The land problems were soon on the way to solution, cotton came to dominate American exports and encouraged the persistence of slavery, and foundations were laid for developing better methods and encouraging farmers to adopt them. At least some of these changes were due to the farm interests of leaders of the new nation.

George Washington and Thomas Jefferson are remembered for their interest in and contributions to agriculture as well as to other fields. Washington experimented with different crops at his home on the Potomac River, Mount Vernon, and fertilized some of his fields with marl and muck from the Potomac River bottom. He developed an excellent strain of wheat by careful seed selection, and, after the Revolution, grew wheat rather than tobacco. Washington was interested in farm implements, especially in improved grain drills. He discussed some experiments along these lines in his diary for April 1786. Washington carried on correspondence with Arthur Young, Sir John Sinclair, and other proponents of farm improvement in England. One of his distinctions was that he was the first American to raise mules. In 1785, the King of Spain sent Washington two jacks and two jennets. One jack died in transit, but the other arrived safely. The next year, General Lafayette sent Washington a jack and two jennets of another breed.

The influence of Thomas Jefferson upon the land policies developed by the Confederation government has been mentioned. He was also interested in farming and sought to make his home estate of Monticello and his other landholdings profitable as well as self-sufficient. This, together with his scientific inclinations, resulted in Monticello's becoming a progressive experimental farm where new machinery, new methods, and new crops were tried out. Jefferson developed a number of farm devices, including a seed drill, a hemp brake, improvements on a threshing machine, and, best known, a design for a mould board for a plow that would turn the soil efficiently. A pioneer in the work of introducing foreign plants into the United States, he sent seeds of upland rice from Italy to America in 1787 and offered to secure plants of the olive tree. Jefferson maintained an extensive correspondence with leading agriculturists at home and abroad. Both in correspondence and otherwise, Jefferson reiterated his belief in agriculture as a way of life and a source of social virtues peculiar to tillers of the soil.

Both Washington and Jefferson were interested in the agricultural societies of their day. These had been preceded by scientific societies, such as the American Philosophical Society, founded in 1769, and the American Academy of Arts and Sciences, founded in 1780. The scientific societies made the encouragement of agricultural experimentation a prominent part of their programs. They were followed by societies devoted entirely to agriculture, the first of which was the Philadelphia Society for Promoting Agriculture, organized in 1785. It was followed very closely by the South Carolina Society for Promoting and Improving Agriculture and Other Rural Concerns, established later in the same year. Other societies were founded in Maine, New Jersey, New York, Maryland, Massachusetts, and elsewhere. These early societies were made up of groups of men of all professions who could afford experimentation and who would seek out and adapt to American conditions the progress made in other countries. The societies awarded premiums, not for definite itemized products that could be produced by the ordinary farmer, but rather for the best solutions of problems of general significance. They probably had little direct influence upon the mass of small farmers, but they were pioneers in agricultural education.

A retired banker and businessman, Elkanah Watson, saw the need for an organization that would reach and encourage the mass of farm people, and he did something about it.

Watson purchased an estate near Pittsfield, Massachusetts. Shortly after moving there in 1807, he acquired a pair of Merino sheep. Since these were the first to be brought into the county, Watson exhibited them in the town square. This excited so much attention that in 1811 Watson organized the Berkshire Agricultural Society, which had for its purpose the sponsoring of an annual cattle show or fair. The Society was made up of neighborhood farmers. The idea spread rapidly, and by 1820, agricultural societies had been formed in all New England counties except those of Rhode Island. The states made grants to the societies, which was one reason for their rapid expansion. The grants were used mainly for paying premiums for products exhibited at fairs. The emphasis was upon greater and better looking crops and fatter animals without regard to production costs or market demands. The exaggerated hopes of many farmers for the benefits to be found in organization were soon disappointed. Little effort was made to preserve the societies when most states discontinued aid following the depression of 1825, and most of them collapsed. Beginning about 1840, the states again made subsidies available, and societies and fairs revived with the establishment of new state and county groups.

While the agricultural societies foreshadowed improvements in farming, the key to the development of American agriculture lay in solving two problems — the method by which the public lands could be transferred to actual farmers and the solving of the "colonial" problem in the settlement of the west. Both of these were resolved by the Continental Congress.

At the beginning of the Revolution, seven states claimed western land. The six states without western claims asserted that the unsettled West was being won by the common blood and treasure of all thirteen states and should thus be their common property. Maryland took the lead in the agitation, calling for the cession of all lands in the West except those that had been granted before the Declaration of Independence. The Continental Congress had complicated the picture by offering to grant land to soldiers. On September 15, 1776, it offered grants to men and officers in the Continental Army ranging from 100 to 500 acres. Land was necessary to redeem the promises, and, late in 1778, Virginia offered to supply all needed for the purpose without charge. This offer, naturally, did not satisfy Maryland.

The first break in the ranks of the landed states came in February, 1780, when New York ceded its tenuous claims. The Continental Congress acclaimed the action, urged other states to do likewise, and on October 10, 1780, declared that the unappropriated lands which should be ceded to the central government should "be disposed of for the common benefit of the United States, and be settled and formed into distinct republican states which should become members of the federal union and have the same rights of sovereignty, freedom and independence as the other states. . . the lands to be granted or settled on such terms and under such regulations as shall hereafter be agreed upon by the United States in Congress assembled. . . ."

On January 2, 1781, the Virginia assembly passed resolutions providing for the cession of its claims north of the Ohio River if the Congress would refuse to recognize private claims. Kentucky, which was also claimed and not included in the cession, achieved separation from Virginia as a new state in 1792 after long and complicated negotiations. Maryland, in spite of the displeasure of those with Indian claims north of the Ohio, agreed on February 2, 1781, to ratify the Articles of Confederation, and formally did so on March 1. There were attempts made in Congress to force Virginia to make its concession without the clause disallowing private claims, but finally, on March 1, 1784, Virginia transferred her lands north of the Ohio with the restriction intact.

The Virginia cession, with its reservation of enough land to reward its soldiers, established a pattern. Massachusetts ceded its claims late in 1784, reserving only an area in western New York which was later compromised with New York and then sold to speculators. Connecticut followed in 1786, retaining its famous "Western Reserve" in northeastern Ohio. North Carolina opened its western claims to speculative sales in 1783, and made its final cession of unsold land in 1790. South Carolina had ceded its claim in 1787. Georgia, hoping to profit from speculative sales, did not concede its claims until 1802.

Almost as soon as the Congress received its first cessions, it was under pressure from a number of interests to fix its procedures for disposing of its land. The more immediate of the reasons advanced for planning with regard to the western claims were: (1) The requests to the government to redeem its promise of land grants to soldiers; (2) The possibility of replenishing the depleted treasury by selling land; (3) The necessity of defending the Northwest against Indian attack; (4)

The need to bind the settlements in Kentucky and Tennessee to the new nation; (5) The need to form a government for the territory; and (6) The pressure of immigration to the West. The result was the Ordinance of 1784, drafted by Thomas Jefferson and others, which never became effective.

Jefferson was also on a committee appointed to draft a plan for surveying and selling the western lands, but was sent as Minister to France before the ordinance was completed, and it was referred to another committee. A new draft was prepared and was passed by Congress as the Ordinance of 1785 on May 20, 1785. The Ordinance provided for the survey of the west into townships, each containing 36 sections of one square mile each, each section comprising 640 acres. One half of the townships were to be sold as such, the alternate ones in sections. In each township, one section was reserved for schools and three for future disposition by Congress. One-third of all gold, silver, and lead was reserved. The lands were to be sold for cash at auction at a minimum price of one dollar per acre, the purchaser paying surveying costs and receiving a deed made out for a specific tract of land. The new system, which permitted the exact fixing of boundaries and led to straight-line roads and fences, provided a considerable contrast to the metes-and-bounds system, where boundaries were fixed by relation to specific natural objects.

Sales under the Ordinance of 1785 were slow even though the requirement of cash was modified in 1787 to provide for one-third cash and the rest in three months. Among reasons for the slow sales were the Indian menace, sales by private land companies, sales by states of land within their boundaries, and the settlement of public lands without the formality of securing title to the lands. This settlement in disregard of law by the settlers, who were called squatters, was a continuation of disregard for land laws demonstrated earlier, and was to color all governmental actions taken on the public domain.

Many government leaders appeared to favor the sale of large blocks of land to colonization or private land companies, and contracts for sales to three such companies were made in 1787 and 1788. The most important sale was to the Ohio Company, headed by General Rufus Putnam and the Reverend Manasseh Cutler, which patented something less than a million acres, according to an authority on land policy. The company was well managed. Its first settlement was at Marietta, Ohio, in 1788.

Another grant, made to speculator John Cleves Symmes, was originally for 1,000,000 acres at 66-⅔ cents an acre. However, he eventually received much less. The most important settlement in his area was Cincinnati.

Meanwhile, it was evident that definite provision must be made for government in the new territories. The Ordinance of 1784 was to go into effect only after all the land-claiming states had ceded their western territories, an event which had not taken place. Too, the West objected to the artificial boundaries of the new states proposed by the Ordinance of 1784, while the East feared that ten new states would dominate the Confederation and that the complete democracy provided by the Ordinance would be a disruptive influence.

Some leading historians credit the Reverend Manesseh Cutler with a major part in bringing about the passage, on July 13, 1787, of the Ordinance of 1787, which replaced the Ordinance of 1784 and established a framework for the government of the West. Regardless of its origin, this Ordinance was perhaps the most important piece of legislation ever passed by any representative assembly in the United States. It put into effect principles previously adopted but never implemented.

Briefly, the Ordinance of 1787 provided: the Northwest Territory should be established as one temporary district which would eventually become at least three but not more than five territories; at first the residents were to be controlled by officials appointed by Congress, but when a territory numbered 5,000 adult males, an elected legislature would share power with a council appointed by the governor and Congress; when a territory numbered 50,000 inhabitants it might frame a constitution and apply for admission into the Union on equal terms with the original thirteen states; a bill of rights similar to that of the Constitution guaranteed territorial settlers certain basic freedoms; slavery was prohibited; and laws permitting the perpetuation of great landed estates were forbidden.

The genius of the Ordianance lay in its provision for the admission of states on equal terms with the original states. No longer were there to be colonials without political privileges; settlers could move into the new areas secure in their personal liberties and secure in the eventual exercise of their political rights. The West was now bound to the nation by the strongest of all possible ties — that of equal rights.

The Act of 1800 was the first law under which a great deal of land was sold directly to settlers and the first in which the frontiersmen's viewpoint was evident. Sponsored by William Henry Harrison, the measure became law on May 10, 1800. While the minimum price of two dollars per acre was retained, land was to be sold at local offices, mostly in tracts of 320 acres. The credit provisions were liberalized in that one-fourth of the price was to be paid in 40 days, another fourth within two years, and the remainder within four years after sale. Interest was to be paid at the rate of 6 percent. Thus, the settler could buy 320 acres of the public domain for $160 down.

The Act of 1800 was modified on March 26, 1804, by the elimination of interest until payments became delinquent and by providing for the sale of land in quarter-sections. Between 1806 and 1820, a dozen laws providing an extension of the time in which payments could be made were passed. No further reduction in the size of tracts to be sold was made until 1816, when a law provided that certain sections of townships might be sold in 80-acre plots.

Following the precedent set in the Revolutionary War, Congress seized upon grants of public lands as suitable rewards for soldiers in 1811 and 1812. In increasing the size of the regular army in 1811, Congress offered, in addition to extra pay, a bounty of 160 acres of land to men completing five years of service, or less if deemed proper by the government. Officers were not included but received retroactive bounties in 1855. Eventually, grants made on account of the War of 1812 amounted to four and one-half million acres.

The evident need for revision of the credit system and the pressure for making land still easier to acquire, engendered in part by the Panic of 1819, led to changes in the basic laws by a law passed on April 2, 1820. This act abandoned the credit system, reduced the minimum price of land to $1.25 per acre, and provided for the sale of tracts as small as 80 acres. The cash requirement and financial difficulties resulting from the Panic of 1819 led to greatly reduced sales for a number of years as compared with the period 1800–1819.

Although the Act of 1820 abolished credit sales, many settlers who had made purchases under the Acts of 1800 and 1804 were still in difficulty in spite of previous relief acts. The act passed on March 2, 1821, was particularly favorable to such debtors. Under its terms, the purchaser delinquent in his payments could relinquish part of his land in payment of the balance due, could continue payments in accordance with the original contract but with the price reduced to $1.25 per acre, or could extend payments for from four to eight years, depending upon the amount still due.

Additional relief acts, with somewhat similar terms, were passed from time to time.

Many debates and laws concerning the public domain were centered upon the problem of transferring land to the individual settler, either for cash or in payment for services in the armed forces. At the same time, many Congressional leaders urged liberal grants of land for various types of internal improvements and other purposes. Grants to aid in improving the national transportation system were to become among the most important grants for internal improvements.

The decisions of the Continental Congress between 1780 and 1787 to make land readily available to farmers and to bring the western areas into the union as equals of the original 13 states as soon as a certain population was present, meant a rapid expansion to the West. This, in turn, meant that labor would continue scarce in relation to land. The westward movement also saw continuing conflicts with the Indians. A series of treaties provided for first one cession and then another by the various tribes. In some instances, as in the Treaty of Holston with the Cherokees in 1791, the United States agreed to supply the tribe with farm implements and agricultural instruction.

At the time of the Revolution it seemed that slave labor might be coming to an end, even though farm labor was still scarce in relation to land. Tobacco, rice, and indigo were becoming less profitable, with a consequent decline in the demand for slaves. Many leaders of the new nation had spoken out against slavery, and both the Ordinance of 1787 and the new Constitution, which became effective in 1787, had limited the expansion of slavery. Foreign visitors and Americans wrote of its evils. But the invention of the cotton gin in 1793 by Eli Whitney, a young graduate of Yale University, made cotton the dominant crop in the South. Plantation agriculture, with its dependence upon slave labor and upon one crop, was established throughout the region. The French traveler, Michaux, noted in 1803 that settlers in such new areas as West Tennessee immediately turned to cotton. In the areas where cotton did not become established, such as Virginia, Maryland, and Kentucky, tobacco continued as the staple crop.

Other implements and machines invented during this period hastened the opening of western lands and cut the need for hand labor at critical points in the farm year. In 1819, Jethro Wood patented an iron plow with interchangeable parts, and within a few years thousands were in use. Neither the traditional wooden nor the new cast iron plows would turn the sticky soil of the prairies. In 1833, a Chicago blacksmith, John Lane, covered the wooden moldboards of the traditional plow with saw-steel. In 1837, John Deere of Grand Detour, Illinois, began manufacturing a one-piece share and moldboard of saw steel or of highly-polished wrought iron. Within ten years, Deere was manufacturing 10,000 of these plows a year at a plant in Moline, Illinois.

Harvesting was the critical point in wheat production. The mechanical reaper replaced much hand power at this point and permitted the rapid expansion of acreage. A practical machine was patented by Obed Hussey in 1833, but the one which would come to dominate the market was patented by Cyrus H. McCromick in 1834. Another useful machine, a thresher for separating grain from straw and chaff, was patented by Hiram and John Pitts of Winthrop, Maine, in 1837.

Agricultural leaders recognized that the continued cultivation of tobacco and, later, of cotton in the South and of wheat and corn in the North could lead to serious soil deterioration. In 1809, John Taylor of Virginia reported to the Philadelphia Society for Promoting Agriculture on experiments with gypsum and manure. The next year, Richard Peters of Pennsylvania urged the liberal use of lime. Later, Edmund Ruffin of Virginia carried out careful experiments in the use of marl, essentially clay mixed with carbonate of lime, often in the form of fossil shells. Ruffin, the first American soil scientist, discussed his work in an "Essay on Calcareous Manures," published in the *American Agriculturist* of December 28, 1821. He continued his experiments and expanded his essay into a book, the third edition of which was published in 1842.

One plantation operator, John Tayloe, used moveable pens for livestock and thus made use of the manure. However, most farmers neglected manure as well as their livestock. At the same time, some efforts were made, particularly by gentlemen farmers, to improve the quality of domestic animals. Merino sheep became popular in the early 1800's. During the War of 1812, with an increased demand for wool and further importation cut off, a virtual craze for the breed developed. Even though the boom collapsed, Merinos had been widely distributed and resulted in improved flocks.

Mule raising, begun by George Washington, spread rapidly. More by chance than by design, a new breed of general-purpose horses, the Morgan, originated in Vermont in the early 1800's. The breed derived from one stallion, Justin Morgan, foaled probably in 1789, of uncertain parentage.

Cattle, as described by European travellers, were generally red in color and of indifferent size and quality. Importations of improved cattle from England of which a record exists were first made in 1783. Matthew Patton of the south branch of the Potomac and H.D. Gough of Maryland imported Bakewell Longhorns and what were probably early examples of Shorthorns. In 1817, Henry Clay brought Herefords into Kentucky at the same time his friend Lewis Sanders was importing Shorthorns and Longhorns. In 1822, John Hare Powel of Pennsylvania began importing improved Shorthorns and soon built up an outstanding herd. A group of Ohio breeders, organized as the Ohio Company for Importing English Cattle began importing Durham cattle in 1833. They soon became popular for improving beef animals.

The adoption of better breeds of livestock was urged by the agricultural press, which had its beginnings during this period. While a few issues of a journal called *Agricultural Museum* appeared in Georgetown, D.C., in 1810–11, the *American Farmer,* established by John Stuart Skinner in Baltimore in 1819, became the first farm periodical to achieve prominence and relative permanence. Skinner and other editors advocated adopting superior crops as well as superior livestock. For example, during 1821 Skinner publicized and helped distribute a Chilean wheat, later known as white club.

Jefferson's earlier concern with Italian rice, as well as Skinner's with wheat, reflected a theme running through much of the literature on improving agriculture. In 1813, John Lorain reported to the Philadelphia Society for Promoting Agriculture that he had improved his corn crop by crossing gourdseed and flint varieties. This foreshadowed later corn breeding, which eventually resulted in hybrid corn.

The Federal government took action in 1819 by a Treasury Circular to encourage United States consuls abroad to send valuable plants and seeds to American customs collectors for distribution. There were no results, according to President John Quincy Adams, because no expenditures were authorized. This led to a new circular in 1827, with instructions for transmitting the seeds and plants, but again no funds were appropriated. However, Henry Perrine, consul at Campeche, Mexico, became interested and began sending plants and information on them to the Secretary of the Treasury. Eventually, Perrine petitioned Congress for a grant of land in Florida to establish a plant introduction garden, and received such a grant in 1838. However, it was soon destroyed in an Indian raid.

In 1837, Henry L. Ellsworth, Commissioner of Patents, recommended that a program for collecting and distributing seeds and plants be established, and began some activities along this line. Congress responded by appropriating $1,000 from the Patent Office fund on March 3, 1839. In a discussion of the 1840 Census, the first to collect agricultural data, Ellsworth emphasized the need to export surplus American farm products in order to attain a balance of trade. Thus began the work which led to the Department of Agriculture.

George Washington, Farmer

Excerpts From Washington's Farm Diaries

From Everett E. Edwards, ed., *Washington, Jefferson, Lincoln and Agriculture* (Washington, 1937), pp. 9–16, 27–36.

Monday, [April] 14. [1760] Fine warm day, Wind So'ly and clear till the Even'g when it clouded; No Fish were to be catchd to day neither.

Mixd my Composts in a box with ten Apartments in the following manner, viz: in No. 1 is three pecks of the Earth brought from below the Hill out of the 46 Acre Field without any mixture; in No. 2 is two pecks of the said Earth and one of Marle taken out of the said Field, which Marle seemed a little Inclinable to Sand;

3. Has 2 Pecks of sd. Earth and 1 of Riverside Sand;
4. Has a Peck of Horse Dung;
5. Has mud taken out of the Creek;
6. Has Cow Dung;
7. Marle from the Gullys on the Hill side, wch. seeemd to be purer than the other;
8. Sheep Dung;
9. Black Mould taken out of the Pocoson on the Creek side;
10. Clay got just below the Garden.

All mixd with the same quantity and sort of Earth in the most effectual manner by reducing the whole to a tolerable degree of fineness and ju[m]bling them well together in a Cloth.

In each of these divisions were planted three grains of Wheat, 3 of Oats and as many of Barley — all at equal distances in Rows, and of equal depth (done by a machine made for the purpose).

The Wheat Rows are next the Numberd side, the Oats in the Middle and the Barley on that side next the upper part of the Garden.

Two or three hours after sowing in this manner, and about an hour before Sunset I waterd them all equally alike with water that had been standing in a Tub abt. two hours exposed to the Sun. . . .

Finishd Harrowing the Clover Field, and began reharrowing of it. Got a new harrow made of smaller, and closer Tinings for Harrowing in Grain — the other being more proper for preparing the Ground for sowing.

Cook Jack's plow was stopd he being employd in setting the Lime Kiln.

Thursday, 1st. Got over early in the Morning and reachd home before Dinner time, and upon inquiry found that my Clover Field was finishd sowing and Rolling the Saturday I left home — as was sowing of my Lucerne: and that on the [   ] they began sowing the last field of Oats and finishd it the 25th.

That in box No. 6, two grains of wheat appeard on the 20th, one an Inch high; on the 22d a grain of Wheat in No 7 and 9 appeard; on the 23 after a good deal of Rain the Night before some Stalks appeard in Nos. 2. 3. 4. 5. and 6, but the Ground was so hard bakd by the drying winds when I came home that it was difficult to say which Nos. lookd most thriving. However in

No. 1 there was nothing come up;
     2........2 Oats.............1 barley;
     3........1 Oat..............2 barley;
     4........1 Oat;
     5........1 Wheat 2 Oats;
     6........1 Do.   3 Do........1 Do.
     7........1 Do.   2 Do........2 Do.
     8........1 Do.   1 Do.
     9........2 Do.   3 Do........2 Do.
     10.............................1 Do.

The two Grains in No. 8 were I think rather the strongest, but upon the whole No. 9 was the best.

Washington's careful attention to his orchard is recorded in the following entry in his Diary for March 1765.

5th. Grafted 15 English Mulberrys on wild Mulberry Stocks on the side of the Hill near the Spring Path. Note, the Stocks were very Milkey.

6th. Grafted 10 Carnation Cherrys on growing Stocks in the Garden — viz. 5 of them in and about the Mint Bed, 3 under the Marella Cherry tree, 1 on a Stock in the middle of the border of the last square, and just above the 2d. fall (Note, this Graft is on the Northernmost fork) of Do.; on the Westernmost one is a Bullock Heart, and on the Easternmost one is a May Cherry out of the Cherry Walk. 1 other on a Stock just above the 2d. Gate. Note this is on the Northernmost prong; the other Graft on the said stock is of the May Cherry in the Cherry Walk.

15. Grafted 6 Early May duke Cherrys on the Nursery, begin'g at that end of the first Row next to the Lane — the Row next the Quarter is meant; at the end of this a Stake drove in.

15. Also Grafted joining to these in the same Row 6 of the latter May dukes — which are all the Cherrys in the Row. Also Grafted 7 Bullock Heart Cherrys in the last Row.

30. Grafted 48 Pears which stand as follows, viz. in the 3d Row beging. at the end next the Cherry Walk are 12 Spanish Pears; next to these are 8 Early June Pears; then 10 latter Bergamy; then 8 Black Pear of Worcester; and lastly 10 Early Bergamy. Note, all these Pears came from Colo. Mason's: and between each sort a Stick is drove down. The Rows are counted from the end of the Quarter.

30. This day also I grafted 39 New Town Pippins, which compleat the 5th Row and which Row are all of this kind of Fruit now.

30. The 6th Row is compleated with Grafts of the Maryland Red Strick, which are all of this sort of Fruit, and contains [   ] trees; so is the 8th Row of this Apple, also 54 in number and 20 in the 9th Row beginning next the Cherry Walk.

30. The 7th. Row has 25 Graffs of the Gloster white Apple which compleats this Row with that sort of Fruit.

Saturday, [April] 8th. [1786] Rid a little after Sun rise to Muddy [hole], to try my drill plow again which, with the alteration of the harrow yesterday, I find will fully answer my expectation, and that it drops the grains thicker, or thinner in proportion to the quantity of seed in the Barrel. The less there is in it the faster it issues from the holes. The weight of a quantity in the barrel, occasions (I presume) a pressure on the holes that do not admit of a free discharge of the Seed through them, whereas a small quantity (sufficient at all times to cover the bottom of the barrel) is, in a manner sifted through them by the revolution of the Barrel.

I sowed with the barrel to day, in drills, about 3 pints of a white well looking Oat, brought from Carolina last year by G. A. Washington in 7 rows running from the path leading from the Overseer's Ho. to the Quarter to the West fence of the field, which the ground was in the best order. Afterwards I sowed in such other parts of the adjoining ground as could at any rate be worked the common Oat of the Eastern shore (after picking out the Wild onion); but in truth nothing but the late Season could warrant sowing in ground so wet.

None of the ground in wch. these Oats were sown had received any improvement from Manure, but all of it had been twice plowed, and then listed, after which the harrow had gone over it twice before the seed harrowing. This, had it not been for the frequent rains, etca., which has fallen would have put the ground in fine order.

Transplanted as many of the large Magnolio into the Grove at the No. end of the Ho. as made the number there.

Also transplanted from the same box, 9 of the live Oak, viz. 4 in the bends of the lawn before the House, and five on the East of the grove (within the yard) at the No. end of the House.

Plowed up my last years turnip patch (at home) to sow Orchard grass seeds in.

No fish caught to day. . . .

Monday, 10th. Began my brick work to day, first taking away the foundations of the Garden Houses as they were first placed, and repairing the damages in the Walls occasioned by their removal; and also began to put up my pallisades (on the Wall).

Compleated Sowing with 20 quarts the drilled Oats in the ground intended for experiments at Muddy hole; which amounted to 38 rows, ten feet apart (including the parts of rows sowed on Saturday last). In the afternoon I began to sow Barley, but finding there were too many Seeds discharged from the Barrel, notwithstanding I stopped every other hole, I discontinued the sowing until another Barrel with smaller holes cd. be prepared. The ground, in which these Oats have been sowed and in which the Barley seeding had commenced, has been plowed, cross plowed, listed (as it is called, that is 3 furrow ridges) and twice harrowed before the drill plow was put into it; with this the furrow is made and the seed harrowed in with the manure afterwds.

Began also to sow the Siberian Wheat which I had obtained from Baltimore by means of Colo. Tilghman, at the Ferry Plantation in the ground laid apart there for experiments. This was done upon ground which, sometime ago, had been marked off by furrows 8 feet apart, in which, a second furrow had been run to deepen them. 4 furrows were then plowed to these, which made the whole 5 furrow Ridges. These being done sometime ago, and by frequent rains prevented sowing at the time intended had got hard, I therefore before the seed was sowed, split these Ridges again, by running twice in the same furrow. After wch. I harrowed the ridges, and where the ground was lumpy run my spiked Roller with the Harrow at the tale over it, wch. I found very efficacious in breaking the clods and pulverising the earth; and wd. have done it perfectly if there had not been too much moisture remaining of the late rains; after this harrowing and rolling where necessary, I sowed the Wheat with my drill plow on the reduced ridges in rows 8 feet apart. But I should have observed that, after the ridges were split by the furrow in the middle, and before the furrows were closed again by the harrow, I sprinkled a little dung in them. Finding the barrel discharged the Wheat too fast, I did, after sowing 9 of the shortest (for we began at the furthest corner of the field) rows, I stopped every other hole in the barrel, and in this manner sowed 5 rows more, and still thinking the seed too liberally bestowed, I stopped 2, and left one hole open, alternately, by which 4 out of 12 holes only, discharged seeds; and this, as I had taken the strap of leather off, seemed to give seed enough (though not so regular as were to be wished) to the ground.

Friday, [January] 22d. [1790] Exercised on horseback in the forenoon.

Called in my ride on the Baron de Polnitz, to see the operation of his (Winlaw's) threshing machine. The effect was, the heads of the wheat being seperated from the straw, as much of the first was run through the mill in 15 minutes as made half a bushel of clean wheat — allowing 8 working hours in the 24, this would yield 16 bushels pr. day. Two boys are sufficient to turn the wheel, feed the mill, and remove the threshed grain after it has passed through it. Two men were unable, by winnowing, to clean the wheat as it passed through the mill, but a common Dutch fan, with the usual attendance, would be *more* than sufficient to do it. The grain passes through without bruising and is well separated from the chaff. Women, or boys of 12 or 14 years of age, are fully adequate to the management of the mill or threshing machine. Upon the whole, it appears to be an easier, more expeditious, and much cleaner way of getting out grain than by the usual mode of threshing; and vastly to be preferred to treading, which is hurtful to horses, filthy to the wheat, and not more expeditious, considering the numbers that are employed in the process from the time the head is begun to be formed until the grain has passed finally through the fan.

Washington to Arthur Young, From Mount Vernon
August 6, 1786

I have had the honour to receive your letter of the seventh of January, from Bradfield-Hall, in Suffolk, and thank you for the favour of opening a correspondence, the advantages of which will be so much in my favour.

Agriculture has ever been amongst the most favourite amusements of my life, though I never possessed much skill in the art; and nine years total inattention to it, has added nothing to a knowledge which is best understood from practice; but with the means you have been so obliging as to furnish me, I shall return to it (though rather late in the day) with hope and confidence.

The system of agriculture, if the epithet of system can be applied to it, which is in use in this part of the United States, is as unproductive to the practitioners as it is ruinous to the land-holders. Yet it is pertinaciously adhered to. To forsake it, to pursue a course of husbandry which is altogether different and new to the gazing multitude, ever averse to novelty in matters of this sort, and much attached to their old customs, requires resolution; and without a good practical guide, may be dangerous; because, of the many volumes which have been written on this subject, few of them are founded on experimental knowledge — are verbose, contradictory, and bewildering. Your Annals shall be this guide. The plan on which they are published, gives them a reputation which inspires confidence; and for the favour of sending them to me, I pray you to accept my very best acknowledgments. To continue them, will add much to the obligation.

To evince with what avidity, and with how little reserve, I embrace the polite and friendly offer you have made me, of supplying me with "men, cattle, tools, seeds, or any thing else that may add to my rural amusement," I will give you, Sir, the trouble of providing, and sending to the care of Wakelin Welch, Esq. of London, merchant, the following articles:

Two of the simplest and best-constructed ploughs for land which is neither very heavy nor sandy. To be drawn by two horses — to have spare shares and coulters — and a mould on which to form new irons when the old ones are worn out, or will require repairing.

I shall take the liberty in this place to observe, that some years ago, from a description, or recommendation of what was then called the Rotherham, or patent plough, I sent to England for one of them; and till it began to wear, and was ruined by a bungling country smith, that no plough could have done better work, or appeared to have gone easier with two horses; but for want of a mould, which I had neglected to order with the plough, it became useless after the irons which came in with it were much worn.

A little of the best kind of cabbage-seeds, for field culture.

Twenty pounds of the best turnip-seeds, for ditto.

Ten bushels of sainfoin-seeds.

Eight bushels of the winter vetches

Two bushels of rye-grass seeds.

Fifty pounds of hop clover-seeds.

And, if it is decided, for much has been said for and against it, that burnet, as an early food, is valuable, I should be glad of a bushel of this seed also. Red clover-seeds are to be had on easy terms in this country, but if there are any other kinds of grass-seeds, not included in the above, that you may think valuable, especially for early feeding or cutting, you would oblige me by adding a small quantity of the seeds, to put me in stock. Early grasses, unless a species can be found that will stand a hot sun, and oftentimes severe droughts in the summer months, without much expense of cultivation, would suit our climate best.

You see, Sir, that without ceremony, I avail myself of your kind offer; but if you should find in the course of our correspondence, that I am likely to become troublesome, you can easily check me. Inclosed I give you an order on Wakelin Welch, Esq. for the cost of such things as you may have the goodness to send me. I do not at this time ask for any other implements of husbandry than the ploughs; but when I have read your Annals (for they are but just come to hand) I may request more. In the meanwhile, permit me to ask what a good ploughman might be had for: annual wages, to be found (being a single man) in board, washing, and lodging? The writers upon husbandry estimate the hire of labourers so differently in England, that it is not easy to discover from them, whether one of the class I am speaking of would cost eight or eighteen pounds a year. A good ploughman at low wages, would come very opportunely with the ploughs here requested.

By means of the application I made to my friend Mr. Fairfax, of Bath, and through the medium of Mr. Rack, a bailiff is sent to me, who, if he is acquainted with the best courses of cropping, will answer my purposes as a director or superintendent of my farms. He has the appearance of a plain honest farmer; — is industrious; — and from the character given of him by a Mr. Peacy, with whom he has lived many years, has understanding in the management of stock, and of most matters for which he is employed. How far his abilities may be equal to a pretty extensive concern, is questionable. And what is still worse, he has come over with improper ideas; for instead of preparing his mind to meet a ruinous course of cropping, exhausted lands, and numberless inconveniences into which we had been thrown by an eight years war, he seems to have expected that he was coming to well organized farms, and that he was to have met ploughs, harrows, and all the other implements of husbandry, in as high taste as the best farming counties in England could have exhibited them. How far his fortitude will enable him to encounter these disappointments, or his patience and perseverance will carry him towards the work of reform, remains to be decided.

With great esteem, I have the honour to be, Sir, your most obedient, humble servant,

G. WASHINGTON.

Washington to Sir John Sinclair, from Philadelphia
July 20, 1794.

* * *

I have read with peculiar pleasure and approbation, the work you patronise, so much to your own honor and the utility of the public. — Such a general view of the agriculture in the several counties of Great Britain is extremely interesting; and cannot fail of being very beneficial to the agricultural concern of your Country and to those of every other wherein they are read, and must entitle you to their warmest thanks for having set such a plan on foot, and for prosecuting it with the zeal & intelligence you do. —

* * *

I know of no pursuit in which more real & important service can be rendered to any Country, than by improving its agriculture — its breed of useful animals — and other branches of a husbandman's cares; — nor can I conceive any plan more conducive to this end than the one you have introduced for bringing to view the actual state of them, in all parts of the Kingdom; — by which good & bad habits are exhibited in a manner too plain to be misconceived; for the accounts given to the British board of Agriculture, appear in general, to be drawn up in a masterly manner, so as fully to answer the expectations formed in the excellent plan *wch* produced them; affording at the same-time a fund of information useful in political economy — serviceable in all countries.

Commons — Tithes — Tenantry (of which we feel nothing in this country) are in the list of impediments I perceive, to perfection in English farming — and taxes are heavy deduction from the profit thereof — Of these we have none, or so light as hardly to be felt. — Your system of Agriculture, it must be confessed, is in a stile superior & of course much more expensive than ours, but when the balance at the end of the year is struck, by deducting the taxes, poor rates, and incidental charges of every kind, from the produce of the land, in the two Countries, no doubt can remain in which Scale it is to be found.

It will be sometime I fear, before an Agricultural Society with Congressional aids will be established in this Country; — we must walk as other countries have done before we can run, Smaller Societies must prepare the way for greater, but with the light before us, I hope we shall not be so slow in maturation as older nations have been. — An attempt, as you will perceive by the enclosed outlines of a plan, is making to establish a State Society in Pennsylvania for agricultural improvements. — If it succeeds, it will be a step in the ladder, at present it is too much in embryo to decide on the result. —

Our domestic animals, as well as our agriculture, are inferior to yours in point of size but this does not proceed from any defect in the stamina of them, but to deficient care in providing for their support; experience having abundantly evinced that where our pasture areas well improved as the soil & climate will admit; — where a competent store of wholesome provender is laid up and proper care used in serving it, that our horses, black cattle, Sheep &c. are not inferior to the best of their respective kinds which have been imported from England. — Nor is the wool of our Sheep inferior to that of the *common* sort with you: — as a proof — after the Peace of Paris in 1783, and my return to the occupations of a farmer, I paid particular attention to my breed of sheep (of which I usually kept about seven or eight hundred). — By this attention, at the shearing of 1789, the fleeces yielded me the average quantity of 5¼ £ of wool; — a fleece of which, promiscuously taken, I sent to Mr. Arthar Young, who put it, for examination, into the hands of Manufacturers. — These pronounced it to be equal in quality to the Kentish Wool. — In this same year, i.e. 1789 I was again called from home, and have not had it in my power since to pay any attention to my farm; — The consequence of which is, that my Sheep, at the last shearing, yielded me not more than 2½ £ — This is not a single instance of the difference between care and neglect. — Nor is the difference between good & bad management confined to that species of Stock; for we find that good pastures and proper attention, can & does, fill our markets with beef of seven, eight & more hundred weight the four quarters; whereas from 450 to 500 (especially in the States South of this where less attention hitherto has been paid to grass) may be found about the average weight. — In this market, some Bullocks were killed in the months of March & April last, the weights of *wch*, as taken from the accounts which were published at the time, you will find in a paper enclosed. — These were pampered steers, but from 800 to a thousand, the four quarters, is no uncommon weight.

Your general history of Sheep, with observations thereon, and the proper mode of managing them, will be an interesting work when compleated; and with the information, & accuracy I am persuaded it will be executed, under your auspices, must be extremely desirable. — The climate of this Country, particularly that of the middle states is congenial to this species of animal; but want of attention to them in most farmers, added to the obstacles which prevent the importation of a better herd, by men who would be at the expence, contributes not a little to the present inferiority we experience. . . .

Both Mr. Adams and Mr. Jefferson had the perusal of the papers which accompanied your note of the 11th of Sep.

With great respect and esteem I have the honor to be Sir, Your Obed Servt.

G. WASHINGTON.

Washington to Jefferson, From Mount Vernon
October 4, 1795

* * *

I am much pleased with the account you have given of the succory. This, like all other things of the sort with me, since my absence from home, has come to nothing; for neither my overseers nor manager will attend properly to any thing but the crops they have usually cultivated; and, in spite of all I can say, if there is the smallest discretionary power allowed them, they will fill the land with Indian corn, although even to themselves there are the most obvious traces of its baneful effects. I am resolved, however, as soon as it shall be in my power to attend a little more closely to my own concerns, to make this crop yield in a degree to other grain, to pulses, and to grasses. I am beginning again with chiccory, from a handful of seed given me by Mr. Strickland, which, though flourishing at present, has no appearance of seeding this year. Lucerne has not succeeded better with me than with you; but I will give it another and a fairer trial before it is abandoned altogether. Clover, when I can dress lots well, succeeds with me to my full expectation, but not on the fields in rotation, although I have been at much cost in seeding them. This has greatly disconcerted the system of rotation on which I had decided.

I wish you may succeed in getting good seed of the winter vetch. I have often imported it, but the seed never vegetated, or in so small a proportion, as to be destroyed by weeds. I believe it would be an acquisition, if it was once introduced properly in our farms. The Albany pea, which is the same as the field pea of Europe, I have tried, and found it will grow well; but is subject to the same bug which perforates the garden pea, and eats out the kernel. So it will happen, I fear, with the pea you propose to import. I had great expectation from a green dressing with buckwheat, as a preparatory fallow for a crop of wheat, but it has not answered my expectation yet. I ascribe this, however, more to mismanagement in the times of seeding and ploughing in, than any defect of the system. The first ought to be so ordered, in point of time, as to meet a convenient season for ploughing it in, while the plant is in its most succulent state. But this has never been done on my farms, and consequently has drawn as much from, as it has given to the earth. It has always appeared to me that there were two modes in which buckwheat might be used advantageously as a manure. One, to sow early, and, as soon as a sufficiency of seed is ripened, to stock the ground a second time, to turn the whole in, and when the succeeding growth is getting in full bloom, to turn that in also, before the seed begins to ripen; and, when the fermentation and putrefaction ceases, to sow the ground in that state, and plough in the wheat. The other mode is, to sow the buckwheat so late, as that it shall be generally about a foot high at the usual seeding of wheat; then turn it in, and sow thereon immediately, as on a clover lay, harrowing in the seed lightly to avoid disturbing the buried buckwheat. I have never tried the latter method, but see no reason against its succeeding. The other, as I observed above, I have prosecuted, but the buckwheat has always stood too long, and consequently had got too dry and sticky to answer the end of a succulent plant.

But of all the improving and ameliorating crops, none in my opinion is equal to potatoes, on stiff and hard bound land, as mine is. I am satisfied, from a variety of instances, that on such land a crop of potatoes is equal to an ordinary dressing. In no instance have I failed of good wheat, oats, or clover, that followed potatoes; and I conceive they give the soil a darker hue. I shall thank you for the result of your proposed experiment relative to the winter vetch and pea when they are made.

I am sorry to hear of the depredations committed by the weevil in your parts; it is a great calamity at all times, and this year, when the demand for wheat is so great, and the price so high, must be a mortifying one to the farmers. The rains have been very general, and more abundant since the 1st of August, than ever happened in a summer within the memory of man. Scarcely a mill-dam, or bridge, between this and Philadelphia, was able to resist them, and some were carried off a second and third time.

Mrs. Washington is thankful for your kind remembrance of her, and unites with me in best wishes for you. With very great esteem and regard, I am, dear Sir, &c.

Last Annual Message to Congress, December 7, 1796

It will not be doubted that with reference either to individual or national welfare agriculture is of primary importance. In proportion as nations advance in population and other circumstances of maturity this truth becomes more apparent, and renders the cultivation of the soil more and more an object of public patronage. Institutions for promoting it grow up, supported by the public purse; and to what object can it be dedicated with greater propriety? Among the means which have been employed to this end none have been attended with greater success than the establishment of boards (composed of proper characters) charged with collecting and diffusing information, and enabled by premiums and small pecuniary aids to encourage and assist a spirit of discovery and improvement. This species of establishment contributes doubly to the increase of improvement by stimulating to enterprise and experiment, and by drawing to a common center the results everywhere of individual skill and observation, and spreading them thence over the whole nation. Experience accordingly has shewn that they are very cheap instruments of immense national benefits.

Washington to William Strickland, from Mount Vernon, July 15, 1797

I have been honored with yours of the 30th of May and 5th of September of last year. As the first was in part an answer to a letter I took the liberty of writing to you, and the latter arrived in the middle of an important session of Congress . . . . I postponed, from the pressure of business occasioned thereby, the acknowledgment of all private letters, which did not require immediate answers, until I should be seated under my own vine and fig-tree, where I supposed I should have abundant leisure to discharge all my epistolary obligations.

In this, however, I have hitherto found myself mistaken; for at no period have I been more closely employed in repairing the ravages of an eight years' absence. Engaging workmen of different sorts, providing and looking after them, together with the necessary attention to my farms, have occupied all my time since I have been at home.

I was far from entertaining sanguine hopes of success in my attempt to procure tenants from Great Britain; but, being desirous of rendering the evening of my life as tranquil and free from care as the nature of things would admit, I was willing to make the experiment.

Your observation, with respect to occupiers and proprietors of land has great weight, and, being congenial with my own ideas on the subject, was one reason, though I did not believe it would be so considered, why I offered my farms to be let. Instances have occurred, and do occur daily, to prove that capitalists from Europe have injured themselves by precipitate purchases of free-hold estates, immediately upon their arrival in this country, while others have lessened their means in exploring States and places in search of locations; whereas, if on advantageous terms they could have been first seated as tenants, they would have had time and opportunities to become holders of land, and for making advantageous purchases. But it is so natural for man to wish to be the absolute lord and master of what he holds in occupancy, that his true interest is often made to yield to a false ambition. Among these, the emigrants from the New England States may be classed, and this will account, in part, for their migration to the westward. Conviction of these things having left little hope of obtaining such tenants as would answer my purposes, I have had it in contemplation, ever since I returned home, to turn my farms to grazing principally, as fast as I can cover the fields sufficiently with grass. Labor, and of course expense, will be considerably diminished by this change, the net profit as great, and my attention less divided, whilst the fields will be improving.

Your strictures on the agriculture of this country are but too just. It is indeed wretched; but a leading, if not the primary, cause of its being so is, that, instead of improving a little ground well, we attempt much and do it ill. A half, a third, or even a fourth of what we mangle, well wrought and properly dressed, would produce more than the whole under our system of management; yet such is the force of habit, that we cannot depart from it. The consequence of which is, that we ruin the lands that are already cleared, and either cut down more wood, if we have it, or emigrate into the Western country. I have endeavoured, both in a public and private character, to encourage the establishment of boards of agriculture in this country, but hitherto in vain; and what is still more extraordinary, and scarcely to be believed, I have endeavoured ineffectually to discard the pernicious practice just mentioned from my own estate; but, in my absence, pretexts of one kind or another have always been paramount to orders. Since the first establishment of the National Board of Agriculture in Great Britain, I have considered it as one of the most valuable institutions of modern times; and, conducted with so much ability and zeal, as it appears to be under the auspices of Sir John Sinclair, it must be productive of great advantages to the nation, and to mankind in general.

My system of agriculture is what you have described, and I am persuaded, were I to proceed on a large scale, would be improved by the alteration you have proposed. At the same time I must observe, that I have not found oats so great an exhauster, as they are represented to be; but in my system they follow wheat too closely to be proper, and the rotation will undergo a change in this, and perhaps in some other respects.

The vetches of Europe have not succeeded with me; our frosts in winter, and droughts in summer, are too severe for them. How far the mountain or wild pea would answer as a substitute, by cultivation, is difficult to decide, because I believe no trial has been made of it, and because its spontaneous growth is in rich lands only. That it is nutritious in a great degree, in its wild state, admits of no doubt.

Spring barley, such as we grow in this country, has thriven no better with me than vetches. The result of an experiment, made with a little of the true sort, will be interesting. The field peas of England (different kinds) I have more than once tried, but not with encouragement to proceed; for, among other discouragements, they are perforated by a bug, which eats out the kernel. From the cultivation of the common black-eye peas, I have more hope, and am trying them this year, both as a crop, and for ploughing in as a manure; but the severe drought, under which we labor at present, may render the experiment inconclusive. It has, in a manner, destroyed my oats, and threatens to destroy my Indian corn.

The practice of ploughing in buckwheat twice in the season as a fertilizer is not new to me. It is what I have practised, or, I ought rather to have said, attempted to practise, the last two or three years; but, like most things else in my absence, it has been so badly executed, that is, the turning in of the plants has been so ill timed, as to give no result. I am not discouraged, however, by these failures; for, if pulverizing the soil, by fallowing and turning in vegetable substances for manure, is a proper preparation for the crop that is to follow, there can be no question, that a double portion of the latter, without an increase of the ploughing, must be highly beneficial. I am in the act of making another experiment of this sort, and shall myself attend to the operation, which, however, may again prove abortive, from the cause I have mentioned, namely, the drought.

The lightness of our oats is attributed, more than it ought to be, to the unfitness of the climate of the middle States. That this may be the case in part, and nearer the seaboard in a greater degree, I will not controvert; but it is a well-known fact, that no country produces better oats than those that grow on the Allegany Mountains, immediately westward of us. I have heard it affirmed, that they weigh upwards of fifty pounds the Winchester bushel. This may be occasioned by the fertility of the soil, and the attraction of moisture by the mountains; but another reason, and a powerful one too, may be assigned for the inferiority of ours, namely, that we are not choice in our seeds, and do not change them as we ought.

The seeds you were so obliging as to give me shared the same fate that Colonel Wadsworth's did, and as I believe seeds from England generally will do, if they are put into the hold of the vessel. For this reason, I always made it a point whilst I was in the habit of importing seeds, to request my merchants and the masters of vessels, by which they were sent, to keep them from the heat thereof.

You make a distinction, and no doubt a just one, between what in England is called barley, and *big*, or *bere*. If there be none of the true barley in this country, it is not for us, without experience, to pronounce upon the growth of it; and therefore, as noticed in a former part of this letter, it might be interesting to ascertain, whether our climate and soil would produce it to advantage. No doubt, as your observations while you were in the United States appear to have been extensive and accurate, it did not escape you, that both winter and spring barley are cultivated among us. The latter is considered as an uncertain crop south of New York, and I have found it so on my farms. Of the former I have not made sufficient trial to hazard an opinion of success. About Philadelphia it succeeds well.

The Eastern Shore bean, as it is denominated here, has obtained a higher reputation than it deserves; and, like most things unnaturally puffed, sinks into disrepute. Ten or more years ago, led away by exaggerated accounts of its fertilizing quality, I was induced to give a very high price for some of the seed; and, attending to the growth in all its stages, I found that my own fields, which had been uncultivated for two or three years, abounded with the same plants, without perceiving any of those advantages, which had been attributed to them.

I am not surprised that our mode of fencing should be disgusting to a European eye. Happy would it have been for us, if it had appeared so in our own eyes; for no sort of fencing is more expensive or wasteful of timber. I have been endeavouring for years to substitute live fences in place of them; but my long absence from home has in this, as in everything else, frustrated all my plans, that required time and particular attention to effect them. I shall now, although it is too late in the day for me to see the result, begin in good earnest to ditch and hedge; the latter I am attempting with various things, but believe none will be found better than cedar, although I have several kinds of white thorn growing spontaneously on my own grounds.

Rollers I have been in the constant use of for many years, in the way you mention, and find considerable benefit in passing them over my winter grain in the spring, as soon as the ground will admit a hoof on it. I use them also on spring grain and grass seeds, after sowing and sometimes before, to reduce the clods when the ground is rough. My clover generally is sown with spring grain; but, where the ground is not too stiff and binding, it succeeds very well on wheat. Sown on a light snow in February, or the beginning of March, it sinks with the snow and takes good root. And orchard grass, of all others, is in my opinion the best mixture with clover; it blooms precisely at the same time, rises quick again after cutting, stands thick, yields well, and both horses and cattle are fond of it, green or in hay. Alone, unless it is sown very thick, it is apt to form tussocks. If of this, or any other seeds I can procure, you should be in want, I shall have great pleasure in furnishing them.

* * *

For the detailed account of your observations on the husbandry of these United States, and your reflections thereon, I feel myself much obliged, and shall at all times be thankful for any suggestions on agricultural subjects, which you may find leisure and inclination to favor me with, as the remainder of my life, which in the common course of things, now in my sixty-sixth year, cannot be of long continuance, will be devoted wholly to rural and agricultural pursuits.

For the trouble you took in going to Hull, to see if any of the emigrants, who were on the point of sailing from thence to America, would answer my purposes as tenants; and for your very kind and friendly offer of rendering me services, I pray you to accept my sincere thanks, and an assurance of the esteem and regard with which I am, Sir, &c.

Thomas Jefferson, Agrarian

Excerpts From Jefferson's *Notes on Virginia*

From Everett E. Edwards, *Jefferson and Agriculture* (U.S. Department of Agriculture, Agricultural History Series No. 7 (Washington, 1943), pp. 23–25, 30–34, 46–48, 66–72.

The political economists of Europe have established it as a principle, that every State should endeavor to manufacture for itself; and this principle, like many others, we transfer to America, without calculating the difference of circumstance which should often produce a difference of result. In Europe the lands are either cultivated, or locked up against the cultivator. Manufacture must therefore be resorted to of necessity not of choice, to support the surplus of their people. But we have an immensity of land courting the industry of the husbandman. Is it best then that all our citizens should be employed in its improvement, or that one half should be called off from that to exercise manufactures and handicraft arts for the other? Those who labor in the earth are the chosen people of God, if ever He had a chosen people, whose breasts He has made His peculiar deposit for substantial and genuine virtue. It is the focus in which He keeps alive that sacred fire, which otherwise might escape from the face of the earth. Corruption of morals in the mass of cultivators is a phenomenon of which no age nor nation has furnished an example. It is the mark set on those, who, not looking up to heaven, to their own soil and industry, as does the husbandman, for their subsistence, depend for it on casualties and caprice of customers. Dependence begets subservience and venality, suffocates the germ of virtue, and prepares fit tools for the designs of ambition. This, the natural progress and consequence of the arts, has sometimes perhaps been retarded by accidental circumstances; but, generally speaking, the proportion which the aggregate of the other classes of citizens bears in any State to that of its husbandmen, is the proportion of its unsound to its healthy parts, and is a good enough barometer whereby to measure its degree of corruption.

Jefferson to John Jay, from Paris, August 23, 1785

. . . The present [letter] . . . is occasioned by the question proposed in yours of June the 14th; "whether it would be useful to us, to carry all our own productions, or none?"

Were we perfectly free to decide this question, I should reason as follows. We have now lands enough to employ an infinite number of people in their cultivation. Cultivators of the earth are the most valuable citizens. They are the most vigorous, the most independent, the most virtuous, and they are tied to their country, and wedded to its liberty and interests, by the most lasting bonds. As long, therefore, as they can find employment in this line, I would not convert them into mariners, artisans, or anything else. But our citizens will find employment in this line, till their numbers, and of course their productions, become too great for the demand, both internal and foreign. This is not the case as yet, and probably will not be for a considerable time. As soon as it is, the surplus of hands must be turned to something else. I should then, perhaps, wish to turn them to the sea in preference to manufactures; because, comparing the characters of the two classes, I find the former the most valuable citizens. . . .

But what will be the consequence? Frequent wars without a doubt. Their property will be violated on the sea, and in foreign ports, their persons will be insulted, imprisoned, &c., for pretended debts, contracts, crimes, contraband, &c., &c. These insults must be resented, even if we had no feelings, yet to prevent their eternal repetition; or in other words, our commerce on the ocean and in other countries, must be paid for by frequent war. . . . I hope our land office will rid us of our debts, and that our first attention then, will be, to the beginning a naval force of some sort. This alone can countenance our people as carriers on the water, and I suppose them to be determined to continue such.

*　*　*

Jefferson to William Drayton, from Paris, July 30, 1787

Sir — Having observed that the consumption of rice in this country, and particularly in this capital, was very great, I thought it my duty to inform myself from what markets they draw their supplies, in what proportion from ours, and whether it might not be practicable to increase that proportion. This city being little concerned in foreign commerce, it is difficult to obtain information on particular branches of it in the detail. I addressed myself to the retailers of rice, and from them received a mixture of truth and error, which I was unable to sift apart in the first moment. Continuing, however, my inquiries, they produced at length this result: that the dealers here were in the habit of selling two qualities of rice, that of Carolina, with which they were supplied chiefly from England, and that of Piedmont; that the Carolina rice was long, slender, white and transparent, answers well when prepared with milk, sugar, &c., but not so well when prepared *au gras;* that that of Piedmont was shorter, thicker, and less white, but that it presented its form better when dressed *au gras*, was better tasted, and therefore preferred by good judges for those purposes; that the consumption of rice, in this form, was much the most considerable, but that the superior beauty of the Carolina rice, seducing the eye of those purchasers who are attached to appearances, the demand for it was upon the whole as great as for that of Piedmont. They supposed this difference of quality to proceed from a difference of management; that the Carolina rice was husked with an instrument that broke it more, and that less pains were taken to separate the broken from the unbroken grains, imagining that it was the broken grains which dissolved in oily preparations; that the Carolina rice costs somewhat less than that of Piedmont; but that being obliged to sort the whole grains from the broken, in order to satisfy the taste of their customers, they ask and receive as much for the first quality of Carolina, when sorted, as for the rice of Piedmont; but the second and third qualities, obtained by sorting, are sold much cheaper. The objection to the Carolina rice then, being, that it crumbles in certain forms of preparation, and this supposed to be the effect of a less perfect machine for husking, I flattered myself I should be able to learn what might be the machine of Piedmont, when I should arrive at Marseilles, to which place I was to go in the course of a tour through the seaport towns of this country. At Marseilles, however, they differed as much in account of the machines, as at Paris they had differed about other circumstances. Some said it was husked between mill-stones, others between rubbers of wood in the form of mill-stones, others of cork. They concurred in one fact, however, that the machine might be seen by me, immediately on crossing the Alps. This would be an affair of three weeks. I crossed them and went through the rice country from Vercelli to Pavia, about sixty miles. I found the machine to be absolutely the same with that used in Carolina, as well as I could recollect a description which Mr. E. Rutledge had given me of it. It is on the plan of a powder mill. In some of them, indeed, they arm each pestle with an iron tooth, consisting of nine spikes hooked together, which I do not remember in the description of Mr. Rutledge. I therefore had a tooth made, which I have the honor of forwarding you with this letter; observing, at the same time, that as many of their machines are without teeth as with them, and of course, that the advantage is not very palpable. It seems to follow, then, that the rice of Lombardy (for though called Piedmont rice, it does not grow in that county but in Lombardy) is of a different species from that of Carolina; different in form, in color and in quality. We know that in Asia they have several distinct species of this grain. Monsieur Poivre, a former Governor of the Isle of France, in travelling through several countries of Asia, observed with particular attention the objects of their agriculture, and he tells us, that in Cochin-China they cultivate six several kinds of rice, which he describes, three of them requiring water, and three growing on highlands. The rice of Carolina is said to have come from Madagascar, and De Poivre tells us, it is the white rice which is cultivated there. This favors the probability of its being of a different species originally, from that of Piedmont; and time, culture and climate may have made it still more different. Under this idea, I thought it would be well to furnish you with some of the Piedmont rice, unhusked, but was told it was contrary to the laws to export it in that form. I took such measures as I could, however, to have a quantity brought out, and lest these should fail, I brought, myself, a few pounds. A part of this I have addressed to you by the way of London; a part comes with this letter; and I shall send another parcel by some other conveyance, to prevent the danger of miscarriage. Any one of them arriving safe, may serve to put in seed, should the society think it an object. This seed too, coming from Vercelli, where the best rice is supposed to grow, is more to be depended on than what may be sent me hereafter. There is a rice from the Levant, which is considered as of a quality still different, and some think it superior to that of Piedmont. The troubles which have existed in that country for several years back, have intercepted it from the European market, so that it is become almost unknown. I procured a bag of it, however, at Marseilles, and another of the best rice of Lombardy, which are on their way to this place, and when arrived, I will forward you a quantity of each, sufficient to enable you to judge of their qualities when prepared for the table. I have also taken measures to have a quantity of it brought from the Levant, unhusked. If I succeed, it shall be forwarded in like manner. I should think it certainly advantageous to cultivate, in Carolina and Georgia, the two qualities demanded at market; because the progress of culture, with us, may soon get beyond the demand for the white rice; and because too, there is often a brisk demand for the one quality, when the market is glutted with the other. I should hope there would be no danger of losing the species of white rice, by a confusion with the other. This would be a real misfortune, as I should not hesitate to pronounce the white, upon the whole, the most precious of the two, for us.

The dry rice of Cochin-China has the reputation of being the whitest to the eye, best flavored to the taste, and most productive. It seems then to unite the good qualities of both the others known to us. Could it supplant them, it would be a great happiness, as it would enable us to get rid of those ponds of stagnant water, so fatal to human health and life. But such is the force of habit, and caprice of taste, that we could not be sure beforehand it would produce this effect. The experiment, however, is worth trying, should it only end in producing a third quality, and increasing the demand. I will endeavor to procure some to be brought from Cochin-China. The event, however, will be uncertain and distant.

I was induced, in the course of my journey through the south of France, to pay very particular attention to the objects of their culture, because the resemblance of their climate to that of the southern parts of the United States, authorizes us to presume we may adopt any of their articles of culture, which we would wish for. We should not wish for their wines, though they are good and abundant. The culture of the vine is not desirable in lands capable of producing anything else. It is a species of gambling, and of desperate gambling too, wherein, whether you make much or nothing, you are equally ruined. The middling crop alone is the saving point, and that the seasons seldom hit. Accordingly, we see much wretchedness among this class of cultivators. Wine, too, is so cheap in these countries, that a laborer with us, employed in the culture of any other article, may exchange it for wine, more and better than he could raise himself. It is a resource for a country, the whole of whose good soil is otherwise employed, and which still has some barren spots, and surplus of population to employ on them. There the vine is good, because it is something in the place of nothing. It may become a resource to us at a still earlier period; when the increase of population shall increase our productions beyond the demand for them, both at home and abroad. Instead of going on to make an useless surplus of them, we may employ our supernumerary hands on the vine. But that period is not yet arrived.

The almond tree is also so precarious, that none can depend for subsistence on its produce, but persons of capital.

The caper, though a more tender plant, is more certain in its produce, because a mound of earth of the size of a cucumber hill, thrown over the plant in the fall, protects it effectually against the cold of winter. When the danger of frost is over in the spring, they uncover it, and begin its culture. There is a great deal of this in the neighborhood of Toulon. The plants are set about eight feet apart, and yield, one year with another, about two pounds of caper each, worth on the spot sixpence sterling per pound. They require little culture, and this may be performed either with the plough or hoe. The principal work is the gathering of the fruit as it forms. Every plant must be picked every other day, from the last of June till the middle of October. But this is the work of women and children. This plant does well in any kind of soil which is dry, or even in walls where there is no soil, and it lasts the life of a man. Toulon would be the proper port to apply for them. I must observe, that the preceding details cannot be relied on with the fullest certainty, because, in the canton where this plant is cultivated, the inhabitants speak no written language, but a medley, which I could understand but very imperfectly.

The fig and mulberry are so well known in America, that nothing need be said of them. Their culture, too, is by women and children, and therefore earnestly to be desired in countries where there are slaves. In these, the women and children are often employed in labors disproportioned to their sex and age. By presenting to the master objects of culture, easier and equally beneficial, all temptation to misemploy them would be removed, and the lot of this tender part of our species be much softened. By varying, too, the articles of culture, we multiply the chances for making something, and disarm the seasons in a proportionable degree, of their calamitous effects.

The olive is a tree the least known in America, and yet the most worthy of being known. Of all the gifts of heaven to man, it is next to the most precious, if it be not the most precious. Perhaps it may claim a preference even to bread, because there is such an infinitude of vegetables, which it renders a proper and comfortable nourishment. In passing the Alps at the Col de Tende, where they are mere masses of rock, wherever there happens to be a little soil, there are a number of olive trees, and a village supported by them. Take away these trees, and the same ground in corn would not support a single family. A pound of oil, which can be bought for three or four pence sterling, is equivalent to many pounds of flesh, by the quantity of vegetables it will prepare, and render fit and comfortable food. Without this tree, the country of Provence and territory of Genoa would not support one-half, perhaps not one-third, their present inhabitants. The nature of the soil is of little consequence if it be dry. The trees are planted from fifteen to twenty feet apart, and when tolerably good, will yield fifteen or twenty pounds of oil yearly, one with another. There are trees which yield much more. They begin to render good crops at twenty years old, and last till killed by cold, which happens at some time or other, even in their best positions in France. But they put out again from their roots. In Italy, I am told, they have trees two hundred years old. They afford an easy but constant employment through the year, and require so little nourishment, that if the soil be fit for any other production, it may be cultivated among the olive trees without injuring them. The northern limits of this tree are the mountains of the Cevennes, from about the meridian of Carcassonne to the Rhone, and from thence, the Alps and Apennines as far as Genoa, I know, and how much farther I am not informed. The shelter of these mountains may be considered as equivalent to a degree and a-half of latitude, at least, because westward of the commencement of the Cevennes, there are no olive trees in 43½° or even 43° of latitude, whereas, we find them *now* on the Rhone at Pierrelatte, in 44½°, and *formerly* they were at Tains, above the mouth of the Isere, in 45°, sheltered by the near approach of the Cevennes and Alps, which only leave there a passage for the Rhone. Whether such a shelter exists or not in the States of South Carolina and Georgia, I know not. But this we may say, either that it exists or that it is not necessary there, because we know that they produce the orange in open air; and wherever the orange will stand at all, experience shows that the olive will stand well, being a hardier tree. Notwithstanding the great quantities of oil made in France, they have not enough for their own consumption, and therefore import from other countries. This is an article, the consumption of which will always keep pace with its production. Raise it, and it begets it own demand. Little is carried to America, because Europe has it not to spare. We, therefore, have not learned the use of it. But cover the southern States with it, and every man will become a consumer of oil, within whose reach it can be brought in point of price. If the memory of those persons is held in great respect in South Carolina who introduced there the culture of rice, a plant which sows life and death with almost equal hand, what obligations would be due to him who should introduce the olive tree, and set the example of its culture! Were the owner of slaves to view it only as the means of bettering their condition, how much would he better that by planting one of those trees for every slave he possessed! Having been myself an eye witness to the blessings which this tree sheds on the poor, I never had my wishes so kindled for the introduction of any article of new culture into our own country. South Carolina and Georgia appear to me to be the States, wherein its success, in favorable positions at least, could not be doubted, and I flattered myself it would come within the views of the society for agriculture to begin the experiments which are to prove its practicability. Carcassonne is the place from which the plants may be most certainly and cheaply obtained. They can be sent from thence by water to Bordeaux, where they may be embarked on vessels bound for Charleston. There is too little intercourse between Charleston and Marseilles to propose this as the port of exportation. I offer my services to the society for the obtaining and forwarding any number of plants which may be desired.

Before I quit the subject of climates, and the plants adapted to them, I will add, as a matter of curiosity, and of some utility too, that my journey through the southern parts of France, and the territory of Genoa, but still more the crossing of the Alps, enabled me to form a scale of the tenderer plants, and to arrange them according to their different powers of resisting cold. In passing the Alps at the Col de Tende, we cross three very high mountains successively. In ascending, we lose these plants, one after another, as we rise, and find them again in the contrary order as we descend on the other side; and this is repeated three times. Their order, proceeding from the tenderest to the hardiest, is as follows: caper, orange, palm, aloe, olive, pomegranate, walnut, fig, almond. But this must be understood of the plant only; for as to the fruit, the order is somewhat different.

The caper, for example, is the tenderest plant, yet, being so easily protected, it is among the most certain in its fruit. The almond, the hardiest, loses its fruit the oftenest, on account of its forwardness. The palm, hardier than the caper and orange, never produces perfect fruit here.

I had the honor of sending you, the last year, some seeds of the sulla of Malta, or Spanish St. Foin. Lest they should have miscarried, I now pack with the rice a cannister of the same kind of seed, raised by myself. By Colonel Franks, in the month of February last, I sent a parcel of acorns of the cork oak, which I desired him to ask the favor of the Delegates of South Carolina in Congress to forward to you.

I have the honor to be, with sentiments of the most perfect esteem and respect, Sir, your most obedient, and most humble servant. —

Jefferson to President Washington, From Philadelphia, June 28, 1793

DEAR SIR, — I should have taken time ere this to have considered the observations of Mr. Young, could I at this place have done it in such a way as would satisfy either him or myself. When I wrote the notes of the last year, I had never before thought of calculating what were the profits of a capital invested in Virginia agriculture. . . . Mr. Young must not pronounce too hastily on the impossibility of an annual production of £750 worth of wheat coupled with a cattle product of £125. My object was to state the produce of a *good* farm, under *good* husbandry as practised in my part of the country. Manure does not enter into this, because we can buy an acre of new land cheaper than we can manure an old acre. Good husbandry with us consists in abandoning Indian corn and tobacco, tending small grain, some red clover, following, and endeavoring to have, while the lands are at rest, a spontaneous cover of white clover. I do not present this as a culture judicious in itself, but as *good* in comparison with what most people there pursue. Mr. Young has never had an opportunity of seeing how slowly the fertility of the *original soil* is exhausted. With moderate management of it, I can affirm that the James river lowgrounds with the cultivation of small grain, will never be exhausted; because we know that under that cultivation we must now and then take them down with Indian corn, or they become, as they were originally, too rich to bring wheat. The highlands, where I live, have been cultivated about sixty years. The culture was tobacco and Indian corn as long as they would bring enough to pay the labor. Then they were turned out. After four or five years rest they would bring good corn again, and in double that time perhaps good tobacco. Then they would be exhausted by a second series of tobacco and corn. Latterly we have begun to cultivate small grain; and excluding Indian corn, and following, such of them as were originally good, soon rise up to fifteen or twenty bushels the acre. We allow that every laborer will manage ten acres of wheat, except at harvest. I have no doubt but the coupling cattle and sheep with this would prodigiously improve the produce. This improvement Mr. Young will be better able to calculate than anybody else. I am so well satisfied of it myself, that having engaged a good farmer from the head of Elk, (the style of farming there you know well,) I mean in a farm of about 500 acres of cleared land and with a dozen laborers to try the plan of wheat, rye, potatoes, clover, with a mixture of some Indian corn with the potatoes, and to push the number of sheep. This last hint I have taken from Mr. Young's letters which you have been so kind as to communicate to me. I have never before considered with due attention the profit from that animal. I shall not be able to put the farm into that form exactly the ensuing autumn, but against another I hope I shall, and I shall attend with precision to the measures of the ground and of the product, which may perhaps give you something hereafter to communicate to Mr. Young which may gratify him, but I will furnish the ensuing winter what was desired in Mr. Young's letter of Jan. 17, 1793. I have the honor to be, with great and sincere esteem, dear Sir, your most obedient humble servant.

Jefferson to President Washington, From Monticello, May 14, 1794

. . . I find on a more minute examination of my lands than the short visits heretofore made to them permitted, that a ten years' abandonment of them to the ravages of overseers, has brought on them a degree of degradation far beyond what I had expected. As this obliges me to adopt a milder course of cropping, so I find that they have enabled me to do it, by having opened a great deal of lands during my absence. I have therefore determined on a division of my farm into six fields, to be put under this rotation: first year, wheat; second, corn, potatoes, peas; third, rye or wheat, according to circumstances; fourth and fifth, clover where the fields will bring it, and buckwheat dressings where they will not; sixth, folding, and buckwheat dressings. But it will take me from three to six years to get this plan underway. I am not yet satisfied that my acquisition of overseers from the head of Elk has been a happy one, or that much will be done this year towards rescuing my plantations from their wretched condition. Time, patience and perseverance must be the remedy; and the maxim of your letter, "slow and sure," is not less a good one in agriculture than in politics. . . . With every wish for your health and happiness, and my most friendly respects for Mrs. Washington, I have the honor to be, dear Sir, your most obedient, and most humble servant.

Jefferson to President Washington, Monticello, June 19, 1796

I . . . talk to you of my peas and clover. As to the latter article, I have great encouragement from the friendly nature of our soil. I think I have had, both the last and present year, as good clover from common grounds, which had brought several crops of wheat and corn without ever having been manured, as I ever saw on the lots around Philadelphia. I verily believe that a yield of thirty-four acres, sowed on wheat April was twelvemonth, has given me a ton to the acre at its first cutting this spring. The stalks extended, measured three and a half feet long very commonly. Another field, a year older, and which yielded as well the last year, has sensibly fallen off this year. My exhausted fields bring a clover not high enough for hay, but I hope to make seed from it. Such as these, however, I shall hereafter put into peas in the broadcast, proposing that one of my sowings or wheat shall be after two years of clover, and the other after two years of peas. I am trying the white boiling pea of Europe (the Albany pea) this year, till I can get the hog pea of England, which is the most productive of all. But the true winter vetch is what we want extremely. I have tried this year the Carolina drill. It is absolutely perfect. Nothing can be more simple, nor perform its office more perfectly for a single row. I shall try to make one to sow four rows at a time of wheat or peas, at twelve inches distance. I have one of the Scotch threshing machines nearly finished. It is copied exactly from a model Mr. Pinckney sent me, only that I have put the whole works (except the horse wheel) into a single frame, movable from one field to another on the two axles of a wagon. It will be ready in time for the harvest which is coming on, which will give it a full trial. Our wheat and rye are generally fine, and the prices talked of bid fair to indemnify us for the poor crops of the two last years.

* * *

Jefferson to Sir John Sinclair, From Washington, June 30, 1803

Dear Sir, — It is so long since I have had the pleasure of writing to you, that it would be vain to look back to dates to connect the old and the new. Yet I ought not to pass over my acknowledgments to you for various publications received from time to time, and with great satisfaction and thankfulness. I send you a small one in return, the work of a very unlettered farmer, yet valuable, as it relates plain facts of importance to farmers. You will discover that Mr. Binns is an enthusiast for the use of gypsum. But there are two facts which prove he has a right to be so: 1. He began poor, and has made himself tolerably rich by his farming alone. 2. The county of Loudon, in which he lives, had been so exhausted and wasted by bad husbandry, that it began to depopulate, the inhabitants going Southwardly in quest of better lands. Binns' success has stopped that emigration. It is now becoming one of the most productive counties of the State of Virginia, and the price given for the lands is multiplied manifold. . . .

I hope your agricultural institution goes on with success. I consider you as the author of all the good it shall do. A better idea has never been carried into practice. Our agricultural society has at length formed itself. Like our American Philosophical Society, it is voluntary, and unconnected with the public, and is precisely an execution of the plan I formerly sketched to you. Some State societies have been formed heretofore; the others will do the same. Each State society names two of its members of Congress to be their members in the Central society, which is of course together during the sessions of Congress. They are to select matter from the proceedings of the State societies, and to publish it; so that their publications may be called *l'esprit des societes d'agriculture*, &c. The Central society was formed the last winter only, so that it will be some time before they get under way. Mr. Madison, the Secretary of State, was elected their President.

Recollecting with great satisfaction our friendly intercourse while I was in Europe, I nourish the hope it still preserves a place in your mind; and with my salutations, I pray you to accept assurances of my constant attachment and high respect.

Jefferson to Jean Batiste Say, Monticello, March 2, 1815

Dear Sir, — Your letter of June 15th came to hand in December, and it is not till the ratification of our peace, that a safe conveyance for an answer could be obtained. . . . The question proposed in my letter of February 1st, 1804, has since become quite a "question viseuse." I had then persuaded myself that a nation, distant as we are from the contentions of Europe, avoiding all offences to other powers, and not over-hasty in resenting offence from them, doing justice to all, faithfully fulfilling the duties of neutrality, performing all offices of amity, and administering to their interests by the benefits of our commerce, that such a nation, I say, might expect to live in peace, and consider itself merely as a member of the great family of mankind; that in such case it might devote itself to whatever it could best produce, secure of a peaceable exchange of surplus for what could be more advantageously furnished by others, as takes place between one county and another of France. But experience has shown that continued peace depends not merely on our own justice and prudence, but on that of others also; that when forced into war, the interception of exchanges which must be made across a wide ocean, becomes a powerful weapon in the hands of an enemy domineering over that element, and to the other distresses of war adds the want of all those necessaries for which we have permitted ourselves to be dependent on others, even arms and clothing. This fact, therefore, solves the question by reducing it to its ultimate form, whether profit or preservation is the first interest of a State? We are consequently become manufacturers to a degree incredible to those who do not see it, and who only consider the short period of time during which we have been driven to them by the suicidal policy of England. The prohibiting duties we lay on all articles of foreign manufacture which prudence requires us to establish at home, with the patriotic determination of every good citizen to use no foreign article which can be made within ourselves, without regard to difference of price, secures us against a relapse into foreign dependency. And this circumstance may be worthy of your consideration, should you continue in the disposition to emigrate to this country. Your manufactory of cotton, on a moderate scale combined with a farm, might be preferable to either singly, and the one or the other might become principal, as experience should recommend. Cotton ready spun is in ready demand, and if woven, still more so.

I will proceed now to answer the inquiries which respect your views of removal; and I am glad that, in looking over our map, your eye has been attracted by the village of Charlottesville, because I am better acquainted with that than any other portion of the United States, being within three or four miles of the place of my birth and residence. It is a portion of country which certainly possesses great advantages. Its soil is equal in natural fertility to any high lands I have ever seen; it is red and hilly, very like much of the country of Champagne and Burgundy, on the route of Sens, Vermanton, Vitteaux, Dijon, and along the Cote to Chagny, excellently adapted to wheat, maize, and clover; like all mountainous countries it is perfectly healthy, liable to no agues and fevers, or to any particular epidemic, as is evidenced by the robust constitution of its inhabitants, and their numerous families. As many instances of nonagenaires exist habitually in this neighborhood as in the same degree of population anywhere. Its temperature may be considered as a medium of that of the United States. . . . On an average of seven years I have found our snows amount in the whole to fifteen inches depth, and to cover the ground fifteen days; these, with the rains, give us four feet of water in the year. The garden pea, which we are now sowing, comes to table about the 12th of May; strawberries and cherries about the same time; asparagus the 1st of April. The artichoke stands the winter without cover; lettuce and endive with a slight one of bushes, and often without any; and the fig, protected by a little straw, begins to ripen in July; if unprotected, not till the 1st of September. There is navigation for boats of six tons from Charlottesville to Richmond, the nearest tidewater, and principal market for our produce. The country is what we call well inhabited, there being in our county, Albemarle, of about seven hundred and fifty square miles, about twenty thousand inhabitants, or twenty-seven to a square mile, of whom, however, one half are people of color, either slaves or free. The society is much better than is common in country situations; perhaps there is not a better *country* society in the United States. But do not imagine this a Parisian or an academical society. It consists of plain, honest, and rational neighbors, some of them well informed and men of reading, all superintending their farms, hospitable and friendly, and speaking nothing but English. The manners of every nation are the standard of orthodoxy within itself. But these standards being arbitrary, reasonable people in all allow free toleration for the manners, as for the religion of others. Our culture is of wheat for market, and of maize, oats, peas, and clover, for the support of the farm. We reckon it a good distribution to divide a farm into three fields, putting one into wheat, half a one into maize, the other half into oats or peas, and the third into clover, and to tend the fields successively in this rotation. Some woodland in addition, is always necessary to furnish fuel, fences, and timber for constructions. Our best farmers (such as Mr. Randolph, my son-in-law) get from ten to twenty bushels of wheat to the acre; our worst (such as myself) from six to eighteen, with little or more manuring. The bushel of wheat is worth in common times about one dollar. The common produce of maize is from ten to twenty bushels, worth half a dollar the bushel, which is of a cubic foot and a quarter, or, more exactly, of two thousand one hundred and seventy-eight cubic inches. From these data you may judge best for yourself of the size of the farm which would suit your family; bearing in mind, that while you can be furnished by the farm itself for consumption, with every article it is adapted to produce, the sale of your wheat at market is to furnish the fund for all other necessary articles. I will add that both soil and climate are admirably adapted to the vine, which is the abundant natural production of our forests, and that you cannot bring a more valuable laborer than one acquainted with both its culture and manipulation into wine.

Your only inquiry now unanswered is, the price of these lands. To answer this with precision, would require details too long for a letter; the fact being, that we have no metallic measure of values at present, while we are overwhelmed with bank paper. The depreciation of this swells nominal prices, without furnishing any stable index of real value. . . . You may judge that, in this state of things, the holders of bank notes will give free prices for lands, and that were I to tell you simply the present prices of lands in this medium, it would give you no idea on which you could calculate. But I will state to you the progressive prices which have been paid for particular parcels of land for some years back, which may enable you to distinguish between the real increase of value regularly produced by our advancement in population, wealth, and skill, and the bloated value arising from the present disordered and dropsical state of our medium. There are two tracts of land adjoining me, and another not far off, all of excellent quality, which happen to have been sold at different epochs as follows:

One was sold in 1793 for $4 an acre, in 1812, at $10, and is now rated $16. The 2d was sold in 1786 for $5⅓ an acre, in 1803, at $10, and is now rated $20.
The 3d was sold in 1797 for $7 an acre, in 1811, at $16, and is now rated $20.

On the whole, however, I suppose we may estimate that the steady annual rise of our lands is in a geometrical ratio of 5 per cent.; that were our medium now in a wholesome state, they might be estimated at from twelve to fifteen dollars the acre; and I may add, I believe with correctness, that there is not any part of the Atlantic States where lands of equal quality and advantages can be had as cheap. When sold with a dwelling-house on them, little additional is generally asked for the house. These buildings are generally of wooden materials, and of indifferent structure and accommodation. Most of the hired labor here is people of color, either slaves or free. An able-bodied man has sixty dollars a year, and is clothed and fed by the employer; a woman half that. White laborers may be had, but they are less subordinate, their wages higher, and their nourishment much more expensive. A good horse for the plough costs fifty or sixty dollars. A draught ox twenty to twenty-five dollars. A milch cow fifteen to eighteen dollars. A sheep two dollars. Beef is about five cents, mutton and pork seven cents the pound. A turkey or goose fifty cents apiece, a chicken eight and one-third cents; a dozen eggs the same. Fresh butter twenty to twenty-five cents the pound. And, to render as full as I can the information which may enable you to calculate for yourself, I enclose you a Philadelphia price-current, giving the prices in regular times of most of the articles of produce or manufacture, foreign and domestic.

That it may be for the benefit of your children and their descendants to remove to a country where, for enterprise and talents, so many avenues are open to fortune and fame, I have little doubt But I should be afraid to affirm that, at your time of life, and with habits formed on the state of society in France, a change for one so entirely different would be for your personal happiness. Fearful therefore to persuade, I shall add with sincere truth, that I shall very highly estimate the addition of such a neighbor to our society, and that there is no service within my power which I shall not render with pleasure and promptitude. With this assurance be pleased to accept that of my great esteem and respect . . . .

Jefferson to Tristram Dalton, From Monticello, May 2, 1817

A method of ploughing our hillsides horizontally, introduced into this most hilly part of our country by Col. T. M. Randolph, my son in law, may be worth mentioning to you. He has practiced it a dozen or 15 years, and it's advantages were so immediately observed that it has already become very general, and has entirely changed and renovated the face of our country. every rain, before that, while it gave a temporary refreshment, did permanent evil by carrying off our soil and fields were no sooner cleared than washed, at present we may say that we lose none of our soil, the rain not absorbed at the moment of its fall being retained in the hollows between the beds until it can be absorbed. our practice is when we first enter on this process, with a rafter level of 10.f. span, to lay off guide lines conducted horizontally around the hill or valley from one end to the other of the field, and about 30. yards apart. the steps of the level on the ground are marked by a stroke of a hoe, and immediately followed by a plough to preserve the trace. a man or a lad, with a level, and two small boys, the one with sticks, the other with the hoe, will do an acre of this in an hour, and when once done it is forever done. we generally level a field the year it is put into Indian corn, laying it into beds of 6.f. wide with a large water furrow between the beds, until all the fields have been once levelled. the intermediate furrows are run by the eye of the ploughman governed by these guide lines. the inequalities of the declivity in the hill will vary in places the distance of the guide lines and occasion gores which are thrown into short beds. as in ploughing very steep hillsides horizontally a common plough can scarcely throw the furrow up hill, Colo. Randolph has contrived a very simple alteration of the share, which throws the furrow down hill both going and coming. It is as if two shares were welded together at their straight side and at a right angle with each other. this turns on its bar as on a pivot, so as to lay either share horizontal, when the other becoming vertical acts as a mouldboard. this is done by the ploughman in an instant by a single motion of the hand at the end of every furrow . . . horizontal and deep ploughing with the use of plaster and clover are but beginning to be used here will, as we believe, restore this part of our country to its original fertility, which was exceeded by no upland in the state.

Land Policies

Toward a Land Policy, 1780

From "Resolution of October 10, 1780," U.S. Continental Congress, *Journals of the Continental Congress, 1774–1789* (Washington, 1910), XVIII, p. 915.

Resolved, That the unappropriated lands that may be ceded or relinquished to the United States, by any particular states, pursuant to the recommendation of Congress of the 6 day of September last, shall be disposed of for the common benefit of the United States and be settled and formed into distinct republican states, which shall become members of the federal union, and have the same rights of sovereignty, freedom and independence, as the other states: that each state which shall be so formed shall contain a suitable extent of territory, not less than one hundred nor more than one hundred and fifty miles square, or as near thereto as circumstances will admit:

That the necessary and reasonable expences which any particular state shall have incurred since the commencement of the present war, in subduing any of the British posts, or in maintaining forts or garrisons within and for the defence, or in acquiring any part of the territory that may be ceded or relinquished to the United States, shall be reimbursed;

That the said lands shall be granted and settled at such times and under such regulations as shall hereafter be agreed on by the United States in Congress assembled, or any nine or more of them.

Arguments on Land Disposal, 1785

From Edmund C. Burnett, ed., *Letters of Members of the Continental Congress*, (Washington, 1936) VIII, pp. 117–19.

William Grayson to George Washington.

New York, May [8], 1785.

DEAR SIR,

I have received your letter of the 25th of Aprill, for which I am much obliged to you; I am sorry for the melancholy occasion which has induc'd you to leave Mount Vernon, and for the affliction which the loss of such near relations must involve Mrs. Washington in.

The ordinance for disposing of the Western territory has been under consideration ever since I wrote you last and has underwent several alterations, the most considerable of which is that one half of the land is to be sold by sections or lots, and the other half by entire Townships; and the dimension of each township is reduc'd to six miles; I *now* expect the Ordinance will be completed in a few days, it being the opinion of most gentlemen that it is better to pass it in it's present form nearly, than to delay it much longer and incur the risque of losing the country altogether. As soon as it is finished I shall do myself the honor to inclose you a copy, and though it will be far from being the best that could be made, yet I verily believe it is the best that under present circumstances can be procured: There have appeared so many interfering interests, most of them imaginary, so many ill founded jealousies and suspicions throughout the whole, that I am only surpris'd the ordinance is not more exceptionable; indeed if the importunities of the public creditors, and the reluctance to pay them by taxation either direct or implied had not been so great I am satisfied no land Ordinance could have been procured, except under such disadvantages as would in a great degree have excluded the idea of actual settlements within any short length of time; This is not strange when we reflect that several of the States are averse to new votes from that part of the Continent and that some of them are now disposing of their own vacant lands, and of course wish to have their particular debts paid and their own countries settled in the first instance before there is any interference from any other quarter. With respect to the different places of sale, it is certainly open to the objections you mention, but it was absolutely necessary to accede to the measure, before we could advance a single step. Since the receipt of your letter I have hinted to some of the members the propriety of altering this part, but find that the idea of allowing the Citizens of each State an equal chance of trying the good lands at their own doors, was one of the strongest reasons with them for consenting to the ordinance. As to the individual states interfering in the sale, it is guarded against; and in case the loan Officer who is responsible only to Congress cannot dispose of the land in a limited time, it is to be return'd to the Treasury board: With respect to the fractional parts of Townships, the Ordinance has now provided for all cases which can occur, except with respect to the Pensylvania line. The Course of the new State from the Ohio will be due North, and the dispute with Pensylvania will be open to discussion hereafter. I am sorry to observe that throughout this measure, there has been a necessity for sacrificing one's own opinion to that of other people for the purpose of getting forward. There has never been above ten States on the floor and nine of these were necessary to concur in one sentiment, least they should refuse to vote for the Ordinance on it's passage. The price is fix'd at a dollar the acre liquidated certificates, that is the land is not to be sold under that; The reason for establishing this sum was that a part of the house were for half a dollar, and another part for two dollars and others for intermediate sums between the two extremes, so that ultimately this was agreed upon as a central ground. If it is too high (which I am afraid is the case) it may hereafter be corrected by a resolution.

I still mean to move for some amendments which I think will not only advance the sale, but increase the facility of purchasing to foreigners, though from present appearances I own I have but little hopes of success.

After this affair is over, the requisition for the current year will be brought forward. The article of 30,000 dollars for the erection of foederal buildings at Trenton I have already objected to, and shall continue to oppose by every means in my power, as I look upon the measure to be fundamentally wrong, and I am in hopes nine States cannot be found to vote for it; should those in opposition to the measure be able to put off the execution for the present year it is to be expected that the Southern States will open their eyes to their true interests and view this subject in a different light. What I at present fear is, that failing to get this article allowed in the requisition they will attempt to draw the money from Holland by a vote of seven States inasmuch as a hundred thousand dollars were voted at Trenton for that purpose although no particular fund was assigned. I own this matter has giv'n me some disgust, as I see an intemperate ardor to carry it into execution before the sense of the Union is known; and I have no doubt that some gentlemen have come into Congress expressly for that purpose.

I take the liberty of introducing Mr. St Greave a delegate from North Carolina a gentleman of great worth, who is travelling through our State to his own Country: He will be very happy to communicate to you the news of this place. I inclose you the report of a Comm'ee for altering the first paragraph of the 9th article of the confederation, which embraces objects of great magnitude, and about which there is a great difference of sentiment.

The Ordinance of 1785

From Clarence E. Carter, ed., *The Territorial Papers of the United States* (Washington, 1934), II, pp. 12–15.

Be it ordained by the United States in Congress assembled, that the territory ceded by individual States to the United States, which has been purchased of the Indian inhabitants, shall be disposed of in the following manner:

A surveyor from each state shall be appointed by Congress, or a committee of the States, who shall take an Oath for the faithful discharge of his duty, before the Geographer of the United States, who is hereby empowered and directed to administer the same; and the like oath shall be administered to each chain carrier, by the surveyor under whom he acts.

The Geographer, under whose direction the surveyors shall act, shall occasionally form such regulations for their conduct, as he shall deem necessary; and shall have authority to suspend them for misconduct in Office, and shall make report of the same to Congress, or to the Committee of the States; and he shall make report in case of sickness, death, or resignation of any surveyor.

The Surveyors, as they are respectively qualified, shall proceed to divide the said territory into townships of six miles square, by lines running due north and south, and others crossing these at right angles, as near as may be, unless where the boundaries of the late Indian purchases may render the same impracticable, and then they shall depart from this rule no farther than such particular circumstances may require; and each surveyor shall be allowed and paid at the rate of two dollars for every mile, in length, he shall run, including the wages of chain carriers, markers, and every other expense attending the same.

The first line, running north and south as aforesaid, shall begin on the river Ohio, at a point that shall be found to be due north from the western termination of a line, which has been run as the southern boundary of the state of Pennsylvania; and the first line, running east and west, shall begin at the same point, and shall extend throughout the whole territory. Provided, that nothing herein shall be construed, as fixing the western boundary of the state of Pennsylvania. The geographer shall designate the townships, or fractional parts of townships, by numbers progressively from south to north; always beginning each range with number one; and the ranges shall be distinguished by their progressive numbers to the westward. The first range, extending from the Ohio to the lake Erie, being marked number one. The Geographer shall personally attend to the running of the first east and west line; and shall take the latitude of the extremes of the first north and south line, and of the mouths of the principal rivers.

The lines shall be measured with a chain; shall be plainly marked by chaps on the trees, and exactly described on a plat; whereon shall be noted by the surveyor, at their proper distances, all mines, salt springs, salt licks and mill seats, that shall come to his knowledge, and all water courses, mountains and other remarkable and permanent things, over and near which such lines shall pass, and also the quality of the lands.

The plats of the townships respectively, shall be marked by subdivisions into lots of one mile square, or 640 acres, in the same direction as the external lines, and numbered from 1 to 36; always beginning the succeeding range of the lots with the number next to that with which the preceding one concluded. And where, from the causes before mentioned, only a fractional part of a township shall be surveyed, the lots, protracted thereon, shall bear the same numbers as if the township had been entire. And the surveyors, in running the external lines of the townships, shall, at the interval of every mile, mark corners for the lots which are adjacent, always designating the same in a different manner from those of the townships.

The geographer and surveyors shall pay the utmost attention to the variation of the magnetic needle; and shall run and note all lines by the true meridian, certifying, with every plat, what was the variation at the times of running the lines thereon noted.

As soon as seven ranges of townships, and fractional parts of townships, in the direction from south to north, shall have been surveyed, the geographer shall transmit plats thereof to the board of treasury, who shall record the same, with the report, in well bound books to be kept for that purpose. And the geographer shall make similar returns, from time to time, of every seven ranges as they may be surveyed. The Secretary at War shall have recourse thereto, and shall take by lot therefrom, a number of townships, and fractional parts of townships, as well from those to be sold entire as from those to be sold in lots, as will be equal to one seventh part of the whole of such seven ranges, as nearly as may be, for the use of the late continental army; and he shall make a similar draught, from time to time, until a sufficient quantity is drawn to satisfy the same, to be applied in manner hereinafter directed. The board of treasury shall, from time to time, cause the remaining numbers, as well those to be sold entire, as those to be sold in lots, to be drawn for, in the name of the thirteen states respectively, according to the quotas in the last preceding requisition on all the states; provided, that in case more land than its proportion is allotted for sale, in any state, at any distribution, a deduction be made therefor at the next.

The board of treasury shall transmit a copy of the original plats, previously noting thereon, the townships, and fractional parts of townships, which shall have fallen to the several states, by the distribution aforesaid, to the Commissioners of the loan office of the several states, who, after giving notice of not less than two nor more than six months, by causing advertisements to be posted up at the court houses, or other noted places in every county, and to be inserted in one newspaper, published in the states of their residence respectively, shall proceed to sell the townships, or fractional parts of townships, at public vendue, in the following manner, viz: The township, or fractional part of a township, N 1, in the first range, shall be sold entire; and N 2, in the same range, by lots; and thus in alternate order through the whole of the first range. The township, or fractional part of a township, N 1, in the second range, shall be sold by lots; and N 2, in the same range, entire; and so in alternate order through the whole of the second range; and the third range shall be sold in the same manner as the first, and the fourth in the same manner as the second, and thus alternately throughout all the ranges; provided, that none of the lands, within the said territory, be sold under the price of one dollar the acre, to be paid in specie, or loan office certificates, reduced to specie value, by the scale of depreciation, or certificates of liquidated debts of the United States, including interest, besides the expense of the survey and other charges thereon, which are hereby rated at thirty six dollars the township, in specie, or certificates as aforesaid, and so in the same proportion for a fractional part of a township, or of a lot, to be paid at the time of sales; on failure of which payment, the said lands shall again be offered for sale.

There shall be reserved for the United States out of every township, the four lots, being numbered 8, 11, 26, 29, and out of every fractional part of a township, so many lots of the same numbers as shall be found thereon, for future sale. There shall be reserved the lot N 16, of every township, for the maintenance of public schools, within the said township; also one third part of all gold, silver, lead and copper mines, to be sold, or otherwise disposed of as Congress shall hereafter direct. . . .

The commissioners of the loan offices respectively, shall transmit to the board of treasury every three months, an account of the townships, fractional parts of townships, and lots committed to their charge; specifying therein the names of the persons to whom sold, and the sums of money or certificates received for the same; and shall cause all certificates by them received, to be struck through with a circular punch; and they shall be duly charged in the books of the treasury, with the amount of the moneys or certificates, distinguishing the same, by them received as aforesaid.

If any township, or fractional part of a township or lot, remains unsold for eighteen months after the plat shall have been received, by the commissioners of the loan office, the same shall be returned to the board of treasury, and shall be sold in such manner as Congress may hereafter direct.

And whereas Congress, by their resolutions of September 16 and 18 in the year 1776, and the 12th of August, 1780, stipulated grants of land to certain officers and soldiers of the late continental army, and by the resolution of the 22d September, 1780, stipulated grants of land to certain officers in the hospital department of the late continental army; for complying therefore with such engagements, Be it ordained, That the secretary at war, from the returns in his office, or such other sufficient evidence as the nature of the case may admit, determine who are the objects of the above resolutions and engagements, and the quantity of land to which such persons or their representatives are respectively entitled, and cause the townships, or fractional parts of townships, herein-before reserved for the use of the late continental army, to be drawn for in such manner as he shall deem expedient, to answer the purpose of an impartial distribution. He shall, from time to time, transmit certificates to the commissioners of the loan offices of the different states, to the lines of which the military claimants have respectively belonged, specifying the name and rank of the party, the terms of his engagement and time of his service, and the division, brigade, regiment or company to which he belonged, the quantity of land he is entitled to, and the township, or fractional part of a township, and range out of which his portion is to be taken.

The commissioners of the loan offices shall execute deeds for such undivided proportions in manner and form herein before-mentioned, varying only in such a degree as to make the same conformable to the certificate from the Secretary at War.

Where any military claimants of bounty in lands shall not have belonged to the line of any particular state, similar certificates shall be sent to the board of treasury, who shall execute deeds to the parties for the same.

The Secretary at War, from the proper returns, shall transmit to the board of treasury, a certificate, specifying the name and rank of the several claimants of the hospital department of the late continental army, together with the quantity of land each claimant is entitled to, and the township, or fractional part of a township, and range out of which his portion is to be taken; and thereupon the board of treasury shall proceed to execute deeds to such claimants.

The board of treasury, and the commissioners of the loan offices in the states, shall, within 18 months, return receipts to the secretary at war, for all deeds which have been delivered, as also all the original deeds which remain in their hands for want of applicants, having been first recorded; which deeds so returned, shall be preserved in the office, until the parties or their representatives require the same.

And be it further Ordained, That three townships adjacent to lake Erie be reserved, to be hereafter disposed of by Congress, for the use of the officers, men, and others, refugees from Canada, and the refugees from Nova Scotia, who are or may be entitled to grants of land under resolutions of Congress now existing, or which may hereafter be made respecting them, and for such other purposes as Congress may hereafter direct.

And be it further Ordained, That the towns of Gnadenhutten, Schoenbrun and Salem, on the Muskingum, and so much of the lands adjoining to the said towns, with the buildings and improvements thereon, shall be reserved for the sole use of the Christian Indians, who were formerly settled there, or the remains of that society, as may, in the judgment of the Geographer, be sufficient for them to cultivate.

Saving and reserving always, to all officers and soldiers entitled to lands on the northwest side of the Ohio, by donation or bounty from the commonwealth of Virginia, and to all persons claiming under them, all rights to which they are so entitled, under the deed of cession executed by the delegates for the state of Virginia, on the first day of March, 1784, and the act of Congress accepting the same: and to the end, that the said rights may be fully and effectually secured, according to the true intent and meaning of the said deed of cession and act aforesaid, Be it Ordained, that no part of the land included between the rivers called little Miami and Sciota, on the northwest side of the river Ohio, be sold, or in any manner alienated, until there shall first have been laid off and appropriated for the said Officers and Soldiers, and persons claiming under them, the lands they are entitled to, agreeably to the said deed of cession and act of Congress accepting the same.

Selling the Western Lands, 1786

From Edmund C. Burnett, ed., *Letters of Members of the Continental Congress* (Washington, 1936), VIII, pp. 380–82.

Rufus King to Elbridge Gerry.

New York, June 4, [*1786.*]

MY DEAR FRIEND,

I have long entertained doubts concerning the line of Conduct, which Congress ought to pursue relative to the Territory of the U.S. Northwest of the Ohio, and am every day more confirmed in the opinion that no paper engagements, or stipulations, can be formed which will insure a desirable connection between the Atlantic States, and those which will be erected to the Northwestward of the Apalachian or Alleghany Mountains, provided the Mississippi is immediately opened. the pursuits and interests of the people on the two sides, will be so different, and probably so opposite, that an entire separation must eventually ensue. this consequence appears to me so obvious, that I very much doubt whether the U.S. would ever receive a penny of revenue, from the Inhabitants who may settle the Western Territory. should there be an uninterrupted use of the Mississippi at this time by the Citizens of the U.S. I should consider every emigrant to that country from the Atlantic States, as forever lost to the Confederacy. Perhaps, I am in error, but when they have no interest in an Union, inconvenient to them in many points, I can discover no principle, which will attach them to such a connexion. I know not what advantages the inhabitants of the Western Territory would acquire by becoming members [of the Confedera]cy. they will want no protection, their locat[ion], would sufficiently secure them from all foreign hostility; [the] exchange of Merchandize, or Commerce, would not be across the Apalachian Mountains, but wholly confined to the Mississippi. If these conjectures are just, in true policy ought the U.S. to be very assiduous to encourage their Citizens to become Settlers of the count[r]y beyond the Apalachian? the object of Congress appears hitherto to have been a sale of this country for the sinking of the domestic debt; the immediate extinguishment of this debt is certainly a very important consideration, but it has its price. I suppose that a treaty could be formed between Spain and the U.S. upon principles of exact reciprocity. So that the Citizens of the latter, might introduce into the European and African dominions of the former; all sorts of goods and merchandize upon the same terms on which the subjects of Spain could introduce the same articles. and on the other hand, that the subjects of Spain, might import into any of the U.S. all sorts of goods and merchandize, upon the same terms, as the Citizens of the U.S. would introduce the same.

I suppose farther, that the Treaty should stipulate that all the Masts, Spars, Timber etc. etc. wanted for the National Marine of Spain, should be purchased and paid for in the U.S. with specie; provided the quality of the Materials equalled that of the same articles in other Countries.

I suppose yet farther that the Phi[lippine Islands] be opened to the American Ships and [in] consequence, the gold and silver of Acapul[co placed within] their reach. Add to the foregoing an article [*torn*], not to relinquish the right to the free navigation of the Mississippi, but *"stipulating that the U.S. should forbear to use the Navigation of the Mississippi for 20 or 25 years."* Would not such a treaty be of Vast importance to the Atlantic States, particularly to the Western division of them? Would not the Fish, Flour and other products of the U.S. acquire thereby a manifest superiority in Spain over similar commodities of any other country? Would not the conventional forbearance of the use of the Mississippi implicate most strongly the right of the U.S. independent of the Convention or Treaty? If these queries are answered in the affirmative, what objection is there on the part of the U.S. to conclude such a Treaty? This question brings into View, the plan of extinguishing the domestic debt by the Sale of the Western Territory, the System whereby it is proposed to govern the people, who shall settle Westward of the Apalachian Mountains and within the U.S. and the ability of the U.S., at this time to contend with Spain in vindicating their right to the free use and navigation of the Mississippi. I am very sensible that the popular opinion throughout the U.S. is in favor of the free Navigation of the Mississippi, and the reasons must be strong, and important, which [could be offered to] oppose this opinion. I am also pretty [well satisfied] that the free navigation of that River, will [some day] be of Vast importance to the inhabitants [within] the Territories of the U.S. yet admitting what will not be denied, that Spain will on no condition agree that any people, except those of our Nation, shall navigate the Mississippi, are the U.S. in a Condition to assert their right? If you answer this question as I should, (*believing as I do* that a War with Spain, France, or England would terminate in the loss of the Fisheries, and the restriction of boundaries, perhaps by Kennebeck on one part and the Apalachian Mountains on the other) is there any substantial objection against an Article in a Treaty with Spain relative to the Mississippi such as is alluded to? It is a consent to what we cannot alter, considering other benefits to be obtained — it must be wisdom then to consent.

But how will this article affect the Sale of the Western Territory? The answer which the delegates of Virginia (all of whom are probably deeply interested in the Ohio and Kentucky lands) would give, is, that the Value of the Country West of the Alleghany Mountains depends in a high degree upon the opening of the Mississippi. Admit the fact; it is desired that the U.S. [should under these] circumstances open that River to thei[r navigation.] If so, the Value placed upon these lands [which] depends upon the opening of the Mississippi, is an ideal value, at this time. With those therefore who do not wish to involve the U.S. in a War against policy and sound reason, this objection is of little consideration. the lands perhaps will not produce so much under the present circumstances of the Mississippi as they would if the river was open. but to all persons desirous of becoming settlers, they will sell for a reasonable price, and go a good way in extinguishing the domestic debt. But how will such an article affect the intercourse between the Inhabitants of the Western Territory, and those of the Atlantic States? In my judgment very favorably. If the former are cut off for a time, from any connections except with the old States, across the Mountains, I should not despair that a government might be instituted, so connecting them with the Atlantic States, as would be highly [beneficial to them both] and promise a considerable [trade].

My dear friend, after all, these are but speculative opinions, and I am very doubtful of them, when a variety of influential motives, which seem to promise well for my country, authorizes my assent. I alluded to this subject in my last letter to you; I wish for your counsel; I wish the New England States were here. I pray you to read these remarks with candor, and in confidence. If I had taken time and care to have expressed my sentiments on this subject, I would have requested you to have communicated this paper to your friend Gov. Bowdoin, and prayed through you his advice. I shall be brought to a decision on this question. Congress must determine. If Spain don't conclude a Treaty with the U.S., I think they will endeavour to guard against the Evils they fear from us, by an intimate connexion with Great Britain. I am of a committee now in conference with the Secretary of Foreign Affairs on this subject. Spain should form a treaty with us, in preference to any other nation, and there is no nation with whom the U.S. could form more beneficial treaties than with Spain and Portugal. Spain will not give up the Mississippi. But I will not add, I write in great haste and in full confidence. If you are at Boston and can consult Mr. Bowdoin, I should thank you to do it. I intended to have written to him relative to the Barbary treaties, but have not been able to find time.

ADIEU YOURS AFFECTIONATELY,

R. KING

Problems of Western Settlement, 1787

From Edmund C. Burnett, ed., *Letters of Members of the Continental Congress* (Washington, 1936), VIII, pp. 588–89.

James Madison to Edmund Randolph.

New York April 22, 1787.

MY DEAR SIR,

I have the pleasure of your's of the 11 inst. acknowledging mine of the 2d.

Cong's are deliberating on the plan most eligible for disposing of the western territory not yet surveyed. Some alterations will probably be made in the ordinance on that subject, in which the idea of townships will not be altogether abandoned, but rendered less expensive. An Act passed yesterday providing for the sale of the surveyed lands, under the direction of the Treasury board. The price is to be one dollar at the lowest the sale is to be duly advertised in all the States, but the office is to be opened and held *where Cong's shall sit*. The original plan of distributing the sale through all the States was certainly objectionable. To confine it to one place, and that so remote as N.Y. is both from the center of the Union, and the premises in question, can not be less so.

The inhabitants of the Illinois complain of the land-jobbers, particularly Pentecost and Clarke, who are purchasing titles among them. Those of St. Vincents complain of the defect of criminal and civil justice among them, as well as of military protection. These matters are before Cong's and are found to be infinitely embarrassing. . . .

James Madison to Thomas Jefferson.

April 23, 1787.

DEAR SIR,

. . . Congress have agreed to Mr. Jays report on the treaty of peace and to an address which accompanies it. Copies of both will no doubt be sent you from his department. The Legislature of this State which was sitting at the time and on whose account the acts of Congress were hurried through, has adjou[r]ned till Jany. next without deciding on them. This is an ominous example to the other States, and must weaken much the claim on Great Britain of an execution of the Treaty on her part as promised in case of proper steps being taken on ours. Virginia we foresee will be among the foremost in seizing pretexts for evading the injunctions of Congress. S. Carolina is not less infected with the same spirit. The present deliberations of Congress turn on 1. the sale of the western lands, 2. the government of the Western settlements within the federal domain, 3. the final settlement of the accounts between the Union and its members, 4. the [treaty with Spain].

1. Between six and seven hundred thousand acres have been surveyed in townships under the land ordinance, and are to be sold forthwith. The place where Congress sit is fixed for the sale. Its excentricity and remoteness from the premises will I apprehend give disgust. On the most eligible plan of selling the unsurveyed residue Congress are much divided; the Eastern States being strongly attached to that of townships, notwithstanding the expence incident to it; the Southern being equally biassed in favor of indiscriminate locations, notwithstanding the many objections agst. that mode. The dispute will probably terminate in some kind of compromise, if one can be hit upon.

2. The government of the settlements on the Illinois and Wabash is a subject very perplexing in itself; and rendered more so by our ignorance of many circumstances on which a right judgment depends. The inhabitants at those places, claim protection agst. the savages, and some provision for both criminal and civil justice. It appears also that land jobbers are among them who are likely to multiply litigations among individuals, and by collusive purchases of spurious titles, to defraud the United States. . . .

Large Land Sales, 1787

From: Edmund C. Burnett, ed., *Letters of Members of the Continental Congress* (Washington, 1936), VIII, p. 659.

William Grayson to James Monroe.

N York 22nd, Octob. 1787.

DEAR SIR,

I have received your favor, and delivered the enclosure to Miss Kortright; the Capt. being on a trip to the township. Congress four days since made a contract with Royal Flint and associates of N. York for three millions of acres on the Wabash on nearly the same terms as that of Cutler and Sarjeant. I believe I informed you that Judge Symms of Jersey had contracted for two millions between the Great and the little Miami. The whole of the contracts will when fully complied with, amount to an extinguishment of six Millions Doll's of the domestic debt, and congress *now* looking upon the Western country in its true light, *i e,* as a most valuable fund for the total extinctionment of the domestic debt, have directed the treasury board to continue the sales on nearly the same terms and principles as those already made. A very considerable emigration will take effect from the five Eastermost States. A Brigade files off from Massachusets immed'ly and which is to be followed by much more considerable ones next spring and fall. A Doctr. Gano a Baptist preacher in this town will carry out (it is said) his whole congregation amounting to five hundred. Symms is beating up for volunteers in the Jerseys, as is the case with Parsons in Connecticut, and Varnum in Rhode Island; these two last are appointed Judges in the Western country.

Congress have authorized St. Clair now Governor of the Western territory to hold a treaty with the Indians next spring if necessary and 14,000 dollars are appropriated for this purpose. a treaty is also directed with the Cherokees and Creeks and 6,000 dollars are appropriated for it. . . .

The Brittish packett arrived last night: but we have not yet received the dispatches from the Secy. for foreign affairs: The news of the day is that the affairs of Holland are in as bad a train as ever: I shall set out from this in about ten days and it is possible I may have the pleasure of seeing you in Richmond. . . . .

The Ordinance of 1787

From Clarence E. Carter, ed., *The Territorial Papers of the United States* (Washington, 1934), II, pp. 39–49.

An Ordiance for Ascertaining the Mode of Disposing of Lands in the Western Territory

Be it ordained by the United States in Congress assembled, That the said territory, for the purposes of temporary government, be one district, subject, however, to be divided into two districts, as future circumstances may, in the opinion of Congress, make it expedient.

Be it ordained by the authority aforesaid, That the estates, both of resident and non-resident proprietors in the said territory, dying intestate, shall descend to, and be distributed among, their children, and the descendants of a deceased child, in equal parts; the descendants of a deceased child or grandchild to take the share of their deceased parent in equal parts among them: And where there shall be no children or descendants, then in equal parts to the next of kin in equal degree; and, among collaterals, the children of a deceased brother or sister of the intestate shall have, in equal parts among them, their deceased parents' share; and there shall, in no case, be a distinction between kindred of the whole and half-blood; saving, in all cases, to the widow of the intestate her third part of the real estate for life, and one-third part of the personal estate; and this law, relative to descents and dower, shall remain in full force until altered by the legislature of the district. And, until the governor and judges shall adopt laws as hereinafter mentioned, estates in the said territory may be devised or bequeathed by wills in writing, signed and sealed by him or her, in whom the estate may be (being of full age,) and attested by three witnesses; and real estates may be conveyed by lease and release, or bargain and sale, signed, sealed, and delivered by the person, being of full age, in whom the estate may be, and attested by two witnesses, provided such wills be duly proved, and such conveyances be acknowledged, or the execution thereof duly proved, and be recorded within one year after proper magistrates, courts, and registers shall be appointed for that purpose; and personal property may be transferred by delivery; saving, however to the French and Canadian inhabitants, and other settlers of the Kaskaskias, St. Vincents, and the neighboring villages who have heretofore professed themselves citizens of Virginia, their laws and customs now in force among them, relative to the descent and conveyance of property.

Be it ordained by the authority aforesaid, That there shall be appointed, from time to time, by Congress, a governor, whose commission shall continue in force for the term of three years, unless sooner revoked by Congress; he shall reside in the district, and have a freehold estate therein in 1000 acres of land, while in the exercise of his office.

There shall be appointed, from time to time, by Congress, a secretary, whose commission shall continue in force for four years unless sooner revoked; he shall reside in the district, and have a freehold estate therein in 500 acres of land, while in the exercise of his office; it shall be his duty to keep and preserve the acts and laws passed by the legislature, and the public records of the district, and the proceedings of the governor in his Executive department; and transmit authentic copies of such acts and proceedings, every six months, to the Secretary of Congress: There shall also be appointed a court to consist of three judges, any two of whom to form a court, who shall have a common law jurisdiction, and reside in the district, and have each therein a freehold estate in 500 acres of land while in the exercise of their offices; and their commissions shall continue in force during good behavior.

The governor and judges, or a majority of them, shall adopt and publish in the district such laws of the original States, criminal and civil, as may be necessary and best suited to the circumstances of the district, and report them to Congress from time to time: which laws shall be in force in the district until the organization of the General Assembly therein, unless disapproved of by Congress; but, afterwards, the legislature shall have authority to alter them as they shall think fit.

The governor, for the time being, shall be commander-in-chief of the militia, appoint and commission all officers in the same below the rank of general officers; all general officers shall be appointed and commissioned by Congress.

Previous to the organization of the General Assembly, the governor shall appoint such magistrates and other civil officers, in each county or township, as he shall find necessary for the preservation of the peace and good order in the same: After the General Assembly shall be organized, the powers and duties of the magistrates and other civil officers, shall be regulated and defined by the said assembly; but all magistrates and other civil officers, not herein otherwise directed, shall, during the continuance of this temporary government, be appointed by the governor.

For the prevention of crimes and injuries, the laws to be adopted or made shall have force in all parts of the district, and for the execution of process, criminal and civil, the governor shall make proper divisions thereof; and he shall proceed, from time to time, as circumstances may require, to lay out the parts of the district in which the Indian titles shall have been extinguished, into counties and townships, subject, however, to such alterations as may thereafter be made by the legislature.

So soon as there shall be 5000 free male inhabitants of full age in the district, upon giving proof thereof to the governor, they shall receive authority, with time and place, to elect representatives from their counties or townships to represent them in the General Assembly: *Provided,* That, for every 500 free male inhabitants, there shall be one representative, and so on progressively with the number of free male inhabitants, shall the right of representation increase, until the number of representatives shall amount to 25; after which, the number and proportion of representatives shall be regulated by the legislature: *Provided,* That no person be eligible or qualified to act as a representative unless he shall have been a citizen of one of the United States three years, and be a resident in the district, or unless he shall have resided in the district three years; and, in either case, shall likewise hold in his own right, in fee simple, 200 acres of land within the same: *Provided, also,* That a freehold in 50 acres of land in the district, having been a citizen of one of the States, and being resident in the district, or the like freehold and two years residence in the district, shall be necessary to qualify a man as an elector of a representative.

The representatives thus elected, shall serve for the term of two years; and, in case of the death of a representative, or removal from office, the governor shall issue a writ to the county or township for which he was a member, to elect another in his stead, to serve for the residue of the term.

The General Assembly, or Legislature, shall consist of the governor, legislative council, and a house of representatives. The legislative council shall consist of five members, to continue in office five years, unless sooner removed by Congress; any three of whom to be a quorum: and the members of the council shall be nominated and appointed in the following manner, to wit: As soon as representatives shall be elected, the governor shall appoint a time and place for them to meet together; and, when met, they shall nominate ten persons, residents in the district, and each possessed of a freehold in 500 acres of land, and return their names to Congress; five of whom Congress shall appoint and commission to serve as aforesaid; and, whenever a vacancy shall happen in the council, by death or removal from office, the house of representatives shall nominate two persons, qualified as aforesaid, for each vacancy, and return their names to Congress; one of whom Congress shall appoint and commission for the residue of the term. And every five years, four months at least before the expiration of the time of service of the members of council, the said house shall nominate ten persons, qualified as aforesaid, and return their names to Congress; five of whom Congress shall appoint and commission to serve as members of the council five years, unless sooner removed. And the governor, legislative council, and house of representatives, shall have authority to make laws in all cases, for the good government of the district, not repugnant to the principles and articles in this ordinance established and declared. And all bills, having passed by a majority in the house, and by a majority in the council, shall be referred to the governor for his assent; but no bill, or legislative act whatever, shall be of any force without his assent. The governor shall have power to convene, prorogue, and dissolve the General Assembly, when, in his opinion, it shall be expedient.

The governor, judges, legislative council, secretary, and such other officers as Congress shall appoint in the district, shall take an oath or affirmation of fidelity and of office; the governor before the President of Congress, and all other officers before the governor. As soon as a legislature shall be formed in the district, the council and house assembled in one room, shall have authority, by joint ballot, to elect a delegate to Congress, who shall have a seat in Congress, with a right of debating but not of voting during this temporary government.

And, for extending the fundamental principles of civil and religious liberty, which form the basis whereon these republics, their laws and constitutions are erected; to fix and establish those principles as the basis of all laws, constitutions, and governments, which forever hereafter shall be formed in the said territory: to provide also for the establishment of States, and permanent government therein, and for their admission to a share in the federal councils on an equal footing with the original States, at as early periods as may be consistent with the general interest:

It is hereby ordained and declared by the authority aforesaid, That the following articles shall be considered as articles of compact between the original States and the people and States in the said territory and forever remain unalterable, unless by common consent, to wit:

Art. 1st. No person, demeaning himself in a peaceable and orderly manner, shall ever be molested on account of his mode of worship or religious sentiments, in the said territory.

Art. 2d. The inhabitants of the said territory shall always be entitled to the benefits of the writ of *habeas corpus*, and of the trial by jury; of a proportionate representation of the people in the legislature; and of judicial proceedings according to the course of the common law. All persons shall be bailable, unless for capital offences, where the proof shall be evident or the presumption great. All fines shall be moderate; and no cruel or unusual punishments shall be inflicted. No man shall be deprived of his liberty or property, but by the judgment of his peers or the law of the land; and, should the public exigencies make it necessary, for the common preservation, to take any person's property, or to demand his particular services, full compensation shall be made for the same. And, in the just preservation of rights and property, it is understood and declared, that no law ought ever to be made, or have force in the said territory, that shall, in any manner whatever, interfere with or affect private contracts or engagements, *bona fide,* and without fraud, previously formed.

Art. 3d. Religion, morality, and knowledge, being necessary to good government and the happiness of mankind, schools and the means of education shall forever be encouraged. The utmost good faith shall always be observed towards the Indians; their lands and property shall never be taken from them without their consent; and, in their property, rights, and liberty, they shall never be invaded or disturbed, unless in just and lawful wars authorized by Congress; but laws founded in justice and humanity, shall, from time to time, be made for preventing wrongs being done to them, and for preserving peace and friendship with them.

Art. 4th. The said territory, and the States which may be formed therein, shall forever remain a part of this confederacy of the United States of America, subject to the Articles of Confederation, and to such alterations therein as shall be constitutionally made; and to all the acts and ordinances of the United States in Congress assembled, conformable thereto. The inhabitants and settlers in the said territory shall be subject to pay a part of the federal debts contracted or to be contracted, and a proportional part of the expenses of government, to be apportioned on them by Congress according to the same common rule and measure by which apportionments thereof shall be made on the other States; and the taxes, for paying their proportion, shall be laid and levied by the authority and direction of the legislatures of the district or districts, or new States, as in the original States, within the time agreed upon by the United States in Congress assembled. The legislatures of those districts or new States, shall never interfere with the primary disposal of the soil by the United States in Congress assembled, nor with any regulations Congress may find necessary for securing the title in such soil to the *bona fide* purchasers. No tax shall be imposed on lands the property of the United States; and, in no case, shall non-resident proprietors be taxed higher than residents. The navigable waters leading into the Mississippi and St. Lawrence, and the carrying places between the same, shall be common highways, and forever free, as well to the inhabitants of the said territory as to the citizens of the United States, and those of any other States that may be admitted into the Confederacy, without any tax, impost, or duty, therefor.

Art. 5th. There shall be formed in the said territory, not less than three nor more than five States; and the boundaries of the States, as soon as Virginia shall alter her act of cession, and consent to the same, shall become fixed and established as follows, to wit: The Western State in the said territory, shall be bounded by the Mississippi, the Ohio, and Wabash rivers; a direct line drawn from the Wabash and Post St. Vincent's, due North, to the territorial line between the United States and Canada; and, by the said territorial line, to the Lake of the Woods and Mississippi. The middle State shall be bounded by the said direct line, the Wabash from Post Vincent's, to the Ohio; by the Ohio, by a direct line, drawn due North from the mouth of the Great Miami, to the said territorial line, and by the said territorial line. The Eastern State shall be bounded by the last mentioned direct line, the Ohio, Pennsylvania, and the said territorial line: *Provided, however,* and it is further understood and declared, that the boundaries of these three States shall be subject so far to be altered, that, if Congress shall hereafter find it expedient, they shall have authority to form one or two States in that part of the said territory which lies North of an East and West line drawn through the Southerly bend or extreme of lake Michigan. And, whenever any of the said States shall have 60,000 free inhabitants therein, such State shall be admitted, by its delegates, into the Congress of the United States, on an equal footing with the original States in all respects whatever, and shall be at liberty to form a permanent constitution and State government: *Provided,* the constitution and government so to be formed, shall be republican, and in conformity to the principles contained in these articles; and, so far as it can be consistent with the general interest of the confederacy, such admission shall be allowed at an earlier period, and when there may be a less number of free inhabitants in the State than 60.000.

Art. 6th. There shall be neither slavery nor involuntary servitude in the said territory, otherwise than in the punishment of crimes, whereof the party shall have been duly convicted: *Provided, always,* That any person escaping into the same, from whom labor or service is lawfully claimed in any one of the original States, such fugitive may be lawfully reclaimed and conveyed to the person claiming his or her labor or service as aforesaid.

Be it ordained by the authority aforesaid, That the resolutions of the 23d of April, 1784, relative to the subject of this ordinance, be, and the same are hereby, repealed and declared null and void.

Land Act of 1800

From "An Act to Amend the Act Entitled An Act providing for the sale of Lands of the United States, in the territory northwest of the Ohio. . . ." May 10, 1800, 2 U.S. Statutes at Large 73.

SECTION 1. *Be it enacted by the Senate and House of Representatives of the United States of America in Congress assembled,* That for the disposal of the lands of the United States, directed to be sold by the act, intituled "An act providing for the sale of the lands of the United States, in the territory northwest of the Ohio, and above the mouth of Kentucky river," there shall be four land offices established in the said territory: one at Cincinnati, for lands below the Little Miami, which have not heretofore been granted; one at Chilicothe, for lands east of the Scioto, south of the lands appropriated for satisfying military bounties to the late army of the United States, and west of the fifteenth range of townships; one at Marietta, for the lands east of the sixteenth range of townships, south of the before mentioned military lands, and south of a line drawn due west from the northwest corner of the first township of the second range, to the said military lands; and one at Steubenville, for the lands north of the last mentioned line, and east or north of the said military lands. Each of the said offices shall be under the direction of an officer, to be called "The Register of the Land Office," who shall be appointed by the President of the United States, by and with the advice and consent of the Senate, and shall give bond to the United States, with approved security, in the sum of ten thousand dollars, for the faithful discharge of the duties of his office; and shall reside at the place where the land office is directed to be kept.

SEC. 2. *And be it further enacted,* That it shall be the duty of the surveyor-general, and he is hereby expressly enjoined, to prepare and transmit to the registers of the several land offices, before the days herein appointed for commencing sales, general plats of the lands hereby directed to be sold at the said offices respectively, and also to forward copies of each of the said plats to the Secretary of the Treasury.

SEC. 3. *And be it further enacted,* That the surveyor-general shall cause the townships west of the Muskingum, which by the above-mentioned act are directed to be sold in quarter townships, to be subdivided into half sections of three hundred and twenty acres each, as nearly as may be, by running parallel lines through the same from east to west, and from south to north, at the distance of one mile from each other, and marking corners, at the distance of each half mile on the lines running from east to west, and at the distance of each mile on those running from south to north, and making the marks, notes and descriptions, prescribed to surveyors by the above-mentioned act: And the interior lines of townships intersected by the Muskingum, and of all the townships lying east of that river, which have not been heretofore actually subdivided into sections, shall also be run and marked in the manner prescribed by the said act, for running and marking the interior lines of townships directed to be sold in sections of six hundred and forty acres each. And in all cases where the exterior lines of the townships, thus to be subdivided into sections or half sections, shall exceed or shall not extend six miles, the excess or deficiency shall be specially noted, and added to or deducted from the western and northern ranges of sections or half sections in such township, according as the error may be in running the lines from east to west, or from south to north; the sections and half sections bounded on the northern and western lines of such townships shall be sold as containing only the quantity expressed in the returns and plats respectively, and all others as containing the complete legal quantity. And the President of the United States shall fix the compensation of the deputy surveyors, chain carriers, and axemen: *Provided,* the whole expense of surveying and marking the lines, shall not exceed three dollars for every mile that shall be actually run, surveyed and marked.

SEC. 4. *And be it further enacted,* That the lands thus subdivided (excluding the sections reserved by the above-mentioned act) shall be offered for sale in sections and half sections, subdivided as before directed at the following places and times, that is to say: Those below the Little Miami shall be offered at public vendue, in the town of Cincinnati, on the first Monday of April one thousand eight hundred and one, under the direction of the register of the land office there established, and of either the governor or secretary of the northwestern territory. The lands east of Scioto, south of the military lands, and west of the fifteenth range of townships, shall be offered in like manner for sale at Chilicothe, on the first Monday of May, one thousand eight hundred and one, under the direction of the register of the land office there established, and of either the governor or secretary of the said territory. The lands east of the sixteenth range of townships, south of the military lands and west of the Muskingum, including all the townships intersected by that river, shall be offered for sale in like manner at Marietta, on the last Monday of May, one thousand eight hundred and one, under the direction of the governor or secretary, or surveyor-general of the said territory. The sales shall remain open at each place for three weeks, and no longer. The superintendents shall observe the rules and regulations of the above-mentioned act, in classing and selling fractional with entire sections, and in keeping and transmitting accounts of the sales. All lands, remaining unsold, at the closing of either of the public sales, may be disposed of at private sale by the registers of these respective land offices, in the manner herein after prescribed; and the register of the land office at Steubenville, after the first day of July next, may proceed to sell, at private sale, the lands situate within the district assigned to his direction as herein before described, disposing of the same in sections, and classing fractional with entire sections, according to the provisions and regulations of the above-mentioned act and of this act: And the register of the land office at Marietta, after the said first day of July next, may proceed to sell at private sale, any of the lands within the district assigned to his direction as aforesaid, which are east of the river Muskingum, excluding the townships intersected by that river, disposing of the same in sections, and classing fractional with entire sections as aforesaid.

SEC. 5. *And be it further enacted,* That no lands shall be sold by virtue of this act, at either public or private sale, for less than two dollars per acre, and payment may be made for the same by all purchasers, either in specie, or in evidences of the public debt of the United States, at the rates prescribed by the act, intituled, ''An act to authorize the receipt of evidences of the public debt in payment for the lands of the United States;'' and shall be made in the following manner, and under the following conditions, to wit:

1. At the time of purchase, every purchaser shall, exclusively of the fees hereafter mentioned, pay six dollars for every section, and three dollars for every half section, he may have purchased, for surveying expenses, and deposit one twentieth part of the amount of purchase money, to be forfeited, if within forty days one fourth part of the purchase money, including the said twentieth part, is not paid.

2. One fourth part of the purchase money shall be paid within forty days after the day of sale as aforesaid; another fourth part shall be paid within two years; another fourth part within three years; and another fourth part within four years after the day of sale.

3. Interest, at the rate of six per cent. a year from the day of sale, shall be charged upon each of the three last payments, payable as they respectively become due.

4. A discount at the rate of eight per cent. a year, shall be allowed on any of the three last payments, which shall be paid before the same shall become due, reckoning this discount always upon the sum, which would have been demandable by the United States, on the day appointed for such payment.

5. If the first payment of one fourth part of the purchase money shall not be made within forty days after the sale, the deposit, payment and fees, paid and made by the purchaser, shall be forfeited, and the lands shall and may, from and after the day, when the payment of one fourth part of the purchase money should have been made, be disposed of at private sale, on the same terms and conditions, and in the same manner as the other lands directed by this act to be disposed of at private sale: *Provided,* that the lands which shall have been sold at public sale, and which shall, on account of such failure of payment, revert to the United States, shall not be sold at private sale, for a price less than the price that shall have been offered for the same at public sale.

6. If any tract shall not be completely paid for within one year after the date of the last payment, the tract shall be advertised for sale by the register of the land office within whose district it may lie, in at least five of the most public places in the said district, for at least thirty days before the time of sale. And he shall sell the same at public vendue, during the sitting of the court of quarter sessions of the county in which the land office is kept, for a price not less than the whole arrears due thereon, with the expenses of sale; the surplus, if any, shall be returned to the original purchaser, or to his legal representative; but if the sum due, with interest, be not bidden and paid, then the land shall revert to the United States. All monies paid therefor shall be forfeited, and the register of the land office may proceed to dispose of the same to any purchaser, as in case of other lands at private sale.

Sec. 6. *And be it further enacted,* That all and every the payments, to be made by virtue of the preceding section, shall be made either to the treasurer of the United States, or to such person or officer as shall be appointed by the President of the United States, with the advice and consent of the Senate, receiver of public monies for lands of the United States, at each of the places respectively where the public and private sales of the said lands are to be made; and the said receiver of public monies shall, before he enters upon the duties of his office, give bond with approved security, in the sum of ten thousand dollars, for the faithful discharge of his trust; and it shall be the duty of the said treasurer and receiver of public monies to give receipts for the monies by them received, to the persons respectively paying the same; to transmit within thirty days in case of public sale, and quarterly, in case of private sale, an account of all the public monies by them received, specifying the amount received from each person, and distinguishing the sums received for surveying expenses, and those received for purchase money, to the Secretary of the Treasury, and to the registers of the land office, as the case may be. The said receivers of public monies shall, within three months after receiving the same, transmit the monies by them received to the treasurer of the United States; and the receivers of public monies for the said sales, and also the receivers of public monies for the sales which have taken place at Pittsburg under the act, intituled ''An act providing for the sale of the lands of the United States in the territory northwest of the Ohio, and above the mouth of Kentucky river,'' shall receive one per cent. on the money received, as a compensation for clerk hire, receiving, safe keeping, and transmitting it to the treasury of the United States.

SEC. 7. *And be it further enacted,* That it shall be the duty of the registers of the land offices respectively, to receive and enter on books kept for that purpose only, and on which no blank leaves or space shall be left between the different entries, the applications of any person or persons who may apply for the purchase of any section or half section, and who shall pay him the fee hereafter mentioned, and produce a receipt from the treasurer of the United States, or from the receiver of public monies appointed for that purpose, for three dollars for each half section such person or persons may apply for, and for at least one twentieth part of the purchase money, stating carefully in each entry the date of the application, the date of the receipt to him produced, the amount of monies specified in the said receipt, and the number of the section or half section, township and range applied for. If two or more persons shall apply at the same time for the same tract, the register shall immediately determine by lot, in presence of the parties, which of them shall have preference. He shall file the receipt for monies produced by the party, and give him a copy of his entry, and if required, a copy of the description of the tract, and a copy of the plat of the same, or either of them; and it shall be his duty to inform the party applying for any one tract, whether the same has already been entered, purchased, or paid for, and at his request to give him a copy of the entry or entries concerning the same. He shall, three months after the date of each application, if the party shall not have, within that time, produced to him a receipt of the payment of one fourth part of the purchase money, including the twentieth part above mentioned, enter under its proper date, in the said book of entries, that the payment has not been made, and that the land has reverted to the United States, and he shall make a note of the same in the margin of the book opposite to the original entry. And if the party shall, either at the time of making the original entry, or at any time within three months thereafter, produce a receipt to him, for the fourth part of the purchase money, including the twentieth part aforesaid, he shall file the receipt, make an entry of the same, under its proper date, in the said book of entries, make a note of the same in the margin of the book, opposite to the original entry, and give to the party a certificate, describing the land sold, the sum paid on account, the balance remaining due, the time and times when such balance shall become due, and that if it shall be duly discharged, the purchaser or his assignee or other legal representative, shall be entitled to a patent for the said lands; he shall also, upon any subsequent payment being made, and a receipt from the receiver being produced to him, file the original receipt, give a receipt for the same to the party, and enter the same to the credit of the party, in a book kept for that purpose, in which he shall open an account in the name of each purchaser, for each section or half section that may be sold either at public or private sale, and in which he shall charge the party for the whole purchase money, and give him credit for all his payments; making the proper charges and allowances for interest or discount, as the case may be, according to the provisions of the fourth section of this act; and upon the payment being completed and the account finally settled, he shall give a certificate of the same to the party; and on producing to the Secretary of the Treasury, the same final certificate, the President of the United States is hereby authorized to grant a patent for the lands to the said purchaser, his heirs or assigns; and all patents shall be countersigned by the Secretary of State, and recorded in his office.

SEC. 8. *And be it further enacted,* That the registers of the land offices respectively, shall also note on the book of surveys, or original plat transmitted to them, every tract which may be sold, by inserting the letter A on the day when the same is applied for, and the letter P on the day when a receipt for one fourth part of the purchase money is produced to them, and by crossing the said letter A on the day when the land shall revert to the United States, on failure of the payment of one fourth part of the purchase money within three months after the date of application. And the said book of surveys or original plat shall be open at all times, in presence of the register, for the inspection of any individual, applying for the same and paying the proper fee.

SEC. 9. *And be it further enacted,* That it shall be the duty of the registers of the land offices to transmit quarterly to the Secretary of the Treasury, and to the surveyor-general, an account of the several tracts applied for, of the several tracts for which the payment of one fourth part of the purchase money has been made, of the several tracts which have reverted to the United States on failure of the said payment; and also an account of all the payments of monies by them entered, according to the receipts produced to them, specifying the sums of money, the names of the persons paying the same, the names of the officers who have received the same, and the tracts for which the same have been paid.

SEC. 10. *And be it further enacted,* That the registers aforesaid shall be precluded from entering on their books any application for lands in their own name, and in the name of any other person in trust for them; and if any register shall wish to purchase any tract of land, he may do it by application in writing to the surveyor-general, who shall enter the same on books kept for that purpose by him, who shall proceed in respect to such applications, and to any payments made for the same, in the same manner which the registers by this act are directed to follow, in respect to applications made to them for lands by other persons. The registers shall, nevertheless, note on the book of surveys, or original plat, the applications and payments thus by them made, and their right to the pre-emption of any tract shall bear date from the day, when their application for the same shall have been entered by the surveyor-general in his own book. And if any person applying for any tract shall, notwithstanding he shall have received information from the register, that the same has already been applied for by the said register, or by any other person, insist to make the application, it shall be the duty of the register to enter the same, noting in the margin that the same tract is already purchased, but upon application of the party made in writing, and which he shall file, he may and shall at any future time enter under its proper date, that the party withdraws his former application, and applies in lieu thereof for any other tract: *Provided always,* that the party shall never be allowed thus to withdraw his former application, and to apply in lieu thereof for another tract, except when the tract described in his former application shall have been applied for previous to the date of that his former application.

Sec. 11. *And be it further enacted,* That the Secretary of the Treasury shall and may prescribe such further regulations, in the manner of keeping books and accounts, by the several officers in this act mentioned, as to him may appear necessary and proper, in order fully to carry into effect the provisions of this act.

Sec. 12. *And be it further enacted,* That the registers of the land offices, respectively, shall be entitled to receive from the treasury of the United States, one half per cent. on all the monies expressed in the receipts by them filed and entered, and of which they shall have transmitted an account to the Secretary of the Treasury, as directed by this act; and they shall further be entitled to receive, for their own use, from the respective parties, the following fees for services rendered, that is to say; for every original application for land, and a copy of the same, for a section three dollars, for a half section two dollars; for every certificate stating that the first fourth part of the purchase money is paid, twenty-five cents; for every subsequent receipt for monies paid, twenty-five cents; for the final settlement of account and giving the final certificate of the same, one dollar; for every copy, either of an application or of the description of any section or half section, or of the plat of the same, or of any entry made on their books, or of any certificate heretofore given by them, twenty-five cents for each; and for any general inspection of the book of surveys, or general plat, made in their presence, twenty-five cents.

Sec. 13. *And be it further enacted,* That the superintendents of the public sales, to be made by virtue of this act, and the superintendents of the sales which have taken place by virtue of the act, intituled "An act providing for the sale of the lands of the United States in the territory northwest of the river Ohio, and above the mouth of Kentucky river," shall receive five dollars a day for every day whilst engaged in that business; and the accounting officers of the treasury are hereby authorized to allow a reasonable compensation for books, stationery and clerk hire, in settling the accounts of the said superintendents.

Sec. 14. *And be it further enacted,* That the fee to be paid for each patent for half a section shall be four dollars, and for every section five dollars, to be accounted for by the receiver of the same.

Sec. 15. *And be it further enacted,* That the lands of the United States reserved for future disposition, may be let upon leases by the surveyor-general, in sections or half sections, for terms not exceeding seven years, on condition of making such improvements as he shall deem reasonable.

Sec. 16. *And be it further enacted,* That each person who, before the passing of this act, shall have erected, or begun to erect, a grist-mill or saw-mill upon any of the lands herein directed to be sold, shall be entitled to the pre-emption of the section including such mill, at the rate of two dollars per acre: *Provided*, the person or his heirs, claiming such right of pre-emption, shall produce to the register of the land office satisfactory evidence that he or they are entitled thereto, and shall be subject to and comply with the regulations and provisions by this act prescribed for other purchasers.

SEC. 17. *And be it further enacted,* That so much of the act, providing for the sale of the lands of the United States in the territory northwest of the river Ohio, and above the mouth of Kentucky river, as comes within the purview of this act, be, and the same is hereby repealed.

APPROVED, May 10, 1800.

Act for Establishment of Ohio, 1802

From "An Act to Enable the People of the Eastern Division of the Territory Northwest of the River Ohio . . . ." April 30, 1802, U.S. Statutes at Large 173.

Be it enacted by the Senate and House of Representatives of the United States of America in Congress assembled, That the inhabitants of the eastern division of the territory northwest of the river Ohio, be, and they are hereby authorized to form for themselves a constitution and state government, and to assume such name as they shall deem proper, and the said state, when formed, shall be admitted into the Union, upon the same footing with the original states, in all respects whatever.

SEC. 2. *And be it further enacted,* That the said state shall consist of all the territory included within the following boundaries, to wit: bounded on the east by the Pennsylvania line, on the south by the Ohio river, to the mouth of the Great Miami river, on the west by the line drawn due north from the mouth of the Great Miami, aforesaid, and on the north by an east and west line, drawn through the southerly extreme of Lake Michigan, running east after intersecting the due north line aforesaid, from the mouth of the Great Miami, until it shall intersect Lake Erie, or the territorial line, and thence with the same through Lake Erie to the Pennsylvania line, aforesaid: *Provided,* that Congress shall be at liberty at any time hereafter, either to attach all the territory lying east of the line to be drawn due north from the mouth of the Miami, aforesaid, to the territorial line, and north of an east and west line drawn through the southerly extreme of Lake Michigan, running east as aforesaid to Lake Erie, to the aforesaid state, or dispose of it otherwise, in conformity to the fifth article of compact between the original states, and the people and states to be formed in the territory northwest of the river Ohio.

SEC. 3. *And be it further enacted,* That all that part of the territory of the United States, northwest of the river Ohio, heretofore included in the eastern division of said territory, and not included within the boundary herein prescribed for the said state, is hereby attached to, and made a part of the Indiana territory, from and after the formation of the said state, subject nevertheless to be hereafter disposed of by Congress, according to the right reserved in the fifth article of the ordinance aforesaid, and the inhabitants therein shall be entitled to the same privileges and immunities, and subject to the same rules and regulations, in all respects whatever, with all other citizens residing within the Indiana territory.

SEC. 4. *And be it further enacted,* That all male citizens of the United States, who shall have arrived at full age, and resided within the said territory at least one year previous to the day of election, and shall have paid a territorial or county tax, and all persons having in other respects, the legal qualifications to vote for representatives in the general assembly of the territory, be, and they are hereby authorized to choose representatives to form a convention, who shall be apportioned amongst the several counties within the eastern division aforesaid, in a ratio of one representative to every twelve hundred inhabitants of each county, according to the enumeration taken under the authority of the United States, as near as may be, that is to say: from the county of Trumbull, two representatives; from the county of Jefferson, seven representatives, two of the seven to be elected within what is now known by the county of Belmont, taken from Jefferson and Washington counties; from the county of Washington, four representatives; from the county of Ross, seven representatives, two of the seven to be elected in what is now known by Fairfield county, taken from Ross and Washington counties; from the county of Adams, three representatives; from the county of Hamilton, twelve representatives, two of the twelve to be elected in what is now known by Clermont county, taken entirely from Hamilton county; and the elections for the representatives aforesaid, shall take place on the second Tuesday of October next, the time fixed by a law of the territory, intituled ''An act to ascertain the number of free male inhabitants of the age of twenty-one, in the territory of the United States northwest of the river Ohio, and to regulate the elections of representatives for the same,'' for electing representatives to the general assembly, and shall be held and conducted in the same manner as is provided by the aforesaid act, except that the qualifications of electors shall be as herein specified.

SEC. 5. *And be it further enacted,* That the members of the convention, thus duly elected, be, and they are hereby authorized to meet at Chilicothe on the first Monday in November next; which convention, when met, shall first determine by a majority of the whole number elected, whether it be or be not expedient at that time to form a constitution and state government for the people, within the said territory, and if it be determined to be expedient, the convention shall be, and hereby are authorized to form a constitution and state government, or if it be deemed more expedient, the said convention shall provide by ordinance for electing representatives to form a constitution or frame of government; which said representatives shall be chosen in such manner, and in such proportion, and shall meet at such time and place, as shall be prescribed by the said ordinance; and shall form for the people of the said state, a constitution and state government; provided the same shall be republican, and not repugnant to the ordinance of the thirteenth of July, one thousand seven hundred and eighty-seven between the original states and the people and states of the territory northwest of the river Ohio.

SEC. 6. *And be it further enacted,* That until the next general census shall be taken, the said state shall be entitled to one representative in the House of Representatives of the United States.

SEC. 7. *And be it further enacted,* That the following propositions be, and the same are hereby offered to the convention of the eastern state of the said territory, when formed, for their free acceptance or rejection, which, if accepted by the convention, shall be obligatory upon the United States.

First, That the section, number sixteen, in every township, and where such section has been sold, granted or disposed of, other lands equivalent thereto, and most contiguous to the same, shall be granted to the inhabitants of such township, for the use of schools.

Second, That the six miles reservation, including the salt springs, commonly called the Scioto salt springs, the salt springs near the Muskingum river, and in the military tract, with the sections of land which include the same, shall be granted to the said state for the use of the people thereof, the same to be used under such terms and conditions and regulations as the legislature of the said state shall direct: *Provided,* the said legislature shall never sell nor lease the same for a longer period than ten years.

Third, That one twentieth part of the nett proceeds of the lands lying within the said state sold by Congress, from and after the thirtieth day of June next, after deducting all expenses incident to the same, shall be applied to the laying out and making public roads, leading from the navigable waters emptying into the Atlantic, to the Ohio, to the said state, and through the same, such roads to be laid out under the authority of Congress, with the consent of the several states through which the road shall pass: *Provided always,* that the three foregoing propositions herein offered, are on the conditions that the convention of the said state shall provide, by an ordinance irrevocable, without the consent of the United States, that every and each tract of land sold by Congress, from and after the thirtieth day of June next, shall be and remain exempt from any tax laid by order or under authority of the state, whether for state, county, township or any other purpose whatever, for the term of five years from and after the day of sale.

APPROVED, April 30, 1802.

Land Bounties for Soldiers, 1811

From "An Act for Completing the Existing Military Establishment," December 24, 1811, 2 U.S. Statutes at Large 669.

Be it enacted by the Senate and House of Representatives of the United States of America in Congress assembled, That the military establishment, as now authorized by law, be immediately completed.

SEC. 2. *And be it further enacted,* That there be allowed and paid to each effective, able bodied man, recruited or re-enlisted for that service, for the term of five years, unless sooner discharged, the sum of sixteen dollars; but the payment of one half of the said bounty shall be deferred until he shall be mustered and have joined the corps in which he is to serve; and whenever any non-commissioned officer or soldier shall be discharged from the service, who shall have obtained from the commanding officer of his company, battalion or regiment a certificate that he had faithfully performed his duty whilst in service, he shall moreover be allowed and paid, in addition to the aforesaid bounty, three months pay, and one hundred and sixty acres of land; and the heirs and representatives of those non-commissioned officers or soldiers, who may be killed in action, or die in the service of the United States, shall likewise be paid and allowed the said additional bounty of three months' pay, and one hundred and sixty acres of land, to be designated, surveyed and laid off at the public expense, in such manner and upon such terms and conditions, as may be provided by law.

APPROVED, December 24, 1811.

Mississippi Petition to Defer Land Payments, 1812

From: Clarence E. Carter, ed., *Territorial Papers of the United States, The Territory of Mississippi, 1809–1817* (Washington,1938), VI, pp. 330–31.

[*November 9, 1812*]

To the Honorable the Senate and House of Representatives of the United States in Congress Assembled.

The Memorial of the Legislative Council and House of Representatives of the Mississippi territory, in General Assembly convened, respectfully represents that since the sale of public lands in this Territory; by reason of the restrictions which have been imposed on American Commerce, from the inimical course pursued towards neutral trade by the belligerents of Europe and from the present disastrous state of American affairs, and of the World, affecting most materially every discription of persons, as well as every human pursuit, the price of the staple commodity of the country has been so reduced, that the honest purchaser who has indulged the fond hope that from close industry and diligence he might accumulate as much from the profits of his farm as would meet the installments due Government, has not been more than a sufficiency with oeconomy from the sale of his crop to discharge the anual expences of his family and consequently is unable to pay for his lands, on which he has from the sweat of his brow, made considerable improvements, and from which alone he can calculate for the future maintainance of his family. Your Memorialists having confidence in the Wisdom of the Measures of the General Government have Great reason to expect the impartial course and the energetic measures pursued towards the Belligerents of Europe must Yet have the desired effect, to wit, a respect for neutral rights, and a restoration of American Commerce to its original importance, and thereby enable the purchasers of land in a few Years to pay the United States the principal and interest due the same. We therefore the Representatives of the People do pray that Your Honorable body will pass a law extending the time of payments to preemption Claimants and all other purchasers of public Lands in this Territory from the day on which the last payment becomes due, until one year after the termination of the present War, or to any other time which in your wisdom you may appoint, and that the payments thereafter may be by annual installments. And Your Memorialists with due defference to Your Honorable Body do further beg leave to represent, that in as much as the present pressure of the times, renders it altogether impracticable to pay the purchase money at the day it has or may become due, they do therefore conceive that a back interest on such installments, predicated on such failure is oppressive, and therefore respectfully pray Your interference for the relief of such delenquent purchasers.

Attest   Tho. B. Reed, Clerk
   H OF REP.
COUNCIL CHAMBER *Nov*ʳ 9ᵗʰ 1812
Attest   Felix Hughes Secʸ
   to the Legislative Council

COWLES MEAD
Speaker of the House
Representatives.
THOMAS BARNES
President of the
Legislative Council

[*Endorsed*] Memorial of the Legislative council and House of Representatives of the Mississippi territory.

11ᵗʰ Decemʳ 1812. Refᵈ to the Committee of the whole House on the report of the Committee on the Public Lands on sundry petitions from purchasers of Public Lands. &c. Mʳ Poindexter refᵈ commee who ho refᵈ lands

Petition for Land for Free Negroes, 1816

From Clarence E. Carter, ed., *Territorial Papers of the United States, Territory of Louisiana-Missouri, 1815–1821* (Washington, 1951), XV, pp. 218–19.

Christopher McPherson to the Secretary of State

[NA:SD, Misc. Letters:ALS]

Washington City Friday the 13ᵗʰ of Decʳ 1816

THE HONORABLE JAMES MONROE Esquire Secretary of State
   RESPECTED SIR,
   I beg leave for your perusal of the four inclosed introductory papers, which shew my Character —
   The object of my visit from Richmond to this City, is to endeavour to purchase from Government a territory of Country on the Missouri & Mississippi Rivers, reaching to a lake for the Free people of Colour in this happy Union — very many in Virginia & North Carolina are ready to sell their property in order to remove thither immediately — and the thing would doubtless become general. — We want three to Six hundred miles square or thereabouts as low as possible with as long Credit as may be suitable. And we wish to march without Gun or sword, under the protection of the General Government — say perhaps in the ratio of one soldier to ten persons. — I pray you Sir, to be so good as embrace

the earliest opportunity of doing me the favour of a conference on this subject. —

I have the honor to be Sir, With great respect & Esteem Your Most obedient & very Hble Servant

CHRIST[r] M[c]PHERSON

[*Addressed*] The Honorable James Monroe Esquire Secretary of State Washington City

[*Endorsed*] Memorial of C M[c]Pherson C. M. M[c]Pherson 13 Dec[r] 1816. N B I put up at the Washington Hotel — C M[c]P

Act For Public Land Sales, 1820

From "An Act Making Further Provision for the Sale of the Public Lands," April 24, 1820, 3 U.S. Statutes at Large 566.

Be it enacted by the Senate and House of Representatives of the United States of America, in Congress assembled, That from and after the first day of July next, all the public lands of the United States, the sale of which is, or may be authorized by law, shall, when offered at public sale, to the highest bidder, be offered in half quarter sections; and when offered at private sale, may be purchased, at the option of the purchaser, either in entire sections, half sections, quarter sections, or half quarter sections; and in every case of the division of a quarter section, the line for the division thereof shall run north and south, and the corners and contents of half quarter sections which may thereafter be sold, shall be ascertained in the manner, and on the principles directed and prescribed by the second section of an act entitled, "An act concerning the mode of surveying the public lands of the United States," passed on the eleventh day of February, eighteen hundred and five; and fractional sections, containing one hundred and sixty acres, or upwards, shall, in like manner, as nearly as practicable, be subdivided into half quarter sections, under such rules and regulations as may be prescribed by the Secretary of the Treasury; but fractional sections, containing less than one hundred and sixty acres, shall not be divided, but shall be sold entire: *Provided,* That this section shall not be construed to alter any special provision made by law for the sale of land in town lots.

SEC. 2. *And be it further enacted,* That credit shall not be allowed for the purchase money on the sale of any of the public lands which shall be sold after the first day of July next, but every purchaser of land sold at public sale thereafter, shall, on the day of purchase, make complete payment therefor; and the purchaser at private sale shall produce, to the register of the land office, a receipt from the treasurer of the United States, or from the receiver of public moneys of the district, for the amount of the purchase money on any tract, before he shall enter the same at the land office; and if any person, being the highest bidder, at public sale, for a tract of land, shall fail to make payment therefor, on the day on which the same was purchased, the tract shall be again offered at public sale, on the next day of sale, and such person shall not be capable of becoming the purchaser of that or any other tract offered at such public sales.

SEC. 3. *And be it further enacted,* That from and after the first day of July next, the price at which the public lands shall be offered for sale, shall be one dollar and twenty-five cents an acre; and at every public sale, the highest bidder, who shall make payment as aforesaid, shall be the purchaser; but no land shall be sold, either at public or private sale, for a less price than one dollar and twenty-five cents an acre; and all the public lands which shall have been offered at public sale before the first day of July next, and which shall then remain unsold, as well as the lands that shall thereafter be offered at public sale, according to law, and remain unsold at the close of such public sales, shall be subject to be sold at private sale, by entry at the land office, at one dollar and twenty-five cents an acre, to be paid at the time of making such entry as aforesaid; with the exception, however, of the lands which may have reverted to the United States, for failure in payment, and of the heretofore reserved sections for the future disposal of Congress, in the states of Ohio and Indiana, which shall be offered at public sale, as hereinafter directed.

SEC. 4. *And be it further enacted,* That no lands which have reverted, or which shall hereafter revert, and become forfeited to the United States for failure in any manner to make payment, shall, after the first day of July next, be subject to entry at private sale, nor until the same shall have been first offered to the highest bidder at public sale; and all such lands which shall have reverted before the said first day of July next, and which shall then belong to the United States, together with the sections, and parts of sections, heretofore reserved for the future disposal of Congress, which shall, at the time aforesaid, remain unsold, shall be offered at public sale to the highest bidder, who shall make payment therefor, in half quarter sections, at the land office for the respective districts, on such day or days as shall, by proclamation of the President of the United States, be designated for that purpose; and all lands which shall revert and become forfeited for failure of payment after the said first day of July next, shall be offered in like manner at public sale, at such time, or times, as the President shall by his proclamation designate for the purpose: *Provided,* That no such lands shall be sold at any public sales hereby authorized, for a less price than one dollar and twenty-five cents an acre, nor on any other terms than that of cash payment; and all the lands offered at such public sales, and which shall remain unsold at the close thereof, shall be subject to entry at private sale, in the same manner, and at the same price with the other lands sold at private sale, at the respective land offices.

SEC. 5. *And be it further enacted,* That the several public sales authorized by this act, shall, respectively, be kept open for two weeks, and no longer; and the registers of the land office and the receivers of public money shall, each, respectively, be entitled to five dollars for each day's attendance thereon.

SEC. 6. *And be it further enacted,* That, in every case hereafter, where two or more persons shall apply for the purchase, at private sale, of the same tract, at the same time, the register shall determine the preference, by forthwith offering the tract to the highest bidder.

APPROVED, April 24, 1820.

Remonstrance to Congress by Inhabitants of Michigan Territory, 1828

From Clarence E. Carter, ed., *Territorial Papers of the United States, the Territory of Michigan, 1820–1829* (Washington, 1943), II, pp. 1175–76.

[*March 10, 1828*]

To the Honorable the Senate and House of Representatives in Congress assembled,

The Subscribers inhabitants of the Ter. of Michigan respectfully remonstrate to your Honorable Body, against the passage of a Bill, reported by the Committee upon Public Lands, entitled "a Bill to graduate the price of public land &ᶜ". We think the Bill is essentially unjust to those who have already purchased of the United States, and have undergone all the labor expense & deprivation of settling a new country — the lands have then become more valuable; but the first settler finds his property falling upon his hands from the fact that new comers can buy for one fifth of what he has paid when the country was a wilderness. Our Territory will be settled by squatters who are a pest to society. By the provisions of this Bill any one may obtain one eighty acre lot by residing upon it and may purchase a thousand acres in its vicinity. Speculation will be the order of the day, immense quantities of land will be purchased at twenty five cents per acre by companies and individuals, a credit system will be established, and the price raised to two or three dollars. The cash system for the disposal of the public lands has produced the happiest results in establishing a moral population for Michigan. The present price secures to the United States a fair compensation for the expenses attendant upon the acquisition and survey of new lands, and secures to the settler a respectable neighbourhood. Your Honorable Body cannot for a moment imagine, that our society would be benefitted by giving away the Public Lands and thus enticing the dregs of community from the old States. We appeal to your Honᵇˡᵉ Body for protection against a system which we think must drive every respectable man out of this country. If your Honorable Body were disposed to establish a place of refuge for the refuse of society, if you were inclined to draw off from the old States, that portion of the people who are a nuisance to any country, we think it might not be amiss to fix upon some place in the interior of the North Western Territory, and to offer lands within a given district to the settler free of expense; but we beg that Michigan may not be selected for that object. We are far from objecting to a grant of land from the United States to the several states and Territories within whose boundaries these lands may lie; but let that grant be made now; now we have roads and bridges and canals to make; let the first settlers be benefitted by these lands; let them have some compensation for the expense of making roads for those who are to come after them. It must be obvious to your Honorable Body, that this Bill if it pass into a law, must procrastinate the settlement of new countries, as emigrants will wait until land falls to its lowest price. We think that this Bill should be entitled, A Bill for the encouragement of squatters and speculators for the destruction of the moral character of new countries, for a tax upon first settlers and a premium to those who come last, and for the establishment of the old credit system.

Land Sales in Arkansas, 1829

From Clarence E. Carter, ed., *Territorial Papers of the United States, the Territory of Arkansas. 1829–1836* (Washington, 1954), XXI, pp. 109–14.

George Graham to the Secretary of the Treasury

Nov. 21. 1829.

HON. SAM¹ D. INGHAM.

SIR,

I have the honor to submit the usual Annual Report in relation to the affairs of this Office. —

The papers marked A. exhibits the periods to which the Quarterly Accounts of the respective Receivers in Office have been returned and adjusted, with the balances on hand agreeably to the last monthly returns of the Receivers of Public Moneys, respectively.

The paper, marked B. exhibits the quantity of Land sold and the amount of purchase money received, under the Cash and Credit system for the year 1828, and the first six months of 1829, and also the amounts paid into the Treasury in Cash & Stock within those periods respectively, after deducting the Incidental Expenses.

The gross proceeds on account of the Public Lands for the years 1827, 1828, and the first six months of 1829, amount to $4.075.282. and the sums paid into the Treasury and accounted for within the same period is $4.079.043, giving an average annual gross receipt of more than $1.600.000., but as a portion of these receipts was for lands heretofore sold under the credit system, this Amount cannot be assumed as the permanent annual revenue to be derived from the sale of the public Lands.

There is reason, however, to believe that there will be an annual demand for about One million of Acres of the public Lands, which, at the minimum price of $1.^{25}/₁₀₀ per acre will give an annual revenue from this source of $1.250.000. which will probably be increased with the progress of population and improvement. That these causes create a demand for the public lands, equal, if not superior, to that arising from their fertility, when taken in connection with their locality, is evinced by the following facts. That the Cash sales of land in the State of Ohio for the year 1828. where the Lands have generally been in market for thirty years, equal to those in any other State, with the exception of Indiana. That the Cash sales in the Steubenville Land District, where the Lands have been in the market for nearly forty years, and within which there is not more than two or three hundred thousand acres of Public Lands left for sale, and which of course, are of the most inferior quality, amounted in 1828, to $35.000. Whereas the sales in the Piqua District in the same State, where the lands have only been in the market for six or Eight years, and a very small portion of them sold, and within which there are extensive bodies of Land of excellent quality, the sales for the same period, amounted to only $2.000. And in the Territory of Arkansas, where there are immense bodies of land of a very superior quality, subject to entry at the minimum price, the sales within the year 1828. amounted to only $2.000. —

The Act of Congress requiring that the public Lands should be sold only for Cash, and the immense quantity of those Lands, that has been brought into market in the States and Territories, extending from the Northern Lakes to the Gulf of Mexico, have had the effect not only of checking, but of almost entirely putting an end to, the speculations in the purchase of the public lands, and it is beleived that the purchases now made from the public, are almost exclusively for the purpose of occupation and improvement; a result the most desirable from any system that may be devised for the disposition of the public domain. —

On the 4th of July last, the several laws relative to the Lands that had been sold on a credit expired, and all Lands purchased on credit which had not been paid for, or relinquished, reverted to the U.S. And the money heretofore paid on them was forfeited. The amount of money thus paid and forfeited is estimated at $700.000. These forfeitures are confined exclusively to the amounts paid on Lands further credited under the provisions of the Act of the 2^d March 1821. and those supplementary thereto. From 1824, until July last, the parties incurring these forfeitures had the privilege of settling their accounts by relinquishing a portion of the Lands purchased, or by paying for the same at a discount of 37½ pr cent without interest. Under these circumstances it is contemplated to offer for sale the relinquished Lands, and those reverted, as soon after April next as circumstances may make expedient, unless Congress shall otherwise, by law, direct.

When the public Lands were sold on a credit, which in fact extended to five years after the payment of One fourth of the purchase money, it became indispensably necessary to protect the U. States against improper combinations by which the purchaser might permit the Land to be sold for taxes, that an arrangement should be made with the people of the Territories, when those Territories were about to assume the form of States, by which the lands sold by the U.S. should be exempted from taxes for a period equal to that to which the Credit given on them was extended, and accordingly, a compact was entered into with the people of the new States within which the public Lands lie, by which the Lands sold by the U.S. should be exempt from taxes for five years. The credit system being now entirely abolished, one of the reasons for this stipulation has ceased to exist, and as those States complain of such exemption of the lands from taxes as a grievance, it is respectfully suggested whether it would not be advisable to exonerate them from this contract. Altho' the exemption of the Lands sold by the U.S. from taxes for a given period, has, no doubt, a tendency to promote emigration, generally, to the States within which the public Lands lie, yet it is perhaps better that each of those States should have the option of deciding how far such exemption might comport with their own views of their own interests.

With the termination of the credit system all legal objections to the entire consolidation of the Land Offices within the respective States, has ceased, and on that subject, I take leave to refer to the following extract from a report made from this Office and submitted to Congress by the Secretary of the Treasury, in Jany 1825,

"After the Lands in any State have been all surveyed and offered at public sale, and when the debt due for lands heretofore purchased in such State, shall have been liquidated, a further and more extensive reduction of the Land Districts may be made by limiting them, in each State, to a single District and locating the Land Office at the Seat of the Government of the State.'' —

"In any arrangement that may be made for the reduction of the number of Land Districts, the fact that a large debt is due on account of Lands, the payments for which are to be made, at the option of the debtors, either at the Treasury of the U.S., or at the Office where the Lands were purchased, is entitled to much attention, in respect both as to the rights and accommodation of the purchasers; and when it is also taken into consideration that the disbursements of the public money are very limited in nearly all that Section of Country where the public Lands lie, and that, from the force of circumstances, these disbursements will not, probably, be materially increased, it is submitted, how far other considerations, than those arising from the reduction of the Expenses of collecting this branch of the revenue, might make it expedient to reduce the Land Districts to the number proposed in this report, until those circumstances shall have occurred which would justify the reduction of them to a Single district in each State.''

A consolidation of the Land Offices would materially reduce the expenses incident to the collection of this branch of the revenue, as well as those incident to this Office, and should Congress, under all the circumstances belonging to the question, deem such an arrangement expedient, it might be immediately carried into effect in the State of Ohio, where the Lands in each Land District have all been offered at public sale, and gradually be extended to the other States, as the fact that all the lands in each of the Land Districts had been offered at public Sale might occur. —

Suspicions having been excited that extensive frauds had been practised in the Territory of Arkansas under the provisions of the Acts of the 26th of May 1824 and 22^d May 1826. authorizing the claimants to Lands, whose titles were derived from the Spanish Gov^t, to institute proceedings in the Court for the said Territory, to try the validity of such claims, I was authorized by the President to instruct Col^o Isaac T. Preston, the Reg. of the L. Office at New Orleans, who possessed in an eminent degree the qualifications necessary for making such an investigation, to proceed to Little Rock for the purpose of examining the Spanish documents that had been filed, and on which the Court had confirmed a number of claims, and by comparing them with the documents that were known to be genuine to ascertain the extent of the frauds that had been practised. The communication from him, dated the 10th of Oct^o 1829 herewith transmitted, and marked C. exhibits the result of his investigation; from which it appears that 117 Claims, covering 60.000 acres of land, were confirmed by the Court of Arkansas between the 19th & 24th of December 1827, that 7 Claims covering 20.000 Acres of Land were yet pending: and that 188 Claims had been withdrawn, or struck from the docket, because security for the costs had not been given, and that there exists no doubt but that the whole of those claims are founded on forged evidences of title. And it further appears that under an order of Court made on the 9th of October 1828, the original papers, on which these claims were founded, have been withdrawn from the files of the Court, with the exception of those filed in 58 of the cases.

Application has been made to this Office to issue patents on entries made at the Land Offices at Little Rock and Batesville for 84 of the claims confirmed by the Court. For the six first cases presented for entries made at Batesville, patents were issued, but as the number of applications for patents increased, as the patents were all required to be issued to assignees, and as on a comparison of the assignments with each other I was satisfied that they had been manufactured by a few individuals, my previous suspicions as to the frauds committed on the Court were so far confirmed as to justify me in withholding the patents in all the other cases. Two of the cases in which patents have issued were no doubt founded on fraudulent evidences of title.

Measures have been taken to procure, if possible, a revision by the Court of Arkansas, of all those cases which have been confirmed by the Court, and the Registers of the Land Offices at Batesville and Little Rock have been instructed not to permit any entry to be made on their books by virtue of such confirmations, and to give public notice that no patents will be issued on entries heretofore made by virtue of such confirmations, until Congress shall have acted upon the subject.

The powers of the Court of Arkansas in relation to these cases will by the provisions of the Law cease in May next, Some immediate Legislative provision is therefore indispensably necessary, by which the powers of the Court may be continued for the special purpose of acting on the cases now pending and on the bills that may be filed to obtain a revision of the cases heretofore decided by the Court. It may perhaps be deemed more expedient to repeal forthwith the Acts of 1824 and 1826, so far as they respect the Territory of Arkansas, and at the same time to re-enact such provisions as may be necessary to enable the Court to receive, and act upon, the bills that may be filed for a revision of the cases heretofore confirmed, and providing for an appeal on such Bills of revision, from the Court of Arkansas to the Supreme Court of the U.S. — Some further Legislative provisions will be also necessary in relation to the papers that have been withdrawn from the files of the Court, and in relation to the admission of entries and granting of patents so as to protect the U. States from imposition by forged assignments from fictitious confirmees should the Court ultimately refuse to reverse the judgments heretofore rendered.

In relation to the fraudulent Claims in Louisiana alluded to in the communication of Mr Preston it is apprehended that they exist in a greater extent than is stated by him; but acting upon the principle that forgery can form no foundation for a good title, the Surveyor of Louisiana has been instructed not to survey, or receive for record any survey including Lands claimed by virtue of written evidence of title emanating from the Spanish Gov^t and confirmed by Congress, where there was good reason to believe that such written evidence of title had been forged, and to proceed to survey the Lands so claimed as public Lands. So soon therefore as any portion of these Lands shall have been sold by the U. S. the parties claiming under a title purporting to have emanated from the Spanish Gov^t and confirmed by Congress, will have opportunity of obtaining a judicial decision as to the validity of their claim, and also as to its proper location. In almost every one of these cases of the original papers on which the claim was founded have been withdrawn from the Land Office in which they were filed: the expediency therefore of some Legislative provision is suggested, by which the original papers shall be required to be produced in Court in all cases of controversy relative to title, arising between the parties claiming by purchase from the U. States, and those claiming by virtue of an Act of Congress confirming a claim purporting to have been founded on written evidence of title derived from the Spanish Government, and which evidence of title has been withdrawn from the Register's Office

All which is respectfully submitted

GEO GRAHAM

Act to Grant Pre-emption Rights to Settlers, 1830

From 4 U.S. Statutes at Large 420.

Be it enacted by the Senate and House of Representatives of the United States of America, in Congress assembled, That every settler or occupant of the public lands, prior to the passage of this act, who is now in possession, and cultivated any part thereof in the year one thousand eight hundred and twenty-nine, shall be, and he is hereby, authorized to enter, with the register of the land office, for the district in which such lands may lie, by legal subdivisions, any number of acres, not more than one hundred and sixty or a quarter section, to include his improvement, upon paying to the United States the then minimum price of said land: *Provided, however,* That no entry or sale of any land shall be made, under the provisions of this act, which shall have been reserved for the use of the United States, or either of the several states, in which any of the public lands may be situated.

SEC. 2. *And be it further enacted,* That if two or more persons be settled upon the same quarter section, the same may be divided between the two first actual settlers, if, by a north and south, or east and west line, the settlement or improvement of each can be included in a half quarter section; and in such case the said settlers shall each be entitled to a pre-emption of eight acres of land elsewhere in said land district, so as not to interfere with other settlers having a right of preference.

SEC. 3. *And be it further enacted,* That prior to any entries being made under the privileges given by this act, proof of settlement or improvement shall be made to the satisfaction of the register and receiver of the land district in which such lands may lie, agreeably to the rules to be prescribed by the commissioner of the general land office for that purpose, which register and receiver shall each be entitled to receive fifty cents for his services therein. And that all assignments and transfers of the right of pre-emption given by this act, prior to the issuance of patents, shall be null and void.

SEC. 4. *And be it further enacted,* That this act shall not delay the sale of any of the public lands of the United States, beyond the time which has been, or may be, appointed, for that purpose, by the President's proclamation; nor shall any of the provisions of this act be available to any person, or persons, who shall fail to make the proof and payment required before the day appointed for the commencement of the sales of lands including the tract, or tracts, on which the right of pre-emption is claimed; nor shall the right of pre-emption, contemplated by this act, extend to any land, which is reserved from sale by act of Congress, or by order of the President, or which may have been appropriated, for any purpose whatsoever. .

SEC. 5. *And be it further enacted,* That this act shall be and remain in force, for one year from and after its passage.

APPROVED, May 29, 1830.

Land Problems in Arkansas, 1830

From Clarence E. Carter, ed., *Territorial Papers of the United States, The Territory of Arkansas, 1829–1836* (Washington, 1954), XXI, pp. 235–36.

William Ferguson to the President

Arkansas Territory 10ᵗʰ June 1830

GENˡ A. JACKSON Washington City D. C. —
Dʳ GENˡ

Feeling a deep intrust in the perilous ciutation of maney of my nighbours, rendered perilous, by Act (N° 41) of the last session of Congress, entitled an act for the further extending the powers, of the Judges, of the Superior Court, of the Territory of Arkansas — under the act of the 26ᵗʰ day of May 1824 and for Other purposes, (Approved May 8ᵗʰ 1830.) I address you this letter, as an acquaintance, wishing to inform you, as well as the Goverment the ruinous effect — that a review of the Spanish Confirmation that are located, or transfered, would have on the — innosent — honest — and industrious — part of the citizens of this Country — to my own Knowledge and to the Territory generally — *should the Clames prove to be spurious, on a review* — There is to my Knowledge. at this time fifty famillys — settled in the County — on lands entered by virtue of Spanish Confirmation, and all, with hardly an exception — poore honest and industrious people who have paid out all, there little earnings for maney years, for a home, to rase there little Children on. — And should the lands be taken from them by the Goverment — ruin, distress, and want must certenly be the conciquences, — for they are left intirely without — recorse, for in evry instance of my Knowledge, the transactions has been by qut clame, The perches of lands by those individuals — wer in no instant made, for speculation — The perches wer all maid under the hiest faith of the Goverment, eather by perchace of the final decree, of Confirmation, after the U.S. Court — had passed upon it — and all times under the former law, for an appeal had expired — Or of the land after the Clames had been, entered and the U.S. land Officers issued there final cirtiffate of entry — which cirtiffcates, they now hold on the Goverment, with such evidence of title — they all though them selves perfectly — safe, went on to making improvements — never dreaming of any defect in the titles, for there land, and now there allarms are greate they fear. from the Current. reports, since the introduction of the act — refered to — that there, homes there farms, will all become vacant—and subject to the mercless speckulater, to be entered over there heads —and they turned out of house and home—such wil be the consequence — with out the aid of Goverment — for they have all exausted there little means in the first perchace and the opeing of farms, and cannot be able to enter the land again as soone as speckulaters under such a state of facts would it not be adviseable — for the Attorney Gen¹ to instruct — the United States attorney for this Territory — to omit the review of such Clames as has been legally transfered to innosent perchesers at least all Clames that have been intered and the lands sold to innosent persons —

I am with due respect your Obᵗ Servent and frieand

W^mD. Ferguson

If further infermation on the subject would be acceptable — I would be glad to render it — on your request — I would be verry glad of a Letter — from you on the subject Yours &c W. D. F.

[*Addressed*] Gen¹ Andrew Jackson President of the United States Washington City D. C — [*Postmarked*] Greenock. A. Ter 13th June Free

[*Endorsed*] Arkansas Territory 10 June 1830 Letter from Mr Ferguson wishing the provisions of an act appro. 8 May 1830, suspended if possible, in as much as it will prove ruinous to many of the good citizens of Arkansas Territory — Refered to the Commission of the Gen¹ Land Office —

Protecting Squatters' Claims in Indiana, 1837

From Timothy H. Ball, *Lake County Indiana, From 1834 to 1872* (Chicago, 1873), pp. 40–49, 64–65.

Constitution of the Squatters' Union, in Lake County, Indiana

"PREAMBLE. *Whereas,* The settlers upon the public lands in this county, not having any certain prospect of having their rights and claims secured to them by a preëmption law of Congress, and feeling the strong present necessity of their becoming *united* in such a manner as to guard against speculation upon our rights, have met and united together to maintain and support each other, on the 4th of July, 1836; and now firmly convinced of the justness of our cause, do most solemnly pledge ourselves to each other, by the strong ties of interest and brotherly feeling, that we will abide by the several resolutions hereto attached (and to which we will sign our names), in the most faithful manner.

"ARTICLE 1st. *Resolved,* That every person who bears all the dangers and difficulties of settling a new and unimproved country is justly entitled to the privilege heretofore extended to settlers by Congress, to purchase their lands at a dollar and a quarter an acre.

"ARTICLE 2d. *Resolved,* That if Congress should neglect or refuse to pass a law before the land on which we live is offered for sale, which shall secure to us our rights, we will hereafter adopt *such measures as may be necessary* effectually to secure each other in our just claims.

"ARTICLE 3d. *Resolved,* That we will not aid any person to purchase his claim at the land sale, according to this constitution unless he is at the time an actual settler upon government lands, and has complied with all of the requisitions of this Constitution.

"ARTICLE 4th. *Resolved,* That all the settlers in this county, and also in the adjoining unsold lands in Porter county (if they are disposed to join us), shall be considered members of this Union as soon as they sign this Constitution, and entitled to all its advantages, whether present at this meeting or not.

"ARTICLE 5th. *Resolved,* That for the permanent and quiet adjustment of all differences that may arise among the settlers in regard to their claims, that there shall be elected by this meeting, a County Board of three Arbitrators, and also a Register of claims, who also shall perform the duties of clerk to the County Board, Arbitrators, and also the duties of a general corresponding secretary. In all elections, the person having the highest number of votes shall be elected.

"ARTICLE 6th. *Resolved,* That the person who may be elected Register (if he accept the office) shall take an oath or affirmation, that he will faithfully perform all the duties enjoined upon him. He shall forthwith provide himself with a map of the county (which shall be subject to the inspection of every person desiring it), on which he shall mark all claims registered, so that it can be seen what land is claimed and what is not; and also a book in which he shall register every claimant's name, and the number of the land which he claims, when it was first claimed, and when the claimant settled upon it, and the date when registered, where the occupant was from, and any other matter deemed necessary for public information, or that the County Board may order.

"He shall give persons applying all information in his power in regard to claims or vacant land, that shall be calculated to promote the settlement of the county. He shall also reply in the same manner to letters addressed him on the subject (provided the applicant pays his own postage.) He shall attend all the meetings of the County Board, record their proceedings, and perform their orders. When required by a member, stating the object, he shall issue notice to the County or District Board, when, where, and for what purpose they are to meet.

"Fees: For every claim he registers, twenty-five cents; and he shall, if required, give the claimant a certificate stating the number of the land, and when registered. For issuing notice to Arbitrators to meet, 12 cents. For attending their meeting the same fees that are allowed them. For duties of corresponding secretary no fees shall be required.

"ARTICLE 7. *Resolved,* That its hall be the duty of every person, when they sign this Constitution, or as soon thereafter as may be, to apply to the Register to have the land he claims, registered (paying the Register his fees at the same time). Where the claimant now resides upon the land which he claims, his claim shall be considered and held good as soon as registered. Every sale or transfer of titles shall be registered the same as new claims. Any person desirous of claiming any land now unoccupied, shall apply to have the same registered, and if he is a resident of the county at the time he applies, residing with, or upon any claim belonging to any other person, or upon any land that has been floated upon by Indian or preëmption claims, he shall be entitled to hold the claim he registers, while he remains a citizen of the county, provided, he shall within thirty days after registering it, make or cause to be made some prominent improvement upon it, and continue to improve the same to the satisfaction of the County or District Board of Arbitrators. Any non-resident who may hereafter be desirous to join this Union shall first sign the Constitution, and after registering his claim, shall proceed, within thirty days, to occupy it with his family, or else make a durable and permanent improvement, either by building a good cabin for his residence, or by plowing at least four acres, and then if he is not able to continue the occupancy of his claim either personally or by a substitute, he shall apply to the Arbitrators, stating his reasons for necessary absence, whether to move on his family, or whether for other purposes; and they shall certify to him what amount of labor he shall perform or cause to be performed within a given length of time to entitle him to hold his claim while he is absent, or for a certain time, which when done and proved to the Register and entered on record, shall as fully entitle the claimant to his claim as though he resided on it.   *Provided,* the Board shall never grant a certificate to extend his absence one year from the date, unless the claimant has performed at least one hundred dollars worth of labor on his claim, and satisfied the Board fully that he will within that time become an *actual settler* upon it.

"Any member of this Union may also register and improve claims for his absent friends, as above provided, if he can and will satisfy the Board (of the county or district), that the identical person for whom he makes the claim will actually become a settler and reside upon it within the specified time.

"Any person found guilty by the Board of making fraudulent claims for speculating purposes, shall, if a member, forfeit his membership in this Union, and forfeit all right and title to hold the same, and it shall be declared confiscated and shall be sold as provided for all forfeited claims, in Article 9th.

"Every person requiring the services of the Arbitrators shall, if required, secure to them before they are bound to act, one dollar and fifty cents for each day's services, of each and all other necessary expense of magistrate, witnesses, Register, or other unavoidable expense.

"ARTICLE 8th.   *Resolved,* that each congressional township, or any settlement confined in two or more townships containing twenty members, may unite and elect a Board of three Arbitrators, who shall possess the same power to settle disputes (when applied to) within their district that the County Board have. And any member of that district may either submit his case to the District or County Board. The opposite party may object to one or two of the District Board, and call one or two of the County Board, or some disinterested member, to sit in their places, provided he pays the extra expense so occasioned. All decisions of County or District Board shall be final.

"Either of the parties, or the District Board, may require the Register to attend their meetings and record their proceedings. But if he is not present they shall certify their judgment to him immediately, and he shall register it as any other claim.

"Any member may also object to one of the County Board, upon the same terms, and require one of a District Board, or some disinterested member, to sit in his place. The same proceedings shall also take place where one of the Board are interested in the dispute. The District Board may order district meetings, and the County Board county meetings.

"ARTICLE 9th. *Resolved,* That the Board of Arbitrators shall, as soon as may be, take an oath or affirmation before some magistrate, faithfully and impartially to perform all the duties enjoined upon them, not inconsistent with law, and that they will do all acts in their power for the benefit of members of this Union.

"On being duly notified, they shall convene, and if they see proper, they shall make their acts a rule of court before some magistrate, according to the statute provided for arbitrated cases.

"They may require the parties in the case to be tried, to be sworn, or affirmed, and hear arguments of parties or counsel, and finally decide which party is justly entitled to hold the claim, and which party shall pay costs or damages.

It shall be the duty of the County or District Board where the claim is situated, to take possession of any claim confiscated under the provisions of article seven, or any unoccupied non-resident claim, the claimant of which has neglected to occupy or improve the same, according to the terms and within the time specified in the certificate, and sell the same to some other person who will become a settler on it, keeping the money obtained for it in their hands (unless hereafter a treasurer shall be appointed) for a fund to defray any expense that may be deemed necessary to maintain our just rights or advance the interest of the Union. And if a fund so accumulated shall not be required for such purpose, the Board shall use it toward purchasing land for any needy widows, or orphan children, or needy members of this Union.

"Provided that the Board having jurisdiction may extend the time to any claimant holding a certificate from them, or application through the corresponding secretary, if the claimant can give them satisfactory reasons therefor, and they may also, when they have sold a forfeited claim, if they deem it just and reasonable, for good cause thereon, refund to the certificate claimant the amount he had actually expended upon it, and retain in the fund only the overplus that the same sold for.

"Any officer of this Union, or any member, shall be discarded if convicted of gross neglect of duty, or immoral conduct tending to injure the character of the Union.

"ARTICLE 10th. *Resolved,* That every white person capable of transacting business, and making or causing to be made, an improvement on a claim, *with the evident design of becoming a settler thereon,* shall be entitled to be protected in holding a claim on one quarter section, and no more — except, where persons holding claims on the prairie or open barrens, where the Board may decide they have not sufficient timber to support their farm, shall be allowed to divide one quarter section of timber between four such prairie claims.

The Board of Arbitrators may require any person making a claim to take an oath or affirmation that he intends the same for actual settlement, or (if timber) use of his farm. No person settling in thick timber shall be allowed to hold more than eighty acres of timber, but shall be protected in a claim of eighty acres on the prairie.

"ARTICLE 11th. *Resolved,* That before land is offered for sale, that each district shall select a bidder to attend and bid off all claims, in the claimant's name, and that, if necessary, every settler will constantly attend the sale, prepared to aid each other to the full extent of our ability in obtaining every claimant's land at government price.

"ARTICLE 12th. *Resolved,* That after the board of Arbitrators have decided that any individual has obtruded upon another claim, and he refuses to give the legal owner peaceable possession, that we will not deal with, or countenance him as a settler until he makes the proper restitution.

"ARTICLE 13th. *Resolved,* That we will each use our endeavors to advance the rapid settlement of the county, by inviting our friends and acquaintance to join us, under the full assurance that we shall now obtain our rights, and that it is now perfectly as safe to go on improving the public land as though we already had our titles from government.

"ARTICLE 14th. *Resolved,* That a meeting duly called by the County Board may alter and amend this Constitution.

"Lake County, Indiana, July 6, 1836.

"I do certify that the foregoing Constitution, as here recorded, is a true copy from the original draft reported by the committee, and adopted by the meeting, except slight grammatical alterations not varying the true sense of any article.

"Attest. SOLON ROBINSON, Register."

Attached to it are 476 signatures.

A few cases of arbitration occurred in regard to disputed claims. To enter upon land which another had claimed was called "jumping" it; and there were, it seems. a few accidental or intentional "jumpers."

The following extracts from the records will surely be of interest as showing the customs of squatter rule:

"Aug. 12. Notified County Board of Arbitrators to meet August 13, at G. W. Turner's, to decide disputed claim between Sam'l Haviland and John Harrison, on Sec. 13, sw. ½ T. 36, R. 8. Aug. 13. * * They decided that Haviland hold the claim on paying Harrison five dollars for his labor, and that Harrison pay the costs, amounting to four dollars and fifty cents."

Harrison, it is to be supposed, had "jumped" this claim and so was the aggressor.

"1837, March 16. This day an arbitration was held between Denton and Henry Miller and John Reed, who had gone on to Millers' claim and built a cabin, and the Arbitrators decide that Reed shall give up the cabin to Millers, and pay the costs of this arbitration, but that Millers shall pay Reed seventeen dollars for the cabin which he has built."

In some cases the costs were divided equally between the parties.

From the decisions of the arbitrators there seems to have been no appeal, in the nature of the case there could be none; and with the decisions the parties appear to have been satisfied. Ten cases of arbitration are on the records.

While improvements were going on during this busy summer every family needed food. The settlers of 1835 had raised provisions sufficient for themselves; but not even in La Porte county had a supply been raised sufficient to meet the wants of new settlers. And on this account "most of the Lake county settlers had to draw their provisions from the Wabash during the summer of 1836." . . .

In March of this year [1837] that event of so much interest to those early settlers, the sale of United States Lands, took place at La Porte. The sales commenced on the 19th. The squatters of Lake were in large force gathered there. The hardy pioneers, accustomed to frontier life and to depend on their strong arms and trusty rifles; the New Englanders and the Yorkers, almost direct from those centers of culture, and possessing their share of the intelligence and energy of those regions; and the firm, sturdy, solid Germans, like those that of late broke the power of the third Napoleon, — Germans who had just left the despotisms of the Old World and had received their lessons of freedom in the New, amid the wilderness of untrodden Western prairies; all were there, determined that no speculator should bid upon their lands. Some trouble had been anticipated. The principle upon which the squatters insisted was of importance to them. They were probably prepared, — from what I heard in those days of my youth, I am satisfied they were prepared — armed men were among them — to use force, if it should be necessary, to secure the right which each squatter claimed of buying his own quarter section at one dollar and a quarter an acre. They knew that in the wilds of Lake, in the retreats of the Kankakee marsh, no officers of justice would search for them if their mode of enforcing their claim should be called lawless. But there arose no necessity. The impression was strongly made that it would not be safe for a speculator to overbid a squatter, about five hundred of whom had solemnly pledged themselves to each other to abide, in the most faithful manner, by their own assertion of squatters' rights. The moral force employed was sufficient. Solon Robinson was bidder for one township, William Kinnison for another, and A. McDonald for the third. The sale passed off quietly, and the sons of Lake returned peacefully to their homes. But unfortunately for some of them, they had expended their silver and gold in making improvements and amid the sickness, and suffering, and death of 1838, "the wild cat" money was not current at the land office, and now what the speculators could not effect in one way they easily accomplished in another. They offered to loan these men money for entering their claims, on the security of their lands, and charged them twenty, thirty, or more, per cent. And thus, after all their care, considerable tracts of Lake county land came into the hands of non-residents.

Petition to Congress by Inhabitants of Jefferson County, Wisconsin, 1838

From John Porter Bloom, ed., *Territorial Papers of the United States* (Washington, 1969), XXVII, pp. 951–52.

From the inhabitants of Jefferson County

To the Honorable the Senate and House of Representatives of the United States in Congress Assembled

Your Memorialests respectfully represent that the settlement of this country is much retarded by the purchase of Large tracts of land by non residents of the Territory who for the most part will keep it out of the setlers hands for many years. Therefore believing that the true policy of the goverment is to protect the actual settlers rights rather than encourage the speculation, & believing that the passage of an equitable preemption Law will be conducive to this object Your memorialests therefore beg leave to present to the consideration of your honorable body their views in relation to this subject. We recommend the total suspension of the publick sales & private entries of Land except to actual settlers or cultivators of the soil we further recommend that the Settler be restricted to a certain quanty of Land not exceeding one section this subject of vital interest to us as actual settlers we respectfully submit and as in duty bound your memorialests will ever pray

Changes in Farming

Opposition to Slave Labor, 1782

From Michel G. Crevecoeur, *Letters from an American Farmer* (London, 1782), pp. 216–19.

While all is joy, festivity, and happiness, in Charles-Town, would you imagine that scenes of misery overspread in the country? Their ears, by habit, are become deaf, their hearts are hardened; they neither see, hear, nor feel for, the woes of their poor slaves, from whose painful labours all their wealth proceeds. Here the horrors of slavery, the hardship of incessant toils, are unseen; and no one thinks with compassion of those showers of sweat and of tears which from the bodies of Africans daily drop, and moisten the ground they till. The cracks of the whip, urging these miserable beings to excessive labour, are far too distant from the gay capital to be heard. The chosen race eat, drink, and live happy, while the unfortunate one grubs up the ground, raises indigo, or husks the rice: exposed to a sun full as scorching as their native one, without the support of good food, without the cordials of any cheering liquor. This great contrast has often afforded me subjects of the most afflicting meditations. On the one side, behold a people enjoying all that life affords most bewitching and pleasurable, without labour, without fatigue, hardly subjected to the trouble of wishing. With gold, dug from Peruvian mountains, they order vessels to the coasts of Guinea; by virtue of that gold, wars, murders, and devastations, are committed in some harmless, peaceable, African neighbourhood, where dwelt innocent people, who even knew not but that all men were black. The daughter torn from her weeping mother, the child from the wretched parents, the wife from the loving husband; whole families swept away, and brought, through storms and tempests, to this rich metropolis! There, arranged like horses at a fair, they are branded like cattle, and then driven to toil, to starve, and to languish, for a few years, on the different plantations of these citizens. And for whom must they work? For persons they know not, and who have no other power over them than that of violence; no other right than what this accursed metal has given them! Strange order of things! O Nature, where art thou? — Are not these blacks thy children as well as we? On the other side, nothing is to be seen but the most diffusive misery and wretchedness, unrelieved even in thought or wish! Day after day they drudge on without any prospect of ever reaping for themselves; they are obliged to devote their lives, their limbs, their will, and every vital exertion, to swell the wealth of masters, who look not upon them with half the kindness and affection with which they consider their dogs and horses. Kindness and affection are not the portion of those who till the earth, who carry burdens, who convert the logs into useful boards. This reward, simple and natural as one would conceive it, would border on humanity; and planters must have none of it!

If negroes are permitted to become fathers, this fatal indulgence only tends to increase their misery: the poor companions of their scanty pleasures are likewise the companions of their labours; and when, at some critical seasons, they could wish to see them relieved, with tears in their eyes they behold them perhaps doubly oppressed, obliged to bear the burden of nature — a fatal present! — as well as that of unabated tasks. How many have I seen cursing the irresistible propensity, and regretting that, by having tasted of those harmless joys, they had become the authors of double misery to their wives. Like their masters, they are not permitted to partake of those ineffable sensations with which nature inspires the hearts of fathers and mothers; they must repel them all, and become callous and passive. This unnatural state often occasions the most acute, the most pungent, of their afflictions; they have no time, like us, tenderly to rear their helpless off-spring, to nurse them on their knees, to enjoy the delight of being parents. Their paternal fondness is imbittered by considering, that, if their children live, they must live to be slaves like themselves; no time is allowed them to exercise their pious office, the mothers must fasten them on their backs, and, with this double load, follow their husbands in the fields, where they too often hear no other found than that of the voice or whip of the task-master, and the cries of their infants broiling in the sun. These unfortunate creatures cry and weep, like their parents, without a possibility of relief; the very instinct of the brute, so laudable, so irresistible, runs counter here to their master's interest; and, to that god, all the laws of nature must give way. [Thus planters get rich;] so raw, so inexperienced, am I in this mode of life, that, were I to be possessed of a plantation, and my slaves treated as in general they are here, never could I rest in peace.

Agricultural Aid to the Cherokees, 1791

From Clarence E. Carter, ed., *Territorial Papers of the United States, Territory South of the River Ohio, 1790–1796* (Washington, 1936), IV, pp. 60–61, 65–65.

THE TREATY OF HOLSTON

[*July 2, 1791*]

A Treaty of Peace and Friendship made and concluded between the President of the United States of America on the part and behalf of the said States, and the undersigned Chiefs and Warriors of the Cherokee Nation of Indians, on the part and behalf of the said Nation.

The parties being desirous of establishing permanent peace and friendship between the United States and the said Cherokee Nation, and the citizens, and members thereof, and to remove the causes of war, by ascertaining their Limits, and making other necessary, just and friendly arrangements: The President of the United States by William Blount Governor of the Territory of the United States of America south of the river Ohio and Superintendant of indian affairs for the southern District, who is vested with full powers for these purposes, by and with the advice and consent of the Senate of the United States: and the Cherokee Nation by the undersigned Chiefs and Warriors representing the said Nation, have agreed to the following articles, namely.

ARTICLE 1. There shall be perpetual peace and friendship between all the citizens of the United States of America, and all the individuals composing the whole Cherokee Nation of Indians.

ARTICLE 2. The undersigned Chiefs and Warriors for themselves and all parts of the Cherokee Nation, do acknowledge themselves and the said Cherokee Nation to be under the protection of the United States of America, and of no other sovereign whosoever; and they also stipulate that the said Cherokee Nation will not hold any treaty with any foreign power, individual State, or with individuals of any State.

ARTICLE 3. The Cherokee Nation shall deliver to the Governor of the Territory of the United States of America south of the river Ohio, on or before the first day of April next at this place all persons who are now prisoners captured by them from any part of the United States: And the United States shall on or before the same day and at the same place restore to the Cherokees all the prisoners now in captivity which the citizens of the United States have captured from them. . . .

ARTICLE 5. It is stipulated and agreed that the citizens and inhabitants of the United States shall have a free and unmolested use of a road from Washington District to Mero District, and of the navigation of the Tennessee River.

ARTICLE 6. It is agreed on the part of the Cherokees that the United States shall have the sole and exclusive right of regulating their trade.

ARTICLE 7. The United States solemnly guarantee to the Cherokee nation, all their lands not hereby ceded.

ARTICLE 8. If any citizen of the United States, or other person not being an indian, shall settle on any of the Cherokees lands, such person shall forfeit the protection of the United States, and the Cherokees may punish him, or not, as they please.

ARTICLE 9. No citizen or inhabitant of the United States, shall attempt to hunt or destroy the game on the lands of the Cherokees — nor shall any citizen or inhabitant go into the Cherokee Country, without a passport first obtained from the Governor of some one of the United States, or territorial districts, or such other person as the President of the United States may from time to time authorize to grant the same.

ARTICLE 10. If any Cherokee indian, or indians, or person residing among them, or who shall take refuge in their nation shall steal a horse from, or commit a robbery or murder or other capital crime on any citizens or inhabitants of the United States — the Cherokee Nation shall be bound to deliver him or them up to be punished according to the laws of the United States.

*　*　*

ARTICLE 14. That the Cherokee nation may be led to a greater degree of civilization, and to become Herdsmen and cultivators, instead of remaining in a state of hunters the United States will from time to time furnish gratuitously the said nation with useful implements of husbandry. And further to assist the said nation in so desirable a pursuit, and at the same time to establish a certain mode of communication, the United States will send such and so many persons to reside in said nation as they may judge proper not exceeding four in number, who shall qualify themselves to act as interpreters — These persons shall have lands assigned by the Cherokees for cultivation for themselves and their successors in office. But they shall be precluded exercising any kind of Traffic.

ARTICLE 15. All animosities for past grievances shall henceforth cease, and the contracting parties will carry the foregoing treaty into full execution with all good faith and sincerity.

ARTICLE 16. This treaty shall take effect and be obligatory on the contracting parties, as soon as the same shall have been ratified by the President of the United States with the advice and consent of the Senate of the United States.

In witness of all and every thing herein determined between the United States of America, and the whole Cherokee nation, the parties have hereunto set their hands and seals, at the Treaty ground on the bank of the Holston near the mouth of the French Broad within the United States this second day of July in the year of our Lord one thousand seven hundred and ninety one.

Wᵐ BLOUNT [L.S.]

Governor in and over the Territory of the United States of America south of the River Ohio and Superintendant of Indian affairs for the southern District.

Invention of the Cotton Gin, 1793

From Matthew Brown Hammond, ed., "Correspondence of Eli Whitney relative to the Ivention of the Cotton Gin," in *American Historical Review* (October, 1897), III, pp. 99–102, 105–07, 112–13, 126–27.

Eli Whitney to Eli Whitney, Sen'r.

New Haven, Sept. 11th, 1793.

DEAR PARENT,

I received your letter of the 16th of August with peculiar satisfaction and delight. It gave me no small pleasure to hear of your health and was very happy to be informed that your health and that of the family has been so good since I saw you. I have fortunately just heard from you by Mr. Robbinson who says you were well when he left Westboro. When I wrote you last I expected to have been able to come to Westboro' sooner than I now fear will be in my power. I presume, sir, you are desirous to hear how I have spent my time since I left College. This I conceive you have a right to know and that it is my duty to inform you and should have done it before this time; but I thought I could do it better by verbal communication than by writing, and expecting to see you soon, I omitted it. As I now have a safe and direct opportunity to send by Mr. Robbinson, I will give you a sumary account of my southern expedition.

I went from N. York with the family of the late Major General Greene to Georgia. I went immediately with the family to their Plantation about twelve miles from Savannah with an expectation of spending four or five days and then proceed into Carolina to take the school as I have mentioned in former letters. During this time I heard much said of the extreme difficulty of ginning Cotton, that is, seperating it from its seeds. There were a number of very respectable Gentlemen at Mrs. Greene's who all agreed that if a machine could be invented which would clean the cotton with expedition, it would be a great thing both to the Country and to the inventor. I involuntarily happened to be thinking on the subject and struck out a plan of a Machine in my mind, which I communicated to Miller, (who is agent to the Executors of Genl. Greene and resides in the family, a man of respectibility and property) he was pleased with the Plan and said if I would pursue it and try an experiment to see if it would answer, he would be at the whole expense, I should loose nothing but my time, and if I succeeded we would share the profits. Previous to this I found I was like to be disappointed in my school, that is, instead of a hundred, I found I could get only fifty Guineas a year. I however held the refusal of the school untill I tried some experiments. In about ten Days I made a little model, for which I was offered, if I would give up all right and title to it, a Hundred Guineas. I concluded to relinquish my school and turn my attention to perfecting the Machine. I made one before I came away which required the labor of one man to turn it and with which one man will clean ten times as much cotton as he can in any other way before known and also cleanse it much better than in the usual mode. This machine may be turned by water or with a horse, with the greatest ease, and one man and a horse will do more than fifty men with the old machines. It makes the labor fifty times less, without throwing any class of People out of business.

I returned to the Northward for the purpose of having a machine made on a large scale and obtaining a Patent for the invintion. I went to Philadelphia soon after I arrived, made myself acquainted with the steps necessary to obtain a Patent, took several of the steps and the Secretary of State Mr. Jefferson agreed to send the Pattent to me as soon it could be made out — so that I apprehended no difficulty in obtaining the Patent — Since I have been here I have employed several workmen in making machines and as soon as my business is such that I can leave it a few days, I shall come to Westboro'. I think it is probable I shall go to Philadelphia again before I come to Westboro', and when I do come I shall be able to stay but few days. I am certain I can obtain a patent in England. As soon as I have got a Patent in America I shall go with the machine which I am now making, to Georgia, where I shall stay a few weeks to see it at work. From thence I expect to go to England, where I shall probably continue two or three years. How advantageous this business will eventually prove to me, I cannot say. It is generally said by those who know anything about it, that I shall make a Fortune by it. I have no expectation that I shall make an independent fortune by it, but think I had better pursue it than any other business into which I can enter. Something which cannot be foreseen may frustrate my expectations and defeat my Plan; but I am now so sure of success that ten thousand dollars, if I saw the money counted out to me, would not tempt me to give up my right and relinquish the object. I wish you, sir, not to show this letter nor communicate anything of its contents to any body except My Brothers and Sister, *enjoining* it on them to keep the whole a *profound secret*.

Mr. Robbinson came into town yesterday and goes out tomorrow, this has been such a bustling time that I have not had oportunity to say six words to him. I have told him nothing of my business — perhaps he will hear something about it from some body else in town. But only two or three of my friends know what I am about tho' there are many surmises in town — if Mr. Robbinson says anything about it, you can tell him I wrote you concerning it, but wished not to have it mentioned. I have been considerably out of health since I wrote you last; but now feel tollerably well. I should write to my Brothers and Sister but fear I shall not have time — hope they will accept my good wishes for their happiness and excuse me.

With respects to Mama I am, kind Parent, your most obt. Son

ELI WHITNEY, JUNR.

Eli Whitney to Eli Whitney, Sen'r.

New Haven, August 17 th 1794.

HON'D SIR, —

It gives me pleasure that I have it in my power to inform you that I am in perfect health. I left Savannah just three weeks ago. We had a passage of Eight Days to New York, where I spent several days and have been here about a week. I was taken sick with the Georgia fever about the middle of June and confined to my bed ten or twelve days, but had got quite well before I left the Country. There were several very hot Days preceeding my sickness during which I fatigued myself considerable and which was probaply the cause of my illness.

My Machinery was in opperation before I came from Georgia. It answers the purpose well, and is likely to succeed beyond our expectations. My greatest apprehensions at present are, that we shall not be able to get machines made as fast as we shall want them. We have now Eight Hundred Thousand weight of Cotton on hand and the next crop will begin to come in very soon. It will require Machines enough to clean 5 or 6 thousand wt. of clean cotton pr Day to satisfy the demand for next Year. I mean for the crop which comes in this fall. And I expect the crop will be double another year.

Within a few weeks a number of persons (I believe about twenty) have died, in this place with a putrid fever it appears to be very contagious and has excited very considerable apprehensions especially in the country. There are various opinions about the disorder — Many suppose it to be the same that was in Philadelphia last summer. It appears pretty certain that the disorder was imported from the W. Indies where it is very prevailant. There are but very few sick at present and if the weather should continue cool I think it will wholly disappear in a few days.

I am going to N. York this week, where I hope not to be detained long, from thence I expect to return here again. My next journey will be to Westboro' where I hope to meet you in happy circumstances. My respects to Mama. I wish to be affectionately remembered to my Brothers and Sister and [subscribe] myself your most Obt. and Dutiful Son

ELI WHITNEY, JUNR.

Phineas Miller to Eli Whitney

Mulberry Grove, May 11th 1797.

DEAR WHITNEY,
To day for the first time Mrs. Miller appears decidedly to be recovering from a confinement of nearly a month past. This affords a relief to my mind which enables me to sit down for the purpose of detailing to you the present situation of our ginning concern.

A constant attendance during every day of the Courts Session, without having been absent one night from home, had quite worn down my patience and health. I felt a few days of quiet perfectly indispensible to my restoration, which having obtained I feel myself once more in usual spirits.

The event of the first Patent suit after all our exertions made in such a varity of ways has gone against us. The preposterous custom of trying civil causes of this intricacy and magnitude by a common Jury together with the imperfection of the Patent law frustrated all our views and disappointed expectations which had become very sanguine.

We had the Judge with a Party to dine with us twice before the trial came on and got him fully prepared to enter into the merits of the case. We had also got the tide of popular opinion running in our favor and many decided friends who adhered firmly to our cause and interest. Added to all this we got the trial brought on, against every measure they could devise for postponement and found them perfectly unprepared as to a knowledge of the strong grounds of their cause and without a single evidence in their favor. We were on the contrary pretty well prepared and neglected no means to become as much so as possible. An instance of our exertion in this respect I will just mention to you. It happened during the trial that a Paper was required to fix the amount of damages which had been left at this place among my other papers. The necessity for this paper appearing very great, Mr. Nightingale immediately mounted my best horse, in the middle of a very hot day came to this place examined my chest, draws, &c., and after a search of nearly half an hour, laid his hand on the paper, remounted his horse, on his way back met a fresh horse I had sent for his relief; and returned to the Court house in just two hours and forty minutes from the time he had left it — the paper came in time to procure admittance to the evidence we had brought, it being an agreement with Peter Robinson and was necessary to support the testimony of W. Shubert of Waynesborough. So that we had the cause well supported, and brought as much within the law as the nature of the trespass would possibly admit. The judge gave a charge most pointedly in our favor, after which the Defendant himself told an acquaintance of his, that he would give two thousand dollars to be free from the verdict — and yet the jury gave it against us after a consultation of about an hour. And having made this verdict general, no appeal would lie, on Monday morning when the judgment was rendered we applied for a new trial but the judge refused it to us inasmuch as that the Jury might have made up their opinion on the defect of the law which states an aggression to consist of *making devising* and using, or selling, and we could only charge the defendants with using. In a private conversation had with the Judge afterwards, he told me that we could have no hope of protecting our Patent rights without an alteration of the law, which he had no doubt but Congress would make for us, on application. Thus after four years of assiduous labour fatigue and difficulty are we again set afloat by a new and most unexpected obstacle. Our hopes of success are now removed to a period still more distant than before, while our expenses are realized beyond all power of controversy. The actual crisis has now arrived which I have long mentioned as possible, and sometimes almost or indeed quite apprehended as probable. This crisis is our insolvency as a Partnership. At the present time I have no idea that any person would chance our engagements to have the whole benefit of our patent, with all the property of which we have possessed ourselves under it. In this unfortunate dilemma I am however far from despairing, or being in bad spirits, since I do not consider the acquirement of property as the greatest blessing or the loss of it as the greatest misfortune which we can experience. I am even willing to make the same unremitted exertions to our mutual benefit, and still to pursue the fleeting prospect. The honorary engagement which Mr. Nightingale considers himself to have made for the [Junction] of our interests will induce him still to connect himself with our affairs in such a manner as would be more favourable than we could otherwise command. The severe indisposition of Mrs. Miller which has produced so much anxiety and confinement from me has hitherto prevented our fixing the terms of agreement but we shall now shortly set about it. Without such an agreement it will be quite impossible for us to proceed one step further in our concerns. For we have no possible resources to meet the demands which stand against us but from the proceeds of Nightingales and my speculation, and this cannot be diverted from his and my private engagements without a junction of his interest to justify such diversion.

Never indeed was the application of money rendered more distressful than it has been to the support of our unfortunate business. You recollect that the small sum which our necessities compelled me to take from the Estate of Gen. Greene I frequently mentioned the necessity of having returned in the course of the present winter. In this necessity I was not mistaken. My last letter from our Creditors in Charleston which was received some time ago, I enclose to you that you may perceive the exact situation of his claim and that we have very little chance of lenity. At the time when this money was taken for our use, I was under the necessity of acquainting Mr. Rutledge that we would certainly indemnify the Estate for any injury it might sustain in consequence of this diversion of its funds. Then you see our difficulties accumulating on every side in consequence of our repeated and continued disappointments and our prospects of success still keeping at a distance — and hence you will perceive a rational source to have existed for the despondency which you have sometimes seen me disposed to feel as respects the Copartnership concerns. If the greatest of our difficulties that of the bad reputation of our cotton could be but once removed, we could however still hold up our heads against the lawless state in which we have the misfortune to carry on our concerns as well as against the enemies of every kind who have conspired our ruin.

The acquirement of money from the speculating concerns of N. and myself would also greatly assist us, in one or other of these resources I will still rest our anchor of hope.

The family all join me in most affectionate regards to you and kind remembrances to your brother.

Your friend and Partner,

PHIS. MILLER.

Eli Whitney to Josiah Stebbins.

Columbia S. Carolina Dcem 20th 1801.

DEAR STEBBINS,
I have been at this place little more than two weeks attending the Legislature. They closed their sessions at 10 oClock last evening. A few hours previous to their breaking up they voted fifty thousand Dollars to purchase my patent right to the Machine for Cleaning Cotton, 20 Thousand of which is to be paid in hand and the remainder in three annual payments of 10 thousand Dollars each. This is selling the right at a great sacrifice. If a regular course of Law had been pursued, from two to three hundred thousand Dollars would undoubtedly have been recovered. The use of the machine here is amazingly extensive and the value of it beyond all calculation. It may without exaggeration be said to have raised the value of seven eights of all the three Southern States from 50 to 100 pr. Cent. We get but a song for it, in comparison with the worth of the thing, but it is securing something. It will enable M and W to pay all their Debts and divide

something between them. It established a precedent which will be valuable as it respects our collections in other States and I think there is now a fair prospect that I shall in the event realize property enough from the invention to render me comfortable and in some measure independent. Tho' my stay here has been short I have become acquainted with a considerable part of the Legislature and most distinguished characters in the State.

Our old school mate H. D. Ward is one of the Senate. He ranks among the first of his age in point of talents and respectability. Is married, has a handsome property and practice in his profession. He has shown me much polite attention as have also many other of the Citizens. I wish I had time to write you more frequently and more lengthily. I go to Georgia for which place I shall start tomorrow. With best regards to Laura I am &c

ELI WHITNEY.

Eli Whitney to Josiah Stebbins.

South Carolina 9th Feb. 1805.

MY DEAR STEBBINS,

I left New Haven about the 25th Oct. last. I made no considerable Stop till I arrived at Columbia in this State; where I staid about four weeks. From thence I went to the city of Charleston, where I spent three weeks. From thence to Savannah in Georgia where I passed one week, and am now on my return, 95 miles from Savannah and 28 miles from Orangeburg, the residence of our old friend H. D. Ward whom I hope to have the Pleasure of taking by the hand to-morrow. I shall proceed directly on by the way of Camden to Rocky Mount upon the Great Falls of the Cataba River where I shall spend two or three days return to, Camden and from thence make the best of my way back to Connecticut, as fast as the season of the year and the state of the roads will permit. I have traveled the rounds thus far and expect to finish my tour by Land and with the same set of horses with which I started. You will perhaps recollect that three years ago I sold my Patent Right to the State of So. Carolina, that soon afterwards, much like Children and much more like rogues, they refused to make the stipulated payments. The principal object of my present excursion to this Country was to get this business set right; which I have so far effected as to induce the Legislature of this State to recind all their former *Suspending Laws* and *resolutions,* to *agree once more* to pay the sum of 30,000 Dollars which was due and make the necessary appropriations for that purpose. I have as yet however obtained but a small part of this payment. The residue is promised me in July next. Thus you see my *recompense* of *reward* is as the land of Canaan was to the Jews, resting a long while in *promise.* If the Nations with whom I have to contend are not as numerous as those opposed to the Israelites, they are certainly much greater *Heathens,* having their hearts hardened and their understanding blinded, to make, propagate and believe all manner of lies. Verily, Stebbins, I have had much vexation of spirit in this business. I shall spend forty thousand dollars to obtain thirty, and it will all end in vanity at last. A contract had been made with the State of Tennessee which now hangs *suspended.* Two attempts have been made to induce the State of No. Carolina to *recind* their *Contract,* neither of which have succeeded. Thus you see Brother Steb. Sovreign and Independent States warped by *interest* will be *rogues* and misled by Demagogues will be *fools*. They have spent much time, *money* and *Credit,* to avoid giving me a small compensation, for that which to them is worth millions.

I have had less leisure to write you this winter than heretofore, tho' not fewer good wishes for your health and happiness. Impart a double portion of my affectionate regards to Laura and her little ones and be assured that I remain

Yr unaltered and unalterable friend

E. WHITNEY.

On the Culture of Tobacco, 1800

From: William Tatham, *An Historical and Practical Essay on the Culture and Commerce of Tobacco* (London, 1800), pp. 1–2, 9–36, 40–41, 55–62.

PART I. INTRODUCTORY REMARKS.

Having lately seen a few plants of American Tobacco growing casually in a gentleman's garden near London, and perceiving that very little is generally known in England concerning the history and ordinary culture of an article of commerce which has occupied a considerable capital in transatlantic traffic for about two hundred years; and indeed a plant which is peculiarly adapted for *an agricultural comparison of climates;* without entering so far into the subject as to consider it a staple produce of the nation, I beg leave to communicate a few particulars in respect to the history and culture of this luxuriant commodity, which I am enabled to state from authorities, and from what I recollect to have noticed during twenty years residence in Virginia, where it is a principal export.

Botanical Definition.

The botanical account of tobacco is as follows: — "NICOTIANA, the tobacco plant, is a genus of plants of the order of *Monogynia,* belonging to the *pentandria* class, order I, of class V. The calyx is a permanent perianthum, formed of a single leaf, divided into five segments, and of an oval figure. The corolla consists of a single petal, funnel-shaped. The tube is longer than the cup. The limb is patulous, lightly divided into five segments, and folded in five places. The fruit is a capsule of a nearly oval figure. There is a line on each side of it, and it contains two cells, and opens at the top. The receptacles are of a half oval figure, punctuated and affixed to the separating body. The seeds are numerous, kidney-shaped, and rugose.

* * *

Of the Culture of the Crop.

First, of preparing the Tobacco Ground.

There are two distinct and separate methods of preparing the tobacco ground: the one is applicable to the preparation of new and uncultivated lands, such as are in a state of nature, and require to be cleared of the heavy timber and other productions with which Providence has stocked them; and the other method is designed to meliorate and revive lands of good foundation, which have been heretofore cultivated, and, in some measure, exhausted by the calls of agriculture and evaporation.

The process of preparing new lands begins as early in the winter as the housing and managing the antecedent crop will permit, by grubbing the under growth with a mattock; felling the timber with a poll-axe; lopping off the tops, and cutting the bodies into lengths of about eleven feet, which is about the customary length of an American fence rail, in what is called a *worm* or *pannel* fence. During this part of the process the negro women, boys, and weaker labourers, are employed in piling or throwing the brush-wood, roots, and small wood, into heaps to be burned; and after such logs or stocks are selected as are suitable to be malled into rails, make clap-boards, or answer for other more partilar occasions of the planter, the remaining logs are rolled into heaps by means of hand-spikes and *skids*; but the Pennsylvania and German farmers, who are more conversant with animal powers than the Virginians, save much of this labour by the use of a pair of horses with a half sledge, or a pair of truck wheels. The burning of this brush-wood, and the log piles, is a business for all hands after working hours; and as nightly revels are peculiar to the African constitution, this part of the labour proves often a very late employment, which affords many scenes of rustic mirth.

When this process has cleared the land of its various natural incumbrances (to attain which end is very expensive and laborious), the next part of the process is that of the hoe; for the plough is an implement which is rarely used in *new* lands when they are either designed for tobacco or meadow.

There are three kinds of the hoe which are applied to this tillage: the first is what is termed the sprouting hoe, which is a smaller species of mattock that serves to break up any particular hard part of the ground, to grub up any smaller sized grubs which the mattock or grubbing hoe may have omitted, to remove small stones and other partial impediments to the next process.

The *narrow* or *hilling* hoe follows the operation of the sprouting hoe. It is generally from six to eight inches wide, and ten or twelve in the length of the blade, according to the strength of the person who is to use it; the blade is thin, and by means of a moveable wedge which is driven into the eye of the hoe, it can be set more or less *digging* (as it is termed), that is, on a greater or less angle with the helve, at pleasure. In this respect there are few instances where the American blacksmith is not employed to alter the eye of an *English*-made hoe before it is fit for use; the industrious and truly useful merchants of Glasgow have paid more minute attention to this circumstance.

The use of this hoe is to break up the ground and throw it into shape; which is done by chopping the clods until they are sufficiently fine, and then drawing the earth round the foot until it forms a heap round the projected leg of the labourer like a mole hill, and nearly as high as the knee; he then draws out his foot, flattens the top of the hill by a *dab* with the flat part of the hoe, and advances forward to the next hill in the same manner, until the whole piece of ground is prepared. The centre of these hills are in this manner guessed by the eye; and in most instances they approach near to lines of four feet one way, and three feet the other. The planter always endeavours to time this operation so as to tally with the growth of his plants, so that he may be certain by this means to pitch his crop within season.

The third kind of hoe is the *broad* or *weeding* hoe. This is made use of during the cultivation of the crop, to keep it clean from the weeds. It is wide upon the edge, say from ten inches to a foot, or more; of thinner substance than the hilling hoe, not near so deep in the blade, and the eye is formed more bent and shelving than the latter, so that it can be set upon a more acute angle upon the helve at pleasure, by removing the wedge. We shall have occasion to notice the application of this implement under a subsequent head of this paper.

Of the Season for Planting.

The term, *season* for planting, signifies a shower of rain of sufficient quantity to wet the earth to a degree of moisture which may render it safe to draw the young plants from the plant bed, and transplant them into the hills which are prepared for them in the field, as described under the last head; and these seasons generally commence in April, and terminate with what is termed the *long season in May;* which (to make use of an Irishism) very frequently happens in June; and is the opportunity which the planter finds himself necessitated to seize with eagerness for the *pitching* of his crop: a term which comprehends the ultimate opportunity which the spring will afford him for planting a quantity equal to the capacity of the collective power of his labourers when applied in cultivation.

By the time which these *seasons* approach, nature has so ordered vegetation, that the weather has generally enabled the plants (if duly sheltered from the spring frosts, a circumstance to which a planter should always be attentive in selecting his plant patch) to shoot forward in sufficient strength to bear the vicissitude of transplantation.

They are supposed to be equal to meet the imposition of this task when the leaves are about the size of a dollar; but this is more generally the minor magnitude of the leaves; and some will be of course about three or four times that medium dimension.

Thus, when a good shower or season happens at this period of the year, and the field and plants are equally ready for the intended union, the planter hurries to the plant bed, disregarding the teeming element, which is doomed to wet his skin, from the view of a bountiful harvest, and having carefully drawn the largest sizeable plants, he proceeds to the next operation.

Of Planting.

The office of *planting* the tobacco is performed by two or more persons, in the following manner: The first person bears, suspended upon one arm, a large basket full of the plants which have been just drawn and brought from the plant bed to the field, without waiting for an intermission of the shower, although it should rain ever so heavily; such an opportunity indeed, instead of being shunned, is eagerly sought after, and is considered to be the sure and certain means of laying a good foundation, which cherishes the hope of a bounteous return. The person who bears the basket proceeds thus by rows from hill to hill; and upon each hill he takes care to drop one of his plants. Those who follow make a hole in the centre of each hill with their fingers, and having adjusted the tobacco plant in its natural position, they knead the earth round the root with their hands, until it is of a sufficient consistency to sustain the plant against wind and weather. In this condition they leave the field for a few days until the plants shall have formed their radifications; and where any of them shall have casually perished, the ground is followed over again by successive replantings, until the crop is rendered complete.

Of Hoeing the Crop.

The operation of *hoeing* comprehends two distinct functions, viz. that of hilling, and that of weeding; and there are moreover two stages of hilling. The first hilling commences, as heretofore described, in the preparation of the field previous to planting the crop, and it is performed, as before explained, by means of the peculiar implement called a hilling hoe; the second hilling is performed after the crop is planted, with a view to succour and support the plant as it may happen to want strengthening, by giving a firm and permanent foundation to its root; and it may be effected according to the demand of the respective plants by a dexterity in changing the stroke with the weeding hoe, without any necessity to recur to the more appropriate utensil.

The more direct use of the weeding hoe commences with the first growth of the tobacco after transplantation, and never ceases until the plant is nearly ripe, and ready to be *laid by,* as they term the last weeding with the hoe; for he who would have a good crop of tobacco, or of maize, must not be sparing of his labour, but must keep the ground constantly stirring during the whole growth of the crop. And it is a rare instance to see the plough introduced as an assistant, unless it be the *flook plough,* for the purpose of introducing a sowing of wheat for the following year, even while the present crop is growing; and this is frequently practised in fields of maize, and sometimes in fields of tobacco, which may be ranked amongst the best fallow crops, as it leaves the ground perfectly clean and naked, permitting neither grass, weed, nor vegetable, to remain standing in the space which it has occupied.

Of Topping the Plant.

This operation, simply, is that of pinching off with the *thumb nail* the leading stem or sprout of the plant, which would, if left alone, run up to flower and seed; but which, from the more substantial formation of the leaf by the help of the nutritive juices, which are thereby afforded to the lower parts of the plant, and thus absorbed through the ducts and fibres of the leaf, is rendered more weighty, thick, and fit for market. The qualified sense of this term is applicable to certain legal restrictions founded upon long experience, and calculated to compel an amendment in the culture of this staple of the Virginia trade, so that it shall at all times excel in foreign markets, and thus justly merit a superior reputation. I do not exactly recollect the present limitation by law, which has changed I believe with the progress of experience; but the custom is to top the plant to nine, seven, or five leaves, as the quality and soil may seem most likely to bear.

Of the Sucker, and Suckering.

The *sucker* is a superfluous sprout which is wont to make its appearance and shoot forth from the stem or stalk, near to the junction of the leaves with the stem, and about the root of the plant; and if these suckers are permitted to grow, they injure the marketable quality of the tobacco by compelling a division of its nutriment during the act of maturation. The planter is therefore careful to destroy these intruders with the thumb nail, as in the act of *topping,* and this process is termed *suckering.*

This superfluity of vegetation, like that of the top, has been often the subject of legislative care; and the policy of supporting the good name of the Virginia produce has dictated the wisdom of penal laws to maintain her good faith against imposition upon strangers who trade with her. It has been customary in former ages to rear an inferior plant from the sucker which projects from the root after the cutting of an early plant; and thus a *second* crop has been often obtained from the same field by one and the same course of culture; and although this scion is of a sufficient quality for smoking, and might become preferred in the weaker kinds of snuff, it has been (I think very properly) thought eligible to prefer a prohibitory law, to a risk of imposition by means of similitude.

The practice of cultivating *suckers* is on these accounts not only discountenanced as fraudulent, but the constables are strictly enjoined *ex officio* to make diligent search, and to employ the posse commitatus in destroying such crops; a law indeed for which, to the credit of the Virginians, there is seldom occasion; yet some few instances have occurred, within my day, where the constables have very honourably carried it into execution in a manner truly exemplary, and productive of public good.

Of the Worm.

There are several species of the worm, or rather *grub* genus, which prove injurious to the culture of tobacco; some of these attack the root, and some the leaf of the plant; but that which is most destructive, and consequently creates the most employment, is the *horn* worm, or large green tobacco worm. This appears to me to be the same species with that which Catesby has described in the second volume of his Natural History of Carolina, p. 94, under the title of *eruca maxima cornuta,* or the great horned caterpillar.

"This caterpillar," says he, "is about four inches long, besides the head and tail; it consists of ten joints, or rings, of a yellow colour; on the head, which is black, grow four pair of horns, smooth and of a reddish brown towards the bottom, jagged or bearded, and black towards the top; on each of the rings arise short jagged black horns, one standing on the back, and two on each side; below which is a *trachaea* on each side; likewise the horn of the back of the last ring is longest: the flap of the tail is of a bright bay colour. It hath eight feet, and six *papillae.*"

There are, besides this kind, others without horns; all of them of a green colour, so far as I recollect. And this, in Catesby's description, differs in respect to colour; this tobacco worm or *horn* worm, as the planters call it more particularly, being of a pale delicate green; an effect I apprehend which proceeds from the colour of its food when it feeds upon growing tobacco plants. The act of destroying these worms is termed *worming* the tobacco, which is a very nauseous occupation, and takes up much labour. It is performed by picking every thing of this kind off the respective leaves with the hand, and destroying it with the foot.

Of the Term "Firing."

During very rainy seasons, and in some kinds of unfavourable soil, the plant is subject to a malady called *firing*. This is a kind of blight occasioned by the moist state of the atmosphere, and the too moist condition of the plant: I do not recollect whether the opposite extreme does not produce an effect something similar. This injury is much dreaded by the planter, as it spots the leaf with a hard brown spot, which perishes, and becomes so far a loss upon the commodity. I apprehend there are two stages when the plant is, in a certain degree, subject to this evil effect: the first is whilst growing in the field, the latter when hanging in the tobacco house. I know of no other remedy than constant working the ground while the feed is growing, and careful drying by the use of fire in the tobacco house.

Of the Ripening of the Crop.

Much practice is requisite to form a judicious discernment concerning the state and progress of the ripening leaf; yet care must be used to cut up the plant as soon as it is sufficiently ripe to promise a good curable condition, lest the approach of frost should tread upon the heels of the crop-master; for in this case, tobacco will be among the first plants that feel its influence, and the loss to be apprehended in this instance, is not a mere partial damage by nippling, but a total consumption by the destruction of every plant.

I find it difficult to give to strangers a full idea of the ripening of the leaf: it is a point on which I would not trust my own experience without consulting some able crop-master in the neighbourhood; and I believe this is not an uncustomary precaution among those who plant it. So far as I am able to convey an idea, which I find it easier to understand than to express, I should judge of the ripening of the leaf by its thickening sufficiently; by the change of its colour to a more yellowish green; by a certain mellow appearance, and protrusion of the web of the leaf, which I suppose to be occasioned by a contraction of the fibres; and by such other appearances as I might conceive to indicate an ultimate suspension of the vegetative functions.

Of Cutting and Gathering the Crop.

When the crop is adjudged sufficiently ripe to proceed to cutting, this operation is assigned to the best and most judicious hands who are employed in the culture; and these being provided each with a strong sharp knife, proceed along the respective rows of the field to select such plants as appear to be ripe, leaving others to ripen; those which are cut are sliced off near to the ground, and such plants as have thick stalks or stems are sliced down the middle of the stem in order to admit a more free and equal circulation of air through the parts during the process of curing, and to free the plant, as far as possible, from such partial retention of moisture as might have a tendency to ferment, and damage the staple.

The plants are then laid down upon the hill where they grew, with the points of the leaves projecting all the same way, as nearly as possible, so that when the sun has had sufficient effect to render them pliable, they may more easily and uniformly be gathered into *turns* by the gatherers who follow the cutting.

Of Gathering the Crop in.

For the better comprehending the method of gathering the crop, it is necessary to understand the preparation which must be previously made for facilitating this part of the process.

In preparing for gathering the crop of tobacco it is customary to erect a kind of scaffold in various places of the tobacco ground which may happen to offer a convenient situation. This is done by lodging one end of several strong poles upon any log or fence which may be convenient, and resting the other end of such poles upon a transverse pole supported by forks, at about five feet from the ground; or by erecting the whole scaffold upon forks if circumstances require it.

In forming this part of the scaffold in the manner of joists, the poles are placed about four feet asunder from center to center, so that when the sticks which sustain the tobacco plants are prepared they may fill the space advantageously by leaving but little spare room upon the scaffold.

Timber is then split in the manner of laths, into pieces of four feet in length, and about an inch and a half diameter. These are termed the *tobacco sticks;* and their use is to hang the tobacco upon, both by lodging the ends of this stick upon the poles of the scaffold which have been previously prepared in the field, in order to render it sufficiently pliable and in condition to carry into the tobacco-house, to which it is now conveyed by such means as the planter has in his power; and by suspending it in the same way in the house, so that the air may pass through it in the process of curing. Instead of this particular method, those who prefer to do so, lay it a short while in bulk upon poles, logs, &c. in the field, before they convey it under cover.

We must now leave the field to attend to the further process in the tobacco-house, or barn, which will form the next part, or division, of this subject.

PART II. ON THE MANNER OF HOUSING, CURING, AND VENDING TOBACCO IN VIRGINIA.

Of the Tobacco House and its Variety.

The barn which is appropriated to the use of receiving and curing this crop, is not, in the manner of other barns, connected with the farm yard, so that the whole occupation may be rendered snug and compact, and occasion little waste of time by inconsiderate and useless locomotion; but it is constructed to suit the particular occasion in point of size, and is generally erected in, or by the side of, each respective piece of tobacco ground; or sometimes in the woods, upon some hill or particular site which may be convenient to more than one field of tobacco.

The sizes which are most generally built where this kind of culture prevails, are what are called forty feet, and sixty feet tobacco houses, that is, of these lengths respectively, and of a proportionate width; and the plate of the wall, or part which supports the eaves of the roof, is generally elevated from the groundsel about the pitch of twelve feet.

About twelve feet pitch is indeed a good height for the larger crops; because this will allow four feet pitch each to three successive tiers of tobacco, besides those which are hung in the roof; and this distance admits a free circulation of air, and is a good space apart for the process of curing the plant.

There are various methods in use in respect to the construction of tobacco houses, and various materials of which they are constructed; but such are generally found upon the premises as suffice for the occasion. And although these sizes are most prevalent, yet tobacco houses are in many instances built larger or smaller according to the circumstances of the proprietor, or the size of the spot of ground under cultivation.

The most ordinary kinds consist of two square pens built out of logs of six or eight inches thick, and from sixteen to twenty feet long. Out of this material the two pens are formed by notching the logs near their extremities with an axe; so that they are alternately fitted one upon another, until they rise to a competent height; taking care to fit joists in at the respective tiers of four feet space, so that scaffolds may be formed by them similar to those heretofore described to have been erected in the open field, for the purpose of hanging the sticks of tobacco upon, that they may be open to a free circulation of air during this stage of the process. These pens are placed on a line with each other, at the opposite extremes of an oblong square, formed of such a length as to admit of a space between the two pens wide enough for the reception of a cart or waggon. This space, together with the two pens, is covered over with one and the same roof, the frame of which is formed in the same way of the walls by notching the logs as aforesaid, and narrowing up the gable ends to a point at the upper extremity of the house, termed the ridge pole. The remaining part of the fabric consists of a rough cover of thin slabs of wood split first with a mall and wedges, and afterwards riven with an instrument or tool termed a *froe*. The only thing which then remains to be done, is to cut a door into each of the pens, which is done by putting blocks or wedges in betwixt the logs which are to be cut out, and securing the jambs with side pieces pinned on with an auger and wooden pins. The roof is secured by weighting it down with logs; so that neither hammer, nails, brick, or stone, is concerned in the structure; and locks and keys are very rarely deemed necessary.

The second kind of tobacco houses differ somewhat from these, with a view to longer duration. The logs are to this end more choicely selected. The foundation consists of four well hewn groundsels, of about eight by ten inches, levelled and laid upon cross sawed blocks of a larger tree, or upon large stones. The corners are truly measured, and squared diamond-wise, by which means they are more nicely notched in upon each other; the roof is fitted with rafters, footed upon wall plates, and covered with *clap-boards* nailed upon the rafters in the manner of slating. In all other respects this is the same with the last mentioned method; and both are left open for the passage of the air between the logs.

The third kind is laid upon a foundation similar to the second; but instead of logs, the walls are composed of posts and studs, tenoned into the sells, and braced; the top of these are mounted with a wall-plate and joists; upon these come the rafters; and the whole is covered with clap-boards and nails, so as to form one uninterrupted oblong square, with doors, &c. termed, as heretofore, a forty, sixty, or one hundred feet tobacco house, &c.

The fourth species of these differs from the third only in the covering, which is generally of good sawed *feather-edged* plank; in the roof, which is now composed of *shingles*; and in the doors and finishing, which consist of good sawed plank, hinged, &c. Sometimes this kind are underpinned with a brick or stone wall beneath the groundsels; but they have no floors or windows, except a plank or two along the sides to raise upon hinges for sake of air, and occasional light: indeed, if these were constructed with sides similar to the brewery tops in London, I think it would be found advantageous.

In respect to the inside framing of a tobacco house, one description may serve for every kind: they are so contrived as to admit poles in the nature of a scaffold through every part of them, ranging four feet from centre to centre, which is the length of the tobacco stick, as heretofore described; and the lower tiers should be so contrived as to remove away occasionally, in order to pursue other employments at different stages in the process of curing the crop.

Of Preparations for curing the Tobacco Plant.

When the plant has remained long enough exposed to the sun, or open air, after cutting, to become sufficiently pliant to bear handling and removal with conveniency, it must be removed to the tobacco house, which is generally done by manual labour, unless the distance and quantity requires the assistance of a cart. If this part of the process were managed with horses carrying frames upon their back for the conveniency of stowage, in a way similar to that in which grain is conveyed in Spain, it would be found a considerable saving of labour.

It becomes necessary, in the next place to see that suitable ladders and stages are provided, and that there be a sufficient quantity of tobacco sticks, such as have been described heretofore, to answer the full demand of the tobacco house, whatsoever may be its size; time will be otherwise lost in *makeshifts,* or sending for a second supply.

Of Hanging the Crop.

When every thing is thus brought to a point at the tobacco house, the next stage of the process is that termed *hanging* the tobacco. This is done by hanging the plants in rows upon the tobacco sticks with the points down, letting them rest upon the stick by the stem of the lowest leaf, or by the split which is made in the stem when that happens to be divided. In this operation care must be taken to allow a sufficient space between each of the successive plants for the due circulation of air between: perhaps four or five inches apart, in proportion to the bulk of the plant.

When they are thus threaded upon the sticks (either in the tobacco houses, or, sometimes, suspended upon a temporary scaffold near the door, they must be carefully handed up by the means of ladders and planks to answer as stages or platforms, first to the upper tier or collar beams of the house, where the sticks are to be placed with their points resting upon the beams transversely, and the plants hanging down between them.

This process must be repeated tier after tier of the beams, downwards, until the house is filled; taking care to hang the sticks as close to each other as the consideration of admitting air will allow, and without crowding. In this position the plants remain until they are in condition to be taken down for the next process.

Of Smoking the Crop.

From what has been said under the head of hanging the plant, it will be perceived that the air is the principal agent in curing it; but it must be also considered that a want of uniform temperature in the atmosphere calls for the constant care of the crop-master, who generally indeed becomes habitually weather-wise, from the sowing of his plants, until the delivery of his crop to the inspector.

To regulate this effect upon the plants he must take care to be often among them, and when too much moisture is discovered, it is tempered by the help of smoke, which is generated by means of small smothered fires made of old bark, and of rotten wood, kindled about upon various parts of the floor where they may seem to be most needed. In this operation it is necessary that a careful hand should be always near: for the fires must not be permitted to blaze, and burn furiously.

Of Stemming Tobacco.

Stemming tobacco is the act of separating the largest stems or fibres from the web of the leaf with adroitness and facility, so that the plant may be nevertheless capable of package, and fit for a foreign market. It is practised in cases where the malady termed the fire, or other casual misfortune during the growth of the plant, may have rendered it doubtful in the opinion of the planter whether something or other which he may have observed during the growth of his crop, or in the unfavourable temperature of the seasons by which it hath been matured, does not hazard too much in packing the web with a stem which threatens to decay. To avoid the same species of risk, stemming is also practised in cases where the season when it becomes necessary to finish packing for a market is too unfavourable to put up the plant in leaf in the usual method; or when the crop may be partially *out of case*. Hence it is that the inspectors mark in the margin of the tobacco note (which is a certificate whereby crops are bought and sold without ever seeing them) the approximate proportion of the hogshead which is of this quality: for it often happens that only one third, one fourth, half, one fifth, five eighths, &c. may be stemmed tobacco, and the remainder of the hogshead be packed in leaf according to the ordinary custom.

Besides the operation of stemming in the hands of the crop-master, there are instances where this partial process is repeated in the public warehouses; of which I shall treat under a subsequent head.

The operation is performed by taking the leaf in one hand, and the end of the stem in the other, in such a way as to cleave it *with the grain;* and there is an expertness to be acquired by practice, which renders it as easy as to separate the bark of a willow, although those unaccustomed to it find it difficult to stem a single plant.

When the web is thus separated from the stem, it is made up into bundles in the same way as in the leaf, and is laid in bulk for farther process. The stems have been generally thrown away, or burnt with refused tobacco for the purpose of soap ashes; but the introduction of snuff-mills has, within a few years back, found a more economical use for them.

Of the Conveyance to Market.

The conveyance of a crop of tobacco to market, is of five different kinds: 1. By carts and waggons. 2. By rolling in hoops. 3. By rolling in fellies. 4. By canoes. 5. By upland boats.

Conveyance by Carts and Waggons.

This kind of conveyance for tobacco, when it is intended to be carried to market, depends mostly upon the leisure of the planter, and not upon any public establishment; and it is not unusual that a crop lays a considerable time in the barn after it is ready to be taken away, because it is not an easy thing for a planter to be absent from his domestic concerns very often upon a tedious journey. When the season and circumstances permit his absence, and his horses can be spared, and are put in condition to encounter a long and rugged road (which formerly was in few instances less than one hundred miles from the inspection, but which is now somewhat reduced by increasing the number of interior inspections), it is usual for several planters in the same neighbourhood to associate together, and join their force of horses, &c. according to their proportions of tobacco to be conveyed to market, each waggon taking two hogsheads. Thus the party set out upon their annual, or, perhaps, biennial, expedition, taking with them their provisions, liquors, and provender for their cattle; and encamping constantly in the woods until their return, by the side of a good *rousing* fire, which is kindled without ceremony upon any man's land, and with any man's fewel, without inhospitable objections from the proprietor. Those who are in more affluent circumstances, and who have occasion to send often to market, generally keep their own waggons in proportion to the extent of their estates; and there are also waggons to be hired, all of them of the same kind, with narrow wheels, carrying each two hogsheads; and all pursuing the same methods for their accommodation. On their return, each one makes it his business to provide for his family, and for such neighbours as he can conveniently serve, by the conveyance of merchandize as part of their *back loads,* or returning freight. Such as are not taken up in this way, are generally occupied by merchants of the interior country, for the supply of their inland stores; and the heavy articles of salt and iron make a material part of this employment. The rates of waggonage (whereof two thousand pounds weight are usually called a load, though some waggons will carry three thousand pounds) are as follow; viz. for one hundred pounds weight, the distance of one hundred miles, the sum of four shillings Virginia money; equal to three shilling sterling.

For one hogshead of tobacco, the distance of one hundred miles, the sum of two pounds Virginia money; equal to one pound ten shillings sterling.

For such a waggon by the day, every thing being furnished by the waggoner, the sum of twenty shillings Virginia money; equal to fifteen shillings sterling.

For such a waggon by the day, provisions and provender being furnished by the employer, the sum of twelve shillings Virginia money; equal to nine shillings sterling money of Great Britain.

Carts are of course half the rates of waggons.

Conveyance by Rolling in Hoops.

I believe rolling tobacco the distance of many hundred miles, is a mode of conveyance peculiar to Virginia; and for which the early population of that country deserves a very handsome credit. Necessity (that very prolific mother of invention) first suggested the idea of rolling *by hand;* time and experience have led to the introduction of *horses,* and have ripened human skill, in this kind of carriage, to a degree of perfection which merits *the adoption of the mother country,* but which will be better explained under the next head of this subject. The hogsheads, which are designed to be rolled in common hoops, are made closer in the joints than if they were intended for the waggon; and are plentifully hooped with strong hickory hoops (which is the toughest kind of wood) with the bark upon them, which remains for some distance a protection against the stones. Two hickory saplings are affixed to the hogshead, for shafts, by boring an auger-hole through them to receive the gudgeons or pivots, in the manner of a field rolling-stone: and these receive pins of wood, with square tapered points, which are admitted through square mortises made central in the heading, and driven a considerable depth into the solid tobacco. Upon the hind part of these shafts, between the horses and the hogshead, a few light planks are nailed, and a kind of little cart body is constructed of a sufficient size to contain a bag or two of provender, and provision, together with an axe, and such other tools as may be needed upon the road, in case of accident. In this manner they set out to the inspection in companies, very often joining society with the waggons, and always pursuing the same method of encamping. This mode of sleeping in the woods upon such a journey; the red clay lands through which most of the tobacco rollers pass; the continual and unavoidable exposure to dews, muddy roads or dusty ones; and the distances which they travel, contribute to add to their long beards a very savage appearance; and the natural consequence of this mode of living produces rough rustic amusements, and similar dispositions. They have hence become an object of apprehension to strangers, and a terror to the English traveller, whom habit has rendered too often wont to view every other country with the eyes of his own; and who expects to find in all men those gradations of humble distance to which he may happen to have been accustomed. To those, in particular, who approach this (or any other) class of Americans, with an air of self-important consequence, they are readily disposed to shew the worst side; and very often, under the mask of ignorance, play such men many an unlucky prank, and bid them a more unpleasant welcome than even the story of the inhospitable *Scotchman* exhibits in the recent travels of an *Irish* gentleman through that well known place, the northern neck of Virginia. Let a man in a *sulky,* however, (of which they are not over fond, perhaps only from his haughty appearance) only put off his offensive attitude of *incubation,* and accost them like fellow mortals of the same species, and they will be the first to do him a real service. The fact is, that men of great respectability, and plentiful hospitality when at home, think it no disgrace to sally forth upon the concerns of their crop; and in this case they accommodate themselves to manners which bid defiance to difficulty, and answer their ends.

Conveyance by rolling in Fellies.

Rolling in *fellies* is an improvement resulting from experience in the former method of rolling in *hoops,* which in long journies are found to shatter (especially upon stony roads), and very often to damage the contents, or occasion delays for a too frequent refitting of the hogshead. Experience has suggested this, and practice in the expedient has rendered the invention of fellies more perfect. They consist of pieces of wood formed into segments of a circle in the manner of cart wheels; and these, instead of being formed into the rim of a wheel supported by spokes fixed into a nave, are fixed round the circumference of the tobacco hogshead by means of auger holes and wooden pins driven into the bulk of tobacco, through the fellies and the staves of the hogshead. By this means the stones upon the road are greatly avoided, and the hogshead may be safely conveyed to a very considerable distance. This improvement has suggested another, which is now reduced to practice in the conveyance of grain, and which doubtless might be farther employed (if need be) in the conveyance of fluid substances. Wheat and other small grain is now *rolled* in many places in Virginia, in hogsheads which are compactly formed; well hooped with iron; the fellies well shod with iron wheel tire; and iron pins for the gudgeons or axles. There is in the head of each cask a small door or scuttle for receiving and delivering the grain; and I can see no reason why fluids may not be as easily received, conveyed, and delivered, by the help of a cock.

This is certainly a cheap and easy-going vehicle; and, when it is considered that the weight of a cart and its contents is thus completely relieved from the back of a horse, and that one horse alone is equal to a considerable burden, I should suppose it worthy an experiment in many English employments.

A Frenchman Views America, 1803

From Francois Andre Michaux, *Travels to the West of the Alleghany Mountains, Spring 1803,* in Reuben Gold Thwaites, ed., *Travels West of the Alleghanies* (Cleveland, 1904), pp. 134–35, 188–92, 207–08, 243–47, 276–81, 298–99, 303–05.

It is reckoned sixty miles from Philadelphia to Lancaster, where I arrived the same day in the afternoon. The road is kept in good repair by the means of turnpikes, fixed at a regular distance from each other. Nearly the whole of the way the houses are almost close together; every proprietor to his enclosure. Throughout the United States all the land that is cultivated is fenced in, to keep it from the cattle and quadrupeds of every kind that the inhabitants leave the major part of the year in the woods, which in that respect are free. Near towns or villages these enclosures are made with posts, fixed in the ground about twelve feet from each other, containing five mortises, at the distance of eight or nine inches, in which are fitted long spars about four or five inches in diameter, similar to the poles used by builders for making scaffolds. The reason of their enclosing thus is principally through economy, as it takes up but very little wood, which is extremely dear in the environs of the Northern cities; but in the interior of the country, and in the Southern states, the enclosures are made with pieces of wood of equal length, placed one above the other, disposed in a zig-zag form, and supported by their extremities, which cross and interlace each other; the enclosures appear to be about seven feet in height. In the lower part of the Carolines they are made of fir; in the other parts of the country, and throughout the North, they are comprised of oak and walnut-tree; they are said to last about five and twenty years when kept in good repair.

The tract of country we have to cross, before we get to Lancaster, is exceedingly fertile and productive; the fields are covered with wheat, rye, and oats, which is a proof that the soil is better than that between New York and Philadelphia. The inns are very numerous on the road; in almost all of them they speak German. My fellow travellers being continually thirsty, made the stage stop at every inn to drink a glass or two of grog. This beverage, which is generally used in the United States, is a mixture of brandy and water, or rum and water, the proportion of which depends upon the person's taste.

Lancaster is situated in a fertile and well-cultivated plain. The town is built upon a regular plan; the houses, elevated two stories, are all of brick; the two principal streets are paved as at Philadelphia. The population is from four to five thousand inhabitants, almost all of German origin, and various sects; each to his particular church; that of the Roman Catholics is the least numerous. The inhabitants are for the most part armourers, hatters, saddlers, and coopers; the armourers of Lancaster have been long esteemed for the manufacturing of rifle-barrelled guns, the only arms that are used by the inhabitants of the interior part of the country, and the Indian nations that border on the frontiers of the United States. . . .

The banks of the Ohio, although elevated from twenty to sixty feet, scarcely afford any strong substances from Pittsburgh; and except large detached stones of a greyish colour and very soft, that we observed in an extent of ten or twelve miles below Wheeling, the remainder part seems vegetable earth. A few miles before we reached Limestone we began to observe a bank of a chalky nature, the thickness of which being very considerable, left no room to doubt but what it must be of a great extent.

Two kinds of flint, roundish and of a middling size, furnished the bed of the Ohio abundantly, especially as we approached the isles, where they are accumulated by the strength of the current; some of a darkish hue, break easily; others smaller, and in less quantities, are three parts white, and scarcely transparent.

In the Ohio, as well as in the Alleghany, Monongahela, and other rivers in the west, they find in abundance a species of *Mulette* which is from five to six inches in length. They do not eat it, but the mother-o'-pearl which is very thick in it, is used in making buttons. I have seen some at Lexinton which were as beautiful as those they make in Europe. This new species which I brought over with me, has been described by Mr. Bosc, under the name of the *Unio Ohiotensis*.

The Ohio abounds in fish of different kinds; the most common is the cat-fish, or *silurus felis*, which is generally caught with a line, and weighs sometimes a hundred pounds. The first fold of the upper fins of this fish are strong and pointed, similar to those of a perch, which he makes use of to kill others of a lesser size. He swims several inches under the one he wishes to attack, then rising rapidly, he pierces him several times in the belly; this we had an opportunity of observing twice in the course of our navigation. This fish is also taken with a kind of spear.

Till the years 1796 and 1797 the banks of the Ohio were so little populated that they scarcely consisted of thirty families in the space of four hundred miles; but since that epoch a great number of emigrants have come from the mountainous parts of Pennsylvania and Virginia, and settled there; in consequence of which the plantations now are so increased, that they are not farther than two or three miles distant from each other, and when on the river we always had a view of some of them.

The inhabitants on the borders of the Ohio, employ the greatest part of their time in stag and bear hunting, for the sake of the skins, which they dispose of. The taste that they have contracted for this kind of life is prejudicial to the culture of their lands; besides they have scarcely any time to meliorate their new possessions, that usually consist of two or three hundred acres, of which not more than eight or ten are cleared. Nevertheless, the produce that they derive from them, with the milk of their cows, is sufficient for themselves and families, which are always very numerous. The houses that they inhabit are built upon the borders of the river, generally in a pleasant situation, whence they enjoy the most delightful prospects; still their mode of building does not correspond with the beauties of the spot, being nothing but miserable log houses, without windows, and so small that two beds occupy the greatest part of them. Notwithstanding two men may erect and finish, in less than three days, one of these habitations, which, by their diminutive size and sorry appearance, seem rather to belong to a country where timber is very scarce, instead of a place that abounds with forests. The inhabitants on the borders of the Ohio do not hesitate to receive travellers who claim their hospitality; they give them a lodging, that is to say, they permit them to sleep upon the floor wrapped up in their rugs. They are accommodated with bread, Indian corn, dried ham, milk and butter, but seldom any thing else; at the same time the price of provisions is very moderate in this part of the United States, and all through the western country.

No attention is paid by the inhabitants to any thing else but the culture of Indian corn; and although it is brought to no great perfection, the soil being so full of roots, the stems are from ten to twelve feet high, and produce from twenty to thirty-five hundred weight of corn per acre. For the three first years after the ground is cleared, the corn springs up too strong, and scatters before it ears, so that they cannot sow in it for four or five years after, when the ground is cleared of the stumps and roots that were left in at first. The Americans in the interior cultivate corn rather through speculation to send the flour to the sea-ports, than for their own consumption; as nine tenths of them eat no other bread but that made from Indian corn; they make loaves of it from eight to ten pounds, which they bake in ovens, or small cakes baked on a board before the fire. This bread is generally eaten hot, and is not very palatable to those who are not used to it.

The peach is the only fruit tree that they have as yet cultivated, which thrives so rapidly that it produces fruit after the second year.

The price of the best land on the borders of the Ohio did not exceed three piastres per acre; at the same time it is not so dear on the left bank in the States of Virginia and Kentucky, where the settlements are not looked upon as quite so good.

The two banks of the Ohio, properly speaking, not having been inhabited above eight or nine years, nor the borders of the rivers that run into it, the Americans who are settled there, share but very feebly in the commerce that is carried on through the channel of the Mississippi. This commerce consists at present in hams and salted pork, brandies distilled from corn and peaches, butter, hemp, skins and various sorts of flour. They send again cattle to the Atlantic States. Tradespeople who supply themselves at Pittsburgh and Wheeling, and go up and down the river in a canoe, convey them haberdashery goods, and more especially tea and coffee, taking some of their produce in return.

More than half of those who inhabit the borders of the Ohio, are again the first inhabitants, or as they are called in the United States, the *first settlers*, a kind of men who cannot settle upon the soil that they have cleared, and who under pretence of finding a better land, a more wholesome country, a greater abundance of game, push forward, incline perpetually towards the most distant points of the American population, and go and settle in the neighbourhood of the savage nations, whom they brave even in their own country. Their ungenerous mode of treating them stirs up frequent broils, that brings on bloody wars, in which they generally fall victims; rather on account of their being so few in number, than through defect of courage. . . .

There reigns in the United States a public spirit that makes them greedily seize hold of every plan that tends to enrich the country by agriculture and commerce. That of rearing the vine in Kentucky was eagerly embraced. Several individuals united together, and formed a society to put it in execution, and it was decreed that a fund should be established of ten thousand dollars, divided into two hundred shares of fifty dollars each. This fund was very soon accomplished. Mr. Dufour, the chief of a small Swiss colony which seven or eight years before had settled in Kentucky, and who had proposed this undertaking, was deputed to search for a proper soil, to procure vine plants, and to do every thing he might think necessary to insure success. The spot that he has chosen and cleared is on the Kentucky river, about twenty miles from Lexinton. The soil is excellent and the vineyard is planted upon the declivity of a hill exposed to the south, and the base of which is about two hundred fathoms from the river.

Mr. Dufour intended to go to France to procure the vine plants, and with that idea went to New York; but the war, or other causes that I know not, prevented his setting out, and he contented himself with collecting, in this town and Philadelphia, slips of every species that he could find in the possession of individuals that had them in their gardens. After unremitted labour he made a collection of twenty-five different sorts, which he brought to Kentucky, where he employed himself in cultivating them. However the success did not answer the expectation; only four or five various kinds survived, among which were those that he had described by the name of Burgundy and Madeira, but the former is far from being healthy. The grape generally decays before it is ripe. When I saw them the bunches were thin and poor, the berries small, and every thing announced that the vintage of 1802 would not be more abundant than that of the preceding years. The Madeira vines appeared, on the contrary, to give some hopes. Out of a hundred and fifty or two hundred, there was a third loaded with very fine bunches. The whole of these vines do not occupy a space of more than six acres. They are planted and fixed with props similar to those in the environs of Paris. . . .

For some time past the inhabitants of Kentucky have taken to the rearing and training horses; and by this lucrative branch of trade they derive considerable profit, on account of the superfluous quantity of Indian corn, oats, and other forage, of which they are deficient at New Orleans.

Of all the states belonging to the union, Virginia is said to have the finest coach and saddle-horses, and those they have in this country proceed originally from them, the greatest part of which was brought by the emigrants who came from Virginia to settle in this state. The number of horses, now very considerable, increases daily. Almost all the inhabitants employ themselves in training and meliorating the breed of these animals; and so great a degree of importance is attached to the melioration, that the owners of fine stallions charge from fifteen to twenty dollars for the covering of a mare. These stallions come from Virginia, and, as I have been told, some were at different times imported from England. The horses that proceed from them have slim legs, a well-proportioned head, and are elegantly formed. With draught-horses it is quite different. The inhabitants pay no attention with respect to improving this breed; in consequence of which they are small, wretched in appearance, and similar to those made use of by the peasantry in France. They appeared to me still worse in Georgia and Upper Carolina. In short, I must say that throughout the United States there is not a single draught-horse that can be in any wise compared with the poorest race of horses that I have seen in England. This is an assertion which many Americans may probably not believe, but still it is correct.

Many individuals profess to treat sick horses, but none of them have any regular notions of the veterinary art; an art which would be so necessary in a breeding country, and which has, within these few years, acquired so high a degree of perfection in England and France.

In Kentucky, as well as in the southern states, the horses are generally fed with Indian corn. Its nutritive quality is esteemed double to that of oats; notwithstanding sometimes they are mixed together. In this state horses are not limited as to food. In most of the plantations the manger is filled with corn, they eat of it when they please, leave the stable to go to grass, and return at pleasure to feed on the Indian wheat. The stables are nothing but log-houses, where the light penetrates on all sides, the interval that separates the trunks of the trees with which they are constructed not being filled up with clay.

The southern states, and in particular South Carolina, are the principal places destined for the sale of Kentucky horses. They are taken there in droves of fifteen, twenty and thirty at a time, in the early part of winter, an epoch when the most business is transacted at Carolina, and when the drivers are in no fear of the yellow fever, of which the inhabitants of the interior have the greatest apprehension. They usually take eighteen or twenty days to go from Lexinton to Charleston. This distance, which is about seven hundred miles, makes a difference of twenty-five or thirty per cent in the price of horses. A fine saddle-horse in Kentucky costs about a hundred and thirty to a hundred and forty dollars.

During my sojourn in this state I had an opportunity of seeing those wild horses that are caught in the plains of New Mexico, and which descend from those that the Spaniards introduced there formerly. To catch them they make use of tame horses that run much swifter, and with which they approach them near enough to halter them. They take them to New Orleans and Natches, where they fetch about fifty dollars. The crews belonging to the boats that return by land to Kentucky frequently purchase some of them. The two that I saw and made a trial of were roan coloured, of a middling size, the head large, and not proportionate with the neck, the limbs thick, and the mane rather full and handsome. These horses have a very unpleasant gait, are capricious, difficult to govern, and even frequently throw the rider and take flight.

The number of horned cattle is very considerable in Kentucky; those who deal in them purchase them lean, and drive them in droves of from two to three hundred to Virginia, along the river Potomack, where they sell them to graziers, who fatten them in order to supply the markets of Baltimore and Philadelphia. The price of a good milch cow is, at Kentucky, from ten to twelve dollars. The milk in a great measure comprises the chief sustenance of the inhabitants. The butter that is not consumed in the country is put into barrels, and exported by the river to the Carribbees.

They bring up very few sheep in these parts; for, although I went upwards of two hundred miles in this state, I saw them only in four plantations. Their flesh is not much esteemed, and their wool is of the same quality as that of the sheep in the eastern states. The most that I ever observed was in Rhode Island.

Of all domestic animals hogs are the most numerous; they are kept by all the inhabitants, several of them feed a hundred and fifty or two hundred. These animals never leave the woods, where they always find a sufficiency of food, especially in autumn and winter. They grow extremely wild, and generally go in herds. Whenever they are surprised, or attacked by a dog or any other animal, they either make their escape, or flock together in the form of a circle to defend themselves. They are of a bulky shape, middling size, and straight eared. Every inhabitant recognizes those that belong to him by the particular manner in which their ears are cut. They stray sometimes in the forests, and do not make their appearance again for several months; they accustom them, notwithstanding, to return every now and then to the plantation, by throwing them Indian corn once or twice a week. It is surprising that in so vast a country, covered with forests, so thinly populated, comparatively to its immense extent, and where there are so few destructive animals, pigs have not increased so far as to grow completely wild.

In all the western states, and even to the east of the Alleghanies, two hundred miles of the sea coast, they are obliged to give salt to the cattle. Were it not for that, the food they give them would never make them look well; in fact, they are so fond of it that they go of their own accord to implore it at the doors of the houses every week or ten days, and spend hours together in licking the trough into which they have scattered a small quantity for them. This want manifests itself most among the horses; but it may be on account of their having it given them more frequently.

Salt provisions form another important article of the Kentucky trade. The quantity exported in the first six months of the year 1802 was seventy-two thousand barrels of dried pork, and two thousand four hundred and eighty-five of salt.

Notwithstanding the superfluity of corn that grows in this part of the country, there is scarcely any of the inhabitants that keep poultry. This branch of domestic economy would not increase their expense, but add a pleasing variety in their food. Two reasons may be assigned for this neglect; the first is, that the use of salt provisions, (a use to which the prevalence of the scurvy among them may be attributed,) renders these delicacies too insipid; the second, that the fields of Indian corn contiguous to the plantations would be exposed to considerable damage, the fences with which they are inclosed being only sufficient to prevent the cattle and pigs from trespassing. . . .

West Tennessea, or Cumberland, being situated under a more southerly latitude than Kentucky, is particularly favourable to the growth of cotton; in consequence of which the inhabitants give themselves up almost entirely to it, and cultivate but little more corn, hemp, and tobacco than what is necessary for their own consumption.

The soil, which is fat and clayey, appears to be a recent dissolving of vegetable substances, and seems, till now, less adapted for the culture of corn than that of Indian wheat. The harvests of this grain are as plentiful as in Kentucky; the blades run up ten or twelve feet high; and the ears, which grow six or seven feet from the earth, are from nine to ten inches in length, and proportionate in size. It is cultivated in the same manner as in other parts of the western country.

The crows, which are a true plague in the Atlantic states, where they ravage, at three different periods, the fields of Indian wheat, which are obliged to be sown again as many times, have not yet made their appearance in Tennessea; but it is very probable that this visit is only deferred, as they do, annually, great damage in Kentucky.

I must also observe here that the grey European rats have not yet penetrated into Cumberland, though they are very numerous in other parts of the country, particularly in those settlements belonging to the whites.

The culture of cotton, infinitely more lucrative than that of corn and tobacco, is, as before observed, the most adhered to in West Tennessea. There is scarcely a single emigrant but what begins to plant his estate with it the third year after his settling in the country. Those who have no negroes cultivate it with the plough, nearly in the same manner as Indian wheat, taking particular care to weed and throw new earth upon it several times in the course of the season. Others lay out their fields in parallel furrows, made with the hoe, from twelve to fifteen inches high. It is computed that one man, who employs himself with this alone, is sufficient to cultivate eight or nine acres, but not to gather in the harvest. A man and a woman, with two or three children, may, notwithstanding, cultivate four acres with the greatest ease, independent of the Indian wheat necessary for their subsistence; and calculating upon a harvest of three hundred and fifty pounds weight per acre, which is very moderate according to the extreme fertility of the soil, they will have, in four acres, a produce of fourteen hundred pounds of cotton. Valuing it at the rate of eighteen dollars per hundred weight, the lowest price to which it had fallen at the epoch of the last peace, when I was in the country, gives two hundred and fifty-two dollars; from which deducting forty dollars for the expenses of culture, they will have a net produce of two hundred and twelve dollars; while the same number of acres, planted with Indian wheat, or sown with corn, would only yield at the rate of fifty bushels per acre; and twenty-five bushels of corn, about fifty dollars, reckoning the Indian wheat at thirteen pence, and the corn at two shillings and two pence per bushel; under the supposition that they can sell it at that price, which is not always the case. This light sketch demonstrates with what facility a poor family may acquire speedily, in West Tennessea, a certain degree of independence, particularly after having been settled five or six years, as they procure the means of purchasing one or two negroes, and of annually increasing their number.

The species of cotton which they cultivate here is somewhat more esteemed than that described by the name of green-seed cotton, in which there is a trifling distinction in point of colour.

The cottons that are manufactured in West Tennessea are exceedingly fine, and superior in quality to those I saw in the course of my travels. The legislature of this state, appreciating the advantage of encouraging this kind of industry, and of diminishing, by that means, the importation of English goods of the same nature, has given, for these two years past, a premium of ten dollars to the female inhabitant who, in every county, presents the best manufactured piece; for in this part, as well as in Kentucky, the higher circles wear, in summer time, as much from patriotism as from economy, dresses made of the cottons manufactured in the country. At the same time they are convinced that it is the only means of preserving the little specie that is in the country, and of preventing its going to England.

The price of the best land does not yet exceed five dollars per acre in the environs of Nasheville, and thirty or forty miles from the town they are not even worth three dollars. They can at that price purchase a plantation completely formed, composed of two to three hundred acres, of which fifteen to twenty are cleared, and a log-house. The taxes in this state are also not so high as in Kentucky.

Among the emigrants that arrive annually from the eastern country at Tennessea there are always some who have not the means of purchasing estates; still there is no difficulty in procuring them at a certain rent; for the speculators who possess many thousand acres are very happy to get tenants for their land, as it induces others to come and settle in the environs; since the speculation of estates in Kentucky and Tennessea is so profitable to the owners, who reside upon the spot, and who, on the arrival of the emigrants, know how to give directions in cultivation, which speedily enhances the value of their possessions.

The conditions imposed upon the renter are to clear and inclose eight or nine acres, to build a log-house, and to pay to the owner eight or ten bushels of Indian wheat for every acre cleared. These contracts are kept up for seven or eight years. The second year after the price of two hundred acres of land belonging to a new settlement of this kind increases nearly thirty per cent.; and this estate is purchased in preference by a new emigrant, who is sure of gathering corn enough for the supplies of his family and cattle.

In this state they are not so famed for rearing horses as in Kentucky; yet the greatest care is taken to improve their breed, by rearing them with those of the latter state, whence they send for the finest mare foals that can be procured.

Although this country abounds with saline springs, none are yet worked, as the scarcity of hands would render the salt dearer than what is imported from the saltpits of St. Genevieve, which supply all Cumberland. It is sold at two dollars per bushel, about sixty pounds weight.

East Tennessea, or Holston, is situated between the loftiest of the Alleghany and Cumberland Mountains. It comprises, in length, an extent of nearly a hundred and forty miles, and differs chiefly from West Tennessea in point of the earth's being not so chalky, and better watered by the small rivers issuing from the adjacent mountains, which cross it in every part. The best land is upon their borders. The remainder of the territory, almost everywhere interspersed with hills, is of a middling quality, and produces nothing but white, red, black, chincapin, and mountain oaks, &c. intermixed with pines; and, as we have before observed, except the *quercus macrocarpa*, the rest never grow, even in the most fertile places.

Indian wheat forms here also one of the principal branches of agriculture; but it very seldom comes up above seven or eight feet high, and a produce of thirty bushels per acre passes for an extraordinary harvest. The nature of the soil, somewhat gravelly, appears more adapted for the culture of wheat, rye, and oats; in consequence of which it is more adhered to than in Cumberland. That of cotton is little noticed, on account of the cold weather, which sets in very early. One may judge, according to this, that Holston is in every point inferior in fertility to Cumberland and Kentucky.

To consume the superfluity of their corn the inhabitants rear a great number of cattle, which they take four or five hundred miles to the seaports belonging to the southern states. They lose very few of these animals by the way, although they have to cross several rivers, and travel through an uninterrupted forest, with this disadvantage, of the cattle being extremely wild. . . .

The low price to which tobacco is fallen in Europe, within these few years, has made them give up the culture of it in this part of the country [South Carolina]. That of green-seed cotton has resumed its place, to the great advantage of the inhabitants, many of whom have since made their fortunes by it. The separation of the seed from the felt that envelopes them is a tedious operation, and which requires many hands, is now simplified by a machine for which the inventor has obtained a patent from the federal government. The legislature of South Carolina paid him, three years since, the sum of a hundred thousand dollars, for all the inhabitants belonging to the state to have the privilege of erecting one. This machine, very simple, and the price of which does not exceed sixty dollars, is put in motion by a horse or by a current of water, and separates from the seed three or four hundred pounds of cotton per day; while by the usual method, a man is not able to separate above thirty pounds. This machine, it is true, has the inconvenience of shortening by haggling it; the wool, on that account, is rather inferior in point of quality, but this inconvenience is, they say, well compensated by the saving of time, and more particularly workmanship.

It is very probable that the various species of fruit trees that we have in France would succeed very well in Upper Carolina. About two hundred miles from the sea-coast the apple trees are magnificent, and in the county of Lincoln several Germans make cyder. But here, as well as in Tennessea, and the greatest part of Kentucky, they cultivate no other but the peach. The other kinds of trees, such as pears, apricots, plumbs, cherries, almonds, mulberries, nuts, and gooseberries, are very little known, except by name. Many of the inhabitants who are independent would be happy to procure some of them, but the distance from the sea-ports renders it very difficult. The major part of the inhabitants do not even cultivate vegetables; and out of twenty there is scarcely one of them that plants a small bed of cabbages; and when they do, it is in the same field as the Indian wheat.

In Upper Carolina the surface of the soil is covered with a kind of grass, which grows in greater abundance as the forests are more open. The woods are also like a common, where the inhabitants turn out their cattle, which they know again by their private mark. Several persons have in their flocks a variety of poll oxen, which are not more esteemed than those of the common species. In the whole course of my travels I never saw any that could be compared to those I have seen in England, which beyond doubt proceeds from the little care that the inhabitants take of them, and from what these animals suffer during the summer, when they are cruelly tormented by an innumerable multitude of ticks and muskitos, and in the winter, through the want of grass, which dries up through the effect of the first frosts. These inconveniences are still more sensible, during the summer, in the low country, through the extreme heat of the climate. The result is, that the cows give but little milk, and are dry at the end of three or four months. In the environs of Philadelphia and New York, where they bestow the same care upon them as in England, they are, on the contrary, as fine, and give as great a quantity of milk. . . .

The best rice plantations are established in the great swamps, that favour the watering of them when convenient. The harvests are abundant there, and the rice that proceeds from them, stripped of its husk, is larger, more transparent, and is sold dearer than that which is in a drier soil, where they have not the means or facility of irrigation. The culture of rice in the southern and maritime part of the United States has greatly diminished within these few years; it has been in a great measure replaced by that of cotton, which affords greater profit to the planters, since they compute a good cotton harvest equivalent to two of rice. The result is, that many rice fields have been transformed into those of cotton, avoiding as much as possible the water penetrating.

The soil most adapted for the culture of cotton is in the isles situate upon the coast. Those which belong to the state of Georgia produce the best of cotton, which is known in the French trade by the name of Georgia cotton, fine wool, and in England by that of Sea Island cotton. The seed of this kind of cotton is of a deep black, and the wool fine and very long. In February 1803 it was sold at Charleston at 1s. 8d. per pound, whilst that which grows in the upper country is not worth above seventeen or eighteen pence. The first is exported to England, and the other goes to France; but what is very remarkable is, that whenever by any circumstance they import these two qualities into our ports, they only admit of a difference of from twelve to fifteen per cent. The cotton planters have particularly to dread the frosts that set in very early, and that frequently do great damage to the crops by freezing one half of the stalks, so that the cotton has not an opportunity to ripen.

In all the plantations they cultivate Indian corn. The best land brings from fifteen to twenty bushels. They plant it, as well as the cotton, about two feet and a half distance, in parallel furrows from fifteen to eighteen inches high. The seed of this kind of Indian corn is round, and very white. When boiled it is preferable to that cultivated in the middle and western states, and in Upper Carolina. The chief part of what they grow is destined to support the negroes nine months in the year; their allowance is about two pounds per day, which they boil in water after having pounded it a little; the other three months they are fed upon yams. They never give them meat. In the other parts of the United States they are better treated, and live nearly upon the same as their masters, without having any set allowance. Indian corn is sold at Charleston for ten shillings per bushel, about fifty-five pounds weight.

Thus rice, long cotton, yams, and Indian wheat, are the only cultures in the maritime part of the southern states; the temperature of the climate, and the nature of the soil, which is too light or too moist, being in no wise favourable for that of wheat or any kind of grain.

Through the whole of the low country the agricultural labours are performed by negro slaves, and the major part of the planters employ them to drag the plough; they conceive the land is better cultivated, and calculate besides that in the course of a year a horse, for food and looking after, costs ten times more than a negro, the annual expense of which does not exceed fifteen dollars.

I shall abstain from any reflexion concerning this, as the opinion of many people is fixed.

The climate of Lower Carolina and Georgia is too warm in summer to be favourable to European fruit-trees, and too cold in winter to suit those of the Carribbees. The fig is the only tree that succeeds tolerably well; again, the figs turn sour a few days after they have acquired the last degree of maturity, which must doubtless be attributed to the constant dampness of the atmosphere.

In the environs of Charleston, and in the isles that border the coast, the orange-trees stand the winter in the open fields, and are seldom damaged by the frosts; but at ten miles distance, in the interior, they freeze every year even with the ground, although those parts of the country are situate under a more southerly latitude than Malta and Tunis. The oranges that they gather in Carolina are not good to eat. Those consumed there come from the island of St. Anastasia, situate opposite St. Augustin, the capital of East Florida; they are sweet, very large, fine skinned, and more esteemed than those brought from the Carribbees. About fifty years ago the seeds were brought from India, and given to an inhabitant of this island, who has so increased them that he has got an orchard of forty acres. I had an opportunity of seeing this beautiful plantation when I was at Florida in 1788.

Origin of the Morgan Horse

From D. C. Linsley, *Morgan Horses* (New York, 1857), pp. 92–95, 131–34, 138–43.

The different accounts that have been circulated in regard to the origin of the Morgan breed of horses, agree that they are descended from a horse called the Justin Morgan, who derived his name from Mr. Justin Morgan, of Randolph, Vt., once his owner. As to the origin of the Justin Morgan, however, they differ widely.

The fact that little or no interest was felt in the subject until after the death of Mr. Morgan, and indeed until after the death of his horse, will account for this diversity.

Almost half a century passed away before any serious effort was made to determine the origin of an animal, whose value was daily more and more appreciated. After the death of Mr. Morgan the horse passed through several hands, and was kept at different places, and when at length serious inquiry was awakened on the subject, it was found that Mr. Morgan had left no written pedigree of his horse, and different reports of what he said in relation to it got into circulation.

We think it may be considered as certain, that during Mr. Morgan's life and until long after his death, very little interest was felt in the question, "What was the exact pedigree of the horse?" When the inquiry became interesting, and discussion arose, different stories were current, and opinions were frequently formed in accordance with previous prejudices or views of the individual forming them, as to the value of different breeds of horses. Some, holding the opinion that no valuable horse could be expected without a great deal of racing blood, sought to make it appear that he was nearly thorough-bred. Others having less faith in the English racer, entertained different opinions, and adopted for their creed stories that ascribed to him a very different origin. No person seemed to take the matter in hand and investigate it thoroughly, until those who might have given the necessary information were gone.

It is not now probable that the blood of the Justin Morgan can ever be exactly and absolutely ascertained. We think, however, it may be considered certain that this unrivalled animal was produced by a cross of the Arabian or thorough-bred with the common stock, but the proportion of each cannot now be determined.

There are no opinions that men maintain so strenuously, and give up so reluctantly, as those which they form and publicly avow upon matters in which they are supposed by others to be particularly well-informed. This is more especially true when these opinions entertained and expressed, relate to matters of practical importance, and not simply to some abstract doctrine.

Thus, suppose a man has studied long upon the subject of rotation of crops, and is thought to understand it thoroughly; if his neighbor acting upon his advice is thereby injured, he will account for the failure in any way, rather than ascribe it to his own want of information, or error of judgment.

So the tradesman much more reluctantly acknowledges an error of judgment in relation to an article of which his very trade presupposes him to be a judge, than the man whose attention has been but momentarily drawn to it. It is to this disinclination to admit they have been in error, coupled with inattention to the proofs that have been offered, that we mainly attribute the disparity that may be noticed in the accounts given by horsemen of the origin of the Morgan horse; for we think all candid persons who will devote but a little time to an examination of the subject, can hardly fail to arrive at one and the same conclusion. So much importance has of late years been attached to this subject, and some excellent horsemen not fully agreeing in relation to it, we have been led to devote a good deal of time and labor to a careful examination of all such evidence as we could discover, and the result of our investigation is (as before stated) the firm conviction that the original Morgan horse was not thorough-bred, but yet, had a large infusion of blood.

Before considering the evidence that has led us to this conclusion, we wish to explain what seems to us should be the character of the evidence that should be deemed amply sufficient to establish this point.

While the friends of these horses have been sometimes ready to declare the original Justin Morgan fully thorough-bred, on the strength of evidence which we think can hardly be deemed sufficient; yet on the other hand, parties interested in denying he had any claim, even to a particle of that noble blood, insist upon proof that shall amount to demonstration.

The pedigree of a stock horse may in one sense be considered property; it is a valuable thing, and one from which the owner may derive great benefit. Now what reason can be assigned for insisting that a man shall adduce stronger evidence to establish his claim to such a pedigree, than is required before a court of justice to secure him in the possession of his house, or his farm; or stronger than would be sufficient to enable his neighbor to turn him out of doors? There can be no reason. But some gentlemen insist that the pedigree of an animal, to be entitled to credit, should be supported by evidence as conclusive as would be required to send the owner to the gallows, if he were indicted for murder. This is certainly most unreasonable and unjust.

* * *

The original, or Justin Morgan, was about fourteen hands high, and weighed about nine hundred and fifty pounds. His color was dark-bay with black legs, mane and tail. He had no white hairs on him. His mane and tail were coarse and heavy, but not so massive as has been sometimes described; the hair of both was straight, and not inclined to curl. His head was good, not extremely small, but lean and bony, the face straight, forehead broad, ears small and very fine, but set rather wide apart. His eyes were medium size, very dark and prominent, with a spirited but pleasant expression, and showed no white round the edge of the lid. His nostrils were very large, the muzzle small, and the lips close and firm. His back and legs were perhaps his most noticeable points. The former was very short; the shoulder-blades and hip bones being very long and oblique, and the loins exceedingly broad and muscular. His body was rather long, round and deep, close ribbed up; chest deep and wide, with the breast-bone projecting a good deal in front. His legs were short, close jointed, thin, but very wide, hard and free from meat, with muscles that were remarkably large for a horse of his size, and this superabundance of muscle exhibited itself at every step. His hair was short, and at almost all seasons soft and glossy. He had a little long hair about the fetlocks, and for two or three inches above the fetlock on the back-side of the legs; the rest of the limbs were entirely free from it. His feet were small but well shaped, and he was in every respect perfectly sound and free from any sort of blemish. He was a very fast walker. In trotting his gait was low and smooth, and his step short and nervous; he was not what in these days would be called fast, and we think it doubtful whether he could trot a mile much if any within four minutes, though it is claimed by many that he could trot it in three.

Although he raised his feet but little, he never stumbled. His proud, bold and fearless style of movement, and his vigorous, untiring action, have, perhaps, never been surpassed. When a rider was on him, he was obedient to the slightest motion of the rein, would walk backwards rapidly under a gentle pressure of the bit, and moved side-ways almost as willingly as he moved forward; in short, was perfectly trained to all the paces and evolutions of a parade horse; and when ridden at military reviews (as was frequently the case), his bold, imposing style, and spirited, nervous action, attracted universal attention and admiration. He was perfectly gentle and kind to handle, and loved to be groomed and caressed, but he disliked to have children about him, and had an inveterate hatred for dogs, if loose always chasing them out of sight the instant he saw them.

When taken out with halter or bridle he was in constant motion, and very playful.

He was a fleet runner at short distances. Running horses short distances for small stakes was very common in Vermont fifty years ago. Eighty rods was very generally the length of the course, which usually commenced at a tavern or grocery, and extended the distance agreed upon, up or down the public road. In these races the horses were started from a "scratch," that is, a mark was drawn across the road in the dirt, and the horses, ranged in a row upon it, went off at "the drop of a hat" or some other signal. It will be observed that the form of the Justin Morgan was not such as in our days is thought best calculated to give the greatest speed for a short distance. Those who believe in long-legged racers will think his legs, body and stride, were all too short, and to them it may perhaps seem surprising that he should be successful, as he invariably was, in such contests. But we think his great muscular development and nervous energy, combined with his small size, gave him a decided advantage in the first start over taller and heavier horses; just as any ordinary horse can distance the finest locomotive in a ten rod race. At all events, the history of racing in this country and in England, proves conclusively, that small horses *may* have great speed. In such a race a horse of great spirit and nervous energy derives a decided advantage from these qualities, especially after being a little accustomed to such struggles. When brought up to the line, his eyes flash and his ears quiver with intense excite-ment, he grinds the bit with his teeth, his hind legs are drawn under him, every muscle of his frame trembles, and swells almost to bursting, and at the given signal he goes off like the springing of a steeltrap. His unvarying success in these short races may perhaps be partly accounted for in this way, though he was undoubt-edly possessed of more than ordinary speed, and was a sharp runner.

Among the many races of this description that he ran, were two in 1796, at Brookfield, Vt., one with a horse called Sweepstakes from Long Island, and the other with a horse called Silver Tail from St. Lawrence Co., New York, both of these he beat with ease. Mr. Morgan (who then owned him) offered to give the owner of Silver Tail two more chances to win the stake, which was fifty dollars, by walking or trotting the horses for it, which was declined. There are many accounts of other races which he ran and won, but these accounts not fully agreeing as to the details, we have not mentioned them.

In harness the Justin Morgan was quiet but full of spirit, an eager and nimble traveller, but patient in bad spots; and although for a long time steadily engaged in the heavy work of a new farm, his owner at that time informs us that he never knew him refuse to draw as often as he was required to, but he pithily adds: "I didn't very often have to ask him but once, for whatever he was hitched to generally had to come the first time trying." This uniform kindness at a pull, was one of the striking characteristics of the horse, and the same trait may be observed in the greater part of his descendants. "Pulling matches" and "pulling bees," were as common in those days as short races, and the "little horse," as he was often called, became quite celebrated for his unvarying willingness to do his best, and for his great power at what is called a "dead lift."

* * *

No man of ordinary judgment could fail to discover his peculiar points of excellence, his oblique shoulders, high crest, fine ear, prominent and sagacious eye, perfect head, large and expanded nostrils, strong loins, long hip, deep and well-spread chest, high withers, short pasterns, strong and sinewy limbs, with all the important muscles far surpassing in size those of any other horse of his weight ever seen in America. The fact that this horse has contributed more than any other animal ever did, to the wealth of the United States, no honest man will deny, but strange to say, in the face of all this, the cry is still heard, *too small, too small*. This reminds us of the man who sold his hen, because she was *too small*, although she daily laid eggs of gold. We rejoice, however, that we live in a day when intelligent men cannot so easily be made the dupes of interested parties. The farming community are thinking and acting with more care and attention than formerly, they are disposed to profit by past experience, they are more close observers of cause and effect, and it is our firm conviction, that the man who is doing most to foster and encourage this principle, is the world's greatest benefactor.

Through life the Justin Morgan was steadily employed in the heavy work incident to the cultivation of a new and mountain country, and was often engaged in similar matches to those just mentioned. Even at the age of fifteen we find him entered at a drawing match that took place at Gen. Butler's tavern, in St. Johnsbury. Some of his opponents are described by persons present as large, heavy horses, yet they were all beaten by the Justin. We mention these facts to show the great muscular development of the horse, and his kind and tractable temper, rather than as an evidence of his value for purposes of heavy draught; for, although the power of an animal in starting a given weight depends more upon his form and muscular development than upon mere size, yet size is indispensable to enable a horse to move off easily upon the road with a heavy load.

The quietness and exceedingly pleasant temper of the Justin Morgan, is strikingly evidenced by the fact that he was often ridden and driven by ladies. A lady of St. Johnsbury once told us she remembered his appearance perfectly, and had repeatedly ridden him, when a girl, to balls and other parties, and spoke with much enthusiasm of his noble appearance, his high spirit, and perfect docility.

It is exceedingly difficult to obtain accurate information respecting the changes in owners that occurred to the horse at different times. To account for this uncertainty, we must consider that his fame has been almost entirely posthumous, that although the champion of his neighborhood, he was little valued, on account of his small size; and it was not until after his death, and his descendants were exhibiting the powers of their sire, in speed, strength, and endurance, in almost every village of Eastern Vermont, that people began to realize they had not properly appreciated him. For this reason, little notice was taken, at that time, of any change of owners, and many persons who very well recollect the horse, recollect nothing of these changes; and those who claim to recollect them, disagree much as to the dates at which his several owners purchased him.

We have made every exertion to collect the most reliable information in relation to this subject, but from the difficulties just mentioned, we fear our chronology may not prove perfectly accurate, but believe it to be in the main correct, and sufficiently so for all practical purposes, as the subject derives its chief importance from the curiosity felt in relation to it, by those interested in his descendants.

As we have before stated, Mr. Morgan used him almost exclusively as a riding horse, though he broke him to harness and occasionally used him in that way. After Mr. Morgan's death, he was sold by the estate to William Rice, of Woodstock, Vt. Mr. Rice used him in the ordinary work of his farm for about two years, or until 1800 or 1801, when Robt. Evans (who had been constantly on the watch for an opportunity to purchase, since he hired him of Mr. Morgan) bought him. Mr. Evans was a poor man, with a large family, and was what is called a great worker. In addition to the work upon his own place, he was constantly undertaking jobs for his neighbors — clearing land, hauling logs, building fence, &c., &c. The "little horse" was Mr. Evans' only team, and of course his labor was very severe. Mr. Evans kept him three or four years, or until 1804, when he was sued for debt; Col. John Goss became his bail, took the horse for security, and finally paid the debt and kept him. Mr. John Goss was not much of a horseman, and therefore took the horse to his brother, David Goss, of St. Johnsbury, who was quite a horseman, and made arrangements with him to keep him for a stock horse. After David had kept him a year, he was so much pleased with him that he exchanged a fine mare with his brother for him, adding cash or other property. The horse in this trade was valued at one hundred dollars. Mr. David Goss kept him seven years, or until 1811, and it was while owned by him that the Hawkins, Fenton, and Sherman horses were sired. Mr. Goss kept him almost constantly at work on his farm, with the exception of about two months in the spring of each year. While his property, although put to hard work, the horse was not overworked or abused, but was properly treated and cared for. David Goss sold him to his son Phillip, and some of his colts about Randolph having grown up and proved valuable, there was some inquiry for the horse in that vicinity, and he was accordingly taken back to that town. This was in 1811. He was now nineteen years old, and those who owned him at different times after this, generally seemed eager to get rid of him, for fear he should die on their hands. Immediately after his return to Randolph, he seems to have been taken care of by Robt. Evans, his former owner, for it was during this year that Bullrush was sired, and he was at that time in the possession of Mr. Evans. Soon after this, or in the autumn of 1811, Phillip Goss sold him to Jacob Sanderson; Sanderson sold him to a Mr. Langmade, who used the old horse hard, considering his age. He worked him some time in a six-horse team, hauling freight from Windsor to Chelsea. Under this treatment he became thin and poor, and was purchased for a trifle by Mr. Chelsea, and shortly after sold by him to Joel Goss, of Claremont, N.H. Mr. Goss kept him one year, and sold him to Mr. Samuel Stone, of Randolph. Mr. Stone kept him two or three years, or until 1819, when he sold him, and he soon after became the property of Levi Bean, who owned him until his death, which happened in the winter of 1821, at the farm of Clifford Bean, situated about three miles south of the village of Chelsea, Vt.

At twenty-nine years of age, no cause need be assigned for his death but the ravages of time and the usual infirmities of years; but old age was not the immediate cause of his death. He was not stabled, but was running loose in an open yard with other horses, and received a kick from one of them in the flank; exposed without shelter to the inclemency of a northern winter, inflammation set in and he died. Before receiving the hurt which caused his death, he was perfectly sound, and entirely free from any description of blemish. His limbs were perfectly smooth, clean, free from any swelling, and perfectly limber and supple.

Those persons who saw him in 1819 and 1820, describe his appearance as remarkably fresh and youthful. Age had not quenched his spirit, nor damped the ardor of his temper; years of severest labor had not sapped his vigor, nor broken his constitution; his eye was still bright, and his step firm and elastic.

However various may be the opinions different persons may entertain respecting the merits of the Justin Morgan, we doubt whether any horse can be instanced, in this or any other country, that has so strikingly impressed upon his descendants, to the fifth and sixth generations, his own striking and valuable characteristics; and it may be safely asserted that the stock of no horse ever bred in this country has proved so generally and largely profitable to the breeders of it. The raising of it has made the fortunes of hundreds of individuals, and added hundreds of thousands, if not millions of dollars, to the wealth of Vermont and New Hampshire.

America's First Agricultural Society

From Philadelphia Society for Promoting Agriculture, *Memoirs* (Philadelphia, 1808), I, pp. i–x.

Preface

The Philadelphia Society for Promoting Agriculture, was formed in the year *one thousand seven hundred and eighty five*, by some citizens, only a few of whom were actually engaged in husbandry, but who were convinced of its necessity; and of the assistance which such an association, properly attended to, would afford to the interests of agriculture. The society continued to meet regularly, for several years; — and published numerous communications from practical men, in the news papers of the day, on various interesting subjects; and thereby contributed to diffuse the knowledge of many improvements in agriculture; the general adoption whereof, has visibly tended to increase the product, and to improve the qualities of the soil of Pennsylvania.

The continuance of a long war with Great Britain had effectually precluded all friendly intercourse, and prevented the receipt of all information from that country, (in a language generally understood here) not only of the improvements in agriculture there existing, but of those in other European countries, wherein the practice and principles of good husbandry are universally attended to. The system generally pursued here at that time, was bad in the extreme. It consisted in a series of exhausting grain crops, with scarcely any interruption, for several years; after which, the land was abandoned to weeds and natural grass, under the fallacious idea of rest; and, when completely worn out, new land was cleared, and the same wretched system pursued. A natural meadow, or one artificially watered, supplied more or less of hay; but where these resources were wanting, the purchase of winter fodder was made from the hard earnings and savings in other products; or the poor animals fed on straw, and the scanty pickings in the fields. — Since the introduction of *red clover*, and other *artificial grasses*, a great and obvious change has taken place; and the most beneficial consequences have followed. The comforts of the farmer are greatly increased, and abundant supplies of summer and winter food for all domestic animals, are furnished. Thus, by the manure obtained, ample means are afforded, of renewing the original strength of the soil. Among other measures tending to produce this happy alteration, the general use of *gypsum* may be mentioned, as one of the most important: for although this substance had been introduced many years before the date of our institution, yet its use was chiefly confined to the vicinity of Philadelphia. The society reflect with patriotic pleasure, upon their agency in diffusing more extensively the knowledge of its effects upon land; and in assisting to dispel the prejudices which unfortunately prevailed against it, by the publication of the communications of practical men, containing the result of their experience with that valuable substance.

Premiums were also proposed and conferred, for the elucidation of subjects upon which information was required, for the adoption of approved systems and modes of European culture, and practices, and for the improvement of certain articles of domestic manufacture. Among the latter, *cheese* may be mentioned; for the best sample of which, and greatest quantity, a gold medal was presented to Mr. *Mathewson of Rhode Island*, in the year 1790; the consequence of this distinction by the society, was a laudable competition among dairy men, and an increased demand, owing to the striking improvement, in the quality of the article, and a rise in price, so as amply to reward, and extend the manufacture, and in a great degree, preclude the necessity of importation. At the present day no occasion exists, for the importation of *cheese* from *Europe*, for general consumption, or as an indispensable supply. Importations on a less scale, continue to be made, but these are in a small proportion to the quantity produced, and manufactured from our own dairies.

After several years of active exertions, the society was unfortunately permitted to fall into a long sleep; but was again revived, in the winter of 1804, and now holds regular meetings. New subjects for premiums have been proposed, as will be seen by the present volume, and have been several months in circulation: numerous communications have been received; from which those now published, are a selection; and some papers before published are added; as being thought worthy of preservation, in our collection. As it is the wish of the society to pursue its labours, with all the zeal due to the importance of the object, for which it was instituted, the communications of all practical agriculturists, upon whose support the usefulness of the Society will in a great measure depend, are earnestly solicited. The example being once set, will be followed by others; and thus, a body of information will be collected; which may essentially benefit the country. The pursuits of the industrious farmer, being more of a practical than a literary nature, he may be induced to think that he is not qualified to give a written account of his improvements, but let not such be backward. The Society are in want of facts, and they care not in what stile of language they are communicated. *Criticism* is missapplied, and out of place, on such occasions. The communications of philosophical and literary characters, on any points contributory to the elucidation of subjects connected with agriculture, will be highly beneficial and gratifying.

Two subjects, in particular, are deemed worthy of great attention, from all concerned in agriculture; and on these the society would gladly receive information: viz. *on the diseases of our domestic animals, and, on new manures*; on both these subjects, very interesting papers will be found in the present volume. A great object in American husbandry, is the improvement of *horned cattle:* the society will therefore receive, with thanks, all information respecting any domestic breeds of neat cattle, sheep, and swine, which have been found to possess peculiar good qualities: and they strongly urge the necessity of preserving, for breed, all those, even of accidental offspring, possessing the desirable and requisite qualities, to entitle them to value and preference. Thus a breed of neat cattle, producing oxen, remarkable for speed of gait and strength, symmetry of form, and gentleness of disposition; and a tendency to fatten quickly, and to increase of flesh and fat, upon those points which recommend them at market, are to be attended to. It is well known, that the diversity in these respects is great, and constitutes the ground of important improvements, by various spirited farmers in *Europe*. And as in many parts of this country, occasional instances of very excellent breeds are to be found, the society think they will render service to the community, by calling the public attention to the subject. It must be acknowledged that the common American oxen fatten well, that they grow to immense sizes; and that as fine samples of beef, are every day to be met with in the markets of *Philadelphia*, as in any other part of the world. But as respects cows, we are much deficient, a circumstance which is the more to be regretted, as probably in no country, does the article of butter, yield greater profit than in the *United States*. Some attempts have indeed been made, to improve our stock, by the importation of bulls and cows, particularly in Maryland and New York; but the public generally, are not yet informed of the success, which has attended the experiment; and whatever may be the result of imported brood animals, the great price at which they must necessarily be held, to remunerate the concerned, for the heavy expences of importation; will prevent the desired benefit from being speedily or generally derived from them. This circumstance ought to operate as an additional reason, for a careful selection of the most valuable animals from our domestic stock, and for the preservation of such others as we may occasionally meet with.

With respect to *sheep*, the objects to be attended to are in part common, with those first noted as to *oxen*. Within a few years, large sizes were chief objects of attention in *England;* but repeated experiments have shown, that they are not so profitable, as those of a moderate size.

The fortunate introduction of the *Spanish, English*, and *Barbary sheep*, all of which are now spreading through the middle States, may be considered as important acquisitions to the agricultural interest. With regard to the *Spanish sheep*, it is found by years of experience, that the cross with *American ewes* produces a healthy, hardy, gentle race, which fatten more speedily than the pure *American* blood; do not loose their wool, when shearing has been neglected beyond the usual time, and do not become diseased when fat. The fine quality of the wool is known to all the world; and what is of great consequence, the weight of *fleece* of the cross with *American* ewes, is evidently increased, when compared with the imported sheep. The same increase takes place in the cross with the English sheep. It may be well to add, that the wool of sheep from the *Spanish* cross, exhibits the most evident marks of improvement; this adds another proof to the many which all parts of the world furnish, that the prejudice respecting the peculiar nature of the climate of *Spain*, being exclusively calculated to produce fine wool, is erroneous.

We owe the introduction of the Barbary mountain sheep, with broad tails, to our gallant countryman, William Eaton, who, when Consul at Tunis, sent them in an armed vessel in the service of the United States, commanded by Henry Geddes, to Timothy Pickering then secretary of state, who presented a fine ram and ewe to the President of our society, from whose disinterested zeal, this valuable breed is now spreading through the State of Pennsylvania, and other States in its immediate vicinity. The wool of those sheep, owing to their health and vigour, does not fall off, like the fleeces of those meagre and degenerate runts, which are too frequent here; it is moreover, in general, of a good staple, and next the skin, peculiarly soft and furry. The weight of the sheep is above mediocrity, but their chief excellence arises from their hardihood, and disposition to fatten speedily; a quality they possess in a remarkable degree, which causes them to be highly valued, both by the grazier and butcher. Hatters, who are acquainted with it, prefer it, for their manufacture, to any other wool. It spins free, and to any fineness. Flossy, fine and well dressed cloth, has been made of it. Those who have worn fleecy stockings, and gloves, of this wool, speak of it with great approbation. Perhaps a cross with the *Merino*, would benefit both.

We possess several valuable breeds of swine; but none, except the *Chinese* and *African* breeds, are distinctly marked. Both these breeds are remarkable for fattening speedily, but their deficiency of flesh, lessens their value, when preserved pure. They both therefore answer best when crossed with our native breeds; as their progeny take on a disposition to diffuse the fat through the flesh, which is also increased; instead of being laid thick on the outside. The Chinese hogs are very prolific, but have thick skins, and therefore not so profitable or delicate for roasters as the African breed, which have remarkably thin skins: these latter will weigh ten pounds at the age of four weeks, and will then bring one dollar twenty five cents at market.

If we have not published all the communications with which we have been favoured, it is not owing to a want of respect for them, or their authors. But our means are yet limited; and our society is only emerging from a state of torpor, into which past circumstances had thrown it. We selected *subjects* rather than *essays;* and risk this recommencement of our well meant endeavours, to promote the happiness and prosperity of our country, with no motive either of personal fame, or interest. Should this attempt be favourably received, and our exertions adequately supported; we have strong hopes, that the usefulness of our association will be extensively experienced. We cannot be disappointed in the satisfaction we feel in having made an effort, to attain a desirable object, however feeble, it may be found, as it respects us in its means, or result. It will at least set an example; and invite men of talents and practical experience, to add to our scanty stock of knowledge, on the important subject of our institution. Those who have enabled us, by their communications, to fulfil our wishes, in the objects we endeavour to attain, merit and receive our sincere acknowledgments; and we are persuaded, have entitled themselves to the grateful attention of those, whose interests they are calculated to promote. It is equally useful to us, to be supplied with information, either new or not generally known here; whether it be obtained by those who impart it, from reading, travel, or original thoughts or practice. As other countries receive the benefits of our labours, in the products supplied to them, through the channels of an extensive and prosperous commerce, it is fit that we should profit by their experience in the arts of cultivation; by which those products will be brought forth more advantageously to us, and beneficially to them, both in quality and abundance. Those who introduce among us, the improvements of foreign countries in agriculture, and the arts and sciences with which it is intimately connected, effect a reciprocity of accommodation. It depends on the good sense and practical attention of our farmers, to adapt them to our climates, soils, habits, and actual circumstances. All foreign practice or improvement, will not suit our situation. We cannot furnish labour, or afford expence, beyond a certain point; but the principles will apply in all countries, and when they are developed, in a plain and intelligible manner, they may, in a greater or less degree, be practiced upon, and fitted to the actual state of things here, so as to produce incalculable and permanent advantages. Although much benefit has been, and will continue to be derived from *European* models, and examples in husbandry; it is with pleasure we observe, that from our own resources, we grow more and more independent of foreign aid. The knowledge of both principles and practice in *agriculture*, is daily increasing; and the general mass of agricultural improvements is evidently advancing throughout our country. Nothing will more conduce to the attainment of the great object of those, who desire to promote this most essential of all arts, than associations to receive and communicate information, on this important subject. Let these be devoted entirely to agricultural enquiries and pursuits; and avoid all topics which are productive of dissension, and calculated to withdraw their attention from the objects of common concern. A small collection of *Books* and *Models*, are attainable at little expence, with some judicious attention in the choice of them. These will be sources of information and useful amusement, as well as cements of union, and means of gaining and diffusing knowledge, auxiliary to practice. A community of interests, may be thus established; mutually supporting and supported, informing and informed; and nothing contributory to the benefit of the whole, will be omitted or lost. Public aid has been so often sought in vain, that private exertions must be redoubled. To this end, a zeal for agricultural knowledge, and practical improvements, must be rendered fashionable, that it may become general and characteristic. Those who seek for personal distinction in our government, and those who from disinterested and virtuous inclinations, perform duties the most honourable to themselves, and beneficial to society; will find the most solid *popularity* and *durable fame*, in measures promotive of the interests (always inseparable from those of commerce and the arts) of agriculturists; who compose the great body of the people. This will shew itself in public improvement; in which the efforts of individuals will be aided and cherished by legislative patronage, and pecuniary support. Our state will then hold its proper rank among our neighbours; and our natural and local advantages, remain no longer inactive. Roads and inland navigation, will be primary objects of legislative attention. The arts of husbandry will be assisted, supported, and rewarded: public men will be popular and eminent, in proportion to the services they render to the leading interests of their country. These, most assuredly, are those of agriculture, and the arts and sciences, all of which are intimately, and indissolubly connected. Our eyes will then be opened, to the sources of wealth and prosperity, which are properly our own; easily attainable, ample, and inexhaustible: and it will no longer be left to the discernment of the intelligent in rival states, to perceive, and take advantage of our culpable blindness, negligence, and mistakes.

Premiums Proposed by the Philadelphia Society

I. The rotation of crops having been found in England constantly to improve the soil instead of exhausting it — and the society being persuaded, that to this management alone is to be attributed the great comparative products of that country — they esteem it of the first importance to America to gain a knowledge of the theory and practice of so admirable a system. — Within the limits of this article, it is impossible to state, with any useful degree of precision, principles, which, after all, must vary with circumstances — but knowing that some farmers, in Pennsylvania and elsewhere, have already made themselves acquainted with this mode of husbandry; and that it is as much the interest, as it is within the power of all to obtain the necessary knowledge — the society, without attempting to lay down any particular directions, offer — For the best experiment of a five years course of crops — a piece of plate, of the value of two hundred dollars, inscribed with the name and the occasion; and for the experiment made of a like course of crops, next in merit — a piece of plate, likewise inscribed, of the value of one hundred dollars.

II. The importance of complete farm or fold-yards, for sheltering and folding cattle — and of the best method of conducting the same, so as to procure the greatest quantities of compost, or mixed dung and manure, from within the farm, induces the society to give for the best design of such a yard, and method of managing it, practicable by common farmers — a gold medal: and for the second best — a silver medal.

III. For the best method of raising hogs, from the pig, in pens or sties, from experience; their sometimes running in a lot or field not totally excluded, if preferred — a gold medal; and for the second best — a silver medal.

IV. For the best method of recovering worn out fields to a more hearty state, within the power of common farmers, without dear or far-fetched manures; but by judicious culture, and the application of materials common to the generality of farms; founded in experience — a gold medal; and for the second best — a silver medal.

V. For the best information, the result of actual experience, for preventing damage to crops by insects; especially the Hessian-fly, the wheat-fly, or fly-weevil, the pea-bug, and the corn chinch-bug or fly — a gold medal; a silver medal for the second best.

VI. For the best comparative experiments on the culture of wheat, by sowing it in the common broad-cast way, by drilling it, and by setting the grain, with a machine, equi-distant; the quantities of seed and produce proportioned to the ground, being noticed — a gold medal; for the second best — a silver medal.

VII. For an account of a vegetable food that may be easily procured and preserved, and that best increases milk in cows and ewes, in March and April, founded on experiment — a gold medal; for the second best — a silver medal.

VIII. For the greatest quantity of ground, not less than one acre, well fenced, producing locust trees, growing in 1791, from seed sown after April 5th, 1785; the trees to be of the sort used for posts and trunnels, and not fewer than 1500 per acre — a gold medal; for the second — a silver medal.

IX. The society believing that very important advantages would be derived from the general use of oxen, instead of horses, in husbandry and other services; and being desirous of facilitating their introduction into all these states; persuaded also, that the comparative value of oxen and cows must very much depend on the qualities of their sires and dams; and that by a careful attention to the subject, an improved breed may be obtained; they propose a gold medal for the best essay, the result of experience, on the breeding, feeding, and management of cattle, for the purpose of rendering them most profitable for the dairy, and for beef, and most docile and useful for the draught; and for the next best — a silver medal.

N. B. Among other things the essay should notice the different breeds of cattle, and their comparative qualities; as their sizes, strength, facility in fattening, quantity of milk, &c.

X. It is a generally received opinion, that horses in a team travel much faster than oxen; yet some European writers on husbandry mention many instances, in which it appeared, not only that oxen would plough as much ground as an equal number of horses, but also travel as fast with a loaded carriage: particularly when, instead of yokes and bows, they were geared in horse-harness, with such variations as were necessary to adapt it to their different shape. To ascertain the powers of oxen in these particulars, and the expence of maintaining them, the society deem matters of very great moment; and are therefore induced to offer a gold medal for the best set of experiments, undertaken with that view; and for the next best, a silver medal. In relating these experiments, it will be proper to describe the age and size of the oxen, their plight, the kinds and quantities of their food, the occasions, manner, and expence of shoeing them; in travelling, the kinds of carriages used, and weight of their loads, and seasons of the year, and the length and quality of the roads: and, in ploughing, the size and fashion of the plough, the quality of the soil, the depth of the furrows, and the quantities ploughed: and, in every operation, the time expended, and number and sorts of hands employed in performing it; with any other circumstances which may more fully elucidate the subject. These experiments will enable the essayist to determine what will be the best form and construction of yokes and bows, and what of ox-harness, to enable oxen, with the best carriage of their bodies and heads, the most ease, and quickest step, to draw the heaviest loads, a description of each of which sort of gears, explained on mechanical principles, must be subjoined to the account of experiments.

XI. For the best method, within the power of common farmers, of recovering old gullied fields to an hearty state, and such uniformity, or evenness of surface, as will again render them fit for tillage; or where the gullies are so deep and numerous as to render such recovery impracticable, for the best method of improving them, by planting trees, or otherwise, so as to yield the improver a reasonable profit for his expences therein, founded on experiment — a gold medal; and for the next best — a silver medal.

XII. For the best cheese, not less than five hundred pounds weight, made on one farm within the United States, and which shall be produced to the society by the first day of January, 1792 — a gold medal — and for the next greatest quantity, not less than two hundred and fifty pounds weight, of equal quality — a silver medal.

XIII. The society believing that the culture of hemp on some of the low rich lands in the neighbourhood of this city, may be attempted with advantage, do hereby offer a gold medal for the greatest quantity of hemp raised within ten miles of the city of Philadelphia. The quantity not to be less than three ton; for the second greatest quantity — a silver medal.

It will be left to the choice of those successful candidates for prizes, who may be entitled to the plate or gold medals, to receive the same either in plate or medals, or the equivalent in money.

The claim of every candidate for a premium is to be accompanied with, and supported by, certificates of respectable persons of competent knowledge of the subject. And it is required, that the matters, for which premiums are offered, be delivered in without names, or any intimation to whom they belong; that each particular thing be marked in what manner the claimant thinks fit; such claimant sending with it a paper sealed up, having on the outside a corresponding mark, and on the inside the claimant's name and address.

Respecting experiments on the products of land, the circumstance of the previous and subsequent state of the ground, particular culture given, general state of the weather, &c. will be proper to be in the account exhibited. Indeed in all experiments and reports of facts, it will be well to particularize the circumstances attending them. It is recommended that reasoning be not mixed with the facts; after stating the latter, the former may be added, and will be acceptable.

Although the society reserve to themselves the power of giving, in every case, either one or the other of the prizes, (or premiums) as the performance shall be adjudged to deserve, or of withholding both, if there be no merit, yet the candidates may be assured, that the society will always judge liberally of their several claims.

Premiums Proposed by the Agricultural Society of Philadelphia for the Year 1806.

1. *Ascertaining the component parts of arable land.*

To the person who shall produce the most satisfactory set of experiments, to ascertain the due proportion of the several component parts of arable land, in one or more of the old counties of this state, by an accurate analysis thereof. A like analysis in detail must also be made of the poorest, medium, and richest soils, in the same county or counties. By a due admixture of these soils, or substances within the reach of common farmers, they are by these experiments, to be enabled to improve, by good tillage, and a course of applicable crops, the poorest or most worn land, with the materials found on their own farms, or those of their neighbours respectively. — Lime, or lime stone, is excluded, its qualities and effects being already well known. But clays, marles, gypsum and sand, or other natural substances, fall within the meaning of this proposal. The crops, so far as consistent with good husbandry, to be the same after improvement as before, and their relative product to be given. All auxiliary, and influencing circumstances to be mentioned; as well as the mode and results of the analysis; and the proportions of the combinations. Artificial manures, after improvement, (lime at this stage may be one) may be used, if the like had been before applied: and all the means and circumstances are to be fairly developed. A piece of plate of the value of one hundred dollars.

The object is, not only to promote experiments calculated to improve farms, out of the materials found upon them; and thus save, or extend the efficacy of artificial manures; but to excite a spirit of exploration for fossils, earths, marle, and clays, applicable to agricultural as well as manufacturing purposes. For subterraneous researches, the society have provided a very complete set of boring instruments, with which those who will use them effectually, may be accommodated.

2. *Trench Ploughing.*

For the greatest quantity and best trench ploughed worn land, not less than five acres. The trenching not less than ten inches deep.

The following mode of trenching is recommended, as being known to be practicable, and easily performed.

1. Provide a light plough, from 12 to 15 inches wide in the hind part of the span or sole, calculated to pare off the sod from 2 to 3 inches deep, according to the depths of the roots of weeds.

2. A strong heavy *Trench Plough*, capable of turning a depth of from 8 to 10 inches of mould, or earth. This must be one or two inches narrower than *The Paring Plough*, or it will cut into the unpared sod. The first is to be drawn by a pair of horses or oxen. The second by two pair of oxen, or strength equivalent. A *Trench* must be first made, with the *Trench Plough* as deep as practicable. The *Paring Plough* must then pare the sod off the next intended furrow, and turn it into the trench. The *Trench Plough* follows, constantly, after the *Paring Plough*. This throws over a body of earth so as to bury all weeds, which are placed too deep for vegetation, and thus, by rotting, become manure. The mould board, of the Trench Plough, should have a thin plate of flexible iron (an old stone-saw the best) screwed on its upper edge, *vertically*, so as to extend the surface and accommodate itself to the curvature of the mould board. — With this auxiliary, the loose earth will be completely thrown into the trench. It is otherwise liable to run over, and choak the Plough. Both Ploughs (the latter the most) require bridles, or clevasses with notches and curvated regulators, to direct and fix both their depth and lateral course. Such are not uncommon. The east Jersey, or low Dutch plan, is the best for the Trench Plough. A Coulter is not much required.

This operation should be performed in the autumn, and the field lay through the winter, to attract from the air, whatever is the food of plants; and to receive the benefits of frequent frosts and thaws. The subsequent ploughing need be no deeper than usual in good tillage. If *limed* the first spring for *Indian Corn*, the better it will produce. A fallow crop *only* should succeed the trenching the first year; and *Corn* admits and requires frequent stirring and exposure of the soil. For the best experiment, *a gold*, and for the second best, *a silver* medal.

3. A course on trench-ploughed ground.

For the best and cleanest course of crops, on not less than five acres of land *trench-ploughed*. The course may be. 1. Indian corn. 2. Legumes. If beans or pease, of a species least subject to the bug; and sown on the fallow of the 2d year, so as to be off in time for a winter crop of wheat or rye. Broad cast of the legumes as a cover, be preferable; though drilling will be highly useful. Potatoes may occupy a part, to be taken off in time for wheat. 3. Clover sown in the winter grain. 4. Clover. This course will be preferred in a competition, unless the society shall be convinced, by the results of another course, that in practice, turns out better. Manure admitted; but the best products, with the least artificial manure, will be preferred. A gold medal for the best; and one of silver, for the second best experiment.

The object of both the above premiums is, to introduce a practice, found very beneficial where it has been fairly tried; and to place the experiments in the hands of spirited and intelligent agriculturists, who will do complete justice to themselves, and the subject recommended to their exertions.

4. Cover of Leguminous Crops.

For the best and greatest crops of beans, pease, or other *legumes*, of the kind before mentioned, sown broad-cast, as covering on fallows, preparatory to winter grain. Not less than five acres, and left clean and fit for wheat. These crops ameliorate, and do not exhaust like all culmiferous plants and those whose seeds produce oil. *Oats* — the worst and most ruinous to succeeding winter crops.

The object is, to introduce the practice of valuable and improving covering crops, in preference to naked fallows, or exhausting covers. A silver medal, or fifty dollars.

5. Destruction of Perennial Weeds.

For the best set of experiments calculated for the destruction of *perennial weeds*. The *daisy* or *May weed, ransted, garlic,* and *St. John's wort*, to be particularly aimed at and noticed. A *botanical account* of the weeds commonly infesting our fields, will highly recommend these experiments; and communications, relative to all or any of those enumerated, will be gratefully received. This account should specially mark the stages of their growth; and periods when they are the most easily destroyed, by the means employed. *Botanizing* for the destruction of weeds, is as necessary and laudable, as it is for the propagation and culture of useful plants. — Nothing promotes the health, increase, and value of the latter, more than expelling the former. *Trench ploughing* is excluded. This has been found to be the surest mode of destroying weeds; especially those with fibrous or bulbous roots. A gold medal.

6. Dairy.

To the person who shall exhibit to the society an account of the profits of the best *dairy*, applied to *butter* or *cheese*. Not less than twenty cows. The greatest proportion of cows kept the longest in profit, and the best. Winter feed (œconomy considered) for carrying the cows productively through the season, enters into the account. The greatest product from an equal number kept without change (except by substitution of well bred heifers raised on the farm) through the year, will have the preference. It is to be understood, that changing cows is not to be admitted, unless full proof, on the annual balance of account, that such practice is comparatively the most productive and profitable, when in competition with one predicated on keeping the same set of cows through the year. The same profits from the permanent dairy (unavoidable casualties allowed) will be preferred. It will be recommendatory of the pretensions of the claimant, if the account be accompanied with experiments, or practical knowledge of the best sizes, description, breed, and ages of dairy cows.

The object is, to induce an attention to the breed and selection of *dairy cows*. Their points and qualities differ from those proper for breeding beef cattle, or for venders of milk. Much depends even with the best stock, on regularity and attention in the dairy women. Unless great care in stripping, and regular periods of milking, are practised, as well as cleanliness in keeping, the best cow will soon cease to be in profit. The quality, and not the quantity of milk is the most important. Nor are the largest the best for the dairy: especially where there are short bites and irregular seasons. A silver medal, or fifty dollars.

7. Live Fences.

For the best experiment on, or practical application of, any species of shrub or tree proper for live fences; and the most economical and practical mode of securing them in their early stages of growth, from injury by cattle or other enemies.

The general idea of European agriculturists has been confined on this subject, to *thorn* or *quick* inclosures. But these may not be found exclusively the best *here*. On *Long Island*, before the revolution, a very able and spirited proprietor of a large estate there, went very extensively into inclosures with quick set, procured not only in this country, but from Europe and elsewhere. He found the thorn, of every description, subject to many casualties and diseases; some of them unknown in Europe. Blights injured a great proportion, after they were in sufficient growth for inclosure without protection. It was not frequent that a sound crop of haws was produced; these being subject to the worm, and other impediments to their perfection. Although it is still desirable, that every attention should be paid to the *hawthorn*, it is not improbable that some other of our native shrubs or trees, may thrive as well, if not better; and equal the thorn in utility. The object therefore is, to promote enquiries and experiments that shall determine this point. The *walnut*, the *apple*, the *honey locust* (*Gleditsia triacanthos*) the *white flowering locust* (*Robinia pseudo-acacia,*) have been tried, on a small scale — Each has its peculiar disadvantages. The white mulberry has also been recommended.

Live fences are of such high importance, in our old settlements, where the timber is daily decreasing, and the expence of inclosure becoming so very serious, that the society cannot sufficiently express their wishes, that some spirited and extensive measures may, without loss of time, be commenced on this momentous subject. The present generation may receive incalculable advantages from successful experiment and practice, in a desideratum so eminently interesting to them. But posterity will bless the memory of those, of whose genius and labours they enjoy the fruits. They will gratefully feel the benefits of durable inclosures, commenced, if even not entirely perfected, in our day: and while they inherit these safe guards to their property, they will perceive the insurmountable difficulties to which they would have been exposed, by a neglect on our part, to establish and provide them.

A gold or silver medal — according to the merit and extent of the experiment or practice.

8. Clearing and Cropping new Lands.

For the best essay, practical and theoretical, founded on experience and facts, as well as calculation and investigation, of the most approved and beneficial mode of clearing and cultivating new settlements, in an unseated, and theretofore uninhabited part of this state, or one in its neighbourhood. A gold medal.

The practice heretofore used of *girdling* trees, can only be justified by the necessity of doing it, through want of labourers, by those who first enter a wilderness. But if lands are inviting, population soon increases, and yet the practice of girdling the timber continues. One part is *girdled* after another, without foresight or precaution. Timber is wantonly, because lavishly and unnecessarily destroyed; and becomes in a few years scarce, where its abundance was at first accounted a burthen. *Culmiferous crops* [plants composed of *straw* and *chaffy husks* for the grain] follow one another in uninterrupted succession, the worst of all bad husbandry. — These are *"stubbled in"* (the phrase of new settlers) till the land is exhausted, and produces nothing but *sorrel* and other execrable vegetation. The timber rots and falls, sometimes dangerously to men and cattle. It is burnt and destroyed, when the field, after a useless waste of time, is cropped again. Fencing, fuel, building, implements, &c. call for timber — but it is distant or gone. The field is choaked with briars, worthless shrubs, and other pests, and its cultivation is generally more expensive than if well cleared originally, and occupied by wholesome and productive crops, either of grain or grass.

Many of us are interested in new lands — and all of us, from public motives, wish to introduce a better stile of clearing and cropping into our new countries. Information from several new settlements (particularly some in the state of New-York) is favourable to a far better plan, of both clearing and cropping. It is, to till less ground cleared perfectly; and crop, according to circumstances, as near as practicably to the rules of good husbandry. Labourers are not there in greater plenty, than elsewhere, in such settlements; and yet the settlers succeed and thrive.

Our object is therefore, to obtain and promulgate every species of information; and thereby be enabled to recommend and encourage better modes of clearing, and a more advantageous, as well as reputable stile of husbandry, in our new countries.

There are in these countries, many intelligent citizens, who may, and it is hoped will assist in both example and investigation. But some of these have not correct ideas on this subject. They conceive that the art of husbandry, for the most part, consists in restoring, or creating fertility, which in new lands is the gift of nature. But the fact is, that fertility without good management, like a savage in power, and subject to no civilized regulation, as often exerts itself mischievously as profitably. It frequently ruins by desultory and misapplied operations. Weeds and other worthless products, are its offspring. These, in many cases, might be prevented, destroyed or converted into benefits, with well directed systems. To instance only the *sorrel* apparently the most mischievous and forbidding. It has been found that with *lime*, it may be made a powerful and efficient auxiliary to profitable crops, and when judiciously applied is known in Europe to be so valuable, that the sorrel is propagated for its uses in husbandry. Limestone is found abundantly in most of our new lands, or at least, in very extensive districts. Careful experiments may point out the mode of liming lands overrun by this apparent pest, so as to destroy its bad qualities, and convert it to salutary and profitable purposes. If this be not now deemed eligible in parts where land is less valuable than *labour*, it will nevertheless be an object e'er long, when the products of land are unattainable, without combinations of labour with ingenuity, good management and appropriate systems of husbandry.

9. Veterinary Essay and Plan.

For the best *essay and plan* for promoting *veterinary* knowledge and instruction, both scientifically and practically, *under the circumstances of our country*. Aid to schools and establishments for this, among other agricultural purposes, ought to be given by the national and state legislatures. But *agriculture*, and the subjects connected with it, have not heretofore been cherished by their patronage. Her young sister, *commerce*, has fortunately fascinated with contributions to revenue, and thereby secured protection and encouragement. But private and individual exertions, for the accomplishment of agricultural objects, must, from necessity, be resorted to, for public benefits derived from this primary source of all the wealth and prosperity we enjoy. — Some of the most worthy and truly respectable governments, and many of the most eminent men, in Europe, have deemed the object here recommended, honourable, politic, and promotive of the public interest and prosperity. While agriculturists are employed in the production of *plants*, their stocks of useful *animals* are abandoned, when diseased, to all the calamities attendant on ignorance of their maladies, or cure. Pretenders and empirics, of the most contemptible characters, prey on the necessities and credulity of those who are compelled to apply to them on this subject.

The essay proposed, should among other requisites, be calculated to rouse the attention of medical professors, to this important branch of neglected knowledge. It should convince them, that they cannot employ themselves, in any part of their studies, in a manner more conducive to *real* respectability of character, than in gaining and promulgating information, so intimately connected with the wealth and political œconomy of their country. This society pledge themselves to distinguish, with some testimony of their gratitude, any medical professor who will assist them in calling the attention of students, to this very interesting subject.

Investigations into *anatomy, diseases* and *remedies*, for the preservation and improvement of animals, on which our subsistence and comforts so materially depend, must assuredly be considered worthy the most patient enquiry, intelligent observation, and professional talents, of the most celebrated among those, who have devoted themselves to medical pursuits. As patriots, it should stimulate their public spirit. — As professional men, nothing can more entitle them to the rewards due to their labours. Who is there among the most respectable of our own citizens, or in the highest grades of society in the old world, who has not deemed it meritorious to promote the interests of agriculture? And is there any branch of that occupation so important, as that now recommended to the notice and enquiry of medical men? If it has held an inferior rank in the classification of science and knowledge, it is entirely owing to the unmerited neglect with which it has been unaccountably treated. It is time it should be rescued from obscurity, and placed among the most commendable and necessary branches of medical education. A gold medal.

10. Domestic or Household Manufactures

For the best and greatest quantity and quality of woolen, cotton or linen fabrics, made in any family, by the members thereof. Weaving, fulling, and dressing, may be done as usual, in the accustomed modes of performing these operations. The object is, to encourage industry in the families of farmers and others, at times when leisure from other occupations permits. Such intervals are too often filled up with dissipation, or suffered to pass away in indolent waste or inattention. The materials being raised or produced on the farm, will entitle to preference in a competition. The breed of sheep, and quality of wool, will be peculiarly recommendatory. A silver medal.

Although the society have principally confined their premiums to honorary distinctions, they will always be ready to commute them for, or add pecuniary reward *to assist* in expensive or difficult experiments. Our *funds* are far below our *zeal*. *But the former* are not of so much moment, as energies excited by emulation, among those who have strong propensities to benefit their country, while they are labouring for themselves. Without the co-operation of our fellow citizens of this description, all our well meant endeavours are vain!

For rules respecting claims — See the laws, art. 14.

RICHARD PETERS, *President*.
JAMES MEASE, M. D. *Secretary*,
No. 192 Chesnut-street — to whom communications may be sent.

Experiments with Gypsum, 1809

From John Taylor, "On Gypsum," in Philadelphia Society for Promoting Agriculture, *Memoirs* (Philadelphia, 1811), II, pp. 51–62.

Port Royal, Virginia, January 30th, 1809.

DEAR SIR,

I have postponed answering your two obliging letters of May last, hitherto, lest the very great pleasure of your correspondence, should seduce me to be troublesome, in the number or length of my letters; and I fear you will allow my apology to be a good one, before you get to the end of this.

Your warning against a reliance on gypsum, and neglecting manure, induces me to give you an idea of my practices respecting both, in hopes of obtaining your corrections. For many years I have enclosed all my arable land at each farm, in one enclosure, and excluded grazing entirely, leaving the whole vegetable matter the land produces, to return, taking a crop of corn and one of wheat, every three or four years. To increase vegetable cover I sow large fields of clover, cutting only a small proportion for seed and for feeding green. These are treated with plaister, and the clover is plowed in dry, when the field comes into culture. It is cheaper to plough it in dry than green, on account of the different seasons of the year for the operations; and however contrary to theory it may be, my experiments have not satisfied me that it is less nutritive to the earth, especially when well clothed with these vegetables until towards the end of the winter. This cover added to the exclusion of the hoof, keeps its pores open longer in winter to the action of the atmosphere, than ploughing itself; and defends it against sun as well as frost, whilst ploughing exposes it to both, my idea is, that this system is fitted for a combination with gypsum, and that such a combination may possibly succeed without the aid of manure. If so, it may be useful towards diminishing the deficiency of that article, for I agree with you, that nothing can be a complete substitute for it. To make the experiment fairly, I have set aside 200 acres, half to be cultivated in corn yearly, half to lie uncultivated and ungrazed, and the whole to receive an annual dressing of three pecks of plaister to the acre. The repetition of the culture being too quick for a perennial plant, I use the bird-foot clover as we commonly call it, to raise clothing for the land, having found that the plaister operated as powerfully on that as on red clover. This grass rises early, dies soon in the summer, abounds in seed so as to set the land thick the following year, affords a good cover, and nourishes a second annual, the crab grass, which springs through it the latter end of the summer, and gives a fresh cover to the earth. This experiment of combining the use of plaister with enclosing has hitherto been very flattering.

As to corn stalks, for about 26 years past I have reduced all mine to food, litter and manure. But my experiments reject the use of cutting boxes, after trying the best for a long time, on account of the expence and inutility of the labour. The expence on a very small farm is not seen; on a very large one, it is felt at once. On mine, the removal of stalks, straw, corn shucks, cobs and tops to the places of consumption, is nearly sufficient for the winter's work. To cut the stalks and straw, would employ the whole labour of the farm. If a good farmer ought to have a vast surplus of dry vegetable matter for litter, beyond what is necessary for food, why should this expence be incurred? Is it not cheaper to feed in waste, and let the waste go for litter? It is with difficulty I reduce this coarse food to manure and apply it in the spring. If the stock is increased and made to eat it, the manure is diminished, and the additional stock is soon killed by the want of a dry warm bed, and a deficiency of summer pasturage.

I find corn stalks gradually became less valuable as food and litter, the longer they stand, therefore I begin to use them as soon as I begin to gather corn, by removing every day the weather will permit, about eight or ten heavy wagon loads, into the stable yard and farm pen; keeping a parcel near each to resort to in less quantity when the weather is bad. Horses and mules thrive better at this crisis, than at any other time of the year. Whether the saccharine juice of the stalk agrees better with them, or whether it is owing to their being able to masticate more of it than the cow, who is chiefly confined to stripping it, they seem to thrive better on this food than horned cattle. Between two and three thousand load of manure is made on the farm I live on, chiefly of corn stalks. It accumulates in the yards until the winter is over, and is never disturbed until the moment it is to be used. This is always in April immediately after the corn, save what is to occupy the land to be manured, is planted. The manure is carried out, spread on land fallowed in winter, that it may separate easily and mix well with the coarse manure; a bushel of plaister to the acre is sown on it after it is spread, and it is ploughed in, all on the same day. I have frequently for experiment, left my manure longer periods to rot, undisturbed — made up into large dunghills — mixed and unmixed with earth — covered and uncovered, and in all have suffered a loss of labour and manure, in proportion to the deviation from my present practice. When manure is suffered to lie to a second year, I think its loss exceeds a moiety. The best instrument for raising and scattering this coarse manure which I have seen, is a hoe, in its eye, shape, helve and dimension, precisely like what is called here a hilling hoe, but having three strong prongs in place of a blade. These prongs pierce the manure by the fall of this forked hoe, it is taken up without stooping, in as large a parcel as the labourer can manage, and shaken into the waggon by suffering the helve of the hoe to fall gently on its top piece.

You ask me the cause of the black heads of wheat in the forward kinds I sent you. They are frequent with us. And the forward is more liable to them, than the later wheat. But in no instance have I known them to produce a material injury to the crop. The infected heads perish young, and communicate no distemper to their neighbours; and the number is never considerable. Like the rust and other disorders of that kind, I suppose it to proceed from repletion. Most of my lands being flat, I have observed that those disorders might be infallibly produced artificially, by graduating moisture alone for their attainment, and trusting to the season for heat; and the remedies I use are, to plough very deep when I sow my wheat; nevertheless covering it shallow, and to lay the land in ridges the width of the corn rows, with a deep and narrow furrow between them. Wheat seems to me to resist these maladies, in proportion to its forwardness, because it is less exposed to the combination of heat with moisture. Early kinds are the resource against the one; draining off rain water by furrows and deep ploughing, seem to me to be the best resource against the other.

A few of the experiments I have made with gypsum are mentioned, to take a chance for adding a fact to your information on that subject.

1803. March 15th. Oats and clover, both just up, plaistered them at one bushel to the acre; three weeks after, plaistered them again with the same quantity. Upon both occasions left the richest portion of the plat unplaistered. This only produced one third, both of oats and clover, of the plaistered land.

April. Mixed or rotted a bushel of plaister with as much seed corn, keeping it wet whilst planting. With such rotted seed planted a field of 40 acres, except eight rows through the centre which were unplaistered. The land poor. The inferiority of these eight rows was visible, from the moment the corn was up, to its being gathered.

1804. April. Rolled the seed corn of two hundred acres in like manner, leaving eight rows across the field, so as to intersect with flat, hilly, sandy, stiff, rich and poor land. Their inferiority was so visible, that from an eminence in the field, a stranger could point out the eight rows from the time the corn was three inches high, until it was all in tassel. In this the eight rows were a week later than the plaistered corn. The plaistered corn stood the best, was forwardest, and produced the greatest crop. Its fodder dried about ten days sooner.

1805. April. Plaistered as above the seed corn of 30 acres of rich *moist* land, leaving eight rows. Corn injured by too much rain. No difference between the eight rows and the rest.

May 7th. Replanted my corn on the high land, which had been much destroyed by mice, moles and birds, mixing two quarts of tar well, with one bushel of seed corn, and then plaistering it as above. The best remedy I ever tried against those evils, and the plaister as usual, accelerated and benefited the corn.

April 25th. Plaistered three bushels on three acres of clover just up, sown alone on land half manured with coarse manure. A good crop.

May 9th. Seven bushels on seven acres of forward wheat and clover. Wheat heading; land thin; the clover exceeded what such land had usually produced. No benefit to the wheat.

May 10th. Six bushels on six acres of very bad clover sown last spring. cover just beginning to bloom. The season became moist, and it improved into a fine crop.

May 10th. Last spring I left an unplaistered strip of 20 feet wide quite across a field of clover. It was all cut except this strip, which was so bad as not to be worth cutting. This spring on this day (clover beginning to bloom) the strip was still much inferior to the adjoining clover, which was good. I plaistered it at a bushel to the acre, leaving the rest of the field unplaistered. It equalled the adjoined clover in one month.

May 16th. Sowed 23 bushels on 23 acres of corn in a large field. Ploughed in part immediately, harrowed in part, and left part on the surface ten days before it was worked in. Corn four inches high. Weather moist. No difference between the three divisions. The seed of the whole field had been rolled. These 23 acres exceeded the adjoining corn 25 per cent: its blades and tops also dried sooner.

June 15th. Plaistered at three bushels to the acre a strip of goose grass or English grass — no effect on land or grass.

August 10th. Sowed 50 acres of thin sandy land in corn at the time, in clover, and 40 bushels of plaister on the seed, harrowing both lightly in. A moderate shower in four days, succeeded by a severe drought. Clover sprouted and chiefly perished. A good cover of bird-foot clover followed land so visibly improved, that a stranger could mark the line of the plaistering by the growth. That and the adjoining land in corn in 1808. The difference visible in favour of the plaistered land.

September 17th, to the 5th, of October. Sowed 88 bushels of yellow latter bearded wheat; 171 of forward, mixing half a bushel of plaister with one of wheat, a little wetted. One bushel of forward, and three pecks of latter wheat were sown to an acre. All among corn. Two slips of 30 feet each were left across the field, in which unplaistered wheat was sown. Where the land was sandy, the unplaistered wheat was best, owing to the great growth of bird-foot clover among the plaistered. This discovered the effect of gypsum on that annual grass. Where this grass did not appear, there was no difference between the plaistered and unplaistered wheat. From the spring of 1806 to this time, the unplaistered slips have been distinctly marked, by a vast inferiority of the weeds and grass naturally produced.

November 23d. Sowed three bushels of plaister on one and an half acres of wheat, left unplaistered for the purpose in the field last mentioned, on the surface. Weather moist. No effect on the wheat, on the ground, or in the growth to this day, though the plaister was of the same kind with that used in the last experiment.

1806, March and April. Sowed 200 acres of clover with plaister, at different times when the weather was dry, moist, windy and still, part at three pecks — a bushel and five pecks to the acre, leaving a slip of 20 feet wide across a field, to ascertain the goodness of the plaister, which was of a hard white kind, that hitherto used being soft and streaked. The clover in this strip was bad, on each side of it, fine. No apparent difference was produced by weather, quantity, or times of sowing. The whole crop far surpassed in goodness whatever such lands had produced before, except the slip, as to which Pharaoh's dream seemed reversed.

April and May. Rolled all my seed corn as usual, leaving slips unplaistered. An excessive drought. No difference between these slips and the rest of the field. The following year when that grass grew, tufts of luxuriant bird-foot clover, designated the exact spots where the plaistered corn had been planted.

April 23d. Sowed 16 bushels of plaister on eight acres of oats and clover, just up, intending to have a great crop, and leaving a slip. Land naturally fine and highly manured. Drought as above, excessive. Oats bad. No difference between the slip and the rest. Clover killed. Land ploughed up in September and put in wheat. Clover sown in 1807 on the wheat. A heavy crop of wheat, clover plaistered in March 1808, at a bushel to the acre; crop very great. No inferiority in the slip unplaistered in 1806.

1807, March 1st, to 12th. Sowed clover seed on one hundred acres in wheat, and 80 bushels of plaister the sowers of the latter following those of the former. Left a strip of 20 feet. Weather dry, moist, windy or calm, and for two days of the sowing a snow two inches or less, deep, on the ground. Land stiff, rich, poor or sandy, and of several intermediate qualities. The clover came up better than any I ever sowed on the surface, the strip was a little, and but a little inferior to the adjoining clover, which I attribute to its receiving some plaister from the effect of a high wind. The whole field received three pecks to the acre in 1808, and was the best piece of high land grass of the size I ever saw. The wheat received no benefit.

March 10th. Sowed 40 bushels of plaister on 60 acres of poor land, cultivated in corn (Indian) last year, and well set with bird-foot clover, leaving an unplaistered slip. Weather dry and windy. Effect vast. Strip visible to an inch, as far off as you could distinguish grass. The bird-foot clover died, and a crop of crab grass shot up through it, and furnished a second cover to the land.

1807 and 1808. In these two years all my corn ground as it was broken up or listed has been plaistered broadcast, with from three pecks to a bushel to the acre, and directly ploughed in, and both the seed corn and seed wheat have been rolled bushel to bushel. In both, the crops have greatly exceeded what the fields have ever before produced. That cultivated last year has doubled any former product. But they have been aided in spots with manure, and the years were uncommonly fruitful. All the manure carried out in these two years has been sprinkled with plaister when spread before being ploughed in, and several fields of the bird-foot clover have been plaistered. The results conform to those already mentioned.

1808, February. Plaistered four ridges of highland meadow oat at a bushel to the acre. No effect.

Some of the inferences I make from these experiments are, that gypsum should be worked into the earth; that half a bushel or less to an acre, worked in, will improve land considerably; that drought can defeat its effects upon corn, but not upon the land, if it is covered; that the weather is of no consequence at the moment it is sown, though the subsequent season is of great; that it may vastly improve red clover even as late as May; that it increases the effects of coarse manure; that a quantity less than half a bushel to an acre, is in some cases as effectual, as a much larger one; that excessive moisture or excessive drought destroys its effect; that its effect is more likely to be destroyed, when sprinkled on the surface, than when mixed with the earth; that sowing it broadcast among Indian corn after it is up, may improve the crop 25 per cent: that sown in June it may not improve English grass; that sown in August and covered, it may improve the land, though drought succeeds; that sown on wheat in November, it may neither benefit the wheat nor land; that about three pecks to the acre immediately sprinkled on clover seed sown on the surface, may cause it to come up, live, and thrive better; that a similar quantity sown on the surface in March may treble the burden of bird-foot clover; that sown broadcast from the 1st of January in breaking up or listing corn ground, the same quantity will probably add considerably to the crop; and that it may not improve the high land meadow oat if sown in February.

I have witheld experiments tending to prove the utility of combining enclosing with the use of gypsum, because they are yet defective; and some others, on account of the length of this letter.

If my poor experiments can in the least degree advance the laudable design of your institution, I shall be always willing both to communicate them, and that you should either select extracts, or suppress them as you please. I expect this year to complete a project for draining 200 or 300 acres of land, subject to tide water, muskrats, and a creek having two mills on it above. It is a considerable work for a farmer, and has been conducted at very little expence. Would a circumstantial account of it be agreeable, should it succeed?

I have been obliged to use the common names of several grasses, from an ignorance of the botanical. Some of them have not I believe been named by the adepts in that science, and I have no botanical vocabulary to look into for the others. Such names I know fluctuate and are often different, in different districts; if those I have used should be unintelligible, I will upon knowing it, try to explain them. I am with great respect, Sir,

YOUR MOST OBEDIENT SERVANT,
JOHN TAYLOR.

DR. JAMES MEASE.

Virginia Farming, 1809

From John Tayloe, "On Virginia Husbandry," in Philadelphia Society for Promoting Agriculture, *Memoirs* (Philadelphia, 1811), II, pp. 100–02.

Read August, 1809

DEAR SIR,

Agreeably to your request I embrace my first leisure of acknowledging your favour of the 22d. of February, and replying thereto as the various subjects occur. First, you mention plaister of Paris, of which I do not make general use, particularly on my low lands, where I have not found it to succeed. I sometimes use it on my highlands, where it answers tolerably well particularly with clover, though I do not cultivate this crop upon a large scale, yet I have some at each of my farms for the purpose of feeding it, when half cured, to my horses, and other work team, through the summer.

My general rotation of crops is corn and wheat, the latter succeeding the former, on the same field, the size of which varies of course according to the size of the farm, for some of the fields or shifts as they are termed here are four hundred acres, whilst others are no more than one hundred upon the different farms, the number of shifts which is generally three, depends in some measure on circumstances and cultivation, as also depends the kind of plough; of late I have been in the habit of making mixed crops, corn, wheat, tobacco, cotton, oats, rye, pease, beans, &c. I seed from three pecks to a bushel of wheat to an acre, and reap from ten to fifteen bushels, and my corn ground produces from three to six barrels per acre, though this again is variable. according to soil and seasons. I have never yet made any accurate estimate of the expence of timber fences, meaning sawed post and railing, which I have had for some years back, and I am highly in favour of, for though they come high in the beginning, yet I think them the cheapest in the end, as I suppose with tolerable care, they would last fifteen or twenty years. The staking and wattleing is also an expensive fence, but looks neat, and is of considerable duration, say from six to ten years, when well done with trimmed cedar brush, or cedar poles interwoven on the stakes; which last kind of fence I have of late been in the habit of making.

The cedar succeeds tolerable well here, though we have not yet any live fences in this vicinage. The stock on my farms are, cattle, sheep, and hogs, though the former succeed tolerably well, I think the latter does best. As I generally kill on my estate, from fifty to sixty thousand pounds of pork annually. The hogs are penned, and fed on corn and vegetables, for six or eight weeks before the killing season. We have an abundance of native manure, in our low ground-marshes, yet such is the routine of my cropping, the extent of the farms, and certain hands appointed to each, I cannot find leisure or means to collect it. I make no artificial manure, except what is made by my cattle in farm yards, which I keep highly littered with straw, marsh hay, corn stalks, &c. through the winter, and spring, and during the summer I have moveable pens, in which I put my cattle at night; these I generally place on my light lands, by which they shortly become equal to those of superior native quality. Our pastures are not sufficiently luxuriant here to make grazing for market an object; yet I have always tolerable good grass beef in the fall, which is rarely sold, but distributed among my overseers, and people: that which comes to my own table is stalled for a few months, and fed with corn-fodder, (corn blades and tops,) clover and vegetables. We have but few instances of the hollow horn here, though immediately on the south side of Rappahannock, there is a distemper the nature and origin of which is not yet known, and proves very fatal to cattle. On my plantations generally, I work horses, oxen and mules; the latter, which I greatly prefer, are by far the most numerous, the oxen that are used for heavy burthens, are worked with a common yoke and bow; the few that plough work in a collar, and are geared some what like horses. — I do not drill my grain, but generally cover it with the plough, followed by the hand hoe, to make a finish; it is generally put in, in September amongst the corn as it stands on the field. I fallow my land in the fall. When I plant my corn in the spring, I plough deep or shallow, though I prefer the former, according to the soil. — Orchards succeed tolerably well here, though I think the peach preferable to the apple for produce. — I have no particular defence for either, except to have the trees looked over, and pruned once a year. I have no doubt but Colonel Taylor's mode of recovering his land, by its own nuisances may be a good one; but then he can raise very little or no stock.

MOST RESPECTFULLY I REMAIN
YOUR OBEDIENT SERVANT,
JOHN TAYLOE.

June 5th, 1809.
RICHARD PETERS ESQ.

Use of Lime in Pennsylvania, 1810

From Richard Peters "On Liming Land," in Philadelphia Society for Promoting Agriculture, *Memoirs* (Philadelphia, 1811), II, pp. 272–86.

Read August 14th, 1810.

The relation to the society in which I am placed by them, impels me to bring to their notice, many subjects, which I should not otherwise conceive myself bound to discuss. When no attention appears to be paid by others, to an important point, I venture to supply, however inadequately, the deficiency. This must be my apology for so often troubling them with my thoughts on topics, to which others, if so inclined, could do more ample justice. We have not a solitary communication upon the practice of *liming lands;* though carried to very great extent in our state. In no country is LIME in more abundance; nor can it be of better quality. Chemical and theoretical accounts of it, may be found in many books. There are good writers on its properties, as they apply to agriculture. But we find in those writers, many positions and remarks, both theoretical and practical, unsuitable to the climate and circumstances of this country. It is more a topic for curiosity than practical use, with common farmers, to enter into its composition chemically; though to those who turn to profitable account such inquiries, they are indeed highly beneficial. I leave all theories; — and wish to call the attention of the society to its practical uses. It would be very desirable to fall on means to acquire every information from those who can, from experience, give it; and multitudes of our fellow citizens have the capacity to afford the fullest satisfaction, in every point of practice required. It is more necessary to excite the inclinations of many of them to communicate their practical knowledge, than it is to give them instruction; as this substance has, in many districts of our country, now become one applied in common course. It is annually becoming dearer to the farmers in old settlements; and especially in the vicinity of the city and large towns; owing to the demand of this material for buildings, and the scarcity of timber for fuel. From twelve to fourteen cords of wood are consumed, in burning one kiln of lime of six hundred bushels. The quarries are inexhaustible. No other fuel equals wood for this purpose; as we may see by comparison of ours with the lime of our coal countries where it is tinged and discoloured; and vitiated by the sulphur of the fossil coal used in its calcination. It would be on these accounts desirable to offer a premium, or, in some way, to encourage improvements in the construction of lime kilns; to same the consumption of fuel. Coked coal might, where timber is exhausted, supply the place of wood.

The quantity, per acre, proper for soils of different textures is by no means fixed; either here or in Europe. I have been surprised, by what I have myself seen, and more by the accounts I have read in European books, at the great, and to us incredible, quantities of lime allowed by Europeans to an acre. Ours is the statute acre of 160 perches. The common computed acres of Europe differ in contents; so that it is difficult always to understand what is meant by writers, even in England, by the term *acre*. But the lime of Europe, applied in the quantity of 160 bushels to a statute acre of 160 perches, at one dressing, must either be of inferior strength and quality to ours; or there must be a vast difference in the effects of climate. As to soils of most countries, they are much alike. There is not, on our globe, better nor worse land, with all the intermediate gradations, than can be found here. It is composed of all the varieties of materials, generally found in soils of other countries; though no accurate analysis, of quantities and proportions of these materials, has been made. It would be highly useful, that geological explorations and inquiries should be more generally prosecuted throughout our country. Every farmer should analyze his own soil; that he may be the better enabled to cultivate to advantage, by knowing its texture, and applying the manure the most suitable to it. THREE HUNDRED BUSHELS (*Winchester*) of lime, have been, at one time, spread on an acre, in England! Half of that quantity, laid on at once here, would ruin any acre of land within my knowledge. I mean a worn acre taken up for amelioration and recovery. Land reduced to sterility, by bad farming and over-cropping, is like the stomach of an animal in a state of debility. It must be recovered by gentle means applied repeatedly, and at proper intervals. Too much food is as destructive to the animal, as over-liming is to the impoverished land; whatever may have been the original stamina of the one, or qualities of the other. It is essential to know what quantity per acre, is advantageous and proper *here;* and the most beneficial modes of using it. When I began, in the early part of my life, to lay on lime, I was advised that the lime would spend itself as much if no culture were carried on, as it would by the severest cropping. I soon found that limed lands required as much care and good management, as others. The lime may sink, or part with its qualities; but severe cropping and bad systems, injure limed, as much as other soils; and, I think, leave them in a state more difficult to recover. I have myself experienced this, when I overlimed or overfarmed, from want of proper information. I therefore suspect, that the lands said, in many places, to be *lime-sick*, must have been badly managed, and overworked. And yet in some parts of our country wherein they have discontinued the use of lime, after having long applied it (perhaps in too great quantities) there are good farmers. I believe land requires a change, after a certain time, of manure as well as of crop; though either may be, after proper intervals and with suitable auxiliaries, again introduced with equal benefit. I do not know enough of facts, relating to *lime-sick* lands, to give an opinion: and this is one among other reasons, why I wish the society to promote inquiries.

I believe it is generally known and agreed, that the poorer the land, either naturally or by wearing, the less lime it will bear. So that 25 bushels will benefit, where 50 would injure. *Lime*, being in itself no manure, must find, in the earth, or in the air, something to act upon, or co-operate with. And, that it may have constant communication with the atmosphere, it should be kept near the surface; both in its first application, and by deepening the ploughings to bring it up when it sinks. I have made much use of it, in every way, and in great quantities, for a long course of years. My soil is various; but generally a kindly loam, mixed with *mica* (isinglass) and in parts sand, as well as clay. The surface is of every description, as to exposure, hill and vale. It had been much worn in some parts; and I have cleared off, from time to time, a considerable portion of the timber. So that I have had all kinds of soil to operate upon. I have generally begun with 40 bushels (sometimes 50, and often 30 and 35) to the acre. I prefer it to be laid in half bushel-heaps, and water-slaked. But I frequently cover these little heaps with earth, and leave it thus to slake; closing the cracks carefully, as they appear. Sometimes I leave it through the winter, in large heaps of 40 or 50 bushels (accordingly as I determine the quantity per acre) well protected by earth and sod. I choose, when practicable, to spread it in the autumn; and either plough or harrow it in. The next season I take only a summer crop. Indian corn I think the best; as its culture mixes the lime most effectually with the soil. I have most frequently put lime on in the spring; and I have cropped fields with winter-grain, when limed in the same season. I have sometimes succeeded with *rye;* but when *wheat* was sown on land fresh limed, I have invariably suffered by mildew, smut, rust, or blight. I scarcely remember an instance to the contrary. The crop is retarded in its maturation, by the lime: and though it shews a deep verdure, and large heads; the former is as deceptive as the blush of a hectic; and the latter seldom, or never fill. Yet in some European books, I see it recommended, to plough or harrow in the grain and lime together. I have never approved of dunging the ground at the time of liming; having made comparative experiments.

My course has been, to lime, — take a summer crop, — fall-plough, — and, the next year, an open fallow, or a covering, but inexhausting, spring-crop, preparatory to dunging for wheat. In this course I have invariably had success; and therefore prefer it to any other. I have, when the field came in course again (in three or four years) limed; and thus repeated the applications to 120, and in one field, to 160 bushels to the acre; including all repetitions of liming, at different, and distant, periods. I have known 80 bushels to the acre (put on, at once, on such land as mine) injure the field for several years; or until recovered by dung, or green manures ploughed in. It is said that clay will bear the heaviest liming. So that climate, strength and quality of lime, (I mean stone-lime, as ours generally is) differ widely in this, from those of other countries.

Farmers of what are called *strong lands*, have told me, that eighty bushels per acre, on the first application, were but a moderate allowance. But I have ever believed that it would have been better to lay on forty or fifty bushels in the beginning; and increase, by repetitions, after proper intervals. Strong lands are precipitated into debility by over stimulation, as strong men, or other animals, are enfeebled by excess, or over exertion. Some of our strongest lands are now thus reduced to a situation to be no longer benefitted by lime. In lime-stone countries, where lime is obtained on easy terms, I have known it spread without rule, or attention to exact quantity. I always predicted, that repentance would, one day, follow when too late, this agricultural enormity.

In *Europe*, lime is heavily spread on a tough old grass lay; and it meliorates the grass, so as to render it highly palatable to cattle, and hastens their fatting. It lies thus, twelve months (having been put on in the autumn) and the field is then ploughed, and taken up for a course of crops; preparatory to being laid down again in grass. In this way, it is alleged, and it seems reasonable, that land will bear the heaviest liming; especially if it be a strong clay; though it is known to benefit lighter soils the most.

Lime on clay has never succeeded with me, to any profitable extent. The idea of its durably warming cold clayey land, is unfounded. Heat is disengaged, when water or moisture solidifies, while lime is slaking: but it becomes shortly thereafter, a cold substance. Its particles are too small and fine to keep asunder those of the clay; and such things as produce this effect are the only proper auxiliaries for clay land. Gravel, sand, shells, unburnt limestone, are better than lime. In clay ridged and drained, and kept dry and friable, lime may be serviceable.

I have spread lime on a clover lay, and suffered it to remain on the surface, through a winter; then ploughed; and the lime being well incorporated by heavy drags or harrows, I have found it a very advantageous mode. I have always preferred, in this and every other mode of application, laying on the lime, and mixing it thoroughly with the soil by frequent stirrings, without dung. I have repeatedly observed, that fresh lime and stable manure, put on together, are by no means so efficacious, as when the latter is applied in the season succeeding the liming: green manures, with fresh lime, do better. Lime, like salt, in very *small quantities* is *septic*, and may with dung be useful: but in the quantities usually applied, it must be injurious, on chemical principles, and in fact, to both the land and dung; which latter *flatters* in its effect on the crop, compared to one with lime alone; as dung will always shew itself in a greater or less degree; but it will shew and act most effectually, when it is not neutralized or consumed by fresh lime.

The varieties of our lime, as to strength or composition, for either masonry or agricultural uses, have been very little attended to. It would be important that some simple test or trial of the qualities of lime, should be established and promulgated. I know that there is in practical result, a great difference in the effects of equal quantities in bulk, measured, or weighed; and the lightest is commonly the best. This I supposed was owing to its being better burnt; so as to have less core. It is but recently, that this subject has been, in Europe, minutely examined. Some kinds of lime have been found, *there*, so composed, as to be prejudicial to agricultural operations. *Here* lime differs widely in effects, on land or crops; so as to require greater quantities of one kind, than of another. I have found it so, in mortar.

It will be perceived, that I have avoided, (as much as possible,) technical disquisition. If I have not mentioned any thing new to experienced farmers, or others acquainted with the subject, I have, at least, endeavoured to set an example; so as to invite their communicating what they know to be instructive. Putting a subject *in requisition*, always rouses attention; and draws forth useful facts; and discussion conveying instruction, which would otherwise remain hidden, or confined to the knowledge of a few individuals. What may be familiar to experienced agriculturists, is nevertheless highly acceptable, and essentially instructive, to those who want practical knowledge.

RICHARD PETERS.

2d July 1810.
To the Philadelphia Society for promoting Agriculture.

Merino Sheep, 1811

From James Mease, "Observations on Sheep," *Archives of Useful Knowledge* (Philadelphia, 1811), I, pp. 102–06.

Until the present year, the only original imported stock rams and ewes, in the United States, were those of Mr. Dupont, Mr. Delessert, Col. Humphreys', and the Author's from Spain; Mr. Livingston's from France, and Mr. Muller's from Hesse Cassel. Of these, the two first mentioned arrived in 1801. Col. Humphreys' and Mr. Livingston's in 1802, the Author's in 1803; and Mr. Muller's in 1807. During the present year, several have been imported into the United States from Spain and Lisbon. The flock of Col. Humphreys' was the largest, he having imported upwards of one hundred sheep, from which and their descendants, many pieces of cloth have been made, and sold from seven to twelve dollars per yard. The particular history of Mr. DuPont's fine ram is as follows:

Don Pedro was imported into the United States, in the year 1801, and is believed to be the first full-blooded Merino ram introduced into North America.

Mr. Dupont de Nemours, then in France, had persuaded Mr. Delessert, a banker of Paris, to send to this country some of those valuable sheep, and he having been at the head of a commission appointed by the French government to select in Spain, 4000 Merino sheep out of the number of 6000, which, by the treaty of Basle, the Spanish government had stipulated to present to France; it is natural to suppose that those which he selected for his own flock, were among the best. Four fine young ram lambs were accordingly shipped, two were intended for Mr. Delessert's farm, called Rosendale, situated near Kingston, on the Hudson river; one was intended for Mr. Dupont de Nemours, who was at that time settled in the vicinity of New-York, and the other was to be presented to Mr. Thomas Jefferson. Mr. Dupont embarked in the ship Benjamin Franklin, on board of which ship the four lambs were shipped, and was unfortunately detained upwards of twenty days in England; his subsequent passage to the United States was long and boisterous, in consequence of which three of the sheep died, and it was with the greatest difficulty that Mr. Dupont preserved the fourth. The ship arrived at Philadelphia on the 16th of July, 1801.

In 1801, Pedro tupped nine Ewes at Mr. Dupont's place near New York; he was then sent to Mr. Delessert's farm, and served a large flock during the years 1802, 3, and 4. In the course of 1805, Mr. Delessert having determined to rent his farm, and to sell all his stock, the progeny of Pedro were sold at public sale, at reduced prices, to the neighbouring farmers, who had no idea of the treasure which was offered to them; being unacquainted with that breed of sheep, they neglected those valuable animals, great numbers of which have perished in their hands, or were sold to butchers; the rest would probably have shared the same fate, had not chancellor Livingston become acquainted with the existence of those sheep, and purchased at advanced prices some of the ewes, which he put to his fine Merino rams of the Rambouillet stock. Pedro, like the rest of the flock of the Rosendale farm, was sold at vendue, and Mr. Dupont's agent bought him for 60 dollars.

In July 1805, Pedro was removed to E. I. Dupont's farm, situated in the state of Delaware, near the borough of Wilmington. That gentleman had a very small flock at that time, but was anxious to see that valuable breed propagated in the country, and with a view to attain that end, he offered the farmers of his neighborhood the use of his ram, gratis; they could not be prevailed upon to think much of what was offered to them free of cost; the consequence was, that very few ewes were sent to Pedro during three seasons, and only by way of experiment.

In 1808, however, Mr. Dupont, with a view of increasing his own flock, purchased from the farmers, his neighbours, as many half or three-quarter blooded ewes of Pedro's breed as he was able to collect, which measure raised his character among the farmers. Since that time Pedro has served every year, from 60 to 80 ewes; the vicinity of Wilmington will therefore be supplied with a large stock of fine woolled sheep, and as Mr. Dupont & Co. are erecting works for the purpose, cloth of any fineness may be made.

Pedro is now ten years old; but very strong and active; he is stout, short, and woolly, and of much better form than Merinos commonly are; and even better than that of a ram figured in a superb engraving lately received by the Agricultural

Society of Philadelphia from Paris. His horns are large and spiral; his legs short, and he weighs 138 pounds; his fleece carefully washed in cold water, weighs 8½ pounds, is extremely fine, the staple is 1¾ inches long, and lying very thick and close upon his body; it is entirely free from loose coarse hairs, called jarr. Every part of his fleece, moreover, is nearly of equal fineness, even the wool of the hind legs and thighs, which is long and coarse upon many Merino sheep, is short and fine upon Pedro. This point, which in the case of wool so valuable as that of Merino sheep, is of great consequence, will be transmitted to his progeny, and proves the value of stock derived from him.

Crossbreeding Corn, 1813

From Philadelphia Society for Promoting Agriculture, *Memoirs* (Philadelphia, 1814), III, pp. 303–25.

Philipsburg, October 25, 1813.

DEAR SIR,

I promised to write you on the early planting of Indian corn, and the power which it possesses to withstand frost, in the earlier stages of its growth, and shall now fulfil that engagement; but conceive it will be useful to commence with remarks on the general properties of this plant.

The prevailing opinion is, that the cultivation of this plant is preeminently injurious to the soil, this opinion has been produced by a bad system of management, originating from a thin population, deficiency in capital, and an abundance of fres.., uncultivated lands, for those causes have retarded the progress of all improvements in husbandry here.

The origin of this error is obvious to me; viz. Maize being a very powerful plant, is capable of contending with an impoverished soil; and when tolerably cultivated, will remunerate the planter on grounds incapable of producing crops of almost any other description, equally valuable, and the farmer abuses this plant, because it continues faithful until the last dying gasp of the soil, which his avarice has destroyed. But it is strange, that men of observation and reflection, viewing the effect without sufficiently examining the cause, have adopted the opinion of the exhausting properties of the corn plant. It is generally agreed, that potatoes, turnips, and cabbages, are meliorating crops; notwithstanding it is known, that neither of those crops will grow on a thin soil, without the assistance of manure, and that if they are continued year after year, (or grown in rotation with others,) without manure, the soil will soon become incapable of producing either of those plants, so as to pay for the gathering of them. This furnishes a striking contrast between those justly esteemed meliorating crops, and the corn plant; for after the former cannot be grown with tolerable advantage, crops of maize, sufficiently productive to induce the continuation of them, may be introduced on the same injured soil, without additional manure: but the fact is, that all the crops that will admit perfect horse hoeing or hand hoeing, may be justly considered meliorating, and also some others, which do not admit this after cultivation, but form a shade sufficiently close to destroy weeds. But it has been well established, that some of the most luxuriant of those crops are great exhausters of the soil; yet they are justly in high estimation; for it being known to every farmer, that they cannot be grown profitably on thin soils, without manure, he readily submits to this necessity; but finding that Indian corn will grow year after year, without manure, he continues the destructive practice of cultivating it in that way, until his fields are ruined, and without due consideration, blames the corn plant for exhausting his soil; an opinion in which he is confirmed from observing the poor crops of wheat he gathers from those fields which have been sown with that grain, among his corn, after the ground had been nearly exhausted, and often overrun with weeds; but being seldom disappointed in obtaining good crops of this or other small grain, after potatoes or turnips, he conceives those plants are exclusively friendly to the soil: they are indeed friendly; for the cultivator is compelled to use them kindly, or lose the labour bestowed on them.

The product of Indian corn being vastly more abundant than any other corn crop, it will demand a due proportion of nutriment from the soil, but will also return much back to it again, even when the grain is not expended on the farm; for the fodder from one acre of luxuriant maize, is fully equal in quantity and nutriment, to one ton of good hay, and the stalks more than equal in value to one ton of straw, for littering the cattle yard; and but few, if any other summer fallow crop, is capable of returning back to the soil, half this amount, unless the most nutritive and most valuable proportion of the crop is added to the amount; and if the grain and fodder of a crop of maize are applied to the fattening of cattle, it will be found much more profitable than turnips and potatoes from the same quantity, and equal quality of land, notwithstanding a statement directly opposite to this assertion has been published, by a very respectable writer on agriculture: but that statement will be found replete with error throughout; for here again, Indian corn, without manure, produced 15 bushels per acre; potatoes manured, produced 200 bushels per acre, the expense of ten acres of potatoes manured, rated at 36 dollars and 60 cents, when it actually cost nearly double that sum to remove them from the ground, and secure them properly.

That Indian corn does not exhaust the soil more than potatoes, appears clear to me, from some mixed crops of those plants, which were grown on my farm. After the ground had been equally manured, one row of corn occupied the same quantity of soil as was assigned to two rows of potatoes, throughout the fields: the plough was too frequently used in the cultivation of the potatoe crop, and sufficiently in the cultivation of the corn, of consequence, a communication between the corn roots with the potatoes was too effectually cut off. Those crops were removed in the fall, and the grounds immediately sown with wheat, and no perceptible difference was ever discovered in the crop growing either where the corn or potatoes had grown, even where the ground had been but once ploughed for seeding, and that too in the same direction the corn and potatoes had stood. The crops of wheat were abundant, except in one field, which being struck with the mildew throughout, yielded only about twenty bushels per acre, of light ordinary grain; this crop you saw, previous to this disaster, and admired it much, and you have now in your possession an abstract copy of the accounts kept for the whole of my crops for 1810, and will find a neat clear profit of about four hundred dollars, from eight and an half acres of wheat, grown after a mixed crop of corn and potatoes. In that statement, not only every expenditure on this crop has been deducted, but also a ground rent of nine dollars per acre, and interest on the capital employed in the cultivation of it. No manure was applied to any of these crops, after the mixed crops were removed.

The lateral roots of Indian corn take a horizontal course, filling the ploughed ground in every direction. The finger-like or perpendicular roots dip deep, far below the range of plants in general. I have seen them traced two feet below the level of the ridge in which they grew, with no better implement than a grubbing hoe, in the hands of an awkward workman; how much further proper investigation might have traced them, I cannot determine: but I am satisfied, that they were numerous, and more capable of running deep and collecting nutriment, than they would have been, had nature formed them into one tap root of equal length and dimensions: this opinion is further confirmed, from the well known properties of this plant to withstand drought, of which a striking instance has occurred in my little field, planted here on a high and dry ridge; for no rain fell on it from the 30th of July until the 1st of September, during which, moisture is particularly wanted in this climate, to fill the grain; yet I have never grown better ears. One just shelled measures one pint and near one gill, but if it had been cribbed, till dried perfectly, might not measure more than one pint.

The prop roots of this plant are wisely constructed to support the great additional height and weight acquired, soon after their appearance, from the shooting of the tassel and ears; for they are stout, and very numerous, and answer the double purpose of props and gathering nutriment; which the plant requires in much greater abundance, than at any other time, for filling the ears. I have seen fields very luxuriant, previous to this very trying crisis, which immediately after, became pale and sickly, and proved unproductive; which is the natural result of crowding the soil with more plants than it is capable of perfecting.

The traces of numerous original corns, are evidently seen in our fields and gardens. Only five of those are in general use for field planting. First, the big yellow and white, in size and form very much alike, the cobs, long and thick, the grains are large, firm, and without indenture; but their size consists principally in their width, which is greatest near the cob, and from this point, the rotundity of their outside surface forms a very considerable vacuity between the rows; and a large circumference, being filled with a few rows of very wide grains, which are short even at their deepest points, covers an extensive surface, without introducing measurement in proportion to the size of the cobs. Secondly, the little yellow and white, resembling each other in size, form and texture, producing still harder grains, more compactly arranged, but not sufficiently so, which, with shorter and slimmer cobs, render them less productive than the two first mentioned; but they ripen early, and are considered safer crops in high latitudes. Thirdly, the gourd-seed. The cob of this is neither so long or thick as the large, solid corns; but the grains are very long, forming a compact round, from the cob to the outside surface of the ear, and gradually taper to a point, where they join into the cob; of course it is vastly more productive than any other known original corn, but ripens late, and the grains are too soft and open for exportation, unless kiln dried. This variety, so far as my observation goes, is invariably white; for although I have frequently heard of a solid yellow gourd-seed corn, yet on investigation, nothing more has appeared than a mixture of the hard yellow corns, with the white gourd-seed. If such an original, firm yellow corn, equally productive with the white gourd-seed, could be procured, it would be invaluable; but by forming a judicious mixture with the gourd-seed and the flinty corn, a variety may be introduced, yielding at least one third more per acre, on equal soil, than any of the solid corns are capable of producing, and equally useful and saleable for exportation. But this mixture should be with the yellow corns; that colour being greatly preferred by shippers, and is most productive, it having the longest and thickest cobs, and would at least compensate for shortening the grain of the original gourd-seed; provided a sufficiency is introduced to lengthen the grain so far as will be consistent with retaining solidity, and that yellow tinge required to make it saleable.

The proportion of the big yellow with the gourd seed, may be determined by the length and thickness of the cob, with wide grains; and of the little yellow, by shorter, slimmer cobs, with narrower grains, of a brighter yellow tinge, and an increased number of rows, in proportion to the width of the grains, and size of the cob; which sometimes amount to upwards of thirty, and seldom less than twenty rows. This sort is firmer and handsomer than that formed with the big yellow, and better calculated for high latitudes, but not so productive. I am induced to believe, that by combining those three, the most valuable properties of each, might be concentered in one variety, in a much greater degree, than at first sight appears, by selecting the ears with the greatest number of rows, compactly formed on the cob, and ripening early. The experiment is easy, and the prospect promising; for in the present unimproved state of that plant, from ninety, to more than one hundred bushels of shelled corn, per acre, have been frequently obtained, and when the seed has been improved, and the arrangement and cultivation of the plant better understood, it does not appear unreasonable to suppose, that one bushel per perch, (160 bushels per acre,) will be obtained with equal facility.

Mr. Stevens's crop of 118 bushels, published in your edition of the Domestic Encyclopedia, is the largest I have noticed; he introduced 26,880 plants per acre. A large ear shells one pint of corn; moderate sized ears will average more than half a pint; if each plant in his field had produced only half a pint of shelled corn, his crop would have yielded at the rate of about 200 bushels per acre, of consequence the number of plants, or their arrangement, or, perhaps, both those causes combined, injured his crop. And the same cause produced similar effects in my crop of 91 bushels, grown in the year 1811, in double rows, on ridges half a perch asunder. The number of plants not exceeding 20,000 per acre, were too much crowded, and leaned out from the ridges into the intervals, to procure the air, until the tops of the plants standing on opposite ridges, intersected, and those which got undermost in the scuffle were entirely barren; and those which predominated, had the shortest ears which had ever been grown before by me; but still the greatest crop of corn. If the same number of plants had been introduced on the same ground, in single rows, at half the distance, it appears reasonable to suppose, that the crop would have been much more productive, but whether so many can be arranged on an acre, so that they will produce large and perfect ears, is unknown to me. Those crops are merely introduced to show, that large increase does not always determine judicious management of the crop; and that if cultivators would carefully observe the progress of their crops, and publish their errors with the publication of the management and the result; those errors, instead of being copied and perpetuated by others, would be avoided, and the knowledge of agriculture rapidly increase.

I introduced the greatest crop of maize that has been published, with another large crop, to show that great increase has been obtained, from very injurious arrangements of the plants, and that much greater might be acquired, if we knew how to arrange the full complement which the soil and climate are capable of perfecting. But it is far better to introduce too few, than too many plants, for in the first case, a good crop is insured, while in the latter, the soil is uselessly exhausted, and the crop is rendered precarious.

Corn should be planted early, more especially in high latitudes, where the farmer's expectation is frequently blasted by an early frost, or a cool spring and summer procrastinating the growth and maturity of the plants; and this evil is increased, if he cultivates the larger and more productive corns, not only as it affects his crops, but also as it establishes the erroneous opinion, that they cannot be grown. But I had been prepared to encounter those false opinions, having planted, my crop much sooner than my neighbours, and observed that maize, in the early stages of its growth, was not subject to such injury from frost, as was generally supposed. For though its tops were frequently turned yellow, and sometimes severely scorched, the roots were established early in the soil, and the plants pressed forward with great rapidity; and no question with me, derived at least as great advantage from early planting, as barley, or any other spring crops. And it frequently occurs, that the weather is more favourable for vegetation in April, than in May, at the usual time of planting; and the ground, early in the spring, is less subject to become dry and crusty, which often prevents the plants from penetrating the soil. The Indians' ideas of planting, are suited to their limited observation; but we see those trees which are generally in blossom at the time of planting corn, are sometimes in full bloom, in February and March. I have not been long enough here to determine how much later spring generally commences here than with you: but believe it cannot be less than two weeks. My corn, and that of some of my neighbours, was planted the last day of April and the first day of May. The weather was mild, and continued so until the 13th of May, when some white frost appeared on the fence rails and bridges. On the 14th some of the corn came up; on the 16th considerable white frost, and ice were also seen in an earthen pot, standing on boards in my yard; on the 17th, the ground was frozen in the bottoms, and ice in troughs, and the same occurred every morning until the 22d, during which the corn continued coming up, and few, if any of the plants were destroyed, although some of them were killed level with the ground. The tops of early planted potatoes were killed, but the crop was abundant. Beans fared worse, being incapable of a general recovery. The older settlers say, this was the severest frosty time, they have seen here at the same season of the year. On the 19th of August there was a considerable white frost, with ice. The upper blades of the corn were wrinkled from twelve to fifteen inches from their points, and hung downward, but generally retained a healthy, rusty, weatherbeaten, green. From this, injury was expected, but none appeared in the crops. The next frost occurred on the 6th of October, and finding the corn plant debilitated by age, it scorched it excessively; and another on the 11th killed it effectually. I have never seen better ears of corn, than were grown in those two crops, when for forty miles or more round, we hear of no good crops, not even on the Susquehannah and other rich bottoms, justly famed for growing corn; except a field grown by a neighbour, which was planted from the 11th to the 13th of May, and that part of his crop seeded with the same kind of corn grown by me, to wit, highly impregnated with the gourd-seed, was not perfectly matured. This general failure in crops of corn, has been commonly attributed to a great redundancy of wet, during the spring and first summer months, and an unusual drought in the latter part of the summer. But the fertile fields mentioned above, were subjected to the same, therefore those causes could have had no general effects elsewhere, except those arising from late planting, which the difficulty of the seasons have rendered more conspicuous. From these facts, I infer, that early planting of corn, more particularly in high latitudes, would augment the general crop to an immense extent.

A grass lay is vastly preferable for corn. The very minute division of the soil, by its innumerable roots, with their fermentation and decomposition, furnish considerable heat and nutriment, and introduce cavities in due proportion to the number and size of those roots, keeping the soil open and mellow, for the ready admission of the corn roots. Stable, or other animal manure, should be spread regularly over the lay. But it is seldom properly done, in the usual random mode of dropping the heaps. The field should be staked on one side, and each end; the distance of one perch is convenient for spreading, and will readily determine the number of loads per acre, with the size of the field. The dropping of the heaps, commencing along the stakes on the side of the field, with the stakes at the ends, insures regularity. If patches appear materially thinner than the soil in general, they should be marked, and additional manure spread over them. The sod, with the manure, should be regularly turned under, with one deep ploughing, or by trench ploughing, which is far preferable; for it will furnish a surface readily pulverized with the harrow, and bury the seeds of most weeds beneath the power of vegetation, and provide larger scope for the roots of the corn. It is said the skim coulter plough, will effect this purpose equally well, and with much less labour and expense. If obstacles prevent trench ploughing, the furrow-slice must be compressed with a roller, and the ground harrowed lengthwise the furrows, sufficiently to close the seams between them, which greatly prevents the growth of the grass. If the cultivator wishes to introduce lime, it may be spread now, with great advantage, and mixed by the harrow with the upper surface of the soil, without injuring the animal manure by its caustic properties: for the cultivation of the crop, and those succeeding it, if readily and effectually executed, will keep the lime near the surface, until it is sunk by its own gravity. Wide intervals, or right angles may be formed, sufficiently correct, with the plough, if they have been regularly staked out, but crossing those furrows, for narrow distances, on the rows, and striking the smaller angles, require mathematical proportion, to admit ready and effectual cultivation. The four furrow wheat drill in your repository, will furnish simple principles for forming a very cheap implement, to effect this purpose.

For dropping the seed, cut an elder stalk of about one inch diameter, just below a limb, which will form a handle; cut and hollow it out above, to hold the number of grains planted, paring the upper edge thin: with this and a small basket, a little boy or girl may drop three times more than a man in the usual way, and more correctly. If two plants are designed at each angle, eight or ten grains should be planted; for numerous birds, and cut-worms, will have their share, and some of the seed will rot, and many that vegetate will not get through the soil. The extra expense of seed is scarcely worth consideration; while replanting is very expensive. Where the planting fails, it will be more profitable to introduce three or four bunch beans, or a potatoe, for neither require hilling, if planted at a proper depth, and the latter will certainly be more profitable than replanted corn. The seed should be covered no deeper, than one inch, with the loosest, moistest, and best mould at hand; and no clods, or other obstacles introduced; moderately patting this covering with the hoe, to secure the moisture. If the soil should form a crust over the seed, near the time of its coming up, the clusters must be daily examined, and when the points of the plants turn downward, they should be liberated immediately, by carefully breaking the crust. I have saved one field of corn by this practice, which must have perished without assistance. This may be readily done, with a small rake, the teeth formed with nails, whose edges have been previously blunted: but a three forked hook, in the form of those used by gardeners, is preferable.

Farmers have been told, and might have seen, the very superior properties of a grass lay for corn, but many who do know it, have not appreciated the advantage, or they would not grow other crops, for the express purpose of subduing the grass, previously to the cultivation of the corn. The cultivation of corn should commence, so soon as the first leaves expand; otherwise grass and weeds will take the lead; hence much labour is saved, by commencing their destruction before they are established. On the proper or improper management, at this critical time, the prosperity of the crop greatly depends; for, if the plants now scarcely three inches high, are pulled up from a free open soil, the roots will be found more than twelve inches long, besides what remain in the ground. Where superficial roots abound, the shovel plough should be used, for the cultivation of the crop; but if those do not exist, a hoe-harrow is much better, and greatly superior to a mould board plough. This circumscribes the roots of the plants, forms ridges and furrows calculated to turn off the rains, and by turning up the grass sods, with the seeds of weeds, causes much useless labour from the growth of both; stops fermentation and decomposition, forming channels from which the heat, moisture and nutriment already accumulated, escape, exposing the manure to be washed away by rain, and exhaled by the sun. The hoe-harrow will more effectually pulverize, and clean the soil without producing any of those injurious effects, and with half the labour, provided a triangular harrow, with tines and handles, follows, with hand hoers after it, to eradicate weeds and grass near the plants, and uncovering and setting up the plants at the same time. The hoe-harrows should have handles, and be formed to close or expand behind, to suit the cultivation of the plants, in the different stages of their growth, with three triangular hoes, supported by coulters, sharp in front and point, for the smaller sizes, and five for the larger, formed and fixed effectually to cut all the ground to the extent of their spread: if wheels were added, they would ensure any desirable depth, with perfect regularity and ease.

Although the open texture of a grass lay admits very early cultivation, after rain it may be too soon done, and cause serious injury.

If the corn has been planted at right angles, the harrows should operate, in the first cultivation, through both directions of the field; first lengthwise the ploughing, and progress until the soil is well pulverized, and the weeds and grass within their reach, entirely destroyed. This will leave little for the hand hoers to do, and the field will be properly cultivated. The supernumerary plants must be pulled up, so soon as the seed at the root disappears, but while a vestige of it remains, squirrels will pull up the plants.

The suckers growing at the joint, in contact with the ground, and the next joint above, should be pulled off so soon and often as they appear in sufficient size to admit the operation. When suffered to become numerous, and large, the nutriment required for their support, with the large and numerous wounds inflicted at once, injure the plants excessively, at the time they require every possible assistance from nature and art. In the second cultivation, the harrows should stop a few days after they have effectually destroyed the grass and weeds, in one direction of the field, allowing time for the mangled and misplaced roots to form a sufficient establishment, to support the plants after they have been subjected to the same unavoidable injury, by the second harrowing. This precaution must be observed if a third cultivation should be found necessary; this may in some measure depend on the seasons, but principally on whether the lay has been ploughed sufficiently deep, and the grass turned under. If this has been done, and the execution of the first and second cultivation effectually performed, the farmer will generally controul the seasons, and sometimes derive advantage from occurrences which prove very injurious to the crops of his less provident neighbours.

To ensure good crops of corn, the cultivation with the harrows must be accomplished, before the tassels and prop roots appear, for after this, the plants requiring all the nutriment which nature and art can supply; their roots should remain perfectly at rest. Still a harrow with blunt or worn tines, may be advantageously used in wide intervals, which favour the growth of weeds, longer than narrow. But this harrow can have little effect among weeds already established, and seldom does much good when contending with grass; therefore the farmer should see the absolute necessity of pressing on a cultivation, to give an early and safe rest to his crop, more especially in narrow intervals,where the size of the plants will soon render the introduction of the harrow impracticable. For although their shade will smother many more weeds than when the intervals are wide, still far too many will remain for the hand hoers, either to pluck or cut off, if the previous cultivation has been procrastinated, or imperfectly executed.

The plants should not be cut off or topped and stripped, until the grain has attained a tolerable solidity, and the milk has entirely disappeared, unless the plants have been killed by frost; in that case they should be immediately cut off, or topped and stripped, to admit the sun and air to the ears. Some say the latter is the surest practice in high latitudes, where the sun is less powerful, and where more cloudy and dripping weather prevails; but others think the former safe, if the plants are set up in small heaps; and that it is the best way, my experience will not authorize a decision. But I do know, from actual and well tried experiments, and regular accounts, that cutting off the plants saves the fodder better, with one half the labour. With an old sithe, crooked at the point, and rags wrapped round the heel for a handle, a man will cut off at least as much more corn, as he could with a hoe, and with greater ease to himself. While he grasps the plants with one hand, he cuts them off at one stroke with the other, and by poising their buts with the point of the sithe, lays them regular with great facility, for the carters. A wooden handle would improve this instrument. The plants may be set up in heaps of six or eight feet diameter at bottom. After the corn has been husked, three or four of those heaps, set up in one, and banded with the stalks, preserves the fodder better than stacking, where the moisture accumulated in the stalks, introduces mould: much labour will be saved by setting up the corn in the field where it was grown, if the fields are not cultivated, until the ensuing spring, in that case the difference in labour, compared with topping and stripping, will be still far greater. Hilling at any time is destructive, it cuts and confines the roots within a narrow compass, and turns off the rains; exhausts the plants by compelling them to form fresh sets of roots, at the expense of those already established, converting the prop roots into common roots, and forming other prop roots from the joints above, which seldom get established in time to support the plants.

I should have mentioned before, that a grass lay is more especially necessary for corn, when it is planted on a flat retentive soil; for unless the ground is very much disposed to wet, it will preclude the necessity of ridging, and the lay forms a tough surface for hauling the manure, which eases the draught. Each furrow-slice forms an effectual underdrain, more especially if the field has not been pastured or mown in the fall, and the furrow-slice is only moderately compressed with the roller, and if a man with a shovel and mattock, follows in the furrows, leveling the heights, removing such roots, stones, &c. as will obstruct the course of the water, forming cuts across the head lands, at convenient distances, for its escape. Tens of thousands of wheat, and other grain fields, are annually injured to an immense extent, through inattention, or to save this trivial labour. The water is stopped in the fields by the obstructions in the furrows, and backed on them at the head lands, or finds a passage over the declivities, sweeping the soil and plants before it, forming innumerable gullies, none of which would appear, if the furrows were opened in proper directions.

Mixed crops of maize are profitable, and believing my crops of this description, have been more productive than any of the same sort, which have been published; the errors in management, with observations on them, may be beneficial. Accident in my crop first published, led me to expect considerable advantage from ridging for corn; but have since found that it was an increased quantity of soil and manure, introduced by the ploughing, that rendered the plants on the ridges which had been accidentally formed, better than the rest of the field, and not, as was supposed, the concentering a double quantity of manure under them; and that ridges produced artificial droughts, without any perceptible advantage from them. I also believed, and continue in the same opinion, that wide intervals, admitting large scope for sun and air, permits the introduction of numerous corn plants with safety, in the rows, and that this is the principal cause why corn is more productive in proportion to the soil occupied by it, when mixed with low growing plants, more especially, if the cultivation of the crops, cuts off the communication between the corn plant and its neighbours, which has been the case with my crops, till a trivial experiment made last summer; and till then, close planting on the rows had been entirely overdone by me. My mixed crop of corn and potatoes, for 1810, published in the second volume of the memoirs of your society, did not discover the impropriety of such close planting, for the failure in corn plants, reduced their number very considerably. But more experience and observation have taught me, that the potatoes planted in that crop, were much too thick, and if two single, instead of two double rows, had been introduced, and planted deeper, the crop would have been much more productive.

My mixed crop of corn and barley, for 1811, clearly discovered the error of too many corn plants. The barley was sown at the rate of three bushels per acre, on six feet beds, and the corn planted on ridges of the same width; the produce in barley at the rate of 36 bushels per acre, and of shelled corn, 138½ bushels per acre. The corn in the ridges was certainly too thick, about 64 plants within the length of one perch, planted triangularly, in double rows. Many plants were entirely barren, nubbins numerous, and the ears generally very short, and badly filled. Although one of the best ridges, husked and shelled under my own inspection, measured at the rate of 152½ bushels per acre, and another ridge, ordered in the same way, at the rate of 149½ bushels per acre, still, every person who saw the crop matured, joined in opinion with me; that half the plants would have produced much more corn. What might have happened, if the soil had not been generally very thin, previous to manuring, for those crops, or if a sufficient quantity of manure had been introduced to supply this defect, I cannot determine; but where the soil was good, evident marks of the injudicious practice, of this close, hedge row planting appeared.

Last spring I planted on one side of my garden, Indian corn, with intervals of six feet, two plants eighteen inches asunder in the rows; the soil appeared good, and the plants were dressed with leach ashes. They were stout, and the ears well filled, although the soil is stiff and disposed to bake, and the latter part of the summer was dry. The only cultivation of this crop was with the hand hoe, until narrow strips were broke up deeper, for the introduction of cabbage between the rows. This mode of planting introduces 9680 corn plants per acre, with an additional produce from one row of potatoes between the rows; or, the large white kidney bunch bean may be introduced, provided neither are hilled or hoed up, for by this practice, the roots of the corn plants will be greatly circumscribed, and compelled to take directions so opposite to nature, that they could not prosper.

If potatoes are planted between the corn rows, the lay should be well turned, eight or nine inches deep. This will introduce no extra labour; for that depth, though not absolutely indispensable, is requisite for the corn crop. A narrow strip, for planting the potatoes, should be well pulverized with a small hoe harrow, without turning up the grass. The ridges formed by opening the furrows, should be turned back for covering the seed. If potatoes are planted at a proper depth, I do not believe that hilling up that plant is so advantageous, as most farmers imagine. It is certain, they too frequently injure the produce by this practice; and their management of it is frequently inconsistent throughout, for they provide a light covering of long manure for planting, and during the cultivation of the crop, heap up on the ridges an enormous weight of mould, and frequently poor cold clay, destroying numbers of the plants, which in a potatoe crop are frequently far behind in vegetation and greatly injure others of the same description; cutting the wire or fruit roots, and oppressing the bulbs which escape, with a weight that prevents due expansion. And this is not all the evil produced, for trenches are formed below the level of the roots, which greatly increases the artificial droughts. — The largest crop of potatoes I ever saw grown but one, was obtained from a stubble field without manure, planted in every furrow, and cultivated with the tined harrow alone, although this mangled some of the plants greatly, and injured others in a less degree. For several years my early potatoes for family use, have been planted in the Irish way, on beds, and have ever been productive and good. This year some of them weighed one pound, which were larger than any early potatoes grown by me before.

YOUR'S, WITH RESPECT,
JOHN LORAIN.

An Englishman Views American Farming

From William Cobbett, *A Year's Residence in the United States of America* (London, 1819), I, pp. iii, 15–17, 26–29, 41–42, 63–65; II, pp. 318–23, 325–26; III, pp. 465–66, 471–76, 478–81, 494, 504–08.

In the course of the THREE PARTS, of which this work will consist, each part making a small volume, every thing which appears to me useful to persons intending to come to this country shall be communicated; but, more especially that which may be useful to *farmers;* because, as to such matters, I have ample experience. Indeed, this is the *main thing;* for . . . this is really and truly a *country of farmers.* Here, Governors, Legislators, Presidents, all are farmers. A farmer here is not the poor dependent wretch that a Yeomanry-Cavalry man is, or that a Treason-Jury man is. A farmer here depends on nobody but *himself* and on his own proper means; and, if he be not at his ease, and even rich, it must be his own fault.

* * *

1817

May 13. Warm, fine day. Saw, in the garden, lettuces, onions, carrots, and parsnips, just come up out of the ground. . . .

15. Warm and fair. The farmers are beginning to plant their *Indian Corn.* . . .

18. Warm and fine. Grass pushes on. Saw some Lucerne in a warm spot, 8 inches high.

19. Rain all day. Grass grows apace. People plant potatoes.

20. Fine and warm. A good cow sells, with a calf by her side, for 45 dollars. A steer, two years old, 20 dollars. A working ox, five years old, 40 dollars.

21. Fine and warm day; but the morning and evening coldish. The cherry-trees in full bloom, and the pear-trees nearly the same. Oats, sown in April, up, and look extremely fine. . . .

July 17. Fine hot day. Harvest of wheat, rye, oats and barley, half done. But, indeed, what is it to do when the weather does so much! . . .

July 24. Fine hot day. Harvest (for *grain*) nearly over. The main part of the *wheat*, &c. is put into *Barns*, which are very large and commodious. Some they put into small *ricks*, or *stacks*, out in the fields, and there they stand, *without any thatching*, till they are wanted to be taken in during the winter, and, sometimes they remain out for a whole year. Nothing can prove more clearly than this fact, the great difference between this climate and that of England, where, as every body knows, such stacks would be mere heaps of muck by January, if they were not, long and long before that time, carried clean off the farm by the wind. The crop is sometimes *threshed* out in the field by the feet of horses, as in the South of France. It is sometimes carried into the barn's floor, where three or four horses, or oxen, going *abreast*, trample out the grain as the sheaves, or swarths, are brought in. And this explains to us the humane precept of MOSES, "not to *muzzle* the ox "as he *treadeth out the grain*," which we country people in England cannot make out. I used to be puzzled, too, in the story of RUTH, to imagine how BOAZ could be busy amongst his threshers in the height of harvest. — The weather is so fine, and the grain so dry, that, when the wheat and rye are threshed by the flail, the sheaves are barely untied, laid upon the floor, receive a few raps, and are then tied up, clean threshed for straw, without the order of the straws being in the least changed! The ears and butts retain their places in the sheaf, and the band that tied the sheaf before ties it again. The straw is as bright as burnished gold. Not a speck in it. These facts will speak volumes to an English farmer, who will see with what ease work must be done in such a country. . . .

Oct. 24. Same weather precisely. Finished Buckwheat threshing and win-nowing. The men have been away at a horse-race; so that it has laid out in the field, partly threshed and partly not, for five days. If rain had come, it would have been of no consequence. All would have been dry again directly afterwards. What a stew a man would be in, in England, if he had his grain lying about out of doors in this way! The *cost* of threshing and winnowing 60 bushels was 7 dollars, 1*l*. 11*s*. 6*d*. English money, that is to say, 4*s*. *a quarter*, or eight Winchester bushels. But, then, the *carting* was next to nothing. Therefore, though the labourers had *a dollar a day each*, the expense, upon the whole, was not so great as it would have been in England. So much does the climate do! . . .

1818

Feb. 16. A hard frost. — Lancaster is a pretty place. No *fine* buildings; but no *mean* ones. Nothing *splendid* and nothing *beggarly*. The people of this town seem to have had the prayer of HAGAR granted them: "Give me, O Lord, neither *poverty* "nor *riches*." Here are none of those poor, wretched habitations, which sicken the sight at the *outskirts* of cities and towns in England; those abodes of the poor creatures, who have been reduced to beggary by the cruel extortions of the rich and powerful. And, this remark applies to *all* the towns of America that I have ever seen. This is a fine part of America. *Big Barns*, and modest dwelling houses. Barns of *stone*, a *hundred feet* long and *forty wide*, with two floors, and raised roads to go into them, so that the waggons go into the *first floor up-stairs*. Below are stables, stalls, pens, and all sorts of conveniences. Up-stairs are rooms for threshed corn and grain; for tackle, for meal, for all sorts of things. In the front (South) of the barn is the cattle yard. These are very fine buildings. And, then, all about them looks so comfortable, and gives such manifest proofs of ease, plenty, and happiness! Such is the country of WILLIAM PENN'S settling! It is a curious thing to observe the *farm-houses* in this country. They consist, almost without exception, of a considerably large and a very neat house, with sash windows, and of a *small house*, which seems to have been *tacked on* to the large one; and, the proportion they bear to each other, in point of dimensions, is, as nearly as possible, the proportion of size between a *Cow* and *her Calf*, the latter a month old. But, as to the *cause*, the process has been the opposite of this instance of the works of nature, for, it is *the large house which has grown out of the small one*. The father, or grandfather, while he was toiling for his children, lived in the small house, constructed chiefly by himself, and consisting of rude materials. The means, accumulated in the small house, enabled a son to rear the large one; and, though, when *pride* enters the door, the small house is sometimes demolished, few sons in America have the folly or want of feeling to commit such acts of filial ingratitude, and of real self-abasement. For, what inheritance so valuable and so honourable can a son enjoy as the proofs of his father's industry and virtue? The progress of wealth and ease and enjoyment, evinced by this regular increase of the size of the farmers' dwellings, is a spectacle, at once pleasing, in a very high degree, in itself; and, in the same degree, it speaks the praise of the system of government, under which it has taken place. . . .

Prices of Land, Labour, Food and Raiment.

310. LAND is of various prices, of course. But, as I am, in this Chapter, addressing myself to *English Farmers*, I am not speaking of the price either of land in the *wildernesses*, or of land in the immediate vicinage of great cities. The wilderness price is two or three dollars an acre: the city price four or five hundred. The land at the same distance from New York that Chelsea is from London, is of higher price than the land at Chelsea. The surprizing growth of these cities, and the brilliant prospect before them, give value to every thing that is situated in or near them.

311. It is my intention, however, to speak only of *farming land*. This, too, is, of course, affected in its value by the circumstance of distance from market; but, the reader will make his own calculations as to this matter. A farm, then, on this Island, any where not nearer than thirty miles of, and not more distant than sixty miles from, New York, with a good farm-house, barn, stables, sheds, and styes; the land fenced into fields with posts and rails, the wood-land being

in the proportion of one to ten of the arable land, and there being on the farm a pretty good orchard; such a farm, if the land be in a good state, and of an average quality, is worth *sixty dollars an acre,* or *thirteen pounds sterling;* of course, a farm of a hundred acres would cost one thousand three hundred pounds. The rich lands on the *necks* and *bays*, where there are *meadows* and surprizingly productive orchards, and where there is *water carriage*, are worth, in some cases, three times this price. But, what I have said will be sufficient to enable the reader to form a pretty correct judgment on the subject. In New Jersey, in Pennsylvania, every where the price differs with the circumstances of water carriage, quality of land, and distance from market.

312. When I say a good farm-house, I mean a house *a great deal better* than the *general run* of farm-houses in England. More neatly finished on the inside. More in a *parlour* sort of style; though *round about* the house, things do not look so neat and tight as in England. Even in Pennsylvania, and amongst the Quakers too, there is a sort of out-of-doors slovenliness, which is never hardly seen in England. You see bits of wood, timber, boards, chips, lying about, here and there, and pigs and cattle trampling about in a sort of confusion, which would make an English farmer fret himself to death; but which is here seen with great placidness. The out-buildings, except the barns, and except in the finest counties of Pennsylvania, are not so numerous, or so capacious, as in England, in proportion to the size of the farms. The reason is, that the *weather is so dry*. Cattle need not covering a twentieth part so much as in England, except hogs, who must be *warm* as well as dry. However, these share with the rest, and very little covering they get.

313. *Labour* is the great article of expence upon a farm; yet it is not nearly so great as in England, in proportion to the amount of the produce of a farm, especially if the poor-rates be, in both cases, included. However, speaking of the positive wages, a *good* farm-labourer has *twenty-five pounds sterling a year* and his board and lodging; and a *good* day-labourer has, upon an average, *a dollar a day*. A woman servant, in a farm-house, has from forty to fifty dollars a year, or eleven pounds sterling. These are the average of the wages throughout the country. But, then, mind, the farmer has nothing (for, really, it is not worth mentioning) to pay in *poor-rates;* which in England, must always be added to the wages that a farmer pays; and, sometimes, they far exceed the wages.

314. It is, too, of importance to know, *what sort* of labourers these Americans are; for, though a labourer is a labourer, still there is some difference in them; and, these Americans are *the best that I ever saw*. They mow *four acres* of *oats, wheat, rye,* or *barley* in a day, and, with a cradle, lay it so smooth in the swarths, that it is tied up in sheaves with the greatest neatness and ease. They mow *two acres and a half of grass* in a day, and they do the work well. And the crops, upon an average, are all, except the wheat, *as heavy* as in England. The English farmer will want nothing more than these facts to convince him, that the labour, after all, is not so *very dear*.

315. The causes of these performances, so far beyond those in England, is first, the men are *tall* and well built; they are *bony* rather than *fleshy;* and they *live*, as to food, as well as man can live. And, secondly, they have been *educated* to do much in a day. The farmer here generally is at the *head* of his

"*boys,*" as they, in the kind language of the country, are called. Here is the best of examples. My old and beloved friend, Mr. JAMES PAUL, used, at the age of nearly *sixty* to go at *the head of his mowers*, though his fine farm was his own, and though he might, in other respects, be called a rich man; and, I have heard, that Mr. ELIAS HICKS, the famous Quaker Preacher, who lives about nine miles from this spot, has this year, at *seventy* years of age, cradled down four acres of rye in a day. I wish some of the *preachers* of other descriptions, especially our fat parsons in England, would think a little of this, and would betake themselves to "work with their "hands the things which be good, that they "may have to give to him who needeth," and not go on any longer gormandizing and swilling upon the labour of those who need.

316. Besides the great quantity of work performed by the American labourer, his *skill*, the *versatility* of his talent, is a great thing. Every man can use an *ax*, a *saw*, and a *hammer*. Scarcely one who cannot do any job at rough carpentering, and mend a plough or a waggon. Very few indeed, who cannot kill and dress pigs and sheep, and many of them Oxen and Calves. Every farmer is a *neat* butcher; a butcher for *market;* and, of course, "the boys" must learn. This is a great convenience. It makes you so independent as to a main part of the means of housekeeping. All are *ploughmen*. In short, a good labourer here, can do *any thing* that is to be done upon a farm.

317. The operations necessary in miniature cultivation they are very awkward at. The *gardens are ploughed* in general. An American labourer uses a *spade* in a very awkward manner. They *poke the earth about* as if they had no eyes; and toil and muck themselves half to death to dig as much ground in a day as a Surrey man would dig in about an hour of hard work. *Banking, hedging*, they know nothing about. They have no idea of the use of a *bill-hook*, which is so adroitly used in the coppices of Hampshire and Sussex. An *ax* is their tool, and with that tool, at *cutting down* trees or *cutting them up*, they will do *ten times* as much in a day as any other men that I ever saw. Set one of these men on upon a wood of timber trees, and his slaughter will astonish you. A neighbour of mine tells a story of an Irishman, who promised he could *do any thing*, and whom, therefore, to begin with, the employer sent into the wood to cut down a load of wood to burn. He staid a long while away with the team, and the farmer went to him fearing some accident had happened. "What are you about all this time?" said the farmer. The man was hacking away at a hickory tree, but had not got it half down; and that was all he had done. An American, black or white, would have had half a dozen trees cut down, cut up into lengths, put upon the carriage, and brought home, in the time. . . .

320. An American labourer is not regulated, as to time, by *clocks* and *watches*. The *sun*, who seldom hides his face, tells him when to begin in the morning and when to leave off at night. He has a dollar, a *whole dollar* for his work; but then it is the work of a *whole day*. Here is no dispute about *hours*. "*Hours* were "made for *slaves*," is an old saying; and, really, they seem here to act upon it as a practical maxim. This is a *great thing* in agricultural affairs. It prevents so many disputes. It removes so great a cause of disagreement. The American labourers, like the tavern-keepers, are never *servile*, but always *civil*. Neither *boobishness* nor *meanness* mark their character. They never *creep* and

fawn, and are never *rude*. Employed about your house as day-labourers, they never come to interlope for victuals or drink. They have no idea of such a thing: Their pride would restrain them if their plenty did not; and, thus would it be with all labourers, in all countries, were they left to enjoy the fair produce of their labour. Full pocket or empty pocket, these American labourers are always the *same men:* no saucy cunning in the one case, and no base crawling in the other. This, too, arises from the free institutions of government. A man has a voice *because he is a man*, and not because he is the *possessor of money*. And, shall I *never* see our English labourers in this happy state? . . .

890. We sold our ark, and its produce formed a deduction from our expences, which, with that deduction, amounted to 14 dollars each, including every thing, for the journey from Pittsburgh to this place, which is upwards of 500 miles. I could not but remark the price of fuel here; 2 dollars a cord for Hickory; a cord is 8 feet by 4, and 4 deep, and the wood, the best in the world; it burns much like green Ash, but gives more heat. This, which is of course the highest price for fuel in this part of the country, is only about a fifth of what it is at Philadelphia.

891. *June 16th*. — Left Cincinnati for Louisville with seven other persons, in a skiff about 20 feet long and 5 feet wide.

892. *June 17th*. — Stopped at VEVAY, a very neat and beautiful place, about 70 miles above the falls of the Ohio. Our visit here was principally to see the mode used, as well as what progress was made, in the cultivation of the vine, and I had a double curiosity, never having as yet seen a vineyard. These vineyards are cultivated entirely by a small settlement of Swiss, of about a dozen families, who have been here about ten years. They first settled on the Kentucky river, but did not succeed there. They plant the vines in rows, attached to stakes like espaliers, and they plough between with a one-horse plough. The grapes, which are of the sorts of claret and madeira, look very fine and luxuriant, and will be ripe in about the middle of September. The soil and climate both appear to be quite congenial to the growth of the vine: the former rich and the latter warm. The north west wind, when it blows, is very cold, but the south, south east and south west winds, which are always warm, are prevalent. The heat, in the middle of the summer, I understand, is very great, being generally above 85 degrees, and sometimes above 100 degrees. Each of these families has a farm as well as a vineyard, so that they supply themselves with almost every necessary and have their wine all clear profit. Their produce will this year be probably not less than 5000 gallons; we bought 2 gallons of it at a dollar each, as good as I would wish to drink. Thus it is that the tyrants of Europe create vineyards in this new country! . . .

906. *June 28th*. — It is, in fact, nothing but a bed of very soft and rich land, and only wants draining to be made productive. We soon after came to the banks of the great Wabash, which is here about half a mile broad, and as the ferry-boat was crossing over with us I amused myself by washing my dirty boots. Before we mounted again we happened to meet with a neighbour of Mr. Birkbeck's, who was returning home; we accompanied him, and soon entered into the prairie lands, up to our horses' bellies in fine grass. These prairies, which are surrounded with lofty woods, put me in mind of immense noblemen's parks

in England. Some of those we passed over are called *wet prairies*, but, they are dry at this time of the year; and, as they are none of them flat, they need but very simple draining to carry off the water all the year round. Our horses were very much tormented with flies, some as large as the English horse-fly and some as large as the wasp; these flies infest the prairies that are unimproved about three months in the year, but go away altogether as soon as cultivation begins.

907. Mr. Birkbeck's settlement is situated between the two Wabashes, and is about ten miles from the nearest navigable water; we arrived there about sun-set, and met with a welcome which amply repaid us for our day's toil. We found that gentleman with his two sons perfectly healthy and in high spirits: his daughters were at Henderson (a town in Kentucky, on the Ohio) on a visit. At present his habitation is a cabin, the building of which cost only 20 dollars; this little hutch is near the spot where he is about to build his house, which he intends to have in the most eligible situation in the prairie for convenience to fuel and for shelter in winter, as well as for breezes in summer, and will, when that is completed, make one of its appurtenances. I like this plan of keeping the old log-house; it reminds the grand children and their children's children of what their ancestor has done for their sake.

908. Few settlers had as yet joined Mr. Birkbeck; that is to say, settlers likely to become *"society;"* he has labourers enough near him, either in his own houses or on land of their own joining his estate. He was in daily expectation of his friends Mr. Flower's family, however, with a large party besides; they had just landed at Shawnee Town, about 20 miles distant. Mr. Birkbeck informs me he has made entry of a large tract of land, lying, part of it, all the way from his residence to the great Wabash; this he will re-sell again in lots to any of his friends, they taking as much of it and wherever they choose (provided it be no more than they can cultivate), at an advance which I think very fair and liberal.

909. The whole of his operations had been directed hitherto (and wisely in my opinion) to building, fencing, and other important preparations. He had done nothing in the cultivating way but make a good garden, which supplies him with the only things that he cannot purchase, and, at present, perhaps, with more economy than he could grow them. He is within twenty miles of Harmony, in Indiana, where he gets his flour and all other necessaries (the produce of the country), and therefore employs himself much better in making barns and houses and mills for the reception and disposal of his crops, and fences to preserve them while growing, *before he grows them,* than to *get the crops first.* I have heard it observed that *any* American settler, even without a dollar in his pocket, would have *had something growing by this time.* Very true! I do not question that at all; for, the very first care of a settler without a dollar in his pocket is to get something to eat, and, he would consequently set to work scratching up the earth, fully confident that after a long summering upon wild flesh (without salt, perhaps) his own belly would stand him for barn, if his jaws would not for mill. But the case is very different with Mr. Birkbeck, and at present he has need for no other provision for winter but about a three hundredth part of his fine grass turned into hay, which will keep his necessary horses and cows; besides which he has nothing that eats but such pigs as live upon the waste, and a couple of fine young

deer (which would weigh, they say when full grown, 200 lbs. dead weight), that his youngest son is rearing up as pets.

910. I very much admire Mr. Birkbeck's mode of *fencing*. He makes a ditch 4 feet wide at top, sloping to 1 foot wide at bottom, and 4 feet deep. With the earth that comes out of the ditch he makes a bank on one side, which is turfed towards the ditch. Then a long pole is put up from the bottom of the ditch to 2 feet above the bank; this is crossed by a short pole from the other side, and then a rail is laid along between the forks. The banks were growing beautifully, and looked altogether very neat as well as formidable; though a live hedge (which he intends to have) instead of dead poles and rails, upon top, would make the fence far more effectual as well as handsomer. I am always surprized, until I reflect how universally and to what a degree, farming is neglected in this country, that this mode of fencing is not adopted in cultivated districts, especially where the land is wet, or lies low; for, there it answers a double purpose, being as effectual a drain as it is a fence.

911. I was rather disappointed, or sorry, at any rate, not to find near Mr. Birkbeck's any of the means for machinery or of the materials for manufactures, such as the water-falls, and the minerals and mines, which are possessed in such abundance by the states of Ohio and Kentucky, and by some parts of Pennsylvania. Some of these, however, he may yet find. Good water he has, at any rate. He showed me a well 25 feet deep, bored partly through hard substances near the bottom, that was nearly overflowing with water of excellent quality. . . .

913. On coming within the precincts of the Harmonites we found ourselves at the side of the Wabash again; the river on our right hand, and their lands on our left. Our road now lay across a field of Indian corn, of, at the very least, a mile in width, and bordering the town on the side we entered; I wanted nothing more than to behold this immense field of most beautiful corn to be at once convinced of all I had heard of the industry of this society of Germans, and I found, on proceeding a little farther, that the progress they had made exceeded all my idea of it.

914. The town is methodically laid out in a situation well chosen in all respects; the houses are good and clean, and have, each one, a nice garden well stocked with all vegetables and tastily ornamented with flowers. I observe that these people are very fond of flowers, by the bye; the cultivation of them, and musick, are their chief amusements. I am sorry to see this, as it is to me a strong symptom of simplicity and ignorance, if not a badge of their German slavery. Perhaps the pains they take with them is the cause of their flowers being finer than any I have hitherto seen in America, but, most probably, the climate here is more favourable. Having refreshed ourselves at the Tavern, where we found every thing we wanted for ourselves and our horses, and all very clean and nice, besides many good things we did not expect, such as beer, porter, and even wine, all made within the Society, and very good indeed, we then went out to see the people at their harvest, which was just begun. There were 150 men and women all reaping in the same field of wheat. A beautiful sight! The crop was very fine, and the field, extending to about two miles in length, and from half a mile to a mile in width, was all open to one view, the sun shining on it from the West, and the reapers advancing regularly over it.

915. At sun-set all the people came in, from the fields, work-shops, mills, manufactories, and from all their labours. This being their evening for prayer during the week, the Church bell called them out again, in about 15 minutes, to attend a lecture from their High Priest and Law-giver, Mr. George Rapp. We went to hear the lecture, or, rather, to see the performance, for, it being all performed in German, we could understand not a word. The people were all collected in a twinkling, the men at one end of the Church and the women at the other; it looked something like a Quaker Meeting, except that there was not a single little child in the place. Here they were kept by their Pastor a couple of hours, after which they returned home to bed. This is the quantum of Church-service they perform during the week; but on Sundays they are in Church nearly the whole of the time from getting up to going to bed. When it happens that Mr. Rapp cannot attend, either by indisposition or other accident, the Society still meet as usual, and the *elders* (certain of the most trusty and discreet, whom the Pastor selects as a sort of assistants in his divine commission) converse on religious subjects.

916. Return to the Tavern to sleep; a good comfortable house, well kept by decent people, and the master himself, who is very intelligent and obliging, is one of the very few at Harmony who can speak English. Our beds were as good as those stretched upon by the most highly pensioned and placed Boroughmongers, and our sleep, I hope, much better than the tyrants ever get, in spite of all their dungeons and gags.

917. *July 2nd.* — Early in the morning, took a look at the manufacturing establishment, accompanied by our Tavern-keeper. I find great attention is paid to this branch of their affairs. Their principle is, not to be content with the profit upon the manual labour of *raising* the article, but also to have the benefit of the machine in preparing it for *use*. I agree with them perfectly, and only wish the subject was as well understood all over the United States as it is at Harmony. It is to their skill in this way that they owe their great prosperity; if they had been nothing but farmers, they would be now at Harmony in Pennsylvania, poor cultivators, getting a bare subsistence, instead of having doubled their property two or three times over, by which they have been able to move here and select one of the choicest spots in the country. . . .

931. *July 10th.* — Leave Frankfort, and come through a district of fine land, very well watered, to Lexington; stop at Mr. Keen's tavern. Had the good fortune to meet Mr. Clay, who carried us to his house, about a mile in the country. It is a beautiful residence, situated near the centre of a very fine farm, which is just cleared and is coming into excellent cultivation. I approve of Mr. Clay's method very much, especially in laying down pasture. He clears away all the brush or underwood, leaving timber enough to afford a sufficiency of shade to the grass, which does not thrive here exposed to the sun, as in England and other such climates. By this means he has as fine grass and clover as can possibly grow. . . .

950. I will here subjoin a list of the prices at Zanesville, of provisions, stock, stores, labour, &c., just as I have it from a resident, whom I can rely upon.

	Dolls.	Cents.	Dolls.	Cents.
Flour (superfine), per barrel of 196 lbs. from	5	.to	5	75
Beef, per 100 lbs.	4	.—	4	25
Pork (prime), per 100 lbs.	4	50—	5	.
Salt, per bushel of 50 lbs.	2	25		
Potatoes, per bushel	.	25—	.	31½
Turnips, ditto	.	20		
Wheat, do. of 60 lbs. to 66 lbs.	.	75		
Indian Corn, ditto, shelled	.	33⅓—	.	50
Oats, ditto	.	25—	.	33⅓
Rye, ditto	.	50		
Barley ditto	.	75		
Turkeys, of from 12 lbs. to 20 lbs. each	.	37½—	.	50
Fowls	.	12½—	.	18¾
Live Hogs, per 100 lbs. live weight	3	.—	5	.
Cows (the best)	18	.—	25	.
Yoke of Oxen, ditto	50	.—	75	.
Sheep	2	50		
Hay, per ton, delivered	9	.—	10	.
Straw, fetch it and have it.				
Manure, ditto, ditto.				
Coals, per bushel, delivered	.	8		
Butter, per lb. avoirdupois	.	12½—	.	18¾
Cheese, ditto, ditto	.	12½—	.	25
Loaf Sugar	.	50		
Raw ditto	.	31¼		
Domestic Raw ditto	.	18¾		
Merino Wool, per lb. avoirdupois, washed	1	.		
Three-quarter Merino ditto	.	75		
Common Wool	.	50		
Bricks, per 1000, delivered	6	.to	7	.
Lime, per bushel, ditto	.	18¾		
Sand, in abundance on the banks of the river.				
Glass is sold in boxes, containing 100 square feet; of the common size there are 180 panes in a box, when the price is	14	.		
The price rises in proportion to the size of the panes.				
Oak planks, 1 inch thick, per 100 square feet, at the saw-mill.	1	50		
Poplar, the same.				
White Lead, per 100 lbs. delivered	17	.		
Red ditto	17	.		
Litharge	15	.		
Pig Lead	9	50		
Swedish Iron (the best, in bars)	14	.		

	Dolls.	Cents.	Dolls.	Cents.
Juniatta, ditto, ditto	14	.		
Mr. Dillon's ditto, ditto	12	50		
Castings at Mr. Dillon's Foundery, per ton	120	.		
Ditto, for machinery, ditto, per lb.	.	8		
Potash, per ton	180	.		
Pearl Ashes, ditto	200	.		
Stone masons and bricklayers, per day, and board and lodging	1	50		
Plasterers, by the square yard, they finding themselves in board and lodging and in lime, sand, laths and every thing they use	.	18¾		
Carpenters, by the day, who find themselves and bring their tools	1	25		
Blacksmiths, by the month, and found in board, lodging and tools	30	.to	40	.
Millwrights, per day, finding themselves	1	50 −	2	.
Tailors, per week, finding themselves and working 14 or 15 hours a day	7	. −	9	.
Shoemakers, the same.				
Glazier's charge for putting in each pane of glass 8 in. by 10 in. with their own putty and laying on the first coat of paint	.	4 to	.	5
Labourers, per annum, and found	100	. −	120	
The charge of carriage for 100 lbs. weight from Baltimore to Zanesville	10	.		
Ditto for ditto by steamboat from New Orleans to Shippingport, and thence by boats, to Zanesville, about	6	50		
Peaches, as fine as can grow, per bushel	.	12½ −	.	25

Apples and Pears proportionably cheaper; sometimes given away, in the country.

951. Prices are much about the same at Steubenville; if any difference, rather lower. If bought in a quantity, some of the articles enumerated might be had a good deal lower. Labour, no doubt, if a job of some length were offered, might be got somewhat cheaper, here.

Importation of Hereford and Shorthorn Cattle, 1822

From Henry Clay, "English Cattle," *American Farmer* (Baltimore, 1822), IV, p. 223.

Ashland, 19th August, 1822.

DEAR SIR,

Your letter under date the 11th ultimo, reached Lexington, whilst I was absent from home, on a visit to one of our watering places, and on that account there has arisen some delay in my answer. I received the portraits by mail, shortly after I addressed you on the subject of them.

You request some account of my imported English Cattle, which I give with great pleasure. In 1816, I wrote to my friend, Peter Irving at Liverpool, the brother of our distinguished countryman, Washington Irving, and requested him to purchase for me two pair of English Cattle, one of the beef and the other of the milk breed. I gave him a carte blanche, both as to price and races. He took great pains to satisfy himself of the best breeds in the kingdom, by a resort to all the means of information at his command; and he became convinced that the Hereford reds was the best, as combining in themselves the three great qualities, beef, milk and draught, which it is desirable that cattle should possess. — Accordingly he caused to be selected and purchased for me, in January 1817, two pair of that breed of cattle. Two of them were two year-olds, and the other two yearlings; the sum of their cost was L 105 sterling. They were shipped at Liverpool, in March 1817, on board the same ship, which imported the English Cattle, for some gentlemen in this neighbourhood, an account of which you have already published, but that was altogether accidental, there having been no concert whatever between those gentlemen and me, in making our respective importations. They were received by Messrs. McDonald and Ridgely, in April or May 1817, and I immediately upon hearing of their landing in the United States, despatched a messenger for them. He brought them about one hundred and fifty miles from Baltimore into the State of Virginia, and owing to the great heat of the season and the wearing away of their hoofs, he was obliged to leave them there to rest, until the weather became cooler; so that it was late in August before they reached Kentucky. One of the bulls died on his way from Baltimore to Virginia, from over feeding on red clover. Estimating the first cost of the cattle, and all incidental charges from the time they were purchased in England, until I received them in Kentucky, and charging the three survivors with what was lost by the death of their companion, those three have come to me at five hundred dollars each. My cattle are very beautiful, fine form, symmetry and color. — They are all without exception, a deep red, white faces, white under the belly, at the tip end of the tail and on one or more of the feet. As I have generally parted with the young, I am not able yet to pronounce, with certainty, whether they will realize the high expectations, which were entertained of them by my friend, Mr. Irving; but I believe he has not been deceived as to their qualities.

We have been for some years breeding in this State, very extensively, from some English Cattle, imported forty or forty-five years ago, by, I believe, Mr. Gough, of Maryland, and Mr. Miller of Virginia. This race of cattle attains a larger size, than any of those of the late importations which we have made. The latter do not, however, want size, and the circumstance of their being smaller than the descendants of the old importation is abundantly compensated by their having less bone, being greatly superior in symmetry, and distributing their flesh on the carcase much better, so as to produce more meat on the good parts. Animals which are ill made, are difficult to keep in good order, because they cannot struggle for subsistence as well as those which are well built. I may, at some future period, inform you how my cattle turn out.

With great respect,

I AM YOUR OBEDIENT SERVANT,

H. CLAY

Another Englishman Views American Farming, 1818

From James Flint, *Letters from America* (Edinburgh, 1822), as quoted in R. G. Thwaites, ed., *Early Western Travels* (Cleveland, 1904), IX, pp. 82, 96–97, 122–23, 139–40, 174–76, 178–81, 226, 232–36.

Produce, in the higher parts of Pennsylvania, may be stated at the rates of from twenty to twenty-five bushels of wheat, and from twenty-five to thirty bushels of Indian corn, per acre. These quantities are raised under slovenly management, and without much labour. A farmer expressed his contentment under existing circumstances; a dollar a bushel for wheat (he said) is a fair price, where the farmer pays neither rent nor taxes to the government. His farm, for example, pays four or five dollars a-year, for the support of the state and county officers.

Labourers receive a dollar per day, and can find board for two dollars a-week. Mechanics, in most cases, earn more. Where health is enjoyed, in this place, poverty bespeaks indolence, or want of economy. . . .

[Ohio] I saw some people thrashing buck wheat: they had dug a hollow in the field, about twenty feet in diameter, and six or eight inches in depth. In this the grain was thrashed by the flail, and the straw thrown aside to rot in the field. The wheat is cleared of the chaff by two persons fanning it with a sheet, while a third lets it fall before the wind.

Indian corn is separated from the husks or leaves that cover the ear, by the hands. In the evenings neighbours convene for this purpose. Apples are also pared for preservation in a similar way. These are commonly convivial meetings, and are well attended by young people of both sexes.

A respectable English family put ashore with a leaky boat, almost in the act of sinking. They had run foul of a log in a ripple. The craft, called family boats, are square arks, nine or ten feet wide, and varying in length as occasion may require. They are roofed all over, except a small portion of the fore part, where two persons row. At the back end, a person steers with an oar, protruded through a hole, and a small fire-place is built of brick. Such boats are so formed as to carry all the necessaries of new settlers. The plough, and the body of the waggon, are frequently to be seen lying on the roof; and the wheels hung over the sides. The bottom is made of strong plank, not liable to be stove in, except where the water is in rapid motion; and the whole fabric is exempt from the danger of upsetting, except in violent gales of wind. Family boats cost from thirty to fifty dollars at Pittsburg. A great proportion of the families to be seen, are from the northern parts of New York, and Pennsylvania, also from the state Vermont, and other parts. They have descended the Allegany, a river that I have not hitherto mentioned as a thoroughfare of travellers. . . .

In this part of Ohio State, first and second rate lands sell at four or five dollars per acre. The richest ground is in bottoms: the hilly has many [98] parts not accessible to the plough. Buildings are most commonly erected on rising grounds. Such situations are believed to be most salubrious, and abound most in good springs.

Farming establishments are small. Most cultivators do every thing for themselves, even to the fabrication of their agricultural implements. Few hire others permanently, it being difficult and expensive to keep labourers for any great length of time. They are not *servants*, all are *hired hands:* Females are averse to dairy, or menial employments. The daughters of the most numerous families continue with their parents. There is only one way of removing them. This disposition is said to prevail over almost the whole of the United States. A manufacturer at Philadelphia told me, that he had no difficulty in finding females to be employed in his work-shop; but a girl for house-work he could not procure for less than twice the manufacturing wages. Some of the children of the more necessitous families are bound out to labour for other people. The Scotch family, recently mentioned, have a boy and a girl living with them in this way. The indenture of the boy expires when he is twenty-one years of age; that of the girl at eighteen. They are clothed and educated at the expense of the employer. The boy, at the expiry of his contract, is to have a horse and saddle, of value at least 100 dollars; and the girl at the end of her engagement, is to have a bedding of clothes. It is said, that a law of the State of Ohio, forbids females to live in the houses of unmarried men.

The utensils used in agriculture are not numerous. The plough is short, clumsy, and not calculated to make either deep or neat furrows. The harrow is triangular; and is yoked with one of its angles forward, that it may be less apt to take hold [99] of the stumps of trees in the way. Light articles are carried on horseback, heavy ones by a coarse sledge, by a cart, or by a waggon. The smaller implements are the axe, the pick-axe, and the cradle-scythe; by far the most commendable of back wood apparatus. . . .

Dec. 15. [1818] Last night a man took the Sheriff of Fayette county [Kentucky] aside, on pretence of business, and immediately commenced an attack on him. The officer of the law drew a dirk, and wounded the assailant.

I note down the prices of live stock, labour, some of the necessaries of life, &c.

	Dollars	Cents
Price of a young male negro, arrived at puberty,	800	—
Hire of ditto per annum, with provisions and clothes,	100 to 150	—
Price of a young female ditto,	600	—
Hire of ditto, per annum, with provisions and clothes,	120	—
Price of a work-horse, from	100 to 120	—
Price of a fine saddle horse,	200 to 300	—
Hire of a four horse team and driver, without provisions,	4	—
Hire of a saddle horse per day,	1	—
Mechanics per day, with board,	¾ to 1	—
Labourers per day, with board,	½ to	75
Wheat per bushel,	1 to	75
Rye,		50
Corn, (Maize)		37½
Oats,	—	33⅓
Potatoes,	—	33⅓
Flour per 100 lbs.	3	—
Beef, per pound, from		5 to 6
Pork, ditto, from		4 to 5
Mutton, ditto, from	—	3 to 4
Turkeys, from 50 cents	to 1	—
Hens and Ducks,		12½
Eggs per dozen,	—	12½
Butter per lb.	—	25
Cheese, ditto,	—	18
Whisky per gallon,	—	40
Tobacco, per 100 lbs.	5	—
Hemp,	8	—
Wool, per lb		33⅓

The indolence and disorderly conduct of slaves, together with their frequent elopements, occasion much uneasiness to their holders. It is not uncommon to hear the master, in ill humour, say that he wishes there was not a slave in the country; but the man who is tenacious of this sort of stock, or who purchases it at a high price, will always find it difficult to convince other people, that his pretensions to humanity towards slaves are in earnest. Some say that the fault is with the British, who first introduced them. Others reprobate the practice; but affirm that, while the laws of the country permit it, and while slaves must be somewhere, *we may have them as well as our neighbours;* and there are a few who vindicate both principle and practice, by declaring, that the negro is a being of an inferior species formed for servitude: and allege that slave-keeping has the divine sanction, as in the case of the Jews. . . .

In exchanging Britain for the United States, the emigrant may reasonably expect to have it in his power to purchase good unimproved land, and to bring it into a rude state of cultivation, with less capital unquestionably, than that employed in renting an equal proportion of good ground at home. He will not be burdened by an excessive taxation, nor with tithes, nor poor's rates; for there are no internal taxes paid to the government, no privileged clergy, and few people who live by the charity of others. His labour and his capital will be more productive, and his accumulation of property more rapid, (good health, industry, and economy, presupposed,) and a stronger hope may be entertained, that extreme poverty or want may be kept at a distance. After residing five years in the country, he may become an elector of those who have the power of making laws and imposing taxes.

The inconveniences or difficulties which attend removing, are upon no account to be overlooked. The man who undervalues these is only holding disappointments in reserve for himself. He must part with friends, and every acquaintance to whom he is attached, a case that he may, perhaps, not fully understand, till he acts his part in it. A voyage and a long journey must be submitted to. He must breathe a new air, and bear transitions and extremes of climate, unknown to him before. His European tinge of complexion must soon vanish from his face, to return no more. A search for the new home will require his serious attention, a diversity of situations may soon be heard of, but it is not easy to visit or compare many of them. Nor is the emigrant, on his first arrival, an adequate judge of the soil of America. In a dilemma of this kind advice is necessary. This is easily procured every where; but it deserves attention to know, whether the informant is interested in the advice he gives. Land dealers, and others, naturally commend tracts of land which they are desirous to sell. The people of the neighbourhood have also an interest in the settling of neighbouring lands, knowing, that by every augmentation of population, the value of their own property is increased. On several occasions I have met with men who attempted to conceal local disadvantages, and defects in point of salubrity, that were self-evident. I do not recollect of having heard more than two persons acknowledge, that they lived in an unhealthy situation. In the high country of Pennsylvania, I was told that Pittsburg is an unhealthy place. At Pittsburg, I heard that Marietta and Steubenville are very subject to sickness. At these places, the people contrast their healthy situation with Chillicothe, which, I was told, is very unhealthy. At Chillicothe, the climate of Cincinnati is deprecated; and at Cincinnati, many people seem willing to transfer the evil to the falls of the Ohio. At this place the truth is partially admitted; but it is affirmed that the Illinois country, and down the Mississippi are very unhealthy. The cautious will always look to the views and character of the man who would direct them, and will occasionally rely on their own judgments.

In the public land-offices, maps of the new lands are kept. Sections of a square mile, and quarter sections of 160 acres, are laid down. The squares entered are marked A. P. meaning advance paid. This advance is half a dollar per acre, or one-fourth of the price. Lands, when first put to sale, are offered by public auction, and are set up at two dollars per acre. If no one offers that price, they are exhibited on the land-office map, and may be sold at that rate at any subsequent time. New settlers, who are sufficiently skilled in the quality of the soil, are in no danger from land-office transactions. Besides the land-offices for the sale of national property, there are agents who sell on account of individuals. . . .

The sixteenth section of each township is reserved for the support of a school.

Lands entered at the public sales, or at the Register's office, are payable, one fourth part of the price at the time of purchase; one fourth at the expiry of two years; one fourth at three years, and the remaining fourth at four years. By law, lands not fully paid at the end of five years, are forfeited to the government, but examples are not wanting of States petitioning Congress for indulgence on this point, and obtaining it. For money paid in advance at the land office a discount of eight per cent. per annum is allowed, till instalments to the amount of the payment become due. For failures in the payment of instalments, interest at six per cent is taken till paid. The most skilful speculators usually pay only a fourth part of the price at entry, conceiving that they can derive a much greater profit than eight per cent. per annum from the increasing value of property, and occasionally from renting it out to others. Where judicious selections are made, they calculate rightly.

The land system now adopted in the United States is admirable in regard of ingenuity, simplicity, and liberality. A slight attention to the map of a district, will enable any one to know at once the relative situation of any section that he may afterwards hear mentioned, and its direct distance in measured miles. There can be no necessity for giving names to farms or estates, as the designation of the particular township, and the number of the section is sufficient, and has, besides, the singular convenience of conveying accurate information as to where it is situated. By the new arrangement the boundaries of possessions are most securely fixed, and freed alike from the inconvenience of rivers changing their course, and complexity of curved lines. Litigation amongst neighbours as to their landmarks, is in a great measure excluded. The title deed is printed on a piece of parchment of the quarto size. The date, the locality of the purchase, and the purchaser's name, are inserted in writing, and the instrument is subscribed by the President of the United States, and the agent of the general land office. It is delivered to the buyer free of all expense, and may be transferred by him to another person without using stamped paper, and without the intervention of a law practitioner. The business of the land office proceeds on the most moderate principles, and with the strictest regard to justice. The proceeds are applied in defraying the expense of government, and form a resource against taxation. The public lands are in reality the property of the people.

The stranger who would go into the woods to make a selection of lands, ought to take with him an extract from the land office map applying to the part of the country he intends to visit. Without this, he cannot well distinguish entered from unentered grounds. He should also procure the names of the resident people, with the numbers and quarters of the sections they live on, not neglecting to carry with him a pocket-compass, to enable him to follow the blazed lines marked out by the surveyor. *Blaze* is a word signifying a mark cut by a hatchet on the bark of a tree. It is the more necessary for the explorer to be furnished thus, as he may expect to meet with settlers who will not be willing to direct him, but, on the contrary, tell him with the greatest effrontery, that every neighbouring quarter section is already taken up. Squatters, a class of men who take possession without purchasing, are afraid of being turned out, or of having their pastures abridged by new comers. Others, perhaps meditating an enlargement of their property, so soon as funds will permit, wish to hold the adjoining lands in reserve for themselves, and not a few are jealous of the land-dealer, who is not an actual settler, whose grounds lie waste, waiting for that advance on the value of property, which arises from an increasing population. The non-resident proprietor is injurious to a neighbourhood, in respect of his not bearing any part of the expense of making roads, while other people are frequently under the necessity of making them through his lands for their own convenience. On excursions of this kind, the prudent will always be cautious of explaining their views, particularly as to the spot chosen for purchase, and without loss of time they should return to the land-office and make entry.

The new abode being fixed, the settler may be surrounded by strangers. Polite and obliging behaviour with circumspection in every transaction, become him in this new situation. . . .

Agriculture languishes — farmers cannot find profit in hiring labourers. The increase of produce in the United States is greater than any increase of consumpt that may be pointed out elsewhere. To increase the quantity of provisions, then, without enlarging the numbers of those who eat them, will be only diminishing the price farther. Land in these circumstances can be of no value to the capitalist who would employ his funds in farming. The spare capital of farmers is here chiefly laid out in the purchase of lands. . . .

All who have paid attention to the progress of new settlements, agree in stating, that the first possession of the woods in America, was taken by a class of hunters, commonly called backwoodsmen. These, in some instances, purchased the soil from the government, and in others, placed themselves on the public lands without permission. Many of them, indeed, settled new territories before the ground was surveyed, and before public sales commenced. Formerly preemption rights were given to these squatters; but the irregularities and complicacy that the practice introduced into the business of the land-office, have caused its being given up, and squatters are now obliged to make way for regular purchasers. The improvements of a backwoodsman are usually confined to building a rude log cabin, clearing and fencing a small piece of ground for raising Indian corn. A horse, a cow, a few hogs, and some poultry, comprise his live-stock; and his farther operations are performed with his rifle. The formation of a settlement in his neighbourhood is hurtful to the success of his favourite pursuit, and is the signal for his removing into more remote parts of the wilderness. In the case of his owning the land on which he has settled, he is contented to sell it at a low price, and his establishment, though trifling, adds much to the comfort of his successor. The next class of settlers differ from the former in having considerably less dependence on the killing of game, in remaining in the midst of a growing population, and in devoting themselves more to agriculture. A man of this class proceeds on a small capital; he either enlarges the clearings begun in the woods by his backwoodsmen predecessor, or establishes himself on a new site. On his arrival in a settlement, the neighbours unite in assisting him to erect a cabin for the reception of his family. Some of them cut down the trees, others drag them to the spot with oxen, and the rest build up the logs. In this way a house is commonly reared in one day. For this well-timed assistance no immediate payment is made, and he acquits himself by working to his neighbours. It is not in his power to hire labourers, and must depend therefore on his own exertions. If his family is numerous and industrious, his progress is greatly accelerated. He does not clear away the forests by dint of labour, but girdles the trees. By the second summer after this operation is performed, the foliage is completely destroyed, and his crops are not injured by the shade. He plants an orchard, which thrives and bears abundantly under every sort of neglect. His live-stock soon becomes much more numerous than that of his backwood predecessor; but, as his cattle have to shift for themselves in the woods, where grass is scanty, they are small and lean. He does not sow grass seeds to succeed his crops, so that his land, which ought to be pasturage, is overgrown with weeds. The neglect of sowing grass-seeds deprives him of hay; and he has no fodder laid up for the winter except the blades of Indian corn, which are much withered, and do not appear to be nutritious food. The poor animals are forced to range the forests in winter, where they can scarcely procure any thing which is green, except the buds of underwood on which they browse. — Trees are sometimes cut down that the cattle may eat the buds. Want of shelter in the winter completes the sum of misery. Hogs suffer famine during the droughts of summer, and the frosts and snows of winter; but they become fat by feeding on the acorns and beech nuts which strow the ground in autumn. Horses are not exempted from their share in these common sufferings, with the addition of labour, which most of them are not very able to undergo. This second rate class of farmers are to be seen in the markets of towns, retailing vegetables, fruits, poultry, and dairy produce. One of them came lately into this place on horseback, with ten pounds of butter to sell; but as he could not obtain a price to his mind, he crossed the river to Louisville market. In going and returning he must have paid twenty-five cents to the ferry-man — a considerable expense, when it is considered that he had travelled twelve miles with his little cargo. Another, who lives at the distance of eight miles from this place, brought a barrel of whisky, containing about thirty-three gallons. He employed neither horse nor vehicle in the transportation, but rolled the cask along the road, which, by the by, is none of the smoothest. Incidents of this kind may, perhaps, cause you to suppose that the condition of the second rate settler is similar to that of subtenants in the north of Scotland, or in Ireland; but the high price of labour in America explains the apparent parity. Men perform offices for themselves that, in Britain, would be done by hiring others. The American farmer, it must be observed, is commonly the proprietor of the land he occupies; and, in the *hauteur* of independence, is not surpassed by the proudest freeholders of Britain. The settler of the grade under consideration, is only able to bring a small portion of his land into cultivation, his success, therefore, does not so much depend on the quantity of produce which he raises, as on the gradual increase in the value of his property. When the neighbourhood becomes more populous, he in general has it in his power to sell his property at a high price, and to remove to a new settlement, where he can purchase a more extensive tract of land, or commence farming on a larger scale than formerly. The next occupier is a capitalist, who immediately builds a larger barn than the former, and then a brick or a frame house. He either pulls down the dwelling of his predecessor, or converts it into a stable. He erects better fences, and enlarges the quantity of cultivated land; sows down pasture fields, introduces an improved stock of horses, cattle, sheep, and these probably of the Merino breed. He fattens cattle for the market, and perhaps erects a flour-mill, or a saw-mill, or a distillery. Farmers of this description are frequently partners in the banks; members of the State assembly, or of Congress, or Justices of the Peace. The condition of the people has necessarily some relation to the age and prosperity of the settlements in which they live.

In Pennsylvania, for instance the most extensive farmers are prevalent. In the earliest settled parts of Ohio and Kentucky, the first and second rate farmers are most numerous, and are mixed together. In Indiana, backwoodsmen and second rate settlers predominate. The three conditions of settlers described, are not to be understood as uniformly distinct; for there are intermediate stages, from which individuals of one class pass, as it were, into another. The first invaders of the forest frequently become farmers of the second order; and there are examples of individuals acting their parts in all the three gradations.

In the district of Jeffersonville, [Ohio] there has been an apparent interruption of the prosperity of the settlers. Upwards of two hundred quarter sections of land are by law forfeited to the government, for non-payment of part of the purchase money due more than a year ago. A year's indulgence was granted by Congress, but unless farther accommodation is immediately allowed, the lands will soon be offered a second time for sale. Settlers seeing the danger of losing their possessions, are now offering to transfer their rights for less sums than have already been paid; it being still in the power of purchasers to retain the lands on paying up the arrears due in the land office. This marks the difficulty that individuals at present have, in procuring small sums of money, in this particular district. . . .

On the 29th of June, [1820], wheat harvest was commenced on several farms to the west of Madison [Ohio]. Oats, at that time, were headed out and luxuriant; but the heat of the climate is uniformly unfavourable to the ripening of this kind of crop. Its weight, relative to measure, is usually about half of that of good grain in the better parts of Britain. The growth of Indian corn is this season luxuriant. The only injury it has suffered arises from squirrels that gathered a considerable quantity of the seed in many fields. Squirrels are not so excessively numerous in the uninhabited woods as in the vicinity of cultivated fields. Potatoes are small and of a bad quality. At Jeffersonville, [Indiana] so early as the 29th of May last, new potatoes were in the market. Turnips (so far as I have observed) do not grow to a large size, nor are they raised in large quantities. Flax, in every field that I have seen, was a short crop, with strong stems, and tops too much forked. Probably thicker sowing would improve its quality. Hemp grows with great luxuriance. The orchards are abundantly productive, and yield apples of the largest size; but little care is taken in selecting or ingrafting from varieties of the best flavour. Small crab apples are the most acid, and produce the finest cider. Pears are scarcely to be seen. Peaches of the best and worst qualities are to be met with. The trees bear on the third summer after the seed is sown, and although no attention is paid to the rearing them, the fruit is excessively plentiful, and is sometimes sold at twenty-five cents (1s. 1½d. English) per bushel. Last year I weighed a peach, and found its weight to be eleven ounces, and I observed in a newspaper about the same time, an account of one of the extraordinary weight of fourteen ounces. A rancid sort of spirit is distilled from them, known here by the name of peach brandy. Cherries are small. The earliest this season at Cincinnati, were ripe on the 22d of May. Wild cherry trees grow to a great height in the woods; the timber is of a red colour, and is used in making tables, bureaus, &c. and forms a tolerable substitute for mahogany.

Ornamental gardening is a pursuit little attended to, and perhaps will not soon be generally exhibited. The soil of the best land being soft, the torrents of rain which almost instantaneously deluge the surface convert it into a paste of a very unsightly appearance. Where the ground has even a slight declivity, it is liable to have deep ruts washed in it. Low walks and other hollows, are often filled with the soil carried down from higher parts of the ground. The severity of the winter is another obstacle; it being difficult to preserve some perennial and biennial plants, or to procure culinary vegetables in the spring. The stock of cultivated flower roots is very small, and these not well selected. Gooseberries and currants are scarce and small. Cucumbers, melons, and a variety of products that require artificial heat in Britain, grow here vigorously in the open air.

Beginnings of the Farm Press, 1819

From *American Farmer* (Baltimore, 1819), I, p. 5.

Baltimore: Friday, April 2, 1819.

To the Public.

It was observed, by a man proverbial for his wisdom, that "KNOWLEDGE IS POWER:" and there is perhaps no pursuit in life, wherein the truth of that saying is more frequently exemplified, than in the various conditions. and fortunes of those, who live, *by the cultivation of the soil.*

How often does it happen that young men, coming by inheritance to ample fortunes, patient of labour, and anxious to accumulate, yet become every day more involved, and presently behold their last acre struck off under the sheriff's hammer; all for want of *skill in "the management of their resources;"* for want of that "knowledge which is power." Much, it is true, depends upon *industry*, but of what avail is mere passive industry, without judgment to apply it? like a fine horse, spirited, vigorous and full of animation, yet if he be blind, leave him without a guide, and he will soon throw himself over the cliff or plunge into the dock.

The great aim, and the chief pride, of the *"American Farmer,"* will be, to collect information from every source, on every branch of Husbandry, thus to enable the reader to study the various systems which experience has proved to be the best, under given circumstances; and in short, to put him in possession of that knowledge and skill in the exercise of his means, without which the best farm and the most ample materials, will remain but as so much *dead capital* in the hands of their proprietor.

Besides articles on the main subject of the paper, it will present original and selected essays and extracts calculated for amusement or instruction, and a *substantial* detail of passing occurrences — and, finally, it will contain a faithful account of the actual prices of all those *principal* articles, which the people of the country generally have to buy, or to sell, in the Baltimore market.

But, as the Editor is aware that "to *promise* is most courtly and fashion-able," he will therefore only add, that the *American Farmer* will be conducted on broad and liberal principles, containing nothing indecorous or personally offensive to the feelings or character of any sect or individual. And further, that if at the end of the year, any subscriber should think he has not received his "penny'orth," he shall be at liberty to withdraw, and his subscription money shall be repaid to him on demand.

Terms.

The price of subscription is $4 per annum; payable $2, half yearly, in advance.

Local Agricultural Societies, 1820

From Elkanah Watson, *History of the Rise, Progress, and existing Condition of the Western Canals . . . and Agricultural Societies* (Albany, 1820), pp. 115–37, 184–87.

In June, 1807, I purchased the elegant seat of Henry Van Schaick, Esquire, in Pittsfield, Massachusetts, thirty-six miles from Albany. The farm, containing two hundred and fifty acres, was much exhausted, — yet from its high state of improvement, abounding with excellent fruit, and the convenience and elegance of the buildings, also a circular pond of twenty acres, and lying within one mile of a beautiful village, — I was induced, at the age of fifty, to hazard my own, and my family's happiness, on the experiment of seeking *"rural felicity,"* — a life I had for twenty years sighed to enjoy; my only mistake was, I commenced too late. In consequence, I soon found my error, — my habits being settled for city life: to retreat was impossible, — to labour in person, *was impossible*. To fill up the void in an active mind, led me first to conceive the idea of an Agricultural Society on a plan different from all others.

With a view of writing this book, in conformity to the wishes of a respectable friend in North-Carolina, I have resorted to all my documents, to trace its commencement, and progress, — they shall be faithfully narrated.

In the fall of 1807, I procured the first pair of Merino sheep, that had appeared in Berkshire, if not in the state. They were the first I had ever seen: although defective in the grade I was led to expect, yet, as all who examined their wool, were delighted with its texture and fineness, *I was induced to notify an exhibition under the great elm tree in the public square, in Pittsfield, of these two sheep, on a certain day. Many farmers, and even women, were excited by curiosity to attend this first novel, and humble exhibition. It was by this lucky accident, I reasoned thus, — If two animals are capable of exciting so much attention, what would be the effect on a larger scale, with larger animals? The farmers present responded to my remarks with approbation. — We became acquainted, by this little incident; and from that moment, to the present, agricultural societies, cattle shows, and all in connexion therewith, have predominated in my mind, greatly to the injury of my private affairs.*

The winter following, I addressed, (under the signature of Projector,) the farmers of Berkshire, with a view to the spread of the Merino sheep; which I considered as invaluable; especially in the hilly countries of New-England. In these first essays, the following *extracts* were an introduction to the subject of establishing an Agricultural society.

"The most certain and direct road to effect this great object, appears to me, will be the organization of an Agricultural society, which will ultimately embrace all the respectable farmers of the county, who will bring to this common fund, like bees to the hive, their stock of experience, for the good of the whole; whereas, for the want of such a central point of communication, how much useful information is necessarily swallowed up in the grave; which, of right, should be the common property of mankind," &c.

The public mind was gradually maturing to embrace the prize in full view.

The breed of swine in Berkshire, was of the most unprofitable kind; — long legs, — tall, — lank sides, — large bones, requiring more to fat them than their value.

In 1808, I procured from Dutchess county, in the state of New-York, a pair of small boned, — short legged, grass fed pigs, so called. The old breed gradually disappeared, and the community have gained largely by the exchange. In the same year, I purchased a young bull at Cherry-Valley, of the celebrated English stock, with a view of ameliorating the breed of cattle. The winter of 1809, I introduced into my pond, in connexion with a few friends, the first pickerel which had been seen in Berkshire, with the exception of a commencement at Stockbridge two years previous. Their prolific increase in all the ponds, and streams, now affords an essential item of delicious, and increasing food for the inhabitants, which cost them nothing but the trouble of catching, and eating them.

I stood alone, the butt of ridicule, till the 1st of August, 1810, when I wrote an appeal to the public, thus:

BERKSHIRE CATTLE SHOW.
"The multiplication of useful animals is a common blessing to mankind" WASHINGTON.

TO FARMERS.
The subscribers take the liberty to address you on a momentous subject, which, in all probability, will materially affect the agricultural interest of this county."

After several other remarks, it concludes:

"In a hope of being instrumental in commencing a plan so useful in its consequences, we propose to exhibit in the square in the village of Pittsfield, on the 1st October next, from 9 to 3 o'clock, bulls," &c. &c. "It is hoped this essay will not be confined to the present year, but will lead to permanent annual cattle shows, and that an incorporated agricultural society will emanate from these meetings, which will hereafter be possessed of funds sufficient to award premiums," &c.

Signed, 1st August, 1810, Samuel H. Wheeler, and twenty-six farmers, among whom I include myself.

In consequence of this first step, on the 1st October, 1810, I find the following notice in the Pittsfield Sun.

The first Berkshire cattle show was exhibited with considerable eclat on Monday last. This laudable measure cannot fail to be highly beneficial, considering its novelty in this part of the world, and that many had their doubts, and even a dread of being held up for the finger of scorn to point at. The display of fine animals, and the number exhibited, exceeded the most sanguine hopes of its promoters, and a large collection of people participated in the display: the weather was delightful. The ice is now broke, — all squeamish feelings buried, — and a general satisfaction evinced. It will now be impossible to arrest its course; we have every thing to hope, and to expect, the year ensuing.

A committee of fourteen respectable farmers, from different parts of the county, was appointed to take preparatory measures for a real exhibition in October, 1811. As I had thus far taken the lead in every thing, by common consent, I was placed at the head of a procession of farmers, marching round the square, without motive, or object; having returned from whence we started, and to separate with some eclat, I stepped in front, gave three cheers, in which they all united, — we then parted, well pleased with the day, and with each other.

The following winter we were incorporated into an Agricultural Society, with ample powers to do good, — but no funds. The persons named in the law, met to organize. I was chosen president, and devoted myself in preparation for a splendid exhibition, the following 24th of September; in pursuance of which, on the 1st August of the same year, in a public notice, in my official capacity, I stated:

We take the liberty to recommend to farmers, to select and prepare prime animals for the exhibition, also for manufacturers to exhibit their best cloth, &c. for inspection or sale. All members of the society are requested to appear in American manufactures. Innocent recreations will be permitted, but every thing tending to immorality, will be discountenanced," &c. &c.

On the 24th September, 1811, we were blessed with another fine day; and the village of Pittsfield was literally crowded with people, at an early hour, by estimation, three or four thousand. Domestic animals were also seen coming from every quarter.

By the request of the committee, I opened the proceedings in the town-house, with an address. Having never spoke in public, and feeling the awkwardness of my situation, — standing before the multitude, (I had been principally instrumental in assembling,) as a visionary projector, — it was with infinite difficulty I could command my nerves, to commence and proceed in my address. It began thus:

On this first occasion of our meeting as an agricultural society, under the sanction of law, and the universal approbation of the community, it has become a duty in the honourable station you have thought proper to assign to me, to address you. Having spent the greater part of my life in cities, (with the exception of the last four years,) it would have been more proper, to have placed a more experienced farmer at the head of an institution, which promises such important benefits to the community at large; especially if conducted with prudence and wisdom; but since you have conferred on me such a distinguished honour, I will endeavour to make good by my zeal and exertions, what I am deficient in experience, in the honourable profession of a farmer.

We all owe to our families, and to our country, the use of such talents as our common Benefactor has been pleased to bestow upon us; and as life is short, with minds disposed to be useful, we may each in our respective stations, contribute to the high destinies which evidently await our favoured land, when we of this generation will be mouldering in the dust," &c. — At the close thus, — "Although the spirit of party has found its way into every other public object in this county, I can solemnly declare, no shade of its deadly poison has as yet, to my knowledge, entered into this institution, — on the contrary, every voice proclaims, in substance, " 'We will strangle the monster at every approach.' Each party professes to be the political disciples of the departed Washington, and yet both stand as political antipodes to each other. Let us then, one and all, for the honour and happiness of ourselves and our country, cultivate a more liberal spirit of political charity towards each other, and leave every man responsible only to his God, and the laws of the land, for the particular cast of his mind, as well in politics as in religion. It is as ridiculous as impertinent, for a man to quarrel with another for not thinking as he thinks, as for not looking as he looks. Two centuries ago Europe was deluged in blood on the score of religious intolerance.

"In these enlightened days, as we are pleased to call them, we look back with astonishment, and disgust, at the folly of man in those days. Will not our descendants have equal cause to regret the folly of the present age? How can Americans boast of our freedom, when we are all combined to enslave each other's opinions, — whereas the freedom of the mind is the most powerful attribute of freemen, and the most valuable prerogative of human nature. Let us then, with one voice and one heart, join hand in hand, on this auspicious day, like a band of affectionate American brothers, intent only on the welfare and happiness of our common country. Should we, on the contrary, suffer the intollerant spirit of party to enter our borders, in any shape, it will prostrate our fair and honourable hopes, and render all our exertions of no avail."

I have noticed these leading features of the address, as it was instrumental in paving the way to the existing system. The rest was a review of the actual state of agriculture in Berkshire: its defects, and a comparative view with that of England, and other European countries.

The procession was immediately formed, — it was splendid, novel, and imposing, beyond any thing of the kind, ever exhibited in America. It cost me an infinity of trouble, and some cash, but it resulted in exciting general attention in the Northern States; and placing our society on elevated ground. In this procession were sixty yoke of prime oxen, connected by chains, drawing a plough, held by two of the oldest men in the county, — a band of music, — the society carrying appropriate ensigns, and each member carrying a badge of wheat in his hat, a stage drawn by oxen, having a broadcloth loom and a spinning jenny, both in operation by English artists, as it moved on, — mechanics, with an appropriate flag, and another stage filled with American manufactures. Four marshals on grey horses, headed by sheriff Larned, conducted the procession, which extended about half a mile. The pens were handsomely filled with many excellent animals. Twelve premiums, amounting only to seventy dollars, were awarded to the most meritorious. Nothing was yet offered for agriculture, — the best farmer, — or domestic manufactures.

The grand difficulty of procuring funds to enable the society to give a powerful impulse on the public mind, was the most serious obstacle we had to combat. From the farmers we could only calculate upon the annual dollar, and as the members did not exceed one hundred and fifty, and as five hundred dollars was indispensable to produce the effect in contemplation, it was evident the difference must be supplied by private donations; or, the society must sink into contempt, and be abandoned. Under this deep apprehension, and encouraged by some gentlemen from Boston, who were present at the recent exhibition; I determined to proceed there, at my own expense, as the society had no funds. I spent one month soliciting charity; although our efforts were highly applauded, and I was greatly honoured in the legislature by personal attentions, yet all my exertions were unavailing. I found myself pursuing an *ignis fatuus*, — and although flattered by some with encouragement, yet I had to abandon the pursuit, much mortified and humbled with this abortive, begging expedition; and returned to Pittsfield, after expending about one hundred and fifty dollars.

The exhibition of 1812, was distinguished by a great increase of premiums, amounting to two hundred and eight dollars, of which fifty dollars was awarded to me, for the best piece of superfine broadcloth, made of the down of my wool, which I gave up to the artists; the residue was awarded on animals. This cloth made a great noise in America. The president of the United States, and several distinguished characters were clothed with it.

Satisfied of the propriety of solemnizing these occasions in the church, by intermixing religious exercises, with appropriate addresses; and the delivery of premiums, and as peculiarly proper in devout acknowledgement for the blessings of the year; being also impressed with a belief that this measure would tend to give popularity to the society, among the graver class of the community; we suggested our wishes to several of the clergy, who were present, soliciting their co-operation in our views. They were also shy; probably considering our measures the bubble of a moment, and that they would make themselves ridiculous. In explaining our views to the Rev. Mr. Porter, of Roxbury, he ascended the pulpit preparatory to the address, and the delivery of our premiums, and favoured us with an animated, pastoral prayer, which was, probably, the first that was ever made on a similar occasion.

The preceding section furnishes an historical review of the successful establishment of the Berkshire agricultural society. The next feature will be to exhibit the means by which the female part of the community were induced, indirectly, to identify themselves with the society.

The grand secret, in all our operations, was to trace the windings of the human heart, and to produce *effects* from every step. It was a great object to excite the females to a spirit of emulation; we were satisfied no measures would lead to that result with so much certainty, as premiums on domestic manufactures, and closing the second *"Farmer's Holiday,"* in *"innocent festivity,"* by an agricultural ball. Also, to unite them in singing pastoral odes at the church. These measures gave us access to hearts naturally warm, and patriotic; we found no difficulty, once in possession of the field, to give their excited minds a direction which will promote their own happiness, and independence; and save millions to our country, as we shall see in the sequel. All this was effected in 1813, as well as the organization of a viewing committee of agriculture. The successful establishment of these two prominent features of the institution, may be considered of primary importance. Since that period, the females have taken a conspicuous lead, in promoting domestic manufactures; they have also added to the popularity of the society. All the other societies, who have adopted the Berkshire system, have wisely commenced in the offset, by uniting the females in their measures. To this source, therefore, may be attributed the increase, and improvement of domestic manufactures, wherever agricultural societies are established.

The vast effects which will grow out of this system, when these societies shall become general, is beyond the reach of figures, in the view of arresting our colonial degradation and dependence on foreign countries, especially for articles of clothing. Perhaps the neat gain to the nation may equal the benefit which agriculture will derive from these institutions. We found it, however, a very difficult, and a very delicate matter to open the way, and induce the women to assemble even in a private room; although a time was set apart, and seven valuable premiums of silver plate were exclusively *devoted to them*, to be awarded on domestic manufactures.

The manner this was effected, forms a memorable, and a curious epoch in the history of this society. At the exhibition in October, the ladies of the county were invited to appear the 12th of January, 1813, with the fruits of their industry, and receive their premiums; — the day arrived, — a long room was prepared at Merrick's tavern, in Pittsfield. Many excellent articles of domestic manufactures, (especially woollens and linens,) were exhibited to a considerable extent; but no female was seen to claim premiums; — this was the crisis, and I was extremely agitated, lest the experiment should fail, on which such important national results were suspended.

Such was their timidity, and dread of being laughed at; (which is a peculiar trait in people residing in the country,) — none dare be the first to support a new project. To break down this folly, we had to resort to a manœuvre, which, in one hour, succeeded to our wishes. I left the hall, proceeded one mile, and with no little difficulty prevailed on my good wife to accompany me to a private room in the house of exhibition.

I then despatched messengers to the ladies of the village in every direction, that she waited for them at the cloth show; they poured forth, — farmers' wives, who were lying in wait to watch the movements of the waters, also issued forth, and the hall was speedily filled with female spectators, and candidates for premiums. I can safely say this was one of the most grateful moments of my life. I immediately rose in the rear of the table, on which the glittering premiums were displayed, and as soon as the awards were determined by the committee, I addressed them as follows: —

LADIES OF BERKSHIRE,
    You have been invited by the society to attend on this auspicious occasion. As its organ, it is made my duty to address a few words to you, cn their behalf. It is as novel, as it is interesting, to find the female portion of the community, called to a public exhibition, on any occasion. Nevertheless, I trust the interesting events of the day will convince you all of the propriety of the measure, by the honourable competition it will necessarily excite, to excel each other in laudable efforts to promote the best interests of our common country, by the rapid increase, both in quantity and quality, of home made domestic manufactures.

    Your worthy example, by your attendance here, will not fail to produce the most beneficial, and extensive good effects. The labours of the society, and your efforts, will go hand in hand; mutually supporting and animating each other. We feel a confidence, that by great efforts, we shall be enabled at the next annual exhibition, to extend some of our awards to females *exclusively*. When we revert our eyes to the elegant display of beautiful cloths, the work of your own hands, now open to our view, our confidence in this hope is much increased. We notice with regret so few candidates for premiums among the ladies, who have honoured us with their attendance Probably, the novelty, in this first essay, has withheld many; but as the charm is dissolved, and all apprehensions of impropriety removed: hereafter no one can receive a premium without appearing in person, and proofs will be required of the articles offered, being wholly manufactured in families. No person to produce more than one piece of cloth.

The residue of the address was remarks of a general nature. Although the legislature were deaf to our loud and repeated calls for help, and blind to the inroads we were making, to overcome a wretched system of husbandry, which pervaded the whole state; although every step we took was leading the community forward by a silent impulse, eventually to see their own good, — yet the exhibition of 1813, was distinguished by sixty-three premiums, amounting to four hundred dollars; — seventeen on agriculture, — fifteen on domestic manufactures, exclusively for females, — eleven ditto, for men, — twenty on domestic animals.

The women came forth with spirit and animation; producing a fine display of domestic manufactures. Our funds were principally obtained from the liberality of a *few individuals*, partly in *Boston*. In my address, on this occasion, in a crowded and spacious church, I observed:

If the measures of our society have been instrumental in promoting the progress of the invaluable objects alluded to, as also ameliorating the breed of domestic animals, it is much to be regretted, that we are deprived of any permanent resources, to sustain a continuation of our efforts. It is with regret I also add, that several of our members are deficient in the discharge of a trifling annual sum, in support of measures of primary magnitude.

No individual will condescend from year to year, to the humiliating, and painful alternative, of soliciting *charity*, especially beyond the county, in support of measures calculated for the general good, — and if encouraged, they will eventually pervade the United States. The legislature of this state will not listen to our repeated supplications, and we have had no other resource but begging.

If then, gentlemen, we are to remain from year to year, unsupported by legislative patronage; we must, I say, reluctant to ourselves, and probably to the immense loss of this immediate community, abandon our pursuits; and leave the progress of improvements in agriculture to grope their way in the dark, with slow and heavy steps as heretofore, — instead of being led forward with rapid strides, by the attentive hand of an organized society of farmers.

Presuming that the wonderful increase of our domestic manufactures, this day displayed to our view, — the marked amelioration in our domestic animals, — the introduction of new and valuable articles of agriculture, — will so far establish the vast importance of this society, as to *command* the patronage of the legislature; — we recommend to every portion of the community to prepare for a grand exhibition the ensuing year, in that hope; especially as they cannot fail to be the gainers, even if disappointed. Should our hopes be illusive, and no further premiums be given, it is contemplated, in that unfortunate event, to establish, on the basis of this institution, regular annual fairs, to be held in this village, for the interchange and sale of animals, and domestic manufactures, from every part of the United States.

On the second evening the young men, by our request, got up a splendid agricultural ball, being the first on this occasion, and many farmers' daughters graced the floor. The officers of the society, and many respectable visitors, attended to give countenance to the measure. This was a proud day for Berkshire. Our country being now in the midst of war, the measures of our society had already so extended domestic manufactures, as to afford large supplies of clothing for our armies. Having exhausted all my powers in soliciting charity in taverns, in travelling, in cities, to support our arduous measures; I wrote urgently to my respectable friend, John Adams, formerly president of the United States, and now president of the Old Massachusetts society, — intreating his influence to procure us aid from their ample funds, no part of which, (by the admission of one of their officers to me,) could be given away in premiums previous to the year 1811, because they did not pursue any efficient measure to excite emulation. They depended too much on types, and too little on personal efforts. Our plan, on the contrary, was in the first instance to seize the bull by the horns, and lead the way by great exertions, into the heart of the community. This we effected, by touching a string which never fails, (if properly directed,) to vibrate in unison with all, viz. — self-love, — self-interest, combined with a natural love of country. I received from that respectable man the following very singular reply.

Quincy, 16th September, 1812.
You will get no aid from Boston, — commerce, literature, theology, are all against *you;* nay, medicine, history, and university, and universal politics might be added. — I cannot, I will not be more explicit.

This is a strange and an inexplicable paragraph; to be thus anathematized, as it *then* appeared to me, by such a host of enemies, in the midst of my ardent efforts, to touch the vital interests of dear America, was to me unaccountable, and unjust in the extreme. It led me to pause, and ask myself, Am I in a visionary dream, or a fit subject for Bedlam?

Mr. Adams was, however, mistaken. The legislature of Massachusetts extended their arms to embrace and patronize the society, in Janaury, 1817. Five years after this severe attack, and one year after I left the state for ever, I requested from this great man an explanation, — it follows: —

When I said to you, that 'so many interests, and respectable societies, were against *you,*' I meant, that at that time, an *anglomania* and an *antigallican* enthusiasm were prevalent, and triumphant in this quarter; and that you, as well as I, had given offence, by our approbation of the war.

In the summer of 1811, a very important feature was added to this society, and as it will, in all probability, be productive of extensive usefulness to the nation, and to posterity; and, considering its novelty, it deserves a particular notice in this place.

The executive committee, of which I was *ex officio* chairman, held frequent meetings to devise measures to ascertain the merits of articles of agriculture offered for premium; as also the best farmer in the county. We found such a diversity of opinion, and so much perplexity, in coming to a result, to guard against frauds, if not perjury; that we were on the point of abandoning the system, as impracticable, when a fortunate idea struck Ebenezer Center, Esq. one of the committee, viz. — to appoint a viewing committee of agriculture, to inspect crops standing in the field, in the month of July, when nine distinguished farmers, or a majority, could come to a satisfactory decision by the *eye;* I considered the project impracticable, but a majority fortunately decided otherwise: and we determined on an experiment. The successful result I detailed in my first address to the Otsego agricultural society, in October, 1817, thus: —

Their duty is to examine, and award on crops standing in the fields, in the month of July, in every part of the county, where candidates for premiums on articles of agriculture, shall have previously sent in their names, residence, and articles offered. The week following, say, middle of July, the committee assemble, and with a map of the county, and a list of candidates, under their eyes, they settle their route: proceeding from town, to town; and from farm, to farm; until they have faithfully accomplished this pleasing and interesting duty, which commonly requires a week. But no member of these committees, has ever yet complained that the time was too long, or the duty too severe; on the contrary, they all express themselves delighted with these agricultural tours. The route, and the time of their arrival at the respective places, being well known, they are received by the whole community with open arms, especially by the candidates. They are frequently attended by a train of anxious and curious spectators.

"To see a group of the most respectable farmers, (as if under the solemnity of an oath,) critically inspecting in the midst of fields of grain, grass, vegetables, &c. also the state of the orchards, buildings, fences, and farming utensils, and to witness the anxious candidate for premiums, attentively seizing every lisp favourable to his husbandry, or probable success, is more exhilarating to the pride of patriotism, than to view the gorgeous pageantry of palaces, and their pampered tenants, decorated in GOLD.

Both which scenes have passed in review under my eyes. To contemplate such an interesting group, in the midst of free and independent Americans, — in a community of enlightened, high minded farmers, the true lords of the soil, we have all reason to join in solemn adoration to the Eternal Fountain of all good, to prostrate ourselves in humble gratitude, that our lot is cast in the only free country on this terrestrial globe, in a country, compared to all others, in the midst of a political millennium, — to kiss the American soil and call it truly blessed."

* * *

The annual exhibition in October, 1814, was peculiarly interesting; being then in the height of a disastrous war, and witnessing the interesting, and useful effects of the viewing committee of agriculture.

On this occasion, I addressed the society in the large church in Pittsfield, commencing with my resignation as president, and giving a general review of our history, and the prospect in anticipation; — exhorting to perseverance, and predicting, that their successful career would eventually lighten up a flame of patriotism, which will pervade the United States. The celebrated General Rial was present on this occasion, with several other British officers, prisoners of war. The times justified severe animadversions, on the conduct of the British in their savage mode of warfare, and in applying the terms, *"modern Goths and Vandals."* They were offended, — and although my doors had been thus far open to receive them with hospitality, they avoided my house. I always regretted the circumstance, but as I had previously apprised the general of my intended remarks, he had no just cause of complaint.

Anxious to establish town committees throughout the county, to be the organs of the executive committee, in the several towns, — and considering this the last wheel to be placed in the machinery of our society; although the plan was coldly met in every direction, as impracticable and unnecessary; — yet, I proceeded on a tour though the northern towns of the county, in the cold month of November, 1815, alone and unauthorized, but I took the responsibility upon my own shoulders, and selected proper characters to be recommended for town committees, and examined critically the situation of the farms, buildings, &c.

Having suffered extremely with the cold on the mountains, it cost me a dangerous fit of sickness, immediately after my return. This measure, and digesting and preparing a new code of by-laws for the society, (which were adopted,) terminated my career in organizing and zealously supporting the Berkshire agricultural society. At several periods I recommended an abandonment of the society, having frequently found myself alone at appointed meetings. I was discouraged for the want of funds, and the beggarly situation I was placed in to solicit charity. But no one would listen to *"give up the ship,"* — on the contrary, *ca-ira* was the tune, — posterity will duly appreciate the fortunate decision. . . .

Epitome of the Berkshire Plan.

Having so often repeated the words Berkshire system, I shall now conclude by unfolding that plan to the view of presidents of agricultural societies, and which I ardently hope they will adopt in preference to any other untried plan that may be offered. I hold out to them this tender of my services, as a legacy, acquired by many years experience, viz. a short epitome of the pith and marrow of the Berkshire system.

1. Let your premiums be announced yearly, in the month of November, so as to give the entire community, male and female, a full year in preparation, — who shall excel?

2. To subdivide your premiums in equal portions, — on agriculture, — on domestic animals, — on domestic manufactures; — also premiums in anticipation for the best farm; — the best orchard of fruit, and forest trees, grasses, &c. &c. three or four years thereafter, — to commence the succeeding year.

3. Preparatory measures in June for a viewing committee of nine respectable farmers, to make the tour of the county in July, in the northern states, to examine crops and farms offered for premium. This has a wonderful effect, — see page 134.

4. To have all your premiums in silver plate, with appropriate devices on the article.

5. To provide for each member handsome diplomas, and for each annual premium neat certificates of *"honourable testimony."*

6. To occupy at least two days in your annual exhibitions. As the time approaches your executive committee to hold frequent meetings, that nothing be neglected. Have neat *permanent* pens; — also proper places prepared for the reception of domestic manufactures; and another for such implements of husbandry, or agricultural products, as shall be offered. Two appropriate flags are also essential, — one bearing agricultural emblems, the other those of manufactures, — to be displayed at the respective places of exhibition the first day, and in the procession the second day. Let the morning of each day be ushered in with the ringing of bells, firing of cannon, and the display of farmers' flags, to rouse the dormant energies of the community.

The First Day's Exhibition. — To be occupied by the several committees of awards in examining animals, domestic manufactures, products of agriculture, and implements of husbandry.

In the afternoon, ploughing matches on the Schoharie plan, — see page 170.

In the evening the several committees to be occupied in preparing their reports, and filling up printed certificates of the premiums, to be furnished by the secretary.

Second Day. — A procession to form at eleven o'clock, and proceed to a church; — for the manner of proceeding, see pages 143 — 149, and 156, — Should the state be districted, (E), it will add much to the general effect.

Pastoral Balls. — These exhibitions generally close in *innocent festivity*, by a PASTORAL BALL. This part of the business has been much reprehended by Quakers and others; although the object is very imperfectly understood. The society take no part in them, — they are the spontaneous work of young men, and it is very natural for young people to indulge occasionally in innocent recreation.

It is also intended as a mark of respect to a day consecrated to promote the individual interests of each member of the community. Men of comprehensive and liberal views, encourage this annual jubilee, principally in a view of promoting domestic manufactures.

This effect begins to unfold in many places, — the times favour the object, — and females are seen at these pastoral balls, ornamented in the work of their own hands, and the native products of our land.

At Hartford, Rensselaer, Schoharie, Otsego, Oneida, Jefferson, Orange, and other counties — evidences of a taste for domestic manufactures have been evinced on these occasions, especially at Hartford. Whenever this shall become general, and the patriotic example of Montgomery county, shall prevail, a re-action will soon commence in our pecuniary affairs; and instead of our country being drained to the last drop, in the purchase of foreign gewgaws, and a universal gloom overspreading the nation, the balance of trade will revert in our favour; the precious metals will, in consequence, flow in upon us, — spreading in a thousand channels into every section of our country.

Having now exhausted a subject, which has engaged my zealous attention for many years, — I feel grateful for having been permitted, by various measures, to give currency to the plan of calling forth female efforts, in connection with agricultural societies, — and to have prompted a laudable spirit of patriotism, a love of country, which is spreading over the entire nation.

Plant Introduction, 1819

From *Niles' Register* (Baltimore, 1819), XVI, pp. 168–69.

Treasury Circular.

Treasury department, March 26, 1819.

SIR,

— In order to ensure uniformity in the execution of the act, of the 20th of April, 1818, supplementary to the collection laws, and more especially to enforce the provisions of the 8th, 13th and 21st sections thereof, the consuls of the U. States residing in foreign states, are informed that,

1st. In all cases where, by the municipal laws of the country in which they exercise their functions, they are restrained from administering oaths, the verification required by the 8th section of the act may be made, in the presence of the resident consul, before any magistrate duly authorised to administer oaths; and such consul shall certify not only the official character of the officer, and that the oath was administered in his presence, but that the person to whom it was administered is of respectable character, and who, according to the provisions of the said act, ought to verify the said invoice.

2d. Where merchandise is purchased for a commercial house in the U. States, by a partner residing abroad, the invoice ought to be verified by such partner, under the 8th section of the act.

3d. All cases embraced by the 13th section of the act are subject to the addition of fifty per cent. the failure to produce invoices duly verified being, in contemplation of the act, equivalent to merchandise fraudulently invoiced at twenty-five per cent. below its appraised value.

4th. Difficulties have occurred under the 21st section of the act relative to discounts. It has been represented to this department, that the most of the discounts which appear on the invoices of merchandise, especially from England, are not made for prompt payment, nor ultimately depend upon any condition of that nature. It is asserted that the true price of the merchandise is ascertained only by deducting the discounts from the invoice prices, and that where discounts are allowed for prompt payment, or, upon a future contingency, they are entered distinct from the common discounts, above described. You are therefore requested to state to this department the general custom in this regard within your consulate, and, as far as depends upon you, to endeavor to have the articles invoiced at their true value, so that no discount may appear thereon, except what may be made and allowed in the payment made for the same within the term of the said section:

5th. You are requested to cause the discount allowed upon such invoices as may be verified before you to be entered upon each invoice, and not upon the summary or recapitulation of several invoices, as is sometimes practised. The continuation of that practice may be productive of inconvenience to the parties, and is at all times calculated to excite suspicions of unfair dealing.

6th. You are lastly requested to furnish this department with semi-annual statements of the articles, the growth or manufacture of the United States, which are entered in the ports within your consulate, and the foreign merchandise which is shipped therefrom to the United States in American vessels; showing, as nearly as practicable, the comparative value of the exports and imports. Conjectural estimates of the foreign shipping employed in the same trade, and of the value of the imports and exports, laden on board such vessels, will be acceptable.

The introduction of useful plants, not before cultivated, or such as are of superior quality to those which have been previously introduced, is an object of great importance to every civilized state, but more particularly to one recently organized, in which the progress of improvements of every kind has not to contend with ancient and deep rooted prejudices. The introduction of such inventions, the results of the labor and science of other nations, is still more important, especially to the U. States, whose institutions secure to the importer no exclusive advantage from their introduction. Your attention is respectfully solicited to these important subjects.

The collectors of the different ports of the United States will cheerfully co-operate with you in this interesting and beneficent undertaking, and become the distributors of the collections of plants and seeds which may be consigned by you to their care. It will greatly facilitate the distribution, if the articles shall be sent directly to those sections of the union where the soil and climate are adapted to their culture.

At present, no expense can be authorised, in relation to these objects. Should the result of these suggestions answer my expectations, it is possible that the attention of the national legislature may be attracted to the subject, and that some provision may be made, especially in relation to useful inventions.

I have the honor to be, very respectfully, sir, your most obed't serv't,

WM. H. CRAWFORD.

The Cast Iron Plow

From *Plough Boy* (Albany, 1820), II, p. 123.

The above is a correct land-side delineation of the improved patent plough of Mr. Jethro Wood, of Cayuga county in this state. The wooden parts are the beam and the handles. The share, the mould-board, the brace that passes up through the beam, and a part of the land side, are all cast in one piece. The remainder of the land side, extending back to the heel of the plough, is cast separately, and is joined to the other in a substantial manner. The edge of the share is also cast separately, and is fastened on in a way that appears to be equally simple and effectual. The purchaser of the plough must provide himself with a suitable number of these cast edges, in order that when one becomes too dull for breaking up sward-land another may be substituted. — For cross ploughing, &c. they may be used long after they become too much blunted for breaking up. Cast iron being much harder than steel, lasts proportionately longer, and remains sufficiently sharp for perhaps three times the length that steel will endure. The depth that the plough is intended to run is determined by placing the chain higher, or lower, in the different notches of the clevis, as in the drawing is exhibited; and the increasing or decreasing the width of furrow is regulated by the upright holt that fixes the clevis to the centre, or to either side of the beam, by holes made in the end of it for the purpose. The coulter is of wrought iron. The characteristics of this plough are simplicity, strength, and durability, and we are fully convinced of its value as applicable, in general, to the lands of this country. The fact is, our lands are as yet too greatly diversified, as well in material, as otherwise — some smooth, and some abounding with stony matter — some level, and some of quite irregular suface — to designate any specific formation of the plough that is most exactly fitted for all, but perhaps the one in question is quite as well adapted, generally speaking, as any other that has been presented to the public. Most of our farmers want a plough that takes a wide furrow, for the greater dispatch of business. The plough taking a wide and shallow furrow, throws it over flat; that taking a narrower, and deeper, lays the one furrow lapping on the other. In some situations the former, and in others the latter furrow is probably most advisable. We would, on the whole, suggest to Mr. Wood, the propriety of having some of his ploughs constructed with a narrower bottom, say 8 or 9 inches wide, in order that purchasers may all be accommodated to their liking, or suited in regard to the particular sort of land they cultivate — The narrower the plough the easier it will run, and with this view the Scotch plough is only 7 inches in width at the bottom. We would suggest another improvement. Instead of having the coulter of wrought iron, and placed where it is, let one, of cast iron, be properly set in the curve, between the point of the share and the beam, or it may extend up through the beam, with the brace, if necessary. In either case a groove is to be made to let in the coulter to its proper place, where it is to be fastened, the method of doing which is unnecessary to describe. When one coulter becomes too much blunted for cutting sward, another is to be put in. Where there is no sward the sharpness of the coulter will not be found material. The improvements here suggested would probably render the plough of Mr. Wood the most perfect of the kind. At all events we have no hesitation in cordially recommending the plough as we find it. We are rather induced to believe that for all sorts of lands, and considering all its good qualities, it has no superior in this country.   J. N.

Wheat From Chile, 1821

From *American Farmer* (Baltimore, 1821), III, p. 271.

Meridian Hill, Nov. 12 [1821]

GENTLEMEN,

— In consequence of the notice in your paper some time since, that I had some Chili wheat to distribute. I have received letters from many gentlemen in different parts of our country, who are desirous of obtaining it, but who are ignorant as to the quantity I have to spare, the price they are to pay for it, and the advantage it possesses over the wheat of the country. May I, therefore, ask of you to inform them that I have now only about 4 or 5 bushels remaining; that I have given my servants orders to give to any one who may send for it, one quart; that all that is required for it is the satisfaction of distributing it: and that the advantage it has over common wheat, is its greater productiveness and less liability to shatter out in the field.

Those who wish to make a trial of it, can obtain a quart by sending for it, so long as I have so much on hand.

With great esteem, your obedient Servant,

D. PORTER.

P. S. When Judge Bland came from Chili, I sowed a small wine-glass full of the wheat he brought with him on 1500 square feet of ground, and reaped therefrom one half bushel and three quarts of grain. The next year I obtained upwards of 25 bushels, which I had ground into flour, the best that has been in my family since I have been in the District.

Teaching Agriculture to Indians, 1822

From Clarence E. Carter, ed., *Territorial Papers of the United States, The Territory of Michigan, 1820–1829* (Washington, 1943), XI, p. 221.

Governor Cass to the Secretary of War

Detroit January 23ʳᵈ 1822.

SIR,

By one of the stipulations in the Treaty of Saguina it is provided, that such an amount shall be expended for the purchase of cattle, farming utensils and seed corn and for employing persons in teaching agriculture to the Chippewa's, parties to that treaty, as the President of the United States might direct.

In the report which I had the honour to make to you of the progress and result of that negociation, I stated that the amount, which it was expected, would be appropriated for these objects was from fifteen hundred to two thousand five Hundred Dollars.

The appropriation for the Indian Department was so limited during the past year, that nothing could be expended from that fund for the fulfilment of this engagement. And as there was no specifick sum appropriated to meet this object, there has been nothing applied for that purpose.

I have recieved from the Saguina indians a representation upon this subject, and a request, that the treaty stipulation may be performed. They are anxious to become Argriculturists, and where this wish is expressed every principle of policy and humanity independent of positive conventional stipulations, dictates a compliance with it.

I take the liberty therefore of suggesting the propriety of a specifick appropriation to meet these objects.

Very Respectfully, Sir, I have the honor to be Your Obᵗ Servᵗ

LEW CASS.

Settlement in Indiana, 1824

From Sandford C. Cox, *Recollections of the Early Settlement of the Wabash Valley* (Lafayette, Ind., 1860), pp. 17–20, 47.

Crawfordsville, Ind., Dec. 24, 1824.

The land sales commenced here to-day, and the town is full of strangers. The eastern and southern portions of the State are strongly represented, as well as Ohio, Kentucky, Tennessee, and Pennsylvania.

There is but little bidding against each other. The settlers, or "squatters," as they are called by speculators, have arranged matters among themselves to their general satisfaction. If, upon comparing numbers, it appears that two are after the same tract of land, one asks the other what he will take to not bid against him. If neither will consent to be bought off, they then retire, and cast lots, and the lucky one enters the tract at Congress price — $1,25 per acre — and the other enters the second choice on his list.

If a speculator makes a bid, or shows a disposition to take a settler's claim from him, he soon sees the white of a score of eyes snapping at him, and at the first opportunity he crawfishes out of the crowd.

The settlers tell foreign capitalists to hold on till they enter the tracts of land they have settled upon, and that they may then pitch in — that there will be land enough — more than enough, for them all.

The land is sold in tiers of townships, beginning at the southern part of the district and continuing north until all has been offered at public sale. Then private entries can be made at $1,25 per acre, of any that has been thus publicly offered. This rule, adopted by the officers, insures great regularity in the sale; but it will keep many here for several days, who desire to purchase land in the northern portion of the district.

A few days of public sale have sufficed to relieve hundreds of their cash, but they secured their land, which will serve as a basis for their future wealth and prosperity, if they and their families use proper industry and economy, sure as "time's gentle progress makes a calf an ox."

Peter Weaver, Isaac Shelby, and Jehu Stanley stopped with us two or three nights during the sale. We were glad to see and entertain these old White Water neighbors, although we live in a cabin twelve by sixteen, and there are seven of us in the family, yet we made room for them, by covering the floor with beds — no uncommon occurrence in backwoods life. They all succeeded in getting the land they wanted without opposition. Weaver purchased at the lower end of the Wea prairie, Shelby west of the river opposite, Stanley on the north side of the Wabash, above the mouth of Indian creek, and my father on the north side of the Wea prairie. . . .

It is a stirring, crowding time here, truly, and men are busy hunting up cousins and old acquaintances whom they have not seen for many long years. If men have ever been to the same mill, or voted at the same election precinct, though at different times, it is sufficient for them to scrape an acquaintance upon. But after all, there is a genuine backwoods, log-cabin hospitality, which is free from the affected cant, and polished deception of conventional life.

Society here at present seems almost entirely free from the taint of aristocracy — the only premonitory symptoms of that disease, most prevalent generally in old settled communities, were manifested last week, when John I. Foster bought a new pair of silver plated spurs, and T. N. Catterlin was seen walking up street with a pair of curiously embroidered gloves on his hands.

After the public sales, the accessions to the population of Crawfordsville and the surrounding country were constant and rapid.

Fresh arrivals of movers were the constant topics of conversation. New log cabins widened the limits of the town, and spread over the circumjacent country.

The reader may be curious to know how the people spent their time, and what they followed for a livelihood in those early times, in the dense forest that surrounded Crawfordsville.

I will answer for the School Master, for I was there myself. We cleared land, rolled logs, and burned brush, blazed out paths from one neighbor's cabin to another, and from one settlement to another — made and used handmills and hominy mortars — hunted deer, turkies, otter, and raccoons — caught fish, dug ginseng, hunted bees, and the like, and — lived on the fat of the land.

We read of a land of "corn and wine," and another "flowing with milk and honey;" but I rather think, in a temporal point of view, taking into the account the richness of the soil, timber, stone, wild game, and other advantages, that the Sugar creek country would come up to, if not surpass, any of them.

I once cut cord wood at 31¼ cents per cord (and walked a mile and a half, night and morning), where the first frame college was built, near Nathaniel Dunn's, northwest of town.

Prov. Curry, the lawyer, would sometimes come down and help for an hour or two at a time, by way of amusement, as there was but little or no law business in the town or country at that time.

Reader, what would you think of going from six to eight miles to help roll logs, or raise a cabin? Or from ten to thirty miles to mill, and wait three or four days and nights for your grist? — as many had to do in the first settlement of this country. Such things were of frequent occurrence then, and there was but little grumbling about it. It was a grand sight to see the log heaps and brush piles burning in the night on a clearing of ten or fifteen acres — a Democratic torch-light procession, or a midnight march of the Sons of Malta, with their Grand Isacusus in the centre, bearing the Grand Jewel of the Order, would be nowhere in comparison with the log heaps and brush piles in a blaze!

But it may be asked, had you any social amusements, or manly past-times to recreate and enliven the dwellers in the wilderness? We had. In the social line we had our meetings and our singing schools, sugar boilings and weddings — which were as good as ever came off in any country, new or old — and if our youngsters did not "trip the light fantastic toe" under a professor of the terpsichorean art, or expert French dancing master, they had many a good hoe-down on puncheon floors, and were not annoyed by bad whisky. And as for manly sports, requiring mettle and muscle, there were lots of wild hogs running in the cat-tail swamps on Lye creek and Mill creek, and among them many large boars, that Ossian's heroes, and Homer's model soldiers, such as Achilles, Hector, and Ajax, would have delighted to have given chase to.

The boys and men of those days had quite as much sport, and made more money and health by their hunting excursions, than our city gents do now-a-days, playing chess by telegraph, where the players are more than seventy miles apart. . . .

I can well recollect when we used to wonder if the youngest of us would ever live to see the day when the whole of the Wea plain would be purchased and cultivated; and our neighbors on the Shawnee, Wild Cat, and Nine Mile prairies were as short-sighted as we were, for they talked of the everlasting range they would have for their cattle and horses on those prairies — of the wild game and fish that would be sufficient for them, and their sons, and their sons' sons. But those prairies, for more than fifteen years past, have been like so many cultivated gardens, and as for venison, wild turkies and fish, they are now mostly brought from the Kankakee and the lakes.

Lucerne in New Jersey, 1823

From John Patrick, "On the Cultivation of Lucerne," Philadelphia Society for Promoting Agriculture, *Memoirs* (Philadelphia, 1826), V, pp. 173–74.

Perth Amboy, N.J. July 10th, 1823.

Having been for eight or ten years past, in the successful practice of cultivating Lucerne [alfalfa], I think it may promote the interests of agriculture, to offer you a few remarks on that subject.

This article I have found by experience, to be not only the most convenient, but also the most profitable of all grasses. It vegetates quicker in the spring than any other grass; it resists the effects of droughts; it may be cut four or five times in, the course of the season; and it will endure from ten to twelve years without renewing. Of all the grasses it is the most profitable for soiling. I am fully of opinion, that one acre properly got in, would be more than sufficient to maintain at least six head of cattle, from 1st May, until the frosts set in; for before it can be cut down in this way, the first part of it will again be ready for the scythe. English writers have recommended the drill system for this grass, but in this climate I have found this plan not to answer. The proper mode is to put the land in good order; and to sow it broad cast during the month of April, or the early part of May. Fall sowing will not answer, as when sowed so late, it, like clover, is found not to resist the effects of the frosts. It may be sown either by itself, or with spring rye, or barley, or with oats, but in the last case, the oats would require to be cut green, and before the seeds form, and by this means an early feed for cattle would be obtained without impoverishing the soil. But the mode I would most confidently recommend, would be, to sow with the Lucerne about half a bushel of winter rye to the acre. The effect of this is, that the rye, which vegetates quickly, serves as a protection to the young grass, against the effects of the scorching sun, and by the time the grass attains sufficient strength to protect itself, the rye withers, and apparently dies. It will however again come forth in the spring, and mixed with the Lucerne, will add much to the quantity of fodder, and prove a most excellent feed for cattle. The rye will admit of being cut green in this way, (before getting into seed) two or three times with the Lucerne, before it decays. The quantity of seed I recommend, is at the rate of from 15 to 20 lbs. to the acre.

The kind of soil most suitable for this culture is a *dry* mellow loam, but a sandy or clay loam will also answer, *provided they are not wet*. In a favourable season, the Lucerne may be cut the ensuing fall, after the first season. You may generally begin to cut it green for cattle by the first of May, which saves the young pasture, and is in every respect a great convenience, as hogs and every description of animals devour it with equal avidity. It produces a great quantity of seed, and is much more easily obtained than clover. The second and third crops are the most productive of seed.

JOHN PATRICK.

A Maryland Plantation, 1826–1837

From George Cooke, Excerpts from *Dairy, 1826–1841* (Hazelwood, Ellicott City, Maryland), Manuscript in National Agricultural Library.

20 Nov. 1826 — Finished hauling in my corn. In the evening husking, ploughing corn ground in the field about the quarter, stable, etc.

21 Nov. — Husked and lofted all my corn. Filled the corn house, 30 ft. long, loft wide and 13 ft. high. Besides feeding my pen hogs, horses, etc. from the 1st of this month, I also used about 15 barrels in the month of October for bread, etc. Ploughing for corn next year.

22 Nov. — Hauling wood to house and quarters, repairing the fence about the stables, putting husks away and doing sundry jobs. Ploughing for corn. Cold in the morning but turned out a fine day.

23 Nov. — Getting out wheat, hauling wood to house, etc. Fine morning, but clouded up and blew hard and cold, a little snow.

27 Nov. — Cleaning up wheat, ploughing for corn. Blind ditching. Wind west and cloudy. Went to Baltimore.

28 Nov. — Ploughing and ditching as yesterday. Grubbing.

29 Nov. — Ploughing and grubbing for tobacco, etc.

30 Nov. — Hauling wheat to Ellicott City. Very indifferent wheat but 53 cents per bushel. Grubbing and hauling wood to Morgans. Warm fine day.

1 Dec. — Treading out wheat. A fine day.

2 Dec. — Sent two loads of wheat to Ellicott City. Grubbing.

4 Dec. — Cloudy morning and commenced snowing and continued most of the day. The ground covered to the depth of two inches. Ploughing for corn and grubbing.

5 Dec. — A fine day. Finished ploughing field adjourning the quarter. Grubbing. Hauling manure in garden, etc.

6 Dec. — Ploughing a piece of ground in the garden and grubbing a piece of meadow land. Snow all gone.

12 Dec. — Much milder. Went to Baltimore but at night. Finished ploughing 2nd years tobacco ground. Grubbing and prepared to kill hogs tomorrow.

13 Dec. — Killed my hogs, 37 in number, most all of them not over 10 months. They are very fat and young meat and weigh 4325 lbs. A fine clear day.

14 Dec. — Cut up and salted my hogs killed yesterday. Hauling stone off the new field intended for corn next year.

25 Dec. — Christmas Day. Fine day, a holiday.

28 Dec. — Very cold. Holiday ends, filling ice house. My waggoner, Clem, not at home.

1 Jan. 1827 — Very cold. Filling ice house.

4 Jan. — Moderate compared with the previous 7 days. Finished filling both ice houses. Clem not come home.

11 Jan. — Cloudy and damp, cutting corn stalks off and hauling them in.

12 Jan. — Raining. Stripping tobacco.

15 Jan. — Fine day. Hauling oats from meadow, cutting corn stocks down.

16 Jan. — Hauling oats and cutting stocks.

17 Jan. — Treading out oats. Find they turn out so bad I shall feed them away as hay. They were all flooded in the spring and laying flat they did not fill well, but the 2nd growth of straw and grass being cut green it makes excellent hay. Very very cold, too cold almost to move.

16 Jan. — Put two large wagon loads of oats hay in stable loft. Getting out clover seed. Killed a mutton. Very cold tho not so cold as yesterday which I think was the coldest day we have had this winter.

19 Jan. — Cutting wood, clearing land for tobacco, top dressing wheat.

23 Jan. — Cloudy, not as cold as yesterday . . . . Killed a mutton. Hanging up my meat to smoke.

30 Jan. — Cloudy and raining most of the day. . . . All hands stripping tobacco.

31 Jan. — Rained in the morning but cleared off in the evening. Stripped out house no. 1 of tobacco.

1 Feb. — Clear and much colder than yesterday. Clearing land for tobacco back of the quarter, hauling wood to tobacco houses for next summer. Clouding up at night, wind west.

8 Feb. — Finished stripping tobacco. Clearing land. Clear and warm.

9 Feb. — Clearing land for tobacco. Putting all the tobacco in house no. 1.

19 Feb. — Put hay in stable loft. Employed in threshing the clover seed to send to mill. A fine day.

20 Feb. — Sent my waggon loaded with clover seed in the cups to the mill at Brookville. . . .

21 Feb. — A light shower of rain last night. Cleared off and a fine day. Make a tobacco bed. The waggon gone with another load of clover to mill. Grubbing, etc.

22 Feb. — A fine day and warm. Finished putting brush on tobacco bed. Grubbing, etc. The waggon returned from Brookville with 5 bushels clover seed. Went to Baltimore and returned at night. Dined with A. S. at General Ridgelys.

23 Feb. — A fine day. Burnt and sowed the tobacco bed. Grubbing, etc. Sowing clover seed.

1 Mar. — Hauling wood to Blacksmiths, mauling rails, digging up the pavement in front of the house intending to alter the road. Raining all day.

31 Mar. — A fine warm day. Returned from Baltimore. Ploughing and hauling out manure. Sowed 2 tobacco beds, making 7 sowed, the plants are up in the 1st bed sown. Had to leave my grey filley on the road being lame.

11 Apr. — Ploughing in manure on corn ground and laying off for planting, burning brush, etc.

12 Apr. — Planted the quarter field in corn, commenced at sunrise and finished 3 o'clock p.m. Harrowing the tobacco house field for planting, clearing off new ground for tobacco. Warm day tho blowing very hard in the evening. 2½ bushels corn planted in quarter field.

20 Apr. — Returned from Annapolis, found my overseer had finished planting corn, having planted a little over 8 bushels. Clearing off tobacco ground, etc.

23 Apr. — Hauled 45 bushels of Plaister from Ellicott City. Run a fence in front of apple orchard to enclose it for yearlings.

27 Apr. — Sowed plaister this morning and hauled another load of 42 bushels from Ellicott City. Some rain in the evening and cold. Clearing off new ground for tobacco.

15 May — Harrowing and ploughing tobacco ground in quarter field. Replanting corn. A little rain at times.

21 May — Raining all day. Planted upwards of 27000 tobacco plants. My grey mare Kate had a horse colt last night. Sire the thorobred horse Mark Antony.

23 May — Harrowing and hoeing corn. Also harrowing tobacco ground. Shearing sheep. Clear and hot.

24 May — Finished shearing sheep. Harrowing corn, etc. Gone to Baltimore, to attend a meeting of the agricultural trustees at Mr. Smiths.

29 May — Cutting potatoes and planted some. Light sprinkle of rain. Hauling manure on potato ground.

30 May — Clear and hot. Planted 30 bushels of potatoes.

1 June — We planted tobacco until breakfast. Wind fresh from NW and very cold. Ploughing corn with double shovel ploughs in quarter field and rounding it with hoes. Also ploughing new meadow.

16 June — Harrowing corn in new field. Weeding out the 1st planting of tobacco.

19 June — Sent a load of 3 hogsheads tobacco to Baltimore and waggon returned at night with 2 barrels pork and 1 barrel whiskey. Ploughing corn and weeding out tobacco. Cool and pleasant.

28 June — Working corn, cutting grass. Commenced cutting rye.

30 June — Finished my rye by dinner, ploughing corn. Stacked some clover hay that was in cocks.

2 July — Bound and set up all the rye. Finished ploughing corn until after I shall cut my wheat. Ploughed my potatoes. Clear and very hot.

3 July — All hands cutting wheat.

7 July — Finished harvest by dinner time. Cloudy and light rain in the evening. Gave the hands holiday this evening.

17 July — Stacking hay in meadow, ploughing fallow and weeding out tobacco.

19 July — I intend going to Washington County to Mr. Dales with Mrs. Cooke.

9 Aug. — Returned last night from Washington County and found my farming operations going on very regularly and the work of the farm forward. Harrowing wheat fallow and chopping grass in corn and thrashing rye. Rain much wanted.

14 Aug. — Hauling out manure and crop ploughing fallow. Worming tobacco. Clear and warm.

16 Aug. — Topping and worming tobacco. Finished crop ploughing the lot in front of orchard. In the evening grubbing in large field next to Weems. Several light showers of rain during the evening.

28 Aug. — A fine day. Wind W. Making cider, worming tobacco, etc. Sent the light waggon to Baltimore.

30 Aug. — Cutting tops and pulling blades. Slow work. The corn so much down. Fine weather.

5 Sept. — Cutting tobacco and filling log tobacco house.

8 Sept. — Firing tobacco in log house. Hauling in fodder and putting in fodder house.

17 Sept. — All the people employed in housing tobacco. Cloudy part of the day.

19 Sept. — Firing tobacco, cutting down corn in quarter field, cross ploughing fallow. A fine day, at night cool. Wind NW.

24 Sept. — A fine day. Seeding rye, cutting off corn and shocking it, firing tobacco.

26 Sept. — Finished seeding rye in quarter field. Having sown 21 bushels. Cutting down corn in new field.

28 Sept. — Got from Mr. Clarke 65 bushels red chaff wheat for seed and 14 bushels blue stem from Mr. Godfrey. In the evening I began to sow the blue stem, setting up corn cut down yesterday, hauling out manure on wheat fallow. Cold and clear.

6 Oct. — Seeding wheat and cutting off corn and shocking it. Got 35 bushels of blue stem wheat from Mr. Williams. Gathering apples.

10 Oct. — Killed a beef. Picking apples and making cider.

18 Oct. — Seeding wheat, digging potatoes, thrashing rye. A fine day.

19 Oct. — A fine warm day. Finished seeding wheat except the potato ground. Digging potatoes. A fine crop of blue ones. The white ones not dug. The fallow in front of the house to the South took 21½ bushels blue stem and 5 of red chaff. Total 26½ bushels.

23 Oct. — Cleared off with wind hard at NW. Set the ploughs to breaking up corn ground for next year. Digging potatoes and put others away. Intend going to Baltimore to attend the cattle show.

7 Nov. — Rained all last night and sometime this morning. Stripping tobacco.

8 Nov. — Hauling in corn and husking. Ground too wet to plough.

22 Nov. — The trustees of the Agricultural Society dined here today. Very cold and blustering.

24 Nov., 1836 — Digging what potatoes we have in the ground. Ploughing corn ground. Cold and clear.

30 Nov. — Fine day. Put my potatoes in the cellar under the office. I had them in the barn. Measures 150 bushels and sent 10 bushels to Unger. Milder than for some time.

10 Dec. — Rained hard last night and cleared off warm this morning tho the wind is high. Hauling wood to house and quarters. Miss Whitier, a governess for our children, came out from Baltimore this morning. She is from Massachusetts.

14 Dec. — Killed my hogs. 28 in number weighed 3200 lbs. All young.

15 Dec. — Cutting up and salting my meat. Hauling hay to stable and wood to house.

28 Dec. — The holidays at an end. All hands at work except Ben who went Weems at West River to see his wife. Having my waggon horses shod . . . Husking corn in barn. . . .

1 Jan, 1837 — Sunday. Ben came home last night from West River. Cloudy . . .

12 Jan. — Clear and more mild. Hauled all the ice I could get off the river, the ice having sunk. Got 21 waggon loads of fine ice about 5 inches thick. Hauled hay to out horses and husking corn in barn.

13 Jan. — Sent a load of corn to mill to be ground in the cob for horse feed. Husking corn in barn. At night snowing.

15 Feb. — Hauled hay to out horses, wood to house and quarters. Cutting and mauling rails. Wind SW and cloudy. My ewes are lambing and have lost 3 lambs.

2 Mar. — Clear and cold. Hauling wood, etc. Went to Washington in company with George Ellicott. Found the crowd of people immense to witness the inauguration of Mr. Van Buren.

16 Mar. — Thrashed out and cleaned up 45 bushels oats and 4 bushels of rye. W. C. Ogle came here last night.

17 Mar. — Put my clover seed hay in barn and ran it thro the wheat machine, which gets the heads off very well.

21 Mar. — Ploughing corn ground and grubbing the same. Hauling manure for early potatoes. Rained most of this evening and at night Got a barrel of flour and herrings.

11 Apr. — More mild and clear. Sowing oats in the field at the bottom of the garden. Putting it in with the cultivator, which do it very nicely. Alexander sick. Layed up with the pain in his hip since Saturday.

28 Apr. — A fine day planting corn, part of which I soaked and rolled in tar water and ashes. I shall see if the birds disturb the corn where it is not tared more than the tared.

29 Apr. — Laying off corn ground adjourning the quarter where I had sown rye and finished planting the same.

27 May — Hauled 8 barrels of plaister from the railroad depot at Ellicott City. Planting corn. Fine day.

31 May — Sowing plaister. Ploughing corn and fallowing with the hoes. At 6 p.m. a light rain. My neighbour Mr. Larkin Dorsey died this evening aged 59 years.

2 June — Hauling out manure for potatoes. Working corn. Warm day and clear. Finished shearing sheep and marking the lambs. Went to the funeral of Mr. Larkin Dorsey.

12 June — Ploughing corn with shovel ploughs. Planting potatoes.

5 July — Commenced cutting clover, it is late tho a good crop, ploughing corn. I plaistered my clover late and it is about right now to cut.

17 July — Fine day. Very hot. Cutting rye. A small crop tho the grain is good. Ploughed most of my rye up, it looked so bad, made a mistake by it, as that which is standing is good.

18 July — Cutting wheat.

29 July — Cocking hay. Fine day and warm. Ben went this morning to West River to see his wife at Mr. Weems.

1 Aug. — Put up all the hay we had cut making 3 stacks of timothy in lot in front of the barn. Commenced mowing timothy in front of the orchard. A good crop. Ploughing corn.

11 Aug. — Cutting oats until about 11 o'clock a.m. when we were stopped by rain. Should have finished the field by dinner having 5 cradlers at work. Most of oats are in the swarth. Raining hard all the evening.

28 Aug. — Thrashed out and cleaned up 4 loads of oats in the barn. Counts 112 bushels besides what had been put away before, making about 140 bushels.

31 Aug. — Thrashed out 20 bushels of wheat. Ploughing fallow for wheat.

1 Sept. — Sold 20 bushels of wheat at $1.55. Ploughing fallow and grubbing.

4 Sept. — Fine day. Thrashed out 30 bushels of wheat. Ploughing fallow. Sold the wheat at $1.50.

2 Oct. — Seeding wheat, sowing the red chaff. Fine day.

18 Oct. — Ploughing ground for rye, cutting off the corn and setting it up.

28 Oct. — Clear. Thrashing out buckwheat.

11 Nov. — Sent 11 bushels wheat and 6 bushels of buckwheat to McCann's Mill for family use. Ploughing corn ground.

2 Dec. — Fine day. Ploughing corn ground. Hauling wood to house and quarters.

Seeds From Abroad, 1827

From Charles Francis Adams, ed., *Memoirs of John Quincy Adams . . . Portions of His Diary . . .* (Philadelphia, 1875), VII, pp. 257–58.

April 12, 1827. Mr. Rush came to mention the result of Dr. Mease's visit. He was with him all day yesterday, and departed this morning on his return to Philadelphia, with many of the documents and books collected respecting silk and the silk-worm. I spoke again to Mr. Rush respecting the project of procuring seeds and plants from foreign countries which may be usefully naturalized in this country. There was a letter from Mr. Crawford to our Consuls abroad on this subject, but which has produced no result, because it authorized no expense. We discussed this matter at some length, and must return to it again. I thought we might venture upon some small expense to collect certain specific seeds or plants and have them planted in the garden of the Columbian Institute.

13th. I have already been tempted by the prevailing warm weather to bathe in the Potomac, but have been deterred by the catarrh still hanging upon me, and by the warnings of physicians, whose doctrines are not in harmony with my experience. I took, however, for this morning's walk the direction to the river, and visited the rock whence I most frequently go into the river. It is yet adapted to the purpose; but all trace of the old sycamore-tree, which was near it, and blew down the winter before last, is gone. There is yet one standing a little below, but it is undermined with every high tide, and must be soon overthrown. The borders of the river are strewed with dead herring and shad, and the waters are not so high as usual at this season.

The French Minister, the Baron de Mareuil, came at one o'clock, and was introduced by Mr. Clay. He is going upon a tour to the interior of New York, perhaps into Canada, and thence to Boston — intending to return to embark the 1st of July at New York for Havre. He said he was going upon a leave of absence, not knowing whether he should return here or be otherwise disposed of.

I assured him of my good wishes, and that it would give us pleasure to see him here again. And I desired him to assure his Government of the continued earnest desire entertained by us to be upon terms of harmony with France.

He spoke in equal terms of courtesy, and expressed himself grateful for the treatment he had received in this country. These were words of course, and without much cordiality, or more sincerity than there is in the closing salutation of a letter.

Durham Cattle in Ohio

From Ohio State Board of Agriculture, *Annual Report* (Columbus, 1857), pp. 301–02, 305.

Ever since the organization of Ohio as a State, the Miami and Scioto valleys have been regarded as *the stock* region of the State. As early as 1808, cattle were taken from the Scioto valley to the eastern markets. There is no doubt that some of the descendents of the Virginia importation of 1783 found their way into the Scioto valley. It is highly probable that some of the best cattle in Southern Ohio owed paternity to Kentucky, and were the descendents of the bulls Phito and Shaker, which were imported into Kentucky in 1803.

The first importation made into Ohio from England direct was made in 1834, under the auspices of the Ohio breeding company.

History of the Ohio Company for Importing English Cattle.

On the 2d November, A. D. 1833, Governor Allen Trimble, George Renick, Esq., and General Duncan McArthur, citizens of the State of Ohio, for the purpose of promoting the interests of agriculture, and of introducing an improved breed of cattle into the State, formed a company, and they, together with the subscribers hereafter named to the written articles of their association, contributed the amount of money necessary to import from England some of the best improved cattle of that country.

The sum of $9,200 was very soon subscribed for that purpose, in ninety-two shares of $100 each; and after making the necessary preliminary inquiries and arrangements, the company appointed Felix Renick, Esq., of Ross county, Ohio, their agent for the purchase and importation of said cattle.

Mr. Felix Renick was accompanied by Messrs. Edwin J. Harness and Josiah Renick, of Ohio, as his assistants, and they left Chillicothe for England on the 30th January, 1834. . . .

Mr. Felix Renick, and his assistants, Messrs. E. J. Harness and Josiah Renick, proceeded to England, and made a careful examination of much of the improved stock of that country, purchased from some of the most celebrated and successful breeders of cattle in England about nineteen at various prices, selected from the herds of Mr. Bates, the Duke of Leeds, the Earl of Carlisle, Mr. Whittaker, Mr. Paley, Mr. Mason, Mr. Ashcroft, and others, consisting of bulls and cows, of the *thorough-bred short horned Durham stock*. They brought these to Ohio, and returned in time to exhibit them at the Agricultural Society of Ross county, on the 31st day of October, 1834.

This stock of English cattle was kept together, under the care of an agent, by said company, and they *increased the number, by additional importations from England, until the 20th day of October, A. D.1836*; when the cattle imported, as well as the natural increase thereof since the 31st October, 1834, were sold at public auction, under regulations adopted by the company. . . .

Very great benefits have resulted to the country by the introduction of this improved English Durham stock into the State of Ohio by this company.

An improved breed of cattle throughout the State has resulted from crossing the English stock with the common stock existing at that time; and a very fine, large and thrifty race of cattle in many parts of Ohio has been bred by this laudable enterprise. Some of their full blood bulls and cows have been sold to farmers of the adjoining States; and thus the benefits of their importations have contributed largely to improve the stock of cattle in the western country.

Mr. George Renick, of Ross county, has bred, from a portion of the stock imported by said company, and the common cows of Ohio, a very fine race of cattle; and for the last six years, as he states, he has annually sold about 50 or 60; the average weight of which, at from three to four years old, was about 1,000 pounds net. Some of them weighed as much as 3,000 pounds, and one (older) as high as 3,400 pounds, gross.

Ex-Governor Allen Trimble, of Highland county; Doctor Arthur Watts, of Ross county; M. L. Sullivant, Esq., of Franklin county — all well known as amongst the most successful farmers and stock growers in Ohio — besides many others of this company — have contributed largely, by their skill and enterprise, to increase and diffuse the improved breed of cattle, resulting from the importations of the company, into every part of this State.

Doctor Watts, at the agricultural exhibition in Ross county, in 1849, exhibited eight two-years old steers, averaging 1,526 pounds each; and at the State agricultural fair at Cincinnati, held in 1850, he exhibited, amongst other cattle, one four years old steer, (full-blood Durham,) weighing 2,550 pounds, gross; and one three years old steer, weighing 2,220 pounds, gross. These weights are given to show the enormous weight which this Durham stock of English cattle attain at an early age when bred by skilful and intelligent farmers; and they show, also, the great value of breeding from this stock to those who are engaged in furnishing the beef markets of our country.

Invention of the Grain Reaper, 1834

From C. H. McCormick, "Cyrus H. M'Cormick's Improved Reaping Machine," *Mechanics' Magazine* (Oct. 11, 1834), IV, pp. 209–10.

To the Editor of the Mechanics' Magazine:

DEAR SIR,

I send you a drawing and description of my Reaping Machine, agreeably to your request.

References — A, the platform; B, the tongue; C, cross-bar; D, hinder end of the tongue; ee, projections in front; F, broad piece on each side; G, circular brace; H, diagonal brace; I, upright post; J, upright reel post; K, braces to upright; L, projection to regulate the width of swarth; M, main wheel roughened; N, band and cog wheel of 30 teeth; O, band; p, small bevel wheel of 9 teeth; Q, do. of 27 teeth; r, do. of 9 teeth; s, double crank; T, cutter; V, vibrating bar of wood, with bent teeth; U, reel pulley; W, reel; X, wheel of 15 inches diameter; Y, reel post.

The platform A is of plank, made fast to a frame of wood, for receiving the grain when cut, and holding it until enough has been collected for a sheaf, or more. The projections in front, ee, are two pieces of the platform frame, extending about 1¼ feet in front, and one or more feet apart. On each outside of these pieces is to be secured a broad piece of wood, as at F, by screw bolts, as at 11, passing through them and the projection of the frame. From the end of the outer broad piece, nearest the platform, rises a circular brace, G, projecting forward, and secured to the reel-post, I, by a moveable screw bolt. About nine inches in front of the screw bolts, at 1 1, are two other *moveable* screw bolts, as at 2, passing through both broad pieces and the ends of both projections, allowing for a rise or fall in adjusting the height of cutting; and at about the same distance, further on, is to play an axis of a wheel to be hung between said pieces. Near each end of this axis is secured an arm with two screw bolts, as at 3 3, one of which is moveable, as will be seen; projecting before the wheel, where the tongue is made fast between them by means of two screw bolts passing through all at D. H is a diagonal brace. On the opposite side of the machine is another reel-post, Y, connected near the top with a piece, K, on each side, with a moveable screw bolt, and extending, one to the end of a piece, L, which is attached to the outside of the platform, and divides the grain to be cut, from that to be left standing, the other to the hind end of the platform. T is an upright post, secured to the braces of G and H, at 7, by a moveable bolt, bracing the reel-post Y by means of a piece, Z, passing diagonally over the reel. 5 5 is a strip of cloth about as high as the grain, for the purpose of keeping entirely separate the grain to be cut from that to be left. On the axis, hung between the hind pieces, is a wheel, M, of about two feet diameter, having the circumference curved with teeth to hold to the ground by. N is a cog wheel on the same axis, which serves also for a band wheel, on which and the pulley U the band O works. The cog wheel p working into the cog wheel N, has another cog wheel, Q, on its axle, which works into another small pinion, as at r, attached to the double crank s. These cranks are in a right line, projecting on opposite sides of the axis and in a line with the front edge of the platform. The lower of these works the cutter T, along the front edge of the platform, and the upper one the vibrating bar V, counter to each other. The cutter is a long blade of steel, with an edge like that of a reap-hook, and is supported on the under side by stationary pieces of wood at suitable distances apart. This blade is attached to the frame piece, below the edge of the platform, by means of moveable tongues or slips of metal; the bolt securing it to said frame-piece acting as a pivot, and that through the blade likewise, so that the motion is described in part of a circle. The vibrating bar is of wood, of the same length, and secured in the same manner, above the cutter, with iron teeth made fast in it, at about 2 inches apart, extending before the edge of the cutter, and bent round under it. This vibrating bar has been and may be made stationary, with bent teeth supporting the stalks on each side of the cutter, thereby dispensing with the upper crank; but the other is much preferable, as it reduces the friction and liability to wear materially, by dividing the motion necessary for one between the two, and counterbalancing each other.

In the upper end of each reel-post is a groove, or long mortice, to receive the end of the axis of the reel, which rests on an adjusting pin, subject to be moved higher or lower, to suit grain of different heights — rye, wheat, or oats, & c. The reel W is composed of two or more cross arms at each end of the axle, projecting about 3 feet each way, and connected at their ends by a thin board of about nine inches in width, which, by the arrangement of the arms, runs in a somewhat spiral direction along the axis (though it might be parallel), the right end bearing up first on the grain. This reel, by the motion given by the strap O as the horses advance, bears the stalks upon the cutter, and when separated lands them on the platform A, which advancing till a sufficient quantity is collected, is discharged as often as may be required by a hand with a rake at the right end of the platform. On the left end of the platform is a wheel, H, of about 15 inches diameter, that may be raised or lowered as the cutting may require, corresponding with the opposite side. The point of the tongue is secured to its place by passing through a pin, 6, that is fastened to the hames of each horse by means of leather straps.

I have made some alterations on the drawing, which I think you will readily understand. Two horses were not used to the machine until the last harvest; the necessary changes of which were only described to the draughtsman, and were not all understood. I directed that it should not exceed 5¼ inches, though I think it does one way. The wheel H I think has a wrong direction.

VERY RESPECTFULLY, YOURS, &c.

C. H. M'CORMICK.

Experiments with Tropical Plants in Florida, 1838

From U.S. Congress, 25th, 2d Sess., House of Representatives, Committee on Agriculture, *Dr. Henry Perrine — Tropical Plants*, House Report 564 to Accompany H. R. No. 553 (Washington, 1838), pp. 1–16, 18–22, 53–54, 78–79.

Report (February 17, 1838)

The Committee on Agriculture, to which was referred the memorial of Doctor Henry Perrine, late consul at Campeachy, asking a grant of land in the southern extremity of East Florida, for the encouragement of the growth of new and important agricultural products, exotic vegetables, and tropical plants, have had the same under consideration, and report:

The memorialist sets forth in his petition that, in the year 1827, while he was acting consul of the United States at Campeachy and Tabasco, he was officially instructed, by the circular of the Treasury Department of the 6th of September of that year, to aid the desires of the General Government to introduce into the United States all such foreign trees and plants, of whatever nature, as may give promise, under proper cultivation, of flourishing and becoming useful. That, in obedience to said circular, the time, labor, and funds of the memorialist were devoted to that purpose, by which he obtained and transmitted to this country much useful information concerning various valuable plants which may be successfully domesticated in the United States; and that, in discharging these governmental duties, he was obliged to sacrifice all opportunities of making money by his profession, or by mercantile pursuits, which, in the same period and region, had furnished fortunes to his unofficial countrymen; and, as funds had not been appropriated by Government to promote the objects of the said Treasury circular, the perquisites of his consulate did not defray one-third of his personal expenses; and, as a reward for his sacrifices, and a premium to encourage the introduction and culture of tropical plants in the United States, he asks of Congress a township of land in the southern extremity of East Florida, which, in soil, climate, and geographical position, for the enterprise, he thinks affords the most favorable location, by means of gardens or nurseries, to contain all tropical vegetables of utility or ornament, which, after due seasoning, may thence be gradually transplanted and acclimated throughout the Territory; and thus be ultimately extended over the adjoining States on the Gulf of Mexico and the Atlantic ocean.

The memorialist enumerates, among the exotics which may be introduced and naturalized in this country successfully, the beautiful and extensively useful family of the palms, the agaves, the shrubs for chocolate, coffee, and tea; the logwood, fustic, cochineal, and other dyes of Mexico, Guatemala, and Brazil; the cinnamon, pimento, ginger, and other spices of the East and West Indies; the mahogany, cedar, ebony, and other precious woods of all parts of the world; the bananas, anonas, mangoes, and other delicious fruits; the Peruvian bark, sarsaparilla, canella, and innumerable salutary medicines for the removal of disease. And that the extensive cultivation of a single species of agave Sisalana will alone furnish a profitable staple to the planters of the South, and a cheap material to the manufacturers of the North; which will supply many wants of our merchant vessels, of our navy, and of our citizens in general.

From some specimens of these plants and vegetables which were exhibited to the committee by the memorialist, they entertain the opinion that, if proper encouragement be given to their introduction and cultivation, they may conduce greatly to promote the agricultural and manufacturing interests of the United States, and the welfare of the people.

The memorialist represents that a tropical climate extends into southern Florida — which opinion seems to be well established by the annexed meteorological table of observations made at Indian Key, during the year 1836; that many valuable vegetables of the tropics do actually propagate themselves in the worst soils and situations in the sun and in the shade of every tropical region, where they arrive either by accident or by design; and that for other profitable plants of the tropics, which require human skill and care, *moisture* is the equivalent to manure; and that tropical cultivation essentially consists in appropriate irrigation, which, in such a climate, goes far to counterbalance the sterility of the soil.

The memorialist being a man of science and untiring industry, and having familiarized himself with botanical studies, and devoted much of his time and pecuniary resources for the last ten or twelve years to the accomplishment of his favorite object, as stated in his memorial, afford, in the estimation of the committee, a sufficient guaranty of his faithful compliance with the terms of the proposed grant of land.

The committee also are induced to believe that the lands asked for by the memorialist are of but little value, and, if applied and improved as by him proposed, the public lands in their vicinity might be enhanced in value, and thereby no pecuniary loss would be sustained by the Government. They have, therefore, unanimously agreed to report a bill setting apart, for this object, one township of the public land south of the twenty-sixth degree of north latitude, in East Florida, upon condition of its occupancy and successful cultivation within a limited period, and under certain restrictions and conditions, as set forth in said bill.

The committee also annex a copy of a report of the Committee on Agriculture in this case, made during the first session of the twenty-second Congress, and the accompanying documents, as a part of this report; which bill and report were not then acted on.

April 26, 1832

The Committee on Agriculture, to which were referred the memorial of Dr. Henry Perrine, consul of the United States at Campeachy, and a resolution of the Legislative Council of the Territory of Florida, recommending a grant of land in that Territory for the encouragement of the growth of new and important agricultural products, exotic vegetables, and tropical plants, have had the same under consideration, and report:

The plan and object of the memorialist, Dr. Perrine, are explained in his petition, hereto annexed, and made a part of this report, (No. 1.) Dr. Perrine has been for some years the American consul at Campeachy and Tabasco. Being a man of science and industry, he has devoted a great portion of his time, for the last four years, in the collection of the most rare and valuable tropical plants, medicinal trees and fruits, dyewoods, and other productions of the Mexican States, in which they abound. From the testimonials exhibited to the committee, they are satisfied that, from the extensive acquaintance of this gentleman in that country, and the high estimation in which he is held by the public authorities of Mexico, he has it in his power to obtain and import to this country the most useful and valuable acquisitions to our agriculture.

The committee do not deem it necessary to offer any remark on the subject of the practicability of such an enterprise. The history of the world shows that, in all ages and countries, trees, vegetables, grains, and plants have been successfully transplanted and domesticated from one country to another. As a general principle of action for the National Legislature, it is better to abstain from any legislation upon such subjects, leaving them to individual enterprise and exertion. There are some cases, however, in which it would be unwise and impolitic not to furnish some facilities in aid of our enterprising fellow-citizens. The Greeks and Romans obtained, at the public expense, a number of grains, vegetables, and plants, from Africa; and all the modern states of Europe have made it one of the leading considerations of national policy to promote new acquisitions to the agriculture as well as to the commerce of the country.

The United States have acquired eighteen or twenty millions of acres of land by the late treaty with Spain, now almost entirely uninhabited, the largest portion of which is incapable of producing any article now cultivated in the United States. This immense tract of land on the borders of our Union must lie unemployed and useless for many years, without some experiment such as Dr. Perrine proposes.

The committee have, therefore, determined to report a bill, setting apart for this object one township of the public lands, to be granted to him and his associates, upon condition of its occupancy and successful cultivation. The committee annex to this report a number of documents and letters explanatory of the object, and showing the importance of the proposed experiment, (numbered from 2 to 7.)

The committee also annex a copy of the Treasury circular to this report, (No. 8,) and translations from a recent work on Cuba, made by the delegate from Florida, showing the extent and value of the productions of that island, and the great importance of the introduction of the same articles, as far as practicable, into this country.

No. 1. To the Senate and House of Representatives of the United States in Congress assembled:

The memorial of Henry Perrine, doctor of medicine, and American consul for
    Campeachy and the adjacent ports in Mexico, respectfully sheweth:

That your memorialist is a native American citizen, whose official district, including the peninsula of Yucatan and State of Tabasco, embraces a section of the Mexican territory which is the most prolific in tropical vegetables of great value to agriculture, manufactures, and commerce.

That, by the circular of the Treasury Department of the 6th September, 1827, your memorialist was officially invoked to aid the desires of the General Government to introduce into the United States all such foreign trees and "plants, of whatever nature, as may give promise, under proper cultivation, of flourishing and becoming useful."

That, in obedience to said circular, the time, labor, and funds of your memorialist were thenceforward devoted to observation and inquiry, amid the difficulties and the dangers incident to the nature of the climate and the face of the country, and the jealousies and the restrictions interposed by the character of the inhabitants and the despotism of the authorities.

That, fortunately, the *profession* of your memorialist afforded him the only means of purchasing favor among all ranks of a semi-barbarous people; and that, hence, a gratuitous and politic distribution of his medical services enabled him to conquer the otherwise insuperable obstacles to the progress of every inquiring foreigner, so far as to acquire much useful intelligence, not obtainable in any other way, concerning various valuable plants which may be successfully domesticated in the United States.

That, as a necessary consequence of thus discharging this governmental task, your memorialist was obliged to sacrifice all opportunities of making money either by professional or mercantile pursuits, which, in the same period and region, have furnished fortunes to his unofficial countrymen; and as funds were not appropriated by Government to promote the objects of its circular, and as the perquisites of his consulate did not defray one-third of his personal expenses, his unaided individual labors have hitherto been of comparatively little practical utility to his country, in consequence of the difficulties, disappointments, and expenses connected with the collection and transmission of living vegetables.

That, hence, when your memorialist became hopeless of the General Government's engaging directly in the important enterprise of domesticating tropical plants in the United States, he, in a letter to the Secretary of the Treasury of the 8th November last, respectfully suggested the propriety of forming an incorporated company in Florida for that purpose; that, hence, the Governor of that Territory, in his message to the Legislative Council of the 2d ultimo, recommended an act of incorporation for your memorialist and his associates; and that, on the — ult., the Tropical Plant Company was instituted by a law, which names both associates and trustees among the most distinguished residents of that peninsula.

That your memorialist now most respectfully asks of Congress a township of land in the southern extremity of East Florida, which can never be of any value either to Government or to your memorialist, without previous heavy expenditures in improvements either upon or around it, but which he is willing to accept as an equivalent for his past sacrifices, or for his future services, as a premium "to encourage the introduction and promote the culture of" tropical plants; since this grant, with his special intelligence, will enable him to secure that leading and permanent participation in this important enterprise, which he believes essential to the speedy, spirited, and persevering progress of the present or any other association, to its zealous, liberal, and patriotic measures, to its ultimate accomplishment of the greatest possible national advantages, and to the consequent elevation of his name to the list of benefactors of his country.

That, in the opinion of your memorialist, the southern extremity of the peninsula of Florida, in soil, climate, and geographical position, affords the only suitable location for the commencement of the aforesaid enterprise, by means of a garden or nursery to contain all tropical vegetables of utility or ornament, which, after due seasoning, may thence be gradually transplanted and acclimated throughout the Territory, and thus be ultimately extended over the adjoining States on the Gulf of Mexico and the Atlantic ocean; and that, in this way, your memorialist firmly believes almost every valuable tropical vegetable may be finally domesticated in all our Southern States, with the great encouragement afforded by the general fact *that most articles of culture flourish best at the more temperate margins of their native zone.*

That, hence, your memorialist confidently anticipates the naturalization of all exotics whose qualities may render them desirable denizens of our free and industrious republic; among which he may name the very beautiful and extensively useful family of the palms, whose diversified products embrace every thing that is essential to the subsistence and comfort of man; the liliaceous order, including the agaves, in his estimation, rank next in their manifold utility to the human race; the shrubs for chocolate, coffee, and tea, which have become articles of necessity in civilized life; the logwood, fustic, cochineal, and other dyes of Mexico, Guatemala, and Brazil; the cinnamon, pimento, ginger, and other spices of the East and West Indies; the mahogany, cedar, ebony, and other precious woods of all parts of the world; the bananas, anonas, mangoes, and numerous delicious fruits for the enjoyment of health; the Peruvian bark, sarsaparilla, canella, and innumerable salutary medicines for the removal of disease.

That, in the opinion of your memorialist, the domestication of the species of a single genus of tropical plants will cause a great revolution in the agriculture of the Southern States, which will not only effectually relieve their present embarrassments, but will also give a productive value to their ruined fields and most steril districts; and that the extensive cultivation of a single species (the agave Sisalana) alone, will furnish a profitable staple to the planters of the South, and a cheap material to the manufacturers of the North; which will supply many wants of our merchant vessels, of our navy, and of our citizens in general; augment our coasting trade and our foreign commerce; and thus contribute greatly to the prosperity and perpetuity of the Union.

That your memorialist, therefore, most respectfully trusts that, either as an equivalent for his past sacrifices, as a reward for his communicated information, as an encouragement for his future services, or as a consideration for all combined, the prayer of this memorial will be granted: especially as the land in that section of Florida will not otherwise, in many years, be of any productive importance to Government; as the settlement of the tract will, from its location, be attended with numerous privations and expenses, which cannot be compensated by a gift of the soil; and as the adjacent territory, in consequence of this very settlement, must speedily acquire a value which will furnish a profitable revenue to the United States

That, to explain more fully the views and expectations of your memorialist, he refers to his various communications on file in the Treasury Department; to the annexed extract of his letter to the Secretary of the Treasury, published in the Globe of the 19th November last; and to the adjoining extract of his letter to Doctor Howell, of Princeton, New Jersey, published in the Telegraph of the 17th ultimo; also, to the subjoined manuscript testimony, (marked A,) and to the file of corroborating Spanish official documents (marked B.)

And your memorialist, &c.

HENRY PERRINE

NEW YORK, *February 6, 1832*

Tropical Plants

Extract from the message of the Governor of Florida, in the Floridian of January 2

"Hundreds of the vegetable productions of tropical climates, of great value, and some in such common use as to be considered articles of necessity, and which we now import at high cost, could be easily cultivated in any part of our Territory. Many, too tropical to flourish in West or Middle Florida, could be reared under the more genial climate of the southern part of the peninsula. The southern part of this continent, and South America and China, abound in trees, plants, herbs, and roots, possessing the most valuable properties, the use of which has been confined to the places of their production, but which could as well be produced and enjoyed by our own citizens. I herewith transmit to the Council, and respectfully invite their attention to an extract of an official letter from H. Perrine, United States consul for Campeachy, to the Secretary of the Treasury, in relation to this subject, which has been published in the newspapers, and from which I have taken it. Other documents worthy of attention are also herewith sent to the Council. It will be noticed that Mr. Perrine is desirous that an act of the Council should be passed, incorporating himself and his associates into a company for the cultivation of tropical exotics; and he proposes to establish the plantation of the company on the southern part of the peninsula. This enterprise should not be classed with the inflated visionary projects of which Florida has been so prolific, and the failure of which has created so much distrust of all novel undertakings. If those who embark in it should not find it a source of gain, and should, after trial, abandon it, the benefits resulting to the country from the introduction of the many valuable foreign products they will have brought among us must be of considerable importance, and should induce us to render every encouragement and aid in our power to promote the success of the undertaking; and although Mr. Perrine has made no direct application, I earnestly recommend the granting of a charter as he wishes, and the bestowment upon the company of as many privileges as is compatible with the public interests. The National Legislature, it is to be hoped, will afford aid to so laudable an enterprise, and one which, if successful, promises to be of national benefit, by a grant of land sufficient for their use, or otherwise. I esteem it, however, of paramount consequence that an interest should be excited among the agriculturists throughout the Territory, in relation to the introduction and adoption of foreign products. The tea-plant, those trees and plants from which are procured the olive, ginger, pepper, cloves, cinnamon, pimento, nutmeg, and cocoa, and many other articles of daily use in our families, could, it is believed, with care and attention, be successfully cultivated in most parts of our Territory; but the practicability of every article mentioned being readily produced by those planters favorably located, as it regards climate, cannot be questioned. The production of these articles, if only sufficient for our own domestic consumption, would be of immense advantage to our citizens; and if experience should prove that Florida might, in a few years, be looked to by our fellow-citizens of the States for such products, the benefits resulting to the Territory would be incalculable.

From the Washington Telegraph of January 17

We publish to-day a highly interesting extract from a letter of Dr. H. Perrine, United States consul at Campeachy, upon the introduction of tropical productions into the Southern States. It is time that the inhabitants of that region should be looking to some new products in the Southern States to avert the evils of the present oppressive system upon their industry and resources.

Extract of a letter from Dr. H. Perrine

"I was much interested by the memoir of Doctor Mease on the materials for thread, twine, and cordage, which appeared in the October number of Silliman's Journal. By thus directing the attention of his countrymen towards those foreign plants which produce these materials, he has rendered an acceptable service to the public. The imperfections or errors which exist in that communication are his misfortune, not his fault. The observations of transient travellers on such subjects must necessarily be very limited and very superficial; and the reports of the natives are still less to be trusted. The attention of such men as Humboldt, Bullock, Poinsett, and Warden, was occupied by too many objects to acquire minute intelligence on all; hence the inaccuracies in at least one portion of the compilation made by Doctor Mease. A residence of several years in a tropical climate has enabled me to obtain a personal acquaintance with some of its valuable fibrous plants. Of the highest importance, in my estimation, are those species of the liliaceous tribe which are there prized on account of the quantity or quality of fibres obtained from the interior of their fresh leaves. Many of them, at first ranked under the aloes, were subsequently gratified with the title of agaves; and some of them are merely waiting for the aid of a botanist to become an independent genus. The agave Americana is still called by travellers the American aloe; and Doctor Mease, with them, has been misled to suppose that this plant produces the Sisal hemp, and the pita a much finer material: but the agave Americana is dedicated to a very different production — the celebrated drink called 'pulque,' derived from the sap of its stem; and hence maguey de pulque is its common name in Mexico. A direct tax on the consumption of this beverage forms an important item in the revenue of that country. 'The entry duties paid in the three cities of Mexico, Tolusa, and Puebla, amounted, in 1793, to the sum of 817,739 piastres.' Humboldt was correct in affirming of the maguey de pulque, 'that its cultivation has real advantages over the cultivation of maize, grain, and potatoes; that it is neither affected by drought nor hail, nor the excessive cold which prevails in winter on the higher Cordilleras of Mexico; that it grows in the most arid grounds, and frequently on banks of rocks hardly covered with vegetable earth; and that it is one of the most useful of all the productions with which nature has supplied the mountaineers of equinoctial America.' But it is not true that the same plant produces the very fine, very strong, and very long fibres, known by the name of pita, from which the most beautiful sewing thread is made; nor does it furnish those coarser fibres for twine and cordage, resembling manilla, but denominated Sisal hemp. If tropical *hemp* be an admissible term for the latter, the former may be honored with the distinction of tropical *flax*. The ixtla, whose *thin* leaves afford the *pita,* grows wild in the shade of the fertile forests of Tabasco. The sosquil o henequen, whose *thick* leaves yield the Sisal hemp, is cultivated in the sun of the steril plains of Yucatan: the stem of neither supplies the drink which constitutes the principal value of the agave Americana; nevertheless, a variety of the maguey de pulque does grow on the tropical shores of the Gulf of Mexico, from which the highland soldiers have occasionally extracted their favorite beverage. Some of the cultivated magueys, brought from a plantation on the mounains to the garden of a gentleman in Campeachy, are there flourishing, notwithstanding the difference in climate, and have produced shoots, which were by me transmitted to New Orleans. Humboldt says that this plant has become wild since the sixteenth century throughout all the south of Europe, the Canary islands, and the coast of Africa; and this fact supports my decided opinion that all the valuable species of the same genus may be successfully cultivated in our Southern States.

"Two varieties of that species, which I take the liberty to christen agave Sisalana, have long been cultivated in the vicinity of Merida, on an extensive scale. Different quantities and qualities of fibres are obtained from several kinds of 'sosquil,' which grow spontaneously through the whole peninsula of Yucatan; but the planters give the preference to the sacqui and yaxqui of the natives, or the whitish and greenish 'henequen.' The young plants are placed about twelve Spanish feet apart, and during the first two years some labor is employed to destroy the weeds between them. In the third year, the cutting of the lower rows of leaves is commenced, and every four months this operation is repeated. Each robust plant will thus give about seventy-five leaves annually, from which are extracted about seven pounds and a half of fibres, and will continue yielding these crops from five to ten years in succession; it is, however, generally cut down as soon as one of the shoots from its roots has grown sufficiently to supply its place: its other offspring are previously removed to form new plantations. The hardiness of the shoots may be inferred from the fact that they are exposed to the sun fifteen or twenty days 'to cicatrize their wounds,' as a necessary preparation for replanting. The simplicity of their cultivation may be conceived from the statement that there is not a hoe, nor a spade, nor a harrow, nor a plough, employed in the agriculture of all Yucatan. The facility of extracting the fibres from their leaves is shown by the rudeness of the instruments which are used by natives for that purpose: a triangular stick of hard wood, with sharp edges, from eight to twelve inches long, and from one to two inches thick, is with them an equivalent to the shaving-knife of the curriers, by which they scrape away from each side of the leaf, on a board resting against the breast, the cuticule and pulpy substance that covers the fibres. Another mode of accomplishing the same object is, by pressing the sharp semilunar extremity of a long flat stick against any fixed surface upon a narrow longitudinal strip of the leaf, which is then drawn through by the unemployed hand. The length, weight, strength, and other qualities of the fibres, as well as the labor of separating them, vary with the magnitude, age, and position of the leaves; but, when extracted, a few hours' exposure to the sun completes the preparation of the Sisal hemp for manufactures and commerce.

"The above brief sketch will show that the bales of exported Sisal hemp may contain materials of very different qualities; and that hence the opinions of its merits expressed by our merchants, our manufacturers, and our scientific men, must vary with the parcels that fall into their hands. The fibres of a single cultivated variety of the agave Sisalana might be assorted like cotton for the foreign market, with denominations and prices corresponding to their relative value; but the collectors for exportation, unconscious of the true interests of themselves or their country, not merely mingle the whole products of both the sacqui and the yaxqui, but add inferior qualities obtained from wild varieties of the same, and even of different species; and injure still further the reputation of this staple abroad, by including the worst proceeds of its imperfect dressings."

Notwithstanding all these disadvantages, the cultivation of Sisal hemp is of the highest importance to the people of Yucatan, *as it is the only article of agriculture* which supplies them with raw materials and domestic manufactures for foreign trade: it has long formed a principal portion of the exports from Sisal to Havana, in the shape of twine, cordage, bagging, &c., for the planters of Cuba. Its ropes and cables have been used in the shipping of various nations; and entire cargoes of the raw material have been transported to the ports, and wrought in the factories, of Europe and of the United States.

As the agave Sisalana is so important an object of cultivation in the peninsula of Yucatan, how much more important would it be to the peninsula of Florida?

Ignorance, and indolence, and ineptitude exist on one side of the American Mediterranean; intelligence, and industry, and ingenuity on the other. Insuperable are the obstacles to enterprise in the nominal republic of the United States of Mexico; multiplied are the encouragements to improvement in the genuine republic of the United States of America!

A. Port of Tabasco, in the District of the United States Consulate at Campeachy, June 6, 1831

Sir: As you are the only native American citizen long resident in Tabasco, and well qualified to answer the present note, do me the favor to tell me what sum of money I should have very probably gained since my arrival here, in June, 1827, if I had dedicated myself exclusively to the interested exercise of my profession, instead of generally practising it gratuitously, with the hope of promoting the usefulness of this consulate, and the inquiry after plants suitable for the United States.

I am, sir, very respectfully, your obedient servant,

H. PERRINE

Ezekiel P. Johnson, *M. D.*

San Juan Bautista, June 6, 1831.

Sir: In reply to your note of this morning, I answer, that if, instead of having gratuitously practised your profession in the families of the persons employed under the local and general Governments, and devoting so large a portion of your time to the collection of seeds and plants useful to our country, (and adapted to our soil and climate,) you had applied yourself to the pay practice of your profession during your residence in the Mexican territory, you would, in my opinion, have been now able to return to the United States with ten or twelve thousand dollars as the reward of your industry.

I remain, sir, very respectfully, your obedient servant,

E. P. JOHNSON

H. Perrine, *M.D., U.S. Consul*

New York, February 3, 1832

I, George Clark, an American citizen, now residing in the city of New York, do hereby certify that, during the years 1829 and 1830, I was engaged in mercantile business in the State of Tabasco, in the district of the United States consulate for Campeachy. That my own personal observations, and the statements of the inhabitants of the country, whether natives or foreigners, corroborate the facts and opinions expressed in the above copy of an original letter from Doctor E. P. Johnson, which I have seen; that the facilities for making money in the consular district of Doctor H. Perrine were abundant, both in professional and mercantile pursuits, which is proved by the fortunes which have been made, since 1827, by his unofficial countrymen; that, within three years of that time, a young merchant gained, by merchandise, upwards of $50,000, according to his own assertions, which were confirmed by my observations; that a young physician, by his practice alone, gained, in my estimation, at least from three to four thousand dollars a year; that the other physicians and merchants made proportionably enormous gains; that Doctor H. Perrine, from the superiority of his medical reputation, and the privileges of his official situation, enjoyed advantages for practice and merchandise not possessed by any of his countrymen: and that, nevertheless, he sacrificed his splendid opportunities for acquiring money through his profession and office, by devoting his medical services gratuitously to the poor and powerful, for the purpose of obtaining useful intelligence concerning the plants of the country, which could not have been purchased in any other way.

GEORGE CLARK

No. 2

Whereas the present Legislative Council, with a view of encouraging the cultivation of useful foreign and tropical plants, have passed a law incorporating the "Tropical Plant Company of Florida:" and whereas it is believed that a grant of land by Congress, somewhere in the southern part of the peninsula of Florida, would be greatly conducive to the public good, and promote the views of said company, and might be made without detriment to the public interest:

Be it, therefore, resolved, That our delegate in Congress be, and he is hereby, requested to use his endeavors to procure the passage of a law making such grant to the said company, for the objects aforesaid, as may best comport with the public good.

Resolved, further, That a copy of this preamble and resolution be immediately forwarded to him.

A true copy.

Test: JOHN K. CAMPBELL, *Clerk.*

No. 3. *Lyceum of Natural History, New York, February 10, 1832*

DEAR SIR: The subject of your paper on the agave Sisalana, read before the Lyceum on January 9th, 1832, has been duly examined by their committee, who gave in their report at the last sitting of the society; which report, by order of the same, I now transmit.

Extract from the minutes

"Feb. 9, 1832. — Mr. Halsey, of the committee to whom were referred the papers read before the Lyceum, on January 9th, by Dr. Perrine, American consul to Yucatan, on the subject of Sisal hemp, (agave Sisalana,) and other tropical plants, gave in their report, accompanied with the following resolutions, which were unanimously accepted.

"At the sitting of the Lyceum, January 9, 1832, Dr. Perrine, American consul to Yucatan, read some observations on the culture, &c. of the Sisal hemp, and other tropical plants, from which materials are furnished of extensive use in commerce and in the arts.

"*Resolved*, That the Lyceum concur in the views of the subject given by Dr. Perrine; and conceiving the great national benefit which might be derived from the introduction and general culture of those plants in such sections of the Union as would be suitable to their growth, they consider the proposition of Dr. Perrine as particularly meriting the patronage of Government, and unite with him in recommending that a grant of land be made by Congress for the purpose of conducting his experiments on the same.

"On motion, it was unanimously

"*Resolved*, That the recording secretary be instructed to transmit to Dr. Perrine a copy of the report and accompanying resolutions on the subject of his paper."

Respectfully yours,

L. D. GALE,
Recording Secretary of the Lyceum

To Dr. PERRINE,
  American Consul to Yucatan

No. 4. New York, February 21, 1831

SIR: Should the memorial of my friend Dr. Perrine, United States consul for Campeachy, relative to the domestication of tropical plants, come before your committee, you will perceive that I have borne my humble testimony to this services and sacrifices in the important enterprise therein set forth.

My residence for several years in a tropical climate, by giving me a knowledge of its productions, has enabled me to appreciate so highly Dr. Perrine's plans, that I have consented, if it should be thought proper, to have my name inserted in the act of the Legislative Council of Florida as one of the trustees of the Tropical Plant Company.

A full, fair, and friendly investigation of his memorial and accompanying documents will, I hope, satisfy you that a grant to him of the land for which he petitions will result in important benefits to the southern portion of our country. If so, permit me to request you to lend your aid to bring the application to a favorable issue. Without the grant of land, he will be unable to continue any measures for the speedy and complete success of this truly national enterprise, which has hitherto been kept up by his individual and unaided exertions.

As I possess no interest in the result of his application to Congress, I feel the greater confidence in addressing you in behalf of the memorial; and I am convinced that no association will execute his liberal views for the public good, unless legislative aid be extended to forward the enterprise.

I am, very respectfully, your most obedient servant,

E. P. JOHNSON

HON. ERASTUS ROOT

No. 5. Pensacola, December 29, 1831

DEAR SIR: The establishment of every new branch of industry, whether it be by the introduction of a new manufacture or of a new article of culture, I consider a positive and permanent addition to national wealth. For this reason, I viewed the letter of our consul at Campeachy, which you had the goodness to enclose to me, on the subject of the Campeachy or Sisal hemp, as being well worthy of attention. The use of what is commonly called the *grass rope* has been extending itself in the United States in a surprising manner of late years. Nearly all our steamboats on the Western waters use no other, and it is getting rapidly into use for hawsers, pulley-ropes, rugs, and even for the running rigging of our merchant ships; indeed, it is impossible at present to say to what extent its uses able than common hemp, and may be applied to many purposes to better advantage than that article. I do not know what quantity is imported in the manufactured or unmanufactured state; but this is certainly a subject well worthy of the attention of the patriotic statesman, as I have ascertained, in the most satisfactory manner, that *the plant is a native of Florida, and of the Southern States in the same latitude*.

From the exact similarity of the ropes made by the Florida Indians, I was satisfied that it was made from the same plant which was pointed out by Spanish gentlemen here as the *pita* cultivated in Campeachy. I have lately seen a person from that country, to whom I exhibited the plant, and he positively assures me that it is the same. It is of the palm family, resembles the bouquet palmetto; the leaves are softer and more pliant, but it has a sharp point or needle like it. My friend and fellow-traveller, Mr. Nutall, can give you its botanical name and character. The plant requires considerable space — say at least five feet square; is placed in hills or squares, like Indian corn, and will occupy the whole by its leaves and side shoots. I should think about a thousand plants might be placed on an acre, producing at least a pound of hemp to each; and if the culture should be successful, it must afford immense profits to those who will first engage in it, as it will require few hands. The plant being perennial, the plantation, once made, will last for years.

The pita grows in great abundance, even in the poor sandy pine lands; but, on the thin oak and hickory ridges, where the soil, although sandy, is more fertile, it grows in perfection. The new settlers now begin to use it for domestic purposes; they rot it, by throwing it into a pond or stream of water for a month, when the fibre is separated from the bark with great ease. As it is easily transplanted, I have no doubt a sufficient quantity can be obtained at once, in its wild state, to make a respectable field, almost anywhere, in the space of a mile or two. I know, from actual experiment, that it can be transplanted without difficulty; that it will bear to be cut once a year; and will, in that period, again attain the former size, or even greater.

Would it not be well to make some more minute inquiries of our consul as to the mode of cultivating this valuable plant? It certainly deserves to be encouraged by our Government; at least, the necessary information might be procured for the benefit of those enterprising individuals who may be disposed to engage in it.

With sentiments of respect, I am your most obedient servant,

H. M. BRACKENRIDGE

HON. J. M. WHITE

No. 6. Plants of Mexico Extract of a letter from Henry Perrine, Esq., United States consul for Campeachy, to the Secretary of the Treasury, dated

NEW YORK, *November 8, 1831*

"SIR: The Treasury circular of the 6th of September, 1827, relative to the introduction of useful exotics into the United States, addressed to a portion of the American consuls, directed the inquiries of the subscriber (United States consul for Campeachy) particularly to the logwood tree, which abounds in the peninsula of Yucatan. His first communication to the Department, dated 1st

January, 1829, contained a brief sketch of the information then obtained, and endeavored to attract its attention towards the *fibrous* plants of that region, as subjects of much greater and more immediate utility. Belonging to the natural family liliacea, the species of the genus agave appeared to him of transcendant importance. In his communication of January 1, 1830, he again expressed that opinion, which has been confirmed by his subsequent observations in Tabasco, and sustained by all the intelligence acquired since his arrival in this city, on the 13th August last. The utility of these plants has been noticed in almost every work on Mexico, from the conquest to the present day. According to Clavigero, the name of Mexico, "quiere decir en el centro del maguey, ó pita, ó aloe Americano." (Agave Americano of Humboldt.) A reference to the same author shows the manifold utility of these plants to the ancient Mexicans. Some species furnish themselves protecting enclosures, and afford impassable hedges to other objects of cultivation. From the juice of others are extracted honey, sugar, vinegar, *pulque*, and ardent spirits. The pulque de maguey is the celebrated substitute for beer, cider, and wine — preferred, even by foreigners, to every other liquor. From the trunk and thickest portion of the leaves, roasted in the earth, an agreeable food is obtained. The sap is applied externally to indolent sores and tumors, and a preparation is used internally for urinary and other diseases. The stalks serve for the beams, and the leaves for the roofs of huts. The thorns answer for lancets, awls, needles, arrow-heads, and other cutting and penetrating instruments. The fibrous substance of the leaves is, however, the most important gift of the agave genus to Mexico. According to the species, the fibre varies in quality from the coarsest hemp to the finest flax, and may be employed as a superior substitute for both. From it the Mexicans fabricated their thread and cordage, mats and bagging, shoes and clothing, and webs, equivalent to cambric and canvass; the hammocks in which they are born, repose, and die; and the paper on which they painted their histories, and with which they adorned and adored their gods. The value of all the agaves is enhanced by their indifference to soil, climate, and season; by the simplicity of their cultivation; and by the facility of extracting and preparing their products. It is not, therefore, surprising that the ancient Mexicans used some part or preparation of these plants in their civil, military, and religious ceremonies, at marriages and deaths; nor that they perpetuated an allusion to their properties in the name of their capital.

"Humboldt, Poinsett, Warden, and other foreigners, seem to consider the *fine fibres*, called pita, a product of the same plant that produces the pulque. But the maguey, from which the drink is obtained, is a totally different species, and furnishes fibres of the coarsest texture. The pita plant, like the cacao, grows best in the shade, and its leaves are long, narrow, and slender; the fibres of one of which accompany this communication. From this species is probably fabricated, in China, that beautiful glossy, fine, and strong stuff, known here by the name of grass cloth. The Sisal hemp of commerce is obtained from two varieties of another species of the agave, which have long been cultivated in the vicinity of Merida. To this species the subscriber refers the Manilla hemp, although he has not hitherto been able to obtain any satisfactory account of the plant which produces it, from our scientific and mercantile men, nor from books. Captain John White, of the navy yard at Charleston, states his impression that it is obtained from the bark of a species of palm tree, but its texture does not warrant that opinion; and Captain Morril, recently from Manilla, confirms the inference above expressed by the subscriber.

"Doctor Hernandez describes nineteen species of agave as indigenous to Mexico, which vary more in the character of their interior substance than in the form and color of their leaves; and, among these, the precise variety which produces the Manilla hemp will probably be found. Some species are prized for the beauty of their flowers, and some for the odor, and others for the flavor of their fruit; but the subscriber limits his recommendation to Government of those which are most valuable on account of the quantity and quality of their fibres. Their cultivation in the United States, he still believes, will form an era in our agricultural and manufacturing prosperity, as distinguished as the invention of the cotton-gin. He has shown, in his former communications, the almost insuperable obstacles to his introducing these plants, unaided and alone; but he had, nevertheless, the satisfaction to be apprized by Mr. Gordon, the collector of New Orleans, that the young Sisal hemp varieties, sent by him to that city, had arrived in a thriving and vigorous condition. The subscriber believes that an act of the Legislative Council of Florida, incorporating himself and associates into a company for the cultivation of tropical exotics, will be necessary to accomplish the views of Government, as manifested in the Treasury circular of the 6th of September, 1827.

"He believes that he has information in his possession sufficient to attract capital to the enterprise. Once formed, the company might, in the same vessel, bring other useful exotics, although their pecuniary interests would confine them to the fibrous plants. Acclimated in the southern extremity of East Florida, they would gradually extend up to the adjoining States on the Gulf of Mexico and the Atlantic ocean. The seed of the logwood would probably be coveted for hedges, on account of its beauty, novelty, and utility. The arnatto would be propagated, were it merely as an ornamental tree. The India rubber would be sought at least as a curiosity, and so would the pimento.

"Ginger and turmeric present the stimulus of immediate profit. The Nankin and the tree cotton would find a genial climate. The tropical shrubs, whose leaves are a substitute for indigo, and the tree whose fruit serves for soap, would likewise there find a home. The 'ramon' would accompany them, to furnish, with its leaves and tender branches, the food for domestic animals in the driest seasons, and, with its fruit, a subsistence for the human family in times of scarcity of corn. In short, every useful tropical plant would likely be introduced by an incorporated company for cultivating the fibrous species at the southern extremity of East Florida."

The above, with some corrections, is from the Globe of the 19th instant.

We are personally acquainted with our enterprising fellow-Jerseyman, Dr. Perrine, and we heartily wish that he may be ultimately recompensed for the time, labor, and money, which he has sacrificed, and is still devoting to the introduction of useful plants into the United States. We understand that he believes the cochineal plant and insect may be successfully reared in East Florida, inasmuch as the experiment has succeeded in Spain; and that he is possessed of all the details relative to the management of both. The stingless bees of Yucatan, introduced by him, (of which he has deposited a hive in Peale's museum at New

York,) will be an invaluable acquisition to our Southern and Southwestern States, and may be gradually propagated throughout the Union. It is to be lamented that Government does not possess the power to appropriate funds to aid its agents in the duties of this class, imposed upon them by the Treasury circular. It is still more lamentable that our consulates, especially in Spanish America, are not salaried, and filled with scientific men, whose pursuits would be useful to their country; and that they are too generally obtained as a mere speculation, to aid the mercantile business of the possessor, to the injury of his countrymen in the same trade, and to the degradation of his office in the eyes of foreigners. — *New Brunswick Times*.

* * *

No. 8. Circular to a portion of the Consuls of the United States Treasury Department, September 6, 1827

SIR: The President is desirous of causing to be introduced into the United States all such trees and plants from other countries, not heretofore known in the United States, as may give promise, under proper cultivation, of flourishing and becoming useful, as well as superior varieties of such as are already cultivated here. To this end, I have his directions to address myself to you, invoking your aid to give effect to the plan that he has in view. Forest trees useful for timber; grain of any description; fruit trees; vegetables for the table; esculent roots; and, in short, plants of whatever nature, whether useful as food for man or the domestic animals, or for purposes connected with manufactures or any of the useful arts, fall within the scope of the plan proposed. A specification of some of them to be had in the country where you reside, and believed to fall under one or other of the above heads, is given at the foot of this letter, as samples merely; it not being intended to exclude others of which you may yourself have knowledge, or be able, on inquiry, to obtain knowledge. With any that you may have it in your power to send, it will be desirable to send such notices of their cultivation and natural history as may be attainable in the country to which they are indigenous; and the following questions are amongst those that will indicate the particulars concerning which information may be sought:

1. The latitude and soil in which the plant most flourishes?
2. What are the seasons of its bloom and maturity, and what the term of its duration?
3. In what manner it is propagated — by roots, seeds, buds, grafts, layers, or how? and how cultivated? and are there any unusual circumstances attending its cultivation?
4. Is it affected by frost in countries where frost prevails?
5. The native or popular name of the plant, and (where known) its botanical name and character?
6. The elevation of the place of its growth above the level of the sea?
7. Is there, in the agricultural literature of the country, any special treatise or dissertation upon its culture? If so, let it be stated.
8. Is there any insect particularly habituated to it?
9. Lastly — its use, whether for food, medicine, or the arts?

In removing seeds or plants from remote places across the ocean, or otherwise, great care is often necessary to be observed in the manner of putting them up and conveying them. To aid your efforts in this respect, upon the present occasion, a paper of directions has been prepared, and is herewith transmitted.

The President will hope for your attention to the objects of this communication, as far as circumstances will allow; and it is not doubted but that your own public feelings will impart to your endeavors under it a zeal proportioned to the beneficial results to which the communication looks. It is proper to add that no expense can, at present, be authorized in relation to it. It is possible, however, that Congress may not be indisposed to provide a small fund for it. The seeds, plants, cuttings, or whatever other germinating substance you may transmit, must be addressed to the Treasury Department, and sent to the collector of the port to which the vessel conveying them is destined, or where she may arrive, accompanied by a letter of advice to the Department. The Secretary of the Navy has instructed the commanders of such of the public vessels of the United States as may ever touch at your port to lend you their assistance towards giving effect to the objects of this communication, as you will perceive by the copy of his letter of instructions, which is herewith enclosed for your information. It is believed, also, that the masters of the merchant vessels of the United States will generally be willing — such is their well-known public spirit — to lend their gratuitous co-operation towards effecting the objects proposed.

I remain, respectfully, your most obedient servant,

RICHARD RUSH

Directions for putting up and transmitting seeds and plants; accompanying the letter of the Secretary of the Treasury of Sept. 6, 1827

With a view to the transmission of seeds from distant countries, the first object of care is to obtain seeds that are fully ripe, and in a sound and healthy state. To this, the strictest attention should be paid; otherwise, all the care and trouble that may be bestowed on them will have been wasted on objects utterly useless.

Those seeds that are not dry when gathered, should be rendered so by exposure to the air, in the shade.

When dry, the seeds should be put into paper bags. Common brown paper has been found to answer well for making such bags. But, as the mode of manufacturing that paper varies in different countries, the precaution should be used of putting a portion of the seeds in other kinds of paper. Those that most effectually exclude air and moisture are believed to be the best for that purpose. It would be proper, also, to enclose some of the seeds in paper or cloth that has been steeped in melted bees-wax. It has been recommended that seeds collected in a moist country or season be packed in charcoal.

After being put up according to any of these modes, the seeds should be enclosed in a box, which should be covered with pitch, to prevent them from damp, insects, and mice. During the voyage they should be kept in a cool, airy, and dry situation — not in the hold of the ship.

The oily seeds soonest lose their germinating faculty. They should be put in a box with sandy earth, in the following manner: First, about two inches of earth at the bottom; into this the seeds should be placed at distances proportionate to their size; on these another layer of earth about an inch thick, and then another layer of seeds; and so on, with alternate layers of earth and seeds, until the box is filled within about a foot of the top, which space should be filled with sand; taking care that the earth and sand be well put in, that the seeds may not get out of place. The box should then be covered with a close net-work of cord well pitched, or with split hoops or laths well pitched, so as to admit the air without exposing the contents of the box to be disturbed by mice or accident. The seeds thus put up will germinate during their passage, and will be in a state to be planted immediately on their arrival.

Although some seeds with a hard shell, such as nuts, peaches, plums, &c., do not come up until a long time after they are sown, it would be proper, when the kernel is oily, to follow the method just pointed out, that they may not turn rancid on the passage. This precaution is also useful for the family of laurels (laurineae) and that of myrtles, (myrti,) especially when they have to cross the equatorial seas.

To guard against the casualties to which seeds in a germinating state may be exposed during a long voyage, and as another means of ensuring the success of seeds of the kinds here recommended to be put into boxes with earth, it would be well also to enclose some of them (each seed separately) in a coat of bees-wax, and afterwards pack them in a box covered with pitch.

In many cases it will be necessary to transmit roots. Where roots are to be transmitted, fibrous roots should be dealt with in the manner herein recommended for young plants. Bulbous and tuberous roots should be put into boxes, in the same manner as has already been recommended for oleaginous seeds; except that, instead of earth, dry sand, as free as possible from earthy particles, should be used. Some of the bulbous and tuberous roots, instead of being packed in sand, may be wrapped in paper, and put in boxes covered with net-work or laths. Roots should not be put in the same box with seeds.

Where the seeds of plants cannot be successfully transmitted, they may be sown in boxes, and sent in a vegetating state. Where more than one kind is sown in the same box, they should be kept distinct by laths, fastened in it crosswise on a level with the surface of the ground in which they are sown; and when different soils are required, it will be necessary to make separate compartments in the box. In either case, they should be properly marked, and referred to in the descriptive notes which accompany them.

When plants cannot be propagated from seeds, with a certainty of their possessing the same qualities which long culture or other causes may have given them, they may be sent in a growing state. For this purpose, they should be taken up when young. Those, however, who are acquainted with their cultivation in the countries where they grow, will know at what age they may be safely and advantageously removed. They may be transplanted direct into the boxes in which they are to be conveyed; or, where that cannot be conveniently done, they may be taken up with a ball of earth about the roots, and the roots of each surrounded with wet moss, carefully tied about it, to keep the earth moist. They may afterwards be put into a box, and each plant secured by laths fastened crosswise above the roots, and the interstices between the roots filled with wet moss. The same methods may be observed with young grafted or budded fruit trees.

Where the time will permit, it is desirable that the roots of the plants be well established in the boxes in which they are transplanted. Herbaceous plants require only a short time for this; but, for plants of a woody texture, two or three months is sometimes necessary.

Boxes, for the conveyance of plants, or of seeds that are sown, may be made about two feet broad, two feet deep, and four feet long, with small holes in the bottom, covered with a shell or piece of tile, or other similar substance, for letting off any superfluous water. There should be a layer of wet moss of two or three inches deep at the bottom; or, if that cannot be had, some very rotten wood or decayed leaves; and upon that about twelve inches depth of fresh loamy earth, into which the plants that are to be transplanted should be set. The surface of the earth should be covered with a thin layer of moss, cut small, which should be occasionally washed in fresh water during the voyage, both to keep the surface moist, and to wash off mouldiness, or any saline particles that may be on it.

When the boxes are about to be put on board the ship, hoops of wood should be fastened to the sides in such a manner that, arching over the box, they may cover the highest of the plants; and over these should be stretched a net-work of pitched cord, so as to protect the plants from external injury, and prevent the earth from being disturbed by mice or other vermin.

To each box should be fastened a canvass cover, made to go entirely over it, but so constructed as to be easily put on or off, as may be necessary to protect the plants from the salt water, or winds, and sometimes from the sunshine. Strong handles should be fixed to the boxes, that they may be conveniently moved.

During the voyage the plants should be kept in a light, airy situation; without which, they will perish. They should not be exposed to severe winds, nor to cold, nor, for a long time, to too hot a sunshine, nor to the spray of the salt water. To prevent injury from the saline particles with which the air is oftentimes charged at sea, (especially when the waves have white frothy curls upon them,) and which, on evaporation, close up the pores of the plants and destroy them, it will be proper, when they have been exposed to them, to wash off the salt particles by sprinkling the leaves with fresh water.

The plants and seeds that are sown will occasionally require watering on the voyage, for which purpose rain water is best. If, in any special case, particular instructions on this point, or upon any other connected with the management of the plants during the voyage, be necessary, they should be made known to those having charge of the plants. But, after all, much will depend upon the judicious care of those to whom the plants may be confided during the voyage.

Plants of the succulent kind, and particularly of the cactus family, should not be planted in earth, but in mixture of dry sand, old lime rubbish, and vegetable mould, in about equal parts, and should not be watered.

It may not be necessary, in every case, to observe all the precautions here recommended in regard to the putting up and transmission of seeds; but it is believed that there will be the risk in departing from them, in proportion to the distance of the country from which the seeds are to be brought, and to the difference of its latitude, or of the latitudes through which they will pass on the voyage. It is not intended, however, by these instructions, to exclude the adoption of any other modes of putting up and transmitting seeds and plants which are in use in any particular place, and which have been found successful, especially if more simple. And it is recommended that not only the aid of competent persons be accepted in procuring and putting up the seeds and plants, but that they be invited to offer any suggestions in regard to the treatment of the plants during the voyage, and their cultivation and use afterwards.

CIRCULAR

NAVY DEPARTMENT

SIR: I have to call your attention to the enclosed copy of a communication from the Treasury Department to the consuls of the United States at various posts; and to desire that the objects of that communication may be promoted by you, on all occasions, as far as may be in your power.

The Executive takes a deep interest in this matter; and, by particular attention to it, you will probably confer a lasting benefit to the country.

The letter of the Secretary of the Treasury is so full and satisfactory, that no farther explanations seem necessary on my part.

You will be pleased to report to the Department what you do in execution of this object, and return the papers to the Department when you are detached from the vessel which you now command.

I am, respectfully, &c.

SAM. L. SOUTHARD

Consulate U.S.A., Campeachy, February 20, 1834

SIR: As an appendix to his communication of the 1st instant, the subscriber avails himself of the only statistical data in his power to *demonstrate* the greatly superior productiveness of slave labor in the United States over slave labor in the West Indies.

British West India colonies, 692,700 slaves, 427,392,000 pounds of sugar, and 19,769,500 pounds of coffee exported.

Spanish island of Cuba, 286,942 slaves, 162,703,425 pounds of sugar, and 42,971,625 pounds of coffee exported.

Louisiana, 109,631 slaves, 70,000,000 pounds of sugar, and 72,000,000 pounds of cotton exported.

Now, admitting for a moment that the culture of cotton is merely equal to the culture of sugar and coffee, as 109,631 slaves produce 142 millions of pounds of sugar and cotton in Louisiana, in the same proportion, 692,700 slaves should produce 897 millions of pounds of sugar and coffee in the British West India islands: and in the same manner 286,942 slaves should produce 371 millions of pounds of sugar and coffee in Cuba. But the former do produce only 447 millions, and the latter only 205 millions; together 692 millions, instead of the 1,268 millions which they should produce in proportion to Louisiana. But the truth is, that the relative value of labor of the production of cotton is at least fifty per cent. more than the value or labor of the production either of sugar or coffee; and hence, the combined 979,642 slaves of British West India islands and of Cuba, should yield 1,590 millions, instead of 672 millions of sugar and coffee, every year, for exportation; or, in other words, with an equal number of slaves Louisiana would supply the consumption of the world!

To obtain the details of the relative productiveness of a single negro, the following estimates are presented of a sugar plantation in Louisiana, and of a sugar plantation in Cuba, each assumed to yield annually 400,000 pounds of sugar.

The first are contained in the report of the agricultural committee of Baton Rouge to the Secretary of the Treasury against the reduction of duties on imported sugar, and must hence be presumed to present the most unfavorable aspect of the cultivation of sugar in Louisiana. The second is taken from pages 108–9 of the Statistical History of Cuba, by Dr. Ramon de la Sagra, who presents the most favorable aspect of the cultivation, in general, of the staples of that island. The first diminishes the average product of an acre, in Louisiana, to 1,000 pounds of sugar; the second exaggerates the average product of an acre, in Cuba, to 2,038 pounds of sugar; although he had previously admitted that Humboldt was correct in limiting it to 1,116 pounds the acre, or 1,500 arrobas the caballeria.

The Louisiana plantation is stated 1,200 acres = $50,000; improvements = $50,000; negroes 80, at $600 each, = $48,000; total 148,000 dollars.

The Cuba plantation is allowed only 30 caballerias, or 981 acres, = $54,000; improvements = $65,490; negroes, 90, at $400 each, = $36,000; total 155,490 dollars.

Of the Louisiana plantation, *one-third*, or 400 acres, is cultivated; giving to each negro five acres, and 5,000 pounds product in sugar.

Of the Cuba plantation, *one-sixth*, or 196/2/10 acres, is cultivated; giving to each negro 2/18/100 acres, and 4,444/4/9 pounds product in sugar, i.e. 555 5/9 pounds *less*.

The proportion of the annual expenses of the whole plantation is, for the negro in Louisiana, only 105 dollars; while for the negro in Cuba in ascends to 151 48/100 dollars; i.e. 46/48/100 dollars *more*.

Hence, although the slave in Cuba may cost 50 per cent. *less*, and the ground he works may produce upwards of 100 per cent. *more*, the slave in Louisiana, both positively in sugar and negatively in money, may gain for his master upwards of 100 per cent. more!

Without reference to the *price* of the sugar, or of the coffee, or of the cotton, it may, in the same way, be shown that on *inferior soils* even our slave labor will create much greater quantities at much less expense! But when we admit the soil and climate to be *equally* productive, how infinitely superior are the products of American *skill*, capital, and *economy*, combined; and when we still further contemplate the *greater* productiveness of most articles of tropical culture, acclimated within our territory, we may safely anticipate that, within twenty years, the southernmost sections of our Union will yield every tropical staple for the consumption of even the torrid zone itself.

I have the honor to be, very respectfully, sir, your humble and obedient servant,

HENRY PERRINE.

To the Hon. LOUIS MCLANE,
    Secretary of State of the U.S.A., Washington City.

Propagation of Fibrous-leaved Plants.

To the Honorable the Committee on Agriculture of the House of Representatives of the United States. Washington, D.C., February 3, 1838.

The introduction of fibrous-leaved plants into tropical Florida, their propagation throughout the steril soils of all our Southern and Southwestern States, the production of their *fibrous foliage*, and the preparation of their *foliaceous fibres*, by small cultivators and family manufacturers, are topics which have occupied my heart, head, and hands, during the last ten years.

My unshaken opinions of the immense importance of the *endogenous* plants, whose *living leaves* yield *textile fibres*, have been expressed in numerous communications to the Government and to the people of these United States; but as many sheets remain on the files of the Departments in Washington. and numerous letters have not yet appeared in the periodicals of agriculture, I now attempt to present a very brief abridgment of the contents of them all, for the consideration of your committee and of Congress. To excite the attention of my readers towards some details of some species of fibrous-leaved plants, I premise a few general statements applicable to the whole.

General Statements.

The fibrous-leaved plants are all hardy, productive, perennial plants, which profitably propagate themselves on sandy, stony, and swampy surfaces, in the sun and in the shade. Their fibrous leaves, produced in any soil and situation, with the least care or cultivation, may be cut in any weather and at every season of the year. These freshly-cut leaves may be immediately manufactured into excellent paper, so cheaply, that it will become as important an auxiliary to popular education as the printing-press itself. These living perennial leaves will yield their fibrous contents in the shortest possible time, with the simplest possible preparations, as the foliaceous fibres are extracted from the green leaves by simple scraping only; and immediately after this mechanical separation, these parallel longitudinal fibres are ready for baling and exportation, or for manufacture. These fresh foliaceous fibres have so much individual strength, length, and elasticity, that they may be instantly wrought, untwisted, into very cheap forms and fabrics, for which the unspun cortical fibres of hemp and flax are entirely unserviceable. Moreover, these foliaceous fibres are so much cheaper, lighter, stronger, longer, more elastic, and more durable than cortical fibres, that they can be spun into thread, twine, and cordage, and can be woven into webs, muslins, or cloths, finer than cambric and coarser than canvass, which will become superior substitutes for similar manufactures of flax and hemp in the general consumption of mankind. Furthermore, many of said fibrous-leaved plants form excellent hedges for themselves and for other objects of cultivation, and the entire *leaves* of many species constitute the best materials for the simplest manufactures of the cheapest possible matting, baling, bagging, and other envelopes of merchandise, for the really domestic manufactures, or farm, family, and female manufactures of hats, bonnets, baskets, and other articles, by an innocent, independent, and rural population. A still more important consideration attending the propagation of fibrous-leaved plants in the poorest soils, will be found in the fact, that, whether the staple desired be fibrous foliage for domestic manufactures, or foliaceous fibres for foreign exports, at least three-fourths of all the requisite labor can be accomplished much more cheaply by horse power than by human power. Moreover, as these perennial plants combine the merits of yielding the greatest possible produce with the least possible labor, in the poorest possible soils, their introduction will be an equivalent to the direct addition of absolute fertility to our hitherto most steril districts, and of positive wealth to our hitherto poorest population.

Hence proceed my convictions that foliaceous fibres may be more profitably produced in the refuse lands of Carolina and Georgia, than cortical or capsular fibres in the richest sections of Ohio and of Louisiana; that hence, even the ruined fields of the Southern States will yield greater prosperity in the production of foliaceous fibres alone, than was ever obtained from their virgin loams by the cultivation of capsular fibres, notwithstanding cotton at present constitutes a great proportion of the whole exports of the United States.

Hence, also, my belief that, as the narcotic leaves of one native plant of Yucatan (which *did* take its name from the then dependent province of Tabasco) do actually afford an annual exportation of many millions in one staple of the South, so the fibrous leaves of another native plant of Yucatan, (which *may* take its name from the actual exporting port of Sisal,) will more probably afford an annual exportation of ten times as many millions of dollars in another staple of the South; and that this new staple will be still more important than all her old staples combined, not merely for the amount, value, and profit of the product itself, but also for the character of the lands, of the labor, and of the population it will employ.

Hence, also, my opinions that the propagation of fibrous-leaved plants in the actually worthless sands and swamps of the Southern States, will form a still more distinguished era in their agricultural prosperity than the invention of the cotton-gin; that the production of fibrous foliage and foliaceous fibres will create still more beneficial revolutions in the commerce and manufactures of all civilized nations than has yet been effected by the cultivation of the capsular fibres called cotton; and that, therefore, their introduction to the intelligent industry of our free institutions should be effectually favored by the statesmen of our nation, and by the philanthropists of the world.

I have the honor to be, gentlemen, very respectfully, your obedient servant,

HENRY PERRINE.

Beginnings of Government Activities for Agriculture, 1838

From U.S. Patent Office, *Annual Report*, 1839, pp. 57–59.

Letter from the Commissioner of Patents to the chairman of the Committee on Patents, in relation to the collection and distribution of seeds and plants.

Patent Office, January 22, 1839.

SIR:

I have the honor to acknowledge the receipt, this morning, of a letter of the 21st instant, from the honorable chairman of the Committee on Patents, requesting the communication of any information relative to the collection and distribution of seeds and plants; also, the practicability of obtaining agricultural statistics, with the addition of any suggestions deemed important in relation to these subjects; and hasten to reply: That, in the discharge of official duties, I could not fail to notice facts deeply connected with the subject of agriculture, and so far as I was able, without the neglect of primary obligations, to give all the adventitious aid in my power to this important branch of national industry.

The beneficial effects attending the selection of a few seeds are attested by many members of Congress, through whose kindness the Commissioner has been able to distribute them.

Numerous letters give assurance of the improvement of the corn crop in the Southern and Western States, by introducing a prolific variety. The successful application of the principles by which this prolific quality was attained, has awakened the attention of agriculturists to this important subject, and inspired them with the hope of improving other seeds and plants by the same method. The sexuality of plants, and the practicability of crossing the same, are no longer matters of doubt; and great improvements may be anticipated from new experiments.

Planters in the rich valley of the Mississippi are confident that the "Baden corn" will increase their crop of maize fifty per cent.; some estimate it much higher; and all agree that the introduction of this seed will increase their agricultural products several millions of dollars annually, in that valley alone.

The distribution of early varieties of maize has not been less beneficial to the Northern sections of the Union. Good crops of maize have been raised the past year on the confines of Canada. The wheat distributed from the Patent Office has been successfully tested; and many large fields will be planted the ensuing season, for the first time, with the Italian or Siberian spring wheat.

Not being authorized to incur expense; the Commissioner has ventured to invite the transmission of seeds, gratuitously, from distant parts, trusting that Congress will, at least, authorize their distribution.

Arrangements could be made for the exhibition of different kinds of grain, exotic and indigenous, in the new Patent Office. Can it be doubted that samples of fifty varieties of Indian corn and one hundred and fifty varieties of wheat, with numerous other grains, showing the plant itself, the weight of seed, the usual product, and the latitude where produced, will be both interesting and useful, and afford much valuable information?

From the exploring expedition large contributions are expected; and the cheefulness with which the commander of our national vessels, and also the diplomatic corps, have offered to aid in the matter, justifies the expectation of great advantage from the proposed effort.

The export of agricultural products from the United States is officially estimated by the honorable Secretary of the Treasury at eighty millions of dollars annually. The home consumption must be far greater.

An increase, by a better selection of seeds, only ten per cent., must secure a gain of upwards of twenty millions of dollars each year.

I rejoice that agricultural statistics have been deemed worthy of an inquiry by the honorable committee.

Other enlightened nations have ranked such information amongst the most important, in providing for the public wants, guarding against speculation, and as a means of estimating the probable state of exchange, so far as it is affected by a surplus or scarcity of crops.

The benefit of national statistics, as well as the practicability of obtaining the same, may be estimated from the success attending the efforts of individual States, and particularly of Massachusetts. The wise and liberal policy of the Legislature, in collecting similar statistics, has met the cordial approbation of all her citizens.

It cannot have escaped notice how many fears were entertained in the Middle and Southern States on the failure of the crop of maize the last season. The timorous and cautious made forced sales of their stock, which they feared would otherwise perish during the winter. Monopolists commenced their speculations; corn rose to $1.50 and $2 per bushel; at the same time, beyond the mountains, the crop of maize, in many places, was seldom better. Such has been the abundance on the waters of the Missouri river, that Government has been able to make large contracts for corn at 12½ and 15 cents per bushel.

The Commissioner of Patents would cheerfully, if desired, collect, as far as possible, agricultural statistics from different sections of the United States, and present the same to Congress, with his annual report. Such statistics might be useful to the Government in their financial estimates, and certainly would be to the citizens generally.

From the Patent Office, allow me to say, much encouragement, it is believed, can be given to agriculture without a neglect of present duties.

A new era seems, indeed, to dawn upon this long-neglected branch of national industry. The introduction of new labor-saving machinery; the preparation of flax for spinning on common cotton machinery, without rotting; the extraction of sugar from the beet, in a method recently discovered, so simple and expeditious as to enable every family to supply itself with one of the necessaries of life; the culture and manufacture of silk, thereby saving in imports twenty millions of dollars annually; all conspire to make this epoch one of the brightest in the history of our republic.

It must be mortifying to an American to reflect that the importation of flax seed, from the Indies, has been among the most lucrative commerce the present season. With land admirably adapted to the culture of this vegetable, which once constituted so great a portion of domestic manufacture, and afforded a considerable staple of export, it must astonish, at least the citizens of the old world, that we now depend upon cultivators of the soil beyond the Cape of Good Hope for supplies of linseed!

I cannot omit to notice that the efforts of Government to obtain, through her navy, foreign seeds and plants, have failed, from the inability of collectors in the different ports to distribute the objects transmitted. Not will the patriotic designs of the nation be accomplished, until some depot shall be established to receive, classify, and disseminate the generous contributions daily offered. The new Patent Office building will afford room for the reception and exhibition of seeds, and a small requisition on the patent fund (expressly constituted to promote the arts, and which cannot be diverted without special legislation) will enable the Commissioner to find a remuneration from expenses already becoming onerous to himself individually.

With the highest respect, I am yours, obediently,

HENRY L. ELLSWORTH.
Commissioner of Patents.

Hon. ISAAC FLETCHER,
Chairman Committee on Patents, H. R.

American Agriculture, 1840

From U.S. Patent Office, *Annual Report*, 1840, pp. 4–5, 68–83.

These annual statistics will, it is hoped, guard against monopoly or an exorbitant price. Facilities of transportation are multiplying daily; and the fertility and diversity of the soil ensure abundance, extraordinaries excepted. Improvements of only ten per cent. on the seeds planted will add annually fifteen to twenty millions of dollars in value. The plan of making a complete collection of agricultural implements used, both in this and foreign countries, and the introduction of foreign seeds, are steadily pursued.

It will also be the object of the Commissioner to collect, as opportunity offers, the minerals of this country which are applied to the manufactures and arts. Many of the best materials of this description now imported have been discovered in this country; and their use is only neglected from ignorance of their existence among us. The development of mind and matter only leads to true independence. By knowing our resources, we shall learn to trust them.

The value of the agricultural products almost exceeds belief. If the application of the sciences be yet further made to husbandry, what vast improvements may be anticipated! To allude to but a single branch of this subject. Agricultural chemistry is at length a popular and useful study. Instead of groping along with experiments, to prove what crops lands will bear to best advantage, an immediate and direct analysis of the soil shows at once its adaptation for a particular manure or crop. Some late attempts to improve soils have entirely failed, because the very article, transported at considerable expense to enrich them, was already there in too great abundance. By the aid of chemistry, the West will soon find one of their greatest articles of export to be oil, both for burning and for the manufactures. So successful have been late experiments, that pork (if the lean part is excepted) is converted into stearine for candles, a substitute for spermaceti, as well as into the oil before mentioned. The process is simple and cheap, and the oil is equal to any in use.

Late improvements, also, have enabled experimenters to obtain sufficient oil from corn meal to make this profitable, especially when the residuum is distilled, or, what is far more desirable, fed out to stock. The mode is by fermentation, and the oil which rises to the top is skimmed off, and ready for burning without further process of manufacture. The quantity obtained is 10 gallons in 100 bushels of meal. Corn may be estimated as worth 15 cents per bushel for the oil alone, where oil is worth $1.50 per gallon. The extent of the present manufacture of this corn oil may be conjectured from the desire of a single company to obtain the privilege of supplying the light-houses on the upper lakes with this article. If from meal and pork the country can thus be supplied with oil for burning and for machinery and manufactures, chemistry is indeed already applied most beneficially to aid husbandry.

A new mode of raising corn trebles the saccharine quality of the stalk, and, with attention, it is confidently expected that 1,000 pounds of sugar per acre may be obtained. Complete success has attended the experiments on this subject in Delaware, and leave no room to doubt the fact that, if the stalk is permitted to mature, without suffering the ear to form, the saccharine matter (three times as great as in beets, and equal to cane) will amply repay the cost of manufacture into sugar. This plan has heretofore been suggested by German chemists, but the process has not been successfully introduced into the United States, until Mr. Webb's experiments at Wilmington, the last season. With him the whole was doubtless original, and certainly highly meritorious; and, though he may not be able to obtain a patent, as the first original inventor, it is hoped his services may be secured to perfect his discoveries. It may be foreign to descend to further particulars in an annual report. A minute account of these experiments can be furnished, if desired. Specimens of the oil, candles, and sugar, are deposited in the National Gallery.

May I be permitted to remark that the formation of a National Agricultural Society has enkindled bright anticipations of improvement. The propitious time seems to have come for agriculture, that long neglected branch of industry, to present her claims. A munificent bequest is placed at the disposal of Congress, and a share of this, with private patronage, would enable this association to undertake, and, it is confidently believed, accomplish much good.

A recurrence to past events will show the great importance of having annually published the amount of agricultural products, and the places where either a surplus or a deficiency exists. While Indian corn, for instance, can be purchased on the Western waters for $1 (now much less) per barrel of 196 pounds, and the transportation, via New Orleans, to New York, does not exceed $1.50 more, the price of meal need never exceed from 80 cents to $1 per bushel in the Atlantic cities. The aid of the National Agricultural Society, in obtaining and diffusing such information, will very essentially increase the utility of the plan before referred to, of acquiring the agricultural statistics of the country, as well as other subsidiary means for the improvement of national industry.

I will only add that, if the statistics now given are deemed important, as they doubtless may prove, to aid the Government in making their contracts for supplies, in estimating the state of the domestic exchanges, which depend so essentially on local crops, and in guarding the public generally against the grasping power of speculation and monopoly, a single clerk, whose services might be remunerated from the patent fund, to which it will be recollected more than $8,000 has been added by the receipts of the past year, would accomplish this desirable object. The census of population and statistics, now taken once in ten years, might, in the interval, thus be annually obtained sufficiently accurate for practical purposes.

All which is respectfully submitted.

HENRY L. ELLSWORTH

Hon. JOHN WHITE,
Speaker of the House of Representatives

* * *

The crops of 1839, on which the Census statistics are founded, were, as appears from the notices of that year, very abundant in relation to nearly every product throughout the whole country; indeed, unusually so, compared with the years preceding. Tobacco may be considered an exception; it is described to have been generally a short crop.

The crops of the succeeding year are likewise characterized as abundant. The success which had attended industry in 1839 stimulated many to enter upon a larger cultivation of the various articles produced, while the stagnation of other branches of business drew to the same pursuit a new addition to the laboring force of the population.

Similar causes operated also to a considerable extent the past year. In 1841, the season may be said to have been less favorable in many respects than in the two preceding ones; but the increase of the laboring force, and the amount of soil cultivated, render the aggregate somewhat larger. Had the season been equally favorable, we might probably have rated the increase considerably higher, as the annual average increase of the grains, with potatoes, according to the annual increase of our population, is about 30 millions of bushels. Portions of the country suffered much from a long drought during the last summer, which affected unfavorably the crops more particularly liable to feel its influence, especially grain, corn, and potatoes. In other parts, also, various changes of the weather in the summer and autumn lessened the amount of their staple products below what might have been gathered, had the season proved favorable. Still, there has been no decisive failure, on the whole, in any State, so as to render importation necessary, without the means of payment in some equivalent domestic product, as has been the case in some former years, when large importations were made to supply the deficiency, at cash prices. In the year 1837 not less than 3,921,259 bushels of wheat were imported into the United States. We have now a large surplus of this and other agricultural products for exportation, were a market opened to receive them.

A glance at the specific crops is all that can be given. Some notice of this kind seems necessary, and may be highly useful to those who wish to embrace, in a narrow compass, the results of the agricultural industry of our country:

WHEAT. — This is one of the great staple products of several States, the soil of which seems, by a happy combination, to be peculiarly fitted for its culture. Silicious earth, as well as lime, appears to form a requisite of the soil to adapt it for raising wheat to the greatest advantage, and the want of this has been suggested as a reason for its not proving so successful of cultivation in some portions of our country. Of the great wheat-growing States, during the past year, it may be remarked that, in New York, Pennsylvania, Virginia, and the Southern States, this crop seems not to have repaid so increased an harvest as was promised early in the season. Large quantities of seed were sown, and the expectation was deemed warranted of an unusually abundant increase. But the appearance of the chinch bug and other causes destroyed these hopes. In the northern part of Kentucky the crop "did not exceed one-third of an ordinary one." In some of the States, as in New Jersey, Ohio, Indiana, Michigan, and Illinois, the quantity raised was large, and the grain of a fine quality. The prospect of another year at the West, if we may judge at so early a period, is for an increased crop, as in some fertile sections more than double the usual amount is said to have been sown. The present open winter, however, may prove injurious, and these sanguine expectations not be realized. Indeed, the wheat and rye, as well as other grain crops, are in parts of the country becoming more uncertain, and, without more attention to the variety and culture, many kinds of grain must probably be still more confined to particular sections. Of all the States, Ohio stands foremost in the production of wheat, as she is also peculiarly fitted for all the grains, and the sustaining of a dense population. About one-sixth of the whole amount of the wheat crop of the country is raised by this State. To this succeed, in their order, Pennsylvania, New York, Virginia, Indiana, Tennessee, Kentucky, Illinois, Maryland, Michigan, and North Carolina. In some of the States a bounty is paid on the raising of wheat, which has operated as an inducement to the cultivation of this crop. The amount thus paid out of the State Treasury, in Massachusetts, for two years, was more than $18,000; the bounty was two dollars for every fifteen bushels, and five cents for every bushel above this quantity. Similar inducements might, no doubt, stimulate to still greater improvements and success in this and other products of the soil.

The value of this crop in our country is so universally felt, that its importance will be at once acknowledged. The whole aggregate amount of wheat raised is 91,642,957 bushels, which is nearly equal to that of Great Britain, the wheat crop of which does not annually exceed 100,000,000 of bushels. The supply demanded at home, as an article of food, cannot be less than eight or ten millions, and has been estimated as high as twelve million of barrels of flour, equal to about forty to sixty millions bushels of wheat. The number of flouring mills reported by the last census is 4,364, and the number of barrels of flour 7,404,562. Large quantities of wheat also are used for seed, and for food of the domestic animals, as well as for the purposes of manufacture. The allowance in Great Britain for seed, in the grains in general, as appears from McCulloch, is about one-seventh of the whole amount raised. Probably a much less proportion may be admitted in this country. Wheat is also used in the production of, and as a substitute for, starch. The cotton manufactories of this country are said to consume annually 100,000 barrels of flour for this and similar purposes; and in Lowell alone, 800,000 pounds of starch, and 3,000 barrels of flour, are said to be used in conducting the mills, bleachery and prints, &c., in the manufactories.

Could the immense surplus amount of this crop, in the West, find access to the ports of Great Britain, as the means of communication are daily becoming more easy and shorter in point of time, it would contribute much to enrich that grain producing section of our country.

BARLEY. — Comparatively little of this grain is raised in this country, with the exception of New York. Maine, Ohio, Pennsylvania, Michigan, Massachusetts, New Hampshire, and Illinois, rank next as producers of this crop. As it is raised principally to supply malt for the brewery, and small quantities of it only are used for the food of animals, or for bread, no great increase in this product is to be anticipated. The crop of 1841 appears to have been somewhat less than the usual one in proportion to the population.

OATS. — This grain in several of the States is evidently deemed an important object of cultivation, and large quantities of it are annually produced. As compared with wheat, it has the precedence in all of them, with the exception of Maine, Maryland, Ohio, and Georgia. New York takes the lead in the amount raised. Then follows, very closely, Pennsylvania; then Ohio, Virginia, Indiana, Tennessee, and Kentucky. It is a favorite crop, too, in the New England States. The crop of oats, in 1841, is believed to have been somewhat below a full one, and may therefore be considered as not having been so successful as some others, although large quantities of the seed were sown in the States where they are most abundantly cultivated. The consumption of oats in this country is confined particularly to the feeding of horses; but in some parts of Europe this article is used, to a considerable extent, as one of the bread stuffs. It enters, to a limited degree, into our articles of exportation, but it is not easy to form any exact estimate of the different appropriations of this crop, at home or abroad.

RYE. — This species of grain is mostly confined to a few States. The proportion which it bears to the other grains is probably greater in the New England States than in any other section of our country. There it likewise, to some extent, forms an article of food for the people. Pennsylvania, New York, New Jersey, Virginia, Kentucky, Ohio, and Connecticut, may be ranked as the chief producers of this crop; at least, these are among the States where it bears the greatest relative proportion to the other important crops. In 1841 it experienced, in some degree, similar vicissitudes with the other grains, and must likewise be estimated as below the increased crop which a more favorable season would probably have produced. The product of this crop is extensively used in many parts of our country for distillation, although the quantity thus applied has probably materially lessened within the few years past, and will doubtless hereafter undergo a still greater reduction.

BUCKWHEAT. — This must be reckoned among the crops of minor interest in our country. With the exception of New York, Pennsylvania, New Jersey, Ohio, Connecticut, Virginia, Vermont, Michigan, and New Hampshire, very little attention seems to be given to the culture of this grain. In England it is principally cultivated, that it may be cut in a green state as fodder for cattle, and the seed is used to feed poultry. In this country it is also applied in a similar manner; and is sometimes ploughed in, as a means of enriching the soil. To a limited extent, the grain is further used as an article of food. The crop of 1841 may be considered as, on the whole, above an average one. This may in part be attributed to the fact that when some of the other and earlier crops failed, resort was had to buckwheat, as a later crop, more extensively than is usual. It is a happy feature in the adaptation of our climate, that the varieties of products are so great as to enable the agriculturist often thus to supply the deficiency in an earlier crop, by greater attention to a later one. There was more buckwheat sown than is commonly the case, and the yield was such as to compensate for the labor and cost of culture.

MAIZE OR INDIAN CORN. — Tennessee, Kentucky, Ohio, Virginia, and Indiana, are, in their order, the greatest producers of this kind of crop. In Illinois, North Carolina, Georgia, Alabama, Missouri, Pennsylvania, South Carolina, New York, Maryland, Arkansas, and the New England States, it appears to be a very favorite crop. In New England, especially, the aggregate is greater than in any of the grains except oats. More diversity seems to have existed in this crop, in different parts of the country, the past year, than with most of the other products of the soil; and hence it is much more difficult to form a satisfactory general estimate. In some sections the notices are very favorable, and speak of "good crops," as in portions of New England; of "a more than average yield," as in New Jersey; of being "abundant," as in parts of Georgia; or, "on the whole, a good crop," as in Missouri; "on the whole, a tolerable one," as in Kentucky. In others, the language is of "a short crop," as in Maryland; or "cut off," as in North Carolina; or "below an average," as in Virginia. On the whole, however, from the best estimate which can be made, it is believed to have equalled, if it did not exceed, an average crop. The improvement continually making in the quality of the seed (and this remark is likewise applicable, in various degrees, to other products) augurs well for the productiveness of this indigenous crop, as it has been found that new varieties are susceptible of being used to great advantage. Considered as an article of food for man, and also for the domestic animals, it takes a high rank. No inconsiderable quantities have likewise been consumed in distillation; and the article of kiln-dried meal, for exportation, is yet destined, it is believed, to be of no small account to the corn-growing sections of our country. It will command a good price, and find a ready market in the ports which are open to its reception. But the importance of this crop will doubtless soon be felt in the new application of it to the manufacture of sugar from the stalk, and of oil from the meal. Below will be found some comparisons and deductions on this subject, and a view of the true policy of our country in relation to it and to agricultural industry generally.

POTATOES. — The Tabular View shows, that in quite a number of States the amount of potatoes raised is very great. New York, Maine, Pennsylvania,

Vermont, New Hampshire, Ohio, Massachusetts, and Connecticut, are the great potato-growing States; more than two-thirds of the whole crop are raised by these States. Two kinds, the common Irish and the sweet potato, as they are called, with the numerous varieties, are embraced in our Agricultural Statistics. When it is recollected that this product of our soil forms a principal article of vegetable food among so large a class of our population, its value will at once be seen. The best common or Irish potatoes, as an article of food for the table, are produced in the higher northern latitudes of our country, as they seem to require a colder and moister soil than corn and the grains generally. It is on their peculiar adaptation in this respect, that Ireland, Nova Scotia, and parts of Canada, are so peculiarly successful in the raising and perfecting of the common or Irish potatoes. It is estimated that, in Great Britain, an acre of potatoes will feed more than double the number of individuals than can be fed from an acre of wheat. It is also asserted that, whenever the laboring class is mainly dependent on potatoes, wages will be reduced to a minimum. If this be true, the advantage of our laboring classes over those of Great Britain, in this respect, is very great. The failure of a crop of potatoes, too, where it is so much the main dependence, must produce great distress and starvation. Such is now the case in Ireland and parts of England and Scotland. Another disadvantage of relying on this crop as a chief article of food for the people is, that it does not admit of being stored up as it is, or converted into some other form for future years, as do wheat and corn. Potatoes also enter largely into the supply of food for the domestic animals; besides which, considerable quantities are used for the purpose of the manufacture of starch, of molasses, and distillation. New varieties, which have been introduced within a few years past, have excited much attention, and many of them have been found to answer a good purpose. Increased improvement, and with yet more successful results in this respect, may be anticipated.

The crop of potatoes in 1841 suffered considerably in many parts of the country, and, perhaps, came nearer to a failure than has been known for some years. In portions of New England and New York this was particularly the case. In other sections, however, if a correct judgment may be formed from the notices of the crop, there appears to have been a more than average increase. In proportion to her population, Vermont may be considered foremost in the cultivation of potatoes. The sweet potato is raised with some success for market as far north as New Jersey, though the quality of the article is not equal to that which is produced in the more southern latitudes. As the climate of the West, compared with that of the Atlantic border, varies perhaps nearly several degrees within the same parallels of latitude, it may be supposed that this variety of the potato can be cultivated even as high up as Wiskonsan or Iowa, in favorable seasons, with tolerable success.

HAY. — This product was remarkably successful during the past year in particular sections of our country, in others less so. In Maine, and in the New England States generally, there was more than an average yield. In New York, which ranks highest in the Tabular View, it was lighter than usual. In New Jersey, and the middle States generally, it was considered "good;" in the more Southern and Southwestern ones, little, comparatively, is cultivated. In the Northwestern States it appears to have been about an average crop. The extensive prairies of the West admit of being covered with luxuriant crops of grass, of better varieties; and when this is done they will prove far more valuable, both for the purposes of stock, and also in raising hay for the Southern market at New Orleans, which is already supplied, to some extent, with this product, brought down the Mississippi, from Indiana, Ohio, and Illinois, as well as by the Atlantic coast, from the New England States and New York. Hay is also an article of export, in some quantities, to the West Indies.

FLAX AND HEMP. — More difficulty has been found in forming an estimate of these two articles than any other embraced in the Tabular View. They are combined in the Census statistics, and the amount is sometimes given in tons, sometimes in pounds, so that it is not easy always to discriminate between them. More than half of the whole combined amount must probably be allotted to flax, as but little hemp, comparatively, is known to be raised. Flaxseed is used for the manufacture of linseed oil, considerable quantities of which are annually imported into this country for various purposes. The oil cake, remaining after the oil is expressed, is a well-known article in use, mingled with the food of horses and other animals.

In these articles of flax and hemp combined, if the Recapitulation of the Census statistics is correct, Virginia is in advance of all the other States; then follow Missouri, North Carolina, Ohio, Kentucky, Indiana, Tennessee, Pennsylvania, New Jersey, Illinois, New York, and other States. It is believed, however, that some of the amounts, as returned by the marshals, should rather have been credited to pounds for flax than to tons, as more nearly corresponding to the actual condition of the crops in our country. Kentucky probably ranks the highest with respect to the production of hemp. The crop of 1840 was a great failure, and that of the past year also suffered much from the dry weather. There is not so much attention paid to the culture of this article as its importance demands; yet there is every ground of encouragement for increased enterprise in the production of hemp, from the supply required in our own country. The difficulty most in the way of its success, hitherto, has been the neglect, either from ignorance, inexperience, or some other cause, properly to prepare it for use by the best process of water-rotting. The agriculturists of our country seem, in this respect, to have too soon yielded to discouragement. The desirableness of some new and satisfactory results on this subject will be seen from the fact that it is stated the annual consumption of hemp in our navy amounts to nearly two thousand tons; besides which, the demand for the rest of our shipping is not less than about eleven thousand tons more; making an aggregate of nearly thirteen thousand tons — the price of which is put at from $220 to $250, and by some even as high as $280 per ton, together with other and inferior qualities, which are used to supply the deficiency of the better article. Our hemp, it is further stated, on high authority, when properly water-rotted, proves, by actual experiment, to be one-fourth stronger than Russia hemp, to take five feet more run, and to spin twelve pounds more to the four hundred pounds. When so much is felt and said on the increase of our navy prospectively, it is an object worthy of attention to secure, if possible, the production of hemp in our own country, adequate to all our demands. The introduction, too, of gunny bags, and of Scotch and Russia bagging, and iron hoops for cotton, renders this direction of the hemp product more necessary and important. It is hoped that some process of water-rotting, which will prove at once both cheap and satisfactory, may yet be discovered by the inventive genius of our countrymen, who are not wont to be discouraged at any slight obstacles.

TOBACCO. — The crop of 1839, in this article, on which the Census statistics are founded, is deemed, as appears from the notices on this subject, to have been a short one, and below the average. The crop of the past year was much more favorable — beyond an average; indeed, it is described in some of the journals as "large."

Virginia, Kentucky, Tennessee, North Carolina, and Maryland, are the great tobacco-growing States. An advance in this product is likewise in steady progress in Missouri, where the crop of 1841 is estimated at nearly 12,000 hogsheads, and for 1842 it is expected that as many as 20,000 may be raised. Some singular changes are going forward with regard to this great staple of several of the States. Reference is here intended to the increasing disposition evinced, as well as the success thus far attending the effort, to cultivate tobacco in some of the Northern and Northwestern States. The tobacco produced in Illinois has been pronounced by competent judges from the tobacco-growing States, and who have there been engaged in the culture of this article, to be superior, both in quality and the amount produced per acre, to what is the average yield of the soils heretofore deemed best adapted to this purpose. In Connecticut, also, the attention devoted to it has been rewarded with much success; 100,000 pounds are noticed as the product of a single farm of not more than fifty acres. It is, indeed, affirmed that tobacco can be raised in Indiana, Ohio, Kentucky, and Tennessee, at a larger profit than even wheat or Indian corn. Considerable quantities, also, were raised in 1841 in Pennsylvania and Massachusetts, where it may probably become an object of increased attention. The agriculturists of these States, if they engage in the production of this crop, will do so with some peculiar advantages. They are accustomed to vary their crops, and to provide means for enriching their soils. Tobacco, as it is well known, is an exhausting crop, especially so when it is raised successive years on the same portions of soil. The extraordinary crops of tobacco which have heretofore been obtained have, indeed, enriched the former proprietors, but the present generation now find themselves, in too many instances, in the possession of vast fields, once fertile, that are now almost or wholly barren, from an inattention to the rotation of crops. The difficulty of cultivating a worn-out soil has induced, and will continue to induce, the emigration of the most enterprising to new lands, where they will bear in mind the lessons that dear-bought experience has taught them. It is a provision of Nature herself, that there must be a suitable rotation of crops; and all history sanctions the conclusion, that the continued cultivation of any specific crop, without an adequate supply of the means of restoration from year to year, must eventually and inevitably terminate in impoverishing its possessors, and entailing on them the necessity of removal from their native homes, if they would not sink in degradation. Had a variety and rotation of crops been resorted to on the lands now so left, the countries suffering by such a course had been far more rich and prosperous.

The value of tobacco exported in different forms in 1839 was $10,449,155, and the amount of tobacco exported in 1840 was about 144,000,000 of pounds. The greater part of this goes to England, France, Holland, and Germany.

COTTON. — This, it is well known, is the great staple product of several States, as well as the great article of our exports, the price of which, in the foreign market, has been more relied on than any thing else to influence favorably the exchanges of this country with Great Britain and Europe generally. The cotton crop of the United States is more than one-half of the crop of the whole world. In 1834, the amount was but about 450,000,000 of pounds; the annual average now may be estimated at 100,000,000 of pounds more; the value of it for export at about $62,000,000. The rise and progress of this crop, since the invention of Whitney's cotton gin, has been unexampled in the history of agricultural products. In the year 1783, eight bales of cotton were seized on board of an American brig, at the Liverpool custom-house, because it was not believed that so much cotton could have been sent at one time from the United States! The cotton crop of 1841, compared with that of 1839 and 1840, was probably less, by from 500,000 to 600,000 bales. In the early part of the last cotton-growing season, an average crop was confidently anticipated; but this hopeful prospect was not realized. In portions of the cotton-producing States, as in parts of Georgia, however, the crop was greater than usual; and in Arkansas it has been estimated at a gain, over that of 1839, of 33⅓ per cent.; but probably, owing to its having suffered from the boll worm, it should be set down at 20 or 25 per cent. A similar advance is expected in future years, among other causes, from the great increase of population by immigration. Mississippi, Georgia, Louisiana, and Alabama, South Carolina, and North Carolina, are, in their order, the great cotton-growing States. An important fact deserves notice here, on account of the relation which the cotton crop bears to other crops. Whenever (to whatever cause it may be owing) the price of cotton is low, the attention of cultivators, the next year, is more particularly diverted from cotton to the culture of corn, and other branches of agriculture, in the cotton-producing States. As cotton is now so low, and so little in demand in the foreign market, unless a market be created at home it must necessarily become an object of less attention to the planters; and it cannot be expected that the agricultural products of the West will find so ready a sale in the Southern market as in some former years. Other countries, too, as India, Egypt, and other parts of Africa, Brazil, and Texas, are now coming more decidedly into competition with the cotton-growing interest of our country; so that an increase of this product from those countries, and a corresponding depression in ours, are to be expected. The amount of India cotton imported into England in 1840 was 76,703,295 pounds; almost equal to the whole cotton crop of North Carolina and South Carolina, or to that of Alabama, for the past year, and nearly double the amount produced by Tennessee, Arkansas, and Florida, combined; being, also, an increase on the importation of cotton from India, the preceding year, of 30,000,000 of pounds, and, in amount, nearly one-sixth of the whole quantity imported during the same year from the United States. From the report of the Chamber of Commerce of Bombay, it appears that, from the 1st of June, 1840, to the 1st of June, 1841, the imports of cotton into Bombay amounted to 174,212,755 pounds; and the whole India cotton crop is estimated, on good authority, at 190,000,000 of pounds. This is a larger quantity than America produced up to 1826, and more than was consumed by England in the same year, and nearly one-third of the whole estimated crop of the United States in 1841. From these facts, it is evident that it is becoming more and more the settled policy of England to encourage the production of cotton in India, while it is equally certain that a foreign market cannot be relied on for our cotton, to the same extent as it has hitherto been. An English authority, speaking of the decline of England and of her manufactures, as having commenced a downward progress, in accounting for this decline, attributes the distress in Leeds, and other places, to the landholders, who, by excluding the foreign bread stuffs, have driven foreigners to manufacture in self-defence. This decline, not being confined merely to her old staple of woollens, must, too, operate in the reduction and diminution of cotton exported from this country. The following statement confirms the position now taken:

"In 1824, Great Britain exported to all foreign countries, including the British possessions, of cloths, &c., 567,317 pieces; in 1828, 566,596 pieces; in 1830, 440,360 pieces; and in 1840, only 250,962 pieces. During the same year last named, (1840,) the total manufactured in only one district in Belgium and Prussia, all within a day's journey of each other, was 333,245 pieces; so that, in one district only, there was made more than was exported by Britain to all the world, by 76,233 pieces."

RICE. — This product is cultivated to comparatively a very little extent in the United States, except in South Carolina and Georgia. In the former of these, it is an object of no small attention, and ranks second only to cotton. It forms a considerable article of export from this country to Europe. England, however, imports annually large quantities of rice from India. The crop of rice in 1841 is said to have been, on the whole, a very good one, equal, if not superior, to the usual average.

SILK COCOONS. — Notwithstanding the disappointment of many who, since the year 1839, engaged in the culture of the *morus multicaulis* and other varieties of the mulberry, and the raising of silkworms, there has been, on the whole, a steady increase in the attention devoted to this branch of industry. This may be, in part, attributed to the ease of cultivation, both as to time and labor required, and in no small degree, also, to the fact that, in twelve of the States, a special bounty is paid for the production of cocoons, or of the raw silk. Several of these promise much hereafter in this product, if a reliance can be placed on the estimates given in the various journals more particularly devoted to the record of the production of silk. There seems, at least, no ground for abandoning the enterprise, so successfully begun, of aiming to supply our home consumption of this important article of our imports. In Massachusetts, Connecticut, New York, Pennsylvania, Delaware, Tennessee, and Ohio, there has been quite an increase above the amount of 1839. The quantity of raw silk manufactured in this country the past year is estimated at more than 30,000 pounds. The machinery possessed for reeling, spinning, and weaving silk, in the production of ribbons, vestings, damask, &c., admit of its being carried to great perfection, as may be seen by the beautiful specimens of various kinds deposited in the National Gallery at the Patent Office. The amount of silk stuffs brought into this country in some single years, from foreign countries, is estimated at more in value than $20,000,000. The silk manufactured in France in 1840 amounted to $25,000,000; that of Prussia to more than $4,500,000. Should one person in a hundred of the population of the United States produce annually 100 pounds of silk, the quantity would be nearly

18,000,000 pounds, which, at $5 per pound, (and much of it might command a higher price,) would amount to nearly $90,000,000 — nearly $30,000,000 above our whole cotton export, nine times the value of our tobacco exports, and nearly five or six times the average value of our imports of silk. That such a productiveness is not incredible, as at first sight it may seem, may be evident from the fact, that the Lombard Venetian kingdom, of a little more than 4,000,000 of population, exported in one year 6,132,950 pounds of raw silk; which is a larger estimate, by at least one-half, for each producer, than the supposition just made as to our own country. Another fact, too, shows both the feasibility and the importance of the cultivation of this product. The climate of our country, from its Southern border even up to 44 degrees of north latitude, is suited to the culture of silk. It needs only a rational and unflinching devotion to this object, to place our country soon among the greatest silk-producing countries of the world.

SUGAR. — Louisiana is the greatest sugar district of our country. The crop of 1841 appears to have been injured by the early frosts; the amount, therefore, was not so great as that of 1839, by nearly one-third.

The progress of the sugar manufacture and the gain upon our imports has been rapid. In 1839 the import of sugars was 195,231,273 pounds, at an expense of at least $10,000,000; in 1840, about 120,000,000 pounds, at an expense of more than $6,000,000. A portion of this was undoubtedly exported, but most of it remained for home consumption. More than 30,000,000 pounds of sugar, also, from the maple and the beet root were produced in 1841, in the Northern, Middle, and Western States; and, should the production of cornstalk sugar succeed, as it now promises to do, this article must contribute greatly to lessen the amount of imported sugars. Indeed, such has been the manufacture of the sugar from the cane for the last five years, that were it to advance in the same ratio for the five to come, it would be unnecessary to import any more sugar for our home consumption. Some further remarks on this particular topic will be found below, in connexion with the subject of cornstalk sugar.

WINE. — North Carolina, Pennsylvania, Virginia, Ohio, and Indiana, rank highest, in their order, in the production of wine. In Maryland, Georgia, Louisiana, Maine, and Kentucky, some thousands of gallons are likewise produced. Two acres in Pennsylvania, cultivated by some Germans, have the past autumn yielded 1,500 gallons of the pure juice of the grape, and paid a nett profit of more than $1,000. Still, the quantity produced is small. The cultivation of both the native and foreign grape, as a fruit for the table, seems to be an object of increasing interest in particular sections of our country; but any very decided advances in this product are scarcely to be expected.

It has thus been attempted to give at least a bird's eye view of the articles enumerated in the Tabular Statistics. There are also a variety of other products which might, perhaps, have been included in the agricultural statistics. These are hops, peas, beans, beets, turnips, and other roots and vegetables; the products of the dairy, of the orchard, and of the bee-hive; wool, live stock, and poultry. Many interesting comparisons in relation to some of the above might be formed from the Census statistics, such as would exhibit in a striking manner the resources our country possesses in the products of her soil and the labor of her hardy yeomanry; but it has been deemed best to omit them in the present report, merely subjoining the Census statistics on these particular articles to the Tabular View. Yet, in estimating the home supply for the sustenance and comfort both of man and beast, these too should always be taken into the account, as a very important item deserving notice.

The whole of the summary now given, with the rapid glance taken at the various products presents our country as one richly favored of Heaven in climate and soil, and abounding in agricultural wealth. Probably no country can be found on the face of the globe, exhibiting a more desirable variety of the products of the soil, contributing to the sustenance and comfort of its inhabitants. From the Gulf of Mexico to our Northern boundary, from the Atlantic to the far West, the peculiarities of climate, soil, and products, are great and valuable. Yet these advantages admit of being increased more than an hundred fold. The whole aggregate of the bread stuffs, corn, and potatoes, is 624,518,510 bushels, which, estimating our present population at 17,835,217, is about 35⅔ bushels for each inhabitant; and, allowing 10 bushels to each person — man, woman, and child — (which is double the usual annual allowance as estimated in Europe,) and we have a surplus product, for seed, food of stock, the purposes of manufacture, and exportation, of not less than 446,166,340 bushels; from which, if we deduct one-tenth of the whole amount of the crops for seed, it leaves for food of stock, for manufactures, and exportation, a surplus of at least 370,653,627 bushels. Including oats, the aggregate amount of the crops of grain, corn, and potatoes, is equal to nearly 755,200,000 bushels, or 42⅓ bushels to each inhabitant. The number of persons employed in agriculture, according to the census of 1840, was 3,717,756. This, it is presumed, refers to the male free white adult population.

The articles of CORN OIL and corn for SUGAR, together with oil from LARD and the castor bean, &c., deserve more than a passing notice. They are destined, it is believed, to call forth increased enterprise among the agriculturists of our country:

CORN OIL is produced from corn meal by fermentation, with the aid of barley malt. It has been produced and used for some time past in certain distilleries, by skimming off the oil as it rises on the meal in fermentation in the mash tub. It has, however, lately become the subject of particular attention, as an article of manufacture, and with success. The meal, after it has been used for the production of this oil, it is said, will make better and harder pork, when fed out to swine, than before. The oil is of a good quality, of a yellowish color, and burns well. Further clarification, it is probable, may render it as colorless as the best sperm oil. Whether or not this may be the case, the ease with which it is made offers strong inducements to engage in the production of this article.

But a more important object in the production of Indian corn is doubtless the manufacture of SUGAR from the stalk. In this point of view, it possesses some very decided advantages over the cane. The juice of the cornstalk by Beaumé's saccharometer, reaches to 10° of saccharine matter, which, in quality, is more than three times that of beet, five times that of maple, and fully equals, if it does not even exceed, that of the ordinary sugar cane in the United States. By plucking off the ears of corn from the stalk as they begin to form, the saccharine matter, which usually goes to the production of the ear, is retained in the stalk; so that the quantity it yields is thus greatly increased. One thousand pounds of sugar, it is believed, can easily be produced from an acre of corn. Should this fact seem incredible, reference need only be made to the weight of fifty bushels of corn in the ear, which the juice so retained in the stalk would have ripened, had not the ear, when just forming, been plucked away. Sixty pounds may be considered a fair estimate, in weight, of a bushel of ripened corn; and, at this rate 3,000 pounds of ripened corn will be the weight of the produce of one acre. Nearly the whole of the saccharine part of this remains in the stalk, besides what would have existed there without such a removal of the ear. It is plain, therefore, that the sanguine conclusions of experimenters the past year have not been drawn from insufficient data. Besides, it has been ascertained, by trial, that corn, on being sown broadcast, (and so requiring but little labor, comparatively, in its cultivation,) will produce five pounds per square foot, equal to 108 tons to the acre for fodder in a green state; and it is highly probable that, when subjected to the treatment necessary to prepare the stalk, as above described, in the best manner for the manufacture of sugar, a not less amount of crop may be produced. Should this prove to be the case, one thousand weight of sugar per acre might be far too low an estimate. Experiments on a small scale have proved that *six* quarts of the juice, obtained from the cornstalk sown broadcast, yielded one quart of crystallized sirup, which is equal to 16 per cent; while for one quart of sirup it takes thirty-two quarts of the sap of maple.

Again, the cornstalk requires only one-fifth the pressure of the sugar cane, and the mill or press for the purpose is very simple and cheap in its construction, so that quite an article of expense will thereby be saved, as the cost of machinery in the manufacture of sugar from the cane is great. Only a small portion of the cane, also, in this country, where it is an exotic, ordinarily yields saccharine matter, while the whole of the cornstalk, the very top only excepted, can be used.

Further, while cane requires at least eighteen months, and sedulous cultivation and much hard labor, to bring it to maturity, the sowing and ripening of the cornstalk may be performed, for the purpose of producing sugar, with ease, within 70 to 90 days; thus allowing not less than two crops in a season in many parts of our country. The stalk remaining, after being pressed, also furnishes a valuable feed for cattle, enough, it is said, with the leaves, to pay for the whole expense of its culture. Should it be proved, by further experiments, that the stalk, after being dried and laid up, can, by steaming, be subjected to the press without any essential loss of the saccharine principle, as is the case with the beet in France, so that the manufacture of the sugar can be reserved till late in the autumn, this will still more enhance the value of this product for the purpose. It may also be true that, as in the case of the beet, no animal carbon may be needed, but a little lime water will answer for the purpose of clarification; after which, the juice may be boiled in a common kettle, though the improved method of using vacuum pans will prove more profitable when the sugar is made on a large scale.

Corn, too, is indigenous, and can be raised in all the States of the Union, while the cane is almost confined to one, and even in that the average amount of sugar produced, in ordinary crops, is but 900 or 1,000 pounds to the acre; not much beyond one-third of the product in Cuba and other tropical situations, where it is indigenous to the soil. The investment in the sugar manufactories from the cane in this country has, it is believed, paid a poorer return than almost any other agricultural product. The laudable enterprise of introducing into the United States the culture of the cane and the manufacture of sugar from the same, has, it is probable, been hardly remunerated, though individual planters, on some locations, have occasionally enriched themselves. The amount of power required, with the cost of the machinery and the means of cultivation, will ever place this branch of industry beyond the reach of persons of moderate resources, while the apparatus and means necessary for the production of corn and other crops lie within the ability of many.

Should the manufacture of sugar from the cornstalk prove as successful as it now promises, enough might soon be produced to supply our entire home consumption, towards which, as has been mentioned, at least 120,000,000 pounds of foreign sugars are annually imported, and a surplus might be had for exportation. In Europe, already, more than 150,000,000 pounds of sugar are annually manufactured from the beet, which possesses but one-third of the saccharine matter that the cornstalk does; and there are not less than 500 beet sugar manufactories in France alone. By this manufacture of sugar at the West, the whole amount of freight and cost of transportation on imported sugar might also be saved — a sum nearly equal, it is probable, to the first cost of the article at the seaport; so that the price of sugar is at least doubled, if not almost trebled, to the consumer at a distance, when so imported. Not less than 6,000,000 pounds of sugar, it is said, are annually imported, for home consumption, in the single city of Cincinnati.

OIL AND STEARINE FROM LARD AND THE CASTOR BEAN, &c. — These two are articles which will hereafter attract much attention in many parts of our country. The use of LARD instead of oil, for lamps of a peculiar construction, has been heretofore attempted with good success, as an article of economy. It has even been adopted in the light-houses in Canada, on the lakes, and is said to burn longer, and free from smoke, while the cost of the article is stated to be but about one-third the cost of sperm oil. But it has now been discovered that oil equal to sperm can be easily extracted from lard, at great advantage, and that it is superior to lard for burning, without the necessity of a copper-tubed lamp. Eight pounds of lard equal in weight one gallon of sperm oil. The whole of this is converted into oil and stearine, an article of which candles that are a good substitute for spermaceti can be made. Allowing, then, for the value of the stearine above the oil, and it may be safely calculated, that when lard is six cents per pound, as it is now but four or five cents at the West, a gallon of oil can be afforded there for fifty cents; since the candles from the stearine will sell for from twenty-five to thirty cents per pound.

Stearine for this purpose has also recently been obtained from castor oil, the product of the *palma christi*, or castor bean, a plant successfully cultivated in portions of our country.

Oil, it is well known, is an article of large consumption in our country. The amount of sperm oil from our whale fisheries, for the year 1841, was 4,965,754 gallons; of whale and fish oil, 6,362,661 gallons — making a sum total of 11,328,415 gallons. The amount for 1840 did not vary much from the same. The amount of sperm and whale oil exported in 1840 was 4,955,486 gallons, leaving for home consumption 6,372,929 gallons. In the year 1840 there was also exported from this country 853,938 pounds of spermaceti candles. From these statements, which do not include linseed, olive, and other oils, it will be seen that the encouragement for the manufacture of oil and stearine, from corn meal, and lard, and the castor bean, is very great. Large quantities of oil for dressing cloths, oiling machinery, &c., are required in the manufactories. In the factories of Lowell, simply, not less than 78,689 gallons are thus needed.

Oil, too, enters largely into the composition of soap; and should it be found, as perhaps by experiment it may be, that the corn meal and lard oils are not liable to the objection which, it is said, attends the use of whale oil in this respect, the demand for this purpose may be of importance to the producers of this article.

It is not improbable that, by further experiments, an oil may be obtained from the cotton seed, of such an excellent quality as to make what is now almost a total loss an article of great value. The Germans at the West are said to obtain oil in some quantities from the seed of the pumpkin; and the seeds of the sunflower, and rape seed, it is well known, have been used to advantage for the same purpose.

While Great Britain and other foreign countries have steadily pursued a policy designed and obviously tending to exclude our agricultural products from their trade, it becomes an object of no small consequence to us to evince, as the foregoing statistics have done, how much wealth we possess in our surplus products of wheat, and various other articles of food, together with the prospective increase of these and other products suited to call out the enterprise and industry of our people, and which, on a fair reciprocity with foreign nations, might greatly contribute to develope and enlarge the resources of our country. Should protective duties abroad continue to exclude our surplus products, the channels of present industry must be diverted to meet the emergency. It may be well for us to learn what makes us truly independent, and also happy. Extravagance in communities, as well as in individuals, leads to inevitable embarrassment. Credit may, indeed, be used for a while as a palliative, but the only effectual remedy is retrenchment and economy. When a constant drain of the precious metals is pressing us to meet the expenditures of our people for foreign imports, and when foreign nations encourage a home policy, by prohibitory duties on our products, it becomes a serious question with us how far and in what directions the industry now expended in raising a surplus beyond our own wants can be diverted to other objects of enterprise. To decide a question of such magnitude and interest, reference must obviously be had to the articles imported, to determine what can be raised or produced in our own country; and possibly it may be found that most of the leading articles, either of necessity or luxury, thus supplied, can be raised and perfected to advantage by the labor and skill of our own inhabitants. The remedy thus lies within our own power. Our true policy is to give variety and stability to our productive industry. Extraordinary prices in particular crops inevitably lead to dangerous extremes in the culture of the same, to the neglect of the usual and necessary articles of produce. Cupidity soon urges even the agriculturist into a spirit of speculation, which too often terminates in great embarrassment, and sometimes in utter ruin. The credulity of Americans is proverbial; and this has, to some extent, been illustrated in the almost universal mania that attended the *morus multicaulis* speculation: a single sprout sold for one dollar, when millions might be produced in one season. Incredulity, likewise, is sometimes yet more injurious to a community, as this shuts out all the light which science pours in, and rests contented with following the beaten path of traditionary leaders. Happy would it be for our country if the spirit of investigation and severe experiment should induce effort to test principles, without diverting it from those channels of industry that will assuredly bring the comforts of life. The balance of trade against us, resulting from our improvidence, can no longer be settled, or, rather, as it might be said, postponed by the remittance of State securities, which seem to have run a brief career, leaving still a vast debt, that can only be honestly cancelled by much *hard work*.

Notwithstanding all this, the daily importation of goods (including many articles of luxury) goes forward to a truly alarming extent; TWO-THIRDS OF WHICH ARE ON FOREIGN ACCOUNT, TO BE PAID FOR IN SPECIE OR ITS EQUIVALENT! Without the admitted means of liquidating the balances against us in foreign countries, we seem still madly bent on increasing them. Eleven and a half millions of dollars in specie were shipped from the single port of New York within the fifteen months preceding January, 1842; and with such a drain going on continually, every dollar of specie in the United States will soon be insufficient to meet our liabilities abroad. Stern necessity, however, will, ere long, extend her laws over us, compelling us to limit our expenditures to the actual income, and to effect exchanges of our agricultural products, either at home or abroad, for the products of mechanical skill and industry. This would be the case, even were the amount of our surplus product likely to be lessened.

Yet there is no reason to apprehend that our surplus products will be diminished. On the contrary, the stoppage of numerous canals, railroads, and other works of internal improvement by the States, will dismiss many laborers, who will resort to agriculture and kindred pursuits; so that the amount of products raised will probably exceed those of former years. The extensive tracts, too, of our unoccupied soil invite emigration to our shores; and when we consider the present extreme distress in portions of the manufacturing districts of Great Britain, we are doubtless to expect a large increase of our population in future years from this cause. It is stated, on high authority, that as many as 20,000 persons die annually in Great Britain, from the want of sufficient and wholesome food. Let the fact of our vast surplus product of the bread stuffs and other articles of food become known abroad, and is it not reasonable to look for increasing additions to the emigration from Europe to this country? — especially since the distance is now, as it were, so much shortened, that a voyage may be compassed in 12 or 15 days. A line of steam packets, too, is in contemplation, to run from Bremen to one of our ports, with the design principally of conveying emigrants, which, no doubt, will prove the means of bringing to us a hardy, industrious German population, most of whom will probably engage in agriculture. With these additions to her laboring force, our growing country, if she be true to herself, offers an unwonted scope for exertion. The diversities of her climate the varieties of her soil, her peculiar combination of population, her mineral, animal, agricultural, mechanical, and commercial wealth, developed as they may be by a rightful regard to her necessities, might thus place her at last in a situation as enviable for her political and moral influence, as for the physical energies she had called into life and action. Our republic needs, indeed, only to prove her own strength, and wisely direct her energies, to become, more than she has ever been, the point on which the eye of all Europe is fixed, as a home of plenty for the destitute, and a field where enterprise reaps its sure and appropriate reward.

The First American Agricultural Revolution and Its Setting, 1840-1870

Introduction

Between 1840 and 1870 American farming saw the development of new institutions and new technologies, and, in the later part of the period, the first American agricultural revolution. This revolution was marked by the adoption of more effective machines and methods, a change from human power to horse power, and, particularly in the North since the South had a highly-developed commercial agriculture, the move from self-sufficient farming to production for market.

In 1840, the farm population made up 9 million persons of a national total of 17 million. Farm workers were 69 percent of all persons gainfully employed. Estimates for 1800 show a farm population of 4.3 million persons out of a total of 5.3 million, with farm workers perhaps 85 percent of all persons gainfully employed. According to a widely-used series of estimates, 807 million pounds live weight of cattle and calves were sold to the nonfarm population or consumed on the farm in 1800, compared with 2,759 million pounds in 1840. Wheat is estimated at 22 million bushels in 1800 and 72.5 million in 1840; corn at 25.6 million bushels in 1800 and 76.5 million in 1840; and cotton at 30 million pounds in 1800 and 690 million in 1840. Total production per farm worker had increased slightly, but the overall increases in total production had resulted mainly from the increase in the number of workers, the increase in the number of acres under cultivation, and the opening of new, fertile land. The area of the United States had doubled between 1800 and 1840, and was to increase again by nearly two-thirds between 1840 and 1850.

Land policies for the nation had been set by the Ordinances of 1785 and 1787. Subsequent modifications, with few exceptions, had made it easier for settlers and speculators to acquire land. The trend continued. The Preemption Act of 1841 allowed a person who had settled on and developed a piece of land prior to its being offered for sale to purchase it at the minimum price. The Graduation Act of 1854 was another law urged by Westerners. It made price reductions on Federal land in proportion to the time land had been on the market.

The culmination of the drive to get western lands into the hands of actual settlers came when President Lincoln signed the Homestead Act of May 20, 1862. Generally, the act gave 160 acres of the public domain to anyone who was the head of a family or over 21 years of age. Title or a patent to the land was issued after the settler had resided on it for 5 years and had made improvements on it, or had resided on the land for six months, made suitable improvements, and paid $1.25 per acre. The passage of the act was controversial and so were its results.

Many land grants were made to states for one purpose or another. In 1850, wet, marshy, and inundated public lands were granted to the states in which they were located in order that the states might reclaim them. In the same year, Congress granted land to the states through which they would pass for the construction of the Illinois Central and the Mobile and Ohio Railroads. This set a precedent for later grants directly to railroads. The first of these, and one of the most important, was the Pacific Railway Act of 1862. These extensive grants paid much of the cost of building the Union Pacific and Central Pacific roads (the eastern and western parts of the same line). The line was completed in 1869.

The Morrill Land Grant College Act of 1862 granted to each state 30,000 acres of public land for each senator and representative in Congress for the purpose of endowing at least one college in each state where agriculture and the mechanical arts should be taught. Some states, such as Michigan and Pennsylvania in 1855, Maryland in 1856, and Iowa in 1858 had established state agricultural colleges. The movement for Federal aid had been underway some twenty years. It was given form during the 1850's by Jonathan B. Turner of Illinois, and became law through the efforts of Senator Justin S. Morrill of Vermont.

The Homestead Act, Pacific Railway Act, and Morrill Land Grant Act were parts of an agricultural reform promised by the Republican Party. The fourth act, also in 1862, established the Department of Agriculture, headed by a Commissioner responsible to the President. A Department had been urged for many years. Its establishment resulted from pressure by the United States Agricultural Society, while proposals by Thomas G. Clemson of South Carolina were influential. After its establishment, such farm periodicals as the *Country Gentleman* and the *American Agriculturist* published many articles on objectives for it and for the new agricultural colleges. Thus, in 1862, institutions had been provided which would help change American agriculture.

All of the reform acts, with the exception of the transcontinental railroad bill, had been opposed by majorities within the Southern Congressional delegations and did not become law until after the start of the Civil War. Between 1840 and 1860, the South increased cotton production from 1.3 million to 3.8 million bales. The slave population had increased from 2.4 million to 3.8 million persons. The cotton plantation, operated by slave labor, dominated the economic, political and social life of the South, as evidenced in the accounts by such travelers as Frederick Law Olmsted. At the same time, some agricultural reformers were urging crop diversification, soil conservation, and greater emphasis on livestock. Such reformers as Edmund Ruffin, James H. Hammond, and C. W. Howard, urged diversification and better utilization of the soil. Howard particularly urged the adoption of grasses suited to the South and to livestock. However, as cotton growing became more important, livestock raising declined in importance. In South Carolina, one of the great Southern centers for range livestock at the turn of the century, the number of cattle declined from 572,608 in 1840 to 506,776 in 1860. The total number in the nation nearly doubled in the same period. The old South as a whole had 22 percent of the nation's livestock in 1840 and 13 percent in 1860. At the same time, as Lewis Sanders pointed out in 1850, improved breeds of livestock were spreading in the South.

Dairying grew in importance in New England and the Middle Atlantic states, and some Middle Atlantic farmers fattened cattle for the city markets. Many of the cattle were driven east from the Ohio country. The cattle raising industry continued to move west to Illinois, Iowa, and Missouri and then to the west coast. There were large herds in California and Texas, but the cattle were long-legged and long-horned. Thus, drives such as that by Thomas Rebar became more frequent. After the Civil War, the long drives from Texas, first to Missouri and then to Kansas railheads and northern pastures began. As cattle raising moved west, hog raising became important in the Old Northwest. In 1840, that region had one-fifth of the nation's hogs and in 1860, one-fourth.

The settlement of Utah by members of the Church of Jesus Christ of Latter-day Saints or Mormons saw the beginning of irrigation — aside from that carried on by Indians in the Southwest — in western agriculture. The movement west by the Mormons began in 1846, when 16,000 persons, 2,000 wagons, and 30,000 head of stock started from Nanvoo, Illinois. The first company arrived at Salt Lake in July 1847 and began irrigation agriculture at once.

At the outbreak of the Civil War in 1861, an efficient commercial agriculture, based upon cotton grown by slave labor, charactertized the South. The North had areas of specialized agriculture, notably dairying as pointed out by Willard, but farming was dominated by small landholders carrying on general farming. New machinery had been invented, the virtues of liming in many areas had been demonstrated, commercial fertilizer was available, and irrigation was being practiced in Utah. The westward movement was continuing unabated, opening new lands to livestock, wheat, and other crops. And, after decades of agitation, the Federal Government was poised to pass the four major agricultural reform bills discussed earlier.

The difficulties in the way of Southern agricultural progress during the Civil War may be seen in the official records of the Confederacy, farm journals — until they were forced to suspend publication because of lack of paper, and accounts by men and women of the South. Persuading planters to grow food for the armed forces instead of cotton for Great Britain was difficult. Farming methods changed but little, and even though labor became very scarce, there was no shift to labor-saving machinery because machinery was not available. The constant drains of men into the army left many sections, particularly those with small farms, almost without workers. There were some Southern planters and farmers, often building upon progress made before the war, who made substantial contributions to food production during the conflict. They, however, were not numerous enough to change the overall picture of a substantial decline in production.

Northern farmers, on the other hand, turned to machinery to replace men in the armed forces. In 1863, the Ohio State Board of Agriculture reported: "machinery and improved implements have been employed to a much greater extent during the years of rebellion than ever before. . . . Without drills, corn-planters, reapers and mowers, horse-rakes, hay elevators, and threshing machines, it would have been impossible to have seeded and gathered the crops of 1863 with the implements in use forty or fifty years ago, by the same laborers that really performed the labor in 1863.'' Throughout most of the northern United States, the same situation prevailed.

Major changes from hand methods to animal power and from near subsistence to commercial agriculture were taking place as a result of labor shortages and of wartime and foreign demand and consequent high prices. Many farmers adopted new machines and methods after the demand for farm products and economic incentives for change had become obvious, so that the impact of the war was more noticeable at its end and the years immediately following than in earlier years. In any case, the Civil War, at least in the north and west, served as a catalyst for the first American agricultural revolution.

After the war, the South faced the problem of restoring its agriculture without the use of slave labor. Since cotton was in demand in both England and the North, cotton culture, based first upon wage and then upon share contracts with the former slaves, was quickly established. Some of the steps looking to diversification, particularly livestock farming, which had been advancing before the war, were abandoned, even though farm journals continued to advocate better practices.

Concern over the farm situation in the South led Isaac Newton, Commissioner of the Department of Agriculture, to send Oliver Hudson Kelley of Minnesota on an inspection trip. As a result, Kelley organized the Patrons of Husbandry, better known as the Grange, in 1867. It became the first nationwide organization to reach a large number of farmers. State and local grangers became active in cooperative buying and selling. Although the Grange itself claimed to be nonpolitical, state groups made their influence felt, particularly in urging legislation to regulate railroads. The high point in early Grange membership was reached about 1875. After that, most of its economic activities were discontinued, and it became stabilized as a social and fraternal order with a continuing educational program.

The Grange urged its members, through lectures and demonstrations, to adopt improved agricultural practices. It joined other institutions, including the new Department of Agriculture and agricultural colleges and the farm press, in such activities. Discussions ranged from the steam plow, to improved breeds of hogs and fowls, to the desirability of growing sugar beets, and to the profitability of growing opium poppies.

In 1870, the farm population was 18.3 million of a total of 38.5 million. Farm workers were 53 percent of all persons gainfully employed. And, because of the first American agricultural revolution, these farmers, whether they knew it or not, were in the midst of a transition from a self-sufficient to a market-oriented commercial agriculture.

Land Policies

Land Grants for Internal Improvements and Individual Rights of Preemption, 1841

From 5 Stat. 453.

Be it enacted by the Senate and House of Representatives of the United States of America in Congress assembled, That from and after the thirty-first day of December, in the year of our Lord one thousand eight hundred and forty-one, there be allowed and paid to each of the States of Ohio, Indiana, Illinois, Alabama, Missouri, Mississippi, Louisiana, Arkansas, and Michigan, over and above what each of the said States is entitled to by the terms of the compacts entered into between them and the United States, upon their admission into the Union, the sum of ten per centum upon the nett proceeds of the sales of the public lands, which, subsequent to the day aforesaid, shall be made within the limits of each of said States respectively: *Provided,* That the sum so allowed to the said States, respectively, shall be in no wise affected or diminished on account of any sums which have been heretofore, or shall be hereafter, applied to the construction or continuance of the Cumberland road, but that the disbursements for the said road shall remain, as heretofore, chargeable on the two per centum fund provided for by compacts with several of the said States.

SEC. 2. *And be it further enacted,* That after deducting the said ten per centum, and what, by the compacts aforesaid, has heretofore been allowed to the States aforesaid, the residue of the nett proceeds, which nett proceeds shall be ascertained by deducting from the gross proceeds all the expenditures of the year for the following objects: salaries and expenses on account of the General Land Office; expenses for surveying public lands; salaries and expenses in the surveyor general's offices; salaries, commissions, and allowances to the registers and receivers; the five per centum to new States, of all the public lands of the United States, wherever situated, which shall be sold subsequent to the said thirty-first day of December, shall be divided among the twenty-six States of the Union and the District of Columbia, and the Territories of Wisconsin, Iowa, and Florida, according to their respective federal representative population as ascertained by the last census, to be applied by the Legislatures of the said States to such purposes as the said Legislatures may direct: *Provided,* That the distributive share to which the District of Columbia shall be entitled, shall be applied to free schools, or education in some other form, as Congress may direct: *And provided, also,* That nothing herein contained shall be construed to the prejudice of future applications for a reduction of the price of the public lands, or to the prejudice of applications for a transfer of the public lands, on reasonable terms, to the States within which they lie, or to make such future disposition of the public lands, or any part thereof, as Congress may deem expedient.

SEC 3. *And be it further enacted,* That the several sums of money received in the Treasury as the nett proceeds of the sales of the public lands shall be paid at the Treasury half yearly on the first day of January and July in each year, during the operation of this act, to such person or persons as the respective Legislatures of the said States and Territories, or the Governors thereof, in case the Legislatures shall have made no such appointment, shall authorize and direct to receive the same.

SEC. 4. *And be it further enacted,* That any sum of money, which at any time may become due, and payable to any State of the Union, or to the District of Columbia, by virtue of this act, as the portion of the said State or District, of the proceeds of the sales of the public lands, shall be first applied to the payment of any debt, due, and payable from the said State or District, to the United States: *Provided,* That this shall not be construed to extend to the sums deposited with the States under the act of Congress of twenty-third June, eighteen hundred and thirty-six, entitled "an act to regulate the deposites of the public money," nor to any sums apparently due to the United States as balances of debts growing out of the transactions of the Revolutionary war.

SEC. 5. *And be it further enacted,* That this act shall continue and be in force until otherwise provided by law, unless the United States shall become involved in war with any foreign Power, in which event, from the commencement of hostilities, this act shall be suspended during the continuance of such war: *Provided, nevertheless,* That if, prior to the expiration of this act, any new State or States shall be admitted into the Union, there be assigned to such new State or States, the proportion of the proceeds accruing after their admission into the Union, to which such State or States may be entitled, uponthe principles of this act, together with what such State or States may be entitled to by virtue of compacts to be made on their admission into the Union.

SEC. 6. *And be it further enacted,* That there shall be annually appropriated for completing the surveys of said lands, a sum not less than one hundred and fifty thousand dollars; and the minimum price at which the public lands are now sold at private sale shall not be increased, unless Congress shall think proper to grant alternate sections along the line of any canal or other internal improvement, and at the same time to increase the minimum price of the sections reserved; and in case the same shall be increased by law, except as aforesaid, at any time during the operation of this act, then so much of this act as provides that the nett proceeds of the sales of the public lands shall be distributed among the several States, shall, from and after the increase of the minimum price thereof, cease and become utterly null and of no effect, any thing in this act to the contrary notwithstanding: *Provided,* That if, at any time during the existence of this act, there shall be an imposition of duties on imports inconsistent with the provisions of the act of March second one thousand eight hundred and thirty-three, entitled, "An act to modify the act of the fourteenth of July one thousand eight hundred and thirty-two, and all other acts imposing duties on imports," and beyond the rate of duty fixed by that act, to wit: twenty per cent. on the value of such imports, or any of them, then the distribution provided in this act shall be suspended and shall so continue until this cause of its suspension shall be removed, and when removed, if not prevented by other provisions of this act, such distribution shall be resumed.

SEC. 7. *And be it further enacted,* That the Secretary of the Treasury may continue any land district in which is situated the seat of government of any one of the States, and may continue the land office in such district, notwithstanding the quantity of land unsold in such district may not amount to one hundred thousand acres, when, in his opinion, such continuance may be required by public convenience, or in order to close the land system in such State at a convenient point, under the provisions of the act on that subject, approved twelfth June, one thousand eight hundred and forty.

SEC. 8. *And be it further enacted,* That there shall be granted to each State specified in the first section of this act five hundred thousand acres of land for purposes of internal improvement: *Provided,* that to each of the said States which has already received grants for said purposes, there is hereby granted no more than a quantity of land which shall, together with the amount such State has already received as aforesaid, make five hundred thousand acres, the selections in all of the said States, to be made within their limits respectively in such manner as the Legislatures thereof shall direct; and located in parcels conformably to sectional divisions and subdivisions, of not less than three hundred and twenty acres in any one location, on any public land except such as is or may be reserved from sale by any law of Congress or proclamation of the President of the United States, which said locations may be made at any time after the lands of the United States in said States respectively, shall have been surveyed according to existing laws. And there shall be and hereby is, granted to each new State that shall be hereafter admitted into the Union, upon such admission, so much land as, including such quantity as may have been granted to such State before its admission, and while under a Territorial Government, for purposes of internal improvement as aforesaid, as shall make five hundred thousand acres of land, to be selected and located as aforesaid.

SEC. 9. *And be it further enacted,* That the lands herein granted to the States above named shall not be disposed of at a price less than one dollar and twenty-five cents per acre, until otherwise authorized by a law of the United States; and the nett proceeds of the sales of said lands shall be faithfully applied to objects of internal improvement within the States aforesaid, respectively, namely: Roads, railways, bridges, canals and improvement of water-courses, and draining of swamps; and such roads, railways, canals, bridges and water-courses, when made or improved, shall be free for the transportation of the United States mail, and munitions of war, and for the passage of their troops, without the payment of any toll whatever.

SEC. 10. *And be it further enacted,* That from and after the passage of this act, every person being the head of a family, or widow, or single man, over the age of twenty-one years, and being a citizen of the United States, or having filed his declaration of intention to become a citizen, as required by the naturalization laws, who since the first day of June, A.D. eighteen hundred and forty, has made or shall hereafter make a settlement in person on the public lands to which the Indian title had been at the time of such settlement extinguished, and which has been, or shall have been, surveyed prior thereto, and who shall inhabit and improve the same, and who has or shall erect a dwelling thereon, shall be, and is hereby, authorized to enter with the register of the land office for the district in which such land may lie, by legal subdivisions, any number of acres not exceeding one hundred and sixty, or a quarter section of land, to include the residence of such claimant, upon paying to the United States the minimum price of such land, subject, however, to the following limitations and exceptions: No person shall be entitled to more than one preemptive right by virtue of this act; no person who is the proprietor of three hundred and twenty acres of land in any State or Territory of the United States, and no person who shall quit or abandon his residence on his own land to reside on the public land in the same State or Territory, shall acquire any right of pre-emption under this act; no lands included in any reservation, by any treaty, law, or proclamation of the President of the United States, or reserved for salines, or for other purposes; no lands reserved for the support of schools, nor the lands acquired by either of the two last treaties with the Miami tribe of Indians in the State of Indiana, or which may be acquired of the Wyandot tribe of Indians in the State of Ohio, or other Indian reservation to which the title has been or may be extinguished by the United States at any time during the operation of this act; no sections of land reserved to the United States alternate to other sections granted to any of the States for the construction of any canal, railroad, or other public improvement; no sections or fractions of sections included within the limits of any incorporated town; no portions of the public lands which have been selected as the site for a city or town; no parcel or lot of land actually settled and occupied for the purposes of trade and not agriculture; and no lands on which are situated any known salines or mines, shall be liable to entry under and by virtue of the provisions of this act. And so much of the proviso of the act of twenty-second of June, eighteen hundred and thirty-eight, or any order of the President of the United States, as directs certain reservations to be made in favor of certain claims under the treaty of Dancing-rabbit creek, be, and the same is hereby, repealed: *Provided*, That such repeal shall not affect any title to any tract of land secured in virtue of said treaty.

SEC. 11. *And be it further enacted*, That when two or more persons shall have settled on the same quarter section of land, the right of pre-emption shall be in him or her who made the first settlement, provided such persons shall conform to the other provisions of this act; and all questions as to the right of pre-emption arising between different settlers shall be settled by the register and receiver of the district within which the land is situated, subject to an appeal to and a revision by the Secretary of the Treasury of the United States.

SEC. 12. *And be it further enacted*, That prior to any entries being made under and by virtue of the provisions of this act, proof of the settlement and improvement thereby required, shall be made to the satisfaction of the register and receiver of the land district in which such lands may lie, agreeably to such rules as shall be prescribed by the Secretary of the Treasury, who shall each be entitled to receive fifty cents from each applicant for his services, to be rendered as aforesaid; and all assignments and transfers of the right hereby secured, prior to the issuing of the patent, shall be null and void.

SEC. 13. *And be it further enacted,* That before any person claiming the benefit of this act shall be allowed to enter such lands, he or she shall make oath before the receiver or register of the land district in which the land is situated, (who are hereby authorized to administer the same,) that he or she has never had the benefit of any right of pre-emption under this act; that he or she is not the owner of three hundred and twenty acres of land in any State or Territory of the United States, nor hath he or she settled upon and improved said land to sell the same on speculation, but in good faith to appropriate it to his or her own exclusive use or benefit; and that he or she has not, directly or indirectly, made any agreement or contract, in any way or manner, with any person or persons whatsoever, by which the title which he or she might acquire from the Government of the United States, should enure in whole or in part, to the benefit of any person except himself or herself; and if any person taking such oath shall swear falsely in the premises, he or she shall be subject to all the pains and penalties of perjury, and shall forfeit the money which he or she may have paid for said land, and all right and title to the same; and any grant or conveyance which he or she may have made, except in the hands of bona fide purchasers, for a valuable consideration, shall be null and void. And it shall be the duty of the officer administering such oath to file a certificate thereof in the public land office of such district, and to transmit a duplicate copy to the General Land Office, either of which shall be good and sufficient evidence that such oath was administered according to law.

SEC. 14. *And be it further enacted,* That this act shall not delay the sale of any of the public lands of the United States beyond the time which has been, or may be, appointed by the proclamation of the President, nor shall the provisions of this act be available to any person or persons who shall fail to make the proof and payment, and file the affidavit required before the day appointed for the commencement of the sales as aforesaid.

SEC. 15. *And be it further enacted,* That whenever any person has settled or shall settle and improve a tract of land, subject at the time of settlement to private entry, and shall intend to purchase the same under the provisions of this act, such person shall in the first case, within three months after the passage of the same, and in the last within thirty days next after the date of such settlement, file with the register of the proper district a written statement, describing the land settled upon, and declaring the intention of such person to claim the same under the provisions of this act; and shall, where such settlement is already made, within twelve months after the passage of this act, and where it shall hereafter be made, within the same period after the date of such settlement, make the proof, affidavit and payment herein required; and if he or she shall fail to file such written statement as aforesaid, or shall fail to make such affidavit, proof, and payment, within the twelve months aforesaid, the tract of land so settled and improved shall be subject to the entry of any other purchaser.

SEC. 16. *And be it further enacted,* That the two per cent. of the nett proceeds of the lands sold, or that may hereafter be sold, by the United States in the State of Mississippi, since the first day of December, eighteen hundred and seventeen, and by the act entitled "An act to enable the people of the western part of the Mississippi Territory to form a constitution and State government, and for the admission of such State into the Union on an equal footing with the original States," and all acts supplemental thereto reserved for the making of a road or roads leading to said

State, be, and the same is hereby relinquished to the State of Mississippi, payable in two equal instalments; the first to be paid on the first of May, eighteen hundred and forty-two, and the other on the first of May, eighteen hundred and forty-three, so far as the same may then have accrued, and quarterly, as the same may accrue, after said period: *Provided,* That the Legislature of said State shall first pass an act, declaring their acceptance of said relinquishment in full of said fund, accrued and accruing, and also embracing a provision, to be unalterable without the consent of Congress, that the whole of said two per cent. fund shall be faithfully applied to the construction of a railroad, leading from Brandon, in the State of Mississippi, to the eastern boundary of said State, in the direction, as near as may be, of the towns of Selma, Cahaba, and Montgomery, in the State of Alabama.

SEC. 17. *And be it further enacted,* That the two per cent. of the nett proceeds of the lands sold by the United States, in the State of Alabama, since the first day of September, eighteen hundred and nineteen, and reserved by the act entitled "An act to enable the people of the Alabama Territory to form a constitution and State government, and for the admission of such State into the Union on an equal footing with the original States," for the making of a road or roads leading to the said State, be, and the same is hereby, relinquished to the said State of Alabama, payable in two equal instalments, the first to be paid on the first day of May, eighteen hundred and forty-two, and the other on the first day of May, eighteen hundred and forty-three, so far as the same may then have accrued, and quarterly, as the same may thereafter accrue. *Provided,* That the Legislature of said State shall first pass an act, declaring their acceptance of said relinquishment, and also embracing a provision, to be unalterable without the consent of Congress, that the whole of said two per cent. fund shall be faithfully applied, under the direction of the Legislature of Alabama, to the connection, by some means of internal improvement, of the navigable waters of the bay of Mobile with the Tennessee river, and to the construction of a continuous line of internal improvements from a point on the Chattahoochie river, opposite West Point, in Georgia, across the State of Alabama, in a direction to Jackson in the State of Mississippi.

APPROVED, September 4, 1841.

Women and Public Land in Florida, 1843

From Clarence Edwin Carter, ed., *Territorial Papers* of the United States, Vol. XXVI, *The Territory of Florida, 1839–1845* (Washington, 1962), pp. 648–49.

GENERAL LAND OFFICE *May* 1st 1843.

W. H. SIMMONS Esqr Register St Augustine Florida.

SIR: I am in receipt of your letter of the 15th Ulto in which you state that "many Applications having been made to" your "Office by Females, for Permits to take up land under the occupation Act, — and the phraseology of the law being somewhat obscure, so as to leave it doubtful whether its privileges were designed to be extended to this class of emigrants," you request directions on the subject.

You state further that "the District Attorney, Mr Douglas, to whom" you "referred in the first instance, inclines to the Opinion that the expressions 'any Person, being the head of a family'" were "intended to include this class of Applicants, — who in all instances, have either sons or slaves, Capable of labor, and therefore of bearing Arms; but does not feel altogether satisfied of the correctness of this interpretation," and at his recommendation you "apply to the Department for its decision and directions on this subject."

That 1st Stipulation in the Act is "That Said Settler shall obtain from the Register of the Land Office, in the District in which he proposes to settle, a permit describing, as particularly as may be practicable, the place where *his* or *her settlement* is intended to be made;" &c. Again — In the 2nd Section of the Said Act, after referring to cases of Settlement of the same quarter Section by two or more settlers, and providing that the right to location shall be determined by priority of Settlement, the following language is employed; "and the subsequent *Settler* or *settlers* shall be permitted to locate the quantity, *he — she —* or *they* may be entitled to elsewhere within the same township upon vacant public lands."

The words underscored in red ink in the foregoing, (not italicized however in the Act) go to Sustain the Correctness of the inclination of the District Attorney's Opinion as quoted by you, Viz. — that "the expressions 'any Person being the head of a family'" were "intended to include this class of Applicants" [Females] "Who, in all instances, have either Sons or Slaves, Capable of labor, and therefore of bearing Arms." — Indeed the underscored words place the matter, in my Opinion, beyond doubt, and you will therefore govern yourself Accordingly, *taking care that the testimony is Satisfactory in every Application, to bring the case fully up* to the *requirements* of this *interpretation*.

Very respectfully Your Obt Servt

Tho: H. Blake Commissioner.

Land Grants to Men Enlisting in the Army, 1847.

From 9 Stat. 123.

Be it enacted by the Senate and House of Representatives of the United States of America, in Congress assembled, That in addition to the present military establishment of the United States there shall be raised and organized, under the direction of the President, for and during the war with Mexico, one regiment of dragoons and nine regiments of infantry each to be composed of the same number and rank of commissioned and non-commissioned officers, buglers, musicians, and privates, &c., as are provided for a regiment of dragoons and infantry, respectively, under existing laws, and who shall receive the same pay, rations, and allowances according to their respective grades, and be subject to the same regulations, and to the rules and articles of war: *Provided,* That it shall be lawful for the President of the United States alone to appoint such of the commissioned officers, authorized by this act, below the grade of field officers, as may not be appointed during the present session: *Provided,* That one or more of the regiments of infantry authorized to be raised by this section may, at the discretion of the President, be organized and equipped as voltigeurs, and as foot-riflemen, and be provided with a rocket and mountain howitzer battery.

SEC. 2. *And be it further enacted,* That, during the continuance of the war with Mexico, the term of enlistment of the men to be recruited for the regiments authorized by this act, shall be during the war, unless sooner discharged.

SEC. 3. *And be it further enacted,* That the President of the United States be, and he is hereby, authorized, by and with the advice and consent of the Senate, to appoint one additional major to each of the regiments of dragoons, artillery, infantry, and riflemen in the army of the United States, who shall be taken from the captains of the army.

SEC. 4. *And be it further enacted,* That to each of the regiments of dragoons, artillery, infantry, and riflemen, there shall be allowed a regimental quartermaster, to be taken from the subalterns of the line, who shall be allowed ten dollars additional pay per month, and forage for two horses.

SEC. 5. *And be it further enacted,* That the said officers, musicians, and privates, authorized by this act, shall immediately be discharged from the service of the United States at the close of the war with Mexico.

SEC. 6. *And be it further enacted,* That it shall and may be lawful for the President of the United States, by and with the advice and consent of the Senate, to appoint one surgeon and two assistant surgeons to each regiment raised under this act.

SEC. 7. *And be it further enacted,* That, during the war with Mexico, it shall be lawful for the officers composing the councils of administration of the several regiments constituting a brigade, either regular or volunteer, in the service of the United States, to employ some proper person to officiate as chaplain to such brigade; and the person so employed, shall upon the certificate of the commander of the brigade, receive for his services seven hundred and fifty dollars, one ration, and forage for one horse, per annum: *Provided,* That the chaplains now attached to the regular army, and stationed at different military posts, may, at the discretion of the Secretary of War, be required to repair to the army in Mexico, whenever a majority of the men at the posts where they are respectively stationed shall have left them for service in the field; and should any of said chaplains refuse, or decline to do this, when ordered so to do by the adjutant-general, the office of such chaplain shall be deemed vacant, and the pay and emoluments thereof be stopped.

SEC. 8. *And be it further enacted,* That the President be, and he is hereby authorized, by and with the advice and consent of the Senate, to appoint two additional surgeons and twelve additional assistant surgeons in the regular army of the United States, subject to the provisions of an act entitled, "An Act to increase and regulate the Pay of the Surgeons and Assistant Surgeons of the Army," approved June thirtieth, eighteen hundred and thirty-four; and that the officers whose appointment is authorized by this section, shall receive the pay and allowances of officers of the same grades respectively; and that the rank of the officers of the medical department of the army shall be arranged upon the same basis which at present determines the amount of their pay and emoluments: *Provided,* That the medical officers shall not in virtue of such rank be entitled to command in the line or other staff departments of the army.

SEC. 9. *And be it further enacted,* That each non-commissioned officer, musician, or private, enlisted or to be enlisted in the regular army, or regularly mustered in any volunteer company for a period of not less than twelve months, who has served or may serve during the present war with Mexico, and who shall receive an honorable discharge, or who shall have been killed, or died of wounds received or sickness incurred in the course of such service, or who shall have been discharged before the expiration of his term of service in consequence of wounds received or sickness incurred in the course of such service, shall be entitled to receive a certificate or warrant from the war department for the quantity of one hundred and sixty acres, and which may be located by the warrantee, or his heirs at law at any land office of the United States, in one body, and in conformity to the legal subdivisions of the public lands, upon any of the public lands in such district then subject to private entry; and upon the return of such certificate or warrant, with evidence of the location thereof having been legally made, to the General Land Office, a patent shall be issued therefor. That in the event of the death of any such non-commissioned officer, musician, or private, during service, or after his discharge, and before the issuing of a certificate or warrant as aforesaid, the said certificate or warrant shall be issued in favor, and inure to the benefit, of his family or relatives, according to the following rules: first, to the widow and to his children; second, his father; third, his mother. And in the event of his children being minors, then the legally-constituted guardian of such minor children shall, in conjunction with such of the children, if any, as may be of full age, upon being duly authorized by the orphans' or other court having probate jurisdiction, have power to sell and dispose of such certificate or warrant for the benefit of those interested. And all sales, mortgages, powers, or other instruments of writing, going to affect the title or claim to any such bounty right, made or executed prior to the issue of such warrant or certificate, shall be null and void to all intents and purposes whatsoever, nor shall such claim to bounty right be in any wise affected by, or charged with, or subject to, the payment of any debt or claim incurred by the soldier prior to the issuing of such certificate or warrant: *Provided,* that no land warrant issued under the provisions of this act shall be laid upon any lands of the United States to which there shall be a preëmption right, or upon which there shall be an actual settlement and cultivation: *Provided, further,* That every such non-commissioned officer, musician, and private, who may be entitled, under the provisions of this act, to receive a certificate or warrant for one hundred and sixty acres of land, shall be allowed the option to receive such certificate or warrant, or a treasury scrip for one hundred dollars; and such scrip, whenever it is preferred, shall be issued by the Secretary of the Treasury to such person or persons as would be authorized to receive such certificates or warrants for lands; said scrip to bear an interest of six per cent. per annum, payable semi-annually, redeemable at the pleasure of the government. And that each private, non-commissioned officer, and musician, who shall have been received into the service of the United States, since the commencement of the war with Mexico, for less than twelve months, and shall have served for such term or until honorably discharged, shall be entitled to receive a warrant for forty acres of land, which may be subject to private entry, or twenty-five dollars in scrip, if preferred; and in the event of the death of such volunteer during his term of service, or after an honorable discharge, but before the passage of this act, then the warrant for such land or scrip, shall issue to the wife, child, or children, if there be any, and, if none, then to the father, and, if there be no father, then to the mother of such deceased volunteer: *Provided,* That nothing contained in this section shall be construed to give bounty land to such volunteers as were accepted into service, and discharged without being marched to the seat of war.

SEC. 10. *And be it further enacted,* That it shall and may be lawful for the President, by and with the advice and consent of the Senate, to appoint, from the officers of the army, four quartermasters of the rank of major, and ten assistant quartermasters with the rank of captain.

APPROVED, February 11, 1847.

Land Grant for Illinois Central Railroad

From 9 Statutes at Large 466.

Be it enacted by the Senate and House of Representatives of the United States of America in Congress assembled, That the right of way through the public lands be, and the same is hereby, granted to the State of Illinois for the construction of a railroad from the southern terminus of the Illinois and Michigan Canal to a point at or near the junction of the Ohio and Mississippi Rivers, with a branch of the same to Chicago, on Lake Michigan, and another via the town of Galena in said State, to Dubuque in the State of Iowa, with the right also to take necessary materials of earth, stones, timber, etc., for the construction thereof: *Provided,* That the right of way shall not exceed one hundred feet on each side of the length thereof, and a copy of the survey of said road and branches, made under the direction of the legislature, shall be forwarded to the proper local land offices respectively, and to the general land office at Washington city, within ninety days after the completion of the same.

SEC. 2. *And be it further enacted,* That there be, and is hereby, granted to the State of Illlinois, for the purpose of aiding in making the railroad and branches aforesaid, every alternate section of land designated by even numbers, for six sections in width on each side of said road and branches; but in case it shall appear that the United States have, when the line or route of said road and branches is definitely fixed by the authority aforesaid, sold any part of any section hereby granted, or that the right of preëmption has attached to the same, then it shall be lawful for any agent or agents to be appointed by the governor of said State, to select, subject to the approval aforesaid, from the lands of the United States most contiguous to the tier of sections above specified, so much land in alternate sections, or parts of sections, as shall be equal to such lands as the United States have sold, or to which the right of preëmption has attached as aforesaid, which lands, being equal in quantity to one half of six sections in width on each side of said road and branches, the State of Illinois shall have and hold to and for the use and purpose aforesaid: *Provided,* That the lands to be so located shall in no case be further than fifteen miles from the line of the road: *And further provided,* The construction of said road shall be commenced at its southern terminus, at or near the junction of the Ohio and Mississippi Rivers, and its northern terminus upon the Illinois and Michigan Canal simultaneously, and continued from each of said points until completed, when said branch roads shall be constructed, according to the survey and location thereof: *Provided further*, That the lands hereby granted shall be applied in the construction of said road and branches respectively, in quantities corresponding with the grant for each, and shall be disposed of only as the work progresses, and shall be applied to no other purpose whatsoever: *And provided further,* That any and all lands reserved to the United States by the act entitled "An Act to grant a quantity of land to the State of Illinois, for the purpose of aiding in opening a canal to connect the waters of the Illinois River with those of Lake Michigan, approved March second, eighteen hundred and twenty-seven, be, and the same are hereby, reserved to the United States from the operations of this act.

SEC. 3. *And be it further enacted,* That the sections and parts of sections of land which, by such grant, shall remain to the United States, within six miles on each side of said road and branches, shall not be sold for less than double the minimum price of the public lands when sold.

SEC. 4. *And be it further enacted,* That the said lands hereby granted to the said State shall be subject to the disposal of the legislature thereof, for the purposes aforesaid and no other; and the said railroad and branches shall be and remain a public highway, for the use of the government of the United States, free from toll or other charge upon the transportation of any property or troops of the United States.

SEC. 5. *And be it further enacted,* That if the said railroad shall not be completed within ten years, the said State of Illinois shall be bound to pay to the United States the amount which may be received upon the sale of any part of said lands by said State, the title to the purchasers under said State remaining valid; and the title to the residue of said lands shall reinvest in the United States, to have and hold the same in the same manner as if this act had not been passed.

SEC. 6. *And be it further enacted,* That the United States mail shall at all times be transported on the said railroad under the direction of the Post-Office Department, at such price as the Congress may by law direct.

SEC. 7. *And be it further enacted,* That in order to aid in the continuation of said Central Railroad from the mouth of the Ohio River to the city of Mobile, all the rights, privileges, and liabilities hereinbefore conferred on the State of Illinois shall be granted to the States of Alabama and Mississippi respectively, for the purpose of aiding in the construction of a railroad from said city of Mobile to a point near the mouth of the Ohio River, and that public lands of the United States, to the same extent in proportion to the length of the road, on the same terms, limitations, and restrictions in every respect, shall be, and is hereby, granted to said States of Alabama and Mississippi respectively.

APPROVED, September 20, 1850.

Tenancy in South Carolina, 1852

From *S. C. Reporter*, reprinted in *Valley Farmer* (St. Louis, May, 1852), IV, no. 5, p. 172.

Letting Farms

A writer in an exchange paper says, 'it is a common remark that to let a farm to a tenant is in nine cases out of ten to destroy its fertility,' and attributes this result to the fact, that farms are usually let from year to year — so that the tenant has no inducement to improve it from the resources at hand. This is no doubt true, and scarcely any farm can maintain its fertility under such a system. But there is another reason still. Landlords are apt to make too hard a bargain with their tenants — they grasp at too much, and by doing so, are sure to lose a good tenant at the end of the year. The man who hires or takes a farm must have a chance to live, or he will not remain. We have had a little experience in this matter and can speak from personal observation. In 1843 the editor of this paper, dazzled with the glowing descriptions of the farmer's life, purchased a farm of one hundred acres in the westerly part of Hopkinton, and resided there three years.

It was naturally a good soil, but had for several years previous been rented to yearly tenants, every one of whom had skinned it to the utmost without replenishing it at all. The hay for the most part had beed carried away, and no manure returned, until it had become almost entirely unproductive — in farmer's parlance, it was pretty much 'run down.'

The three years we occupied it, brought it up to a tolerable state of productiveness. At the end of that time it was let at the halves as it is termed, to a good, industrious, intelligent, tenant, who has remained there ever since and remains there still; and the farm has all the time instead of growing worse, been growing better; so that at the present time, it is not exceeded in productiveness by any farm in its neighborhood. Taking it with the understanding that he could have it as long as he pleased, he has managed it according, and has collected and saved his manure, gathered up compost, and made all the preparations for the next year that he would if he owned it. The stock was all left upon the place, and he is interested in its growth for he has half of the increase, with other privileges and advantages, which makes the situation satisfactory to him, and for his interest to remain. When, had he been screwed down as many tenant are, he would no doubt have left the farm at the end the first year. Whilst the tenant is thus doing well, we are receiving a good interest on the investment, in the shape of butter, cheese, pork, potatoes, beans, corn, rye &c., all of which we have in abundance, for family use and some to sell, and of the best quality. Nobody knows how to make better butter than the woman on that farm. The result of our experience then is, that if you would have a good tenant you must first be a good landlord; you must not only give the tenant a chance to live, but you must make it for his interest to take care of you interest.

Ohio Land for Sale

From *Ohio Cultivator* (Columbus, December 15, 1853), IX, p. 380.

THE MARKETS

Cultivator Office, December 13, 1853.

Most kinds of farmers' products are a shade lower than at our last report. Two causes have operated to produce this result. 1st, the unfavorable advices from abroad on account of the unsettled state of European affairs; and 2d, the unusually pleasant weather which has seemed to extend the Indian Summer far into the domain of winter.

New York, December 10. — *Flour and Meal*. — Under the influence of unfavorable advices from Europe, Western State Flour has declined. The arrivals are light, and good shipping brands not plenty. The better grades are heavy at the close. Sales of Western Canal at $6.75@$6.81¼ for common to straight State; $6.81¼@$6.94 for common to good Ohio, and mixed to straight brands.

Grain. — There is a little variation in Wheat market; prime parcels are not plenty and these are in request; prime white Genesee $1.78; good white Southern $1.66; good white Pennsylvania $1.73; red Ohio on private terms at $1.53. Rye is better and in request — $1.03. Barley is steady; good six rowed Wisconsin 84½c. Oats in fair request at 50@53c. for State and Western and 47@49c. for Jersey. Corn is a shade firmer, but not active; 78@80c. for Southern white and yellow, $80@81½c. for Western mixed, 81@82½c. for Southern and Jersey yellow and, and 81c. for round yellow.

Provisions. — A fair demand for Pork for the local trade, and the market is more steady; $13.50@$13.62½ for Mess and $11.12½@ $11.25 for Prime. Beef is in moderate request at $8.50@$11 for country Mess, $13.50 for repacked Chicago, $5.75 for country Prime, and $15.50@$16.50 for extra Chicago. Beef Hams are selling at $13@$15. Prime Mess is firm; sales of 200 tcs. common Ohio at $20. Butter is in fair demand at 10@13c. for Ohio and 16@19c. for State dairies. Cheese is hardly so firm; sales of Ohio at 7½c. and State dairies at 9@9¾.

Cincinnati, December 13. — *Flour*. — Prices slightly lower, in lots at $5.22½@$5.25 per bbl. Buckwheat Flour has declined to $2.50 per cwt. *Grain*. — Wheat remains firm at $1, with a good demand. Corn is taken freely by distillers, at 37½c. Oats have improved — sales from Canal, in lots at 41@41½ per bu. Barley is in good demand at 50@55c. Rye 60@65c. We notice sales of prime small White Beans at $1.15 per bu. *Seed*. — Clover is in good demand and firm at $5.45@$5.50 per bu. for prime new crop. The demand to fill orders is fully equal to the supply. Timothy is dull at $1.50@$2.50. Flax $1.40 per bu. *Hogs* $4.25@$4.35. *Beef Cattle,* $4.50@$6 per cwt. *Sheep* $2@$4 per head. Cheese, W. R., 9; English Dairy 11½@12c. Butter, choice, 16@17. Eggs 13@14.

More Land for Sale. — I now offer for the first time *all* my real estate in Ohio, including my Home Farm containing five thousand acres, west of and adjoining the city of Columbus, and extending west to the five mile stone, on the National

Road Said Farm is capable of being sub-divided to advantage into farms of fifty acres or more, having on much of it fine durable running water, springs, timber and cleared land. The soil of the best Scioto bottom and upland.

Also, on both sides of the National Road, and Columbus and Xenia Railroad, between the five and six and a half mile stone, about two thousand acres. It can also be sub-divided into convenient farms and lots of any size.

Also, twenty-three hundred acres, on Big Darby, the geographical centre of which is about two miles from the Columbus and Xenia Railroad. The nearest points within one hundred rods of the National Road, and ten and a half miles west from Columbus. It is washed for three miles on its western boundary by Big Darby, a fine, bold mill stream having forty feet fall from its upper to its lower corner; having fine sites for mills and manufactories. It is covered with a fine growth of splendid white oak, burr oak, black walnut, blue ash, hickory, elm, &c. The soil is mostly first rate upland. The timber will pay thirty dollars per acre clear of expense of manufacturing and sale, with ordinary management. I will sell none less than the whole tract. Price twenty dollars per acre, one-half in hand, the balance in one year, with interest from date of sale.

Also, a tract of two thousand acres in Brown township, twelve miles northwest from Columbus, near the Columbus, Piqua and Indiana Railroad. The soil is extraordinary. The timber is worth thirty dollars per acre, clear of expense of manufacturing and sale. It is surrounded on all sides by finely improved farms. Price twenty dollars per acre, as above. I wish to sell the tract entire. I expect to erect thereon, and have in operation, a good Saw Mill, by first of December next.

Also, divers and sundry patches in different parts of the county, in tracts of two, three and four hundred acres each. Some choice pieces for fifteen dollars per acre.

Also, one thousand acres in Logan county, on Rush Creek, a never-failing spring stream, running its entire length, three-fourths of a mile south of the Big Spring Station, on the Marion and Bellefontaine Railroad. Soil extra. Timber tolerable. Price twelve dollars per acre.

Also, eight hundred acres in the vicinity of Ridgway, one-half mile from Railroad Station. Soil extra, and good timber, such as black walnut, cherry, oak, ash and sycamore. Price twelve dollars per acre.

Also, six hundred acres, opposite, and only separated by the Scioto river from Kenton, the flourishing county seat of Hardin county. Price twenty dollars per acre.

The above are the largest tracts I have, not yet sold. I have also several thousand acres in smaller tracts, too tedious to describe, in different parts of Logan and Hardin counties.

I am willing and desirous to sell all or any on reasonable terms and time. Land buyers will do well to give me a call at my office, in Franklinton, one mile west from the State House, Columbus, Ohio, where I can give particulars, and, if necessary, send a man to show the land.

The lands near Columbus will be sold in city lots and out lots, to suit buyers.

M. L. SULLIVANT.

Columbus, December 15, 1853–2t.

OHIO CULTIVATOR FOR 1854

THE TENTH VOLUME of this popular Journal will commence on the first of January, 1854. The circulation has been steadily increasing from year to year until it has reached TEN THOUSAND COPIES, giving evidence that it is truly esteemed as

THE LABORER'S FRIEND!

Our aim is the *diffusion of useful knowledge* among the Industrial classes of society. We labor for the FARMER, the MECHANIC, and the DOMESTIC CIRCLE, all of whom we shall endeavor to address in the language of

PLAIN AND PRACTICAL TRUTH,

Suited to the wants and means of every-day life. Our information is gathered in a great measure from personal observation in all parts of the country, by which means we become familiar with the various wants and capabilities of different sections, and are better able to suggest means of improvement. Besides this we have correspondence from all quarters, giving facts and experience in the sway of Crops, Farm Stock and Domestic management, which must insure for our paper among the intelligent working classes, a

WELCOME TO THE FIRESIDE!

The past year has been one of plenty and prosperity to the Farmer, and he can well afford to bestow the small sum necessary to secure for himself and family the reading of a good and reliable Agricultural paper, and thus give encouragement to an agency which has contributed in no small degree to the general prosperity of the State. To this end we shall continue to devote our best energies, and believing that there are yet many thousands of families who would gladly welcome the *Ohio Cultivator* to their generous firesides, if it was once introduced to their notice, we invoke the kind offices of our friends in this behalf, feeling assured that in so doing they will confer an equal favor upon their friends and the community at large. Our Publication is established upon a firm basis, having outlived all the storms that have assailed it, and is now without a rival of its class, in all this region, while we allow ourselves to believe, that in adaptation to the demands of Western Farmers, it has no superior any where. We shall spare no reasonable labor or expense to make it worthy of the confidence and support of all our old friends and the thousands of new ones which we hope to enrol for the coming year.

THE TERMS of the *Cultivator* will be as heretofore, viz: Single subscriptions $1 a year. 4 copies for $3. 9 copies for $6 — always in advance; and all subscriptions to begin with the year.

As no Traveling Agents are employed, we rely upon the good will of Postmasters and our local friends, to make up Clubs, in their several districts; and to remunerate such as interest themselves for us, we offer the following

PREMIUMS — To persons who send us nine subscribers and $6, we will send the complete volume for the past year, or any previous volume, if preferred, in printed cover, *postage paid*. The person sending us the largest number of subscribers for the coming year, will be entitled to a complete set of the *Cultivator* from its commencement — 9 volumes — making a good *Farmer's Library*. For the second largest number, 8 volumes; the third 7 volumes; the fourth 6, &c.

Packages of Seeds will also be sent as premiums, to those who signify that they prefer them — nine or more varieties, free of postage to eac person who sends a Club of nine or more subscribers.

Postage Stamps can in all cases be sent to us for fractions of a dollar.
Address,   BATEHAM & HARRIS,
Columbus, Ohio.

Homestead Act, 1862

From 12 Stat. 392.

Be it enacted by the Senate and House of Representatives of the United States of America in Congress assembled, That any person who is the head of a family, or who has arrived at the age of twenty-one years, and is a citizen of the United States, or who shall have filed his declaration of intention to become such, as required by the naturalization laws of the United States, and who has never borne arms against the United States Government or given aid and comfort to its enemies, shall, from and after the first January, eighteen hundred and sixty-three, be entitled to enter one quarter section or a less quantity of unappropriated public lands, upon which said person may have filed a preëmption claim, or which may, at the time the application is made, be subject to preëmption at one dollar and twenty-five cents, or less, per acre; or eighty acres or less of such unappropriated lands, at two dollars and fifty cents per acre, to be located in a body, in comformity to the legal subdivisions of the public lands, and after the same shall have been surveyed: *Provided,* That any person owning and residing on land may, under the provisions of this act, enter other land lying contiguous to his or her said land, which shall not, with the land so already owned and occupied, exceed in the aggregate one hundred and sixty acres.

SEC 2. *And be it further enacted,* That the person applying for the benefit of this act shall, upon application to the register of the land office in which he or she is about to make such entry, make affidavit before the said register or receiver that he or she is the head of a family, or is twenty-one years or more of age, or shall have performed service in the army or navy of the United States, and that he has never borne arms against the Government of the United States or given aid and comfort to its enemies, and that such application is made for his or her exclusive use and benefit, and that said entry is made for the purpose of actual settlement and cultivation, and not either directly or indirectly for the use or benefit of any other person or persons whomsoever; and upon filing the said affidavit with the register or receiver, and on payment of ten dollars, he or she shall thereupon be permitted to enter the quantity of land specified: *Provided, however,* That no certificate shall be given or patent issued therefor until the expiration of five years from the date of such entry; and if, at the expiration of such time, or at any time within two years thereafter, the person making such entry; or, if he be dead, his widow; or in case of her death, his heirs or devisee; or in case of a widow making such entry, her heirs or devisee, in case of her death; shall prove by two credible witnesses that he, she, or they have resided upon or cultivated the same for the term of five years immediately succeeding the time of filing the affidavit aforesaid, and shall make affidavit that no part of said land has been alienated, and that he has borne true allegiance to the Government of the United States; then, in such case, he, she, or they, if at that time a citizen of the United States, shall be entitled to a patent, as in other cases provided for by law: *And provided, further,* That in case of the death of both father and mother, leaving an infant child, or children, under twenty-one years of age, the right and fee shall enure to the benfit of said infant child or children; and the executor, administrator, or guardian may, at any time within two years after the death of the surviving parent, and in accordance with the laws of the State in which such children for the time being have their domicil, sell said land for the benefit of said infants, but for no other purpose; and the purchaser shall acquire the absolute title by the purchase, and be entitled to a patent from the United States, on payment of the office fees and sum of money herein specified.

SEC. 3. *And be it further enacted,* That the register of the land office shall note all such applications on the tract books and plats of his office, and keep a register of all such entries, and make return thereof to the General Land Office, together with the proof upon which they have been founded.

SEC. 4. *And be it further enacted,* That no lands acquired under the provisions of this act shall in any event become liable to the satisfaction of any debt or debts contracted prior to the issuing of the patent therefor.

SEC. 5. *And be it further enacted,* That if, at any time after the filing of the affidavit, as required in the second section of this act, and before the expiration of the five years aforesaid, it shall be proven, after due notice to the settler, to the satisfaction of the register of the land office, that the person having filed such affidavit shall have actually changed his or her residence, or abandoned the said land for more than six months at any time, then and in that event the land so entered shall revert to the government.

SEC. 6. *And be it further enacted,* That no individual shall be permitted to acquire title to more than one quarter section under the provisions of this act; and that the Commissioner of the General Land Office is hereby required to prepare and issue such rules and regulations, consistent with this act, as shall be necessary and proper to carry its provisions into effect; and that the registers and receivers of the several land offices shall be entitled to receive the same compensation for any lands entered under the provisions of this act that they are now entitled to receive when the same quantity of land is entered with money, one half to be paid by the person making the application at the time of so doing, and the other half on the issue of the certificate by the person to whom it may be issued; but this shall not be construed to enlarge the maximum of compensation now prescribed by law for any register or receiver: *Provided,* That nothing contained in this act shall be so construed as to impair or interefere in any manner whatever with existing preëmption rights: *And provided, further,* That all persons who may have filed their applications for a preëmption right prior to the passage of this act, shall be entitled to all privileges of this act: *Provided, further,* That no person who has served, or may hereafter serve, for a period of not less than fourteen days in the army or navy of the United States, either regular or volunteer, under the laws thereof, during the existence of an actual war, domestic or foreign, shall be deprived of the benefits of this act on account of not having attained the age of twenty-one years.

SEC. 7. *And be it further enacted,* That the fifth section of the act entitled "An act in addition to an act more effectually to provide for the punishment of certain crimes against the United States, and for other purposes," approved the third of March, in the year eighteen hundred and fifty-seven, shall extend to all oaths, affirmations, and affidavits, required or authorized by this act.

SEC. 8. *And be it further enacted,* That nothing in this act shall be so construed as to prevent any person who has availed him or herself of the benefits of the first section of this act, from paying the minimum price, or the price to which the same may have graduated, for the quantity of land so entered at any time before the expiration of the five years, and obtaining a patent therefor from the government, as in other cases provided by law, on making proof of settlement and cultivation as provided by existing laws granting preëmption rights.

APPROVED, May 20, 1862.

Comments on the Homestead Act

From Country Gentleman (Albany, May 22, 1862), IXX, p. 337.

THE HOMESTEAD BILL having passed both houses of Congress, will probably have received the signature of the President, and become a law before this comes before our readers. It purports to grant to actual settlers, at a nominal price, parcels of public lands. It is called a *homestead* bill, as if a piece of wild land could with any propriety, be considered a homestead for man. A home it may be, to birds, beasts, and reptiles, but how different from a home for a civilized family — who need society, friendship and love. It may pass *land* to the *landless,* but a *home* is to be wrought out by human labor and skill. It is a misnomer to call a parcel of wild land a *farm* and more inappropriate to call it a *homestead.* With much labor and care a farm may be made of it; and with neighbors enough, it may become a homestead. It is no grant of Congress. It *was* and is the property of the people, and Congress could only make needful rules to regulate it. If a free homestead law is valid, it is because it regulates in a needful and proper manner, the public land. The truth cannot be too strongly impressed on the mind of our youth, that *wild land* has no money value except in particular localities, where value has been given to it by labor expended in the neighborhood and reflected on it. Nine-tenths of the farms in the lake States — a region of more merit than any other on the Continent — can now be bought for a sum less than the cost which has been expended in their improvements — leaving nothing for the first cost of the land in a state of nature. The right to appropriate wild land, at a nominal price, accorded by the act of Congress, will in nine cases out of ten, be barren of benefit. If it shall induce additional immigration, especially from Germany, as is not unlikely, it may do good. Western Europe could spare a million a year with little loss there and great gain here. Our rich lands are hungry for tillage. "The poor German in the pleasure which the ownership of land gives him will submit cheerfully to the privations and small returns of a new farm."

Land Grant to Build Union Pacific Railroad

From 12 Statutes at Large 489.

Be it enacted by the Senate and House of Representatives of the United States of America in Congress assembled, That Walter S. Burgess, William P. Blodget, Benjamin H. Cheever, Charles Fosdick Fletcher, of Rhode Island; Augustus Brewster, Henry P. Haven, Cornelius S. Bushnell, Henry Hammond, of Connecticut; Isaac Sherman, Dean Richmond, Royal Phelps, William H. Ferry, Henry A. Paddock, Lewis Stancliff, Charles A. Secor, Samuel R. Campbell, Alfred E. Tilton, John Anderson, Azariah Boody, John S. Kennedy, H. Carver, Joseph Field, Benjamin F. Camp, Orville W. Childs, Alexander J. Bergen, Ben. Holliday, D. N. Barney, S. De Witt Bloodgood, William H. Grant, Thomas W. Olcott, Samuel B. Ruggles, James B. Wilson, of New York; Ephraim Marsh, Charles M. Harker, of New Jersey; John Edgar Thompson, Benjamin Haywood, Joseph H. Scranton, Joseph Harrison, George W. Cass, John H. Bryant, Daniel J. Morell, Thomas M. Howe, William F. Johnson, Robert Finney, John A. Green, E. R. Myre, Charles F. Wells, junior, of Pennsylvania; Noah L. Wilson, Amasa Stone, William H. Clement, S. S. L'Hommedieu, John Brough, William Dennison, Jacob Blickinsderfer, of Ohio; William M. McPherson, R. W. Wells, Willard P. Hall, Armstrong Beatty, John Corby, of Missouri; S. J. Hensley, Peter Donahue, C. P. Huntington, T. D. Judah, James Bailey, James T. Ryan, Charles Hosmer, Charles Marsh, D. O. Mills, Samuel Bell, Louis McLane, George W. Mowe, Charles McLaughlin, Timothy Dame, John R. Robinson, of California; John Atchison and John D. Winters, of the Territory of Nevada; John D. Campbell, R. N. Rice, Charles A. Trowbridge, and Ransom Gardner, Charles W. Penny, Charles T. Gorham, William McConnell, of Michigan; William F. Coolbaugh, Lucius H. Langworthy, Hugh T. Reid, Hoyt Sherman, Lyman Cook, Samuel R. Curtis, Lewis A. Thomas, Platt Smith, of Iowa; William B. Ogden, Charles G. Hammond, Henry Farnum, Amos C. Babcock, W. Seldon Gale, Nehemiah Bushnell and Lorenzo Bull, of Illinois; William H. Swift, Samuel T. Dana, John Bertram, Franklin S. Stevens, Edward R. Tinker, of Massachusetts; Franklin Gorin, Laban J. Bradford, and John

T. Levis, of Kentucky; James Dunning, John M. Wood, Edwin Noyes, Joseph Eaton, of Maine; Henry H. Baxter, George W. Collamer, Henry Keyes, Thomas H. Canfield, of Vermont; William S. Ladd, A. M. Berry, Benjamin F. Harding, of Oregon; William Bunn, junior, John Catlin, Levi Sterling, John Thompson, Elihu L. Phillips, Walter D. McIndoe, T. B. Stoddard, E. H. Brodhead, A. H. Virgin, of Wisconsin; Charles Paine, Thomas A. Morris, David C. Branham, Samuel Hanna, Jonas Votaw, Jesse L. Williams, Isaac C. Elston, of Indiana; Thomas Swan, Chauncey Brooks, Edward Wilkins, of Maryland; Francis R. E. Cornell, David Blakely, A. D. Seward, Henry A. Swift, Dwight Woodbury, John McKusick, John R. Jones, of Minnesota; Joseph A. Gilmore, Charles W. Woodman, of New Hampshire; W. H. Grimes, J. C. Stone, Chester Thomas, John Kerr, Werter R. Davis, Luther C. Challiss, Josiah Miller, of Kansas; Gilbert C. Monell and Augustus Kountz, T. M. Marquette, William H. Taylor, Alvin Saunders, of Nebraska; John Evans, of Colorado; together with five commissioners to be appointed by the Secretary of the Interior, and all persons who shall or may be associated with them, and their successors, are hereby created and erected into a body corporate and politic in deed and in law, by the name, style, and title of "The Union Pacific Railroad Company;" and by that name shall have perpetual succession, and shall be able to sue and to be sued, plead and be impleaded, defend and be defended, in all courts of law and equity within the United States, and may make and have a common seal; and the said corporation is hereby authorized and empowered to lay out, locate, construct, furnish, maintain, and enjoy a continuous railroad and telegraph, with the appurtenances, from a point on the one hundredth meridian of longitude west from Greenwich, between the south margin of the valley of the Republican River and the north margin of the valley of the Platte River, in the Territory of Nebraska, to the western boundary of Nevada Territory, upon the route and terms hereinafter provided, and is hereby vested with all the powers, privileges, and immunities necessary to carry into effect the purposes of this act as herein set forth. The capital stock of said company shall consist of one hundred thousand shares of one thousand dollars each, which shall be subscribed for and held in not more than two hundred shares by any one person, and shall be transferable in such manner as the by-laws of said corporation shall provide. The persons hereinbefore named, together with those to be appointed by the Secretary of the Interior, are hereby constituted and appointed commissioners, and such body shall be called the Board of Commissioners of the Union Pacific Railroad and Telegraph Company, and twenty-five shall constitute a quorum for the transaction of business. The first meeting of said board shall be held at Chicago at such time as the commissioners from Illinois herein named shall appoint, nor more than three not less than one month after the passage of this act, notice of which shall be given by them to the other commissioners by depositing a call thereof in the post office at Chicago, post paid, to their address at least forty days before said meeting, and also by publishing said notice in one daily newspaper in each of the cities of Chicago and Saint Louis. Said board shall organize by the choice from its number of a president, secretary, and treasurer, and they shall require from said treasurer such bonds as may be deemed proper, and may from time to time increase the amount thereof as they may deem proper. It shall be the duty of said board of commissioners to open books, or cause books to be opened, at such times and in such principal cities in the United States as they or a quorum of them shall determine, to receive subscriptions to the capital stock of said corporation, and a cash payment of ten per centum on all subscriptions, and to receipt therefor. So soon as two thousand shares shall be in good faith subscribed for, and ten dollars per share actually paid into the treasury of the company, the said president and secretary of said board of commissioners shall appoint a time and place for the first meeting of the subscribers to the stock of said company, and shall give notice thereof in at least one newspaper in each State in which subscription books have been opened at least thirty days previous to the day of meeting, and such subscribers as shall attend the meeting so called, either in person or by proxy, shall then and there elect by ballot not less than thirteen directors for said corporation; and in such election each share of said capital shall entitle the owner thereof to one vote. The president and secretary of the board of commissioners shall act as inspectors of said election, and shall certify under their hands the names of the directors elected at said meeting; and the said commissioners, treasurer, and secretary shall then deliver over to said directors all the properties, subscription books and other books in their possession, and thereupon the duties of said commissioners and the officers previously appointed by them shall cease and determine forever, and thereafter the stockholders shall constitute said body politic and corporate. At the time of the first and each triennial election of directors by the stockholders two additional directors shall be appointed by the President of the United States, who shall act with the body of directors, and *to* be denominated directors, on the part of the government; any vacancy happening in the government directors at any time may be filled by the President of the United States. The directors to be appointed by the President shall not be stockholders in the Union Pacific Railroad Company. The directors so chosen shall, as soon as may be after their election, elect from their own number a president and vice-president, and shall also elect a treasurer and secretary. No person shall be a director in said company unless he shall be a bona fide owner of at least five shares of stock in the said company, execpt the two directors to be appointed by the President as aforesaid. Said company, at any regular meeting of the stockholders called for that purpose, shall have power to make by-laws, rules, and regulations as they shall deem needful and proper, touching the disposition of the stock, property, estate, and effects of the company, not inconsistent herewith, the transfer of shares, the term of office, duties, and conduct of their officers and servants, and all matters whatsoever which may appertain to the concerns of said company; and the said board of directors shall have power to appoint such engineers, agents, and subordinates as may from time to time be necessary to carry into effect the object of this act, and to do all acts and things touching the location and construction of said road and telegraph. Said directors may require payment of subscriptions to the capital stock, after due notice, at such times and in such proportions as they shall deem necessary to complete the railroad and telegraph within the time in this act prescribed. Said president, vice-president, and directors shall hold their office for three years, and until their successors are duly elected and qualified, or for such less time as the by-laws of the corporation may prescribe; and a majority of said directors shall constitute a quorum for the transaction of business. The secretary and treasurer shall give such bonds, with such security, as the said board shall from time to time require, and shall hold their offices at the will and pleasure of the directors. Annual meetings of the stockholders of the said corporation, for the choice of officers (when they are to be chosen) and for the transaction of annual business, shall be holden at such time and place and upon such notice as may be prescribed in the by-laws.

SEC. 2. *And be it further enacted,* That the right of way through the public lands be, and the same is hereby, granted to said company for the construction of said railroad and telegraph line; and the right, power, and authority is hereby given to said company to take from the public lands adjacent to the line of said road, earth, stone, timber, and other materials for the construction thereof; said right of way is granted to said railroad to the extent of two hundred feet in width on each side of said railroad where it may pass over the public lands, includig all necessary grounds for stations, buildings, workshops, and depots, machine shops, switches, side tracks, turntables, and water stations. The United States shall extinguish as rapidly as may be the Indian titles to all lands falling under the operation of this act and required for the said right of way and grants hereinafter made.

SEC. 3. *And be it further enacted,* That there be, and is hereby, granted to the said company, for the purpose of aiding in the construction of said railroad and telegraph line, and to secure the safe and speedy transportation of the mails, troops, munitions of war, and public stores thereon, every alternate section of public land, designated by odd numbers, to the amount of 10 alternate sections per mile on each side of said railroad, on the line thereof, and within the limits of 20 miles on each side of said road, not sold, reserved or otherwise disposed of by the United States, and to which a preëmption or homestead claim may not have attached, at the time the line of said road is definitely fixed: *Provided,* That all mineral lands shall be excepted from the operation of this act; but where the same shall contain timber, the timber thereon is hereby granted to said company. And all such lands, so granted by this section, which shall not be sold or disposed of by said company within three years after the entire road shall have been completed, shall be subject to settlement and preëmption, like other lands, at a price not exceeding one dollar and twenty-five cents per acre, to be paid to said company.

SEC. 4. *And be it further enacted,* That whenever said company shall have completed forty consecutive miles of any portion of said railroad and telegraph line, ready for the service contemplated by this act, and supplied with all necessary drains, culverts, viaducts, crossings, sidings, bridges, turnouts, watering places, depots, equipments, furniture, and all other appurtenances of a first class railroad, the rails and all the other iron used in the construction and equipment of said road to be American manufacture of the best quality, the President of the United States shall appoint three commissioners to examine the same and report to him in relation thereto; and if it shall appear to him that forty consecutive miles of said railroad and telegraph line have been completed and equipped in all respects as required by this act, then, upon certificate of said commissioners to that effect, patents shall issue conveying the right and title to said lands to said company, on each side of the road as far as the same is completed, to the amount aforesaid; and patents shall in like manner issue as each forty miles of said railroad and telegraph line are completed, upon certificate of said commissioners. Any vacancies occurring in said board of commissioners by death, resignation, or otherwise, shall be filled by the President of the United States: *Provided, however,* That no such commissioners shall be appointed by the President of the United States unless there shall be presented to him a statement, verified on oath by the president of said company, that such forty miles have been completed, in the manner required by this act, and setting forth with certainty the points where such forty miles begin and where the same end; which oath shall be taken before a judge of a court of record.

SEC. 5. *And be it further enacted,* That for the purposes herein mentioned the Secretary of the Treasury shall, upon the certificate in writing of said commissioners of the completion and equipment of forty consecutive miles of said railroad and telegraph, in accordance with the provisions of this act, issue to said company bonds of the United States of one thousand dollars each, payable in thirty years after date, bearing six per centum per annum interest, (said interest payable semi-annually,) which interest may be paid in United States treasury notes or any other money or currency which the United States have or shall declare lawful money and a legal tender, to the amount of sixteen of said bonds per mile for such section of forty miles; and to secure the repayment to the United States, as whereinafter provided, of the amount of said bonds so issued and delivered to said company, together with all interest thereon which shall have been paid by the United States, the issue of said bonds and delivery to the company shall ipso facto constitute a first mortgage on the whole line of the railroad and telegraph, together with the rolling stock, fixtures and property of every kind and description, and in consideration of which said bonds may be issued; and on the refusal or failure of said company to redeem said bonds, or any part of them, when required so to do by the Secretary of the Treasury, in accordance with the provisions of this act, the said road, with all the rights, functions, immunities, and appurtenances thereunto belonging, and also all lands granted to the said company by the United States, which, at the time of said default, shall remain the ownership of the said company, may be taken possession of by the Secretary of the Treasury, for the use and benefit of the United States: *Provided,* This section shall not apply to that part of any road now constructed.

SEC. 6. *And be it further enacted,* That the grants aforesaid are made upon condition that said company shall pay said bonds at maturity, and shall keep said railroad and telegraph line in repair and use, and shall at all times transmit despatches over said telegraph line, and transport mails, troops, and munitions of war, supplies, and public stores upon said railroad for the government, whenever required to do so by any department thereof, and that the government shall at all times have the preference in the use of the same for all the purposes aforesaid, (at fair and reasonable rates of compensation, not to exceed the amounts paid by private parties for the same kind of service;) and all compensation for services rendered for the government shall be applied to the payment of said bonds and interest until the whole amount is fully paid. Said company may also pay the United States, wholly or in part, in the same or other bonds, treasury notes, or other evidences of debt against the United States, to be allowed at par; and after said road is completed, until said bonds and interest are paid, at least five per centum of the net earnings of said road shall also be annually applied to the payment thereof.

SEC. 7. *And be it further enacted,* That said company shall file their assent to this act, under the seal of said company, in the Department of the Interior, within one year after the passage of this act, and shall complete said railroad and telegraph from the point of beginning as herein provided, to the western boundary of Nevada Territory before the first day of July, one thousand eight hundred and seventy-four:

Provided, That within two years after the passage of this act said company shall designate the general route of said road, as near as may be, and shall file a map of the same in the Department of the Interior, whereupon the Secretary of the Interior shall cause the lands within fifteen miles of said designated route or routes to be withdrawn from preëmption, private entry, and sale; and when any portion of said route shall be finally located, the Secretary of the Interior shall cause the said lands hereinbefore granted to be surveyed and set off as fast as may be necessary for the purposes herein named: *Provided,* That in fixing the point of connnection of the main trunk with the eastern connections, it shall be fixed at the most practicable point for the construction of the Iowa and Missouri branches, as hereinafter provided.

SEC. 8. *And be it further enacted,* That the line of said railroad and telegraph shal commence at a point on the one hundredth meridian of longitude west from Greenwich, between the south margin of the valley of the Republican River and the north margin of the valley of the Platte River, in the Territory of Nebraska, at a point to be fixed by the President of the United States, after actual surveys; thence running westerly upon the most direct, central, and practicable route, through the territories of the United States, to the western boundary of the Territory of Nevada, there to meet and connect with the line of the Central Pacific Railroad Company of California.

SEC. 9. *And be it further enacted,* That the Leavenworth, Pawnee, and Western Railroad Company of Kansas are hereby authorized to construct a railroad and telegraph line, from the Missouri River, at the mouth of the Kansas River, on the south side thereof, so as to connect with the Pacific railroad of Missouri, to the aforesaid point, on the one hundredth meridian of longitude west from Greenwich, as herein provided, upon the same terms and conditions in all respects as are provided in this act for the construction of the railroad and telegraph line first mentioned, and to meet and connect with the same at the meridian of longitude aforesaid; and in case the general route or line of road from the Missouri River to the Rocky Mountains should be so located as to require a departure northwardly from the proposed line of said Kansas railroad before it reaches the meridian of longitude aforesaid, the location of said Kansas road shall be made so as to conform thereto; and said railroad through Kansas shall be so located between the mouth of the Kansas River, as aforesaid, and the aforesaid point, on the one hundredth meridian of longitude, that the several railroads from Missouri and Iowa, herein authorized to connect with the same, can make connection within the limits prescribed in this act, provided the same can be done without deviating from the general direction of the whole line to the Pacific coast. The route in Kansas, west of the meridian of Fort Riley, to the aforesaid point, on the one hundredth meridian of longitude, to be subject to the approval of the President of the United States, and to be determined by him on actual survey. And said Kansas company may proceed to build said railroad to the aforesaid point, on the one hundredth meridian of longitude west from Greenwich, in the territory of Nebraska. The Central Pacific Railroad Company of California, a corporation existing under the laws of the State of California, are hereby authorized to construct a railroad and telegraph line from the Pacific coast, at or near San Francisco, or the navigable waters of the Sacramento River, to the eastern boundary of California, upon the same terms and conditions, in all respects, as are contained in this act for the construction of said railroad and telegraph line first mentioned, and to meet and connect with the first mentioned railroad and telegraph line on the eastern boundary of California. Each of said companies shall file their acceptance of the conditions of this act in the Department of the Interior within six months after the passage of this act.

SEC. 10. *And be it further enacted,* That the said company chartered by the State of Kansas shall complete one hundred miles of their said road, commencing at the mouth of the Kansas River as aforesaid, within two years after filing their assent to the conditions of this act, as herein provided, and one hundred miles per year thereafter until the whole is completed; and the said Central Pacific Railroad Company of California shall complete fifty miles of their said road within two years after filing their assent to the provisions of this act, as herein provided, and fifty miles per year thereafter until the whole is completed; and after completing their roads, respectively, said companies, or either of them, may unite upon equal terms with the first-named company in constructing so much of said railroad and telegraph line and branch railroads and telegraph lines in this act hereinafter mentioned, through the Territories from the State of California to the Missouri River, as shall then remain to be constructed, on the same terms and conditions as provided in this act in relation to the said Union Pacific Railroad Company. And the Hannibal and St. Joseph Railroad, the Pacific Railroad Company of Missouri, and the first-named company, or either of them, on filing their assent to this act, as aforesaid, may unite upon equal terms, under this act, with the said Kansas company, in constructing said railroad and telegraph, to said meridian of longitude, with the consent of the said State of Kansas; and in case said first-named company shall complete their line to the eastern boundary of California before it is completed across said State by the Central Pacific Railroad Company of California, said first-named company is hereby authorized to continue in constructing the same through California, with the consent of said State, upon the terms mentioned in this act, until said roads shall meet and connect, and the whole line of said railroad and telegraph is completed; and the Central Pacific Railroad Company of California, after completing its road across said State, is authorized to continue the construction of said railroad and telegraph through the Territories of the United States to the Missouri River, including the branch roads specified in this act, upon the routes hereinbefore and hereinafter indicated, on the terms and conditions provided in this act in relation to the said Union Pacific Railroad Company, until said roads shall meet and connect, and the whole line of said railroad and branches and telegraph is completed.

SEC. 11. *And be it further enacted,* That for three hundred miles of said road most mountainous and difficult of construction, to wit: one hundred and fifty miles westwardly from the eastern base of the Rocky Mountains, and one hundred and fifty miles eastwardly from the western base of the Sierra Nevada mountains, said points to be fixed by the President of the United States, the bonds to be issued to aid in the construction thereof, shall be treble the number per mile hereinbefore provided, and the same shall be issued, and the lands herein granted be set apart, upon the construction of every twenty miles thereof, upon the certificate of the commissioners as aforesaid that twenty consecutive miles of the same are completed; and between the sections last named of one hundred and fifty miles each, the bonds to be issued to aid in the construction thereof shall be double the number per mile first mentioned, and the same shall be issued, and the lands herein granted be set apart, upon the construction of every twenty miles thereof, upon the certificate of the commissioners as aforesaid that twenty consecutive miles of the same are completed: *Provided,* That no more than fifty thousand of said bonds shall be issued under this act to aid in constructing the main line of said railroad and telegraph.

SEC. 12. *And be it further enacted,* That whenever the route of said railroad shall cross the boundary of any State or Territory, or said meridian of longitude, the two companies meeting or uniting there shall agree upon its location at that point, with reference to the most direct and practicable through route, and in case of difference between them as to said location the President of the United States shall determine the said location; the companies named in each State and Territory to locate the road across the same between the points so agreed upon, except as herein provided. The track upon the entire line of railroad and branches shall be of uniform width, to be determined by the President of the United States, so that, when completed, cars can be run from the Missouri River to the Pacific coast; the grades and curves shall not exceed the maximum grades and curves of the Baltimore and Ohio railroad; the whole line of said railroad and branches and telegraph shall be operated and used for all purposes of communication, travel, and transportation, so far as the public and government are concerned, as one connected, continuous line; and the companies herein named in Missouri, Kansas, and California, filing their assent to the provisions of this act, shall receive and transport all iron rails, chairs, spikes, ties, timber, and all materials required for constructing and furnishing said first-mentioned line between the aforesaid point, on the one hundredth meridian of longitude and western boundary of Nevada Territory, whenever the same is required by said first-named company, at cost, over that portion of the roads of said companies constructed under the provisions of this act.

SEC. 13. *And be it further enacted,* That the Hannibal and Saint Joseph Railroad Company of Missouri may extend its roads from Saint Joseph, via Atchison, to connect and unite with the road through Kansas, upon filing its assent to the provisions of this act, upon the same terms and conditions, in all respects, for one hundred miles in length next to the Missouri River, as are provided in this act for the construction of the railroad and telegraph line first mentioned, and may for this purpose, use any railroad charter which has been or may be granted by the legislature of Kansas; *Provided,* That if actual survey render it desireable, the said company may construct their road, with the consent of the Kansas legislature, on the most direct and practicable route west from St. Joseph, Missouri, so as to connect and unite with the road leading from the western boundary of Iowa at any point east of the one hundredth meridian of west longitude, or with the main trunk road at said point; but in no event shall lands or bonds be given to said company, as herein directed, to aid in the construction of their said road for a greater distance than one hundred miles. And the Leavenworth, Pawnee, and Western Railroad Company of Kansas may construct their road from Leavenworth to unite with the road through Kansas.

SEC. 14. *And be it further enacted,* That the said Union Pacific Railroad Company is hereby authorized and required to construct a single line of railroad and telegraph from a point on the western boundary of the State of Iowa, to be fixed by the President of the United States, upon the most direct and practicable route, to be subject to his approval, so as to form a connection with the lines of said company at some point on the one hundredth meridian of longitude aforesaid, from the point of commencement on the western boundary of the State of Iowa, upon the same terms and conditions, in all respects, as are contained in this act for the construction of the said railroad and telegraph first mentioned; and the said Union Pacific Railroad Company shall complete one hundred miles of the road and telegraph in this section provided for, in two years after filing their assent to the conditions of this act, as by the terms of this act required, and at the rate of one hundred miles per year thereafter, until the whole is completed: *Provided,* That a failure upon the part of said company to make said connection in the time aforesaid, and to perform the obligations imposed on said company by this section and to operate said road in the same manner as the main line shall be operated, shall forfeit to the government of the United States all the rights, privileges, and franchises granted to and conferred upon said company by this act. And whenever there shall be a line of railroad completed through Minnesota or Iowa to Sioux City, then the said Pacific Railroad Company is hereby authorized and required to construct a railroad and telegraph from said Sioux City upon the most direct and practicable route to a point on, and so as to connect with, the branch railroad and telegraph in this section hereinbefore mentioned, or with the said Union Pacific Railroad, said point of junction to be fixed by the President of the United States, not further west than the one hundredth meridian of longitude aforesaid, and on the same terms and conditions as provided in this act for the construction of the Union Pacific Railroad as aforesaid, and to complete the same at the rate of one hundred miles per year; and should said company fail to comply with the requirements of this act in relation to the said Sioux City railroad and telegraph, the said company shall suffer the same forfeitures prescribed in relation to the Iowa branch railroad and telegraph hereinbefore mentioned.

SEC. 15. *And be it further enacted,* That any other railroad company now incorporated, or hereafter to be incorporated, shall have the right to connect their road with the road and branches provided for by this act, at such places and upon such just and equitable terms as the President of the United States may prescribe. Wherever the word company is used in this act it shall be construed to embrace the words their associates, successors, and assigns, the same as if the words had been properly added thereto.

SEC. 16. *And be it further enacted,* That at any time after the passage of this act all of the railroad companies named herein, and assenting hereto, or any two or more of them, are authorized to form themselves into one consolidated company; notice of such consolidation, in writing, shall be filed in the Department of the Interior, and such consolidated company shall thereafter proceed to construct said railroad and branches and telegraph line upon the terms and conditions provided in this act.

SEC. 17. *And be it further enacted,* That in case said company or companies shall fail to comply with the terms and conditions of this act, by not completing said road and telegraph and branches within a reasonable time, or by not keeping the same in repair and use, but shall permit the same, for an unreasonable time, to remain unfinished, or out of repair, and unfit for use, Congress may pass any act to insure the speedy completion of said road and branches, or put the same in repair and use, and may direct the income of said railroad and telegraph line to be thereafter devoted to the use of the United States, to repay all such expenditures caused by the default and neglect of such company or companies: *Provided,* That if said roads are not completed, so as to form a continuous line of railroad, ready for use, from the Missouri River to the navigable waters of the Sacramento River, in California, by the first day of July, eighteen hundred and seventy-six, the whole of all of said railroads before mentioned and to be constructed under the provisions of this act,together with all their furniture, fixtures, rolling stock, machine shops, lands, tenements, and herediataments, and property of every kind and character, shall be forefeited to and be taken possession of by the United States: *Provided,* That of the bonds of the United States in this act provided to be delivered for any and all parts of the roads to be constructed east of the one hundredth meridian of west longitude from Greenwich, and for any part of the road west of the west foot of the Sierra Nevada mountain, there shall be reserved of each part and instalment twenty-five per centum, to be and remain in the United States treasury, undelivered, until said road and all parts thereof provided for in this act are entirely completed; and all the bonds provided to be delivered for the said road, between the two points aforesaid, there shall be reserved out of each instalment fifteen per centum, to be and remain in the treasury until the whole of the road provided for in this act is fully completed; and if the said road or any part thereof shall fail of completion at the time limited therefor in this act, then and in that case the said part of said bonds so reserved shall be forfeited to the United States.

SEC. 18. *And be it further enacted,* That whenever it appears that the net earnings of the entire road and telegraph, including the amount allowed for services rendered for the United States, after deducting all expenditures, including repairs, and the furnishing, running, and managing of said road, shall exceed ten per centum upon its cost, exclusive of the five per centum to be paid to the United States, Congress may reduce the rates of fare thereon, if unreasonable in amount, and may fix and establish the same by law. And the better to accomplish the object of this act, namely, to promote the public interest and welfare by the construction of said railroad and telegraph line, and keeping the same in working order, and to secure to the government at all times (but particularly in time of war) the use and benefits of the same for postal, military and other purposes, Congress may, at any time, having due regard for the rights of said companies named herein, add to, alter, amend, or repeal this act.

SEC. 19. *And be it further enacted,* That the several railroad companies herein named are authorized to enter into an arrangement with the Pacific Telegraph Company, the Overland Telegraph Company, and the California State Telegraph Company, so that the present line of telegraph between the Missouri River and San Francisco may be moved upon or along the line of said railroad and branches as fast as said roads and branches are built; and if said arrangement be entered into, and the transfer of said telegraph line be made in accordance therewith to the line of said railroad and branches, such transfer shall, for all purposes of this act, be held and considered a fulfilment on the part of said railroad companies of the provisions of this act in regard to the construction of said line of telegraph. And, in case of disagreement, said telegraph companies are authorized to remove their line of telegraph along and upon the line of railroad herein contemplated without prejudice to the rights of said railroad companies named herein.

SEC. 20. *And be it further enacted,* That the corporation hereby created and the roads connected therewith, under the provisions of this act, shall make to the Secretary of the Treasury an annual report wherein shall be set forth —

First. The names of the stockholders and their places of residence, so far as the same can be ascertained;

Second. The names and residences of the directors, and all other officers of the company;

Third. The amount of stock subscribed, and the amount thereof actually paid in;

Fourth. A description of the lines of road surveyed, of the lines thereof fixed upon for the construction of the road, and the cost of such surveys;

Fifth. The amount received from passengers on the road;

Sixth. The amount received for freight thereon;

Seventh. A statement of the expense of said road and its fixtures;

Eighth. A statement of the indebtedness of said company, setting forth the various kinds thereof. Which report shall be sworn to by the president of the said company, and shall be presented to the Secretary of the Treasury on or before the first day of July in each year.

APPROVED, July 1, 1862.

Homestead Law in Operation, 1863

From *American Agriculturist* (New York, July, 1863), XXII, no. 7, p. 207.

The following extracts from a familiar letter written by a subscriber to the *Agriculturist*, indicate how great a benefit has been conferred by the Homestead Law upon thousands who need only a *start* in life to become the independent possessors of a home and the means of support. Let those who from untoward circumstances find it impractible or difficult to make their way by other means, find encouragement in the example here shown. The writer says: ''I failed up in the mercantile business, had nothing left but a span of horses and some household furniture and a few dollars in money, with which I started for Nebraska to take a homestead under the new law. I arrived here in March with just $5 left, took a claim, put up a log house, and went to work. I have 160 acres of splendid land which will make me a good farm, which only cost me $13; and five years' residence on it secures me the title by paying $2 more at the end of that time. I have got 10 acres of land broken up and a good garden started, and am greatly indebted for the latter to volumes of 20 and 21, of the *Agriculturist*, which I bought on the way out where I stopped over night. I have read them thoroughly, and come to the conclusion that I cannot get along without the paper. I have no experience in farming, and when I want information on any point I refer to the paper and am almost sure to find it — consequently I send the dollar for the present year, which I got by working out by the day, and this is the very best investment I can make. I wish some thousands of the hard-worked clerks and mechanics in the city, that have families to support, could know what a chance there is here for them to secure a home and a sure competency. I have been through the mill, and truly can say that I am happier and better contended here in my log house, with the prospect before me of securing an attractive home for myself and children, than I ever was when in successful pursuit of a mercantile business. Here is ample room for thousands — produce of all kinds is high and commands cash at any time. The soil is light loam with a slight intermixture of sand, the country is healthy, plenty of good water to be had by digging 10 to 20 feet, to say nothing of creeks. My claim is on the great military road from Omaha to the mines, upon which hundreds of teams pass daily, laden with stores of every desciption. The middle branch of the Pacific Road (when built) will pass near here, and right here in the valley of the Platte River are thousands of acres waiting for somebody to take them in possession — "to tickle with a hoe, that they may laugh a harvest." Do tell the poor hard-working drudges that barely eke out a scanty subsistance, that here they could be lords of the soil and soon gain an independence.

House Speech on Homesteads for Former Slaves, 1864

From *Congressional Globe*, 38th Congress, 1st Session, pp. 2252–53.

William HIGBY (Rep., Cal.): Mr. Speaker, the attack on this bill from the opposite side of the House has been against the general policy of the confiscation law. That discussion is in reference to a policy already established by law. It is a foregone conclusion and beyond our action, and therefore, to my mind, it is wholly unnecessary to occupy so much time on that point.

Mr. Speaker, the main feature which seems to create a great deal of disturbance here is that one which gives to a man with a dark skin the same compensation for his services which is given to a man with a white skin. Justice to the one class is injustice to another. That is, if a man does a dollar's worth of work for me, if he is a white man I must give him a dollar, but if he is a black man I must pay him less, or else I am raising him to a level with the white man.

Sir, this bill, if it becomes a law, takes away no rights from the soldier, whether he be black or white. He may take lands under the homestead law in other parts of the Union; but the bill provides that if he wishes to take them in the insurrectionary districts, he must take them in accordance with the provisions of this bill, if it becomes a law. It takes away no rights under existing laws, but simply confines him to the terms contained in this bill, provided they see fit to take lands in the insurrectionary districts.

Sir, some gentlemen are very uneasy. The question of equality arises to disturb them. They say if you give them freedom you give them an equal chance for acquiring property; that you then propose to make them voters, and then, of course, follows amalgamation and all the evils their imaginations can conjure up upon this subject.

Some gentlemen have spoken very learnedly and very exclusively here upon this question, when they represent States in which negroes every year go to the polls and vote. How many black men are there to-day in the State of New York who are entitled to go to the polls and vote? Gentlemen pretend to claim that if you secure to these men the rights conferred in this bill, if you guaranty to them pay for their services, you therefore give them all the privileges which the Constitution and the Government bestow upon white citizens. Does it follow that if a black man is paid justly for his services, that he therefore has all other rights which are extended to white men? For that reason I asked my colleague on the committee, when he was being questioned, how many centuries into the future his intentions extended. "Sufficient unto the day is the evil thereof."

We are trying to remedy in some small degree the terrible evil which we now have on our hands. We can see the consequences of this evil in our streets this very day in the thousands of maimed and mutilated soldiers who are pouring through them, and are thronging our hospitals all around us. They are the fruits of slavery. Now, there are thousands of colored men in the State of New York, my native State, who, if there were an election to-day, would have the right to go to the polls and vote. In my earliest childhood I can remember to have seen black men go to the polls and vote. It is true a qualification was imposed — a property qualification. Now, if the Government will tolerate the black man under any circumstances whatever as citizens or inhabitants only, it must encourage the idea of their acquiring property that they may not become paupers and pensioners upon the Government.

The only question is whether such a bill as this should follow those which have already been passed. And I will answer the gentleman from Ohio [Mr. PENDLETON] who asked my colleague upon the committee so many questions as to whether there were any lands now in a condition to be disposed of under the bill if it becomes a law. I would have Congress pass this law if there was not an inch of land in possession of Government under previous laws. We have reason to believe there will be thousands and millions of acres, and possibly within the next six months. I do not know how soon or how late we may get possession of these lands, nor is that the business of Congress. But it is wise, it is proper, it is just, that a provision should be made for the disposition of such lands, should any come into the possession of the Government of the United States under either of the two acts named in this bill. These two laws, as I have said, are now in existence, making provision for the forfeiture of these lands, and are wise provisions, too; and this third one, containing provisions for the disposition of these lands, is something which needs and demands the action of Congress. I hope it will become a law. I move the previous question.

Education and Experimentation

Against a National Agricultural Society, 1841

From *Farmers' Register* (Petersburg, Va., Apr. 30, 1841), IX, pp. 248–50.

For the Farmers' Register.

To the Farmers and Planters of Virginia

April 8th, 1841.

Friends and brethren — Will you permit one of your own fraternity, without deeming him obtrusive, to solicit your attention to a subject deeply interesting, not only to our own class, but to every other in the community, since *their* prosperity is so intimately connected with and dependent upon *ours,* that neither can permanently prosper unless *we* do so — at least in the aggregate. The subject to which I allude is, the establishment, at the city of Washington, of a National Society of Agriculture.

To such of you as are conversant with the history of this vital art, both in our own and other countries, it is needless to dwell long on the well known fact, that there is not now a civilized country upon earth, *except our own,* but what has had, for years past, either such a society, or some similar institution. It is equally well known, that the improvement of each nation in all the various branches of husbandry, has been almost stationary for centuries, before the establishment of such institutions, and rapid thereafter, beyond what any one could have imagined to be possible. Even in China, a country which we, in our self-imputed wisdom, deem almost barbarous, agriculture has always been fostered by the government, and held in the highest honor, ever since there was any authentic history of the country. Are *our* people and country so entirely different from all others in the world, that *we* can prosper without any resort to the means which every other civilized nation has deemed essential to their welfare? I confidently think not; and, with your permission, I will proceed to offer a few reasons to prove that no country whatever is more in need of a national society of agriculture than our own, if indeed there be any that require it so much.

In addition to those general arguments in favor of such an institution, which apply to every country, there are some peculiarly applicable to our own, that seem to me unanswerable and which I beg leave respectfully to state. Certain causes which have a strong tendency to destroy our heretofore happy union have long been operating among us; and, I deeply regret to say, have manifestly been on the increase for some years past, so that it is now quite common to hear men talking familiarly of disunion, whereas it was once considered a sort of treason even to speak of it as a possible event. But this most happy state of things no longer exists. Demons in human shape, whose inmost souls are cankered with lawless ambition, and reckless fanatics, with too little sense to perceive the fatal tendency of their opinions and actions, have been indefatigably engaged in disseminating sectional jealousies and animosities throughout the United States; and the success of these diabolical efforts has been far greater than could well have been anticipated. They have, in fact, most fearfully disturbed that harmony and good will which might have bound us inseparably together, as long as time shall last, had the same efforts been made to strengthen, as have been made to sever those ties of friendship and brotherly love, upon the preservation of which the peace, the prosperity, and the happiness of the American people most unquestionably depend. To annihilate, if possible, these baneful elements of discord, or at least to neutralize them, *ought* to be the paramount duty of our national legislature. But, alas! the members themselves, or rather a considerable portion of them are deeply infected with the deadly poison of disunion, and, of course, appear to have no other object in meeting but to aggravate all the causes of sectional animosities and dissensions, by making party questions and quarrels of almost every subject of discussion that comes before them. This has so often happened of late years, that it is by no means uncommon in these times to hear Congress Hall stigmatised with that most disgraceful, but not unjustly bestowed, nickname, *"bear-garden."* All the consequences of such shameful conduct are bad enough; but the worst of them is, that the hostile feelings thus generated by those misrepresentatives of the people, are carried back by them, or transmitted through their vile party-newspapers, even to the extremities of our union, and multitudes of the people thereof have become so deeply infected with them, as to look upon each other as little better than natural enemies.

If this deadly poison has not already reached the heart's core of our body politic, I beseech you to consider well, whether there is any thing better calculated to stop, or, at least, to mitigate its fatal progress, than a national society of agriculture. This would annually collect, from the remotest states of our confederacy, many of the yeomanry of the land — the very bone and sinew of our country — who, being drawn together by a common interest, and without any possible cause of quarrel, would very soon learn that they had been most grossly misrepresented to each other, and would part, after each meeting, with a desire continually increasing to meet again. All would learn something new to them by such intercommunication; and the professional benefits which each would derive from it would so increase their mutual good will, would so warm their hearts towards each other, that the fraternal regard of the farmers and planters of our country — who fortunately yet constitute a vast majority of our whole population — would continue to "grow with their growth, and strengthen with their strength," in defiance of all the attempts which could be made to destroy it.

And the incarnate devils, who are now laboring so hard to accomplish objects which, if attained, would inevitably dissolve our union, would soon fall into that utter contempt and detestation which their infernal purposes most justly deserve.

And now, my friends, if you approve of the foregoing suggestions, will you not give to the friends of a national society of agriculture some cause to hope for your co-operation in the attempt to establish one at the city of Washington? A meeting for the purpose will probably be held during the extra session of congress, of which due notice will be given, if there is a probability of effecting it. And if only *one* person would attend from each congressional district, or *one or two* from each of our agricultural societies, (to the members of which I particularly appeal,) I have no doubt that a large assemblage, friendly to the object, might easily be convened; for a similar appeal will be made to the agriculturists of the other states, many of whom have already manifested a strong desire to form such a national institution. Thousands of our citizens, I believe, would hail it, not only as the harbinger of rapid improvement in all the different branches of husbandry, but as the peacemaker that would finally exterminate all sectional jealousies and animosities; every element of popular dissension; and would unite in one perpetual league of concord and amity all the different states of our union.

Possibly I may ascribe to a national society of agriculture more power than it could possibly acquire, more extensive influence than it could ever possess. But this self-deception, if indeed it be one, can do no imaginable harm either to my country or myself. I will, therefore, continue to cherish it, and will still hope to witness the trial of such a society before I die. Should such trial be made, and fail, I will then, but *not until then,* acknowledge my error.

On this subject, of a national society of agriculture, and the duty of congress to promote some such establishment, there are some arguments so powerful and conclusive in a late address of Chilton Allan, Esq., the patriotic president of the State Agricultural Society of Kentucky, that I could wish to see them republished in every agricultural paper in the United States; for I have read nothing so well calculated to carry conviction to every mind. The man who could read them, and remain unconvinced, must have an intellect very differently constituted from any thing that I can imagine.

Before I conclude, I must beg our friend Ruffin to tell us what *he* thinks of the foregoing project. A few favorable words from him would greatly encourage the efforts of many others, as well as the hopes and exertions of your, and his old friend,

JAMES M. GARNETT.

P. S. If any of the editors of our political journals are friendly to the establishment of a national society of agriculture, I hereby respectfully ask them to republish this communication, or to give us something of their own, which I should much prefer.

Editor's Reply

Our esteemed correspondent rates our influence and recommendation at much too high a value; but, whether worth any thing or not, our best wishes go for the success of his plan and proposal. But we confess our want of confidence, nay, our despair, as to our government, state or federal, doing any thing for agriculture. Further, we cannot believe our own class, the agricultural interest, could have enough of zeal, public spirit (or even enlightened self-interest,) and energy to perform their part in the great and important work proposed. It might be easy enough to assemble enough individuals at Washington to form a sufficiently numerous "National Agricultural Society;" but, we fear, it would be impossible to induce the proper men to go, and especially from the remote parts of the United States. Moreover, if a body, as well constituted as could be reasonably hoped for, could be assembled for this purpose in Washington, we doubt whether the novel attractions and political excitement of the place would not divert the attention of many of the most disinterested and independent members from their designed labors; and taking the whole body, there would probably be more exertion made by members of the society in using the opportunity for seeking office, or other private benefits to themselves individually from the public purse, than to promote the interest of agriculture and the common weal. If the individuals would not so act, they would form a rare exception to the general course of things in the corrupt political atmosphere of the city of Washington.

We have not examined the question of the constitutional power of the general government to aid the agricultural interest and improvement in this mode, and therefore do not mean to express an opinion thereon. We would readily publish the argument of the subject, from the address of Chilton Allan, esq., but have mislaid our copy. But it is not any constitutional obstacle that will prevent the action of congress for this beneficial object. If there were no such objection, any favored individual could more easily obtain money for some useless job or merely nominal public service, and given solely for his private emolument than the same amount would be appropriated for the most important services to agriculture. Thus, as one example among hundreds, that scientific quack and empty pretender, Featherstonhaugh, though a foreigner, obtained from congress for two years $5000 a year, upon the pretence of geological surveys in the North-Western territories, of which the plan was doubtless devised by himself, and solely for his private benefit. The service was performed by his making a pleasure excursion in a couple of summer months, and the writing a report thereof, (to be published at the public expense,) which report consisted principally of loose generalities, and was more like an introductory lecture of a professor of geology, than presenting precise results of laborious and accurate field investigation. This fat sop thus dispensed by favor, also enabled the recipient to assume and usurp impudently the title of "Geologist of the United States;" and the reputation thereby stolen no doubt helped him subsequently to the very important appointment by the the English government to survey the disputed boundary line, and by which he has been enabled to go far towards embroiling the two governments in war. It is to such applicants as these, who are sufficiently urgent and patient, and truckling enough to the money dispensers, that appropriations are readily made by government — and not to such public objects as the encouragement and promotion of agriculture. The Smithsonian fund will be wasted by congress in jobs for private benefit, just as has already been done with a very large part of the great general education fund bequeathed to the city of Philadelphia.

The Importance of Farming, 1842

From *American Agriculturist* (New York, April, 1842), I, pp. 1–3.

To Our Readers

In introducing ourselves to the public as the conductors of a new Agricultural Journal, we should do injustice to their good sense, not to confess frankly, a diffidence in entering upon a career, which is in the extent and weight of its obligations, new, arduous, and responsible.

It is but a few years since the first agricultural periodical issued from an American press, to flicker for a short period before the listless gaze of public apathy, and then expire for want of adequate support. In like manner several papers of conceded merit sucessively rose and sunk, until at last, from their unrequited efforts, the spirit of our people was aroused to the importance of agricultural information, and the public mind had become awake to a partial appreciation of the necessity of scientific enquiry as connected with the cultivation of our soil, and the general improvement of our farming system. In accordance with the public demand, several useful and well sustained papers have sprung into existence within the last few years, which from the intellegence they have imparted on agricultural subjects throughout the country, have added millions to our national wealth and prosperity.

We too, have come into this ample field as laborers. In doing so, we arrogate to ourselves no superiority of intellect, or higher sources of knowledge than are possessed by our contemporaries and fellow Journalists. We seek only to enlarge the boundaries of that high order of intellegence, which we know the pursuits of an enlightened system of husbandry require, and we trust we shall so demean ourselves in conducting our work as not only to merit the approbation of that public whom we address, but inculcate also, that respect for this noble profession which it so eminently deserves.

The pursuit of agriculture in its broadest sense, it need hardly be observed, constitutes the basis of our national virtue and national wealth. Yet important as it is in the accomplishment of these great objects, and in its truly elevated and dignified character, it has been, and still is, in its real merits, estimated below that of other professions in our land.

The pecuniary troubles, which, within the last few years, have so extensively visited our country, have been produced *almost exclusively* by the neglect of this indispensable basis, on which all other pursuits should rest; and the disastrous fate which other occupations have met, constituting as they have done, a too enlarged super-structure for our agricultural foundation, has taught the present generation, lessons of wisdom which they will not soon forget. We must now begin, once more, to build our social system and make our *foundation sufficiently broad and strong,* and if our intemperate ambition and misguided enterprise induce us to enlarge the edifice beyond the extent of a perfectly adequate support, we must inevitably expect to see it "topple down headlong." There are principles and laws in the social system, as firm and unyielding as in the mechanical, and whenever they are disregarded, we must look for its partial derangement, or entire subversion. That shrewd observer of men and things, Baron Rothschild, foresaw with a noon-day clearness, the tendency of our system in the recent heyday of our fictitious prosperity, when he refused the tempting offer of an American loan at a large premium, for says the sagacious banker, "a nation that ought from its pursuits, to furnish a surplus of produce, can find money neither for principal or interest while it has to buy its bread in Europe."

We need to have the occupation of farming made more popular and attractive; it should occupy a higher niche in popular estimation and in the scale of national employments; it should command not only the cold respect and distant admiration of our active and enterprising business men, but their warmest regard and cordial participation. When this is the state of public opinion and our professional and mercantile and speculating pursuits, are disincumbered of their legions of supernumeraries, our mechanics and manufacturers sustained in their meritorious efforts to supply our country with her own fabrics; then may we expect that real and substantial return of prosperity, which we may otherwise look for in vain.

To aid in directing and stimulating the efforts of our enterprising and patriotic countrymen in consumating so desirable a result, our time and talents will be devoted, and we shall feel amply compensated if we shall succeed in contributing, even in a remote degree, to the accomplishment of an object so dear to every American heart. . . .

Our location being in the heart of a Commercial Metropolis, may cause some apprehesion of a want of that entire devotion to Agricultural matters, which must characterize a Journal that is expected will be in the highest degree beneficial to the Farming community. In answer to any such well suggested objection, we would say, that we have maturely weighed the pros and cons of our position, and it was not till after the fullest deliberation, and the opinion of many of our friends whose judgment we would rather confide in than our own in this matter, that we have decided to take our stand here. We have in this great Commercial emporium, access to information and means for procuring as well as distributing it, that cannot be found elsewhere. More Farmers and Planters resort here than to any other city of the Union, and the facility of procuring back numbers and volumes, as well as those just issued, without the expense of postage, is no inconsiderable item. And our own opportunities for looking over crops and herbs, and visiting our subcribers is greater than it would be in any other place. And to those who know us, we need not assure them, that though we are personally in the city, *our hearts are in the country* as well as our interests. We shall not prove recreant to our trust.

On Agricultural Colleges, 1844

From *American Agriculturist* (New York, February, 1844), III, pp. 52–54

The 19th century presents the singular anomaly, of an age, skilful to a degree beyond any that has preceded it, in all the arts that minister to the comforts and luxuries of man, with the single exception of that art, which is alone the base and support of all others — the art of an enlightened agriculture. All the elegancies of life too, and the refinements of intellectual culture, the useful and recondite sciences, literature, poetry, music, painting, and sculpture, have been patronised, illustrated, and studied, under every advantage, and have thus been pushed far toward their maximum of improvement, yet is the foundation of this varied and beautiful superstructure, the only portion of the edifice which is destitute of strength, order, symmetry, or design. And if we look back through the history of the ancients, reaching, according to the most approved chronology, much farther than 6,000 years, we find no record from which we can learn that any branch of the world's ancestry has been wiser, in this respect, than their descendants of the present day.

We shall not attempt to account for this gross and most inexcusable neglect, beyond the effect of that principle, which may almost be taken as an axiom in human conduct, that man's exertions are withheld, just in the ratio of the Deity's munificence. Supreme Benevolence has wisely provided for the success of the humblest efforts of unenlightened reason, in its struggles to procure from the earth the elements of subsistence; and on the very threshold of this success, have all human efforts been arrested. Content with having achieved the bare means of existence, the human mind has been stayed in this vast field of inquiry; and has turned away from it, if not with loathing, at least with indifference, and with a keen and delighted relish to other and less important and less praiseworthy objects of ambition. Whence comes this lack of reason, this short-sightedness to our own best interests? We must acknowledge ourselves incompetent to give the answer, and we gladly assign the solution of this difficult problem to our modern philosophers who are so worthily busying themselves with "the law of progress;" from them alone must light come, if it come at all.

Whatever the cause may be, certain it is, that the world has hitherto taken but the initiatory steps in the art of agriculture; and this broad land, like the western hemisphere in the days of Columbus, remains a terra incognita, an unexplored continent, inviting the most intellegent research, and ready to repay its explorers, with the highest rewards. It may be true, indeed, that portions of this goodly land have been heretofore discovered by the Northmen of preceding times; and even inhabited by a refined race of Aztelans possessing a high degree of culture; yet to the present race of man, no chart or history has been bequeathed, to point out its location or well-defined boundaries.Whatever discoveries may have been made in this great art in the early ages of the world, by the Egyptians, or other early civilized nations, who possibly, may have inherited from the antediluvians, a science and practice far beyond any thus far reached by successive generations — it is certain, that modern inquirers must re-discover it for themselves, if they wish now to have it in possession.

We would not be ungrateful for the worthy and efficient service, rendered, since the commencement of the present century, by the devoted sons of genius, who have given a portion of their time to the elucidation of the principles of agriculture, and who have begun a systematic investigation of the laws of nature, that needs only to be followed up, rigidly and unremittingly, to result in all the benefits which may be fairly demanded at their hands. But, we ask, what has been the success in this all-important pursuit, that will compare with the improvements in the mechanic arts, as shown in the application of steam, machinery for the manufacture of the different fabrics from wool, cotton, silk, the metals; and the various other new and important aids rendered to the useful occupations of the present day? With the facilities afforded by the above inventions, one person can now do as much, as could have been accomplished by twenty, without them, only 40 years ago. Can any approximation to such improvement be shown in the cultivation of the soil? We speak not of the mechanical instruments of the farm, which have measurably, and perhaps to the extent which could reasonably have been expected, participated in the modern progress of improvement.

Our meaning is much broader and deeper, and includes the whole science of agriculture, in all its varied phases and relations. We look to, and demand for agriculture, that enlarged and liberal measure of discovery, which will enable the human race to provide sustenance for its thousand millions of inhabitants, now covering the face of the earth, destined, probably, hereafter to be indefinitely augmented; with an approximation to that certainty and success, that attends human labor in the other departments of life. We prepare our land and sow it to wheat, or plant it in corn; and after much doubt and uncertainty, reap from the first an average, in these United States, probably, not exceeding 14 bushels; and gather from the last, not more than 20 bushels per acre. Yet we have seen under favorable circumstances, that the farmer has yielded 80 bushels, and the latter over 180 bushels per acre. We claim, that abating somewhat for the accidents of seasons, unusual droughts, humidity, or frosts; or perchance, the destruction following upon the eccentricities of the elements, as a hail-storm, or whirlwind, on an ungarnered crop, we might look for the highest results from every well-directed effort, with the same confidence that we now look to the attainment of any given speed from a steamboat, after providing it with a suitable model, engine, and fuel; or the weaving a definite number of yards by a powerloom, properly constructed, and moved by the requisite force. To accomplish thus much, we have but to place our soil, and seed, and culture, in the same precise conditions, that have once been so successful; and yet how seldom is this achieved, even on the same field, and under the same direction as may have been before employed.

If we look beyond the discoveries hitherto applied, and bring to the science of agriculture such analogies as are appropriate to the subject, as shown from the progress of human invention in other departments of enterprise, we may reasonably expect developments in aid of this object, which would now be considered as perfectly Utopian. What brilliant results may yet crown the researches of the devotee of agricultural science, and what green and enduring wreaths of glory are destined to circle the brow of genius, who may hereafter successfully explore this hitherto almost untrodden wastes. And how the comforts of this world, and its means of subsistence will be multiplied, when all the aids to its cultivation are rendered, which mankind have a right to demand.

We have then our deficiencies for the present and past, and our hopes for the future pointed out. Where are the remedies for the former, and the proper and reliable foundations for the latter? First and mainly, it may be answered, in bringing the right minds to the just and full consideration of this subject; and secondly, and as a necessary sequence to the former, the application of the requisite amount of funds, which shall secure genius of the highest caste, under all the circumstances of advantage, essential to its fullest effect.

Briefly, and in a form that all may comprehend, we say; we want an agricultural institution, founded and arranged on the best principle which can be dictated by enlightened experience, sound judgment, and a shrewd common sense; and so guarded, as to be unassailable by the corruptions of party, and beyond the reach of any hostile innovations of the fickle multitude; *and such an institution should be endowed with a permanent fund of one third, to half a million of dollars.* In this institution, we would place a chemist and geologist; an anatomist and physiologist; a botanist; an entomologist; and a practical agriculturist, who should give embodiment and effect to the suggestions of science, and run each out to a clear, distinct and definite result. These professors should be such as the choicest spirits of the age could afford; surrounded with all necessary assistants, books. . . .

With such an institution, how long would it be, ere the tyro in agriculture could go to it, with the same certainty of receiving the requisite information, that the mariner now does in consulting his chart and compass? The slow and dangerous coasting, amid shoals and breakers, that now mark out his benighted course, would at once give way to bolder movements, and more direct and certain success. Thus guarded, thus endowed, and thus filled, such an institution would revolutionize the practice of agriculture within the present age, and more than double the products of the earth, with the same labor and expense now devoted to them. Is this not an object worthy the legislation of statesmen, or the munificence of intelligent and patriotic individuals?

But with legislatures constituted as at the present day, we can not, probably, look to a single one of our 26 state governments, for the object desired; and as for Congress, nothing can be hoped from that quarter. From $10,000,000 to $12,000,000 is the amount of our annual *peace* appropriation for war; and this preparation for human butchery, is all legitimate and proper; but an appropriation of one twentieth of this amount, to feed the hungry, and clothe the naked, and carry comfort and consolation to the diseased and destitute, the aged and infirm, and afford thrift and abundance to all, would, in the opinion of our strict constructionists, rend our constitution to tatters. Verily the extremes of human wisdom and folly, like the continued extension of the arch of a circle, finally meet in the same point.

Hopeless then, as may be the realization of the desired aid from any present legislation, we have to expect it, if at all, from individual bounty alone. Here indeed is a glorious field for immortality, for one sufficiently enlightened to grasp it, and the man who shall have the good sense and liberality, to found the first Agricultural College on the enlarged and munificent plan proposed, will secure a fame for all coming time, before whose brightness that of an Alexander or Napoleon would become dim, or distinguishable only by its intensity of darkness.

We must confess our hopes in the beneficial results of the present efforts in the cause of agriculture — our inquiries and discussions — our treatises and periodicals — our agricultural premiums and shows — come up to this extent, and scarcely more: they are awakening the public mind to a sense of its deficiencies; they are discovering the vacuum which yet remains to be filled. They are the crepuscular light which heralds the coming morn, but they are not the glorious effulgence of the king of day. But his approach is indicated beyond the possibility of doubt; and ere long the world will be in the full enjoyment of his benignant rays.

On The Advancement of Agriculture, 1846

From: *American Agriculturist* (New York, April, 1846), V, pp. 106–07.

Agricultural Colleges and Schools.

To Legislators throughout the United States — or rather to their constituents, as the former are merely their servants — their waiting echo.

The establishment of Agricultural Schools and Colleges by our legislative bodies has been repeatedly urged in these columns, but hitherto, like many other important things, without success. Although hopeless of securing any present aid for these most praiseworthy objects, by our National or State Legislatures, we yet deem it incumbent upon us, as conductors of a public journal, whose sole object is the advancement of the agricultural classes, to reiterate and re-urge this question. If we cannot for the present induce any favorable action from those who are delegated to enact laws, we can bring the subject to the attentive consideration of those who select their representatives for this purpose. We thus hope to enlighten public sentiment on this all-important matter; and if *the people* are once awakened to its importance, they will see to it, that their representatives do not long continue to neglect their interests. If they *will not lead* in a measure of such vital consequence to this, the largest interests of the community, they *can be driven* into it when the people have become aroused.

It is somewhat strange, and entirely unaccountable on any other principle than narrowness of views, discreditable to the age, or the utter subserviency of our leading men to the behests of party, that they cannot take this single step in advance of the practice of past ages, and assume the responsibility of maturing and carrying out a measure fraught with so much benefit to the country at large, as would result from the establishment of one or more institutions, which will bring to the minds of adults as well as youth, the great principles and the most approved practice of Agricultural Science. Should our law-makers vote an amount perfectly adequate to the purchase of a suitable experimental farm and the erection of proper buildings, apparatus, &c., and engage some able men to carry out the objects of the undertaking, can it be doubted that the farmers of the great State of New York would not most fully sustain it? Are they for ever to remain the hewers of wood and drawers of water to every other class in the community, and see thousands annually devoted to the higher branches of education in other professions, and they not be allowed to receive a meager percentage of this outlay, for the necessary improvement of their own profession and interests? True, they can participate, in common with others, in the higher walks of academical and collegiate education, provided by the munificence of the State; yet they will find there is little to fit them for their own peculiar sphere.

The discovery of a new world of agricultural science has burst upon this age; and order and design are found to govern, by the exactest principles and laws, every one of nature's operations. Many of these principles and laws have been detected by the ablest scientific explorers of Europe, such as Davy, Chaptal, Boussingault, Liebig, Johnston, and others. These discoveries, and what are destined to succeed them, will give to agricultural pursuits a precision and advantage, similar to what followed the discovery of the magnet in maritime affairs. The farmer has, from time immemorial, been groping in the dark; he know only what experience revealed to him; and even from this he often drew false conclusions, from not being able to comprehend all the premises, and the most ordinary operations of nature. With well known, indeviating principles with which to work, the Agriculturalist could push boldly into the ocean of experiments, and calculate, with uneering certainty, his latitude and longitude, and the precise distance he was from any given point, instead of slowly coasting along dangerous, dreary coasts, in continual fear of shipwreck. The difference between the practice of a farmer of the last century and of one in the age to come, will not be less than the difference between a voyage from Liverpool to Boston by the Cunard line of steamers, and a coasting voyage from the same point, by the Scottish coast, the Okrneys and Shetland isles, Nova Zembla, the Polar ice, Greenland, and the north-eastern coast of America, in the ancient craft of the Carthaginians or freebooting Danes and Swedes.

Talk of the agricultural intelligence of this age! Why, it is merely this; some few intellectual men of other countries — scarcely any of our own, — have just overstepped that horizon of darkness, which has hitherto hedged in the world, and made a few preliminary discoveries! What a poor amount of agricultural knowledge is this! It is in the spirit of the age, and should be peculiarly that of this country, which boasts of its intelligence, to carry out by every means in its power, so laudable, so intelligent, and withal *so money-making a scheme,* — for *"money"* is the talismanic word we are forced to use. But so little light has hitherto penetrated among our agriculturists, that five-sixths of the most intelligent of them will tell you, that "the new of the moon is the time for this, the first quarter for that, her fulling for another, and her waning for something else." Not even a *Farmer's Almanac* will sell, without a mystic figure installed as the key of nature, surrounded by the Zodiac and its signs, whose converging rays indicate the hidden secrets of nature, and expose the whole cycle of her operations! Astrology, that has been abandoned by the world at large for two hundred years or more, is good enough to reveal the mysteries of the farmer's art, embracing, as it does, almost all the laws of nature! Out upon the *twaddle,* and more unmeaning *gibberish* than nursery maids deal out to nurslings, when they tell the farmers, "their dear constituents," — "the bone and sinew of the land" — "the most enlighted class of the country," — "the expectancy and rose of the fair State." and other holiday and juggling terms, with which they are smothered, adding that "*they* do not require any assistance from art, they have only to plow and delve, and cast in their seed, and nature will do the rest — they need but pay their taxes for the support of the State, and others will take care that government and the professions are well looked after.

Were our own views carried out, we would appropriate at once, half a million of dollars for the founding of an Agricultural College and experimental farm, the interest of which should for ever be devoted to the employment of the ablest professors the world affords, whose whole genius and attainments should be devoted to discoveries in this art, and in teaching them to our most intelligent youth. We would invite the Liebigs, and Boussingaults, and Johnstons, and Bakewells, to occupy chairs in the institution, *at salaries which would command their acceptance;* and as soon as others could be appropriately filled by American genius, who should be pressed into service to the full extent of the demand. Minor establishments should receive encouragement and support, and every pecuniary aid which could facilitate the discovery and dissemination of agricultural science and art, should be freely and liberally granted. We should then see the beginning of the end of the shameful neglect of the agricultural class; we should be able to console ourselves with the reflection, that we had at least made whatever effort was in our power, to accomplish the greatest good to the greatest number.

What say you, farmers, to this proposition? Shall anything be done or not? If anything is to be accomplished, you will have to make the first move. You must *command* your delegates to give you from their loaded coffers, some small part of the means *you* have so liberally provided for them, that you may be able, from its judicious expenditure, to supply still more. You have only to set about this in earnest, and the object is already accomplished.

The above was written for our March number, but unfortunately crowded out. By reference to the proceedings of the American Agricultural Association, it will be seen that one of our citizens has generously offered the free use of his farm for five years, for the benefit of an Agricultural School. This farm is in the finest possible order, and one of the best in this vicinity. Its buildings also are very complete, and nearly new. We hope others will be stimulated to follow this munificent example. If public bodies will not move, let private bodies do so, and the former will soon emulate their example. It is painful to think of the wealth which is annually lavished on vanity and folly in this country, which might, if the owners would but will it, be devoted to the glorious cause of the advancement of the science and practice of agriculture.

Agricultural Education

From *Prairie Farmer* (Chicago, September, 1847), VII, pp. 266–67.

For several years, many of our newspapers and agricultural clubs have had much to say of the establishment of agricultural schools, and in one or two instances have thrown themselves upon the subject as a sort of hobby. We have rather stood aloof from the discussion, because we have not very clearly seen the aim at which they were driving, and have been under the impression that those who were most eager in the business, did not themselves see their way with entire plainness.

If we have understood the ideas generally put forth, they have been to set up single schools, in which agriculture, theoretical and practical, should be the whole study. The models for these schools are to be found in Great Britain, or on the continent of Europe. If this be the object aimed at, we are free to say, that we regard the plan with little favor. Nothing, it seems to us, could sooner put such a conceit out of the head of Americans, than the perusal of Mr. Colman's account of these schools. The idea of making a man a farmer, and so educating him that *he can be nothing else,* may do for that country; but it will never do here.

The facility with which men here change their occupations, though productive of some evils, and many times abused, is a peculiarity belonging to our country, which it is not desirable to change, until there is less change in other things. In England, a man works all his life at one branch of business; and if he is thrown out of that, he is as helpless as a child of ten years. We do not desire to see men thus educated here, whether the education be agricultural or in any thing else. The duties of an American citizen require that he shall possess a fair knowledge of a great many things. Particular and thorough knowledge of one kind, and no knowledge of any thing else, may give him power in that direction; but it exposes the individual and society to perils which ought not to be risked. We do not want lawyers, who know nothing but law; nor ministers of the gospel who know nothing but theology; nor physicians who know nothing but physic. Their chief force may and ought to be expended in the line of their professions; but they must be properly instructed in other departments of knowledge; that the mind and character may be balanced; otherwise they may become dangerous members of society, if by any chance they are thrown out of their sphere. Nor do we want farmers to be educated to know nothing but farming. What is wanted of all professions and pursuits is, that they know *one thing intimately, and others well.*

Our system of education has hitherto been mainly right. It has been to teach children the principles of things, to instruct them in the art of thinking. Hence in our higher schools and colleges, the aim is not so much to perfect the student in the minutiae of a single branch of knowledge, or to impart a *great amount* of knowledge, as to instruct him in the elements of many branches, and qualify him to go forward himself, in any direction he pleases, after he leaves the institution. In short, the object is to enable the student to educate himself. Foreigners often amuse themselves at the number of studies to which our pupils attend; and some among us declaim about it, mainly because it differs from the plans of other countries; but we maintain, that though not free from evils, the *plan* is right, and adapted to our circumstances; and that it would be folly to substitute for it any other.

It is possible that we do not do full justice to the views of those of whom we speak; but that their plan is to establish what may be called agricultural *colleges* — institutions to which young men may repair after finishing the ordinary routine of studies, for the study of agriculture, both as an art and a science — something like our law and medical schools. If this be the plan, we are not prepared to say but that it may at some future time, be desirable. But matters in this country are not ripe for it yet, as will be plainly developed by the failure of all such projects for years to come.

An obstacle, sufficient to defeat this plan is, that the *science* of agriculture is not yet sufficiently perfect, to stand by itself as a system of university instruction.

Its truths are often too much in the dark to be taught to young men, as scientific truths are usually taught, in the higher schools.

Then there are not at present teachers to man such institutions; but as we might be treading on tender toes to enlarge here, we will be content with the mere statement of the fact.

But not to deal merely in objections — there is manifestly a better plan — one which was recommended in this journal several years ago.

The country is already, in a measure, supplied with schools. Common schools exist in some shape and perfection in all the States. Then there are private and select schools, academies, seminaries, and high schools; and still further, colleges and universities. Now why cannot the study of agriculture, in its different branches, be introduced into all these schools? We maintain that it can.

It must be remembered that agriculture is no single study, like arithmetic, or mental philosophy; but that it is based upon and divided into a great number of branches of study, varied and diverse in character. There is perhaps no division of education which consists in maintaining an acquaintance with things, which does not bear upon it. Those elementary branches, necessary to educate the farmer, are necessary to educate every other man; and hence up to the higher and more abstruse of its scientific principles, all should be educated together.

The studies which belong to an agricultural education are, arithmetic, geometry, natural philosophy, algebra, mensuration, chemistry, geology, mineralogy, physiology of vegetables and animals, and political and domestic economy. To these, at some stage of the course, others might be added — such as would impart a through acquaintance with the domestic animals, practical and particular husbandry, and analytic chemistry.

Is there any difficulty in attaching any or all of these branches to our present school systems, according to their various grades? Is not the proper place for them to be taught in our schools, academies and colleges? It is hard to believe that there can be two opinions held by candid persons acquainted with the principles and details of youthful instruction. Yale and Cambridge have already departments attached for instruction in scientific agriculture, and others will undoubtly follow.

It may be advisable, and undoubtedly will be found so, to modify in the higher schools the course of instruction for those who are to become farmers. A college course of study might easily be formed, leaving out principally, or wholly, the dead languages, which consume much time, and substitute those which teach instead a knowledge of *things* — leaving it optional with the student, at his entrance, to pursue the present, or the new course. The two courses much of the way would be indentical.

Should the students of the new class increase to warrant it, colleges of that sort might be established for them as demanded; and these, having grown out of the demand, would be permanent. It is in general much easier and safer to reform and remodel old institutions, than to revolutionize for new ones. All sweeping and radical changes in education we look upon not only as undesirable but chimerical.

There is no rational doubt, that a general introduction into our courses of education, of such studies as bear upon agriculture, is demanded by the general interests of society, and will be demanded by society itself, so soon as the case is made clear to its apprehension.

Agricultural Science, 1850

From Ezra Graves, "Agricultural Science," in New York State Agricultural Society, *Transactions*, 1850, (Albany, 1851), X, pp. 237–39.

The human mind, like the human body, requires a change of exercise and employment, to fit and prepare it for usefulness to the world, and make it agreeable to those with whom it mingles, and useful to those whose happiness and welfare depend upon its exercise; and he who fails to cultivate his reasoning powers for any other purpose than merely to subserve his own private notions, or his own cupidity and selfishness, deserves not his habitation in a land of freedom, and in a country and climate so happily calculated to invigorate and strengthen every effort of the human intellect, to dispense joy, diffuse knowledge, and to cast into the circle of a broad and extended acquaintance a halo of light, a beacon of intelligence, and the oil-can of tranquility and satisfaction. And while our common schools, academies and colleges are shadowing forth their benign influence, and giving to scholastic genius a brillant and enchanting view of the last and most elevated round in the ladder of fame, the world is becoming satisfied that the cultivation of the soil upon which God placed his creature man is an elevated, and important science, the beauties of which are not seen or enjoyed by the freshman in the class of Agriculture.

It may be asked, what is meant by agricultural science? and how is the science to be acquired, and who are the teachers? I mean, by agricultural science, an analysis of the soil, and a comparison of its yearly products; enabling the careful learner always to fit and prepare his ground to meet the demands of the seed that he sows, or plants, and qualifying him to select the suitable field for the particular species of grain that he intends to raise with such prudence, foresight and experience, that he will be amply compensated for his toil in an abundant crop. And how is this science to be acquired? I answer, would it be difficult for the agriculturist, or horticulturist, to determine what his lands were best calculated to produce, who should sow three kinds of grain in the same field, the grounds of which are equally well fitted, if he watched with care the growth, ripening and harvesting of each kind; and then estimated the quantity and the value of each, in proportion to the labor and expense incurred upon each? There are but few crops in this country but what pay the cost of growing them, and very many pay from fifty to a hundred per cent., and not unfrequently even more than this. If the farmer can make the experiment successfully upon one field, I see no good reason why he cannot upon his whole farm. These experiments made, and entered in a journal, which should be kept by every farmer, he would be able at all times, by referring to his journal to ascertain the year, and the crop that he had upon a particular part of his farm, and upon every field of his farm; thus enabling him to fit and prepare his land for that kind of crop which pays the best, with the least labor and expense. And is this knowledge not worth acquiring? And allow me to ask how this labor-lessening and money-saving knowledge is to be acquired by the common farmer, with his common education, in any other way? The agricultural chemist, who by chemical appliances, can analyze the soil, and determine its properties, aided by botanical researches, can determine the different properties of the different productions in the vegetable kingdom, and the soils that contain the ingredients congenial to the growth of the plants, grains and herbage, which he wishes to produce. And this knowledge of which I have spoken is to be desired, because it is always reliable, for the earth is like its Divine Creator, ever ready to yield its blessings to those who justly merit them; and the husbandman who pays the soil for its benevolence to him, by judiciously manuring, and skillfully fitting it, and adapting his seed to the proper field, will never look upon his cultivated fields and charge them with having extorted from him labor and toil, without returning a liberal equivalent. And who that has toiled and sweat upon lands made barren by his repeated drafts upon her, without feeding her with even the stalks of her own productions, has not learned the sorry lesson, at the expense of his personal health and comfort, as well as by the chagrin and mortification with which he is met, by finding the productions of his farm too scanty to fill a barn of small capacity, or granary of limited dimensions.

On An Agricultural Bureau, 1852

From *Praire Farmer* (Chicago, December, 1852), XII, pp. 535–36.

Once or twice we have alluded to this important question. We have said little, because we expected nothing, until after the business of President making. Well, that has been accomplished, and the dominant party in Congress has been successful, and will soon hold the reins of Government, and continue to control National Legislation. We are measureably satisfied with this result, because, personally we have little to do with politics, and less with parties, and in our editorial capacity, nothing at all. But here is a question of great national moment, and one in which our readers are specifically interested; and we should doubtless be neglecting a specific duty, were we to remain comparatively silent much longer.

The proposed measure is essentially a national and democratic one, and that party is in the majority, and its leaders have no longer an excuse for neglecting to act upon it: for they will not now "strengthen the hands of the present administration," by doing justice to the Agriculturalist. — And it is to be hoped, too, that sectional feelings will yield to the general good.

As a class we are three or four fifths of the people; and all admit that ours is the great interest of the nation; and we are represented in the machinery of our Government by a petty clerkship, located "in a cellar room" of the Patent Office!!

Is there any need of further argument? Does our immense and controlling interest deserve nothing more? If our wise men think we are fitly represented, we have nothing more to say. We shall work on, and hope on, as we have heretofore done; and trust to the all powerful march of progress. Perhaps we can afford to wait; and we certainly feel too indignant, to beg as a boon, what we should claim as a right. Men learned in the statistics of our agriculture are of opinion, that we are not loosing much over twenty or thirty millions per annum, in the value of misdirected labor and unnecessary deterioration in the fertility of our cultivated lands; and as long as we possess new territory and a virgin soil to fall back on, this will never be missed, any more than the hundred million or so, which we might make, and ought to make, over present profits were a right direction given to our whole routine of cultivation.

But seriously; before this paper can reach all of its readers, the working session of Congress will have commenced; and there are good men, and just men in Congress, as well as out of it; and there are those who watch the signs of the times, and shape their policy accordingly; and if we urge this measure now, it will be carried, ere three months have passed over us.

Our press is most respectable, and not without influence; our associations are now counted by the hundred, if not by the thousand. We do not advise Farmers to go before Congress in the ordinary form of petitioners — we have no time for it; and furthermore, we believe it beneath our dignity; as the most numerous class, representing the most important and neglected interest of the nation; and who will say that a just and honest pride hath not its rights?

Still, our associations should memorialize Congress, and our press should speak the sentiments of its supporters. — And we would furthur suggest, that our reading and thinking men, (and they are many) should correspond with the members known to them, and thus gain friends by the weight of opinion, and help to direct the details of legislation, by suggestions drawn from experience and reflection.

With men of science, there is no question in regard to the utility of a department of Agriculture, in our Government; and the only weighty one, urged by politicians, is the fear that a department of the kind would degenerate into a party machine, in the hands of party politicians.

This we can never permit; ours must be an economical WORKING DEPARTMENT — a school of Agricultural Science, and a Bureau of Agricultural statistics, rather than a portion of the Government, following the changing destinies of parties. Hence, we have been in favor of an AGRICULTURAL BUREAU, instead of an unweildy DEPARTMENT, with its Secretary in the Cabinet.

One word more and we have done. The Chief, and the heads of desks, in such a Bureau, should all be selected for their peculiar fitness, and should be able to carry with them the confidence of our reading men and Scientific Agriculturalists, rather than the good will of a party; and then they would be enabled, in addition, to show themselves duly grateful for their places, by so performing their duties as to reflect the more credit on the appointing power, the more useful and satisfactory their labors might prove, to the great interests subserved.

Without some such tacit understanding, we would not give a rye straw for a "Department," while with it we fully believe that a simple Bureau would satisfy the simple desires of the great mass of our Agricultural population.

The Operation of the Agricultural
Bureau of the Patent Office, 1857

From *DeBow's Review* (New Orleans, 1857), XXIII, pp. 77–85.

The astonishing progress which has been made in all the branches of natural science during the last thirty years, is owing in no small degree to the overthrow of those pedantic and intolerant monopolists of science and learning, who, while speaking *ex cathedra*, very seldom recognized any but their own authority, or gave credit to any system, theory, view, and experiment, other than such as emanated from the school to which they belonged, or of which they were regarded the shining lights.

Any one conversant with the progress made in Physics, Chemistry, Geology, and Physiology, knows of the desperate struggle and severe trials, which the modern masters of science had to contend with against the old oligarchy of professors and scholars, before they could succeed in revolutionizing the vast field of applied sciences. These old fashioned gentlemen fell upon Liebig, when he first startled the world with his bold views, daring theories, and ingenious hypothesis, and his contemporaries, in other branches of natural sciences, did not fare better.

However, the revolution has been accomplished; and, at this day, there is in the scientific world of Europe no other authority recognized but what can show cause for the pretence on the very face of its teachings, investigations, or conclusions. Humboldt, Faraday, Agassiz, Liebig, Dumas, themselves, form no exceptions to the rule.

It is our purpose now to present a few critical remarks on the operations of the Agricultural Bureau of the Patent Office, as they are developed in its "Reports."

The man of science, as well as the practical agriculturist, was accustomed to date the advent of the period of a more rational theory and practice in agriculture from the days of "father Thaer," or the latter part of the last or the beginning of the present century; but the compiler of the Agricultural Report frequently rejoices in referring to the sayings of Cato, Pliny, Palladius, Columella, and others of the old Latin writers. The men of our days, to whom we are indebted for the scientific structure and more rational development of the art of culture, must feel some surprise at this. One might just as well, in treating of astronomy, physics, chemistry, and other branches of the natural sciences, resort to Pythagoras, Archimedes, Galen, Hermes, or the old Egyptian magicians, for his principal authorities. Was it the object of the founders of the Agricultural Bureau to make it the archives of all the historical rubbish of the dark ages — the infant tune of science and industry — a source of every description of information however destitute of any real value to the agriculturalist of our days and of our land? While we know that the Romans, Greeks, Egyptians, and other ancient nations had not the means furnished by the very modern sciences of geology, chemistry, and physiology, to look into the nature, the laws, and the conditions of vegetable life, it is indeed quite immaterial to our farmers and planters to learn what Cato, Theophratus, or Pliny may have thought in this regard. But if even worthy of note, it is certainly beyond the proper province of the Agricultural Bureau to convey this kind of knowledge. The object of this Bureau, as we regard it, is to examine into the condition of our national agriculture, and by pointing out its deficiences, and ascertaining and proposing the certain or probable remedies, to devise the means and ways for its general or special amelioration and improvement.

There are undoubtedly many good things contained in all of the ''reports,'' but they are so mixed up with incongruous, contradictory, impractical, trifling, and very often erroneous statements, suggestions, and propositions, that their value is greatly impaired, and not seldom entirely paralyzed. So far we can see in these ''reports'' nothing but cheap and unfair competitor to our statistical and agricultural periodicals, which does not possess the adaptability and practical value for specific regions, of which many of the latter may be proved.

The suggestions of the Commissioner, in the volume for 1855, relating to the system to be adopted in promotion of the objects of the Agricultural Bureau, we are not prepared altogether to contradict, though it may be doubted among other things if the ''assessors'' will be suitable agents for procuring ''*reliable* annual statistics.'' If one remembers the stubborn and persistent opposition which the marshals of the census encountered on account of the peoples' horror against taxation, it naturally suggests itself that such will be still more the case with assessors.

The propriety of the introduction of the valuable meteorological tables of Professor Henry in the Agricultural Report may be questioned. This is going back to elementary and abstract principles, with which the mind of the farmer and planter may not be troubled successfully. In agricultural matters the science of climatology is chiefly destined to aid in establishing the laws for the natural geographical range and distribution of the plants, by showing the analogies of various regions in the character of their seasons, the changes in temperature, of winds, moisture, etc., in the course of the year. Beyond the study of these varying or parallel characteristics of certain agricultural districts, the practical cultivator is not likely to go. But as this science is in its infancy yet, it appears that its exposition could find a more appropriate quarter. The views of Prof. Henry, as set forth in his article, ''Meteorology in its connection with Agriculture,'' no matter how ingenious, are hypothetical to a great extent, and however much praise the author deserves, I am decidedly of the opinion, that the communication is in the wrong place. We can, further, not agree with the Commissioner of Patents, when he calls the chemical analysis of soils and products ''a supplement to these meteorological investigations,'' insisting ''that the full purpose of the latter cannot be carried out without a resort to the former.'' We maintain that any competent judge would rather express the opinion that the objects of chemical analysis of soils, products, and ashes, may be more fully and conclusively secured if mere meteorological observations are once reduced to certain immutable and generally understood laws.

Many of the statements in regard to the results so far obtained from this wholesale practice of distributing ''seeds and cuttings,'' are made on *ex parte* experience, and the anticipations attached to them, to say the least, are, in most cases, sanguine and premature. It is surprising, that the culture of certain trees, shrubs, and herbs is recommended on the score of our yearly importing the useful parts of them to the amount of such and such value from other countries.

Are we to make a China or Japan out of this country? or are we afraid that England, France, or Spain, in case of war with us, would close up all the seas to our merchantmen? When was it ascertained that an import of raw and manufactured materials is injurious to the prosperity of a country, which is already gradually advancing towards a period when foreign countries have of necessity to buy more from it, than they can hope to sell in return? Shall commerce decrease or be arrested, in order to make our land the receptacle of nature's gifts all over the globe, if such a thing were attainable? We certainly do not object to a bed or garden of the "almond," "cork oak," "prune," "liquorice," "vanilla," "box wood," and a thousand such, as matters of taste or ornament here and there, but we deem it exceedingly idle to encourage such experiments *for economical ends,* on a large scale, so long as we have need to impress on the minds of our farmers the necessity of an improvement in the culture of the very first staples, which constitute the wealth of the country. If we dare take it for granted, that a cultivation of the "opium poppy," "palmated rhubarb," "asafoetida," "malabar cinnamon," and similar medicinal plants, would, after all the experiments made in southern Europe and elsewhere, give no satisfaction to the scrutinizing pharmacologist, it must still sound amusing to find included in the list even the Iceland moss, common all over northern and middle Europe and cheap as dirt, and which in all probability grows unnoticed in equal abundance in many districts of our more northern States. Indeed having just taken up the dispensatory of the United States, we learn that the plant is found in the northern latitudes of the old and *new* continents, and on elevated mountains further south. It is also abundant on the mountains and in the sandy plains of New England. What is said of this "Iceland moss" in respect to its cheapness, must be said of the equally recommended "orris root," "quassia," "rhatany," and others which are paraded in the "report." Why not introduce with the same show of reason, all at once, the whole batch of plants found in the materia medica? So far has the compiler of the "report" transgressed the limits of his proper province, that he does not shrink from giving the medical and technical properties, and corresponding applications of some of his pet plants, which properties any one desirous of learning, can be found more correctly and professionally discussed by calling for the Dispensatory at the first drug store in reach. It is new to us that the "quassia" is narcotic, because it kills flies! The *recipes* for making tooth-powder, port wine, tincture, quassia beer, or for cooking chesnuts à la France, would be better fitted for a six-penny "golden book" or "household treasure," than for being made part of the contents of the report of the National Agricultural Bureau!

That we import $30,000 worth of castor-oil, does not prove, that the "palma christi" is not extensively cultivated in some of our States, where happily some enterprising fellow-citizens have grown rich by pressing the viscid liquid from the castor-beans. The import goes to show, what is founded in fact, that the people of the United States consume more castor-oil, than all the world and the rest of mankind together. To have this, our own experience approved by incontestable authority, we refer once more to the pharmacoepia of the United States, which says: "That the ricinus is perhaps in no country more largely cultivated than in the United States."

It is further a fallacious view, that our wheat lands "average twenty bushels to the acre," the yield, according to the last census, as well as to more recent information, being not quite *ten bushels per acre.* And, pray, what American farmer would be led to plant the "Persian walnut," because "the product of each tree will be about one bushel of nuts in twelve or fifteen years after planting?" As to the laudations bestowed on the oil, obtained from the kernel of the Persian walnut, we must respectfully suggest to substitute and use, if needs be, the oil of the indigenous "Juglans," which is abundantly found in the forest of the Canadas and of all the northern, eastern, and western States, and which oil enjoys similar properties with the one from the "Juglans regia."

In another place the "report" maintains, that if we were to cultivate the "opium poppy," and to raise a surplus, it could be sent to China in exchange for tea. An odd proposition, if it be taken into consideration, that, but for the prohibitory laws of China, that plant would there prosper as "hardy" as in Hindoostan. Persia, or Arabia. The English East India Company has besides taken upon itself the disgraceful business of smuggling upon the poor children of the Celestial Empire that wicked drug in large quantities, and we would not advise our human and liberal people to rival John Bull in the execrable traffic.

But there is hardly an item in that portion of the "report" to the introduction of which we could not take exception. If, as another instance, it is presumed, that we might save the cost for the imported Russian and Chinese rhubarb, "if its culture were successfully prosecuted here," we have but to answer, that no competent medical man substitutes, at this day, and after fifty years trial, the French, English, or German rhubarb for the Asiatic article. Above all others, the recommendation of the culture of the "asafoetida plant" is ludicrous and trifling; for, as a medicament, its use is exceedingly limited, and as to its spice or relish, it is certain that our people will not indulge in the taste of the Persians, but adhere their kindred, yet less offensive onion and garlick.

Experiments in planting the "Malabar cardamom" and similar spices, would prove, doubtless, unsuccessful. But if there were any reason to expect satisfactory results, we would recommend in their place the cultivation of the more extensively used and consequently more lucrative Java coffee, Ceylon cinnamon, mace and nutmeg, cloves, ect.; though we should regret to state, that the natural geographical range of all these vegetable tribes is exceedingly limited, and does not, therefore, encourage the hope of a successful experiment.

Now, in order to point out the foregoing mistakes in the Agricultural Report of the Patent Office, we had scarcely need to review more than a portion of the introductory pages.

Space does not at present admit of any further extension of these remarks, though we have advanced but a very short distance into the report. There is one subject, however, upon which something may with great propriety be said, considering the interest it has excited in this country.

The Agricultural branch of the Patent Office claims no little merit for the introduction of the sorghum or Chinese sugar cane, which it is said and thought will destroy the present monopoly in that most necessary article of consumption. At the same time, a statement has appeared, and is going the round of the public press to the following effect:

"CHINESE SUGAR CANE. — At the annual meeting of the Boston Natural History Society, on the 6th, Dr. A.A. Hayes read a paper on the Chinese sugar cane, in which he concludes that the sorghum cultivated in this country does not secrete cane sugar, or true sugar, its saccharine matter being purely glucose in a semi-fluid form. For sweetening properties nearly four pounds of this glucose would be required to equal one of true sugar; but as a raw material for the production of spirit, and as an addition to the forage crop, the plant may be found to have a high economical importance. Prof. John Bacon confirmed the results at which Dr. Hayes had arrived. He was unable to obtain any crystals of sugar cane in the sorghum."

Whether the opinion of Professors Bacon and Hayes be correct or not, it furnishes another instance to demonstrate the deficiency in the operations of the Agricultural Bureau. Before any new species or variety of culture plants is diffused over our rural districts, and the farmers invited to spend their time, labor, and money in experimenting, their value should be tested first by some competent experimenters. If it turns out that the Chinese sugar plant furnishes *grape* instead of *cane* sugar, the Bureau is guilty of a very serious mistake. How, for one moment, could any one overlook that most important question — what kind of sugar is it which the sorghum yields? The Chinese and other people may be content with the quality of the saccharine principle contained in that plant for their economical wants and tastes; or, more probably, may it serve them the same purposes that the sweet fruit juices, raisins, figs, plums, and others do in our households; but all this would not warrant the recommendation of the culture of the plant as a substitute for the sugar cane of our Southern States. We have no objection to let the sorghum pass for a very excellent fodder plant, and hope that further experience will prove that it is that without any deleterious effort on the soil's constituents. But it is more than we can digest if the packages containing the seeds are at once labeled — "good for fodder, green or dry, *and for making sugar*." So far we have only seen that the stalks yielded *syrup*, a fact which by no means warrants the production of a sugar equal to that obtained from the sugar cane.

We are, however, not at all astonished to see endorsed by the Bureau opinions like that from one of its correspondents, who, indeed, did not attempt to make sugar, but has no doubt that it can be made from such a syrup as the one obtained from the sorghum juice, and "more and better sugar, too, than the Louisiana cane does yield." In the report of 1854, the sorghum is introduced in the following terms: "The great object sought in France in the cultivation of this plant is the juice contained in its stalks, which furnishes three important products, viz: sugar, which is identical with that of cane sugar, alcohol, and a fermented drink analogous to cider. The juice, when obtained with care, by depriving the stalk of its outer coating or woody fibre and bark, is nearly colorless, and contains merely sugar and water, producing from ten to sixteen per cent. of the former." Much of the correspondence displayed in the reports reflects no credit on the intelligence of either the author and the endorser; it sounds like clap-trap, and in some instances reminds one of the adage — "tickle me and I'll tickle you."

There is, however, no doubt but that the "sugar millet" is already a favorite with abolitionists, who fervently hope that it will deliver them from the necessity of using the slave-made sugar of Louisiana and Texas. We know the sentiments of some of the correspondents of the office on the sorgho. One, Mr. F. Alunch, from Warren county, Mo., was accustomed to leave his farm last summer and mount the stump in behalf of the woolly horse. He is in earnest with negro free-love, and although he did *not* make sugar from the sorgho, hints upon the propriety of inventing a suitable machine to crush the stalks, being probably unwilling to apply the instrument now in use at the South, because they sometimes crushed negroes!

The truth is the South has so long been accustomed to hear of great discoveries likely to put an end to her monopoly in the great agricultural staples, that
she instinctively doubts when they are mentioned or presented from the usual
sources. The upland rice, the flax cotton, and, in the event, perhaps, the Chinese
sorghum will take their places in that same catagory.

Now that the importation of new cuttings of the sugar cane has so signally
disastrously failed, for whatever reason, the writer of this article will close for the
present with an extract from an address submitted by him to the Southern members
of Congress, and published before the expedition sailed:

"Is it advisable to have the experiment made with new cuttings on a 'liberal and
extensive scale,' what is equivalent with a costly plan, without having previously ascertained
whether that will or will not likely prove to be the remedy! Is it mere child's play to have an
agent sent to 'Venezuela, Guiana, Brazil, East India, Mauritius, or Java,' in order to bring
home new varieties of sugar cane, the planting of all of which may in the end not give the
desired and expected satisfaction, because there was no necessity for them, they did not
constitute the proper remedy to be applied? If either the degeneration of the plant, or the
exhaustion of the soil, or mismanagement and want of rotations, or the scarcity of the
application of manures, or any other is the cause of the unsatisfactory yield of the sugar cane
in Louisiana, it strikes one very forcibly, that before all other contrivances for an amelioration, the remedy should be sought at home.

"How can it be done? Let us select some spots in the sugar region of the United States
where the sugar cane seems healthy, or the crop is satisfactory; a similar course is adopted in
districts where the contrary is the case. If we can manage it to embrace in our research a
dozen cane fields in various situations and of various qualities, so much the better. If not done
before, we have to institute at the same time a series of climatological observations. Next we
have to ascertain the physical and geological condition of the surface of the various cane
districts, and then to enter upon a very careful chemical examination of the surface soil and
sub-soil. Finally, we analyze likewise the ashes of the sugar cane varieties planted on
different soils, and compare the results of the whole series of investigation. We shall,
thereupon, learn to a certainty whether the conditions for satisfactory results from the soils of
those various cane fields, to which the experiment extended, are uniform and equal, or they
are not. If the latter is the case we will be enabled to supply the deficiency without much
difficulty. The soil is brought up to the standard of that of a healthy and prosperous sugar
region, and the same variety of cane planted again. Is the same discrepancy still showing
itself, we are justified to ascribe the cause to other influences besides the deficiency in the
composition of the soil, and the want or excess of moisture.

"Without pursuing such a systematic and well-understood course we remain most of
the time in the dark, and losses are encountered and labor and capital thrown away without
avail.

"Many diseases fostering upon culture plants, as the appearance of pernicious insects, the growth of parasitical vegetations, are very often but the result of the exhaustion and
subsequent inadaptability of the soil, and disappear with the restoration of the latter to its
former standard.

"We must never forget that as little as animals can feed or be healthy on a insalubri-
ous air, on bad water, or insufficient and spoiled food, as little can plants be prevented from
becoming diseased and degenerated whenever a remarkable deficiency or change in their
principle nutriment, *the soil's constituents,* occurs.

N. B. — It was stated at the head of this article that guano was being
distributed from the agricultural rooms of the Patent Office. This guano is the
product of Baker's island, in the Pacific, and under the protection and encouragement of Government, was intended to be brought into competition with, or perhaps
altogether supersede the Peruvian article. The State chemists of Maryland, having

submitted to severe tests the samples received from the Patent Office, pronounce the article to be destitute of any practical value. Could not the office itself have ascertained this fact, making thus an appropriate use of its revenues, instead of holding out deceptive hopes so long to the public? The Maryland chemists say:

"It is evident from the analysis that the American guano from Baker's island is, as to composition and general character, identical with the common Mexican guano of the West Indies. It represents a Mexican guano of excellent quality; inasmuch as its total amount of phosphoric acid (39.11 per cent.) is equal to 85.37 per cent. of bone-phosphate of lime, a per centage which is seldom reached by Mexican guano, and surpassed only by Columbian.

"The almost total absence of ammonia in this guano (like in Mexican) makes it unfit for comparison with the Puruvian guano."

An Agricultural State University

From *American Farmer* (Baltimore, April, 1857), XII, pp. 318–20.

Letter I

We have observed in a recent issue of one of our city papers that a committee, consisting of five members, have been appointed by the National Agricultural Society to present to the consideration of Congress, the practicability and necessity of establishing in each State an Agricultural University. Of the character of the gentlemen who constitute the association, which has delegated these gentlemen to solicit Congress, as well as of the gentlemen who have been delegated, any comment would be superfluous. They are known over the wide extent of the land. The advantages of the proposed measure, it will be observed, unlike many, will be confined to no locality, but will extend to all of the States.

As a Marylander, and one, too, who knows something of the claims of Agricultural interests upon the favorable consideration of Congress — who knows, moreover, by experience, something of farming, a few years of my early life having been spent in that pursuit — I feel it to be a privilege, through your press, to lend my aid in advancing interests so fraught with consequence to our national prosperity.

It was remarked by no less a man than General Jackson, that the "products of the soil and our minerals were among the most productive sources of wealth to our nation." That this is true cannot be questioned. The experience of the nation most abundantly confirms a theory upon the recognition of which depends, in no small degree, our wealth and prosperity as a nation.

Among the means of securing thrift and enterprise to our commercial and mechanical interests, does the culture of land stand prominent? If this be apparent, its tendency, and that, too, its proximate tendency can be none the less so, to wit: the augmenting our national resources. There is no merchant who intelligibly pursues his vocation but knows that the value of the commodity with which he traffics depends upon the *abundance* or *scarcity* of those products which form the *bases* of wealth. Of what value is money unless a corresponding and substantial equivalent be found?

There is no mechanic who pursues his trade understandingly who can do otherwise than admit that its activity depends upon something behind the immediate demands for more houses, dwellings and steam engines, &c.

Of what value is your Baltimore and Ohio Railroad if there be no grain or stock to be conveyed over it? Were the soil of the West undeveloped, need there be any occasion for travelling?

If the first proposition be true, which we are presumptuous enough to believe none will controvert, the sequel of the same can be none the less so, to wit: — the necessity of encouraging those men and those interests which develop the staple products of soil that constitutes one of the bases of our wealth. How shall these interests be encouraged?

The answer, so far as it extends is categorical. Establish State Universities, wherein shall be taught to Farmers some of the great and *fundamental principles* which must be *known* and *observed* in the *culture* of land, and which, because *neglected,* after a series of years, must at last be *stumbled* over and *tested*. This is a day of progress, when the profession and vocations of men must be made *specialities*. If a man is to become a *lawyer* he must be prepared as such. If a surgeon (which though under a category with that of a *physician*,) is nevertheless such a department of that profession as needs special attention in order to eminence, in addition to a scientific knowledge of the human system, his hand must be skilled in all of the uses of the instruments. If a *merchant*, he must pass through an ordeal in the counting house. If a *banker,* he must become an adept at the counter. If a *farmer,* does he need no preparation?

If such a University be established what will be the advantages accruing from the same?

It will afford facilities for the instruction of men in the science and practice of farming. It will secure for each State a reservoir of intelligence. It will elevate a vocation which by some, because not understood, has been looked upon as a subterfuge for the indolent, or as a successful mode of securing a living to the industrious.

Suppose only a few can directly reap the advantages of such an institution? Will not these farmers thereby educated exert an influence over those by whom surrounded, and thereby secure one of the many advantages anticipated by the creation of such an institution?

But, says some practical farmer, when we come to place this land under cultivation, this "book farming" will not suit. When we come to plough up these old lands in our own State, where scarcely a spear of grass can be found, or where if sedge, the symbol of *poverty* is visible, it serves only for a *shelter* to rabbits.

I denounce as much as any one this *book-farming* if carried to an extreme. Being ignorant of the science, did you ever, after the introduction of guano into our State, misapply this manure? How long was it ere the true process in its application was fallen upon? Instead of applying it upon the unproductive soil, in virtue of which an ingredient was placed into the soil necessary to vegetation, we find would-be *economists,* making the application of this great fertilizer upon the land to which had already been applied lime.

Though after this, in some measure efficient, the fact of its greatest efficiency upon unimproved and untilled soil was at last as an accident stumbled upon. This is one illustration of the advantages accruing from a University established for the purposes *above* specified. By means of this, farmers, if not *stubborn* in their adherence to their old and stereotyped processes of farming, will have an institution whence will emanate information necessary to be known to an efficient and successful culture of their lands.

The advantages accruing to commercial, mechanical, or professional interests indirectly, I shall not specify. In my next, I shall atttempt to show that Congress, in view of her immense revenue arising from the sales of public lands, has the ability to establish these Universities. The practicability of the measure and in view of the universality of the claim, it *cannot* be easily rejected by that body.

In conclusion, allow me to say, that the farmer has interests which must be conceded to him, and we too, my fellow-countrymen, have interests involved in his success.

FRIEND OF EDUCATION.

Letter Second

In my first communication, which has already reached the public, I attempted to show, so far as was possible in so small a space allotted to a newspaper article, the important relation sustained by Agriculturists to other departments of the body *politic,* be it mechanical or commercial; I moreover intimated that in proportion as the resources of our soil are developed, just in such proportion do we find other *interests* enhanced, *business* brisk, and *money* abundant. In the same, I alleged that one of the conditions of a thoroughly and properly developed soil was based upon the idea of the land being intelligently tilled — that farmers were not to stumble upon principles necessary to be known and observed in the culture of lands, but must learn and apply them. How are they to do this? Can any more happy expedient be fixed upon than the one proposed by the National Agricultural Society, to wit: the establishing in each State of an agricultural college? It is now my intention to show that *Congress has the ability, in view of the immense revenue accruing from the sale of public lands, to make the appropriation asked for: that the measure is entirely practicable, interfering with no principle which should regulate their action, and that in view of the universality of the claim, cannot be easily rejected.* To any one conversant with the financial condition of our general government, argument would be unnecessary to prove the ability of Congress to do what has been proposed to be asked of that body. The allegation is all that is necessary. The immense amount of funds now in the treasury, and our almost fabulous accessions annually, are glaring facts. She is *able* — the nation's coffers are full. Is such a scheme practicable? In other words, will such a step conflict with the previous policy of the government, or of administrations?

Does not this touch some of the *favorite* theories of your party distinctions? Has not this question of internal improvement been one about which political parties formerly differed? To all of these, and a host of other questions, which are incidental to what we have here affirmed, we cannot now furnish answers. The appropriation proposed to be asked for by the *national* society is for all of the States. It is not for Virginia or for New Jersey — not for Maryland or for Delaware, but for each and every one. It comprises the whole of the States: it is, then, a universal measure. There is one feature in this whole matter which we should not lose sight of, to wit, the claims of the old States for their quota of the benefits ensuing upon the sale of government lands. It is now one of the principles of our general government, and a commendable one too, that so soon as a new territory has become settled to such an extent as to warrant said territory to apply for admission as a State, to admit the same, and to make such an appropriation of public lands to said newly-admitted State as, when judiciously disposed of, or improved, promise princely endowments for the objects appropriated. Do you wish an illustration? Look at the school funds of the new States. This is all proper. It is an evidence of our nation's appreciation of this glorious institution.

The next expedient is the appropriation of each alternate section of the public domain towards the encouraging the building of mammoth railroads. To this we make no objection. The latter opens avenues over which the products of the interior find their way to market; these contribute largely to the settling and developing our new and wide extended territory. The former furnishes the new commonwealths with the means whereby they are enabled to prepare their youthful citizens for becoming useful and intelligent men. But how is this? Are those *old parent* States upon the Atlantic coast, which fought for and achieved our country's independence, in virture of which all of this vast extended territory becomes the common property of us all — how in this, I ask, are they receiving, directly or indirectly, benefits equivalent to those which are secured to the emigrant to the new State? How are we, as one of the old Colonies and States, receiving similar benefits from our common patrimony? 'Tis true, we may have a school fund, such as received its first impulse from a tax upon imported liquors; from the appropriation or purchase of a few hundred acres of land in the old counties of the State in its early history; perhaps a school fund from a tax imposed upon Banks, or from the appointment of the surplus revenue from the general Government in '26. But in any one of these Atlantic States, what is it? Have we any railroads built from public lands to develop those portions of our States remote from markets, except what private individuals or corporations built? Are not our farmers emigrating from the East to the West? Is not our land, in the same ratio of the emigration, depreciating? We want some of the proceeds of the sales of our common territory, and we are entitled to it. We want the farmers of our own States to know that these lands of ours can be made to produce crops far beyond what their most sanguine expectations would have supposed. Such is their seclusion they would have never dreamed that on this side of the mountains, there are lands which can be made to produce as abundantly as those in the remote West. These are in proximity with the great commercial emporiums of our great nation and with the sea. Shall they be vacated by their occupants or their children, who go to seek a more luxuriant soil far removed from their homes and old associations? Is this not the tendency of the times? Is it not palpably so of the farming part of our community? Here our lands, many of them though unproductive, because worked down by the process of farming in vogue with the old farmers, in juxtaposition to market and the sea, and near the old domestic fire-sides of so many of us, are poor, but can be made, by a proper and judicious application of lime, guano, &c., to produce crops more luxuriant than when the land possessed its pristine verdure. These "old fields" can be, and are now being renovated and made to "blossom as the rose." In the remote West lands are cheap and productive, though far removed. Need our lands be unproductive when nature and man have laid their contributions at the disposal of the intelligent farmer? To accomplish this, I then conceive that one of the most eligible means is an appropriation by Congress to each of our States of five hundred thousand dollars, with which an Agricultural School could be established in each State.

FRIEND OF EDUCATION

Education In Rural Districts

From *Southern Cultivator* (Augusta, Ga., July, 1857), XV, pp. 218–19.

In an able and interesting report, made by the President of the Virginia State Agricultural Society, PHILIP ST. GEORGE COCKE, Esq., we find the following earnest appeal in behalf of popular education in the Old Dominion: —

> Seventy thousand of our adult population can neither read nor write! And these, too, are 'bone of our bone, and flesh of our flesh;' they are Virginia's sons and daughters! In the name of humanity! in the name of all that is generous, unselfish and noble in our nature! in the name of country, of Christianity and of God! will the farmers of Virginia any longer permit the existence of this deplorable state of ignorance. If my humble voice could be heard beyond this assembly, I would say to the farmers of Virginia, consider tha your children, aye, the descendants of the richest of your present number, will inevitably in after generations be numbered amongst the poor. Transport yourselves then, in imagination, but thirty, forty, or fifty years into the future, and whilst you yet live, make yourselves the tender and blest fathers of the poor, and shed abroad your hearts and means until every child within the limits of our broad Commonwealth shall, at least, have the advantage of a a Free School education.

We rejoice to see the leading minds in the noble Mother of States so devoted to the cause of educating the poor, who are unable to educate themselves. They are often the descendants of the wealthiest families; and in future, changes from affluence to poverty are likely to be more frequent still, as the fatness of wealth with its indulgencies, and the sharpness of want with its energies, make rich men poor, and poor men rich, as rapidly as the seasons change from Spring to Summer, Summer to Autumn, and Autumn to Winter. But we prefer to let the eloquent Virginian be heard in our editorial columns, rather than our humble selves, in the matter under consideration. He says:

> It is a very remarkable fact, that amongst all the numerous and varied pursuits of man, the very one of those pursuits which has the most intimate, the most extended and often the most recondite connection with all the laws of physical nature, with all science, with all art, in short, with the whole range of knowledge — a pursuit, too, upon which depends the subsistence and the very existence of the human species — upon which is based the well-being, the happiness, the progress and prosperity of individuals, of States and of nations. It is

remarkable, I say, that the pursuit of agriculture should be the *last* and the *least* to benefitted and advanced by all the vast progress that has been made in other departments of skill, knowledge and industry. And why is this? First, the science and art of agriculture having their infinite connections, near and remote, with all knowledge, the general subject is more difficult to be understood and fully known, as it is one of the most extensive and recondite that can engage the human mind; and in the next place, because throughout all history, and in every country, the very men who are most engaged and interested in agriculture, have been precisely those who have been least cultivated and improved by means of scholastic exercises and education suited to their pursuits.

The last remark above quoted hits the nail square on the head. It is no reproach to farmers to lack mental culture, when they had no fair opportunity to attend school so as to acquire a good education. That fact, however, is no good reason why they should not vote for giving all coming generations a better chance to improve the noble faculties of our common nature than they enjoyed. Without some material increase of knowledge, our future progress must be in desolating the land we cultivate, not in making it more fruitful. But hear Mr. COCKE'.

In our Southern States, the entire class of proprietors or cultivators of small landed property, the managers or overseers having in a great measure the more immediate supervision and control of the landed estates of wealthy proprietors, are universally and utterly ignorant of every abstract principle of physical or natural science.

And it is reasonable to believe that the loss to Southern agriculture each year, in consequence of this lamentable state of ignorance, if such loss could be prevented, and could the amount so saved for a single year be appropriated and applied to educational purposes, that it would itself be sufficient richly to endow as many Agricultural Schools and Colleges as are required by our Southern States. When we contemplate the vast amount of ignorance, the total want of education existing amongst the mass of agricultural population of our State, we shall be at no loss to conjecture that the pecuniary loss to Virginia from this cause is immense indeed.

The writer of the above proposes to add three agricultural professorships to the University of the State; which would be a valuable addition to its educational force, although, in our humble judgment, to teach the profession of agriculture properly, it should be divided among not less than six professors or the least number employed to teach the profession of medicine in Colleges. Some of the Agricultural Schools in Europe have thirteen professors; but six in this country would do the work, if duly qualified, in a worthy and effective manner. Public opinion is growing up to demand the advantages which schools devoted to the elevation of tillage and husbandry as an enlightened calling may readily afford, and place within the reach of all. Every friend of improvement in agriculture and in those with whom it is a profession, should speak out on the question. If the president of every agricultural society would take the high ground occupied by the President of the Virginia Society, he would be every where sustained by the Agricultural Press, and our country would soon be in advance of all other nations in both the science and the honors that legitimately appertain to this the greatest interest of mankind. It has something substantial to build upon; something most enduring to uphold the wisdom and virtue devoted to the supply of its manifold wants. Its friends should have *faith,* and work accordingly. We have often wished that we had an efficient Agricultural Society to do what we once hoped that the United States Agricultural Society would achieve. The writer labored more than a year to get that institution organized at the seat of the Federal Government. Our object was not shows of fat oxen or babies, but to reach hundreds of thousands and millions by cheap publications, and thus create that kind of popular sentiment which supports every well-considered effort to increase our agricultural knowledge. To make the human family think, and think to the consummation of a good purpose, they need not a little plain talking to in order to set their best thoughts in motion. We know from personal experience that there is a remarkable affinity between mental rust and *rusticity*. A little more rubbing and scrubbing of the intellect is needful on many a plantation to keep it bright. We find an agricultural library valuable for its daily conversation, its cheap and pleasant instruction, and the interest it awakens in even the dullest routine of the isolated farmers life. As social companions, books conceived by the best minds the world has ever produced, are alike above all praise and all price. It would be an improvement of our home philosophy if it made us think a little more of the soul and less of the body, and not dwarf and peril the former while we push the latter recklessly into an untimely grave.

L.

Founding of Maryland Agricultural College, 1858

From *American Farmer* (September, 1858), XIV, pp. 65–66.

With no ostentatious display, but with sobriety and decorum, on a fair and pleasant day — the twenty-fourth of the last month — and in the presence of a goodly company was laid, by Charles B. Calvert, Esq., the President of the Board of Trustees, the corner stone of the Maryland Agricultural College. The stone was of granite, with a square opening excavated in the centre, in which was deposited by Dr. J. O. Wharton, the Register, a covered box of copper, containing the daily papers of Baltimore and Washington, a copy of the act of incorporation, lists of the officers of the Institution, of the stockholders, and of all those, of every degree, engaged in the construction of the building, specimens of all the coins of the United States of the present day, except those of gold, the county papers of Prince George's county, (in which county the College is situated), specimens of corn, wheat, rye, oats, cloverseed and timothy, in glass vials, sealed, and a specimen of tobacco; and last, but, we flatter ourselves, by no means least, the August number of the American Farmer.

Mr. Calvert preceded the ceremony of laying the corner stone — which was done by himself, with mortar and trowel, handled in very artistic style — by an address, which was everything that the occasion should have called forth —manly, feeling, terse, appropriate and eminently practical. He referred to the unpretending character of the ceremony at that time, and remarked, that it was so because the Trustees wished to avoid the delay that the attempt to have a more imposing celebration would involve; but, he hoped and believed, that when the building was completed, that then, as would indeed be fitting, large delegations, not only of the farmers themselves, but of the farmers' wives and daughters, from every part of the good old State, would be present, to honor by their presence, the commemoration of such a happy consummation. Then would be the true time for the great rejoicing. Meanwhile let each Trustee, and every earnest friend of the Institution, exert every effort to collect the funds that would be required to complete the structure according to the original design; and, though the foundation then in process of being laid in their presence, was for a building one hundred and twenty feet in length, fifty-four feet in width, and five stories high, yet this was but one wing of the noble edifice that it is their purpose to construct. Two or three dollars from each Farmer in the State, would give to the Institution the funds they required to establish, on a firm basis, a College where the farmer would be taught to know and feel that he too, as well as the lawyer and physician, has a *profession*. Other States had undertaken enterprises in some respects similar to this College, but to Maryland was in truth reserved the honor of being the pioneer in founding an Institution that in its plans and purposes was original and American, and not a mere copy of European models. This was not to be only an American edition of an European Manual Labor School — its aims and its destiny, he trusted, were to be far higher, and its results would be felt, if not immediately, yet ultimately, in the increased weight and influence that its impulses would give to the great, but not sufficiently co-operating, Agricultural class.

Mr. Calvert commented with just severity upon the course of the Senate at Washington, the past session, in abolishing their Agricultural Committee, and did not spare a severe criticism upon the inconsistent course of squandering the public lands in reckless donations to undeserving objects, whilst the great interests of Agriculture, and Agricultural educational institutions, were overlooked. Money could be appropriated to West Point, to supply officers for the army, whose occupation was to destroy life — means could be found to support a Naval School, at Annapolis, the duty of whose scholars would be, when called upon, to destroy life — thousands could be appropriated to aid in laying a telegraphic cable upon foreign soil — and all this was constitutional and proper; but when appropriations were sought for objects in aid of the greatest and most valuable of all interests — the Agricultural — the Constitution was immediately appealed to, and the liberties of the land were in jeopardy, if more than a scanty pittance was doled out, barely sufficient to buy a few seeds and specimens of plants, under the auspices of the Patent Office. But let Agricultural Colleges be established, and there would arise a class of men in their midst, whose voice would be heard in the halls of Legislation, and who would secure for the farming interest those rights they are now denied, for want of sufficient advocates. With an earnest appeal to his hearers, to continue with unabated zeal their efforts on behalf of the Institution, Mr. Calvert closed the very excellent address, of which we have given a very faint and imperfect outline.

Of the Trustees we noticed as being present, Col. J. H. Sothoron, of St. Mary's Co., Col. Chas. Carroll, of Howard Co., John Merryman, Esq., of Baltimore County (the President of the State Agricultural Society), W. W. Corcoran, Esq., of Washington, D.C., James T. Earle, Esq., of Queen Anne's Co., (late President of the State Agricultural Society,) Hon. J. Dixon Roman, of Washington Co., and W. T. Mitchell, Esq., of Charles Co.

It was remarked as a singular coincidence, that the day was the anniversary of the somewhat too famous battle of Bladensburg; and it is to be hoped that this anniversary, in its peaceful and happy inauguration of a great undertaking, will be the commencement of a cycle, whose revolution shall completely obliterate the memory of the blood and rapine of the past. After the ceremonies of the day had been concluded, the Trustees were entertained at the hospitable board of Mr. Calvert, and, as he himself was formerly President of the State Agricultural Society, we had the agreeable pleasure of seeing at the same board a President and two Ex-Presidents of our valued association. By a happy accident, the day selected for laying the corner stone of the College, was also the birth-day of Mr. Calvert, and with hearty wishes that his useful and honorable life might be prolonged to another half century, was his health and prosperity drank, in the mantling glass.

The progress of the College building bids fair to be rapid — the workmen employed being numerous and efficient. The situation is superb — on a lofty eminence, but of gentle slope, whence a view is commanded of the surrounding country for many miles. Health, abundance of water, and diversity of soil, accessibility by railroad and by turnpike, proximity to Washington, and to Baltimore and Annapolis, render it one of the very best sites that could have been selected; and, under the fostering charge of Dr. Wharton, the Register of the Institution, who is shortly about to take up a more permanent residence upon the premises, in the building which has been recently fitted up, it cannot fail to give fresh evidences of the successful energy which has hitherto characterised the enterprise.

Presidential Veto of Agricultural Colleges Act, 1859

From J. D. Richardson, ed., *A Compilation of the Messages and Papers of the Presidents* (Washington, 1896), VII, pp. 3074–81.

WASHINGTON CITY, *February 24, 1859.*

To THE HOUSE OF REPRESENTATIVES OF THE UNITED STATES:

I return with my objections to the House of Representatives, in which it originated, the bill entitled "An act donating public lands to the several States and Territories which may provide colleges for the benefit of agriculture and the mechanic arts," presented to me on the 18th instant.

This bill makes a donation to the several States of 20,000 acres of the public lands for each Senator and Representative in the present Congress, and also an additional donation of 20,000 acres for each additional Representative to which any State may be entitled under the census of 1860.

According to a report from the Interior Department, based upon the present number of Senators and Representatives, the lands given to the States amount to 6,060,000 acres, and their value, at the minimum Government price of $1.25 per acre, to $7,575,000.

The object of this gift, as stated by the bill, is "the endowment, support, and maintenance of at least one college [in each State] where the leading object shall be, without excluding other scientific or classical studies, to teach such branches of learning as are related to agriculture and the mechanic arts, as the legislatures of the States may respectively prescribe, in order to promote the liberal and practical education of the industrial classes in the several pursuits and professions in life."

As there does not appear from the bill to be any beneficiaries in existence to which this endowment can be applied, each State is required "to provide, within five years at least, not less than one college, or the grant to said State shall cease." In that event the "said State shall be bound to pay the United States the amount received of any lands previously sold, and that the title to purchasers under the State shall be valid."

The grant in land itself is confined to such States as have public lands within their limits worth $1.25 per acre in the opinion of the governor. For the remaining States the Secretary of the Interior is directed to issue "land scrip to the amount of their distributive shares in acres under the provisions of this act, said scrip to be sold by said States, and the proceeds thereof applied to the uses and purposes prescribed in this act, and for no other use or purpose whatsoever." The lands are granted and the scrip is to be issued "in sections or subdivisions of sections of not less than one-quarter of a section."

According to an estimate from the Interior Department, the number of acres which will probably be accepted by States having public lands within their own limits will not exceed 580,000 acres (and it may be much less), leaving a balance of 5,480,000 acres to be provided for by scrip. These grants of land and land scrip to each of the thirty-three States are made upon certain conditions, the principal of which is that if the fund shall be lost or diminished on account of unfortunate investments or otherwise the deficiency shall be replaced and made good by the respective States.

I shall now proceed to state my objections to this bill. I deem it to be both inexpedient and unconstitutional.

1. This bill has been passed at a period when we can with great difficulty raise sufficient revenue to sustain the expenses of the Government. Should it become a law the Treasury will be deprived of the whole, or nearly the whole, of our income from the sale of public lands, which for the next fiscal year has been estimated at $5,000,000.

A bare statement of the case will make this evident. The minimum price at which we dispose of our lands is $1.25 per acre. At the present moment, however, the price has been reduced to those who purchase the bounty-land warrants of the old soldiers to 85 cents per acre, and of these warrants there are still outstanding and unlocated, as appears by a report (February 12, 1859) from the General Land Office, the amount of 11,990,391 acres. This has already greatly reduced the current sales by the Government and diminished the revenue from this source. If in addition thirty-three States shall enter the market with their land scrip, the price must be greatly reduced below even 85 cents per acre, as much to the prejudice of the old soldiers who have not already parted with their land warrants as to Government. It is easy to perceive that with this glut of the market Government can sell little or no lands at $1.25 per acre, when the price of bounty-land warrants and scrip shall be reduced to half this sum. This source of revenue will be almost entirely dried up. Under the bill the States may sell their land scrip at any price it may bring. There is no limitation whatever in this respect. Indeed, they must sell for what the scrip will bring, for without this fund they can not proceed to establish their colleges within the five years to which they are limited. It is manifest, therefore, that to the extent to which this bill will prevent the sale of public lands at $1.25 per acre, to that amount it will have precisely the same effect upon the Treasury as if we should impose a tax to create a loan to endow these State colleges.

Surely the present is the most unpropitious moment which could have been selected for the passage of this bill.

2. Waiving for the present the question of constitutional power, what effect will this bill have on the relations established between the Federal and State Governments? The Constitution is a grant to Congress of a few enumerated but most important powers, relating chiefly to war, peace, foreign and domestic commerce, negotiation, and other subjects which can be best or alone exercised beneficially by the common Government. All other powers are reserved to the States and to the people. For the efficient and harmonious working of both, it is necessary that their several spheres of action should be kept distinct from each other. This alone can prevent conflict and mutual injury. Should the time ever arrive when the State governments shall look to the Federal Treasury for the means of supporting themselves and maintaining their systems of education and internal policy, the character of both Governments will be greatly deteriorated. The representatives of the States and of the people, feeling a more immediate interest in obtaining money to lighten the burdens of their constituents than for the promotion of the more distant objects intrusted to the Federal Government, will naturally incline to obtain means from the Federal Government for State purposes. If a question shall arise between an appropriation of land or money to carry into effect the objects of the Federal Government and those of the States, their feelings will be enlisted in favor of the latter. This is human nature; and hence the necessity of keeping the two Governments entirely distinct. The preponderance of this home feeling has been manifested by the passage of the present bill. The establishment of these colleges has prevailed over the pressing wants of the common Treasury. No nation ever had such an inheritance as we possess in the public lands. These ought to be managed with the utmost care, but at the same time with a liberal spirit toward actual settlers.

In the first year of a war with a powerful naval nation the revenue from customs must in a great degree cease. A resort to loans will then become necessary, and these can always be obtained, as our fathers obtained them, on advantageous terms by pledging the public lands as security. In this view of the subject it would be wiser to grant money to the States for domestic purposes than to squander away the public lands and transfer them in large bodies into the hands of speculators.

A successful struggle on the part of the State governments with the General Government for the public lands would deprive the latter of the means of performing its high duties, especially at critical and dangerous periods. Besides, it would operate with equal detriment to the best interests of the States. It would remove the most wholesome of all restraints on legislative bodies — that of being obliged to raise money by taxation from their constituents — and would lead to extravagance, if not to corruption. What is obtained easily and without responsibility will be lavishly expended.

3. This bill, should it become a law, will operate greatly to the injury of the new States. The progress of settlements and the increase of an industrious population owning an interest in the soil they cultivate are the causes which will build them up into great and flourishing commonwealths. Nothing could be more prejudicial to their interests than for wealthy individuals to acquire large tracts of the public land and hold them for speculative purposes. The low price to which this land scrip will probably be reduced will tempt speculators to buy it in large amounts and locate it on the best lands belonging to the Government. The eventual consequence must be that the men who desire to cultivate the soil will be compelled to purchase these very lands at rates much higher than the price at which they could be obtained from the Government.

4. It is extremely doubtful, to say the least, whether this bill would contribute to the advancement of agriculture and the mechanic arts — objects the dignity and value of which can not be too highly appreciated.

The Federal Government, which makes the donation, has confessedly no constitutional power to follow it into the States and enforce the application of the fund to the intended objects. As donors we shall possess no control over our own gift after it shall have passed from our hands. It is true that the State legislatures are required to stipulate that they will faithfully execute the trust in the manner prescribed by the bill. But should they fail to do this, what would be the consequence? The Federal Government has no power, and ought to have no power, to compel the execution of the trust. It would be in as helpless a condition as if, even in this, the time of great need, we were to demand any portion of the many millions of surplus revenue deposited with the States for safekeeping under the act of 1836.

5. This bill will injuriously interfere with existing colleges in the different States, in many of which agriculture is taught as a science and in all of which it ought to be so taught. These institutions of learning have grown up with the growth of the country, under the fostering care of the States and the munificence of individuals, to meet the advancing demands for education. They have proved great blessings to the people. Many, indeed most, of them are poor and sustain themselves with difficulty. What the effect will be on these institutions of creating an indefinite number of rival colleges sustained by the endowment of the Federal Government it is not difficult to determine.

Under this bill it is provided that scientific and classical studies shall not be excluded from them. Indeed, it would be almost impossible to sustain them without such a provision, for no father would incur the expense of sending a son to one of these institutions for the sole purpose of making him a scientific farmer or mechanic. The bill itself negatives this idea, and declares that their object is "to promote the liberal and practical education of the industrial classes in the several pursuits and professions of life." This certainly ought to be the case. In this view of the subject it would be far better, if such an appropriation of land must be made to institutions of learning in the several States, to apply it directly to the establishment of professorships of agriculture and the mechanic arts in existing colleges, without the intervention of the State legislatures. It would be difficult to foresee how these legislatures will manage this fund. Each Representative in Congress for whose district the proportion of 20,000 acres has been granted will probably insist that the proceeds shall be expended within its limits. There will undoubtedly be a struggle between different localities in each State concerning the division of the gift, which may end in disappointing the hopes of the true friends of agriculture. For this state of things we are without remedy. Not so in regard to State colleges. We might grant land to these corporations to establish agricultural and mechanical professorships, and should they fail to comply with the conditions on which they accepted the grant we might enforce specific performance of these before the ordinary courts of justice.

6. But does Congress possess the power under the Constitution to make a donation of public lands to the different States of the Union to provide colleges for the purpose of educating their own people?

I presume the general proposition is undeniable that Congress does not possess the power to appropriate money in the Treasury, raised by taxes on the people of the United States, for the purpose of educating the people of the respective States. It will not be pretended that any such power is to be found among the specific powers granted to Congress nor that "it is necessary and proper for carrying into execution" any one of these powers. Should Congress exercise such a power, this would be to break down the barriers which have been so carefully constructed in the Constitution to separate Federal from State authority. We should then not only "lay and collect taxes, duties, imposts, and excises" for Federal purposes, but for every State purpose which Congress might deem expedient or useful. This would be an actual consolidation of the Federal and State Governments so far as the great taxing and money power is concerned, and constitute a sort of partnership between the two in the Treasury of the United States, equally ruinous to both.

But it is contended that the public lands are placed upon a different footing from money raised by taxation and that the proceeds arising from their sale are not subject to the limitations of the Constitution, but may be appropriated or given away by Congress, at its own discretion, to States, corporations, or individuals for any purpose they may deem expedient.

The advocates of this bill attempt to sustain their position upon the language of the second clause of the third section of the fourth article of the Constitution, which declares that "the Congress shall have power to dispose of and make all needful rules and regulations respecting the territory or other property belonging to the United States." They contend that by a fair interpretation of the words "dispose of" in this clause Congress possesses the power to make this gift of public lands to the States for purposes of education.

It would require clear and strong evidence to induce the belief that the framers of the Constitution, after having limited the powers of Congress to certain precise and specific objects, intended by employing the words "dispose of" to give that body unlimited power over the vast public domain. It would be a strange anomaly, indeed, to have created two funds — the one by taxation, confined to the execution of the enumerated powers delegated to Congress, and the other from the public lands, applicable to all subjects, foreign and domestic, which Congress might designate; that this fund should be "disposed of," not to pay the debts of the United States, nor "to raise and support armies," nor "to provide and maintain a navy," nor to accomplish any one of the other great objects enumerated in the

Constitution, but be diverted from them to pay the debts of the States, to educate their people, and to carry into effect any other measure of their domestic policy. This would be to confer upon Congress a vast and irresponsible authority, utterly at war with the well-known jealousy of Federal power which prevailed at the formation of the Constitution. The natural intendment would be that as the Constitution confined Congress to well-defined specific powers, the funds placed at their command, whether in land or money, should be appropriated to the performance of the duties corresponding with these powers. If not, a Government has been created with all its other powers carefully limited, but without any limitation in respect to the public lands.

But I can not so read the words "dispose of" as to make them embrace the idea of "giving away." The true meaning of words is always to be ascertained by the subject to which they are applied and the known general intent of the lawgiver. Congress is a trustee under the Constitution for the people of the United States to "dispose of" their public lands, and I think I may venture to assert with confidence that no case can be found in which a trustee in the position of Congress has been authorized to *"dispose of"* property by its owner where it has been held that these words authorized such trustee to give away the fund intrusted to his care. No trustee, when called upon to account for the disposition of the property placed under his management before any judicial tribunal, would venture to present such a plea in his defense. The true meaning of these words is clearly stated by Chief Justice Taney in delivering the opinion of the court (19 Howard, p. 436). He says in reference to this clause of the Constitution:

It begins its enumeration of powers by that of disposing; in other words, making sale of the lands or raising money from them, which, as we have already said, was the main object of the cession (from the States), and which is the first thing provided for in the article.

It is unnecessary to refer to the history of the times to establish the known fact that this statement of the Chief Justice is perfectly well founded. That it never was intended by the framers of the Constitution that these lands should be given away by Congress is manifest from the concluding portion of the same clause. By it Congress has power not only "to dispose of" the territory, but of the "other property of the United States." In the language of the Chief Justice (p. 437):

And the same power of making needful rules respecting the territory is in precisely the same language applied to the other property of the United States, associating the power over the territory in this respect with the power over movable or personal property; that is, the ships, arms, or munitions of war which then belonged in common to the State sovereignties.

The question is still clearer in regard to the public lands in the States and Territories within the Louisiana and Florida purchases. These lands were paid for out of the public Treasury from money raised by taxation. Now if Congress had no power to appropriate the money with which these lands were purchased, is it not clear that the power over the lands is equally limited? The mere conversion of this money into land could not confer upon Congress new power over the disposition of land which they had not possessed over money. If it could, then a trustee, by changing the character of the fund intrusted to his care for special objects from money into land, might give the land away or devote it to any purpose he thought proper, however foreign from the trust. The inference is irresistible that this land partakes of the very same character with the money paid for it, and can be devoted to no objects different from those to which the money could have been devoted. If this were not the case, then by the purchase of a new territory from a foreign government out of the public Treasury Congress could enlarge their own powers and appropriate the proceeds of the sales of the land thus purchased, at their own discretion, to other and far different objects from what they could have applied the purchase money which had been raised by taxation.

It has been asserted truly that Congress in numerous instances have granted lands for the purposes of education. These grants have been chiefly, if not exclusively, made to the new States as they successively entered the Union, and consisted at the first of one section and afterwards of two sections of the public land in each township for the use of schools, as well as of additional sections for a State university. Such grants are not, in my opinion, a violation of the Constitution. The United States is a great landed proprietor, and from the very nature of this relation it is both the right and the duty of Congress as their trustee to manage these lands as any other prudent proprietor would manage them for his own best advantage. Now no consideration could be presented of a stronger character to induce the American people to brave the difficulties and hardships of frontier life and to settle upon these lands and to purchase them at a fair price than to give to them and to their children an assurance of the means of education. If any prudent individual had held these lands, he could not have adopted a wiser course to bring them into market and enhance their value than to give a portion of them for purposes of education. As a mere speculation he would pursue this course. No person will contend that donations of land to all the States of the Union for the erection of colleges within the limits of each can be embraced by this principle. It can not be pretended that an agricultural college in New York or Virginia would aid the settlement or facilitate the sale of public lands in Minnesota or California. This can not possibly be embraced within the authority which a prudent proprietor of land would exercise over his own possessions. I purposely avoid any attempt to define what portions of land may be granted, and for what purposes, to improve the value and promote the settlement and sale of the remainder without violating the Constitution. In this case I adopt the rule that "sufficient unto the day is the evil thereof."

JAMES BUCHANAN.

In Support of the Morrill Bill, 1859

From *Cultivator* (Albany, April, 1859), VII, p. 129.

THE AGRICULTURAL COLLEGE BILL — Since our last we have received Mr. BUCHANAN'S Veto of the Bill granting Lands in promotion of the cause of Agricultural Education. This Bill was only passed by Congress after much opposition, in overcoming which the thanks of the public are due to Mr. MORRILL, its originator, and perhaps also to others. The veto of the bill is based upon its being both "inexpedient and unconstitutional" because, 1, it will deprive the Treasury, the President asserts — of nearly the whole revenue it next year counts upon from sales of Public Lands; 2, it interferes with the constitutional relations existing between the General and the State Governments; 3, it will operate to the injury of the new states; 4, its favorable influence for the advancement of agriculture and the mechanic arts is questioned, and, 5, it would interfere seriously with already existing institutions.

We give these positions assumed in the veto message, which do not represent the whole argument it contains, not in order to combat them, but to show in part the grounds on which the designs of Congress in behalf of our immense Agricultural interests, are ostensibly prevented from execution by the exertion of the objective power lodged in the hands of the President. If the same objections were considered of equal force, when the interests of merchants, manufacturers, and speculators are brought up for fostering aid; when steam lines are to be established, tariffs to be adjusted, bounty warrants thrown out in a flood over the country, and vast territories partitioned off section by section to railroad lines, the Farmers of the United States would feel less sensibly a discrimination against their claims, which must strike them, we fear, as partial and one-sided, however it may be fortified by constitutional considerations and legal arguments.

Agricultural Department at Washington, 1859

From *American Agriculturist* (1859), XVII, pp. 103–04.

After sending our last number to press we spent ten days on a visit to our National Capital, partly to gain health and vigor by release from business cares, and partly to witness the congressional proceedings during the last week of the Session. Another object in view was to look into the operations of the so-called agricultural Department, connected with the Patent Office, and supported by the Public Treasury. To prevent any embarrassment, or interference with our investigations, we purposely avoided direct contact with the chief "agricultural clerk," who, though not nominally, yet really holds under his exclusive surveillance, control, and direction, the entire operations of the agricultural department. We passed much time, however, with sundry gentlemen in Washington, who are well informed as to the way things are managed, including sundry members of Congress, members of the Congressional Committee on Agriculture, etc.; and we also had a lengthy personal interview with Commissioner Holt, who is (or ws then) the nominal head of the Agricultural Department. From the information thus gained, in addition to what we had previously known, and from several sources of *future* intellinece opened to us, we propose from time to time to set before the public the defects and wants of the department, with the hope of enlightening our readers, and so far as may be, contributing to improvement in the management of one of the most important departments connected with our General Government.

As now managed, the agricultural operations at Washington are a sham — a shame to us as an agricultural people. Our government might well, and ought to spend at least a million dollars annually in promoting the agricultural and hortical-tural improvement of the country, but without a change in the present organization it would be far better to save the sixty or seventy thousand dollars spent in salaries, in seeds, and in getting up the Annual Reports, and also the hundred and fifty or two hundred thousand dollars more for printing, binding, and distributing these "Reports" (See next page for notes on the last published Report). This view is already taken of the subject by many Members of Congress. This year the appropriation is cut down to the pittance of forty thousand dollars, and several Members stated to us that even this sum would have been withheld, had it not been for the hurried legislation of the closing hours, when it passed, without discussion, as an appendage to the general apropriation bill. We were in the gallery of the House at the time, and noted that it received but a small vote though enough to constitute a majority of those present, giving attention *voting* when this particular apropriation chanced to be passed along with many others. As a member of the House remared to us, "several Representatives who give no attention to the matter of agriculature and know little or nothing on the subject, were afraid to vote against any measure of this kind, lest it should be constued by their constituents in the 'Rural Districts' as an evidence of want of sympathy and interest in the 'bone and sinew.' " We were assured by Members of the *next* Congress, that the entire agricultural department would be abolished next Winter unless a decided change be made in its organization, efficiency and usefulness. Appended to the appropriation was this significant clause:

"*Provided*, That no part of the appropriation shall be used or expended in defraying the expenses of a body of men or delegates assembled in Washington or elsewhere as an agricultural college or 'advisory board of agriculture,' convened under the authority of the Secretary of the Interior, or any other person, under any name, for any object whatever."

This was designed as a direct censure upon the recent enterprize of the "agricultural clerk," (noticed by us in February, p. 35) viz.: the secret calling together of a selected paid coterie of person to whitewash the doings of the department. [In this connection we would inquire why the report of that "Advisory Board of Agriculturists" has never been permitted to see the light though called for by Congress. Rumor says, the "agricultural clerk" caught a Tartar in the report itself, as prepared by them. We call for its publication as originally made by the committee of that body.]

As a further indication of the feeling in Congress, we may add that the Senate refused to print the usual copies of the Agricultural Report. The House, at first also refused to print them, but after the loss of the bill to abolish the franking privilege, several members, who wished to have a supply of electioneering documents, to frank as a "sop" or compliment to their "rural constituents," contrived to get a hasty vote in the House for printing 210,000 copies or some kind of an Agricultural Report — they knew not what, for they only voted upon the title page, and for aught they or we know, it will be as poor a thing as its immediate predecessor.

How the Agricultural Department is Organized.

As every one perhaps understands, the executive government is divided into "Departments," as the Department of the Interior, the Department of War, of the Navy, of the Treasury, and of the Post Office. The head or chief officer of each Department is called the Secretary of that Department, except the P. O. Secretary, who is called Post Master General. These several chief officers are appointed by the President, and they together form his Cabinet.

The Secretary of the Interior has charge of several sub-Departments, such as the Patent Office, Indian Affairs, etc. Under him, is the Commissioner of Patents, who employs a so-called "Agricultural Clerk." All business matters done, and Documents issued relating to agriculture, are in the name of the Commissioners of Patents, who is himself a secondary officer. His attention is, however mostly given to subjects connected with Patents, and his agricultural clerk really manages and controls all matters connected with agriculture.

Mr. Thompson, is the present Secretary of the Interior. Mr. Holt was, until recently, the Commissioner of Patents, but he has just been appointed P. M. General, and the office of Com. of Patents is vacant at the time of this writing.

This tacking agriculture as a sub-department on to still another sub-department, is not only placing it below its proper position in point of importance, but this very fact so depreciates its dignity, that little attention is given to placing at its head a man of acknowledged superior abilities. This is abundantly proved by the fact, that for ten years past, the really important station of "agricultural clerk," or chief manager of agricultural affairs, has been occupied by a man of only ordinary ability (D. Jay Browne, who is the present incumbent of the office).

When we say "ordinary ability," we only repeat what is the general opinion of the great mass of intelligent men in the country, that is of those who take interest in the subject of agriculture. In all agricultural transactions connected with the Patent Office while under his control, there has been shown a lack of system, of valuable research, and of broad comprehensive views, which has brought the department to its present low standard in the estimation of the people at large.

Said Commissioner Holt to us: "I do not understand this apparent opposition, or at least this want of sympathy with a department so deserving of the cordial support of all classes as that of Agriculture." The real cause of it we endeavored to set forth to him, viz.; the want of an efficient man at its head to conduct and guide its affairs in such a manner as to command the respect and confidence of the country. We found that the Commissioner has himself known very little of the criticisms of at least three-fourths of the Agricultural Press, and of the more influential agricultural individuals and societies. His information in this respect has come to him *through* his agricultural clerk, and of course he has been permitted to know only the *favorable* side. [We suggest to our agricultural contemporaries, that hereafter when they have occasion to censure the agricultural operations at Washington, they send a marked copy sealed up and directed personally to the Commissioner of Patents.]

So far as we could learn, whenever anything unfavorable to the agricultural clerk has chanced to come to the Commissioner's notice, it has been promptly attributed by his clerk to the influence of interested seedmen, or to personal aspirations, or personal enmity of editors. While at Washington, we heard for the first time that the last named motive had been attributed to this journal. We beg to say to the Commissioner and to Mr. Browne himself, that there is not the slightest ground for this supposition. We heard (at Washington) for the first time, and at only second hand from Mr. Browne himself, that he had, at some former period, a personal difficulty with one of the *former* publishers of this journal. With that *we* have nothing to do — and care nothing. Until he published his famours ''autobiography'' and sent it over the country under the goverment frank, we did not even know that had so much as set foot in the office of this paper.

The truth is, we do not know Mr. Browne personally, but from the day we read his Book on Manures, and his Book on Trees, we set him down as a man of very moderate ability, native or acquired, and when we heard of him as ''agricultural clerk'' at Washington, we could not but regret that that important station had not been better filled. Still we hoped for the best, and did what we could to uphold him and the department. But after long trial and waiting, the feeble, inefficient character of all that has been done, the wishy-washy reports that have annually emenated from that department, the character of the seeds collected and distributed, and the way the thing has been carried on — all these matters have led us to the irresistible conclusion, that without a change in the administration of affairs, no good will come of the money expended by government in trying to promote the interest of agriculture in our country. This department, even in its present third-rate position, if properly conducted, might be productive of great good, and we earnestly hope that the in-coming Commissioner of Patents, whoever he may be, will place it in such a position, and under such control that we can unite with our cotemporaries in extending to it the strongest sympathy and support. It is in the power of the Commissioner to make this the most efficient, the most popular branch of government. We hope he will apreciate this, and take hold of it with a strong determination to make it what is *should be.*

With these statements we leave the subject now, intending soon to give some specific illustrations of the way things are and have been managed at the Capital. These will in part refer to: *how* the money goes; what salaries are paid and to whom; what special favors are conferred; how the articles for the Patent Office Report are obtained or made up; how seed has been knowingly put up and sent abroad *wrongly labeled;* some of the nonsensical ''offical instructions'' *e. g.* in regard to alligator's blood for orange tree insects, etc.; who helped the clerk to his position and how he is still rewarded for it; etc., etc. The developments will be interesting and instructive.

Agriculture in 1860

From Thomas G. Clemson, ''Preliminary Remarks,'' U.S. Patent Office, *Report on Agriculture,* 1860 (Washington, 1861), pp. 5–6, 11–26.

PRELIMINARY REMARKS

The agriculturists of the United States, to whose enlightened judgment these remarks are addressed, will doubtless appreciate the spirit in which they are offered.

The requirements of the present age, and the permanent importance of the subjects embraced in its operations, demand that the powers of this agency of the Government should be enlarged. This opinion was expressed in the views I had the honor to submit to the Secretary of the Interior at the period of my being called by that functionary to the position of Superintendent of Agricultural Affairs.

A vast majority of the intelligent agriculturists of the country, dissatisfied with the limited functions now exercised by the Government, not only confidently anticipate, but demand an organization at least equal in importance to that of any other department.

No object is more worthy of governmental care; nor is there any field of action in which the satisfactory realization of progress in population, wealth and civilization can be so certainly attained.

All civilized nations have in all times fostered agriculture as a primary and indispensable employment of man. This fostering has been direct in the bestowment of bounties; indirect in the restraints imposed upon foreign competition; and educational and providential in the encouragement of industry and ingenuity through the information and facilities which governments alone are capable of providing with efficiency and to a satisfactory extent. The Agricultural Division of the United States Patent Office has been created as the agent of the Government to give effect to its purposes in the last-named and most beneficent manner; and is, to the common mind, the only visible or appreciable agency for the promotion of this great and essential interest.

The degree of encouragement imparted by our Government by means of import duties, the bestowment of public lands upon actual settlers, and the distribution of the Reports of this Office — to say nothing of the influence and aid of local and general agricultural societies — cannot be estimated. It is, therefore, impracticable to institute a comparison between the benefits conferred by it and those conferred by the governments of other nations upon this branch of industry; yet it may be profitable in this connexion to review briefly the provisions made by several countries of Europe for the promotion of agriculture.

England

There is no special Bureau of Agriculture connected with the British Government; but the statute books are replete with enactments having in view the encouragement of this branch of industry; and the government expenditures therefor are upon a most magnificent scale. A living, active interest in it pervades the whole empire, and finds expression in many forms, giving existence to organized associations, to magazines and journals, to schools of varied grades, and to experimental gardens and farms in many localities. Moreover, the aggregation of wealth in the hands of comparatively few proprietors, and the stability of its tenure, enable them to execute upon enlarged plans many of the enterprises which in other countries, but especially in our own, can only be accomplished through the aid of the General Government.

France

The Agricultural Department at Paris is composed of a Director, three Chief Clerks of Bureaus, three Assistant Chief Clerks of Bureaus, and twenty-seven

Clerks of various grades, who are under the control of a Minister of Agriculture, Commerce, and Public Works. The three Bureaus are thus divided:

I. Bureau of Agricultural and Veterinary Instruction

The organization of the Imperial Schools of Agriculture is governed by a decree dated October 3, 1848. There are three schools, situated at Grignon, Grandjouan, and La Saulsaie. The instruction is both theoretical and practical, the object being to qualify managers, or overseers. The period of study is three years, and the charge 750 francs, or $150, for board and instruction. The officers, professors, and tutors are appointed by the Minister of Agriculture; the cultivators, or practical instructors, by the Director of the schools. The cultivation is carried on at the expense of the State, which also defrays the expense of instruction.

In the farm schools, of which there are fifty-one, the instruction is only practical. The Director, who is appointed by the Minister, personally controls the school and its field operations. He is allowed an annual sum of 175 francs, or $35, for the support of each pupil.

United States of America

The Agricultural Division of the Patent Office comprises as its personnel a superintendent; four clerks, including translators and writers; and a curator or gardener, and assistants; and its average annual expense for the last three years has been about $53,000, including the distribution of plants.

I should be wanting in fidelity to the trust reposed in me were I not earnestly to urge a more efficient encouragement to this great basis of all prosperity. The enlarged organization I have proposed is indispensable to the prosperity of our country; and the consummation of such a creation is an achievement in which man may well be proud to engage.

That the great interest of agriculture should be without suitable representation in the Government appears as an anomaly, and indicates a want of appreciation of the true state of our civilization. The present embryotic organization owes its existence to ideas of expediency expressed in the form of an annual grant to collect and distribute seeds and cuttings and information on their culture. That it should prove inefficient for the accomplishment of great and far-seeing enterprises is necessarily incident to its limited foundation and unstable tenure. The remedy is with the American people and their legislators; and it is confidently believed that, as the members of the great producing family become imbued with these truths, they will manifest their opinions by firm and vigorous action. An adequate organization and corresponding appropriations will be greeted throughout the land with the approving response of millions. A Department established under such auspices for the benefit of the paramount agricultural interest of the country, should be separate and apart from all influences other than those prompted by the highest regard for the public good, unobtrusive in its conduct as in its nature, and having truth for its object. It should endure untrammelled, and free from all partisan considerations. It should know no section, no latitude, no longitude. It should be subservient to no party other than the great party of production.

In the early stages of the formation of this Government it was not to be expected that a Department of Agriculture would be established. In the records of the colonial history of this nation, there are indications that something was done for the encouragement of this pursuit. As the necessity more and more pressed upon the people, sheep, as well as the fruits of the earth, were guarded from their natural enemies by premiums or bounties for the destruction of the depredators, and numerous intimmations were given by way of recommendation calling the attention of farmers to their own interests, intimately associated as these were with the public welfare. The pressure of emergencies during the Revolution was not friendly to the prosecution of measures encouraging to agriculture, although most of the great leaders were cultivators of the soil. Population was then sparse, and the wants of men were easily supplied from their wide acres on the fertile domains of a virgin continent. The necessities of the new Government, too, were those produced by war and revolutionary events, and had to be met at once. Hence the appointment of Secretaries of War, of the Navy, and of the Treasury. For our Foreign relations, a Secretary of State was then needed; and, as other wants became imperative, with the return of peace and the increase of population, a Postmaster General, and new Departments with their Bureaus, a Secretary of the Interior, an Attorney General, a Commissioner of Lands, of Pensions, of Indian Affairs, of Patents, of Customs, and of Public Buildings were called into existence.

Gradually, as the country advanced, and the vast forest fell before the axe of that hardy pioneer, and the broad prairies felt the hand of the immigrant scattering abroad the seed-corn and grain borne further and further from the East to the West, a new spirit of agricultural enterprise started into life; and societies have hence sprung into existence all over the land. Great credit is due to these, and to many an intelligent farmer and planter for their zeal and practical adaptations to secure and extend a higher state of culture, as well as to the periodicals and public journals wholly or partially devoted to this subject and to those arts and sciences intimately connected with its progress and improvement.

The public desires have been naturally directed to the General Government for its fostering aid and such encouragement as it may rightfully give. With its wide domain, its resources of information far and near, and its ability to present to the people accurate knowledge of the best productions of foreign culture, the agriculturists of the country have everywhere felt that they had a claim upon it for all that might be done to promote the efficiency both of associated effort and of personal or individual enterprise.

From this conviction the Agricultural Division of the Patent Office has arisen. At first, a mere clerkship charged with the duty of gathering agricultural statistics and having them printed in the report of the Commissioner of Patents, with various statements of the condition of agriculture, together with the disbursement of a trifling appropriation for the distribution of seeds, it has increased with the demands of the agricultural people to the large proportions of a Bureau. Its Report in 1842, consisting of less than 20 pages, and embraced in the same volume with the Report on Mechanics, has been gradually enlarged, and now forms a separate volume of useful and scientific matter. The distribution of seeds, though in some instances influentially opposed because misunderstood, has grown in favor with intelligent and patriotic agriculturists, and the demand for plants and seeds introduced from foreign countries has become general and urgent, and their propagation has already conferred incalculable benefits upon the country.

These operations are a necessity of the time. They form the connecting link between the hitherto forgotten farmer and the Government. Through the commerce of the country, they are to aid the great interest of agriculture by the purchase and introduction of new varieties of seeds from other countries, and the interchange of those native to it, but neglected, and hence requiring renewal in our climate. Information from every available source, at home and abroad, is collected, and disseminated, as the seeds are, gratuitously. The farmer and the planter are thus encouraged in their experiments, while this Division becomes a means of communication with the governments and peoples of all lands, providing what may be suitable for their soils and climates, strengthening our friendly relations with them, and at the same time using its official power and influence to obtain whatever may advance the agricultural interest of our own country.

At the time when our Government was organized, the art of agriculture was mere empiricism. The natural laws by which a successful practice is governed were not known: for the nature of substances, their true history, and the part they perform in producing and sustaining life, were unknown. It is only within the last quarter of a century that science has vindicated its true position. Without it there is no help for agriculture. All history teaches us that sterility and depopulation, and changes of locality by civilization, are the consequences of governmental neglect to sustain this great branch of industry. At the present period no man can hold himself guiltless who ignores its importance or withholds that fostering aid without which a disappearance of population must surely ensue; and such governmental improvidence and neglect of individuals must lead to wretchedness and death. No nation can prosper without progression in this branch of industry.

It is the duty of the Government to care for its domain — the joint property of the people, and the nation's hope.

All our vast domain is indeed far from being capable of cultivation; but there are means in science, if properly invoked, by which millions of acres that, remaining under their present condition, will forever be unoccupied, might be brought to a state of fertility. Our swamps, now fruitful only of disease, might be made to perform the most important of duties. Our uplands are being rapidly exhausted and abandoned to further waste. We cannot calculate our loss in worn out lands, but if we would form a faint conception of its immensity, let us make an effort to estimate the amount of money that would be required to put each acre in the condition of fertility in which it was when first occupied. But the day is not far distant when thousands of square miles of uplands will cease to be cultivated, and when the population of vast regions will be restricted to the alluvial and the marl formations.

In the North and in the South the soils have been wearing out through the reckless nature of our system of cultivation, and the cry now is, What shall we do to meet this difficulty? Agricultural societies, associations, clubs, and libraries, are everywhere organized; and agriculturists generally, though conscious how much good has been done by the efforts put forth, are yet aware how insufficient in some respects their unaided work is, and turn to the Government for the might of its countenance and coöperation. Wanting aggregated capital and the science which capital can command, they justly demand this coöperation of the Government as a right.

A glance at the various duties thus far embraced and performed in the agricultural agency intrusted to my care may help to form a better judgment of what further development it admits for increased usefulness by the enlargement of its means. With the moderate appropriations heretofore made, it is admitted that vast benefit has already resulted to the country from its operations. Taking these results as the means of estimating what a more liberal patronage by the National Legislature, such as lies within its province to confer, might have accomplished, it may well be doubted that any method of furthering the public welfare promises better to reward the earnest attention of those who have it in their power, if they feel it to be their duty, to promote an interest on which so largely depends the means of sustenance and comfort of our people for generations to come.

The distribution of seeds of ordinary character is not alleged to be necessarily an object of governmental duty; yet, so far as heretofore practiced, it has been attended with undoubted good. New varieties of vegetables have been introduced; valuable plants not known in the country have been naturalized with decided advantage to the entire community; while the crops of cereals have been vastly improved.

A large proportion of the limited appropriation made by Congress for the fiscal year ending June 30, 1860, for the collection of seeds, was disbursed in the purchase of tea seeds, in the construction of propagating houses for their reception, in the expenses attendant on their distribution, and in the preparation of the annual Agricultural Report. It was therefore impossible to procure seeds and cuttings for general use, as before.

It is believed, however, that the money so expended will prove a judicious investment. Thirty-two thousand healthy plants have been disseminated among gentlemen who had expressed a desire to experiment with them; and of that amount fully two-thirds were forwarded to planters residing south of Virginia and Kentucky. There will be eight thousand more rooted cuttings for distribution this winter; and it is contemplated to continue their propagation to a large amount each year in order to supply the continued demand, and thus insure a fair trial of the tea plant in our country.

It is confidently hoped that by substituting machinery and steam power for the tedious and laborious Chinese mode of preparation exclusively by hand, tea may be extensively manufactured here, and even become an article of export, especially as the necessary care and labor in the process is better suited to the weakly and young, who are unfitted for the culture of cotton and tobacco, and the heavier duties of the plantation.

In Assam there is a variety of the tea plant said to be of hardier growth than the usual Chinese, and its leaves bring a higher price in the English market. It may be advisable at an early day to procure a supply of the seed to ascertain their adaptedness to our climate and soil.

The appropriation made by Congress for the fiscal year ending June 30, 1861, will enable the Agricultural Division to extend its operations and usefulness by procuring and propagating various economical, medicinal, and useful plants, which it has been difficult heretofore to introduce with any prospect of success. By judicious changes the propagating houses connected with the garden of this Office have been better adapted to the general purposes in view. The garden, as far as it would admit of it, has been laid out with taste, and when planted with useful and ornamental shrubs and trees will present an attractive appearance. If enlarged and properly located, it may become an ornament to the capital, an object of utility to the country, and of general interest to the world; but the ground at present appropriated to propagation is not suited, either in extent, in position, or in the nature of the soil. Other grounds should be selected for the purpose, for in no other manner can an equal amount of money be so advantageously employed.

The introduction of foreign plants and their acclimation are not more important than that many indigenous plants, now uncultivated, should be subdued and applied to our use. But neither of these projects can be satisfactorily consummated without skill and knowledge. The procurement of valuable foreign plants and cuttings is not unfrequently attended with much trouble and expense; yet their indiscriminate distribution, often among persons not acquainted with the requirements for their successful acclimation, has been, and must continue to be, attended with the loss of time and of money, and with the impairment of the public confidence in enterprises otherwise full of promise and beneficence. Had the recently imported tea seeds been distributed, on their arrival upon this continent, the experiment would have entirely failed, as in former instances, with respect to the cork acorns, for example, wherein, with but few exceptions, the whole distribution was lost, while those planted in the Propagating Garden, with scarcely an exception, have been developed into vigorous plants. With suitable grounds and houses, and apposite appurtenances, the successful introduction of a single specimen will usually insure any desired number of plants, and the object is thus gained forever. These necessary adjuncts, while advancing the agriculture of the entire country, may also be made economically subservient to supplying all the ornamental trees requisite for the public grounds of the District of Columbia, and of the entire country.

Considering the nature of the duties which should fitly devolve upon this Department, it would seem eminently proper that the administration of these public grounds should be intrusted to its care and attention. It is, indeed, a matter worthy of consideration whether the best interests of the country would not be promoted by bringing all the public lands under the control and immediate supervision of such a Department. There are questions of serious import connected with the management of these lands which may not longer be postponed without sinister and momentous detriment to the future welfare of the country. As is elsewhere intimated in these remarks, the mineral wealth of the nation, so intimately connected with its agricultural prosperity, demands the highest and most intelligent offices of the Government, guided by the lights of science. These great national resources cannot be neglected with impunity; they demand, more than an other portion of the nation's dependencies, other cares than they have received; and this interest could not be placed in better keeping than under a Department of Agriculture of ample capacity and power.

Through our consular and diplomatic agents it is possible to introduce into the United States every ornamental and useful plant, and every animal, bird, and fish, valuable for its special qualities.

A large sum has been applied to the purchase of seeds abroad. The Superintendent was instructed to proceed to Europe to make the necessary selections, to obtain them on the most advantageous terms, to ascertain the safest manner of transporting them to this country, and to procure such information relative to agricultural matters as would be likely to interest and benefit our country.

The seeds, cuttings, &c., so obtained, will be forwarded from their respective localities as soon as they shall be in a condition to bear transportation. Seeds have been procured from Poland, Algiers, and the borders of the Black sea, where the climates appear adapted to the production of the grains most sought after by our farmers.

The Department has also given great attention and encouragement to the cultivation of the native grape, and the manufacture of wine therefrom; and there can be no doubt that its exertions have had considerable influence in causing many intelligent persons to engage in this important branch of industry, from the success of which, it is believed, improved physical and mental health, as well as increased wealth to our people, may be expected. It is, indeed, certain that the native grape may be introduced and cultivated with success in those regions which have become exhausted by over-cropping with other objects of culture.

Cuttings of superior native vines have been received from cultivators and amateurs. These will be propagated with care, and will afford interesting opportunities for making experiments in hybridizing with foreign or other unacclimated varieties.

An assortment of seeds and cuttings collected by Rev. Dr. Barclay in the Holy Land have likewise been received, and a portion of them distributed. The last shipment contained some articles which will be sent to the Southern States in time for their early spring sowing.

The effort to procure alive a few swarms of the Italian Bee, "Apis Ligustica," has been unsuccessful, owing to inattention to the instructions given by the agent of the Office. It is expected that the loss will be repaired.

Not many years ago the cotton plant was little better than a mere weed. It now vivifies the commerce of the world. The silk worm is an introduction of immense emolument to France. New varieties have been acclimated in that country which feed on the oak, the ailanthus, and the palma christi; and it is expected that their product can be manufactured at prices so low as to bring it into use in making sails for vessels, and for other common purposes.

The importance of domesticating the buffalo was brought to the attention of the Agricultural Division of this Office by the Hon. E. Thayer, Chairman of the Committee on Public Lands of the House of Representatives, in a letter to the Secretary of the Interior, accompanying a communication from Colonel Daniel Ruggles, U.S.A., together with a proposed resolution. The following is an extract from the reply submitted to Congress upon the subject:

"It is beyond a doubt that the buffalo, which once roamed over this entire continent, has gradually disappeared from his former haunts, and is now restricted to the most distant prairies of the Northwest, and the gorges of the Rocky Mountains. Judging from the past, and the active causes leading to inevitable extermination, the continuance of this animal, without the aid of domestication, is certainly questionable.

"The buffalo has ranged as far north as Slave Lake, in latitude 63° to 64°, and as low as 33° in New Mexico; and it would be rash to say that any part of this continent has been unfrequented by him.

"The buffalo is too well known to require any special description of its appearance or habits. In its osteology there is a marked difference from that of the ox species as found in every part of the world, the buffalo having fifteen ribs on each side, while the common ox has only thirteen. The civilized man, equally with the savage, appears to have pursued a course of wholesale slaughter, more for pleasure than for the satisfaction of his wants.

"Perhaps no animal with which we are acquainted possesses such remarkable properties or qualities. His migratory habits and fitness for great extremes of heat and cold are the results of 'natural selection and the struggle for existence' for untold centuries, by which he has arrived at a vigor of constitution, fleetness, and muscular strength rarely, if at all, met with in the ox tribe. These are qualities of great value which cannot be disregarded, and particularly when we consider the direct and indirect advantages that judicious crossings of domestic animals have bestowed upon civilization to an extent not to be calculated.

"A full-grown male buffalo will weigh from 1,200 to 2,000 pounds, and even more. In winter, his whole body is covered with long, shaggy hair, mixed with much wool: on the forehead this hair is a foot long. The Indians work the wool into cloth, gloves, stockings, &c., which are very strong, and look as well as those made from the best sheep's wool. The fleece of a single animal has been found, according to Pennant, to weigh as much as eight pounds.

"The dressed buffalo robe is esteemed everywhere on this continent and in Europe. It is used by the Indians in lieu of blankets for clothing, and as a covering to their habitations. In the North and Northwest it is an indispensable accompaniment to the traveller.

"The flesh has been extolled by those who have eaten of it, and bears the same relation to beef that venison does to mutton; and all concur in their praises of the hump as at once rich, tender, savory, and never cloying.

"The hoofs and horns are converted into cups, spoons, powder-flasks, &c., while the bones, independently of other uses, when broken and boiled, yield an oil or marrow which is used for culinary purposes; and one animal has given as much as 150 pounds of tallow.

"For military purposes, the buffalo, if domesticated, would appear to be particularly adapted, perhaps more so than any other animal, not excepting the camel. His great endurance, fleetness and strength would make him efficient as a beast of burden and for draught; and when no longer needed, he could be slaughtered for food.

"If, by crossing the buffalo upon our domestic stock, we could gain the qualities of fleetness, strength of constitution, and muscular vigor, with the chances for properties not to be calculated in advance, such results should not be undervalued. Neither should we lose sight of the possibility of the cross proving free from those diseases and epidemics which occasionally make such havoc among our domestic cattle. An indigenous race may be expected to possess, in this respect, special qualities which would render a cross with it highly advantageous.

"The domestication of the buffalo has been accomplished to a limited extent in more than one instance; and its feasibility is placed beyond a doubt by the experience of Mr. R. Wickliffe, of Kentucky, who has bred and crossed them with our native cattle. Any experiments, to be reliable, should extend over a series of years, and would be attended with considerable expense; and it is believed that an enterprise of such great magnitude demands the constant and vigilant attention of intelligent agents, to whom it should be committed as a special charge."

In the progress of arrangements for the further extension and development of the Agricultural Division, it is deemed advisable to commend to the consideration of the Agricultural Societies of the country a more intimate union and a more decided coöperation on their part with the General Government in the great work of agricultural improvement.

Our country is vast, its climate is varied, its soils are diversified, and its products are of many kinds. This coöperation, therefore, cannot fail to be productive of salutary results; and since the Agricultural Societies are composed of intelligent persons in the respective communities, who can readily understand the advantages of this means of promoting their own prosperity and the general welfare, application is made to them in this behalf.

Though the life of a nation may be counted by centuries, its present existence is always dependent upon the annual production of the soil. Without fertility, neither population nor civilization can abide. It is vain to expect that lands once worn out may be recuperated at will. Many who now live have seen individuals enriching themselves by exhausting the soil at the expense of the nation, and passing to their progeny untenable estates.

The land is for the good of all. Fertile soils were given to the nation as a trust, and are dispensed by the nation to the people to be used, but not abused. Fertility, the nation's endowment and hope, should hence be maintained; and, in order that the public at large may be impressed with these truths, and that the agricultural portion of the people may be brought into active coöperation with the Government, the aid of Agricultural Societies is thus invited in procuring agricultural statistics and reliable information upon subjects affecting agriculture, by which the whole community may be benefitted and civilization advanced.

This object may probably be best attained by the adoption of a system for the guidance of individual members of Societies in collecting facts in relation to every branch of interest to the farmer and the planter, and in reporting them at stated intervals — quarterly would be most judicious — to each State Society for its information, for publication if desired, and especially for transmission to this Office for elaboration and subsequent use. A summary of the condition of agriculture in every part of our country may, in this manner, be obtained, of material benefit to all, from the nature and reliability of its facts and the medium and regularity of their publication.

We have now a population exceeding 30,000,000, and an area of land of more than 3,000,000 square miles. It is the duty of the government to care for this immense property, and to prevent exhaustion of the soil and depopulation. This can be done by diffusing agricultural knowledge, and by procuring new plants, and bringing into notice and successful cultivation those which may be unknown or uncultivated; by the introduction of new animals, valuable for their wool, their flesh, or other qualities; of birds, useful for their eggs, feathers, or flesh; of fish that do not naturally inhabit our rivers, or which may have ceased to exist. To effect these great ends, it is not necessary to spend a large amount of money. It may be done without risk, and at a much less cost than is supposed by many.

The duties to be performed by this agency of the Government are onerous and responsible, and would be still more so if the views submitted in this paper were carried out. Among these present and contemplated duties I would mention the following:

1. An organized correspondence with the Agricultural Societies of the United States, and with the learned societies of the civilized world, would elicit correct statistical information which could not be collected in any other manner, and which would be of untold interest and advantage to our country and the world.

2. The publication of a Report on the subject of Agriculture, in which information could be authoritatively presented and diffused, would be of the greatest value.

3. The study of unknown indigenous plants for familiar cultivation in our own country, many of which may doubtless prove an addition of the greatest importance to our wealth.

4. Entomological investigations into the nature and history of the predatory insects which have proved so injurious to our crops of cereals, fruits, &c., and also to timber.

5. Questions of the highest moment and variety, connected with agriculture, requiring chemical aid and investigations in the field as well as the laboratory.

6. Familiar examples of special modes of culture, such as irrigation, might be put into operation and opened to the examination and study of the public, who would thus have ocular demonstrations of the methods of renovating lands, of keeping them in a constant state of fertility, and of producing crops which cannot be obtained in any other way without further outlay than by the use of water. Thousands of acres in the South, now waste and entirely unproductive, might by such means be brought to produce large crops of grass, which cannot be grown in our southern climate as the lands are now cultivated. That which is looked upon as impracticable would thus become feasible and profitable by means of irrigation.

7. The stocking of our rivers with fish such as do not live in them is a matter of great interest, and can only be carried out by the Government. We may judge of its importance when we understand that one million brood of salmon, without special attention or care, will in two years produce ten millions of pounds of the most healthful food. This subject has not only attracted the attention of European governments, but it has been repeatedly carried into successful operation there, and, upon a limited scale, in a section of this country.

These, with other subjects, necessarily come under the consideration of the Agricultural Division.

A celebrated statesman has remarked that "Agriculture feeds us; in a great measure it clothes us; without it we could not have manufactures, and we should not have commerce; they will stand together; but they will stand together like pillars, the largest in the centre, and that is agriculture."

By calling in the aid of science, and by the introduction and acclimation of new and valuable plants and animals, new sources of wealth are created and new industrial occupations are opened up for an increasing population, while our home markets are improved and we become less dependent upon other countries. By organizing with distant nations a system of exchange of the most useful and best developed productions within the limits of their respective climates, the Agricultural Division will be performing a most important duty, and one which cannot fail to be highly beneficial to our country. Foreign governments have expressed a willingness to promote this object. Many of our diplomatic and consular agents are equally desirous of showing their appreciation of the importance of this work, and have proved it by forwarding, together with various useful seeds and plants, interesting information which can be advantageously laid before our people.

The duties of a chemist in connexion with this division are most important. We should be able to give information upon all questions of general interest relating to agriculture, as connected with the sciences, such as the analysis of soils and of all other substances, the effects of geological formations on soils, and the composition of the divers mineral substances. In the yearly Report it is important to give a summary of the advancement of this science connected with agriculture, and for this he should have the acquirements which would enable him to discriminate with accuracy. But there are many other branches of knowledge, most of them having a direct influence upon the agricultural prosperity of the country, which would fall naturally into the class of duties assigned to his Office, including metallurgy and the development of the mineral wealth of the country, subjects much neglected to the manifest detriment of the prosperity of the country. Other nations consider them of such consequence that they have public schools to instruct in these branches, under the immediate jurisdiction of the government. If we cannot do this, much good may yet be done in indicating what has failed and what has succeeded in other countries, and thus making the time and attention bestowed on them there subservient to our advantage.

In many ways the government would receive direct benefit from the establishment of an efficient chemical laboratory. In the selection of the materials of which the public buildings are constructed grave errors have been made. The Executive Mansion, the old or central portion of the Capitol, of the Patent Office, and of the Treasury Department are instances in point, their walls now decomposing and disintegrating, and requiring the constant use of paint to preserve them from destruction. This would have been avoided had there been an officer of the Government competent and authorized to indicate the defects and advantages of the different materials. Such considerations always precede the creation of public buildings in Europe. A material may answer for one construction and be entirely unsuited to another, or it may be useless in any edifice. Not only in regard to the public buildings, but to all other works, is advice of this nature important. Certain rocks undergo decomposition in contact with salt water. It is therefore necessary that science should aid in designating those proper to be used in works in which durability is so great a consideration. The durability of the timbers of our vessels of war, and the appropriateness of the paints applied for their preservation, have engaged the attention of the Government and called into requisition the judgment of enlightened men; but surely there would be advantage in having a responsible authority at hand to consult on such subjects as each successive case is presented, rather than confide in less responsible yet more interested parties. The adaptedness of various kinds of iron to certain uses is also an important subject within the range of duties committed to such an officer. It is true that the architects, the civil engineers, the naval constructors and engineers, and the engineers of our army have all approved themselves well in the judgment of the world; but it may be affirmed with equal confidence that the suggestions herein made, and attempted to be enforced by these remarks, will meet with a peculiarly approving response from the intelligent and accomplished officers of the classes named, all of whom fully appreciate the disadvantages to which the Government is subjected from the want of a properly constituted and appointed laboratory of chemistry and metallurgy. But the primary necessity for such an establishment is in its connexion with agriculture. No true assistance can be given to the farmer in which the results of chemistry do not bear a part. The extraordinary progress which agriculture has made within a quarter of a century is due to it. The prospects for advancement in that vast interest are greater now than ever before, and their realization will doubtless lead civilization to the goal most ardently desired by the majority of mankind.

I have already shown that the want of a proper understanding of the nature of the soil, and of the arts based upon its cultivation, is the cause of the melancholy spoliation of so much of our land in this country and throughout the world. As the science of chemistry takes cognizance of the properties of all substances, and of their action upon each other, its varied and multiplied connexions with all that appertains to agriculture is without limit.

Particular attention has been paid to the management of this Division, and the improvement of the work performed. Large masses of letters and documents, the accumulation of years, are being classified and arranged for reference and use. Books better suited to its requirements have been opened, and the work systematized and simplified through the skilful labor of assistants, whose capabilities and fidelity it is alike my duty and my pleasure to commend.

How far it is accordant with the true interests of the Patent Office and with the rights of inventors to continue the administration of agricultural affairs under the aegis of that Office; how far it accords with verity to hold the Commissioner of Patents to responsibility before the country and the world for the performance of duties of which in the nature of things he cannot be cognizant, for the expression of opinions he cannot have matured, and for the promulgation of scientific discoveries in fields his accustomed pursuits have seldom or never permitted him to traverse, are regarded as proper and important inquiries at the present era, without respect to the converse of the several propositions implied, namely, that operations that are *sui generis* should not be embarrassed by incongruous alliances, but that the labor and responsibility involved, and, it may even be added, the honor of all creditable achievement, should fall upon the real agents in their consummation.

THOMAS G. CLEMSON,
Superintendent of Agricultural Affairs of the United States.

Massachusetts Agricultural College, 1861

From *American Agriculturist* (January, 1861), XXVI, p. 6.

It is often a good thing that events happen which put a dead lock upon the proceedings of people in power, or those holding important trusts, and force them to *stop and think*. This has occurred to the Trustees of the Mass. Agricultural College, which body, unwieldy from its numbers and slow to act — its members disagreeing among themselves — has finally come to the point of stopping to think. Their president and also the president of the College, Hon. H. F. French, has resigned, Prof. Chadbourne of William's College has been appointed president; a first rate farmer, Levi Stockbridge, of Hadley, has been appointed farm-manager, and things look now as if beginning anew with more moderate ideas, the Institution might gain a healthy *growth*. A mistake too often made, is in attempting to create a great institution in a short time. Strength, vigor and sound vitality come with a gradual growth — as witness Rome and the Oak. — In the beginnings of our agricultural colleges and similar institutions, the error fallen into, has been in considering a grand building the most important thing. Set a number of earnest men, capable of teaching agriculture, down upon a good farm, with a good large house and barns upon it, and the co-operation of a good farmer; put up a few temporary buildings, if need be, for lecture rooms now, and perhaps for stables by and by; give the faculty a little money to spend upon books, apparatus and fitting up; let them know that they shall have more as fast as they can show results; let all permanent improvements be made with a view to the future; and leave the Faculty as unhampered in regard to matters of instruction and discipline as possible, and success of the most gratifying character will be almost certain in any State of the Union. The grand establishment, with all desirable surroundings, will come in good time, with that practical fitness of things to ends, which comes with gradual healthy growth.

Report of the Commissioner of Agriculture

The opening of a new session of Congress last month was the occasion of the message from the President and reports from the Departments, and the report of "The Honorable" (what are titles worth?) Isaac Newton, Commissioner of Agriculture, comes also among them. This document has been extensively published by the daily press of the country and we will here only call attention to the fact, that nothing of any importance has been done, with the exception of collecting and publishing statistics in regard to growing crops and the prospects, the whoesale distribution of seeds, (amounting to 992,000 packages, and of plants, (34,000,) with the reported "interesting and suggestive" operations of the experimental farm, where grains, etc., received from all over the world are tested. These, with the publication of the Annual Report, has cost $162,600.43, but how much real good is accomplished? Mr. Newton has sent away his best men. The chemist has been kept at work analysing copper, iron and silver ores, testing rhubarb wine and such things (for Mr. Newton's friends). The museum has grown, as it should under the care of Mr. Glover; the garden improves year by year under Mr. Saunders; the farm is, we presume, in good hands, and the statistical department seems to have careful thought and labor. Mr. Newton has nothing to do with these things, except to make himself a nuisance, and interfere with their better operation, as he has with the able chemist, Dr. Erni who, for faithfulness, has lost his position. Mr. Newton has kept his own place, though notoriously inefficient and a disgrace to the country, which has such an illiterate, thick-headed man in so responsible a place simply by the dilligent use of *means* — fruits to this Senator, flowers to the wife of another, delicacies to the White House, a sinecure clerk-ship to the lazy cousin of some one of influence, and so on. — Does agricultural education receive a thought from the Commissioner? Do the causes of the diseases which are so destructive to our animals? Hog-cholera has been among us for years, and is as yet uninvestigated; the Spanish fever threatens great damage to the cattle of our Western States; abortion in cows is on the increase at the East, and glanders and farcy are destroying thousands of horses. Why no word about these things? Does he concern himself with agricultural immigration, and the occupation of United States land under the Homestead Act? The instruction of immigrants? The projects of planting trees upon the prairies? The encouragement of emigration southward, to restock and cultivate the Southern States? and many such like things? Not he. — In his concluding sentence, Mr. Newton says, ''he cannot repress the conviction that a new era is dawning upon the agriculture of our country.'' — May his own speedy retirement from office give force to the conviction.

Lincoln Calls for a Department of Agriculture, 1861

From Everett E. Edward, ed., *Washington, Jefferson, Lincoln and Agriculture* (Washington, 1937), pp. 88–89.

From Lincoln's First Annual Message to Congress

Agriculture, confessedly the largest interest of the nation, has not a department nor a bureau, but a clerkship only, assigned to it in the Government. While it is fortunate that this great interest is so independent in its nature as to not have demanded and extorted more from the Government, I respectfully ask Congress to consider whether something more can not be given voluntarily with general advantage.

Annual reports exhibiting the condition of our agriculture, commerce, and manufactures would present a fund of information of great practical value to the country. While I make no suggestions as to details, I venture the opinion that an agricultural and statistical bureau might profitably be organized.

Pennsylvania Farm School, 1862

From *Country Gentleman* (Albany, Jan. 2, 1862), IXX, p. 16.

We are indebted to Dr. Evan Pugh, president of the Farmer's High School of Pennsylvania, for a copy of its Third Annual Catalogue, just issued. We are glad to be able to present a brief outline of its affairs, and to know that they now occupy a more hopeful position than ever before.

During the past three years the Pennsylvania Farm School has been going on under circumstances of great difficulty, owing to the unfinished state of the College buildings. But an appropriation of last winter by the State Legislature of $50,000, has enabled the Trustees to advance in the work of completing the buildings, so that they will be entirely finished early next summer.

The main College building, we are told, is the largest edifice devoted to agricultural instruction in the world. It is, with the basement, six stories high, and covers an area of 19,200 square feet. It contains 165 dormitories 10 by 18 feet square, and 9 to 11 feet high, affording ample room for 330 students. The building is also well supplied with commodious rooms for museums, scientific collections, lecture rooms and laboratories for chemical and philosophical study and experimentation.

The cost of construction is estimated at $121,000. Other property belonging to the institution, including a farm of 400 acres, makes the entire property of the school worth about $178,000.

The Farm School has been in operation for three years, and from the commencement has been well patronized. Heretofore it has been found necessary to exclude students from other States, in order to make room for those from Pennsylvania, but the enlarged capacity of the building will now allow students from all States to enter its classes.

The course of instruction is intended to be thorough in regard to the natural sciences in general, and especially so in regard to those having bearing upon agriculture. Any student having a knowledge of the ordinary elementary branches of an English education can enter in classes and graduate after a four years course of study. The *first year* is devoted to a review and more complete study of the English branches. During the *second* the student is conducted into the elementary branches of the natural sciences, and the *third* and *fourth years* are mainly devoted to the latter. The mathematical course is about as thorough as that usually followed in other colleges, the scientific course is much more thorough than in literary colleges generally, while no attention at all is given to other languages than the English. It is the design of its friends to make the course as thorough and complete as that of the best European Agricultural Colleges, with such differences from them as the differences between American and European institutions generally require. Students who complete the course and pass satisfactory examinations and prepare dissertations approved by the Faculty, take the degree of Bachelor of Scientific and Practical Agriculture, *B. S. A.*

The college has just sent forth its first graduates, the class embracing 11 students. The Catalogue contains the titles and a general summary of the subjects of their graduating dissertations. The subjects are of an agricultural or manufacturing character, treated of with the aid of science. Artificial manures, plant ashes, slags of iron furnaces, iron ores, limestones and soils are submitted to chemical analyses, and the results given. One dissertation is devoted to the graminaceous plants in the neighborhood of the Farm School. The course combines manual labor with study. Each student performs three hours labor daily and after three years experience the Faculty speak with full confidence as to the practicability of combining manual labor with study. All the work of the farm, garden and nursery is performed by students, all of whom are required to work; by this means the terms of admission are kept down at the very low rates of $100 per session of ten months.

The next session will open on Wednesday, the 19th of February, and close on the 18th of December following.

Persons wishing to obtain farther particulars should address the President of the Institution, Dr. E. Pugh, Farm School P. O., Penn.

On a Department of Agriculture, 1862

From *Country Gentleman* (Albany, March 13, 1862), XIX, pp. 176–77.

HON OWEN LOVEJOY sends us the Report of the Congressional Committee on Agriculture, of which he is chairman, submitted in the House of Representatives, Feb. 11th, in favor of the "establishment of an Agricultural Department or Bureau."

This action was specially recommended in President Lincoln's last annual Message; and the Secretary of the Interior, in his report then presented, also urged "the establishment of a bureau of agriculture and statistics, the need whereof," he remarks, "is not only realized by the heads of departments, but is felt by every intelligent legislator. . . . One of the objects contemplated by Congress in the appropriations for the promotion of agriculture was the 'collection of agricultural statistics.'. . . Annual reports made under the direction of such a bureau, *setting forth the condition of our agriculture, manufactures, and commerce,* with well digested statements relative to similar facts in foreign countries, which the present rapid intercommunication enables us to obtain often in advance of their publication abroad, would prove the most valuable repertories of interesting and important information, the absence at which often occasions incalculable loss to the material interests of the country."

The Report of the Committee on Agriculture seconds these suggestions, and argues the importance of the Bureau proposed, very fairly, but not with as much force as we should have been glad too see the question discussed, nor apparently with as clear an idea of the whole scope and uses of such a department as was entertained by the Secretary himself.

Thus, in answer to the supposed objection, "Why not have a minister of commerce, of manufactures, as well as a minister of agriculture?" It is justly answered that "commercial and manufacturing interests, being locally limited and centralized, can easily combine and make themselves felt in the halls of legislation, and in the executive departments of the government," while farmers are lacking in similar combinations to press their interests; and thus "New-York and Lowell have often more immediate influence in directing and moulding national legislation than all the farming interests in the country." But the committee might also and still more forcibly as it seems to us, have urged the importance of such a bureau, from the benefits it will render to our commerce and manufactures, quite as much as from those which our agriculture may expect from its establishment. For its true design should be to include — as a department of *"Agriculture and Statistics"* — the whole productive capacities of the country in every department of industry. It is this which the Report of the Secretary of the Interior, as above quoted and italicized by us, distinctly advocates; and, in such a bureau. Agriculture justly takes the first and most prominent place because, upon *its* condition, and upon the facts revealed by its statistics, so large a share of both our commercial and manufacturing prosperity is mainly dependent. The interests of manufacturers, of merchants, and of farmers, are, we fully believe, in the long run far more likely to be harmonious and coincident with each other, than they are to clash, and it thus results, not only that such a bureau, properly managed, is of itself partially a bureau of commerce and manufacturers, but also that where it *is* specially promotive of the interests of Agriculture, it also exerts a secondary effect to the advantage of those other pursuits which can never be most prosperous except in the prosperity of the Farmer, and which depend upon him entirely for the resources of their subsistence, and, very largely, for the purchase of their goods.

A Department of Agriculture and Statistics,'' we do need. It should receive condense and circulate the Statistics now obtained, in States which already collect them, and secure their collection in other States. It should be an active, living, *working* department. It should deal with the great facts of constant practice; and leave theoretical investigations mainly to those better fitted to carry them on wisely and well. It should not be turned into a hospital for rejected office seekers in other departments; not into a seed and flower establishment for the supply of what any seedsman or florist will be glad to sell us; nor into a publication office of collated extracts and re-hashed essays, presenting nothing that has not been printed before in other forms, and nowhere serving to carry forward one step the real agricultural knowledge of the day. If our Government proposes in earnest thus to extend its helping hand to our industrial interests, we trust the new bureau will be committed to hands not notoriously incompetent; that it will be raised above the level of petty political influences; that it will publish reports to which we may turn with a reasonable hope of eliciting here and there a new fact, or getting now and then some ray of light, however feeble, in channels before obscure. If space permitted we should be glad to treat this subject at greater length, — we have avoided it hitherto, in despair that what any true friend of Agriculture might say would receive even the most curosy attention, and now we dare scarcely entertain much hope that the new bureau, if indeed it be created, will be placed upon such a basis as to secure the confidence of the reasoning and thinking Farmers of the country.

Act Establishing the Department of Agriculture

From 12 Statutes at Large 387.

Be it enacted by the Senate and House of Representatives of the United States of America in Congress assembled. That there is hereby established at the seat of the Government of the United States a Department of Agriculture, the general designs and duties of which shall be to acquire and to diffuse among the people of the United States useful information on subjects connected with agriculture in the most general and comprehensive sense of that word, and to procure, propagate, and distribute among the people new and valuable seeds and plants.

Sec. 2. *And be it further enacted,* That there shall be appointed by the President, by and with the advice and consent of the Senate, a "Commissioner of Agriculture," who shall be the chief executive officer of the Department of Agriculture, who shall hold his office by a tenure similar to that of other civil officers appointed by the President, and who shall receive for his compensation a salary of three thousand dollars per annum.

Sec. 3. *And be it further enacted,* That it shall be the duty of the Commissioner of Agriculture to acquire and preserve in his Department all information concerning agriculture which he can obtain by means of books and correspondence, and by practical and scientific experiments, (accurate records of which experiments shall be kept in his office,) by the collection of statistics, and by any other appropriate means within his power; to collect, as he may be able, new and valuable seeds and plants; to test, by cultivation, the value of such of them as may require such tests; to propagate such as may be worthy of propagation, and to distribute them among agriculturists. He shall annually make a general report in writing of his acts to the President and to Congress, in which he may recommend the publication of papers forming parts of or accompanying his report, which report shall also contain an account of all moneys received and expended by him. He shall also make special reports on particular subjects whenever required to do so by the President or either House of Congress, or when he shall think the subject in his charge requires it. He shall receive and have charge of all the property of the agricultural division of the Patent Office in the Department of the Interior, including the fixtures and property of the propagating garden. He shall direct and superintend the expenditure of all money appropriated by Congress to the Department, and render accounts thereof, and also of all money heretofore appropriated for agriculture and remaining unexpended. And said Commissioner may send and receive through the mails, free of charge, all communications and other matter pertaining to the business of his Department, not exceeding in weight thirty-two ounces.

Sec. 4. *And be it further enacted,* That the Commissioner of Agriculture shall appoint a chief clerk, with a salary of two thousand dollars, who in all cases during the necessary absence of the Commisioner, or when the said principal office shall become vacant, shall perform the duties of Commissioner, and he shall appoint such other employees as Congress may from time to time provide, with salaries corresponding to the salaries of similar officers in other Departments of the Government; and he shall, as Congress may from time to time provide employ other persons, for such time as their services may be needed, including chemists, botanists, entomologists, and other persons skilled in the natural sciences pertaining to agriculture. And the said Commissioner, and every other person to be appointed in the said Department, shall, before he enters upon the duties of his office or appointment, make oath or affirmation truly and faithfully to execute the trust committed to him. And the said Commissioner and the chief clerk shall also, before entering upon their duties, severally give bonds to the Treasurer of the United States, the former in the sum of *of* ten thousand dollars, and the latter in the sum of five thousand dollars, conditional to render a true and faithful account to him or his successor in office, quarter yearly accounts of all moneys which shall be by them received by virtue of the said office, with sureties to be approved as sufficient by the Solicitor of the Treasury; which bonds shall be filed in the office of the First Comptroller of the Treasury, to be by him put in suit upon any breach of the conditions thereof.

Approved, May 15, 1862.

The First Morrill Act, 1862

From 12 Statutes at Large 503.

Be it enacted by the Senate and House of Representatives of the United States of America, in Congress assembled, That there be granted to the several States, for the purposes hereinafter mentioned, an amount of public land, to be apportioned to each State a quantity equal to thirty thousand acres for each Senator and Representative in Congress to which the States are respectively entitled by the apportionment under the census of 1860; *Provided,* That no mineral lands shall be selected or purchased under the provisions of this act.

Sec. 2. *And be it further enacted,* That the land aforesaid, after being surveyed, shall be apportioned to the several States in sections or subdivisions of sections, not less than one-quarter of a section; and wherever there are public lands in a State, subject to sale at private entry at one dollar and twenty-five cents per acre, the quantity to which said State shall be entitled shall be selected from such lands, within the limits of such State; and the Secretary of the Interior is hereby directed to issue to each of the States, in which there is not the quantity of public lands subject to sale at private entry, at one dollar and twenty-five cents per acre, to which said State may be entitled under the provisions of this act, land scrip to the amount in acres for the deficiency of its distributive share; said scrip to be sold by said States, and the proceeds thereof applied to the uses and purposes prescribed in this act, and for no other purpose whatsoever: *Provided,* That in no case shall any State to which land scrip may thus be issued be allowed to locate the same within the limits of any other State or of any territory of the United States; but their assignees may thus locate said land scrip upon any of the unappropriated lands of the United States subject to sale at private entry, at one dollar and twenty-five cents, or less, an acre: *And provided further,* That not more than one million acres shall be located by such assignees in any one of the States: *And provided further,* That no such location shall be made before one year from the passage of this act.

Sec. 3. *And be it further enacted,* That all the expenses of management, superintendence, and taxes from date of selection of said lands, previous to their sales, and all expenses incurred in the management and disbursement of moneys which may be received therefrom, shall be paid by the States to which they may belong, out of the treasury of said States, so that the entire proceeds of the sale of said lands shall be applied, without any diminution whatever, to the purposes hereinafter mentioned.

Sec. 4 (as amended April 13, 1926, 44 Stat. L. 247). That all moneys derived from the sale of lands aforesaid by the States to which lands are apportioned and from the sales of land scrip hereinbefore provided for shall be invested in bonds of the United States or of the States or some other safe bonds; or the same may be invested by the States having no State bonds in any manner after the legislatures of such States shall have assented thereto and engaged that such funds shall yield a fair and reasonable rate of return, to be fixed by the State legislatures, and that the principal thereof shall forever remain unimpaired: *Provided,* That the moneys so invested or loaned shall constitute a perpetual fund, the capital of which shall remain forever undiminished (except so far as may be provided in section 5 of this act), and the interest of which shall be inviolably appropriated, by each State which may take and claim the benefit of this act, to the endowment, support, and maintenance of at least one college where the leading object shall be, without excluding other scientific and classical studies and including military tactics, to teach such branches of learning as are related to agriculture and the mechanic arts, in such manner as the legislatures of the States may respectively prescribe, in order to promote the liberal and practical education of the industrial classes in the several pursuits and professions in life.

Sec. 5. *And be it further enacted,* That the grant of land and land scrip hereby authorized shall be made on the following conditions, to which, as well as to the provisions hereinbefore contained, the previous assent of the several States shall be signified by legislative acts:

First. If any portion of the fund invested, as provided by the foregoing section, or any portion of the interest thereon, shall, by any action or contingency, be diminished or lost, it shall be replaced by the State to which it belongs, so that the capital of the fund shall remain forever undiminished; and the annual interest shall be regularly applied without diminution to the purposes mentioned in the fourth section of this act, except that a sum, not exceeding 10 per centum upon the amount received by any State under the provisions of this act, may be expended for the purchase of lands for sites or experimental farms, whenever authorized by the respective legislatures of said States;

Second. No portion of said fund, nor the interest thereon, shall be applied, directly or indirectly, under any pretense whatever, to the purchase, erection, preservation, or repair of any building or buildings;

Third. Any State which may take and claim the benefit of the provisions of this act shall provide, within five years, at least not less than one college, as prescribed in the fourth section of this act, or the grant to such State shall cease; and said State shall be bound to pay the United States the amount received of any lands previously sold, and that the title to purchasers under the State shall be valid;

Fourth. An annual report shall be made regarding the progress of each college, recording any improvements and experiments made, with their costs and results, and such other matters, including State industrial and economical statistics, as may be supposed useful; one copy of which shall be transmitted by mail free, by each, to all the other colleges which may be endowed under the provisions of this act, and also one copy to the Secretary of the Interior;

Fifth. When lands shall be selected from those which have been raised to double the minimum price in consequence of railroad grants, they shall be computed to the States at the maximum price, and the number of acres proportionally diminished;

Sixth. No State, while in a condition of rebellion or insurrection against the Government of the United States, shall be entitled to the benefit of this act;

Seventh. No state shall be entitled to the benefits of this act unless it shall express its acceptance thereof by its legislature within two years from the date of its approval by the President.

Sec. 6. *And be it further enacted,* That land scrip issued under the provisions of this act shall not be subject to location until after the first day of January, 1863.

Sec. 7. *And be it further enacted,* That land officers shall receive the same fees for locating land scrip issued under the provisions of this act as is now allowed for the location of military bounty land warrants under existing laws: *Provided,* That maximum compensation shall not be thereby increased.

Sec. 8. *And be it further enacted,* That the governors of the several States to which scrip shall be issued under this act shall be required to report annually to Congress all sales made of such scrip until the whole shall be disposed of, the amount received for the same, and what appropriation has been made of the proceeds.

Approved, July 2, 1862. (12 Stat. 503.)

Lincoln on the Department of Agriculture, 1862

From Everett E. Edwards, ed., *Washington, Jefferson, Lincoln and Agriculture* (Washington, 1937), p. 89.

From Lincoln's Second Annual Message to Congress

December 1, 1862

To carry out the provisions of the act of Congress of the 15th of May last, I have caused the Department of Agriculture of the United States to be organized.

The Commissioner informs me that within the period of a few months this Department has established an extensive system of correspondence and exchanges, both at home and abroad, which promises to effect highly beneficial results in the development of a correct knowledge of recent improvements in agriculture, in the introduction of new products, and in the collection of the agricultural statistics of the different States.

Also, that it will soon be prepared to distribute largely seeds, cereals, plants, and cuttings, and has already published and liberally diffused much valuable information in anticipation of a more elaborate report, which will in due time be furnished, embracing some valuable tests in chemical science now in progress in the laboratory.

The creation of this Department was for the more immediate benefit of a large class of our most valuable citizens, and I trust that the liberal basis upon which it has been organized will not only meet your approbation, but that it will realize at no distant day all the fondest anticipations of its most sanguine friends and become the fruitful source of advantage to all our people.

Objectives for the Department of Agriculture, 1863

From *Country Gentleman* (Albany, January 8, 1863), XXI, no. 2, pp. 25–26.

It may be proper to illustrate what has been said with reference to the ground on which this Department should stand, and the general objects it should have in view, by one or two instances in point, before proceeding farther.

I. *In its relations to the Government, it should act as the representative of the Farmers of the country*. Thus when a measure is proposed like the Riciprocity Treaty with the British Provinces, manufacturers whom it will benefit, are sure to be represented by committees, and to impress every argument in their favor upon the public officers and members of Congress concerned. The farmer, on the other hand, has no centralized organization in permanent session, to weigh the influence upon his interests of the movement under way; and it cannot be expected that our statesmen, with the best intentions in the world should look at these interests in every light, with *one side* of the question perhaps strongly urged upon their notice, and no one really ready to investigate the other side as thoroughly as it ought to be examined, or to advocate its claims, if found likely to be injuriously affected, with similar earnestness. We have always been of the opinion, that the Reciprocity Treaty referred to, was a measure tending to build up the agricultural interests of Canada at the expense of our own, and only to the benefit of some of our manufacturers and forwarders, and the cities deriving increased prosperity from their trade. It was our desire to have investigated the subject at the time of the conclusion of the treaty, but no private individual could well devote the time nor probably indeed obtain access to the necessary channels of information, to present effectively such objections as might be found to exist; and so, like many other things of the kind, it was allowed to pass by. What could a Department of Agriculture have better taken up for thorough investigation, so as to insure such provisions in the Treaty as would have protected the Farmer of the United States? Or if, in its present form, as some perhaps may claim, the Treaty is not injurious to him, would not the government at least have had a compilation of facts, settling the question for all future time, and proving the correctness and justice of its policy, when the continuance or the abrogation of the Treaty comes up for discussion? Are we not now "all at sea" in this matter, for the lack of a Department with some one at its head capable of discerning the bearing, and prompt of his own accord to enter into the close examination of just such points? We take it if a measure is brought before Congress, affecting in any way the revenue of the country, the Secretary of the Treasury will not wait for an invitation to scrutinize it; and that, if he finds it likely to result in harm, he will speedily seek some means of impressing his views upon the attention of committees and of members generally.

II. This Department should be in effect the representative of our Agriculture *with other governments as well as with our own*. We mean by this, not only that it can thus do much to overcome the misrepresentations of our productions and productiveness. (referred to in our last number,) which frequently appear in foreign journals, but also that our Ministers abroad should be able to derive from its reports or from correspondence with it, those facts and arguments most likely to prove attractive to emigrants and to lead to the investment of capital here. The different colonies of Great Britain are constantly exerting themselves in this direction, through every influence they can bring to bear upon the population of Great Britain herself, and of the Continent as well. A case illustrative of the means thus employed, is the visit to America in 1850, of Prof. J. F. W. JOHNSTON, the noted scientific and agricultural author, who printed a two-volume work on his return, proving (to the astonishment of everybody not in the secret of his having made the journey in the pay of some Colonial Emigration or Steamship company,) that the English possessions on this continent, from Novia Scotia all the way through to Vancouver's Island we believe, are far preferable to the settler, in climate, in fertility, in institutions and in every possible respect, to the adjacent or any other lands belonging to the United States! Now our Patent Office Reports have quoted *usque ad nauseam* the investigations of prominent foreign agriculturists and men of science; and as they thus testify to the very high rank in which the opinions of these authorities should be held, it is barely possible that the Department of Agriculture could not do a wiser thing than to invite some such person as LIEBIG in Germany, BARRAL or LAVERGNE in France, VOELCKER or JOHN C. MORTON in England, to visit us and examine our Agriculture, not only in the light of its statistics, but also in that of careful and extended personal observation — with the understanding that the fruits of such research should be published here in the Report of the Department, for our own instruction, so far as they might tend to that result, as well as abroad, in other forms, with the view of placing a knowledge of our agricultural resources before other nations.

III. In our remarks thus far we have been disposed to assume that a Department of the Government is organized primarily with some higher object than to teach the working farmer how to hoe corn, or to assure the studious farmer that three parts of hydrogen to one of nitrogen sufficiently condensed, constitute what is called ammonia. These are both important in their bearings upon the production of the country, but knowledge with reference to them finds other vehicles of dissemination. We have heard of gigantic hammers, which can crack an egg shell just as perfectly and precisely as they will forge into desired shape immense masses of solid metal: but we are not aware that any were ever erected simply for the former purpose — unless possibly the framers of the Law organizing the Agricultural Department are open to such an imputation.

We do not think, however, that this can have been their intent or aim. Without the text of the law before us, we judge from the newspaper summary published soon after its enactment, that it defines "the designs and duties" of the department as being "to acquire and diffuse among the people useful information connected with agriculture; also to procure, propagate and distribute seeds and plants." But the subsequent clauses appear to leave so much within the descretion of the Commissioner, that this definition may be made to cover all that we have indicated or propose to suggest as to the sphere of his departmental action, and almost anything else that he may secure appropriations for from the custodians of the public money.

1. And at the head of the classes of "useful information," which the Department should "acquire and diffuse among the people" — and this, under our classification, is the third, and possibly the most important of the chief objects it should have in view — we rank the *Collection of Agricultural Statistics*. If there are agencies which can be brought into active operation to accomplish this end by the actual enumeration annually of the products and livestock of the country, of course this is much to be preferred. It has been, we believe, already a subject of consideration with the Department, whether this may not be effected through the collectors or assessors of the Internal Revenue. No one would be more pleased to see this measure carried through successfully than ourselves; but should it fail of enactment, owing to the opposition which has manifested itself heretofore in several of the States when a corresponding law has been before State Legislatures, we may suggest that approximative estimates may be reached very much more exactly than was the case when the experiment was tried in 1842–9, by extending the correspondence of the Department, and securing the assistance of a larger and better qualified corps of reporters, who shall contribute from different parts of every county, if necessary, throughout the Union, regular statements of the condition and prospects of the crops, and, as soon after harvest as possible, a comparative calculation of its results by the side of years in which actual returns have been taken. There would therefore be involved both the amount of the crop obtained, and all the various influences affecting that amount; and a summary of the whole, as condensed as might be consistent with the preservation of all its principal features, would present an agricultural view and history of the year which we could obtain from no other source.

There are many who think, — and the present Commissioner of Patents, Mr. HOLLOWAY, after a careful examination of the subject, is among the number* — that Agriculture, Manufactures and Commerce should be united under the general supervision of the same Department; and this was, in the main, our suggestion last year in advocating a "bureau of agriculture and statistics." And how instructive might that Report be rendered which should trace the bearings of the crops of the year upon the exports and imports and mercantile prosperity of the country; and the bearings of public events — the construction of new lines of internal intercourse — the introduction of improvements — the changes of relation between different districts and between our own and other countries — upon the course of our Agriculture! Wheat growing has *not* been abandoned, for instance, in our Eastern States, as a whole — as so many have asserted, because their soils have been impoverished, but because they can be more profitably turned to account in other ways; and yet this single fact, which should exercise a potent influence upon the internal migration of our people, as well as upon our agricultural reputation abroad, has never been set forth in a sufficiently authoritative form to command the attention it deserves.

2. And this leads us to remark secondly, that it would be within the province of the Department *to investigate and survey the present condition of American Agriculture*. It has been already argued that a point of great importance is to render each year's report a transcript of the agriculture of the country for that particular year. And, to accomplish this end — to form "a base of operations"

from which to estimate and understand the changes which take place, for the better or the worse as the case may be, — nothing could answer so well as an exhaustive inquiry into the farming of the country, carried on from year to year, and either from one State to another, or from one district comprising analogous soils, climate and productions, to another marked by different characteristics in these respects. One State, or one such district, thoroughly reported upon by a competent person employed for the purpose, who should devote his time wholly to observation and investigation until qualified to write understandingly and correctly, — would be a marked feature in each volume, requiring but brief *addenda* in subsequent volumes, until at length, we might have a yearly Report that would speak for itself as a chart of our actual farm practice, its besetting sins, and the foundation of common sense and truth on which as a whole it probably rests.

3. Cognate to the last named branch of investigation is *the thorough discussion of particular subjects*. To relieve the proposed surveys of our present systems of farming, from the burden of much that would possess a bearing upon them, more or loss direct, and which could be referred to if separately printed either in the same or other volumes, there is a large class of subjects which will reward especial research. There is, for example, the actual change which this or that particular improvement — say the introduction of Short Horned Cattle, or of Merino Sheep — has effected, in its pecuniary bearings upon our farming — the enlarged value and increased returns thence derived — basing every statement upon actual investigation into the statistics involved, aand not upon guess-work, wherever there are census or transportation or market returns which throw light upon the question. It is not to place the history of our improvements on record, that such an inquiry should be alone conducted, but because it would present the most powerful of all arguments for the extension of exertion in directions in which *it is found to pay*, and because new and unexpected suggestions might thus be developed as to farther efforts. Under this head, moreover, are to be classed the discussion of principal crops, either with reference to their culture throughout the country at large, or in this or that important locality, — and of particular processes, such as the manufacture of beef on the Western prairies, or in the stalls of our Eastern farmers, or the production of mutton for farm use and for city butchers, and so on. Of this kind have been many of the articles in the Patent Office Reports already published, and some of them drawn up most creditably to their authors; but the great error has been a proneness to make these essays very full and exhaustive on points which have been previously treated at length by foreign and home writers, and quite lean and barren where original research or the eliciting of any new information is involved.

4. *Scientific Investigation*. What we have thus far adduced has purposely been of an immediately practical nature. We cannot entertain any very brilliant anticipations from the establishment of governmental laboratories. Where the necessity arises for chemical research on any particular question, that research had better be delegated to some competent chemist away from Washington. We should have no faith in the thoroughness or value of any investigation (worth taxing the resources of the government to accomplish) carried on in the midst of dozens of others under a department bureau.

Not but what there are points enough in which scientific investigation may appropriately receive the encouragement if not the direct supervision of the Department of Agriculture. Take, as a single example, the experiments in the laboratory and in the field, which might be instituted with reference to our great crop, Indian corn, not only to lead to a better understanding of the action of fertilizers, in promoting its growth here at the east where we manure for it so largely, but also to test whether it is within the reach of science to provide any safeguard against exhaustion on soils where it is now continuously cultivated with little or no manure — cheap enough to lead to its general adoption. Or, to go into matters more purely scientific, any of the great questions to which Mr. LAWES and Dr. GILBERT have devoted years of unceasing labor and expenditure, are of fully as great interest to the American as to the English farmer, but they are beyond the reach of the private purse in this country, and, if undertaken, it can only be with the aid of such a Department.

5. In other branches of Science, *Entomology* for example, there are investigations which might be appropriately conducted. But here again it seems to us preferable on the score both of efficiency and economy, to commit them to competent naturalists to carry on, rather than to endeavor to assemble a corps of second rate assistants amidst the temptations and expenses of life at the Capitol.

6. *The Condition of Foreign Agriculture*. There may be much that is useful derived from occasional investigations into the course of agriculture abroad by competent persons, but this point we shall not enlarge upon, as it suggests itself only too readily, and has been frequently acted upon in the past.

7. *The Introduction of New Plants*. It is very well that the Department of Agriculture should be authorized to correspond with the representatives of the country abroad, or perhaps occasionally send agents to other countries competent to select seeds or plants for trial here. But the continued importation of large quantities of common seeds, as already indicated, appears to us useless, and, if not conducted with greater care than has been the case heretofore, absolutely injurious from the risk to which we are exposed of introducing both noxious insects and noxious weeds.

In the midst of the pressing engagements of the New-Year season, we have not been able to condense and elaborate, as we should have liked, this hurried and very imperfect outline of the duties and opportunities of an Agricultural Department. Those, however, who will take the pains to follow out the suggestions that have been given, to the full results, to which, as we believe, they may be made to lead, will perhaps derive from their own conclusions, if not from ours, a somewhat enlarged idea of the responsibilities of the Department, and of the expectations in which thinking men are entitled to indulge, if it shall be properly managed. Above all things we would have the notion done away with, that a book of certain size must be published every year, whether it is filled with appropriate and valuable articles or not. We would have the Commissioner lay before Congress annually an exact account of just what he and his assistants have been about for the promotion of the ends to which they were appointed to office. But in the "accompanying documents" we would have the rule of the Royal Agricultural Society of England, with reference to its prize essays, strictly adhered to, that all the information they contain, "shall be founded *on experience or observation*, and not on simple reference to books or other sources." And we would have no bureau established in or about the Department, simply as an ornamental appendage to make up a certain printed programme.

The motions of inquiry already brought forward in the present session of Congress, are sufficient to prove that there is little public confidence at present — whether in the necessity of the existence of the Department, or in the capacity of those who have it in charge — we do not undertake to say. And these articles have been prepared partly with the view of giving voice as nearly as we can, to the hopes which the farmers of the country may justly base upon the organization of such a branch of the general government, and partly in order to show that it has a sphere wide enough and important enough, in which to exert its energies, if Congress will prolong its existence, and insist upon the effective prosecution of the ends it is designed to accomplish.

Goals for the Department of Agriculture, 1863

From *American Agriculturist* (New York, January, 1863), XXII, no. 1, p. 8.

What Our Agricultural Bureau Ought to Do.

When the new Agricultural Bureau was provided for by Act of Congress, we had some hopes that good would come out of it. That the General Government should do something — should do much — to foster and develop the greatest interest of our country, its agriculture, is too evident to require argument. That no change for the worse could be made upon the system pursued during several years past, seems almost equally evident. The appointment of a head to the new department being a matter of so much importance, we tried to indicate to the President that in the selection of the Commissioner he should not be guided by his kind hearted feelings, by family considerations, or by importunity, but appoint the best man, the one of the most comprehensive views, of activity, experience, administrative talent, and enterprise. How far he was guided by such considerations we do not pretend to say. The appointment being made, we determined to judge of it by the result produced. So far we have waited and are still waiting to see what will be done. Any real good accomplished we stand ready to approve. Whatever hints we may offer to the gentlemen in charge of the Bureau, are therefore given in the kindest spirit. The Department belongs to us, in common with every other person interested in the agriculture of the country.

What ought it to do? First, we say, that it should aim at investigation, at the collection of information and statistics *which can not be attempted by individuals*. To illustrate: The distribution of seeds, excepting those of rare and costly character, can be done by individuals. Our Agricultural Department at Washington, *has* been mainly a free government seed store, largely devoted to collecting, at public expense, a great number of seeds, mostly common, and of good, bad, and indifferent quality. These, together with an annual volume of little value, have been distributed at random by members of Congress as political instrumentalities. It is well for the Government Bureau to collect rare and new seeds from other countries, and *test* their utility in different parts of our own country. But this should be only incidental.

Here are a few of the things we would propose for the attention of the Agricultural Bureau:

I. — The opening of a comprehensive and systematic correspondence with leading, reliable, and intelligent cultivators, at least one in each county in the United States, after the plan of M'Killop's commercial agency in this city. At that agency one can learn, on the instant, the exact status, the financial condition of any business man in the entire country. Such a system of government correspondence would enable the Agricultural Bureau to gather prompt information on any topic of general interest.

II. — The collection of accurate early information from the whole country in regard to the amount, condition and prospects of the growing crops. This information to be gathered frequently during the growing season, say from May to September, and the general result to be published for the guidance of both farmers and commercial men, and to be given to the public at once — not a year afterwards when of no particular value. The special announcement by telegraph, that "full returns to the Agricultural Bureau indicate a given amount of wheat or corn growing, and that the prospects at a given date indicated an average or a deficient or a surplus yield," would be hailed by all classes as something tangible and useful.

III. — A thorough discussion, *founded on comprehensive and general information*, of two or three leading crops, each year. To illustrate: Grass, or the forage crop, is the most important one of the country. Could not the Bureau of Agriculture, with its facilities, set on foot *and carry out* an investigation which would tell us definitely: what are the peculiar characteristics of the Blue Grass regions of Kentucky, and into what other portions of the country that grass might be introduced with advantage; what kind of grass proves to be the best for prairie soils in the different localities, and why; and the same of clay soils, loams, bottom lands, etc., in the various climates, and at different elevations; the relative value of timothy, clover, lucerne, red-top, etc., for growing cattle, working animals, dairy purposes, also for horses, sheep, etc. The information should not be an Essay for the Report, at so many dollars per column, by one man, founded on his own limited observation, but it should embrace the results of a collection of reliable information from the whole country. Let the whole force of the Department be concentrated upon one, two, or three crops a year, according to its facilities for doing it thoroughly.

IV. — The introduction and testing of new seeds and plants. The present system is wholly wrong. It is worse than useless to collect a great mass of seeds, and scatter them broad-cast over the land, at the caprice of Congressmen who use them at random as electioneering or political appliances. Let the Department secure a moderate supply of several new seeds and put a portion into the hands of a few persons of known skill and enterprise, in a sufficient number of localities to make the experiment general for the whole country, and let careful returns of the results be obtained and published. A hundred parcels of seed thus tested, would furnish more information than a million parcels scattered promiscuously. One or two hundred specimens of a new plant thoroughly tried in as many localities, would be amply sufficient to test its value, and the results obtained from their careful trial in judicious hands, and under specific instructions, would be decisive.

The above are a few suggestions we would offer to the managers of the new Bureau of Agriculture. We may add others hereafter.

Goals for Agricultural Colleges, 1863

From "Agricultural Education," *Country Gentleman* (Albany, September 17, 1863), XXII, p. 187.

EDS. CO. GENT. — Congress having made liberal grants to each of the States to endow Agricultural Colleges, it becomes a matter of interest to inquire what kind of institution an agricultural college ought to be.

If an agricultural college is to be a college for the purpose of teaching agriculture, it must fail; because agriculture is neither a science nor an art, but a handicraft or trade. As well establish colleges to teach shoe-making, or house-painting, or cotton-spinning, as to teach agriculture. Scientific agriculture, as it is called, cannot be said to exist as a science. It is aimed at and hoped for, but until analytical chemistry has been carried to a much greater perfection than at present, it must continue to be among the things hoped for, and not one of the things to be taught in colleges. Analytical chemistry discovers no difference between the components of cotton and sugar. Even in the common analysis of water, one of the most eminent chemists of New-England now confesses that the method hitherto pursued has been all wrong. If scientific agriculture is based upon the idea that by a knowledge of the component parts of vegetable products, added to a knowledge of the component parts of soils and manures, a given vegetable product may be obtained, very much as a cook makes a pudding by compounding the articles according to her receipt, the idea may be a very pleasing one, but while the fact is that chemists are as yet able to analyze vegetable products only in the rudest and most elementary manner, the idea cannot be carried into practice. What good farmer ever derived any benefit from an analysis of the soil? Is it not admitted that chemists can detect no difference between some of the most fertile and some of the most barren? Something may be taught of botany and the physiology of plants, but if agricultural colleges are to graduate a parcel of young men with a smattering of chemistry, a touch of botany, and an inkling of vegetable physiology, who think themselves good farmers, agricultural colleges will be a nuisance, because they will increase the great defect of American education, superficial knowledge. Scientific agriculture stands to day with phrenology, and biology and magnetism. It is an undeveloped theory, not a science. Of practical sciences those only can be taught which admit of accumulated knowledge of facts leading to theories, which again are proved by the facts. But the known facts of agriculture are of the simplest kind, and discovered themselves for the most part while Adam delved. The theories of scientific agriculture are not yet proved by the facts.

Agricultural colleges then must simply be high schools for farmers. What makes a good farmer? The same qualities which make a good mechanic or man of business — intelligence, judgment, and industry. Can a school teach these to its pupils? To a certain extent, and indirectly, it can; but as it is the object of all schools to do so, your object and means will be the same as those of other good schools. If you wish to teach young farmers to know when they know a thing, and when they do not, you will not put them through a course of agricultural chemistry, for the result would probably be a persuasion that they knew something of that of which they knew nothing at all.

If then the noble endowments of Congress are to result in anything but a delusion and a snare, let those who are to direct the organization of these colleges pitch their profession low, and the results will be higher. If the colleges turn out well drilled lads, thoroughly grounded in an English education, knowing something of surveying, book keeping, and mechanics, with such lessons in farming as they may learn by example and practice on a good farm, it will be well. Such boys will have a better education than George Washington. But if they graduate youths who think they know something of vegetable physiology, agricultural chemistry, and the theories of Liebig, they will merely produce a considerable number of badly educated men, who are worse than uneducated men, because they use their common sense less, and are more conceited.

D.

The Agricultural Department Reaches Minnesota

From O. H. Kelley, "The Agricultural Report in the North-West," *Country Gentleman* (Albany, December 17, 1863), XXII, p. 401.

The Agricultural Report in the North-West.

MESSRS. EDITORS. — I have just examined the *Report of the Commissioner of Agriculture* for 1862 — a work of some 600 pages. Its value to any intelligent farmer is at least five dollars — in fact I doubt if that amount would purchase the same quantity of strictly reliable information which this Report contains.

On the second page I notice the resolution requiring one hundred thousand copies for the use of the House of Representatives, and twenty thousand only for the Department. I presume by this, that they also have a peculiar value for political purposes. So large a demand by the House shows very plainly that the Department has at least friends among its members as well as with the people at large. In a late Eastern paper I notice that the Department is called the "*Great National Seed Store*," — probably the writer of that article has the impression that the gratuitous distribution of seeds injures his business, if he is in that line. On the distribution of books and seeds in the State of Minnesota, I wish to make a few remarks, presuming the same will apply to other Western States.

Within a circuit of ten miles of where I write, I can count one hundred farmers — mostly eastern men, consequently favored with at least a common school education — as near as I can ascertain about five of the number generally secure a copy of the Annual Report, and probably twenty-five receive one variety of seeds each, the latter I have had the pleasure of distributing. I find that the recipients of the Reports generally read them faithfully, and that they are valued for their contents. The seeds are usually of some new varieties to us, and with but few exceptions are planted and properly cared for, and now for the results. I am called upon every season for Reports and Seeds, not that the farmers wish to save the money the seeds would cost at the stores, for I have frequently been offered pay for those I have distributed, but because those received from the Agricultural Department during the past four years have proved in nearly every instance to be pure and of choice varieties, which were not offered by the seedsmen. Where the crops are Annuals, the farmers invariably liberally distribute the seeds they sowed, and the consequence is that we have good varieties, and of the best qualities, particularly of vegetables. Of Biennials but few secure seed for further use, and then the call comes upon the seedsman for a supply, so that instead of the Department injuring their business by the gratuitous distribution, it is rapidly increasing it, as well as introducing to our notice plants many of us were before ignorant of.

I presume the introduction of "King Philip corn" in this vicinity alone, has been of more value to us than all our taxes have amounted to since the first package was received. If it were possible, we would like to have every one of our farmers receive at least one paper of seeds annually, and have the number of reports somewhat increased. I take the liberty to suggest that as the Department now has upwards of two thousand regular correspondents, who are all agriculturists and interested in its welfare, there be a much larger number of reports placed in the Commissioner's hands for distribution among the class of men who will be directly benefited, and who will treasure them. Those who reside in cities, and have libraries of their own or the privilege of public libraries, and not directly interested in agriculture, might look upon the publication of these reports as a waste of money by government; in fact, I have heard the expression many times even in this State. But I know of no one move that will make the delegation from our State in Congress more popular, than to favor at the present session an appropriation large enough to place in the possession of every farmer in the Western States a copy of the Report for 1862; and I venture to say if some of our members would examine some of the libraries of their constituents during the next political canvass, they would find them to consist in many instances of the Bible and the Report for the past year, which would be good proof that they had done something towards the diffusion of knowledge. Until the past year, we have not had that interest shown in our welfare as a class which we have deserved, but now the establishment of an Agricultural Bureau has begun to show us very plainly our strength, and the labors of the Commissioner and his assistants thus far, has been attended with the most pleasing and valuable results to us in the West, and every intelligent farmer will expect Western members to be very liberal in their sentiments towards our pet, unless they desire to lose popularity at home. O. H. KELLEY. *Itasca, Minn.*, Dec. 1, 1863.

Farming in Wisconsin, 1864

From Wisconsin State Agricultural Society, *Annual Report, 1864* (Madison, Wis., 1868), VII, pp. 61.

In compliance with the law, I have the honor herewith to transmit the Fiscal Report of the Wisconsin State Agricultural Society for the year ending Dec. 14th, 1864.

After the lapse of three years, during which there was held no general exhibition of the industry of the State, early in the spring of 1864, the Society again rallied its forces and renewed this portion of its regular work. The sequel has shown that the public were prepared for this action on the part of the Society; for notwithstanding the still unsettled condition of the country and the unfavorableness of the season, agriculturally considered, the Exhibition, as such, was creditable to the enterprise of our people, the attendance numerous, and the cash receipts larger than at any previous one since the organization of the State.

The Fair afforded a most satisfactory demonstration of the wonderful energy and growing zeal with which the industrial classes have pushed forward in their own legitimate work, practically unheeding the tumult of war, only just beyond the borders of their State. There were the inventors of implements and machinery with the products of their genius in the form of numberless improvements — there the producers of grain and breeders of stock, with indubitable proofs of successful effort for the attainment of results that must give to themselves larger profits and to the nation greater assurance of increased means for the early extinguishment of its immense burden of debt — there the fruit-growers, with surprising evidence of what resolution and perseverence may accomplish even against the odds of an adverse, if not perverse, climate and the voice of almost universal complaint coming up to them from every quarter of the State — there an army of resolute mechanics and manufacturers, offering to the over-tasked husbandman the means of multiplying his productive energies.

The State of Wisconsin has reason to be proud of her population. No more industrious, enterprising, determined and heroic people live in the world.

We cannot conclude this brief Report without again asking your attention to the importance of prompt action in the matter of the proposed College of Agriculture and Mechanic Arts, provided for in the Congressional act of 1862. Lands have been located, the net proceeds of which, if judiciously managed, should, at least, equal the sum of $300,000. This, with such aid as the State ought to be willing to give, will constitute a foundation for the beginning of an educational work, in the interest of the industrial arts, which must ultimately result in great good to both people and State.

If it were possible, on a fair basis, to connect the proposed College with the State University, that would undoubtedly be the most economical and best disposition to make of the question; but as this is believed to be impracticable, the incorporation of a separate institution seems to be the only alternative.

The bill for the incorporation of the "State Agricultural College of Wisconsin," which passed the Senate by so large a majority, last year, appeared to us to meet the conditions of the Congressional Act and to provide for the educational wants of the industrial classes in a very satisfactory manner, and it is hoped that a similar measure may receive the approval of both branches of the present legislature.

The period allotted to the State for full compliance with the act donating the lands is passing, and we are unable to see any substantial reason why provision should not be made at once for an advantageous disposition of the lands and the early establishment of the Wisconsin College of Agriculture and the Mechanic Arts.

Agricultural Colleges, 1865

From U.S. Department of Agriculture, *Annual Report*, (Washington, 1865) pp. 140–52, 185–86.

The Act of Congress

Congress by an act entitled "An act donating public lands to the several States and Territories which may provide colleges for the benefit of agriculture and the mechanic arts," approved July 2, 1862, granted to each State, for such purposes, an amount of public land equal to thirty thousand acres for each senator and representative in Congress, to which the States are respectively entitled by the apportionment under the census of 1860.

The subject of agricultural schools and colleges has long attracted the attention both of our people and legislators, and many attempts, most of which have proved failures, have been made to establish such institutions. The disposition to expend money in large and expensive buildings, and to indulge the American propensity to own all the lands that join us, induced Congress, in the act referred to, to fix judicious restraints upon the States accepting its grant. To the careless observer, a college, is, chiefly a group of magnificent buildings, with pleasant surroundings of lawns and trees, where students are expected some how to gain an education, however starved and pinched may be the internal organization, including the corps of professors and teachers.

Seeing how many institutions have been ruined or contracted in their usefulness by extravagance in the external management of their affairs, and especially by indulgence in architectural display, Congress wisely provides "that all the expenses of management, superintendence, and taxes, from the date of the selection of said lands previous to their sales, and all expenses incurred in the management and disbursement of the moneys which may be received therefrom, shall be paid by the States to which they may belong, out of the treasury of said States, so that the entire proceeds of the sale of said lands shall be applied, without any diminution whatever, to the purposes hereinafter mentioned;" and that "no portion of said fund, nor the interest thereon, shall be applied directly or indirectly, under any pretence whatever, to the purchase, erection, preservation or repair of any building or buildings."

To guard against loss of the fund by improvident investment, the act provides that all moneys derived from the lands granted shall be invested in stocks of the United States, or of the States, or some other safe stocks yielding not less than five per cent.; and that if any portion of the fund, or of the interest thereon, shall be lost or diminished, it shall be replaced by the State, so that the capital shall remain forever undiminished, except that a sum not exceeding ten per cent, upon the amount received by any State under the act may be applied to the purchase of lands for sites or experimental farms, whenever authorized by the legislature.

The general object and character of the colleges thus to be established is briefly stated in the fourth section of the act, which provides that the interest of the fund shall be inviolably appropriated by each State which may take and claim the benefit of the act, "to the endowment, support, and maintenance of at least one college, where the leading object shall be without excluding other scientific and classical studies, and including military tactics, to teach such branches of learning as are related to agriculture and the mechanic arts, in such manner as the legislatures of the States may respectively prescribe, in order to promote the liberal and practical education of the industrial classes in the several pursuits and professions of life."

Our object being to discuss the subject in such a way as to aid those who are engaged in organizing colleges under the grant of Congress, it is important to ascertain at the outset what limitations are prescribed by the act. The grant was made by Congress to all the States, and it was then, and still is, impossible to devise a defined plan to be adopted by all. The New England States, with their thoroughly organized system of common schools, require different colleges from the southern States, where no such system is known, or the new States of the west, where society has hardly begun to crystallize into towns or villages. Great latitude was therefore left to the several States in establishing their respective institutions under the act.

Certain marked features, however, remain prescribed by the act of Congress, which good faith, if not the power of the law, requires each State to incorporate into every institution benefited by its grant. "The leading object" of the college shall be, says the act, "to teach such branches of learning as are related to agricultural and mechanics arts;" and the title of the act expresses the same general object — to "provide colleges for the benefit of agriculture and the mechanic arts."

These fundamental provisions call for the establishment of institutions different from our ordinary colleges, which can in no fair sense be said to be maintained for the benefit of agriculture and the mechanic arts, or to teach especially such branches of learning as are related to agriculture and the mechanic arts. In a loose and general sense, all learning may be said to benefit agriculture and the mechanic arts, and to be related to them; but the colleges maintained by the grant of Congress are required to be distinctively and essentially of this character. It is therefore a fraud on the act for a State to transfer the bounty of Congress to existing literary institutions without requiring them, at least, to establish a regular course of study in such branches of learning as are distinctively related to agriculture and mechanic arts.

We find nothing in the act to limit the colleges established under it to the mere practical teachings of agriculture and mechanics; but, on the contrary, the idea, so far as developed, is of colleges of the grandest scope, where, "without excluding other scientific and classical studies," the branches of learning related to agriculture and the mechanic arts are to be taught "in such manner as the legislatures of the States may respectively require, in order to prosecute the liberal and practical education of the industrial classes, in the several pursuits and professions of life." Several points should be noted in the language of the act just quoted.

"Liberal" as well as "practical" education is provided for, and education in the several pursuits and "professions" of life.

The grand idea which seems to underlie the whole act, and which, no doubt, was prominent in the minds of the framers of it, is the elevation of the laboring classes. This is clearly expressed in the language already cited, giving the grand object, which is "to promote the liberal and practical education of the industrial classes," &c. The "industrial classes" are ordinarily those engaged in agriculture and the mechanic arts. To raise them to equality in education with the classes more favored by fortune, is the first care of a republican government. The rich may educate their own children, but the government should take care that the poor are not neglected. Already colleges exist in most of the States, where youth, a majority of whom are from the wealthier families, are educated for the professions. Colleges to teach the branches of learning relating to agriculture and the mechanic arts offer peculiar attractions to the industrial classes, and it is desirable to bear in mind, in their organization, the fact that these classes have not, usually, large means at their command, and that institutions for their benefit must furnish the means of education at moderate expense.

Again, it is clear that, although the primary object is the education of the industrial classes, it is not intended so to conduct their education as to confine to them to any class, in their after life. The object is rather to offer to the industrial classes such facilities for education as they are most likely to use, to give them instruction in the branches relating to agriculture and mechanics, to offer them instruction in "scientific and classical studies," and finally to prepare them, by a "liberal and practical education," not only for farmers and mechanics, but for success "in the several pursuits and professions of life."

Whether Independent or Connected with Other Institutions

Whether the college, to be established under the grant of Congress, shall be an institution independent, or whether it shall be, to greater or less extent, connected with existing colleges, is a question raised in every State where the subject has been discussed.

Assuming that a union with an existing institution is consistent with the act of Congress, let us consider the expediency of such a union. The question is attracting much attention in Europe, particularly in Germany, where, as in this country, scholars, and especially officers of universities and colleges, generally advocate such a connexion. The principal arguments in favor of a union, so far as relates to this country, may be arranged under a few heads:

1. The great cost of buildings for lecture and recitation rooms, halls, libraries, laboratories, and many other accommodations, may, for the most part, be saved, since in all our colleges there is accommodation for many more students than now attend.

2. Existing institutions, too, are already supplied with musuems of natural history, geology, comparative anatomy, and the like, and with libraries for general

reading and scientific works, all of which may be available to a larger number of students. It requires a long period of time as well as a large amount of money to form such collections, and without them an agricultural college could not be expected to maintain a position of dignity or usefulness.

3. Existing institutions have organized corps of professors, many of them (as of chemistry, physics, botany, physiology, mathemtaics, ethics) the same that would be necessary in the agricultural college, and those, with slight addition to their labors or numbers, could give instruction to the students in agriculture and mathematics.

4. The great leading minds of the country are already engaged and attached to existing institutions, and it will be found impossible to organize new colleges with competent professors.

5. The union of the highest education in the sciences, and in their application, is impracticable; and true education consists in the apprehension of principles and in general discipline, rather than in practical arts, which may be readily learned afterwards.

6. That knowledge is advanced by the devotion of thoroughly trained minds to special branches of science, whereby discoveries are made and actual additions to the sum of human knowledge are published to the world. The Smithsonian Institution at Washington and the Museum of Comparative Zoology at Cambridge, Massachusetts, are illustrations of this special mode of study, and all the higher universities and colleges, incidentally at least, to some extent, adopt the same method.

The reasons in favor of independent agricultural colleges, and the answers to the foregoing arguments in favor of a connexion, may be thus stated:

1. Admitting the great value of libraries and museums already formed, as well as the economy of using buildings already built, it is fair to suggest that funds for the erection of new buildings, and for libraries and collections, can usually be raised by local subscriptions or by contributions, in aid of an agricultural college, from persons who would give no aid to an existing institution. There is a deep interest among farmers and mechanics in the success of colleges adapted to their practical wants, which is of more value than all that the older colleges can offer.

2. It is no disrespect to existing institutions to maintain that no one of them has within itself a corps of instructors competent to manage an agricultural college. Wedded to their own approved and time-honored theories, almost unanimously distrusting the possibility of a union of manual labor and study, accustomed to instruct mainly in theory, unfamiliar with practical agriculture, believing that Latin and Greek furnish the best discipline for the youthful mind, the regular professors in existing colleges are peculiarly unfit to develop or execute a new and peculiar plan of education. The agricultural college, thus controlled, would of necessity sink into a subordinate branch of the university, and fail of all its purposes. In an independent institution, under a government devoted to its peculiar objects, the professors of the other colleges might be procured to deliver courses of lectures in their several departments, and thus their learning may be made available to the new college. Nearly all college professors have periods of leisure which they devote to lectures abroad, and such interchange would be mutually beneficial.

3. The arguments in favor of a union, based upon the incompatibility of the study of abstract and applied science, and upon the idea that the advancement of

knowledge rather than its diffusion is the chief object in view, are founded in a misapprehension of the intention of Congress as shown in the act.

The manifest object of the act is, as has been already shown to furnish a more practical education for the industrial classes than other colleges afford; and if such education is incompatible with the theories of existing institutions, there can be no union between the two systems. Again, the new colleges are designed to educate boys, and not to advance the knowledge of learned professors. Their first object is to diffuse knowledge already existing, to teach their pupils what is already known to the best farmers, the best mechanics, and to the professors of the various sciences — boys between the ages of sixteen and twenty-one, incapable of receiving education beyond this. It is not expected of them that they should make discoveries in science, or enlarge the boundaries of human knowledge. Let us train them in body, in mind, in taste, in morals, developing each capacity harmoniously, to make them perfect men, robust and manly, with knowledge of men, of business, of practical affairs — observers and lovers of nature as well as students of books — and so prepare them for contact with the world "in the several pursuits and professions of life."

4. The essence of republicanism is equality and freedom from caste. They who advocate union do not, in general propose to annex farms to the existing colleges, and so do not intend to make manual labor a part of their system. They thus avoid the divison of their students into classes of scholars and laborers, by sacrificing the advantages which, in another place, we claim for manual labor. So far, they doubtless do well; for the harmonious operation of any system in which a part of the pupils should be required to perform farm labor, in the costume adapted to it, and another part should be exempt from labor, would be impossible; and the case would be even more manifestly hopeless were the attempt made to introduce a class of laboring boys into the ranks of an established institution, where the older classes had, by the natural course of their education, imbibed the common prejudice against manual labor.

The customs of students in old institutions seem fully as strong as the authority of the faculty. The attempts to abolish *fagging* in England, and *hazing* in America, in the universities, has taxed the utmost power of the authorities, with only partial success. This love of power and assumption of superiority seems to be one of the innate depravities of students, and the experiment of introducing a new class, to be known distinctly as agricultural or mechanical, even without the requirement of labor, would not be found without its embarrassments. If again, the agricultural students are not distinguished from the rest, we have remaining only our old colleges, and our whole plan of agricultural colleges is destroyed.

The opinion of Dr. Hitchcock himself, president of Amherst College, was decidedly in favor of an independent institution. His reasons why mere agricultural professorships are insufficient, and in favor of independent agricultural colleges, are briefly as follows: 1. Because lectures upon such subjects attract but few of the students of colleges, most of whom are looking forward to professional life; 2. Because the two classes of students who would thus be brought together would have too little sympathy to act in concert and as equals in the same university; 3. Because, without such concert and sympathy, one or other of the classes of students would feel no pride in the institution, and without such *esprit de corps* it could not prosper; 4. Because the field is wide enough to require such establishments. The principles of agriculture are based upon a large part of the physical sciences. No man can understand the *principles* of farming who is not more or less acquainted with chemistry, anatomy, physiology, botany, mineralogy, geology, meteorology, and zoology; and then the practical part requires an extensive acquaintance with various branches of mathematics and natural philosophy. 5. Because it demands extensive collections of various kinds in order to elucidate the principles of husbandry; enough, indeed, to belong to any scientific institution, and too many to form a mere subordinate branch of some institution with a different object in view. 6. Because the number of instructors must be so large that they could not conveniently form an adjunct to some other institution.

Manual Labor

Whether students in agricultural colleges shall be required to perform manual labor, is a question which everywhere excites discussion, and which deserves most careful consideration. Learned professors, and indeed nearly all who have been engaged in education in our academies and colleges conducted on the ordinary principles, doubt the success of combining labor with study.

Manual labor schools were a few years ago much advocated. The idea upon which they were based was, that students by laboring a part of their time might defray the expenses of their own education. It was supposed that four or six hours' labor daily, well applied on the farm or in the workshop, might not only pay the board but the tuition of the pupil, and all his incidental expenses. The difficulty, however, was not in the theory, but in its development. A single faithful industrious young man in a farmer's family might no doubt, by even four hours' daily labor, pay for more than his board; and perhaps a practical farmer might take into his family a small class of such youths, and teach them practical agriculture, and receive fair compensation for their support and his assistance to them, by their labor for a third or half the time. The farmer would invest in the enterprise only the supplies for his table and some additional house-room. His teachings would be given without loss of time from his business, and he would make no expenditures for apparatus, or for buildings for lecture and recitation rooms.

Suppose now that this same farmer undertakes to enlarge his plan of usefulness, and, instead of his small class, to educate two hundred boys in agriculture, not only practical but scientific; to teach them not only manual labor and commence farming by example, but to give them a regular course of education in chemistry, physics and engineering, natural history, comparative anatomy and physiology, including veterinary surgery; to instruct them in French and German, and, generally, to give them, in the language of the act of Congress, a "liberal and practical education in the several pursuits and professions of life." How can he do it? He must erect large and expensive buildings, with halls and lecture-rooms, and museums and laboratories; he must employ professors learned in the several departments, who must be paid enough at least to support them; he must provide his students with rooms for lodgings and for study, and make provision for their board; he must expect only the average amount of intelligence, industry, and fidelity in his pupils; and he must provide for the profitable employment of his two hundred boys, in all seasons — summer and winter, rain and sunshine. If he finds his own time occupied on the farm, he must employ some discreet educated person to take general charge of his establishment, to organize classes, conduct correspondence, listen to the complaints and requests of the students — in short, to *preside* over the enterprise, which it may be perceived has grown from a farmer's family into an agricultural college, with a president, professors, the usual expensive buildings, and our farmer himself as farm superintendent. The main object now being to educate two hundred boys, and not merely to farm profitably for the farmer's benefit, incidentally teaching a half dozen young men, the result of the scheme pecuniarily is entirely changed.

Thoughtful men might have foreseen what experiment proved, that manual labor schools as such — schools where the pupil's labor was to pay all his expenses and those of the school — must fail. It is difficult enough for the average of men to succeed in life when they devote themselves to one object, and give to it all their energies; but when they undertake a grand project like education, and expect that an incidental adjunct like a system of half-time labor shall maintain it, their disappointment is sure.

When we consider, further, that the men who have undertaken to establish manual labor schools have not been usually of the class called practical, but rather of the enthusiastic and philanthropic order — educated rather in theory than otherwise, it would seem strange, indeed, if they should be able even to conduct fairly experiments involving farms and workshops, buying and selling, and all the complicated machinery of education and self-support combined.

The failure of manual labor schools furnishes no argument against manual labor in agricultural colleges, but tends to prove only that such labor cannot be expected to be very profitable as a matter of dollars and cents, however profitable it may be as a part of a system of education.

It should be distinctly understood by the public, by legislators, and by all connected with these institutions, that the principal object is the education of the pupil, and that this object is kept in view in his hours of labor as well as in his hours of study. Profit and education may be quite inconsistent in many instances. The young man will earn more for the institution if placed in the employment which he best understands, and kept there regularly through his course. His education will be best promoted, on the other hand, by allowing him to engage in those branches of labor of which he has no knowledge.

It is frequently said by advocates for manual labor that three or four hours' labor a day ought to support the pupil. The same persons, if you ask them, will say that the pupils should be taught to perform with their hands every process of farm labor. Let the farmer test this matter by applying the question to his own case. How much would it profit him, if he has a fine dairy stock of twenty cows, to have them milked for a fortnight by twenty boys who never had milked a cow before: How much richer would he be to set twenty boys, who never mowed a swath before, into grass fields to mow for him a week, and furnish them scythes: Ask similar questions as to all farm operations, fencing, cutting timber, planning sawing, tending stock — as to gardening, pruning, grafting, budding, transplanting, and we shall see that unskilled labor of boys can be of little value; especially when they are employed in large numbers, so that they cannot be constantly superintended and watched as a farmer would do with his own family.

It is important to organize our colleges with the right idea upon this point. If legislators and trustees assume that student labor must be profitable and productive, and insist that it shall be made so, they compel their officers to sacrifice the prime object of their instructions, or to disappoint the expectations of the public. The writer visited the agricultural colleges of Pennsylvania and Michigan in June, 1865, and carefully investigated this subject at both institutions. He believes that the views already expressed will be fully confirmed by the testimony of the officers of those colleges. In another place we shall have occasion more particularly to refer to the systems there in operation. Manual labor should be required of every student, because in no other way than by actual practice can a man learn the proper use of implements. The processes of husbandry can no more be learned by study, than one can learn by study how to ride, or skate, or swim. A four years' course of lectures without practice would never teach a youth to mow or plough, or to plant trees, or graft or bud them. No man can safely go into the market to buy or sell live stock, seeds, manures, or any product of the farm, without practical and daily familiarity with such kinds of property.

Again: no person without a thorough knowledge of the processes of husbandry is fit to direct labor. The relations between proprietor and laborer are very delicate in this country. The laborer is intelligent, and knows when he is fairly treated, and will soon learn whether his employer is entitled to respect. Many gentlemen purchase farms, and entirely fail in their hopes of enjoyment of rural life because they do not know what a fair days's work is. They are unreasonable in their demands, and find fault with the poor fellow who has done a hard day's work, and the laborer feels that his best efforts are unappreciated, and ceases from his honest endeavors.

To encourage men of wealth of all pursuits and professions to create and occupy tasteful homes in the country is a legitimate object of agricultural education, and this can only be done by teaching the proprietors themselves the practical details of the farm, or by educating a class who shall correspond to the land stewards of England, who are competent to take full charge, for a fair salary, of large estates.

Almost every merchant, shipmaster, and manufacturer looks anxiously forward to the time when, bidding adieu to the peculiar cares of his own occupation, he may retire with a competence, perhaps to his paternal acres in the interior, perhaps to some elegant suburban residence, and devote his declining years to the peaceful pursuits of agriculture. The long-expected day arrives, and "with sweet dreams of peace" the rural home is secured. Field is added to field, and costly barns and stables are erected. Extravagant prices are paid for Short-horns, and Jersey's, and Devons, as caprice or the casual suggestion of friends may dictate; magnificent operations in draining and subsoiling, in planting orchards and vineyards, are commenced. Guano and phosphates, bonedust and poudrette, are purchased and applied to hasten nature's tardy operations. Heneries and duck ponds are constructed, and stocked with fowls of wonderful names and pedigree. The dairy, with its never-failing spring, with the thousand appliances recommended in modern treatises, is elaborately furnished. Oxen and horses, ploughs and harrows, carts, harness, hay-cutters, root-cutters, mowers and reapers, with an endless variety of small tools, all of the most costly description, are added to the working capital, and cheerfully paid for, with the certainty that by and by the harvests will bring a rich return, and the proprietor will rejoice in his successful experiment in scientific farming.

A very few years, however, are sufficient to reverse this pleasing picture. The "hired men" are unfaithful and indolent; the fancy cows break into the cornfields or young clover, and are ruined; the drains are obstructed by the frosts of the first winter; the apple and peach orchards and vines yield no fruit; the poultry cannot keep enough feathers to cover their nakedness, and much less can they afford any eggs; the potatoes rot; the horses fall lame unaccountably, and, to cap the climax of misery, the kitchen help goes suddenly off, and the "angel in the house" either takes refuge in a fit of illness or finds relief in tears, with an occasional reminder of "I told you so." Scientific farming is pronounced a humbug, and our disappointed but worthy citizen suddenly sells out at a sacrifice, and returns to his city home "a sadder and a wiser man." Such cases are constantly occurring, and they not only bring disappointment to the parties themselves, but discouragement to all who would fain believe that agriculture may be made, at the same time, a rational amusement and a safe and profitable business. These men fail because they know nothing of practical agriculture themselves, and because we have no class competent to take charge for them of their agricultural affairs. Manual labor should be required in agricultural colleges, because the cultivators of the soil are usually the owners of it, and because conveniences, as well as the theory of our government, requires that the head and the hand shall be united in the same person; and a great proportion of students will have occasion to labor on their own farms. A course of study of several years without labor would unfit them for actual work, both physically and mentally. We deem it important, too, that labor at these colleges be compulsory upon all. The idea has been suggested of leaving the matter optional with the student, and allowing those who work compensation. The objection to this is obvious. We desire, as a prominent object, to do away with caste, and especially with all distinctions founded upon an exemption from labor. Interest in the work of the farm can only be maintained by constant association of work and study, by constantly testing in the field the theories of the school-room. The idea that labor is degrading is already (though not designedly) fostered by setting apart, in our ordinary colleges, an educated class, who are not workers, and who from superior education occupy high positions. If we would dignify labor, we must combine and associate it with intellect and culture of the mind and taste, and in our agricultural colleges allow no divorce between what God has joined together — the mind and the body.

In the agricultural colleges of Michigan and Pennsylvania three hours' daily labor is required of each student. In the Michigan college, after detailing a sufficient number to take care of the stock and to attend to minor affairs, the students are divided into three equal classes, one of which works in the gardens, under the charge of the professor botany and horticulture; while the other two work in the field, under the professor of physiology and practical agriculture. At the end of a certain term the class from the garden is put into the field, and one of the other classes is put into the gardens, new details being made for the care of stock.

At the agricultural college of Pennsylvania the time allotted to labor is the same. The students labor, however, under the farm superintendent, and not, as in Michigan, under the professors. It seems to us that this is the true system. It is objected that the professors cannot have time to spend with their pupils in the field; that they need their whole time in their studies and laboratories. This is the old reason urged in a new form against combining manual labor and study. The professors of practical agriculture and horticulture, and of botany, surely should be able to find useful topics of instruction in the field, and in our battles for the dignity of labor we cannot afford to yield the point so far as to set apart an aristocracy of intellect in our own professors, by position and education above manual labor. We need the eye or the master in the field. We should hardly expect young men to submit patiently to the direction and supervision of such a man as we are at present likely to employ as farm and superintendent, and there are manifest advantages in having the labor of the pupils directed by their professors — illustrating in the field the lessons of the lecture-room, and, with the students, conducting to definite results experiments in the many vexed questions of practical agriculture.

The only objection to manual labor by students is in the supposed incompatibility of physical and mental labor. We admit that severe long-continued daily labor in the field is inconsistent with the close and absorbing pursuits of science and art, but we maintain that two or three hours of the light labor in which students of a college would participate may be healthful for body and mind.

Mr. Colman, in his reports upon the agriculture of Europe, in speaking of manual labor in such schools, says: "There can be no doubt that a man will perform more intellectual labor who devotes a portion, and not a small portion, of every day to healthful physical exertion, than the man who, neglecting such exertion, abandons himself in his study exclusively to his books. I am quite aware that many occupations of a mechanical or a commercial nature may so occupy the mind as to unfit it for scientific pursuits; but agricultural labors, quiet in their nature and carried on in the open air, when pursued with moderation, so far from fatiguing, refresh and invigorate the mind and prepare it for the more successful application to pursuits exclusively intellectual."

Is a Farm Necessary?

Whether a farm is a necessary adjunct of an agricultural college, depends very much upon whether manual labor by the students is an essential element of their education, and whether the college is to be connected with another institution or be independent. If we adopt the theory that practice and study cannot profitably be pursued at the same time, we have no occasion for a farm. Connecticut has granted her land script fund to Yale College, which has established a "course in agriculture" in the Sheffield scientific school, which will be given at large in this paper.

It is proposed in this place to call attention to two or three points, having a bearing upon the topic under consideration.

The circular says: "The details of farming cannot be learned advantageously in an agricultural school. They are only to be acquired during a long apprenticeship on the farm. No young man is well prepared to attend an agricultural school who is not practically familiar with most of the ordinary operations of farming.''

To this it may be fairly objected, that it practically excludes from the course all but the sons of farmers, for ''the comparatively high standard of admission'' prescribed is such as would not be often attained by boys who should be sent from home into farmers' families to learn practical agriculture. More than one-half of all the pupils who have thus far attended the agricultural college of Pennsylvania are other than farmers' sons — the most of them from the cities and large towns. We apprehend that such will be the case in most of these colleges in the old States, and it is desirable that it should be so. The circulation from city to country, from merchandise and the professions to agriculture, and in the next generation back to the city, so in accordance with the spirit of our institutions, and healthful to the community, promotes harmony and equality, and checks all tendency to caste.

Each position in life seems hardest and least desirable to him who fills it. The city boy sees in agriculture only visions of bliss in the country such as he has enjoyed there in his holidays, while the farmer's son regards the farm only as a place for hard work, and envies the position of the merchant and the lawyer. The parents sympathize with these views, and the sons as often as otherwise seek a different business from that of the fathers. It will not be contended that these colleges are designed exclusively for the benefit of the sons of farmers, although this is sometimes thoughtlessly assumed.

A college in this country which should not open its doors as readily to the sons of the poorest mechanic, the wealthiest merchant, the lawyer, the doctor, and the minister, as to the sons of the farmer, would occupy a position at variance with our common school system and our fundamental principles of government.

Whether the details of farming can be advantageously learned in an agricultural school depends upon the appointments of the school, the capacity of the teachers, and the apparatus provided. With an extensive farm, stocked and furnished with specimens of the various breeds of cattle, horses, sheep, and swine, and with such other animals as may be newly introduced, and with the best variety of farm implements — a farm where the ordinary as well as experimental processes of husbandy were conducted, would certainly furnish every facility for learning the details of farming. Whether, as at Yale, the agricultural warehouses and neighboring gardens and farms can, to some extent, supply the place of a farm, must depend much on location. In Michigan and Pennsylvania the agricultural colleges are too far away from any such collections or examples of good husbandry to be aided by them, and we suppose this may be the case in other States. As was said of the labor of students, so it may be said of the farm — it should be regarded as part of the apparatus of the college, and not as a source of profit. The farm that should be chiefly for experiment and educational farming is never pecuniarily profitable, however profitable it may be for education. Experiments which fail, so far as money is concerned, may be as valuable as those which succeed. A beacon or a buoy is often as valuable to the mariner as a compass, and it is as important to the farmer to know what to avoid as what to pursue. A ''model farm'' is connected with most of the agricultural schools abroad and the director is required to farm it to a profit; and this is for example to the surrounding farmers, to convince them, by actual observation, that good farming is profitable. This is more important in Ireland or France, where the occupants of land are less intelligent than with us, where each farmer knows pretty well the capacity of his own farm. The objections to it are, that by farming for profit merely we lay aside experiments and pursue the established course of the neighborhood, and we must employ the students in what they already best understand, instead of teaching them what they need learn. The idea of a *model* farm is such a farm as may serve for a model for surrounding farmers in its extent, its arrangement of buildings, its live stock, and its course and processes of husbandry. Inasmuch as in most of our States there is so great a variety of soil — wet and dry, clay, sand, and loam — it would be difficult to make any one farm a model for others. But an experimental farm should be of sufficient extent to embrace a variety of soils, and in its various products illustrate something for the benefit of all the farms of the State.

We have carefully examined the authorities upon the question of the expediency of having land connected with an agricultural college, and this question is closely connected, practically, as we have seen, with that of the independent organization of the college, or its connexion with a university.

Mr. Flint, in his report already cited, refers to the latter question, and says that volumes have been written upon it, and that in Germany it is still warmly discussed, the larger party taking ground in favor of a union, and he cites Liebig among the number.

This controversy is also referred to by Mr. Klippart, in his excellent address before the agricultural convention of Ohio, and he gives a conversation between himself and Baron Liebig, in which the baron says: "You want to teach agricultural *science* in the same manner that medical science is taught — that is, by series of lectures delivered by competent professors. You must not trouble yourself about teaching practical agriculture. The several lecturers on the several branches of agriculture can make excursions of one or two days every week, into different parts of the State, and can see and examine the operations on the best farms in the State. In this way they will learn what the present system and practice is with the best farmers; many improvements in the manual part of farming will thus suggest themselves to the students, which they can put into practice themselves. But you must teach the science of agriculture as purely, that is, with as little reference to application, as the science of geometry or trigonometry is taught.   *   *   *   But you do want 'experimental stations.' Let the object of these experiments be to obtain the greatest crops at the least expense, without impairing the fertility of the soil.   *   *   One centrally located institution, to teach pure agricultural science, is as much as you need (in Ohio) until your population has at least doubled; but if you can afford it, you should have an experimental station in each county.   *   *   You will not require a great amount of land — a few hundred acres is all-sufficient for all manner of experiments."

The argument of Liebig is evidently not against having experimental farms, but against a system of mere model farms, with schools of mere practical agriculture, where science is not taught, but where the processes of culture are learned by rote. Further on he is quoted thus: "The agricultural department to a college, without an experimental station, is simply nonsense.   *   *   The object of an agricultural college is not simply to teach what is already known, but to teach a better system of farming. How will you do this? Certainly not by employing a practical farmer to manage a model farm for you; for he knows only what is practical generally, and his superior ability will consist simply in his better *management* over other ordinary farmers. This will be teaching financiering and not agriculture. The only method by which you can possibly advance and develop agriculture, is by experiments; that is the only plan, for there is no branch of industry so completely built up by experiments as agriculture. * * So far as cattle-breeding is concerned, *all* of that can be taught at the college proper. A few of each kind of cattle, horses, sheep, and swine will be sufficient. You must not calculate that the experimental farm will, in any sense, be a source of revenue to the finances of the institution, for while some experiments may show considerable net profit, others will show a corresponding loss.''

It seems quite unnecessary that Americans should enter into the controversies which have grown up in Europe. However it may be abroad, there is no obstacle to establishing a college in each of the larger States in America, which shall, in due time, combine all the advantages claimed for both high and low schools in Europe. We assume that, in this country, our eollege is to be established for the admission, not of ignorant laborers or illiterate boys, but of youth who have had the early advantages of good schools, and who are advanced enough in common branches to enter intelligently upon courses of scientific study.

Although literary colleges already exist, they are not generally so rich in libraries, museums, buildings and funds, nor do they so engross the talent and time of scientific teachers that they may not soon be rivalled by our new agricultural colleges. To these new colleges may be attached experimental farms, where science may be illustrated and tested by practice, and where that familiar acquaintance with soils, implements and processes, andswith animals, their habits, laws of breeding and uses, and that manual dexterity with tools may be attained, which cannot otherwise be acquired by those not bred upon a farm.

Nearly every agricultural institution in Europe, high or low, has connected with it, in some way, an experimental farm. Hohenheim, the most celebrated agricultural school in the world, has nearly 800 acres in a farm and about 5,000 acres in forest. Its three independent schools are on the same estate and under the same roof, but the different classes cannot meet in the same room or field. The institute is for "young gentlemen," and the school of practical farming for the sons of peasants. The latter work nearly all the time, while the former are not obliged to labor, though they are instructed (it is said) partly "by actual practice."

In France and Ireland, as will be seen, farms are attached to all the schools of agriculture, and so it is with nearly all those in the Germanic states.

Dr. Hitchcock gives a list of 352 agricultural schools existing in Europe in 1850, and he remarks "with very few exceptions, (I do not recollect any save the University of Edinburg,) a farm of at least a few acres of land is connected with the school." And it may be added, in conclusion, that the opinion of this eminent friend of agricultural education is decidedly in favor, not only of independent colleges, but of having connected with them farms of at least 100 or 200 acres.

Plan for Half-Year Institutions

The grant of Congress being proportioned to the number of senators and representatives from the respective States, gives to the smaller States but a small fund for the maintenance of a college, and such States may prudently inquire whether some modification of a plan adapted to the larger States may not, in their own case, be expedient. The annual expense of maintaining an institution of high rank as a college in this country is probably not well understood. To enable those who are considering the matter of establishing colleges to count the cost more accurately, we give a table by the late lamented president of the agricultural college of Pennsylvania, Dr. Pugh, which, although imperfect, is of great interest.

Table showing the educational resources of the more prominent American colleges.

Colleges.	No. of professors.	No. of students.	Amount of endowment.	Annual expenses.	No. of volumes in library.
Bowdoin College	18	181	$182,000	....	30,595
Dartmouth College	20	307 $13,000	217,667	$17,907	35,402
Harvard University	56	833 $68,000	1,613,884	153,431	149,000
Amherst College	17	229	590,000	18,500	30,000
Brown University	12	202	220,000	36,000	37,000
Yale College	40	617	....	78,000	75,000
Columbia College	43	689 52,000	1,650,666	79,269	18,000
University, city of New York	36	488	250,000	14,011	10,000
New York Free Academy	25	916 42,000	....	52,590	10,000
Union College	17	276 19,400	658,000	30,000	18,000
Rochester University	11	160 10,950	123,224	13,408	7,000
Vassar Female College	..	....	408,000	....	....
Princeton College	13	221	....	....	22,000
University of Pennsylvania	28	642	306,654	26,844	8,000
Philadelphia High School	19	502	....	23,430	....
Girard College	13	400	2,000,000	85,000	....
University of Michigan	27	286	600,000	40,000	8,000
University of Illinois	38	....	427,625	....	....
Georgetown College	26	225	....	36,000	22,000
St. Louis University	26	350	....	....	25,000

Georgetown College has around it 200 acres of ground in a high state of cultivation, this too, independently of a large vegetable and botanical garden, a greenhouse, and observatory containing many valuable astronomical instruments.

The grant to New Hampshire is but 150,000 acres, which, at the price at which the scrip has been sold during the last year in the market, (about eighty cents per acre,) would give but $120,000, the interest of which, at six per cent., would be $7,200, a sum entirely inadequate to pay a corps of professors, even if the farm, buildings, library, museums, apparatus and furniture were supplied by private subscription.

New Hampshire has in Dartmouth College, at Hanover, in a strictly rural district, an excellent literary college, with a scientific school. The amount of about $18,000 is now annually paid for expenses of all kinds in that institution, as appears in the above table. The idea has been suggested, and certainly deserves consideration, whether in that State, where a large majority of the people are engaged in agriculture, and where farmers' sons must form the greater part of the students, a half-year system of study for agricultural students may not be expedient and best. A majority of the literary students of the college are usually away, engaged in teaching the district schools in winter, leaving the professors in comparative leisure.

By connecting the agricultural college with Dartmouth, a few professors in the requisite agricultural departments might be added, and agricultural pupils might, during the winter months, attend to lectures and recitations, and in summer return to their homes, or find employment wherever they could best practice the theories learned in winter. "Study," (says an officer of Dartmouth, in a letter now before us discussing this plan,) "say, from November 1 to May 1; then send home the boys, each with half a dozen practical problems about soils, fertilizers, crops, &c., to be wrought out experimentally, and results noted and reported at the beginning of the next term. This would turn the whole State into an agricultural farm, and make all the farmers who had boys here, or whose neighbors were thus favored, both teachers and pupils. In the warm months our leading professors could lecture in different parts of the State, thus diffusing knowledge and awakening interest."

KANSAS

The Kansas State Agricultural College, formerly the Bluemont College, "opened under the auspices of the State in September, 1863," (says the superintendent of public instruction,) "and has been doing a great and good work in the education of teachers, and in training young men and women for active business life, and also in fitting them to graduate from the highest course of a first-class collegiate institution." A president and four professors are employed, and the number of students was 113, as shown by the catalogue of 1865. The ages of the students range from 9 to 27 years, there being a large preparatory class. The college is at Manhattan, and has 80 acres of land, a college building, and the foundation of a library. The annual expenses are estimated at only $4,000 a year. A boarding-house is about to be erected and the institution, now in its infancy, has large prospective means. It is believed to be the only agricultural college where females are instructed. We have not at hand any definite programme of its course of study.

MAINE

After much discussion, the agricultural college of Maine has been located at Orono, and is to be conducted as an independent institution. No buildings have yet been erected, and no plan of organization has been published.

So far as can be learned, no other agricultural colleges than those above noticed have yet been established. The Maryland Agricultural College, established as early as 1857, and still in operation, has a farm attached, but is rather a school of general education than of agriculture distinctively.

The Mechanic Arts

The act of Congress provides that colleges maintained by its provisions shall teach, not only such branches of learning as are related to agriculture, but such as are related to the mechanic arts.

Massachusetts has granted the income of about one-third of her fund to the Institute of Technology, where the mechanic arts receive special attention, and her agricultural college is therefore regarded as released from obligation to teach the mechanic arts, further than they are essential to agriculture.

A good water-power, with shops of various kinds, or steam or caloric power for want of water, are greatly to be desired connected with every agricultural college. The act of Congress calls for earnest attention to the department for instruction in the branches related to the mechanic arts, which seem to have been nearly overlooked. It is hoped that the subject may receive due consideration in the organization and progress of these institutions.

We close our paper with the following conclusions:

1. Public sentiment and the public good require a more practical course of education than our literary colleges afford, with more attention to modern and less to ancient languages.

2. Colleges established under the act of Congress should "teach such branches of learning as are related to agriculture and the mechanic arts," both scientifically and practically, so as to prepare their students to labor and to teach in the highest branches of their respective pursuits.

3. If the means are sufficient, these colleges should be independent, and not united with existing colleges.

4. But one such college in a State should be established. Experimental farms or stations, or subordinate schools, may be organized in counties or districts.

5. Manual labor for practice and education is essential to education in agriculture, and should be required of all students in colleges which have farms attached.

6. Farms for experiment, illustration, and practice, with live stock and farm implements, are essential to strictly agricultural colleges.

7. Where means for independent institutions are wanting, a half-year system of study in winter, and labor at home or on an experimental farm in summer, is practicable.

8. The promotion of equality, and the dignity of labor, being principal objects in our government, we find no models for our agricultural colleges in the aristocratic communities of Europe.

Agricultural Colleges and Experiment Farms

From *American Agriculturist* (New York, May, 1870), XXVII, p. 170.

A pamphlet issued by the Trustees of the Agricultural College of Penn., reporting the results at their *three* experimental farms, has been received. The report presents a mass of results through which the inquirer must wade for hours to learn much. Yet there is a good deal of value in it. We have the highest opinion of the good which well managed, experimental farms may be to the community — and can well see that great benefit might be derived by the students at an agricultural college, if they could watch and help conduct the experiments; — but these farms seem to be conducted entirely separate from the college, and without any idea of giving instruction to the pupils, according to the provisions of the Agricultural College Act. This may be all right, and whether it is or not, do let us have well digested statements of the experiments, showing what they were undertaken to prove, and a classification of the results, showing clearly what they teach. If we criticise unfavorably those institutions, which, accepting the bounty of the government, teach agriculture without practical demonstration, what shall we say of Agricultural Colleges ''running'' farms 150 miles from their students?

The Civil War

King Cotton, 1861

From *National Intelligencer* (Washington August 27, 1861), XLIX, p. 3.

The Charleston Mercury continues to put faith in the imperial sway of King Cotton. In a recent number it holds the following language with regard to the dependence of foreign nations and their "coercion" by this staple:

"We can make the foreign nations who require our cotton our friends — nay, our allies against the United States, to put an end to the war which interferes with their necessities and welfare. Neither to Great Britain nor France would a war with the United States be one half as disastrous as the deprivation of the cotton of the Confederate States. The commerce of the United States is nothing to them. On cotton, therefore, more than on diplomacy; on cotton more than on fighting on sea or land, do we rely for coercing the recognition by foreign nations of the independence of the Confederate States and the termination of the war."

The Confederacy and Food Supplies, 1862

From *The War of the Rebellion — Official Records of the Union and Confederate Armies*, Series 4 (Washington, 1900), I, pp. 870–79.

CONFEDERATE STATES OF AMERICA, SUBSISTENCE DEPT.,
Richmond, Va., January 18, 1862.

HON. J. P. BENJAMIN,
Secretary of War Confederate States of America:

SIR: I have the honor to acknowledge the receipt from you of a copy of the resolution of Congress dated 11th of January, 1862, to wit.*

The current indispensable business of this office and the comprehensive nature of the resolution has caused delay. The papers herewith submitted, being "Abstract of purchases and returns of provisions," "Abstract of engagements of salt beef and bacon,'' marked B; copy "Contract of Wilson & Armstrong,'' marked C; copy "Contract of Haxall, Crenshaw & Co.,'' marked D; and Maj. F. G. Ruffin's report, marked E, will, as far as practicable, fulfill its requirements. This Bureau has been conducted on the principle that the essential duties of its chief are to ascertain and to anticipate the present and future wants of the commissariat in general and particular; thereon to apply to the Secretary of War for the means, and to appoint the agents deemed most competent to accomplish these objects; then to effect them. On commencing the organization in Montgomery it was known that supplies, especially of salt meats, could not be obtained to an adequate extent except in the enemy's country. Accordingly appropriate steps were taken to reach them. The stores of bacon and pork thus acquired, at a cost to the Government of much less than one-half the current rates, are still being issued. In considering the question of a future supply of salt meats the inapplicability of the rules of purchasing prescribed by the Army Regulations was as obvious as the difference between peace and extensive fields of supply on the one hand and revolution on the other. Under existing circumstances an absolute deficiency of pork in the Confederacy added to the contrast. To meet this vital necessity and the competition that was inevitable only one way was open; that was to anticipate its operations and contend with its development. This was determined on, and arrangements were initiated early in July. In the middle of August certain papers were sent to me with directions to return them with my remarks in relation to their respective contents. After disposing of the subjects included I closed my communication with this paragraph:

> The real evil is ahead. There are not hogs in the Confederacy sufficient for the Army and the larger force of plantation negroes. Hence competition must be anticipated by arranging for the purchase of the animals and getting the salt to cure them. Furthermore, beeves must be provided for the coming spring. Cattle must be collected from Texas before the rains set in, and be herded in ranging grounds convenient to the Mississippi. I am arranging for these matters.

Though not — so far as packing and curing was involved — in accordance with the rules prescribed by the regulations for purchasing supplies, the War Department concurred in these views, and they have been prosecuted vigorously, to make use of the whole season and withdraw the products as soon as possible from the hostile front along which they have been prepared.

The abstract of contracts and the statement of Major Ruffin thereon, from the data which we had collected, exhibit the nature of the operation. The responsibility of recommending it and the expenditure of such large sums — the products being necessarily laid so near the enemy's lines — has been heavy and is the best guarantee that every plan of meeting the necessity that was on us had been carefully considered. There appeared no alternative. The existing establishments and the experts best adapted to securing the end had to be employed in the interests of the Government, or to be its antagonists. Among the agents there was one exceptional case, whose action having long ceased, and the entire results of it been taken off the hands of the Government before its extent was known, his name does not appear, but the correspondence is on file in the department. Serving without compensation, and not intended for any specific action, he was not appointed by the War Department as other agents are. The following is in relation to him: To begin operations in

Kentucky I proposed to a gentleman of that State, of large fortune and above all suspicion, to make a tour and urge the farmers of Southern feelings to hold their hogs for the Government. On his making favorable reports I asked him to see about arrangements for a packing-house for receiving the hogs and curing the meat. On the 23d of October he reported that he had made a conditional contract for the packing establishment at Clarksville at a rent of $8,000. The alternative presented to me was to permit competition in an important position and lose results hoped for from Kentucky, or accept. The rent seemed extortionate, and I paused for information.

In the meantime Mr. R. T. Wilson, with whom contracts had been made to purchase and cure meat in Kentucky and Tennessee, came to Richmond shortly after this proposition was presented to me and agreed to go to Clarksville and hire the house himself, believing that it was important to secure it for his operations in Kentucky, accepting on the private account of himself and company the action of this preliminary agent.

It subsequently appeared that before this was done the conditional contract had been disputed and $10,000 absolutely demanded by the owners, and had been yielded. But the matter had passed from the hands of this department. Whether better could have been done or not thus becomes immaterial, but I must defend this disinterested agent. General A. S. Johnston seems to find the securing of that establishment and the packing arrangement of this department opportune, for he has directed the products at Clarksville to be transported to Bowling Green, and ordered the agents from Nashville to slaughter and cure from 5,000 to 8,000 hogs at the latter place as a reserve for his army.

I proceed to another subject. All subsistence stores that are allowed to the Army have to the fullest capacity of our country been obtained, and no essential supplies have ever failed to be ready for transportation when and where required, timely notice having been given to this department. It is known to the War Department that from the time I came to Richmond I urged the opinion that the railroads would be found unequal to the demands that soon would be made on them and that subsistence stores must sometimes fail to reach their destination. This terminated my duty, but not my apprehensions. Frequently I have had occasion to make such representations. To illustrate it, sugar is now needed in the camps, and there are invoices here of thirteen different shipments from New Orleans of near 1,400 hogs heads on the way, starting from November 27 onward, and stores have been on the road from Nashville from one to three months. On the 12th fifty barrels of pork arrived that were shipped on the 22d of August, 1861. In this communication I referred previously to papers sent to me in the month of August, with directions to make remarks thereon. I advert again to them.

One of those papers contrasted the prices of flour at Manassas with that sent from Richmond, leading to the inference that purchasers here could be so foolish or so criminal as to pay 75 per cent more for equal grades. The facts are, that the flour sent from here was worth from $1.50 to $2 more than similar grades of country flour. This difference is fixed by market prices. Second. That the flour though sent in the month of July to Manassas, and according to department rule invoiced at actual cost, had been bought in the month of May, long before I came to Richmond and without the possibility of my knowing anything of it. The resolution of Congress covers all provisions that have been purchased. A large class of accounts have been contracted by officers conducting troops from places of rendezvous and enrollment. Third. Other commands for which preparations of cold provisions had been made have, by delay on the roads, been provided on those occasions by special purchases absolutely necessary, generally economical. None otherwise have been observed. Another class of accounts have been tavern bills, which have in almost every instance been rejected, and none paid but extreme cases of a very special nature, from which no precedent could be deduced. It may safely be affirmed that troops thus rapidly assembled from remote points could not have cost less for subsistence en route.

Referring to the papers accompanying,

I am, sir, very respectfully, your obedient servant,

L. B. NORTHROP,
Commissary-General of Subsistence.

[Sub-inclosure.]

CONFEDERATE STATES OF AMERICA, SUBSISTENCE DEPT.,
Richmond, January — , 1862.

Col. L. B. NORTHROP,
Commissary-General of Subsistence:

SIR: By your instructions I proceed to submit the following report on the purchases and contracts made by this department:

Salt Meats

The supply of salted meats was that which the department felt most solicitous to secure. Provisions of that kind have been heretofore mainly sent to the South from States now foreign to us, or the seat of active hostilities. Reliance on that was out of the question after the amount that could be got early in the war had been obtained. In the packing season of 1860–'61 upward of 3,000,000 head of hogs were packed at the various porkeries of the United States, besides those packed by farmers at home; of which less than 20,000 were packed at regular establishments south of the lines of our armies. Of this whole number experts estimate that the product of about 1,200,000 hogs was imported in the early part of the last year from beyond our present lines into what is now the Southern Confederacy. This was accomplished, and to the extent of a bountiful supply, by the action of State authorities in some cases, by the enterprise of private parties, and by this department through agencies of its own. Of this number it is estimated that about 300,000 hogs, in their bacon equivalent, have been consumed by our State and Confederate armies since the commencement of hostilities.

Tennessee then became the main reliance for a supply for the future use of the Army, which, together with the accessible portions of Kentucky, had been so ravaged by hog cholera and injured by short corn crops for three years preceding the year just closed that the number slaughtered at the porkeries had dwindled from 200,000 head to less than 20,000. It was into this field, just recovering from these disasters, and almost the sole resource of the Army, the planters, and the inhabitants of cities, that this department had to enter as a purchaser — dubious of a sufficiency, but assured of a heavy and active competition. If, when the price of hogs was only 6 to 7 cents per pound gross in the South, it had been the custom of many planters to buy the live-stock from the drovers and put up their own supplies, drovers would of course take hogs to them when the price was double, and supplies thus diverted could never come into army consumption. But besides this loss, what would have gone into commercial hands would also have been open to the planters' bids, and must have been lost to the Government or secured at exorbitant prices. Now, if the usual mode had been adopted of obtaining supplies by bid and contract, it is obvious that, as each speculator or packer could operate most profitably on a theater of scant supply, and contracts under that system could not have been awarded to all, those who failed to get contracts would have made as much, if not more money, by speculating against the Government than by working for it. This state of things would have wrought the double effect of raising prices upon the Government and preventing its full supply, and the latter would have been disaster, if not ruin. To prevent this it was necessary to combine all the packers in the interest of the Government, and to accomplish that it was necessary to offer them a fair and liberal compensation, placing all upon one footing. Such compensation, it was clear, they would have at any rate, and in most cases without the outlay of capital in buildings and fixtures which their undertaking for the Government would require. This compensation, though liberal, was not exorbitant, and in view of the uncertainties of the times was not more than ought to have been offered. It was paid in kind and in a class of products perishable in their nature, for which the Government had no use, and by the sale of which, on a large scale, it could have made little or nothing. It will be fully understood by reference to the contract with Wilson & Armstrong, herewith inclosed, marked C, and given as a sample of all contracts on the same subject-matter.

The only danger would be that under the stimulus of gain the contractors might compete with each other and so run up prices. This however, was partly prevented by the practical difficulties of the undertaking, such as the impossibility of obtaining cooperage for unusually large quantities of lard and the risk of preserving offal. It was further endeavored to be provided against by districting the country, as far as an imperfect knowledge of its agriculture would permit assigning each contractor to certain limits. But even if such competition has existed, its effect has certainly been to secure more bacon to the Government and at lower prices than the opposite system could possibly have brought; and the main object was to secure a full supply cost, however important, being secondary to that. That object accomplished. The number secured is about 250,000 head of unusually large hogs, including some 20,000 which have been obtained from Kentucky within the Federal lines by Government agents, acting under the instructions of this department, at much risk and with occasional losses to those undertaking it. The above number is increased by others obtained at other points and mostly on similar terms, as may be seen by reference to the abstract of purchases of hogs and beeves herewith transmitted.

At Thoroughfare, in the rear of Manassas Junction, a packing establishment has been put in operation on Government account and the same has been done in Richmond, and at each of those points every product is saved to the Government, because it either finds ready sale or prompt and grateful consumption by the Army. The management of this important work has been intrusted to agents under official bonds, or under contracts secured by bonds. To these the money has been advanced as needed, when it could be obtained from the Treasury, and while no greater advance has been made than necessary no greater risk has been incurred than the usual confidence reposed in officers of the Army disbursing large sums of money. It is due to the patriotism of these agents to say that in several cases when funds were scarce they have freely advanced both their funds and credit to their respective trusts. Thus Wilson & Johnson and Wilson & Armstrong have advanced at various times about $520,000, J. H. Craigmiles and J. F. Cummings, respectively, $878,878 and $2,204,977, and all these have strained their credit to its utmost tension to ease the strain upon the Treasury; others have assumed obligations of the same character. These agents were severally instructed to set the price, first at 6, subsequently at 8 cents per pound, and if these prices would not secure the hogs, then to give such as would do it, but in no instance to go higher than was absolutely necessary. Under these circumstances prices have ranged from 6½ to 11½ cents, the latter in a few cases, and of late they have receded from these figures. These prices have been high, but the rate of rise is not greater than in other kinds of produce of prime necessity and scant supply, and not as great if the greater scarcity be considered. They could not have been kept lower except by a military order prohibiting exportation. Such an order was applied for, but refused, to the great enhancement of prices. In other cases it became necessary to get military authority to enforce contracts against numerous parties to whom higher prices had been conceded on contracts for lower rates, but who were only emboldened by such concessions to make still higher and more dishonest demands. The compensation of these agents has been in kind — as when they assumed the character of agent and contractor — or in money alone, which has occurred in two cases, and in both these the rate is $500 per month for the time of service, being a much less sum than could have been made by the same parties operating on private account.

If frauds have been committed under this system they have not been heard of except in rumors, which, upon investigation, have either failed or declined to assume a proper accusatory form, or in the hints and insinuations of scandal or slander. The whole course of the department in obtaining its supply of hogs has been guided by the policy disclosed in the above. To complete the supply of salted meat, beef has also been engaged, to be packed at different places, stated in the accompanying paper, marked B, at prices adapted to the various localities. The establishments at which this work is done are all under the charge of agents (or, in cases where the amount packed was deemed too small to justify the appointment of an agent, it is intrusted to the parties themselves) who are under bonds to furnish a merchantable article. Further contracts will be made, or existing contracts extended, so as to insure enough beef to subsist the troops until the returning summer shall again afford an abundance of fresh beef. The different agents and contractors have been instructed to put as much of this beef as was practicable into barrels or tierces, but it has been impossible to put it all into such packages. Cooperage is scarce and high, and enough coopers cannot be had at any price to make the requisite number of barrels. To meet this difficulty the packers who are convenient to the different forces have been instructed to use boxes in which the beef, after having been brined in the usual way, is salted down and directed to be distributed for speedy consumption. The plan has now been tried sufficiently to prove its efficacy, and if timely transportation can be had there is no doubt of its success. The barreled beef will be kept for later consumption and moved as fast as prepared, and when transportation can be had, either direct to its destination or to secure depots for future consumption and distribution. The bacon will be reserved, as far as can be, for summer and fall supply. The price paid for this beef has varied, according to locality and the condition of the animal, from 3¼ to 4½ cents gross per pound, averaging less than 4 cents, and will go higher as the season advances. The contractors and agents have been instructed in their purchases to consult economy up to the limit of fair market rates, and never to exceed them. The compensation has been the fifth quarter, as it is called, which was the usual butchers' profit from time immemorial. In the case of R. A. Porter, of Louisiana, it is different, because he had to erect, upon short notice, an immense establishment, and had to furnish his own salt and cooperage, and his hides, requiring a larger amount of salt to preserve them, would yet bring less than those nearer to manufacturers. His compensation, therefore, has been apparently increased by 1½ cents per hundredweight gross. Still the beef cured by him will not amount to 10 cents per pound laid down at Memphis, a far less sum than it can be had for in the general market. The department has establishments of its own of this kind at Richmond and Thoroughfare, erected for the same reasons and conducted on the same principles as for hogs at the same places.

Fresh Beef

Whenever it has been practicable the commanders of the different forces have not been interfered with in obtaining fresh beef in their several bounds. As a general rule, local officers can make such purchases as well as this department, and with more satisfaction to the generals. It is so difficult to control commissaries who are under the special orders of such officers that it has not been attempted except when deemed advisable in special cases, and it has been compelled to let these purchases rest upon the administrative responsibility by the commanders. All that has been done in such cases has been approve the contracts made by such commissaries as have submitted them to this department. As in some cases this has not been done and the purchases made have only appeared through returns of commissaries to this office, a statement of such contracts as have been submitted is not given, since it might mislead as to the quantities provisions that have been or are being purchased. But where the commands had to be supplied from a distance, or where, from the vast size of the force or the probable conflict of purchase among the commissaries of different commands, difficulties might arise, a different course has been pursued; officers or agents of the department have been appointed or detailed, and they have been instructed to furnish by direct purchase and through such sub-agencies as they might deem necessary, in districts geographically prescribed and where they were the sole purchasers, the proper number of beef cattle at their appointed depots. General instructions have been given to all these parties to discourage speculation by refusing to buy at second-hand whenever practicable, but no minute instructions were necessary in these cases from the character of the officers and agents and their thorough acquaintance with their duties. Wherever that has been done the price of cattle has been kept at a moderate rate and arrangements have been made with more or less completeness and celerity, as the exigencies of the case or of public business would permit to save to the Government all or a very large share of the profits of the fifth quarter. The prices in these cases have varied from 3 to 4½ cents per pound gross. The agents thus specially employed are Maj. B. P. Noland, of Loudoun, for the district that feeds Manassas and Fredericksburg; Mr. William M. Tate, of Augusta, for the district that feeds Richmond and the Peninsula; Mr. C. L. Snyder, of Roanoke County, commissary agent, and Mr. T. J. Higginbotham, of Tazewell County, for Southwestern Virginia; Mr. William Falconer, of Petersburg, for Norfolk City, and Mr. R. T. Wilson, of Loudon, Tenn. In one or more cases some of these gentlemen have found it necessary to employ sub-agents.

Flour

Want of money has prevented such contracts for flour as would have secured contracts for the whole year when wheat was low. In the absence of funds it was necessary to attempt some plan by which after supplying its immediate wants, the department might regulate its future prices. All that it could do in that direction has been to adopt a system by which its purchases could be arranged for present and future supply on a basis which would be safe for the Government and ought to be satisfactory to the seller. That basis was simply the application of the universally accepted commercial law that the price of any article not at a ruling market must be the price of that market less cost and charges. As our ports were all blockaded different flour marts were assumed as the points of sale and the deduction scaled by the distance of the seller from his usual market, and all such markets were generously put upon the footing of the best, though there had always been a very considerable difference among them. In addition to this it was also determined to adopt in such contracts as might be made a sliding scale by which flour should rise with any anticipated rise in the price of wheat, thus guaranteeing the contractor against loss and guarding the Government against applications for relief — a most fruitful source of corruption. The price of flour under this policy was fixed upon the price of wheat at $1 per bushel, at which the Government stipulated it should commence in this market. If this course has produced discontent it was because it was not understood, or because parties who had wheat to sell could not comprehend that a very abundant article must rate low in the market, whilst articles of as much relative consumption but of absolute scarcity should command far greater prices. This policy of the department has been somewhat interrupted by speculation, though that is now believed to be subsiding, but it was mainly thwarted by the want of money and transportation, with which at command it could have made large purchases before the rise in flour took place.

The only large contract the department has made has been with Messrs. Haxall, Crenshaw & Co., a copy of which is herewith transmitted, marked D, and the officers and agents of this department have been instructed to observe its principles in their own similar transactions. As this contract has been the ground of much unjust animadversion upon the department and the contractors, it may not be amiss to state, in justice to the propriety of its selection and their liberality, that where they had an admitted right to a compensation of $6.76 per barrel they voluntarily remitted 26 cents per barrel, or $6,500 of their claim.

The only agents to purchase flour that it has been thought necessary to appoint are Mr. James M. Ranson, of the county of Jefferson, Va., and a party (whose name is known because it has been very recently determined on, and has been intrusted, for special reasons, to Maj. B. P. Noland) for the county of Loudoun. Both these parties have received or will receive instructions from the post commissary at Manassas, to whom full authority has been given in the premises. The other purchases of flour have all been made through regular commissaries.

The amount of flour purchased up to this time will be found in the abstract of purchases, sent herewith. A resolution passed by Congress at its last session directed the erection of bakeries to furnish "well-baked bread" to troops in the field, or in lieu of that, that contracts might be made for the supply of such bread. Such bakeries have been erected wherever practicable or where the Army Regulations did not provide for the case. But it was found necessary to procure a bakery in which hard bread should be prepared, an ample supply of that being represented as indispensable; and though these representations were not concurred in, yet it was deemed proper to meet this requisition, and accordingly, it being impossible to contract for any large amount of bread, a bakery was purchased and put in operation with complete machinery. Its full working capacity is 140 barrels of flour or 280 barrels of hard bread per day of twenty-four hours. Since the bakery has been in operation the wants of the Army for hard bread have decreased and there is now on hand here a stock of 2,700 barrels of hard bread and 330,000 pounds at Manassas. This bread, made of superior flour, is cheaper by 12½ per cent. than it could have been bought from outside parties making a very inferior article.

Sugar and molasses are purchased and only await transportation to be furnished in full supply at all the camps. These articles are had at New Orleans from first hand. Rice is being purchased under agencies which are now in the course of completion, and it is hoped bonds can be used exclusively in payment. In addition to the quantities of salt reported, contracts have been made with Messrs. Stuart, Buchanan & Co., at the salt-works near Saltville, in Southwest Virginia, to secure an ample quantity of salt for army supplies and for packing purposes. The first of these stipulates for 10,000 bushels of salt per month, at 75 cents per bushel of fifty pounds; the second for 22,000 bushels per month, at the same price, plus the cost of bags or barrels at the option of the manufacturers. Besides this about 40,000 bushels were purchased at Nashville at $3 per bushel, to be used in packing pork and beef, transportation from the salt-works being impossible. Besides the above-recited purchases or arrangements to purchase supplies of all kinds, local commissaries at various places have made and are making similar purchases by the direction or under the sanction of their respective commanders, whose duty it is, by Army Regulations, to give proper supervision in the case.

It will be observed that this department has not been conducted on the system of contracts by bids. That system as a whole is not approved of, though in some cases and in favorable circumstances it may be advantageously blended with the system of purchases directly or through agencies. But if it had been the best, as a general rule it could not have obtained in the administration of this department in the circumstances which have surrounded it.

Very respectfully, your obedient servant,

FRANK G. RUFFIN,
Major and Commissary of Subsistence.

Exhibit B

Statement showing the actual and contingent number of beeves and hogs contracted to be slaughtered, and of bacon and fresh beef contracted to be purchased for and by the Commissary Department for the subsistence of the C. S. Army.

Williams & Lancaster, Bristol, Tenn., 12,000 hogs; T. J. Bretlow, Southampton, Va., 10,000 hogs; D. Morris & Co., Morristown, Tenn., 25,000 hogs; Wilson & Armstrong, Nashville, Clarksville, Bowling Green, and Patriot, 66,000 hogs; Wilson & Johnson, Loudon, Sweet Water, and State Line, 14,000 hogs; Government account, Thoroughfare, Va., 10,000 hogs; Government account, Richmond, Va., 1,300 hogs (beef is packed at both these places); John Blacknall, Oxford, N.C., 500 hogs; Cummings, Gilkeson & Co., Nashville, Tenn., 35,000 hogs, 6,000 beeves; Cummings & Waterhouse, Shelbyville, Tenn., 35,000 hogs, 25,000 beeves; Chandler & Co., Chattanooga, Tenn., 25,000 hogs, 2,000 beeves; J. H. Craigmiles, Cleveland, Tenn., 1,000 hogs; H. B. Henegar, Charleston, 1,000 hogs; J. M. Toole, Maryville, 1,200 hogs, John Grant, Muddy Creek, 2,000 hogs; C. M. McGehee, Knoxville, 10,000 hogs; R. A. Porter, Alexandria, Va., 20,000 beeves; C. L. Snyder, Salem, Va., 2,000 to 3,000 beeves; D. Morris & Co., Morristown, Tenn., 500 to 2,500 beeves; Wilson & Johnson, Loudon, Tenn., 1,000 to 2,000 beeves; Wilson & Armstrong, Nashville and Clarksville, Tenn., 15,000 to 20,000 beeves; A. Cone, Bulloch County, Ga., contingent.

Contracts for Bacon

George W. White, 1,000,000 pounds, or a sufficiency for Fort Smith and Fayetteville, Ark., 15 cents per pound; George W. White, 500,000 to 1,000,000 pounds, to be delivered at San Antonio, Austin, Navasota, and Jefferson, Tex., 15 cents per pound; John G. Todd, 450,000 to 650,000 pounds, Galveston, Houston, or Columbus, Tex., 15 cents per pound; P. C. Pendleton, 100,000 pounds or more, 17½ cents per pound; A. Cone, bacon contingent quantity 17½ cents per pound; A. Cone, pickled pork, contingent quantity, 13 cents per pound; A. Cone, bulk meat, contingent quantity, 15 cents per pound.

Contracts for Beef on the Hoof

George W. White, all required by the troops in West Arkansas, Cherokee, Creek, and Choctaw Nations, and as far north as Springfield, Mo., at 6½ cents net per pound; George W. White, all required by the troops for the coast of Texas, Corpus Christi, and all points east of it, at 6 cents net per pound; R. A. Harvard, in Confederate bonds at par, 8,000 to 10,000 pounds, Prairie Home, La., at 6⅓ cents net per pound; Price Williams, 3,000 pounds, Mobile, Ala., at 9 cents net per pound.

Merinos and Their Wool

From *Southern Cultivator* (Augusta, Ga., July–August 1862), XX, p. 137.

EDITOR SOUTHERN CULTIVATOR — As there is much interest being felt, just now, in the wool-growing business at the South, I send you an account of the yield of my little flock, including everything that we sheared. Twenty-seven Rams and Ram lambs (seventeen of the latter, three of which were very small and late,) yielded 245¾ lbs. being an average of nine pounds and a small fraction. Our breeding ram's fleece weighed 15½ lbs. Some two year olds weighed 12, 11½, 11 and so on. Two large French Merinos 10 lb. each. Some ram lambs yielded 11½, 11,10, and so on. Thirty-one ewes, all but five being lambs, (including an old ewe thirteen years old, a sickly one and one lamb of last summer) yielded 185 lbs. an average of six pounds, which was almost half washed, by repeatedly crossing water in the pasture — the rams had not been wet.

These are all full blooded Merinos, and most of them Spanish and small, showing their wonderful capacity for producing wool. The French have more size and style than the Spanish Merinos, but where wool is the main object, I prefer them to the French, though if you want both wool and mutton in equal degree, the French are the sheep, the only material difference being in size; but when compared with other kinds, I would prefer the most inferior Merino I ever saw to the best of any other kind, and all the Southern people have to do to find it out to their entire satisfaction, is to try them.

Most Respectfully Yours,

JAMES W. WATTS.

Cartersville, Ga., May 1862.

Advice to the South, 1862

From *Southern Cultivator* (Augusta, Ga., July–August, 1862.), XX, pp. 142–43.

The contest in which we are engaged, must produce a revolution in our industrial pursuits, certainly so far as the plantation States are concerned. Heretofore we have had but one money crop. We have sold Cotton and little else. It has been supposed to be good economy to buy almost every thing that we have consumed. This state of things is brought to a sudden close. We are glad of it. Loss of life is indeed to be deplored. But so far as expenditure is concerned, if this war shall cost us millions, it will have been money well spent, if it have the effect of so developing our resources as to render us commercially an independent people. It would seem that nothing short of this violent convulsion in public affairs could bring our people to reason. To talk or write of permanent and not temporary and fallacious prosperity; to remonstrate against the excoriation and denudation of our soil; to utter warnings against the time when the exhausted earth should cease to bring forth her accustomed products, was all throwing straws against the wind, so long as Cotton bore a full price, and that price could more conveniently buy the necessaries of life than we could make them ourselves.

The keen-sighted Yankee has been smitten with blindness. His silly blockade not only closes his own market for the present, but for all future time. He will compel us to see what we should have known before, that the great bulk of the articles with which he has supplied us can be made more cheaply by ourselves — a thousand times more cheaply, when we take into account the rapid improvement of our soil, which must be consequent upon a diversified industry.

Let us look around us. This change in our pursuits must be promptly made. We have no time for delay. There are articles of prime necessity which we have been in the habit of getting from the North, which we can get no longer from that quarter. What are they? Can we make them ourselves? If so, what is the best way of doing it? These are important questions, and deserve prompt and thoughtful attention.

So far, the promise of Grain is abundant. Unless some accident should happen to the Corn crop, we may dismiss apprehensions about bread.

The same cannot be said of Meat. Hogs have been terribly thinned out by disease. The scarcity of food for several past winters has greatly diminished the number of Cattle. Dogs have been suffered to commit such ravages among Sheep, that if this county (Bartow) is a fair sample of the South, the number of Sheep has been seriously diminished.

The supply of meat demands grave attention. No sow pigs should be spayed, at least until after they have brought one litter of pigs. By adopting this course, for this season, the deficiency could be made up. The hogs for next winter should be pushed from the time that they have gleaned the harvest fields. In fattening Hogs, it should be borne in mind that Corn cut up and thrown to Hogs, just after it is in roasting-ear goes much further than when it is matured. They will then eat all the blades and a good deal of the stalk. They fatten much faster then than in cold weather. Where Corn is cut off at that early stage the Pea crop is much increased. By letting the Hogs get a good start in the harvest fields, pushing them with green Corn in a field or large shady lot, and afterwards giving them the run of the Pea fields, small Hogs may be made to attain a heavy weight. By these and similar precautions, we may obtain a supply of bacon. We shall not obtain it unless extraordinary effort is made. No heifer calves and no ewe lambs should be killed. Why do not our planters cure mutton hams? Why do they not make corned beef? Both are excellent articles of food for white and black, and both are much less expensive than bacon, though not so convenient for use.

In this connection, although not an article of food, it may be proper to advert to a kindred subject. *We should for the present stop raising mules* and raise horses alone. The epidemic among horses has destroyed great numbers of them. There are now scarcely enough in the country for farm and plantation purposes. Thus far, in the conduct of the war, we have had but little need for cavalry. The points threatened heretofore have been chiefly on the sea coast. As the war assumes the form of invasion, dragoons will be increasingly demanded. Where are the horses to come from, and yet keep up the necessary operations of the farm? In the county in which we live (Bartow) which is a farming county, we learn that there has been difficulty in obtaining horses of the right kind for a single cavalry company. The deficiency, where there used to be an over-supply, arises from the two causes mentioned — the epidemic and the excessive raising of mules. Brood mares are now difficult to get. If the practice of mule raising be not arrested, we shall suffer serious inconvenience from it.

Heretofore our Hay has been brought almost solely from the North. Planters have usually made enough fodder for their own consumption. But not enough to supply the towns and cities. This is an indispensable article. Great quantities will be needed for army purposes, as well as for city supplies. There are some expedients to which we can resort for present purposes — our permanent reliance must be upon meadows.

As soon as small grain can be got out of the field, Peas can be drilled by running furrows three feet apart and covering with another furrow, and then breaking out the middles. As soon as the Peas begin to ripen, they should be cut with a scythe. On rich land, it is better to sow Corn at the same distance and cut and cure it. An acre of Corn Fodder cut, will make more Hay than an acre of Peas, but requires much richer land. Both can be packed into a bale in the common Cotton screw, and can be shipped to market in neat and portable packages.

For Hay for future use, Lucerne should be sowed about September first. By sowing it so early, it will give a heavy cutting next year. Remember that it should be sown only on the richest of upland, broken at least ten inches deep — the deeper the better. Lucerne seems to require but two things at the South — that the land on which it is sowed should be rich and dry. It thrives as well on the sandy soil of our islands as it does on the clay lands of the middle country.

On almost every plantation, there are either river or creek bottoms, or branch lands, which are too wet for cultivation. Planter or farmer — these are the most valuable parts of your plantation. Begin on them as soon as you lay by your crop, or as soon as they are dry enough to allow you to enter them. Cut down the timber upon them — grub them carefully — burn everything that can be burned, and sow early in the fall with Herds Grass and White Clover. Other grasses might be mentioned, but the seed cannot probably be obtained. Herds Grass can be gotten in quantity from Tennessee. You need not drain the land for Herds Grass no matter how wet it is in the winter, provided the water is not deep and stagnant. Don't plow such a piece of ground — you will tear it up into tussocks. Scratch it with an iron-toothed harrow, then sow the seed — if it is raining, it need not be covered. If it is dry, harrow it again or brush it in. Your first crop will be mainly swamp grass. That should be cut very early and will make tolerable hay — the Herds Grass and Clover will finally take possession. It seems that only those things which were meant to be cut with the scythe, can stand cutting — the whole tribe of annual weeds and grasses disappear before it.

What are we to do in the cities for Butter? The question is applicable not only to cities but the interior towns and villages. Northern butter has been constantly brought in and was in fact a reliance, in Atlanta and other towns as far removed from the coast. So far as we know, there is not a dairy farm in the State of Georgia. Why should we not make butter enough for the supply of our demand. There is no good reason for it. There is a mistake on this subject. It is supposed that there is little profit in butter making. Nothing can be more erroneous. The profit is large, where it is made a business of by proper persons.

We had an opportunity once to look carefully into the farm accounts of a Scotch nobleman's estate. Nothing could be more exact and accurate. No merchants books could be better kept. These books showed a result which surprised us. The land of the estate, 600 acres, was valued at $500 per acre. It averaged in Wheat from 30 to 40 bushels per acre — the usual price of Wheat was about $2 per bushel. A dairy of 100 Ayrshire cows was kept on the farm. The butter was sent daily to Glasgow, 30 miles distant, by rail and was sold at 12 cents per lb. To our surprise, the books showed that there was more profit in Butter at 12 cts. per lb., than in Wheat at $2 per bushel, when an acre averaged upwards of thirty bushels. This comparative result so astonished us, that we went over the accounts repeatedly, in connection with the educated and very sensible steward of the estate, and with the same results. It was estimated that the sour milk fed to hogs and the manure of both cows and hogs, was more than equivalent to the expense of the dairy, and the butter was clear. There is a large profit in a well managed dairy.

But we do not advise large Cotton planters to undertake making butter beyond their own consumption. It has been one of our errors to jump at conclusions too rapidly. Because silk growing was very profitable under certain circumstances and in several countries, It was rashly concluded that it would be profitable every where, if climate allowed it. Hence the *Morus Multicaulis* fever and its results. If the attempt had been judiciously made, it would have been profitable and will be so still, whenever it is thus made.

The persons who can make money by making Butter, are small planters or farmers, living near towns or on railroads. More will depend on the farmer's wife than on himself. A good milker is as important as good feed or a good cow. A careless negro will soon make the best cow unprofitable. Her errors are those which the eye of a man, especially a Southern man, will not readily detect. Unless he has a stirring, active wife, it will be best for him to let a dairy alone. If he has such a one, the sooner he begins the better. With butter at 50 cents per lb., as it is now and as it is likely to be for some time, the profit on a considerable dairy would be enormous. But in order to do this, a person must be prepared for the business. There must be a plenty of succulent cow feed, a proper succession of it, a good dairy house, &c. When we speak of a Cotton plantation, we know exactly what is meant. When we speak of a dairy farm, our understanding of the term should be equally explicit. A dairy farm is one in which all other occupations are subordinate and conductive to this leading pursuit.

Away from towns and railroads, making cheese would be more profitable than making butter, because in our warm climate it is the more readily kept and transported than butter. We have no practical difficulties in cheese making at the South greater than those which exist elsewhere. We have rarely eaten better cheese than that made in Georgia — certainly no Northern cheese.

To planters, this is a subject of interest in a form in which they have not regarded it. Where cheese is generally made, it is a cheap substitute for meat among laboring men. It is very hearty food, and much relished by negroes. A cheese room, and a cheese press, should be as indispensable a part of the plantation fixtures as a ginhouse. It will be so regarded before long.

Where are we to get our clothing from? We make Cotton enough. But our Wool crop does not begin to supply us. As a first step towards this, let us begin with the dogs. There are more than one million persons in Georgia — if we allow one dog to every ten persons, that will give us 100,000 dogs in the State. A dog eats as much as a hog — it costs as much to keep him. One hundred thousand hogs would feed a large army. But they prevent many times their own number of sheep. Can any one doubt if there were no dogs at large that the Wool crop of the State would soon nearly, if not quite, equal its Cotton crop? We suggest to our readers to bring this subject to the attention of the Grand Juries of their counties, praying them to present these dogs as a nuisance, which the public good requires should be abated, and call on our Legislatures to impose such a tax as will reduce their number to reasonable limits. It is a sin and a shame that these, for the most part, worthless animals, should not only consume so much food, but be an effectual barrier against the introduction of an important national industry.

But if we had the Wool, we have not the requisite mills to supply us with clothing. This is a bad time to attempt costly enterprises. Until these mills can be created, we must go back to the old hand-loom. Let us put them up at once. Let no one be ashamed of home-made stuff — rather let him glory in it. It is a credit to a man to have a wife and daughters who are smart enough to clothe him and them-selvs from the producs of his own farm.

What are we to do for Shoes? We shall feel this difficulty before next winter. We have not a sufficiency of hides, and if we had the hides we have not the tan-yards.

These must be rapidly increased. It is fortunate that by the new process hides can be so quickly converted into good leather — only fifteen days. It is equally fortunate, in the scarcity of bark, that the noxious May weed has been found to possess admirable tanning properties, dispensing entirely with bark. Last fall we were presented with a pair of boots tanned by this new and rapid process. We have worn them some eight months, have been constantly on foot and in the field, and we have never had any leather which were better. The kind donors were Messrs. HUNT, of Mt. Zion, Hancock Co., Ga., who own the patent for Georgia, and to whom inquiries might be addressed. The rapidity with which this process is completed, renders it a subject of importance in the present juncture of affairs.

We shall experience trouble in regard to shoes both for laborers and soldiers, unless increased attention is paid to the subject.

So much for food and clothing. In regard to drinks, it is Utopian to expect in these days that a large portion of our population will be content with cold water. We must act not in view of what men ought to do, but what they will inevitably do. The consumption of coffee at the South is enormous. It is used excessively by our laboring white population, three times daily, at each meal. It has been thought best to say nothing of this, for fear that economy should lead them to adopt something cheaper but more hurtful. If the war prevents them from procuring their favorite beverage, Coffee, it is to be apprehended that they, both men and women, will adopt the Scotch plan and use whiskey.

To prevent this, we must press on the cultivation of the Grape. The wines made from our grapes are comparatively harmless. They can be made cheaper than Coffee even at ordinary times. He who plants a vineyard is then, in our opinion, a social benefactor. A few acres in vines, on every plantation, would supply the country with a cheap and healthy beverage, in place of the poisonous liquors which now infest it. To the vineyard, in suitable localities (that is on rich bottom lands,) might be added the culture of Hops, with a view to the manufacture of malt drinks. Hops, when successful, are the most profitable crop that grows out of the ground — the produce of an acre not unfrequently selling for $2,500 in one year — as a drawback it is expensive and liable to casualties, perhaps not more so than Cotton or Rice.

This topic is worthy the attention of the humane and philanthropic. The most strenuous efforts should be made to prevent, by a harmless substitute, any increased use of ardent spirits; It is possible to find such a substitute, which at the same time will largely remunerate him who provides it.

Negro Labor in the South, 1862

From *The War of the Rebellion — Official Records of the Union and Confederate Armies,* Series 4 (Washington, 1900), II, pp. 34–35.

DEAR GENERAL:

. . . While writing I will refer to another matter that is creating some unpleasant feeling in our State. General Mercer is impressing negroes to complete the fortifications at Savannah, and is going to the plantations, where our planters give up their cotton crops to raise corn and provisions for the Army and country, and he goes just at the time when they are saving their fodder and when all their hands are required. Our planters very naturally say that we ought to take the negroes working upon railroads, accustomed therefore to such work, and besides the railroads can wait. Corn and fodder cannot wait. In addition to this, the offer has been made to General Mercer to do the whole work by contract at less expense to the Government. I mention this matter because it is creating much bad feeling. Our people are willing to make any and all sacrifices, but they like to see reason and common sense in the officials of Government. My health has improved much slower than I had hoped, but I shall still return to camp next week and try it, anyhow.

Your friend,

HOWELL COBB.

Advice to Jefferson Davis, 1862

From *The War of the Rebellion — Official Records of the Union and Confederate Armies*, Series, (Washington, 1900), II, pp. 39–40.

RICHMOND, VA., *August 6, 1862.*

His Excellency JEFFERSON DAVIS,
Richmond, va.:

The policy of Northern leaders in the war for the subjugation of the Southern people has been to take our chief sea-coast cities, so as to cut off all supplies from foreign countries, get possession of the border States of Kentucky, Missouri, and Tennessee, which are the great grain-growing States, properly belonging to the Confederacy; cut the railway connections between Virginia and the cotton States, and cut the cotton region in two divisions by getting full possession of the Mississippi River. By getting possession of the sea-coast cities on the one side and the principal grain-growing region on the other; by separating the cotton region of the Confederacy from Virginia and cutting it into two separate divisions; by commanding completely the Mississippi River, they expected to starve the people into subjection, or crush out one division after another by the great advantage they would possess in concentrating heavy forces upon any given section or division. The lull brought upon the people of the Confederate States by their great success during the first six months of the contest has enabled their persevering enemy to half succeed in their well-laid schemes for the complete subjugation of the Southern people. The late victories of the Confederate forces, and the repulses which the Northern troops have met with lately, have stirred up the Northern Government and people to such exertions as will in their opinion complete our subjugation at no distant day. The object of first magnitude, under existing circumstances, upon our part, is to get possession of Western Virginia, Kentucky, and Tennessee. By securing a firm foothold in these States and arming the people loyal to our cause, all the land forces within the limits of the Confederate States proper belonging to the enemy, and not protected by the sea or inland navigation too wide to be commanded by cannon, may be taken or driven beyond our limits; the Mississippi River and all the railway connections we have lost may be regained. The shortest way, then, to clear our coast of the invaders (provided a majority of the people of the Northwest could first be brought to favor an honorable peace) would be to plant an army of sufficient power to completely control all direct communication between the people of the Northwest and those of the commercial and manufacturing States of the Northeast, by selecting and holding a military line from the most suitable point in Kentucky or Western Virginia, to a point above Lake Erie, where the communication between the lakes could be commanded with artillery on land. As the people of the Confederate States could under such circumstances live much better than those of the Northeast, we might reasonably expect them to give up the contest as hopeless.

Next in importance to obtaining full possession of the Border States and Tennessee is the holding of Charleston, Savannah, and Mobile, and the regaining of

Norfolk and New Orleans. Without extraordinary exertions in a very short time on the part of our Government and people to add greatly to the effective defenses of the first three named cities, they must fall before the great force which the enemy is now energetically preparing for their possession or destruction. The heavy guns which the enemy are now preparing to arm their fleet of iron-clad steamers with will be able to batter down any of the forts as they were last season. Unless the forts are greatly strengthened and guns placed in them sufficiently powerful to disable their iron-clad steamers, we should not calculate on being able to hold those cities. The possession of these important places in addition to that of New Orleans on the part of the enemy would make the war at least one year longer than it would be, provided we hold them. Through these cities we would receive many necessary supplies from abroad. If we lose them, then the war must go on entirely within ourselves. The timely completion or procurement of the most substantial iron-clad steamers or gun-boats, with as heavy ordnance as the enemy are preparing, is one of our great necessities for coast defense. With the present progress of the work upon those at Charleston, they are not likely to be finished in time to be of any service in defending that city. The proposed plan of flanking Washington, Philadelphia, New York, and Boston was communicated verbally to you last March through Colonel Withers, of Jackson, Miss.

The foregoing views are very respectfully submitted for your consideration.

Very respectfully, yours, &c.,

J. B. GLADNEY.

Salt From Old Brine

From *Southern Cultivator* (Augusta, Ga., September–October, 1862), XX, p. 164.

We continue to publish everything having a practical bearing on the manufacture of Salt. The Raleigh (N.C.) *Standard,* of June 21st., says: "In answer to our note, Mr. E. EMMONS, Jr. has kindly furnished us the following easy method of obtaining salt from pickle and the earth in smoke-houses. We advise our readers to preserve the article. The information given is important at any time, and especially so at the present period:"

To purify Salt and Brine from Fish and Meat Barrels.

The solid salt, if there is any in the barrels, should be scooped out and drained, and the draining returned to the brine. Boil the brine down to a solid. This, together with the salt already removed from the brine, must be heated to a dull red heat, or sufficient to char the organic matter contained in it; if it cakes in burning, it should be stirred to bring all parts in contact with the heat. Then dissolve in clean water, using no more water than is necessary for the purpose. This impure solution must be carefully strained through a fine cloth — a bag made of Canton flannel is the best. If it does not come through a second or third time without washing the strainer, the strained brine must be boiled down again. As the evaporation progresses, salt will be formed at the bottom of the pan or kettle, and as this retards the evaporation, it can be ladled out and drained, the drainings returned to the kettle and the salt spread out on clean vessels to dry, while the boiling must bee evaporation progresses, salt will be formed at the bottom of the pan or kettle, and as this retards the evaporation, it can be ladled out and drained, the drainings returned to the kettle and the salt spread out on clean vessels to dry, while the boiling must be continued until the water is nearly evaporated, when the salt may be removed and dried. In this way, Salt equal to the finest table-salt may be made from the most impure brine. A saturated solution of salt contains about one fourth, by weight, of salt; consequently, a gallon of brine should yield one and a half to two pounds of salt. This process could be advantageously employed in families, even with salt at the old price. Salt from springs and that leached from earths, decomposed sand stone, &c., containing organic matter, of which there are several localities in Chatham county and other parts of the State, may be purified by the process given above for brines.

Earth from smoke-houses may be leached and treated in the same manner. The process is perfectly simple; the only care required is that the burning is sufficient, (salt is not injured by heat,) and the straining neatly conducted, and all the vessels used, kept clean.

To the Planters and People of Georgia

From *Southern Cultivator* (Augusta, Ga., September–October, 1862.), XX, p. 181.

The undersigned has been requested by the Executive Committee of the Georgia Relief and Hospital Association to address you a few sentences on the importance of providing, during what remains of our present summer, abundance of dried fruit and vegetables for our troops in the field.

Many of our people, without any such reminder, have been diligently engaged in this work. But there are many still, who are blessed with quantities of fruits and vegetables, and yet, from inadvertence, or engrossing occupations, have given no attention to this subject. To such I more particularly appeal — in the name of our brave boys whose breasts are so freely presented as bulwarks for the protection of our homes and their comforts; in the name of those humane attentions which they so well merit at our hands — and most earnestly entreat that they will, even at the cost of a little personal inconvenience, give us their aid in endeavoring to secure this supply.

The crop of peaches is now abundant; summer apples are to be obtained in many places; fall and winter apples will soon be matured in considerable quantities throughout all the upper regions of our State; tomatoes, okra, peppers, cushaws, pumpkins, and some similar vegetables, are now, or soon will be, in great abundance, all over the State. All these can be expediously, economically, and safely dried or preserved. They can be put up in bags or boxes (the latter are preferable,) and, through our Association, or otherwise, they can be forwarded to our soldiers as fast as prepared. They will prove very great comforts indeed to those of our brave fellows who may be threatened, or suffering, with scurvy, from a prolonged salt meat diet. They will be very precious as a light and refreshing diet to others who are debilitated by disease, and have no food fit for their systems. They will preserve many such in their places of duty. They will restore others to strength and service. And they will save others again from prostration and death. Let the fathers and mothers who have sons, and the sisters who have brothers in our camps, think of this; and if anything can add to the self-sacrificing energy already manifested in our State, I am sure this thought will; and that we will receive such an enthusiastic response to this call as has never failed to gladden our hearts, to bless and to brighten our labors in all similar efforts made by this committee whilst endeavoring to administer the charity entrusted to our charge.

Papers in all parts of the State interested in the cause, (and who are not,) will please copy.   E. STARNES,

For Ex. Com. Ga. Relief and Hospital Association.

Augusta, Ga.

Orders for Confederate Soldiers, 1862

From *The War of the Rebellion — Official Records of the Union and Confederate Armies*, Series 4 (Washington, 1900), II, pp. 234–35.

GENERAL ORDERS,   ADJT. AND INSP. GENERAL'S OFFICE,
    No. 104.

Richmond, December 13, 1862.

I. Encampments of troops near towns and villages must be avoided where it is not indispensable. Whenever it is so, a sufficient force for guards and outposts must be selected from the best disciplined troops, and assigned as a garrison, &c.; and officers and men will not be permitted to enter the town or village except on written permission of the commanding officer.

II. Arms must not be carried from the camp, nor will mounted men in camps be permitted to ride their horses, except upon duty.

III. Private property is invariably to be respected, and must not be taken or used, except when indispensable for the public service, and then only by orders of competent authority, and in the manner pointed out in the Army Regulations and orders of the Quartermaster and Commissary Departments.

IV. The reckless destruction of fencing, wood, and other property of the citizens, which has occurred in so many instances, cannot be too strongly condemned. Commanders of troops of whatever grade should, by the exercise of diligence and strict discipline, endeavor to prevent such results, entailing, as they will, poverty upon individuals and useless expense on the Government. Fencing ought not to be disturbed where it can possibly be avoided; and when wood is necessary for the public use, that which is least valuable must be selected with as little waste as practicable.

V. A careful observance of these orders is enjoined on the Army as of the first importance to the public interests. All violations of them are directed to be reported to the proper authority for such punishment as may be requisite.

By order:

S. COOPER,
Adjutant and Inspector General.

Farming in New York, 1863

From New York State Agricultural Society, *Transactions, 1862* (Albany, 1863), XXII, pp. 3–4.

The Executive Committee of the New York State Agricultural Society most respectfully report:

The year 1862 has been of unsurpassed interest in the history of our country, and agriculture, as well as the various other interests, has been more or less affected by the terrible conflict which has been in progress and still continues.

In our own State, it is a matter of rejoicing, that, although more than one hundred thousand men have been withdrawn from our fields and workshops, to sustain the free institutions of our country, our farmers have been enabled to carry on their usual operations, and a kind Providence has blessed us with remunerating crops, so that we are enabled to meet the demands which are made upon us.

The increased attention, which necessity has forced upon us, to substitute machinery for hand labor, in order to supply the deficiency of laborers, has enabled the farmers to carry on their operations in a manner that has enabled them in the main to obtain the usual returns from the culture of the soil. Our mechanics have felt the pressure of the times, still, the demands for machinery and implements have been quite as extensive as was anticipated. A rigid economy, which has generally been adopted, will doubtless enable them to meet the demands upon them, and to prepare in season a full supply for the demands of the ensuing season.

Early in the season the question whether, in the existing state of affairs in our country, it was desirable to hold the usual annual exhibition, was discussed. The Executive Committee deemed it best to have the usual exhibition, believing that the farmers and mechanics of the State would not only expect this, but that they would sustain the Society; and the result of the exhibition has most signally shown that this was the only decision that would have met the wishes of the people. The Monroe county Agricultural Society and the citizens of Rochester, with a liberality and energy worthy of the highest praise, entered upon the work of preparing grounds and erections for the Society with a determination that everything necessary to meet the demands of the Society should be provided, and the testimony of the thousands in attendance at the Annual Fair of their entire success, must have been in the highest degree gratifying. Of the results of the annual exhibition we shall more particulary refer in another part of this report.

It was stated in the report of our predecessors, that "It is highly probable that the result of the present difficulties in our country will cause a change in the agriculture in our State in many respects." It was stated that "Our farmers were already turning their attention to the rearing of sheep, both for wool and mutton;" and it is evident, from the best exhibition of sheep we have ever had at any of our fairs, that in this respect increased attention had been given, and the result it is believed, if judiciously pursued, will be most gratifying. It is hoped, however, that this will be prudently managed, so that there will not be a reaction, which might produce serious losses as in former years. It was also stated, that "The investigations making in relation to flax as a substitute for cotton, was attracting much attention." This also has been realized. The legislature at its last session appropriated $2,000, which was placed in charge of the Society, for the purpose of investigating the improved machinery and process of manufacture, which were claimed to have been made. The Executive Committee appointed a competent committee: Messrs. Samuel Campbell, of New York Mills, Mr. A. Wild, of Albany, J. Stanton Gould, of Hudson. A committee from the Rhode Island Society, Ex-Governor Dyer and Albert Briggs, Esq., attended with them and aided them in their examinations, and favored them with their experience in like investigations in Rhode Island. It is ascertained, as their report will show, that enough has been secured to make flax a very profitable crop to the farmer and to the owners of machinery now in operation; and the market is such as to work up all that is now raised; and the prospect is not without favorable indications, that we may reach the great object in view — flax-cotton a substitute for cotton in the finer fabrics of our country. On the coarser fabrics and with wool it can be used, and is in use profitably now. We hope the Legislature will continue the appropriation, so that the Society can pursue the investigation, if practicable, to a successful conclusion. A little aid now, when these pioneers are struggling for success, would be of more importance to them than a much larger sum when the obstacles now in their way shall be overcome. . . .

View of a Southern Planter, 1863

From W. H. Russell, *My Diary North and South* (London, 1863), I, p. 382–83.

After breakfast, the Governor drove out by the ever-silent levée for some miles, passing estate after estate, where grove nodded to grove, each alley saw its brother. One could form no idea, from the small limited frontage of these plantations, that the proprietors were men of many thousands a year, because the estates extend on an average for three or four miles back to the forest. The absence of human beings on the road was a feature which impressed one more and more. But for the tall chimneys of the factories and the sugar-houses, one might believe that these villas had been erected by some pleasure-loving people who had all fled from the river banks for fear of pestilence. The gangs of negroes at work were hidden in the deep corn, and their quarters were silent and deserted. We met but one planter, in his gig, until we arrived at the estate of Monsieur Potier, the Governor's brother-in-law. The proprietor was at home, and received us very kindly, though suffering from the effects of a recent domestic calamity. He is a grave, earnest man, with a face like Jerome Bonaparte, and a most devout Catholic; and any man more unfit to live in any sort of community with New England Puritans one cannot well conceive; for equal intensity of purpose and sincerity of conviction on their part could only lead them to mortal strife. His house was like a French château erected under tropical influences, and he led us through a handsome garden laid out with hothouses, conservatories, orange-trees, and date-palms, and ponds full of the magnificent Victoria Regia in flower. We visited his refining factories and mills, but the heat from the boilers, which seemed too much even for the all-but-naked negroes who were at work, did not tempt us to make a very long sojourn inside. The ebony faces and polished black backs of the slaves were streaming with perspiration as they toiled over boilers, vat, and centrifugal driers. The good refiner was not gaining much money at present, for sugar has been rapidly falling in New Orleans, and the 300,000 barrels produced annually in the South will fall short in the yield of profits, which on an average may be taken at 11*l.* a hogshead, without counting the molasses for the planter. With a most perfect faith in States Rights, he seemed to combine either indifference or ignorance in respect to the power and determination of the North to resist secession to the last. All the planters hereabouts have sown an unusual quantity of Indian corn, to have food for the negroes if the war lasts, without any distress from inland or sea blockade. The absurdity of supposing that a blockade can injure them in the way of supply is a favourite theme to descant upon. They may find out, however, that it is no contemptible means of warfare.

Effects of the War in Virginia, 1863

From Catherine Cooper Hopley, *Life in the South* (London, 1863), II, pp. 115–17, 120–22.

As the days grew shorter and cooler, the blockade began to be felt more severely in its effects. Such trifling articles of household requirements as matches, soap, candles, starch, glue, &c., were becoming exceedingly scarce. All of them could be produced at home; but it seemed no one's business to begin. Southern extravagance and affluence had never thought of saving grease for soap, any more than rags for paper, or hides for leather. The bugbear of "raising the blockade" impeded speculation; and by slow degrees, and in scanty quantities, these things found their way into the market.

We had already begun to drink rye mixed with our coffee, though indeed it was scarcely apparent to the taste; and Mrs. McGee, who entered heart and soul into the all-important supplies for the table, amused us highly with her adventures in search of eatables. Tea was then three dollars and a-half per lb., and began to be used as an occasional treat. An abundance of delicious milk, however, was always at hand; and this is so usual a beverage, that the tea was not much missed. Coffee had risen from ten cents to seventy-five cents a pound. Salt, as a great favour to "such a good customer," was purchased at three dollars and a-half per sack. There happened to be a supply at Warrenton just then, because at the same time it was selling at twenty-two dollars per sack at Lynchburg, and eight, ten, or eighteen dollars in different other places, according to the supply on hand. These unequal supplies gave rise to very inconvenient speculations. Once when the pupils and other persons had learned that a good supply of certain articles of clothing were to be found at such a store, they wrote home to their parents for money for the purchases, all resolving to avail themselves immediately, on the principle of "first come first served;" and such was, in fact, the case, for by the time the funds arrived, the disappointing intelligence was received that a merchant from Richmond had been up and "bought the whole stock!"

* * *

In spite of the dearness of provisions, our table was always well supplied. The heads of the establishment had no idea of diminishing the comforts of their inmates to increase their own profits or save their purse-strings. The only anxiety seemed to be that they should not find enough to eat; and Mrs. McGee, after engaging a carriage several weeks beforehand, went on a foraging expedition around the country, to try and persuade certain farmers to reserve their poultry, eggs, and butter for the use of the College.

Even those rides of hers belonged to the war. Her carriage broke down, and the horses were worn out, and, poor lady, at one time she had a narrow escape from passing the night in a mud hole by the road side, into which the carriage sank, and the united efforts of driver and horses could not drag it from its bed. Other travellers finally lent their aid, and released both lady and horses.

The prices then paid for country articles were, for butter, from 40 to 60 cents per pound, instead of 12 or 15 cents; eggs from 40 to 60 cents per dozen; a turkey, the usual price for which had been one dollar, was now two dollars and a half; and chickens one dollar and a half per pair instead of thirty cents. "I declare it is too ridiculous of those negroes," said Mrs. McGee; "it costs them nothing in the world to feed their chickens, and they are pretending that the 'hard times' compel them to double their price: what do they know of hard times, I wonder!" Wood for burning was eight or ten dollars a cord, or load, instead of three dollars. "The wood waggons have grown wonderfully short this year," said our indefatigable *ménagère;* "the cords are not two thirds their usual length, but more than three times their usual price." The difficulty of transportation, and the want of men to cut it, caused this price in a country so covered with timber, that at other times persons might procure as much as they chose, and be almost thanked for doing it. We heard of much suffering among the poor as soon as fires became necessary, and much inconvenience was felt by others who either could find no stoves, or no dwellings, and where families accustomed to spacious houses were crowded together to share their fires and apartments, and assist each other in "enduring all things" in the same cheerful contented hopeful spirit, which was a marvel to contemplate. . . .

Hints for Southern Farmers, 1863

From *Southern Cultivator* (Augusta, Ga., March–April, 1863), XXI, p. 47.

March and *April* are busy months on the plantation, and not a moment must be lost from the field, until all our crops are put into the ground in the best manner. Let us, as often before since the beginning of this war, urge upon our readers the *necessity* of devoting their whole land and force and time and thought and strength to the making of an OVERWHELMING AND SUPER ABUNDANT CROP OF PROVISIONS. For the present, let *Cotton* alone, "severely." Do not plant a single seed more than will be necessary to produce cotton for domestic use, and make seed for next year. Turn your entire force, now, into the making of Corn! — the greatest of all our food-producing plants! As an intelligent Arkansas correspondent once said: "If you have plenty of Corn, you have *everything.*" So we repeat, again and again — "Plant Corn early — plant late — plant hill-side and bottom — plant hammock and sandy land, and prairie, that your garners may be filled to overflowing, and that there may be PLENTY IN ALL THE LAND!"

Also, be sure to plant extra full crops of *Sweet Potatoes;* Sow *Spring Oats,* early *Cow Peas,* in the drill, for fodder; *Lucerne,* in the drill, &c. Plant the *Chinese Sugar Cane,* also, for syrup and forage, and do not abandon a really valuable plant because it requires a little care and labor on your part. Syrup is now three or four dollars per gallon, and it will become dearer and dearer until the war is over. Plant, also, *immediately,* the true Sugar Cane, if you can obtain seed; and for culture, &c., see Jan. & Feb. number.

In preparing for your regular *Corn Crop,* plow or subsoil your land 10 to 12 inches deep, manure heavily and plant early. Do not lose a moment after the danger of late frost is over, and plant a *greater amount of land* and in *better style* than ever before.

Do not neglect the sowing of plenty of *Corn* in *drills* for fodder. It is a great help to your stock all through the summer months, and it is the sole food for many working animals in the West Indies and elsewhere, for a good part of the year. Plant, also, all the *Egyptian Millet* you can, in three foot drills, land very deep and rich. It is one of the very best plants we have for feeding green; and on good land may be cut at least half a dozen times in the season.

Sweet Potatoes should now be bedded out, so as to secure an abundant supply of "draws." No crop cultivated in the South is more worthy of attention than the Sweet Potato. It is one of the most valuable esculents for man or beast, and every planter should have full "banks" at the setting in of winter, so that he may have plenty to "sell and to keep." Two Dollars ($2) per bushel and "upwards," is *now* the price of Sweet Potatoes in the Augusta market, and "scarce at that." The Hayti (white Yams,) the Yellow Yams, and the Red "Brimstone" (so called) are all fine, productive varieties. See the excellent hints of Mr. Nelson on Sweet and Irish potatoes, in present number.

Irish Potatoes should be planted in drills three feet apart and covered with a thick layer of pine straw, or leaves, as heretofore directed, or, scatter manure in the bottom of a deep trench, drop the setts upon this, and cover up with earth, drawing it to the stems as they grow.

THE VEGETABLE GARDEN. — If you have Cabbage plants that have been kept over winter, set them out now. Sow more Cabbage seed to head in the summer; Flat Dutch is the best. Thin out Turnips as soon as they have four leaves, leaving them at the distance of 6 inches apart and sow more Turnip seed; Early White Dutch and Red Top Dutch are the best for spring use. If you did not sow Onion seed (black,) last month, do it at once; they will come into use in the latter part of the summer, when all that were raised from the sets or buttons are gone. If you did sow black Onion seed in September, it can now be transplanted, Sow Carrots, Beets, (Extra Early are the finest,) Parsnips, Salsify, Lettuce, Radishes, Thyme, Parsley and Rape (for early greens.) Plant all in rows 15 inches apart. Sow, also, a little spot with Celery and protect it from the sun. When Cherry trees are in bloom, plant Snap Beans; and when Apple trees are in flower plant Squashes (Scallop is the best) in hills 3 feet apart; also, Cucumbers and Muskmelons 6 feet apart; the Nutmeg and Citron Melon are very fine and the earliest Beechwood Melon is very superior, but a little later. All vines are greatly benifitted by guano, phosphate, or poultry manure. At the same time, also sow Okra, Tomatoes and Egg Plants. Asparagus will now begin to sprout; don't suffer any to run up to seed, but cut all down. Cabbages which have been set out and are starting to grow, should, once a week, have a watering of liquid manure — a shovelful of chicken-manure, dissolved in ten gallons of water, will make an excellent fertilizer of this kind, a little of which may be occasionally applied.

All vegetables, that already have a start, should have a good hoeing by the latter part of this month.

Plant a full crop of *English Peas,* as heretofore directed.

THE ORCHARD AND FRUIT GARDEN. — As soon as the trees are beginning to bloom, hang up a number of wide-mouthed bottles, half filled with molasses-water, in your trees — you will catch a great number of insects, and thus prevent them from doing injury to your fruit.

THE FLOWER GARDEN. — Propagate Dahlias as soon as you can see the sprouts or buds; with a sharp knife split the stem right through, leaving a piece of the stem and one or two buds to each piece; plant them so deep as to be covered with at least 4 inches of soil. Tie up all your flowering plants to stakes; the wood of the China tree, when splintered out, furnishes the best and most durable stakes, where Cypress cannot be had. If annual flower seed has not been sown yet, it should be done at once. Recollect, that fine seeds will only need to be covered *slightly.* If covered deeply, they will not sprout.

The War and Agriculture

From William Bacon, "The War and Agriculture," *Country Gentleman* (Albany, Mar. 26, 1863), XXI, pp. 210–11.

In the early history of our country, it constituted much of the farmer's independence that he raised and manufactured so many of the articles essential to household comfort. All his provisions nearly were the products of his soil. His clothing also, originated from his fields, whether wrought from the fleeces of his flock, or acquired by the more laborious operation of transforming his flax into fine linen. Then, machinery was in its simplest forms, hardly showing the infancy of the manhood power it now possesses. The extreme known west lay in the immediate neighborhood. The fertile lands of the great valley had not been explored, of corse the rich harvest, with which they have since blessed the world were not in anticipation. The occupied south was of limited area, and its productions of more limited amount. The farmer must do all he possible could within his family, to secure to them a respectable maintainance.

But national enterprise, aided by the blessing of Heaven, caused this day of old things to pass away, and herald a morning of brighter and more agreeable events. The South expanded like the rose, and the North found the sweetness springing from her soil more easily attainable than that drawn from her own maples. Her cotton-fields offered to supply the whole country and the world with a fabric much cheaper than had ever been realized. The cotton-gin and other machinery were brought into requisition to aid in the proposal. The west exulted in its more than abundant harvests, and said to the south, raise cotton and sugar, and to the east, build your manufactories high and strong, and fill them with operatives, and we will feed you both, and England to boot. And so . . . Agriculture, Manufactures, Commerce, the Arts, every thing was invited to flourish, and did flourish under this growing, spreading, deep-rooting state of things.

War, viewed in its most favorable light, even if it is clad in the armor of justice, is a fearful calamity. If between different nations, it for the time carries a blighting against itself, and arrayed with hostile intent, it becomes more dreadful; and when a nation like ours, so beautifully united by nature with all the elements of power, greatness, and goodness, unlooses the fiery demon, there must be a tremendous breaking up, for a time at least, of some of the elements of its glory.

For two years our fair and beautiful country, on which Heaven has been lavish with its smiles has been embroiled in the bitterness of civil war. The effect of this war are daily becoming more manifest. They are of wide extent, and universal in consequences. It is only in relation to our agriculture, however, that we propose to consider it, and in doing this, we do not claim to attempt showing that any one part of section of the grand whole is suffering any more than all sections.

The value of a farm, and the strength of a country lies in its productive resources. However great these resources may be, they are unavailing where there is not power to bring them out. This power lies in the physical and intellectual strength that may be brought into action to effect the object. How much of this power and strength has been called from the field of agriculture, as well as from other departments of life, in the last two years, others can estimate. They will find a very material difference in the effective force to cultivate and gather in the harvests. This want of force must, to a great extent, then diminish the amount of harvest. We know of no remedy for this, and it can be only a partial one, but for those who remain at the home to labor more diligently to make up the deficiency created by loss of the labor of those who have exchanged the quiet of the farm for the turmoil of the tented field.

Our system of agriculture is now called to change to meet new circumstances. The products of the South have been so far kept from us, as to place those that do reach us, at prices unknown for many years, if ever in the history of the country. The present exorbitant prices of cotton have caused an advance, though not proportionate, on wools, and the whole cloth-affording material is placed above a real for a necessary value. Very likely cotton will be cultivated in some parts of the West with appreciable success. So will the sorghum. This will give us new resources to some extent, but it will be by diminishing the old. If Illinois can raise cotton, and does it, and if other Western States, with New Jersey and Pennsylvania, go into the culture of sorghum and tobacco, as they no doubt may very successfully do, then the territory heretofore occupied with corn and wheat will be diminished, the price increased, and the sections of country that have depended on them for these articles must suffer, unless they too change their policy, and raise more of them, as they did in olden time. But not Illinois, nor all the North capable of producing the article, can raise cotton enough for a northern supply. Yet all the North can raise flax, (we rejoice that it is to be extensively proved the coming season,) and machinery can be set to work for its manufacture, and we have no doubt a linen can be manufactured so cheap as almost to compete with cotton in its cheapest days.

So we conclude that this scathing war will be productive of two results, at least, to the northern farmer. It will, from the very necessity of the case, lead him to a more varied system of husbandry, by inducing him as his ancestors did in olden time, so far as it is in his power, to raise all he can, and in variety, for the consumption of his family. Many a field, on which it has not been supposed that wheat would grow for the last half century, will show a golden harvest. Exhausted lands will be enriched, and waste places reclaimed to meet the increasing demands of the productions they can be made to give.

This varied system of agriculture will call for intellectual as well as physical action in the management of the soil. The farmer must think more intensely and observe more closely as his employment becomes varied.

By leading us into a more varied use of agriculture, this war will show the farmer wherein his true independence lies, to wit, by his raising from his own soil such productions as he has heretofore purchased from abroad. It may have been a matter of economy for him so to purchase, in years gone by, but now, what was formerly comparatively cheap, can be obtained only at extravagant prices, and with the unexpected revulsion that has overtaken us, how poorly prrepared we are to meet these prices.

Let the agricultural public, then, change its policy to meet the exigency of the times, and may we learn from the lesson before us, in all future time to seek out and depend upon our own resources, rather than upon other countries or distant sections of our own.

Richmond, Mass., March 3, 1863.   WILLIAM BACON.

Governor Brown on Behalf of Poor Soldiers, 1863

From Herbert Fielder, *A Sketch of the Life and Times and Speeches of Joseph E. Brown* (Springfield, Mass. 1883), pp. 278–81.

EXECUTIVE DEPARTMENT,
MILLEDGEVILLE, April 6th, 1863.

To the General Assembly:–

The armies of the Confederate States are composed, in a great degree, of poor men and non-slaveholders, who have but little property at stake upon the issue. The rights and liberties of themselves and of their posterity are, however, involved; and with hearts full of patriotism, they have nobly and promptly responded to their country's call, and now stand a living fortification between their homes and the armed legions of the Abolition Government. Upon their labor their families at home have depended for support, as they have no slaves to work for them. They receive from the Government but *eleven dollars per month,* in depreciated currency, which, at the present high prices, will purchase very little of the necessaries of life. The consequence is, that the wives of thousands of them are now obliged to work daily in the field to make bread — much of the time without shoes to their feet, or even comfortable clothes for themselves or their little children. Many are living upon bread alone, and feel the most painful apprehensions lest the time may come when enough even of this cannot be afforded them. In the midst of all the privations and sufferings of themselves and their families, the loyalty of those brave men to the Government cannot be questioned, and their gallantry shines more conspicuously upon each successive battle field. Freemen have never, in any age of the world, made greater sacrifices in freedom's cause, or deserved more of their country or of posterity.

While the poor have made and are still making these sacrifices, and submitting to these privations to sustain our noble cause and transmit the rich blessings of civil and religious liberty and national independence to posterity, many of the rich have freely given up their property, endured the hardships and privations of military service, and died gallantly upon the battle field. It must be admitted, however, that a large proportion of the wealthy class of people have avoided the fevers of the camp and the dangers of the battle field, and have remained at home in comparative ease and comfort with their families.

If the enrolling officer under the conscript act has summoned them to camp, they have claimed exemption to control their slaves, or they have responded with their money, and hired poor men to take their places as substitutes. The operation of this act has been grossly unjust and unequal between the two classes. When the poor man is ordered to camp by the enrolling officer, he has no money with which to employ a substitute, and he is compelled to leave all the endearments of home and go. The money of the rich protects them. If the substitution principle had not been recognized, and the act had compelled the rich and the poor to serve alike, it would have been much more just.

Again, there is a class of rich speculators who remain at home preying like vultures upon the vitals of society, determined to make money at every hazard, who turn a deaf ear to the cries of soldiers' families, and are prepared to immolate even our armies and sacrifice our liberties upon the altar of mammon. If laws are passed against extortion, they find means of evading them. If the necessaries of life can be monopolized and sold to the poor at famine prices, they are ready to engage in it. If contributions are asked to clothe the naked soldier or feed his hungry children, they close their purses and turn away. Neither the dictates of humanity, the love of country, the laws of man, nor the fear of God, seem to control or influence their actions. To make money and accumulate wealth is their highest ambition, and seems to be the only object of their lives. The pockets of these men can be reached in but one way, and that is by the tax gatherer; and, as they grow rich upon the calamities of the country, it is the duty of patriotic statesmen and legislators to see that this is done, and that the burdens of the war are, at least to some extent, equalized in this way. They should be compelled to divide their ill-gotten gains with the soldiers who fight our battles; both they and the wealthy of the country, not engaged as they are, should be taxed to contribute to the wants of the families of those who sacrifice all to protect our lives, our liberties and our property.

I consider it but an act of simple justice, for the reasons already stated, that the wages of our private soldiers be raised to twenty dollars per month, and that of non-commissioned officers in like proportion, and that the wealth of the country be taxed to raise the money. I therefore recommend the passage of a joint resolution by the Legislature of this State, requesting our Senators and Representatives in Congress to bring this question before that body, and to do all they can, both by their influence and their vote, to secure the passage of an act for that purpose, and to assess a tax sufficient to raise the money to pay the increased sum. This would enable each soldier to do something to contribute to the comfort of his family while he is fighting the battles of his country at the expense of his comfort and the hazard of his life.

I respectfully but earnestly urge upon you the justice and importance of favorable consideration and prompt action upon this recommendation.

Let the hearts of our suffering soldiers from Georgia be cheered by the intelligence that the Legislature of their State has determined to see that justice is done them, and that the wants of themselves and their families are supplied, and their arms will be nerved with new vigor when uplifted to strike for the graves of their sires, the homes of their families, the liberties of their posterity, and the independence and glory of the Republic.

JOSEPH E. BROWN.

The Legislature, by joint resolution, promptly united with the Governor in the justice and importance of the matter, and appealed in vain to the Confederate Congress to adopt it.

The efforts to increase the pay of soldiers was met by the Government with the presentation of the reason that it was ruinous to increase their pay, because the currency was already depreciated, and the depreciation would be increased with the increase of its volume thus to be rendered necessary. The troops in the army could not appreciate the reason while they could see it violated by the Government through its agents in every expenditure, except that of paying the soldiers in service. Agents were advancing prices rapidly for all government supplies, and competitors in market with ready government cash for all the necessaries of life; in many instances the agent and officers buying on speculation and selling to the Government at large and fabulous profits. The soldiers could see that the expansion was rapid; that the prices advanced by Government officers and agents, in order to carry out their speculations, excluded their families from the markets at any lower rates. They therefore could not see the justice of the Government that kept them in service whether they were willing or not under the conscript system, and refused to increase their pay. There were thousands of men in the service who did not regard the matter and would have remained voluntarily in service without any pay at all. But the effect was very injurious upon thousands of others who desired the money for their families at home.

Destruction in the South, 1863

From *The War of the Rebellion — Official Records of the Union and Confederate Armies* (Washington, 1889), XXIII, Pt. 1, Series 1, pp. 246–50.

CORINTH, MISS., *May 5,* 1863.

SIR: I have the honor to submit the following report of the expedition up the Tuscumbia Valley to Courtland, Ala. . . .

On my return, I burned all provisions, produce, and forage, all mills and tan-yards, and destroyed everything that would in any way aid the enemy. I took stock of all kinds that I could find, and rendered the valley so destitute that it cannot be occupied by the Confederates, except provisions and forage are transported to them. I also destroyed telegraph and railroad between Tuscumbia and Decatur, and all the ferries between Savannah and Courtland. . . .

The expedition, so far, can be summed up as having accomplished the object for which it started, the infantry having marched 250 miles and the cavalry some 400, and fought six successful engagements, driving the enemy, 3,000 strong, from Bear Creek to Decatur, taking the towns of Tuscumbia and Florence, with a loss not to exceed 100, including 3 officers. Destroyed 1,500,000 bushels of corn, besides large quantities of oats, rye, and fodder, and 500,000 pounds of bacon. Captured 150 prisoners, 1,000 head of horses and mules, and an equal number of cattle, hogs, and sheep; also 100 bales of cotton, besides keeping the whole command in meat for three weeks. Destroyed the railroad from Tuscumbia to Decatur; also some 60 flat-boats and ferries in the Tennessee River, thereby preventing Van Dorn, in his move, from crossing to my rear; also destroyed five tan-yards and six flouring mills.

It has rendered desolate one of the best granaries of the South, preventing them from raising another crop this year, and taking away from them some 1,500 negroes.

We found large quantities of shelled corn, all ready for shipment, also bacon, and gave it to the flames.

I cannot speak too highly of the conduct of the officers and troops under my command. They were guilty of but one disobedience of orders — in burning some houses between Town Creek and Tuscumbia; on the discovery of which I issued orders to shoot any man detected in the act. After that nothing was burned except by my order.

I am, very respectfully, your obedient servant,

G. M. DODGE,
Brigadier-General.

Capt. S. WAIT,
Assistant Adjutant-General, Left Wing, Sixteenth Corps.

Prices of Provisions, 1863

From *Southern Cultivator* (Augusta, Ga., July–August, 1863), XXI, p. 99.

A large meeting of the citizens of Jefferson county, Ga., was held at Louisville, May 12, for the purpose of adopting measures that would best supply the wants of our patriotic soldiers in the field:

The committee appointed to assess and fix the prices of supplies and forage for the army agree upon the following rates: Bacon, 50 cts. per lb.; corn, $1.50 per bushel; wheat, $3.00 per bushel; flour, $8.00 per hundred; oats, $1.00 per bushel; oats in sheaf, $2.00 per hundred; rye, $3.00 per bushel; wool, $2.00 per lb.; leather, $1.50 per lb.; sugar cane syrup, $2.00 per gallon; Chinese cane syrup, $1.50 per gallon, and peas, $1.50 per bushel. This report was unanimously received. On motion, it was resolved that these proceedings be sent to both of the Augusta papers for publication. The meeting then adjourned. . . .

The undersigned, Quartermaster at Augusta, Ga., hereby notifies the Wool growers, that he is prepared to purchase, for the use of the Confederate Government, their Wool at the price fixed by the patriotic people of Jefferson, and trusts that the people of the several counties will call similar meetings to that held in Jefferson, and unanimously resolve to strike a death blow to speculators, who are a curse to our country, and resolve to sell their surplus Wool to the Government of their choice at the liberal price fixed by the good citizens of Jefferson.

J. T. WINNEMORE,
Major and Quartermaster.

Impressments in the South, 1863

From *The War of the Rebellion — Official Records of the Union and Confederate Armies* (Washington 1900), Series 4, II, pp. 943–44.

EXECUTIVE DEPARTMENT,
Milledgeville, Ga., *November 9, 1863.*

Hon. JAMES A. SEDDON,
Secretary of War, Richmond, Va.:

SIR: I have the honor to acknowledge the receipt of your letter of the 31st ultimo upon the subject of impressment. Deeply regretting, but feeling assured of the truth of your statements in reference to the embarrassments that do now and for some time past have attended the operations of the commissaries of the Army, I have felt it my duty to aid and encourage them in procuring supplies and to sustain them with all my influence in making legal impressments. There have been so many outrages committed in this State under the guise of making impressments for the Army by unauthorized persons, who have resorted to it as a convenient mode of stealing and robbing from peaceful and unoffending and in many cases unsuspecting citizens, and so many irregularities and acts of partiality, injustice, and oppression committed by some of those who are authorized to make impressments, stripping some of nearly all their provisions and stock, in violation of the act of Congress, and refusing to grant to the owners the rights provided for them in the act, that I have felt it to be my duty to interpose in behalf of common justice and right, and if possible to force lawless persons to abandon this mode of robbery, and legally authorized impressing agents to discharge their duties in subordination to the laws of the country and the acts of Congress.

I assure you that I have no disposition whatever to interfere with the legal execution of the laws regulating impressment, but, on the contrary, to aid and encourage it. It is due to the people that the burdens of supporting the Army should be, as nearly as practicable, equally distributed, and it is grossly unjust that agents indisposed to perform their duties faithfully should be tolerated in going into some sections and neighborhoods and taking from the people all they have to subsist upon, in violation of law, denying them the right given by law of an arbitration, which the act of Congress entitles them to, and at the same time leaving other persons and neighborhoods and sections untouched. I have called the attention of the General Assembly of this State, now in session, to the subject, and have recommended the enactment of penal laws to punish those who are unauthorized and who resort to this method of committing robberies upon the people, and also of authorized agents who violate the laws of Congress, under which they are appointed, and which laws are wisely intended to restrain and regulate them. There is, as I before stated, not the slightest intention on my part or of the Legislature, so far as I know or have reason to believe, to interfere with impressments that are legal and conducted in accordance with the laws of Congress.

While this is true I feel it to be my duty, in this our great time of need and difficulty in supplying the Army with provisions, to endeavor to impress upon your mind the absolute importance and necessity of a change of policy on the part of the Government upon the subject of the compensation allowed to owners for articles purchased or impressed for the use of the Army. The effect of the present system of low prices and inadequate compensation, imposing as it does the burdens of supplying the Army upon the producing classes, levying contributions upon them in every case to the extent of the difference in the price paid by the Government and the market value, a burden in which other tax-paying classes do not share, is to withhold the supplies from the market and cause them to be secreted and concealed from the Government agents. This result has inaugurated the system of supplying the Army by impressment instead of by purchase, which is contrary to the true policy of the Government and against the injunctions in the act of Congress which forbids impressments until after there is a refusal to sell. By this system and under its baneful operations to the difficulty of procuring supplies and the danger of suffering in the Army for the want of food is added the evil spirit, bordering already in many cases upon open disloyalty, which it engenders among the people.

The evil increases and must of necessity continue to increase so long as the Government persists in taking the produce of the people at rates so far below the market price and in distributing the operations of impressing agents so unequally in the different sections of the country. So far as I am able to understand the spirit of our people, they are willing and ready to furnish to the Government whatever they can possibly spare and to give the preference to the Government over all other consumers, but not so when they see the burdens so unequally imposed.

I therefore urge upon your early consideration the necessity of a change of policy and the propriety of paying the market price for all articles purchased, which will supersede the necessity in most cases of making impressments at all and restore quiet to the disturbed communities of this State.

If this change cannot be made under the law as it now exists, I would through you urge upon the President the importance of procuring a speedy change of the law.

I am, very respectfully, your obedient servant.

JOSEPH E. BROWN.

More on Impressment, 1863

From *The War of the Rebellion — Official Records of the Union and Confederate Armies* (Washington, 1900), Series 4, II, p. 1066.

EXECUTIVE DEPARTMENT, NORTH CAROLINA,
Raleigh, *December 29, 1863.*

Hon. JAMES A. SEDDON,
Secretary of War:

DEAR SIR: I have the honor herewith to transmit to you certain resolutions of the General Assembly of this State, passed at the recent extra session, upon the subject of illegal impressments and the scarcity of provisions, to which I invite your early attention.

There is great reason to believe that the supply of provisions is very limited, and I earnestly request that the Government will impress as small quantities as possible within our borders. Impressing agents in many instances act in such manner as to create great dissatisfaction among our people, and I sincerely hope that you will look to their conduct, and issue such instructions as will protect citizens from illegal and unjust annoyance. These agents sometimes assume the right to judge of the quantity which the citizen needs for the use of his family, and impress what they regard as the surplus, thus leaving him without an adequate supply. This crying evil and injustice should be corrected without delay.

Many military officers also, in violation of the law of Congress, assume the right of impressment. This evil cannot longer be tolerated, and I invoke your aid in its suppression.

With sentiments of great respect, your obedient servant,

Z. B. VANCE.

[Inclosure.]

Resolutions in relation to impressment.

Whereas, this Legislature has been informed that certain persons, claiming to be officers and soldiers in the military service of the Confederate States, have been and are now going through portions of the State making impressments of corn, pork, and other articles of food essential to the support of the inhabitants of the said localities in addition to the tithes demanded by the Government, and before the tithes are collected; and whereas, it is the duty of the authorities of the State to protect its citizens in the enjoyment of their constitutional rights and privileges, as well as the subject to render obedience to its constitutional and legal requirements: Therefore,

Resolved, That His Excellency the Governor is hereby requested to correspond with the authorities of the Confederate Government in regard to such impressments, and ascertain whether they are directed and countenanced by them, and to make an effort to prevail upon said authorities to put a stop to all such illegal proceedings and public nuisances.

Resolved further, That in consideration of the scarcity of provision, after the tithes shall be withdrawn from the State, His Excellency the Governor be respectfully requested to use every honorable means to keep in the State the balance of provisions; otherwise great and almost universal suffering must ensue.

Read three times and ratified in General Assembly this the 12th day of December, A. D. 1863.

R. S. DONNELL,
Secretary of House of Commons.
GILES MEBANE,
Secretary of Senate.

Agriculture in 1863

From U.S. Department of Agriculture, *Annual Report,* 1863, (Washington, 1864) pp. 3–6.

DEPARTMENT OF AGRICULTURE,
Washington, D.C., *January 1, 1864.*

SIR: I have the honor to submit my second Annual Report. Although the year just closed has been a year of war on the part of the republic over a wider field and on a grander scale than any recorded in history, yet, strange as it may appear, the great interests of agriculture have not materially suffered in the loyal States. With the exception of some fruitless incursions along the border, and the invasion of Pennsylvania and defeat of the insurgents on the now historic field of Gettysburg, the loyal people have everywhere enjoyed a ''broad and quiet land,'' with abundant health and prosperity, while a wider territory has been cultivated and a larger yield realized, except where drought and frost interfered, than during any previous year. Notwithstanding there have been over one million men employed in the army and navy, withdrawn chiefly from the producing classes, and liberally fed, clothed, and paid by the government, yet the yield of the great staples of agriculture for 1863, as compared with the previous year, has been as follows, viz:

	1862.	1863.
Wheat, in bushels	169,993,500	191,068,239
Oats . . . . do	172,520,997	174,858,167
Corn . . . . do	586,704,474	449,163,894
Hay, in tons	20,000,000	18,500,000
Tobacco, in pounds	208,807,078	258,462,413
Wool . . . . do	63,524,172	79,405,215

The comparison, with the exception of corn and hay injured by drought and frost, is even more favorable for 1863, if instituted in regard to the general products of the farm. This wonderful fact of history — a young republic, carrying on a gigantic war on its own territory and coasts, and at the same time not only feeding itself and foreign nations, but furnishing vast quantities of raw materials for commerce and manufactures — proves that we are essentially an agricultural people — that three years of war have not, as yet, seriously disturbed, but rather increased industrial pursuits, and that the withdrawal of agricultural labor and the loss of life by disease and battle have been more than compensated by machinery and maturing youth at home, and by the increased influx of immigration from abroad. In spite of the vast influence of the enemies of free institutions in Europe, brought to bear on the masses of her people against our republic, notwithstanding the flame of civil war still rages within our borders, yet the tide of immigration was never stronger, healthier, or more promising. While some as adventurers seek this western world for military fame, stimulated by our large bounties and the chances of promotion, or to fight, sincerely, the battles of freedom and equality, the greater part come to labor, to enjoy independence and quiet, and to make happy though humble homes for themselves and their children.

According to the report of the New York Commissioners of Immigration, the number of immigrants arriving at that port during the eleven months ending November 30, of the year 1863, was 146,519, against 76,306 during 1862. This proportional increase holds good in respect to the other great ports of our country, independent of the large number of persons from Canada and other portions of America. To an intelligent mind, and especially to every American, the causes of this increasing influx of foreign population, even during a period of war, are very evident and gratifying. I shall simply indicate some of these causes, without discussing them at length.

In the first place the present rebellion is being understood abroad in its true light — as a revolt against democratic institutions, the rights of labor and human nature, and that the triumph of the government guarantees to immigration its great rewards of peace, prosperity, and freedom.

In the second place none but actual citizens, or those who have legally declared their intention of becoming such, are liable to military service; while at the same time such aliens enjoy nearly all the advantages of citizenship. However manifestly unjust this immunity is to those who defend and support the state, yet it stands forth as a noble proof of the generosity of the republic towards all who, in foreign lands, oppressed and poor, desire to better their condition.

Again, the political, religious, and social institutions of the United States, such as the elective franchise; freedom of speech, of the press, and of worship; the separation of church and state; the tenure of the soil and of other property; the honorable condition and remuneration of labor; the cheapness of education, of food, and raiment; the equality of all citizens in the courts, and the open and fair field in the race of public and private life; all these, and more, are justly attractive to the masses in Europe when properly explained and understood. And no better service could be rendered the government and the great cause of agriculture than the preparation and dissemination of correct knowledge throughout Europe respecting the United States as the country of immigration.

A fourth cause, attracting at this present time so many honest sons of toil from the farms and workshops of Europe to America, is the demand for general labor, especially agricultural, owing to the vast number of our citizens now in the army and navy, or who have perished in the defence of their country.

And lastly, as the more general and positive cause of all, may be stated the cheapness and fertility of some lands, and the cheapness and capability of other lands, in connexion with the provisions of the homestead act, by which the government offers, gratis, to each actual settler one hundred and sixty acres of the best unappropriated land. Besides this free grant, the government offers vast tracts of land at $1 25 per acre, while improved or exhausted farms, in all parts of our country, may be purchased on the most favorable terms. In this connexion there are several topics worthy of most serious consideration, and to which I desire briefly to call attention.

Whatever, owing to the war and the march of events, may be the future condition of land and labor in the rebellious States, or the legal decisions of the courts, arising out of confiscation, litigation, or the demands of the military service, yet a great change must gradually take place, not only in the tenure of the soil and its modes of culture, but in the people themselves and their institutions. Much of the land will gradually pass out of the hands of its present proprietors, either by purchase, the decision of the courts, or by the force of circumstances. Many estates will be divided into smaller farms and occupied by the humbler classes in the south, whites and freedmen, and by industrious and enterprising settlers from the other States and from Europe. The old fallacy, so long inculcated by politicians and accepted by the people like many other fallacies respecting the south, that none but negroes can toil there, will be thoroughly exploded during the present generation. Once divide there the vast estates and elevate labor to its true dignity, by hiring instead of owning it, and I venture the prediction that in less than ten years after the close of the war, over a million of the industrial classes, native and foreign, will have settled in the sunny south, making it teem with new beauty, progress, and wealth. The tides of immigration which now flow across the sea, and sweep west and northwestward with such irresistible power, bearing and leaving in their course the rich deposits of industry and art, of prosperity and life, will then divide at the Alleghanies and equally enrich the hills, the valleys, and savannas of the south.

The great laws governing the flow of population are as palpable as those governing the physical world; and these laws should be studied and heeded by our legislators if they desire to populate and develop equally every part of our country. Men who have been oppressed in the Old World, and have yet manhood enough left to seek a free life in the New, will not settle in the mild latitudes of the south, where labor is legally degraded, but go, though it be to the forests and winter snows of the northwest, where labor is honorable in all.

Now, in respect to the south, with its magnificent zones of climate and naturally fertile soil, there is no question but that her agricultural products and general prosperity will be vastly increased by the new condition of things imposed upon her by the rebellion. Terrible as is the ordeal, time, moderation, freedom, and industry will be the great healers and rectifiers; so that it shall be seen that even war offers its compensations as well as peace. Plantations that now contain from three to five thousand acres of land will be divided into farms of from three to five hundred acres, which can be more easily and better tilled, and made far more productive. While the south will continue to grow the great staples, such as cotton, sugar, and rice, many other semi-tropical productions may be introduced, of equal value and more easily cultivated, together with all the cereals, grasses, fruits, and vegetables of the temperate zone.

The half has not been told or tested in regard to the capabilities of the southern States. In times past, all their available capital, skill, and labor having been devoted to the cultivation of the great staples, no special attention was given to other crops of equal value, more developing to the country and more conducive to the comfort of the people. With smaller farms and intelligent and interested labor, the following, among other articles, might be introduced and successfully cultivated in the south: the tea and coffee plants; the opium poppy, the vanilla, ginger, castor bean, assafoetida, wax and quassia plants, silk cocoons, gum arabic, mastic and camphor trees, the Chinese yam, the sweet chestnuts, the earth and other almonds of southern Europe, the Persian walnut, the cork and gall-nut oak, the arrow, licorice, and orris roots, various valuable hemps and grasses, the prune, fig, date, pomegranate, olive, tamarind, guava, nectarine, shaddock, pineapple, and pistache, Iceland moss, the cochineal, indigo, dyer's madder, frankincense, balsam, Egyptian senna, and various other productions which we now purchase abroad at an annual expense of many millions of dollars.

The articles above enumerated form but a small portion of the possible productions of the south; while she is known to be capable of yielding not only the great staples of commerce and manufactures, but an abundance of almost every kind of food if sufficient attention was paid to its cultivation. It is a notorious fact that in the palmiest days of the south, except on the tables of the wealthier classes, the diet of the great body of the people, in variety of meats, vegetables, and fruits, was generally poor and often produced elsewhere, simply because tillage was otherwise directed or totally neglected.

Although discussion on the subject of the exhausted an abandoned lands of the south and the best modes of reclaiming them would be, just now, a most valuable contribution to the agricultural needs and changing condition of the country, yet I have not, at present, the data necessary for such an undertaking.

War and Agriculture, 1864

From Judge French, "War and Agriculture," *Country Gentleman* (Albany, Jan. 14, 21, 1864), XXIII, pp. 25–26, 41–42.

Pursuits far different from those of agriculture have occupied my time for the year now past, the first year of many in which I have not regularly contributed to some periodical. It is said that you may put into a vessel full of water a quantity of sugar, without running it over, and that after saturating the water with sugar, a quantity of salt may be dissolved, and still the water will remain of the same measure. So with employment; one may devote to his profession as much time daily as his health will allow, and he may also give to his favorite amusement of agriculture or horticulture, either in the field or study, a fair proportion of his remaining hours, with no additional wear upon his system, and even with positive advantage. One idea, whether in morals, in politics or business; the constant devotion of any human mind to one special object, always cramps and narrows the man. In spite of the wise old sayings to the contrary, I believe a man may as readily understand two things as one. I believe in a rotation of crops in the mind as well as the soil, and that a good sharp business-like merchant or lawyer may pursue a science or art along with his daily pursuit with advantage every way. Let me be understood. It is not that a farmer, after a hard day's work in hoeing or haying can, as some have imagined, sit down and be a student in the evening. Severe physical labor, producing fatigue, unfits the mind for hard study during such fatigue, but physical labor in an agreeable occupation rests the over-tasked intellect, and so the lighter and voluntary attention to literary or scientific subjects, brings recreation and rest together to one whose energies have been throughout the day taxed and strained in a particular direction.

Retrospect at 1863

The changes affecting the interests of agriculture and kindred pursuits during the past year are worthy of the careful study of all who are making or intend to make the culture of the earth, or the growth of live stock the business of life. The war and its results are as much the business of the farmer as of the soldier. The great questions of the demand and supply of provisions, of horses, of clothing, all immediately affect the producer. The withdrawal and diversion of labor from the soil, by the employment of our sons and brothers in the army or in service connected with the army, comes home at once to our farms and our firesides. The breaking up of the great system of involuntary service at the south, the sale of estates for taxes, the desolation of large portions of the States which have been the scenes of active army operations, all are opening new fields for the ambition of our young northern farmers, and offering problems difficult of solution to the land owners of the whole country. Of these great changes it is our duty to take thought, early and carefully, that we may so direct our agricultural energies as to produce the best results both for ourselves and our country. The war, its objects, progress and results, are really the only topics that can long engage the attention of any audience. The minister who does not pray for the soldier, the lecturer or political orator who does not talk of the war, and the editor who does not write of it, is not in sympathy with the great heart

of the nation, which now beats to martial music. Dividing the subject into convenient heads, let us notice some of the prominent effects of the war upon agricultural
production.

Agriculture our Strength

When the rebellion broke out, and without even waiting for the arrival in
England of our new minister, Mr. Adams, the British government at once joined
with France in acknowledging the Confederates as belligerents, and there is no
doubt that throughout the first two years of the war, both the governments referred
to, expected, if they did not also intend, that the rebellion would be successful. The
Southern idea that cotton was king, and that in some way, cotton must be supplied
from America to Europe, took possession of the leading minds of the British
government, and of the only mind of any importance in France, that of the emperor.
Further on, when we come to speak of cotton, we shall see how terrible was the
pressure upon the manufacturing interests of those countries, how distressing the
suffering of the manufacturing population. Much, however, as France and England
needed cotton to employ their laborers, they needed bread far more to feed them.

Mr. Caird, who is a member of parliament, and of the highest authority in
agricultural statistics, estimates that Great Britain, in addition to all she produces on
her own soil, buys and imports the enormous amount of three hundred thousand
bushels of corn, which includes all kinds of grain, a day, throughout the year. This,
Mr. Caird gives, as the average import, in ordinary years. But, fortunately for our
country, the crops in Great Britain in 1860 and 1861, were short, while in our
Northern States they were abundant.

The parliamentary returns of Great Britain for the year 1861, show that in
that year she imported of wheat, flour, and Indian corn, alone, one hundred and six
millions of bushels, 48 per cent. of which, or nearly one half, came from this
country.

The New York trade tables show that for the year ending Sept. 1, 1862, we
exported to Europe more than fifty-two million bushels of wheat, flour and Indian
corn, equal to one hundred and forty-two thousand bushels every day in the year!

Again, the British board of trade reports, that in the ten months ending
October 31, 1862, Great Britain received of the United States, produce amounting
in value to eighty-seven and a half million dollars, which is more than $290,000 per
day for all that time, and the amount for the same period in 1861 was not six
millions less. The English relief committees estimate that two shillings, about a half
dollar a week, is about what is necessary to support their starving population. Now
$290,000 a day is little more than two millions a week, enough to support on the
above estimate, four million of people.

Our contributions of food to England were not charity, but merely trade, but
in 1861 and 1862, there was really no market in the world where England could
have procured her supply of food, had war suddenly broken out between that
country and America. Her statesmen are wise and far-sighted, and it seems manifest, when we remember the feeling existing in that government in favor of the
South, that nothing but the apprehension of a want of food, occasioned by a rupture
with the United States, prevented England from interfering with the blockade, and

so producing war, for the sake of obtaining cotton. Now, better counsels prevail there, and it is sincerely hoped that the peace which was preserved through motives of policy, may ever in future be maintained through mutual recognition of the rights of nations and the rights of man.

Cotton

Prior to 1861, of all the cotton manufactured in England, it was for several years found that 85 per cent. came from the United States. It is estimated that in 1860 only 12½ per cent. of the vast quantity of a thousand million pounds of cotton used in England was supplied from all other countries, 87¼ per cent. being imported from the United States.

This enormous supply was suddenly almost entirely cut off by our civil war, and the blockade of the Southern ports. The exports of manufactured cottons from Great Britain in 1860, were about two hundred and fifty million dollars. No statement of the importations of cotton into England in 1861 is at hand, but it is estimated that the whole amount from all sources in 1862, was but four-tenths of the usual quantity, and of this only 4½ per cent., instead of 85 per cent. as formerly, came from this country.

The effect of the sudden failure of the supply of cotton to English manufacturers, may be imagined from the foregoing statement. There is no doubt that more real distress has been produced in England by our civil war, so far as regards the necessaries of life, than anywhere in the Northern States. By the reports of the Central Executive Relief Committee of Manchester, England, it appears that about 300,000 of the best work-people of that country, were in April, 1863, after sixteen months idleness, still out of employment. These were cotton operatives.

The number receiving relief from committees and parishes at that time, was a little above 360,000, with no reasonable prospect of immediate reduction. That the British nation should have been roused almost to desperation at such a condition of affairs, coming upon its people with no fault of their own, is not strange. That the suffering class, the operatives themselves, should instinctively have sympathised with the North, while their government sympathised with the South, is both strange and true.

Leaving England to take care of her own troubles, and test the various plans of relief for her starving people, of which the most prominent are employment on public works, emigration and charitable support till cotton is again supplied, let us return to our inquiry into the effect of these changes in the cotton trade and culture upon American agriculture. At 60 cents a pound, which cotton has of late commanded, and which is about five times its price in the years preceding the war, and by great exertions of British capitalists the supply of cotton from other countries has greatly increased, and no doubt a sufficient quantity may be produced at present prices to supply the world. But some day, we trust not far distant, peace will return to our shores, and the best cotton growing country in the world will again be open to cultivation. Most of us believe that slavery has already received at the hands of its friends, its death wound, but whether it live or die, the southern country must be open to cultivation by sombody. Under sales for taxes, as in the Sea Islands, by confiscation, as in Louisiana, by the death of the former owners in battle, and by the dissipation of their property, which was principally in slaves and lands, and burdened with debt, it is plain that to a large extent the cotton lands must come under a new ownership. Whether free labor can at once be organized so as to produce cotton at old prices, is not the question, but that at double or triple those prices, northern men with white free labor even, can make fortunes in cotton growing, cannot be doubtful.

There is no doubt that in 1860, cotton manufacturers had already exceeded the demand, and that without our war, there must have been distress in the British manufacturing districts, and cotton must have borne a very low price. When again we compete with the world in cotton-growing, we shall commence with high prices and a fair field, and if we cannot compete with the rest of the world, it will be because the South was right, and slave labor is more profitable than free labor.

Already Massachusetts men and Massachusetts capital are going to the Sea Islands, purchasing at nominal prices the most valuable cotton lands in the world, organizing the freedmen into families and villages, and paying in the first year's experiment all expenses, and the price of their land. They who believe our government can protect its citizens in their occupation there, may consider those lands open to them to enter upon. They who believe that Mr. Jefferson Davis and his friends will soon establish their dominion with its peculiar institution, over the South, had much better remain under such vines and fig trees as at present shelter them at the North. . . .

Looking at the war as a fact merely, and endeavoring not to go off into any discussion of its origin or the conduct of it, and especially avoiding anything like political discussion, let us continue the inquiry how the present condition of affairs affects agriculture and its prospects in the long future.

In a former article, we endeavored to show that the true strength of the country is in its agriculture. The spectacle of a great nation, suddenly calling into the field a million of men, maintaining such an army with an expenditure beyond precedent, diverting labor from its accustomed channels in the arts of peace to those of war, building a navy almost in a year, and yet being able at the same time to supply the starving millions of the old world with food, could only be furnished by a country whose argicultural resources are almost beyond computation.

Land our True Wealth

Emerson has well said that "all nobility rests on the possession and use of land." In Europe, this is an acknowledged truth. In this country, where land is so abundant and cheap, we hardly realize its true value. When we call land *real* estate, we use no figure of speech. It is almost the only property upon which through a series of years one can safely count, and in troublous times like the present, the farmer who labors with his own hands on his own land, is the most secure man in the country. The owner of bank and railroad stocks can form no estimate of his property, till he looks in the morning paper for the sales of the day before, and when he sees, by the daily increasing prices, a sure indication of a coming crisis, he looks anxiously round for some piece of *real* estate at old prices, for sure investment. But the farm has risen too in price and so have all its products, and the farmer wisely holds on to it, till the storm passes by.

Although it may not enable him, like a shoddy contractor, to load his wife and daughters with unaccustomed jewels, or to drive 2.40 horses, it is his secure and quiet home, which, well and worthily occupied, makes him the peer of the best in the land. Therefore, whatever has been, or may be, said in this discussion, of improved prospects at the South or West, let men who have established homes already, especially if they are past their youth, consider wisely before they abandon sure competency in their own native or adopted State, for the chances, however flattering, of a new country.

Effect on Improvements

While prices of produce of all kinds, have greatly increased, yet such has been the diversion of labor from its accustomed fields, that the price of farm labor has perhaps increased still more. The field of labor has been deserted for the field of battle. The farmer, who depended for help mainly on his boys, has seen them, one after another, shoulder their muskets for the army. The effect of this has been to check, in a great measure, all substantial improvements on the farm. The new barn which was planned, must wait for its erection till after the war. The field, which the son had laid out for draining, must go as it is until he returns again, and so the permanent changes contemplated upon the old homestead, are all omitted for the present. But the grass still grows rank in the meadow, though the war trumpet sounds, and the fields are white with the harvest though the reapers be few. A man too old, or a boy too young for a soldier, can drive the horses with the mower or reaper, or the horse-rake, and perform the work of all who are gone.

And so it has come to pass, that the makers of these farm implements have hardly been able to meet the increased demand.

Throughout New England, the use of mowing machines has suddenly increased ten-fold. Many farmers who keep but a single horse, find him worth a half dozen men, at what used to be the hand-work of mowing. So far, it may be safely said, that while the war has prevented extensive and permanent improvements upon the farm, it has, according to the old adage, that "necessity is the mother of invention," greatly hastened the introduction of useful machines, and of useful processes in farm labor.

Sheep Husbandry

For some years before the rebellion, it had become manifest to observing men, that all over the country, on the prairies and on the small farms of New England, sheep husbandry would in future prove a most profitable branch of agriculture. Those who were familiar with English farming, and knew that its whole system depends upon the sheep, as the enricher of the soil, and that their system was a permanent and improving system, while ours, if indeed we have any system, is temporary and exhausting, they especially, were advocates for the introduction of sheep upon our old farms.

It was argued too, that breeds of mutton sheep, such as are approved in England, instead of the fine wools common in this country, might be found most profitable, especially rear large markets, where the freight, which is so much greater in proportion to value upon the meat than upon wool, would be of small account.

All these considerations have still their weight, but the war, with the high price of cotton, has added new reasons for increased attention to the subject. We have heretofore used cotton because it was cheaper than wool. Now, taking all things into account, with about equal prices per pound of the raw material, woolen fabrics are even cheaper than cotton. The clothing of soldiers is almost exclusively of wool, and it is so, because it is the only material that sufficiently protects those exposed to great hardships, and to sudden alternations of heat and cold. To fit out a million of men, and maintain half that number constantly in the field, has consumed vast quantities of wool.

It may be assumed that cotton will not for many years be reduced to its former low price, and it may safely be assumed too, that woolen fabrics, after having been thoroughly introduced into general use, will always be preferred to cotton or linen, except for special uses.

We are rapidly finding out too, in this country, what has long been known abroad, especially in England, that mutton of really good quality is not inferior as food, to any other meat. Before the war, Boston used to get some of her best mutton from Kentucky, and the British steamers furnished the Astor House and Fifth Avenue Hotel with choice South-Down. It is easily demonstrated that a pound of good mutton can be produced generally, cheaper than the like quantity of beef. The sheep pays large interest annually, both in the fleece and in lambs, while the bullock takes a three years credit before he repays at all.

The report of the Department of Agriculture for 1862, contains a capital article, covering 43 pages upon the "Condition and Prospects of Sheep Husbandry in the United States."

The author's name does not appear, but for a guess, we will attribute it to Mr. Grinnell, the chief clerk of that department.

It is well considered and exhaustive, giving the statistics of sheep husbandry in the States, the imports and exports of wool and woolen fabrics, the comparative prices of wool in different years, since 1824, the wool supply of Great Britain, with well considered remarks upon the different breeds of sheep and their management, and the profits of raising mutton, wool and lambs. In short, could the reader have access to the article in question, we should recommend him not to spend much time in reading our remarks on the subject, which have only the merit of brevity, which is imposed upon writers for weekly papers.

One fact is worthy of special attention, namely, that long wools in the fleece command about the same price now as short wools. Without presuming to even intimate an opinion in favor of any particular breed, which we well know would call down wrath and indignation from the breeders of every other sort, it may be said in general, that the long wools or middle wools are more hardy, more prolific, better nurses, and bear a heavier fleece than the fine wools. The improved Leicester or Dishley sheep, which in England is prized because of its early maturity under the most luxurious feeding with turnips and oil-cake, is an exception to this rule. An English friend in a letter just received, says that his flock of nine hundred improved Lincolns averaged 9½ pounds per fleece at the last shearing, which is perhaps greater than any equal number of any breed ever yielded in our country.

Again, it is said that the silky lustre which belongs to some wool, and which is now much prized in ladies' dress goods, depends not on fineness of fibre, but is found in greatest perfection in certain breeds of long wools. It is a quality so valuable, that it occasions a difference of 25 per cent. over other wools of equal fineness.

On the whole, we believe that the prospects of sheep husbandry were never so good as at the present time, whether upon the rough old New England farms, which demand a renewal that no other treatment can give, or upon the vast prairies where guarding and slight protection are the only requisites to abundant success.

Flax-Cotton

With a hundred million dollars invested in cotton-mills in this country, and with cotton at five times its usual price, and no probability that it can for many years to come return to its old prices, there is every inducement for pursuing the experiment of cottonizing flax, or in other words, of preparing flax to be spun on cotton machinery, as a substitute for cotton. The great obstacle is, not in the difficulty of producing the raw material, but in its preparation. Cotton fibre is some six or seven eighths of an inch in length, while flax fibre is from two to three inches long, and one of the great difficulties seems to be found in cutting or pulling apart the flax fibre, so that it may be assimilated to cotton. Congress has appropriated $20,000 to be expended by the Agricultural Department in the investigation of this subject, and various individuals claim that they are progressing satisfactorily in raising up this rival to King Cotton. From all that we can learn, no very definite or satisfactory success has yet been attained, and in conclusion we should be inclined for the present, to adopt the language of the Rhode Island Society for the Encouragement of Domestic Industry, which reported on this subject through a committee in 1861, that "the encouragements for ultimate success are too strong to allow the investigation to rest here."

Sorghum Molasses and Sugar

If every white person in this country now consumes about forty pounds of sugar a year, and "this is not an approximation to the amount it would be if a cheap and home-made article were accessible to all," as is said in the last Agricultural Report, it becomes us all to consider well where this vast amount of "sweetening" is to be procured. Of the billion pounds of sugar used in this country, we have heretofore produced less than one-third, and the general impression has been, that the Louisiana planters could never fairly compete with importers, without a protective tariff. What may hereafter be done under a system of free labor, remains to be seen. In the meantime, sugar is very dear, and notwithstanding the flattering accounts published from time to time, none of the sorghum plants seem to yield the right sort of sugar in sufficient amount to promise us a supply for a long time to come. Chemists tell us that the sugar of these plants is partly grape-sugar, which cannot be practically chrystalized into the form of sugar, as that of the proper sugar cane may be, and how much cane sugar can be extracted from any of the new plants, does not appear in any publication to which we have access.

It is, however, already demonstrated, that molasses or syrup may be produced in any quantities, from the sorghum and imphee plants. The preliminary report of the Commissioner of Agriculture for 1862, at p. 22, says, "the yield of syrup in 1862 in the Western States has exceeded 40,000,000 of gallons."

Under the head of Agricultural Statistics however, at p. 552, we find it estimated that the crop of 1863 will be at least 25 per cent. greater than that of 1862," and a little lower on the same page, "the yield of 1863, with a favorable season, will not be under twenty millions of gallons!" In one place it is called syrup, in the other molasses, but no doubt the same product is meant.

This kind of inaccuracy is very annoying in a public document of this character, but enough is known to indicate clearly that our Middle and Western States may hereafter produce at home an abundant supply of molasses for our whole country, and probably at prices at which they may compete with the world, at least in our own markets.

Of the effect of the war upon the habits of our people, already too restless to remain long in any position, of the probable rush of emigration from the North toward the South as that country is opened to free labor, of the tide of labor as already setting from the old world to our shores, of all that lies before us in this great sacrifice which we are now making in defence of popular liberty, we might pertinently speak in an agricultural paper, for all that concerns the country concerns the farmer. We are content, however, to leave our subject here, if we have succeeded in illustrating the idea that agriculture, though reckoned among the arts of peace, is that which to us affords our strength and chief defence in war.

Supplies In The South, 1864

From *The War of the Rebellion — Official Records of the Union and Confederate Armies* (Washington, 1891), Series 1, XXXIV, Pt. 2, pp. 902–03.

ALEXANDRIA, LA., *January 21, 1864.*
BRIGADIER-GENERAL BOGGS, *Chief of Staff:*

GENERAL: The question of subsistence and supplies for the troops under my command assumes an importance to require prompt action, and I respectfully invite the attention of the lieutenant-general commanding to the subject. The incursion of the enemy in this region of the country, occurring at the most critical period of the growing corn crop, diminished largely the yield. More corn was made than we could have expected, but the supply has been at last exhausted, and the condition of the river prevents the transportation from above. This deficiency is artificially increased by the schedule of prices fixed for this portion of the district. Corn readily sells for $3 per barrel, while the price allowed by the Government is only about half that amount. To avoid loss planters declare they have no more corn than is necessary for their own use, and the Government is now compelled to resort to impressment even for bread corn. This practice alienates the affections of the people, debauches the troops, and ultimately destroys its own capacity to produce results. The planters will hide much of their produce or remove it beyond our reach, and will assuredly in future plant no more than they themselves require.

These remarks regarding the effect of the schedule of prices fixed by orders apply to other necessary articles, notably so to leather. By the exertions of Major Brent a considerable supply of leather has been obtained and a small work-shop established. Without this there would not be a single battery in the field. The price paid for the leather as fixed by the schedule was less than half the market value, and the tanners declare that they will put no more sides in vat, as their leather is liable to seizure by the Government, and I know of no law to compel a man to tan against his will. Much relief would be afforded by abrogating at once the schedule of prices and letting the Government pay in the market the full current value of its purchases. My experience here but confirms my experience of the past.

The effort to force a depreciated currency on an unwilling people has never succeeded, and I hope the whole system will be abrogated. Although our hopes were excited by the glowing accounts of the abundance of the Texas harvest, ten days' rations of flour is all the troops have received during about seven months. Last spring at Natchitoches I had a conversation with the present chief quarter-master of this department on this subject of supplies, suggesting that light flats might be prepared to transport supplies whenever the river was too low for steam-boat navigation. Perhaps this method has been tried. Not one ounce of salt provisions has been received this fall or winter for the troops under my command. Beef-cattle are not fit for food at this late period of the year, and their condition is unusually bad now, owing to the severity of the weather. The supply of this bad food has become very uncertain. The officer in Texas charged with the purchase of cattle for the Government utterly failed to furnish them.

Under instruction from department headquarters, I have recently sent an officer into Texas to try and procure an adequate supply. Mean time necessity has compelled me to impress all the cattle I could find on the prairies, much to the discontent of the inhabitants, who have since, as far as they were able, driven their cattle to the enemy. As intimately connected with the subject of the supply and quality of beef-cattle, I beg to inclose copy of an extract received from Brigadier-General Mouton. I trust this officer is mistaken supposing that there was an intention to censure his conduct, or that the doctrine has been established that nothing less than the starvation of his troops will justify a commander in the field in diverting food from the object contemplated by the chief commissary of the department. The censure of our superiors, when, as is the case in this department, they are entitled to all respect as well as prompt obedience, is exceedingly painful to officers striving to do their whole duty.

To obviate this, I most respectfully ask for full instructions regarding the rights and duties of officers of the general departmental staff and of those command-ing troops in the field. A recent decision in this matter has so overturned the preconceived opinions of my life and shown me to be so much in error that I am entirely without rule to guide my own action or that of my subordinate officers. I most earnestly hope that this question of supplies will meet the early attention of the lieutenant-general commanding, for if some relief is not granted we must withdraw from this whole lower region. I also beg to inclose a copy of a report from Colonel Vincent, commanding on the Teche, which will exhibit to the lieutenant-general commanding the condition of affairs in that quarter. It may be added that the last communication from Franklin village was signed by Brigadier-General Emory, commanding U.S. forces.

Respectfully, your obedient servant,

R. TAYLOR,
Major-General.

Life In The Confederate States

From *Southern Literary Messenger* (Richmond June, 1864), XXXVIII, p. 382.

The wife of a General in the Confederate service, writing to her friend in Europe, says: "There are many little things in which our daily life is changed — many luxuries cut off from the table which we have forgotten to miss. Our mode of procuring necessaries is very different and far more complicated. The condition of our currency has brought about many curious results; for instance, I have just procured leather for our negro shoes by exchanging tallow for it, of which we had a great quantity from some fat beeves fattened and killed upon the place. I am now bargaining with a factory up the country to exchange pork and lard with them for blocks of yarn to weave negro clothes; and not only negro clothing I have woven, I am now dying thread to weave home-spun for myself and daughters. I am ravelling up or having ravelled all the old scraps of fine worsteds and dark silks to spin thread for gloves for the General and staff, which gloves I am to knit. These home-knit gloves and these home-spun dresses will look much neater and nicer than you would suppose. My daughters and I, being in want of under garments, I sent a quantity of lard to the Macon factory, and received in return fine unbleached calico, — a pound of lard paying for a yard of cloth. They will not sell cloth for money. This unbleached calico my daughters and self are now making up for ourselves. You see some foresight is necessary to provide for the necessaries of life. If I were to describe all the cutting and altering of old things to make them new, which now perpetually goes on, I should far outstep the limits of a letter — perhaps I have done so already, but I thought this sketch would amuse you, and give you some idea of our Confederate ways and means of living and doing. At Christmas I sent presents to my relations in Savannah, and instead of the elegant trifles I used to give at that season I bestowed as follows: — several bushels of meal, peas, bacon, butter, lard, eggs, sausages, soap (home-made), rope, string, and a coarse basket! all of which articles, I am assured, were most warmly welcomed, and more acceptable than jewels and silks would have been. To all these we are so familiarised that we laugh at these changes in our ways of life, and keep our regrets for graver things. The photographs of your children I was so happy to see. You would have smiled to have heard my daughters divining the present fashion from the style of dress in the likenesses. You must know that, amid all the woes of the Southern Confederacy, her women still feel their utter ignorance of the fashions whenever they have a new dress to make up or an old one to renovate. I imagine that when our intercourse with the rest of mankind is revived, we shall present a singular aspect; but what we have lost in external appearance, I trust we shall have gained in sublimer virtues and more important qualities.''

Illinois Agriculture In 1864

From Illinois State Agricultural Society, *Transactions*, 1861–64 (Springfield, Ill., 1865), pp. 11–13.

Conforming to an established custom, the Executive Board of the Illinois State Agricultural Society respectfully submit the following brief review of the condition of the industry of the State, and of the operations of the organization they have the honor to represent, for the four years last past.

The period covered by this report has been the most eventful in the history of the State and the nation. Our labors as officers of the society commenced just at the dawn of the great rebellion in January 1861, were concluded some three months prior to the final capture or dispersion of the armed insurgent forces, and were therefore performed under an ever-present sense of threatened danger and disturbance to the interests it was made our special duty to guard.

As indicating our views of the best policy to be pursued by the agriculturists of the State, we quote the concluding paragraphs of an appeal which was widely disseminated with copies of the society's premium list in the early spring of 1861:

"One word more to the farmers of Illinois. Domestic troubles which, in the history of the world, have always been attended with neglect of agricultural pursuits, and generally followed by famine, seem to threaten us as a nation.

"Upon the farmer and his labor must at last depend the production of the prime necessities of human life; and whether the present national disturbances shall be of long or short duration, your duty as tillers of the soil is plain.

"There is now, in one portion of the State, what, in times of quiet, would be regarded as a large surplus of breadstuffs, and in another portion the prospect is good for a heavy crop of wheat the coming harvest; but notwithstanding these things, unless the proper efforts are put forth and continued to save the surplus now in your granaries, and to sow and plant and produce still more, that surplus will soon disappear. Let us exhort you to till this year every productive acre of your soil. Let no excitement, no interest in the stirring events of the day interrupt the operations of the farm. You know not what a year may bring forth. Your market is certain, and all history is a lie if it shall not be remunerative.

"We urge you then to strain every nerve; your interest financially cannot fail to be promoted by it, while your country and the cause of humanity alike demand it.''

The inventors and manufacturers were also assured by the same circular that "the Mechanical Department, Class F, had been entirely and throughout reconstructed, both as to the objects and to the amounts offered in premiums. The artisans of the State will find their wants and interests have been carefully consulted." In short, so far as the management of this society could effect it, the industry of the State was placed upon a war basis at the earliest possible moment after civil war became a verity, and throughout our entire term everything within the means at our command which could be done to stimulate production or economise labor has been done.

We have no cause to be dissatisfied with the results. From 1861 to 1864, inclusive, there were actually mustered into the various branches of the Federal military service, and credited to this State one hundred and ninety-eight thousand five hundred and eighteen (198,518) able-bodied men, and this out of an entire population in 1860, of one million seven hundred and nineteen thousand four hundred and ninety six, (1,719,496) or more than eleven per cent. of the whole. The muster rolls in the Adjutant General's office show that by far the largest number of these soldiers were drawn from the farms, and that the workshops of the State furnished almost all the remainder; this could hardly be otherwise in a State where, as in this, the industrial interests so largely preponderate over all others, and is mentioned in this connection simply to illustrate the extent to which our early apprehensions were realized, and that it may be held in remembrance while considering the following report of progress made during the same period in the several branches of industry referred to.

Diary Of A Confederate War Clerk, 1864

From J. B. Jones, *A Rebel War Clerk's Diary* (Philadelphia, 1866) II, pp. 173, 188–89, 277.

March 18th. [*1864*] — Bright and warmer, but windy.

Letters received at the department to-day, from Georgia, show than only one-eighth of the capacity of the railroads have been used for the subsistence of the army. The rogues among the multitude of quartermasters have made fortunes themselves, and almost ruined the country. It appears that there is abundance of grain and meat in the country, if it were only equally distributed among the consumers. It is to be hoped the rogues will now be excluded from the railroads.

The belief prevails that Gen. Lee's army is in motion. It may be a feint, to prevent reinforcements from being sent to Grant.

My daughter's cat is staggering to-day, for want of animal food. Sometimes I fancy I stagger myself. We do not average two ounces of meat daily; and some do not get any for several days together. Meal is $50 per bushel. I saw adamantine candles sell at auction to-day (box) at $10 per pound; tallow, $6.50. Bacon brought $7.75 per pound by the 100 pounds. . .

April 19th — Cloudy and cold.

We have no authentic war news, but are on the tip-toe of expectation. The city is in some commotion on a rumor that the non-combating population will be required to leave, to avoid transportation of food to the city. Corn is selling at $1.25 per bushel in Georgia and Alabama; here, at $40 — such is the deplorable condition of the railroads, or rather of the management of them. Col. Northrop, Commissary-

General, said to-day that Gen. Lee and the Secretary of War were responsible for the precarious state of affairs, in not taking all the means of transportation for the use of the army; and that our fate was suspended by a hair. . .

The famine is becoming more terrible daily; and soon no salary will suffice to support one's family.

The 1st and 2d Auditors and their clerks (several hundred, male and female) have been ordered to proceed to Montgomery, Ala. Perhaps the government will soon remove thither entirely. This is ill-timed, as the enemy will accept it as an indication of an abandonment of the capital; and many of our people will regard it as a preliminary to the evacuation of Richmond. It is more the effect of extortion and high prices, than apprehension of the city being taken by the enemy.

September 3, 1864

The President has called upon the Governor of Alabama for the entire militia of the State, to be mustered into the service for the defense of the States. It is dated September 1st, and will include all exempted by the Conscription Bureau as *farmers*. Every farm has its exempted or detailed man under bonds to supply meat, etc.

Laying Waste, 1864

From *The War of the Rebellion — Official Records of the Union and Confederate Armies* (Washington, 1893), XLIII, Series 1 Pt. 1, pp. 916–17.

CITY POINT, VA., *August 26, 1864 — 2.30 p.m.*

(Received 12.10 a.m. 27th.)

Major-General SHERIDAN,
      Halltown, Va.:

I telegraphed you that I had good reason for believing that Fitz Lee had been ordered back here. I now think it likely that all troops will be ordered back from the Valley except what they believe to be the minimum number to detain you. My reason for supposing this is based upon the fact that yielding up the Weldon road seems to be a blow to the enemy he cannot stand. I think I do not overstate the loss of the enemy in the last two weeks at 10,000 killed and wounded. We have lost heavily, mostly in captured, when the enemy gained temporary advantages. Watch closely, and if you find this theory correct push with all vigor. Give the enemy no rest, and if it is possible to follow to the Virginia Central road, follow that far. Do all the damage to railroads and crops you can. Carry off stock of all descriptions, and negroes, so as to prevent further planting. If the war is to last another year, we want the Shenandoah Valley to remain a barren waste.

U. S. GRANT,
Lieutenant-General.

Feeding the Confederate Army

From *Southern Cultivator* (Athens, Ga., January, 1865), XXIII, pp. 19–20.

If the enemy leave Georgia entirely unmolested the current year, which they are by no means likely to do, her production of food will not be over one half what it was in 1864, and probably less than one third what it was in 1863. The best wheat soils in the State were last fall, at seed time, within the enemy's lines, and very few inhabitants or hands, were left to occupy and cultivate them. In other parts of the State, the movements of the enemy, or of our own troops — the taking into service of men over forty-five — the enrollment by medical boards of great numbers of infirm men, hitherto excused from duty, who could, in some degree, (although unable to labor) exercise an oversight over their hands and plantations — the revoking of agricultural details — the placing at that critical time the militia up to the age of sixty in the field, instead of leaving them upon their farms — must result in an alarming deficiency in the next crop of wheat.

Even those planters whose age exempts them from service, we regret to find, have sown less wheat than usual. With the latter, the impressment of their crop, year after year, at a price less than it cost to grow it, and payment for the same in a constantly depreciating currency, has had the legitimate effect of such a course. The consequence is, that none but the very sanguine and very patriotic, have sown much more than they will require for home consumption, and even that was sown generally too late. Unfortunately, the impressment system, is now beginning to ripen its natural and legitimate fruits. Lack of food for the army itself, and starvation for all other men producers appears imminent. If we look at the supply of meat, the prospects is not much more encouraging. Our stock of neat cattle is getting greatly reduced, and not more than half the quantity of beef can be furnished in 1865, that was killed in 1864. With hogs, the numbers are reduced somewhat by cholera, but the quantity of pork made, depends upon how much corn is now grown. Upon the corn crop, then, more than ever, must be our dependence for food the current year. Yet, with the most strenuous exertion, it will prove a deficient one. Thousands of acres of fertile land have no fences left upon them. Other plantations have had the best horse or mule taken off by Sherman's army or "Wheeler's cavalry," or by both. Others still have horses and teams left, but the directing mind is absent, and the negroes, left to themselves, will not make their own subsistence for the year.

What then is to be done? The first thing should be, to stop at once the impressment of food at prices below the market value. Let the Government continue to be a preferred customer, but let it pay the full value of the products it requires. Look at the impressment prices quoted in this number, and tell us if production can be kept up at any such rates? Remember, too, that of late they are seldom paid for, even in currency, but in certificates of quartermasters, &c, for which the planter is often glad to get half the face of them in the currency, depreciated as it is. Is this the "just compensation" contemplated by the Constitution, in the case of private property taken for public use? The next thing to do is should be to send back every farmer that is detailed for light duty, and encourage him to make all the food possible, by paying the full value of it. A third way to improve the supply of provisions is instead of impressing more horses, to dismount about two-thirds of the cavalry. It is pretty certain our independence is not to be gained by men on horseback. Send, therefore, their horses back into the corn fields. In this way, the horses will make, instead of consuming food; and their riders can be kept in camp, where they will be of some service, instead of being in every roasting-ear patch, collard garden, chicken roost, and hog sty in the Confederacy. A cavalry horse will consume about as much as seven men, and so far they have been of but little service. On the farm, they would enable thirty or forty additional acres to be worked, and increase, instead of exhausting our resources.

If the present suicidal plan of obtaining supplies is persisted in — if care is not taken to protect the producer, both from pillage by the troops, and in his undoubted right to the full market value of his product — production will continue to diminish, until nothing is made beyond what is of absolute personal necessity to the grower. We shall soon be destitute of flour, meat, corn, salt, and the upshot will be a choice between starvation, subjugation, submission and expatriation — all of which, by a little wise forethought, could have been, and can still be, avoided. Even now, it is rumored that another more sweeping order is out for the impressment of horses, than which a more suicidal measure could not be taken, unless it be to put all negroes into the army, and feed the people with Confederate notes — the most considerable product we have made of late years.

We speak earnestly. If this war for independence fails; if, with our magnificient resources in money, means and gallant hearts, with which the contest commenced, our country is subjugated, and we endure a national extinction, suicide will be the manner of our political death. Our case, though critical is far from desperate, if, by a Fabian policy, our armies are saved, and if, in every possible way, the growth of food and the production of stores is properly encouraged. This war cannot be kept up without food. Even the scarcity, to a certain degree, of the single article of salt, for a very short time would compel us, if it could not otherwise be obtained, to make terms with the enemy, or cut our way out of the country. A soldier's entire wages now, will not more than supply his family with salt and necessary medicines at present rates.

Subsistence for Troops, 1865

From *The War of the Rebellion*, Series 1, XLVI, Pt. 2, p. 1297.

BUREAU OF SUBSISTENCE,
Richmond, Va., March 10, 1865.

General I. M. St. John, *Commissary-General:*
GENERAL: The following memoradum is based upon the lines of communication in Virginia being at once restored and protected, and independent of voluntary contributions; also upon the ability of the Treasury to meet the requisitions of this

Department, and the Quartermaster General to provide the necessary transportation for the stores when purchased and collected. With gold or United States currency to operate with (in sections of the country where Confederate money is not current or acceptable), I am satisfied the estimate made could be doubled — in other words, the subsitence necessary for the troops operating in Virginia and North Carolina is only limited by the amount of specie and Confederate money available for its purchase.

The crops south of North Carolina, in Georgia, Alabama, and Mississippi, were never so large, and with the railroad communications restored could at once be made available, although not included in the estimate.

Very respectfully, your obedient servant,

S. B. FRENCH,
Major and Commissary of Subsistence.

From North Carolina: Rations bread, 7,500,000; rations meat, 6,000,000. From Virginia: Rations bread, 5,000,000; rations meat, 5,500,000. Total — rations bread, 12,500,000; rations meat, 11,500,000. With the loss of the Central railroad and the James River Canal, we must deduct 2,000,000 rations meat and 3,000,000 rations bread.

N. B. — With East Tennessee in our possession and protected as far as Morristown, and local transportation to haul the supplies, 15,000,000 rations of bread and 5,000,000 rations of meat may be added to the above.

Agriculture In Ohio, 1865

From Ohio State Board of Agriculture, *Twentieth Annual Report, 1865*, (Columbus, 1866), Pt. 2, pp. 247–48.

After four years of the most unexampled rebellion to be found in the records of civilized man, it may be well to take a survey of the condition in which the war found and left our industrial resources; more especially that of agriculture — for in all wars heretofore the agricultural interest suffered more severely than any other productive or industrial pursuit.

During the war just closed Ohio promptly furnished all the men and horses required of her to conduct the war to a successful issue. In a State like that of Ohio, where all are engaged in some active pursuit, and the greater portion of these pursuits are productive industry, the withdrawal of any considerable number of the adult population must necessarily affect the industrial relations and products. However much the mechanical interest were effected, certain it is that the agricultural interests were not effected to that extent which, from the number of men sent into the field, would at first sight appear almost inevitable. The area cultivated during the war does not vary materially from the ordinary annual average; but, whilst the acreage presents such a favorable account, the attention to live stock has very manifestly been sadly neglected. The assessors' returns show a diminution of horses in numbers of less than 10 per cent. between 1861 and 1865; but in cattle, the most important staple of our live stock, the diminution has almost reached 35 per cent., or an annual diminution of about 9 per cent., or about 175,000 head of cattle annually. If the war had not taken place, there is little doubt that both horses and cattle would have increased just as much as they decreased. Sheep, on the contrary, nearly doubled in numbers during the war; but this increase was caused by the interruption of cotton growing, and the unprecedented price of it, as well as the great demand for the middle and coarser grades of wool for army clothing.

The census of 1860 show that in Ohio we had —

Males.	Whites.	Colored.
Between 15 and 20 years of age	127,972	1,997
Between 20 and 30 years of age	200,493	3,208
Between 30 and 40 years of age	144,587	2,209
Between 40 and 50 years of age	100,224	1,500
	573,276	9,014
Grand total		582,290

The census report further shows that 638,130 of the population were engaged in recognized and specific avocations; but many females are included in this list, such as tailoresses, seamstresses, milliners, mantua makers, etc.; but we find that there were —

Farmers	223,485
Farm laborers	76,484
	299,969

Or about one-half of the white male population between 15 and 50 years of age, or what is about the same thing, one-half of the white male population between 20 and 70 years of age were engaged in farming. The white male population between 50 and 70 years of age is somewhat less than that between 15 and 20.

According to the Adjutant General's report, Ohio furnished 372,353 men for and during the war, or an aggregate of more than half of the male population, both white and colored, between the ages of 15 and 70, as enumerated in 1860. In these heavy drafts upon the male population for military purposes the agricultural portion were by no means exempt; and it would not be surprising to learn that the area cultivated was considerably below the normal annual average.

Changes in Farming

Northern Opinions of Southern Agriculture, 1841

From *Farmers' Register* (Petersburg, Va., Mar. 31, 1841.), IX, pp. 168–169.

To the Editor of the Farmers' Register.

Shrewsbury, N.J., 2 mo. 19, 1841.

ESTEEMED FRIEND, — In reading the editorial article on a common objection to agricultural periodicals, as contained in the number of the Register just received, I was sorry to notice thy *opinion,* (for such I must accept it, although stated as a fact,) viz.: that nearly all the farmers north of ''Mason and Dixon's line'' hold every thing in and about southern agriculture, and agricultural opinions, in supreme contempt.

This is, indeed, a grave charge, and did we not verily believe such an opinion of us to be erroneous, we should be inexcusable. The privilege that I have had of associating with some of the distinguished agriculturists in at least four of the states north of said line, emboldens me to *deny it.*

Thou surely cannot be aware of the esteem in which Ruffin's 'Essay on Calcareous Manures' is held with us, to say nothing of the compliments and praises (which thou calls 'unsubstantial food') offered to 'Arator' and other southern agricultural writers.

I have estimated as a prominent advantage attending farming, that of a freedom in good measure from that petty and narrow-minded jealousy of the success of others in the same line which so commonly attaches to most other kinds of business.

I am proud to say, that the disposition *''to do good and communicate,''* as relates to successful modes of improving our practice, is, and I hope will always be, a leading feature of our *craft.*

I pray that national, much more sectional feelings, may never produce ''supreme contempt'' for the honest opinions or practice of one tiller of God's earth, wherever his lot may be cast.

Commending thee to a reconsideration of thy remarks, and more full inquiry as to our lack of fraternal regards, and if we are, indeed, alienated from our southern brother farmers, bear with us, we entreat it of thee, and with our folly as patiently as may be, and trust to kindness, persuasion and forbearance, to win us back to the bonds of good fellowship.

ROBERT WHITE, Jr.

We receive the friendly and well-meant rebuke of our correspondent, in a better spirit than that which prompted the hasty and *too general* expression of ours, which he censures. If the words were to be taken *literally* and strictly, and without any allowance for the manner in which they were brought in, we would admit, to the fullest extent, that all the censure of our friend was deserved. But while confessing the wrong of not having mentioned, or referred to as existing, the exceptions to the rule which we stated, we must still maintain our opinion as *generally,* though certainly not *universally,* and, without any exception, true. We could present sundry striking proofs of the *very general* disregard by northern agriculturists, and agricultural journals, of southern agriculture, (even the most improved and admirable,) and agricultural opinions, which would justify our strong expression of this general opinion amounting to evincing "supreme contempt." But the evidences are not only uncalled for, but would be objectionable for very obvious reasons. But though, in the rapidity of uttering a mere incidental and parenthetical remark, we stated the general proposition, without referring to the exceptions, it certainly was not with any intention of denying or concealing them, or because of undervaluing them. On the contrary, our past volumes have exhibited, among the most valued of their contents, the communications of several northern farmers, and enlightened friends of agriculture; and their good opinion of the work we have the honor to conduct, (which is principally an abstract and exposition of southern agriculture,) has been abundantly proved in their continued support of it, and also in the excellent contributions of some of them to its contents. And the good opinion entertained by even these few individuals, enlightened and judicious as they are, of our own labors, and their continued support, are so highly valued as to go very far to compensate for the neglect or disregard of all others of the great northern agricultural community. Besides our present correspondent, with whom we have but recently been acquainted, (and connected as reader and publisher,) we may be permitted to mention, among the few, but highly valued exceptions to the cause of neglect or contempt of southern agriculture and agricultural publications, too *sweepingly* expressed, the names of Walker, Hulme, and Vanuxem, of Pennsylvania, of Beekman and Wadsworth of New York, and of Colman of Massachusetts, who have given us highly prized evidences of their favor and approbation, and by whom the readers of the Farmers' Register, as such, have been instructed, or otherwise greatly benefited.

But still, "the exception proves the rule." While we value very highly the marks of approbation, and the long continued favor of the very few northern agriculturists or others who have bestowed them on our labors, or on southern agriculture, the existence of these few facts, and also of the few northern farmers who have visited, seen and appreciated the admirable farming, and enlightened views of improvement to be seen in some parts of Virginia, serve but to make more striking and remarkable the general opinion, (or total want of all opinion,) entertained in the north, of southern agricultural practices and opinions. It is very true that our agricultural practices, in ninety-nine cases in the hundred, are wretched and abominable; but such is also the case even in the northern states, and still better farming countries. And we of the south at least receive readily the instruction and lights of the north, (for we derive almost all our reading from the north — agricultural, and on every other subject,) while the very existence of southern publications is scarcely known to the great mass of northern agriculturists. But enough — and we will say no more than to repeat to our correspondent and friend the assurance that we highly value the few cases of exceptions to which he belongs, and that it was very far from an intention to count the approbation of such persons as "unsubstantial food." The application for which this expression was designed, was to high (and sometimes greatly exaggerated) praises of our labors and our publication, from persons who have never given any other evidence of valuing either — and who have not in any other manner aided our exertions, or attempted to advance their great objects, whether as literary contributors, as subscribers, or by using their influence to extend the knowledge and circulation of the work. Such praises, we cannot but rate as "unsubstantial food," and count them not much more as evidence of the interest felt by, and the sincerity of the utterers, than are the newspaper puffs of many periodicals, which can be supplied in any quantity, and of any degree of *strength,* according to order, and either upon *purchase,* or *exchange,* if not as *alms* to the begging publisher. In our case, however, we have never been indebted for praise to any of these means; for, however common the practice, or made legitimate by long usage of the trade, we never have begged, bought, exchanged, or otherwise bargained for any puff or praise, from any quarter, and we too heartily despise the practice, ever to resort to it. Whenever puffing shall be essential to the support of any publication of ours, it will sink at once. — ED. F. R.

Hussey's Reaper, 1841

From *Farmer's Register* (Petersburg, Va., Mar. 31, July 31, 1841), IX, pp. 129, 434–35.

I see, by your last number, you decline recommending Hussey's reaper, for the good and sufficient reasons, that you had not seen it work, and those who have tried it differ in opinion. Letters recommendatory in our country, both of men and things, are often too easily obtained, and I approve your caution.

A few of Mr. Hussey's first invention have been in use in this, and the adjoining county of Talbot, for two years, and I understand those who used, approve them. When I went into his shop last year to get one, I was offered his first and second invention; I preferred the latter, on account of the machinery being less complicated, and I was in some degree influenced by learning that a gentleman of your state, whose intelligence I estimated by his writings in the Register, had preferred it.

I have not, as yet, from my own experience, been able to decide on the value of the reaper, though I must confess my expectations are much in favor. It broke early in harvest, and I did not get it well repaired till towards the close. I then, for parts of two days, cut some strong wheat, afterwards my oat crop, and also my clover for seed, much to my satisfaction. Mr. Hussey's reaper may still be considered in experiment. An efficient wheat cutter is a great disideratum in agriculture; and, perhaps, the best means of improvement will be found in exposing its imperfections. The defects which I ascertained were, the part on which the lever plays (which I call the perch) could not sustain the impulse of the lever, being made of cast iron, and the knives could not be set to cut higher than nine inches, which cut more straw than was useful, and also grass among the wheat; this imposed unnecessary labor on the horses, and a heavy strain on the machine. The first defect I corrected by substituting a perch of wrought iron. Mr. Hussey was here towards the close of harvest, and I pointed out to him the second defect; at his request, I sent the reaper to Baltimore last fall; he has returned it within a few days, and it may now be set to cut high or low.

An implement, which promises an important improvement in agriculture, is a matter of as much interest to farmers, as a measure of national policy to politicians, and ought to be freely discussed, and will generally be sooner decided. I observe, from some late movements in congress, that a national bank and a protective tariff, are still considered open questions, though they have been under discussion for forty years.   RUSTICUS.

Queen Ann's Co., Md., 18th Feb., 1841.

To the Editor of the Farmers' Register.

Agreeably to the arrangement made between Mr. Hussey and yourself, with a view of testing the merits of his reaping machines fully in a long harvest, be made his appearance here on the 27th ult., having been preceded some days by two of his reapers, one of either sort; and having received due notice of the probable time of our commencing that interesting operation to the farmer, the securing of our crop of wheat. He accordingly put them in operation the next day, but from the unskilfulness of the hands, and from the horses not being accustomed to the work, and probably from the greater friction in using new machines, the work was so badly done, and the loss of time so great, that I more than once regretted having given my consent to make trial of them here. It is proper to observe that the first day the experiment was made on ten-foot beds, where the furrows were quite deep; and in crossing the beds the horses were a good deal jostled, and in going with the beds, or diagonally across them, as was done over part of the ground, the wheels occasionally would run into the furrows, and in both cases the machines were prevented from operating successfully. The next day they continued to work on the same unfavorable ground, and though there was a manifest improvement in the work done, and in the quantity of it, I was still very much dissatisfied. The third day, however, we removed them to a more favorable site, where the beds were wide, the furrows shallow, and the wheat heavy, and I very soon became convinced that Mr. Hussey's reaper did not deserve to be classed with the humbugs of the day. By this time the horses and hands employed had become better trained, and the work was beautifully done — better indeed than I ever saw done by the most expert cradler and binder, "with every appliance and means to boot" to enable them to do the work well. Less wheat was left on the ground traversed by the machines, either standing or cut, than I ever observed in any wheat field before.

I wish I could speak as strongly in favor of the reaper as a time-saving machine, but the truth obliges me to say that I cannot. Still I think that it will save time; but the question is, how much? A very difficult question it is, too, and by no means so easily solved as might at the first glance be imagined. Indeed so much depends on the locality, the length of the rows and the heaviness of the crop, (the reaper operating to most advantage in heavy wheat,) that the time saved is constantly varying; and to approximate the truth, therefore, is as much as can be expected. Something, indeed, a good deal, depends upon the fact, whether good cradlers have to be stopped in order to run the machine; good policy, however, would always suggest the propriety of stopping the worst.

It is not enough to ascertain the number of binders required to run the machine, in order to determine precisely the time saved. Say 8 hands are required for this purpose in heavy wheat, and where the rows are pretty long, and such situations are the most favorable to the reaper, and six where the wheat is lightest and the rows short, and a good deal of time consequently lost in turning. Are six cradlers saved in the former case, and four in the latter, estimating the driver and raker, who ought to be good and efficient hands, as of equal value with cradlers? Certainly not; and for this reason. The reaper cannot be started as long as there is any dew on the wheat in the morning, nor can it operate after much has fallen in the evening. At such times the hands that attend the machine have to be employed in some other way; and moving from one kind of work to another is always attended with more or less loss of time. Nor is this all. In shocking wheat after the machine, some loss of time is also incurred. Where we use the cradles, the binders follow immediately behind them, and then come the pickers-up, followed by the shockers, and the whole work goes on together. The reaper, however, when operating in long rows, as it must do to work to advantage, scatters the work so much, leaving it in long narrow strings, that shockers cannot find constant employment in following it. We have found it necessary, therefore, to stop a part of our cradles, once a day, in order to bring up the shocking after the machines, which certainly occasions some loss of time. Still I think on the whole that the securing of our crop has been somewhat expedited by the use of these machines; and if binders could have been hired to operate them, without stopping the cradles for the purpose, our harvest would have been very materially shortened; and the loss of wheat would unquestionably have been much less.

It would add greatly to the value of these machines, if the ingenious inventor, Mr. Hussey, could devise some way to make them cut damp straw; so that they might be kept at work all day. Whatever Mr. Hussey has not accomplished, however, is, I am sure, owing to the intrinsic difficulty of making the improvement desired; for the wonder with me is not that he has achieved no more, but that he has done so much.

The reaper compares most advantageously with cradles in cutting heavy wheat that stands well, cutting it quite as rapidly as it would a lighter crop, which the cradles would not do; or in cutting fallow wheat that inclines altogether one way. The fallow wheat, however, must be cut the way it inclines, the knife going under it, and it is laid beautifully, as it falls from the machine, for the binders; but the machine must go back without cutting. I am not of opinion that the reaper will answer in all situations, or will even supersede the use of the cradle altogether; but I incline to think that it may be used to great advantage in securing parts of almost every large crop; at least on level land.

After timing these machines repeatedly, I have not been as yet able to get either of them to cut more than an acre per hour, and, by the way, that is quite expeditious work in heavy wheat. Before trying the reaper, I had supposed that good scythemen would average more than 2 acres a day in good wheat, but I am now convinced that this is quite as much as can be done. My overseer, Mr. Adams, who superintended the machines, and is quite a judicious man, entertains the belief that 1½ acres might be accomplished by the reaper in an hour, with fast horses and superior driving. It is probable, too, that the experience of another season might enable us to effect more than we have yet done. But still I doubt if an acre and a half an hour can ever be counted on for many consecutive hours.

An observant gentleman of Charles City, and a practical farmer too, who has one of these machines which he worked last year, informed me recently that it would cut down sixteen acres of wheat a day, or would do the work of eight cradles. The testimony of this gentleman is every way entitled to credit, and justice to Mr. Hussey seems to require that it should be mentioned. I presume of course that some allowance was made for the time lost in the morning and evening, when the straw was damp.

In removing the machines, we employed for the most part the same horses and hands. When they can be conveniently changed, so as to lose no time in feeding, the amount of work will no doubt be much greater; as full two hours are allowed at dinner time in harvest. When this plan is adopted, the horses and hands intended to work the latter half of the day, must be fed only.

I have heretofore mentioned that the two machines sent me by Mr. Hussey are of different kinds. The one has two large wheels and the other one, and I give a decided preference to the former. It is an easier draught and performs better in every respect.

Before taking leave of the subject, I would respectfully suggest to those who are making trial of these machines for the first time, that they ought not to be disheartened, if they fail to perform very perfectly for the first day or two. Whatever difference of opinion may exist in regard to their saving time, there certainly can be none as to the very superior style, in which they save wheat, when properly managed.

With regard to the durability of these machines, I can say no more than that neither gave out during our harvest, and that they are to appearance quite strong.

WM. B. HARRISON.

Brandon, July 12th, 1841.

The Adaptation of Particular Wheats to Particular Localities and Patent Machines, 1842

From William Carmichael, "The Adaptation of Particular Wheats to Particular Localities," *Farmers' Register* (Petersburg, Va., February 28, 1842), X, pp. 89–90.

In a late number of the Register, Mr. Carter suggested the benefit to be derived from the change of seed wheat. On a former occasion, I endeavored to draw attention to this subject, and finding that it has now attracted some notice, I shall give an account of the wheat most successfully grown here. From my early recollection till the year 1798, when the Hessian fly first made its appearance in this district of country, the white Washington was the general crop. Its destructive ravages induced many farmers to try other varieties; but it was soon found that this pestilent insect made little distinction. The late Col. Edward Hays, of Talbot, who was much devoted to agriculture, and fond of experiments, for some years abandoned the growth of Washington wheat: he afterwards got seed of me; and said, at a later period, he had tried many varieties, but had found none so valuable, and productive, as the Washington wheat. The grain is white, and makes the best family flour; the straw is soft, and both horses and cattle prefer it to any other wheat straw. It only grows well in salt-water districts. If carried into the interior it degenerates, loses its color, and becomes chaffy. For the last fifty years to which my recollection runs, this has been the principal crop on the farm where I now reside, till last year. About six years ago, supposing that changing seed with a neighbor might be beneficial, I got a few bushels of the same sort of wheat. Though he was ignorant of it, his wheat contained a small portion of smut, which soon infected all my white wheats; and after some fruitless attempts to clean it, I was under the necessity of abandoning its growth. Some of my neighbors, who have suffered in the same way, are now raising seed picked by hand, and as soon as I get pure seed, I shall sow it again. I have supposed this wheat would suit on some of the salt-water districts in Virginia. About thirty years ago, a wheat was introduced here, and is known by the name of the red chaff beard, I apprehend the same mentioned by Mr. Carter. It seems adapted to a variety of soils, and is more generally grown in this county than any other. It ripens early, though sown late; and suffers less by the frost, especially on low, wet grounds. For these properties, it has been part of my crop for many years, though it sells at a much inferior price in the Baltimore market. When my white wheat was infected with smut, though this was thrashed by the same machine, cleaned with the same fan, and stored in the same granary, it did not contract the disease. An experienced farmer, to whom I mentioned this circumstance, informed me that his white and red wheat had been equally infected.

A farmer always incurs risk by changing seed. When I abandoned the growth of Washington wheat, I purchased white flint from a gentleman in the neighborhood, who got the seed from New York in 1836, and had grown it exclusively on his farm. He had no suspicion of smut; but upon careful examination, I found it to contain a small portion, and declined sowing it. I then procured Pennsylvania white wheat, which has been lately introduced here: it does not ripen early, and I do not think produces equal to the Washington, or red chaff.

I last summer saw some Chester county farmers in Washington, who stated that they grew there a wheat which they called the Mediterranean, which, though sown the last of August, would resist the Hessian fly. Dr. Darlington, a gentleman of science and intelligence of West Chester, confirmed their statement; and I purchased a few bushels, which I have sown for experiment. I have also sown a small quantity of white wheat which I procured from Owego county. It has a fine white grain, and much resembles our Washington white wheat. . . .

The community stands much indebted to Mr. Garnett, for his persevering efforts in the cause of agriculture; and I am pleased to find that an agricultural convention has met in the city of Washington, which has framed a constitution, and organized a society; though, I much fear, it will languish for want of funds. Agriculture, though the paramount interest of the country, and extending into every section, it seems is not so connected with *the general welfare* as to be entitled to the countenance of the general government. Perhaps one of the causes is, that no demands have been made. If I was a member of this society, I would bring the subject before them at their next meeting, to see if something could not be made out of this clause of the constitution. The manufacturers, though for the most part a sectional interest, placed themselves under its fostering wings, where they would still have remained, but for the ardent and determined spirits of the south. I am not a latitudinarian in construction; but there are many problems in agriculture not solved, many discoveries not assayed, for want of means, and I think an appropriation to these objects would be more for the general welfare, than sending ships on discovery to unknown seas, even if they should be so fortunate as to find an island within the Arctic circle.

Among other new discoveries, the wheat drill was presented last year. The patentees are Pennsylvanians, from Chester county. They stated that, in their richest lands, one and a half bushel had been found sufficient seed; on my improved grounds I generally sow two to the acre. I had a few acres put in at the rate they proposed, alternating spaces, with the plough, in which I sowed two to the acre. The next harvest will test the accuracy of their statement. I kept the time of the work done by the drill, for two successive hours, and found it would plant one acre to the hour, with two men and two horses, without fatigue to either. It deposites the wheat and covers it by the same operation; and none can be found on the surface. The saving in labor and the saving of seed are no small considerations. I was much pleased with the work done by the drill. It is somewhat complicated, and I have not sufficient information in mechanics to determine on its stability, and whether it can sustain the rough usage of rough hands. . . .

Cotton in South Carolina, 1842

From James H. Hammond, "Anniversary Oration of the Agricultural Society of South Carolina," *Farmers' Register* (Petersburg, Va., May 31, 1842), X, pp. 246–48.

The extraordinary progress of the growth of cotton has, it is well known, occasioned a series of calculations, at various periods, from the time of the invention of the saw-gin, to prove that it must soon shoot far ahead of the consumption, and that competition would in a short time confine its culture to a few favored spots. Contrary, however, to all reasonable expectation, consumption has advanced with the same wonderful pace as production; and relying on the frequent failure of past predictions, many are even yet sanguine that it will continue to do so for a long period to come. I very much fear that they are at last mistaken — that what has been prophecy for such a time, is shortly to be fact; and that the present depression of the cotton market is neither accidental nor temporary, but the result of natural causes, and likely to be permanent.

The rapid increase of consumption has been owing to causes which can be explained, and to which limits may be assigned. A spirit of improvement in machinery for manufacturing it had sprung up in England, some twenty years before cotton began to be cultivated in this country. The hand-frame, cylinder cards, woof and warp machine, spinning jenney, and power loom, followed each other in quick succession, and Watt soon taught the whole to move by steam. The quality of goods was in consequence very much improved, and prices also reduced, so as to recommend them at once to all classes of consumers. The raw material then furnished, as might be supposed, fell far short on the augmented demand, and production was in turn vastly stimulated. The supply increased from America with great rapidity after Whitney's gin came into general use, but new inventions in machinery continued, and as even the home market had not yet been overflowed, and the demand still kept ahead of it, until the general pacification of Europe and the world opened fresh markets to this new style of goods, which could never apparently be stocked. Peace, also, made money abundant. New and immense investments were made in manufactories, and cotton mills arose all over England. It was of no small consequence to the result that England was the seat of this great manufacturing revolution. Her incalculable wealth, her great naval, superiority, and her vast territorial possessions, scattered over every quarter of the globe, gave her the command of a commerce unknown before in the annals of the world. And during a period of profound peace, longer than mankind have enjoyed since the death of the elder Antonine — and happily not yet terminated — this great nation has devoted herself to almost the sole purpose of opening new markets and filling them with cotton goods. To all these circumstances is owing the rapid stride of consumption, which has thus far out-stripped all calculation. But these causes have run their cycle; their utmost effects have been fully felt; all the markets now accessible to cotton manufactures are kept not only stocked but glutted; and although peace should continue and improvements in machinery go on, and the power of England still remain unbroken, none of which, to say the least, are certain, it is impossible for consumption to increase in any thing like the ratio it has hitherto done.

That it will increase to some extent, in every given series of years, is perhaps certain. I believe that cotton goods must undoubtedly drive linen from the almost entire monopoly which it yet enjoys in the domestic uses of the continent; that they have, sooner or later, to clothe the naked barbarians of Africa, as well as the silk-robed myriads of the Chinese empire: to ascend the Euphrates; to break more effectually through the barriers of the Bosphorus, and penetrate to all the nations of the Black Sea, the Volga and the Obe; while every birth in a large portion of the old world, and in the remotest civilized corner of the new, creates a fresh demand. But all this must be the work of time. Popular prejudices must be broken down; the policy and the agriculture of nations now devoted to growing rival commodities must be revolutionized, and manufactures must spring up and gain the ascendency where poverty and ignorance and despotism now flourish. Years, perhaps centuries, must elapse before all this can be accomplished; and peace and commerce must, for all the time, hold the world subject to their benignant influence.

It might be thought that the great increase in the sales of raw cotton during the last year (1840) argues an equal increase of consumption. But it must be remembered that the sales of 1838 were nearly equal to those of the last: that this year they have again fallen off to a fearful degree, receding to a point below those of 1839, and not much in advance of those of 1837, and that in spite of the greatest falling off in the crop ever known in one year, there has been an actual increase of the surplus on hand in Liverpool and Havre. The large sales of the last year were in fact owing to the great fall of prices, and the reduced sales of this year, to the trifling increase of them in the early part of the season, thus showing, that in the present state of the markets, the smallest advance instantly checks consumption, which would not be the case, if the supply had not, to say the least, fully reached the limits of the immediate demand. The disastrous condition of the manufacturers during the present year, proves also, that the purchases of the last, even by the trade, were speculative, and did not indicate the true ratio of consumption. Indeed it is now generally conceded, that ever since 1834, notwithstanding the great appearance of manufacturing prosperity, more goods have been turned off then were required for consumption, and have accumulated all over the world, to an extent not so easily perceived, but not less fatal in effect than the accumulation of the raw material in the great markets of England and France.

But while consumption is thus fluctuating — giving the clearest evidence that it has approached a point beyond which it cannot advance, except with the slow march of time and the mighty changes to which I have alluded, production is not only going forward with gigantic step, but yearly developing a capacity which proves that it has yet scarcely passed its infancy, and has been creeping lazily along, compared with what it can do. If we are to believe the English, the experiment of growing improved cotton in their Indian possessions, under the direction of Americans — from our seed, and prepared with our gins, is likely to succeed. Already has some of the new made cotton been sold in Liverpool at a price higher than that of our best New Orlean's brands. I do not, however, feel very deep apprehensions from this quarter. The sample sent was probably most of it from the little Isle of Bourbon, where the finest Sea Island cotton, next to our own, is known to grow. With a soil impoverished by 2000 years or more of cultivation; with a climate in which it rains continually for half the year, and for the remainder of it never rains, so that during one period cotton will not grow at all, and during the other must depend on dews and laborious irrigation; almost without animal power; with an idle and feeble race of laborers; paralyzed by absurd social forms; and subjected to the most unprofitable as well as the most wretched system of slavery; with all these drawbacks, I cannot believe that India will be able to compete with us. It is idle to talk of her doing so with other sections of our country. To Egypt — notwithstanding the temporary effect of the galvanic energies of Mehemet Ali — and to Western Asia, belong nearly all the disadvantages of India, without the benefit of English capital and English enterprise. Brazil and other parts of South America might become more formidable rivals, but their institutions are too unsettled; and their population too motley and uncivilized, if they had no other impediments, to give us serious alarm.

But the cotton growers of South Carolina need not look abroad for competition. It is much nearer home. It is our own kith and kin—the hardy and industrious and enterprising vanguard of civilization—that have levelled the gigantic forests of the south and southwest, and furrowed the rich bottoms through which pour the tributaries of the Gulf of Mexico, from the Suwanee to the Sabine, and that have but recently rescued from a slothful race the fertile empire stretching beyond the Sabine to the Rio Grande—who are destined at no distant day, to supply the foreign markets of the world with this inestimable staple. They have just overcome all the incipient difficulties of the enterprise, and are now prepared to put forth, on the finest soil and in the most favorable climate of the earth, an energy which must inevitably crush all serious competition.

A few facts will show their progress and our own. From 1789 to 1811, the production of cotton in the United States had increased from nothing to 90 millions of pounds, of which but 2 millions were grown in the gulf states, not counting Georgia and Tennessee amont them. By 1826, these states had risen by slow degrees to 135 millions in a crop of 368. In 1834, the whole crop had increased to 457 millions, of which they produced 252 millions, or rather more than half. And in 1839, when the crop amounted to 830 millions, the same states grew 600 millions, having in the last five years doubled their production and made 100 millions more, while the rest of the cotton growing states in the aggregate had actually, for thirteen years previous, made no increase at all. The world is accustomed, especially of late, to speak with astonishment of the unparalleled progress of the growth of cotton in this country. Yet for fifteen years now the whole of that progress has been made in the gulf states. Notwithstanding the high prices, we have been stationary. Is not this a startling fact? It seems to me of itself to settle the question, and to leave no doubt that these states, having now but fairly prepared themselves for it, so soon as the check on consumption shall place in strict competition all the cotton growers of the world, and reduce prices to their lowest point, the cultivation of this staple must be confined almost entirely to these fertile regions.

Now, I believe, prices have already reached, if indeed they have not gone below, the lowest point at which we can profitably grow cotton made in South Carolina, and I may include a large part of Georgia also, does not exceed 1200 lbs. per hand. The average expenses per hand cannot be less than $35. When every thing is taken into account for which money is paid or labor abstracted from the field to make, I doubt whether the most judicious planter is able to reduce them lower. If, then, cotton sells at an average of 10 cents net, on the plantation, to do which it must bring in the seaports 11 or 12 according to the distance of transportation, the clear profits for each hand in these states will be $85. And this includes the rent of land. If cotton sells at 8 cents net, on the plantation, the clear profit, rent included, will be only $61. If this estimate be correct, it will be seen at once that cotton cannot be profitably grown here at 8 cents per lb. Yet this is at least as much as we shall realize for the present crop, and more, I believe, than we can safely anticipate hereafter.

If any one doubts the permanency or abundance of the future supply from other quarters at 8 cents per lb. or even less, let him look a moment at the profits of the planters in the gulf states to which I have alluded. The vast land speculations incident to new and fertile countries, and the fact that everything has been conducted there on the credit system, have, it is true, involved those states in great embarrassment. But this does not affect the productive capacity of the soil. The average crop per hand there cannot be rated at less than 2000 lbs. The expenses per hand I will estimate at $40, although when compelled to be economical, I see no reason why they should exceed our own. At 10 cents net, for his cotton, the gulf planter will make $160, clea, against $85 at the same prices here, and at 8 cents, $120 against $61; which shows that, at prices ruinous to us, he will realize a handsome interest on his investment. But while I have estimated our production fully as high as truth will permit, I am satisfied I have underrated that of some of these states, perhaps all. On their best lands 300 lbs. per hand is not at all an extraordinary crop, and more is often made. The planters themselves, though great advocates of short crops at certain seasons of the year, would scarcely be willing to estimate their average crops at less than 2500 lbs. At these rates of production, and even 6 cents a pound net, they will make the very fair profit of $110 to 140 per hand; and unless cotton is for ever to baffle all the laws of trade, it is certain that prices must ere long range about and possibly below that point.

A result so fatal to us could only be arrested by the want of sufficient land or labor in these fruitful countries. But of this there is no prospect. Both may be already found or soon placed there in ample abundance, to supply not only the whole quota furnished by the United States, but all that is now furnished by every cotton growing region, for the foreign markets of the world. A slight examination will show the fact conclusively. The crop of the world for 1839, the last of which I have seen a full return, and the largest perhaps ever grown, supplied the markets of the United States and Europe with a little less than 1000 millions of pounds. At the rate of 250 lbs. to the acre, it would require but 4 millions of acres to grow it all. The four states bordering on the gulf—Louisiana, Mississippi, Alabama and Florida—not to include that almost equally fertile section of Georgia between the Flint and Chattahoochie rivers—contain 130 millions of acres. If then only one acre can be found in 32 capable of producing 250 lbs. they alone can supply the present demand of all the foreign markets of the world, and of our own also. It is difficult to compute the entire crop of the whole world, including that portion which is con-sumed at home in other countries than our own, as well as that which goes abroad, but according to the most extravagant estimate which has been made—excepting however the very veracious returns of our late census takers—it cannot exceed 1500 millions of lbs. At the rate per acre mentioned, it would require but 6 millions of acres, or one in 21 of those contained in these states, to yield the whole amount, and supply the entire demand not only of the foreign markets, but of the whole human family at the present moment.

In this estimate I have said nothing of the magnificent wilderness which joins us on our south-western border. Texas alone contains 150 million acres. The climate has not been fairly tested, and like that of other regions approaching the tropics, its vicissitudes may prove too great for complete success in the cultivation of cotton. As yet, however, those best acquainted with it regard it as the most favorable of all others, while a large proportion of its soil is undoubtedly as fertile as any which has ever yielded its fruits to the hand of man. Already it is swarming with the adventurous offspring of the great Anglo-Saxon family and offering the most formidable competition to even the bountiful bottoms of the Mississippi and Red River. When sufficient labor shall have found its way thither, as it is now rapidly doing, no one can venture to assign a limit to either the extent or cheapness of its production of cotton. There is little question, however, that out of the 280 millions of acres, embraced in this country and our own gulf states, land enough will be found in considerable bodies of such quality as to grow, at 6 cents a pound, or less, cotton sufficient to supply the progressive demand of mankind in all time to come. For were they at this moment to be civilized by some supernatural influence, with every avenue of commerce thrown wide open, and every article competing with it drawn from the market to the full extent that probably they ever can be, the 800 millions who inhabit all the earth would scarcely require more; 8,000 millions of pounds. This they might produce on 30 millions of acres, or 1 acre in 9. And when it is remembered that one half of France and one third of England, barren as they were by nature, are now in actual tillage, it cannot be extravagant to suppose one ninth of these prolific soils well adapted to the growth of cotton.

Nor will there be any deficiency of laborers. While labor, it is true, can be made available only to a very limited extent. But there are in the United States and Texas upwards of two millions and a half of slaves, 1,500,000 of whom may be rated as efficient hands. Those who grow cotton make their own provisions, or can do it. To meet the remaining agricultural and domestic, as well as all other demands for slave labor, will scarcely require more than half these hands. So that the other half, or 750,000, may be employed in cotton culture. These, at even 2000 lbs. each, will produce 1500 millions of pounds, which is precisely the amount at which I have computed the crop of the whole world. But the present demand of all the foreign markets and our own can be supplied by 500,000 properly located; and there is little question that such a location will be speedily effected. This done, we shall have an actual surplus of slave labor on our hands.

If, then, the consumption of short staple cotton has reached such a point that the least advance on the present low prices immediately checks it, and one at which it seems scarcely possible, at any prices, to maintain it, if at the same time the production of this kind of cotton is increasing every where over the world, and especially in the countries bordering on the Gulf of Mexico, both in and out of the United States, who, though latest to begin its culture, possess such advantages of soil and climate as to have already far outrun all competition, and, having conquered most of the difficulties incident to new countries and new enterprises, do now furnish six-tenths of the demand of both the American and European markets, and are capable of supplying, almost immediately, the whole, at such prices as are utterly ruinous to us, if all these things be true, as I have endeavored to show they are, the conclusion is irresistible, that the planters of South Carolina will be speedily compelled almost if not altogether to abandon its longer cultivation."

Virginia Agriculture, 1842

From Edmund Ruffin, "Report to the State Board of Agriculture," *Farmer's Register* (Petersburg, Va., June 30, 1842), X, pp. 257–66.

On the most important recent improvements of agriculture in lower Virginia—and the most important defects yet remaining.

The time can yet be remembered by many persons now living, when the improvement of the general or average rate of product of land by its fertilization was scarcely attempted, or even thought of, in the lower and middle regions of Virginia. The efforts to increase the amount of the products and profits of agriculture, were not wanting then, more than nos. But it was by adding to the cultivated surface by new clearings of wood-land, and by thus substituting fresh and richer ground for such as was worn-out and totally worthless. The manuring was almost limited to the partial saving and use of the excrement of the live-stock; no more vegetable matter, as litter, being used, or thought worth the labor of using, than would serve to absorb and and retain the animal juices. Perhaps only the more ignorant cultivators believed that vegetable matter was not manure of itself; but of the more intelligent, but few would have deemed that the doubling the ordinary small amount of litter in their farm yards and stables, would not have been of more labor than profit, even for lands the best adapted to be benefited by such manures. With very few exceptions, there was no more litter used than necessary to prevent the cattle from suffering by the wetness and filth of their pens and beds; and to serve to soak up partially the fluid matter, of which otherwise the waste was manifest. The straw of the wheat and oat crops, and the stalks of the tobacco, were necessarily brought to the barns to be separated; and, with greater or less waste, these materials were in some manner necessarily converted to manure. But the much larger quantity of corn-stalks were left in the field, to be afterwards burnt out of the way of the next ensuing corp; and still less was any other supply of coarse litter sought or desired. Even this small extent of manure-making was a great improvement on the older practice. Thomas Adams, an observant and respectable old farmer of Prince George, who died about 20 years ago, told me that he remembered the first beginning, in that county, of keeping up the work horses at night, on litter, for the purpose of increasing the manure, instead of turning them out to graze in the pastures. The practice was introduced by a new overseer on Old Town farm, then one of the "quarters" of the head of the Center family; and it attracted much more of notice for censure, than for applause or imitation. The comment of one of the neighborn was, that "whenever *he* should be obliged to make manure for his land, he should think it full time for him to move to the *back woods*,"—the name by which the more western settlements were then called.

With a few honorable exceptions, very few of the wealthy land holders, attended personally or regularly to the management of their farms, previous to the last 30 years. Still, some few excellent cultivators, and judicious managers of their farms, were profiting by pursuing a different course. Such were Philip Tabb of Gloucester, Fielding Lewis and John Minge of Charles City, and John Taylor of Caroline. The writings of the latter subsequently gave an impulse which may be considered as being the commencement of our improving state of agriculture, and with which the following recital will set out.

The 'Arator' of John Taylor, was published in a volume in 1813, and soon acquired a popularity, and a ready acceptance of its worst errors as well as its most valuable truths, which no other agricultural publication has obtained. The two most important improvements which the author thus made generally known, were both in regard to the using of vegetable matters as manure.

I. The making of farm-yard and stable manure, in which the vegetable materials are in very large proportion to the animal matter

The first of these lessons, and great improvements on previous general opinion and practice, was the almost unheard of plan of using as litter all the vegetable offal of the crops, no matter how excessive the proportion might be to that of the animal excrement; or, in teaching a reliance on the fertilizing properties of vegetable matter as manure to soil, and food for plants, without regard to the amount of admixture of animal matter. And this, however universally understood now, so much, indeed, that the former state of general opinion can scarcely now be realized, was then totally new doctrine to nineteen in twenty of those who to it soon became converts. And the improved practice, thus brought about, was made the more extensive by another then novel lesson taught by Taylor, that the fermentation of manure before application was unnecessary and wasteful. Before that time, the little farm-yard manure which was made, was left exposed first on the yard, and next heaped, and in both states wasting, and the use lost, for a whole year before being used for the next crop of tobacco. The advantages and disadvantages of the different modes of preparing farm-yard and stable manure, of the different conditions when applied, will be again adverted to under a subsequent head, when the still newer practice of top-dressing on grass will be considered.

II. Manuring by "enclosing" or non-grazing, or excluding stock from fields when in grass, or not cultivated

The next great improvement, and innovation on previous opinions and practice, was the plan of manuring fields by leaving on them, to die and decay, their own vegetable growth, whether of artificial grass, or, in its absence, of the weeds and other natural growth, whenever tillage crops were no on the field. This, though it be Nature's own and universal mode of restoring fertility to ompoverished lands, and therefore is offered to the observation of all persons, had not been practised before any where; nor had even the theoretical truth on which the practice rests, been received or considered. The plan may be claimed as an improvement entirely Virginian, in large and systematic use; and the first promulgation and extension of the practice was due to the mind and the labors of John Taylor, and especially to the publication of his 'Arator.' It is true, that this mode of fertilization is not suited to lands already in a productive state, nor where animal products are very profitable, or indispensable—nor where putrescent manures can be obtained by purchase, or from other foreign sources, more cheaply than to surrender the grazing of entire fields, however poor and bare of grass. But the system was, nevertheless, admirably suited to the general condition of eastern Virginia—nearly every farm of which presented much of exhausted, and miserably poor fields, producing, when grazed, very scanty sustenance to a half-starved stock of cattle, which, even if 100 in number, if of well kept cattle would have yielded more, in milk, meat, and labor, and nearly as much in manure. In fact, the gross product of the then large stocks of cattle on the farms of eastern Virginia, estimated at fair market prices, did not compensate their owners for the cost of keeping, them, without including the enomous attendant expense of the fences dividing the several fields of each farm, and which were rendered necessary solely that nearly starved cattle might roam over the almost naked pasture fields. To adopt some other mode of keeping a much smaller stock, was of itself a great saving; and by so doing, all the natural growth of the fields, when resting from tillage, served to manure the ground on which it grew—giving back, in the large proportion formed from the air and water, perhaps twice as much of enriching matter as they had drawn from the soil. And however scanty might be the cover of weeds to the acre, the whole amount on the field would be greater than the owner could possibly have supplied to his farm in prepared vegetable manure; and moreover, it was already spread over the whole surface, and applied in the best manner, as well as produced without labor or cost. If the several fields of a farmer (under the then usual circumstances) were 100 acres each, and three in number, (the rotation, corn, wheat or oats, and rest,) while he could barely manure 10 acres of his corn field from the farm-yard and stable, and by naked summer cow-pens also, he could add thereto the slight but costless manuring of the 100 acres to which the rotation gave rest that year. Suppose this cover to the 100 acres to be worth but half as much as the other manure, (in the earliest practice, and the then most impoverished condition of the land,) the 100 acres so treated would at least receive the value of 5 other acres manured; and this would be adding at once 50 percent to the previous amount of manuring done, and that without any increased expense, or loss of other products.

But, great as was the improvement undoubtedly made by these two modes of vegetable manuring, there was much waste of effort, and much loss subsequently suffered, by applying improperly in practice the doctrines and instruction of Taylor, which were true in theory and of which the application would have been profitable under other circumstances. It was necessary to learn, (what is now much better and generally understood, but what Taylor neither taught nor seemed to know,) that naturally poor land cannot be durably or profitably enriched by vegetable or putrescent manures, so as to be raised and kept above its original state of fertility. And on such lands, therefore, after a certain slight degree of improvement obtained, even the cheap manuring methods of Taylor were almost thrown away. And on better original soils, after being much recruited and enriched by such means, there are enormous and yearly increasing evils produced, in the troublesome weeds encouraged to grow, and the myriads of insects sustained by their growth, and which become depredators on the succeeding cultivated crop. Even the ploughing under the rank weeds on moderately rich land became a difficult operation; and, like the also difficult ploughing under of coarse unrotted manure from the farm-yard, was often very troublesome to tillage, and sometimes even injurious to the first tilled crop. But these objections do not lessen the merit of Taylor's "enclosing system," for a vast extent of impoverished fields, for the restoration and improvement of which, that system was not only profitable, but indispensable. The greatly increased and extended benefits of these and other applications of vegetable manures, were, however yet to be shown by the subsequent introduction of the knowledge and use of calcareous manures.

III. Marling, or manuring from beds of fossil shells

This mode of fertilization, now so general through all the marl region of lower Virginia, was not practised except on three on four detached farms, and that to but small extent, before 1802. Some few and generally small experimental applications of marl had indded been made, by different individuals, from 15 to as far back as 45 years before; but which applications, from total misconception of the true mode of action of calcareous manures, had been deemed failures; and without exception, of course, had been abandoned by the experimenters, as worthless; and the experiments had been almost forgotten, until again brought to notice, after the much later and fully successful introduction of the practice.

Henley Taylor and Archer Hankins, two plain and illiterate farmers, and near neighbors in James City county, were the earliest *successful* and continuing appliers of marl in Virginia. But at what time they began, and which of them was the first, I have not been able to learn, though visiting Mr. Hankins' farm for that purpose as well as to see his marling, and making inquiries of him, personally, in 1833. Mr. Taylor had then been long dead, and his improvements said to be almost lost, by the exhausting cultivation of the then occupant of his land. Mr. Hankins was unable to say when he and his neighbor began to try marl. He was only certain that it was before 1816. Yet, though these farms are within 12 or 15 miles of Williamsburg, to which place I had made visits once a year or oftener, yet I never heard an intimation of their having begun such practice, until some time after my own first trials in 1818. At that time, when led to the use, as I was, altogether by theoretical views, and by reasoning on the supposed constitution of the manure, it would have been to me the most acceptable and beneficial information to have heard that any other person in Virginia had already proved practically the value of marling. The slow progress of the knowledge of the mere fact of marl having been successfully used before that time, was a strong illustration of the then almost total want of communication among farmers, as well as of their general apathy and ignorance, in regard to the means of improving their lands.

Much earlier than the commencement of marling in James City, the practice had been commenced, (in 1805) in Talbot county, Md., by Mr. Singleton. His account of his practice is in the 4th volume of the 'Memories of the Philadelphia Agricultural Society,' dated December 31, 1817, and first published some time in 1818. But successful as was his practice, and also that of Mr. Taylor and Mr. Hankins in connexion with much worse farming, it is certain that neither of these individuals had the least idea of the true action of marl; and they were indebted to their good fortune, more than to any exercise of reasoning, that they received profitable returns, and did no injury by marling. They all three applied their putrescent manures with the marl. But though this was the safest and most beneficial plan, the thus uniting them prevented the separate action and value of putrescent and calcareous manures being known, compared, and duly appreciated.

My own application of marl, on Coggins Point farm, Prince George county, which in 1818 extended only to 15 acres, (of which but 3 or 4 were under the crop of that year,) by 1821 had been increased to above 80 acres a year, and so continued until nearly all the then arable land on that farm requiring it, (more than 600 acres,) had been covered. In 1821, my earliest publication on the subject was made. Though the facts and reasoning thus made known by that time were beginning to attract much notice, and to induce many persons to begin to marl, still it was some years later before incredulity and ridicule had generally given place to full confidence in the value of the improvement. Even at this time, when nearly 25 years of my own experience of marling and its benefits have passed, and the results are open to public notice and scrutiny, not half the persons who could marl are engaged at it, or are marling to but little purpose; and of all who are using marl, nineteen in twenty are proceeding injudiciously, without regard to the mode of operation of the manure, and therefore are either doing harm or losing profit, almost as often, though in less degree, as doing good. At this time, however, there are scarcely any persons, however negligent in practice who do not fully admit the great value and certain profit of applying marl whenever it is found.

But with all the existing neglect of using this means of fertilization, and with all the still worse ignorance of or inattention to its manner of operating, there never has been a new improvement in agriculture more rapidly extended, or with such beneficial and profitable results. In Prince George county, there is not one farmer having marl on or near his land, who has not applied it to greater or less extent, and always with more or less profit — and, in most cases, largely as well as profitably. In James City county there has been perhaps the next largest as well as the oldest practice. In York county, as in James City, some of the most valuable and profitable improvements by marling have been made. And some of the farms of both counties, adjoining Williamsburg, and having the benefit of putrescent town manures, show more strikingly than any others known, the remarkable power of calcareous manure to fix the putrescent in the soil, and make them more efficient and far more durable. In Surry, Isle of Wight, Nansemond, Charles City, New Kent, Hanover, King William, King and Queen, Gloucester, and Middlesex, counties in the middle of the marl region of Virginia, marl has been already extensively applied, and the profits therefrom are annually increasing. And in other surrounding counties, worse supplied with marl, the practice has been carried on in proportion to the facilities, and to the more scanty experience and degree of information on the subject. It would be a most important statistical fact, if it could be ascertained how much land in Virginia has already been marled. The quantity however is very great; and all the land marled has been thereby increased in *net* product, on the general average, fully 8 bushels of corn or oats, or 4 bushels of wheat — and the land increased in intrinsic value fully 200 per cent. on its previous value or market price. Where the marling has been judiciously conducted, these rates of increase have been more than doubled. From these data, might be calculated something like the already prodigiously increased values and products due solely to marling, and which will be still more increasing from year to year. If not already reached, the result *will* soon be reached, of new value to the amount of millions of dollars having been thus created. It is not designed in this hasty sketch to enter into minute details of results, nor to prescribe rules for practice, both of which have been given in other publications. The purpose here is but to state improvements and describe results in general.

It required the improvement by marling, on originally poor or middling soils, (or liming, which in final and general results is the same thing,) to render as generally available the best and otherwise but rarely found benefits, of the two kinds of vegetable manuring recommended by Taylor. When such soils have been made calcareous, by marling or liming, then, and not until then, all the benefits, present and future, that his readers might have been induced to expect, may be confidently counted upon. In my own earlier practice — and Taylor had no greater admirer, or more implicit follower — I found my farm-yard manurings on acid soils scarcely to pay the expense of application, and to leave no trace of the effect after a very short time. And land allowed to receive for its support all its vegetable growth (of weeds and natural grass) of two and a half years in every four, and the products in corn having been measured and compared, showed no certain increase in more than twenty years of such mild treatment. Since, on the same fields, farm-yard manures, in every mode of preparation and application, always tell well, both in early effect and in duration. And even the leaves raked up on wood-land, spread immediately and without any preparation, as top-dressing on clover, always produce most manifest improvement, and are believed to give more *net* profit than any application of the much richer farm yard manure, per acre, made on like land before it was marled. This utilizing and fixing of other manures, and the fitting land to produce clover, which effects of marling are in addition to all the *direct* benefit produced, would alone serve to give a new face to the agriculture of the country. Whatever may be done by clover, and almost everything that can be done to profit by vegetable manures, on the much larger proportion of the lands of lower Virginia, will be due to the application of marl or lime.

IV. Liming

The kindred improvement by liming began to be extensively practised on some of the best James river lands, where no marl was found, soon after the use of the latter began to extend. Who may have made the earliest and small applications of lime is not known, nor is it at all important. The earlier profitable use of lime in Pennsylvania, and the much earlier and more extended use in Britain, were known to every well-informed or reading farmer. Such a one was Fielding Lewis of Charles City, as well as a most attentive, judicious, and successful practical cultivator and improver. He is believed to have been the earliest considerable limer, and the one who obtained the most manifest profits therefrom, and whose example had most effect in spreading the practice. Some of his disciples and followers have since, in greater rapidity and wider extent of operations, far surpassed their teacher and leader — to whom, however, they award the highest meed of praise for bringing into use, and establishing this great benefit to the agriculture of lower Virginia. Nearly all the best soils on James river are comparatively of low level, as if of ancient alluvial formation, and have no marl with which the neighboring higher and poorer lands are mostly supplied. Of such rich lands, are the farms of Weyanoke, Sandy Point, Westover, and Shirley, &c., in Charles City, and Brandon (Upper and Lower,) in Prince George — and on all these lands, as well as some other, lime has been largely applied. The use is extending to the lands on all the tide waters of the state; and it has recently received a new impulse from the low price at which northern stone-lime is now brought and sold. It is ready slaked, and the vessels are loaded in bulk. The lime is sold on James river at 10 cents the bushel, and even may be contracted for at 8 cents, from vessels that come for cargoes of wood, and would come empty but for bringing lime. The greater lightness and cheaper transportation of lime, will enable it to be applied where marl could not be carried with profit; and with the two, there will be but little of lower Virginia which may not be profitably improved by calcareous manures.

V. Introduction of clover, and its application directly as manure

As has been already observed, incidentally, the clover culture of Virginia is due to the previous use of calcareous manures. Without enough of lime, or some combination of lime, to constitute a really good soil, and still more on sandy lands, as most are in lower Virginia, it is almost impossible to raise clover at all, and always unprofitable to attempt it in field culture, and as part of a regular rotation. It is true, that on some of the best and richest natural soils, such as were just above named, and on the repeatedly and long manured house lots of worse original soils, some good clover had been made. But even there, the products were precarious. Field culture of clover was out of all question, and failed wherever attempted in lower Virginia, except on the best and very peculiar natural soils. It may even be doubted whether it was any where so productive, and so certain, as to be a profitable and regular manuring crop of the rotation. But after marling or liming, clover grew much better on the few places on which only it could be raised at all before, and it also became a crop as sure as any other, and *naturalized* as it were, on the before foreign and altogether unsuitable soils. It is true that marl or lime, alone, will not produce heavy crops of clover on very light or other very poor land. But even there, a good and healthy "stand" of clover is readily obtained, the plants taking and keeping possession of the ground, and are fit to be raised to a heavy crop by other auxiliary means — such as putrescent manures, and also gypsum, and perhaps green-sand. For according to the present feeble and uncertain lights on both the latter substances, the balance of testimony, furnished by my own experience and the operations of other persons in lower Virginia, is in favor of their acting well as manures on clover generally on naturally calcareous or marled and limed land, as well as of their having no useful effect an acid soils.

Clover is a most valuable addition to the comforts and minor profits of a farm, and also as an indirect fertilizer, by increasing the supply of animal manure. But its great benefit is different from, and superior to all these — and is to be secured by the crop being turned into the soil, and thus serving directly as manure for the crop of wheat to be sown thereupon. This process of culture, and mode of manuring, though not original in lower Virginia, but derived from the upper country, is believed to be peculiar to the United States. For though it be called by the same name, (fallow,) as is the best preparation for wheat in Britain, the two processes, and their designed and actual effects, are altogether different. In Britain, the object is to cleanse a field which had became foul with the grass and weeds which had been encouraged to grow and take possession of the land, by the moist climate, and the want of enough hoed crops (like our Indian corn,) in the British rotations. And the object is effected by repeated ploughings and cross ploughings, harrowings, raking and hand-picking of root weeds, continued at intervals through a whole year, during which the land is kept as naked as possible. On the contrary, in the perfection of what is so improperly called "fallow" for wheat in Virginia, a heavy growth of clover is raised, and the second crop, if not also the first, (as is the practice of some of the best farmers on James river,) is left ungrazed and unmown, to be well turned under in August or September, by good ploughs drawn by 3 or 4 horses; then to harrow the new surface, and close the seams between the furrow slices; and without more stirring of the earth, (if possible,) to sow wheat thereupon, at the proper time, and cover the seed with harrows if soft, or such other shallow cutting implement, as the degree of hardness of the surface may require. The object is to have but one, and that a good ploughing; (unless the quantity of wire-grass or blue-grass compel a subsequent shallow ploughing;) and this restriction to one ploughing is not only to save labor, but to secure the clover manure below, and out of the way, and to keep a clean surface above, of which the tilth and mellowness are maintained by the gradual yielding and sinking of the mass of clover turned under.

Such a coat of clover is an excellent manuring for the wheat succeeding, and moreover has been obtained at small cost, as its broad leaves drew its principal nourishment from the atmosphere. The smothering growth, also, of thickly set and rank clover, serves to keep the land clean of almost every thing but the clover itself, which is ploughed under deep enough to be effectually put out of the way of the succeeding wheat. The total and sudden change, also, of the condition of the field, by the ploughing and complete burying of the clover, and the leaving for some time a perfectly bare surface, must serve to starve and destroy nearly all the depredating insects that fed on the clover, and which, if not thus destroyed would continue to feed on the wheat. Thus *wheat on clover,* if well managed in all respects, is an admirably clean crop, both from weeds and from insects, and is heavily manured by the clover, which experience has proved is well adapted as a manure for wheat. Hence, it is not strange that the very best crops of wheat that can be obtained from the land, are raised by this mode of manuring with clover.

In speaking above of the practice of ploughing in clover to precede wheat, as being peculiarly American, it was only meant in reference to it as in regular and profitable use, and forming a part of the ordinary system of rotation and culture on some farms. The like explanation and understanding should be applied to several other things spoken of in this piece, as new improvements here. It is certain that the ploughing under of various green crops, as manure, and especially clover, has been sometimes done in Britain in modern times, and still more formerly in Greece and Italy, judging from the recommendations of the practice by ancient writers. "But," says Sir John Sinclair, "notwithstanding the recommendations of the ancients — notwithstanding the great need of manure — and the excessive price of every extraneous article of that description — this practice is now in a great measure abandoned. Peas, buck-wheat, clover, and other juicy plants, when ploughed in, yield no doubt a portion of manure; but they do little more than *restore* the nourishment they had abstracted from the soil; whilst it is extremely difficult to plough down rank vegetables in a growing state, so as to cover them in a complete manner. The farmers of Scotland are, therefore, of opinion that the only way of securing the highest degree of benefit from green crops, is to cut and bring them into a state of putrefaction, before they be applied to the soil." (General Report, &c., of Scotland, vol. ii. p. 530.) That, in the very different circumstances of Britain, the ploughing in of green clover may be bad economy, may be admitted, without affecting the propriety of the practice in Virginia. The like admission may apply to sundry other processes which are proper and economical and preferable here, and which would be the reverse in Britain, in Flanders, and still more in China. But the distinguished and able agriculturist above quoted was altogether wrong in his main theoretical ground of objection, viz.: that crops ploughed under only *restore* to the soil as much as they had previously abstracted. He forgot, or was not aware of the much larger portion of their substance which most plants, and especially clover and other broad-leaved plants, draw from the atmosphere and form into substantial vegetable matter, from the elements of air and water. But for this, the most bountiful provision of nature for the productive power of the soil, the draught therefrom on every farm, would regularly exceed the returns by manuring, &c., by the whole amount of crop sold off; and certain and rapid exhaustion, would necessarily be the consequence, to every part of the cultivated face of the earth, not regularly and abundantly supplied with putrescent manures from abroad. Yet all know that but a small proportion of the products of a field being returned as manure, will serve to maintain, and under favorable condition of soil, &c., may even greatly increase its productive power.

Perhaps there is no improving process or general practice, of one country or district, and however peculiar to it, that has not been used in some, and it may be many, isolated cases elsewhere. But, the very fact of such trials, however numerous, being rare and remaining separate, and not becoming general practice even on a few farms, is enough to prove that the processes or plans were not found profitable — and that such practical operations were but exceptions to the general rules of approved procedure. If there is any one of the modes of enriching by putrescent matters here spoken of, which might be supposed altogether new, it would be the plan introduced by Taylor, of manuring land by its own vegetable cover, wholly or in part, left to decay on the fields. Yet even this was advocated in Scotland, by Lord Belhaven, in 1707, and his publication is quoted by Sinclair, for the purpose of strongly condemning the opinion. The passage from Lord Belhaven is as follows: "I must say, by-the-by, that if in East Lothian they did not leave a higher stubble [when reaping grain,] than in other places of the kingdom, they would be in worse condition than they are, though bad enough. A good crop of corn makes a good stubble; and a good stubble is the equallest enriching that can be given." (The Country Man's Rudiments, p. 23.) On which Sinclair remarks, "Such sentiments are now condemned by the practical farmer, whose object it is to cut and to bring home the whole, or nearly the whole straw; under an impression that, *unless it be brought into a putrefied state in the fold yard, little advantage, as a manure, can be obtained from it.'* (General Report, &c., of Scotland, vol. ii. p. 510.)

VI. Top dressing with putrecent manures, and especially on clover

After allowing all merit due to the improvement to be made by collecting, and to the greatest possible extent, more vegetable materials for farmyard and stable manure — and also for the greater benefit of applying such manures on lands marled or limed, there would still remain a powerful obstacle to the extension of manuring, in the enormous amount of labor required, on Taylor's proposed plan, to carry out the manure in the spring, and plough under for corn before planting — to say nothing of the difficulty of tillage, and of danger to the young plants in dry weather, by the presence of 50 large ox cart loads (4 oxen for each,) per acre, the quantity of which he speaks. In so short a time, and in a busy part of the year, it is impossible, with the ordinary force of the farm, to carry out, spread, and turn in proper manner, all the manure that could be made from all the offal of a grain farm — independent of the leaves and other materials which ought to be used. If, to lessen these difficulties, the manure be earlier heaped and fermented, then indeed it is made more manageable and more speedily operative; and the time for applying it is extended. But, on the other hand, the *whole* labor is increased, and the manure loses much of its fertilizing principles by fermentation. A third choice remains, which is to leave in the farm-yard, without heaping, and undisturbed, either all the manure, or all that cannot be conveniently carried out before planting the corn land intended to be so manured. The balance, or the whole so left, rots slowly through the summer, and may be carried out in September on the clover land, to be ploughed immediately for wheat; or otherwise kept until the next spring, to be ploughed under for the corn crop. In the first case, there is lost the use of the manure for six months, and in the second for a whole year; and in both, during the delay, the manure must lose much of the gaseous products of decomposition, which there is nothing growing to receive and secure. I have tried and suffered loss by all these three courses; and was fully sensible that in all, either the waste of manure, or the loss of its use by delay was very considerable.

The plan of top-dressing is an admirable means of overcoming most of these difficulties — and indeed all the important difficulties, if there be no *greater* waste of the fertilizing principles of manure spread on the surface of the soil, and no less benefit to the growing crop, than if the manure had been buried by the plough. Both these points are disputed, and there have been no careful or conclusive experiments made to decide the question. But from my general experience and observation, and now general practice for 12 years, I am satisfied that the evil of the greater waste of manure or of its effects, if indeed there be any greater, is not to be compared to the great saving of expense, found in applying the manure as top-dressing on clover, in March and April, or even somewhat later. When thus applied, and spread evenly and immediately, as it should be, the first rain carries into the earth all the soluble and active parts, and they are absorbed immediately by the roots of the clover, the growth of which will show the benefit in a few days. The coarse parts of the manure, are soon covered over and kept shaded and moist, by the growth of the clover, and thereby the rotting is both hastened, the products preserved better from waste; and every rain carries to the roots, and plus them to immediate use, as food for the growing plants. Thus, a good growth of clover is made even on inferior land, and perhaps by food that might have mostly gone to waste, if the manure had been fermented and made hot in the heap — or more slowly, by lying spread during all summer over the cattle yard. And this clover manure, being more than half derived from the atmosphere, would be worth to the succeeding crop of wheat or corn, more than double the prepared manure it fed upon and consumed. The saving of labor also is great. The mode of application permits the manure to be carried out in any state, from the time of the earliest beginning of the spring's growth of clover, to the first of May; and, if not before so used, even if put out at any time in summer, would do as well, except that the benefit of helping the earlier growth would have missed. But as to summer-made stable manure, which is so generally injured by fire-fanging, if bulked, or wasted, if spread open in the stable yard, it is decidedly better to spread it on clover, during all summer.

The best application of top-dressing is to clover, because every portion of the manure so used by the clover, is converted to (probably) double the value of the manure in clover, ready to be again used by the wheat the same year. But though clover, as a broad-leaved leguminous plant, (or of the pea tribe,) draws more of its nourishment from the atmosphere, and gives more of manure to the next crop, than any other grass, still, if clover were not at hand, the top-dressing should be given to any grass or any weeds that stood thick enough soon to cover and shade the manure, and to increase the growth of which would be beneficial to the land, and to the succeeding crops.

But there is another material for top-dressing to which there can be no objection on the score of supposed waste, or because any thing better or more economical could be done with it, and which application is highly beneficial. This is leaves raked up in wood-land, and laid on young clover, in winter or spring. Valuable as is this material, on account of its abundant quantity, and usual cheapness of supply, it is very poor in quality compared to straw; and would perhaps not be worth the labor of double hauling and twice handling, required to carry it through the farm-yard to prepare it for manure. And dry or unrotted leaves would be very troublesome to plough under, and if thick might do more harm than good. But when spread on clover, not too thickly immediately from the woods, there is little trouble, no possibility of injury, and a decided improvement to the clover, and of course to the next wheat crop, and to the land. Wheat straw, spread in July, or later, on the young clover sown the year before, is also a valuable manure; and may be applied most advantageously on the poorest spots, by bringing back to the field return loads of straw, when carting the wheat from the field to the thrashing machine.

Another, and perhaps the best mode of dressing by pine leaves, (which are best, because they will not be blown away by high winds,) is on wheat, as soon as possible after it is sown. This is too busy a time for much of such manuring to be then done. But it will be especially beneficial to such extent as it can be done. When so applied, the leaves, merely as covering, protect the wheat from being killed by the winter's cold, and from violent alternations of temperature. And they do the same for the clover sown afterwards, and also protect the young plants by shade and moisture. And besides these merely mechanical modes of action, the leaves serve as well as manure, both to the wheat and to the clover, as they could in any other mode of application. An important objection however to this mode may exist if the leaves serve to shelter and protect chinch bugs, as coarse farm-yard litter does, if thus laid on wheat, when those destructive insects are numerous.

Two other modes of top-dressing have been practised and approved by good farmers, in different parts of Virginia. The first is of well-rotted manure, applied on wheat during winter and spring. It was practised by Mr. Richard Sampson, and his account of it may be seen at page 58 vol. i. of Farmers' Register. The practice is doubtless good when manure so rotted is on hand, and the most convenient application is to the wheat field. The other mode, which was in use on some of the Rappahannock farms, was to apply the coarse unrotted manure on the surface of the corn-field, after the corn had been planted, or had come up. I have also practised this plan when much more convenient to do so; and would care but little to choose between this top-dressing and ploughing the manure under, before planting the corn. But neither to wheat or corn ought the top-dressing to be given, if the clover land is as convenient, and as much needing the manuring.

VII. Rotation of crops

One of the best indications of the agriculture of any region being in an improving state, though this improvement be not yet much extended, is the regard recently paid to the proper principles of rotations of crops. And if no regard be had to this part of husbandry, there will not be good farming, no matter what may be the goodness of the soul, or the industry and skill of the cultivator. If judged by this test, the agriculture of lower Virginia is making good progress towards an improved state, though yet very far below the point aimed at. There is no rotation universally admitted to be the best, even under the same and most favorable cirucmstances. But almost every intelligent farmer at least exerts his reasoning faculties on this subject, and adopts such rotation as his judgment makes out to be the best for his particular circumstances. Still, no one of even the best farmers and most successful improvers, seems to be entirely satisfied with his own system of rotation. And this very discontent serves to prove that each one is striving to find out and to pursue what is the best course, without being bigoted to even that which he so far prefers. These remarks however apply only to a few of the best farmers; who, however, as in all cases when agreed themselves as to the best courses, will serve as exemplars and guides to all others who can or will profit by any instruction or example.

The Eastern Shore two-field rotations, of 1st, corn, 2d, oats, followed the same year by volunteer Magothy bay bean, serving as a manure crop for the corn of the next (the third) year, however scourging, without great attention to manuring, is a rotation sustained by reason, as well as by experience.

The old three-shift rotation, common on the greater number of farms which have any rotation at all, or where there is any thing deserving the name of farming, is 1st, corn, 2d, wheat or oats, 3d, the fild at rest from tillage, but under close grazing. This is a most exhausting course. But the same frame-work, altered only by the land being limed, regularly sown in clover on the wheat, and the clover grazed only slightly and late in autumn, and the greater part turned in for manure, is a very improving, cleansing, and altogether excellent rotation for light loams, or good corn lands. This is the rotation on the lighter of the two Upper Brandon farms, belonging to Mr. Wm. B. Harrison.

The rotation most approved, however, and most practised by the best farmers on James river, and also on the lighter soils of the Messrs. Wickham, on the Pamunkey, is the four-field rotation, of 1, corn, 2, wheat, with clover sown, 3, clover, turned in for manure to 4, wheat. This seems to be a very severe rotation, and is even so considered by some of the best farmers who pursue the plan. However, this rotation, with all the objections to it, both theoretical and practical, is connected with and serves to exhibit most admirable and profitable farming, and also improvement of land such as may well compare, in these respects, with any in this country. Such a rotation *demands,* and will not do without, a good soil, good manuring, good clover, (and therefore marling or liming, or a naturally calcareous soil,) good teams and good ploughing, and in every respect good farming. He who can meet all these requisitions will succeed admirably, both as to profit and improvement, with the four-field rotation; but if he falter in the work, he will be soon obstructed by insuperable difficulties, and be defeated, and driven from this course of cultivation.

VIII. Implements of husbandry and tillage

Great improvements have been made within the last 25 years in agricultural utensils and machinery. Before that time, two-horse ploughs were rarely used, and only on the few richest and best cultivated farms, for breaking lands generally, and on the rich low-grounds and perhaps a few acres more of lots, on some other farms of middling grade, as to fertility and management. On the far greater number of farms there was neither a two-horse plough, nor a mould-board plough for a single horse. Ninety-nine acres in the hundred were broken up by one-horse ploughs; and half of the whole quantity with the trowel-hoe, or fluke-hoe plough, having cutting wings to the share on both sides alike, and no mould-board. The ploughing was rarely deeper than three inches, (often less,) and even that depth, on the naturally poor lands, often reached the barren subsoil. Harrows, and the advantages of harrowing, to get rough ploughed land in good order, which no farmer can now dispense with, were scarcely known. On most farms, the only approach to a harrow was the ''drag'' with large wooden teeth, used to smooth newly sown wheat land, after the seed had been covered by the trowel-hoe plough. The wheat was almost universally trodden out of the straw by horses (sometimes aided by oxen) on the earth, and out of doors. Thrashing machines, (and those very inferior, on the old Scotch plan,) were not on half a dozen farms on James river and perhaps not thrice as many in all eastern Virginia.

It can now scarcely be conceived how land could be broken up, and left in tolerable order by such ploughs and ploughing. Nor could it be done on even the poorest land, if suffered to be at rest long enough for its surface to be clothed with grass, or the earlier and more scant growth of weeds to remain. But most of the farmers who broke up their corn fields with trowel-hoe ploughs, secured the feasibility of the process by cultivating the land in corn every other year, and grazing as close as possible the intervening years. Of course there was no turf, and scarcely a living root, on poor land. Very many did not break up the land at all, at first, but merely ran a furrow with the trowel-hoe ploughs along each of the former corn-rows, checked by similar cross-rows, then planted, and ploughed the intervals at leisure some 6 to 8 weeks after, when the corn was several inches high. Many persons objected altogether to using mould-board ploughs, believing that turning over the furrow slices would exhaust the soil by greater exposure to sun and air.

Since, perhaps the error has been in the opposite direction. Many improving farmers have been even too ready to buy new and highly priced ploughs and other implements; and there are but few who have not had to throw aside, as worthless, at least as many kinds of new implements as they have retained. However, ploughs of good construction are now in general use. Two-horse ploughs are common wherever there is any farming better than the worst; and three-horse, and even four-horse ploughs are used where the natural depth or subsequent improvement of soil requires ploughing 8 or more inches deep. Good thrashing machines, either fixed or portable, of several kinds, are in general use, except on very small farms; and many a large wheat farmer would now abandon the culture of wheat, rather than to have to return to treading out every crop, as formerly.

The processes of tillage have improved with the implements for tillage; and imperfect as both yet are, these two improvements are now more advanced than the more important improvements of the fertility of the soil, and the preservation and economy of the means and resources for fertilization.

Defects of agriculture still remaining

What has been said of improvements made, will prevent the necessity of treating of the remaining defects of agriculture at much length, copious as may be that subject. The most important defects may be despatched in a few words, as they are either the omitting, or negligently and imperfectly executing, any such of the foregoing named improvements, as may be suitable and wanting to the particular circumstances of each farmer. But there are also minor branches of these several general subjects, and also other distinct subjects, of neglect or omission, which remain to be briefly considered.

In regard to the application of marl — although it has been more universally or generally and extensively adopted and practised than any other as new and as great innovation in agriculture, and although the unusual rapidity with which the new practice has spread, is as remarkable as the extended space of operations, still very few of those farmers who have adopted this mode of improvement have operated properly, or to half the profit that even their actual expense therein, if properly directed, would have returned. The far greater number of persons commence and continue the always heavy business of marling, without providing for it any additional force, or suspending any of the previous cultivation and full employment on their farms. Few had before been able to perform properly all the ordinary labors required; and yet they count on marling their whole farms with no more labor than was before engaged. They marl only "at leisure times;" and for want of regular and continued operations, very little can be done at all, and that little at double cost. But suppose that there were no such disadvantages, but only the slower rate of work, and that the owner of 600 acres to be marled, could cover 10 acres a-year at odd jobs and leisure times, or 100 acres by merely suspending a third of his cultivation, and devoting the labor so released to regular and continued marling. In the first case 36 years would be required to marl the whole surface, and in the latter only six years. In the latter case there would be no diminution of product by the lessened tillage, even in the first year; for the two-thirds of the first field marled would yield; in the first year, more (and usually much more) than the whole without marling. And the increased rate of product of the farm, at the end of the 6 years serving to marl the whole, would more than pay for thrice the labor usually required for the marling. The mere difference to the farmer of having his land all marled, say in 1842, or not until 1852, would be the loss of the increased product for ten years. And on most of the lands of lower Virginia, that increased product, from marling alone, would be equal to half the gross product, (or 100 per cent. on the original product,) and would amount to more in 10 years than the previous fee-simple value of the whole farm. He who has determined to marl his land, and yet delays to do it, is thereby submitting, during the delay, to a loss in every crop fully equal to all that is made on the unmarled land in tillage.

But putting aside this enormous waste of value, incurred by mere delay of operations, the marling actually done, (and at double cost, from irregular working and want of method,) does not yield half the available returns, because of ignorance or carelessness in the application, and the subsequent treatment of the land. Frequently, the quantity applied is unsuited to the nature and wants of the soil. Sometimes, from variation in the strength of the marl, and the proprietor being ignorant of the quality, or, more often, from irregular spreading, injury is done, and loss sustained, by there being too much marl on some parts, and too little on other parts, of the same field, and sometimes even of the same acre. Again, if the time and manner of application be never so judicious, and the improved fertility be never so great at first, half the surely available future profit will be lost, if the subsequent management be as improper as it usually is in almost all cases. Under ordinary tillage and management, and a rotation merely *not* exhausting, (say the three-shift, without clover and without grazing, and very little manuring,) marling may, and probably will, give a *permanent* increase to the crops of between 50 and 100 per cent. But if clover be sown, and partly or wholly ploughed under as manure, and if the *increased* supply of materials for farm-yard manure be all used, (which also will be due entirely to marl,) these means will in a few years double the direct effects of the marl alone. The whole permanently increased product, thus to be made by the most judicious marling (and the same, it is presumed, of liming,) and subsequent operations, on the ordinary acid soil of lower Virginia, might be safely counted as four-fold the previous gross products; and a ten-fold increase of the net profit, and also of the intrinsic value of the land. Yet the loss, or failure to obtain, the much larger proportion of this increase of value is borne by nearly all of those who apply marl, because they will not attend to the properties, and the mode of operation of the manure which they are using.

Having begun to marl without any guide whatever, or source of proper instruction, except my own theoretical views of soils and manures, I have myself committed more or less of every serious error above referred to, and suffered something from every kind of loss; as well as having realized enough of every benefit promised, to know what greater and more speedy general profit a proper conduct throughout would surely produce. My mistakes and losses, necessarily caused by the ignorance and inexperience of an uninstructed novice, and pioneer in a totally new business, may serve to guard from the same all followers; and they, being so guarded, many enjoy all the benefits of marling, unalloyed by any of the risks and losses of ignorant beginners. But it is a subject of regret, and of great public as well as private loss, that the actual results are very different. It is as if each new beginner were resolved to learn nothing from the previous experience of others, but prefers to owe all knowledge to his own errors and losses.

In regard to putrescent manures, there are very great and very general defects of management. On acid soils there is excuse for neglect of manuring, for it scarcely pays the cost. But on well constituted natural soils, and on naturally acid and bad soils, after being marled or limed, too much labor cannot well be given to collecting all the farm materials for, and applying putrescent manures. But even of the quantity of materials operated on, in the farm-yard and stables, there is great waste of enriching principles permitted, in various ways, but more especially by fermentation, whether the manure be heaped, or not heaped, or carried out on the fields before rotting. And such waste must take place, if decomposition of the manure proceeds, and its gaseous products be set free, when there are no growing plants present, or not enough of them, to absorb and put to use the fleeting fertilizing principles.

The neglect of proper drainage is another great defect of our agriculture, affecting not only the profits, but the health of the lower country. On almost every farm there is something of this defect. But in the low and level lands of the south-eastern counties, the whole region suffers from excessive wetness, which might be easily removed by a proper system of drainage and of tillage suited to that object. Princess Anne and Norfolk counties might be raised from their present most wretched agricultural condition, to their proper grade of fine and productive and valuable farming lands; and even the rich lands of Elizabeth City, and the now fertile and productive farms of the more carefully cultivated and beautiful low lands of Gloucester, might yet be greatly improved in value, by more thorough and general drainage.

A great defect on almost every farm to some extent, and entirely on most land, is the neglect of grass crops for hay. Even the small extent of clover husbandry, admirable and profitable as it is, scarcely forms an exception to this defect; for the clover is turned in for manure, and rarely converted to hay. On some of the best farms, and where the benefits of clover-manure have been best known by experience, as Westover and Brandon, the proprietors deem this use so much the more profitable, that they would consider the conversion of any considerable portion of the clover crop to hay as a wasteful abstraction from the product of the field under the succeeding crops. But without undervaluing the worth of clover as manure, I doubt whether these excellent farmers and improvers do not in this respect adhere too closely to their system. Before forming the seeds, which is the exhausting part of the growth, the clover has drawn nearly all its nourishment from air and water, and taken very little from the enriching ingredients of the soil. If then the first crop were mown for hay, before the seeds had been formed, there would be the following benefits derived, at very slight cost to the land: 1st, the hay, for farm consumption, 2nd, the cutting off by the scythe and destroying all annuals not then having seeded — and the keeping down of all perennial weeds, shrubs, and bushes; 3d, the having a clean second or summer growth of clover, to furnish seed if desired, or otherwise thus made free from the smothering and sometimes destructive cover of the dead first or spring crop. Of course, the mowing should be on clover of good size, and where the coarse manure had not been applied.

But putting hay from clover aside, there is general neglect of natural meadows, and still more of artificial grasses on the low and moist spots, which would be very profitable under permanent grass, and are worth but little as arable land. Yet hay, either for sale if near to market, or for consumption on the farm where needed, is perhaps the most valuable crop of suitable land, and certainly the crop most cheaply made. For want of attention to grass husbandry, our towns continue to be supplied with hay imported from New England — and our corn crops are stripped of the leaves for fodder, though the cost of labor so employed, and the injury to the grain, must amount to much more than the value of the fodder so obtained.

A region in which hay is scarcely made at all, and grass so much neglected, must necessarily also show poor live stock, and small profits therefrom. To both these bad courses we have been seduced by the quantity of cattle-forage furnished in the offal (shucks and stalks) of our large corn crops. But all that even this bountiful supply permits, is that the cattle may live, gradually becoming poorer, through winter and spring, and yielding through the year little or no net profit, except in furnishing supplies necessary to the comfort of the family, in milk, butter, and meat, and of manure to the land.

We raise not enough hogs, for the home supply of what is the principal animal food of the country; though this great defect is rapidly becoming less. We raise very few horses for labor. Sheep are not to be found at all on most farms, owing in a great degree to the risks they would incur from dogs. If every useless and worthless dog in lower Virginia were a sheep or a hog, the expense of maintenance would be less, and the returns to agriculture greater by many thousands of dollars, besides the rendering the raising of sheep in general safe and profitable, instead of being as hazardous as if the country were infested by wolves instead of by worthless dogs.

On Calcareous Manures, 1842

From Edmund Ruffin, "Essay on Calcareous Manures" (Petersburg, Va., June, 1842), pp. 19–23, 57–63.

With some exceptions to every general character, the tide-water district of Virginia may be described as generally level, sandy, poor, and free from any fixed rock, or any other than stones rounded apparently by the attrition of water. On much the greater part of the lands, no stone of any kind is to be found of larger size than gravel. Pines of different kinds form the greater part of a heavy cover to the silicious soils in their virgin state, and mix considerably with oaks and other growth of clay land. Both these kinds of soil, after being exhausted of their little fertility by cultivation, and "turned out" to recruit, are soon covered by young pines which grow with vigor and luxuriance. This general description applies more particularly to the *ridges* which separate the *slopes* on different streams. The ridge lands are always level, and very poor — sometimes clayey, more generally sandy, but stiffer than would be inferred from the proportion of silicious earth they contain, which is caused by the fineness of its particles. Whortleberry bushes, as well as pines, are abundant on ridge lands — and numerous shallow basins are found, which are ponds of rain water in winter, but dry in summer. None of this large proportion of our lands has paid the expense of clearing and cultivation, and much the greater part still remains under its native growth. Enough, however, has been cleared and cultivated in every neighborhood to prove its utter worthlessness under common management. The soils of ridge lands vary between sandy loam and clayey loam. It is difficult to estimate their general product under cultivation; but judging from my own experience of such soils, the product may be from five bushels of corn, or as much of wheat, to the acre on the most clayey soils, to twelve bushels of corn, and less than three of wheat, on the most sandy — if wheat were there attempted to be made.

The *slopes* extend from the ridges to the streams, or to the alluvial bottoms, and include the whole interval between neighboring branches of the same stream. This class of soils forms another great body of lands, of a higher grade of fertility, though still far from valuable. It is generally more sandy than the poorer ridge land, and when long cultivated is more or less deprived of its soil, by the washing of rains, on every slight declivity.

The washing away of three or four inches in depth exposes a sterile subsoil, (or forms a "gall,") which continues thenceforth bare of all vegetation. A greater declivity of the surface serves to form gullies several feet in depth, the earth carried from which, covers and injures the adjacent lower land. Most of this kind of land has been cleared and greatly exhausted. Its virgin growth is often more of oak, hickory, and dogwood, than pine; but when turned out of cultivation, an unmixed growth of pine follows. Land of this kind in general has very little durability. Its best usual product of corn may be, for a few crops, eighteen or twenty bushels — and even as much as twenty-five bushels, from the highest grade. Wheat is seldom a productive or profitable crop on the slopes, the soil being generally too sandy. When such soils as these are called rich or valuable (as most persons would describe them,) those terms must be considered as only comparative; and such an application of them proves that truly fertile and valuable soils are very scarce in lower Virginia.

The only very rich and durable soils below the falls of our rivers are narrow strips of high-land along their banks, and the low-lands formed by the alluvion of the numerous smaller streams which water our country. These alluvial bottoms, though highly productive, are lessened in value by being generally too sandy, and by the damage they suffer from being often inundated by floods of rain. The best high-land soils seldom extend more than half a mile from the river's edge — sometimes not fifty yards. These irregular margins are composed of loams of various qualities, but all highly valuable; and the best soils are scarcely to be surpassed in their original fertility, and durability under severe tillage. Their nature and peculiarities will be again adverted to, and more fully described hereafter.

The simple statement of the general course of tillage to which this part of the country has been subjected is sufficient to prove that great impoverishment of the soil has been the inevitable consequence. The small portion of rich river margins, was soon all cleared, and was tilled without cessation for many years. The clearing of the slopes was next commenced, and is not yet entirely completed. On these soils, the succession of crops was less rapid, or, from necessity, tillage was sooner suspended. If not rich enough for tobacco when first cleared, (or as soon as it ceased to be so,) land of this kind was planted in corn two or three years in succession, and afterwards every second year. The intermediate year between the crops of corn, the field was "rested" under a crop of wheat, if it would produce four or five bushels to the acre. If the sandiness, or exhausted condition of the soil, denied even this small product of wheat, that crop was probably not attempted; and, instead of it, the field was exposed to close grazing, from the time of gathering one crop of corn, to that of preparing to plant another. No manure was applied, except on the tobacco lots; and this succession of a grain crop every year, and afterwards every second year, was kept up as long as the field would produce five bushels of corn to the acre. When reduced below that product, and to still more below the necessary expense of cultivation, the land was turned out to recover under a new growth of pines. After twenty or thirty years, according to the convenience of the owner, the same land would be again cleared, and put under similar scourging tillage, which, however, would then much sooner end, as before, in exhaustion. Such a general system is not yet every where abandoned; and many years have not passed, since such was the usual course on almost every farm.

How much our country has been impoverished during the last fifty years, connot be determined by any satisfactory testimony. But, however we may differ on this head, there are but a few who will not concur in the opinion, that [up to 1831] our system of cultivation has been every year lessening the productive power of our lands in general — and that no one county, no neighborhood, and but few particular farms, have been at all enriched, since their first settlement and cultivation. Yet many of our farming operations have been much improved and made more productive. Driven by necessity, proprietors direct more personal attention to their farms — better implements of husbandry are used — every process is more perfectly performed — and, whether well or ill directed, a spirit of inquiry and enterprise has been awakened, which before had no existence.

Throughout the country below the falls of the river, and perhaps thirty miles above, if the best land be excluded, say one tenth, the remaining nine tenths will not yield an average product of ten bushels of corn to the acre; though that grain is best suited to our soils in general, and far exceeds in quantity all other kinds raised. Of course, the product of a large proportion of the land would fall below this average. Such crops, in very many cases, cannot remunerate the cultivator. If our remaining wood-land could be at once brought into cultivation, the *gross* product of the country would be greatly increased, but the *net* product very probably diminished; as the general poverty of these lands would cause more expense than profit to accompany their cultivation under the usual system. Yet every year we are using all our exertions to clear wood-land, and in fact seldom increase either net or gross products — because nearly as much old exhausted land is turned out of cultivation as is substituted by the newly cleared. Sound calculations of profit and loss, would induce us even greatly to reduce the extent of our present cultivation, in lower Virginia, by turning out and leaving waste, (if not to be improved,) every acre that yields less than the total cost of its tillage.

No political truth is better established than that the population of every country will increase, or diminish, according to its regular supply of food. We know from the census of 1830, compared with those of 1820 and 1810, that our population is nearly stationary, and in some counties is actually lessening; and therefore it is certain that [to 1830] our agriculture is not increasing the amount of food, or the means of purchasing food — with all the assistance of the new land annually brought under culture. In these circumstances, a surplus population, with all its deplorable consequences, is only prevented by the great current of emigration which is continually flowing westward. No matter who emigrates, or with what motive — the enterprising or wealthy citizen who leaves us to seek richer lands and greater profits, and the slave sold and carried away on account of his owner's poverty — all concur in producing the same result, though with very different degrees of benefit to those who remain. If this great and continued drain from our population was stopped, and our agriculture was not improved, want and misery would work to produce the same results. Births would diminish, and deaths would increase; and hunger and disease, operating here as in other countries, would keep down population to that number that the average products of our agricultural and other productive labor could feed, and supply with the other necessary means for living.

A stranger to our situation and habits might well oppose to my statements the very reasonable objection, that no man would, or could, long pursue a system of cultivation of which the returns fell short of his expenses, including rent of land, hire of labor, interest on the necessary capital, &c. Very true; if he had to pay those expenses out of his profits, he would soon be driven from his farm to a jail. But we own our land, our laborers, and stock; and though the calculation of net profit, or of loss, is precisely the same, yet we are not ruined by making only two per cent. on our capital, provided we can manage to live on that income. If we live on still less, we are actually growing richer, (by laying up a part of our two percent.,) not withstanding the most clearly proved regular loss on our farming.

Our condition has been so gradually growing worse, that we are either not aware of the extent of the evil, or are in a great measure reconciled by custom to profitless labor. No hope for a better state of things can be entertained, until we shake off this apathy — this excess of contentment, which makes no effort to avoid existing evils. I have endeavored to expose what is worst in our situation as farmers; if it should have the effect of arousing any of my countrymen to a sense of the absolute necessity of some improvement, to avoid ultimate ruin, I hope also to point out to some of their number, if not to all, that the means for certain and highly profitable improvements are completely within their reach.

The cultivators of eastern Virginia derive a portion of their income from a source quite distinct from their tillage — and which, though it often forces them to persist in their profitless farming, yet also, in some measure, conceals, and is generally supposed to compensate for its losses. This source of income is, the breeding and selling of slaves; of which, (though the discussion of this point will not be undertaken here,) I cannot concur in the general opinion that it is also a source of profit.

It is not meant to convey the idea that any person undertakes as a regular business the breeding of slaves with a view to their sale; but whether it is so intended or not, all of us, without exception, are acting some part in aid of a general system, which taken altogether is precisely what I have named. No man is so inhuman as to breed and raise slaves, to sell off a certain proportion regularly, as a western drover does with his herds of cattle. But sooner or latter the general result is the same. Sales may be made voluntarily, or by the sheriff — they may be met by the first owner, or delayed until the succession of his heirs — or the misfortune of being sold may fall on one parcel of slaves, instead of another; but all these are but different ways of arriving at the same general and inevitable result. With plenty of wholesome though coarse food, and under such mild treatment as our slaves usually experience, they have every inducement and facility to increase their numbers with all possible rapidity, without any opposing check, either prudential, moral, or physical. These several checks to the increase of population operate more or less on all free persons, whether rich or poor; and slaves, situated as ours are, perhaps are placed in the only possible circumstances in which no restraint whatever obstructs the propagation and increase of the race. From the general existence of this state of circumstances, the particular effects may be naturally deduced; and facts completely accord with what these circumstances promise. A gang of slaves on a farm will often increase to four times their original number, in thirty or forty years. If a farmer is only able to feed and maintain his slaves, their increase in value may double the whole of his capital originally vested in farming, before he closes the term of an ordinary life. But few farms are able to support this increasing expense, and also furnish the necessary supplies to the family of the owner; whence very many owners of large estates, in lands and negroes, are throughout their lives too poor to enjoy the comforts of wealth, or to encounter the expenses necessary to improve their unprofitable farming. A man so situated, may be said to be a slave to his own slaves. If the owner is industrious and frugal, he may be able to support the increasing number of his slaves, and to bequeath them undiminished to his children. But the income of few persons increases as fast as their slaves; and if not, the consequence must be, that some of them will be sold, that the others may be supported; and the sale of more is perhaps afterwards compelled, to pay debts incurred in striving to put off that dreaded alternative. The slave first almost starves his master, and at last is eaten by him — at least he is exchanged for his value in food. The sale of slaves is always a severe trial to their owner. Obstacles are opposed to it, not only by sentiments of humanity and of regard for those who have passed their lives in his service — but every feeling he has of false shame comes to aid; and such sales are generally postponed until compelled by creditors, and are carried into effect by the sheriff, or by the administrator of the debtor. But when the sale finally takes place, its magnitude makes up for all previous delays. Do what we will, the surplus slaves *must* be sent out of a country which is not able to feed them; and these causes continue to supply the immense numbers that are annually sold and carried away from lower Virginia, without even producing the political benefit of lessening the actual number remaining. Nothing can check this forced emigration of blacks, and the voluntary emigration of whites, except increased production of food, obtained by enriching our lands, and the consequent increase of farming profits. No effect will more certainly follow its cause than this — that whenever our land is so improved as to produce its present amount of population. The improving farmer who adds one hundred bushels of corn to the previous product of his country, also effectually adds, and permanently, to its population, as many persons as his increase of product will feed and support. . . .

We have seen from the proof furnished by the analysis of wood ashes, that even poor acid soils contain a little salt of lime, and therefore must have been slightly calcareous at some former time. But such small proportions of calcareous earth were soon equalled, and then exceeded, by the formation of vegetable acid, before much productiveness was caused. The soil being thus changed, the plants suitable to calcareous soils died off, and gave place to others which produce, as well as feed and thrive on, acidity. Still, however, even these plants furnish abundant supplies of vegetable matter, sufficient to enrich the land in the highest degree; but the antiseptic power of the acid prevents the leaves from rotting for years, and even then the soil has no power to profit by their products. Though continually wasted, the vegetable matter is continually again forming, and always present in abundance; but must remain almost useless to the soil, until the accompanying acidity shall be destroyed.

It may well be doubted whether any soil destitute of lime in every form would not necessarily be a perfect barren, incapable of producing a spire of grass. No soil thus destitute is known, as the plants of all soils show in their ashes the presence of some lime. But it is probable that our sub-soils, which, when left naked by the washing away of the soil, are so generally and totally barren, are made so by their being entirely destitute of lime in any form. There is a natural process regularly and at all times working to deprive the sub-soil of all lime, unless the soil is abundantly supplied. What constitutes soil, and makes the strong and plain mark of separation and distinction between the more or less fertile soil and the absolutely sterile sub-soil beneath? The most obvious cause for this difference which might be stated, is the dropping of the dead vegetable matter on the surface; but this is not sufficient alone to produce the effects, though it may be so when aided by another cause of more power. When the most barren surface earth was formed or deposited by any of the natural agents to which such effects are attributed by geologists, it seems reasonable to suppose that the surface was no richer than any lower part of the whole upper stratum so deposited. If, then, a very minute proportion of lime had been equally distributed through the body of poor earth to any depth that the roots of trees could penetrate, it would follow that the roots would, in the course of time, take up all the lime, as all of it would be wanting for the support of the trees; and their death and decay would afterwards leave.

Nearly all the wood-land now remaining in lower Virginia, and also much of the land which has long been arable, is rendered unproductive by acidity, and successive generations have toiled on such land, almost without remuneration, and without suspecting that their worst virgin land was then richer than their manured lots appear to be. The cultivator of such soil, who knows not its peculiar disease, has no other prospect than a gradual decrease of his always scanty crops. But if the evil is once understood, and the means of its removal is within his reach, he has reason to rejoice that his soil was so constituted as to be preserved from the effects of the improvidence of his forefathers, who would have worn out any land not almost indestructible. The presence of acid, by restraining the productive powers of the soil, has in a great measure saved it from exhaustion; and after a course of cropping which would have utterly ruined soils much better constituted, the powers of our acid land remain not greatly impaired, though dormant, and ready to be called into action by merely being relieved of its acid quality. A few crops will reduce a new acid field to so low a rate of product, that it scarcely will pay for its cultivation; but no great change is afterwards caused, by continuing scourging tillage and grazing, for fifty years longer. Thus our acid soils have two remarkable and opposite qualities, both proceeding from the same cause: they can neither be enriched by manure, nor impoverished by cultivation, to any great extent. Qualities so remarkable deserve all our powers of investigation; yet their very frequency seems to have caused them to be overlooked; and our writers on agriculture have continued to urge those who seek improvement to apply precepts drawn from English authors, to soils which are totally different from all those for which their instructions were intended. . . .

It has already been made evident that the presence of calcareous earth in a natural soil causes great and durable fertility. But it still remains to be determined, to what properties of this earth its peculiar fertilizing effects are to be attributed.

Chemistry has taught that silicious earth, in any state of division, attracts but slightly, if at all, any of the parts of putrescent animal and vegetable matters. But even if any slight attraction really exists when this earth is minutely divided for experiment in the laboratory of the chemist, it cannot be exerted by silicious sand in the usual form in which nature gives it to soils; that is, in particles comparatively coarse, loose, and open, and yet each particle impenetrable to any liquid, or gaseous fluid, that might be passing through the vacancies. Hence, silicious earth can have no power, chemical or mechanical, either to attract enriching manures, or to preserve them when actually placed in contact and intermixed with them; and soils in which the qualities of this earth greatly predominate, must give out freely all enriching matters which they may have received, not only to a growing crop, but to the sun, air, and water, so as soon to lose the whole. No portion of putrescent matter can remain longer than the completion of its decomposition; and if not arrested during this process, by the roots of living plants, all will escape in the form of *gas* (the latest products of decomposition,) into the air, without leaving a trace of lasting improvement. With a knowledge of these properties, we need not resort to the common opinion that manure *sinks* through sandy soils, to account for its rapid and total disappearance.

Aluminous earth, by its closeness, mechanically excludes those agents of decomposition, heat, air and moisture, which sand so freely admits; and therefore clay soils, in which this earth predominates, give out manure much more slowly than sand, whether for waste or for use. The practical effect of this is universally understood — that clay soils retain manure much longer than sand, but require much heavier applications to show as much effect early, or at once. But as this means of retaining manure is altogether mechanical, it serves only to delay both its use and its waste. Aluminous earth also exerts some chemical power in attracting and combining with putrescent manures, but too weakly to enable a clay soil to become rich by natural means. For though clays are able to exert more force than sand in holding manures, their closeness also acts to deny admittance beneath the surface to the enriching matters furnished by the growth and decay of plants. And therefore, before being brought into cultivation, a poor clay soil would derive scarcely any benefit from its small power of combining chemically with putrescent matters. If then it is considered how small is the power of both silicious and aluminous earths to receive and retain putrescent manures, it will cease to cause surprise that such soils cannot be thus enriched, with profit, if at all. It would indeed be strange and unaccountable, if earths and soils thus constituted *could be* enriched by putrescent manures alone.

Davy states that both aluminous and calcareous earth will combine with any vegetable extract, so as to render it less soluble, (and consequently not subject to the waste that would otherwise take place,) and hence "that the soils which contain most alumina and carbonate of lime, are those which act with the greatest chemical energy in preserving manures." Here is high authority for calcareous earth possessing the power which my argument demands, but not in so great a degree as I think it deserves. Davy apparently places both earths in this respect on the same footing, and allows to aluminous soils retentive powers equal to the calcareous. But though he gives evidence (from chemical experiments) of this power in both earths, he does not seem to have investigated the difference of their forces. Nor could he deem it very important, holding the opinion which he elsewhere expresses, that calcareous earth acts "merely by forming a useful earthy ingredient in the soil," and consequently attributing to it no remarkable chemical effects as a manure. I shall offer some reasons for believing that the powers of attracting and retaining manure, possessed by these two earths, differ greatly in their degrees of force.

Our aluminous and calcareous soils, through the whole of their virgin state, have had equal means of receiving vegetable matter; and if their powers for retaining it were nearly equal, so would be their acquired fertility. Instead of this, while the calcareous soils have been raised to the highest condition, many of the tracts of clay soil remain the poorest and most worthless. It is true that the one labored under acidity, from which the other was free. But if we suppose nine-tenths of the vegetable matter to have been rendered useless by that poisonous quality, the remaining tenth, applied for so long a time, would have made fertile any soil that had the power to retain the enriching matter.

Many kinds of shells are partly composed of gelatinous animal matter, which, I suppose, must be chemically combined with the calcareous earth, and by that means only is preserved from the putrefaction and waste that would otherwise certainly and speedily take place. Indeed, the large proportion of animal matter which thus helps to constitute shells, instead of making them more perishable, serves to increase their firmness and solidity. When long exposure, as in fossil shells, has destroyed all animal matter, the texture of the calcareous substance is greatly weakened. A simple experiment will serve to separate, and make manifest to the eye, the animal matter which is thus combined with and preserved by the calcareous earth. If a fresh-water muscle-shell is kept for some days immersed in a weak mixture of muriatic acid and water, all the calcareous part will be gradually dissolved, leaving the animal matter so entire, as to appear still to be a whole shell — but which, when lifted from the fluid which supports it, will prove to be entirely a flaccid, gelatinous, and putrescent substance, without a particle of calcareous matter being left. Yet this substance, which is so highly putrescent when alone, would have been preserved in combination with calcareous matter, in the shell, for many years, if exposed to the usual changes of air and moisture; and if secured from such changes, would be almost imperishable.

Calcareous earth has power to preserve those animal matters which are most liable to waste, and which give to the sense of smell full evidence when they are escaping. Of this, a striking example is furnished by an experiment which was made with care and attention. The carcass of a cow, that was killed by accident in May, was laid on the surface of the earth, and covered with about seventy bushels of finely divided fossil shells and earth, (mostly silicious,) their proportions being as thirty-six of calcareous, to sixty-four of silicious earth. After the rains had settled the heap, it was only six inches thick over the highest part of the carcass. The process of putrefaction was so slow, that several weeks passed before it was over; nor was it ever so violent as to throw off any effluvia that the calcareous earth did not intercept in its escape, so that no offensive smell was ever perceived. In October, the whole heap was carried out and applied to one-sixth of an acre of wheat — and the effect produced far exceeded that of the calcareous manure alone, which was applied at the same rate on the surrounding land. No such power as this experiment indicated (and which I have since repeated in various modes, and always with like results) will be obtained, or expected from clay.

Quick-lime is used to prevent the escape of offensive effluvia from animal matter; but its operation is entirely different from that of calcareous earth. The former effects its object by "eating" or decomposing the animal substance, (and nearly destroying it as manure,) before putrefaction begins. The operation of calcareous earth is to moderate and retard, but not to prevent putrefaction; not to destroy the animal matter, but to preserve it effectually, by forming new combinations with the products of putrefaction. This important operation will be treated of more fully in a subsequent chapter.

The power of calcareous earth to combine with and retain putrescent manure, implies the power of fixing them in any soil to which both are applied. The same power will be equally exerted if the putrescent manure is applied to a soil which had previously been made calcareous, whether by nature, or by art. When a chemical combination is formed between the two kinds of manure, the one is necessarily as much fixed in the soil as the other. Neither air, sun or rain, can then waste the putrescent manure, because neither can take it from the calcareous earth, with which it is chemically combined. Nothing can effect the separation of the parts of this compound manure, except the attractive power of growing plants — which, as all experience shows, will draw their food from this combination as fast as they require it, and as easily as from sand. The means then by which calcareous earth acts as an improving manure are, *completely preserving putrescent manures from waste, and yielding them freely for use*. These particular benefits, however great they may be, cannot be seen very quickly after a soil is made calcareous, but will increase with time, and, with the means for obtaining vegetable matters, until their accumulation is equal to the soil's power of retention. The kind, or the source, of enriching manure, does not alter the process described. The natural growth of the soil, left to die and rot, or other putrescent manures collected and applied, would alike be seized by the calcareous earth, and fixed in the soil.

This, the most important and valuable operation of calcareous earth, then gives nothing to the soil; but only secures other manures, and gives *them* wholly to the soil. In this respect, the action of calcareous earth in fixing manures in soils, is precisely like that of *mordants* in "setting" or fixing colors on cloth. When alum, for example, is used by the dyer for this purpose, it adds not the slightest tinge of itself — but it holds to the cloth, and also to the otherwise fleeting dye, and thus fixes them permanently together. Without the mordant, the color might have been equally vivid, but would be lost by the first wetting of the cloth.

Thus, reasoning *a priori,* from that chemical power possessed by calcareous earth which is wanting to both sandy and clayey earths, would lead to the conclusion that calcareous earth serves to combine putrescent matters with the soil in general; and the known results of fertility being therein so fixed, might serve for the like proof, even without the other course of reasoning. There is still another proof of this combination being formed, which is obtained by a chemical process, but which is so simple that no chemical science is requisite to make the trial.

If a specimen of any naturally poor soil, after being dried and reduced to powder, be agitated in a vessel of water, (as a common drinking glass,) and then allowed to stand still, the coarser silicious sand will subside first, the finer sand next, and last the clay. In this manner, and by pouring off the lighter parts, before their subsidence, it is very easy to separate with sufficient accuracy the sand from the clay. But if a specimen of a good rich *neutral soil* be tried in that manner, it will be found that the fine sand and the clay and putrescent matter hold together so closely that they cannot be separated by mere agitation in water. Then take another sample of the same soil, and pour to it a small quantity of diluted muriatic acid; and though no effervescence is produced, (the lime not being in the form of carbonate,) the acid will take away the lime, or destroy its combination with the other earths, so that the sand and the clay may then be separated by agitation in water, as perfectly and easily as in the case of the poorest soils. This difference between good and bad soils, (whether light or stiff,) or those naturally rich and those naturally poor, cannot escape the observation of the young experimenter; and the cause can be no other than what I have supposed. This then serves as the third mode of proof of the important position, that calcareous earth (or lime in some other form) not only combines with vegetable and animal matters, but also serves (as a connecting link) to combine these matters with the sand and clay of the soil.

The next most valuable property of calcareous manures for the improvement of soil is their *power of neutralizing acids,* which has already been incidentally brought forward in the preceding chapter. According to the views already presented, our poorest cultivated soils contain more vegetable matter than they can beneficially use; and when first cleared, they have it in great excess. So antiseptic is the acid quality of poor wood-land, that before the crop of leaves of one year can entirely rot, two or three others will have fallen; and there are always enough, at any one time, to greatly enrich the soil, if the leaves could be rotted and fixed in it at once.

This alleged antiseptic effort of vegetable acid in our soils receives strong support from the facts established with regard to *peat soils,* in which vegetable acids have been discovered by chemical analysis; and though the peat or moss soils of Britain differ entirely from any soils in eastern Virginia, (except that of the great Dismal Swamp, almost the only peat bog known,) still some facts relating to the former class may throw light on the properties of our own soils, different as they may be. Not only does vegetable matter remain without putrefaction in peat soils and bogs, and serve to increase their depth by regular accretions from the successive annual growths, but even the bodies of beasts and men have been found unchanged under peat, many years after they had been covered.* It is well known that the leaves of trees rot very quickly on the rich lime-stone soils of the western states, while the successive crops of several years' growth, in the different stages of their slow decomposition, may be always found on the acid woodland of lower Virginia.

The presence of acid in soils, by preventing or retarding putrefaction, keeps the vegetable matter inert, and even hurtful on cultivated land; and the crops are still further injured by taking up this poisonous acid with their nutriment. A sufficient quantity of calcareous earth, mixed with such a soil will immediately neutralize the acid, and destroy its powers; and the soil, released from its baneful influence, will be rendered capable, for the first time, of using the fertility which it really possessed. The benefit thus produced is almost immediate; but though the soil will show a new vigor in its earliest vegetation, and may even double its first crop, yet no part of that increased product is due to the direct operation of the calcareous manure, but merely to the removal of acidity. The calcareous earth, in such a case, has not made the soil richer in the slightest degree, but has merely permitted it to bring into use the fertility it had before, and which was concealed by the acid character of the soil. It will be a dangerous error for the farmer to suppose that calcareous earth can enrich soil by direct means. It destroys the worst foe of productiveness, and uses to the greatest advantage the fertilizing powers of other manures; but of itself it gives no fertility to soils, nor does it furnish the least food to growing plants.*

These two kinds of action are by far the most powerful of the means possessed by calcareous earth for fertilizing soils. It has another however of great importance — or rather two others, which may be best described together as the *power of altering the texture and absorbency* of soils.

At first it may seem impossible that the same manure can produce such opposite effects on soils as to lessen the faults of being either too sandy or too clayey — and the evils occasioned by both the want and the excess of moisture. Contradictory as this may appear, it is strictly true as to calcareous earth. In common with clay, calcareous earth possesses the power of making sandy soils more close and firm — and in common with sand, the power of making clay soils lighter. When sand and clay thus alter the textures of soils, their operation is altogether mechanical; but calcareous earth must have some chemical action also in producing such effects, as its power is far greater than that of either sand or clay. A very great quantity of clay would be required to stiffen a sandy soil perceptibly, and still more sand would be necessary to make a clay soil much lighter — so that the cost of such improvement would generally exceed the benefit obtained. Much greater effects on the texture of soils are derived from much less quantities of calcareous earth, besides obtaining the more valuable operation of its other powers.

Every substance that is open enough for air to enter, and the particles of which are not absolutely impenetrable, must absorb moisture from the atmosphere. Aluminous earth, reduced to an impalpable powder, has strong absorbing powers. But this is not the form in which such soils can act — and a close and solid clay will scarcely admit the passage of air or water, and therefore cannot absorb much moisture except by its surface. Through sandy soils, the air passes freely; but most of its particles are impenetrable by moisture, and therefore these soils are also extremely deficient in absorbent power. Calcareous earth, by rendering clay more open to the entrance of air, and closing partially the too open pores of sandy soils, increases the absorbent powers of both. To increase that power in any soil, is to enable it to draw supplies of moisture from the air, in the driest weather, and to resist more strongly the waste by evaporation of light rains. A calcareous soil will so quickly absorb a hasty shower of rain as to appear to have received less than adjoining land of different character; and yet if observed in summer, when under tillage, some days after a rain, and when other adjacent land appears dry on the surface, the part made calcareous will still show the moisture to be yet remaining, by its darker color. All the effects from this power of calcareous manures may be observed within a few years after their application — though none of them so strongly marked, as they are on lands made calcareous by nature, and in which time has aided and perfected the operation. These soils present great variety in their proportions of sand and clay; yet the most clayey is friable enough, and the most sandy firm and close enough, to be considered soils of good texture; and they resist the extremes of both wet and dry seasons, better than any other soils whatever. Time, and the increase of vegetable matter, will bring those qualities to the same perfection in soils made calcareous by artificial means, as they are in soils made calcareous by nature.

The subsequent gradual accumulation of vegetable or other putrescent matter in the soil, by the combining or fixing power of calcareous earth, must have yet another beneficial effect on vegetation. The soil is thereby made darker in color, and it consequently is made warmer, by more freely absorbing the rays of the sun.

Additional and practical proofs of all the powers of calcareous earth will be furnished, when its use and effects as manure will be stated. I am persuaded, however, that enough has already been said both to establish and account for the different capacities of soils for improvement by putrescent manures. If the power of fixing manures in soils has been correctly ascribed to calcareous earth, that alone is enough to show that soils containing that ingredient, in sufficient quantity, must become rich; and that aluminous and silicious earths mixed in any proportions, and even with vegetable or other putrescent matter added, can never form other than a sterile soil.

On Louisiana Sugar, 1846

From *De Bow's Review* (New Orleans, 1846), II, pp. 322–28, 330, 332–40, 344–45.

Dear Sir: — I yield to your request that I should give you the result of my studies on the cultivation and manufacture of sugar in this State. I do it the more cheerfully as I indulge a hope of eliciting communications from others on a subject so vitally interesting to Louisiana, and in this way that any errors into which I may fall from want of experience or defective information will be pointed out and corrected. The subject is vast: volumes have been written upon it, and any survey however general must make my communication extend over more space than you may be able to accord it in your pages. If so, hesitate not to retrench any portion that you may deem least likely to afford interest and information.

To a person accustomed to regard the bountiful returns which nature yields to man's labor in the cultivation of other crops, no fact strikes with more surprise than the small comparative return obtained from the cane. The seed seldom yields more than four-fold, hardly ever more than five-fold. The very smallest quantity of cane required for planting one hundred acres is twenty acres of the finest cane, and if, as too frequently is practised, the smallest and poorest cane is saved for planting, it is necessary to put up thirty, forty, and sometimes even fifty acres of cane in order to plant one hundred acres. If in the cultivation of the cane like that of the grains, it were necessary to plant the entire field each year, the large portion of each crop required for seed would form a very serious draw-back, and in some instances might even cause the abandonment of the culture. But fortunately the cane is not an annual plant. Each year fresh shoots spring from the stubble which remains after cutting the crop; the cane *rattoons,* as it is termed. In the West Indies where no frosts interfere with this natural re-production, it is said that the cane rattoons sometimes for a period of eighteen or twenty years, although I am inclined to believe this an exaggeration and that it is, in general, necessary to re-plant every ten or twelve years. — In Louisiana as a general rule, the fields are divided as near as may be into three equal parts, one of which is planted each year, so that in a plantation with six hundred acres of cane in cultivation, two hundred acres are plant cane, two hundred acres are rattoons of the first year, and two hundred acres rattoons of the second year. After a field of cane has thus yielded three crops, it is usual to plough up the stubble, and plant afresh, and if we take this as a general rule, and assume as an average that one acre of cane will suffice for planting four acres, it results that the yield of the seed is twelve-fold, or in other words, that one twelfth of each crop must be reserved for planting the next.

In giving an account of the cultivation, I shall commence by describing the process of laying by from each crop the seed for the next. Just before commencing the gathering of the crop, usually about the 1st of October, the planter selects the cane intended for seed. And here if I may be allowed without presumption to say so, a general and fatal error prevails. Most planters have not the courage to *sacrifice,* as they term it, their best and finest cane for seed. Selecting the fields of the oldest rattoons where the plant is sparsest and smallest, they act in direct opposition to those principles of nature which both theory and experience have established for guides in re-production. In both the animal and vegetable systems all agree in the general maxim, that like will produce like. In sowing grain, in producing vegetables, in breeding animals, in the whole reproductive system of nature, it has been universally established as a rule, that a healthy and vigorous offspring can be expected only from parents of similar constitution, and in all cases where this principle has been acted on with perseverance, it has not only succeeded in preventing deterioration, but in superinducing progressive development. I cannot but believe that this practice of always selecting the poorest plants for seed, was one of the main reasons which caused that fine variety of cane called the *Creole,* to degenerate to such an extent that in late years it has been almost entirely banished from our fields. In some instances the planters have pushed the "penny wise, pound foolish" system to such an extent as actually to reserve no cane for plant, but the tops, that is, the green upper joints which are cut from the plant when it is gathered for the mill, and which are not mature enough to afford sugar. I am the more emboldened in making these remarks, as experiment has shewn that in this respect cane is not an anomaly in the vegetable kingdom. A friend who is an experienced and intelligent planter, with sufficient energy of mind to break through the trammels of routine, when in opposition to good sense and sound principle, thoroughly tested this plan of reserving tops for planting in a portion of his field some years ago, and the result was a marked degeneration in the product.

The cane when cut for seed is preserved in *mattrasses* — it is laid in the field in beds of about two feet in height, in layers in such manner that the leaves of each layer over-lap and cover the stalks of the preceding layer, and thus form a protection against the frost: the mattrasses are also laid with their leaves towards the south, so that the north wind cannot lift them in its passage, nor penetrate under them. — In selecting the plant also care should be taken to have in view as much as possible proximity to that part of the field which is to be replanted, and thus to avoid any unnecessary labor in carting the plants long distances when seed time arrives.

Cane may be planted in Louisiana at any time between the first of October and the end of March — but if planted in the fall, care must be taken that the ground be thoroughly drained: otherwise the plant will freeze if the winter be severe, or rot if it be mild. Cane planted in the fall should be planted at least four inches deep to protect it from the frost. Few planters, however, are able to plant before, or during the grinding season. This work is usually commenced immediately after the crop is taken off. The ground is prepared by the plough, and the cane planted in January, February, and March. Much diversity prevails in the mode of planting; formerly, the cane was planted in rows, from three to five feet apart: but recently a very decided change is perceptible, and the cultivators have become convinced that a width of seven or eight feet between the rows, is as little space as ought ever to be left. When cane is planted in narrow rows, the effects of crowding the plant are not visible in the early part of the season, nor are they as pernicious in very dry seasons: but late in the year the narrow rows are found to be shaded the entire day, the access of sun and air is debarred, the cane does not ripen as well, nor are the stalks as heavy, and in fact all the evils attendant upon crowding too much vegetation in too small a space are clearly apparent. The following mode of planting and cultivating the cane has been pursued for a number of years, by the friend alluded to above, and has been attended with signal success. As soon as the ground has been prepared in January, the cane is planted in rows at a distance of eight feet. Three canes are laid in a row at a distance of four inches from each other: care is taken that the cane be so laid as to place the *eyes* from which the plant is propagated on each side of the cane: if the cane is thrown into the row without regard to this point many of them will be so placed that one series of the eyes will rest on the bottom, and the opposite series will be on top: the bottom eyes will thus come out later, the cane will be unequal in the rows, and will present to the eye a strikingly different appearance to that which is planted with the precaution of having the eyes on each side, so that nothing may obstruct the first efforts of the tender shoot in its struggles to reach light and air. The canes are laid straight in the row, the crooked stalks being cut when necessary, so as to make a straight line. The plants thus arranged in the rows are covered with finely pulverized earth to the depth of an inch, but care is taken after the plant is up to supply an additional depth of earth, round the roots at a much earlier period than is usually done, because most planters cover their cane deeper in planting. The advantage of this light covering is to hasten the first vegetation, and force an early start; a matter vitally essential in a cultivation, like that of the cane in Louisiana, which must be forced into maturity within a term, several months shorter than that which it naturally requires. My limits forbid following minutely the whole process of cultivation through the year, there being but little difference in the subsequent management from that followed by most planters, except in one particular, which I shall now point out. When the cane is cut in the fall, a large portion of the produce of the soil remains on the field, as is well known in the tops and leaves of the cane, the ripe portion of the stalk being alone conveyed to the mill. This is called the *trash*, and is placed on the stubble to assist in protecting from the frost, that part of the cane which remains under ground, and from which the rattoons shoot up in the ensuing season. As soon in the spring, as danger of frost is no longer apprehended, the trash is raked off the rows of stubble to allow access to the sun and air, and on nearly all plantations this trash which is a useful and fertilising manure is burnt up, instead of being returned to the earth. One cause of the difficulty of making use of this trash as manure, was the narrowness of the space between the rows under the old system of planting, which left so little room as to make the operation of ploughing in the trash difficult and laborious, but where the rows are eight feet apart, the task is easy. Independently of the considerations to which I shall presently advert, and which derive their force from the chemical constitution of the cane, it is difficult for a person who has not witnessed the results to form an adequate idea of the improvement to a soil that is naturally at all stiff, or clayey from the mere mechanical subdivision of its particles attendant on the decay of the large quantity of this trash left annually in the fields. This system was first put into operation on the plantation of which I am part-owner, last year. The trash on the first ploughing of the rattoons, was covered with the earth turned over from the furrow, which is run alongside of the stubble. At the second ploughing, when it became necessary to turn up the entire space between the rows, the difference in the soil was so perceptible as to create strife amongst the negroes for the preference of ploughing these rows, the subdivision of the soil caused by the decay of the trash, rendering the work much lighter and easier than in others, where from causes not worth detailing, we had been compelled to burn the trash. The advantages of this system are such that in lands which have been thus treated for a term of ten years without repose, I have been assured that the soil far from deteriorating is perceptibly improved in each successive year. The space between the rows not only reposes for three years, but is enriched by an annual increment of the best manure, and when it becomes necessary to replant, the cane is planted in the spaces thus fertilised, and the former rows then become intervening spaces to receive in their turn the benefits of this rich nutriment for the soil.

I referred in support of the advantage derived from the plan of ploughing in the trash to the chemical constitution of the cane as established by organic analysis. Although I am satisfied from reasons which I will give when I come to treat of the manufacture of sugar, that no accurate or satisfactory analysis of the sugar-cane has yet been made, or at least published, still the errors are not such as to affect the results in relation to cultivation.

Sugar-cane is composed of water, woody fibre, and soluble matter, or sugar. In round numbers it may be stated that the proportions are 72 per cent. of water, 10 per cent. of woody fibre, and 18 per cent. of sugar. But sugar itself is shewn by organic analysis to consist entirely of carbon and water, and woody fibre consists principally of the same elements combined with inorganic bases; so that the oxygen and hydrogen found in the sugar-cane, in the state of water, or as constituent elements of the sugar and woody fibre form about nine-tenths of its weight, and are entirely derived from the atmosphere and from water, thus abstracting nothing from the soil. But this is not all. Vegetable Physiologists agree that a very large proportion of the carbon of plants is derived from the air through the action of the leaves, which decomposes the carbonic acid of the atmosphere, and appropriate to the formation of the tissues of the plant, the carbon contained in this acid. For the purposes of the present illustration, it may, therefore, be assumed that not more than about six per cent. of the growth of the cane is derived from the soil, and hence the fact that this crop can be cultivated on the same soil without exhausting it for a long series of years — but it is certain that a system which is constantly abstracting *something* from the earth and never making to it any return, must by degrees impair and eventually destroy the fertility of even the alluvial soil of lower

Louisiana. — Now by ploughing into the land each year, the tops and leaves stripped from the stalks, not only is the soil improved by the mechanical subdivision of its particles above referred to, but it is kept in good tilth by having restored to it not only at least as much carbon as was abstracted from it, but a large portion of the inorganic bases. And if to this the bagasse were added as a manure, we should never hear of a soil being worn out on a sugar plantation in Louisiana. I am aware that it was formerly doubted whether any of the carbon of plants was derived from the soil, but later researches have put this point at rest, and have shewn that a large portion of this element is derived by plants from the carbonic acid evolved from vegetable substances during their decay in the soil, either by its inhalation into the roots, in an aeriform state, or by its first entering into solution into the water found in the soil, and being afterwards absorbed in this form by the roots. The experiment of Sir Humphrey Davy on this point appears conclusive, that eminent chemist having shewn that different plants and grasses grow much more luxuriantly when watered with solutions of sugar, than with common water, the two liquids differing in nothing but the presence of carbon in the former, and its absence in the latter.

Before closing these remarks on the cultivation of the cane, allow me to say something on a point, in comparison with which all others sink into insignificance. In the closing lecture of a series delivered in New Orleans by Professor B. Silliman, jun., on Agricultural Chemistry, he observed that if he were asked by what means the planter of Louisiana could, with certainty, add largely to the product of the soil, he would say, as Demosthenes said of action in its effects on eloquence, drainage, drainage, drainage. The present season has given to nine-tenths of our planters melancholy proof of the truth of this remark, and although the quantity of water which has fallen in this State the present year is altogether unprecedented, yet it is well known that every few years we may expect what is called a wet season, the effects of which on each plantation in the State, are in exact inverse proportion to the extent of its drainage. It is in such seasons that the most striking contrasts are shown between the results of skilful and imperfect cultivation; but it is a great error to suppose that drainage, thorough and perfect drainage, is without its influence in the driest season. In the alluvial soil of our Mississippi river, and the bayous leading out of it, exposed to the action of the water which filters through the banks, and which in the spring of the year is rendered icy cold by the melting of the snows in the northern regions, from which it flows, it is impossible to over-rate the importance of draining. The effect of this low temperature of the water which penetrates into our fields is so great, as perceptibly to retard the spring vegetation, unless means are taken to obviate its effects. In the recent experiments of planting cane in the parish of Rapides, it has been observed that the cane is earlier and more vigorous in its first vegetation, although in a more northern latitude than it is even in our lower river parishes, the soil on the Red river being higher and naturally drier than that on the banks of the Mississippi, and not being exposed to the same deleterious influence of the water percolating the banks of that stream. Now, this very serious injury to the crop is at once obviated by the digging of a deep ditch along the entire front of the field which intercepts the seepage water, and being connected with the drainage canals, carries off this water behind without allowing it to penetrate into the soil, and chill the roots of the plant. . . .

Manufacture of the Sugar

A sugar plantation is incomplete without its workshop, that is, its sugar-house. The owner is manufacturer as well as agriculturist, and the manufacture is one of great delicacy and difficulty. Until within a very few years the process has been of the rudest and most primitive character. A partial extraction of the juice was effected by the simplest and most imperfect machinery: the juice when extracted was tempered with lime which was added empirically without measure or proportion, and with scarce any regard to the varying quality of the juice, and thus tempered was boiled in open kettles over a fire, until evaporation produced a sufficient concentration of the saccharine matter to admit of crystallisation on cooling. The loss to the planter exceeds belief: the sugar-cane treated with care in the laboratory of the chemist yields eighteen per cent. of its weight in pure sugar, whilst in the rude process above described, its yield is scarcely five per cent. Such until a few years ago was the process *universally* used in the plantations of the West Indies and Louisiana, and such is now the process on very many estates, with occasional trifling improvements, none of which suffice to carry the yield beyond one-third of the real quantity of sugar in the cane. . . .

The first and almost insumountable difficulty in obtaining from the cane all its sugar results from the imperfection of the mills used to extract the juice. The cane contains ninety per cent. of juice, and ten per cent. of woody fibre, which is of a spongy consistence. The cane is crushed between cylindrical iron rollers, three in number, placed horizontally and moved by the steam engine. The quantity of juice thus extracted rarely exceeds two thirds of that contained in the cane, so that from this cause alone the planter loses one third of his crop, which remains in the bagasse. All efforts have hitherto proved fruitless to diminish this enormous loss.

Two-thirds of the juice being thus extracted from the cane, its conversion into sugar is attended with further loss. The juice as it runs from the mill is impure. It is impregnated with feculencies, with the dust and earth which have adhered to the cane when cut, with the coloring matter of the rind, much of which is pressed out by the rollers, and with fragments of the fibrous matter, both of the inner and outer part of the stalks; this latter containing inorganic bases, principally silicon. Before commencing the manufacture of the sugar, all careful planters take pains to purify the juice as far as possible by mechanical means. The juice runs into a vat divided into separate compartments, by one or more tissues of iron or copper wire, by which all the grosser impurities are arrested, and the juice thus cleansed is ready for the first operation, which is the defecation or clarification. According to the old system of manufacture in kettles, this defecation was effected by boiling the juice over an open fire, tempering it with lime in variable proportions, and skimming off the scum as it arose to the surface. The loss of juice, in this operation, is four or five per cent. The juice thus defecated was passed from one kettle to another, (the number of kettles being generally four, but sometimes five or even six,) until it reached the last kettle, called the battery, in which it was finally concentrated, till the syrup attained a density of about forty-two degrees of Beaumé's saccharometer, at which point it was ladled out of the battery into large wooden vats, called coolers. It was retained in these coolers till its crystalization, generally about twenty-four hours, at the end of which time it was taken out and placed in hogsheads in the draining room or purgery over a cistern, into which the molasses fell as it drained through holes placed in the bottom of the hogshead. The sugar thus drained was generally ready for market in two or three weeks. I give but a very meagre and hasty outline of the process hitherto pursued, because it is familiar to nearly all your readers; and I shall require several pages in giving some of the details of late improvements. I will merely remark, that this system produces a sugar highly colored, containing a large quantity of molasses, say about fifty gallons to each thousand weight of sugar, and consumes a large quantity of fuel, amounting on an average to two cords and a half of wood per hogshead. This large quantity of molasses is produced as above remarked by the imperfection of the manufacturing process, as none exists naturally in the juice of the cane; and as molasses rarely sells for more than one-third of the price of sugar per pound, the loss suffered by the planter in this item is again very serious.

Such was the mode generally, nay, universally adopted in the manufacture of sugar, till within the last twelve or fifteen years, when an apparatus was introduced into the French colonies, the joint invention of Mr. Degrand and Messrs. Derosne and Cail. This apparatus is now generally called by the name of the latter gentlemen; that is, the Derosne and Cail apparatus. . . .

We will now take the juice as it flows from the mill, after passing through the wire cloth, and examine its treatment, in detail, by the Derosne and Cail apparatus. It first flows into defecators, which are iron kettles with a double bottom, technically called a steam-jacket. The steam from the boilers is conducted by a pipe which is connected with this steam-jacket and which is provided at the opening into the steam-jacket with a cock, by which steam can be admitted or shut off at will. In these defecators, the first operation of cleansing or defecating the juice takes place, and in them the lime is introduced. Different opinions exist as to the proper time of introducing the lime, some mixing it with the juice when cold, and others preferring to await its rise to a temperature of about 150 degrees of fahrenheit. I think the latter plan preferable, and believe it to be also quite essential not to introduce the lime without previous preparation. This is especially important, when our common oyster shell lime, manufactured on plantation, is used, as it almost invariably is combined with a notable proportion of potash, which has a powerful effect in causing sugar to deliquesce. Impurities of a similar kind, but less abundant, are also found in the Thomaston and Western lime, used by most planters. The nature of the action of lime on cane juice is somewhat involved in obscurity. One effect is to saturate a small quantity of acid, which is always found in cane juice, but the quantity which is used with advantage in defecating far exceeds that which is required for destroying this slight acidity. Besides this effect, there is no doubt that the lime has a certain action, whether mechanical or chemical is not fully known, upon the mucilaginous or gummy matters found in the juice, by virtue of which it causes those matters to unite in a thick scum on the surface of the juice when heated.

It has unfortunately been impossible hitherto to discover a fixed rule by which to regulate the proportion of lime required for a given quantity of juice, and indeed this proportion is necessarily variable according to the quality of the juice and the nature of the soil on which the cane is grown. Ripe juices, and juices the product of calcareous soils, require much less lime than those which are extracted from unripe cane, or those produced on lands rich in animal or vegetable manures. In order to attain the proper porportion, and at the same time to avoid mixing with the cane juice any of the impurities that occur in unslacked lime, Mr. Payen advises the following process which recommends itself by its simplicity, and which I detail, because all agree that the defecation is the most important operation in the whole process of the manufacture. The lime should be slacked with care, and in quantities large enough to last for some weeks. It should be slacked by successive additions of warm water, and slowly stirred, so that the water may penetrate every part of it as equally as possible, and should be repeatedly washed, by allowing it to settle and pouring off the water from the top. The potash or other impurities will be dissolved and carried off by the water, and the lime remain pure. In this state if covered with water, it will remain for several weeks without being perceptibly injured by atmospheric action, and the whole mass will be of one quality. When used, it should be mixed with water to an extent sufficient to make a milk of lime marking 13 or 14 degree of Beaumé's saccharometer. a fixed quality and density being thus obtained, it only remains to ascertain by experiment what proportion of this lime, thus prepared, is required for a gallon of juice, and Payen advises the following mode: Prepare six separate equal quantities of lime, say one pennyweight each; then heat a gallon of juice, and when it has reached 150° of Fahrenheit, add one portion of the lime: continue the heat till it almost reaches the boiling point, then withdraw from it a table spoon full of the juice, and filter it through a small filter in a funnel: then add a second portion of lime, replace the juice on the fire and repeat the same operation. Continue till you have added the six portions of lime and have withdrawn six samples of the juice. Place the six samples in their regular order in small phials, and the *first* of them that shows the liquid to be of a clear amber color, contains the proper dose of lime. The subsequent phials containing a larger quantity of lime, will probably show a clear liquid less highly colored, but in these there is an excess of lime which would give a greyish tint to the sugar, and it is an admitted principle that the least quantity of lime that will serve the purpose of defecating, is the best. By this simple test the quantity of lime required will be readily shown: for instance, as there are twenty pennyweights to the ounce, if it be found that two pennyweights give the proper point to a gallon, we know that we require an ounce of the prepared milk of lime for each ten gallons of the juice — and instead of spoiling entire strikes or batteries by deficient or excessive doses of lime, the manufacturer would proceed in perfect confidence as long as the quality of the cane juice remained the same, and it would be easy to repeat the essay when a different quality of juice presented itself from a different part of the field. The juice thus tempered remains in the defecator with the steam under it until it reaches the boiling point, for the purpose of ascertaining which, a thermometer is hung with its bulb plunged in the juice. So soon as 211° of Fahrenheit are marked by the thermometer, the steam is shut off by turning the cock. On no account must the thermometer be allowed to pass 212°, which is the boiling point, because ebullition then commences, the effect of which is to break the scum that has formed on the surface, and by stirring the juice to mix the scum with it, and thus destroy the whole operation — at 211° or 212°, it will be found that the impurities of the juice have arisen to the surface, forming a thick scum of considerable consistency. After shutting off the steam a cock is opened under the bottom of the defecator and the juice is drawn off clear, the scum gradually sinking, and as soon as the juice ceases to flow clear, the cock is turned so as to arrest the flow into the juice pipe and open another orifice in a different pipe which carries off the scum. The great superiority of this mode of defecation over that in the open kettles, is palpable. The perfect control which the manufacturer has over the heat applied to the juice, enables him to arrest it at a given point and thus prevent ebullition, which, in the open kettles, is constantly going on; the scums in the latter as they arise are only partially removed by the skimming paddles, and by the continual motion which the ebullition imparts to the fluid, some of the impurities become so mixed up with the juice as to make it impossible to separate them.

The juice thus defecated flows through a pipe placed under the defecator, and which carries it to the filters. The filters used in the Derosne and Cail apparatus are called the Dumont filters, that being the name of the inventor, and their use forms perhaps the greatest improvement in the manufacture of sugar that the present century has produced, not even excepting the vacuum pan of Howard. These filters are iron cisterns nearly cylindrical: are six feet in height, five in diameter at the top, and four and a half at the bottom. They are filled nearly to the top with animal charcoal, or bone black in coarse grains about the size of cannon powder. This bone black is the carbonaceous substance into which bones are converted by calcination in close vessels. It possesses the extraordinary property of appropriating to itself the coloring matter of nearly all fluids that are filtered through it, and so powerful is its agency in this respect that in testing the qualities of some bone black offered me for sale, a dark colored claret was so completely discolored in a single filtration through a depth of twelve inches of the black as to be undistinguishable by the eye from the purest spring water. Another property possessed by this singular substance is that of abstracting from syrup any excess of lime that may remain after the defecation, and in addition to these two inappreciable advantages in the manufacture of sugar, it increases the crystalisation to an extent that is scarcely credible, amounting according to some experiments to eighteen or twenty per cent. The introduction of this powerful auxiliary has created a complete revolution in the process of manufacturing and refining the best sugar in France, and the result in Louisiana must inevitably be the same. The only draw-back to its use was its cost, because formerly it was thrown away as soon as repeated filtrations had saturated the black with the coloring matter and impurities of the syrup to such an extent as to deprive it of its efficacy; but the discovery of a mode of renovating, or as it is technically termed *revivifying,* the bone black has obviated this difficulty by enabling the manufacturer to use the same black for an indefinite length of time with but little loss in quantity or quality. The process of revivification is simple and not expensive, but the length of this article prevents my describing it in detail. The cane juice in passing through the filters is purified, brightened, and flows from a cock at the bottom ready to undergo the next process which is that of evaporating the water which it contains.

The evaporation is conducted by a very ingenious process, the invention of Mr. Degrand, and calculated particularly with a view to economise the quantity of cold water required to condense the exhaust steam from the vacuum pan used to boil the syrup up to the crystalising point, and which will be subsequently described. It is impossible to give an intelligible explanation of this part of the process without a plan of the condenser, but my object is to state the mode of manufacture, not the mechanism of the apparatus. With this view it will suffice to state that the juice is made to fall over a steam pipe, through which the exhaust steam from the vacuum pan returns to the boilers, and that a double effect is thus produced: the juice by falling in a shower over the hot steam pipes, is concentrated to 15 or 16 degrees of the saccharometer instead of 8 or 9, and at the same time serves to condense the exhaust steam which is pumped back in the state of hot water into the boilers. The economy of fuel is here very great, as none of the heat of the steam which boils the vacuum pan is lost, all either serving to evaporate the juice or being returned to the boilers.

The cane juice has now become a syrup of a density of 15 degrees, and is immediately conducted through a pipe into the vacuum pan in which it is concentrated to a density of 28 degrees. From the vacuum pan it again passes over the filters in order to effect a further discoloration, and is collected into a reservoir from which it is returned into the vacuum pan where it is finally concentrated to the point of crystalisation. This vacuum pan, its theory, its action on the syrup and its advantages are matters of very great interest to the planter and require some development. I must be excused if in explaining them I am compelled to state a few familiar general principles of physical science in such manner as to make the subject intelligible to those whose attention has never been directed to these matters.

It is known to all that if heat be applied to water until the thermometer marks 212°, vapor will be formed and the water will all pass off in steam if the heat be continued for a sufficient length of time. This is the evaporating point of water in the open air. It is equally well known that the atmosphere of our earth presses on all objects with a weight which is calculated to be equivalent to fifteen pounds per square inch of surface. The tendency of water to evaporate into steam is therefore repressed in the open air by a weight of fifteen pounds on every square inch of its surface, and it has been found that if this pressure be withdrawn the water will evaporate at a much lower temperature than 212°, and the same principle applies to other liquids. If, therefore, an air tight iron pan be made, and if a vacuum be formed in this pan by withdrawing the air by means of an air pump, water introduced into this pan would boil at a temperature diminishing in proportion to the diminution of the pressure of the air. It is difficult to say what would be the lowest temperature at which it could be made to boil because a perfect vacuum is not attainable by any means yet invented, but a vacuum can readily be produced by the air pump, in which water would boil at a temperature of 120°. A vacuum pan for making sugar then is an iron vessel, now generally made cylindrical, air-tight, connected by a pipe with an air pump worked by the steam engine, whereby the air is withdrawn from the pan to an extent sufficient to diminish the pressure of the atmosphere so far as to enable us to boil the syrup at a temperature varying from 130 to 160 degrees, instead of 235 or 240 degrees, which is the boiling point of syrup in the open air when concentrated to the density of 42° or 43° of the saccharometer. The vacuum pan is heated by means of a steam jacket or steam pipes, or both, and it is the steam which has served for this purpose that in escaping passes into the condenser mentioned above, and serves to evaporate the cane juice and is then returned in the form of hot water to the boilers, to be again converted into steam and renew the same round of service.

Such are the outlines of the system introduced into the manufacture of sugar by Messrs. Degrand and Derosne & Cail. . . .

In concluding, may I not be allowed to congratulate your readers on the prospects of permanent prosperity in this the most important branch in our State industry, and largest source of State wealth. A fortunate concurrence of circumstances has rendered harmless the reduction in the protective duty which had been levied in favor of this very extensive manufacture. The sudden and unexpected repeal by Great Britain, of that provision in her laws, which discriminated between sugar grown by slave labor and by free labor, has destroyed the barrier of prohibition which prevented the import into that country of the Cuba sugars; and the still further prospective reduction in the English duties secures us against a competition which must have ruined two thirds of our planters. The largely increased consumption which must inevitably result in Great Britain from the reduction of prices consequent on the diminution of the duty, will suffice to absorb so great a portion of the Cuba crop as to leave to our State almost the exclusive supply of the home market. The extent to which the production of sugar can be carried in Louisiana is appreciated but by few, but those who reflect on the subject and who feel an interest in all that concerns the prosperity of our State, foresee with exultation the day not far distant when boundless tracts now covered by the primeval forest shall teem with plenteous harvests of the cane; when nearly every plantation shall be a manufactory of refined sugar, supplying not only the wants of our own country, but forming a large item in our annual exports; when in a word the industry and enterprise of our population shall succeed in developing to their full extent the resources which a bounteous Providence has lavished on this favored land.

I remain with great regard
your friend and servant,
J. P. B.

On the "Working Man's Cottage," 1846

From A. J. Downing, "On Simple Rural Cottages," *Horticulturist* (Albany, 1846), I, pp. 105–10.

On Simple Rural Cottages

The simple rural cottage, or *the Working Man's Cottage*, deserves some serious consideration, and we wish to call the attention of our readers to it at this moment. The pretty suburban cottage, and the ornamented villa, are no longer vague and rudimentary ideas in the minds of our people. The last five years have produced in the environs of all our principal towns, in the Eastern and Middle States, some specimens of tasteful dwellings of this class, that would be considered beautiful examples of rural architecture in any part of the world. Our attention has been called to at least a dozen examples lately, of rural edifices, altogether charming and in the best taste.

In some parts of the country, the inhabitants of the suburbs of towns appear, indeed, almost to have a mania on the subject of ornamental cottages. Weary of the unfitness and the uncouthness of the previous models, and inspired with some notions of rural Gothic, they have seized it with a kind of frenzy, and carpenters, distracted with *verge-boards* and *gables,* have, in some cases, made sad work of the picturesque. Here and there we see a really good and well-proportioned ornamental dwelling. But almost in the immediate neighborhood of it, soon spring up tasteless and meagre imitations, the absurdity of whose effect borders upon a caricature.

Notwithstanding this deplorably bad taste, rural architecture is making a progress in the United States that is really wonderful. Among the many failures in cottages, there are some very successful attempts, and every rural dwelling, really well designed and executed, has a strong and positive effect upon the good taste of the whole country.

There is, perhaps, a more intuitive judgment — we mean a natural and instinctive one — in the popular mind, regarding architecture, than any other one of the fine arts. We have known many men, who could not themselves design a good common gate, who yet felt truly, and at a glance, the beauty of a well-proportioned and tasteful house, and the deformity of one whose proportions and details were bad. Why then are there so many failures in building ornamental cottages?

We imagine the answer to this lies plainly in the fact, that the most erroneous notions prevail respecting the proper use of DECORATION in rural architecture.

It is the most common belief and practice, with those whose taste is merely borrowed, and not founded upon any clearly defined principles, that it is only necessary to adopt the *ornaments* of a certain building, or a certain style of building, to produce the best effect of the style or building in question. But so far is this from being the true mode of attaining this result, that in every case where it is adopted, as we perceive at a glance, the result is altogether unsatisfactory.

Ten years ago the mock-Grecian fashion was at its height. Perhaps nothing is more truly beautiful than the pure and classical Greek temple — so perfect in its proportions, so chaste and harmonious in its decorations. It is certainly not the best style for a country house; but still we have seen a few specimens in this country, of really beautiful villas, in this style — where the proportions of the whole, and the admirable completeness of all the parts, executed on a fitting scale, produced emotions of the highest pleasure.

But, alas! no sooner were there a few specimens of the classical style in the country, than the Greek temple mania became an epidemic. Churches, banks, and court-houses, one could very well bear to see *Vitruvianized*. Their simple uses and respectable size bore well the honors which the destiny of the day forced upon them. But to see the five orders applied to every other building, from the rich merchant's mansion to the smallest and meanest of all edifices, was a spectacle which made even the warmest admirers of Vitruvius sad, and would have made a true Greek believe that the gods who preside over beauty and harmony, had for ever abandoned the new world!

But the Greek temple disease has passed its crisis. The people have survived it. Some few buildings of simple forms, and convenient arrangements, that stood here and there over the country, uttering silent rebukes, perhaps had something to do with bringing us to just notions of fitness and propriety. Many of the perishable wooden porticoes have fallen down; many more will soon do so; and many have been pulled down, and replaced by less pretending piazzas or verandas.

Yet we are now obliged to confess that we see strong symptoms manifesting themselves of a second disease, which is to disturb the architectural growth of our people. We feel that we shall not be able to avert it, but perhaps, by exhibiting a *diagnosis* of the symptoms, we may prevent its extending so widely as it might otherwise do.

We allude to the mania just springing up for a kind of *spurious* rural Gothic cottage. It is nothing more than a miserable wooden thing, tricked out with flimsy verge-boards, and unmeaning gables. It has nothing of the true character of the cottage it seeks to imitate. It bears the same relation to it that a child's toy-house does to a real and substantial habitation.

If we inquire into the cause of these architectural abortions, either Grecian or Gothic, we shall find that they always arise from a poverty of ideas on the subject of *style* in architecture. The novice in architecture always supposes, when he builds a common house, and decorates it with the showiest *ornaments* of a certain style, that he has erected an edifice in that style. He deludes himself in the same manner as the schoolboy who, with his gaudy paper cap and tin sword, imagines himself a great general. We build a miserable shed, make one of its ends a portico with Ionic columns, and call it a temple in the Greek style. At the same time, it has none of the proportions, nothing of the size, solidity, and perfection of details, and probably few or none of the remaining decorations of that style.

So too, we now see erected a wooden cottage of a few feet in length, *gothicized* by the introduction of three or four pointed windows, little gables enough for a residence of the first class, and a profusion of thin, scolloped verge-boards, looking more like card ornaments, than the solid, heavy, carved decorations proper to the style imitated.

Let those who wish to avoid such exhibitions of bad taste, recur to some just and correct principles on this subject.

One of the soundest maxims ever laid down on this subject, by our lamented friend Loudon, (who understood its principles as well as any one that ever wrote on this subject), was the following: *"Nothing should be introduced into any cottage design, however ornamental it may appear, that is at variance with propriety, comfort, or sound workmanship."*

The chiefest objection that we make to these over-decorated cottages of very small size, (which we have now in view,) is that the introduction of so much ornament is evidently a violation of the principles of *propriety*.

It cannot be denied by the least reflective mind, that there are several classes of dwelling-houses in every country. The mansion of the wealthy proprietor, which is filled with pictures and statues, ought certainly to have a superior architectural character to the cottage of the industrious workingman, who is just able to furnish a comfortable home for his family. While the first is allowed to display even an ornate style of building, which his means will enable him to complete and render some-what perfect — the other cannot adopt the same ornaments without rendering a cottage, which might be agreeable and pleasing, from its fitness and genuine simplicity, offensive and distasteful through its ambitious, borrowed decorations.

By adopting such ornaments they must therefore violate propriety, because, architecturally, it is not fitting that the humble cottage should wear the decorations of a superior dwelling, any more than that the plain workingman should wear the same diamonds that represent the superfluous wealth of his neighbor. In a cottage of the smallest size, it is evident, also, that, if its tenant is the owner, he must make some sacrifice of comfort to produce effect; and he waives the principle which demands sound workmanship, since to adopt any highly ornamental style, the possessor of small means is obliged to make those ornaments flimsy and meagre, which ought to be substantial and carefully executed.

Do we then intend to say, that the humble cottage must be left bald and tasteless? By no means. We desire to see every rural dwelling in America tasteful. When the intelligence of our active-minded people has been turned in this direction long enough, we are confident that this country will more abound in beautiful rural dwellings than any other part of the world. But we wish to see the workingman's cottage made tasteful in a simple and fit manner. We wish to see him eschew all ornaments that are inappropriate and unbecoming, and give it a simple and pleasing character by the use of truthful means.

For the cottage of this class, we would then entirely reject all attempts at columns or verge-boards. If the owner can afford it, we would, by all means, have a veranda (piazza), however small; for we consider that feature one affording the greatest comfort. If the cottage is of wood, we would even build it with strong *rough* boards, painting and sanding the same.

We would, first of all, give our cottage the best *proportions*. It should not be too narrow; it should not be too high. These are the two prevailing faults with us. After giving it an agreeable proportion — which is the highest source of all material beauty — we would give it something more of character as well as comfort, by extending the roof. Nothing is pleasanter to the eye than the shadow afforded by a projecting eave. It is nearly impossible that a house should be quite ugly, with an amply projecting roof: as it is difficult to render a simple one pleasing, when it is narrow and *pinched* about the eaves.

After this, we would bestow a little character by a bold and simple dressing, or facing, about the windows and doors. The chimneys may next be attended to. Let them be less clumsy and heavy, if possible, than usual.

This would be character enough for the simplest class of cottages. We would rather aim to render them striking and expressive by a good outline, and a few simple details, than by the imitation of the ornaments of a more complete and highly finished style of building. . . .

Last, though not least, this mode of building cottages is well adapted to our country. The material — wood — is one which must, yet for some years, be the only one used for small cottages. The projecting eaves partially shelter the building from our hot sun and violent storms; and the few simple details, which may be said to confer something of an ornamental character, as the rafter brackets and window dressings, are such as obviously grow out of the primary conveniences of the house — the necessity of a roof for shelter, and the necessity of windows for light.

Common narrow *siding*, (*i.e.* the thin *clap-boarding* in general use,) we would not employ for the exterior of this class of cottages — nor, indeed, for any simple rural buildings. What we greatly prefer, are good strong and sound boards, from ten to fourteen inches wide, and one to one and a fourth inches thick. These should be tongued and grooved so as to make a close joint, and nailed to the frame of the house in a *vertical* manner. The joint should be covered on the outside with a narrow strip of inch board, from two to three inches wide. . . .

We first pointed out this mode of covering, in our "Cottage Residences." A great number of gentlemen have since adopted it, and all express themselves highly gratified with it. It is by far the most expressive and agreeable mode of building in wood for the country; it is stronger, equally cheap and much more durable than the thin *siding;* and it has a character of strength and permanence, which, to our eye, narrow and thin boards never can have. When *filled in* with cheap soft brick, it also makes a very warm house.

The rafters of these two cottages are stout joists, placed two feet apart, which are allowed to extend beyond the house two feet, to answer the purpose of *brackets,* for the projecting eaves.

The window dressings, which should have a bold and simple character, and made by nailing on the weather boarding stout strips four inches wide, of plank one inch, and a half in thickness. The coping piece, is of the same thickness, and six to eight inches wide, supported by a couple of pieces of joists, nailed under it for brackets.

We have tried the effect of this kind of exterior, using *unplaned* boards, to which we have given two good coats of paint, *sanding* the second coat. The effect we think much more agreeable — because it is in better keeping with a rustic cottage, than when the more expensive mode of using planed boards is resorted to.

Some time ago, we ventured to record our objections to *white* as a universal color for country houses. We have had great satisfaction, since that time, in seeing a gradual improvement taking place with respect to this matter. Neutral tints are, with the best taste, now every where preferred to strong glaring colors. Cottages of this class, we would always paint some soft and pleasing shade of drab or fawn color. These are tints which, on the whole, harmonize best with the surrounding hues of the country itself.

These two little designs are intended for the simplest cottages, to cost from two to five hundred dollars. Our readers will not understand us as offering them as complete models of a workingman's cottage. They are only partial examples of our views and taste in this matter. We shall continue the subject, from time to time, with various other examples.

Louisiana Agriculture, 1874

From *De Bow's Review* (New Orleans, 1847), III, pp. 412–19.

The present and Future Products of Louisiana, and the Means for Augmenting Them

In the few remarks I propose submitting to the readers of your valuable publication, my object will be rather to elicit than to impart information. A residence of a few weeks only in a State whose climate, soil, and topography is so entirely new to me, and even this brief period having been almost entirely absorbed in business pursuits, has afforded me little opportunity of acquiring sufficient knowledge of the products of this State to become an instructor.

In some occasional numbers of your Review that have come under my notice, I have been gratified to perceive many valuable practical articles, particularly on the subject of agriculture, which are calculated to develope the great and hitherto untasked resources of this State. The history of the production of former times, our experience of the present, and a wise forecast of the future, are all essential for determining the most successful policy for the reward of enterprize, labor, and capital hereafter. From the accumulated facts of the past, and the data they afford on the subject of climate, soil, and production, in connexion with the probable future wants of the commercial world, a system of practical agriculture for this State may be educed which will be subject to little risk or fluctuation, while it is attended with large and measurably certain profit.

A national board of agriculture, comprising great intelligence, sagacity, and judgment, which should have the whole subject of American production, agriculture, manufactures, and commerce, before it, could do more to indicate the true policy for each section to pursue, than can be acquired in any other way. This was the favorite plan of our illustrious Washington, and has been seduously cherished and ably advocated by many of our most intelligent statesmen since; and it would seem reasonable to the simple minded and practical citizen that while our peace establishment requires $8,000,000 for the support of the army and navy, and our war appropriations for redressing real or fancied injuries, come up to $40,000,000, a few thousand might be reasonably asked for the developement of our natural resources, in which, after all, consist our only safe reliance and strength.

Next to a national board, and in its absence, one organized by the State Legislature is the most important, and where this is unprovided by the proper authority, individual associations, comprising representations from every interest within the State, is the only means left for acheiving this object. Such a one, I am happy to know, exists here, yet without that spirit and general interest which is essential to secure the requisite success. The means are entirely within the reach of the planters whenever they choose to call them forth, and it is to be hoped that they will not long permit them to remain undeveloped.

No country of equal extent on the face of the globe seems to possess such a prodigal affluence, such an unstinted measure of agricultural wealth as the alluvial portions of Louisiana. With an area of deltal formation of several thousand square miles, which no combinations of earths or organic materials for the highest production of vegetable fertility ever surpassed; with wide-spread luxuriant prairies, and its rolling productive uplands, every acre of this State seems teeming with the elements of vegetation, the foundation of future wealth, and the sustenance of future millions. And every section of it is accessible within a convenient distance by navigable waters, or admits of easy construction of roads. Even the waters which pervade and border the State would furnish sufficient food for a population larger than now inhabits it. With a climate generally mild and healthful, and with such redundancy of resources for the support of life and the acquisition of wealth, it would seem almost superfluous to suggest the means or the motives for the attainment of either. Actual want or suffering under such circumstances cannot exist, but that absence of individual prosperity is often to be found that creates a morbid restlessness under present exigencies, and induces efforts for its alleviation in the removal to some fancied El Dorado in the yet unexplored wilderness. Such would do well to consider that there is scarcely an acre either of land or water in Louisiana that cannot be put to some profitable use, and that, too, near a market whose commerce reaching to every part of the habitable globe, renders surfeits or over supply absolutely impossible. Let us consider these products somewhat in detail.

Sugar may undoubtedly be assumed as the leading staple of the State at the present moment. In 1845, there were produced here from nine hundred and fifty five sugar-mills 207,337,000 lbs. of sugar, and about 9,330,000 gallons of molasses, amounting together to near $15,000,000. It is estimated there will be during the present year, 1,240 mills, which, at the same ratio, will carry production up to about $19,000,000 in this article alone. Accidents, mismanagement, and unforeseen casualties from the elements and the season, will probably lessen the quantity, yet it is certainly within reason to assert that scientific and careful cultivation, the use of better machinery, the general application of well established chemical principles in the manufacture of the cane, would swell the amount far beyond the assumed maximum.

The extension of cane cultivation is undoubtedly advancing more rapidly at the present moment than at any former period. Each succeeding year witnesses the extension over new territory. It is descending on both banks of the river near to its mouth; it is climbing still higher upwards on the main stream and its tributaries, and it is fast occupying every one of its innumerable bayous or outlets, while more thorough ditching and especially the adoption of draining wheels is rapidly bringing into use larger portions of tillable land in the rear, and making all far more productive. The last we conceive to be one of the most efficient means for reclaiming vast bodies of land for the future cultivation of the cane. Still further means for the augmentation of the crop are to be found in much deeper and more thorough ploughing; the use of the sub-soil plough; manuring with the bagasse and trash buried between the furrows; and a proper rotation with the cow pea or other green or vegetable fertilizers.

Cotton may be ranked next in the order of the staples of this State. But a few years since this was the leading product; but while it has been reclaiming new territory and advancing in quantity, in much of the old, the greater profit afforded by the cane has enabled the latter to usurp many of the plantations hitherto exclusively devoted to the former. In the cultivation of this leading export of America, much improvement has been witnessed within the few past years, and although excessive rain or drought, the army worm or catterpillar, blight, mildew, or rust, occasionally disappoints the hopes of the planter, yet a closer study of the habits and diseases of the plant, a careful selection of seed, the introduction of new and improved varieties, as the mastodon and others, and a nicer and more careful cultivation are all aiding to swell the aggregate of the cotton fields.

Maize, or Indian corn, ranks next among the products of the State, though what is raised within it enters to a small extent only in the exchange of commerce. It is generally consumed on the plantations where it is produced, and its value is absorbed to swell the exports of the two former staples. If viewed, however, as it is, an indispensable article of food for the laborer, the working animals, swine and poultry, it assumes a vast importance among the leading objects of attention, and much beyond the measure assigned to it at the current rates in dollars and cents. Besides the large expense of preparing, sacking, and sending to market, there to pay additional sums in freight, drayage, storage, *rattage* and commissions, there is a corresponding expense of purchasing, freight, drayage, etc., in bringing it back to the plantation for consumption. All these several items must first be subtracted before we can get at the relative value of corn raised on a remote plantation or the one where it is consumed. If we go a step further, and consider its presence or absence in our granaries, as involving the question of sustenance or starvation, of life or death, of which we have at the present moment so terrible an example in Europe, we shall hereafter place a higher value on this article than we have hitherto done since the early settlement of the country. What has occurred elsewhere may occur here, and exemption from any particular calamity hitherto is no guarantee against its presence hereafter. The cultivation of maize on nearly every plantation within this State to the extent at least of its own consumption, ought to be considered a fundamental principle in its management.

Although not equally adapted to the highest production of corn as some of the choice lands between the great chain of northern lakes and the south line of Tennessee, yet where well drained and properly treated, the delta of Louisiana every where gives a remunerating crop of corn, and the lighter soils of the uplands requires but a judicious system of tillage to make a fair return in this crop for the labor and expense bestowed upon them.

If considered in an economical or domestic point of view, Indian corn throughout the valley of the Mississippi, is the most profitable crop that can be raised, as one man's labor will produce more human and animal food than in the cultivation of any other one product. With the best ploughs, a planting machine, cultivators and harrows, one person can easily plant, cultivate and harvest fifteen acres with four months labor, that will produce an average of forty bushels per acre, a quantity sufficient to sustain the existence of forty or fifty persons for an entire year. It is also like the cane, subject to fewer accidents or maladies than any other crop. Nothing but frost, excessive moisture, poverty of the soil, or negligent management, will prevent a good crop. The first may be always avoided by a late planting; thorough drainage effectually removes injurious moisture; deep ploughing and fine pulverization, especially the use of the sub-soil plough, will mitigate, if it does not wholly obviate the effects of drought, and rotation of crops and occasional application of green manures, if others are deficient, will be sufficient to prevent exhaustion.

Rice at one time formed an important staple of the State, and is now produced in quantities far greater than is generally supposed, yet to an extent much less than the soil, climate, and value of the article will justify. The rice lands of the Carolinas and Georgia are considered among their most valuable, the best being worth five hundred dollars per acre, while the best cotton lands will not command more than fifty. Why should they not be of equal value here? The want of skilful management, we fear, must be the only answer. Next to maize, rice is capable of affording the largest amount of food to man. In localities precisely suited to it, this capacity even much exceeds its rival, and nowhere is it believed it can be raised more advantageously than in this State. Immense bodies of the swamps and low-lands throughout the delta, are easily susceptible of being every way fitted for the highest and most profitable production of this grain. Suitable dykes or levees, proper ditches, both for draining and flooding the fields, with the addition of draining wheels, where their presence is necessary, is all that is necessary to secure millions of acres for this object that are now solely tenanted by every worthless specimen of the amphibious, vegetable, and animal creation. Rice may also be advantageously grown upon the uplands, and even the highest pine soils will yield enough to make it an object of attention. But in such great care is requisite to prevent exhaustion, which is scarcely possible on the rich alluvial bottoms that can be properly flooded, as the turbid water that overspreads the fields comes to the support of the crop charged with every necessary ingredient of vegetable nutrition.

One reason why rice has not hitherto been made an object of greater attention here is the want of proper machines for planting and preparing for market. Those of the latest and best construction have been for some time used in the Atlantic States, and may now be had in this city. With these at command, with a soil, climate, and the facilities for irrigation so entirely adapted to the purpose, there is no good reason why rice should not again become one of the most important branches of agricultural attention in Louisiana.

Indigo was the leading product of this State a century since, yet now it is scarcely cultivated. Two millions of acres of the most fertile cotton lands within the State are every way adapted to its profitable growth. Its culture here was gradually abandoned for the greater profits afforded by other articles, particularly sugar and cotton. The demand for it from the extension of our manufacturers is annually increasing its consumption in this country, and the application of the latest chemical science to its maceration and preparation for market, would undoubtedly render this an object well worthy of attention at the present time. Tobacco may be raised here of the finest quality and to an unlimited extent. If production be combined with its manufacture for the supply of this and other markets, few objects would better repay the labor and capital invested than this. The choicest qualities of leaf are produced on this soil which are scarcely surpassed by the best brands of the Havana.

Madder, woad, weld, saffron, sumach, etc., used primarily for dyes, and already in large demand by the northern manufacturers in this country, can be raised here with decided profit. The first is also a valuable food for cattle; saffron is used medicinally, and the astringent properties of the sumach renders it a substitute for the tannin of the oak and hemlock, where they do not exist. Roots and almost every species of culinary vegetable can be raised to the full extent of the wants of the inhabitants, and the sweet potato may be grown for exportation with decided profit. Of fruits, the orange and the fig thrive remarkably within the State, and the former may be exported, and with equal advantage to the planter as any other crop. The peach, the apricot, and nectarine produce largely and of the finest quality, when properly treated. The plum and the apple, the olive, the lemon, the lime, etc., may, with proper attention be reared on such soils and such localities as are suited to their habits and characteristics. The wild mulberry grows spontaneously in the forests of this State, and the *morus multicaulis* and other varieties of the Italian succeed admirably on the drier soils. This ensures success for the silk worm wherever introduced, as it may undoubtedly be hereafter, on the uplands which decided advantage. The equable temperature and condensed nutritive foliage afforded by such localities, and they sweep around a larger portion of the northern part of the State, will undoubtedly produce healthy silk worms, and as heavy, valuable cocoons as are yielded in any part of the world.

These are a few among the many objects that should arrest the attention of the intelligent and enterprizing agriculturalists before seeking them further south and west, and still more remote from the seaboard, where for a coming century at least he must look for the most profitable market for his products.

The false ambition for large plantations, and operations and achievements beyond the legitimate means of the owner, has been and still continues to be the bane of citizens of our new States. This policy may result in giving to the few, large landed estates, yet really less pecuniary income than would result to the shrewd manager, where a denser population existed, and more aggregate and active wealth circulated among the mass, the necessary result of a greater and more *intense* production. In looking over some of the plantations of this region, where large masses of land are either wholly or partially unsubdued, and the remainder admits of much higher cultivation, one cannot but be forcibly impressed with the consideration that the old maxim *"divide and conquer,"* has a much more pregnant and salutary bearing on the welfare of the human race than was ever assigned to it by the ambitious Roman. "A little land well tilled," while vastly more beneficial to the State and the middle property classes, is perhaps, of equal or even greater advantage to the opulent, than the present system of overgrown and half cultivated plantations. A division of labor and a variety in the objects of agricultural pursuits, are equally essential to call into profitable action the various traits of human character, the attainments of the greatest good to the greatest number, and the full development of the vast agricultural resources of this great State.

The foregoing embrace a few hints which may be successfully and almost indefinitely extended by your more experienced correspondents, for the more effectual and profitable augmentation of the present and future products of Louisiana.

Very respectfully,
R. L. A.

Potato Production, 1848

From U.S. Patent Office, *Annual Report*, 1847 (Washington, 1848), pp. 134–37.

Potatoes

The time was when this crop was numbered among our most successful ones. Affording as it does, not merely a favorite esculent for the table, but an excellent fodder stuff, it was a means on which the farmer relied to supply him with sustenance for various animals, with no little confidence. It was likewise assuming a place among our home manufactures for the starch which it afforded. But within a few years there has been a sorrowful change, and throughout almost the whole extant of the country where the common potato is cultivated, instead of ascertaining the amount of the crop, our attention is rather demanded to learn the amount of the loss suffered. It is only in this way that we can hope to arrive at any reliable conclusions respecting the productiveness of this crop. We have as a data to fall back upon the returns of the census, and the estimates of two or three years subsequent, when, happily, the destroying plague had not appeared. As we shall devote some time to the subject of the disease, we shall only run over the estimates reported to us somewhat rapidly, embodying most of our information under the former branch of the subject.

The loss of this crop falls heavily on the state of Maine and the New England states generally. Maine has been celebrated for her fine potatoes. Her climate or soil, and perhaps both, have seemed peculiarly adapted to this plant. In her prosperous years of cultivation, large quantities have been raised for export, besides what were necessary for the home consumption. This year, however, this source of her commerce and wealth has been greatly cut off, and in some sections hardly enough have been raised for her own population. In some of the instances reported to us, we are told that the crop of 1847 is fifty per cent. less than that of 1845, and ninety per cent. less than that of 1843. The average number of bushels also is estimated at not more than twenty bushels to the acre. In some other parts of the state the evil may not have been so severely felt, but the aggregate crop of the state is greatly reduced from the fair average one.

In New Hampshire and Vermont, with the other states of New England, the estimate of the loss varies. Perhaps the amount of loss is not, on the whole, so great in Connecticut as in the other states. In some instances we have understood that very fine potatoes, and of excellent quality, have been raised there. For some reason or other, the more northern sections of the country seem to have suffered most; twenty-five and thirty per cent. decrease from the crop of 1845 are the common estimate, where the amount per acre is set from one hundred to one hundred and eighty bushels.

The loss on the crop of 1847, in the state of New York, is variously estimated. In some counties, from apprehension of the evil, less were planted, while in others, more contiguous to the markets, there was a large extent of ground devoted to this vegetable. The early appearance of the crop, both here and in New England, was most promising. The vines flourished, and the tubers apparently set so as to encourage the hope of the farmer, that his potatoes would escape, and his labors be at last again rewarded. But in the latter part of August they told another tale — the blight came, and the tubers already formed began to decay. In proportion as the quantity planted had been large, was the evil. In some places the loss is estimated at twenty-five, in others thirty, fifty, and up to sixty per cent. At all events, there was in the aggregate a great reduction from an average crop. The acreable product also ranges, according to the estimates furnished, at from thirty up to sixty, eighty and one hundred bushels.

In the potato crop of New Jersey likewise, and Pennsylvania, similar deductions are required. In these states there has been rather more attention recently paid to the sweet potato, in consequence of the evil by which the common potato has been rendered less productive. In the reports from this state, as well as in some others, the crop is estimated as it actually was, and the loss by the rot then given. There would seem to be reason for supposing, that viewed in the light of productiveness simply, there may have been an increase in some parts of the state, but the rot has been very severely felt. The disease appears to have been felt for the first time, this year, in particular portions of this state. While the loss is variously estimated, from ten up to fifty per cent., yet we are inclined to believe, in the aggregate it did not cause as great a reduction as in some of the states we have already mentioned.

The potato rot seems likewise to have been felt to a considerable extent among the common potato, when cultivated in Maryland. It has not heretofore prevailed with very great severity in this state, but it has been gradually extending its attacks further to the south and west every year.

In the western parts of Virginia, owing to the scarcity of seed, there were less potatoes planted; the rot also made its appearance, and very considerably lessened the crop. It is thought, however, in some places to have abated. Not a great many are ordinarily raised, and chiefly for table use; they are not resorted to for the food of cattle. The average product per acre is by some estimated at one hundred bushels. The loss is estimated at one-third. Indeed, the general reports from the state represent the crop as seriously affected; though in the eastern counties it is stated that the productiveness was good, but they sustained loss after they had matured. We should judge that the crop of sweet potatoes was a less one than that of 1845, as the drought diminished the amount produced. The sweet potato crop of South Carolina is represented as having been a very fine one, in some instances probably double or three times that of 1845. The general average of the state, per acre, is given at about fifty bushels, though in the vicinity of Charleston it is said to be one hundred, and in other districts one hundred and fifty bushels per acre. The Irish potato, where raised, likewise were fine.

The advices received as to the sweet potato crop in Georgia and Alabama and other states, are likewise favorable; twenty-five per cent., one-third, one-half, are mentioned as the probable gain on the crop of potatoes for 1845. The season promoted their growth, and the average product, in different places varied from fifty up to one hundred and fifty or two hundred bushels per acre. The common potato is occasionally used, and no complaint is made as to its success this year. In Louisiana it is planted in January.

In Tennessee and Kentucky the potato crop is given as varying from ten to twenty and twenty-five per cent. better than that of 1845. Both kinds of potatoes are raised here, so that it is difficult to discriminate. No mention, however, is made of the rot; the average product per acre is given as from fifty to one hundred and two hundred bushels.

The Ohio potato crop suffered in particular sections of the state, to some considerable extent. The evil, however, was less in others, and instances are mentioned of a decided increase. The sweet potato is raised here, but chiefly however the common or Irish kind.

The loss does not appear to have been so great in Indiana, Illinois and Missouri as in Ohio, and the states from the east; in some sections it is estimated at a loss from the rot, of 10 to 25 per cent. In others there was an evident gain, and there may have been so in the whole aggregate crop. The disease was felt somewhat this year in Michigan, Wisconsin and Iowa, but the report on the whole is judged a favorable one.

It will be recollected that there was a great falling off of the crop in 1845, so that taking this as the basis of comparison for the present year, we must probably reckon it considerably under the average crop. Regarding the entire crop of the United States, we doubt whether the Irish potato crop will reach to the returns of the census, although there has evidently been an increasing cultivation from other causes. The sweet potato crop we believe may have gained on that of the census; and taking both together we believe that it is above the crop of 1845, but not equal to the census returns.

Improvements in Farm Tools, 1848

From *De Bow's Review* (New Orleans, 1848), VI, pp. 131–34.

THERE has been within a year or two back established in New-Orleans, an Agricultural Warehouse, more complete than any other in the Southern country, and equal to the best in the Northern States. This establishment is in the charge of R. L. Allen, brother of the Editor of the *American Agriculturist*. One is surprised on examination to discover how numerous and ingenious are the conveniences for assisting the labors of man in his natural and necessary duty of tilling the earth and working its products. We took a few notes during some occasional visits to the establishment, and present them to the reader.

PLOWS. — Of these the variety is so great as to preclude a description of each kind. From a draught requiring four to six horses, in heavy ground, to a plow easily drawn by one horse or mule, the intermediate styles afford a wide scope for selection. Many are of iron, composed of an admixture of several kinds, which produces a metal of great strength and durability, increasing at least one hundred per cent. more service in those parts so soon worn out in other plows. Of 195 premiums offered by agricultural societies in Massachusetts, New-York and Vermont, for the best plows, the manufacturers of the class to which we refer obtained 158 as the reward to their skill. We noticed the Side Hill or Swivel Plows, of which are five different sizes, so constructed that the mould-board can be instantly changed from one side to the other, which enables the operator to perform the work horizontally upon side hills, going back and forth on the same side, and turning all the furrow-slices with great accuracy, downward.

There are four sizes of the *Sub-soil Plow,* to be used immediately after the team which turns up the surface-soil, and in the same furrow. This is of great advantage to the crops, both in dry and wet land. In the former, the sub-soil being deeply broken up and well pulverized, the moisture is retained much longer than it otherwise would be, and the roots of plants can descend much lower and wider for their food; while in the latter, the excess of moisture filters below, and is readily carried off. This plow will break and pulverize the soil any required depth to eighteen inches.

The *Three-Share Plow* spreads two feet six inches wide, and is used for plowing in wheat, rye, and other grain, after sowing; and taking so many furrows at a time, it gets over the ground very rapidly. It is also highly useful as a cultivator, doing the work of three small plows with the same force.

The *Paring Plow* is used for paring turf-lands, preparatory to burning. The share is thin and flat, made of wrought iron, steel edged. It has a lock-coulter in the centre, and short coulters on the outside edge of each wing of the share, cutting the turf, as it moves along, into two strips about one foot wide, and as deep as may be required.

Mr. Allen having made himself acquainted with southern soils, and the best modes of culture, has constructed a series of plows expressly for this region, combining the advantages of the best northern models, without being so cumbersome. In addition to those specified, he has various other plows suited to every locality in North and South America, and the West India Islands, among which are steel points and shares, steel and wrought iron mould-boards, and every species of castings for plantation or farming labor.

CULTIVATORS. — Of these are several varieties. The Cultivator is a great labor-saving implement for stirring the earth between the rows of corn and other crops. It is well adapted for mixing manures in the soil, and pulverizing it after plowing. It leaves the soil much lighter, and in better condition to receive the seed than when the harrow only is used. It is useful for covering grain broad-cast, and buries it at a more suitable and uniform depth than the plow, and in one fourth the time; and much more perfectly than the harrow. There are various forms of teeth, some enter and stir the soil deep; others are broad and flat, to skim the surface and cut up the weeds; others narrow, acting as scarifiers, — and all fit the same size and form of mortice. The farmer, by purchasing different forms of teeth, can use them in the same frame-work. The *Universal Cultivator,* which may be easily repaired by blacksmiths, is made to expand from two to five feet.

Langdon's Cultivator is, in reality, a plow with a light, wide, flat share, sharp at the edges, and coulters on the mould-board. It is used for running between the rows of different crops, to cut up the weeds and loosen the soil. It is also an excellent implement for digging potatoes. The shares can be detached and wings added, which converts it into a double mould-board plow. It is recommended only for light soils free from stones.

The *Cotton Sweep Cultivator* is made expressly to take the place of the cotton sweep, besides doing much additional work. It has very sharp teeth so arranged as to cut up all grass and weeds, at the same time finely pulverizing the soil; and can be expanded or contracted to suit any width of row. It works so lightly that one mule can draw it.

The *Hand Cultivator* is made entirely of iron, except the handle, and expands from ten to eighteen inches. It is a very useful implement in garden culture, and is often used in fields. It cuts up and leaves the weeds exposed, and stirs the earth thoroughly. The operator, with his hands behind him, clenches the cross handle, and walks easily forward between the rows, and performs the work better and faster than several men with hoes, leaving the ground well pulverized and the weeds destroyed. This being expeditious, it can often be used to advantage.

ROLLERS. — These implements are fast coming into general use. They crush all sods and lumps that remain on the top of the ground after the harrow has passed, and force down small stones level with the surface. They render the field smooth for the cradle, scythe and rake, press the earth close about the seeds, and secure a more certain and quick germination. Their greatest benefits are realized when used on such light, sandy and porous soils as are not sufficiently compact to hold the roots of plants firmly, and retain suitable moisture. On such lands they are invaluable, and in all cases their use has greatly increased the product. Much benefit is undoubtedly found in compressing the surface of such soils by preventing the escape of those gasses from the manure, so essential to vegetation, and which are so rapidly extracted by the sun and winds. The rollers in highest estimation are made of iron, 18 or 24 inches in diameter, in separate sections, each one foot long, placed on a wrought iron arbor, on which they turn independently of each other; — thus turning without much friction and leaving the ground smooth. They are generally used with three to six sections. If four only are required, shafts may be substituted for the tongue, and drawn by one horse, or both may be used alternately according to the team. The box is attached to receive stones, etc., picked up on the field, and giving weight to the roller as the work may require.

HARROWS. — Of these are many kinds and sizes, from the one horse up to the large four horse harrows. The *Triangular Folding, Scotch and Geddes* are most approved. The latter, in two triangles, is superior to the square, as it draws from one point with a regular, not a straggling motion, and of course is easier for the team. Either part is easily lifted when in motion, to let off any trash that may have collected among the teeth.

Sufficient attention is not paid to harrowing. It is the next most important operation after plowing. The harrow should run from four to six inches deep, cutting up all the lumps, and leaving the ground in a finely pulverized state.

SEED-SOWERS, &c. — These machines are quite ingenious and labor-saving in their contrivance. The *Improved Brush* has been long in use in this country and in England, and is found to be the only one that plants all the variously formed small seeds rapidly and with precision. It is easily arranged to plant a greater or less quantity as may be required.

Bachelder's Corn-Planter deposits the seed at any distance in drills or otherwise, from a hopper above the beam; and as the horse moves along, the share below opens the furrow; arms moving horizontally drop the corn through a tube conducting it to the bottom of the drill. A triangular iron follows to remove all lumps and stones, and a roller to compress the earth over the seed. The machine requires a small horse to draw it, and with a boy to drive, will plant from ten to twelve acres per day, according to the width of the rows.

The *Horse-Drill* will plant wheat, rye, Indian corn, oats, peas, beans, ruta-bagas, &c., and can be regulated to drop any required quantity on an acre. The drills can be thrown in or out of gear separately, so as to plant a field of any shape without seeding any part twice. They are so arranged as to operate equally well on all kinds of land, hilly and rough, as well as level and smooth. A man with two horses can put in from 10 to 12 acres with wheat in a day; and with one horse he can plant 20 acres with corn per day.

COTTON SEED PLANTER. — Both the corn-planter and horse-drill above described are easily adapted to sowing cotton seed, and can be made to do the work of six or eight hands, and much more perfectly.

HARVESTING IMPLEMENTS, &c. — The *Reaping Machine* can reap fifteen acres of wheat in a day, and will cut the grain as smooth and clean as can be done with a sickle or scythe. It has low wheels, drawn by a pair of horses, and cuts a swath five feet wide, with twenty knives, working horizontally, which require sharpening only once a day. A man sitting on the side of the platform with a rake pushes off the grain as fast as it is cut. A field of oats or barley may be cut as neatly and expeditiously as one of wheat or rye.

Railway Horse Power and Thrasher is adapted to one horse, which, with the aid of two men and a boy, can thrash at the rate of 75 to 100 bushels of wheat, or 100 to 150 bushels of oats in a day.

The Grain Cradle, Grass, Lawn, Bramble, or Bush Scythes are implements which admirably answer the purposes intended. They are the most approved in the market, and are made of the best cast and German double-refined steel.

The *Revolving Hay Rake,* with a horse, man and boy, will clean from 15 to 20 acres per day. It can be used to good advantage, even on rough ground.

Rice Thrashers, Fanning Mills, Corn Shellers, Smut Machines, Corn and Cob Crushers, Burr Stone Mills, Rice Hullers, Straw Cutters, Machines for raising water by horse power, &c. We enumerate, at random, a few more of the articles or inventions found in the warehouse of Mr. Allen.

Egg Hatching Machine is constructed of tin or other materials, with the brooding-chamber surrounded by water, warmed to a suitable degree of temperature by a spirit-lamp, which may be constantly kept burning for less than ten cents a day. The apparatus is so simple in its construction and management, that a child can superintend its operation, with two hours' time in a day, and requires no attention during the night, after ten o'clock. It is made of three different sizes, to contain from 200 to 600 eggs at once.

The variety of corn mills in this establishment exceeds any we have before seen, from the little hand steel mill, and such as combine hand and horse power, to the largest sizes of the best French burr stone. Also, corn-shellers equally various, — there being more than a dozen kinds to shell by hand, and four to shell by horse or steam power, some of the latter being capable of turning out 1500 bushels per day. To these may be added straw cutters of twelve or fifteen styles, working in all kinds of ways, by hand and horse power, yet all efficient in lessening the mastication of the animal, and thereby increasing the value of the fodder, whether corn stalks or shucks, straw or hay.

The large furnace kettle, which is always set and ready for use, in doors or out, is of great convenience and utility, economizing fuel and labor. These are of all sizes.

Having referred to the corn and cob-crushers, we may remark, that it is a vast labor saving machine for the stomach of the animal, and in connexion with the straw-cutter, is capable of saving at least one third of the food, while it augments the working capabilities of the beast. The food must be reduced, ground or divided, before it can be acted upon by the gastric juices of the stomach, and assimilated to the system. Now, a great deal of this may be accomplished by these machines, and to much more advantage than by mastication; and when the animal has received his rations in this prepared form, he more readily digests them, and is sooner refreshed and invigorated for his work.

The Water Ram, a modification of the invention of Montgolfier in the last century, for raising water to any required height, with a fall not exceeding three feet, we conceive to be an instrument of great value for various purposes. We noticed pumps and engines of several descriptions, adapted to almost every conceivable use, all of which are the best suited to some one or more peculiar situations or objects. Also, a variety of Hoes, of domestic manufacture, far superior to any ever imported, and which Mr. Allen informed us, he had made especially for such discriminating planters as can appreciate an article which will pay for itself by extra service in every week's use. They are made, not only in an improved form, but of a fine quality of metal, and will do work much better, faster, and with much more ease to the hand than the common hoe.

The garden implements and tools for shrubbery form a beautiful collection. Many are delicate and ingenious, fit for the hands of a lady, uniting amusement with healthful exercise. Spades, shovels, scythes, forks, and every other instrument needed in rural occupations, may be had at Mr. Allen's warehouse.

But we cannot further particularize. We have accomplished our object, and what we deem a duty to our agricultural readers, by calling their attention to the first establishment of this kind ever undertaken south or west of Baltimore, and one probably more varied and complete than can be found elsewhere in the United States. When we reflect that nearly all the agricultural improvements of the present day consist in the superiority of the implements, and the economy and perfection of labor thereby secured, we cannot too earnestly recommend the adoption of such implements as are most likely to attain this result.

That policy is best which brings most reward with less sacrifice to the operator. Of late, the mechanic arts have wonderfully multiplied the comforts and reduced the labor of man. Prejudice against anything new, because it happens to excel old models, is no longer tolerated by the intelligent. There is the same reason why the best tools should be used in producing from the soil, as that the carpenter or other handicraftsman should employ instruments adapted to the work he has to accomplish. We shall conclude our remarks on this subject by a quotation which we consider eminently applicable:

Furius Cresinus, an emancipated Roman slave, having obtained from his very small estate much larger crops than his more wealthy neighbors from their vast domains, they became so envious that they charged him with employing enchantment to attract into his grounds the produce of their fields. Having been summoned by Spurius Albinus, and being fearful of condemnation, he introduced into the forum, as the tribunes prepared to vote, his robust and well clad family, and his agricultural implements, — his heavy mattocks, his ingeniously constructed plows, and his well-fed oxen, and then exclaimed, — Behold! Roman citizens, my magic; but I am still unable to show you, or to bring into the market place y *studies,* my constant *vigilance,* my fatiguing *labor.* Scarcely had he concluded, when he was absolved by public acclamation.

It is in enterprise, study, unremitting study, vigilance and industry, more than in *money,* that the mystery of great crops and successful husbandry consists.

To succeed in this "enchantment" of full crops, let parties within reach of New Orleans call on Mr. Allen, who will cheerfully assist them in the work. His collection cannot fail to interest any visiter, — producing new and enlarged ideas, with increased love of nature. Such is the tendency of all associations connected with rural life.

Sugar Farming, 1848

From Charles L. Fleischmann, "Report on Sugar Cane and Its Culture," U.S. Patent Office, *Annual Report*, 1848, pp. 274–77, 281–86.

To the Hon. EDMUND BURKE,
  Commissioner of Patents.

SIR: Agreeably to your suggestion that I should collect during my stay in Louisiana, all such information as would be of interest to the sugar planters of the United States, I set out for that State in the middle of October, and arrived in New Orleans, at the beginning of the rolling season. I lost no time in accomplishing my object and visited several plantations, with a view of becoming acquainted with the mode of cultivating the cane, and the improvements for boiling sugar. But, as my investigations progressed, I found that there existed a great many conflicting views, in regard to cane culture, arising from the different kinds of soil, higher or lower locations, and the influence of climate.

It was necessary, therefore, in order to arrive at correct statements, that I should visit other portions of the State, as for instance, Bayou Lafourche, Red River, and all those parishes where the sugar is now and has been heretofore grown, and also those where the planters intend to raise the cane instead of cotton. To accomplish all this, the time allowed me was too short, and I came to the conclusion to transmit to you only a part of the collected facts and drawings and to continue my investigations, if so authorized, during another rolling season.

It is a subject of too great importance to be slightly treated, especially at this moment, when so many cotton planters within the region suitable for the culture of sugar cane, intend to quit the growing of cotton and plant the cane.

The cane has been successfully cultivated for several years in the higher latitudes of Louisiana, and as the profits, arising therefrom, are greater than those from the cultivation of cotton, the production of sugar will necessarily increase in a few years to such an extent, as not only to supply our home consumption, but leave a surplus for exportation. In that portion of the States of Louisiana and Mississippi, where the cane can be raised with advantage, nearly all the cotton plantations will be diverted to the cultivation of sugar. And as the cotton crop in that region amounts to 300,000 bales, we can assume that the same force employed for raising that quantity of cotton, will produce nearly 250,000 hogsheads of sugar. Thus doubling the present crop, without bringing into the calculation the sugar which Texas, Alabama, Georgia, Florida, and the other sugar growing States of the Union will produce. It will thus be seen that it will take only a few years to increase the sugar crop of the United States at least to three times the amount of what it now is. All the facts emanating from a source like that of the United States Patent office, at a moment like the present, will be of incalculable importance.

With regard to the collection of facts, having relation to the apparatus used in the boiling of sugar, I had less difficulty to contend with; although it is a subject of great moment to the planter, to enable him to produce a sugar which will bear a comparison with the best Havana, in grain, purity, and whiteness, obtained cheap, with the least amount of fuel and labor. There is no exaggeration in saying, that there is no sugar growing country, where all the modern improvements have been more fairly tested and adopted than in Louisiana, and where such perfect boiling apparatus is used, fulfilling all the conditions that science and experience have pointed out as necessary for obtaining a pure and perfect crystaline sugar, combined with the utmost economy of fuel.

The success of these improved modes is due to the enterprise and high intelligence of the Louisiana planters, who spare no expense to carry this important branch of agriculture and manufacture to its highest perfection. They have succeeded in making, *strictly from the cane juice,* sugar of absolute chemical purity, combining perfection of crystal and color. "This is, indeed, a proud triumph," says Professor McCulloch, in his valuable report to Congress. "In the whole range of the chemical arts, I am not aware of another instance where so perfect a result is in like manner immediately attained."

What was supposed impossible, has been accomplished by the Louisiana planter, notwithstanding the obstacles of the late maturity of the cane, early frosts, and other incidents occurring there, which casualties are unknown to the sugar planter of the tropical regions. But not only in the raising of cane and the manufacture of sugar does the Louisiana planter excel. He deserves also commendation for the manner in which he has embellished his country. His leisure hours are devoted to the beautifying of his estates, thus rendering the margin of the Mississippi a continuation of beautiful villages, surrounded by tropical plants and trees.

I cannot describe the delight I felt when I first entered the State of Louisiana. Its river, the creator of this rich alluvial territory, after having tossed and rolled its mighty waters against the wild shores of the upper country, carrying away and building up, inundating vast tracks and leaving everywhere traces of its destructive sway, begins at once to slacken its current and keep its turbid stream within the bounds of fertile banks, gliding majestically through highly cultivated plains covered with the graceful sugar cane, the uniformity of which is continually diversified by beautiful dwellings, gardens, and the towering chimneys of the sugar-houses, the handsome fronts of which stand forth in the picturesque back ground of the forest, forming an everchanging scene.

The traveller who floats in one of the gigantic palaces of the southwest, can from the high deck behold with delight the enchanting scenery the whole day long, and look with regret on the setting sun, which, gradually withdrawing behind the dark outline of the cypress forest, leaves this lovely country reposing under the dark mantle of night. Not less beautiful and well cultivated are the shores of the great bayous and tributaries crossing the State in all directions. I invariably met with that far-famed hospitable welcome peculiarly characteristic of the southern gentleman and planter.

I consider it my duty to mention, especially as such, Mr. Theodore Packwood, Mr. Benjamin, Mr. McMaster, and Colonel Mennsel White, senator from Plaquemine. At the plantation of this last gentleman I was seized with a violent fever. Owing to the kind attentions I received I was in a few days sufficiently restored to enable me to examine his beautiful estate, on which utility and taste are everywhere blended. His large and, in fact, splendid sugar-house, placed in the midst of extensive cane fields, vies in neatness with any in the State. His superior draining machine, large canals, and innumerable ditches, deserve the highest praise as also his garden, adorned with a great variety of roses, all then in bloom, with sweet myrtle, orange, plantain, oleander, aloe, and many other tropical shrubs and plants, as well as with live oak and other native forest trees, the intermingling of which is novel and curious for those born in the north. This valuable plantation extends a considerable distance up and down the coast, enclosed by a neat fence, lined with the hardy sour orange, its golden fruit adding to its picturesque appearance. To these gentlemen and their families I shall always feel most highly indebted; and, indeed, to every gentleman I called on, since they all gave me, with most condescending kindness, every information I desired favoring me with the perusal of their memorandum books, and extracts therefrom.

The sources from which I drew my information were various, but I stand principally indebted to Mr. Theodore Packwood, of Scarsdale, to Mr. Benjamin, of Bellechasse, to Colonel White, Deer Range, Messrs. Morgan, and Wilkinson, and Urquhart.

I found also a great deal of valuable matter in Professor De Bow's Commercial Review, a work of sterling merit, containing a rich source of information for those interested in the statistics of our extensive country.

Mr. De Bow most obligingly tendered me the free use of his office and library, and I have availed myself of his kind, disinterested, and occasional assistance.

I have made copious quotations from the notices on sugar, furnished to him by the ablest planters. I feel myself excusable for having done so, as Mr. De Bow's work is new and scarcely known among those who live at a distance from the Mississippi.

I would strongly recommend every one interested in sugar growing, to peruse the Commercial Review, in order to keep pace with the progress of the Louisiana planter, who seeks information not only in his own country, but also in the West Indies and in Europe, the summer months being spent by him in journeying through these countries, and on his return in the fall he makes use of what he has acquired in his travels, attending to his business with a care and interest that surprises even our active business men of the north. Such is the Louisiana planter, so far as I have had an opportunity of becoming acquainted with him.

There remains still much to be done. But the desire of improvement is aroused, the proofs of successful endeavors attained by some, must have convinced the less energetic, and there is no doubt, in a short time, the progress in the culture and the manufacture of sugar, will be rapid and general throughout the State of Louisiana.

A book which ought to be in the hands of every one engaged in the manufacture of sugar, is Professor McCulloh's able report in relation to the chemical nature of saccharine substances, and the art of manufacturing sugar, published by Congress in 1847. It contains all the valuable improvements up to that period, that had been made in this branch of industry.

I must crave some indulgence, as I had only a few weeks to visit the principal plantations, collect notes, and make drawings, but I hope that during another grinding season, I shall be able to complete a fuller report.

Through the disinterested kindness of Professor Corda, from Prague, in Bohemia, well known through his works upon vegetable physiology, and his numerous microscopic analyses, whom I met in New Orleans, in his scientific tour through the United States, I am enabled to lay before you beautiful and almost invaluable drawings representing the cane as seen, highly magnified; with a description of its structure, both of which will be found in their proper places.

Baron Von Humboldt most earnestly recommends Professor Corda to the American nation in a general letter of introduction, a copy of which I here annex, to show the interest this eminent savant yet takes, in his advanced age, in the progress of science.

I have also, with your permission, compiled from the records of the Patent Office, improvements patented in the manufacture of sugar, with illustrations.

It is highly important that the planter should be acquainted with the various improvements made in this branch of industry, in order to avail himself of them when found useful.

The publications of extracts from the scientific journals of England, France, and Germany, on this subject, in the official report of the Patent Office, would also be of great importance to the sugar growing interest, a branch of agriculture and industry which every year becomes more extensive, and in which many millions of dollars are invested.

I am, sir, with great respect, your obedient servant,

CH. L. FLEISCHMANN.

Cultivation, &c., of Sugar Cane

Varieties of cane cultivated in Louisiana. — The planters of Louisiana cultivate five different kinds of cane; the Bourbon, the green ribbon, the red ribbon, the Otaheite, and the Creole cane.

The *Bourbon cane,* is very extensively cultivated. I found it almost the only kind of cane raised on some plantations; it has a good coating of silica, which forms a strong protection against the cold; the dark purple color of its cortex increases the absorption of light, and accelerates its maturity. It is thought a hardy cane, rattoons well, and yields good sugar; it has large eyes, which resemble those of the red ribbon, and somewhat the eye of the Creole, and withstands the influence of a slight frost.

I was told by a planter that, when red ribbon cane is planted in new rich land, it loses its stripes and becomes altogether purple; but that, when the same cane is planted in old dry land, the ribbon appears again. Others think that, when the eyes of the red ribbon cane are situated upon the purple stripe, the cane produced from such eyes is an entirely purple cane. The leaves, and the whole appearance of the plant, resembles that of the red ribbon.

The *green ribbon* is undoubtedly a different species of cane, not only from its light yellow color and delicate green stripes — whence its name is derived — but also from the difference of shape and formation of the eye, which is small, elongated, and delicate in its structure, resembling that of the Otaheite. The cortex is less strong than that of the Bourbon, it yields well, but is much more easily affected by frost than the former.

Next to the Bourbon, the red ribbon is the most extensively cultivated in Louisiana. It is a beautiful cane, and its purple stripes vary from one inch to a line in width. Like the Bourbon, it has a strong coating of silica, which makes it more hardy and capable of resisting slight frost. Its eyes are in shape and size like those of the Bourbon, and are less affected by the inclemency of the weather than the green ribbon, Otaheite, or Creole cane. It rattoons well, yields well, and the juice from the ripe cane is rich sugar.

The Otaheite cane has large joints, but grows less high, and its cortex is less thick, than in the former species; its eyes are of a very delicate structure; this cane does not rattoon well, which must be ascribed to its delicate eyes. It is easily affected by the frost, in consequence of which very little of this kind of cane is cultivated, although its juice is rich and yields very abundantly.

The Creole cane has, in former years, been very extensively cultivated in Louisiana; but the Bourbon and red ribbon, from their hardy nature, have nearly every where superceded it. It is only planted in small patches for eating; and in the neighborhood of the city of New Orleans, it is raised for the markets. Its cortex is easily crushed, and yields a very rich and delicious juice, from which a superior kind of sugar is produced. Its eyes are rather small, but still of a larger size than those of the green ribbon or Otaheite, and resemble somewhat those of the Bourbon and red ribbon. I have been informed that, in the parish of Plaquemines, the planters begin to cultivate it again more extensively. I have seen there, on Mr. Morgan's plantation, who always cultivated it, considerably large fields planted in that cane. This cane grows short, its leaves are straight, and do not droop like those of the Bourbon or red ribbon.

The joints of this cane, like that of the Otaheite, are often covered with a black matter, which comes off when rubbed, and appears to be a kind of fungus.

Among the Bourbon and red ribbon cane, there grows sometimes a large cane of whitish color, with very large joints, containing a great deal of juice, but of a disagreeable taste. This is a diseased cane, caused by long wet spells; its whole appearance is coarse and unnatural, something like the diseased limb of a man. Colonel White planted that cane in his garden, and remarks, in his memorandum, kindly extended to me, "I also took the white sickly, looking cane, frequently found among the purple and red ribbon, the bark of which is softer, and the cane itself more juicy. It grew well and large, and, from having air and sun, produced a yellowish, slightly purple, looking cane, which I continued to plant until I got about five acres. I then abandoned it as worthless."

I have been told of different varieties of cane produced, by planting ribbon and Otaheite cane in the same field; but, as the cane never flowers, how could it be possible to produce a different variety?

The length of the ripe joints vary; those of the Bourbon and red ribbon have joints varying from four to nine inches in length. The cane cut for grinding measures from three to five feet in length. I saw some over eight feet high, and with from twenty-four to twenty-eight joints; but they are a rare instance. The eyes are situated above the joints, and alternate in a straight line on each side of the cane. The knots, or ring, of the joints are of a light yellow color, and have two or three rows of minute points, from which the roots project; below the joint the cane is surrounded with a ring of greyish substance, which Mr. Avequin analysed, and found it to be a substance similar to wax, and called it, from this resemblance, *cerosie*.

This film of wax-like substance covers the whole length of the joint. I observed it to be abundant on the Bourbon and ribbon cane. Mr. Avequin states, that cane which is much covered with this kind of wax yields little sugar, while those species of cane which have little of the waxy coating are richer in sugar. Mr. Avequin presented me with two specimens of the cerosie. It is a substance, in appearance and touch, very much like wax of a dirty greenish color, exceedingly brittle, requires a higher degree of temperature to melt it than sperm, tallow, or beeswax, and when melted together with them, it separates from them in cooling.

Mr. Avequin showed me some cane arrows which he had brought with him from the West Indies; they were cut from the last joint of the cane and measured from five to seven feet in length, the cane itself being about six to eight feet high. The spike is about one and a half foot long, bearing very small delicate grains; the glumes are provided with a fine silk-like plume, to facilitate the winds to carry the seeds of this useful plant great distances, showing that nature has intended it should propagate from the seed; still there exist a great diversity of opinion with regard to the power of reproduction through the seeds of cane.

The spike is composed of a great number of delicate branches. . . .

Even in the West Indies, where the climate is so favorable to the growth of cane, it arrows very sparingly. Mr. Avequin told me that he there saw in a whole field sometimes only few plants with arrows, and that the seed from the cane never sprouts; he collected great quantities of their delicate seeds, analysed them and found them to contain all the constituent parts of seed belonging to the graminaceous tribe. . . .

Time of planting. — The planting time of the cane in Louisiana, is in the fall, and immediately after the grinding is over. Few planters commence planting as early as October; but few have time to begin planting before or during the grinding season.

They are fully occupied at that period, laying up their seed-cane, and in making all the necessary preparations for grinding and boiling.

Cane planted in the fall, gives generally a fine crop, but it requires well drained fields, for if the ground be wet and the weather mild, the cane will rot; furthermore, it must be well covered with earth, at least to the height of four to six inches, to protect it from the frost, in case of a severe winter. Cane well planted in dry land, comes up early in the commencement of the warm season, and outgrows cane planted early in the spring; for this has to contend against wet or dry weather.

Distance of planting cane. — There exists a great difference of opinion among the Louisiana planters, with regard to the distance that cane should be planted apart. Many still adhere to the old mode of planting, that is, in rows from three to five feet, while others plant it, with great advantage, eight feet apart, or at such distance, that the carts and cattle straddle the rows in carting cane from the field, without injury to the rattoon.

As in Louisiana, the cane must reach its full growth in nine months, the planter carefully endeavors to expose it as much as possible to the influence of the sun and free circulation of the heated air.

I have seen cane planted at eight feet, which was so luxuriant in its growth, that the rays of the sun could scarcely penetrate, although it was a field planted with cane for twenty successive years, and had only the year previous a crop of Indian corn and peas on it; that one year's rest, wide planting, and proper culture, gave it such a vigorous growth, as I never saw in any agricultural produce. . . .

I think some of the Louisiana planters look too much to the number of acres planted in cane, and not enough to a more perfect culture, whereby they would increase the size of the cane, and improve the quality and add to the quantity of the sugar.

Seed cane. — The Louisiana planters use generally for cane seed, the riper part of the stalk. Some cut the cane in the middle to use the tops for planting, and bring the lower joints to the mill, some again, use the green tops alone for planting.

It is with cane as with all other plants, imperfect seed produces a poor plant and bad fruit. The planter cannot expect that seed-cane, with delicate imperfect eyes, short thin joints, will produce a cane like one of vigorous growth, with perfectly well developed eyes and a great deal of juice which supports the young shoot, until its roots are strong enough to obtain nourishment from the soil.

The young sprout from poor cane is less able to support the inclemency of the climate and is more liable to disease.

In the West Indies, we are told, the few upper joints of the plant nearest the leaves, commonly designated as the "cane tops," are used for cane seed. In the West Indies, where the cane arrives to perfect maturity, where every joint is ripe, and every eye well developed, the top joints may answer, but in Louisiana, where the cane is never entirely matured, where it must be cut before the upper joints are formed, the tops are not fit for seed, and the result must necessarily be a bad one. All the planting of cane that I have seen, was done with cane from five to twelve joints or more in length.

Some planters select for seed the oldest or poorest rattoon cane; this cane is of small growth, has a few short joints, and bad eyes, and appears almost to be a different species of cane from the plant cane which has been raised upon well prepared and thorough drained soil, or after Indian corn or peas. These planters save all their fine cane for the mill and forget that by this kind of economy they reduce their crop for the following year, and that it will affect the sugar, both in quantity and quality.

"I cannot believe," says the writer of that able article on sugar, in De Bow's Reviews, "that this practice of always selecting the poorest plant cane for seed, was one of the main reasons which caused that fine variety of cane called *creole,* to degenerate to such an extent, that in late years it has been almost entirely banished from our fields."

Mode of planting. — Previous to planting, the land is ploughed, well harrowed, and the rows marked with a two horse plough, after which the double moulded plough follows, which opens a clear furrow ready for planting.

In this furrow three canes are laid straight, in rows four inches apart. If we examine the drawing of the various species of cane, we shall observe that near every joint is an eye which is always opposite to the eye of the next joint above or below, so that in every cane there are two rows of eyes in regular order and in a straight line.

It is, therefore, very important in planting to lay the cane in such manner in the furrow that both rows of eyes be free to grow their young shoots at the same time. This mode of planting cane requires attention and much labor, because the cane must be refixed sometimes with earth to keep it in its proper position, to prevent its turning when the earth is thrown upon it, and few planters have hands enough to spare them for extra labor, just at that time. In case the cane is thrown promiscuously in the furrow, and some of the cuttings have to lay with one row of eyes facing the ground, the young shoots will have to twist and force their way up to the light, and their growth is necessarily retarded and unequal.

The cane, when thus laid in straight lines, three thick, two joints of the top, and two of the butt always brought together and overlapping, must be carefully covered with the hoe; the depth of the earth thrown over it varies according to the season; in the fall, as mentioned above, it requires from four to six inches, according to a more southern or northern latitude of the sugar region. For cane planted in the spring, two inches of well pulverized soil is sufficient, in order to hasten the development of the shoots, and accelerate the growth of the cane, that it may reach its maturity before the cold season sets in. In the tropical region the planter has nothing to fear from frost, and he gives his cane as much time as it naturally requires to come to maturity.

The Louisiana planter labors under great disadvantage in this respect, and his principal aim must be to obtain fine large canes in as short a period as possible; and this depends not only on early planting, but also on a careful preparation of the soil, good seed cane, and proper cultivation during its vegetation.

On Manure. — The Mississippi valley, and especially the lower portion, which embraces the State of Louisiana, has been the depository of enormous beds of fertile soil since time immemorial, composed of the minutest particles of organic and inorganic matter, brought together down from the source of that father of rivers, rolling its muddy waters over various rocky beds, washing the extensive limestone, coal and chalk formations in its current southward, mixing with it in its course vegetable as well as animal matter, and adding thereto the deposites of its numerous tributaries, which have their sources among the primitive masses. The rapid current rolls on for thousands of miles, and deposites gradually its more bulky and heavy particles as it winds its way down to the lower portion of that valley, where its bed becomes deep and its current more gentle, forming on either side of its channel the inexhaustible treasures it holds in suspension.

The Drovers and Butchers, 1848

From William Renick, *Memoirs, Correspondence, and Reminiscences* (Circleville, Ohio, 1880), pp. 26–31.

MOUNT OVAL, NEAR CIRCLEVILLE, O., Dec. 25, 1848.

Mr. A. M'Makin, American Courier, Philadelphia.   DEAR SIR:
The following article appeared in your paper of the 16th inst., under the head of "Destroying a Conspiracy:" "An effort is about to be made by the victualers, with the aid of the public, to destroy a combination of drovers and middlemen, or

forestallers, who conspire to raise the price of butcher's meat to an unfair and oppressive height. They may count on our aid to effect this public good.''

It is evident, from your manifest willingness to participate in the correction of the alleged abuse referred to in the above article, that you have taken as Gospel truth all you have heard in relation thereto, on one side of the subject. Now, would it not be as well (before a final decision) to hear what may be said upon the other side, that you may be the better enabled to judge more correctly for yourself, whether or not the alleged abuse really does exist? — for it is not at all likely that either you or a large majority of the Philadelphia public are practically acquainted with the subject under consideration, and therefore must form your opinion from arguments advanced from each and both sides of the question. I say that the alleged ''conspiracy'' does not exist, and if you will allow me a brief space in your widely circulated paper for my remarks, I will endeavor to prove to you, and through you, to the Philadelphia public, that it cannot, to any considerable extent, take place, and that the drovers have been wrongfully accused. Although I am well aware of the strong natural influence with which I will have to contend (I mean the pocket interest), yet I think the large body of your community — actuated by a spirit of reciprocity, and the still more noble sentiment of ''do as you would be done by,'' — will, when their minds are disabused from any imposition being practised upon them, sustain the old adage, ''live and let live,'' at all events, I trust you and they will give my remarks an impartial consideration. I will be brief as possible, and confine myself entirely to the western branch of the business — my long experience enabling me to speak with confidence, so far as *it* is concerned, for I myself have been engaged for more than twenty-five years, in occasionally furnishing the butchers of the Eastern cities with beef cattle, and also for the whole of that period been engaged in the business of raising, grazing, and feeding cattle for the supply of those markets — and for the better understanding of the whole matter, I will commence with the beginning of the business of furnishing beef (which, I take it for granted, is the principal article complained of) for your market, and the probable cost of production.

In the first place, cattle are bred and raised until they are four years old, generally for, say $24 per head, or six dollars per head per annum, which, I think, you will say is reasonable enough, though the best of the improved breed will command that price, and even more, when three years old. Then the animal must be put to better keeping, and will consume all the grass that will grow on two acres of the best land in an ordinary season; put that down at $6, and you cannot call that charge extravagant. Then in five months more he will consume at least seventy-five bushels of corn, or one-half bushel per day; put that down at $15, or twenty cents per bushel; can you complain of that charge? Total so far, $45, and the animal will weigh about 900 pounds. So much for the cost of production in Ohio. But yet I have allowed nothing for interest of capital, nor for loss of cattle by murrain, which is at least three or four per cent. per annum, but that must come off the farmer.

Now we will see wherein the western drover is at fault. He buys the cattle at the above rates — that is, the best prime cattle, that will weigh 900 pounds, for $45, or $5 per hundred pounds, and then he is at an outlay for driving of $11 to Philadelphia, or $13 to New York, provided the cattle are driven on grain, and two-thirds that price if driven on early grass. This includes all expenses except the interest of money, which will add one dollar more, making the whole cost to the drover $57 to Philadelphia, and $59 to New York, and the cattle will weigh, on an average, from 750 to 775 pounds, at Philadelphia, and near a quarter less at New York.

Now, admit the cattle to carry to market, on an average, the highest weight here named, which they will not do, and allow the drover to sell them at the highest prices that the Philadelphia butchers have ever paid for the last ten years, except perhaps for a single market day or two, at most, and even then where is the margin for the drover's extravagant profit? There is none. The fact is, that for many years it has seldom happened that your butchers have been willing to pay a price that would remunerate the western drover for *superior cattle,* and the consequence has been that seven-eights of the best Ohio and Kentucky cattle have passed your doors, to reach a better and a more remunerating market. I do not say that the people of Pennsylvania cannot produce good beef at a less cost than Ohio and Kentucky, but I do not see how they can do it, although they are measurably clear of the great cost of driving, and also the great loss of weight, yet they pay more for their store cattle, and the food wherewith the cattle are made fat is worth more than double the estimated cost of ours. So much for the actual cost of *good* beef delivered at market.

Now for the "conspiracy," or combination. How is that possible, with a consumption among the Eastern cities, large and small, of more than 200,000 cattle annually, and with 15,000 or 20,000 head always ready for market, scattered over a vast extent of territory, and at all times more or less of them starting to market, without the knowledge of other parties? Would not the conspirators have to buy up all the cattle within reach of the market? And could they do that? Would there not, in spite of them, be a sufficient number untouched by the drovers, ready and waiting to step in the moment they would get the price up, to take advantage of the high market? Remember that fat cattle, after being driven to market, are a perishable article — they cannot be held out of market like a barrel of pork or flour, they must be sold on arrival, or very soon thereafter, be they few or many; the expense of keeping is too great, and they lose in weight too fast after the drive, to hold any length of time.

Now, sir, if this is all true, will you not be ready to conclude that the butchers are mistaken? I know they are; but still I do not say that they do not believe what they assert, for a large majority of them are perhaps, but little, if any better acquainted with the actual cost of production, than other citizens generally of Philadelphia. They, and we all know that it often happens that the supply of beef cattle at hand is not equal to the demand; in that case, of course, the drover will put up the price if he can, just as any other dealer in any other article would, under similar circumstances; but then again it more frequently happens that the supply is greater than the demand; then the drover in his turn must submit to take the best price he can get. I admit that the monopolizing of the market for a short period has often been attempted, but which has nearly as often resulted in a failure. But there is no sane man acquainted with the business that would not at once denounce anything like a permanent monopoly as literally impracticable; the competition is too great. You might as well tell me of a "conspiracy" of the merchants of Market street, to raise the price of their goods to their Western customers, or a combination of the shoemakers of Lynn, for the purpose of raising the price of boots and shoes, either of which perhaps would be as easily effected as the "conspiracy" complained of. There is no branch of business with which I am acquainted that is so precarious, or that is not in the aggregate more profitable, than that of driving cattle from the West to the Eastern markets.

In the foregoing remarks in relation to the cost of the production of beef for your markets, I allude *only* to the *best article*. Inferior beef can be, and is, produced at a considerable less cost. But are you willing to be put off with the inferior article? I think not; but I assure you that the best beef cannot be afforded to you (without a change of times) at a less price than you have obtained it for the past two years, without a sacrifice on the part of the producer, drover, or butcher of their just and moderate profits; and I am yet to be made to believe that the people of Philadelphia are unwilling to pay such a price for their beef as will allow all concerned a fair and moderate profit. They must bear in mind that good beef cannot be produced, to any extent, on low-priced, or inferior land. It requires the best of land to produce it. For that reason the production is principally confined to comparatively small and detached portions of the United States. All the best beef furnished from Ohio and Kentucky is produced on land that will command from $30 to $50 per acre for other purposes than the making of beef.

Compare the prices of the different articles of food in the United States, relatively, with those of other countries, and you will find the disparity greater in the article of beef than that of most any other thing, and it is in your favor. And again I call upon your people to take into consideration the fact, that good beef cattle cannot be reared at former prices — the prices of 1824–25. At that time, we of the West were almost without resources; we were obliged to take the offered cash price for any article we had to dispose of, whether remunerative or not — there was no alternative. We were compelled to have some money wherewith to procure the necessaries of life, pay taxes, etc., and nothing but cattle and hogs would at all times command the cash. I myself have seen wheat offered at twenty-five cents per bushel, and the cash refused even at that low price. But thanks to the shade of Clinton, the opening of the New York and Ohio canals put a new face on our affairs, and the completion of each successive important canal and railroad has added to our resources and diversified our pursuits, the result of which (taken in connection with the late foreign demand) has been an advance in price of from fifty to one hundred per cent. in most articles of produce, and a consequent rise, though a less porportionate one, in the article of beef. With these remarks I will leave the matter to your consideration.

Mowing Machines, 1849

From *Prairie Farmer* (Chicago, October, 1849), IX, p. 320.

Now that the demand for machines for cutting grain is tolerably well supplied a new one begins to be heard. We must now mow our grass by machinery. The attempt to supply this want is not new. It has been attempted often and with about the same amount of success up to a late period.

A machine for mowing was exhibited at Buffalo last Autumn; but its moving wheel was of such construction that we did not think it worth mention. It has, however, been since improved; gearing having been substituted, and as we learn, operates up to a certain point of perfection.

Mr. Hussey also offers his reaper as a mowing machine, and we learn that it can also be made to work up to the same point.

Mr. McCormick has also endeavored to adapt his reaper to cutting grass, with what success we do not learn.

Mr. Charles M. Grey, the cradle maker in this city, has also constructed a machine for mowing; which we saw in operation a few days since. The implement is the first one constructed, and is imperfect in some respects; but sufficed to show what may easily be done, and what it will be difficult to do.

First then, to mow with machinery, there *must be a smooth bottom*. Grass does not grow like wheat, three or four feet in height; so that it can be cut higher or lower as the case may be. From one foot, or less, to eighteen inches in about its height. Hence an inch or two at bottom will make some considerable difference in the amount of hay obtained from an acre of ground. If the ground be uneven a cutting implement, five or six feet long, cannot get all the hay.

Second. There is no difficulty in constructing a machine to cut timothy or clover, particularly if it stands well; and there is as yet about all that has been done by mowing machines. Ketchum's, Hussey's and Mr. Gray's present machine will all cut these grasses tolerably well. This is something; but we want a machine that will cut the fine grasses, such as grow thick and close at the bottom.

Third. This is not easy to be done, and no machine is yet invented which will do it successfully. To this point Mr. Grey is now bending his energies. The construction of his implement is such that he can get close enough to the gound, provided it be smooth; but the difficulty to be overcome is the clogging of the knives, *"Hic opus est"* in Latin — tough business, in English. We hope Mr. Grey will succeed. He is an ingenious mechanic, and there is no man whom we should be pleased to see the inventor of a good mower sooner than him.

Cotton Growing, 1849

From U.S. Patent Office, *Annual Report, Agriculture*, 1849, pp. 307–16.

Cotton

Various causes operated to reduce the crop of cotton grown in 1849, below that of 1848. Prominent among these, were severe frosts about the middle of April, in all the cotton-growing states, which destroyed the young plants, and left many planters without seed to repair their losses by planting anew. An excess of rain fell in the months of May, June, and July, which was followed by an ususual drought, soon after, in large districts. The immense quantity of rain that fell did additional damage, by causing rivers to overflow their banks and inundate vast areas of bottom land, to the destruction of cotton and other crops. These disasters were attended by the prevalence of cholera and other diseases of a fatal character, which induced the temporary abandonment of many rich and promising cotton fields. Insects did no inconsiderable injury, as usual, in all the cotton-growing States.

Had it not been for the industry that prompted the planting of more acres than were ever before devoted to this important staple, the supply of this year would be less by some 300,000 bales than it now is.

If the year 1850 shall prove favorable to the growth of cotton, and no serious misfortune befall the cultivators, a harvest of 3,000,000 bags may reasonably be expected. But no crop grown in the United States is more liable to casualties; and not till it is fairly gathered and housed can one feel safe from loss. The receipts up to the latest dates are as follows: (From the Augusta Chronicle and Sentinel of March 6, 1850.)

	1850	1849
Savannah, Feb. 26	244,505	237,365
Mobile, Feb. 22	263,302	392,114
New Orleans, Feb. 27	597,147	716,460
Charleston, Feb. 28	263,180	302,493
Florida, Feb. 20	118,128	107,251
Texas, Feb. 20	16,795	16,296
North Carolina, Feb. 16	5,919	2,158
Virginia, Feb. 1	5,275	5,780
	1,514,251	1,779,817
Decrease at New Orleans	119,313	
Mobile	128,812	
Charleston	39,313	
Virginia	505	287,943
Increase at Florida	10,877	
Texas	499	
Savannah	7,240	
North Carolina	3,761	22,377
Total decrease		265,566

Stock on hand.	1849–50.	1848–49.
New Orleans, Feb. 27	229,376	256,643
Mobile, Feb. 22	129,562	167,325
Florida, Feb. 20	48,716	46,548
Texas, Feb. 20	1,750	800
Charleston, Feb. 28	62,950	99,168
Savannah, Feb. 26	62,465	46,879
North Carolina, Feb. 16	550	200
Virginia, Feb. 1	750	650
	536,119	578,213
Decrease in stock		42,101
Stock in New York, Feb. 19	88,952	64,561

Exports.	1849-50.	1847-48.
Great Britain	437,651	673,417
France	148,694	153,731
Other foreign ports	81,417	139,265
Total foreign exports	667,762	966,413
Decrease in foreign exports		298,651
Shipments to Northern ports	473,816	425,814
Increase to the North		48,002

We omit quotations, only remarking that the sales made were about 11½ and 11⅝ for good middling: 11¾ and 11⅞ for middling fair, and 12⅛ and 12½ for fair to fully fair and choice cottons.

We learn that contracts have been made for crops of nankeen cotton at 14 cents, for the next season.

The receipts of cotton at Augusta and Hamburg, up to the 1st of March, reach 208,628 against 226,260 bales last year, showing a deficiency in our receipts of 17,532 bales up to the 1st inst., and a deficiency of receipts in the month of February of 20,019 bales. The shipments so far this season reach 159,820 bales against 196,159 bales last year, and the stock in store 62,527 against 56,654 bales at same date. The total of receipts at the receiving points now reaches 1,514,251 against 1,779,817 bales last year, showing a decrease of 265,566 bales. The falling off is very heavy at Mobile and New Orleans, and the deficiency at other points is also on the increase.

The estimates lately received from New Orleans put down 850,000, and our calculation is, that the Atlantic will not exceed, if it reaches, 700,000 bales.

The foreign exports show a considerable falling off, and the stock in the Southern seaports shows a deficiency of 42,101 bales.

As the crop grown in a year is never all sent to market within any definite time after it is ginned and packed, (some always holding back for better prices,) nothing short of an actual and well-taken census can determine the quantity made in 12 months. The time approaches when some steps will be taken by each State to learn the amount of agricultural products annually called into existence within its limits. As has often been suggested by the writer and others, this can be done with little inconvenience by the assessors or collectors of taxes, in each county or parish, every year.

Every farmer can tell the amount of his last crops: and every one should know all that any dealer or consumer knows about what is annually grown and needed of the articles which he produces. Knowledge of this kind is exceedingly valuable to all classes, but peculiarly so to that great agricultural class who both feed and clothe the world.

Planters and farmers should regard the commercial world as members of one family, the extent of whose wants ought to be studied with nice discrimination, to avoid the errors of producing too much of some necessaries, and too little of others. There is considerable danger that planters will devote too much attention to cotton culture, and too little to growing grain, making meat, and the improvement of lands. The tide never rises so high as not to ebb and leave a wide beach and many a wreck behind.

The cotton grown in one State being often sent to market in another, as from North Carolina and Georgia to Charleston, and from Arkansas, Tennessee, and Mississippi to New Orleans, there are no reliable data to determine the quantity grown in any State. As the United States census will soon be taken, perhaps as early as this report will be printed and distributed, an estimate of the product by States will hardly be worth the expense of publishing in 130,000 copies. The aggregate is likely to be not far from 2,100,000 bales. The average cost of producing 100 lbs. of good cotton is a matter of general interest, and has elicited not a little discussion. Deriving his information from large planters, Mr. Solon Robinson thus estimates the profits and expenses of growing this staple: —

(From the National Intelligencer)

"What does it cost a pound to grow cotton? This is a question of vast importance to the United States. Who can answer it? Not one in ten of those that make it their staple crop, I venture to say; for cotton planters are as careless in this respect as though they were conducting a business of cents and dimes, instead of dollars and eagles.

I therefore propose to give you an extract from my notes, which I have been taking during my extensive agricultural tour the past winter and spring, not only to show the character of the information that I have been gathering, but in the hope that it may induce others to come out and give more and better information, or point out any errors in my statements.

The cost of making 331,136 pounds of cotton last year upon one of the best plantations of South Carolina was $17,894.48, or a fraction over five cents and four mills a pound, including freight and commission, as well as interest upon a fair valuation of property.

The cost, exclusive of freight and commission and including interest, of making 128,000 pounds upon the 'cane-brake lands of Alabama' last year was $6,676.80, a fraction over five cents and two mills a pound. This is considered the richest cotton land in the world, and although the crop was called a small one, it was probably about an average one. The field hands upon this place numbered seventy-five, counting all over twelve years old, which gives a fraction less than four and one-third bales to each. Now this crop has to be hauled over about twenty-five miles of the worst roads in the world, when wet, as they usually are at the time the crop is ready to go to market, and then down the difficult and dangerous navigation of the Tombigbee river.

I am satisfied that these two crops give a better showing than three-fourths of the cotton crops of the United States. My own opinion is, that whenever cotton is below six cents it does not pay interest upon the capital invested, except perhaps in some few instances.

Below I give a table of items of expense upon the first plantation mentioned. This is owned by Colonel J. M. Williams, of Society Hill, and lies upon what is called the swamp-lands of the Pedee river. These items are necessary to show that I have not stated the expenses too high.

The capital consists of 4200 acres of land, (2700 in cultivation,) at $15........$ 63,000.00
254 slaves at $350 each, average old and young.................................... 89,900.00
60 mules and mares, and 1 jack, and 1 stud, average $60........................ 3,720,00
200 head of cattle, at $10 .... 2,000.00
500 head ofhogs, at $2.. 1,000.00
23 carts and 6 wagons 520.00
60 bull-tongue ploughs, 60 shaving ploughs, 25 turning do.,
   15 drill do., 15 harrows, at an average of $1.50 each 262.00
All other plantation tools estimated worth.. 1,000.00

Total...$161,000.00

Cash expenses:
Interest is only counted on the first five items, $158,620, at 7 per cent......... 11,103.00
3980 yards Dundee bagging, at 16 cents, (5 yards to a bale,).................... 636.80
3184 lbs. of rope, at 6 cents... 191.04
Taxes on 254 slaves, at 76 cents... 193.04
Taxes on land.. 70.00
Three overseers' wages 900.00
Medical attendance, $1.25 per head.. 317.00
Bill of yearly supply of iron, average 100.00
Ploughs and other tools purchased, annual average.............................. 100.00
200 pairs of shoes, $1.75; annual supply of hats, $100 .... 275.00
Bill of cotton and woolen cloth 810.00
100 cotton comforters, in lieu of bed blankets 125.00
100 oil-cloth capotes (New York cost,)... 87.50
20 small woolen blankets for infants... 25.00
Calico dress and handkerchief for each woman and girl,
   (extra of other clothing,)... 82.00
Christmas presents, given in lieu of ''negro crop'' 175.00
50 sacks of salt.. 80.00
Annual average outlay for iron and wood-work for carts and wagons 100.00
lime and plaster bought last year.. 194.00
Annual average outlay for gin, bitts. &c 80.00
400 gallons molasses 100.00
3 kegs of tobacco, $60; 2 bbls. of flour, $10.................................... 70.00
5/6 of a cent a pound on cotton for freight commission 2,069.60

                17,879.48

The crop of cotton at six cents will amount to 19,868.16
   Colonel Williams has also credited this place with the
     additional items drawn from it.
13,500 lbs. of bacon taken for home, place, and factory........................... 675.00
Beef and butter for ditto and sales.. 500.00
1100 bushels of corn and meal for ditto and sales................................ 550.00
80 cords of tan bark for his tan-yard.. 480.00
Charges to others for blacksmith work.. 100.00
Mutton and wool for home use and sales.. 125.00

                $22,298.16

Profits over and above interest and expenses upon this total are $4,403.68.

Counting cotton only at six cents, profits are $1973.68; counting it at seven cents, ($23,179.52,) and profits are $5285.04. It is proper to state that part of the crop was sold at seven cents, and it may average that.

Now it must be borne in mind that this is one of the best plantations, as well in soil as management, and that this was an extraordinary good crop. It must also be assumed that the land will continue to maintain its fertility and value, and that the same hands will keep the building in repair, as no allowance is made in the expense account for such repairs, or there will be aloss under that head.

Most of the corn and meal credited comes from a toll mill on the place. All the cloth and shoes are manufactured by Colonel Williams, but upon a distinct place.

The place mentioned in Alabama belongs to Robert Montague, Esq., of Marengo county.

The items of valuation are:

1100 acres of land, at $25	$27,500.00
120 slaves, at $400	48,000.00
4 wagons	400.00
5 yoke of oxen, at $30	150.00
30 mules and horses, at $75	2,250.00
4000 bushels of corn on hand for plantation use, at 35 cents	1,400.00
Fodder and oats on hand for plantation use	200.00
40 head of cattle, at $5 do	200.00
70 head of sheep, at $2 do	140.00
250 head of hogs do	600.00
20,000 lbs. bacon and pork for plantation use	1,000.00
Ploughs and all other tools do	500.00
	$82,240.00
Interest on capital at 7 per cent	5,756.80
Cash expenses, taxes, average	100.00
Blankets, hats, and shoes, (other clothing all home-made,)	250.00
Medical bill, average not exceeding	40.00
500 lbs. of iron, $30; hoes, spades, &c., $30	60.00
Average outlay for mules over what are raised	100.00
Average expense yearly for machinery repairs	20.00
Bagging and rope	350.00
	$6,676.80

This crop (28,000 lbs.) at six cents net will leave a balance of $1004.20, which is just about enough to pay the owner common wages of an overseer, which business he attends to himself.

Now, while there may be a few better places, there are thousands not near as good in all the cotton-growing region.

I am, most respectfully, &c.
SOLON ROBINSON.

Washington, *June* 4, 1849.''

Both of the above estimates are defective and erroneous, but they supply data which will aid in eliciting the truth. No allowance is made in either case for the increase in number and value of the slaves in the course of a year. If this is not equal to 7 per cent. on an average, it is to 3½ per cent., and sometimes reaches 8 or 10 per cent. The calculation is faulty in estimating a great deal more land than is planted, at $15 per acre in one instance, and $25 in the other. The latter price is three times larger than the average of cotton plantations at the South.

Although there are branches of agriculture more profitable than making cotton at six cents a pound, that is not an ill-rewarded business, if skillfully conducted. The way to secure better prices is to be fully prepared to encounter returns smaller than six cents. It is plain that if planters will generally make their own meat, raise their own mules and horses, produce wool enough to pay for all the woolen cloth they need on the plantation, and manage to have a few hundred pounds of butter to sell after supplying their own table, there will be little danger of overstocking the markets with cotton and depressing the price below 8 or 10 cents a pound. On the other hand, while two-thirds of a crop will bring more money than a full one, the sum received for cotton will be nearly so much clear surplus over and above all contingent expenses for the current year. There are three cardinal objects to be pursued in farming or planting:

1st. To make a comfortable and independent living.

2d. To keep the soil one cultivates constantly improving, so that all the crops grown may be produced at less cost.

3d. To lay up property or capital beyond the attainment of the objects above indicated.

It would be easy to write an essay on each of these points; but instead of this, it is believed that a better service will be done to the cotton-growing interest to copy from the Southern Cultivator, some practical remarks on the preparation of seed and land in cotton culture, from a gentleman of large experience, and well known as a writer on rural topics: —

Remarks on the Cultivation of Cotton

By Dr. M. W. Philips, of Mississippi

No. 1. Preparation of Land. — In writing out the detailed plan I pursue in the cultivation of cotton, I must begin, I suppose, on the 1st of January, so as to carry your readers regularly through. I will endeavor not to be tedious, yet I cannot possibly be minute without at least being tiresome to somebody — as there is always somebody who already knows every thing.

For ten years past I have thrashed down all cotton-stalks, cut down all corn-stalks, and turned them under as well as possible with a turning-plough. When planting cotton after corn, I strive to break up the land with two-horse ploughs — what I term flushing, that is, breaking up 30 to 50 feet beds. Last year I broke up every acre of land I planted with two-horse ploughs, whether planted in cotton, corn, oats, or potatoes.

If any land has been in cotton, I generally open out water-furrows, deep, with a shovel-plough; to this I throw two furrows, one on each side, with one or two-horse turning-ploughs. Thus the land remains until a day or two before I wish to plant, when I have the baulk broken out, thus having fresh earth to plant upon, and yet firm earth for the seed to be planted in. There will be a narrow ridge of earth not covered by the fresh earth, but I invariably run an iron-tooth harrow along the ridge so as to break clods, and rake off pieces of stalk, and to leave the ridge fresh; if once running the harrow will not do, I run it twice.

The opener then follows and opens out a furrow, say one-half inch is deep enough, and narrow; if this furrow could be as straight as a bee line, and half an inch wide, I would esteem it better, if upon level land. The seed is scattered thinly and regularly, then covered with a board or block; I would prefer a roller. As to distance, this depends upon quality, age, and locality of land, rich and fresh land requiring greater distance; and I am inclined to think that the same quality of land north of say 33°, will tend more to longer joints than does cotton about 31° to 33°, and particularly Western lands, these lands tending to short joints, and greater yield to height of cotton. I do not plant any land that requires rows to be over five and a half feet, even to grow 15 to 20 cwt. of cotton per acre. There is sometimes, I am sure, much loss by too sparse planting. I desire to have the plants meet in the rows by the first of August, and should it after this date lap in row, the crop will not be materially injured. I find the new varieties, as sugar-loaf and cluster, to require less distance both ways than does the Mexican. When I planted my crop with Mexican — Petit Gulf — I gave 5 to 5½ feet by 2 to 3 feet on my best land.

For four years I have grown sugar-loaf, and 4½ feet by 18 to 24 inches, preferring about 18 inches. Upon second quality of land I reduce distance to 4 feet less, by 18 inches. Upon this department of planting (the preparation) I use more time and labor than is usual, being careful to break up deep, throw out into beds all the land, leaving no unploughed ridges; the ridges I endeavor to pulverize well, and do not run ploughs unless land will pulverize, thinking ploughing may be done too early and land injured by being ploughed wet. My object in ploughing, say 3 furrows, early, is to permit the foundation of ridges to settle somewhat, as seed germinates freer and grows off better than upon light earth. I break out the residue as late as planting time, so that the plant will start before or with the grass and weeds. I prefer never more than a bushel of seed per acre, because solitary stalks are not injured by cold weather when scraped out as when grown in a hot-bed.

I have been asked how I plant seed when I buy. I reply I wet the seed thoroughly with salt and water, and sometimes use brine made by steeping stable manure in salt and water for 10 days before wanted, until fermentation has ensued. The seeds are then dried off with ashes, or lime, or plaster; I prefer the two latter, as the seeds are white, and the master can see that care in dropping is practiced by hands. These seeds are dropped at the required distance, and are covered with the foot, by brushing a little earth upon the seeds and pressing them into the earth with the foot. I would prefer a seed-planter, but could not make the one I tried drop regular. Five to ten seeds in a place are ample. I have dropped only one, and two, and three; when I did this myself, I failed not in a stand.

When a good ridge, clean of clods and litter, hand can scrape more; the labor of planting carefully, and time seemingly lost in this, as well as of dropping seed, is fully regained in the scraping. I have cultivated for ten years 9 to 10 acres of cotton, and 8 to 9 of corn, besides potatoes, oats, &c. This could not have been done, but by doing all work well; time is saved by good ploughing and neat planting.

No. 2. Preparation of Land and Planting. — Last night, I gave you the preparation and planting of the cotton crop; yet I could not, in length of one article, give more than a rapid survey. I prefer short articles, and yet it is best to be particular, even minute, though there is even here an objection, for a writer should leave something for his readers to think of. When I plant oat-land, land that was the year previous at rest, or corn-land, I invariably break up into large beds, size according to width of rows to be planted, so as to throw water-furrow of the flushing as a water-furrow of the row. When four feet rows, I run off land thirty-two feet, and keep furrows as straight as possible on level land. I then lay off rows, always with a shovel-plough, and then two furrows as before. Sometimes I open out the water-furrow of old rows, as deep as two mules can draw a shovel-plough; bed up to this entire, then open out a new water-furrow deep, and reverse two furrows with a one-horse plough. I am satisfied that there is no land I plant bit what is materially benefited by breaking up with a two-horse plough; thus all trash, grass, seed, &c., is well buried below the one-horse plough furrow. I use a piece of wood two or three feet long, running level on the land, the front end shod with iron, for the purpose of opening out furrows for planting seed. My object is to make a clean, straight furrow, and impact the loose earth. This stick of wood is rounded below, and fastened to a shovel-plough stock.

The straighter the row on level land, or the more regular on rolling land, if circling be practiced, the closer can the scraper be run, thus giving less labor to hoe hands, and if cotton seed be scattered very regular, so as to give a stand, no stalks touching, the hoe hand can thin out faster, and thus save time. If I were able to plant my cotton crop with the neatness and order with which Col. Wade Hampton plants his crop, I believe I could cultivate an acre or two more per hand. Being in company with him in 1847, on a steamboat, we discussed the subject of planting for hours, and he assured me that all his furrows were opened out for planting with the corner of the hoe, narrow and straight. If I could drop seed in a furrow only an inch wide and quite straight, I think I could manage two acres of scraping per day to each full hand; I regard planting a crop, if done in the best manner, more in the light of half cultivated, than many would believe. I have scraped three acres in a day. I can dirt easily four acres per horse, and can with the solid sweeps break out four to nine acres per horse, owing to whether rows be four or five feet wide; thus, besides the earthing furrow, it requires one or two to sweep out the middle. But land has to be put in good order, and seed planted in order. This matter has called for many a line from my pen in the different papers I have written for, and I must be pardoned for thus dwelling so long. It is really no interest of mine whether planters cultivate well or ill; whether they can cultivate a fair crop easily or not, I cannot be benefited; yet as a citizen of this beautiful world, as a sojourner in this southern clime, I feel an abiding interest in the welfare of my fellows. Therefore, I say, if planters will devote more care and attention in tilling their lands, and in putting in their crops in a good manner, they will be able to make more, and yet spare their servants and their beasts much labor in the cultivation.

Look at the garden. Take one bed and trench it, spade up two apades deep, reversing the soil even, what will be the result? But suppose the first spit be laid one side, then the second spit well and finely dug up, the first returned reversed, or thoroughly mixed, will not that bed be more or less moist all the year? And if there is any chance for water to pass off, will it not be fit to work, after a rain, sooner than any part of the garden? and must it not, of necessity, produce better?

I admit a planter cannot plant so great a crop, but he will need much less to make an equal crop.

The misfortune is, the body of cotton planters want a large crop, and will not be at the expense of team and tools. Would they not ridicule the carpenter, who, instead of getting tools to tongue and groove his flooring, would attempt to rabbet each side of plank, or to dig grooves, and then dig for a tongue with a chisel? And yet, though not quite so absurd, planters so act. What difference in cost in twenty years, if a planter buys six shovels, six one-horse turning-ploughs, three two-horse turning-ploughs, six scrapers, six harrows, or to buy all turn-ploughs? These same ploughs will last by changing, those not used to be taken care of, as long as the same number of one kind, and for all work. "Think ye, and judge ye."

Mr. Simeon Oliver, of Hernando, Miss., writes us that the average yield of seed cotton is 1000 lbs per acre in that county, and the cost of production about 6 cents a pound. The *sugar-loaf cluster,* or Prout variety, yields best.

Mr. E. A. Holt, of Montgomery, (good authority,) estimates the crop of Alabama for 1849 at 400,000 bales. The Mexican or Petit Gulf is regarded as the best variety.

Dr. David L. White, of Florida, writes, that the average yield per acre there is 800 lbs., and per hand from 3 to 4 bales. Petit Gulf is preferred; of the upland class of plants, seed of long staple obtained from South Carolina.

From St. Francis Co., Arkansas, as Mr. J. W. Calvert writes, they gather about 1000 lbs. of seed cotton from an acre, in some instances much more. Varieties grown, Mastodon, Mexican, &c. He prefers Mastodon for their land, and Mexican for rich soils. No manure except cotton seed is used.

In south-western Texas, cotton crops are quite uncertain and irregular. Mr. Pryor Lea writes, that the worm does great injury; but aside from these casualties, a bale of cotton per acre and ten per hand are a common return.

In the north of Alabama, as Mr. James Williams, of Jackson Co., informs us, the crop is very light. He regards cotton at 6 cents a pound as a better crop than corn or wheat.

Mr. Thomas A. Heard, of Clark Co., Ark., estimates the cost of growing a bag of cotton at $25, where, he says, the average is 1200 lbs. per acre; some lands go as high as 2500 lbs., which would turn out nearly two bales, or 800 lbs.

Mr. Samuel S. Graham, of Coosa Co., Alabama, says that cotton will grow on land too poor for wheat. He estimates the crop of the State at 450,000 bales of 500 lbs. each. Cost of production, five cents a pound; average yield, 150 lbs. of clean cotton per acre.

Lands cultivated in cotton are extremely liable to injury by washing rains. The warmer a climate, the more water the air will hold in the form of an invisible vapor, or of clouds, and the more sudden and voluminous the fall of showers. A large portion of the water thus precipitated upon the mellow soil runs off, and carries with it, even where it forms no gulleys, much of the fine particles of mould, loam, and clay. In this connection, it is important to bear in mind that it is the smallest fragments of rocks and decaying plants and insects, which dissolve in water and enter the roots of cultivated vegetables to nourish them. Hence, the organic and inorganic matter borne along in a muddy stream (and all streams that flow over ploughed ground are muddy) robs the soil of its fertility vastly more than would the removal of an equal weight of coarse sand, gravel, or compact clay. One valuable means of preventing the wash of hoed fields is to seed them in autumn as early as practicable, with some small grain, such as rye, barley, wheat, or oats. These plants will not only serve to prevent the washing of the ground, but they will imbibe volatile and involatile food from the soil, which would be lost without their presence. In addition to this, the winters are warm enough at the South for small grains to grow and draw largely upon the atmosphere for their organic elements. These crops, consumed by sheep or other animals, will yield valuable meat and not a little equally valuable manure. So soon as cotton and corn cease to grow, and peas are ripe when planted in cornfields, rye or some winter grain should be sown for the benefit of the land. The winter is the time to renovate the soil at the South. Particular pains should be taken in saving cotton seed, and the droppings from all domestic animals in stables and yards. Manure has been found in so cold a climate as that of Scotland to lose three-fourths of its fertilizing power in a few months, if permitted to lie out in an open yard and exposed to rains and sunshine.

Observations of a Cattle Expert, 1850

From Lewis Saunders, "History of Kentucky Cattle" *Southern Planter* (Richmond, Jan., 1850), X, pp. 19–25.

History of Kentucky Cattle

The first emigration to Kentucky — "the *dark and bloody ground*," the hunting grounds of the Southern and of the Northern Indians — with the view of permanent occupancy, of holding the country at all hazards, by men determined to overcome the tomahawk and scalping knife, by the use of the rifle, took place in 1775–6. The country then belonged to Virginia; a large proportion of the first settlers were from that State; next from Pennsylvania, then North Carolina, Maryland, New Jersey, &c. It is presumed that the emigrants brought with them domestic animals, such as were then in common use. H. Marshall, speaking of Gen. Ben. Logan, in his history of Kentucky, Vol. I., says, "In the fall of the year 1775, Col. Logan removed his cattle and the remainder of his slaves to his camp," (near where Danville now stands.) Horses and cattle were subsisted in the summer, in the *range*, consisting of a great variety of nutritive native grasses, including the buffalo clover and the wild pea vines, luxuriant beyond description; and in the winter, in the *cane brakes*.

It seems to me that the general characteristics of the cattle of the United States, at the commencement of the present century, were very similar to those of Devonshire, Dorsetshire, and Somersetshire, in England, as represented in prints of cattle in those counties in the last century. I have observed the cattle of Virginia, Maryland, Pennsylvania, New Jersey, New York, and the New England States; they seem to have had a common origin.

The first improvement in the breed of cattle in Kentucky was made by Mr. Matthew Patton and his family, to whom the country is much indebted for the introduction of several valuable animals. . . .

I have heard it estimated that the introduction of the Patton cattle increased the weight of the four year-old bullocks twenty-five to thirty per cent., besides improving the quantity and quality of the milk. This was a great gain.

The next marked improvement in the breed of cattle, was brought about by the importation of some animals direct from England in 1817. At that period and for many years previous, I lived in Lexington. My pursuits were otherwise directed than to agriculture; but I had early inbibed a fondness for fine stock, particularly horses and cattle. I admired good fruits, and gave some attention to their culture. For several years I was in receipt of a variety of English publications, on agricultural subjects and agricultural improvements, from which I got a glance of what was going on, in some respects, in the old country. It astonished me greatly to see the enormous prices paid for animals of particular breeds. First, the Long-horns, brought to a high state of perfection by the justly celebrated Bakewell, Princep, Munday and Fowler. Toward the close of the last century, they were at the height of their popularity. . . .

Much time was required, combining capital, skill, and untiring perseverance, to bring this breed to such a high state of perfection. Notwithstanding all this, it was suffered to run out, almost to disappear in the course of a few years. About the time that the Long-Horns were held in such high estimation, commenced the improvement of the Short-Horns. Skilful breeders, with Charles Colling at their head, brought this breed to a very high state of perfection. Their value was at the height in 1810. In this year, public sale took place. The list of animals sold and the very high prices paid for each have often been published. Countess, out of Lady, four years old, brought four hundred guineas. Comet, six years old, brought one thousand guineas. He was bought by four farmers.

It seemed to me that if four farmers were willing to pay five thousand dollars for a bull, there was a value in that breed that we were unapprised of, and that I would endeavor to procure it. I made up an order for six bulls and six cows. My views were then more inclined for a good milking than for a beef breed. The weight of authorities, given by the writers on the subject of cattle, at the close of the last and the commencement of the present century, were in favor of the *Holderness* breed as the best for milk, and the *Teeswater* and *Durham* as having the handsomest and most perfect forms. I settled on these breeds. In frequent conversations with Capt. William Smith, about the contemplated importation, he strongly urged me to include the *Long-Horns;* he had witnessed the marked improvements made by the use of old Mr. Patton's first Long-Horn bull, and he was extremely anxious to have a bull of that breed. I had great respect for him as a man, and, confiding in his judgment, two pairs of the Long-Horns were added to the list. The order was forwarded in the fall of the year 1816, to Buchanan, Smith & Co., Liverpool, with instructions to cause selections to be made of the best young animals for breeders, all to be two years old in the following spring.

First, a bull and heifer of the Holderness breed, to be procured from that district in Yorkshire. Next, two bulls and two heifers of the Teeswater breed, to be procured on the River Tees, in the county of Durham. Then a bull and heifer of the

Durham breed, and two bulls and two heifers of the Long-Horn breed. A minute description was given, particularizing each breed — no limit as to price. If the money sent was not sufficient to put that number on board ship, they were to be reduced, so as to have the best animals that could be had for breeders.

Buchanan, Smith & Co., employed Mr. Etchers, of Liverpool, to go into the different districts to make the selections and purchases, and he seems to have executed the order with much ability. The following is the invoice:

"Cattle shipped on board the Mohawk for Baltimore, consigned to Messrs. Rollins & McBlair, merchants there:

"No. 1. A bull from Mr. Clement, Winston, on the river Tees, got by Mr. Constable's bull, brother to Comet.

"2. A bull of the Holderness breed, of Mr. Scott, out of a cow that gave thirty-four quarts of milk per day — large breed.

"3. A breed from Mr. Reed, Westholm, by his own old bull.

"4. A bull from the Holderness breed from Mr. Humphreys, got by Mr. Wase's bull, of Ingleton.

"5. A bull of the Long-Horn breed, from Mr. Jackson Kendall, out of a cow that won the premium.

"6. A bull of the Long-Horn breed, from Mr. Ewartson, of Crosby Hall — is of a very fat breed.

"7. A heifer from Mr. Wilson, Staindrop, Durham breed.

"8, 9, 10. Three heifers from Mr. Shipman, on the river Tees— his own breed.

"11, 12. Two heifers of the Long-Horned breed, from Mr. Ewartson, Crosby Hall — of Westmoreland breed."

The Mohawk arrived in Baltimore in May, 1817. The cattle were safely landed, in good condition. Great pains had been taken in procuring comfortable accommodations for them in the ship, and an experienced herdsman employed to feed and take care of them on the voyage. On arrival, they were taken in charge by my friend, Mr. John Hollins, who caused them to be put in the best pasture and particularly cared for.

After the cattle had been shipped, and before their arrival at Baltimore, I sold to Capt. William Smith one-third of the concern, and to Dr. William H. Tegarden another third; reserving to myself one-third only. A suitable agent was sent to Baltimore for them, and they were brought to Kentucky at the joint risk and expense of the three parties. On their arrival at Lexington they were divided. There fell to my lot:

Bull No. 1, which I named Tecumseh.
    "   No. 2, named Sam Martin.
    "   No. 8,   "   Mrs. Motte.
    "   No. 10, "   Georgiana.
Capt. Smith's lot:
Bull No. 5, which he named Bright.
    "   No. 7, named the Durham cow.
    "   No. 9,   "   "   Teeswater cow.

Dr. Tegarden's lot:

Bull No. 4, which he named Comet.
" No. 6, named Rising Sun.
" No. 12, Long-Horn cow.

No. 10 died in Maryland; No. 3 (bull) became lame on the travel out to Kentucky, and was left on the way. He was afterwards received and sold by the Company to Captain Fowler, who sold him to Gen. Fletcher, of Bath county, Kentucky, where he died.

When the division took place, Captain Smith evinced great anxiety to own the largest Long-Horn bull; Dr. Tegarden preferred No. 4, and, as neither of them were my favorites, I cheerfully yielded; and in consequence they gave me choice of the cows. I selected one of the Teeswater heifers, and named her Mrs. Motte. It was a very pleasing occurrence to have each party highly gratified with receiving the very animals he preferred.

The narration of a pertinent coincident will not, I think, be deemed ill-placed.

Mr. H. Clay, being in England in 1816, having always had fondness for fine horses and for other fine stock, concluded to send home some fine cattle. At this time, the Herefords were great favorites at Smithfield. Either from Mr. Clay's own taste, or from the recommendation of others, he selected that stock, purchased a cow, a young bull, and heifer of that breed, and sent them to Liverpool, to be shipped to the United States. It so happened that they were put on board the Mohawk, the same ship with my cattle, and they arrived together at Baltimore, were there placed in the same pasture, and the agent that was sent for my cattle brought out Mr. Clay's to Kentucky.

Although Mr. C. and myself at that period resided in the same city and had been personal and political friends from the time of his coming to Kentucky, in 1798, till March, 1825, and our social and personal relations have been unchanged for fifty years — yet neither Mr. C. nor myself had the slightest knowledge or intimation of the intention or views of the other in regard to *importing foreign cattle.*

Mr. Clay at one time had a good stock of horses. He bred the dam of Woodpecker, one of our best race horses, and he proved to be a good stallion. His flock of sheep were celebrated for the fineness of their fleece.

But, having introduced the Herefords, I may as well finish them.

At this time (1817), Mr. Isaac Cunningham owned the largest and best grass-farm in Kentucky — the identical farm settled by old Mr. Mat. Patton, the father of the Patton family, who introduced the Patton cattle. Mr. C. was wealthy, had a good stock of Patton cows and had been in the habit of selling his young ones for breeders. Mr. Clay's good judgment led him to place his *Herefords* in the hands of Mr. Cunningham. Notwithstanding all these advantages, Herefords made no impression; in a very few years they were unknown as a breed in Kentucky, and at this day a part blooded one is rarely to be met with.

As to the Long-Horns, although there were two bulls and two cows imported, the breed has nearly run out. Captain Smith kept them up for a while, but, as he died soon after they were introduced, his stock was neglected. The Rising Sun left a good stock in Clarke and Bourbon counties, and for a while they were very popular with the feeders in those counties; but they have gradually yielded to the Short-Horns. A mixture of Long-Horn blood, in a remote degree, is deemed by many feeders of great value (and that is my opinion.) The hide is thick, the hair is long, and very closely set; they are of very hardy constitution, well adapting them to *our* mode of feeding. Cattle are not housed or sheltered, but fed out in the fields taking the weather as it comes. The Short-Horns have thin hides, fine short hair, and do not stand exposure to the weather so well.

The importation of 1817 (alluding to which it seems that the Long-Horns and the Herefords are to be omitted.) The young ones were much sought for throughout Kentucky and parts of Ohio, and were all sold for breeders. *Tecumseh* and *Sam Martin* were the principal instruments used in effecting this great improvement. Mrs. Motte, the Durham cow, and the Teeswater cow, were excellent breeders. The Durham cow was equal to the best milk cow I ever saw. Napoleon was her best bull calf. Mrs. Motte was the neatest, the finest animal of the importation.

A year or two previous to 1831, I observed that my young cattle were not up to the mark of improvement that I wished to see progressing, but were rather falling back. The only remedy that I then thought and still believe necessary to arrest this downward tendency and to give a fair prospect of improvement, was the *introduction of remote blood.*

Col. John Hare Powell, of Philadelphia, imported a number of animals of the improved Short-Horn breed, several years subsequent to 1817. He ordered his selections from the best herds in England, with great particularity as to pedigree, form and milking qualities, and without stint as to price. My attention was directed to this stock, to procure a cross on the Short-Horns of 1817.

In the spring of the year 1831, I procured of Mr. Barnitz, of York, Pennsylvania, a young bull and three young cows of Col. Powell's stock. In several points, their forms were better than those of 1817. The cross was very beneficial to me.

Some few years afterward, David Sutton, of Lexington, introduced several animals of Col. Powell's stock.

Then other gentlemen imported cattle from Philadelphia, and from other parts of the United States and from England; so that we have had a number of bulls and cows of the best known breeds in England and in the United States. From this basis, intelligent gentlemen, with abundant capital and great skill, have continued to improve, by judicious crossing, until we have arrived at a high state of perfection as to *form* and *early disposition* to take on *fat,* points most desired of all others by the grazier and the feeder.

Notwithstanding that Col. Powell's stock were drawn from the best milking families in England, their descendants did not prove with us to be as good milkers as the stock of 1817, nor were they so healthy.

The dairy is but a secondary consideration with a Kentucky farmer — beef is more profitable, and as the great object of all pursuits is *money,* the one putting most in the purse will be pursued. For a dairy of cows where there is a demand, selling milk is most profitable; next cheese, if the climate suits; last, making butter. A Kentucky farmer in general has no demand for milk. Cheese can be made here as well as any where else, but it costs too much labor to save it. Some writers say that it ought not to be relied on as a business south of forty degrees. Butter could be made, of the best quality and in quantities, but it seems that the farmers prefer taking only as much milk from the cows as supplies their families with milk and butter, giving the remainder to the calves. From these considerations it would seem that the *breed* of cattle bringing most money from the butcher at two and three years old will have the preference with the grazier and the feeder, they using nine-tenths of the cattle bred in the State.

It will be seen from what has been stated that great attention has been given to the breeding of cattle in this State for more than fifty years, and the course pursued has been to procure the best known breeds to cross with; so that we now have excellent breed for the grazier and the feeder — forms approaching nearer and nearer to perfection, and an aptitude to take on fat at an early age. But, in obtaining these grand objects, *perfect forms* and *early maturity,* so much desired by the grazier and the feeder, we have *sacrificed,* mainly, the *milking qualities.*

"Whatever be the breed, there are certain conformations which are indispensable to the thriving and valuable ox or cow. — If there is one part of the frame, the form of which, more than of another, renders the animal valuable, it is the *chest.* There must be room enough for the heart to beat and the lungs to play, or sufficient blood for the purposes of nutriment and of strength will not be circulated — nor will it thoroughly undergo that vital change which is essential to the proper discharge of every function. Look, therefore, first of all, to the wide and deep girth about the heart and lungs; we must have both. The proportion in which the one or the other may preponderate, may depend on the services we require from the animal; we can excuse a slight degree of flatness of the sides, for he will be lighter in the fore hand, and more active; but the grazier must have *breadth* as well as *depth.* And not only about the heart and lungs, but over the whole of the ribs must we have both length and roundness — the *hooped* as well as the deep barrel is essential; there must be room for the capacious paunch, room for the materials from which the blood is to be provided. The beast should also be ribbed home; there should be little space between the ribs and the hips. This seems to be indispensable in the ox, as it regards a good healthy constitution, and a propensity to fatten; but a largeness and drooping of the belly is excusable in a cow, or rather, notwithstanding it diminishes the beauty of the animal, it leaves room for the udder; and, if it is also accompanied by swelling milk veins, it generally indicates her value in the dairy."

The introduction of the Patton stock into Kentucky effected as much benefit to us in the improvement of our cattle, in a little more than twenty, as was effected in England in more than *sixty years.* . . .

Twenty fat cows were sold in the early part of this month, by one drover, at Cincinnati, the average weight of which was over one thousand pounds, the four quarters. These cows were Kentucky bred. All but three had produced calves.

I expected to receive authentic data, to state the average *age* and *weight* of the *four quarters* of cattle slaughtered at Louisville and at Cincinnati for three periods. Though promised, the paper has not yet come to hand.

In 1832, I took to New Orleans three bullocks, produced by a cross of cows of the Patton and Miller stock, by bulls of the importation of 1817.

No. 1. Red, 6 y'rs old, live weight,   3,448 lbs.
No. 2. Red, same age,   3,274 "
No. 3. Brindle, 4 years old,   2,868 "

I sold these three animals together, at auction, for the sum of nine hundred and twenty-five dollars.

I was at the New York State agricultural exhibition at Saratoga, in September, 1847. I very attentively examined the cattle stock there shown. The oxen are better than were generally to be met with in Kentucky; all others not so good.

The Ayrshire cattle may be classed with our half-blooded Durhams, from common cows.

We can derive no benefit from a cross of Devon blood. The diminutive size and ill-forms of the Alderneys would exclude them from our pastures.

Our climate is favorable for breeding and rearing cattle. They are free from any marked disease. I have never known an epidemic among them.

It is the custom with some farmers, as soon as the corn is in the roasting ear, to cut it up, giving stalk and all to hogs. — The hogs masticate the stalk — suck and swallow all the juice, throwing out the remaining fibrous matter, which soon becomes dry. Cattle are very fond of this refuse stuff; but when taken in quantities, it causes a derangement of the maniplus, for which no remedy has yet been discovered. At first the animal becomes restless, and is feverish. Soon after it begins to rub its head up and down a post or any thing it can rub against — manifesting the greatest pain and misery. It continues rubbing until it dies. [We suppose this to be what is called the "mad itch." — EDS.] I have seen several so affected, and, after the rubbing commenced, I never knew of one that was cured. Upon opening the animal, it is found that the maniplus is entirely deranged, dry, and hard, mortification having in some instances already commenced. The only remedy is to keep your cattle from the place where green corn stalks have been fed to hogs.

Cattle of Ohio and Indiana are not so healthy as are the cattle of Kentucky. I was told by a Cincinnati butcher, who supplies with beef a portion of the Jews of that city, that he was compelled to procure his cattle for these people from Kentucky. The *Priest* sticks the animal, which is dressed in his presence by the butcher. Upon opening the animal, if any imperfection of the intestines is visible, such as blisters on the liver, &c., the priest remarks, "this one may do for the Christians, but will not do for the Jews — you must bring up another." The cattle of Kentucky have no blemish; the intestines are in a perfectly healthy condition; so we, only, can supply the Cincinnati Jews with beef.

I was informed by Dr. Watts, of Chilicothe, a gentleman of intelligence and great enterprise, who feeds and grazes on a large scale, that he would pay five per cent. more for Kentucky raised cattle for either purpose than he would for Ohio or Indiana cattle. He considered the risk of life that per cent. in favor of the cattle of Kentucky.

There are three epochs in the history of Kentucky cattle:

First. The introduction of the Patton cattle, say in the year 1790, and some years afterwards, the Miller stock of the like blood. These were generally diffused throughout the State, improving our stock twenty-five to thirty per cent. in a period of twenty-five years.

Second. The importation of 1817, which gave us finer forms and an aptitude to take on fat at an earlier age, adding twenty-five to thirty per cent. upon the Patton improvement, in a period of less than twenty years.

Third. The numerous importations made into Kentucky and into Ohio, from 1831 to 1836, from which has arisen our present superior breed. To keep up this breed as it now is, requires sound judgment and unceasing vigilance, or a decline must follow.

I recommend to the breeders in Kentucky to import at least half a dozen young bulls from the Netherlands, Holland, or Northern Germany at once — and renew such an importation every five or six years for twenty years. Then to draw their young bulls from the best stocks to be found in England.

I do not think it is desirable to have a very large breed; but *form* and *early maturity* are not for a month to be lost sight of. A skilful breeder endeavors to shape the animal so as to carry most flesh on the valuable points, to have the *loin and hind quarters* much the heaviest, as these parts bring to the butcher the most money.

LEWIS SANDERS.

GRASS HILLS, KY., *December,* 1848.

Small Threshing Machines, 1850

From *Prairie Farmer* (Chicago, January 1850), X, p. 21.

Wheeler's machine for threshing, has been pretty extensively introduced into Illinois, Iowa, and Wisconsin during the past season; and we have taken some pains to learn how far it has succeeded. The result is, that no doubt remains of the superiority of the principle involved in its operation over any sweep machine known to us. We have it from several persons using them. that with one of these implements, two horses and four or five men have done as much work, day by day, as threshers with eight horses and ten men operating along side of them. One man informs us, that he threshed one hundred and sixty bushels in a day, of wheat yielding only eight bushels per acre. This we call large threshing. Another informs us that his wheat yielded from seven to nine bushels per acre, with straw enough to produce in common seasons twenty or twenty-five bushels. — With the same span of horses working every day, and his machine elevated sixteen inches, his ordinary threshing was eighty bushels per day. Others in the same circumstances, by hard driving, or changing horses, have threshed from one hundred to one hundred and fifty, and in one case, as we have said, one hundred and sixty. That the machine in years when there is a fair yield will fully come up to those run with eight horses we do not believe; but that it will pull hard after them we have little doubt, from the representations made to us. The great saving effected is apparent on inspection. In the starting of a sweep machine the eight horses are put under full way, and are obliged to do their utmost before any power at all is applied, or even before the band is attached to the cylinder. To what purpose is all this power put forth? To *turn the gearing* and nothing else. In a tread machine of the sort above named, the power is applied directly to the cylinder, saw, or whatever else is to be worked; none being lost in working ponderous gearing — and the friction being but little, nearly the whole force of the team is applied to the work to be done.

It is not to be denied that a tread power is somewhat straining upon the ancles of the horse; and there are sometimes animals which cannot work upon them. They are, however, few, and the comparatively low elevation at which this power runs will soon enable a horse to be accustomed to it, so as to work easily.

The only objection we have heard to Wheeler's machine is in relation to its cleaning. It is provided with a separator which divides the straw from the chaff, and for further cleaning uses a fanning mill, driven by the same power, at the same time. The objections to this are, that it requires an extra man or even two; and does not, after all, clean the grain fit for market. In this respect it does not come up to the promise of the advertisements, and of course fails to satisfy expectations. A fixture, beneath the separator perhaps, for blowing out the chaff, so that the grain might be stored and cleaned afterwards, would fully meet all wants. There are also some defects in manufacture; such as the wheel which drives the mill being too small; the part of the rod to which the band wheel is attached being round instead of square, which might easily be remedied.

For several years we have been urging the want on the part of the farmers, *of a small horse power,* which each one could own, keep and use, with his own help, without being obliged to call on all his neighborhood to assist him in getting out his grain. This implement meets that want, so far as we can judge.

But Wheeler's is not the only machine of the sort in use. We saw in Milwaukie county, the past season, a one horse power, which had driven out of its reach, all larger machines. It was patented, we believe, by one Smith, and was worked by one horse and three men; and would thresh in ordinary seasons, about one hundred bushels per day. — It was run with less ease than Wheeler's, requiring a sharper elevation, and necessitating a change of horses every half hour, or so. But the cost of threshing was not over one-half that of the ordinary mode.

Wheeler's horse power is so small and compact — so easily worked and so durable, that we have no doubt it will be in demand when known. The Prairie Farmer is now printed on a press run by one of the single horse powers, which, at an elevation of about two inches, give force enough to work a half dozen printing presses. As it is it drives it in a steady and satisfactory manner, and with any speed desired. This power is equally useful in sawing wood or for any thing else where a horse power is required.

On Fruit Culture, 1851

From A. J. Downing, *Rural Essays* (New York, 1854), pp. 435 –41.

By far the most important branch of horticulture at the present moment in this country, is the cultivation of Fruit. The soil and climate of the United States are, on the whole, as favorable to the production of hardy fruits as those of any other country — and our northern States, owing to the warmth of the summer and the clearness of the atmosphere, are far more prolific of fine fruits than the north of Europe. The American farmer south of the Mohawk, has the finest peaches for the trouble of planting and gathering — while in England they are luxuries only within the reach of men of fortune, and even in Paris, they can only be ripened upon walls. By late reports of the markets of London, Paris, and New-York, we find that the latter city is far more abundantly supplied with fruit than either of the former — though finer specimens of almost any fruit may be found *at very high prices,* at all times, in London and Paris, than in New-York. The fruit-grower abroad, depends upon extra size, beauty, and scarcity for his remuneration, and asks, sometimes, a guinea a dozen for peaches, while the orchardist of New-York will sell you a dozen *baskets* for the same money. The result is, that while you may more easily find superb fruit in London and Paris than in New-York — if you can afford to pay for it — you know that not one man in a hundred tastes peaches in a season, on the other side of the water, while during the month of September, they are the daily food of our whole population.

Within the last five years, the planting of orchards has, in the United States, been carried to an extent never known before. In the northern half of the Union, apple-trees, in orchards, have been planted by thousands and hundreds of thousands; in almost every State. The rapid communication established by means of railroads and steamboats in all parts of the country, has operated most favorably on all the lighter branches of agriculture, and so many farmers have found their orchards the most profitable, because least expensive part of their farms, that orcharding has become in some parts of the West, almost an absolute distinct species of husbandry. Dried apples are a large article of export from one part of the country to another, and the shipment of American apples of the finest quality to England, is now a regular and profitable branch of commerce. No apple that is sent from any part of the Continent will command more than half the price in Covent Garden market, that is readily paid for the Newtown pippin.

The pear succeeds admirably in many parts of the United States — but it also fails as a market fruit in many others — and, though large orchards have been planted in various parts of the country, we do not think the result, as yet, warrants the belief that the orchard culture of pears will be profitable generally. In certain deep soils — abounding with lime, potash, and phosphates, naturally, as in central New-York, the finest pears grow and bear like apples, and produce very large profits to their cultivators. Mr. Pardee's communication on this subject, in a former number, shows how largely the pear is grown as an orchard fruit in the State of New-York, and how profitable a branch of culture it has already become.

In the main, however, we believe the experience of the last five years has led most cultivators — particularly those not in a region naturally favorable in its soil — to look upon a pear as a tree rather to be confined to the fruit-garden than the orchard; as a tree not so hardy as the apple, but sufficiently hardy to give its finest fruit, provided the soil is deep, and the aspect one not too much exposed to violent changes of temperature. As the pear-tree (in its finer varieties) is more delicate in its bark than any other fruit-tree excepting the apricot, the best cultivators now agree as to the utility of sheathing the stem from the action of the sun all the year round — either by keeping the branches low and thick, so as to shade the trunk and principal limbs — the best mode — or by sheathing the stems with straw — thus preserving a uniform temperature. In all soils and climates naturally unfavorable to the pear, the culture of this tree is far easier upon the quince stock than upon the pear stock; and this, added to compactness and economy of space for small gardens, has trebled the demand for dwarf pears within the last half-dozen years. The finest pears that make their appearance in our markets, are still the White Doyenne (or Virgalieu), and the Bartlett. In Philadelphia the Seckel is abundant, but of late years the fruit is small and inferior, for want of the high culture and manuring which this pear demands.

If we except the neighborhood of Rochester and a part of central New-York (probably the future Belgium of America, as regards the production of pears), the best fruit of this kind yet produced in the United States is still to be found in the neighborhood of Boston. Neither climate nor soil are naturally favorable there, but the great pomological knowledge and skill of the amateur and professional cultivators of Massachusetts, have enabled them to make finer shows of pears, both as regards quality and variety, than have been seen in any part of the world. And this leads us to observe that the very facility with which fruit is cultivated in America — consisting for the most part only in planting the trees, and gathering the crop — leads us into an error as to the standard of size and flavor attainable generally. One half the number of trees well cultivated, manured, pruned, and properly cared for, annually, would give a larger product of really delicious and handsome fruit, than is now obtained from double the number of trees, and thrice the area of ground. The difficulty usually lies in the want of knowledge, and the high price of labor. But the horticultural societies in all parts of the country, are gradually raising the criterion of excellence among amateurs, and the double and treble prices paid lately by confectioners for finely-grown specimens, over the market value of ordinary fruit, are opening the eyes of market growers to the pecuniary advantages of high cultivation.

Perhaps the greatest advance in fruit-growing of the last half-dozen years, is in the culture of foreign grapes. So long as it was believed that our climate, which is warm enough to give us the finest melons in abundance, is also sufficient to produce the foreign grape in perfection, endless experiments were tried in the open garden. But as all these experiments were unsatisfactory or fruitless, not only at the North but at the South — it has finally come to be admitted that the difficulty lies in the variableness, rather than the want of heat, in the United States. This once conceded, our horticulturists have turned their attention to vineries for raising this delicious fruit under glass — and at the present time, so much have both private and market vineries increased, the finest Hamburgh, Chasselas, and Muscat grapes, may be had in abundance at moderate prices, in the markets of Boston, New-York, and Philadelphia. For a September crop of the finest foreign grapes, the heat of the sun accumulated in one of the so-called cold vineries (i.e. a vinery without artificial heat, and the regular temperature insured by the vinery itself) is amply sufficient. A cold vinery is constructed at so moderate a cost, that it is now fast becoming the appendage of every good garden, and some of our wealthiest amateurs, taking advantage of our bright and sunny climate, have grapes on their tables from April to Christmas — the earlier crops forced — the late ones slightly retarded in cold vineries. From all that we saw of the best private gardens in England, last summer, we are confident that we raise foreign grapes under glass in the United States, of higher flavor, and at far less trouble, than they are usually produced in England. Indeed, we have seen excellent Black Hamburghs grown in a large pit made by covering the vines trained on a high board fence, with the common sash of a large hot-bed.

On the Ohio, the native grapes — especially the Catawba — have risen to a kind of national importance. The numerous vineries which border that river, particularly about Cincinnati, have begun to yield abundant vintages of pure light wine, which takes rank with foreign wine of established reputation, and commands a high price in the market. Now that the Ohio is certain to give us Hock and Claret, what we hear of the grapes and wine of Texas and New Mexico, leads us to believe that the future vineyards of New World Sherry and Madeira may spring up in that quarter of our widely extended country.

New Jersey, so long famous for her prolific peach orchards, begins to show the effects of a careless system of culture. Every year, the natural elements of the soil needful to the production of the finest peaches, are becoming scarcer and scarcer, and nothing but deeper cultivation, and a closer attention to the inorganic necessities of vegetable growth, will enable the orchardists of that State long to hold their ground in the production of good fruit. At the present moment, the peaches of Cincinnati and Rochester are far superior, both in beauty and flavor, to those of the New-York market — though in quantity the latter beats the world. The consequence is, that we shall soon find the peaches of Lake Ontario outselling those of Long Island and New Jersey in the same market, unless the orchardists of the latter State abandon *Malagatunes* and the *yellows,* and shallow ploughing.

The fruit that most completely baffles general cultivation in the United States, is the plum. It is a tree that grows and blossoms well enough in all parts of the country, but almost every where it has for its companion the curculio, the most destructive and the least vulnerable of all enemies to fruit. In certain parts of the Hudson, of central New-York, and at the West, where the soil is a stiff, fat clay, the curculio finds such poor quarters in the soil, and the tree thrives so well, that the fruit is most delicious. But in light, sandy soils, its culture is only an aggravation to the gardener. In such sites, here and there only a tree escapes, which stands in some pavement or some walk for ever hard by the pressure of constant passing. No method has proved effectual but placing the trees in the midst of the pig and poultry yard; and notwithstanding the numerous remedies that have been proposed in our pages since the commencement of this work, this proves the only one that has not failed more frequently than it has succeeded.

The multiplication of insects seems more rapid, if possible, than that of gardens and orchards in this country. Every where the culture of fruit appears, at first sight, the easiest possible matter, and really would be, were it not for some insect pest that stands ready to devour and destroy. In countries where the labor of women and children is applied, at the rate of a few cents a day, to the extermination of insects, it is comparatively easy to keep the latter under control. But nobody can afford to catch the curculios and other beetles at the price of a dollar a day for labor. The entomologists ought, therefore, to explain to us some natural laws which have been violated to bring upon us such an insect scourge; or at least point out to us some cheap way of calling in nature to our aid, in getting rid of the vagrants. For our own part, we fully believe that it is to the gradual decrease of small birds — partly from the destruction of our forests, but mainly from the absence of laws against that vagabond race of unfledged sportsmen who shoot sparrows when they ought to be planting corn — that this inordinate increase of insects is to be attributed. Nature intended the small birds to be maintained by the destruction of insects, and if the former are wantonly destroyed, our crops, both of the field and gardens, must pay the penalty. If the boys must indulge their spirit of liberty by shooting *something* innocent, it would be better for us husbandmen and gardeners to subscribe and get some French masters of the arts of domestic sports, to teach them how to bring their light artillery to bear upon bull-frogs. It would be a gain to the whole agricultural community, of more national importance than the preservation of the larger birds by the game laws.

We may be expected to say a word or two here respecting the result of the last five years on pomology in the United States. The facts are so well known that it seems hardly necessary. There has never been a period on either side of the Atlantic, when so much attention has been paid to fruit and fruit culture. The rapid increase of nurseries, the enormous sales of fruit-trees, the publication and dissemination of work after work upon fruits and fruit culture, abundantly prove this assertion. The Pomological Congress which held its third session last year in Cincinnati, and which meets again this autumn in Philadelphia, has done much, and will do more towards generalizing our pomological knowledge for the country generally. During the last ten years, almost every fine fruit known in Europe has been introduced, and most of them have been proved in this country. The result, on the whole, has been below the expectation; a few very fine sorts admirably adapted to the country; a great number of indifferent quality; many absolutely worthless. This, naturally, makes pomologists and fruit-growers less anxious about the novelties of the nurseries abroad, and more desirous of originating first-rate varieties at home. The best lessons learned from the discussions in the Pomological Congress — where the experience of the most practical fruit-growers of the country is brought out — is, that for every State, or every distinct district of country, there must be found or produced its improved indigenous varieties of fruit — varieties born on the soil, inured to the climate, and therefore best adapted to that given locality. So that after gathering a few kernels of wheat out of bushels of chaff, American horticulturists feel, at the present moment, as if the best promise of future excellence, either in fruits or practical skill, lay in applying all our knowledge and power to the study of our own soil and climate, and in helping nature to perform the problem of successful cultivation, by hints drawn from the facts immediately around us.

Thoreau on Farming, 1851

From Bradford Torrey, ed., *The Writings of Henery David Thoreau, Journal, III, Sept. 16, 1851–April 30, 1852* (Cambridge, Mass., 1906), pp. 41–43.

Minott was telling me to-day [Oct. 4, 1851] that he used to know a man in Lincoln who had no floor to his barn, but waited till the ground froze, then swept it clean in his barn and threshed his grain on it. He also used to see men threshing their buckwheat in the field where it grew, having just taken off the surface down to a hard-pan.

Minott used the word ''gavel'' to describe a parcel of stalks cast on the ground to dry. His are good old English words, and I am always sure to find them in the dictionary, though I never heard them before in my life.

I was admiring his corn-stalks disposed about the barn to dry, over or astride the braces and the timbers, of such a fresh, clean, and handsome green, retaining their strength and nutritive properties so, unlike the gross and careless husbandry of speculating, money-making farmers, who suffer their stalks to remain out till they are dry and dingy and black as chips.

Minott is, perhaps, the most poetical farmer — who most realizes to me the poetry of the farmer's life — that I know. He does nothing with haste and drudgery, but as if he loved it. He makes the most of his labor, and takes infinite satisfaction in every part of it. He is not looking forward to the sale of his crops or any pecuniary profit, but he is paid by the constant satisfaction which his labor yields him. He has not too much land to trouble him, — too much work to do, — no hired man nor boy, — but simply to amuse himself and live. He cares not so much to raise a large crop as to do his work well. He knows every pin and nail in his barn. If another linter is to be floored, he lets no hired man rob him of that amusement, but he goes slowly to the woods and, at his leisure, selects a pitch pine tree, cuts it, and hauls it or gets it hauled to the mill; and so he knows the history of his barn floor.

Farming is an amusement which has lasted him longer than gunning or fishing. He is never in a hurry to get his garden planted and yet [it] is always planted soon enough, and none in the town is kept so beautifully clean.

He always prophesies a failure of the crops, and yet is satisfied with what he gets. His barn floor is fastened down with oak pins, and he prefers them to iron spikes, which he says will rust and give way. He handles and amuses himself with every ear of his corn crop as much as a child with its playthings, and so his small crop goes a great way. He might well cry if it were carried to market. The seed of weeds is no longer in his soil.

He loves to walk in a swamp in windy weather and hear the wind groan through the pines. He keeps a cat in his barn to catch the mice. He indulges in no luxury of food or dress or furniture, yet he is not penurious but merely simple. If his sister dies before him, he may have to go to the almshouse in his old age; yet he is not poor, for he does not want riches. He gets out of each manipulation in the farmers' operations a fund of entertainment which the speculating drudge hardly knows. With never-failing rheumatism and trembling hands, he seems yet to enjoy perennial health. Though he never reads a book, — since he has finished the ''Naval Monument,'' — he speaks the best of English.

Breaking Wisconsin Prairie, 1852

From J. Milton May, "Breaking Prairie," Wisconsin State Agricultural Society, *Transactions*, 1851 (Madison, Wis., 1852), I, pp. 243–46.

The work of Breaking Prairie is very justly considered of the first importance by the settler in a Prairie country. Indeed it is not uncommon to find a quarter or a half section of land, broken and sown with fall or winter wheat, which has attained a thrifty growth before it is enclosed.

Necessity, and a prudent forecast on the part of the pioneer, indicate the importance of growing a crop as early as possible, for with his grounds broken and sown, the long winter will afford ample time to procure his fencing materials, even though his timbered land is a half dozen of miles distant from the farm he is making; and the following season finds a field well enclosed, with the appearance of having been under cultivation a half century.

In the early settlements of the prairie country the obstacles in the way of rapidly and easily *breaking prairie* were somewhat numerous and formidable. Some of the principal ones may be mentioned in this connexion: — First, the tenacity and strength of the prairie sward, arising from the ten thousand wire-like fibrous roots, interlaced and interwoven in every conceivable manner. — Second, the *red root,* so called. This is a large bulbous mass of wood or root, gnarled and hard, very much resembling cherry timber in color and density. When in a live state, it sends up annually a twig or shoot similar to the willow, which is destroyed by the prairie fires, so that no tree or shrub is formed, while the root continues to grow, and attains a diameter of six, eight, and sometimes twelve inches. These roots are found usually in a given neighborhood, while other and large sections of country are entirely free from them. — Third, in locations where adjacent improvements prescribe limits to the annual burning of the prairie and in the neighborhood of the groves, hazel bushes spring up, forming a thicket which are called "hazel roughs" by those who break prairie.

Formerly, to overcome these obstacles and make any considerable progress in the work of "breaking," four or six yoke of oxen and two men were found necessary, but ingenuity and enterprize have wrought a great change in this important department of labor. Instead of the heavy, uncouth, and unmanageable wooden ploughs, with iron or steel points, formerly used, various kinds of improved breaking ploughs are brought into requisition, reducing the cost of breaking to one-third or one-half the former price.

Among these improved ploughs may be noticed a singular and unique looking one called the *skeleton plough,* which turns the furrow with a gradual motion, by means of small iron rods instead of a mould board, thus reducing the friction and making the draft very light. Sometimes wheels are attached and so arranged that the plough is self-holding, and the depth of the furrow regulated by a lever. This arrangement dispenses with the labor of one man.

In later years, small ploughs that can be used by one man and a span of horses are considered the most valuable. One description of these small breaking ploughs has for its foundation a thin plate of wrought iron or steel, combining lightness with strength. The coulter is composed of a single plate of steel about three feet long, pointed at both ends, and fastened by a single bolt near its centre, to the plough post, about six inches below the beam. When one end is dulled by use, the other is presented by reversing its position.

The share and coulter of breaking ploughs should be of the best steel, such as is used for edge tools, the angle of the share relatively with the landside should be about thirty degrees, and the angle of the coulter about twenty degrees.

If good materials is used, the above angles (substantially) observed, and the plough otherwise constructed with due regard to proportion, an implement will be produced that will cleave the sward easily, and dispose of the furrow with the utmost precision, and at a cost not exceeding that of breaking meadow or pasture sward land, in the eastern or middle States. It is not uncommon for a man, with one span of horses, to break one and one-half or two acres of prairie per day, during the season.

The width of furrow, the depth most desirable, and the best season of the year for breaking prairie, are questions in regard to which a variety of opinions have existed. Now, however, amongst practical men the narrow furrow is generally preferred, as it is more readily pulverized under the harrow, in preparing the ground for seeding, while the best depth, undoubtedly, is two and a half or three inches. A greater thickness of the furrow slice will tend to preclude the atmospheric and chemical influences necessary to the decomposition of the vegetable portion of the sward, and, as a consequence, the ground will be unfit for fall seeding.

If a less thickness than two and a half inches is made, there is danger that grass shoots and weeds will penetrate through the thin sod, and a healthy growth of vegetation promoted, instead of a subjugation of the land to agricultural purposes.

The months of May and June are the best for breaking prairie; and should the amount of work to be done be insufficient to require these entire months, the time intervening from the twentieth of May to the twentieth of June, without doubt is the most appropriate, although many persons commence earlier, and continue later, than the time here indicated.

Frequently a crop of corn is raised on the sod by "chopping in" the seed corn with a sharp hoe or axe, or by dropping the seed along the edge of the third or fourth furrow, and then covering it by the succeeding one, and often ten to twenty bushels per acre is raised in this manner.

An experiment was made by a practical farmer during the last season which presents a new phase in corn raising on newly broken prairie. An ordinary furrow was turned, and immediately after another plough running deeper cut a second furrow, from the same place, which was turned directly upon the first one. This process was continued, until the entire field was subsoiled in the same manner. On the upper furrows the corn was planted, and from the absence of the vegetating portion of the grass roots decomposition was more rapidly promoted, and this sod soon fell in pieces under the cultivator in dressing the corn. This experiment resulted in a very heavy crop.

Does not this successful innovation indicate advantages, (not only in growing corn, but in a greater broken depth of soil for subsequent cultivation), that are worthy the practical consideration of agriculturists?

Will not this experiment materially aid in advancing this important branch — prairie farming?

The disparity in time and expense of "making a farm," in a heavily timbered country, or on a well chosen prairie, is greater than would at first seem apparent. True it is, that most of us have listened with delight to the "loud sounding axe," as with "redoubling strokes on strokes" the forest denizens were laid low with a crash that was right musical, as the echo reverberated amongst the hills; — but consider then the burning and clearing off the timber, at a cost of from five to twenty dollars per acre, with the stumps remaining, as a memorial for hard labor, or a quarter of a century — contrasted with two to three dollars per acre for breaking prairie, which is as free from obstructions as though cultivated an hundred years, and which suffers by the comparison?

"But the item of fencing! Of enclosing these lands!" Based on careful estimates the conclusion is reached, that fencing a thousand acres, in the ordinary manner, is more expensive in a timbered than a prairie country; — besides the prairie country can furnish examples where the *first crop* has paid for the soil one dollar twenty-five cents per acre, besides breaking, fencing, seed, harvesting, and marketing the crop, with a surplus remaining of some forty cents per acre.

At the risk of being considered extravagant the opinion is hazarded, that had the advantages of Wisconsin, Illinois, and Iowa, been as fully understood thirty years ago as they now are, and the facilities for reaching these States been as great as at present, many portions of the eastern and middle States would have remained an unbroken forest, valued chiefly for its game and timber.

Settling In Oregon, 1853

From "Letter from Oregon," *Ohio Cultivator*, July 15, 1853, IX p. 214.

PACIFIC FARM, Oregon Territory, *May, 1853.*

EDS. OHIO CULTIVATOR:

— I have wandered off so far from home and acquaintance, that I sometimes feel quite lonely during my leisure hours. This evening, as I sat musing upon the scenes and events of other years, and trying to picture in my imagination how every thing looked at the old "Evergreen Farm," among the many things that presented themselves, was the *Ohio Cultivator,* which was a welcome visitor at our house for so many years, that it almost seems like one of the family. This broke up my reverie, and I got my pen immediately to dispatch an order for the *Cultivator.*

Last fall, after my arrival, I traveled over the greater part of the Territory, in search of a place to settle upon, as my future home. During my travels I saw the largest wheat, oats, potatoes, beets, onions, turnips, &c., the fattest cattle, horses and sheep, and the best beef that I ever saw in my life. I also listened with great satisfaction (but some doubt) to stories of producing 58@60 bushels of wheat per acre; and on the Columbia bottoms, harvesting 800 bushels of potatoes per acre, and selling them at $2,50 per bushel! The price was doubtless true, as wheat has been worth $5 per bushel since; and now potatoes command $3. And I am well satisfied, that in this climate and soil, by proper culture, such yields may be easily obtained.

After seeing and hearing all of these, and many more remarkable things, together with the scenery of the country, which is varied and beautiful, I became perfectly in love with Oregon. But about a month of constant rain and snow in December, rather cooled my flame. This was of but short duration, however, as January and February were warm and pleasant, with bright sunny days, and frosty nights, making the winter, as a whole, more mild, pleasant, and agreeable, than the winters of Southern Ohio.

The spring has been early, warm, and genial; and vegetation advances with surprising alacrity. Farmers have good reasons to expect abundant harvests.

I have selected a situation upon the bank of Lewis & Clark's river, 6 miles from Astoria, and 3 miles from the Ocean. This river was named in honor of Lewis & Clark, who encamped here, during the winter that they spent in this country. My garden is now the very spot that was then their camping ground. Some of the remains of their old houses may yet be seen. The river at this place is 120 yards wide, and 2 fathoms deep at low tide — the tide rises and falls 6 to 8 ft.

Three miles west of here, is the Clatsop beach, one of the most beautiful beaches on the Western coast. To this place we resort, for sea bathing, gathering clams, &c. My land is all heavily timbered, except the "tide land." These are low lands along the river, that are overflown a few times a year during the highest tides. They yield abundant crops of grass.

Hemlock, spruce, fir, white and yellow pine, and American arbor vitae, are our chief varieties of timber. The trees are tall, straight, and handsome. Many of them are 10 feet in diameter, and from 250 to 300 feet in height. This is as large as I dare speak of, through fear of having my veracity questioned: so I will not mention a few trees in this vicinity, that are from 15 to 20 feet in diameter!

Lumber is our chief article of commerce, and commands high prices in California.

A great portion of this territory is admirably adapted to grazing. I believe that I have seen as fat beef taken off of the prairie, as any stall-fed beef that I ever saw in Ohio. We have no fine, blooded stock here; but a spirit of improvement is manifesting itself among our farmers; and it will doubtless soon be introduced. Cows are worth from $80 to $125 per head; sheep $8@$10, each; and American mares, from $150 to $250 per head.

Agriculture is yet in the very cradle of infancy. Her beauty and strength are yet in embryo. But could a few hundred copies of the *Ohio Cultivator,* and kindred works, be distributed among her nurseries, methinks her puny arms would soon unfold to mighty length, and she wax strong and vigorous. We have the territory, climate, and soil, for a great Agricultural State; and all that is now wanting, is *means* to develop it.

YOURS, TRULY,
P. W. GILLETT.

Cattle Industry In 1853

From U.S. Patent Office, *Annual Report, Agriculture,* 1853, pp. 11–12, 23–24.

Statement of Thomas W. Sampson, of Ashland Farm, Rocheport, Boone county, Missouri.

The California trade has produced quite a revolution in the cattle trade in Missouri. They have advanced at least 200 per cent. in price, and decreased in numbers in about the same proportion. The consequence is, that the rearing of cattle is about the most profitable business our farmers can enter into. Of the common stock of the county, calves are worth at present from $5 to $8; yearlings, from $10 to $12; two-year-olds, from $15 to $18; three-year-olds, from $20 to $25; and four-year-olds, from $30 to $45 each.

Cattle are generally grazed on blue grass pasture through the spring and summer months, by such farmers as have it in sufficient quantities; others drive to the prairies in April, and have them herded on the prairie grass until about the middle or last of October. The prairie grass is very fine, and fattens equal to blue grass. The cost of herding is about 50 cents per head. Cattle are usually wintered on what we call "stock fodder," which is the corn cut up whilst it is yet green, and shocked up sixteen hills square, or two hundred and sixty-five hills of corn, and on an average about six hundred and forty stalks of fodder to the hill. The corn is shucked and the husks left on the fodder stalk. One shock a day will keep about eight head of young cattle through the winter, and is worth on an average about eight cents a shock. Counting from the first of November to the first of April, one hundred and fifty days, we have a cost of $1.50 a head for the first year, 50 cents a head for herding in summer, and 20 cents a head for incidental expenses — salt, driving out, &c. We have a cost of $2.20 for the first year; $2.50 for the second year, and $3 for the last or third year.

Our stock of cattle has been very much improved within a few years past, by the introduction of pure-blooded short-horns from England, New York, Ohio, and Kentucky; much the largest number of which have been brought from the latter State. The cattle all over the county are getting more or less mixed, and sell proportionally higher in accordance with the amount of Durham blood. We have some very fine imported Durham bulls; no other improved breeds of cattle have been introduced in this section. Milch cows of the common stock and one-fourth Durham are worth from $20 to $40, while those that are from one-half to three-fourths blooded are from $40 to $100. Full-blooded Durhams are worth from $100 to $500, according to quality, size, and appearance, pedigree, &c. I am breeding my cows, about twenty in number, this season, to three different bulls, all pure-blooded Durhams brought out from Kentucky.

Statement of Micajah Burnett, of the United Society of Shakers, Pleasant Hill, Mercer county, Kentucky

The dairy business, though much favored by soil and climate, has not been put forward with a view to export its products. Every farmer liberally supplies his own family with milk and butter, and has enough to spare to supply the home demand. The business, however, is receiving increased attention, and may, with proper care and management, be made a source of great profit when we have better facilities for getting to market. The Durham cows, which have been gradually increasing in number since the year 1817, are, by those most familiar or conversant with them, acknowledged to be better adapted to the dairy than those of any other breed ever introduced into this section.

The dairy cows of this society, about one hundred and fifty in number, being all Durhams — some thorough-bred, others with a slight strain of the Patton breed — are unsurpassed in the quantity and richness of their milk. They will give for the first three or four months after calving, while on grass alone, from 30 to 68 pounds of milk daily; and from experiments made in the early part of October, 18 pounds of milk will make a pound of butter. Some of these cows do not go dry during the year; others are dry a longer or shorter time, but rarely exceeding eight weeks.

We never soil our milch cows, but believe it much better to have them run at large in the pastures day and night, except in the winter season, when they are kept in the stable, tied each in her own proper stall, from five o'clock in the evening until eight in the morning. They are driven up, however, and stabled morning and evening during the balance of the year while they are milked, and then turned out to range. From early in the spring until late in the fall they subsist on blue-grass and clover pastures, and are only fed during the remainder of the year, morning and evening, with sheaf-oats, cut fine, made wet, and mixed with meal or "ship-stuffs"; also, occasionally with slops from the kitchen, pumpkins, small potatoes cooked for the purpose, and hay at will during the night. In this way they are kept fat or in fine order from year to year.

Good dairy cows, mixed from one-half to seven-eighths Durham, are worth from $40 to $50 each; common stock, from $15 to $25. Butter is worth from 12½ to 15 cents per pound; cheese 10 cents. The price of milk, from its abundance, scarcely nominal. The Kendall churn is reputed the best of any proved by the society.

Frederick Law Olmsted Visits the Seaboard Slave States, 1853–54

From Frederick Law Olmstead, *A Journey in the Seaboard Slave States, 1853–1854*, I, pp. 6–13, 43–47, 310–12; II, pp. 94–110.

WASHINGTON, Dec. 14th [1853]. Called on Mr. C., whose fine farm, from its vicinity to Washington, and its excellent management, as well as from the hospitable habits of its owner, has a national reputation. It is some two thousand acres in extent, and situated just without the District, in Maryland.

The residence is in the midst of the farm, a quarter of a mile from the high road — the private approach being judiciously carried through large pastures which are divided only by slight, but close and well-secured, wire fences. The mansion is of brick, and, as seen through the surrounding trees, has somewhat the look of an old French chateau. The kept grounds are very limited, and in simple but quiet taste; being surrounded only by wires, they merge, in effect, into the pastures. There is a fountain, an ornamental dove-cote, and ice-house, and the approach road, nicely gravelled and rolled, comes up to the door with a fine sweep.

I had dismounted and was standing before the door, when I heard myself loudly hailed from a distance.

"Ef yer wants to see Master, sah, he's down thar — to the new stable."

I could see no one; and when I was tired of holding my horse, I mounted, and rode on in search of the new stable. I found it without difficulty; and in it Mr. and Mrs. C. With them were a number of servants, one of whom now took my horse with alacrity. I was taken at once to look at a very fine herd of cows, and afterwards led upon a tramp over the farm, and did not get back to the house till dinner time.

The new stable is most admirably contrived for convenience, labor-saving, and economy of space. (Full and accurate descriptions of it, with illustrations, have been given in several agricultural journals.) The cows are mainly thorough-bred Shorthorns, with a few imported Ayrshires and Alderneys, and some small black "natives." I have seldom seen a better lot of milkers; they are kept in good condition, are brisk and healthy, docile and kind, soft and pliant of skin, and give milk up to the very eve of calving; milking being never interrupted for a day. Near the time of calving the milk is given to the calves and pigs. The object is to obtain milk only, which is never converted into butter or cheese, but sent immediately to town, and for this the Shorthorns are found to be the most profitable breed. Mr. C. believes that, for butter, the little Alderneys, from the peculiar richness of their milk, would be the most valuable. He is, probably, mistaken, though I remember that in Ireland the little black Kerry cow was found fully equal to the Ayrshire for butter, though giving much less milk.

There are extensive bottom lands on the farm, subject to be flooded in freshets, on which the cows are mainly pastured in summer. Indian corn is largely sown for fodder, and, during the driest season, the cows are regularly *soiled* with it. These bottom lands were entirely covered with heavy wood, until, a few years since, Mr. C. erected a steam saw-mill, and has lately been rapidly clearing them, and floating off the sawed timber to market by means of a small stream that runs through the farm.

The low land is much of it drained, underdrains being made of rough boards of any desired width nailed together, so that a section is represented by the inverted letter A. Such covered drains have lasted here twenty years without failing yet, but have only been tried where the flow of water was constant throughout the year.

The water collected by the drains can be, much of it, drawn into a reservoir, from which it is forced by a pump, driven by horse-power, to the market-garden, where it is distributed from several fountain-heads, by means of hose, and is found of great value, especially for celery. The celery trenches are arranged in concentric circles, the water-head being in the centre. The water-closets and all the drainage of the house are turned to good account in the same way. Mr. C. contemplates extending his water-pipes to some of his meadow lands. Wheat and hay are the chief crops sold off the farm, and the amount of them produced is yearly increasing.

The two most interesting points of husbandry, to me, were the 'large and profitable use of guano and bones, and the great extent of turnip culture. Crops of one thousand and twelve hundred bushels of ruta baga to the acre have been frequent, and this year the whole crop of the farm is reckoned to be over thirty thousand bushels; all to be fed out to the neat stock between this time and the next pasture season. The soil is generally a red, stiff loam, with an occasional stratum of coarse gravel, and, therefore, not the most favorable for turnip culture. The seed is always imported, Mr. C.'s experience, in this respect, agreeing with my own: — the ruta baga undoubtedly degenerates in our climate. Bones, guano, and ashes are used in connection with yard-dung for manure. The seed is sown from the middle to the last of July in drills, but not in ridges, in the English way. In both these respects, also, Mr. C. confirms the conclusions I have arrived at in the climate of New York; namely, that ridges are best dispensed with, and that it is better to sow in the latter part of July than in June, as has been generally recommended in our books and periodicals. Last year, turnips sown on the 20th July were larger and finer than others, sown on the same ground, on my farm, about the first of the month. This year I sowed in August, and, by forcing with superphosphate — home manufactured — and guano, obtained a fine crop; but the season was unusually favorable.

Mr. C. always secures a supply of turnips that will allow him to give at least one bushel a day to every cow while in winter quarters. The turnips are sliced, slightly salted, and commonly mixed with fodder and meal. Mr. C. finds that salting the sliced turnip, twelve hours before it is fed, effectually prevents its communicating any taste to the milk. This, so far as I know, is an original discovery of his, and is one of great value to dairymen. In certain English dairies the same result is obtained, where the cows are fed on cabbages, by the expensive process of heating the milk to a certain temperature and then adding saltpetre.

The wheat crop of this district has been immensely increased, by the use of guano, during the last four years. On this farm it has been largely used for five years; and land that had not been cultivated for forty years, and which bore only broom-sedge — a thin, worthless grass — by the application of two hundred weight of Peruvian guano, now yields thirty bushels of wheat to an acre.

Mr. C.'s practice of applying guano differs, in some particulars, from that commonly adopted here. After a deep ploughing of land intended for wheat, he sows the seed and guano at the same time, and harrows both in. The common custom here is to plough in the guano, six or seven inches deep, in preparing the ground for wheat. I believe Mr. C.'s plan is the best. I have myself used guano on a variety of soils for several years with great success for wheat, and I may mention the practice I have adopted from the outset, and with which I am well satisfied. It strikes between the two systems I have mentioned, and I think is philosophically right. After preparing the ground with plough and harrow, I sow wheat and guano together, and plough them in with a gang-plough which covers to a depth, on an average, of three inches.

Clover seed is sowed in the spring following the wheat-sowing, and the year after the wheat is taken off, this — on the old sterile hills — grows luxuriantly, knee-high. It is left alone for two years, neither mown nor pastured; there it grows and there it lies, keeping the ground moist and shady, and improving it on the Gurney principle.

Mr. C. then manures with dung, bones, and guano, and with another crop of wheat lays this land down to grass. What the ultimate effect of this system will be, it is yet too early to say — but Mr. C. is pursuing it with great confidence.

Mr. C. is a large hereditary owner of slaves, which for ordinary field and stable-work, constitute his laboring force. He has employed several Irishmen for ditching, and for this work, and this alone, he thought he could use them to better advantage than negroes. He would not think of using Irishmen for common farm-labor, and made light of their coming in competition with slaves. Negroes at hoeing and any steady field-work, he assured me, would "do two to their one;" but his main objection to employing Irishmen was derived from his experience of their unfaithfulness — they were dishonest, would not obey explicit directions about their work, and required more personal supervision than negroes. From what he had heard and seen of Germans, he supposed they did better than Irish. He mentioned that there were several Germans who had come here as laboring men, and worked for wages several years, who had now got possession of small farms, and were reputed to be getting rich. He was disinclined to converse on the topic of slavery, and I, therefore, made no inquiries about the condition and habits of his negroes, or his management of them. They seemed to live in small and rude log-cabins, scattered in *different* parts of the farm. Those I saw at work appeared to me to move very slowly and awkwardly, as did also those engaged in the stable. These, also, were very stupid and dilatory in executing any orders given to them, so that Mr. C. would frequently take the duty off their hands into his own, rather than wait for them, or make them correct their blunders: they were much, in these respects, like what our farmers call *dumb Paddies* — that is, Irishmen who do not readily understand the English language, and who are still weak and stiff from the effects of the emigrating voyage. At the entrance-gate was a porter's lodge, and, as I approached, I saw a black face peeping at me from it, but, both when I entered and left, I was obliged to dismount and open the gate myself.

Altogether, it struck me — slaves coming here as they naturally did in direct comparison with free laborers, as commonly employed on my own and my neighbors' farms, in exactly similar duties — that they must be very difficult to direct efficiently, and that it must be very irksome and trying to one's patience, to have to superintend their labor.

* * *

This morning I visited a farm, some account of which will give a good idea of the more advanced mode of agriculture in Eastern Virginia. It is situated on the bank of James River, and has ready access, by water or land-carriage, to the town of Richmond.

The soil of the greater part is a red, plastic, clayey loam, of a medium or low fertility, with a large intermixture of small quartz pebbles. On the river bank is a tract of low alluvial land, varying from an eighth to a quarter of a mile in breadth. The soil of this is a sandy loam, of the very finest quality in every respect, and it has been discovered, in some places, to be over ten feet in thickness; at which depth the sound trunk of a white oak has been found, showing it to be a recent deposit. I was assured that good crops of corn, wheat, and clover, had been taken from it, without its giving any indications of "wearing out," although no manure, except an occasional dressing of lime, had ever been returned to it. Maize, wheat, and clover for two years, usually occupy the ground, in succession, both on upland and lowland, herd's-grass (red-top of New York), sometimes taking the place of the clover, or being grown with it for hay, in which case the ground remains in sward for several years. Oats are sometimes also introduced, but the yield is said to be very small.

Hay always brings a high price in Richmond, and is usually shipped to that market from the eastward. This year, however, it is but a trifle above New York prices, and the main supply is drawn from this vicinity. I notice that oats, in the straw, are brought, in considerable quantity, to Richmond, for horse-feed, from the surrounding country. It is often pressed in bales, like hay, and sells for about the same price. At present, hay, brought from New York in bales, is selling at $1.25 to $1.50 per cwt.; oats, in straw, the same; oats, by the bushel, 40 to 50 cents; maize, 66 to 70 cents; wheat straw, 75 cents per cwt.; maize leaves ("corn fodder"), 75 cents per cwt.

Wheat, notwithstanding these high prices of forage crops, is considered the most important crop of the farm. The practice is to cut the maize (which is grown on much the same plan as is usual in New York) at the root, stook it in rows upon the field, plough the lands between the rows (one way) and drill in wheat with a horse drilling machine: then remove the stooks of maize into the sown ground, and prepare the intervening lands in like manner. The maize is afterwards husked in the field, at leisure, and carted off, with the stalks, when the ground is frozen. Sometimes the seed-wheat is sown by hand on the fresh-ploughed ground, and harrowed in. In the spring, clover-seed is sown by hand. The wheat is reaped by either Hussey's or M'Cormick's machine, both being used on the farm, but Hussey's rather preferred, as less liable to get out of order, and, if slightly damaged, more readily repaired by the slave blacksmith on the farm.

Lime is frequently applied, commonly at the time of wheat-sowing, at the rate of from twenty-five to fifty bushels an acre. It is brought, by sea, from Haverstraw, New York, at a cost, delivered on the farm, of 7¼ to 7½ cents a bushel. Plaster (sulphate of lime) has been tried, with little or no perceptible effect on the crops.

Dung, largely accumulated from the farm stock, is applied almost exclusively to the maize crops. Guano is also largely used as an application for wheat. After trying greater and less quantities, the proprietor has arrived at the conclusion that 200 lbs. to the acre is most profitable. It will, hereafter, be applied, at that rate, to all the wheat grown upon the farm. It has also been used with advantage for ruta baga. For corn, it was not thought of much value; the greatest advantage had been obtained by applying it to the *poorest land of the farm, some of which was of so small fertility, and at such a distance from the cattle quarters and the river, that it could not be profitably cultivated, and had been at waste for many years. I understand this may be the case with half the land included in the large farms or plantations of this part of the country.* Two hundred weight of Peruvian guano to the acre brought fifteen bushels of wheat; and a good crop of clover was perfectly sure to follow, by which the permanent improvement of the soil could be secured. This the proprietor esteemed to be the greatest benefit he derived from guano, and he is pursuing a regular plan for bringing all his more sterile upland into the system of Convertible husbandry by its aid.

This plan is, to prepare the ground, by fallowing, for wheat; spread two hundred pounds of guano, broadcast, on the harrowed surface, and turn it under, as soon as possible after the sowers, with a "two-shovel plough" (a sort of large two-shared cultivator, which could only be used, I should think, on very light, clean soils), the wheat either being sown and covered with the guano, or, immediately afterwards, drilled in with a horse-machine. In the spring, clover is sown. After the wheat is harvested, the clover is allowed to grow, without being pastured or mown, for twelve months. The ground is then limed, clover ploughed in, and, in October, again guanoed, two hundred weight to the acre, and wheat sown, with clover to follow. The clover may be pastured the following year, but in the year succeeding that, it is allowed to grow unchecked until August, when it is ploughed in, the ground again guanoed, and wheat sown with herd's grass (red-top) and clover, which is to remain, for mowing and pasture, as long as the ground will profitably sustain it.

The labor of this farm was entirely performed by slaves. I did not inquire their number, but I judged there were from twenty to forty. Their "quarters" lined the approach-road to the mansion, and were well-made and comfortable log-cabins, about thirty feet long by twenty wide, and eight feet wall, with a high loft and shingle roof. Each, divided in the middle, and having a brick chimney outside the wall at each end, was intended to be occupied by two families. There were square windows, closed by wooden ports, having a single pane of glass in the centre. The house-servants were neatly dressed, but the field-hands wore very coarse and ragged garments.

* * *

Among American patriots of this period of our history, should always be classed John Taylor, of Caroline county, Virginia, the author of "Arator," and John S. Skinner, who, in 1819, commenced at Baltimore, in Maryland, the publication of the first special agricultural journal in America. Other men, many of whose names are enrolled among those of our national statesmen, were then united with them, in strenuous and concerted exertion, to give a better direction to the labor and agricultural capital of those States.

The convalescence of Virginia agriculture, however, if convalescent it may be considered ever to have been, should more especially be dated from the introduction of lime, as an application, in connection with better tillage, judicious rotations, and more frequent applications of dung and green crops, for the improvement of the land. And for this, Virginia is chiefly indebted to the study, experiments, preaching, and publications of Edmund Ruffin. Mr. Ruffin was, for many years, the editor of the *Virginia Farmers' Register,* but is best known as the author of "A Treatise on Calcareous Manures," than which no work on a similar subject has ever been published in Europe or America based on more scientifically careful investigation, and trusty, personal experience, or of equal practical value to those for whose benefit it was designed.

But, contemporaneously with the invigoration of the planning class, the depression of the tobacco market, and the introduction of these improvements in agriculture which promised so much for the future of the State, there entered a still more potent element into the direction of her destiny. This was occasioned by the increasing profit and extending culture of cotton in the more Southern States, which gave rise to a demand for additional labor, increased the value of slaves, and, the African Slave Trade having been declared piracy, led to a great extension of the internal Slave Trade.

The value of the cotton exported from the United States was:

In 1794,	$ 500,000
1800,	5,000,000
1810,	15,000,000
1820,	22,000,000
1830,	30,000,000
1840,	64,000,000
1850,	72,000,000

Closely corresponding to the increase in the exportation of cotton, was the growth of the demand for labor; and as, in any slave-holding community, experience shows no other labor can be extensively made use of but that of slaves, the value of slaves *for sale* has steadily advanced in Virginia, with the extension of cotton fields over the lands conquered or purchased for that purpose of the Indians in Alabama and Florida; of France, in the valley of the Mississippi; and of Mexico, in Texas.

Rice and its Culture

Although nineteen-twentieths of all the rice raised in the United States is grown within a district of narrow limits, on the sea-coast of the Carolinas and Georgia, the crop forms a not unimportant item among the total productions of the country. The crop of 1849 was supposed to be more than two hundred and fifteen million pounds, and the amount exported was equal, in value, to one-third of all the wheat and flour, and to one-sixth of all the vegetable food, of every kind, sent abroad. The exportation of 1851 was exceeded in value, according to the Patent Office Report, only by that of cotton, flour, and tobacco.

Rice is raised in limited quantity in all of the Southern States, and probably might be in some at the North. . . .

In Louisiana and the Mississippi valley, where the rice culture is, at present, very limited, there are millions of acres of now unproductive wilderness, admirably adapted to its requirements, and here, "it is a well known fact," says a writer in De Bow's *Review*, "that *the rice plantations, both as regards whites and blacks, are more healthy than the sugar and cotton.*" The only restriction, therefore, upon the production of rice to a thousand-fold greater extent than at present, is the cost of labor in the Southern States. . . .

Rice continues to be cultivated extensively on the coast of Georgia and the Carolinas, notwithstanding the high price of labor which Slavery and the demand for cotton has occasioned, only because there are unusual facilities there for forming plantations in which, while the soil is exceedingly rich and easily tilled, and the climate favorable, the ground may be covered at will with water, until nearly all other plants are killed, so as to save much of the labor which would otherwise be necessary in the cultivation of the crop; and which may as readily be drained, when the requirements of the rice itself make it desirable.

Some of the economical advantages thus obtained, might certainly be made available, under other circumstances, for other crops. Luxuriant crops of grain and leguminous plants are sometimes grown upon the rice fields, and I have little doubt that there are many swamps, bordering upon our Northern rivers, which might be converted into irrigated fields with great profit. On this account, I shall describe the rice plantation somewhat elaborately.

A large part of all the country next the coast, fifty miles or more in width, in North and South Carolina and Georgia, is occupied by flat cypress swamps and reedy marshes. That which is not so is sandy, sterile, and overgrown with pines, and only of any value for agriculture where, at depressions of the surface, vegetable mould has been collected by the flow of rain water. The nearer we approach the sea, the more does water predominate, till at length land appears only in islands or capes; this is the so-called Sea Island region. Below all, however, there stretches along the whole coast a low and narrow sand bar — a kind of defensive outwork of the land, seldom inhabited except by lost Indians and runaway negroes, who subsist by hunting and fishing. There are, upon it, several government relief stations and light-houses, far less frequent, alas! than skeleton hulks of old ships, which, half buried — like victims of war — in the sand, give sad evidence of the fury of the sea, and of the firmness with which its onsets are received.

At distant intervals there are shallow breaches, through which the quiet tide twice a day steals in, swelling the neutral lagoons, and damming the outlet of the fresh water streams, till their current is destroyed or turned back, and their flood dispersed far and wide over the debatable land of the cypress swamps.

Then when heavy rains in the interior have swollen the rivers, their eddying currents deposit, all along the edges of the sandy islands and capes of the swamp, the rich freight they have brought from the calcareous or granitic mountains in which they rise, with the organic waste of the great forests through which they flow. With all is mingled the silicious wash of the nearest shore and the rich silt of the salt lagoons, aroused from their bottoms in extraordinary assaults of the ocean.

This is the soil of the rice plantations, which are always formed in such parts of the tidal swamps, adjoining the mainland or the sandy islands, as are left nearly dry at the ebb of the water. The surface must be level, or with only slight inclinations towards the natural drains in which the retiring tide withdraws; and it must be at such a distance from the sea, that there is no taste of salt in the water by which it is flooded, at the rise of tide.

In such a situation, the rice fields are first constructed as follows: Their outline being determined upon, the trees are cut upon it for a space of fifty feet in width; a ditch is then dug at the ebb of the water, the earth thrown out from which soon suffices to prevent the return of ordinary tides, and the laborers are thus permitted to work uninterruptedly. An embankment is then formed, upon the site of the first made ditch, sufficiently thick and high to resist the heaviest floods which can be anticipated. It is usually five feet in height, and fifteen in breadth at the base, and all stumps and roots are removed from the earth of which it is formed, as, in digging the first ditch, they have been from its base. The earth for it is obtained by digging a great ditch fifteen or twenty feet inside of it; and if more is afterwards needed, it is brought from a distance, rather than that the security of the embankment shall be lessened by loosening the ground near its base.

While this embanking has been going on, the trees may have been felled over all the ground within, and, with the underbrush, drawn into piles or rows. At a dry time in the spring, fire is set to the windward side of these, and they are more or less successfully consumed. Often the logs remain, as do always the stumps, encumbering the rice field for many years. Usually, too, the larger trees are only girdled, and their charred or rotting trunks stand for years, rueful corpses of the old forests.

The cleared land is next divided into fields of convenient size, by embankments similar to, but not as large as, the main river embankment, the object of them being only to keep the water that is to be let into one field out of the next, which may not be prepared for it; commonly they are seven or eight feet wide at base and three feet high, with ditches of proportionate size adjoining them; a margin of eight or ten feet being left between the ditches and the embankments. Each field must be provided with a separate trunk and gate, to let in or exclude the water of the river; and if it is a back field, a canal, embanked on either side, is sometimes necessarily made for this purpose. Such a canal is generally made wide enough to admit of the passage of a scow for the transportation of the crop.

These operations being concluded, the cultivation of the land is commenced; but, owing to the withdrawal of shade, the decay of roots and recent vegetable deposit, and the drainage of the water with which the earth has hitherto been saturated, there continues for several years to be a gradual subsidence of the surface, making it necessary to provide more ditches to remove the water, after a flooding of the field, with sufficient rapidity and completeness. These ditches, which are, perhaps, but two feet wide and deep, are dug between the crops, from time to time, until all the fields are divided into rectangular beds of a half or a quarter acre each. Now, when the gates are open, at the fall of tide, any water that is on the beds flows rapidly into these minor drains (or "quarter ditches"), from these into the outside ditches of each field, and from these through the field trunks into the canal, or the main embankment ditch, and from this through the main trunk into the river. The gates in the trunk are made with valves, that are closed by the rise of water in the river, so as not to again admit it. Another set of gates, provided with valves opening the other way, are shut down, and the former are drawn up, when it is wished to admit the water, and to prevent its outflow.

The fields can each be flooded to any height, and the water retained upon them to any length of time desired. The only exceptions to this sometimes occur on those plantations nearest the sea, and those farthest removed from it. On the lower plantations, the tide does not always fall low enough, for a few days at a time, to draw off the water completely; and on the upper ones, it may not always rise high enough to sufficiently flood the fields. The planter must then wait for spring-tides, or for a wind from seaward, that shall "set up" the water in the river.

In times of freshet of the river, too, it will be impossible to drain a greater or less number of the plantations upon it. These circumstances occurring at critical periods of the growth of the rice plant, always have a great effect upon the crop, and are referred to in factors' and brokers' reports, and are often noticed in the commercial newspapers.

There is another circumstance, however, connected with the character of the season for rain, that still more essentially concerns the interests of the rice planters, especially those nearest the ocean. In a very dry season, the rivers being low, the ocean water, impregnated with salt, is carried farther up than usual. Salt is poisonous to the rice plant; while, on the other hand, unless it is flooded from the river, no crop can be made. The longer the drought continues, the greater this difficulty becomes, and the higher up it extends.

An expanse of old rice ground, a nearly perfect plain surface, with its waving, clean, bright verdure, stretching unbroken, except by the straight and parallel lines of ditch and wall, to the horizon's edge before you, bounded on one side by the silver thread of the river, on the other by the dark curtain of the pine forest, is said to be a very beautiful sight. But the new plantation, as I saw it in February, the ground covered thickly with small stumps, and strewn with brands and cinders, and half-burnt logs, with here and there an old trunk still standing, seared and burned, and denuded of foliage, with a company of clumsy and uncouth black women, armed with axes, shovels and hoes, and directed by a stalwart black man, armed with a whip, all slopping about in the black, unctuous mire at the bottom of the ditches, is a very dreary scene.

In preparing the ground for the crop, it is first thoroughly "chopped," as the operation with the thick, clumsy, heavy hoe is appropriately termed. This rudely turns, mixes, and levels the surface, two or three inches in depth. It is repeated as near as possible to the planting time, the soil being made as fine and friable, by crushing the clods, as possible — whence this second hoeing is termed the "mash." From the middle of March to the first of April planting commences, the first operation in which is opening drills, or, as it is termed on the plantation, "trenching." This is done with narrow hoes, the drills or trenches being chopped out about four inches wide, two inches deep, and thirteen inches apart. To guide the trenchers, a few drills are first opened by expert hands, four feet four inches apart, stakes being set to direct them; the common hands then open two between each of these guide rows, measuring the distance only by the eye. The accuracy with which the lines are made straight is said to be astonishing; and this, as well as the ploughing, and many other operations performed by negroes, as I have had occasion to notice with colored laborers at the North, no less than among the slaves, indicates that the race generally has a good "mathematical eye," much more so at least than the Irish.

As fast as the trenches are made, light hands follow, strewing the seed in them. It is sowed very thickly through the breadth of the trenches, so that from two to three bushels of rice are used upon an acre. The seed is lightly covered with hoes as rapidly as possible after it is sowed.

The force employed must always be large enough to complete the sowing of each field on the day it is begun. The outer gate in the trunk is opened as soon as the sowing is finished; and on the next rise of tide the water flows in, fills the ditches, and gradually rises until the whole ground is covered.

This is termed the "sprout flow," and the water is left on the field until the seed sprouts — from a week to a fortnight, according to the warmth of the season. It is then drawn off, and the field is left until the points of the shoots of the young plants appear above ground, when the second flooding is given it, called the "point flow." At this time, the water remains on till all the grass and weeds that have come up with the rice are killed, and until the rice itself is three or four inches in height, and so strong that the birds cannot pull it up. As soon as the ground is sufficiently dry, after the "point flow," the rice is hoed, and a fortnight or three weeks later it is hoed again, remaining dry in the meantime. As soon, after the second hoeing, as the weeds are killed by the sun (or, if rainy weather, immediately, so as to float them off), the field is again flooded, the water being allowed to rise at first well above all the plants, that the weeds and rubbish which will float may drift to the sides of the field, where they are raked out, dried, and burned: the water is then lowered, so that the points of the rice may be seen above it. The rice will be from six inches to one foot in height at this time, and the water remains on at the same height for two or three weeks. The exact time for drawing it off is determined by the appearance of the rice, and is a point requiring an experienced and discreet judgment to decide. This is called the "long flow."

The field is again left to dry, after which it receives a third and a fourth hoeing, and, when it is judged to need it, the water is again let on to a depth that will not quite cover the rice, and now remains on till harvest.

The negroes are employed, until the rice is headed, in wading through it, and collecting and bringing out in baskets any aquatic grasses or volunteer rice that have grown in the trenches. "Volunteer rice" is such as is produced by seed that has remained on the ground during the winter, and is of such inferior quality that, if it is left to be threshed with the crop, it injures its salable value much more than the addition it makes to its quantity is worth.

When the rice has headed, the water is raised still higher, for the purpose of supporting the heavy crop, and to prevent the straw from being tangled or "laid" by the wind, until it is ripe for the sickle.

The system of culture and irrigation which I have described is that most extensively practised; but there are several modifications of it, used to a greater or less extent. One of these is called "planting in the open trench;" in which the seed is prepared by washing it with muddy water, and drying it, so that a slight coating of clay remains upon it, which, after it is sown, is sufficient to prevent its rising out of the trench when the field is flooded. This save the labor of covering it, and, the water being let on at once after the sowing, it is protected from birds. The water remains until the plant has attained a certain size and color (commonly from two to three weeks), when it is withdrawn, and the subsequent culture is the same as I have described, after the second or "point" flow, in the first plan. The "long flow" and the "lay-by flow" are sometimes united, the water being gradually raised, as the plant increases in height, and only drawn off temporarily and partially, to supply its place with fresh, to prevent stagnation, or to admit the negroes to go over the field to collect weeds, etc. When this follows the open trench planting, the rice is flooded during all but perhaps two weeks of its growth, and receives but two instead of four hoeings. Some keep the water on as much as possible, only drawing off for barely the time required for the negroes to hoe it, when necessary to free the crop from weeds. Good planters use these and other modifications of the usual plan, according to the season, each having occasional advantages.

It will be obvious that in each method, the irrigation, by protecting the seed and plants, destroying weeds and vermin, and mechanically sustaining the crop, allows a great deal of labor to be dispensed with, which, with an unirrigated crop, would be desirable. This economy of labor is probably of greater consequence than the excessive moisture afforded the plant. Crops of rice have been grown on ordinarily dry upland, in the interior of the State, quite as large as the average of those of the tidal-swamps, but, of course, with an immensely greater expense in tillage.

I should remark, also, that as moisture can be commanded at pleasure, it is of much less consequence to be particular as to the time of seeding, than it would otherwise be. One field is sowed after another, during a period of two months. The flowings, tillage, and harvest of one may follow that of another, in almost equally prolonged succession. A large plantation of rice may therefore be taken proper care of with a much smaller force of hands than would otherwise be necessary. Many of these advantages, the Northern farmer should not neglect to consider, would be possessed by grass meadows, similarly subject to irrigation.

The rice harvest commences early in September. The water having been all drawn off the field at the previous ebb tide, the negroes reap the rice with sickles, taking three or four rows of it at a cut. The stubble is left about a foot in height, and the rice is laid across the top of it, so that it will dry rapidly. One or two days afterwards it is tied in small sheaves, and then immediately carried to the barn or stack yard. This is often some miles distant; yet the whole crop of many plantations is transported to it on the heads of the laborers. This work, at the hottest season of the year, in the midst of the recently exposed mire of the rice fields, is acknowledged to be exceedingly severe, and must be very hazardous to the health, even of negroes. Overseers, who consider themselves acclimated, and who, perhaps, only spend the day on the plantation, often at this time contract intermittent fever, which, though not in itself immediately dangerous, shatters the constitution, and renders them peculiarly liable to pneumonia, or other complaints which are fatal. When there is a canal running in the rear of the plantation, a part of the transportation of the crop is made by scows; and very recently, a low, broad-wheeled car or truck, which can be drawn by negroes on the embankments, has been introduced, first at the suggestion of a Northerner, to relieve the labor.

The rice is neatly stacked, much as wheat is in Scotland, in round, thatched stacks. Threshing commences immediately after harvest, and on many plantations proceeds very tediously, in the old way of threshing wheat, with flails, by hand, occupying the best of the plantation force for the most of the winter. It is done on an earthen floor, in the open air, and the rice is cleaned by carrying it on the heads of the negroes, by a ladder, up on to a platform, twenty feet from the ground, and pouring it slowly down, so that the wind will drive off the chaff, and leave the grain in a heap, under the platform. But on most large plantations, threshing-machines, much the same as are used with us, driven either by horse-power or by steam-power, have been lately adopted, of course with great economy. Where horse-power is used for threshing, the wind is still often relied upon for removing the chaff, as of old; but where steam-engines are employed, there are often connected with the threshing-mill, very complete separators and fanners, together with elevators and other labor-saving machinery, some of it the best for such purposes that I have ever seen.

After the ordinary threshing and cleaning from chaff, the rice still remains covered with a close, rough husk, which can only be removed by a peculiar machine, that lightly pounds it, so as to crack the husk without breaking the rice. Many of the largest plantations are provided with these mills, but it is now found more profitable (where the expense of procuring them has not been already incurred), to sell the rice "in the rough," as it is termed, before the husk is removed. There are very extensive rice-hulling mills in most large towns in Europe and America. In most of the European States a discriminating duty in favor of rough rice is laid on its importation, to protect these establishments. The real economy of the system is probably to be found in the fact, that rice in the rough bears transportation better than that which is cleaned on the plantation; also, that when fresh cleaned it is brighter and more salable. Rice in the rough is also termed "paddy," an East Indian word, having originally this signification.

The usual crop of rice is from thirty to sixty bushels from an acre, but even as high as one hundred bushels is sometimes obtained. Its weight (in the rough) is from forty-one to forty-nine pounds per bushel. The usual price paid for it (in the rough), in Charleston and Savannah, is from eighty cents to one dollar a bushel.

Planters usually employ their factors — merchants residing in Charleston, Savannah, or Wilmington, the three rice ports — to sell their crop by sample. The purchasers are merchants, or mill-owners, or the agents of foreign rice mills. These factors are also employed by the planters as their general business agents, making the necessary purchase of stores and stock for their plantation and family supply. Their commission is 2½ per cent.

Dairying in Ohio, 1854

From Ohio Board of Agriculture, *Annual Report*, 1854, pp. 203–08.

Agricultural reports hitherto have given us but meager and indefinite statistical information upon the subject of dairy husbandry in the United States. The small extent of territory at present appropriated *strictly* to dairy purposes, is perhaps one reason why, in all our general views of agriculture, the dairy interest scarcely arrests our attention. In the absence of statistical data of the number of acres of land used, and the number of cows kept *exclusively* for dairy purposes, I may briefly enumerate the principal localities, in which either from peculiar adaptation of soil, or for other sufficient reasons, dairying is the leading agricultural interest. In all the States of New England, though considerable butter and cheese is made in the aggregate, yet there are but a few counties where dairying is pursued as a principal or leading business.

In the State of New York we find more extensive districts where this branch of agricultural industry occupies the attention of the people. We may mention the counties of Herkimer, St. Lawrence, Jefferson, a part of Montgomery, of Onedia and of Madison, the county of Chemung, a part of Erie county, and the counties of Alleghany, Cattaraugus and Chetauque, as comprising the principal dairy districts of that State.

In the State of Ohio we have the counties of Ashtabula, Trumbull, Geauga, Cuyahoga, Lorain, Portage and Lake.

This enumeration embraces not all, but the principal districts where dairying may properly be said to be the leading agricultural pursuit of the people.

It is not strange, then, that an interest appropriating to itself so small a portion of our immense territory should fail to attract that attention in the public mind which is requisite to the full development of its resources, and which is necessary to the diffusion of that practical and scientific knowledge so important in the pursuit of any avocation of industry.

It is true that the aggregate dairy products of the U. States are greatly augmented, and probably more than doubled by additions outside of the territory I have indicated, but it is equally true that this is not the result of any *direct* efforts to increase the dairy products of the country. It is merely a surplus necessarily (and no doubt profitably) resulting from other leading branches of agriculture.

Most farmers in all the Northern and Central States, pursue a sort of mixed husbandry. They cultivate a part of their land, keep some sheep, raise a few cattle, &c. As a consequence, they keep more cows than would be strictly necessary to supply the family with milk, cream and butter; and after the calves are weaned, the surplus of milk is made into cheese or butter for market. No doubt in many localities, and in the condition and circumstances of many families, this is a very safe and judicious mode of farming. But, confining myself to the subject of dairying as an independent branch of business, the question naturally arises, can the dairy business be profitably extended? In other words, can the production of butter and cheese be very much increased with the prospect of remunerating prices to the producer? We believe they can, and shall offer a few of the reasons on which our opinion is founded.

First. The price of dairy products, like every other farm crop, will always depend, mainly, upon "supply and demand."

Now, that the consumption (or "demand") of cheese and butter has more than kept pace with the increase of "supply," will be admitted, I think, by any one who will consult the price current tables for the past twenty-five years.

Secondly. The consumption of cheese in particular, will be likely to continue to increase fully equal to any probable increase of supply. The present and prospective high prices of bread and meat favors this idea. Again, the nutricious and healthful properties of cheese as an article of diet, are being more and better appreciated and understood, and will tend to increase the consumption of this convenient and economic article of food.

Finally. That we may safely increase the productions of butter and cheese, is evident from the fact, that if all the cheese at present made in one year in the United States, was to be cut up and equally distributed among all the people of the Union at once, we should hardly get a handful apiece!

But there is another reason why we may be encouraged to increase the dairy business to the full extent of the demand. Dairy products can be produced at, relatively, less cost than cultivated grain crops.

First. Because there is, probably, less exhaustion of the soil in proportion to the value of the product.

Second. There is less liability to disappointment in the amount of product, from the effects of disastrous seasons.

The past summer of 1854 has been the severest test that our dairymen have ever passed through, during my acquaintance on the Western Reserve, embracing a period of twenty-five years; and yet I think that dairy products will average at least two-thirds the usual crop, while the cultivated crops of corn and potatoes, and many of the small grains, fall far below half a crop; and in the central portions of our State, where wheat is relied upon as the principal crop, there are numerous instances of entire failure; and I know of many pretty extensive farmers who did not harvest a single bushel of wheat.

The reason of this difference in the amount of risk, is quite obvious. The dairyman, relying principally upon pasturage and hay for the support of his stock, has little to contend with, except the effects of drouth, and this effect seldom extends through the entire season. It is quite rare that the hay crop, on a well managed farm, is very materially lessened by the effects of drouth, and in the spring months, and generally the fall months, there is plenty of pasturage, so that the dairy crop is never a *failure*.

We feel confident then, in the assertion, that the dairy business may be very much increased in Ohio, with a fair and almost a certain prospect of remunerating prices for the products of our labor.

For the benefit of those who have no practical knowledge of dairying, I append a general statement of the expense and profits of a well managed dairy farm, leaving out minor items on both sides, which would generally balance each other, and premising that I present this subject, not *theoretically,* but in the *practical* manner, pursued by intelligent and skillful husbandmen.

The profitable management of a dairy of 40 cows, will require say 200
    acres of improved land. This, at $20 per acre, would cost $4,000 00
40 good cows, say $25 each... 1,000 00
                                                                   $5,000 00

The sales from these cows will average —
375 lbs. cheese at 7c ....$26 25 and
25 lbs. of butter at 15c........... 3 75
                $30 00 × 40 = ...$1,200 00
The whey and dairy slops will make 2,500 lbs. of pork at 4c. per lb. ... 100 00
    Gross sales $1,300 00
Deduct interest on $5,000.. $300 00
Salt and cloth for capping 20 00
Ten stock hogs $3 per head 30 00
Managing the dairy and cutting hay 300 00
    Leaving a profit of... $650 00

In this statement I have not estimated the value of milk, cream, butter and cheese, consumed by an ordinary family, say of eight persons, but have only stated the actual sales after allowing the family all they require for the year.

In explanation of the above statement, it is proper to say that I have not made full allowance for the cost of tending the dairy and cutting the hay and wintering the cows, though the estimate is but a trifle too low, as I have allowed interest on the whole capital invested, whereas 150 acres of good average land is all that is necessary for the yearly support of the 40 cows, thus leaving 50 acres of land for cultivation, orcharding, &c., and for the support of teams, &c., necessary for carrying on the farming operations.

If the question be asked why I have taken 200 acres of land instead of 150, as the basis of my statements, I answer, that in the *practical* operation of the dairy business, some land for tillage is always desirable, and is in fact necessary to its most profitable management.

To milk 40 cows, requires four good milkers, and if part are females or children, there may be more. After the milking is done, one man can make the cheese and tend it, and have about half the day for other work.

As there will generally be a proportion of male help in most families, this extra quantity of land gives employment between milking hours, and will handsomely support any ordinary family: so that the net sales in the foregoing statement will be clear net profits.

It is readily conceded that this does not show any remarkable profits from the quantity of land, yet when we take into consideration the fact that these results are seldom affected by any of the contingencies which so frequently destroy the cultivated crops, and further that this system may be pursued for many years without a very material deterioration of the soil, we think the results will compare satisfactorily with other branches of husbandry. I would not be understood, however, as expressing the opinion that a very long course of grazing for milking purposes, may not ultimately exhaust the soil of some of its elements of fertility.

I have said nothing of dairies, exclusively for the making of butter, for the reason that the results under favorable circumstances will not be materially different. But as few have the necessary buildings and other conveniencies for making and preserving butter, I have not deemed it prudent to encourage its manufacture upon a large scale at any great distance from a ready market.

The rapid growth of towns and cities, and the constantly increasing demand for good butter, will always offer sufficient inducement for its manufacture, and there is scarcely a prospect that butter of the right quality will be in excess of the demand.

So much has been written by abler pens than my own, upon the best methods of making both butter and cheese, that I shall offer no suggestions upon these subjects, but shall pass briefly to a consideration of the management of pastures and the selection of dairy cows.

Pastures

Long experience has convinced most dairymen that cows do best when allowed the full range of all their pasture. The reasons for this may not be so obvious to a person having little or no experience in the business. I am satisfied, however, that there is economy and utility in the practice.

First. There is less expense in fencing.

Second. A saving of time in driving cows to and from different pastures.

Third. It obviates the difficulty experienced on most farms, of having always a good supply of water in the pasture, few farms being so well supplied as to have plenty of water in any desirable number of pastures for rotation.

Fourthly. The cows will be more healthy, and more uniform in the quantity and *quality* of their milk, and will yield a better average when they have the entire range of pasture; and if this be true, there can be no doubt of the economy of the practice.

Almost every farm furnishes quite a variety of soils, and consequently as great a variety of grasses.

The cow, when left free to choose, always manifests her love of variety in the selection of her food, and she will make the entire range of her pasture almost every day, selecting such food as is most agreeable to her taste, or best adapted to her wants.

Again, dairymen well understand that the quality and flavor of the milk is more or less affected by the kind and quality of the food consumed by the cow. Now if we give the cow the entire range of pasture to pick and cull and mix her food as best may please her, she will not be surfeited one day and half starved the next; she will always find as much tender and succulent food as she will in a change or rotation of pastures, and as a consequence she will give a larger average yield of milk, and it will be of more uniform quality. Both are considerations of much importance to the dairyman.

Selection of Dairy Cows

From what breed of cows shall the dairyman make his selections? is an important question. I am not aware of any extensive stock-grower, in Ohio, of what is technically called "blooded cattle," who has been long improving his stock with the express object in view of developing the milking qualities of the race.

My remarks, then, will be understood as expressing my opinion of the *present condition* of the different breeds or races (if that be a proper distinction), and not what they might or might not be brought to, by a long course of judicious breeding, with this particular object in view. The statement of a few principles will assist us in forming correct conclusions upon this subject.

First. The most profitable cows for the dairyman, are those that will give the best return in proportion to the food consumed, or cost of keeping.

Second. The cost of keeping cows is nearly in proportion to their bulk or live weight.

Third. Cows do not yield milk in proportion to their size or live weight.

If these principles be correct, and as general principles I believe they are, then we should not expect that the Durham stock would take precedence for dairy purposes. It is no disparagement to this truly noble race of cattle, that they cannot fill every place in the different departments of agricultural economy. They have been long bred for a distinct purpose, they have been brought to great perfection, and have nobly realized the object of that purpose. They have acquired what may be termed a fixed constitutional tendency, and it will require long years of skillful breeding to change that constitutional tendency, and develop the best milking qualities of the race, and in doing this we would perhaps destroy much that has been gained by the opposite system of breeding — a consummation, we presume, which no one desires to see.

The Devons

This is a very beautiful breed of cattle, and the objection as to size, certainly does not apply to them. I am not aware, however, that as a race, they have ever been distinguished in this country as very extraordinary milkers. So far as I have been acquainted with this race, I should say that their yield of milk is below the average of what dairymen call good milkers, neither do they continue in milk so long as is desirable; yet I think it may be fairly conceded that the milk is remarkably rich in cream and butter, and no doubt good selections from this breed might give quite satisfactory results in a butter dairy.

Ayrshires

No breed of cows have been more highly extolled than the Ayrshires. It is claimed, I believe, that they have been long bred in Europe, with distinct reference to their milking qualities. If this be true, and the maxim hold good that "like produces like," we would expect the Ayrshires, as a race, to exhibit extraordinary milking properties. It is claimed for most of the importations, that they are superior milkers, but I have seen no published statement where any considerable number of full blood Ayrshires, in a single dairy, have ever been fairly tested in this country.

I have raised half-blood Ayrshires that were very superior milkers, but they were a cross from some of the best cows I ever milked, and the produce might have been equally good milkers, crossed with some other breed.

Whether the Ayrshires will sustain in this country the high reputation they have acquired elsewhere, I think remains to be tested.

The reader no doubt has anticipated by this time, that for the purposes of the dairy, I prefer

The Native Breed

I express the opinion with much confidence, and without fear of successful contradiction, that proper selections from what we call native cows, are the most profitable for dairy purposes. And in this I am sustained by the generally expressed opinion of practical dairymen all over the country. I doubt not, however, that for the "general farmer," who breeds cattle stock for all purposes, that crossing with the improved breeds will be greatly to his benefit. I am aware that some (I might say many), individual cows of what is called blooded stock, have produced extraordinary yields of milk and butter. The most successful trial I have heard of, was by Mr. George Vail, of the State of New York, in 1844, when his 6 short horn cows, in a trial of 30 days, made nearly 1½ lbs. of butter each per day, and one of his best cows, which was milked separate, made 1¾ lbs. per day. I think this is the largest yield we have any record of, from any considerable number of blooded cows in this country. Now these cows had always received the most careful attention, and no doubt were in better condition to give successful results than cows generally are. And yet at the trial of dairy cows in the State of New York, as reported from the State Fair at Syracuse, there were entered nine dairies of native cows, quite a number of which came up to this yield, and several I think exceeded it; and we find many records of individual native cows that have produced more than two pounds of butter per day.

But my purpose is not to draw an invidious distinction between native and blooded stock, they both have their uses, and I trust I shall never fail to appreciate the value of the improved stock of our State. The dairyman, however, should never forget that there is a wide difference in the milking qualities of what is called native stock. In the strictly grain-growing sections, where little hay is made, and little succulent food is provided for cattle, where the young stock are compelled to brouse in brush pastures in summer, and have nothing but straw in winter, we do not find so good a race of milkers as we do in the grazing districts of our country, and we have no reason to expect it, for I believe it is an admitted principle, that the quality of the food exerts quite a degree of influence in developing the characteristics of the growing animal.

I have already extended this article beyond the limits I had assigned myself in its commencement, yet I am unwilling to close without saying a few words to those who may be desirous of engaging in this branch of husbandry: Remember that the success of your undertaking will depend upon the amount of knowledge and skill you may bring into requisition in the prosecution of your occupation. It is a mistaken idea, that any branch of farming can be successfully and profitably pursued in the shades of ignorance. This is especially true of dairy husbandry. This knowledge is not hidden, and this skill is not difficult to acquire. Let it be our aim, then, to give a reputation and a character of excellence to the dairy products of our State, and to place our occupation in that honorable position among the other industrial pursuits of our country, to which, from its importance, it may justly lay claim.

A. KRUM

Seeds and Cuttings, 1855

From Daniel J. Browne, "Report on the Seeds and Cuttings Recently Introduced into the United States," U.S. Patent Office, *Report on Agriculture*, 1854, (Washington, 1855), pp. x –xxxv.

SIR: Agreeably to your suggestion, herewith I furnish such information relative to the nature, origin, culture preparation, and uses of the principal seeds, cuttings, &c., imported or introduced into this country within the last two years, as might prove beneficial or acceptable to a great body of our agriculturists, who may have received them for experiment. I regret that I am not able to report at length on many of the products, as they are quite as important, perhaps, as those which are more fully treated. I beg leave to add, that there are numerous useful products in Europe and other distant parts, that never have found their way into this country, which, I am persuaded, might be cultivated with a fair chance of success. The time for believing that the exclusive possession of any benefit contributes solely to the privilege or prosperity of any particular country or kingdom, has gone by; and the principles of free and universal intercourse and exchange are now conceded to constitute the surest foundation for the happiness of nations. This is so obviously true in matters of this sort, that it cannot for a moment be doubted. Hence it may be inferred that there is ample room for exertion on the part of our general government, as well as of States or individuals, to increase our agricultural and botanical riches, more especially those products which so conspicuously and permanently add to our useful and economical resources.

Among the foreign products which have more recently been imported or introduced, and distributed for experiment, and which appeared to be susceptible of profitable cultivation in this country, I would instance the following:

Cereals and other Plants, Cultivated for their Farinaceous Seeds, Straw, or Haulm.

Turkish Flint Wheat, from Mount Olympus, in Asia; a fall variety, with rather large, long, flinty berries no very dark-colored, and possessing remarkable properties for long keeping in a moist climate, or for transportation by sea without kiln-drying. It has proved itself both hardy and prolific in the Middle States, and its culture deserves to be extended. The spikes are of good length and size, having only a light beard.

Algerian Flint Wheat, from the province of Oran. This variety has a remarkably large berry, rather dark-colored, and weighing 70 pounds to a bushel. From a sample sown in the valley of Virginia, in November last, it yielded at the rate of 35 bushels to the acre, a berry equal in size and weight to the original. The spikes are large, bearing an enormous beard.

Pithusian Flint Wheat, from the island of Iviça; another fall variety, resembling the Algerian, but having larger berries, varying in color from light to dark.

Syrian Spring Wheat, from the "Farm of Abraham," at the foot of Mount Carmel, in the Holy Land. The berry of this variety resembles that of the last preceding, and is reputed to have matured in sixty days after sowing.

Cape Wheat, from the Cape of Good Hope, procured by Commodore Perry, of the Japan Expedition. This is a beautiful light-colored wheat, slightly flinty in its character, and doubtless produces an excellent flour. It probably will do much better at the South than at the North, if sown in autumn, unless it should prove to be a spring wheat. If successful, it will be liable to degenerate, unless the seed is often replenished or changed.

Spanish Spring Wheat, (Trigo candeal,) from Alicante; a beautiful variety, of unsurpassable whiteness, and is reputed to have ripened in less than ninety days after sowing. It will doubtless succeed well as a winter wheat at the South, and a March or spring variety at the North. The berry is rather long, plump, and slightly flinty in its character. The flour is of unrivalled whiteness, and is celebrated in Spain as entering into the composition of candeal bread (pan candeal.)

White Hungarian Wheat, (Blé blanc de Hongrie, of the French,) from the south of France. The spikes of this variety are white, of medium length, very compact, and square-like, terminating abruptly, or not tapering to the extremity; chaff, smooth and thin; spikelets, containing four grains, which are quite large, short and plump, or rounded, white, and slightly transparent. Weight, 66 pounds to the bushel. It is reputed to be about a week longer in ripening than other sorts, but from its superior qualities it well deserves a trial in this country, as a fall or winter wheat at the South, and as a March or spring variety at the North.

Red-chaff White Wheat, from England, having a very large, short, rounded berry, generally soft, but often transparent. It is rather tender, and probably would not succeed as a fall wheat north of Virginia.

White Neapolitan Wheat, (Richelle blanche de Naples,) from the south of France, where it is much cultivated. The spikes are long, but not very compact; terminal spikelets, having short awns from one-fourth of an inch to an inch in length; chaff, delicately tinged with a dull yellow or copper color; grains, large, considerably elongated, and generally of a yellowish-white color. It has the disadvantage of ripening late, and is believed to be too tender for the North. Possibly, it may prove to be a March or spring wheat, if sown early in the Middle States or at the South.

Girling's Prolific Wheat, from England; a very prolific variety, with a large, short, plump, brown berry, but inclined to be soft. Like the Red-chaff White, it is thought to be tender, and unsuited for the Northern States.

White Chilian Wheat, from Santiago de Chili; a beautiful variety, with large, rounded, plump, white grains, resembling those of the Red-chaff White from England, and, like which, it is believed to be too tender for the North.

Saumur Spring Wheat, (Blé de saumur de mars,) originally from the valleys of Anjou, a southeastern department of France, and is a very remarkable variety for fall or winter-sowing. The berry is rather soft, though full, of a reddish color, and much esteemed by farmers for its early maturity, which perfects itself some days before the ordinary sorts. As its name implies, it may also be sown in March, which will add to its value in this country as a spring wheat. If sown in autumn, it probably would succeed in the middle or central range of States.

Early Noé Wheat, (Blé de l'Ile de Noé,) introduced into the central part of France by M. de Noé, and is commonly known there under the name of *Blé bleu.* From its hardy and productive nature, it is gradually superseding the Saumur wheat in the high latitudes of Paris, and is much sought after on account of its precocity. As this wheat and the preceding variety have the property of ripening some days before the common sorts, if they succeed in our climate in this respect, a great point will be attained. A single week thus gained in ripening would often secure the crop from injury by the fly or rust, aside from the advantages to be acquired from an early market. It would probably succeed well as a spring wheat if sown early.

Geja Wheat, from the south of Spain; with a large, moderately long, full berry, of a brown color, rather inclined to be flinty. It probably would be too tender to sow at the North as a fall wheat, although it might succeed if sown early in the spring.

Large Northern Prolific Rye, from Germany; with a large grain, and doubtless will be suited to the Middle, if not to the Northern States.

Spanish Barley, from the south of Spain; with a full, well-filled grain, which promises well.

Common Black Oat, (Avoine noire de Brie,) from France. In the length of the straw, and the form of the panicle, this variety is similar to the Potato oat. The grain is rather large, well filled, and of a shining black color, lighter towards the point. It is very prolific, and about a week earlier than the Potato oat, weighing 42 pounds to the bushel.

Chenailles Oat, (Avoine noire de Chenailles,) from France; resembling the preceding in the character of the grain, but somewhat earlier and of taller growth.

Spanish Oat, from the south of Spain, with light-colored grains, heavy, and well filled with farinaceous matter. It probably would succeed well in the Middle and Southern States for late fall or winter-sowing.

Silver Buckwheat, (Sarrazin argenté,) from France; an esteemed sort, with whitish grains, and employed for the same purposes as the common kind.

White Quinoa, (Chenopodium quinoa,) from France, but originally from Peru, where it is a native. The grains are round, white, and about the size of mustard-seed. The leaves of this plant, before it attains full maturity, are eaten like spinach; but the seeds are the parts most generally used as food, being both nutritious and wholesome, as well as easy of digestion. They are prepared in a variety of ways, but most frequently are boiled in milk or soups, or cooked with sweet peppers and cheese.

This plant is very vigorous, quite insensible to cold, and produces an abundance of seed on a good, light, warm soil. Its culture is simple. If intended for its grain, it may be sown in a sheltered border early in the spring, in order that it can be transplanted before the return of summer heats; or it may be sown in open culture in drills, in the middle of the spring. When the plants become of sufficient size, they are removed and planted at the distance of twenty inches apart, well exposed to the sun. If desired for the leaves only, they may be set nearer to each other, and the stalks cut off at the first gathering, in order to cause them to branch out for a succession of crops. By watering during the summer, should there not be rain, the product of leaves will be incessantly renewed.

Forty Days Maize, (Maïs quarantain,) a dwarf variety from the south of Spain, reputed once to have ripened high up in the Alps in forty days after planting. The object of introducing this grain into the United States was on account of its quick growth, early maturity, and sweet flavor in the green state, as well as the delicacy of the bread made from its meal. Besides, it appears to be well adapted to the high latitudes and elevated valleys in many parts of the country, where most other varieties of corn will not thrive, and with a chance of a successful result in crossing it with the larger sorts, to which it might impart, in a degree, its quality of early ripening, if not its taste.

Indian Millet, or Dourah Corn, (Holcus sorghum,) from St. Martin, in the West Indies; described at length in another part of this volume.

Legumes

Early Long-podded Bean, from England; quite as prolific as the common Long-podded, but considerably earlier. It probably will do well at the South, but of doubtful success north of Virginia.

Long-podded or Butter Bean, from Germany; an esteemed sort for eating in a green state when shelled.

Early Dwarf French Bean, (Haricot flageolet, or Nain hâtif de Laon,) rather long, narrow, and cylindrical in shape, and of a whitish or pale-green color. It is one of the most esteemed varieties in the neighborhood of Paris; very dwarfy and rapid in its growth, and is much employed there as ''snaps,'' or shelled in a green state, and even when dried. From its bushy and dwarfy habit, it will bear close planting — say from two feet to two and a half feet apart.

Pearl Bean, without strings, from Germany; a fine variety, used as ''snaps'' when green, or in a dried state when shelled. It probably will prove a runner.

Pearl or Round Turkey Pea Bean, from Germany; represented as an excellent and prolific sort, with yellow transparent pods.

Mexican Beans, (Frijoles,) two varieties, ''Black'' and ''Reddish,'' treated of in last year's report.

Early May Pea, from England; already known to the market gardeners and seedsmen of the United States.

Early White May Pea, from Germany; represented as an excellent variety for early sowing.

Dwarf Hamburg Cluster Pea, from Germany; the best and earliest of the earliest sorts of that country.

Late Fall Golden Pea, from Germany; well adapted for very late sowing for autumn use, and not affected in its growth by mildew or heat.

Auvergne Pea, from England; a very hardy productive sort, growing to a height of four or five feet, of an excellent quality, and adapted for late sowing for fall use.

Capucine Pea, from Germany; a fine variety to be used in succession.

Champion of England Pea, from England; much esteemed as a second sowing; already well known to American seedsmen, as well as to private growers.

Oregon Pea, described in last year's report, the origin of which is unknown. It greatly resembles, if it is not identical with, the Oleaginous pea, *(Dolichos viridis,)* lately introduced into France, from China, by M. Montigny, French consul at Shanghai, to whom we are already indebted for the Sorgho Sucré and the Chinese Yam.

Japan Pea, also described in last year's report, and has been since cultivated with remarkable success.

Soja Bean, (Soja hispida,) procured by the Japan Expedition; two varieties, the ''White'' and the ''Red-seeded,'' both of which are employed by the Japanese for making *soy,* a kind of black sauce, prepared with the seeds of this plant, wheaten flour, salt, and water. This ''soy,'' or ''soja,'' which is preferred to the *Kitjap* of the Chinese, is used in almost all their dishes instead of common salt. The soy may be made as follows:

Take a gallon of the beans of this plant and boil them until soft; add bruised wheat, one gallon; keep in a warm place for twenty-four hours; then add common salt, one gill, and water, two gallons; and put the whole into a stone jar, and keep it tightly closed for two or three months, frequently shaking it; and then press out the liquor for use.

The seeds of this plant only require to be sown in a warm, sheltered situation at the time of planting Indian corn, and cultivated as any garden bean.

White Lupine, (Lupinus albus,) from the south of Spain, where it is culti-vated to a limited extent for forage, as well as for soiling. It was employed as food by the ancient Romans, and, as with the inhabitants of the present day, was ploughed into the soil as a manure. In Germany, also, it has been found to be one of those plants by which unfruitful, sandy soils may be most speedily brought into a productive state. The superiority of this plant for the purpose of enriching the soil depends upon its deep roots, which descend more than two feet beneath the surface; upon its being little injured by drought, and not liable to be attacked by insects; upon its rapid growth; and upon its large produce in leaves and stems. Even in the north of Germany, it is said to yield, in three and a half to four months, ten or twelve tons of green herbage. It grows in all soils except such as are marly and calcareous; is especially partial to such as have a ferruginous subsoil; and, besides enriching, also opens stiff clays by its strong stems and roots. It abounds in potash, nitrogen, and phosphoric acid, and is considered the best of green manures, being almost equal to farm-yard dung. The seeds are somewhat expensive, and about the size of peas. They should be sown as early in the spring as the season will admit, without injury from frost, and the plants will blossom in three or four months; soon after which, they may be turned into the soil, and succeeded by most of our field or garden crops. Although rather slow to decay, its decomposition may be hastened, if desirable, by the addition of caustic lime.

Yellow Lupine, (Lupinus luteus,) from Germany, where it is extensively cultivated as a green or vegetable manure, to be ploughed under in poor soils. Large crops are also obtained for the seeds, which, when ground or crushed, serve to fatten cattle and swine. Its culture is nearly the same as the preceding.

Garbanzo, (Cicer arietinum,) or chick-pea, from Alicante, in Spain. This is an annual plant, much cultivated in the south of Europe, as well as in Asia and Africa. Cooked whole, it is not easy of digestion; but when eaten in the form of a soup or porridge, it is much esteemed. The famous Parisian dish called *purée aux croûtons,* and the *olla podrida* of Spain, particularly the former, are composed of this pea. In warm countries, it is sown in autumn, and harvested the following summer; but in a more temperate climate, it is sown in spring, and gathered in autumn just before its perfect maturity, in order that it may more readily be cooked.

Gallardon's Large Light-colored Lentil, (Ervum lens,) from the south of France, but much cultivated in the neighborhood of Paris, both in the garden and open field. It is usually sown in lines or hills, but seldom broadcast. It is best adapted to a dry and sandy soil, as on rich land it runs too much to stalks and leaves rather than seeds. In France, it is sown late in March or in the beginning of April. In order that the lentils may be of a better or firmer quality, they are shelled or threshed out only as they are required for use. They may be cooked with bacon or in the form of a porridge or soup, like split peas. The ancient Romans are said to have caused them to germinate before cooking, in order to develop their saccharine qualities.

Plants Cultivated Chiefly for their Tubers or Roots

Potato Seed, (Solanum tuberosum,) from Germany; obtained from the apples, or balls, of the potato haulm. The importance of experimenting with seed and the mode of culture are treated at length in another part of this volume.

Fluke Potato, from England; a superior variety, much esteemed at Liverpool for its flat shape, and fine qualities for domestic use, and for long keeping. It bears late planting, yields well, and has never been known to be much affected by the rot.

In order fairly to test its adaptation to the Middle or Northern States, it would require to be cultivated for several years.

Regent Potato, "The Potato" of London market. It is roundish in shape, of good size, having a yellowish, rough skin, dry, mealy, of excellent flavor, and light-colored within. It matures rather early, keeps well during the winter and spring, and is productive in its yield.

Lapstone Kidney Potato, a fancy variety lately originated in Yorkshire, England, by a shoemaker. From its slight resemblance, in shape, to a lapstone, it has acquired its name. It is rather small, smooth, and light-colored without, and perfectly white and flour-like within, when cooked.

Early White Potato, from England; another fancy variety of small size; finger-shaped, and early to mature.

Chinese Yam, (Dioscorea batatas,) originally from China, but more recently from France, where it is proposed as a substitute for the common potato. It is fully described and treated of at length in another part of this volume.

Earth Almond, or Chufa, (Cyperus esculentus,) from the south of Spain. In addition to what is said of this plant in another part of this volume, in order to remove any prejudice which may exist in supposing that it is identical with the creeping cyperus, *(C. repens,)* or nut-grass, which is found growing wild on the banks of streams, in pastures, and cultivated ground, from New York to Florida and Louisiana, I would state that the latter differs essentially from the chufa in its height, as well as in the size, shape, and color of its spikelets. The roots, also, contain many fibrous branches, often terminating in edible tubers, about the size of a pea, creeping continuously along with, and just below, the surface, and send up numerous suckers, which are regarded by Southern planters as a great scourge to their crops. The chufa is quite different in this respect, only throwing up several stalks from one root, like the common potato, but does not spread.

Considerable attention has long been devoted to the cultivation of the chufa in the south of Spain, where it is stated that more than $400 have been realized from an acre in the short space of five months.

Turnips. — Through the liberality of Messrs. Charlwood & Cummins' extensive seedsmen of London, the office received the following twenty-six varieties of turnip-seed, on condition that they should be sent to every State and Territory in the Union for experiment, with the view of testing their adaptation to the soil and climate; said experiments to be conducted as uniformly as practicable, and the result made known to the public in the form of a report: Skirving's Swede, or Ruta-baga; Rivers' Stubble Swede; Laing's Swede; Green-topped Swede; Dale's Hybrid; Green-topped Six-weeks; Snow Ball; Strap-leaved; Small Yellow Malta; White Globe, or Norfolk White; Green Round, or Norfolk Green; Green Globe, or Green Norfolk; Golden Ball; Red Globe, or Norfolk Red; White Tankard, or Decanter; Green Tankard, or Decanter; Yellow Tankard, or Decanter; Red Tankard, or Decanter; Green-topped Scotch; Purple-topped Scotch; Skirving's Purple-topped Scotch; Early Stone, or Stubble-stone; Yellow Stone; Red-topped Stone; White Dutch; Yellow Dutch.

It may be needless to state that the above-named request has been complied with, so far as this office is concerned, and the seeds distributed with an appropriate circular to all parts of the United States.

There have also been imported from England, in addition to the above, the following sorts, and extensively distributed far and wide: Rivers' Swede; Ashcroft's Swede; Sutton's Green-topped Yellow Hybrid; Sutton's Purple-topped Hybrid; Sutton's Cruicksfield Hybrid; Sutton's Early Six-weeks; Border Imperial; Orange Jelly; Yorkshire Paragon; Sutton's Improved Green Globe; and Lincolnshire Red Globe.

Four varieties were likewise imported from France, namely: Navet long des vertus, Navet de Freneuse, Navet turnip, and Rave d'Auvergne.

Radishes. — Two varieties of radish were imported from England, the "Yellow Turnip," and the "Long Scarlet." From France, there were received the "Large Field Radish" (Raifort champêtre,) "Olive-shaped Radish," (Radis rose demi-long,) "Short Scarlet Radish," (Radis rose demi-long écarlate,) and the "Winter Rose-colored Radish," (Radis rose d'hiver.)

Beets. — Of these, there were imported from England the "London Red," and the "Bassano," the latter a turnip-rooted variety, which originated in Italy, and is already known to American cultivators. There was also introduced from France the "Scaly or Rough Red Beet," (Betterave rouge crapaudine,) and from Germany the "White Silesian Sugar Beet," the latter of which is particularly valuable for feeding to milch cows.

Of the *Mangold Wurzel*, two sorts were imported, the "Large Yellow Globe" from England, and the "Large Long Yellow" (Betterave jaune grosse,) from France. The former is a fine variety which originated in England, growing mostly above the surface of the ground, which renders it particularly fit for shallow soils. The latter holds the first rank in the neighborhood of Paris for feeding milch cows. There is also the "German Yellow," (Betterave jaune d'Allemagne,) introduced into France about thirty years ago, and is now extensively cultivated. From this a sub-variety has been produced, called *Betterave jaune des Barres,* which some consider as a model forage beet, or mangold wurzel. It is of elliptical form, and so little buried in the ground as to be easily torn up by the hand, and is equal in quantity and quality to the varieties named above.

Carrots. — Of carrots, five sorts were imported: "St. James" from England, and the "Short Red," (Carotte rouge courte,) "Long Yellow" (Carotte jaune longue,) "Vosges White" (Carotte blanche des Vosges,) and the "Green-crowned White" (Carotte blanche à collet vert,) from France.

Parsnips. — Of these, only two varieties were introduced, named the "Hollow-crowned," from England, and the "Round Parsnip," (Panais rond,) from France.

Onions. — From France, three varieties of the onion were imported: the "Brunswick Dark Red," (Oignon rouge foncé de Brunswick,) "Cambray," (Oignon de Cambrai,) and the "Early White," (Oignon blanc hatit.)

A variety of *Leek* (Poireau long) was also imported from France.

The *Celeriac or Turnip-rooted Celery* (Apium graveolens rapaceum) has been introduced both from Germany and England. It is mentioned at length in another part of this volume.

Chickory, or Succory, (Chicorée sauvage à café,) from France; principally cultivated for use in salads, and for its roots to roast for mixing with coffee. Mentioned at length in another part of this volume.

Plants, the Leaves of Which are Chiefly Used for Salads, Potage, &c

Early York Cabbage, from England; already known to American cultivators.

Large Ox-heart Cabbage, (Chou coeur de boeuf gros,) from France; one of the best kinds cultivated.

Alsacian or Quintal Cabbage, (Chou d'Alsace ou quintal,) from France, with short thick stalks, and very large heads, having festooned leaves, of a very bright green color. The head of this cabbage grows to an enormous size, on rich, new land.

Large Red Cabbage, (Chou Milan des vertus,) from France; the largest of the Milan varieties, which are noted for growing more open, and in being more delicate and less musky in their flavor. It requires a very rich soil.

London Cauliflower, from England; a superior variety.

Medium Cauliflower, (Chou-fleur demi-dur,) from France; possessing qualities between the fine, tender sorts and those which are coarse and tough.

Broccoli. — There have been imported from England three varieties of broccoli: the "Mammoth," "Imperial White," and the "Purple Cape;" the latter direct from the Cape of Good Hope.

Two varieties of *Kohl-rabi,* or "Turnip-stemmed Cabbage," one from England, and the other from France; described in another part of this volume.

Brussels Sprouts, (Chou de Bruxelles,) from France; producing in the axils of the leaves small heads, resembling those of other cabbages. They are very tender, and much esteemed. By successive sowing, in the Northern and Middle States, from April to June, this excellent vegetable will be fit for use from early in autumn till late in the winter, as it stands frosts better than most other kinds.

Sea Kale, (Crambe maritima,) from England; a notice of which will be found in another part of this volume.

Lettuces. — Of lettuce, three varieties were received from Germany: the "Blood Red," "Spotted" or "Tiger," and the "Asparagus lettuce." There are three kinds, also, from France: the "Large, Brown Slow-growing," (Laitue grosse brune paresseuse,) with greyish-green leaves, marked with pale, brownish spots, having very large and regular heads slightly tinged with red at the top; "Roman Pale-colored Marsh," (Laitue romaine blonde meraîchère,) an esteemed sort, much cultivated in the neighborhood of Paris, forming heads without tying; and the "Gotte" or "Gau" lettuce (Laitue gotte,) a variety suitable for growing under glass for winter use.

Celery. — Of celery there was imported from France one variety, "Early Dwarf," (Celeri court hâtif,) which, from its compactness of growth, does not require to be tied before earthing up, with fine dense heads, and prompt in blanching; two varieties, also, were imported from England, "Seymour's White Solid" and "Cole's Red," both of which are already known to American growers.

True Giant Asparagus, from England.

Lettuce-leaved Spinach, (Epinard à feuille de laitue,) from France; with very large, thick, dark-green leaves, which form themselves into a bunch or head.

Plants Cultivated for their Berries or Fleshy Fruits

Large Yellow-fleshed Pumpkin, or Squash, (Potiron jaune gros,) from France; the fruit of which is very heavy, of a gold yellow within, and grows to an enormous size.

Cassabar Melon, from Asia Minor, which, when pure, is of a sweet, delicious flavor, and may be eaten, even by invalids, with impunity. These seed were reported to be seven years old.

Valencia Melon, (Melon de Valencia,) from the south of Spain; a variety of the Canteleup tribe, celebrated for its delicious, sweet flavor, and preferred to all others in the countries where it grows.

Summer Green-fleshed Melon, (Melon d'été à chair verte,) from France.

Honfleur Melon, (Sucrin de Honfleur,) from France; very large, somewhat long, with thick ribs; having a rather coarse flesh, but full of sweet juice.

Prescott Canteleup, (Cantaloup Prescott,) from France; a variety much cultivated, and the most esteemed of any at Paris. Its color varies from green to a silvery tint, having ribs more or less rough.

White-fleshed Winter Melon, (Melon de invierno con carna blanca,) from the south of Spain.

Winter Melon, (Melon d'hiver,) from the south of France; with a smooth rind, greenish-white, brittle flesh, juicy, and of a delicate flavor. It keeps well as late as the month of February.

Long, Thick, Smooth Cucumber, from Germany; suitable for salting when nearly turning yellow, and after extracting the seeds.

London Short, Prickly Cucumber, from England.

Long White Cucumber, (Concombre blanc long,) from France.

Gherkin Cucumber, (Concombre à cornichon,) from France; producing small green fruits, much prized for pickling.

Long Violet-colored Egg-plant, (Aubergine violette longue,) from France; a small, long-fruited variety, of much better flavor and delicacy than the large oval or round sorts.

Early Red Tomato, (Tomate hâtive,) from France.

Sweet Pepper, (Pimiento dulce,) from the south of Spain.

Plants Cultivated for Fodder, Manure, or for their Uses in Manufactures and the Arts

Sorgho Sucré, (Sorghum saccharatum?) a new gramineous plant from France, the seeds of which were sent to that country some four years since from the north of China, by M. de Montigny, the French consul at Shanghai. A full description of this product will be found in another part of this volume. There is one feature in regard to this plant, perhaps it may be well to state: It would seem that in a tropical climate, while the sugar-cane is perfecting its growth, there might be three crops of the Sorgho obtained from the same ground; and should it prove to be as rich in saccharine matter as has been alleged, a greater amount of sugar would be obtained from a given space than from the cane, besides the advantages of the distribution of the work throughout the year, instead of a press of labor forced upon the planters at one time. There is also another feature in this plant which would seem to be worthy of notice, as a forage plant in the Middle and Western States: If the seeds are sown early in May, two crops of fodder can be raised from the same roots in the season — say, one about the first of August, and the other in October.

Moha de Hongrie, (Panicum germanicum,) from France; an annual, good for forage, green or dry, very productive, of quick growth, and flourishes well on dry soils. The seeds may be sown from May until July. The following extract from the "Bon Jardinier" for 1855 will give some idea how this grass is appreciated in France:

"My first attempts on a large scale did not succeed well; but one of my neighbors, among others, made such good use of it for feeding horses and cows, that I was induced to try it again, which I did with success. I sowed it in 1835, in a dry, calcareous soil, when I remarked in it the quality of great resistance to drought. It remained green and in an excellent state in spite of the high temperature and great drought of that year, even in places which had not been manured. This time, I sowed it in drills, and regularly weeded it; but haricot beans and Panicum italicum, sowed and treated in the same manner, withered and lost their leaves, while the Moha remained green and fresh. The disastrous drought of 1842 gave new proof of the superiority of this plant in this respect. In the middle of a calcareous plain, where everything had perished, the Moha remained unchanged. A considerable portion of the heads were tolerably well filled with seed, and the threshing gave a good though diminished product. When it is intended to perfect the grain the Moha must be sown in May; when it is only wanted as a green forage, it may be sown as late as July, at the rate of ten to twelve pounds to the acre in the first case, and fourteen to sixteen in the latter."

Sainfoin, (Hedysarum onobrychis,) two varieties from France: the "Common" (Sainfoin ordinaire,) and the "Double-bearing," (Sainfoin à deux coupes,) both of which are perennial. The former is best adapted for poor soils, and will not admit of but one crop in a season. The latter is generally cultivated in all parts of France, is more vigorous, hardier, and more productive than the common sort, yielding two cuttings in a year. The farmers consider it necessary to sow it upon good land, lest the plants should deteriorate, as on soils of inferior quality the sowings must be occasionally renewed. As its stalks are thicker and harder, and its seeds larger than those of the other variety, it must be sown more closely — say, at the rate of six or seven bushels to an acre. The sowing is generally done in spring, but sometimes early in autumn. If the seed is a year old, it is not liable to vegetate.

Serradilla, (Ornithopus sativus,) an annual from Germany; employed in Portugal as an artificial forage, in dry, sandy soils, where it affords an early pasturage for cattle. As it is somewhat tender, it probably would only answer for our Southern or, perhaps, the Middle States. From its fine quality and great productiveness, it is desirable to experiment with it as far North as it would be likely to grow, when it would be better to sow it in spring with other grain, in order to obtain in autumn a green crop, or cut it for hay.

Heracleum Sibiricum, (Berce de Sibérie,) a perennial from Germany, producing a very abundant, early green forage. It is sown in autumn and comes up the following spring.

Chilian Clover, or Alfalfa, (Medicago sativa?) from Chili; a perennial variety of lucerne, which succeeds well in our Middle and Southern States. It differs from the common lucerne of Europe only in the color of its flowers, which are purple. It is sown in autumn in drills, in a deep, rich soil, producing good forage for animals, either green or dry, the following summer, and will endure for many years. Deep culture is absolutely necessary, in order to allow the extension of the roots into the earth.

Yellow or Black Trefoil, (Medicago lupulina,) a biennial from England, at present considerably cultivated in the central parts of France. One of its advantages is, that it grows well in dry and inferior soils. Its forage, though less abundant than other trefoils, or clovers, is of fine and good quality, and not dangerous to cattle when eaten green, in producing hoven. It is much more valuable, however, for an early sheep pasture, than to convert it into hay. It may be sown in March or April, like other spring grains.

Cow Grass, or Perennial Clover, (Trifolium medium vel perenne,) from England; usually sown among other grass-seeds for a permanency, but not with the common red clover.

Alsyke or Swedish Clover, (Trifolium hybridum,) from England; believed to have originated in the south of Sweden, where it is particularly abundant. It is best adapted to moist and strong soils, and has the property of self-sowing, when the flowers are left to mature, which will cause it to endure fifteen, twenty or more years. The usual course to pursue is to cut it once a year for hay, afterwards leaving it for pasturage. Its flowers, which put forth in June in great profusion, resemble in shape those of the common white clover, but are larger and of a rosy tint, of a sweet, agreeable odor, and afford an excellent forage for bees. It may be sown with autumn or spring grain; with the latter it is preferable, to prevent winter-killing.

Suckling Red Clover, (Trifolium filiforme,) from England.

Perennial Ray Grass, (Lolium perenne,) two varieties from England, the "Italian" and the "Improved." The former is said to be distinguished from the common ray-grass of England, by its earlier maturity, larger leaves, deeper green color, and by the greater height to which it grows. It is usually sown in autumn, as is the general practice with grass-seeds in the south of Europe. After the field is harrowed, it is sown at the rate of sixteen to eighteen pounds to the acre, and the seed rolled in. In the following autumn, the turf is covered like an old meadow, and the crop of the next year is more than double. It may also be sown in spring. It is eaten greedily by cattle, whether green or dry, and yields fifty per cent. of hay.

The "Improved Ray Grass" possesses several desirable properties, which recommend it to the attention of cultivators, the principal of which are — its adaptation to a great variety of soils; the facility with which it is propagated, by reason of its seeds being produced in abundance, and their uniformity in ripening; and the fibrous structure of its roots, which fits it in an eminent degree for alternate husbandry. Notwithstanding all these good qualities, its culture in the Middle and Southern portions of the Union, at least, should be entered into with caution, from the great heats and summer droughts. Again, at the extreme North there is danger from the winter frost.

Meadow Fescue, (Festuca pratensis,) from England; an excellent perennial grass, either for alternate husbandry or permanent pasture, but more particularly the latter. It is relished well by cattle, horses, and sheep.

Sheep's Fescue, (Festuca ovina,) from England; an admirable perennial grass, well adapted for growing on elevated sheep pastures, where it is well relished by those animals, which prefer it to all other herbage where it exists.

Rough-stalked Meadow Grass, (Poa trivialis,) a valuable perennial grass, from England suitable for mixed pastures, particularly on damp soils, and where partly shaded by trees.

Sweet-scented Vernal Grass, (Anthoxanthum odoratum,) a perennial from England and France, yielding but a scanty herbage, and is not particularly relished by any kind of live stock, perhaps with the exception of sheep. It is remarkable for giving out a pleasant odor during the process of drying. It has been recommended to be sown in sheep pastures for the purpose of improving the mutton, a quality which it is said to possess, and which is founded on the fact that places in which it naturally abounds are said to produce the finest mutton. From its dwarfy growth, and the close sward it forms, it is recommended to be sown on lawns or ornamental grounds.

Burnet Grass, or Pimprenelle, (Poterium sanguisorba,) an annual from France, well suited for pasturage on poor dry soils, whether sandy or calcareous. It may be sown early in the spring.

Goldbackia Torulosa, a new perennial oil plant from Germany, producing an abundance of seed, suitable for making oil. It is said to be hardy, and affords an early pasturage for sheep.

Gold of Pleasure, or Camelina Sativa, (Miagrum sativum,) an annual from France, which produces a finer oil for burning than rape, having a brighter flame, less smoke, and scarcely any smell. It succeeds well on light, shallow, dry soils, and in our Middle and Southern States it probably would produce two crops in a season. Besides the use of the seeds for oil, the stems yield a coarse fibre for making sacks and a rough kind of packing-paper, and the whole plant may be employed for thatching. The culture is similar to that of flax.

Colza, or Rape, (Brassica campestris oleïfera,) two varieties from France, the "Colza froid" and "Colza parapluie." The former is highly recommended, the yield being much greater than the common varieties of rape. It may be distinguished by its luxuriant growth and reddish seeds. The latter, principally cultivated in Normandy, though less productive, has the advantage of throwing out lateral branches, which, falling towards the ground, support the plant and prevent it from lodging in consequence of heavy rains that may happen near the time of maturity. Both varieties may be sown from the middle of July till the end of August, and treated in every respect like other winter rape.

Spurry, (Spergula arvens is,) an annual from Germany and France, where it is much cultivated as a winter pasture for cattle and sheep. Mutton, as also the milk and butter of cows fed with it, are stated by Thaër to be of very superior quality. It is usually sown on stubble-fields after the grain crops have been removed.

But the principal use to which this plant can be applied in this country is as a green manure, on poor, dry, sandy, or worn-out soils. It may be sown either in autumn on the wheat stubble, or after early potatoes, and ploughed under in spring preparatory to the annual crop; or it may be used to replace the naked fallow, which is often hurtful to lands of so light a character. In the latter case, the first sowing may take place in March, the second in May, and the third in July, each crop being ploughed in to the depth of three or four inches, and the new seed then sown and harrowed. When the third crop is ploughed in, the land is ready for a crop of winter grain.

Sand or Sea-side Lyme Grass, (Elymus arenarius,) a perennial from Holland. This grass is not eaten by any of our domestic animals, owing, no doubt, to its excessive hardness and coarseness. Sir Humphrey Davy found, by analyzing the soluble matter afforded by this plant, that it contained one-third of its weight of sugar. Hence it has been called the "Sugar-cane of Great Britain." It has been recommended, however, that the hay made from it be cut like chaff and given to cattle, either alone or mixed with other food.

The purpose for which this plant is generally employed, and for which its creeping, matted roots fit it in an eminent degree, is for binding loose sands, when sown with the Arundo arenaria, to prevent the encroachment of the sea.

Sea Reed, (Arundo arenaria,) from Holland. This plant, like the preceding, is unworthy of cultivation as food for cattle, but can only be employed to advantage in raising a barrier against the encroachment of the ocean.

The object of importing the seeds of these grasses was, to sow them on such parts of our coasts as may be threatened or are suffering injury from the sea, particularly on beaches or sand-hills which are liable to changes from abrasion or drifting winds. The world-renowned dykes of Holland owe much of their strength and durability to the protection afforded by those remarkable plants. With regard to their culture, I have no definite knowledge.

Trees and Shrubs

The Carob Tree, or St. John's Bread (Ceratonia siliqua). — Of all the seeds imported for the purpose of distribution there are none more interesting nor more valuable than those of the carob tree. The pods, when matured, contain a few drops of a substance resembling honey. The tree is unquestionably of Eastern origin, and is supposed to be identical with that upon which St. John fed while in the wilderness. The seeds were procured for the office from Alicante, in Spain. In Murica, Valencia, Catalonia, and other provinces in that country, it abounds, and frequently forms, with the olive and other valuable trees, large forests. It was, without doubt, introduced there by the Moors, who knew its nutritive qualities as a food for their horses, mules, and cattle. They probably brought it from Palestine and Egypt, whence it appears to have originated. In these Spanish provinces, it now grows naturally in every kind of ground, not excepting the driest and most barren spots, where the underlying rock shows itself more frequently than earth. Its roots, twisting in every direction, accommodate themselves to the lightness or depth of the soil; while the trunk, remarkable for its smooth and light-colored bark, attains in sheltered positions a colossal size. The branches, furnished with greyish colored leaves, spread majestically around the trunk, and, when loaded with fruit, hang down quite to the ground in the form of a tent. The fruit ripens rapidly, and such is its abundance and weight that it is necessary at once to gather it. The pods are sweet and rich in sugar, and animals feed on them with avidity, and become quite fat and in good condition for work.

There are several varieties of the tree. The produce is necessarily in proportion to the attention given. It blooms twice a year — about the first of February and the middle of September — and when well watered arrives at a considerable height, and sometimes covers a space of one hundred feet in diameter, bearing upwards of a ton of pods. It will doubtless succeed in the Southern and perhaps in the Middle States.

The Olive (Olea europaea). — Of the olive, it has been said, with much truth: "Olea prima omnium arborum est;" and when we consider its productiveness, longevity, and usefulness, a little enthusiasm on the subject, perhaps, would not be altogether misplaced. The present importation is by no means the first attempt to cultivate this tree in this country, as it had already been introduced into California by the Jesuits one hundred and fifty years before. In about the year 1755, Mr. Henry Laurens, of Charleston, imported from remote parts of the globe a great variety of useful and ornamental productions, among which were olives, capers, limes, ginger, Guinea-grass, the Alpine strawberry, (which bore fruit nine months in the year,) red raspberry, and blue grapes; also directly from the south of France, apples, pears, plums of choice varieties, and the white Chasselas grape, the latter of which bore abundantly. The fruit raised from the olive tree was prepared and pickled, equal to those imported.

In 1769, the olive was introduced into Florida, by a colony of Greeks and Minorcans, brought over by a Dr. Turnbull, an Englishman, who founded a settlement called "New Smyrna."

In 1785, a society was incorporated in South Carolina for the promotion of agriculture. The object was to institute a farm for agricultural experiments, to import and distribute foreign productions suitable to the climate of Charleston, and to direct the attention of agriculturists of the State to economical objects, as well as to reward those persons who should improve the art of husbandry. Among other objects of interest, the society imported and distributed some cuttings of vines and olives. The latter answered well, but the climate near Charleston proved too moist for the grapes. Attempts have been made to propagate the olive from seeds in various parts of the South, but hitherto with little success. This may be attributed to a tendency in the olive to sport into inferior varieties when so planted; but there is every reason to hope that the new importations of cuttings of improved kinds will increase the production in many parts of the South.

Congress, in the year 1817, granted four townships of land in the present State of Alabama, on a long credit, to a company of French emigrants, for the purpose and on the condition of their introducing and cultivating the olive and grape; but the enterprise never was prosecuted to any considerable extent, and it finally fell through, and the lands reverted to the government.

Of the olive stocks and cuttings recently from France, the following varieties were received and distributed in the Carolinas, Georgia, and other States bordering on the Mexican Gulf: Olivier blanquet nain; Olivier vermillion nain; Picholine nain (a variety yielding the kind of olives most celebrated for pickling, and is not very particular in the choice of soil and climate); Olivier verdal nain; Olivier de cruan nain; Olivier de salon (a variety producing a small round fruit, good for oil, and prefers dry, elevated ground); Olivier bouquetier nain; Olivier gros Redonaon; and Olivier violet.

The Fig (Ficus carica). — The fruit of this tree is a great and wholesome luxury, both in a green and in a dried state, and its multiplication in our Southern and Southwestern States cannot fail to be fraught with great advantage. It will grow well upon the poorer and drier soils, provided it is sheltered, and can be propagated with great ease; and such is the goodness and abundance of its fruit, and the number of its varieties, that in some parts of Southern Europe it goes by the name of the "Providence of the Poor." In Spain it grows side by side with the carob and almond trees, and lines the fields and vineyards, its deep-green boughs forming an agreeable shelter from the heat of the sun.

The nature of the soil and its aspect influences considerably the choice and cultivation of the different kinds of figs. The white varieties, for instance, seem to prefer an elevated position and a strong, light soil; while the darker kinds succeed best where the situation is sheltered and low. A very choice sort, the fruit of which is of a deep rose-color, while the trunk of the tree is nearly black, seems to thrive best in low, shady places, provided it be exposed to the rays of the rising sun. It is possible to increase the varieties of the fig *ad infinitum,* either by seed or by the more common method of cuttings, inclined and buried from two to three feet in the earth. In the third year the young tree is pruned, and the head is formed by leaving three branches, which in due time are covered with fruit. Some cultivators graft them in various ways about the time when the sap begins to move. With due attention, the product is greatly improved and increased, although few fruit trees, perhaps, bear so abundantly, considering the little trouble taken with them.

In all countries which may properly be called "fig climates," two crops are produced in a year. The first is from the old wood, and corresponds with the crops of England and the middle portions of the United States; and the second from the wood of the current year, the figs produced by which, in the last-named countries, are never ripened except in hot-houses. In Greece, Syria, and Egypt, a third crop is sometimes produced. The first crop is ripened, in the south of France and in Italy, in May, and the second crop in September.

The only variety of cuttings lately imported from France was the large "White Fig," (Figuier blanc,) which is sufficiently hardy, with slight protection, to withstand the climate of the Middle States.

The Prune (Prunus domestica). — The scions of two varieties of prunes, "Prunier d'Agen," and "Prunier Sainte Catherine," have been imported from France, and distributed principally in the States north of Pennsylvania, and certain districts bordering on the range of the Alleghany mountains, in order to be engrafted upon the common plum. These regions were made choice of in consequence of their being freer from the ravages of the curculio, which is so destructive to the plum tree in other parts as often to cut off the entire crop. It has been estimated that the State of Maine, alone, where this insect is rarely seen, is capable of raising dried prunes sufficient to supply the wants of the whole Union.

The *Prune d' Agen,* which is considered the best for drying, is of good size, of a violet-color, with deep-yellow flesh of a delicious flavor. This variety succeeds best when engrafted upon a wild stock, or when it springs up directly from the root.

The *Prune Sainte Catherine,* in the climate near Paris, is also esteemed as excellent for drying. It likewise furnishes to commerce the well-known "Pruneaux de Tours." The tree is of medium size, about twenty-five feet high, and grows well both as a pyramid and as a standard. The branches are long, slender, and but little ramified; their shape being rather slight. Throughout their whole length there grow a large number of buds, so near to each other that on a branch a yard long there are often produced from fifty to sixty plums. Hence it is easy to conceive the excessive abundance of the crop of a tree thus laden with fruit, the productiveness of which is not equalled by any other kind. This plum is of medium size, obovate or nearly round, divided by a deep suture throughout its length. The stem is slender, about three-fourths of an inch long, curved at its upper part and inserted in a small cavity. The skin is fine, pale yellow, sometimes tinted with red on the sunny side, and lightly covered with a white transparent bloom. The flesh is yellowish, sometimes firm and adhering to the stone, very juicy, sweet, and agreeably flavored. It ripens in the neighborhood of Paris in September and October. This plum, beyond its unrivalled merits for preserving in a dried state, has the advantage of being an excellent dessert fruit when fully mature.

In very warm, dry climates, prunes are prepared by drying on hurdles by solar heat alone; but in France, they place the plums upon round wicker baskets, about two feet in diameter, and two inches deep, putting into an oven heated sufficiently warm to cause the fruit to wrinkle after an exposure of about twelve hours. The oven is again heated, continuing to increase the temperature until the plums become firm, when they are flattened by pressure between the fingers, while under the process of desiccation. Great care is observed to remove the plums from the oven as soon as they arrive at a certain stage of dryness to prevent them from cooking too much. Finally, after the prunes are baked for the last time, the oven is heated as it should be for bread, in which the plums are exposed until they begin to swell and bubble, when they must be taken out. As soon as the temperature of the oven falls to about half-heat, the prunes are put back to remain over night. Then, if properly cooked, they are covered with a beautiful white "bloom." They are then assorted by sizes, and packed in baskets, boxes, or jars, for sale or use.

If it is desirable to make what are called "Pruneaux fourrés," the stones are taken out when they are about half baked, and insert in its place another plum which has also been deprived of its stone, and continue the cooking as above.

Raisin Grape-vines. — Two varieties of small grapes, the "Vigne cheveles," and the "Vigne Corinth," from which are made the Ascalon, Stoneless or Sultana raisins, and the Zante or Corinth currant, imported from France, and principally distributed in the Middle and Western States. The berries are small, often without seeds, with a fine pulp, and of an agreeable flavor. They are much used in a dry state in domestic cookery, and, should they succeed in this country, will add to the many varieties of useful and wholesome fruit already introduced. The English name of "currant" given to the Ribes rubrum, arises evidently from the similarity of that fruit to the small grape of Zante, or the common grocer's "Corinths," or "currants."

The Levant and Grecian Islands supply the largest proportion of dried currants for the markets, and retain their reputation by the general superiority of the fruit they furnish. Spain, Italy, and the southern portions of France, also supply a considerable amount. The method pursued for making these currants varies somewhat with the locality and the variety employed. They are more easily prepared than the larger grapes, which are known in commerce under the name of "raisins." These require to be dipped, in the first stage, into a rather strong ley, made of wood-ashes, sweetened by an addition of aromatic plants, such as thyme, lavender, orange leaves, &c.; but the small grapes here in question are merely gathered a few days after complete maturity, at the moment when it is perceived that the berries are about to fall from the vines. They are then placed upon hurdles of close wicker-work, or upon large sheets, in the sun. When it is perceived that the berries are detaching themselves from the main stalk, although still preserving their stems, the operation is often hastened by striking the bunches slightly with a stick. The stalks are then separated from them by means of a sieve, and the dust and other remains are got rid of by winnowing; after that, they are packed in boxes, where they are pressed in closely, covered with thick paper, and kept in a dry, cool place.

A very important point in the management of all varieties of grape is the mode and season for pruning. No general system or rule will suit. Experience must be the guide as to what will answer best in different climates, soils, and situations. A method which will do well in the North may be destructive to the plant in the South.

The Jujube Plum, (Zizyphus sativa,) a small tree or thorny shrub, from the south of France, bearing a reddish plum about the size of olives, of an oval shape and sweet clammy taste, including a hard oblong stone, pointed at both ends. From this fruit is made the ''Jujube paste'' of the shops. In Italy and Spain it is served up at the table in desserts during the winter season, as a dry sweetmeat. These seeds have principally been distributed in the Middle and Southern States.

Pistachio Nut, (Pistacia vera,) an extremely interesting tree, has been imported, not merely on account of its ornamental character, but because it is useful and produces agreeable nuts. For the twofold reason, a quantity of them has been imported from the southern part of Europe and widely distributed throughout the Middle and Southern sections of the Union. In favorable situations, it will attain a height of fifteen or twenty feet, and frequently, while a mere shrub of five or six years' standing, will bear. Its branches spread out widely, without being numerous; and the trunk is covered with a greyish-colored bark. The inflorescence takes place about April or May. The male flowers, which appear first, shoot from the side of the branches in loose panicles, and are of a greenish tint. The female flowers put forth in clusters in the same manner.

As the pistachio tree is dioecious, it is necessary to plant male and female trees together, or they will not produce. The nuts are of an oval form, about the size of an olive, slightly furrowed, and of a reddish color, containing an oily kernel of a mild and agreeable flavor. It is a native of Persia, Syria, Arabia, and Barbary, and is supposed to have been introduced into Italy in the second century by the Emperor Vitellius; whence it was carried into France, in the southern parts of which it is so far naturalized as really to appear indigenous. Later still, that is in 1770, it was introduced into England, where in sheltered positions it bears without protection from the cold of ordinary winters. The summers there are scarcely warm enough to ripen its nuts. Although severe frost is to be dreaded, it will bear a greater degree of cold than either the olive or the almond, and hence is better adapted to the climate of our Middle and Southern States, where it is thought it could be cultivated with profit. The finest kinds are those known as the Aleppo and Tunis varieties — the former for its large size; the latter, though smaller, for qualities which recommend it to French confectioners, who cover the fruit with sugar and chocolate, and flavor creams and ices with it. A similar pistachio nut is used in France in the preparation of sausages and in seasoning meats. It is considered as a tonic, and as beneficial for coughs and colds. It is frequently eaten raw, but oftener in a dried state, like almonds.

The Cork Tree, (Quercus suber,) from the south of Europe. Much is anticipated from the successful introduction of this product, as the acorns have been distributed throughout the Middle and Southern portions of the Union for experiment, where it is hoped that it will prove to be adapted to the soil and climate.

Should a portion of the present distribution by any untoward circumstance fail to answer expectation, care should be taken by the office to obtain another supply for those who feel an interest in growing this useful tree. Plantations might be established in every favorable locality, so that in due time, the increasing wants of the country for cork may fully be met by the home supply. Therefore, if the introduction should prove successful, the enterprise cannot be regarded otherwise than of national importance.

This tree, under favorable circumstances, grows rapidly, and attains a height of upwards of thirty feet. Indeed, even in England, there are specimens over fifty feet high, with a diameter of more than three feet. In the south of Europe, cork trees are much esteemed, and lands planted with them are considered the most profitable of all that are not irrigated. They seem in general to prefer those localities where gneiss, sandstone, schistose and calcareous rocks abound. The substance so familiarly known to us as "cork," is the epidermis or outer bark, which sometimes acquires a thickness of two or three inches. This is rarely taken off until the tree has arrived at an age of fifteen or twenty years. This operation, which is carried on every six, seven, eight, or nine years, according to circumstances, is generally completed in the months of May and June, while the sap is still active in the tree. Although easy to accomplish, some care is required to avoid injuring the true bark, (Liber,) which lies under the cork. A circular incision is usually made round the foot of the tree, and another near the branches. Longitudinal cuts are then made; and finally, by using the handle of a hatchet as a wedge, the cork is detached from the under bark. The larger branches are treated in a similar manner.

Maté, or Paraguay Tea, (Ilex paraguariensis). — We are indebted for the seeds of this shrub to Lieutenant Page, of the U.S. steamer Water Witch, while engaged in exploring the sources of the Rio de la Plata, in South America. It is worthy of attention of persons living in the Middle and Southern sections of the Union. As a tree, it is highly ornamental; and wherever the Magnolia grandiflora will thrive, there it may be successfully cultivated. The inhabitants of Paraguay, and indeed most of those who use it on the southern part of this continent, attribute to it almost fabulous virtues. It is unquestionably aperient and diuretic, and produces effects very similar to opium; but most of the qualities so zealously attributed to it may, with some reason, be doubted. Like that drug, however, it excites the torpid and languid, while it calms the restless, and induces sleep. Its effects on the constitution, when used immoderately, are similar to those produced by ardent spirits; and when the habit of drinking it is once acquired, it is equally difficult to leave it off. The leaves of the plant are used by infusion, and all classes of persons partake of it, drinking it at all hours of the day at their various meals, rarely indeed beginning to eat before tasting their favorite beverage. Not only is this the case in Paraguay, Uruguay, and the Argentine Republic, but in Peru, Chili, and Ecuador, it is no less esteemed. They drink the tea from the spout of a pot which they call *maté*, adding to it a little burnt sugar, cinnamon, or lemon-juice. The wealthier or more refined class draw it into the mouth through a tin or silver pipe, called *bombilla,* which, being perforated with holes at one end, and inserted in the *maté*, or teapot, enables them to partake of the liquid without swallowing the smaller particles of the pulverized leaves floating on the surface. The quantity of leaves used by a person who is fond of it is about an ounce. The infusion is generally kept at the boiling temperature, but those who are accustomed to it seem to drink it thus without inconvenience. In the mean time, hot water is supplied as fast as it is consumed, every visiter being supplied with his *maté* and pipe. If allowed to stand long, the tea acquires an inky color. The leaves, when fresh, taste somewhat like mallows, or inferior Chinese green tea.

Morocco Dressers' Sumach, (Rhus coriaria,) from the south of Europe. The seeds of this shrub have been imported for experiment in the Middle States, where it is thought it will be adapted to the climate. It usually grows six to eight feet in height, on dry, sandy, or rocky soils, in exposed situations. The branches and leaves are imported into this country, and employed for tanning leather. It is said that they are used in Turkey and Barbary for preparing the Turkey morocco from the skins of sheep and goats. The seeds are sold at Aleppo, where they are eaten to provoke an appetite.

Furze, (Ulex europaeus,) from Brittany, in France; a low prickly shrub, used an an excellent green fodder for cattle, when bruised. It was imported for a hedge-plant in the Middle and Southern States, and is described in another part of this volume.

French Broom, (Genista scoparia,) from France; a low, hardy shrub, growing from three to nine feet in height, with numerous straight, sharp branches, and used as fodder for sheep and for making brooms. It will grow on any dry, meagre or sandy soil, and is well adapted for protecting the sides of the embankments and cuttings of railroads.

It may be remarked that most of the fore-mentioned seeds and cuttings have been, or are to be, placed in the hands of members of Congress, and the secretaries of State and County Agricultural Societies, for distribution in their respective districts, reference having been made to their adaptation to the soil and climate, as well as to the economy of the sections where intended to grow. All of those procured in Europe were obtained from reliable sources, and are believed to be of superior quality and true to their kind, the vitality of which has been tested, as far as practicable, by actual germination under glass or by other means. It is not to be expected, however, that every variety will succeed in all parts of the country, if in any, where the experiments are to be made, as one may have the disadvantages against him incident to a change of soil and climate, as well as from an unfavorable season, which no human power can prevent or avert.

I am, sir, very respectfully, your obedient servant,

Bread Crops, 1855

From "Bread Crops," U.S. Patent Office, *Report on Agriculture*, 1854 (Washington, 1855), pp. 131–33, 143–47.

Statement of Peter Reid, Lake Post Office, Greenwich, Washington county, New York

In this county are cultivated the favorite varieties of Indian corn usually grown in the Northern States. The average yield to the acre, on ordinary soil, well cultivated, is about 45 bushels.

Statement of Joshua Harris, of Welche's Mills, Cabarras county, North Carolina

The present crop of corn is light on account of the drought; but "bottom lands" have yielded pretty well. I had 40 bushels to the acre the present season on such lands. We plant 4½ feet apart each way, and two stalks to the hill. The present price of corn is 75 cents per bushel.

Statement of S. S. G. Franklin, of Cuba, Clinton county, Ohio

Owing to the severe drought last summer, our present crop has been cut short. The average yield to the acre will probably not exceed 25 bushels, though from some good "bottom lands" 75 bushels may be gathered.

The cost of production is 10 cents a bushel. Present price at Wilmington, 56 cents, and at Cincinnati 66 cents per bushel.

We generally break up sward land in the winter, harrow it well in the spring, mark it out into squares 3½ feet each way, and cover the seed by hoes. Our time of planting is from the 1st to the 20th of May. When the corn is up, we keep the ground loose and free from weeds with the cultivator or double-shovel plough. We cut up the stalks in the fall for fodder, and sometimes sow the ground with wheat.

Statement of D. C. M. Evans, of Scio, Harrison county, Ohio

The average yield of corn to the acre in this section is about 40 bushels, although some fields exceed 100 bushels. The cost of production is about 18 cents a bushel. The price of corn is from 62½ to 80 cents.

We cultivate by breaking up sward land in the winter, and harrowing well before planting. We mark the rows out 3½ feet apart each way, and cover by hoes. When the corn is three or four weeks old, the cultivator and double-shovel plough are used to keep the ground to loose and to destroy the weeds. The crop is cut up in the fall, and the ground usually sown with wheat.

Statement of Isaac R. Evans, of Harrisville, Butler county, Pennsylvania

Last season, I treated a field of corn in the following manner: The land had been in clover and other grass for three years previous. In the spring, as soon as the ground was fit to plough, I turned under the sod to the depth of eight or nine inches, and did nothing more to it till planting time, when it was thoroughly harrowed. After planting, it received a slight dressing of compost and coal ashes. The cultivator was then passed twice and the plough once over the ground. The drought was so severe that the yield was not more than half what it otherwise would have been. I harvested 40 bushels from each acre.

The two varieties most cultivated here are what are known as the "Pennsylvania White-cob" and the "Pennsylvania Red-cob." The latter is lighter in the grain, more easily shelled, and ripens a little earlier than the former, but is not so productive.

Statement of E. Shoemaker, near Ebensburg, Cambria county, Pennsylvania

An early and hardy variety of Indian corn has long been a desideratum among the Alleghany mountains, where our summers are generally too short for the successful cultivation of the common kinds known to our farmers.

Towards the last of May, I planted a sample of "Brown" or "Improved King Philip corn," on rather thin, poor land, and at the same time planted, on a similar spot in every respect, an equal quantity of the best variety of corn previously cultivated in this section. To each hill of the "Brown corn," I applied a quantity of a "fertilizer" purchased in Philadelphia, which, however, proved to be worthless, as, was fully ascertained by its application to other crops. To each hill of the other corn I applied a shovelful of manure. The "Improved King Philip" produced decidedly a better crop, both in quantity and quality, than the other variety, and ripened nearly or quite three weeks earlier than any other corn in this county.

Statement of George Buchanan, Samuel Gilliland, James T. Hale, David Duncan, and William P. Fisher, being that portion of their report which relates to Indian corn, addressed to the Centre County Agricultural Society of Pennsylvania

The varieties of corn mostly cultivated in this county are the "Large Yellow," "Squaw," and the "White Gourd-seed," the first two having the preference, as being earliest and most reliable in product.

The preparation of our land for corn is as follows: A stiff clover or Timothy sod is selected; and if low land, it is ploughed in the fall, at least eight inches deep, but if upland, the ploughing is usually deferred till spring, and in that case, all the manure used is hauled out in the winter or early in the spring, and spread upon the land. The ploughing should not be done till the frost is entirely out of the ground; otherwise, the soil will bake and remain hard all summer. The ground is then well harrowed, so as to be of fine tilth, and finally furrowed out, say at distances of 3 by 4 feet. The corn is planted at the intersection of these furrows. Some farmers only furrow one way, 3 feet apart, and then drop two or three grains in each hill at distances of about 20 or 22 inches. The reason assigned for this method is, that more corn can be raised on a given quantity of land than by the other mode. Corn-drills are not in use among us, and hence, we cover with the hoe. When the corn is up, we stir the ground with a light harrow, having a boy to follow and set up such plants as have been disturbed. Immediately after this, we commence with the cultivator, and continue its frequent use till the corn is too large for tilling. We believe that the thorough use of the cultivator will insure a fine crop in favorable seasons, without employing the plough, and leave the land in much better condition for the succeeding crop of wheat, than by its use. At the same time, candor compels us to say, that most of our experienced farmers prefer the plough at the last working of the ground.

Seed corn should be selected from the fairest ears, rejecting the grains on the extreme ends. Before planting, we recommend that the seed be soaked over night in a strong solution of salt and water, and then mixed with equal parts of lime, ashes, and gypsum, sufficient in quantity to enable the corn to be readily dropped. This promotes the early vegetation of the corn, and in some measure prevents the ravages of crows and worms. When the corn is well up, ashes and gypsum are thrown on each hill; and this process is repeated, or the mixture is sown broadcast over the field, at the rate of 10 or 12 quarts to the acre.

Corn is generally husked in the field, between the 15th and 25th of October, and hauled to the cribs as soon after as possible. Some farmers, also cut off the stalks entirely, and then sow the ground to wheat; but this method is followed only to a limited extent.

Corn in the ear is principally used for fattening hogs, and is considered excellent food for horses throughout the year. Corn-meal is also used for fattening cattle and hogs, and as food for horses. The past season, owing to the extreme drought, has not been favorable to a large yield of corn, but in some parts of the county an average crop has been obtained, which is estimated at from 60 to 80 bushels of ears to the acre. Shelled corn is worth, with us, from 55 to 60 cents a bushels.

Statement of James S. Montgomery, of Eagle Lake, Colorado county, Texas

In reference to insects which infest grain crops, I would state that, for several years, I have tested, with complete success, a plan for preserving corn against the ravages of the weevil. It is to store the corn dry and in good condition, in air-tight cribs. My cribs are built of logs, pointed and plastered with clay, and shedded all round. I have known others to pursue the same course, and invariably with complete success. I think storing in large quantities conducive to preservation.

*   *   *

Statement of Samuel J. Fletcher, of Winchester, Clark county, Missouri

My wheat crop of 50 acres, this year, averaged 20 bushels of prime grain to the acre; while, on the same quality of land around me, the hogs were turned into most of the wheat fields, the grain not being worth cutting. I plough deep, put 2 bushels of seed on an acre, harrow thoroughly, and finish with a two-horse roller. Wheat should also be rolled in the spring, when the ground is dry. On stiff lands, harrowing in the spring before using the roller, I have found, increases the yield one-third in quantity, and improves the quality. My crop has not failed for fourteen years.

Last year, I got three varieties of wheat from Baltimore, namely: the "Australian," "Gale's Early Flint," and "White Blue-stem." The first mentioned all died; the second filled well till it reached the milky state; it then shrivelled, and at cutting-time was worthless. The "White Blue-stem" filled and yielded well, 2 bushels of seed producing 38½ bushels, the whole of which I have sown the present year; and the young crop looks remarkably fine. I think this variety will suit our prairies.

Statement of Joshua Harris, of Welche's Mills, Cabarras county, North Carolina

The crop of wheat in this region has been not more than half the ordinary yield. The "Troy" wheat, which I spoke of in the last year's report, yielded well, but does not make good flour, even in the best mills. Most of our farmers sow the "May" wheat, as it succeeds best of any variety among us. It can be sown any time in November, and yields well.

Wheat is harvested from the 1st to the 10th of June. A bushel weighs about 64 pounds. The present price is from $1 25 to $1 50 per bushel. The price of flour, at our nearest railroad market, is $8 50 per barrel.

But it is again assuming a place in the fields of our farmers with fair success. The difficulty in raising wheat has not been altogether on account of the ravages of the weevil and Hessian fly. For, where neither was found, the straw would be imperfect, brittle, and turn black, and the kernel small, shrivelled, and make inferior flour. The same has been the case with oats and rye. It was first observed soon after threshing machines were introduced.

Statement of Henry H. Holt, of Cascade, Kent county, Michigan

Wheat is the most important crop cultivated in this county, and is usually of a fine quality. Its cultivation is increasing, as it is now considered our most certain as well as most profitable crop. For the last two years, it came to maturity before the severe droughts commenced, while other crops were more or less injured thereby. The "Soule White-flint" variety is mostly cultivated, as it generally yields better crops than any other.

A large portion of the land in this county is "oak openings," and much of it new. Consequently, farmers do not see the need of manuring it, therefore, very little manure is used. In preparing for the first crop, which is usually wheat, the timber is "girdled," the fallen trees cut and burned, and the land "broken up" in June. It remains thus until a short time before sowing, when it is worked with a heavy harrow, which loosens the roots, which are cut off by the plough. The loose roots are then collected and burned. The "breaking up" requires a strong team and plough, and costs from $4 to $5 an acre. The first crop of wheat usually pays the expense of "breaking up," sowing the grain, and building the fence.

The timber lands here usually produce the largest crops of wheat, but the "oak openings" are considered the most certain. The yield is from 15 to 30 bushels to the acre; average, 20 bushels. The time of sowing is from the 10th of September to the 1st of October. Wheat sown before the 20th of September has been more or less injured by the Hessian fly, especially the present season. The best time is between the 18th and 25th of September. The quantity of seed usually sown to the acre is about 1½ bushels. The weevil has never been known here. The usual price of wheat, for a number of years past, has been from 63 cents to $1 per bushel; this year, it has been from $1 25 to $1 40 a bushel.

Statement of C. F. Mallory, of Romeo, Macomb county, Michigan

The usual yield of wheat in this region is from 20 to 25 bushels to the acre. Cost of raising, about 50 cents a bushel. Market value, $1 50 a bushel.

Statement of J. D. Yerkes, of Northville, Wayne county, Michigan

The crop of wheat in this vicinity fell considerably short of the average, the present season, in consequence of excessive freezing and exposure last winter.

Statement of S. S. G. Franklin, of Cuba, Clinton county, Ohio

Wheat is our principal crop. The average yield in this county, the present year, could not have been more than 10 bushels to the acre. This will hardly pay expenses. Present price at our nearest market, $1 40 per bushel. The "Mediterranean" and "Rock" wheats are considered the best varieties.

Statement of D. C. M. Evans, of Scio, Harrison county, Ohio

The usual product of wheat in this region is from 12 to 20 bushels to the acre, although as many as 33 bushels are often raised. The time of sowing is between the 1st and 25th of September. Some sow still later. The harvest commences the first week in July, and continues about three weeks. The quantity of seed sown to the acre is from 1 to 2 bushels, generally 1½ bushels. The varieties which have proved most successful are "Soule's Garden," "White-velvet," and "White Blue-stem." Price $1 62½ to $2 per bushel.

Statement of Elias Green, of Wakeman, Huron county, Ohio

But little wheat is raised here, not much more than enough for home consumption. What we have principally to guard against is winter-killing. This can be prevented in a measure by ploughing up the ground into narrow lands, or ridges, of about 12 feet in width, and making cross-drains of sufficient depth to let off the surface water.

The "White Blue-stem" is now the most popular wheat with us. The "Mississippi" variety is earlier, and on that account preferred by many.

The general yield is about 20 bushels to the acre, at an average cost of from 40 to 60 cents a bushel, according to the skill or thoroughness practised in its cultivation. The surest and most common mode of cultivation is to summer fallow the ground, and sow 1½ bushels to the acre from the 1st to the 20th of September.

Statement of Thomas F. Hicks, of Jelloway, Knox county, Ohio

Wheat is the staple crop of this county, all others being subsidiary. It generally follows corn or clover. If corn, the soil should be ploughed deep, harrowed lightly, the seed sown broadcast, and brushed or harrowed in or drill-planted. If clover, and too tall or weedy to be turned under properly with once ploughing, it should be broken up, in May or June, and re-ploughed in the fall; or stirred once or twice through the summer with the harrow, to prevent the weeds from starting. It should then be stirred again thoroughly at sowing time with the cultivator, and the seed sown at the rate of 1¼ bushels to the acre.

There was more wheat sown in this county in the fall of 1853 than there had been for some time before; and there was every prospect of an abundance until within three or four weeks of harvest time, when it was discovered, to the consternation of all, that the red weevil, or wheat midge, was destroying it. A few weevils made their appearance years ago, but not enough to do much injury; but now the little pests have come with force sufficient to take, in many cases, nearly the whole crop. There was not a quarter so much sown this fall as last; and, if the weevil still continues its exterminating war, it is probable that the culture of wheat will be abandoned entirely, for a few years at least, but it is hoped that a remedy will be found. It has been observed that wheat sown very early ripened before the weevil commenced its ravages, and conversely, wheat sown very late did not mature till they had ceased their destructive work. From this, we might infer that by sowing either very early or very late the wheat might escape their depredations, and secure a crop.

Every means should be used to force the early sowing rapidly to maturity, by scattering a small quantity of gypsum over the field.

The varieties most extensively cultivated and universally approved here are the "White Blue-stem," and "Golden-strawed." The "Mediterranean" has been very little cultivated, in consequence of its coarseness and the darkness of its flour; but it was not so much injured by the midge as the other sorts, and it will probably be preferred hereafter on account of its hardihood. The "Golden-strawed" is a few days the earliest, presents a beautiful appearance, and has a plump grain.

Wheat has been worth, since harvest, from $1 50 to $2 per bushel. The "Blue-stem" is worth 2 or 3 cents more a bushel than the other sorts. The usual time of harvesting is from the 1st to the 15th of July. A few reapers are used, but the major part of the wheat is cut with the cradle, bound in sheaves, and set up twelve to a shock.

The cost of producing an acre of wheat, including the rent or value of the land for one year, may be estimated as follows:

Ploughing once	$1 00
Harrowing three times	70
Cradling, shocking, and housing	2 00
Threshing 20 bushels, at 7 cents per bushel	1 14
Conveyance to market	1 00
Seed-wheat, 1¼ bushels, at $1	1 24
Interest on land for one year	2 00
	9 09

Twenty bushels of what in market, at $1	20 00
Value of the straw	1 00
Total value of the crop	21 00
Cost of culture as above	9 09
Profit per acre	11 91

Ten bushels to the acre will pay for production at the present prices; but the permanency of a good price cannot be depended upon; four years ago it was but 50 cents.

Statement of C. Jacobs, of Dayton, Yam Hill county, Oregon

Wheat is the only crop on which we can safely rely. With proper cultivation, we may generally expect a yield of 20 bushels to the acre.

The best mode of cultivating is, to fallow in summer, ploughing twice, and just before sowing plough a third time, and harrow in the seed. The cost of cultivating and harvesting is about $15 an acre, which, deducted from the value of the crop, say $20, leaves $5 profit to the acre.

Statement of George Buchanan, Samuel Gilliland, James T. Hale, David Duncan, and William P. Fisher, being that portion of their report which relates to the cultivation of wheat, addressed to the Centre County Agricultural Society of Pennsylvania

The mode of raising wheat here is to turn under either clover sod or a field that has been in corn the year before; corn-field is preferred. Manure is put on in the spring in an unfermented state before ploughing for corn. This dressing answers well both for the corn crop and wheat that is to follow it. The ground is harrowed once or twice during the summer, then stirred after harvest, and sometimes harrowed again. Seeding commences the first of September, 1½ bushels being sown to the acre. It is either harrowed, cultivated, or drilled in. Clover seed is sown in March or April. Harvest commences from the 1st to the 10th of July. Our soil can be put and kept in a high state of cultivation, with deep ploughing, proper management of barn-yard manure, red clover and gypsum, so as to be made, on an average, to produce 30 bushels to the acre, without the aid of guano or any artificial fertilizers. With our present mode of farming, the yield is perhaps not over 16 or 20 bushels to the acre, in average seasons. Our farms are entirely too large, and too much stock kept on them.

The "White Blue-stem" variety is principally sown. It is fast deteriorating, and must be replaced by some other kind. "Australian," "Mediterranean," and "Golden-stem" are also sown. Our crops this year fall below 10 bushels to the acre.

Our soil is strictly a limestone clay, equal, if not superior, to any other in the State for wheat. The district comprising the valley of the Susquehanna, in the counties of Centre, Union, Clinton, and a part of Blair, and Huntingdon, 70 miles in length by 12 to 20 in width, has the elements of wealth, both mineral and agricultural, as profusely scattered as any other portions of the State; and had we better facilities of getting to market, this region would be second to none. At present, our surplus products have to be transported in wagons, over mountains, from 20 to 30 miles to get to a railroad. The average annual amount of surplus wheat in this valley is over 1,000,000 bushels.

Statement of William Smoot, of Boone Court House, Virginia

Wheat should be sown by the last of August or first of September in this region. By early sowing it escapes the rust.

Driving Cattle to California, 1855

From Letter, Thomas Rebar to James C. Riggin, Apr. 15, 1855. This item is reproduced by permission of The Huntington Library, San Marino, Calif.

Mission of San Jose April 15/55

DEAR JIM

It is now nearly one year Since I bid you & your family farewell when I last seen you, I promised to write to you frequently, which promise I have most shamefully neglect, and for which I have no apology to make, but feel myself censurable for so gross a neglect, at the Same time hoping that Musser has kept you booked up, on all the haps & mishaps of our adventurous trip across the plains, and also Since our arrival here, But the Old proverb is better late than never, So I will endeavour to give ou a few outlines of Our doing Since we left you, you remember we left Tarkio on the 4th of May Crossed the Missouri River on the 9th Struck Platte on the 15th passed Fort Kerney on the 21st We experienced some heavy storms thus far crossed South Platte & struck North Platte the one the 25th, no detension in crossing, Passed Fort Laramie June 4th, crossed North Platte on the 10th over the bridge passed South Pass on the 18th with the exception of a few days we had fine weather to travel but grass generaly short, crossed Green River on the 21st Struck Bear River 25th Soda Springs 29th, East Branch of Raft River on July 4th, Where we feasted on fine trout, and tasted our small Keg of Brandy but nevertheless I felt from home, Struck the Junction of the Salt lake & California road on the 7th the Humboldt on the 14th along this stream for several 100 Miles we had pretty good grass & got along fine, when we lost our big Mule & horse by some disease, hard to tell what the cause was, after which we lost, now and then a cow briste (?), until we struck the desert on the 29th, entered in the Morning, Stoped three times and struck Carson River on the Morning of the 30th about Sunrise, we had a shower while on the desert and Several cool, cloudy, days, before we struck it, which you Know was very favourable to us, we traveled Slow up this river, the cattle dieing off very fast Some days 7 and 8 heads John Musser left us in the Midst of this Mortality and him & Levi took on the horses Spare Mules as fast at convenient, on the 2nd of August, Hiram was a faithful fellow, and we done the best we could clumb the last Summit on the 12th arrived at Hangtown on the 17th, John met us again on the Macosma on the 20th, arrived on the San Joaquin on the 25th, with about 285 head of cattle, out of 360 which we started with, lost 2 mules & the large horse, and let me assure you we done well to what many others did, very few got in with two thirds and many lost half what they started with, we have sold nearly to the amount of 6000 Dollars & have 231 head left besides 4 mules 1 Horse & one waggon, we can occationaly sell a cow at pretty fair prices Say 100 Dols Large work cattle we sold at 150 dols per Yoke, but they would not bring it now, we sold our horses & Mules well, 4 of each at 250 Dols per head, we could buy them again for less money I could buy cows through the country at from 40 to 50 dols per head from people that have to raise money, we intend Keeping our cattle until next winter, there is no selling them any sooner at reasonable fair prices on a/c of the Scarcity Money, times are hard here and every thing low flour brings about nine dols per barrel Potatos one cent per pound, Beef 9 & 10 dols per Hundred by the Carcass, the Supposition here is, that there well be a great deal of stock brought from the lower part of the state this season, and but very few if any across the plains, You will please answer this letter and inform me whether there are any driven from your Country this season or not, and about what Number, We got all the cattle through that we bought of you excepting the Spotted off Steer we Sold the other 3 at 75 dols per head and have the cow yet Brandy died since we sold him, him & Buck were Sick on the way & very much Swollen Backs, one Eye was Swelled Shut for several days Hiram is still here what he intends to get at I do not Know he got but 5 Yoke of his cattle through & sold two yoke, the young Black Bull we got through but none of the Others we started with

With very few exceptions I have enjoyed very good health since I left you, the rest of the boys, that is John Dick Wils & Hiram are well John was sick Several Month during the winter, he intends going home next month, Wils & Dick will buy his interest in the sheep, & take care of his interest in the cattle until they are ready for market he is still very much taken with that Ella of his, Kostenborder is some place about Volcano how he is doing I do not Know, neither do I care he is decidedly a mean man in my estimation Burns & Olymer are on Deer Creek doing pretty well, the rest are at and about Hangtown Louis is at Weaverville in Company with Dicks brother & John Candor, we have had a good deal of rain here of late & consequently fair prospects of a good harvest, we are located about five miles from the Mission of San Jose, in the Coast Lange, have a fine range for our cattle good grass & fine water they are doing well and improveing fast

Now Jim I want you to answer this letter as soon as convenient & any information you ask I will give you if in my power Remember me to all my acquaintances
My Respects to you & your family
This day
your daughter
is one year Old,
I hope the coast
is yet clear

YOURS TRULY
THOS. REBAR

Observations on the Prairie, 1855

From Franklin Langworthy, *Scenery of the Plains, Mountains and Mines*, Reprinted: Paul C. Phillips, ed. (Princeton, N.J., 1932), pp. 195–96.

Question Answered

If gold is so universally diffused through the mining country, and if the soil in the valleys is so fertile, and produce sells at so high a price, — why cannot every man who is prudent and industrious, soon acquire wealth in California?

Answer. — Although gold is quite generally diffused, yet there are comparatively but few places sufficiently rich to pay the expense of obtaining it. Whether the miner is earning anything or not, his expenses are always going on, and five hundred dollars per year, is a moderate bill for a miner's board and expenses. A majority do not earn much to exceed that amount, and thousands even fall short of it, and in consequence are getting deeper in debt. I believe that one-half of the miners would return to their homes, if they had money to pay their passages.

As to farming, it requires some capital to make even a small beginning. Claims on good lands are high. A small outfit for a farmer would stand something as follows:

Claim on forty acres of land,	$400
Yoke of Oxen,	300
One Cow,	200
Plow, Drag, Chains, and other tools,	100
One Sow,	100
Twenty-five dunghill Fowls,	100
Seed,	200
Total,	$1,400

To fence this field, might cost as much as all the foregoing bill, amounting in all, to nearly three thousand dollars. Such a capital would enable a man to commence farming upon a small scale. But how shall a man begin without capital at all?

In the mines it seems to be equally difficult to commence business to advantage without some capital. Without means to purchase stock, one cannot become a partner in any of the large Damming, Tunneling, Quartz-mining, or Water Companies. The only chance is at individual enterprise. By this mode, there are two ways in which to proceed. First, To work on wages for others. Secondly, To find diggings for one's self. Wages are high, but employment fluctuating. A man frequently engages to labor for another at five dollars per day. Before the end of the week, the mine fails, and all the hands are discharged. What was earned here, must all be paid for board, before employment can again be found. Hence there is no certainty of acquiring capital by working for wages. The miner must then either find diggings of his own, or else purchase a claim supposed to be good, of some other person. A claim cannot be purchased for a sum much short of five hundred dollars, that holds out a tolerably fair prospect of yielding a return of five dollars for each day's labor. But when this amount of money has been paid, the claim in a majority of cases, proves to be good for nothing. Perhaps a deception had been practised by the seller. Gold dust has frequently been by men's hands, privately mixed with the earth, and then washed out while some person was present, by which means the claim obtains a high reputation for riches, and sells accordingly. This process the miners call "salting a claim." If the miner searches for diggings of his own, he may strike a fortune the first day, or he may toil faithfully for months and years, and all his labor prove totally unavailing.

The foregoing facts are a sufficient answer to the question, — why every man cannot gain wealth in California. . . .

The Prairies

The word "prairie," simply signifies a meadow; and all such tracts in a state of nature are covered with native grasses of different varieties. These natural pastures are as good as those that have been cultivated, and the "prairie hay" is but little, if any, inferior to timothy and clover. Many persons who have never travelled in the Mississippi Valley, conceive that a prairie is a tract of land perfectly level, and covered with a rank growth of tall grass. This impression has doubtless been made, by reading the common school geographies now used and taught to children in every part of the country. In those books I find a prairie described as before stated. It is time that such a false impression be removed. I will admit, that near the banks of some large rivers, tracts of low, wet land, may be found, producing rank and tall grass. But not one acre of prairie in a hundred is of this character.

The prairie country is by no means a dead level, but is beautifully diversified with hills, and swells of land, of different elevations, similar to countries that are naturally covered with forests. The grass upon them is rather short, and not generally as tall as timothy and clover. The prairie looks some like an old meadow, that has been improved for so great a length of time that the grass is somewhat "thin," being crowded out by various weeds and flowering plants that have sprung up. In the western country, rivers have wide bottoms, or intervals. These are somewhat level. But a small part even of these is covered with the tall grass, and when that is found to be the case, it is only upon the low and sunken spots. The inhabitants and farms are principally to be found upon the high, rolling prairie.

There are numerous springs of good water, and the streams have a rapid descent, and have, in general, pebbled or rocky beds. The soil on the high prairie is very fertile, having originated from the decomposition of successive generations of grass. The vegetable mold, of a dark color, is commonly about two feet in depth. Underneath this lies a thick stratum of clay. Not the least appearance of stone is to be seen upon the high prairie. Yet the whole country is based upon strata of horizontal rocks, and building-stone is found in the sides of ravines, and the banks and bluffs of rivers, and can generally be found in all parts of the prairie country.

The Mississippi Valley is the only real agricultural portion of North America, upon the east side of the Rocky Mountains. If our fathers had landed here, instead of at Plymouth Rock, New England might have been a wilderness to this day.

Philosophy of Prairies

There is no appearance, or even probability, that these prairies were ever cultivated, or ever had timber growing upon them. Upon digging, we find no appearance of the roots of trees, either decayed or sound. The surface of the land is perfectly smooth. If trees had ever grown here, or had the land ever been cultivated, the surface would not have been left thus even. To what philosophic cause can the prairies be attributed? I will frankly express my opinion on this, as I always do on every subject. I think the waves of Old Ocean once rolled over this part of the world. During that long period, the general surface of the land was formed when covered with water. When the waters retired, grass and vegetation sprung up, and has ever since maintained the ascendancy over timber. Since that time, timber has taken root along the water-courses, and has been gradually making encroachments upon the prairies. I regard the grassy prairies as more ancient than the groves.

The horizontal strata of limestone rocks upon which the country rests, is full of petrifactions of marine shells. This very season, a large petrified fish has been found imbedded in a strata of rock in this State. This specimen was discovered on high land, by the workmen in excavating for the railroad. How came fishes and shells imbedded in the rocks, thus far in the interior of the continent, unless the Sea God once held dominion here? Geological science alone enables us to understand anything at all in relation to these and such like mysteries of nature.

Lost Rocks, or Boulders

In various parts of the prairie country are large rocks, called boulders, lying upon the very surface of the ground. These boulders are of a species of rock, entirely different from any others found in the country. By some process in nature, they must have been transported here from some distant region. Their native country was, perhaps, the Ozark Mountains, in Missouri, or they might have come from the base of the Rocky Mountains. But how came they here? By what means did these heavy rocks travel a thousand miles? Answer: — An ocean once extended from here to the Rocky Mountains. These rocks were frozen into huge cakes of ice. The tides broke the ice loose from the shore. The strong westerly winds of winter drove the ice before it, until they grounded in the localities where they are now found. Such is my opinion on this subject. If any one can suggest a more rational hypothesis, I am agreed. These boulders are termed by the inhabitants, "lost rocks."

An Erroneous Impression

In the Eastern States, a prejudice exists against the country lying in the Mississippi Valley, in consequence of the numerous extensive tracts entirely destitute of timber. Some of the prairies in Illinois are of sufficient size for a number of counties, in which but little or no timber can be seen. The question is often asked, can such districts ever be settled and improved? I reply in the affirmative. These prairies can and will be all settled, at no very distant period. The Grand Prairie is even now pervaded by the great Illinois Central Railway. Flourishing towns are starting up as if by magic, at short distances along the entire length of the line. Timber for fuel, fencing and building, is by this means conveyed to every locality where it is required.

On the large prairies the soil is peculiarly fertile, and the expense of fuel and fencing bears no proportion to the cost of clearing heavily timbered land of its natural growth. Wood is at present used for fuel, in all parts of the State, but in some localities the price is as high as three or four dollars per cord. If wood, for fuel, should ever fail, coal will supply the deficiency, and of this article the State contains vast and inexhaustible strata.

Modes of Fencing

Most farms in Illinois, are at present enclosed by rail fences. Many, however, use boards for that purpose, and a few lots are fenced with split or sawed pickets, or palings, and some few farms are enclosed with wire fencing. Some of the best improved and fenced farms in the State, are situated six or eight miles from any growing timber.

View from the Car Windows

Farmers have discovered one considerable advantage in having their land far from the groves. In such places their crops are never disturbed by crows, blackbirds, or squirrels. Hedges, of Osage orange, are now coming into use, and from appearances will in a few years supersede all other modes of fencing in the prairie country. In three or four years, it forms a barrier impenetrable by horses, cattle, sheep, or swine. Farmers who settle on the prairie, far from timber, soon have their buildings surrounded by artificial groves of the most splendid locusts, or orchards of apples and peaches.

No prospect can be more rich and beautiful, than that which may be seen from the windows of the cars, as you roll through oceans of verdure, dotted over as far as your sight can reach, with rural dwellings, embowered among ornamental trees, and surrounded by rich fields, blooming orchards, and flowery gardens. Such is the appearance of Northern Illinois, even now, and it is scarcely a score of years since the wolf, and timid deer, were the undisputed possessors of the whole region. How will this country appear when a century shall have rolled away?

Breaking Prairie

Those who have never visited the prairie country, have heard the alarming fact stated, that the turf is so strong, that five or six yoke of oxen are required to turn over the green-sward. I admit that such teams are frequently used in breaking prairie. In this case however, the plows are of huge size, and cut a furrow from twenty-four to thirty inches wide. With an ordinary plow, cutting a furrow twelve inches in width, I have seen prairie broken up with the aid of one span of horses. The motive in having so heavy a plow, propelled by a team of such strength, is to break a greater number of acres in a day. For example: — A man with a light plow and team will break one acre in a day, whereas, with a heavy plow and strong team, a man can in a day break three acres. The large breaking plows are guided in their proper course by means of two wheels running upon an axletree, and firmly fastened to the forward end of the beam. Being furnished with this apparatus, a driver of the team is all that is required, the plow runs alone without any person to hold it. The surface of the ground being naturally smooth, furrows lie even, like planks jointed and laid in a floor. The edges of the plow-share are always kept sharp and keen, by means of frequent applications of a file which the driver always carries with him for that purpose. When a field is broken up and fenced, it may then be considered as being in the highest state of cultivation. Indeed it is in a better state of improvement, and more perfectly fitted for a crop, than is generally found to be the case in fields once covered with timber, after fifty years of cultivation. The turf being all completely turned over, is soon decomposed and reduced to a fine, dark, rich mould, admirably adapted to the production of wheat, barley, oats, indian corn, potatoes, and all kinds of garden vegetables. Such are a few of the advantages of a prairie country in an agricultural point of view.

Comparative Amount of Labor

Those best qualified to judge, are of opinion that it requires a similar amount of labor to cultivate one acre of land in the Northern States, that it does three acres in the prairie regions of the West. The western farmer has a decided advantage over the eastern, although his produce may sell at prices one-third lower. But the advantage will still be greater when railroads and other facilities of communication shall equalise prices in all parts of the Union. This state of things is rapidly approximating. The Mississippi Valley possesses advantages in the construction of railroads much superior to any other large division of the United States. The great Father of Waters and its innumerable branches, furnishes an extent of inland navigation unprecedented on our globe, amounting to twenty thousand miles. . . .

Slavery Viewed by a European, 1854

From Robert Russell, *North America, Its Agriculture and Climate* (Edinburgh, 1857), pp. 133–37, 140–44.

Baltimore, 20th Nov. 1854. — My feelings on entering a slave State for the first time were not easily described. In traveling through the north and west, little incidents are now and then occurring, which show an entire absence of the class-feeling necessarily arising at home in our dense and more stationary society. The equality which prevails among travellers who observe the ordinary courtesies is striking, and we feel happy that the progressive condition of all classes admits of such a state of things. A few days ago I enjoyed the company of the frank and lively farmers of Ohio; and to-day, when at dinner, I learned that the gentleman on my left was General Scott, the hero of the Mexican war, who was receiving no more attention than I. The abrupt line, therefore, that is drawn betwixt different races appears more unreasonable; and we wonder why labour, so honourable in one State, should in another become associated with a feeling of degradation. That one class is stationary in Baltimore, the mean huts in the suburbs bear sufficient testimony. . . .

I called upon a gentleman in Baltimore who takes considerable interest in the agriculture of Maryland. He was an earnest advocate for slavery, and maintained that it was no drawback to the cultivation of the land. He also spoke with vehemence regarding the conduct of the abolitionists in the north, upon whom he laid the blame of having been the means of putting a stop to the attempts which at one time were made to concert measures for the gradual extinction of slavery. Schemes having this end in view were at one time openly discussed in Maryland; now, however, nothing is thought of but such measures as are best calculated for making slave property secure. But perhaps the truth of the matter is, slaves have now become more valuable. He assured me that regular societies existed in Pennsylvania to aid slaves in making their escape from their owners in Maryland, who were constantly meeting with great losses from this cause. He also stated, however, that the value of land had of late years risen as much in Maryland as it had done in the adjoining free State of Pennsylvania. Guano has been applied to many of the worn-out lands in raising wheat, with highly satisfactory results. The effects of the manure on this crop are most certain when it is applied in autumn at the time the seed is sown.

Maryland is a comparatively fertile State. Its agricultural statistics show us that there is a vast difference in the natural capabilities of land. Perhaps some may have thought that my descriptions of the general poverty of the soil in the New England States is somewhat exaggerated; but its truth will be rendered apparent by comparing their produce with that of Maryland. It is often said that slavery tends to exhaust land and to check the development of its resources, whilst free labour has effects precisely opposite. This opinion is no doubt true to a certain extent; and if it be assumed as a fact that the effects of free and of slave labour are such, it only shows more clearly that good land is not easily exhausted, and poor is not easily enriched. From all the information which I could gather, I do not think that the cultivated land of the State of Maryland is so naturally fertile as the cultivated land of Scotland; yet how much more productive in wheat and tobacco is that Slave State than the whole of the New England States, where there is no lack of industry and activity. The agricultural statistics for 1850 put us in possession of the following figures: —

	Reclaimed Land.	Wheat.	Indian Corn.	Tobacco.
	Acres.	Bushels.	Bushels.	Lbs.
Maryland	2.797.905	4.494.680	11.104.631	21.407.497
New England States	11.147.096	1.090.845	10.176.056	1.485.510

Thus, although the six Free States of New England have about four times more land reclaimed than Maryland, yet their production of wheat is less than a-fourth, and of tobacco less than a fourteenth, of this Slave State.

The comparative value of free and of slave labour is a question upon which there has been much discussion. In going into the Southern States, I was more anxious to inquire into this question than into the physical and moral condition of the slaves, a subject upon which so much has already been written. Indeed, it was one of the principal objects of my curiosity in visiting the United States, to make myself acquainted with the circumstances which favour the institution of slavery, and give it so great a hold on the agricultural and commercial systems of the country. I was amused with the view taken by an influential paper in the north on the comparative economy of free and slave labour, and as a pretty fair specimen of the way in which the merits of the question are often discussed by the northern press, I shall give the concluding paragraph. The article, after commenting upon the recent fall of labour in the north, and its continued rise in the south, sums up: — "Corn-field hands in the south bring 125 to 150 dollars a year, cooks and house-servants 50 to 75 dollars. A woman, and a child eight months old, sold the other day, brought 1310 dollars, which at 7 per cent is 91.70; this, with life insurance at 3 per cent, 39.30. Taxes, doctors' bills, and clothing, with food, must sum up the cost of such a negro to 225 dollars a year. Free women servants with us receive from 5 to 7 dollars a month, which averages 72 a year. Children we would not have if they were given to us. The south, then, cannot stand for any length of time this competition of labour. The cost of negroes must come down, decidedly down. Hence we give the warning, stand from under.''

In Maryland, an able-bodied slave is hired out by his master to work in the fields at from 120 to 150 dollars a year, and of course the person who employs him gives board, which is no doubt greatly inferior to what the free labourer receives. A common Irish labourer employed on the railway near Boston, had a dollar a day throughout the year. He paid 12 dollars a month for board, but had coffee at breakfast, meat at dinner, wheaten bread, butter, cheese, and tea at supper. Farm servants at Burlington, Vermont, had 150 dollars a year, and their meals at their master's table. In the Genesee district, the wages of farm servants were 16 dollars a month for eight months, and 12 dollars a month for the other four. The nominal hire of a slave engaged in agricultural operations is therefore rather less than that of a free labourer. The cost of maintaining the slave is also less than of boarding the freeman, a difference which will so far assist in compensating the planter for the inferior work of the slave. These facts tend to show that slave labour cannot be materially dearer than free, even in those States where the two admit of being fairly compared. This opinion is further borne by the circumstance that after many inquiries among the farmers of the Free State of Ohio and the Slave State of Kentucky, I failed to satisfy myself that there is any difference in the value of land adapted for grazing and raising grain in the two States; for had slave labour been so much inferior to free, as is commonly supposed, it should have lessened the value of property in the Slave State.

But it is generally believed that the chief profits or the Kentucky of Maryland slave owners arise from *breeding* slaves and selling them to the cotton and sugar planters in the Southern States. It appears to me, however, that the breeding of slaves could, under no conceivable circumstances, be profitable on its own account. We must remember that the natural increase of the slave population in the Northern States is about 5 per cent per annum. This increase, therefore, being scarcely equal to the ordinary rate of interest of money, the mere breeding of slaves would not be profitable, though their maintenance did not cost anything. The natural increase of slaves is no doubt a considerable item in the profits of the slave-owner in Maryland and Virginia; yet, excepting in the rice and sugar districts, it is a far larger item in the profits of the slave-owners in the Southern States. On a cotton plantation, the sum invested in slaves bears a much larger proportion to the gross amount of the capital of the planter than where the land is only adapted for wheat and maize. The climate also of the greater portion of the States bordering on the Gulf of Mexico is quite as favourable to the increase of the negroes as that of Kentucky or Virginia. . . .

Delaware, after Maryland, is the most fertile State on the seaboard. It is of small extent, and its soil is only suited for growing the ordinary crops — wheat, maize and oats. Through the subdivision of property, the average size of farms is reduced to 90 acres. In 1850 there were 6053 farms, and 2289 slaves, or scarcely one slave, reckoning old and young, for every two farms. In 1790 there were 8887 slaves, but the numbers have since been gradually decreasing. The owners of slaves in Delaware are therefore in a very small minority, and if the majority willed, the institution might be abolished. This fact shows, however, that the political feeling is not very strong here betwixt those who own slaves and those who do not. The dread lest abolition would only be a transferring of the slaves from the Northern to the Southern States, serves to check the zeal of those who wish to have the system uprooted. A good field negro is worth at present about 1000 dollars, £208: 6s., a price which would subject slave-owners to a severe test were they put to the alternative of manumitting or of selling their slaves. It is affirmed that large numbers of slaves were sold to the southern planters when the legislatures of New York and Pennsylvania set a period for the extinguishing of slavery.

The great upholder of slavery in the Northern States is the cultivation of tobacco, and not the breeding of slaves. Slavery possesses great advantages over free labour in the cultivation and tending of this plant. This does not arise, as Adam Smith supposed, from the raising of tobacco being more profitable than the raising of grain, for if it were so its culture is as open to free as to slave labour, and it would undoubtedly be preferred. Nevertheless, in the present circumstances of the country, free labour cannot successfully compete with slave labour in the production of tobacco; for, among other reasons, slave-owners can always command the quantity as well as quality of labour that are required to raise this crop economically.

On good land, a freeman or a slave can cultivate twenty acres of Indian corn, and as many of wheat. The management of a slave property on which nothing but wheat and Indian corn are raised, is necessarily attended with great disadvantages, because the operations are diffused over a great area, and the superintendence must be more imperfect. But in corn-growing districts, free labourers, or, more strictly speaking, *small proprietors,* have great advantages over slave-owners. A large slave plantation which may have become unprofitable through exhaustion, will not afford a profitable investment for a capitalist to buy and to farm it by employing free labourers. But such a plantation would only afford a subject for free labour *were it divided into small farms,* whose proprietors would cultivate them with their own hands. In present circumstances, this is the only process by which slavery is uprooted, and it takes place more rapidly in poor than in rich land.

Though a slave may, under very favourable circumstances, cultivate twenty acres of wheat and twenty of Indian corn, he cannot manage more than two acres of tobacco. The culture of tobacco, therefore, admits of the concentration of labour, and thus the superintendence and management of a tobacco plantation will be more perfect and less expensive than a corn one. And while slavery can always command labour, it likewise possesses the great advantage of organizing labour. Tobacco cannot be cultivated in the Free States by hiring and employing labourers; it is only cultivated there by small farmers. These circumstances give slave labour great advantages over free in the culture of this crop.

But besides these advantages, slavery admits of having an economical division of labour in the raising and preparing of tobacco for market. Mr. Babbage points out the existence of an important principle in the division of labour when applied in manufactures, that is, "the master manufacturer, by dividing the work to be executed into different processes, each requiring different degrees of skill and force, can purchase exactly that precise quantity of both which is necessary for each process; whereas, if the work were executed by one workman, that person must possess sufficient skill to perform the most difficult, and sufficient strength to execute the most laborious of the operations into which the art is divided." The same principle applies to the organization of slave labour for tobacco culture, though in a different way, inasmuch as both young and old slaves can find suitable employment in the culture and preparation of the crop for market. Worms require to be picked off the plants during their growth, and the leaves are gathered as they become ripe at different periods of the season. These operations can be done as well, and consequently as cheaply, by women or children, as by full grown men. But often a small proprietor in the Free States can command no other labour than his own, which would be greatly misapplied in most of the manual operations connected with tobacco culture; because his team of horses might sometimes be standing in the stable while he was picking worms off the plants, which would render this very costly work. Thus, through the organization and the division of employment which slave labour admits of, it is virtually cheaper than free. . . .

If we compare, however, the counties in Maryland that raise a large quantity of tobacco with those that raise little or none, some interesting results are brought out regarding the relative numbers of free and slave population which particular descriptions of land can support. Prince George county, in the south part of Maryland, having an area of 600 square miles, is bounded on the west by the Potomac, and on east and north-east by the Patuxent. Its soil is well suited to the growth of tobacco; so much so, indeed, that it produces more than any county in the Union. The agricultural statistics of 1850 give its produce at 8,380,851 lbs. of tobacco, 1,590,045 bushels of Indian corn, 231,687 of wheat, and 100,947 lbs. of butter. The population by the last two censuses was as follows: —

	1840.	1850.
Slaves	10,636	11,510
Free	8,903	10,039

On the other hand, the county of Cecil, situated at the head of Chesapeake Bay, having an area of 300 square miles, is not suited to the growth of tobacco. Its chief produce in 1850 was 410,060 bushels of Indian corn, 168,112 of wheat, 208,380 of oats, and 9288 tons of hay. The population during the same periods was —

	1840.	1850.
Slaves	1,352	844
Free	15,880	18,095

The county of Alleghany, also, forming the western extremity of Maryland, and bordering on Pennsylvania and Virginia, has an area of 800 square miles. Being intersected by the Alleghany mountains, its surface is broken and irregular. The soil in the valleys is represented to be fertile, and well adapted for grazing. In 1850 it produced 101,733 bushels of Indian corn, 73,525 of wheat, 163,943 of oats, 231,038 lbs. of butter, and 10,896 tons of hay. The population during the last two censuses was —

	1840.	1850.
Slaves	812	724
Free	14,878	22,045

Thus, in the tobacco-growing county of Prince George, the number of slaves increased about 10 per cent from 1840 to 1850; but in the grain and pastoral counties of Cecil and Alleghany, slavery appears to be undergoing a process of gradual extinction.

A New Reaper, 1857

From *American Farmer* (Baltimore, April, 1857), XIII, p. 123.

The Ottawa (Illinois) Free Trader, gives the following account of a Harvesting Machine, which, if it accomplishes all that is expected of it, will, indeed, *"make a noise in the world"* — so Messrs. Murray & Van Doren may go ahead, and they will be sure to "win the race," or else "kill the horse." The editor says: —

"Our main purpose at present, is to speak of a new Reaper that we saw in operation on the farm of Messrs. Murray & Van Doren, in the town of Farm Ridge. Hitherto our Reaper inventors and manufacturers have at best been able to produce a machine that could cut and rake the ground ready for the binders. While the machines have thus enabled the farmer to overcome the crushing labor of the cradle, and to dispense with a large number of hands (always at this season of the year next to impossible to procure) the Harvester has been of no further advantage to him. In point of expense, a Reaper with two men and four horses, cutting 12 to 14 acres a day, is no cheaper than half a dozen of cradlers would be, who could cut down the same amount of grain in the same time.

"The Reaper of Messrs. Murray & Van Doren, whose first start in the great race for pre-eminence and public favor we witnessed on Tuesday, takes, therefore, an immense stride ahead of any Reaper now in use; it not only aims to cut and rake the grain for the farmer, but it delivers it over to him also bound and stacked. Thus the labor of at least eight men, which is required to bind and place in shocks the wheat cut by an ordinary Harvester, is entirely dispensed with, and the farmer saves the cost of just so many hands.

"We shall not attempt a description of this machine. Suffice it to say, the main driving wheel, the mode of giving motion to the sickle, and of propelling all the machinery about it, is entirely different in principle from any other machine we have witnessed. An advantage about its peculiar construction also is, that all the weight of its gearing is in the centre, so that it is evenly balanced, and there is no side draught, although the horses go in front.

All the essential parts of the machine, we believe, have already been patented. Two or three experimental ones have been made, in a rough way, merely for present trial. As the machine goes into the field, the inventors follow its motions, noting every irregularity or imperfection, and devising means to overcome every difficulty. In this way they are quite confident they will, by the end of the present harvest, have brought their Reaper to such perfection, that by the next season they can challenge the world to an open competition.

"The machine we saw in use was placed in a ten acre lot of poor winter wheat, although it had worked well the day before in stout spring. It was in company and competition with one of Haynes & Hawley's headers, keeping even pace with it handsomely, cutting about the same width, (between six and seven feet,) and while the header tumbled the grain into an awkward and ungainly wagon box or tender that must be kept at its side to catch the grain and then carry it off and stack it, the Murray & Van Doren machine, (dispensing with the two extra teams and half a dozen of hands required by the former,) as carefully and rapidly cut the grain, bound it, and place it in stacks. The stacking apparatus being placed on the opposite side of the machine from the sickle, balances the weight and draught so perfectly, that it requires no extra power to propel it. The stacks are small, amounting to about as much as half a dozen of ordinary shocks, but are so proportioned and symmetrically shaped as not only to allow the grain to season perfectly, but also to be secure from the weather without any further handling or even care.

"The honor of the first conception of the rough outlines of this machine belongs, we believe, conjointly to Messrs. Murray & Van Doren, although they have not refused to accept many useful hints from others. The simpleness and originality of the conception, however, is such, that if it succeeds, as they anticipate, no McCormick, Haynes, or Hussey, we imagine, will have the hardihood to claim that their patents have been infringed.

"We confess, from what we have seen of it, we have confidence in the success of this Reaper; and in that event, we are equally confident, it will make a noise in the world. Messrs. Murray & Van Doren are men of intelligence and abundant means, and when they enter the lists, will either 'win the race or kill the horse.'"

Farming in Maine, 1857

From Maine Board of Agriculture, *Annual Report, 1857* (Augusta, 1858), pp. 7–30.

In viewing my field of labor for the present year, two paths seemed to invite attention; the one regarding the improvement of agriculture as it exists in the older parts of the State, the other having to do with the development of the agricultural resources and capabilities of such portions as are not yet settled. Of the urgent necessity of the former there can be no doubt, and equally certain is it, that such was the principal end aimed at in the establishment of this Board; yet in view of the fact that the State is possessed of a large extent of what is almost an unbroken wilderness, portions of which are reported to be rich and fertile in a high degree, but of which very little definite and reliable information seems to be generally diffused among our citizens, and of one of the signs of the times, to wit; that the tide of emigration westward, which for years past has operated so disastrously in draining our State of brains, money and muscles, is now partially stayed, and so a favorable opportunity presented to invite anew, a candid and critical examination of the inducements offered by a judicious selection of virgin soil within our own borders, I do not feel at liberty to neglect the latter path, and accordingly propose herewith, first, to present such facts and considerations regarding a part of our newer territory, as I have been able to gather by personal observation and by inquiries instituted on the spot, and afterwards, to resume the consideration of agriculture in the State at large, viewed specially with regard to its defects and available modes of improvement.

The portion of our unsettled territory which is believed at this time to present the greatest inducements to immigration, is what is known as the Valley of the Aroostook, together with a tract fifty miles, more or less, south of this, embracing the five easternmost ranges of townships, and which is drained in part by other tributaries of the St. John, but principally by those of the Penobscot.

The greater part of this territory, embracing upwards of two thousand square miles, is what is usually denominated settling land, although lumbering has been largely, and is still, to a considerable extent, carried on in some sections of it; in other portions no more timber now exists than will be needed for building purposes. The land throughout is uniformly good; in some of these townships scarce a lot of one hundred and sixty acres can be found which is not capable of being made a good farm, and but little waste land is believed to exist in any of them.

The surface is more or less undulating, the easterly ranges of townships being less hilly, and more free from stone, than is usual in the State at large; the ranges west of these, as fourth and fifth, are more broken in surface, sometimes hilly, and with frequent boulders and out-crops of limestone, slate, &c. Some townships in the first range are so free from stones that even a sufficiency for wells, cellars, &c., is not always readily obtained.

The soil is various, but consists mainly of a deep rich hazel loam, and is usually underlaid with a substratum of limestone, sometimes, but to less extent, with slate, the depth varying from two to six or more feet. The soil seems to have originated mostly from the decomposition of limestone and slate. I noticed nothing like hard pan, clay or other appearance of retentive subsoil, the water passing readily downward so as to obviate any necessity for underdraining. . . .

Natural Growth

The forest trees of this section are of mixed growth. The sugar maple and yellow birch prevail chiefly, and these attain very great size. They are intermingled with occasional lofty pines, spruce, fir, white cedar, poplar, elm, ash, &c. In the lower lands, the evergreens here named, with some hemlock and abundance of larch or hackmetac, are plenty, but they are by no means confined to wet soils. Although much choice timber has been cut and found a market via the river St. John, considerable wealth of forest yet remains.

Climate

This is the great bugbear, in the minds of many, as an obstacle to successful farming any where in Maine; and Aroostook, being the most northerly county in the State, is often deemed more objectionable for this reason. But while it has its peculiarities, I failed to see cause to deem it the worst.

It is an exceedingly healthy climate. Upon this point, I cannot do better than to quote from the late statistical report on the sickness and mortality in the army of the United States, compiled from the records of the Surgeon General's office, as the testimony of the surgeons stationed in Aroostook in 1844–5, when their reports were made, is both disinterested and conclusive, and reveals a remarkable freedom from pulmonary disease so common in most sections of New England. . . .

The growing season, it is true, is shorter than elsewhere, but the rapidity of growth when once begun, is unparalleled in other parts of New England. Of this, I cannot state from observation, making as I did, my visit at mid-summer, but the uniform testimony of settlers on this point, and the progress actually made towards maturity which I witnessed, was fully satisfactory.

The snow falls early, sometimes as soon as the end of October, and before much frost (sometimes none) is in the ground. There it remains, steadily covering the soil until spring opens, a warm blanket two to four feet deep, with no alternations of freezing and thawing. When it goes off, the transition from winter to summer is almost instantaneous, and the soil *may be worked at once*. Being thus blanketed through the winter and porous enough as before remarked to allow superfluous moisture readily to pass downwards, no time is lost either for the ground to thaw, or to become sufficiently dry and warm to be worked to advantage. The crops are put in with no delay, and once in, they proceed with rapid strides to maturity.

On a farm of Mr. Cary's, at Houlton, I was shown thirty acres of wheat, ten of which were sown April 17th to 20th. This, in the latter part of July, was fully in milk, and past liability to injury from the wheat midge or fly. The rest was sown considerably later and the grain not yet fully formed. Upon this, I was sorry to see that the midge threatened to levy a serious contribution. In the neighborhood of Presque Isle, I was informed that oats, sown as late as June 10th, usually ripened without injury from frost. With regard to the usual period at which frosts occur, it was not ascertained to differ materially, of late years, from other sections. In low grounds frost is often noticeable at an early date, but on the higher lands usually tilled, one sufficient seriously to check vegetation is not expected before "the full of the moon in September," and this period safely passed, not until some weeks, possibly a month later. From the best information I could gather, frosts have not been so early or destructive for the ten years past, as before that time, and when the clearings were generally smaller and afforded little opportunity for circulation of air. The last week in August, 1842, there occurred a frost which did considerable injury, especially to the crops of such of the settlers as had been engaged in spring, in driving timber, and so had deferred their seed time to a period too late for safety. June 4th, 1844, ice made as thick as window glass. In 1845, the last spring frost occurred on the 31st of May. The injury which ensued from the early and late frosts from 1842 to 1846, was, in many cases, of a serious character, and had a very discouraging effect upon immigration. I did not learn of serious injury since that period in any case where crops were put in at the proper season. In some years frost has first occurred in Aroostook several weeks after it appeared in Penobscot county, and I learn that the present year no frost had occurred up to September 26th.

The term, during which cattle required to be fed from winter stores of forage, proved shorter than was anticipated, the autumnal feed being said to be abundant and good *until the snow fell,* and *as soon as this disappeared in spring,* cattle could find plenty of fresh and nutritious grasses. Some of the residents, who had been familiar with agricultural pursuits in other States, assured me that in this regard Aroostook possessed decided advantages over southern Maine or Massachusetts. The value of such pasturage will be readily appreciated by every practical farmer.

A very noticeable peculiarity of the climate of Aroostook, is the exemption hitherto enjoyed from injurious droughts. The settlers informed me that although there had been times when rain would have been acceptable somewhat sooner than it came, yet, that it could not be truthfully said that actual injury had ever ensued for want of it.

Whether, and to what extent, this exemption may be attributed to the existence of primeval forests, and whether it may be expected to continue after the woodman's axe has done its work, may be a matter of some uncertainty; but the probability that they are intimately connected, the one with the other, adds force to the well known and abundant arguments against indiscriminate waste and strip, and in favor of retaining, (or, if preferred in some cases, allowing an immediate second growth of,) sufficient wood for fuel, timber, and especially for *shelter* to their homes, fields, orchards, cattle and crops. The subject of shelter in a climate like ours, is one of such importance that I cannot forbear to express the hope that it may be duly considered and acted upon.

Productions

All the small grains thrive well. Wheat is not so extensively grown as formerly, the fly, or midge, (commonly called weevil,) rust and mildew being found serious drawbacks upon its profitable culture. I was informed by Mr. Nathaniel Blake of Portage Lake, (number thirteen, in sixth range,) that the wheat-fly had never troubled the grain there, and that he usually reaps twenty-five bushels per acre; but this was the only instance of entire exemption found, although at Patten and some other places, injury from the midge had been far less than from rust. Mr. Blake also stated of this locality, that frosts were usually two weeks or more later than at number eleven, some ten or twelve miles south of it. Mr. J. W. Haines, an old settler from Kennebec county, on Letter D, in first range, firmly holds from his own experience, that the fly, though often abundant, rarely injures the crop unless rust, mildew, or some unfavorable atmospheric influence retards the growth of the plant, thus giving the maggot time to commit his ravages, as otherwise, the plant being perfectly healthy and thrifty, the grain fills plumply and is abundant, be the maggots never so plenty.

When successful, twenty to twenty-five bushels is considered a good crop. Instances of much larger yield were narrated. Mr. Haines stated that he had grown forty-one and a half bushels of spring wheat, of sixty pounds to the bushel, to the acre. This was on land which had been in grass for four years and manured for potatoes the year previous; and that his neighbor, Mr. Goss, in adjoining township Letter C, grew last year one hundred and thirty-five bushels of bearded wheat, on five acres, and in another instance, fifty-two and a half bushels of winter wheat on one acre of new land from which no previous crop had been taken. This was mentioned as a very unusual crop, as winter wheat had rarely succeeded so well as spring wheat. Mr. Alfred Cushman of Golden Ridge, (number three, in fifth range,) President of the Penobscot and Aroostook Union Agricultural Society, informed me, that in one instance, from two and a half bushels sowing, he had reaped one hundred and seventy-five bushels, on four acres, one acre of which proving too wet, yielded only about half as much as the rest, thus indicating fifty bushels to the acre, on three acres. Since then he had been less successful, having experienced rust often and mildew occasionally.

On the whole, it seems doubtful whether wheat may be depended on as a staple crop, or that it can be profitably grown to an extent much beyond the amount needed for home consumption. But no drawbacks were found to exist which may prevent large production of the other grains. Oats, barley and rye, with fair treatment, grow luxuriantly, and yield bountifully. It is true, some very small and meagre crops were noticed; one, for instance, of about twenty bushels of oats; but upon inquiry it was ascertained to be the *seventh successive crop* of oats on the same spot, without any application of manure, and that the fifth yielded thirty-five bushels. Who can wonder if such management, or rather gross mismanagement, prevails, that some, even here, become discontented, complain of the climate, take the western fever, and talk of emigrating? And here it may be added, that evidence was abundant that two, three, or even four white crops in immediate succession on the same ground, was by no means so unfrequent as it should be. Several instances came to my knowledge, in which three successive crops of oats on the same land had yielded an average of fifty bushels or upwards per acre. The average production of these grains, *under good treatment,* may be set down as fifty bushels of oats, thirty of barley, and thirty to thirty-five of rye; the *actual average* would however be found at least twenty-five per cent. less than this.

Buckwheat is largely grown, and is probably gaining in estimation, for the remark was frequently made by residents, that although prejudiced hitherto against it, they had at length adopted its culture. The variety grown, is called here, rough buckwheat, and elsewhere known as Indian wheat. The smooth variety proves much less successful and is nearly abandoned. Its yield varies from twenty or thirty to fifty bushels — sometimes considerably more. With good treatment, and on soil in good condition, forty to fifty bushels may be confidently expected. It is usually grown upon the poorest. Its weight is from forty-five to fifty pounds to the bushel, and it yields about one-third of fine flour, which makes excellent bread and cakes, from a third to two-fifths of a coarser description, but very nutritious and highly esteemed for swine and other animals, the remainder being principally hull, is of little or no worth. The value of buckwheat for fattening animals, as compared with Indian corn, was variously estimated at from one-third to two-thirds its value, some deeming a bushel and a half of buckwheat equal to one of corn, others rating a bushel of corn worth three of buckwheat. The more usual estimate was one-half, although some who professed to have proved its value with care, were confident that deducting thirty-three per cent. for the hull, it was equal, weight for weight, to Indian corn, for fattening stock. The market value of Indian corn is usually two and a half times that of buckwheat. . . .

The success attending this grain in Aroostook, and the comparative extent to which it is grown, may be judged of from the fact, that by the census of 1850, this county, containing a little more than a fiftieth part of the population of the State, produced the previous year eighty-six thousand five hundred and twenty-nine bushels, while all the rest of the State produced only eighteen thousand bushels.

Indian corn is not extensively grown in Aroostook county, but its cultivation seems steadily, rather than rapidly, on the increase. By areful selection of early seed, a fair crop is generally secured. As far north as number eleven, fifth range, I saw corn silked out at the end of July. Last year Mr. Bean of letter G, in this vicinity, raised fifty-one bushels of sound corn per acre. The yield is not usually above this, and the actual average probably below forty bushels.

Mr. Cushman, at Golden Ridge, (number three, in fifth range,) some fifty miles south of number eleven, informed me that he had grown at the rate of two hundred and twenty bushels of ears per acre. He had, when I was there, a very promising field of corn, which was planted more closely than I had ever before seen, viz: three feet by eighteen or twenty inches asunder, thus giving more than double the usual number of hills to the acre. Mr. Cushman is doubtless wise in endeavoring to adapt the distance between hills to the anticipated size of the plant, and to get the full benefit of a well prepared plot for corn; but four and a half or five square feet only to the hill, seemed rather close, even for small Canada corn, in Aroostook. I had the pleasure of meeting Mr. Cushman subsequently, at the State Fair in Bangor, early in October, and learned from him that an acre of this had been harvested, and the product proved to be two hundred and ten bushels of ears of sound corn — a very good crop for any where.

Messrs. Gerry, Cushman, and other residents of this vicinity, assured me that they deemed Indian corn a surer crop than wheat. The seed used, is partly the Canadian variety, and partly what is called there, the early Dutton, a twelve rowed sort, originally brought from Massachusetts, and gradually acclimated. By the census returns of 1850, it would appear that the crop of Indian corn exceeded that of wheat, by several thousand bushels; but there is doubtless some serious error in the figures.

Roots

Large crops of these are as easily grown, perhaps more so, as in any other parts of New England. Potatoes are excellent and abundant, the usual crop being from two to three hundred bushels, per acre. In some sections, very little or no injury has ensued from disease, and it was estimated by several persons, that for ten years past, not over a quarter of the crop had ever been lost from the rot in any locality. Turnips to the amount of five hundred bushels to the acre, are grown with no labor beyond brushing in the seed on new land, and perhaps a little thinning out, no hoeing or weeding being bestowed. Much larger crops can be grown with additional labor and care. Carrots are highly esteemed, the crop, with fair treatment, varying from six hundred to twelve hundred bushels per acre. I learned of one crop of eight hundred and sixty-eight bushels, by actual measure, to the acre, where the rust had materially checked the growth. As in other sections of our State, root crops receive far less attention than they deserve.

Grass

No better district for hay, grazing and dairying, can be found in New England, than here. Indeed, I have never seen better in Orange county, New York, nor any where else — and should a person accustomed to much richer pastures than are usual in New England or New York, tramp over some of these, with red clover well up to the knees, and a dense mat of honeysuckle under foot, (the pastures tolerably well stocked too,) he could scarcely fail to deem it a country of rare excellence for grazing and dairying. Nor can I conceive sufficient reason, why Aroostook butter and cheese, may not be profitably exported to large extent, and by the application of proper skill in manufacturing, be made to rival that of Orange county, and command as good a price. Whether it can or not, one thing is sure, a good name must first be established, for the little butter which Maine has sent abroad, has by no means an enviable reputation in Boston market, let the quality of some which is eaten at home, be what it may. As for cheese, Maine now buys hundreds of thousands of pounds annually, and so our farmers might command a sufficient market for a good article at home, for a long time to come.

The propensity to take off successive crops of grain, until the yield seriously diminishes, is so great, that little land is sown to grass until its fertility is very sensibly impaired; hence the low average yield of hay, which does not much exceed a ton per acre, — perhaps it may a little, while with better treatment, an average of two tons might be ha just as easily. The reason alleged, or excuse offered for the practice was, that as it is, they have quite as much hay as they could cure, or store, or use, and more than this would be of no value.

Among the few exports from the Aroostook valley, may be named herds grass and clover seed. Last year, Mr. John Allen, near Presque Isle, offered for premium a crop of two thousand and twenty-four pounds clover seed, grown on seven acres, and which he sold at fifteen cents per pound. He stated the profit on the crop to be one hundred and sixty-three dollars sixty cents, or upwards of twenty-three dollars per acre. I heard of a crop upon ten acres, in another locality, of twenty-five hundred pounds. It is deemed very profitable when the heads "seed well," but this is by no means sure always to occur. It is rarely cut for seed unless promising upwards of one hundred pounds per acre; and sometimes three hundred are realized. In 1850, six hundred and sixty-one bushels clover seed, or forty thousand pounds, and ten hundred and eighty bushels of other grass seeds, were grown in the county. Herds grass, or timothy, usually yields six or seven, and sometimes ten bushels of seed per acre. In one instance, I learned of one hundred and four bushels grown on ten acres.

Fruit

Of the culture of fruit in Aroostook, it may be premature to speak with confidence; but the prospect is strongly in favor of ultimate success. There are a number of nurseries established, principally of the apple, and many trees have been planted out. In the village of Houlton I was told that little success had attended the planting of any other than the Siberian crab apple, which lived and bore well, but that a few miles out they succeeded tolerably well. Here, also, I saw plum trees of choice varieties which had borne abundant crops, quite too heavy indeed for their ultimate good. The appearance of the apple trees in this vicinity indicated a growth of wood too late to become well ripened and hard, and so, unfit to withstand the severity of winter. The circumstance of the roots being for a large part of ordinary winters in a soil above the freezing point, and the tops, at the same time, in a much lower temperature, may also have had an injurious influence. On higher lands, rocky knolls and side hills, especially in fifth range, I found orchards which bid fair to be productive and profitable. Mr. Elisha Brown of number six, in this range, has an orchard of some three hundred trees, many of them planted eight to twelve years, and most of which are succeeding finely. In his earlier attempts, he lost largely by grafting to Baldwin, Greening, Roxbury Russet, and other varieties of good repute further west, but which prove utterly unfit for this climate. Success in orchard culture here, will depend greatly upon a proper selection of varieties, and in this much help may be obtained from the experience of cultivators in the northern parts of Penobscot and Piscataquis counties, where also, some very promising seedlings have originated. It is highly probable that the sorts which in most parts of New England ripen in autumn, will here prove winter, or at least, early winter varieties. Mr. Cushman of Golden Ridge, informed me that he had received the first premium for apples at their Agricultural Fair held in October, on the Red Astrachan, such was its fine quality and good condition. This variety proves throughout Maine to be one of the hardiest, but in the western part of the State, I have never seen it in eating after August; and as with this, so probably with other early varieties, the period of maturity may be considerably later, and in some cases the quality improved, as it is proved to be with the Duchess of Oldenburg, another extremely hardy, early sort. Mr. Cushman's success has been such, that he proposes to plant at least two thousand apple trees on his farm, (with reference to its future division,) to graft them as soon after they attain suitable age, as he can decide in his own mind upon the most profitable varieties for leading sorts in their adaptation to his soil and climate. The treatment he proposes being somewhat original, may be stated. First cut down and burn the original growth; "hand-pile" the logs remaining sufficiently to allow the planting out of the apple trees, and seed down the land to clover at once. The clover to be neither mown or pastured, but left to decay on the ground, year after year. Three or four tons per acre every year, he thinks, will keep the ground in good heart, and what is more, secure the trees from the attacks of mice which are often troublesome on tilled and mown land, as they will hardly care to eat apple tree bark, or wood, while "living in clover," and with plenty of seed to fatten upon.

The smaller fruits, as currants, gooseberries, &c., &c., thrive perfectly well, and yield freely. Mr. Brown had a plantation of barberries which were quite flourishing. English gooseberries were entirely free from mildew, and so far as I could learn, this troublesome affection is entirely unknown in the eastern part of Washington county, and also in the adjoining Province of New Brunswick. It was not ascertained that grapes or pears had been tried. That there are among the early ripening grapes lately introduced to notice, or among the countless seedlings now on trial in all parts of the country, some which are both sufficiently early and hardy to succeed well any where in Maine, no doubt is entertained. Few among us now know the luxury of *good* and *ripe* grapes, or are aware of the impulse given to their culture in the last few years — of the progress which has been made, and the probability that all who will bestow the needful attention, can soon enjoy this delicious fruit. And that there are varieties of pears which will succeed in Aroostook, seems at least probable, from the fact that some are known to have grown well, and borne well, in a still higher latitude.

Markets.

This is a matter of prime importance, and ever to be well considered in estimating the expediency or profit of production. Of what avail is it that lands be never so fertile, if crops, when grown, can find no remunerating sale?

It may be said that agriculture can *exist* without markets, for the laborer can be fed from his own products, and clothed in home manufactured flax and wool, also of his own growth; but such labor is for life, not for profit — a struggle for existence, and such agriculture cannot be deemed a distinct and desirable branch of industry. . . .

The only market now existing in Aroostook for ordinary agricultural productions, is that created by the lumbering operations. This is generally a good one to an extent sufficient to absorb the surplus which the settlers now tilling the soil have to dispose of; but it is by no means a uniform one, varying as it necessarily must, with the fluctuations of that interest (proverbially uncertain) which creates it. Viewed in another aspect, this market can hardly be deemed a particularly desirable one, inasmuch as the manure yielded by the hay, oats, &c., cannot be purchased by the farmer, and returned to the soil to maintain its original fertility. Neither can I deem it a reliable one, for even admitting that the demand for forest productions will be always and uniformly good, the timber crop itself, although it may not be all harvested in one year, nor ten, nor twenty, will not last always; and while it does, the market which it affords will be gradually receding from the tilled lands.

Suppose every township, as soon as stripped of its timber, to be settled with energetic, industrious farmers, and their fields presently to smile with bounteous harvests, what would they be worth? What would be the net proceeds of a thousand acres of oats, yielding a hundred bushels to the acre, if it costs a hundred and fifty miles cartage over an earth road, however good, to market them? *Perhaps a dime per bushel.* And this leads to the inquiry, whether with present facilities for exportation alone, farming can be extensively carried on in Aroostook at a profit; and I hesitate not to say in reply, that it can, *provided* the mode of procedure be adapted to the circumstances, and this, in my opinion, is only by adopting, in the main, low farming, to wit, the growing of cattle and horses, and sheep husbandry. Pity 'tis, that on so fertile a soil, a higher grade may not be as successfully pursued; but it is the part of prudence to be governed in our action by circumstances, never forgetting to control those circumstances as fast and to such extent as may be within our power. I cannot doubt that a vigorous prosecution of dairy business and of wool growing, the yield from both which, on lands both cheap and good, will bear an export charge of one or two cents per pound, without destroying profit, or the growing of horses and lean cattle, possessing locomotive powers of their own, to take themselves to market, may be carried on to advantage. Very probable is it, or at least, so it seemed to me, that fat cattle might be produced, barreled, and find an Atlantic market via the St. John, at a cost enabling the producer successfully to compete with any section.

Such, in its leading features, is the more judicious mode of procedure, wherever land is abundant and cheap, bearing but small ratio to the value of labor, connected with distant markets. But to do even this to the best advantage, so as to reap a handsome profit, requires the outlay of considerable capital, notwithstanding the cheapness of land, first, to construct the needful buildings *comfortably* to house a large number of animals in winter, in order to save food and turn them out in spring in a condition *duly to thrive* on their summer feed; next, to obtain the most desirable breeds, with which to commence the undertaking; buildings also, for storing a sufficiency of winter stores, by no means forgetting an abundant supply of roots, safely stored in well ventilated, yet warm cellars. That this might be done, and the operation yield a satisfactory profit, seems sure enough, but from whence will such come? Capitalists, or those possessing means to enable them to do this, can easily get a living where they now are, in all probability comfortably settled, and in the enjoyment of greater privileges than can be expected in a new country, so that after all, inducement seems but scanty to cause *such* to emigrate and "rough it in the bush."

In estimating the comparative advantages of emigration to the new lands of our own State, and to those of the great west, there are many considerations to be taken into account. Land may be had cheaply in either case, but cheaper here than there, in fact, almost for nothing, as the half dollar per acre, which the State asks, may be chiefly paid in making the settlers' own roads — roads which he would have to make for his own convenience, if not thus paid for by the State. But in going west, cheapness ends with the price of land. The settler may get enough at a dollar and a quarter per acre, *provided* he go far enough from roads and rivers to find such as is not already taken up, but house, barn and fences are as necessary as land, and when he proceeds to their erection, he finds timber can only be obtained at a high price, usually at considerable distance, often having to be carted several days journey. Other building materials, bricks, lime, stone, nails, in fact, all, bear very high rates. Labor, especially mechanical labor, is costly and scarce. These considerations alone, leaving out many others, neither few nor trifling, (as the scarcity of water and its bad quality, when obtained, the agues, prostrating fevers, etc.,) I found to have proved operative in deciding the question in the minds of some residents of Aroostook, who had visited the west for the purpose of personal examination and deliberate balancing of advantages. They came home, content to remain, *fully* satisfied of the superiority of a residence here, and congratulating themselves that they went first to look before selling out, and thus taking a step not so easily or cheaply recalled — satisfied, that although a man possessing abundant means, might there obtain higher rates of interest and find more tempting opportunities for speculation — the man with little besides strong hands and a willing mind had a better prospect here.

It may not be inappropriate to mention here some particulars regarding the practice of agriculture in Aroostook, as noticed while there. It is no more true of these lands, that they may be drawn upon any how and any long, without exhaustion, than of virgin soil elsewhere. Allusion has been made to repeated grain crops in immediate succession. This is a serious departure from judicious practice, and if persevered in, must result, as it has ever done in other places, in barrenness and exhaustion. It is comparatively easy to maintain fertility when once in possession, but how difficult to restore it when lost, thousands and millions of acres over the length and breadth of the land, and the struggles of their owners for a livelihood testify in most emphatic tones.

I saw nothing to prove true what rumor had said: that the settlers esteemed manure to be a nuisance, and carted it to the nearest stream to be rid of it, but with few exceptions, there was less care for its preservation than is desirable. In some instances gratifying evidence appeared that it was properly valued, and on the whole it seemed gaining in estimation, and that the settlers were gradually coming to a just appreciation of its value.

In the neighborhood of Presque Isle, there have been, through the exertions of members of the North Aroostook Agricultural Society, introductions of choice cattle, and a marked improvement has been the result. The Hereford and Durham blood prevails mostly in the crosses observed. Ayrshires were not known as such, but some cattle brought in from the neighboring Province of New Brunswick, and considered natives, bore strong evidence of an infusion of Ayrshire blood. These, including the Devon more lately introduced, are probably the best breeds yet proved for this climate, taking into consideration all their good qualities, and the absence of serious defects, all (except the Durhams, which require good feed and shelter,) being very hardy, easily kept, and valuable. The Ayrshire breed has been somewhat extensively introduced into New Brunswick. Mr. Samuel Gray, of Frederickton, brought with him from Ayrshire some years since, numbers of these to breed from for sale, and had, when I was there, a considerable herd of full blooded animals, including many choice specimens. In other parts of Aroostook, I found no more attention paid to the selection of choice stock than prevails in some other counties in the State, and much less attention paid to stock-growing as a leading branch of agriculture, than it seemed to me there should be.

Sheep husbandry receives little attention in comparison with what might be profitably bestowed upon it. The sheep which I saw there would average decidedly better for mutton than those of the State at large, the flocks having been improved by admixtures from those of Mr. Perley, a well known extensive stockgrower at Woodstock, N.B., near Houlton — and who rears principally, if not wholly, the pure Leicester breed.

The most objectionable feature noticed in connection with the growth of domestic animals, was in regard to swine — these being both too few in number and too bad in quality. Some good hogs there are, others tolerable, but more prevalent were those too nearly resembling the landpikes. These were noticed in even greater purity in some of the more sparsely settled parts of New Brunswick, from whence they may probably have been derived — big-eared, long-legged, long-snouted, slab-sided, thick-skinned, large-boned, ravenous brutes, which look as if they might have originated in a cross between a jackass and an alligator, and from which it would be the height of imprudence for one to contract to furnish mess pork for less than two or three York shillings per pound; certainly, unless he had a term of years to do it in. That such neglect of swine culture should exist, is the more to be regretted, as considerable quantities of pork are annually imported into the county at large addition to first cost from the price paid for transportation. From the best information I could obtain, between eighteen and twenty-five hundred barrels are required every year, and sometimes more than this, as supplies for the lumbering operations; the quantity varying from year to year with the demand for timber. No one acquainted with the subject, of whom inquiry was made, reckoned the amount at less than fifteen hundred in any year, and most stated it to be from two to three thousand in ordinary seasons. So far as could be learned, nearly all of this was brought from abroad, while there can be no doubt whatever, that by judicious management, the whole of it could be grown on the spot, from the refuse of the dairy, with roots, and fattened on buckwheat, with perhaps a little corn, and yield a handsome profit. So good an opportunity of earning thirty to fifty thousand dollars per annum, should by no means be neglected.

Grasses for the South, 1858

From *Southern Cultivator* (Augusta, Ga., Feb., 1858), XVI, p. 42.

EDITORS SOUTHERN CULTIVATOR — Dr. E. Jinkins, of Horse Penn, Miss., I gladly welcome on the Grass question, and hope many others may come out and advocate the grasses for the South. I have, in a small way, petted many grasses and I am so fully convinced that we of the South can and ought to cultivate grasses for economic purposes, that I have bought several kinds.

As to Bermuda, I have seen hay from this grass that the stalks measured 12 to 15 inches, I feel certain, grown in the flats of a large bayou subject to overflow, If all had at command such lands, Bermuda would do for summer pastures and for hay. But lest they have not, I recommend Orchard Grass. Admit it is not a native here, and to show if not so known, it will do, allow me to state to you a fact concerning it. I have a brother-in-law, living about Montgomery, Ala., who is himself a progressive man (I think he got it from the Philips stock about his house); well, he had a remarkable fine native grass, that a friend had presented to him, called *Taylor Grass* — this was sold me. I, of course, wanted to know of it, and he sent out, with my brother, a bunch of it, root and stem. When I received it, I carefully divided it and planted in my garden. No more vigorous grass grows in the South; numbers of gentlemen were taken to see this prodigy. I assure you it was as pretty an object as I ever saw.

Well! it seeded, and, to my regret, all its glory passed away, for it was the *Orchard Grass*. I wrote of this to our friend, Dr. Cloud, months ago, for the sport of taking the wind out of the sails of our friends, who had borrowed plumage — *but he never received it.*

The *Rescue Grass,* another *native*. I believe it to be a native of the United States; am certain it was sent me by my friend C. B. Stewart as a Musquit Grass. I am certain it is a profitable winter grass, on good land; it should be pastured late, otherwise. when left to seed, there are such quantities of seed left on the earth that the succeeding year the pasture is not good, for too many plants; or, as practiced in the "swamp," take off stock 1st of March, and when perhaps 1-4th to 1-3d seed be matured, cut for hay and thresh enough seed for use — enough seed will be left on the ground to re-seed. The seed are good for poultry and hogs — fattening like oats. I shall sow down 6 or 8 additional acres this fall and continue to increase yearly.

Stanford's Wild Oat Grass grew for me from October, 1856, until August, 1858. I gathered seed 2 years from the same plants; have not examined since, having very little, though bought at $15 per bushel.

By the way, these prices tend greatly to retard progress, it is far better that the possessor of such valued articles would *increase their stores* until enough to sell at $2 to $5, or as friend Peabody did — make a fortune at once.

Lucerne I have planted for years — the last time as *Alfalva,* or Chili Clover, and on good land it will pay for soiling or as hay, I almost think equal to Bermuda; it will be green all winter and is the earliest grass for soiling. It is not a grazing grass, though it might, in an emergency, be used as such. It pays admirably for deep culture and high manuring,

The South relies too much on muscle. There is no need to press negroes and mules to nett an equal amount of cash in a life time. Pastures and stock will benefit land and save money. We wear out everything, whereas a change of policy would not.

Yours, &c.,
M. W. PHILIPS.

The New-York State Fair, 1858

From *Cultivator* (Albany, N.Y.), VI, pp. 346–47.

In every feature presenting more or less of interest and merit, and, in most departments, exhibiting a gratifying advance upon its predecessors, the State Fair held last week at Syracuse, was not only an example of the advantages of a central location, but also a most creditable witness to the agricultural improvement of the immediate locality in which it was held. Now that it is over, and we have all returned once more to our usual duties, the general results of the week that is passed can but claim a review; and it is matter of congratulation to the farmers of the State that there is room for so much commendation, and that we are saying words, not of compliment but of "truth and soberness," when we add, that seldom, if ever has such an anniversary been the occasion of so thorough and general satisfaction.

The Exhibition

In *Cattle,* which stand first upon the Premium List the Exhibition as a whole was probably never excelled in this country. The only breed wholly or entirely wanting to complete the list was the Adderney, which had we think not a representative on the grounds, — but in Short-Horns, Devons, Herefords and Ayrshires, the first two especially the turnout was good and large. And no one could have passed the stalls devoted to the Grades, without renewed conviction of the benefit which the importation of improved breeds has been to the country. The different classes of Working Oxen so far as the writer could ascertain were remarkably well filled. Some of the Fat Cattle were of immense frame, and perfectly loaded with flesh. In *Horses* the different classes were generally full — if we except Thorough-breds, which were almost or entirely wanting. The show of Stallions of all work and of Morgans or Black Hawks, is spoken of as very good and there were some excellent matched horses, geldings and mares. Mules and Jacks were rather deficient. In *Sheep*, the show was very good and extensive, and this was also the case in *Swine*. In *Poultry,* the Superintendent assured us he had never seen a better display, including a wide collection of different varieties, and good competition in nearly all.

The Weather and Attendance

The heavy showers of the preceding week had, it was hoped, opened the way for clearer skies during our Exhibition, and Tuesday we began to think the promise good for the three succeeding days. Wednesday was very fine, but the wind shifted at evening to south and east, and Thursday we woke to a rain-storm, every hour of which from early dawn until noon, diminished the attendance by thousands. With the fine receipts up to Wednesday evening, and the evident feeling along all the lines of Railroad centering in Syracuse, as well as among the farmers of Onondaga and adjoining counties, there was every reason to anticipate the largest receipts the Society's treasury had ever known. The result as it proved to be, was, in view of the weather, most encouraging — the total being $10,815 81.

Mowers and Reapers

The collection of these was extensive and valuable. There were several that were new, of which some appeared to possess much merit. Among the newer machines, were those of Bullock & Bros. from Chautauque Co.; J. & G. Lord of Watertown, who exhibited a cam machine, with an ingenious self-raking attachment; Willard & Ross, Vergennes, Vt.; Gore's New-England mower, a one-horse machine; Wheeler's patent from Shourds & Mosher, Cayuga county; the iron mower of H. Marcellus, of Amsterdam, N.Y.; Parkhurst's Buffalo mower; Tyler's patent, from Washington county; J. V. Wemple's, from Fonda; Hubbard's machine, and others. The older and well known mowers and reapers of Allen, Kirby, Wood, Hussey, Ketchum, Miller and Aultman, and others were also on the ground, including the one-horse machines of Kirby and Ketchum. Some of the new machines were in many particulars imitations of the best older ones, with important improvements in some particulars, and complex encumberments in others. Sherwood's reaper and binder, for attaching to any reaping machine, excited much attention. The operator, by its assistance, will bind as fast as the larger machines will cut, and thus save the labor of several men. Annealed wire is used for the bands, and costs but 15 cents per acre.

The costly and complex corn-huskers which were shown last year at Buffalo, have given place to the cheaper and simpler ones exhibited this year. One of the best of these was the simple and compact iron husker of G. Bellows, of Seneca Falls, which cuts the cob, and shoots out the ear with one blow of the lever-handle. A less compact, but equally efficient and perhaps better machine is Gould's patent, also from Seneca Falls. We were rather disappointed with the operation of Perkin's machine, which requires two distinct blows of a mallet to push an ear.

Morrison's patent corn-sheller was one of the best we saw in operation, clearing the cob in a neat and perfect manner, and with ease to the operator — the price $12. Another, known as the "Young America," and invented by J. P. Smith, although hardly as efficient as the former, was greatly superior in compactness, and is offered at $10.

Cahoon's Sowing Machine, both for horse and hand power, was exhibited — the horse machine has proved of great efficiency, and the latter would be also, but for the hard labor it involves to the operator. There were three modifications of the rotating harrow, the amount of merit of which is not yet fully determined. Winegar's gate excited much attention, from the ease of its operation to the driver of carriages without dismounting. The "Parallelogram Gate," of W. Tobey, Naples, N.Y., a new contrivance, is opened and shut by riding on perpendicular levers — it is very simple in construction, but requires further trial to determine its value. There were several modifications of portable fences, the pannels locking together at the ends; some zig-zag and others straight. The exhibitors stated their cost, variously at 45 to 70 cents a rod, but we did not find any that appeared to be faultless — those merely resting on the ground being liable to be upset, and others intended for staking or pinning to the ground being consequently attended with much labor in getting up. It is proper to add, however, that the former may answer well in sheltered valleys, excluded from winds.

Two hand-sawing machines were in operation, one of them from Heth, Hall & Co., and the other from Porter, Kellogg & Co., both of Jefferson county. The former is worked by both hands and feet, the operator standing on a reciprocating platform — the latter by working a lever something like the motion of a pump-handle. The exhibitors claim that the application of the force is so much more efficient than with a common wood saw, that a great increase in effective work is attained, some five cords of wood being cut from logs in a day by one hand. The price is about $20.

Gladding's patent hay fork was shown in partial operation, and is obviously a valuable contrivance. The hay is raised by a horse, and the load cast off and dropped by pulling a cord. The price $10.

We observed several good horizontal horse powers, among them one from Walrath Brothers of Chittenango, N.Y., compact, well made and quiet running; and another from B. & H. Wakely of McLean, N.Y.

The platform scales from Strong & Ross of Brandon, Vermont, appeared to be of excellent manufacture, and they are said to possess great accuracy.

Emery Brothers as usual made an extensive display of their well made agricultural machinery. Among the objects shown by them, was a collection of plows, their dynamometer, which served so valuable a purpose at the Syracuse trial of implements last year, their good and simple sugar-cane mill, a simple and efficient shingle jointing machine, a clover mill, a corn and cob crusher, a horse-fork, corn-sheller, portable grist mill, horse-power churn, portable cider mill, and last and not least, a set of railway horse-powers, thrashers and separators.

The collection of Richard II Pease, also of Albany, comprised Horse Powers, Threshers and Separators, Circular Saw Mills and Saws, a Clover Huller, and Cider Mill, on which several prizes were awarded.

Wheeler, Melick & Co., also exhibited sets of Horse Powers, Threshers, &c., from their extensive Factory.

One of the most admirably made machines on the ground, was the portable steam engine from A. N. & E. D. Wood of Utica. They have recently made a valuable improvement for confining the fire and rendering the engine more safe and secure. So perfect was the construction of this engine, that the least noise could not be perceived from its active working at a distance of two paces, and of course it was perfectly free from oscillation — a difficulty of a formidable character in some other portable engines.

Among the other articles which we can only enumerate, were Starks and Perigo's spoke planing machine; Birdsell's clover thresher and cleaner; Reynold's band cutter and self-feeder for threshing machines, simple and apparently efficient; Spencer's thresher and separator, from Tompkins county, well made and well arranged; Westinghouse's set of excellent railroad horse powers and threshers, circular saws, &c., and Badger's horse powers. R. C. Pratt of Canandaigua exhibited a simpler and improved form of his ditcher, which is reduced in weight and price to about one-half of the former machine, and we should think much easier to handle and manage.

A collection of plows from Walfer Warren of Utica, R. M. Hermance of Syracuse, P. Auld of Utica, Holmes, Stringer & Co., Munnsville, J. & G. Lord & Co., Watertown, and Woodworth, Whitney & Co., Manlius, and cultivators from Sayre & Remington, Utica, J. P. Cramer, Schuylerville, and J. S. & M. Peckham of Utica.

Allen's potato digging plow, and a much more complex digger (costing some $50) from J. E. Hardenburgh, Fultonville, N.Y.

Abraham Lincoln on Agriculture, 1859

From Wisconsin State Agricultural Society, *Transactions, 1858–59* (Madison, 1860), V, pp. 287–99.

Agricultural Fairs are becoming an institution of the country; they are useful in more ways than one; they bring us together, and thereby make us better acquainted, and better friends than we otherwise would be. From the first appearance of man upon the earth, down to very recent times, the words *"stranger"* and *"enemy"* were *quite* or *almost* synonymous. Long after civilized nations had defined robbery and murder as high crimes, and had affixed severe punishments to them, when practiced among and upon their own people respectively, it was deemed no offence, but even meritorious, to rob, and murder, and enslave *strangers,* whether as nations or as individuals. Even yet, this has not totally disappeared. The man of the highest moral cultivation, in spite of all which abstract principle can do, likes him whom he *does* know, much better than him whom he does *not* know. To correct the evils, great and small, which spring from want of sympathy, and from positive enmity, among *strangers,* as nations, or as individuals, is one of the highest functions of civilization. To this end our Agricultural Fairs contribute in no small degree. They render more pleasant, and more strong, and more durable, the bond of social and political union among us. Again, if, as Pope declares, "happiness is our being's end and aim," our Fairs contribute much to that end and aim, as occasions of recreation — as holidays. Constituted as man is, he has positive need of occasional recreation; and whatever can give him this, associated with virtue and advantage, and free from vice and disadvantage, is a positive good. Such recreation our Fairs afford. They are a present pleasure, to be followed by no pain, as a consequence; they are a present pleasure, making the future more pleasant.

But the chief use of Agricultural Fairs is to aid in improving the great calling of *Agriculture,* in all its departments, and minute divisions; to make mutual exchange of agricultural discovery, information and knowledge; so that, at the end, *all* may know everything, which may have been known to but *one,* or to but *few,* at the beginning; to bring together, especially, all which is supposed to not be generally known, because of recent discovery or invention.

And not only to bring together, and to impart all which has been *accidentally* discovered or invented upon ordinary motive; but, by exciting emulation, for premiums, and for the pride and honor of success — of triumph, in some sort — to stimulate that discovery and invention into extraordinary activity. In this, these Fairs are kindred to the patent clause in the Constitution of the United States; and to the department, and practical system, based upon that clause.

One feature, I believe, of every Fair, is a regular *Address*. The Agricultural Society of the young, prosperous, and soon to be, great State of Wisconsin, has done me the high honor of selecting me to make that address upon this occasion — an honor for which I make my profound and grateful acknowledgement.

I presume I am not expected to employ the time assigned me in the mere flattery of the farmers, as a class. My opinion of them is that, in proportion to numbers, they are neither better nor worse than other people. In the nature of things they are more numerous than any other class; and I believe there really are more attempts at flattering them than any other; the reason of which I cannot perceive, unless it be that they can cast more votes than any other. On reflection, I am not quite sure that there is not cause of suspicion against you, in selecting me, in some sort a politician, and in no sort a farmer, to address you.

But farmers, being the most numerous class, it follows that their interest is the largest interest. It also follows that that interest is most worthy of all to be cherished and cultivated — that if there be inevitable conflict between that interest and any other, that other should yield.

Again, I suppose it is not expected of me to impart to you much specific information on Agriculture. You have no reason to believe, and do not believe, that I possess it — if that were what you seek in this address, any one of your own number, or class, would be more able to furnish it.

You, perhaps, do expect me to give some general interest to the occasion; and to make some general suggestions, on practical matters. I shall attempt nothing more. And in such suggestions by me, quite likely very little will be new to you, and a large part of the rest possibly already known to be erroneous.

My first suggestion is an inquiry as to the effect of greater *thoroughness* in all the departments of Agriculture than now prevails in the North-West — perhaps I might say in America. To speak entirely within bounds, it is known that fifty bushels of wheat, or one hundred bushels of Indian corn can be produced from an acre. Less than a year ago I saw it stated that a man, by extraordinary care and labor, had produced of wheat what was equal to two hundred bushels from an acre. But take fifty of wheat, and one hundred of corn, to be the possibility and compare it with the actual crops of the country. — Many years ago I saw it stated in a Patent Office Report that eighteen bushels was the average crop throughout the United States; and this year an intelligent farmer of Illinois, assured me that he did not believe the land harvested in that State this season, had yielded more than an average of eight bushels to the acre; much was cut, and then abandoned as not worth threshing; and much was abandoned as not worth cutting. As to Indian corn, and indeed, most other crops, the case has not been much better. For the last four years I do not believe the ground planted with corn in Illinois, has produced an average of twenty bushels to the acre. It is true, that heretofore we have had better crops, with no better cultivation; but I believe it is also true that the soil has never been pushed up to one-half of its capacity.

What would be the effect upon the farming interest, to push the soil up to something near its full capacity? Unquestionably it will take more labor to produce *fifty* bushels from an acre, than it will to produce *ten* bushels, from the same acre. But it will take more labor to produce fifty bushels from *one* acre, than from *five?* Unquestionably, thorough cultivation will require more labor to the *acre;* but will it require more to the *bushel?* If it should require just as *much* to the bushel, there are some *probable,* and several *certain* advantages in favor of the thorough practice. It is probable it would develop those unknown causes, which of late years have cut down our crops below their former average. It is almost certain, I think, that in the deeper plowing, analysis of the soils, experiments with manures, and varieties of seeds, observance of seasons, and the like, these cases would be found. It is certain that thorough cultivation would spare half, or more than half the cost of land, simply because the same product would be got from half, or from less than half the quantity of land. This proposition is self-evident, and can be made no plainer by repetitions or illustrations. The cost of land is a great item, even in new countries; and constantly grows greater and greater, in comparison with other items, as the country grows older.

It also would spare the making and maintaining of inclosures — the same, whether these inclosures should be hedges, ditches or fences. This again, is a heavy item — heavy at first, and heavy in its continual demand for repairs. I remember once being greatly astonished by an apparently authentic exhibition of the proportion the cost of an inclosure bears to all the other expenses of the farmer; though I cannot remember exactly what that proportion was. Any farmer, if he will, can ascertain it in his own case, for himself.

Again, a great amount of "locomotion" is spared by thorough cultivation. Take fifty bushels of wheat, ready for the harvest, standing upon a *single* acre, and it can be harvested in any of the known ways, with less than half the labor which would be required if it were spread over *five* acres. This would be true, if cut by the old hand sickle; true, to a greater extent, if by the scythe and cradle; and to a still greater extent, if by the machines now in use. These machines are chiefly valuable, as a means of substituting animal power for the power of men in this branch of farm work. In the highest degree of perfection yet reached in applying the horse power to harvesting, fully nine-tenths of the power is expended by the animal in carrying himself and dragging the machine over the field, leaving certainly not more than one-tenth to be applied directly to the only end of the whole operation — the gathering in of the grain, and clipping of the straw. When grain is very thin on the ground, it is always more or less intermingled with weeds, chess and the like, and a large part of the power is expended in cutting these. It is plain that when the crop is very thick upon the ground, a larger proportion of the power is directly applied to gathering in and cutting it; and the smaller, to that which is totally useless as an end. And what I have said of harvesting is true, in a greater or less degree of mowing, plowing, gathering in of crops generally, and, indeed, of almost all farm work.

The effect of thorough cultivation upon the farmer's own mind, and, in reaction through his mind, back upon his business, is perhaps quite equal to any other of its effects. Every man is proud of what he does *well;* and no man is proud of that he does not well. With the former, his heart is in his work; and he will do twice as much of it with less fatigue. The latter performs a little imperfectly, looks at it in disgust, turns from it, and imagines himself exceedingly tired. The little he has done, comes to nothing, for want of finishing.

The man who produces a good full crop will scarcely ever let any part of it go to waste. He will keep up the enclosure about it, and allow neither man nor beast to trespass upon it. He will gather it in due season and store it in perfect security. Thus he labors with satisfaction, and saves himself the whole fruit of his labor. The other, starting with no purpose for a full crop, labors less, and with less satisfaction; allows his fence to fall, and cattle to trespass; gathers not in due season, or not all. Thus the labor he has performed, is wasted away, little by little, till in the end, he derives scarcely anything from it.

The ambition for broad acres leads to poor farming, even with men of energy. I scarcely ever knew a mammoth farm to sustain itself; much less to return a profit upon the outlay. I have more than once known a man to spend a respectable fortune upon one; fail and leave it; and then some man of modest aims, get a small fraction of the ground, and make a good living upon it. Mammoth farms are like tools or weapons, which are too heavy to be handled. Ere long they are thrown aside at a great loss.

The successful application of steam power to farm work, is a *desid ratum* — especially a steam plow. It is not enough that a machine operated by steam, will really plow. To be successful, it must, all things considered, plow *better* than can be done with animal power. It must do all the work as well, and *cheaper;* or more *rapidly*, so as to get through more perfectly *in season;* or in some way afford an advantage over plowing with animals, else it is no success. I have never seen a machine intended for a steam plow. Much praise and admiration are bestowed upon some of them; and they may be, for aught I know, already successful; but I have not perceived the demonstration of it. I have thought a good deal, in an abstract way about a steam plow. That one which shall be so contrived as to apply the larger proportion of its power to the cutting and turning the soil, and the smallest, to the moving itself over the field, will be the best one. A very small stationary engine would draw a large gang of plows through the ground from a short distance to itself; but when it is not stationary, but has to move along like a horse, dragging the plows after it, it must have additional power to carry itself; and the difficulty grows by what is intended to overcome it; for what adds power also adds size, and weight to the machine, thus increasing again, the demand for power. Suppose you should construct the machine so as to cut a succession of short furrows, say a rod in length, transversely to the course the machine is locomoting, something like the shuttle in weaving. In such case the whole machine would move north only the width of a furrow, while in length the furrow would be a rod from east to west. In such case, a very large proportion of the power, would be applied to the actual plowing. But in this, too, there would be difficulty, which would be the getting of the plow *into*, and *out of*, the ground, at the end of all these short furrows.

I believe, however, ingenious men will, if they have not, already, overcome the difficulty I have suggested. But there is still another, about which I am less sanguine. It is the supply of *fuel*, and especially *water*, to make steam. Such supply is clearly practicable, but can the expense of it be borne? Steamboats live upon the water, and find their fuel at stated places. Steam mills, and other stationary steam machinery, have their stationary supplies of fuel and water. Railroad locomotives have their regular wood and water stations. But the steam plow is less fortunate. It does not live upon the water; and if it be once at a water station, it will work away from it, and when it gets away cannot return, without leaving its work, at a great expense of its time and strength. It will occur that a wagon and horse team might be employed to supply it with fuel and water; but this, too, is expensive; and the question recurs, "can the expense be borne?" When this is added to all other expenses, will not plowing cost more than in the old way?

It is to be hoped that the steam plow will be finally successful, and if it shall be, *"thorough cultivation"* — putting the soil to the top of its capacity — producing the largest crop possible from a given quantity of ground — will be most favorable for it. Doing a large amount of work upon a small quantity of ground it will be, as nearly as possible, stationary while working, and as free as possible from locomotion; thus expending its strength as much as possible upon its work, and as little as possible in traveling. Our thanks, and something more substantial than thanks, are due to every man engaged in the effort to produce a successful steam plow. Even the unsuccessful will bring something to light which in the hands of others will contribute to the final success. I have not pointed out difficulties, in order to discourage, but in order that, being seen, they may be the more readily overcome.

The world is agreed that *labor* is the source from which human wants are mainly supplied. There is no dispute upon this point. From this point, however, men immediately diverge. Much disputation is maintained as to the best way of applying and controlling the labor element. By some it is assumed that labor is available only in connection with capital — that nobody labors, unless somebody else owning capital, somehow, by the use of it; induces him to do it. Having assumed this, they proceed to consider whether it is best that capital shall *hire* laborers, and thus induce them to work by their own consent, or *buy* them, and drive them to it, without their consent. Having proceeded so far, they naturally conclude that all laborers are naturally either *hired* laborers or *slaves*. They further assume that whoever is once a *hired* laborer, is fatally fixed in that condition for life; and thence again, that his condition is as bad as, or worse, than that of a slave. This is the *"mud-sill"* theory. But another class of reasoners hold the opinion that there is no *such* relation between capital and labor, as assumed; and that there is no such thing as a freeman being fatally fixed for life, in the condition of a hired laborer, that both these assumptions are false, and all inferences from them groundless. They hold that labor is prior to, and independent of, capital; that, in fact, capital is the fruit of labor, and could never have existed if labor had not first existed — that labor can exist without capital, but that capital could never have existed without labor. Hence they hold that labor is the superior — greatly the superior of capital.

They do not deny that there is, and probably always will be, *a* relation between labor and capital. The error, as they hold, is in assuming that the *whole* labor of the world exists within that relation. A few men own capital; and that few avoid labor themselves, and with their capital, hire or buy another few to labor for them. A large majority belong to neither class — neither work for others, nor have others working for them. — Even in all our slave States, except South Carolina, a majority of the whole people of all colors, are neither slaves nor masters. In these free States, a large majority are neither hirers nor hired. Men, with their families — wives, sons, and daughters — work for themselves, on their farms, in their houses and in their shops, taking the whole product to themselves, and asking no favors of capital on the one hand, nor of hirelings or slaves on the other. It is not forgotton that a considerable number of persons mingle their own labor with capital; that is, labor with their own hands, and also buy slaves or hire freemen to labor for them; but this is only a mixed, and not a distinct class. No principle stated is disturbed by the existence of this mixed class. Again, as has already been said, the opponents of the *"mud-sill"* theory insist that there is not, of necessity, any such thing as the free hired laborer being fixed to that condition for life. There is demonstration for saying this. Many independent men, in this assembly, doubtless a few years ago were hired laborers. And their case is almost if not quite the general rule.

The prudent, penniless beginner in the world, labors for wages awhile, saves a surplus with which to buy tools or land, for himself; then labors on his own account another while, and at length hires another new beginner to help him. This say its advocates, is *free* labor — the just and generous, and prosperous system, which opens the way for all — gives hope to all, and energy, and progress, and improvement of condition to all. If any continue through life in the condition of the hired laborer, it is not the fault of the system, but because of either a dependent nature which prefers it, or improvidence, folly, or singular misfortune. I have said this much about the elements of labor generally; as introductory to the consideration of a new phase which that element is in process of assuming. The old general rule was that *educated* people did not perform manual labor. They managed to eat their bread, leaving the toil of producing it to the uneducated. This was not an insupportable evil to the working bees, so long as the class of drones remained very small. But *now*, especially in these free States, nearly all are educated — quite too nearly all, to leave the labor of the uneducated, in any wise adequate to the support of the whole. It follows from this that henceforth educated people must labor. Otherwise, education itself would become a positive and intolerable evil. No country can sustain, in idleness, more than a small percentage of its numbers. The great majority must labor at something productive. From these premises the problem springs — "How can *labor* and *education* be the most satisfactorily combined?"

By the *"mud-sill"* theory it is assumed that labor and education are incompatible; and any practical combination of them impossible. According to that theory, a blind horse upon a tread mill, is a perfect illustration of what a laborer should be — all the better for being blind, that he could not kick understandingly. According to that theory, the education of laborers, is not only useless, but pernicious and dangerous. In fact, it is, in some sort, deemed a misfortune that laborers should have heads at all. Those same heads are regarded as explosive materials, only to be safely kept in damp places, as far as possible from that peculiar sort of fire which ignites them. A Yankee who could invent a strong *handed* man without a head would receive the everlasting gratitude of the *"mud-sill"* advocates.

But free labor says "no!" Free labor argues, that as the Author of man makes every individual with one head and one pair of hands, it was probably intended that heads and hands should co-operate as friends; and that that particular head, should direct and control that pair of hands. As each man has one mouth to be fed, and one pair of hands to furnish food, it was probably intended that that particular pair of hands should feed that particular mouth — that each head is the natural guardian, director and protector of the hands and mouth inseparably connected with it; and that being so, every head should be cultivated, and improved, by whatever will add to its capacity for performing its charge. In one word free labor insists on universal education.

I have so far stated the opposite theories of *"mud-sill"* and "free labor" without declaring any preference of my own between them. On an occasion like this I ought not to declare any. I suppose, however, I shall not be mistaken, in assuming as a fact, that the people of Wisconsin prefer free labor, with its natural companion, education.

This leads to the further reflection, that no other human occupation opens so wide a field for the profitable and agreeable combination of labor with cultivated thought, as agriculture. I know nothing so pleasant to the mind, as the discovery of anything that is at once *new* and *valuable* — nothing that so lightens and sweetens toil, as the hopeful pursuit of such discovery. And how vast, and how varied a field is agriculture, for such discovery. The mind, already trained to thought, in the country school, or higher school, cannot fail to find there an exhaustless source of enjoyment. Every blade of grass is a study; and to produce two, where there was but one, is, both a profit and a pleasure. And not grass alone; but soils, seeds, and seasons — hedges, ditches, and fences, draining, droughts, and irrigation — plowing, hoeing, and harrowing — reaping, mowing, and threshing — saving crops, pests of crops, diseases of crops, and what will prevent or cure them — implements, utensils, and machines, their relative merits, and to improve them — hogs, horses, and cattle — sheep, goats, and poultry — trees, shrubs, fruits, plants, and flowers — the thousand things of which these are specimens — each a world of study within itself.

In all this, book-learning is available. A capacity, and taste, for reading, gives access to whatever has already been discovered by others. It is the key, or one of the keys, to the already solved problems. And not only so. It gives a relish and facility for successfully pursuing the unsolved ones. The rudiments of science, are available, and highly valuable. Some knowledge of botany assists in dealing with the vegetable world — with all growing crops. Chemistry assists in the analysis of soils, selection, and application of manures, and in numerous other ways. The mechanical branches of natural philosophy, are ready help in almost everything; but especially in reference to implements and machinery.

The thought recurs that education — cultivated thought — can best be combined with agricultural labor, or any labor, on the principle of *thorough* work — that careless, half-performed, slovenly work, makes no place for such combination. And thorough work, again renders sufficient, the smallest quantity of ground to each man. And this again, conforms to what must occur in a world less inclined to wars, and more devoted to the arts of peace than heretofore. Population must increase rapidly — more rapidly than in former times — and ere long the most valuable of all arts, will be the art of deriving a comfortable subsistence from the smallest area of soil. No community whose every member possesses this art, can ever be the victim of oppression in any of its forms. Such community will be alike independent of crowned-kings, money-kings, and land-kings.

But, according to your programme, the awarding of premiums awaits the closing of this address. Considering the deep interest necessarily pertaining to that performance, it would be no wonder if I am already heard with some impatience. I will detain you but a moment longer. Some of you will be successful, and such will need but little philosophy to take them home in cheerful spirits; others will be disappointed, and will be in a less happy mood. To such, let it be said, "Lay it not too much to heart." Let them adopt the maxim, "Better luck next time;" and then, by renewed exertion, make that better luck for themselves.

And by the successful, and unsuccessful, let it be remembered, that while occasions like the present, bring their sober and durable benefits, the exultations and mortifications of them are but temporary; that the victor will soon be vanquished, if he relax in his exertion; and that the vanquished this year, may be victor the next, in spite of all competition.

It is said an Eastern monarch once charged his wise men to invent him a sentence, to be ever in view, and which should be true and appropriate in all times and situations. They presented him the words, *"And this, too, shall pass away."* How much it expresses! How chastening in the hour of pride! How consoling in the depths of affliction! "And this, too, shall pass away." And yet, let us hope, it is not *quite* true. Let us hope, rather, that by the best cultivation of the physical world, beneath and around us, and the intellectual and moral world within us, we shall secure an individual, social, and political prosperity and happiness, whose course shall be onward and upward, and which, while the earth endures, shall not pass away.

Frederick Law Olmsted on the Highlanders

From Frederick Law Olmsted, *A Journey in the Back Country* (New York, 1860), pp. 221–25.

The northernmost cotton-fields, which I observed, were near the Tennessee line. This marks a climactic division of the country, which, however, is determined by change of elevation rather than of latitude. For a week or more afterwards, my course was easterly, through parts of Tennessee and Georgia into North Carolina, then northward, and by various courses back into Tennessee, when, having finally crossed the Apalachians for the fourth time, I came again to the tobacco plantations of the Atlantic slope in Virginia.

Agricultural Notes in the Highlands

The climate of this mountain region appears not to differ very greatly from that of Long Island, southern New Jersey and Pennsylvania. It is perhaps more variable, but the extremes both of heat and cold are less than are reached in those more northern and less elevated regions. The usual crops are the same, those of most consequence being corn, rye, oats, and grass. Fruit is a more precarious crop, from a greater liability to severe frosts after the swelling of buds in the spring. The apple-crop has been thus totally destroyed in the year of my journey, so that in considerable orchards I did not see a single apple. Snow had fallen several inches in depth in April, and a severe freezing night following, even young shoots, which had begun to grow, forest trees, and leaves which had expanded, were withered.

The summer pasture continues about six months. The hills generally afford an excellent range, and the mast is usually good, much being provided by the chestnut, as well as the oak, and smaller nut-bearing trees. The soil of the hills is a rich dark vegetable deposit, and they are cultivated upon very steep slopes. It is said to wash and gully but little, being very absorptive. The valleys, and gaps across the mountain ranges, are closely settled, and all the feasible level ground that I saw in three weeks was fenced, and either under tillage or producing grass for hay. The agricultural management is nearly as bad as possible. Corn, planted without any manure, even by farmers who have large stocks of cattle, is cultivated for a long series of years on the same ground; the usual crop being from twenty to thirty bushels to the acre. Where it fails very materially, it is thought to be a good plan to shift to rye. Rye is sown in July, broadcast, among the growing corn, and incidentally covered with a plow and hoes at the "lay by" cultivation of the corn. It is reaped early in July the following year with cradles, an acre yielding from five to fifteen bushels. The following crop of corn is said to be much the better for the interpolation. Oats, and in the eastern parts, buckwheat, are sowed in fallow land, and the crops appeared to be excellent, but I could learn of never a measurement. Herds-grass (*agrostis vulgaris*), is sown on the valley lands, (rarely on the steep slopes of the mountains,) with oats, and the crop, without any further labor, pays for mowing and making into hay for from four to eight years afterward. Where it becomes mossy, weedy, and thin, it is often improved by harrowing or scarifying with a small "bull-tongue," or coulter-plow, and meadows thus made and occasionally assisted, are considered "permanent." The hay from them soon becomes in large part, however, coarse, weedy and bushy.

Natural meadows are formed on level land in the valleys, which is too wet for cultivation, by felling the timber and cutting up the bushes as close to the ground as practicable, in August. The grass is cut the following year in June, and again in August or September, at which time the new growth of bushes yield to the scythe. The sprouts cease to spring after the second or third year. Clover is a rare crop, but appears well, and is in some localities a spontaneous production. Hay is stored but little in barns, the larger part being stacked in fields. The hay fields are pastured closely, and with evident bad effect, in the spring and autumn.

Horses, mules, cattle, and swine, are raised extensively, and sheep and goats in small numbers, throughout the mountains, and afford almost the only articles of agricultural export. Although the mountains are covered during three months of the winter with snow several inches in depth, and sometimes (though but rarely) to the depth of a foot or more, and the nights, at least, are nearly always freezing, I never saw any sort of shelter prepared for neat stock. In the severest weather they are only fed occasionally, hay or corn being served out upon the ground, but this is not done daily, or as a regular thing, even by the better class of farmers. One of these, who informed me that his neighbor had four hundred head that were never fed at all, and never came off the mountain, in consequence of which "heaps of them" were starved and frozen to death every year, said that he himself gave his stock a feed "every few days," sometimes not oftener than "once in a week or two." The cattle are of course small, coarse, and "raw-boned." They are usually sold to drovers from Tennessee when three years old, and are driven by them to better lowland pastures, and more provident farmers, by whom they are fattened for the New York market. During the past five or six years, in consequence of the increasing competition, the drovers have purchased also the two year olds.

No dairy products are sold. I saw no cheese; but butter of better quality than I found elsewhere at the South, is made by all farmers for their own tables. Mules are raised largely. The mares with foals are usually provided with a pen and shed, and fed with corn, cut oats, (the grain and straw chopped together), and hay, daily during winter. This is done by no means universally, however. In no single case did I find stabling and really comfortable shelter prepared for a stock of mules; as a consequence, the mules are inferior in size and constitution to those of Kentucky, Tennessee and Missouri, and command less prices when driven to the plantations of South Carolina and Georgia — the market for which they are raised.

The business of raising hogs for the same market, which has formerly been a chief source of revenue to the mountain region, has greatly decreased under the competition it has latterly met with from Tennessee and Kentucky. It is now a matter of inferior concern except in certain places where the chestnut mast is remarkably fine. The swine at large in the mountains, look much better than I saw them anywhere else at the South. It is said that they will fatten on the mast alone, and the pork thus made is of superior taste to that made with corn, but lacks firmness. It is the custom to pen the swine and feed them with corn for from three to six weeks before it is intended to kill them. In some parts of the mountains the young swine are killed a great deal by bears. Twenty neighbors, residing within a distance of three miles, being met at a corn-shucking, last winter, at a house in which I spent the night in North Carolina, account had been taken of the number of swine each supposed himself to have lost by this enemy, during the previous two months, and it amounted to three hundred.

Bears, wolves, panthers, and wild-cats are numerous, and all kill young stock of every description. Domestic dogs should also be mentioned among the beasts of prey, as it is the general opinion of the farmers that more sheep are killed by dogs than by all other animals. Sheep raising and wool growing should be, I think, the chief business of the mountains. If provided with food in deep snows, a hardy race of sheep could be wintered on the mountains with comfort. At present no sheep are kept with profit. I have no doubt they might be, were shepherds and dogs kept with them constantly, and were they always folded at night. Eagles are numerous, and prey upon very young lambs and pigs.

Many of the farmers keep small stocks of goats, for the manageable quantity of excellent fresh meat the kids afford them, when killed in summer. Their milk is seldom made use of. They require some feeding in winter, and the new-born kids, no adequate shelter ever being provided for them, are often frozen to death. Goats, in all parts of the South, are more generally kept by farmers than at the North.

The agricultural implements employed are rude and inconvenient. A low sled is used in drawing home the crops of small grain. As it is evident that large loads may be moved with a sled across declivities where it would be impracticable to use a cart or wagon, hill-side farmers, elsewhere, might frequently find it advantageous to adopt the plan.

Importance of Grass Culture to the South

From C. W. Howard, "Grasses for the South," U.S. Patent Office, *Report on Agriculture*, 1860 (Washington, 1861), pp. 224–35, 238–39.

As this Essay is designed to be an inquiry into the adaptation of the South for the cultivation of the grasses and into the different kinds of the grasses best suited to a Southern latitude, many things will be found in it without interest to a Northern reader. In fact, statements will be made which, though true at the South, would be untrue at the North. The purposes of grass culture at the South and the North differ widely from each other. At the North the chief object is hay as a supply for winter. At the South the desideratum is the cultivation of those grasses which will grow during our mild winters, and therefore save us the expense and labor of making hay. Hence it will be seen that recommendations which will hold good in one section will not always hold good in the other.

As this Essay is designed for general readers, it will not be a Botanical Essay. Whenever it is possible, the common popular names will be applied to the grasses considered.

In no part of Christendom, enjoying a good government, and settled by an intelligent population, does land sell at so contemptible a price as in the Plantation States. In Georgia, for instance, land does not command an average price of five dollars per acre. Various causes have been assigned for this low value. It will be instructive to examine them.

The reason generally assigned at the South is the proximity of an abundance of cheap fertile lands at the West. If this be a sound reason at the South, it should also be true at the North, as it is as easy to reach new lands from New York as it is from Georgia. But land is steadily rising in value in New York and other northern States. The proximity of new lands cannot, therefore, be the cause of the low price of land at the South, as it does not produce this result at the North.

It is said, again, that the supply of land is greater than the demand, in consequence of the sparseness of our population; capital seeks its most profitable investment. There is money enough in the Southern States to have given a much higher value to our land. But the truth is that prudent men have found that, under our present system, land will not pay an interest on more than its present price. Hence this capital, instead of being invested in land, is appropriated to the building of railroads, factories, &c. It will also be found that in the Southern States where the white population is least dense the lands are highest in price, and the reverse.

Many persons suppose that it is the form of labor prevalent at the South which diminishes the value of Southern lands. This supposition is worthy of a brief consideration.

The remarks made upon it will not touch the moral or political aspect of Negro Slavery; it will be considered merely as a matter of agricultural interest.

If Negro Slavery diminishes the value of Southern lands, it must produce this result in some one of the following forms.

Before noticing these forms it may be proper to make the general remark that at the South where the negroes are the most numerous the lands bear the highest price, as the rice, Sea Island cotton, and sugar-cane lands. Some of our best rice lands now command from two hundred to three hundred dollars an acre. The reason of this high price will be given hereafter.

Does slave labor affect injuriously the value of Southern lands from its want of constancy? It is the most constant form of labor. The negro has no court-house, no jury, no musters, no mill to attend. He has no provision to buy, and no anxiety or loss of time on this account; food for himself and family is provided. If his family are sick, careful nurses are provided for them. The details of cotton and rice culture could not be conducted with a form of labor less constant.

Is there a deficiency of vigor in slave labor? In all forms of out-of-door bodily and severe labor, to be continued for a length of time, the well-fed negro is more capable than the white man. The regular and almost universal allowance of food upon plantations shows that, as a general rule, the negroes have a sufficiency of hearty and nutritious food.

Is there a deficiency of intelligence in Slave labor? There is less intelligence than among white laborers at the North, in Scotland, and some parts of England; but not less intelligence than exists among the mass of French, Irish, and Belgian laborers. Yet land rates as high in Belgium as in any other part of Europe. The cultivation is also as perfect as can be found elsewhere. It is not so much the intelligence of the laborers as of the controlling and directing mind, which is of the greatest moment in agriculture.

Is there a deficiency of economy in slave labor? The entire expense of a negro laborer on a plantation cannot be put down at more than fifty cents a day. Can any other labor in this country be obtained as cheaply as this? Beyond this, multitudes of men have largely increased their fortunes by the natural increase of their laboring force.

If there be no deficiency in the constancy, vigor, intelligence, or economy of slave labor, it cannot be supposed, with justice, to affect the value unfavorably of Southern land.

In the present excited state of the public mind it is proper to repeat the remark that this brief inquiry is made, not with a view to exciting discussion of a vexed topic, but solely of arriving at the true cause of the low price of Southern land, and of suggesting a remedy. This inquiry could not be conducted without an examination of the character of the labor employed upon the land.

Does the Southern climate affect injuriously the price of Southern lands? It does not; because the lands are of the greatest value [greater than anywhere else in the Union] in those parts of the South which are not sickly, as the rice lands. As a general remark, the climate of the middle belt of the Southern States, including rolling oak and hickory lands, very closely resembles the climate of France, which is considered to be the best climate of Europe for agricultural purposes. In most of this region there are but few days in winter in which the plough need be stopped on account of the frozen state of the earth.

Is there a deficiency in the natural fertility of the Southern soil? No one will pretend to say that the original fertility of the great body of the Southern States was inferior to that of the Middle and Northern States, where land has attained a great comparative value.

Is there a deficiency in the salable value of Southern products of the soil? These products generally command a better price at the South than the North. The most valuable products of the South, cotton and rice, are peculiar to it.

If the low value of landed estate at the South is to be attributed neither to the proximity of cheap Western lands, to slave labor, to defective climate, to sparseness of population, or deficiency in the value of its products, to what is this low value attributable?

The answer is, to the *Defective System of Southern Agriculture*. That system is defective, among others, in the following particulars:

1st. This system is such that the planter scarcely considers his land as a part of his permanent investment. It is rather a part of his current expenses. He buys a wagon and uses it until it is worn out, and then throws it away. He buys a plough or hoe, and treats both in the same way. He buys land, uses it until it is exhausted, and then sells it, as he sells scrap iron, for whatever it will bring. It is with him a perishable or movable property. It is something to be worn out, not improved. The period of its endurance is therefore estimated in the original purchase, and the price is regulated accordingly. If it be very rich level land, that will last a number of years, the purchaser will pay a fair price for it. But if it be rolling land, as is the great bulk of the interior of the Southern States, he considers how much of the tract is washed or worn out, how long the fresh land will last, how much is too broken for cultivation, and in view of these points determines the value of the property. Of course he places a low estimate upon it.

2d. The system of Southern agriculture is such that a very large proportion of the landed estate yields no annual income. A considerable amount is in woodland, yielding nothing but a supply of rails and fuel. This is to a great degree dead capital. A large number of acres on almost every farm in the older parts of the cotton States is worn out and at rest — of course paying no interest. The only paying part of the tract is that which is under the plough. The interest on the land which the planter does not cultivate must be charged to that which he does cultivate, and this brings down the value of the whole property to a very low figure.

3d. The Southern system of agriculture allows to land no value independent of the labor put upon it. The negro is the investment rather than the land. The value of the negro is instantly affected by a change in the price of cotton, while the value of the land which grows the cotton is comparatively unaffected. It is an extraordinary anomaly that perishable labor should take precedence of imperishable land. It is not uncommon to hear young men at the South giving it as a reason for their entering a profession, that while they owned a large body of land they owned but twenty or thirty negroes, and that it would be impossible to make a support with so small a force. When asked how the rest of the world manage who have no negroes, the reply is "our system differs from theirs, ours requires a large amount of labor."

Precisely, and therein it is defective, and until that defect be remedied, land will continue to be comparatively a drug in the market. It is the design of this Essay to show that it is possible to give land a value independent of any costly or complicated annual labor bestowed upon it.

4th. The Southern system of agriculture includes a succession of crops of a most exhausting or otherwise injurious character. These crops are cotton and corn, varied only by small grain. This succession is continued until the land is worn out and turned out to rest.

5th. These crops are not only exhausting and hurtful in consequence of the clean culture they require, but they also require an amount of labor not known elsewhere. If we consider the amount of productive land, that is, the number of acres yielding an annual income, we shall find the amount of labor used on an ordinary Southern plantation to be greater per productive acre than the amount of labor used in the most perfectly cultivated portions of Europe. In the latter every acre produces something, whether in pasture, meadow, or cultivated crops. At the South nothing but the cotton or grain pays. The rest of the plantation is idle.

The causes mentioned are those which have the greatest influence in depreciating the value of Southern land. They are as follows:

The planters buy land as something to be worn out, not improved; they suffer a large portion of their investment in land to remain as dead capital paying no annual income; they pursue a system which allows no value to their land independent of the labor bestowed upon it; they cultivate a succession of crops of an exhausting or otherwise hurtful character; and, lastly, in the cultivation of these crops, they use an amount of labor not known elsewhere in intelligent agriculture.

The people of the Southern States must be in all time chiefly an agricultural people. Their land must be the basis of their wealth. Upon its skilful use must rest their permanent prosperity. Large and remunerative crops at the expense of the land indicate a prosperity which is fallacious. It is living upon capital. That which gives the greatest value to their land most conduces to their permanent prosperity. There is no question of political economy of equal importance to the people of the South as the best means of increasing the value of their landed estate. For instance, there are in the State of Georgia, selected because the writer happens to be most familiar with it, about thirty-seven millions of acres of land. An average increase of value of ten dollars per acre, which would not bring its salable value to fifteen dollars per acre, would give an increase to the property of the State of three hundred and seventy millions of dollars — a sum larger by more than fifty millions of dollars, than the whole value of the personal and real estate of Georgia by the census of 1850 — an increase which would yet leave the land of the State worth less by thirteen dollars per acre than the average value of land in the Middle States.

The remedy proposed for this depreciation of Southern land is *A Change in the System of Agriculture prevalent at the South.*

When a change is spoken of, it is not meant to advise an abandonment of any of our staples. The world requires southern cotton, rice, and sugar. It cannot now dispense with them. A total failure in the cotton crop would create, if not a revolution, a fearful disturbance of affairs in England.

The change proposed is the incorporation into the Southern system of agriculture, of a feature by which crops for the improvement of the soil shall receive as regular atrention as crops for sale.

Prominent among these crops are the artificial grasses. By them, most profitably, sufficient live stock can be raised to keep in a state of progressive improvement the lands cultivated in cotton or the cereals. By their use the whole of the farm becomes productive. There is no dead capital in it. The soil is prevented from washing or exhaustion. An equal area may yield an income with much less labor.

In fact, a portion of the farm yields a fair income without annual labor, save the cost of keeping up the fences.

As a confirmation of the truth of these remarks, if we take the map of Europe, we shall find that land rates in price in proportion to the attention paid to the artificial grasses. It is least in Spain; it rises in France; it is still greater in Belgium, and greatest in Holland, which is almost a continuous meadow.

If we take our own country, the same observation holds good. Passing by the older Northern States, let us compare Kentucky and Georgia. The Kentucky lands, as a general remark, are much better naturally than those of Georgia. But there are some bottom lands in Georgia equal to the best of the Kentucky lands, leaving out the rice lands. We can easily find any of the best bottom lands in Georgia which will command more than fifty dollars an acre, while farms sometimes sell in Kentucky for more than one hundred dollars an acre. Both are slave States. Kentucky is younger than Georgia. She is nearer to the new lands of the West than Georgia. Her climate is not so good as the climate of Georgia. The cause of the difference must be found in the different systems of cultivation adopted by each. Every acre of the Kentucky farm, including woodland, produces something, as all the land not under the plough is in grass, yielding a return without labor. On the contrary, no acre of the Georgia farm produces, except that which is under the plough, and the farm does not contain within itself, under the present system, the means of repairing the damage done by the plough. The cause of the difference, then, between the Georgia and Kentucky land of equal fertility in value is obviously the reason above given.

This is the proper point to allude to the exception afforded by the rice lands and some of the Sea Island lands. Both of these have the means of keeping up their fertility within themselves. If they did not possess these facilities, they would quickly fall into the condition of the rest of the Southern lands. The rice lands can keep up their fertility by flooding with water; the sea islands by using the marsh mud which surrounds them. They are to a degree independent of live stock. The rest of the South must look to live stock as their only permanent reliance for keeping up the fertility of the lands. All experience teaches that there is no way of keeping live stock in sufficient numbers to answer this end but by the aid of the artificial grasses, which follow and do not precede the domestic animals. All grasses are considered as artificial grasses which it is necessary to sow.

Can the Artificial Grasses be Grown in the Cotton States?

A great many things are now done at the South which, a few years since, were deemed impracticable. It was supposed that winter fruits could not be raised at the South. The reason was that the experiments were made with trees suited to a Northern climate. It is now conceded that the Southern apples and pears for winter use quite equal the Northern. It was supposed that wine could not be made at the South, yet Southern wines cannot be exceeded in this country. It was supposed that malt liquors could not be made at the South, yet beer of excellent quality is made here. It was supposed that cheese could not be made at the South, yet as good cheese as any made at the North has been exhibited at our Fairs. It was supposed that wool of the Saxon and merino sheep would deteriorate in the climate of the South, yet at the World's Fair in London, Tennessee wool took the premium over all competitors. We should never conclude that a thing cannot be done because it has not been done.

It is said that the climate of the South will be an effectual barrier to the extensive growth of the artificial grasses. These grasses make up a large class. There are several hundred of them — some belong to a cold climate, and some to a warm climate, some requiring a damp soil, and others a dry soil. There is as much difference in their habits as there is in the habits and wants of fruit trees. Because the fig and pomegranate will not grow in Massachusetts, shall we conclude that no fruit trees of any kind will grow there? And because a Northern artificial grass will not grow in a Southern soil, shall we conclude that no artificial grass will grow there?

The Southern climate has its advantages and disadvantages in grass culture. At the North the danger is that the grass will be frozen out, at the South that it will be burned out. Proper precautions must be taken in both instances. A traveller passing through the Southern States in July, and finding the grass parched under the summer's sun, may, on his reaching the North in the same month, find their meadows looking green and refreshing, and may conclude that grass culture succeeds in the North, but is unsuited to the South. But suppose the same traveller finds in the following March that the land at the North is stiff frozen or covered with snow; that the cattle are all housed and eating costly food; and when he reaches the South in the same month, finds the pastures green and verdant, and the cattle luxuriating in new grass, might he not with the same propriety conclude that the North was a bad grass country, and the South a good one?

The hot suns of the South are against the grass in the summer, but they are very much in its favor in the winter. Almost all the Northern grasses which will live at the South change their habits and become winter grasses. They grow during the winter. It cannot be expected that they should grow during the summer. Everything which the Almighty has made, possessing either animal or vegetable life, requires some period of repose. These grasses, which have so changed their habits as to grow during the winter, must not be grazed during the summer. Nor must they be so closely grazed during the winter that their roots will be left in a denuded state to encounter the hot sun of the summer, which will keep them. They must be allowed to rest from their winter and spring labors under a cover of a portion of their own vegetation.

The ability to use the artificial grasses during the winter is a great advantage of the Southern climate. It amounts to letting the stock mow their own hay, and is a saving of the expense both of mowing and of expensive barns. The cost of one Pennsylvania stone barn would lay down a considerable Southern plantation in the winter grasses. The substitute in the summer is the crab grass, which springs as soon as the grain is cut, and affords a bite of fresh young grass at a time when, in countries destitute of this invaluable product, cattle feed reluctantly on the old grass of the spring.

It may be said by some of the readers of this Essay, "This is fair theory, but we have tried these grasses at the South, and they do not succeed." Perhaps the experiment has been made on poor land. Neither cotton, corn, nor grain would succeed if put on poor land and left without subsequent attention. The great body of the Southern States was once covered with a carpet of nutritious grasses, as Texas now is. They were then natural to grass. Most of these grasses have disappeared from among us. The pea vine once grew on land now too much exhausted for profitable cultivation without manure. The history of a range is as follows: When the stock of the settlers first enter it they attack only the grasses which they like. They continue these attacks from year to year, beginning as soon as the first bud of the spring puts forth. They leave the grasses which they do not like. These flourish while the valuable grasses are destroyed. Hence we may go into a range in the newer parts of the South, and while the ground is covered with grass in August almost knee-high, we shall find the cattle hungry in the midst of apparent plenty.

Cultivation adds to the destruction of the valuable natural grasses. The salts necessary to them are exhausted by it. And if we wished to replace these grasses upon the soil on which they once flourished, it would be absolutely necessary to manure it heavily. The careful observer will occasionally find some of these grasses growing in the older parts of the South, but always in rich places in which they have by some means been secured against the exterminating "hoof and tooth." If it be necessary to manure the land in order to make valuable grasses grow which were once native to it, how much more is it necessary to manure the same land in order to make grasses grow which are foreign to it. The artificial grasses are highly concentrated food. They contain much in a small space. They are composed of the elements which make up the flesh and bones of the animals which eat them. They must previously feed before they can feed these animals. If the grass food is not in the soil, it must be put there. There is no crop on which manuring pays better than on grass lands, as it pays twice — in the profit on the animals fed and in the improvement of the soil. Most of the unsuccessful experiments in grass culture in the South have been unsuccessful because they have been made on poor land.

Good bottom land at the South will generally produce the grasses suited to them without the aid of manure. There is very little cultivated upland in this country which will produce good grass crops without manure. These exceptions are generally found in the West and Northwest. They rarely, if ever, occur at the South.

The question is asked, Will it pay to manure grass lands at the South? If it will pay anywhere, it will certainly do so here. If it will pay to manure a meadow at the North, from which hay is to be cut, cured, carted, and stacked, or housed, much more will it pay at the South, where all this expense can be saved by an advantage of climate. The person who is considering grass culture at the South on upland, must take into the account the cost of manuring as an indispensable preliminary. On open land this cost may be abated and sometimes more than compensated by sowing grain with grass seeds, the increase of the grain crop covering the expense of the manure.

Persons attempting the cultivation of the grasses at the South have sometimes failed, because they did not understand the nature of the perennial grasses. They are accustomed chiefly to the annual grasses, as the crab and crowfoot grasses. These mature rapidly, as they are short-lived. It is a law of nature that those of her products which are designed to last long mature slowly. The planter will recognize an illustration of this remark in the growth of a broom sedge field. The first season the young grass barely makes its appearance, yet it is there. It is several years before it entirely occupies the ground. The same field, if ploughed, would spring up in crab grass, and in two months would be covered with a heavy carpet of grass and annual weeds.

When a field is sown with the artificial grasses at the South, the first season the ground will apparently be occupied almost exclusively by the natural grasses and weeds which always follow the stirring of the soil by the plough. The experimenter, observing this result, concludes that his experiment is a failure and ploughs up his ground.

He should have remembered that these grasses and weeds are annuals; that they follow the plough; that they will not appear the next year, and that they have shaded from the scorching sun the delicate needle-like spears of the young perennial grasses which are hardly visible to his eye. He should have waited until the next year, when he would probably have found a fair stand of the grass sown by him. Again, some experimenters in grass culture, delighted at the succulent appearance of the young grass in the autumn, when everything else has been withered by frost, turn upon it their equally delighted animals. These continue upon it all winter, wet or dry. A blade of grass which appears above ground is instantly bitten. This process is repeated until late in the spring. The summer's sun comes, the naked roots are exposed to it, the grass is killed, and the experimenter declares that the Southern climate is unsuited to any of the artificial grasses.

These failures, whether from the selection of unsuitable grasses, from sowing in poor land, from ignorance of the comparatively slow growth of the perennial grasses, or from overstocking and too close feeding, are all set down to the climate.

The writer has given for more than twenty years a considerable degree of attention to the growth of the artificial grasses at the South. He does not hesitate to give it as his opinion, based upon long observation both in all parts of this country and many European countries, that, for the production of several of the most valuable grasses, the climate of the Southern States possesses advantages which are not exceeded by any other climate whatever — the temperature of the whole year being taken into the account. Our deficiency is much more in fertility of soil than in suitableness of climate. A disadvantage of climate is irremediable by man; a disadvantage of soil may be remedied by skilful culture.

What Are the Grasses Suited to the Southern States?

Failures are often as instructive as successes. To save labor and expense to others, a statement will be made of the results of all the experiments in grasses and forage plants made at Spring Bank Farm.

This farm is situated near Kingston, Cass county, Georgia. It is in what is called the blue limestone formation. The soil is a stiff, red clay, having in it a very small quantity of sand. Except the bottom land, it has been manured where most of the experiments have been made. Latitude between 34 and 35 degrees.

FORAGE PLANTS. — 1. Sainfoin. — From the great value of this plant in the South of Europe it was hoped that it would prove an acquisition to Southern agriculture. Seeds have been sowed on this farm obtained from England, France, and Naples. They have been sowed, at intervals of time, on upland and low ground, on manured and unmanured land, on limed and unlimed land, and in no instance with success. The plants have lived, but their existence has been a sickly and useless one.

2. Burnet or Pimpernel. — This plant does not grow here high enough to mow. It is green all winter, being scarcely touched by our severest frosts. It is not liked by stock during the summer, but is readily eaten during the winter. It is worthy of more extended trial than has been given it here. It grows in tufts or bunches; its blood-red blossoms, with their peculiar round form, (from which it derives its name, "Sanguisorba,") render it an ornamental plant.

3. Lupine. — This was another plant from the South of Europe, from which much was expected. The plant has grown vigorously here, but soon after the formation of the beans it was attacked by an insect, which touched nothing else, and destroyed it. A similar insect has followed several other experiments with this plant in Georgia. Besides being subject to this casualty, it is an annual which diminishes its value even where it can be successfully grown.

4. Vetch. — Experiments have been made with both the English summer and winter vetch. They have grown very well, but do not produce as much as the better sort of our field peas. These are also annual. There is a native vetchling which is propagated with ease, which comes earlier than any other forage plant except lucerne. This is an insignificant plant on poor land, but on rich land grows more than knee-high and makes a fine hay. Its best use is to be sown with winter grasses for pasture; after once being sowed, resowing is not necessary unless it be grazed too severely. It is now becoming generally diffused over this farm.

5. Scabious. — This is a forage plant much valued at the cantons of the Cevennes. Besides being nutritious, it is believed to possess valuable medicinal properties for live stock. It is known in our flower gardens under the popular name of "Mourning Bride." It has not been found here to stand grazing, and does not answer for hay.

6. Chicory. — Arthur Young speaks with enthusiasm of this plant: "I never see this plant," says he, "this excellent plant, without congratulating myself for having travelled with the view of acquiring and spreading useful knowledge." According to him the introduction of this forage plant into England, if a man had done nothing else during his life, would be sufficient to prove that he had not lived

in vain. In the Southern States it certainly does not promise equal benefits. It grows with vigor; stock are fond of it when cut and thrown to them. But it is propagated too slowly, and requires too much culture for the present state of Southern agriculture. Such at least is the conclusion based upon the experiments made with it upon this farm.

7. French crimson clover. — This is a beautiful annual when in blossom, resembling a field of large ripe strawberriès. On rich land it thrives well during winter and spring, and affords early and valuable pasture. As a fertilizer it would be valuable sown with wheat on land already in good heart. The seeds differ from the common red clover in being enveloped in a kind of down, which enables them to be diffused by the wind.

8. Neapolitan clover. — This species closely resembles the preceding. Both the plant and the seeds are somewhat larger. It is also an annual, and its value is diminished on this account.

9. Spurry. — This plant, which has been called the "clover of sandy lands," has been unsuccessfully tried at this place. The growth was meagre and valueless. It is possible that it might thrive on lands containing more sand.

10. Melilot. — Two species of melilot have been experimented upon; one an annual having yellow blossoms, the seeds of which were obtained on the Battery at Charleston, S.C. It grows luxuriantly here, but is rejected by all kinds of live stock. The other kind, with white blossoms, is a biennial, growing on rich land four or five feet high; it is also rejected by stock. It is a singular fact that this foreign plant was in use by the Cherokee Indians before this country was taken possession of by the whites. It was valued by the Indians as a fragrant ingredient in a salve in much repute among them.

11. Narrow-leaved plantain. — This plant is regarded as a pest at the North. In England, on the contrary, it is much valued, particularly in Yorkshire, where it passes under the name of ribwort. Almost all the cloverseed brought to the Southern States from the North contain seeds of the plantain. It will live at the South on the poorest land, but is valuable only on good land. If not grazed during the summer it will afford a considerable amount of winter food. Cattle, horses, and sheep eat it in winter with avidity. Hogs are not fond of it. It is a useful constituent of a winter pasture at the South, and its growth is therefore encouraged on this farm.

12. White clover. — This is an invaluable plant in Southern agriculture. It springs naturally in almost every place on which ashes have been thrown, and which have been left for any length of time, without cultivation. Its benefits are not generally known, because the spots on which it grows are generally thrown open to cattle all the year round. They are very fond of it, and it is therefore rarely suffered to attain its full height. On this farm, if not grazed during the winter and spring, it grows on manured land sufficiently tall to mow. If sowed with tall grasses for hay, it stretches in its efforts to obtain its share of light, and thus gives a heavy cutting near the ground. Its tendency at times to slabber horses is an objection; but this objection is trifling when compared with its many advantages. It is very valuable as a hog and sheep pasture, as it grows during the warm spells in winter in this latitude and below it. It thrives on any land that is rich enough for it, growing as well on the sandy lands near the coast as on the dry lands of the interior. It is of much use as a fertilizer. When a piece of ground has been made rich enough to bear a good coat of

white clover, which is suffered to shed seeds, it becomes as natural to the soil as crab grass. The process of subsequent improvement is easy. It yields little summer pasturage. It possesses an interest to the Southern planter, as it will thrive on sandy soils on which red clover will not live. It should be sowed with some of the grasses. It combines admirably well with Bermuda grass, as the white clover appears as the Bermuda ceases to grow in the autumn. The extensive cultivation of this apparently insignificant though really valuable plant is strongly advised.

13. Red clover. — It is unnecessary to speak of the value of this forage plant. This is universally known. The only question is, Will it grow at the South? Careful inquiry, experiments, and observations have determined the following results in regard to it: It will grow on rich and dry bottom lands in all parts of the South. It will not thrive in a wet subsoil, however rich the surface may be. It will thrive on any of our lands made sufficiently rich and ploughed to a sufficient depth. Deep ploughing is essential, in order to enable its tap root to sink rapidly into the earth. On lands destitute of clay, it is useless to attempt its culture. Clay lands which have been worn by cultivation must be well manured in order to grow red clover successfully. It will die out if pastured much, under our fiery sun, after the month of June. It should be suffered to remain untouched during the summer, and be grazed again as cool weather commences. Red clover possesses a peculiar advantage to the Southern planter, not to cut for hay, but chiefly as a hog pasture in the spring, to last until the stubble fields are open. One of the greatest expenses of the plantation is the cost of meat for the negroes. This meat, during the greater part of the year, is bacon. The most troublesome season of the year in hog raising is the spring. Red clover meets the wants of that season. Every planter should sow enough red clover to graze his hogs at that time. There is scarcely a plantation in which suitable land cannot be found. Of it as a fertilizer in a rotation of crops, it is unnecessary to speak. It grows well at the South in woodland which has been well thinned out, and the ashes from the burned timber and brush carefully scattered.

14. Lucern. — On many accounts, Lucern is one of the most bountiful gifts of Nature to the Southern planter. No grass or forage plant in cultivation at the North will yield nearly as much hay as Lucern at the South. In good seasons, and on land sufficiently rich, it can be cut four or five times during the year. An acre of good Lucern will afford hay and cut green food for five horses the whole year. Ten acres will supply fifty head of plantation horses. This can be cut down in a day with a mowing machine. How unwise in the planter, then, to damage his corn by pulling fodder — that most irksome and senseless work of the plantation. A few acres of Lucern would save him this labor, and the tedious time occupied in pulling fodder could be employed in the improvement of his land. It is useless to attempt the cultivation of Lucern on poor land. It will live, but it will not be profitable. There are certain indispensable requisites in the cultivation of Lucern. The ground must be good upland; it must be made very rich; it cannot be made too rich. If the ground is as carefully prepared for it as an asparagus bed, the Lucerne will spring almost with the rapidity (after cutting) of asparagus. It must be very clean. When the Lucerne is young it is delicate, and may be smothered with the natural weeds and grasses of a foul soil. Land which has been in cotton, worked very late, if made sufficiently rich, is in a good state of preparation for Lucern. The manure put upon it must be free from the seeds of weeds; hence, a mixture of guano and phosphatic manures would

be an excellent application. On this farm, land designed for Lucerne is put in drilled turnips well manured and worked. The turnips are folded with live stock — that is, they are fed on the ground, which thus gets all the solid and liquid excrements of the animals, and becomes very rich, and is also very clean.

Great depth of cultivation is necessary in preparation of the soil for Lucern. If the ground was broken up with a four-horse plough, and in the same furrow a two-horse subsoil plough was run, stirring it eighteen or twenty inches, it would be to the advantage of the subsequent crops of Lucerne.

Ten pounds of seed are required for an acre, sowed broadcast. Drilling is unnecessary if the ground be properly prepared and the Lucern is not pastured. If the preparation has been imperfect, and the Lucerne is to be occasionally pastured, it is better to drill at such a distance as will allow a narrow plough to be passed between the rows when the surface requires stirring.

Either early in autumn or early in February are good seasons for sowing Lucern. The seed should be lightly harrowed in, and then the surface should be rolled. Lucern lasts a great number of years, the roots ultimately becoming as large as a small carrot. It should be top-dressed every third year with some manure free from the seeds of weeds. Ashes are very suitable for it. The Lucerne field should be as near as possible to the stables, as workhorses, during the spring and summer, should be fed with it in a green or wilted state. As Lucern is much earlier than red clover, it will be found a useful adjunct in hog raising. Hogs are very fond of it, and will thrive on it in the spring, when it is cut green and thrown to them. This extended notice of Lucern is given because it is remarkably adapted to our soil and climate, and is, beyond all comparison, the most valuable plant for hay-making and soiling to the Southern planter. It thrives in no part of Europe with greater vigor than it does in the Southern States.

It will be seen from the necessarily brief remarks upon these fourteen different forage plants that the writer has bestowed much attention upon this interesting branch of agriculture. His observations are not taken from books, but are the result of personal examination and experiment. Some of the seeds of these plants have been obtained with a good deal of trouble. When some of these plants, the seeds of which had been obtained from England, failed, it was supposed to be possible that the failure might have arisen from the fact that the seeds were grown in a climate different from our own. A second trial was therefore made with seeds obtained direct from Italy. The result has been given. The whole of them are rejected as being not deserving of attention at the South, except Lucern, red and white clover, and possibly burnet.

THE GRASSES. — The reader should bear in mind that when grasses are represented as growing successfully at the South, it is to be understood that they were sown on land either naturally or artificially rich. He will be misled if he applies the conclusions hereinafter stated to poor land.

15. Blue Grass. — This well-known grass will grow at the South on all lands having a good clay foundation. On extremely rich land, if not grazed during the summer or autumn, it will yield a tolerable burthen of green food during the winter.

Its ordinary growth is low. It almost disappears during the heat of summer. It has a great tendency to become, in common language, "hide-bound." Its chief value is when sown with other grasses which grow in tufts, as orchard grass, as it fills up the intervals between these tufts. It should not be suffered to find its way into meadow land designed for hay. On rich bottom land it will overrun almost any other grass than Bermuda. It rarely grows at the South high enough to be cut, and it therefore converts a meadow into a pasture. It grows best in woodland. It should not be sowed alone, but in connexion with other grasses, and is valuable when thus used.

16. Orchard Grass. — This grass succeeds at the South on lands having a clay subsoil, as low down as the oak and hickory rolling country extends. In the flat sandy lands it is said not to perfect its seeds, and quickly dies out. It is of little use at the South as a hay grass, but possesses great value as a winter pasture. It grows best in the shade, which result its name would indicate. It should not be grazed during the summer. All stock should be taken from it in June and not allowed to return to it until Christmas. It is not among the most permanent of the artificial grasses. Hence it is proper to sow it with red and white clover, when these are used in a rotation, for the improvement of the soil. Orchard grass is proper to be mixed with clover, when the latter is to be cut for hay, as both blossom at the same time. Herds grass and timothy are much later than red clover, and therefore unsuited to be sown with it.

17. Timothy. — On rich bottom lands, well drained, timothy succeeds well, generally, at the South. Recent experiments near Atlanta and Athens, in Georgia, indicate that it will grow satisfactorily on manured upland. There are other grasses which will yield more hay on upland, and several others which will afford better winter pasture. It is advised to confine its use to rich bottom lands and for hay.

18. Tall meadow Oat-grass. — This grass has been introduced into Georgia under several names: the Stanford wild oat, the Snythe grass, the Utah grass, and the Oregon grass. These are the same grass. The seed stems of this grass grow to four or five feet in height. On rich upland it yields a large amount of good hay. On good bottom land the yield is still larger. Its winter growth is heavier than that of any other grass, except the Italian rye-grass in favorable positions. It is somewhat surprising that this grass is not enumerated in the list of Texas grasses appended, as it is certainly a native of the West. The writer has been informed by others that it stands winter grazing at the South well. He cannot speak from his own experience, as he has been unable to obtain seed in sufficient quantities to make an experiment on a scale large enough to render it practically reliable.

If his information be correct, so far as present experiments have gone, it certainly deserves to be placed at the head of winter grasses for the South.

In confirmation of this opinion, it may be instructive to quote a few authorities. Dr. Muhlenberg, and Mr. Taylor, of Virginia, the author of Arator, consider this as the most valuable of the grasses.

"It possesses the advantage," says Judge Buel in the Farmer's Companion, "of early, late, and quick growth, for which the Orchard grass is esteemed, and is well calculated for a pasture grass. We have measured it in June, and when in blossom, (at the time in which it should be cut for hay,) and found the seed stems four and a half feet long.

"The latter math is nearly equal in weight and superior to the seed crop."

Sinclair says: "It thrives best on a strong tenacious clay, and Muhlenberg prefers for it a clover soil."

Dickerson remarks that "it makes a good hay, but is most beneficial when retained in a state of close feeding."

The British Cyclopaedia states that "it affords a greater weight of hay than most grasses. On the Continent, in comparison with common grass, it is found to yield in the proportion of 20 to 2."

Loudon says: "Every animal that eats grass is fond of it, while it makes the best hay and affords the richest pasture. It abounds in the best meadows about Lacock and Chippenham, and has the valuable property of abiding in the same land, while most other grasses are constantly changing."

Mr. Lewis Sanders, of Kentucky, says: "I have been informed by an experienced trainer that the use of hay made of the tall meadow oat-grass, as the fodder fed in training keeps the bowels in a natural condition, dispensing with any use of balls, physic, &c., thereby giving the horse several lengths the advantage of the one weakened by physicking. My own observation and experience in feeding cattle sustain the remarks of 'an experienced trainer,' as to the feeding this species of stock with hay made of the tall meadow oat-grass or orchard grass. Sow the seed on ground well prepared, as it usually is for the reception of timothy or other small seed, fall or spring. As early in the spring as the ground can be prepared is the surest and best time to sow meadow oat-grass or orchard-grass seed. In June of the first year weeds will make their appearance, then, with a keen scythe, mow weeds, grass, and all close to the ground. The next year comes a good crop of seed. As soon as the top seeds are ripe secure it with a cradle or scythe. Immediately after the seed is cut mow closely for hay; then, about the last of August, you have a second crop of hay, yielding more than the first, leaving the best of all pastures for colts and calves."

Mr. George H. Waring, of Habersham county, Georgia, a gentleman in every way reliable, makes an extraordinary statement as to the yield of this grass in hay on a small piece of rich land. He states that the produce of a piece of ground 90 feet by 10, was weighed by him, and the result in dried hay was 210 lbs. This is at the rate of about five tons per acre, which would be an enormous return.

Mr. Gideon Dowse, of Richmond county, states, in the Southern Cultivator, that this grass thrives well near the Richmond Baths, in Richmond county, Georgia. The lands there are naturally poor and very sandy, and would be considered as unfavorable to grass culture as any lands in the Southern States.

On the whole, Southern planters are advised to make careful, judicious, and yet vigorous experiments with the tall meadow oat-grass. The seeds can be obtained from the seedsmen at the North, especially in Philadelphia. The attention of the Southern people was first called to the especial value of this grass in Southern agriculture by Mr. J. R. Stanford, of Clarkesville, Habersham county, Georgia, who has been raising it successfully for a number of years.

19. Randall grass. — The writer has been unable to obtain the seeds of this grass, which is so much valued in parts of Virginia. Experiments made with it by others in this immediate vicinity have not been satisfactory, the hot sun of the past dry summer having been very hurtful to it.

20. Terrel grass or Wild rye. — This native grass obtained the name of Terrel grass from the fact that it was first brought to notice and use by Dr. Terrel, of

Sparta, Hancock county, Georgia, a gentleman whose name will be perpetuated in Georgia as the liberal founder of an agricultural professorship in the Georgia University.

The botanical name of this grass is *Elymus*. There are two species of it — one a swamp and the other an upland growth. They differ in the shape of the seeds and the size of the plant, the swamp plant being the taller of the two. The difference in value is not material. This grass is a native in all the cotton States. It is found in Georgia from the sea-coast to the mountains, only, however, in spots in which it has been inaccessible to live stock.

Since it has been made so frequent a subject of remark in the Southern Cultivator, specimens have been sent to the writer from South Carolina, Alabama, Tennessee, Mississippi, Arkansas, and Texas, with inquiries as to whether this was the Terrel grass. In each instance such proved to be the case, showing that it was originally diffused throughout all these States. It will be found fully described in the accompanying list of Texas grasses. The Terrel grass makes an abundant but coarse hay, which is, however, relished by live stock. Its chief value is for winter pasture, for which purpose it is admirably suited. It will be valuable in proportion to the richness of the land. About one bushel of seed should be sowed to the acre. The difficulty of obtaining the seed is an obstacle to its rapid introduction, as it is not in the hands of the seedsmen.

Almost every observant planter can, however, obtain enough of it on his own land to make a commencement. It may be readily known by the long beards of the seed, closely resembling the beards of rye. No planter in any part of the South who can obtain the seeds of this grass need be without a good winter pasture.

The Terrel grass succeeds admirably in woods pasture; this, indeed, is its natural position.

21. English rye grass. — Experiments with this grass have not been satisfactory. It has lived, but yielded no good result, and has been abandoned.

22. Italian rye grass. — This is the most beautiful of all the grasses. Its winter growth on very rich land is enormously great. Nothing can be more beautiful than the deep glossy verdure of this grass, when surrounding nature bears the desolate appearance of mid-winter. It is, however, capricious as to its duration, being sometimes annual and sometimes perennial. It must be recommended rather as an ornament than a utility.

It may be an object near large towns, where manure is abundant, but will scarcely find its place in the rough usages of the plantation.

In consequence of the extraordinary results stated by Colman as being attained in Europe from this grass, special pains were taken in experimenting with it. The first seeds tried were obtained from England, the second from France, and the third direct from Italy.

There was no material difference in the results. The Italian seed was sown last spring; they came up and grew vigorously, but almost entirely perished during the severe drought of the past summer.

It is worthy of notice that a few Lucern seeds came mixed with the grass seeds.

The Lucern is now growing luxuriantly, while the grass has disappeared.

23. English meadow soft grass. — This grass, which grew tolerably well for two years, and which stood winter grazing satisfactorily, was killed by the drought of the last summer.

24. Feather grass — Paris grass — velvet grass. — All popular names of the same grass. Rejected as unworthy of attention.

25. Deer Park grass. — This is a native of Louisiana. It resembles the Terrel grass, to which, on the whole, it is inferior.

26. Meadow vernal grass. — This grass is highly valued in England, as it springs very early and gives a delightful fragrance to the hay of which it forms a part. It has been long known as a border in southern gardens under the name of "vanilla grass," from its peculiar and agreeable odor. It is an evergreen, grows readily, but is diminutive, and has no merit in which it is not exceeded by other grasses.

26a. English fox-tail grass. — Rejected as unsuited to a Southern climate.

27. Herds grass. — For certain positions and for certain purposes this is an exceedingly valuable grass. It is suited to moist ground; in fact, it grows almost in running water. It requires a stiff, close, and wet soil. It will render valuable lands otherwise useless. It will grow on a pipe clay soil, provided it be not that kind of pipe clay which is very wet in winter and very dry in summer — it should be moist all the year. This grass is recommended as occupying favorably the wettest part of the plantation, from which it will yield a heavy return of valuable hay — quite as valuable as sheaf oats. It will bear being under water a good portion of the winter without injury to it. It will grow well under a partial shade, as in woodland thinned out.

28. Musquit grass. — The variety of musquit under experiment by the writer has been a soft, woolly grass, green during the winter, but not standing grazing as well as some other winter grasses. It is probably the same with No. 2 in the Texas list.

29. Crab and crowfoot grasses. — The latter of these valuable annuals is peculiar to the sandy lands of the South. The former is a universal product. Until a plantation is stocked with permanent grasses, these should be used for making hay. For this purpose it is not prudent to rely on the after growth from the grain fields. Sufficient ground to afford hay enough for the use of the plantation stock should be manured in the spring; the ground should be deeply ploughed, harrowed, and rolled. If it should happen, as is frequently the case, that weeds precede the grass, the weeds will disappear, and the grass alone will come up. Even this process of getting hay is much less laborious than pulling fodder. Still, it is a labor that should be encountered only until permanent meadows are established, which being once laid down last a lifetime.

30. Rescue grass. — This is an annual, and in every way inferior to common rye, when used for winter grazing.

31. Bermuda grass. — Public opinion seems to have changed very much at the South in regard to this grass. It was at one time regarded with terror by the cotton planters. Now many of them are setting it out on their plantations. It is quite certain that no other grass will, in a hot climate, yield as much good grazing during the spring, summer, and autumn as Bermuda grass. It is, however, objectionable, as it requires to be set out by the roots, and when once fixed in the soil it is very

difficult to eradicate. Where plantations are destitute of good pastures, and time and inclination are wanting to put land in sufficient condition to bear other grasses, it would, without question, be good policy to stock the permanent pasture land with Bermuda grass. Care will prevent its entrance into the cotton fields. On sandy lands, in which it is destroyed with comparative ease, it may be questioned whether it would not be judicious to introduce it as part of a regular rotation of crops. Its fertilizing power is certainly very great — perhaps quite as great as that of red clover. It will live on land so poor as to be incapable of supporting other valuable grasses, though its value is in proportion to the fertility of the soil. It seems to be determined that below the mountainous parts of the Southern States, if stock be kept away from Bermuda grass during the summer and autumn, although the ends of the grass may be nipped by frost, that there will be sufficient green grass underneath to feed stock during the winter. This being the case, it must stand unrivalled as a grazing grass in the Southern States, taking into the account the whole year, both summer and winter. On very rich land it grows tall enough to be made into hay, and the hay is of the very best quality. The premium bale of hay at the recent Fair of the Georgia State Agricultural Society was made from Bermuda grass. . . .

IV. — Laying Down Meadow and Pasture Land to Grass

Some plain and simple directions as to these points will conclude this Essay. Northern readers who have been accustomed all their lives to grass culture may consider these directions very unnecessary. But at the South, to the majority of readers, the subject is a new one. The cotton planter has heretofore considered it to be his especial mission to "kill grass." He forgets that while it is a part of his business to kill hogs, sheep, and cattle at the proper time, he yet raises pigs, lambs, and calves for a future supply of food for himself and family; and thus while at the proper time he "kills grass," it is equally necessary to raise young grass in other places to afford future food, directly and indirectly, to his cotton and grain crops, and that in no other way than by grass culture can the fertility of his lands be kept up, except at an expense which is ruinous.

For low ground meadow the best bottom land on the plantation should be selected. If it be very wet, and cannot be drained without too great an expense, herds grass alone should be sowed at the rate of a half bushel to the acre. If it be well-drained bottom land, four quarts red clover, the same of white clover, a peck of herds grass and a peck of timothy per acre should be sowed. The ground should have been thoroughly ploughed and harrowed until all clods are broken down, and the seed should then be lightly bruised or rolled. For upland meadow, Lucerne alone is by all means to be selected. Directions for its cultivation have been given above.

For summer pastures on open land, reliance must be had chiefly on the natural grasses, if the planter is afraid of Bermuda grass. No pasture, however, need be better than crab grass, after the grain crops are removed. The growth of these grasses is a peculiar and inestimable advantage of the Southern States.

For spring, autumn, and winter pastures, on upland, sow a mixture of blue meadow oat, orchard, terrel grass, with red and white clover. For pastures on low land add to the above herds grass; if the ground be very wet, sow herds grass and white clover alone. These will actually dry up wet places.

The chief reliance at the South should, however, be on wood and pastures. All the grasses mentioned above will grow in the shade of trees. The use of woodland for this purpose answers a double end. Land now idle is put to a valuable use, and all the open land of the plantation by the additional stock raised can be kept in a state of improvement.

The preparation of woodland for pastures is as follows: Take out rail timber enough to fence it, grub the land carefully, cut out useless trees, leave especially the most bearing trees, as in a thinned and trampled pasture they will bear fruit almost every year, and thus greatly diminish the cost of hog raising. Burn the logs and brush in as small heaps as possible and scatter the ashes. Wherever a log or brush heap has been burned, the grass sown will take hold with vigor.

As it is impossible to make woodland fine with the plough, great care must be taken in covering grass seeds. Wherever a clod is turned over on seeds they do not vegetate. If it be possible to sow during a snow or drizzling rain, no covering of the seeds is necessary. A short roller is perhaps the best instrument to cover with. If oxen be used, or a very gentle mule, a short roller can be managed in thinned woods. The ground for wood pasture should be ploughed as well as the nature of the case will admit.

No stock of any kind should be allowed to go on the pasture after it is sown until the grass has once gone to seed. This will fill the ground with seeds for future use. If the pasture be designed for winter use no stock should go on it during the summer. If it be designed for summer use no stock should go on it in the winter. The same pasture cannot be used all the year round. Stock should be taken from the winter pastures in very wet weather unless it be sheep. They should not be grazed at any time too severely; too close grazing will destroy the grass; it will produce this effect in any climate. The planter should treat his winter grass pastures as he has been in the habit of treating his rye and barley lots for winter grazing.

A woodland pasture, well stocked with good grasses, will pay the interest annually on more than fifty dollars an acre, while at the same time the valuable timber is preserved for plantation use. No planter who has ever tried such a pasture would part with it for fifty dollars an acre. If it cost him ten dollars per acre to prepare the ground, buy and put in the seeds, &c., how great is his gain!

There must be a revolution in Southern agriculture. The process of exhaustion must be stopped, or it will stop itself. The improvement of the soil must be considered. This is the point at which to begin. The commercial manures answer a special and temporary purpose. The only permanent reliance for valuable improvement of the soil is by means of the manure of the domestic animals. These can be raised to profit only with the aid of the artificial grasses. Our cultivated lands are too much exhausted to produce valuable grasses in their present state. Our woodlands, with a dressing of the ashes produced in thinning them out, will produce these grasses. Thus a sufficient quantity of manure can be obtained to begin a system of improvement which may ultimately comprise the whole plantation.

It is firmly believed that the now idle woodlands of the South can be put in a situation to raise sheep enough to allow a sale of wool equal in value to the present cotton crop, and that this result can be obtained with a positive increase of the cotton crop from the manure of the sheep.

The opinions positively expressed in this Essay are not speculations. They are not mere theories. They are based, so far as is possible, on the practical observations and experiments of a man who has been grieved to witness the gradual impoverishment of the fair domain of the South, who has pondered the subject long and carefully, and who has been brought to the conclusion that our hope of improvement is to be based upon the adoption of a mixed husbandry, of which the cultivation of the artificial grasses is an essential part, and which contemplates by the aid of these grasses the rearing of sufficient liieetock to manure annually all the land in cultivation.

Cattle Disease, 1860

From G. Emerson and A. L. Elwyn, "Cattle Disease, or Pleuro-Pneumonia." U.S. Patent Office, *Report on Agriculture, 1860* (Washington, 1861), pp. 239–41.

SIR: In pursuance of instructions received from you, dated the 19th of June, we visited the State of Massachusetts for the purpose of investigating the "Cattle Disease," or Pleuro-pneumonia, which, during the last year and up to this time, has existed there, and from which much loss has been sustained by some farmers, and a general damage inflicted upon the rural interests throughout the New England States.

The alarm occasioned by the recent spreading of this cattle disease in Massachusetts has extended far and wide over our country, a proof of which was given by the number of commissions sent from various States, by public authority, to investigate this epizootic. Beside those from Maine and other parts of New England, New York, New Jersey, Pennsylvania, Ohio, and Indiana were all represented by deputations with the same object. Governor Banks and other State authorities, with Charles L. Flint, Secretary of the State Board of Agriculture, all contributed every facility we could desire to further our objects, and we should be ungrateful if we did not acknowledge our obligations to them for their assistance in procuring information.

A knowledge of the existence of a similar disease among cattle in some parts of New Jersey and Eastern Pennsylvania — three hundred and four hundred miles to the Southwest — had not reached Massachusetts previous to the time of our visit, and might probably have exerted some influence in modifying legislation upon the subject. But in New Jersey and Eastern Pennsylvania the presence of the cattle disease was for a long time suppressed, nor did it obtain publicity until inquiries were set on foot by the Philadelphia Society for Promoting Agriculture, at its meeting held in the first week of June. This society then appointed us a committee of investigation, and, at the instance of the Hon. Thomas G. Clemson, Chief of the Agricultural Division, we were subsequently charged with instructions from your Bureau of the General Government to proceed to Massachusetts to investigate the cattle disease, and to report to you upon the subject. The very short time allotted for this purpose must be an apology for omissions and imperfections naturally to be expected.

In Massachusetts, where the agricultural interests are based upon its neat stock, the spread among them of a fatal disease, resembling that which for many years past has been devastating the herds of Europe, naturally produced much public excitement and alarm. The legislature, during its regular session, took the subject into serious consideration, and displayed, in their vigorous action, the tone which has animated most of the European governments in their efforts to check the plague or lessen its violence.

Viewing the disease as one brought from a foreign country, and propagated by contagion alone, the Legislature, at its regular session, passed, on the 4th of April, an act providing for the appointment of three commissioners, who were required to take measures for the *extirpation of the disease*. They were authorized and required to visit, without delay, the several places in the Commonwealth where the disease was known or supposed to exist, and empowered to cause all cattle which had been diseased, or had belonged to diseased herds, to be forthwith killed and buried, and the premises where they were kept cleansed and purified; to appraise, in their discretion, the value of the cattle killed which were apparently well, and certify to the governor and council the allowances made to the owners, and give such lawful orders and directions as, in their judgment, the public necessity might require.

The commissioners promptly entered upon their duties, and, from the sickness found in various herds and the number of cattle they had occasion to slaughter, the appropriation of $10,000, placed at their disposal by the legislature, was soon exhausted, with a like amount furnished by contributions from the State Agricultural Society and liberal public-spirited individuals. But it was soon found that the disease had spread itself over a larger extent of territory than it was at first supposed to occupy. More appropriations were required, and the governor summoned an extra session of the Legislature, on the meeting of which a large amount of valuable information relative to the nature and progress of the disease was collected and immediately published.

The commissioners reported that, in compliance with the summary instructions given them, they had slaughtered over eight hundred head of cattle found diseased or suspected. They were released, at the extra session, from the necessity of slaughtering all diseased stock, and, at their option, empowered to carry out a system of isolation. The governor, in his message to the extra session, viewed the disease as one purely contagious, not a disaster affecting Massachusetts or New England alone, but a contagion, which, if allowed to spread without effort to extirpate or restrain it, must ultimately ravage the whole country. Upon the basis furnished by the Census of 1850, he estimates the present property in cattle throughout the Union at no less than six hundred and forty millions of dollars, an interest only second in importance to that of Indian corn. "But these figures," he observes, "very imperfectly represent the interest of the American people in their gigantic industrial product. How far it enters into the employment of the great majority of persons — how many millions are dependent upon it for the luxuries and necessities of life — to what extent it contributes, indirectly, to public health and enjoyment, and how large a part it forms of the sound and reliable business of the country, are considerations which naturally occur to the mind of every intelligent person."

"If," he continues, "we could confine the ravages of the fatal distemper, so unfortunately deposited on our shores, to our own State, it would still be of sufficient importance to demand the earnest attention of the people; but unless extirpated on the instant when it appears, it cannot be so confined. If it spread over our own territory, it must ravage other States; and it becomes a duty of the highest character — one which we owe alike to ourselves, to the honor of Massachusetts, and to the people of the whole country — to make every available and possible effort to restrain its ravages, if extirpation is impossible.

"Admitting, if need be, that it is doubtful whether it partakes more of the character of a contagion or an epidemic — admitting that it may not be in our power to prevent its spreading through the country — nevertheless every citizen of Massachusetts should have it in his power to say that every proper effort had been made by the State to produce that result. I am constrained to express the opinion that all has not yet been done which may be wisely if not successfully performed; and this fact I offer to you as a chief reason for this extraordinary convocation. This would seem to be a measure which the natural comity existing between friendly States would absolutely demand. Such a measure would for a brief period — not necessarily a long one — interfere with the freedom of trade, but it is such an interference as the continued existence or the trade in cattle itself requires. A line may easily be established, beyond which no cattle shall be passed without sufficient assurances that they do not carry contagion with them."

The governor, in his message, recommends the adoption of public regulations to prohibit, so far as it can wisely and properly be done, the exportation from Massachusetts to neighboring States of cattle in which the seeds of disease may possibly exist. In all this vigorous action of the public authorities of Massachusetts we recognize the most laudable efforts to arrest the disease, and, if possible, prevent its extension to other States, at whatever sacrifice.

Admitting that the first alarm may have led to exaggeration of the evil at hand, and that the prevailing impression in regard to the mode in which the disease was solely propagated was an open question, still the motives which inspired the public authorities of Massachusetts do them much honor and merit the commendation of sister States.

The legislature of the neighboring State of Connecticut has since taken up the subject, and recently passed stringent laws involving heavy fines and imprisonment where persons are found transporting cattle from one part of the State to another without proper authority. It remains to be seen, and time will soon show, what results will be gained from measures so promptly taken in New England to arrest the progress of this cattle plague. Meanwhile a similar disease has manifested its presence in New Jersey and in the neighborhood of Philadelphia, where it has caused much destruction in many dairy herds during the past winter and spring, and up to this time. Rumors of its appearance in other parts of the country are given by the press.

Having thus brought to your notice the action of legislative bodies, so far as they have yet proceeded in this country, in regard to the visitation of this new disease in the United States among cattle — one which threatens so much our agricultural interests — we offer the following statement of what we have gathered relative to its history, causes, nature, means of prevention, and remedial treatment:

History of the First Recognition and Diffusion of Pleuro-Pneumonia

The French government, with that zeal which it always shows for the promotion of its public interests, some years since appointed a scientific commission for the investigation of epizöotic pleuro-pneumonia in cattle. From the report of this commission it would appear that this disease was an endemic epizöotic, confined in former times to various isolated regions in the mountains of Piedmont, Switzerland, Franche Compté, Dauphiné, the Vosges, the Pyrenées and Auvergne. With the exception of these places, its inroads upon the agricultural interests were so limited as not seriously to affect the public weal until the year 1789. Whether the French commission may not have gone further back in dating the first appearance of this disease in Germany, Holland, and perhaps other countries, is a question of some interest, which, however, we have not time to investigate.

This formidable disease of neat cattle has of late years prevailed simultaneously over large tracts of Europe and other parts of the Old World, exhibiting everywhere the same leading characteristics. It would, therefore, appear to result from the operation of some general cause, and hence belong to the class of diseases denominated epizöotic. By its wide-spread ravages in many parts of Europe it has ruined many dairymen, breeders, and stock-owners, and exerted a great influence upon the beef-market.

So far as we have carried our investigations we have not found the prevalence of the cattle disease on this side of the Atlantic affected by the character of the country, whether upland or valley, dry or marshy, in proximity to the sea, or distant from it.

In Massachusetts the first cases known were on high land, about seven miles west of Boston, and the most general spread was after a leap of sixty miles to the west, also in a hilly region. We have, however, no idea that it was attracted in this movement by the character of the locality. Nothing unusual has transpired in the conditions of the weather, except the prevalence of drought during the spring months of the present year in New England. But in New Jersey and eastern Pennsylvania the weather has been seasonable, and the winter past of the average temperature.

That the cattle disease has prevailed in New Jersey and in Pennsylvania, in the proximity of Philadelphia, during the past winter and spring, and there prevails at the present time, is certainly known. In the absence of any proof of its introduction here from abroad, we are inclined to ascribe its presence to epizöotic agency. Be this as it may, there is good reason to believe that here, as in Massachusetts, contagion has been active in spreading the disease. Wherever we have met with it, the symptoms and post mortem appearances correspond most strikingly with those described by continental and English authorities.

Whether spreading by contagion or by epizöotic agency, pleuro-pneumonia is usually found to select herds exposed to some influence calculated to impair their health, whilst it spares those where everything contributed to maintain full vigor. . . .